COMPLEX MACROMOLECULAR ARCHITECTURES

COMPLEX MACROMOLECULAR ARCHITECTURES

SYNTHESIS, CHARACTERIZATION, AND SELF-ASSEMBLY

Editors

Nikos Hadjichristidis
University of Athens, Greece

Akira Hirao
Tokyo Institute of Technology, Japan

Yasuyuki Tezuka
Tokyo Institute of Technology, Japan

Filip Du Prez
Ghent University, Belgium

John Wiley & Sons (Asia) Pte Ltd

This edition first published 2011

Registered office
John Wiley & Sons (Asia) Pte Ltd, 1 Fusionopolis Walk, #07-01 Solaris South Tower, Singapore 138628.

For details of our global editorial offices, for customer services and for information about how to apply for permission to reuse the copyright material in this book please see our website at www.wiley.com.

Library of Congress Cataloging-in-Publication Data

Complex macromolecular architectures : synthesis, characterization, and self-assembly / Nikos Hadjichristidis ... [et al.].
p. cm.
Includes bibliographical references and index.
ISBN 978-0-470-82513-6 (hardback)
1. Macromolecules. 2. Polymerization. I. Hadjichristidis, Nikos, 1943-
QD381.C658 2011
547′.7–dc22

2010053511

Print ISBN: 978-0-470-82513-6
ePDF ISBN: 978-0-470-82514-3
oBook ISBN: 978-0-470-82515-0
ePub ISBN: 978-0-470-82827-4

Set in 10/12pt Times by Thomson Digital, Noida, India
Printed and bound in Singapore by Markono Print Media Pte Ltd

Contents

Contributors

Amelia M. Anderson-Wile, Cornell University, Ithaca, NY, USA
Toufic Nabil Aridi, University of Waterloo, Waterloo, Canada
Christopher Barner-Kowollik, Karlsruhe Institute of Technology, Karlsruhe, Germany
Haskell W. Beckham, Georgia Institute of Technology, Atlanta, GA, USA
Erik B. Berda, University of New Hampshire, Durham, NH, USA
Wolfgang H. Binder, Martin-Luther University Halle-Wittenberg, Halle, Germany
Michelle A. Chavis, Cornell University, Ithaca, NY, USA
Wen-Chang Chen, National Taiwan University, Taipei, Taiwan
Geoffrey W. Coates, Cornell University, Ithaca, NY, USA
Alain Deffieux, CNRS, Université de Bordeaux, Pessac, France
Filip Du Prez, Ghent University, Ghent, Belgium
Joseph B. Edson, Cornell University, Ithaca, NY, USA
Claudia Enders, Martin-Luther University Halle-Wittenberg, Halle, Germany
Mario Gauthier, University of Waterloo, Waterloo, Canada
Jessica Gwyther, University of Bristol, Bristol, UK
Katharina Hackethal, Martin-Luther University Halle-Wittenberg, Halle, Germany
Nikos Hadjichristidis, University of Athens, Athens, Greece
Hirokazu Hasegawa, Kyoto University, Kyoto, Japan
Mayumi Hayashi, Tokyo Institute of Technology, Tokyo, Japan
Florian Herbst, Martin-Luther University Halle-Wittenberg, Halle, Germany
Tomoya Higashihara, Tokyo Institute of Technology, Tokyo, Japan
Akira Hirao, Tokyo Institute of Technology, Tokyo, Japan
Richard Hoogenboom, Ghent University, Ghent, Belgium
Hermis Iatrou, University of Athens, Athens, Greece
Andrew J. Inglis, Karlsruhe Institute of Technology, Karlsruhe, Germany
Takashi Ishizone, Tokyo Institute of Technology, Tokyo, Japan
Toyoji Kakuchi, Hokkaido University, Sapporo, Japan
Chi-Ching Kuo, National Taiwan University, Taipei, Taiwan
Mieke Lammens, Ghent University, Ghent, Belgium
Sébastien Lecommandoux, Université de Bordeaux, Pessac, France; CNRS, Pessac, France
Cheng-Liang Liu, National Taiwan University, Taipei, Taiwan
Guojun Liu, Queen's University, Kingston, Ontario, Canada
Jun Luo, University of California, Irvine, CA, USA

Ian Manners, University of Bristol, Bristol, UK
Rachel K. O'Reilly, University of Warwick, Coventry, UK
Christopher K. Ober, Cornell University, Ithaca, NY, USA
Marinos Pitsikalis, University of Athens, Athens, Greece
David A. Rider, University of Bristol, Bristol, UK
Georgios Sakellariou, University of Athens, Athens, Greece
Toshifumi Satoh, Hokkaido University, Sapporo, Japan
Michel Schappacher, CNRS, Université de Bordeaux, Pessac, France
Evan L. Schwartz, Cornell University, Ithaca, NY, USA
Kenneth J. Shea, University of California, Irvine, CA, USA
Kenji Sugiyama, Hosei University, Tokyo, Japan
Yasuyuki Tezuka, Tokyo Institute of Technology, Tokyo, Japan
Michael L. Turner, University of Manchester, Manchester, UK
Mitsuru Ueda, Tokyo Institute of Technology, Tokyo, Japan
Kenneth B. Wagener, University of Florida, Gainesville, FL, USA
George R. Whittell, University of Bristol, Bristol, UK
Ian Wyman, Queen's University, Kingston, Ontario, Canada
Takuya Yamamoto, Tokyo Institute of Technology, Tokyo, Japan
Hideaki Yokoyama, The University of Tokyo, Tokyo, Japan
Hee-Soo Yoo, Tokyo Institute of Technology, Tokyo, Japan
Chin-Yang Yu, University of Manchester, Manchester, UK

Preface

This book is dedicated to materials with complex macromolecular architectures. The book is unique in its focus, since only a few of these structures have, until now, been considered in volumes covering the broader area of macromolecular engineering. In addition, the collection of chapters provides an in-depth study of the self-assembly and potential applications of these cutting-edge materials, providing information not available in any other reference volume.

The structures described include those previously discussed elsewhere (miktoarm stars, combs, grafts, rings, dendritic, hyperbranched and arborescent), as well as newly synthesized complex architectures (multicyclic, hydrogen-bonded complex architectures, structures from living alkene polymerization, ADMET and polyhomologation, as well as topological polymer chemistry).

Each chapter is the outstanding contribution of top scientists, currently active in the field. The content is thorough and scientifically rigorous, yet presented with such clarity that all interested researchers will easily grasp the concepts and will surely be inspired to learn more about these highly complex materials.

The well-defined macromolecular structures presented here are only a few of the myriad of those possible. Imagination, nature and other scientific disciplines (polymer physics, materials science and molecular biology) will lead polymer scientists to novel structures with the ultimate goal of designing and synthesizing complex macromolecular architectures with predetermined properties.

Preface

About the Editors

Nikos Hadjichristidis obtained his BSc at the University of Athens, Greece, his PhD at the University of Liege, Belgium, and his DSc at the University of Athens. He conducted postdoctoral research at the University of Liege and at the National Research Council of Canada. He has been the Director of the Industrial Chemistry Laboratory at the University of Athens since 1994 and the Chairman of the Chemistry Department (1991–1995, 1999–2003 and 2005–2009). He has supervised, until now, 45 PhD and 55 Masters' theses.

He was a Visiting Scientist at the University of Liege, Visiting Research Officer at the NRC of Canada, Visiting Scientist at the University of Akron and a Distinguished Visiting Scientist at the NRC of Canada. He has been a Visiting Professor at Exxon Research and Engineering Co., NJ, since 1984, and at REPSOL-YPF Research Center, Madrid (2000–2009).

He has received the International Award of the Society of Polymer Science, Japan (SPSJ, 2007), the ACS PMSE Cooperative Research Award (2010) and the ACS Rubber Division Chemistry of Thermoplastic Elastomer Award (2011). He was elected as a PMSE Fellow in 2004 and was the "Ralph Milkovich" Memorial Lecturer in 2006 at the University of Akron. He was a member of the Editorial Board of *Macromolecules* (1997–1999), and is currently an Editorial Board member of *Journal of Polymer Science, Polymer Chemistry* and *Progress in Polymer Science*. He is also one of the editors of *European Polymer Journal*.

Nikos Hadjichristidis has dedicated his career primarily to the synthesis of model polymers with complex macromolecular architectures and has published more than 370 papers and 23 reviews in refereed scientific journals, 6 patents, two books (as editor), and is the author of *Block Copolymers: Synthetic Strategies, Physical Properties, and Applications* (2003).

Nikos Hadjichristidis's Web Site: http://www.chem.uoa.gr/polymers/pg_polym_index.htm

Akira Hirao was born in Tokushima, Japan, on December 4, 1947. He received his Doctorate Degree from Tokyo Institute of Technology, Japan in 1975. Then, he joined the group of Professor Charles U. Pittman, Jr. at the University of Alabama, USA, as a postdoctoral fellow, from 1975 to 1977. His professional career at the Tokyo Institute of Technology progressed from Assistant Professor (1977–1983), Associate Professor (1983–1996), to Full Professor (1996–present). He held the position of the Chairman of Polymeric and Organic Materials Department for four terms and was the Vice-Dean of Chemistry Division, undergraduate course from 2004 to 2006. In 2010, he received The Award of the Society of Polymer Science, Japan. He has been working in the area of polymer synthesis since 1975. His research is mainly focused on the following areas: synthesis and living anionic polymerization of new functional monomers, the precise synthesis of architectural polymers such as block copolymers, graft copolymers, comb-like polymers, chain-end- and in-chain-multifunctionalized polymers, star-branched polymers, dendritic hyperbranched polymers, surface control and analysis, as well as molecular self-assemblies and nanostructured materials derived from multiphased polymers. He was a member of the Editorial Board of *Macromolecules*, *Polymer Journal* and is currently a member of *Macromolecular Research*, *European Polymer Journal*, and others. He has published more than 300 refereed scientific journals and several book chapters (as editor) and has given 80 plenary and invited lectures at international conferences on polymer science.

Yasuyuki Tezuka, born in 1953, is a graduate of The University of Tokyo. In 1982, he received his doctorate degree from Ghent University (Belgium) with a thesis on poly(THF)-based telechelic polymers. He then joined Nagaoka University of Technology (Japan) as an assistant professor. In 1991, he was promoted to an associate professor. He moved to the Tokyo Institute of Technology in 1994 and has been a professor since 2003 in the Department of Organic and Polymeric Materials. He received the Tokyo Tech Award of Best Teacher in 2004, the Tokyo Tech Innovative Research Engineering Award in 2009, and The Award of the Society of Polymer Science, Japan (2010). He has served as an Asian Editor of *Reactive and Functional Polymers* since 2006. He was an associate editor of *Polymer Journal*, a publication by The Society of Polymer Science, Japan (2002–2006). He has co-authored three books and published about 170 original papers and review articles. His current research is focused on topological polymer chemistry, in particular, on the design of topologically unique macromolecular architectures by developing a new process of electrostatic self-assembly and covalent fixation, and of novel polymer materials by their topology effects.

Yasuyuki Tezuka's Web Site: http://www.op.titech.ac.jp/lab/tezuka/ytsite/sub0e.html

Filip Du Prez is since 1999 head of the Polymer Chemistry Research Group (www.PCR.UGent.be) at Ghent University in Belgium (PhD in 1996 at Ghent University, Belgium and Lehigh University, USA; post-doc in Montpellier, France), where about 20 researchers are dealing with the design of functional polymer architectures and polymer materials, controlled polymerization techniques and "click" chemistry. In 2008, he had a Visiting Professor position at the Centre for Advanced Macromolecular Design (CAMD) (UNSW, Sydney), hosted by Christopher Barner-Kowollik. A few actual research domains are polymeric nano- and microparticles, self-healing materials, step-growth polymerization in combination with "click" chemistry, polymeric solid supports for ATRP and "click" catalysts, shape-memory polymers and highly branched copolymer structures. He is the author of more than 130 peer-reviewed publications and patents, 8 book chapters and (co-)chairman of 10 (inter)national conferences on polymer chemistry related topics. Since 2008, he has been one of the editors of *European Polymer Journal*. In addition, he is a member of the Editorial Board of *Polymer, Designed Monomers and Polymers, Journal of Macromolecular Science* (*Part A: Pure and Applied Chemistry*) and *Reactive and Functional Polymers*.

Filip Du Prez's Website: http://www.pcr.ugent.be/

Part One

Synthesis

1

Cyclic and Multicyclic Topological Polymers

Takuya Yamamoto and Yasuyuki Tezuka
Department of Organic and Polymeric Materials, Tokyo Institute of Technology, Japan

1.1 Introduction

Tailored design of polymer chain architectures has been an enduring challenge in synthetic polymer chemistry. A remarkable progress has been observed to extend the frontier from conventional linear structures, formed by a successive chain-growth mechanism, toward nonlinear and complex topologies composed of branched and cyclic segments (Adachi and Tezuka, 2009). Consequently, unprecedented opportunities have been created for the functional design of polymer materials by programming their topologies.

The topological form of polymers is classified by the principal geometric parameters, namely their numbers of termini (T) and junctions (J) (Scheme 1.1) (Tezuka, 2008; Tezuka and Oike, 2001; Tezuka and Oike, 2002). Thus, a linear polymer is classified as a form of two termini ($T=2$) and the absence of junctions ($J=0$), while an n-armed, star-shaped polymer as a form of $T=n$ and $J=1$. Likewise, a polymer brush having n side chains can be defined as a form of $T=n+2$ and $J=n$. Accordingly, the branched polymer structure tends to become more complex with the increase of the number of the termini and junctions. It is remarkable, on the other hand, that a single cyclic (ring) polymer is unique in the absence of termini and of junctions ($T=0$, $J=0$). Moreover, various complex topologies including single- or multicyclic units are shown by the form having relatively small T and J numbers to imply their geometrically primitive nature, though their synthesis in polymers has been a formidable challenge. In this chapter, we show recent developments in this area, as well as new insights in polymer topology effects revealed by precisely designed cyclic and multicyclic polymers.

Complex Macromolecular Architectures: Synthesis, Characterization, and Self-Assembly, First Edition.
Edited by Nikos Hadjichristidis, Akira Hirao, Yasuyuki Tezuka and Filip Du Prez.

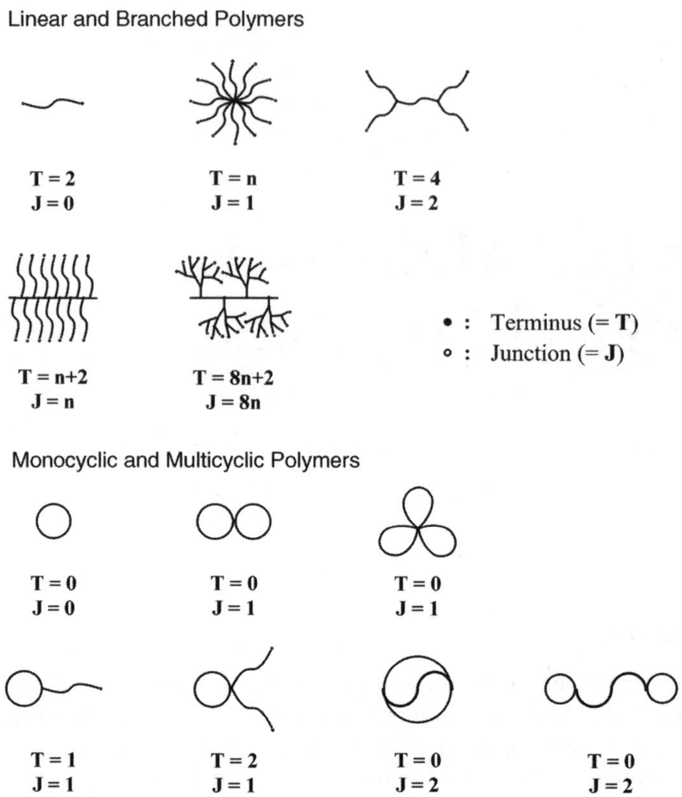

Scheme 1.1 Various polymer topologies with the numbers of termini (T) and junctions (J) (Tezuka, 2008; Tezuka and Oike, 2001; Tezuka and Oike, 2002).

1.2 The Progress on the Synthesis of Ring Polymers

Ring (single-cyclic) polymers are different from their linear and branched counterparts in their static (smaller hydrodynamic volume), dynamic (less entanglement and nonreptational diffusion) as well as thermal (higher T_g) properties. In a recent decade, new synthetic protocols have been developed to produce a variety of ring polymers of guaranteed purity, to provide a basis for the rational polymer materials design that relied on the ring topology (Adachi and Tezuka, 2007).

1.2.1 Ring-Expansion Polymerization

A ring-expansion polymerization proceeds with the successive insertion of monomers into a cyclic initiator to form a ring polymer. The process does not require dilution, in contrast to an alternative and conventional end-to-end polymer cyclization process by linear polymer precursors. Until recently, however, the ring-expansion process has not been considered as a practical means, since a reactive initiator fragment is inherently retained in the ring polymer structure. And it is not an easy task to remove it by keeping the ring topology of the product (Kricheldorf, 2010; Shea *et al.*, 1998).

In order to overcome this drawback, Nishikubo *et al.* employed a cyclic thioester initiator to form a ring polythioether through the insertion of thiirane monomers followed by an intermolecular transesterification (Kudo *et al.*, 2005; Kudo *et al.*, 2008). As another approach, Grubbs *et al.* have introduced a Ru catalyst bearing a cyclic ligand containing a Ru-alkylidene unit, which could initiate the ring-expansion polymerization of cyclic olefins such as 1,5-cyclooctadiene or cyclooctene (Bielawski *et al.*, 2002; Bielawski *et al.*, 2003; Xia *et al.*, 2009). The chain transfer to the Ru-alkylidene unit took place during the chain growth to eliminate the initiator complex, to form a ring polybutadiene or a ring poly(cyclooctene), respectively, which are free of the catalyst fragment (Scheme 1.2). The ring poly(cyclooctene) obtained was then hydrogenated to give a ring polyethylene, showing distinctive properties in comparison with commercial linear polyethylenes (Bielawski *et al.*, 2003). Furthermore, Waymouth *et al.* have employed an N-heterocyclic carbene compound as an initiator for the ring-expansion polymerization of lactones or lactide to form a ring polyesters or polylactide, respectively, having a narrow polydispersity index (Scheme 1.2) (Culkin *et al.*, 2007; Jeong *et al.*, 2009). The control of the molecular weight of ring polymers, free of the catalyst fragment, has been achieved for the first time in these ring-expansion polymerization methods.

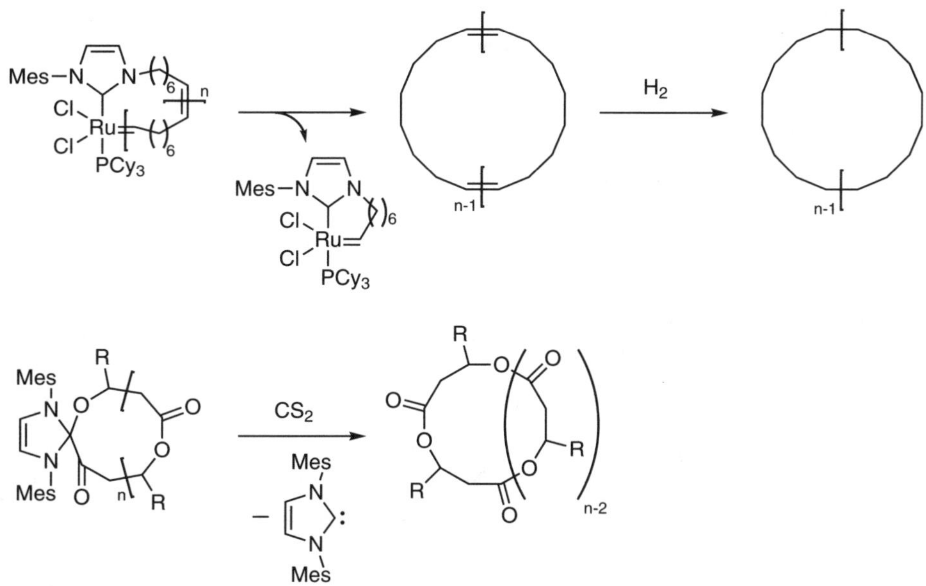

Scheme 1.2 New ring-expansion polymerizations (Bielawski *et al.*, 2002; Bielawski *et al.*, 2003; Culkin *et al.*, 2007; Jeong *et al.*, 2009; Xia *et al.*, 2009).

1.2.2 Cyclization by Telechelic Polymer Precursors

The conventional synthetic protocol for ring polymers has been an end-to-end linking reaction between a linear polymer precursor having reactive groups, that is, telechelics, and an equimolar amount of a bifunctional coupling reagent (Laurent and Grayson, 2009). This bimolecular polymer cyclization should be performed with strictly equimolar amounts of telechelics and coupling reagent, and moreover with dilution to suppress the concurrent chain extension. In consequence, this *bimolecular* process is often considered unattractive in practical use (Scheme 1.3A).

Alternative *unimolecular* processes have subsequently been introduced as a significantly improved means to prepare ring polymers (Scheme 1.3B). As a recent example, Kubo *et al.* have prepared an α-amino-ω-carboxy poly(*t*-butyl acrylate) through a living polymerization using an initiator having a protected amino group, followed by the end-capping reaction to introduce a carboxylic group (Kubo *et al.*, 2003). The resulting asymmetric telechelic precursor underwent, after the deprotection, an intramolecular amidation reaction to give a ring polymer product. By the subsequent hydrolysis, moreover, a ring poly(acrylic acid) was produced as a polymer electrolyte of lower viscosity for potential applications. More recently, Grayson *et al.* have prepared a telechelic polystyrene having an alkyne group and an azide group, which was subjected to the end-to-end polymer cyclization via click chemistry (Laurent and Grayson, 2006). The click process has been recognized as an effective means to prepare other ring polymers, such as poly(*N*-isopropyl acrylamide) and poly(ε-caprolactone) to elucidate the

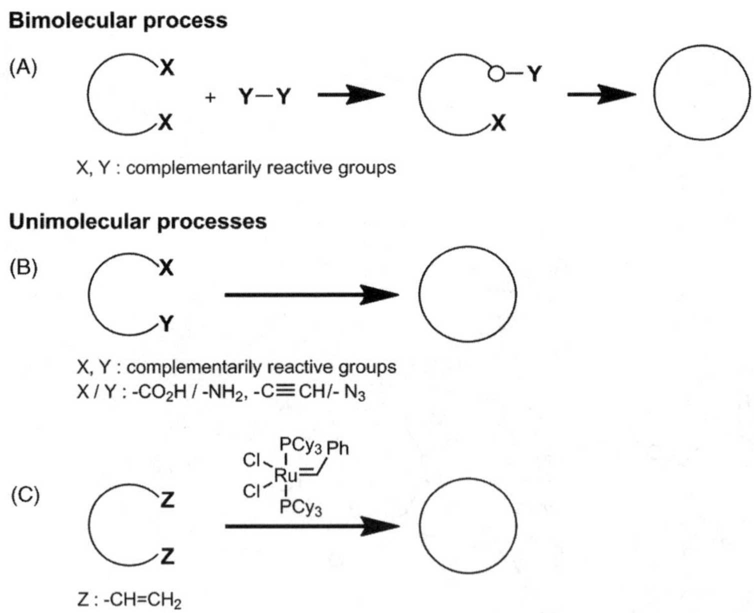

Scheme 1.3 Bimolecular (A) (Laurent and Grayson, 2009), unimolecular (B) (Hoskins and Grayson, 2009; Kubo *et al.*, 2003; Laurent and Grayson, 2006; Qiu *et al.*, 2007; Ye *et al.*, 2008) and (C) end-to-end linking processes for the synthesis of ring polymers (Adachi *et al.*, 2008; Bielawski and Grubbs, 2007; Guidry *et al.*, 2005; Hayashi *et al.*, 2007; Tezuka and Komiya, 2002).

topology effect on phase transition (Ye *et al.*, 2008; Qiu *et al.*, 2007), and on biodegradation (Hoskins and Grayson, 2009), respectively.

An olefin-metathesis process has so far been applied in synthetic polymer chemistry, as in the ring-opening metathesis polymerization (ROMP) and in the acyclic diene metathesis (ADMET) (Bielawski and Grubbs, 2007). Moreover, new functional-group tolerant catalysts have recently been developed and successfully applied for the synthesis of topologically unique molecules like catenanes (Guidry *et al.*, 2005; Chung *et al.*, 2009) as well as ring polymers through the ring-expansion polymerization, as shown in Section 1.2.1 (Bielawski *et al.*, 2002; Bielawski *et al.*, 2003; Xia *et al.*, 2009). We have prepared α,ω-diallyl telechelic poly(tetrahydrofuran) and poly(acrylic ester)s, through the relevant living polymerization followed by the end-capping reaction. The subsequent metathesis condensation, that is, metathesis polymer cyclization (MPC), produced the corresponding ring polymers in high yields even under dilution (Scheme 1.3C) (Tezuka and Komiya, 2002; Hayashi *et al.*, 2007). Moreover, the MPC process has been applied for the practical synthesis of amphiphilic cyclic block copolymers (Adachi *et al.*, 2008).

Another practical means has been developed by combining a classical ring-expansion polymerization process, in which the initiator fragment is included within the ring polymer structure, with an end-to-end polymer cyclization process. Thus, Jérome *et al.* have prepared a ring poly(ε-caprolactone) having short block segments of a photocrosslinkable, olefin-modified monomer units through a living ring-expansion polymerization using a cyclic stannate initiator (Scheme 1.4) (Li *et al.*, 2006). The subsequent irradiation under dilution caused the crosslinking of the two chain-end segments. The subsequent removal of the stannate group produced a chemically stable ring polymer product. Furthermore, a *tadpole* polymer,

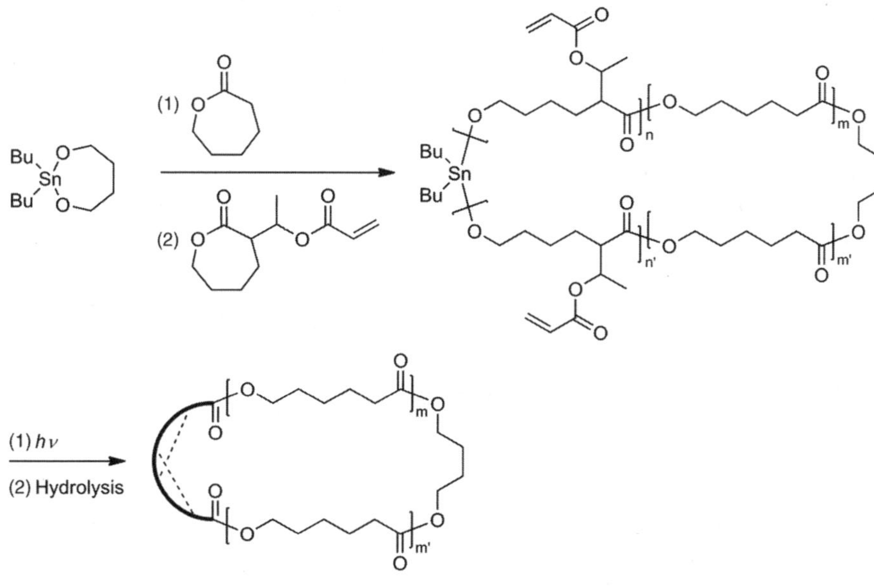

Scheme 1.4 Ring-expansion polymerization and the subsequent covalent linking for the synthesis of ring polymers (Li *et al.*, 2006).

that is, a ring with linear branches, was obtained by the reinitiation of the polymerization after the chain-end linking and the subsequent removal of the stannate group, causing the cleavage of the ring polymer unit to form the two outward branches on a ring polymer unit.

By taking advantage of this methodology, Deffieux *et al.* have prepared a high-molecular weight ring poly(chloroethyl vinyl ether), which was subsequently subjected to the grafting-onto reaction with a living polystyrene, to form a densely grafted large ring polymer product as a nano-object. Notably, the topology of the resulting nano-objects, directly observable by atomic force microscopy (AFM), corresponds to that of the original polymer itself (Schappacher and Deffieux, 2008a; Schappacher and Deffieux, 2008b). This study showed, for the first time, that single as well as multicyclic polymer topologies are indeed formed by synthetic polymers through the polymer cyclization process, as previously shown in DNA systems (Seeman, 2003).

We have developed an *electrostatic self-assembly and covalent fixation* (ESA-CF) process using linear or star telechelic precursors having cyclic ammonium salt groups in combination with small or polymeric plurifunctional carboxylates (Oike *et al.*, 2000; Tezuka, 2003; Tezuka, 2005). Various kinds of telechelic polymers have so far been employed, including poly(tetrahydrofuran) (Oike *et al.*, 2000), poly(ethylene glycol) (Tezuka *et al.*, 2002), polystyrene (Oike *et al.*, 2001a), and poly(dimethylsiloxane) (Tezuka *et al.*, 1997), obtainable through the relevant living polymerization followed by the end-capping technique. The electrostatic polymer self-assemblies have subsequently been converted into covalently connected and prescribed polymer structures, such as star polymers, network polymers, graft copolymers as well as single- and multicyclic polymers by applying the dilution condition.

In particular, telechelics having five- and six-membered cyclic ammonium, that is, *N*-phenylpyrrolidinium and *N*-phenylpiperidinium, respectively, salt groups, are of a particular interest (Adachi *et al.*, 2006). The former undergoes a selective ring-opening reaction to form an amino ester group upon heating with a carboxylate counteranion through the nucleophilic attack exclusively at the N-adjacent ring (*endo*) position. On the other hand, the latter having unstrained six-membered cyclic ammonium salt groups gives a chemically stable, simple ester group through the nucleophilic attack at the alternative N-adjacent (*exo*) position, to eliminate *N*-phenylpiperidine (Scheme 1.5).

It is also notable that cations and anions balance the charges during the formation of the self-assembly between telechelics having cationic groups accompanying carboxylates as

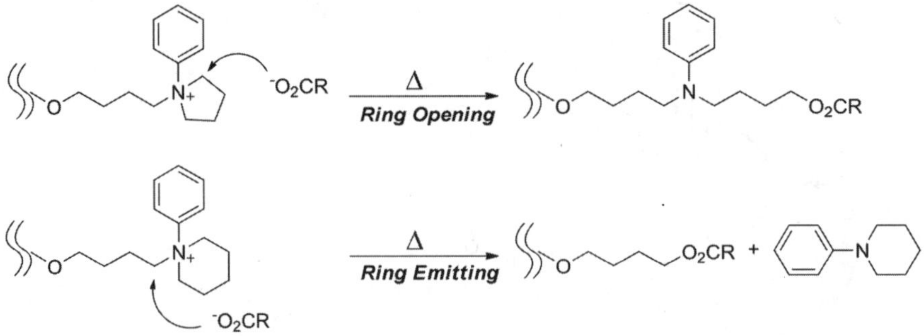

Scheme 1.5 Ring-opening vs. ring-emitting processes for the covalent conversion (Adachi *et al.*, 2006).

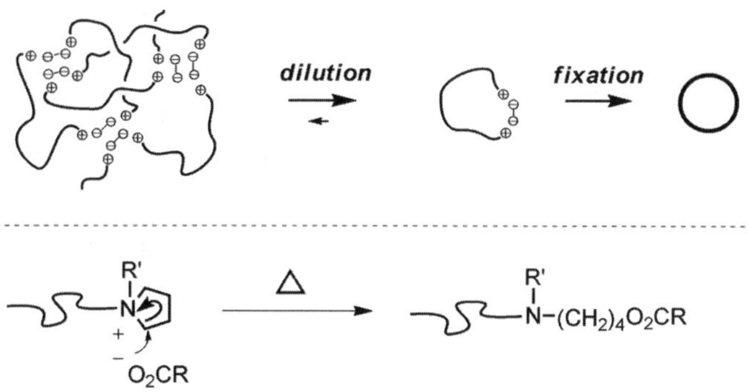

Scheme 1.6 *Electrostatic self-assembly and covalent fixation* (ESA-CF) process by the selective ring-opening reaction of a five-membered cyclic ammonium salt group (Oike *et al.*, 2000).

a counteranion. And by dilution, the self-assembly tends to restructure into that having the smallest number of polymer components by keeping the balance of the charges. Thus, when a linear telechelic precursor having *N*-phenylpyrrolidinium salt groups accompanying a biphenyl dicarboxylate counteranion was heated at the concentration of 1 g/L, the selective ring-opening reaction of the cyclic ammonium groups took place, to produce ring polymer products in high yields (Scheme 1.6) (Oike *et al.*, 2000).

By employing telechelic precursors having such functional groups as hydroxyl and allyl groups at the designated, typically a center, position by using the relevant initiator, or by introducing a functional dicarboxylate counteranion, a variety of ring polymers possessing the prescribed functional groups at the designated positions, that is, *kyklo*-telechelics and cyclic macromonomer, have also been prepared (Oike *et al.*, 2001b). These were further utilized to construct more complex polymer topologies containing ring polymer units (Oike *et al.*, 2001c).

1.3 Functional Ring Polymers and Topology Effects Thereby

Ring polymers having programmed functionalities with guaranteed purity have opened up new research opportunities in the frontier of synthetic polymer chemistry and of designing polymer materials based on their topologies. Recent achievements by our collaborative studies include; (1) the efficient preparation of polymer catenanes using a ring polymer precursor having a hydrogen-bonding unit (Ishikawa *et al.*, 2010), (2) the single-molecule fluorescence spectroscopy study on the polymer diffusion dynamics using a ring polymer having a chromophore unit (Habuchi *et al.*, 2010) and (3) the crystallization dynamics using a *defect-free* ring polymer (Tezuka *et al.*, 2008).

1.3.1 Polymer Catenanes Using a Ring Polymer Precursor Having an H-Bonding Unit

Polymer catenanes are among the most challenging targets in synthetic polymer chemistry, which has been enlightened by recent breakthroughs in supramolecular chemistry (Lehn, 1995). The effective synthesis should address the ambivalent requirements, namely

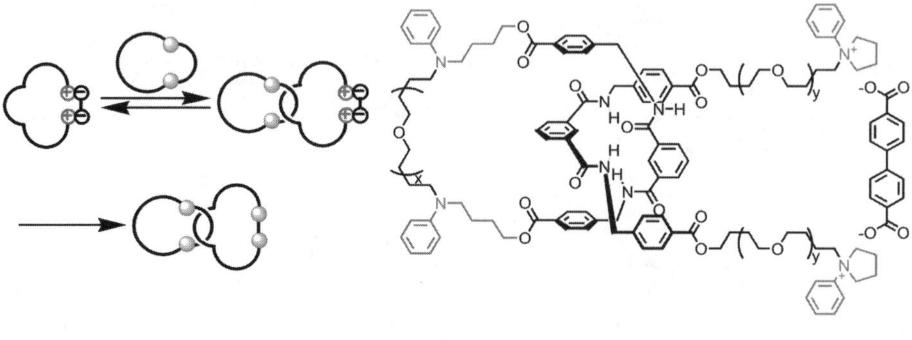

Scheme 1.7 Synthesis of a polymer hetero[2]catenane through the cooperative electrostatic/hydrogen-bonding self-assembly and covalent fixation (Ishikawa *et al.*, 2010).

the association of polymer chains favored in concentrated solution and the unimolecular end-to-end polymer cyclization promoted under dilution. It is notable, moreover, that polymer [2]catenane consisting of the two catenane ring polymer units and a corresponding large single cyclic polymer of the same chain length are regarded as a pair of topological isomers having the identical molar mass and chemical structure units. Consequently, they are not simply distinguishable by any of the NMR or MALDI-TOF MS techniques.

We have achieved effective synthesis of polymer catenanes using a ring polymer precursor having a programmed hydrogen-bonding unit, by taking advantage of the cooperative hydrogen-bonding/electrostatic interaction of polymer precursors (Scheme 1.7) (Ishikawa *et al.*, 2010). In this process, we have first prepared a ring poly(THF) having a hydrogen-bonding, isophthaloyl benzylic amide group. Next, a linear telechelic poly(THF) having cyclic ammonium salt groups and having an isophthaloyl benzylic amide group at the center position was mixed to form the hydrogen-bonding self-assembly. The subsequent ESA-CF process under dilution could promote the polymer cyclization by keeping the hydrogen-bonding interaction. Polymer catenane products could be isolated as an acetone-insoluble fraction due to the distinctive solubility from the corresponding ring polymers. The following MALDI-TOF MS analysis confirmed the formation of polymer *hetero*[2]catenane having two distinguishable ring polymer units together with polymer *homo*[2]catenane concurrently produced during the polymer cyclization process. The isolated yield of polymer catenanes reached as high as 7%, which is significantly higher than all previous reports, indicating that the polymer cyclization with the hydrogen-bonding interaction is indeed effective for the polymer catenane formation.

1.3.2 Single-Molecule Spectroscopy Using a Ring Polymer Having a Chromophore Unit

The dynamics of *linear* or *branched* polymers is known to follow the reptation mechanism, in which the chain ends of polymer molecules play a critically important role. On the contrary, *ring* polymers are unique in the absence of the chain ends and therefore their non-reptation dynamics has been a fundamental issue in both theory and practice in polymer materials science.

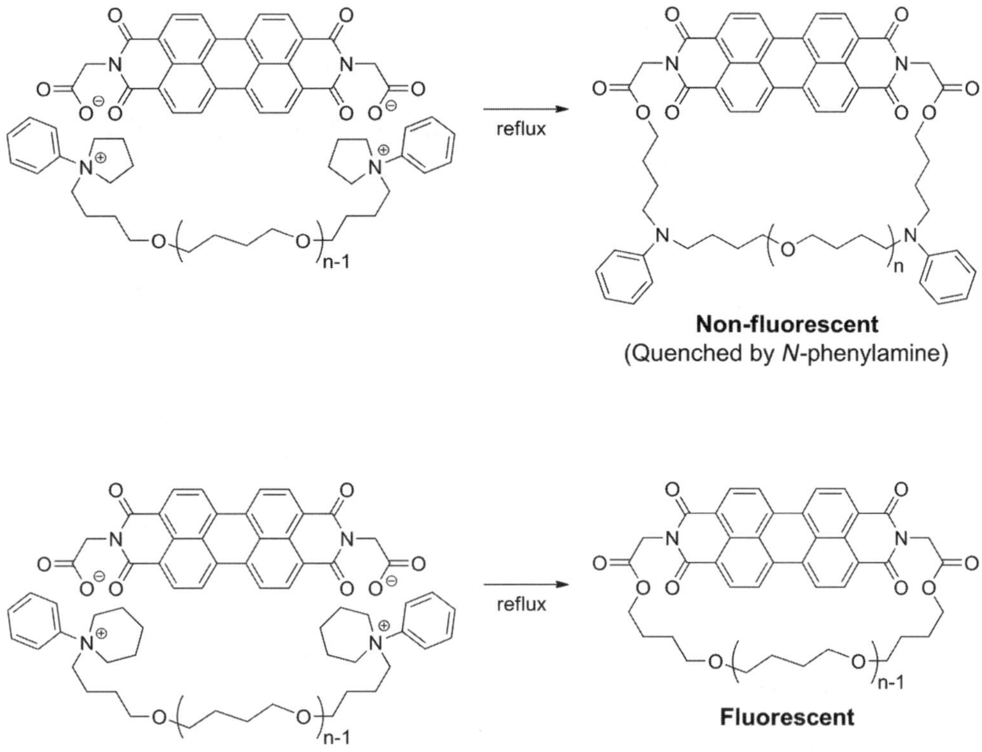

Scheme 1.8 Ring polymer formation through the covalent fixation of *N*-phenyl pyrrolidinium (top) and *N*-phenyl piperidinium end groups (bottom) via ring-opening and ring-emitting reactions, respectively (Habuchi *et al.*, 2010).

We have disclosed the multiple-mode diffusion process by ring polymers in contrast to the single-mode process by linear counterparts, through the single-molecule fluorescence spectroscopy using a newly designed ring polymer having a perylene chromophore (Habuchi *et al.*, 2010). By employing the ESA-CF process, we have prepared a linear telechelic poly(THF) having *N*-phenyl piperidinium (6-membered ring) salt groups accompanying a perylene-dicarboxylate counteranion and subsequently a ring polymer having a perylene unit and having a simple ester linkage by the elimination of *N*-phenylpiperidine as shown in Section 1.2.2. (Scheme 1.8) (Adachi *et al.*, 2006). It is notable that a relevant ring polymer was obtainable by using an alternative telechelics having *N*-phenyl pyrrolidinium (five-membered ring) groups. However, the latter was inapplicable because the photoquenching took place by the *N*-phenylamine group introduced by the ring-opening reaction of the pyrrolidinium salt group.

The single-molecule imaging experiment has been performed with these designed ring poly(THF)s as well as their linear analogs, mixed in a matrix of nonlabeled linear poly(THF) to reveal the real-time motion of a single polymer molecule. Figure 1.1 shows time-course single-molecule fluorescence images of perylene-labeled cyclic and linear poly(THF)s in unlabeled linear poly(THF) matrices. The multimode diffusion mechanism has subsequently been

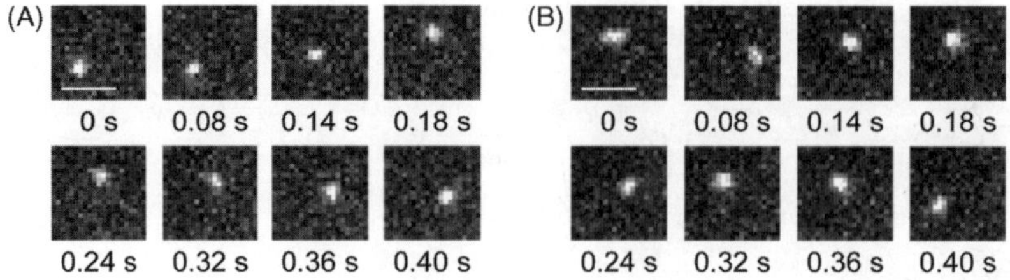

Figure 1.1 Time-course single-molecule fluorescence images of perylene-labeled cyclic (A) and linear (B) poly(THF)s mixed with unlabeled linear poly(THF)s (Habuchi *et al.*, 2010). Scale bar = 2 μm.

disclosed, for the first time, exclusively for ring polymers, corresponding to the diffusion either suppressed by the threading of linear polymers in the matrix, or unrestricted by the absence of the threading.

1.3.3 Crystallization Dynamics Using a Defect-Free *Ring Polymer*

Polymer crystallization involves the diffusion of polymer molecules, and thus should be affected by their topologies. In order to elucidate rigorously this topology effect, we have employed a *defect-free* ring poly(THF) together with its linear analog (Tezuka *et al.*, 2008). First, a poly(THF) having allyl end groups with a controlled molecular weight has been prepared. The following metathesis polymer cyclization (MPC) in the presence of a Grubbs catalyst under dilution produced a ring poly(THF) having a butenoxy linking group. The subsequent hydrogenation afforded a ring poly(THF) consisting exclusively of oxytetramethylene units, that is, a *defect-free* ring poly(THF) (Scheme 1.9). The isothermal crystallization experiments have been performed together with the linear counterpart having ethoxy end groups, to show a noticeable topology effect in the suppressed crystallization kinetics as well as

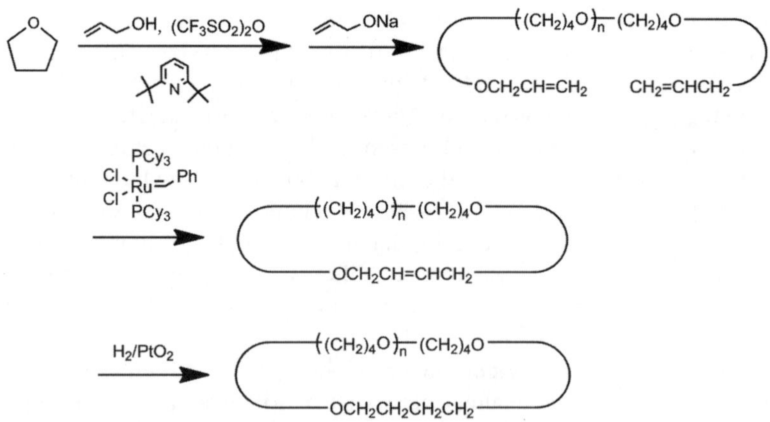

Scheme 1.9 Formation of a *defect-free* ring polymer through the metathesis polymer cyclization (MPC) and subsequent hydrogenation (Tezuka *et al.*, 2008).

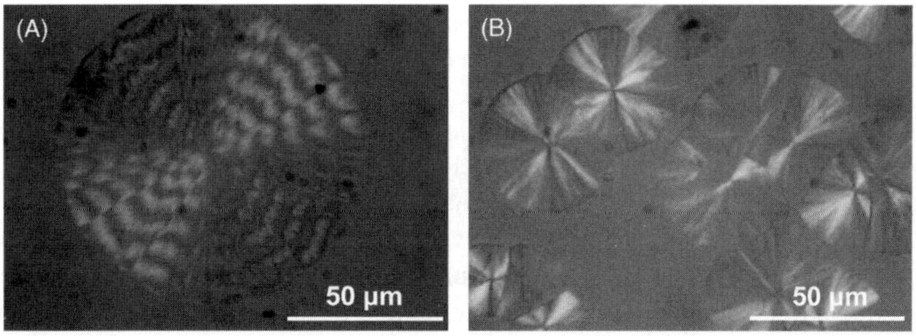

Figure 1.2 Optical micrographs of spherulites of a *defect-free* cyclic poly(THF) (A) and linear poly(THF) having ethoxy termini (B) (Tezuka *et al.*, 2008).

in the unusual spherulite morphology for ring polymers. Figure 1.2 shows optical micrographs of spherulites of the *defect-free* cyclic poly(THF) and linear counterpart having ethoxy termini. A banded structure with concentric ring pitch of approximately 7 μm is observed only for the cyclic polymer.

1.4 New Developments in the Construction of Multicyclic Polymer Topologies

The topology of multiple cyclic constructions is classified into three basic groups, namely *fused*, *spiro*, and *bridged* types, and their combinations. As for the dicyclic topology, three basic constructions are designated as θ- (*fused*), 8- (*spiro*), and manacle- (or two-way paddle) shape (*bridged*), respectively (Scheme 1.10). And the tricyclic topology contains fifteen constructions, including four *fused* types, namely α, β, γ, and δ graphs, respectively, *spiro* types such as a trefoil and a tandem triple ring, and *bridged*-types, such as a three-way paddle-shaped rings. Notably in the tetracyclic topology, a triply fused, $K_{3,3}$ construction, known as a nonplanar graph in topological geometry, is included (Tezuka, 2008; Tezuka and Oike, 2001).

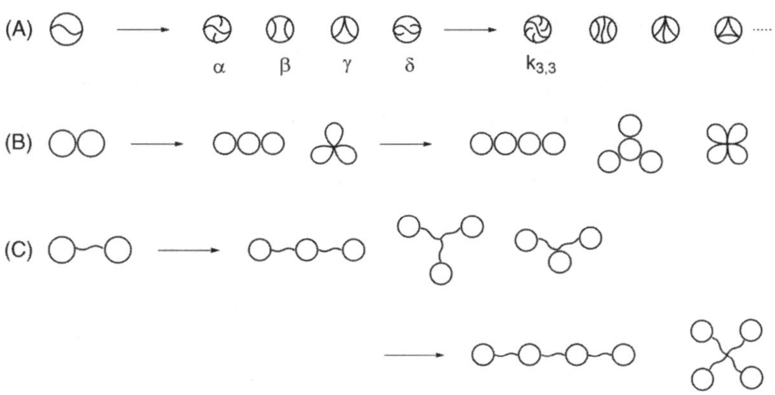

Scheme 1.10 *Fused* (A), *spiro* (B), and *bridged* (C) multicyclic constructions (Tezuka, 2008; Tezuka and Oike, 2001).

As for the separation and the subsequent characterization of polymers of complex topologies, a variety of powerful means has now been developed. In particular, a ring polymer and its linear analog, often a synthetic precursor of the former, as well as a pair of dicyclic polymeric topological isomers, that is, θ- and manacle-shaped polymers, could be separated and eventually fractionated by means of a reversed-phase liquid chromatography (RPC) (Tezuka *et al.*, 2007). Together with the size exclusion chromatography (SEC) to distinguish their hydrodynamic volumes, a set of polymers having different topologies have now been more convincingly assigned. By the RPC analysis under critical conditions, in particular, Takano *et al.* have shown that ring polymer products, obtained through the conventional bimolecular end-to-end polymer cyclization process, still contain a noticeable amount of linear polymers as contaminant (Takano *et al.*, 2007).

In addition, the MALDI-TOF MS technique has now become a powerful tool for the structural characterization of topologically complex polymers by showing the absolute molar mass of up to 10^4 for a polymer component of a single definite DP. Together with a traditional NMR method, the total chemical structure of polymer products including the precise linking structure of ring polymers as well as the end-group structure of linear ones, can be fully determined to prove the topology of polymers.

Upon these developments, topologically significant polymers have now become an attractive challenge in synthetic polymer chemistry to extend the frontier of the field. In this respect, the *electrostatic self-assembly and covalent fixation* (ESA-CF) process (Section 1.2) provides a unique opportunity to construct various complex polymer topologies, either by the direct application or by the combination with the effective polymer linking process such as a metathesis condensation and/or a click coupling processes.

1.4.1 Fused *Multicyclic Polymers*

A three-armed star telechelic precursor having cyclic ammonium salt groups has been prepared to form a self-assembly with one unit of tricarboxylate counteranion under dilution in order to keep the balance of the charges. The subsequent covalent fixation by heating produced a *fused* dicyclic, that is, θ-shaped, polymer effectively (Tezuka *et al.*, 2003a).

The combination of the linear bifunctional telechelic precursor with a tricarboxylate counteranion allows production of a self-assembly under dilution, consisting of the three polymer units and the two counteranions. The subsequent covalent conversion by the ring-opening leads to the mixture of manacle-shaped and θ-shaped polymers (Scheme 1.11) (Oike *et al.*, 2000; Tezuka *et al.*, 2003b). The two polymer products are regarded as topologically distinctive constitutional isomers, and could be resolved and fractionated by means of a preparative RPC technique near the critical condition. The two polymeric isomers are assignable by their different hydrodynamic volumes measured by SEC, as the manacle isomer is presumed to be larger than the θ-counterpart. The assignment was further confirmed through the chemical conversion by the use of a pair of topological isomers including a cleavable unit in one segment component (Tezuka *et al.*, 2007). The relevant pair of manacle- and θ-shaped polymeric isomers were formed also from a self-assembly consisting of two units of a three-armed star telechelic precursors and three units of dicarboxylate counteranions (Tezuka *et al.*, 2004), and from an H-shaped telechelic polymer precursor having four allyl groups via the double-metathesis condensation (Scheme 1.12) (Tezuka and Ohashi, 2005).

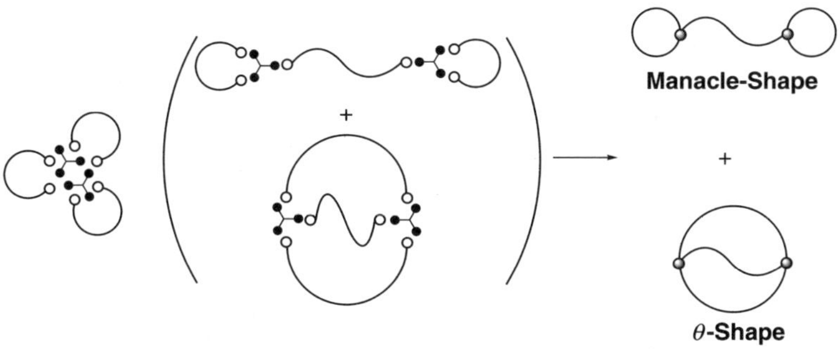

Scheme 1.11 Synthesis of topological polymeric isomers having manacle- and θ-shaped constructions by the ESA-CF process (Oike *et al.*, 2000; Tezuka *et al.*, 2003b).

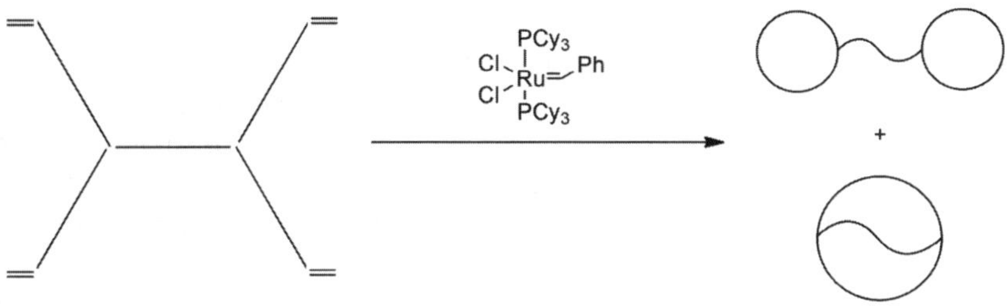

Scheme 1.12 Synthesis of manacle- and θ-shaped topological polymeric isomers through the double metathesis of an H-shaped prepolymer (Tezuka and Ohashi, 2005).

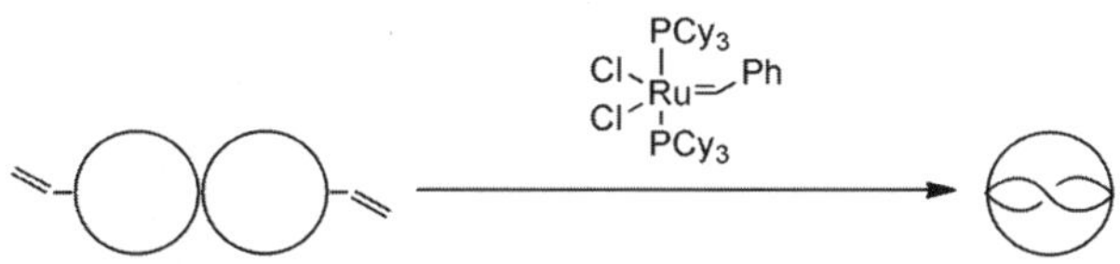

Scheme 1.13 Synthesis of a δ-graph polymer from an 8-shaped prepolymer having two allyl groups (Tezuka and Fujiyama, 2005).

Moreover, a dicyclic 8-shaped polymer precursor having two allyl groups at the opposite positions of the two ring units has been produced through the ESA-CF process by using a linear telechelic precursor having an allyl group at the center position and having cyclic ammonium salt end groups. The subsequent intramolecular metathesis condensation could produce a polymer product having a doubly fused tricyclic topology, that is, δ-graph (Scheme 1.13) (Tezuka and Fujiyama, 2005).

1.4.2 Spiro *Multicyclic Polymers*

As shown in the section above, a self-assembly composed of the two units of a linear telechelic precursor having cyclic ammonium salt groups and one unit of a tetracarboxylate counteranion

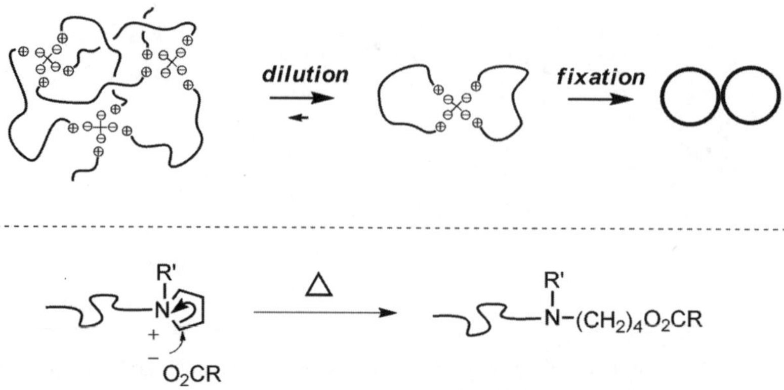

Scheme 1.14 Synthesis of an 8-shaped polymer by the ESA-CF process (Oike *et al.*, 2000).

is formed under dilution by keeping the balance of the charges and with including the smallest number of the polymer components. The subsequent covalent conversion by the ring-opening reaction by heating produced a *spiro*-dicyclic, 8-shaped polymer effectively (Scheme 1.14) (Oike *et al.*, 2000). Alternatively, a four-armed star telechelic precursor having allyl end groups has been prepared and subjected to the MPC process to form an 8-shaped polymer (Hayashi *et al.*, 2008).

Moreover, the ESA-CF and the metathesis condensation process has been combined to produce 8-shaped polymers by three different methods (Scheme 1.15) (Tezuka *et al.*, 2003c). These are: (1) a bimolecular metathesis coupling reaction of the two ring polymers having an allyl group, (2) an intramolecular metathesis condensation of a twin-tailed tadpole polymer

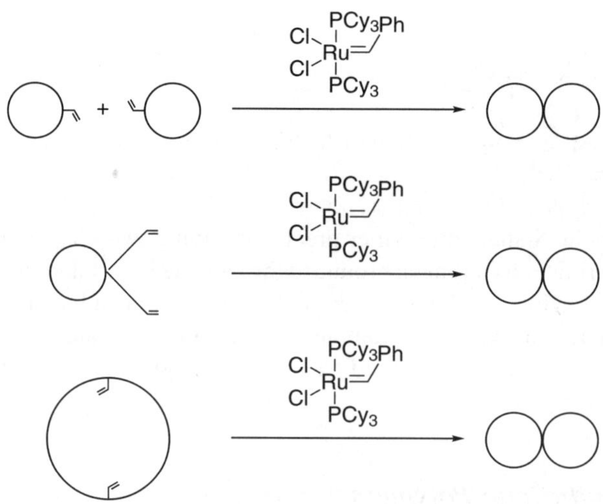

Scheme 1.15 Synthesis of 8-shaped polymers through the metathesis condensation process (Tezuka *et al.*, 2003c).

precursor having allyl end groups, and (3) an intramolecular metathesis condensation of a ring polymer precursor having two allyl groups at the opposite positions.

1.4.3 Bridged *Multicyclic Polymers*

As discussed in the preceding sections, a *bridged*-dicyclic, that is, manacle-shaped (or two-way paddle-shaped), polymer was obtained together with the θ-shaped polymeric isomer by means of the ESA-CF process (Scheme 1.11) (Oike *et al.*, 2000). On the other hand, the selective one-step synthesis of the manacle-shaped polymer has been circumvented due to the restricted geometrical symmetry of this topology. Therefore, an alternative process has been introduced by the combination of the ESA-CF process with a "click" chemistry (Iha *et al.*, 2009; Fournier *et al.*, 2007). By the ESA-CF process, a ring polymer precursors having an alkyne group has been prepared, and subjected to the "click" reaction with either linear or three-armed star telechelic precursors having azide groups. A *bridged*-dicyclic, manacle-shaped (or two-way paddle-shaped) and a *bridged*-tricyclic, three-way paddle-shaped polymers, respectively, were thus produced. Moreover, *topological block copolymers* consisting of ring and linear/branched polymer segments were obtained by using a bifunctional ring polymer precursor having two alkyne groups at the opposite positions (Sugai *et al.*, 2010).

1.5 Conclusions and Perspectives

This review highlighted recent developments in *topological polymer chemistry*. Through intensive research efforts by an increasing number of research groups around the globe, a variety of polymers having unprecedented topologies have now been synthesized and convincingly characterized. Ongoing synthetic challenges will certainly extend the frontier of polymer science and supramolecular science. The future studies will certainly be promoted also by the progress of modern topological geometry and graph theory. Moreover, conceptually novel design of polymer materials will be realized based on the precisely controlled polymer topologies, currently restricted mostly in conventional linear and branched ones.

References

Adachi, K. and Tezuka, Y. (2007) Topological polymer chemistry: Designing unusual macromolecular architectures. *Kobunshi Ronbunshu*, **64**, 709–715.

Adachi, K. and Tezuka, Y. (2009) Topological polymer chemistry in pursuit of elusive polymer ring constructions. *Journal of Synthetic Organic Chemistry, Japan*, **67**, 1136–1143.

Adachi, K., Takasugi, H., and Tezuka, Y. (2006) Telechelics having unstrained cyclic ammonium salt groups for electrostatic polymer self-assembly and ring-emitting covalent fixation. *Macromolecules*, **39**, 5585–5588.

Adachi, K., Honda, S., Hayashi, S., and Tezuka, Y. (2008) ATRP-RCM synthesis of cyclic diblock copolymers. *Macromolecules*, **41**, 7898–7903.

Bielawski, C.W. and Grubbs, R.H. (2007) Living ring-opening metathesis polymerization. *Progress in Polymer Science*, **32**, 1–29.

Bielawski, C.W., Benitez, D., and Grubbs, R.H. (2002) An "endless" route to cyclic polymers. *Science*, **297**, 2041–2044.

Bielawski, C.W., Benitez, D., and Grubbs, R.H. (2003) Synthesis of cyclic polybutadiene via ring-opening metathesis polymerization: The importance of removing trace linear contaminants. *Journal of the American Chemical Society*, **125**, 8424–8425.

Chung, M.-K., White, P.S., Lee, S.J., and Gagne, M.R. (2009) Synthesis of interlocked 56-membered rings by dynamic self-templating. *Angewandte Chemie International Edition*, **48**, 8683–8686.

Culkin, D.A., Jeong, W., Csihony, S. *et al.* (2007) Zwitterionic polymerization of lactide to cyclic poly(lactide) by using N-heterocyclic carbene organocatalysts. *Angewandte Chemie International Edition*, **46**, 2627–2630.

Fournier, D., Hoogenboom, R., and Schubert, U.S. (2007) Clicking polymers: A straightforward approach to novel macromolecular architectures. *Chemical Society Reviews*, **36**, 1369–1380.

Guidry, E.N., Cantrill, S.J., Stoddart, J.F., and Grubbs, R.H. (2005) Magic ring catenation by olefin metathesis. *Organic Letters*, **7**, 2129–2132.

Habuchi, S., Satoh, N., Yamamoto, T. *et al.* (2010) Multimode diffusion of ring polymer molecules revealed by a single-molecule study. *Angewandte Chemie International Edition*, **49**, 1418–1421.

Hayashi, S., Adachi, K., and Tezuka, Y. (2007) An efficient route to cyclic polymers by ATRP-RCM process. *Chemistry Letters*, **36**, 982–983.

Hayashi, S., Adachi, K., and Tezuka, Y. (2008) ATRP-RCM synthesis of 8-shaped poly(methyl acrylate) using a 4-armed star telechelics. *Polymer Journal*, **40**, 572–576.

Hoskins, J.N. and Grayson, S.M. (2009) Synthesis and degradation behavior of cyclic poly(ε-caprolactone). *Macromolecules*, **42**, 6406–6413.

Iha, R.K., Wooley, K.L., Nyström, A.M. *et al.* (2009) Applications of orthogonal "click" chemistries in the synthesis of functional soft materials. *Chemical Reviews*, **109**, 5620–5686.

Ishikawa, K., Yamamoto, T., Asakawa, M., and Tezuka, Y. (2010) Effective synthesis of polymer catenanes by cooperative electrostatic/hydrogen-bonding self-assembly and covalent fixation. *Macromolecules*, **42**, 168–176.

Jeong, W., Shin, E.J., Culkin, D.A. *et al.* (2009) Zwitterionic polymerization: A kinetic strategy for the controlled synthesis of cyclic polylactide. *Journal of the American Chemical Society*, **131**, 4884–4891.

Kricheldorf, H.R. (2010) Cyclic polymers: Synthetic strategies and physical properties. *Journal of Polymer Science: Part A: Polymer Chemistry*, **48**, 251–284.

Kubo, M., Nishigawa, T., Uno, T. *et al.* (2003) Cyclic polyelectrolyte: synthesis of cyclic poly(acrylic acid) and cyclic potassium polyacrylate. *Macromolecules*, **36**, 9264–9266.

Kudo, H., Makino, S., Kameyama, A., and Nishikubo, T. (2005) Synthesis of cyclic polymers: ring-expansion reaction of cyclic *S*-dithioester with thiiranes. *Macromolecules*, **38**, 5964–5969.

Kudo, H., Sato, M., Wakai, R. *et al.* (2008) A novel approach to cyclic polysulfides via the controlled ring-expansion polymerization of cyclic thiourethane with thiiranes. *Macromolecules*, **41**, 521–523.

Laurent, B.A. and Grayson, S.M. (2006) An efficient route to well-defined macrocyclic polymers via "click" cyclization. *Journal of the American Chemical Society*, **128**, 4238–4239.

Laurent, B.A. and Grayson, S.M. (2009) Synthetic approaches for the preparation of cyclic polymers. *Chemical Society Reviews*, **38**, 2202–2213.

Lehn, J.-M. (1995) *Supramolecular Chemistry: Concepts and Perspectives*, Wiley-VCH, Weinheim.

Li, H., Debuigne, A., Jérome, R., and Lecomte, P. (2006) Synthesis of macrocyclic poly(ε-caprolactone) by intramolecular cross-linking of unsaturated end groups of chains precyclic by the initiation. *Angewandte Chemie International Edition*, **45**, 2264–2267.

Oike, H., Imaizumi, H., Mouri, T. *et al.* (2000) Designing unusual polymer topologies by electrostatic self-assembly and covalent fixation. *Journal of the American Chemical Society*, **122**, 9592–9599.

Oike, H., Hamada, M., Eguchi, S. *et al.* (2001a) novel synthesis of single- and double-cyclic polystyrenes by electrostatic self-assembly and covalent fixation with telechelics having cyclic ammonium salt groups. *Macromolecules*, **34**, 2776–2782.

Oike, H., Kobayashi, S., Mouri, T., and Tezuka, Y. (2001b) *Kyklo*-telechelics: Tailored synthesis of cyclic poly(tetrahydrofuran)s having two functional groups at opposite positions. *Macromolecules*, **34**, 2742–2744.

Oike, H., Mouri, T., and Tezuka, Y. (2001c) A cyclic macromonomer designed for a novel polymer network architecture having both covalent and physical linkages. *Macromolecules*, **34**, 6229–6234.

Qiu, X.P., Tanaka, F., and Winnik, F.M. (2007) Temperature-induced phase transition of well-defined cyclic poly(*N*-isopropylacrylamide)s in aqueous solution. *Macromolecules*, **40**, 7069–7071.

Schappacher, M. and Deffieux, A. (2008a) Synthesis of macrocyclic copolymer brushes and their self-assembly into supramolecular tubes. *Science*, **319**, 1512–1515.

Schappacher, M. and Deffieux, A. (2008b) Atomic force microscopy imaging and dilute solution properties of cyclic and linear polystyrene combs. *Journal of the American Chemical Society*, **130**, 14684–14689.

Seeman, N.C. (2003) DNA in a material world. *Nature*, **421**, 427–431.

Shea, K.J., Lee, S.Y., and Busch, B.B. (1998) A New strategy for the synthesis of macrocycles. The polyhomologation of boracyclanes. *The Journal of Organic Chemistry*, **63**, 5746–5747.

Sugai, N., Heguri, H., Ohta, K. *et al.* (2010) Effective click construction of *bridged-* and *spiro-*multicyclic polymer topologies with tailored cyclic prepolymers (*kyklo-*telechelics). *Journal of the American Chemical Society*, **132**, 14790–14802.

Takano, A., Kushida, Y., Aoki, K. *et al.* (2007) HPLC characterization of cyclization reaction product obtained by end-to-end ring closure reaction of a telechelic polystyrene. *Macromolecules*, **40**, 679–681.

Tezuka, Y. (2003) Topological polymer chemistry by electrostatic self-assembly. *Journal of Polymer Science Part A: Polymer Chemistry*, **41**, 2905–2917.

Tezuka, Y. (2005) Topological polymer chemistry by dynamic selection from electrostatic polymer self-assembly. *Chemical Record*, **5**, 17–26.

Tezuka, Y. (2008) Topological polymer chemistry: An insight with Poincaré into nonlinear macromolecular constructions. *Kobunshi*, **57**, 81–85.

Tezuka, Y. and Fujiyama, K. (2005) Construction of polymeric δ-graph: A doubly fused tricyclic topology. *Journal of the American Chemical Society*, **127**, 6266–6270.

Tezuka, Y. and Komiya, R. (2002) Metathesis polymer cyclization with telechelic poly(THF) having allyl groups. *Macromolecules*, **35**, 8667–8669.

Tezuka, Y. and Ohashi, F. (2005) Synthesis of polymeric topological isomers through double metathesis condensation with H-shaped telechelic precursors. *Macromolecular Rapid Communications*, **26**, 608–612.

Tezuka, Y. and Oike, H. (2001) Topological polymer chemistry: Systematic classification of nonlinear polymer topologies. *Journal of the American Chemical Society*, **123**, 11570–11576.

Tezuka, Y. and Oike, H. (2002) Topological polymer chemistry. *Progress in Polymer Science*, **27**, 1069–1122.

Tezuka, Y., Iwase, T., and Shiomi, T. (1997) Tailored synthesis of star and network poly(dimethylsiloxane)s through electrostatic self-assembly and subsequent covalent fixation of telechelics having cyclic onium salt groups. *Macromolecules*, **30**, 5220–5226.

Tezuka, Y., Mori, K., and Oike, H. (2002) Efficient synthesis of cyclic poly(oxyethylene) by electrostatic self-assembly and covalent fixation with telechelic precursor having cyclic ammonium salt groups. *Macromolecules*, **35**, 5707–5711.

Tezuka, Y., Tsuchitani, A., Yoshioka, Y., and Oike, H. (2003a) Synthesis of θ-shaped poly(THF) by electrostatic self-assembly and covalent fixation with three-armed star telechelics having cyclic ammonium salt groups. *Macromolecules*, **36**, 65–70.

Tezuka, Y., Tsuchitani, A., and Oike, H. (2003b) Synthesis of polymeric topological isomers through electrostatic self-assembly and covalent fixation with telechelic poly(THF) having cyclic ammonium salt groups. *Polymer International*, **52**, 1579–1583.

Tezuka, Y., Komiya, R., and Washizuka, M. (2003c) Designing 8-shaped polymer topology by metathesis condensation with cyclic poly(THF) precursors having allyl groups. *Macromolecules*, **36**, 12–17.

Tezuka, Y., Tsuchitani, A., and Oike, H. (2004) Synthesis of polymeric topological isomers through electrostatic self-assembly and covalent fixation with star telechelic precursors. *Macromolecular Rapid Communications*, **25**, 1531–1535.

Tezuka, Y., Takahashi, N., Satoh, T., and Adachi, K. (2007) Synthesis of polymeric topological isomers having θ- and manacle-constructions with olefinic groups at designated positions. *Macromolecules*, **40**, 7910–7918.

Tezuka, Y., Ohtsuka, T., Adachi, K. *et al.* (2008) A defect-free ring polymer: Size-controlled cyclic poly(tetrahydrofuran) consisting exclusively of the monomer unit. *Macromolecular Rapid Communications*, **29**, 1237–1241.

Xia, Y., Boydston, A.J., Gorodetskaya, I.A. *et al.* (2009) Ring-expansion metathesis polymerization: catalyst-dependent polymerization profiles. *Journal of the American Chemical Society*, **131**, 2670–2677.

Ye, J., Xu, J., Hu, J. *et al.* (2008) Comparative study of temperature-induced association of cyclic and linear poly(*N*-isopropylacrylamide) chains in dilute solutions by laser light scattering and stopped-flow temperature jump. *Macromolecules*, **41**, 4416–4422.

2

Ultrarapid Approaches to Mild Macromolecular Conjugation

Andrew J. Inglis and Christopher Barner-Kowollik
Preparative Macromolecular Chemistry, Karlsruhe Institute of Technology (KIT), Engesserstraße 18, Geb. 11.23, 76128 Karlsruhe, Germany

2.1 Introduction

The evolution of the synthetic macromolecular sciences has largely been driven by the increasing need for purposefully designed and precisely engineered materials for use in a wide variety of "high-demand" applications in fields ranging from biomedicine to the nanosciences. In pursuit of meeting such demands, the two major (and fundamental) issues that lie at the core of modern polymer research are controlling macromolecular architecture and chemical functionality. To this end, the relatively recent developments of controlled radical polymerization (CRP) technologies and highly orthogonal and efficient conjugation chemistries (many of which have been described as *click* chemistry) that may be thereafter applied have given us, as chemists, the ability to control these properties with unprecedented precision.

To set matters in a more concrete perspective, it is interesting (and highly instructive) to view the ultimate goal of synthetic polymer science as striving to emulate the exquisite control in architecture and function that is displayed in such biomacromolecules as proteins. Such entities have a very specific function that is directly correlated to these properties. Bearing this in mind, the controlled self-assembly of synthetic polymers into such higher order structures as micelles may be viewed as an initial step along the very long path ahead towards synthetically replicating the complexity and precision of proteins.

Although the use of such processes as atom-transfer radical polymerization (ATRP) (Ouchi *et al.*, 2009; Braunecker and Matyjaszewski, 2007) and reversible addition-fragmentation chain transfer (RAFT) polymerization (Moad *et al.*, 2009; Moad *et al.*, 2008) have been exploited in and of themselves in producing well-defined architectures; it has only been with

Complex Macromolecular Architectures: Synthesis, Characterization, and Self-Assembly, First Edition.
Edited by Nikos Hadjichristidis, Akira Hirao, Yasuyuki Tezuka and Filip Du Prez.

the introduction of the *click* chemistry philosophy (Kolb *et al.*, 2001; Barner-Kowollik and Inglis, 2009; Binder and Sachsenhofer, 2008; Fournier *et al.*, 2007; Lutz, 2007) and the development of the tools with which it may be carried out that we have seen a tremendous burst of creativity in the design and synthesis of functional materials. As such, there has been no shortage of reviews dedicated to this very subject matter. To highlight just a few key examples, the use of highly efficient chemical transformations within the general context of macromolecular engineering has been discussed in the excellent review of Hawker and Wooley (2005) and the perspective article of Sumerlin and Vogt (2010). Meanwhile, Klok and colleagues have specifically focused upon the postpolymerization modification of polymers constituted from monomers bearing chemoselective handles (Gauthier *et al.*, 2009).

The greatest utility that *click* chemistry has brought to macromolecular chemistry is allowing one the freedom to choose from virtually any synthetic technique to generate functional, polymeric intermediates. We now start to view our synthetic polymers as building blocks, as smaller parts of an ultimately larger structure achievable through conjugation.

The concept of macromolecular conjugation may be defined by the following:

- the covalent addition of small molecules to form a larger structure, that is, step-growth polymerization in an arithmetic fashion;
- dendrimer synthesis: covalent addition of small molecules in a geometric fashion;
- functionalization of a polymer with a smaller molecule through a covalent linkage either to a polymer end group or to multiple positions along the polymer backbone;
- polymer–polymer conjugation: the direct coupling of two (or more) polymers or indirect coupling via a small multifunctional linker; and
- conjugation of a polymer to a surface.

It is readily observable upon inspection of the literature that there are a number of tools available that may be used to carry out all the above forms of macromolecular conjugation. However, now that there is an increasing interest in combining synthetic polymer chemistry with biology (and other chemically sensitive systems) within the context of conjugation, there is a concomitant increasing need for highly orthogonal, rapid and mild conjugation chemistries that can be efficiently applied under the conditions generally required for such systems: ambient temperature, absence of toxic catalysts and considerably dilute. Awareness of this burgeoning interest is especially reflected in two recent publications. Lutz (Lutz, 2008) recently highlighted various copper-free azide–alkyne cycloadditions within the context of macromolecular ligation chemistry. In a much wider scope, Schubert and colleagues recently presented a collection of copper-free *click* technologies that extend beyond azide–alkyne chemistry (Becer *et al.*, 2009). Nevertheless, the focus of both of these publications is strictly upon the absence of copper from the discussed reactions and not necessarily upon other conditions.

To this end, the principal aim of this chapter is to highlight those conjugation chemistries which either have been (or have the potential to be) highly useful in the field of mild, ultrarapid macromolecular conjugation. That is not to say that every reaction presented herein may be labeled as a *click* reaction for, as will be shown, some examples fail the orthogonality criterion. Not qualifying as *click* chemistry does not, and should not be inferred to mean useless.

Naturally, what is said to constitute an ultrarapid and mild reaction is rather subjective as it can be very system specific. For example, the use of the copper(I) catalyzed azide–alkyne

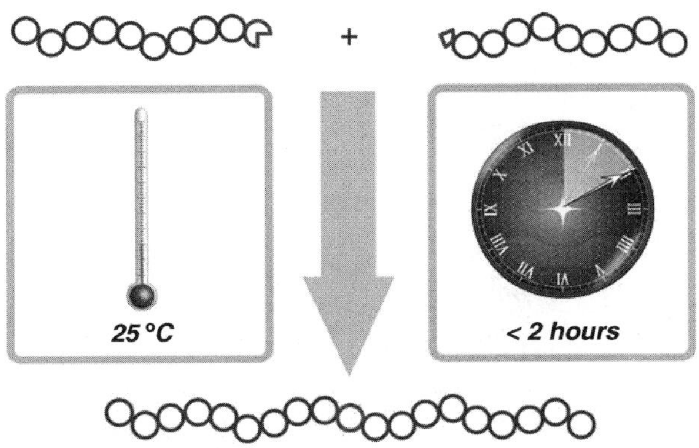

Figure 2.1 Criteria to which the various chemistries discussed are restricted. (Reproduced with permission from A.J. Inglis and C. Barner-Kowollik, "Ultra rapid approaches to mild macromolecular conjugation," *Macromolecular Rapid Communications*, 2010, **31**, 14, 1247–1266. © Wiley-VCH Verlag GmbH & Co. KGaA.)

cycloaddition (CuAAC) may be performed in a couple of hours at room temperature and in benign solvents (such as water), however, the requirement of the cytotoxic copper catalyst precludes its applicability to some biomedical applications, including *in vivo*. Furthermore, an ambient temperature and overnight reaction may be considered sufficiently fast for many applications, such as block copolymer formation, however it may not be suitable for biological labeling. Thus, the application in which the *click* reaction is used dictates what is considered to be fast and mild. Therefore, some guideline must be provided here in order for a critical evaluation of the available technologies to be accessible to a broad readership. For the current purpose, we have included those reactions that may be performed at ambient temperature and be completed within a 2-h time frame (in the absence of transition-metal catalysts), as conceptually depicted in Figure 2.1.

Recently, we (jointly and collaboratively with Prof. Stenzel as well as Dr. Sinnwell) have developed the RAFT-hetero Diels–Alder (HDA) reaction that may be used in place of, or in cooperation with, the CuAAC to achieve a variety of well-defined macromolecular architectures and functional materials. During the continued course of our research in this area, an ultrarapid (<10 min) and ambient temperature variant was designed in 2009. The following will outline the evolution of the RAFT-HDA concept to ultrafast macromolecular conjugation chemistry and critically evaluate the efforts of other research groups in developing alternative chemistries that also meet such specifications. A graphical overview of these chemistries is presented in Figure 2.2.

2.2 RAFT-HDA Chemistry

Diels–Alder reactions have been used in polymer synthesis as a method of producing highly crosslinked structures (Hay *et al.*, 1989; Stork *et al.*, 1999; Watson and Bass, 2000; Nenov *et al.*, 2002) and even as the polymerization mechanism itself (Bailey, 1972; Kamahori *et al.*, 1999). It has only been rather recently that it has been effectively utilized in the modular

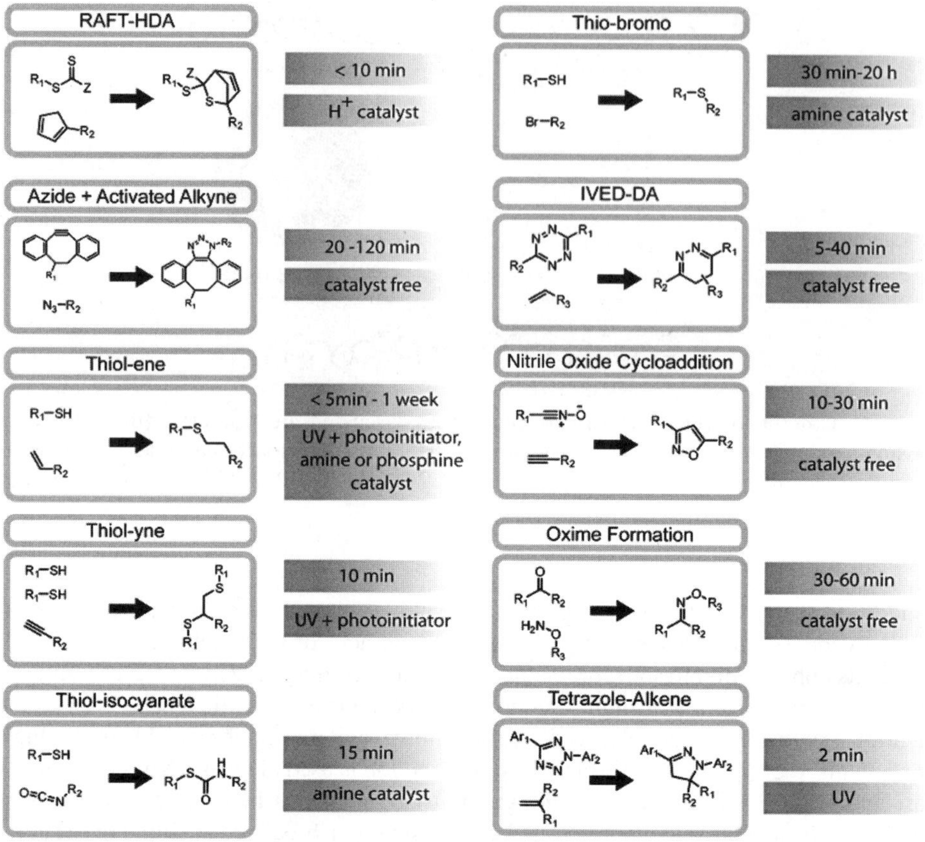

Figure 2.2 Overview of ultrarapid and mild conjugation chemistries. (Reproduced with permission from A.J. Inglis and C. Barner-Kowollik, "Ultra rapid approaches to mild macromolecular conjugation," *Macromolecular Rapid Communications*, 2010, **31**, 14, 1247–1266. © Wiley-VCH Verlag GmbH & Co. KGaA.)

construction of precisely defined polymeric materials. Hizal and Tunca have pioneered the use of anthracene and maleimide functional polymers to generate very well-defined structures ranging from block copolymers (Durmaz *et al.*, 2006a; Durmaz *et al.*, 2007), star polymers (Durmaz *et al.*, 2006b; Dag *et al.*, 2009), graft polymers (Dag *et al.*, 2008) and more complex architectures (Gungor *et al.*, 2009). Although making use of readily available materials and no catalyst, elevated temperatures (>110 °C) and extended reaction times (36–120 h) are severe limitations.

Our introduction to the world of *click* chemistry came in 2006 (jointly and collaboratively with the team of Stenzel) at the University of New South Wales in Sydney, Australia, when we investigated the synthesis of azide/alkyne functional RAFT agents to be used in the modular construction of block copolymers of styrene and vinyl acetate (Quemener *et al.*, 2006). Whilst proving to be highly efficient in the generation of previously unattainable structures, the requirement of first equipping the appropriate controlling agents with complementary *click*

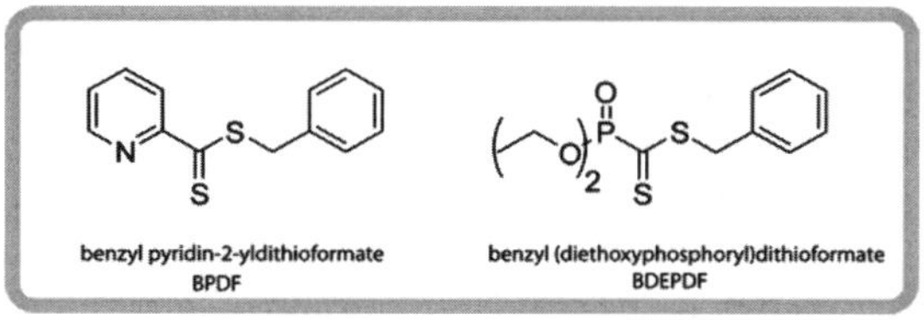

Figure 2.3 Chemical structures of the RAFT agents that may be utilized in RAFT-HDA *click* chemistry. (Reproduced with permission from A.J. Inglis and C. Barner-Kowollik, "Ultra rapid approaches to mild macromolecular conjugation," *Macromolecular Rapid Communications*, 2010, **31**, 14, 1247–1266. © Wiley-VCH Verlag GmbH & Co. KGaA.)

functionalities was somewhat of a drawback. Borne out of a desire to simplify the process, attention was drawn to the thiocarbonyl thio moiety that is inherent to materials formed by the RAFT process in its potential to behave as a dienophile in [4 + 2] cycloadditions.

As will be shown throughout the course of this chapter, a common theme (however, not strictly a requirement) that has emerged within the field is the activation of one of the *click* moieties by an electron-withdrawing group (EWG). In the case of RAFT-HDA chemistry, it was found that dithioesters bearing an EWG on the thiocarbonyl moiety not only may be effectively used as controlling agents in RAFT polymerization, but also serve as highly efficient dienophiles in [4 + 2] cycloadditions. The structures of such dithioesters are presented in Figure 2.3.

As such, initial investigations into this concept saw the highly efficient construction of block copolymers (Sinnwell *et al.*, 2008a), star polymers (Inglis *et al.*, 2008), star-shaped block copolymers (Sinnwell *et al.*, 2008b) and surface modification of polymeric microspheres (Nebhani *et al.*, 2008). In these cases, depending upon the structure of the dienophile, quantitative conversions into the targeted products were achieved at 50 °C within a 2–24 h timeframe. Importantly, and unlike some other reactions that have been touted as macromolecular *click* conjugations, the above systems all proved to be highly efficient (by macromolecular standards) by using 1 : 1 stoichiometry between the respective reactants, a characteristic that becomes highly useful when purification of the *click* products become problematic.

During the initial stages of the investigation, the high convenience of *trans*,*trans*-2,4-hexadien-1-ol (in terms of both commercial availability and chemical accessibility) was put to great utility in achieving the required outcomes. However, soon after a faster and ambient temperature variant of the concept was sought.

A particularly advantageous trait of the Diels–Alder cycloaddition is the generality of its mechanism, meaning that there exists a wide variety of diene–dienophile combinations from which one may make a selection based upon their requirements. Striving for a faster RAFT-HDA reaction, the use of cyclopentadiene was investigated.

2.3 Ultrafast RAFT-HDA Chemistry

Cyclopentadienyl functionality (Cp) in polymers has not received as much attention in the recent literature as, for example, azides and thiols. Nevertheless, the potential for such materials to form tuneable and reversible linkages holds great promise in the design of various uniquely thermoresponsive materials. There is, however, an inherent difficulty in achieving

Cp-functional polymers not because of the high reactivity of the diene, but rather the harsh nature of the most common pathway to functionalized cyclopentadienes, which makes use of sodium cyclopentadienide (NaCp). In much the same way that the inherent halide functionality of polymers formed by ATRP can be readily transformed into azides, quantitative transformation into Cp-functionality can also be achieved. From our experience, NaCp may be effectively used in the preparation of Cp-functional poly(styrene) and poly(ethylene glycol) owing to their lack of potentially crossreactive functionality (Inglis *et al.*, 2009a). Conversely, poly(acrylates) and poly(methacrylates) are incompatible with NaCp due to the multitude of ester functionality. A much milder and highly efficient route to Cp-functional polymers involves the use of nickelocene, as we have recently reported (Inglis *et al.*, 2009b).

With Cp-functional polymers and those prepared by RAFT polymerization in hand, conjugation of the two may be simply performed under ambient conditions of atmosphere and temperature within a few minutes (Figure 2.4a) (Inglis *et al.*, 2009a). In the case of BPDF end groups, the ultrarapid reaction is triggered by protonation of the pyridine ring nitrogen atom. In the case of BDEPDF end groups, no catalyst is required.

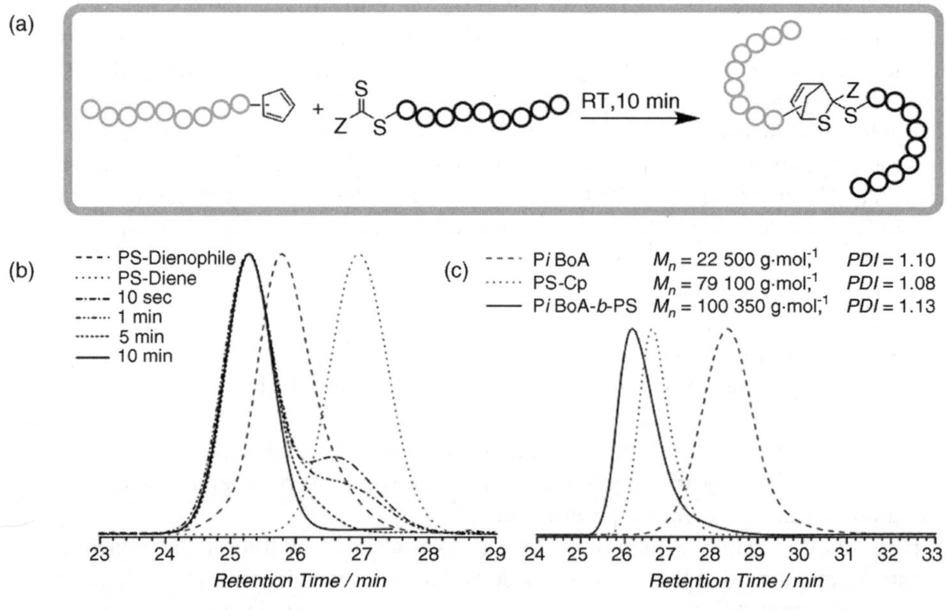

Figure 2.4 (a) Ultrarapid RAFT-HDA click chemistry; (b) kinetic monitoring of block copolymer formation via SEC (Reproduced with permission from A.J. Inglis, S. Sinnwell, M.H. Stenzel and C. Barner-Kowollik, "Ultrafast click conjugation of macromolecular building blocks at ambient temperature," *Angewandte Chemie International Edition*, 2009, **48**, 13, 2411–2414, © Wiley-VCH Verlag GmbH & Co. KGaA.); (c) SEC traces of high molecular weight block copolymer formed via the RAFT-HDA *click* reaction and its precursor homopolymers. (Reproduced with permission from A.J. Inglis, M.H. Stenzel and C. Barner-Kowollik, "Ultra-fast RAFT-HDA click conjugation: An efficient route to high molecular weight block copolymers," *Macromolecular Rapid Communications*, 2009, **30**, 21, 1792–1798, © Wiley-VCH Verlag GmbH & Co. KGaA.)

Using the more efficient BPDF as the example, a series of SEC measurements were conducted to elucidate the performance of block copolymer formation via this technique. Upon inspection of the various SEC traces presented in Figure 2.4b, the majority of the block structure is formed within the first 10 s of reaction, with quantitative conversion achieved within 10 min.

In a further example, the molecular weights of the precursor building blocks were increased to investigate their influence on macromolecular–macromolecular *click* conjugations (Inglis *et al.*, 2009c). As expected, when a PS chain of 80 000 g mol^{-1} molecular weight was reacted with P*i*BoA of 20 000 g mol^{-1}, a 2-h reaction time at ambient temperature was required. However, efficient construction of the required block structure (with molecular weight of over 100 000 g mol^{-1}) was still achieved, representing the largest synthetic block copolymer synthesized in a modular fashion from singly functionalized precursors (Figure 2.4c).

In another strategy, novel RAFT agents bearing highly electron-withdrawing sulfonyl-based Z-groups were synthesized and tested for both their performance in controlled polymerization of various monomers and also as a reactive dienophile (Nebhani *et al.*, 2009a; Nebhani *et al.*, 2009a). Interestingly, the reactivity of such dithioesters is so great that it was shown to undergo a Diels–Alder reaction with styrene and a Michael-type addition with butyl acrylate. As such, these compounds are not suitable for general use as RAFT agents, however isobornyl acrylate was successfully polymerized in a controlled fashion without any side reactions observed. This can be attributed to the rather sterically crowded double bond in such a monomer.

In order to evaluate the performance of these dithioesters in HDA cycloadditions, poly (ethylene glycol) monomethyl ether (PEG) was equipped with an open-chain diene and, in another example, with a cyclopentadienyl endgroup in order for the HDA adducts to be clearly characterized by electrospray mass spectrometry (ESI-MS). The reaction schematic is presented in Figure 2.5a.

In the case where the open-chain diene functional PEG was utilized, 24 h were required to attain completion in a catalyst-free, ambient-temperature reaction. However, use of the more-reactive Cp-functional PEG resulted in complete conversion (under identical reaction conditions) in less than 1 h. Despite this high performance, the HDA adduct partially degrades to form the structures depicted in Figure 2.5b. The degree of degradation can be qualitatively discerned from the ESI-MS spectra presented in Figure 2.5c. It must be noted that residual precursor material observed in the spectra (species **1**) is attributed to the occurrence of a retro-HDA cycloaddition under the conditions of measurement.

The RAFT-HDA *click* strategy has certainly been proven to enable highly efficient macromolecular–macromolecular conjugations to be performed under very mild conditions. It also stands as a very rare example within the field of CRP in that the functionality utilized to mediate controlled polymerization is also the same functionality that serves as the *click* function. As such, the technique may be termed "atom economical" for this very reason. Furthermore, the reversibility of such chemistry may serve as the inspiration for the development of a variety of thermoresponsive materials.

2.4 Cycloadditions with *Strained* or *Activated* Alkynes

Typically, organic azides and terminal alkynes do not react together (within the context of a *click* reaction) at ambient temperature and in the absence of a transition-metal catalyst.

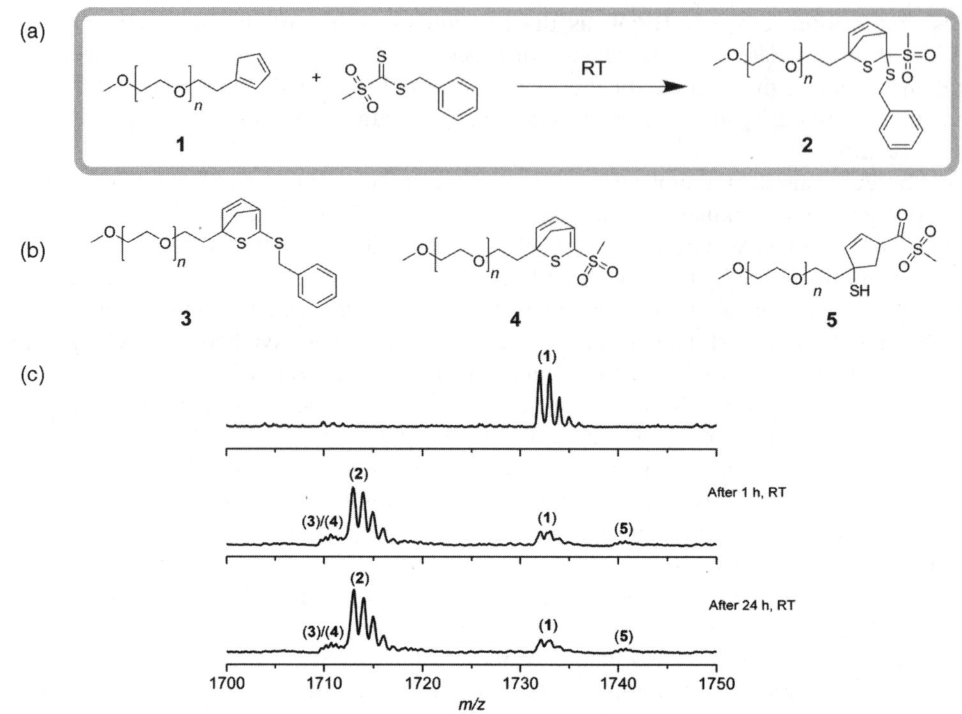

Figure 2.5 (a) RAFT-HDA chemistry utilizing highly electron-withdrawing dithioesters; (b) side products resulting from the fragmentation of the RAFT-HDA adduct with sulfonyl-based dithioesters; (c) ESI-MS monitoring of the RAFT-HDA reaction of cyclopentadienyl-capped PEG and benzyl methylsulfonyldithioformate (Nebhani *et al.*, 2009). It should be noted that the signals for the various species represent polymer chains of differing number of monomer repeat units. (Reprinted from L. Nebhani, S. Sinnwell, C.Y. Lin *et al.*, "Strongly electron deficient sulfonyldithioformate based RAFT agents for hetero Diels–Alder conjugation: Computational design and experimental evaluation," *Journal of Polymer Science: Part A Polymer Chemistry*, **47**, 22, 6053–6071 (Figure 6), © 2009, with the permission of John Wiley and Sons, Inc.)

Atypically however, alkynes may be *activated* by alternative strategies that eschew the use of a catalyst altogether. The manners in which *activation* may be achieved have been either the application of ring strain to an alkyne, the incorporation of an EWG or a combination of both.

One of the earliest examples, if not the first, of a strained-ring promoted cycloaddition between an alkyne and azide was provided by Wittig and Krebs (1961). Neat cyclooctyne was reacted with phenylazide to yield the corresponding triazole as the only product. This has been appropriated by Bertozzi and colleagues for selective modifications of biomacromolecules and living cells (Agard *et al.*, 2004). With their now called "first-generation" cyclooctyne (Figure 2.6a), successful copper-free and ambient-temperature cycloadditions with azides were achieved, although the reaction kinetics were comparable to those of the incumbent Staudinger ligation technology (Köhn and Breinbauer, 2004) and much slower than the CuAAC. Furthermore, the noncatalyzed azide–alkyne cycloadditions (in general) do not show the same regioselectivity as the CuAAC. However, in the vast majority

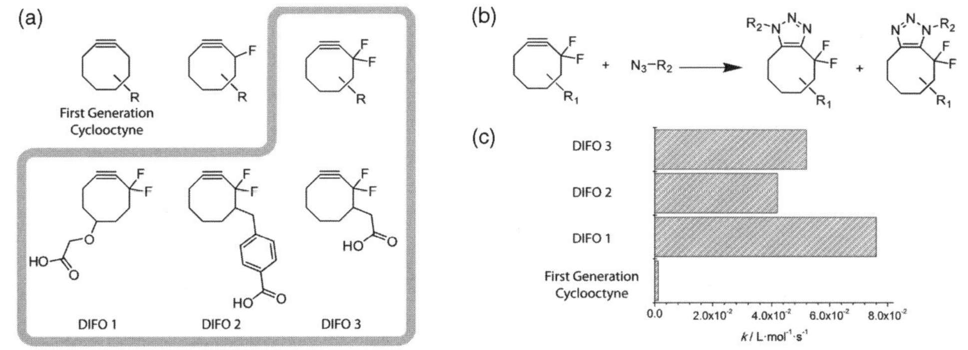

Figure 2.6 (a) Structures of the various cyclooctyne derivatives used in copper-free azide–alkyne cycloadditions; (b) nonregion-specific, copper-free and rapid cycloaddition between azides and DIFO derivatives; and (c) comparison of the second-order rate coefficients for the reactions of selected cyclooctyne derivatives and benzyl azide at room temperature. An azide: alkyne ratio of 10 : 1 was used. (Reproduced with permission from A.J. Inglis and C. Barner-Kowollik, "Ultra rapid approaches to mild macromolecular conjugation," *Macromolecular Rapid Communications*, 2010, **31**, 14, 1247–1266. © Wiley-VCH Verlag GmbH & Co. KGaA.)

of cases, regioselectivity is not a crucial criterion for coupling reactions, least of all in macromolecular chemistry.

A dramatic improvement in reactivity was subsequently achieved through the introduction of EWG (specifically propargylic fluorine atoms) to the cyclooctyne derivative. The monofluorinated variant only provided a roughly twofold increase in reactivity (Agard *et al.*, 2006); however, the difluorinated derivative (named difluoronated cyclooctyne (DIFO) 1) improved upon the reactivity by a further 20-fold increase (approximately) (Baskin *et al.*, 2007). It should be noted that a 10 : 1 ratio of azide to alkyne was utilized in the kinetic experiments. As such, highly efficient labeling of azide-expressed cells was achieved within 1 h of exposure to DIFO under ambient conditions.

Such fluorinated cyclooctynes have proven to be biocompatible and stable in aqueous media in addition to their remarkable reactivity. However, the synthesis of DIFO 1 is very inefficient, achieving under 2% yield in 12 steps. Furthermore, its hydrophobicity was shown to also contribute to nonspecific protein and cell binding. Substantially improving the efficiency of the synthesis of DIFO derivatives, Bertozzi and colleagues reported DIFO 2 and DIFO 3, which were formed in a 6-step, 36% yield and a 7-step, 28% yield strategy, respectively. It is noted that the reactivities of the discussed DIFO derivatives are comparable (Codelli *et al.*, 2008). The structures of these cyclooctyne derivatives, a general reaction schematic and a graphical comparison of their reactivities are presented in Figure 2.6.

The azide-DIFO cycloadditions therefore present a mild and fast conjugation methodology perfectly suited to biochemical applications such as *in vivo* dynamic imaging of cell-surface glycans in live cells. However, the laborious synthetic pathway of the precursor DIFO derivatives still stands in the way of such a system to be more widely investigated and, as of yet, the technique has not been applied to the design and construction of synthetic materials.

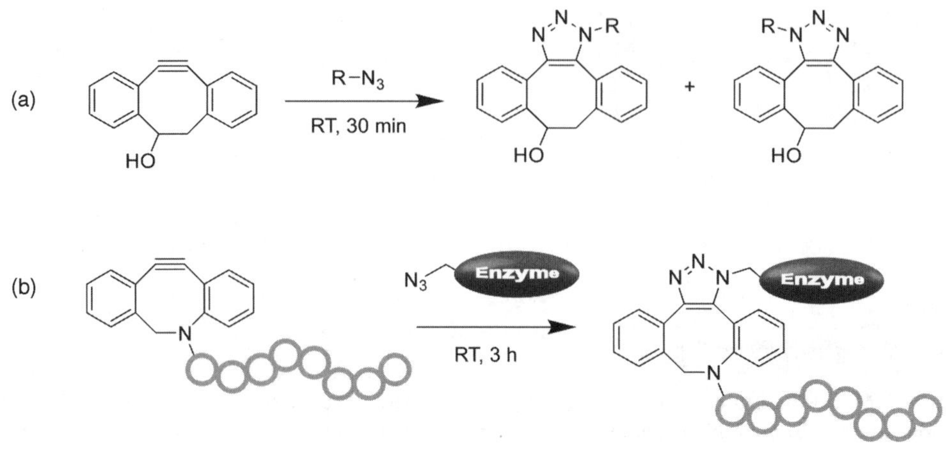

Scheme 2.1 (a) Strained cycloctyne utilized by Boons *et al.* for mild and rapid cycloadditions with azides and; (b) PEGylation of enzymes via a copper-free azide–alkyne cycloaddition. (Reproduced with permission from A.J. Inglis and C. Barner-Kowollik, "Ultra rapid approaches to mild macromolecular conjugation," *Macromolecular Rapid Communications*, 2010, **31**, 14, 1247–1266. © Wiley-VCH Verlag GmbH & Co. KGaA.)

An alternative strategy was reported by Boons and colleagues in which additional ring strain was applied to a cyclooctyne by the introduction of benzyl groups to either side of the reactive center (Scheme 2.1a) (Ning *et al.*, 2008). Such 4-dibenzocyclooctynols show similar reactivity to that of the DIFO derivatives of Bertozzi and are biocompatible. In model studies, various azides and cyclooctyne were mixed in an equimolar ratio and were found to undergo quantitative conversion into the corresponding 1,2,3-triazole within 30 min at ambient temperature. Such cyclooctynes are also claimed to have a long shelf life and there exists the potential to further enhance reaction performance by decorating the aromatic moieties with EWGs. Furthermore, a functional aza-dibenzocyclooctyne was made to great use in the recently reported PEGylation of an enzyme (Scheme 2.1b) (Debets *et al.*, 2010). Taking this into consideration, along with the simpler synthesis when compared to the DIFO derivatives, it is envisaged that such structures will be the ones to push forward the application of fast, copper-free azide–alkyne cycloaddition chemistry to include further synthetic polymer conjugations.

Yet another strategy to boost the reactivity of alkynes towards cycloadditions with azides is through the direct functionalization of the alkyne with an EWG. Ju and colleagues investigated a series of such alkynes (Scheme 2.2) (Li *et al.*, 2004). All reactions with model compounds were performed in aqueous solution at ambient temperature and were performed with a 1: 1 stoichiometry. Depending upon the structure of the alkyne, the reactions were completed in 6–12 h. Although not fulfilling our criteria for <2 h reaction times, it is interesting to note that in one illustrated example, complete regioselectivity for the 1,4-regioisomer of the triazole was observed. Furthermore, and similar to the earlier examples from Bertozzi and Boons, electron-deficient internal alkynes may also undergo efficient cycloadditions with azides. This is noteworthy as the CuAAC may only be used for terminal alkynes.

Accelerating the rate of reaction such that it may be completed within 2 h, Gouin and Kovensky reported a further electron-deficient alkyne (*p*-toluene sulfonyl alkyne) as shown in

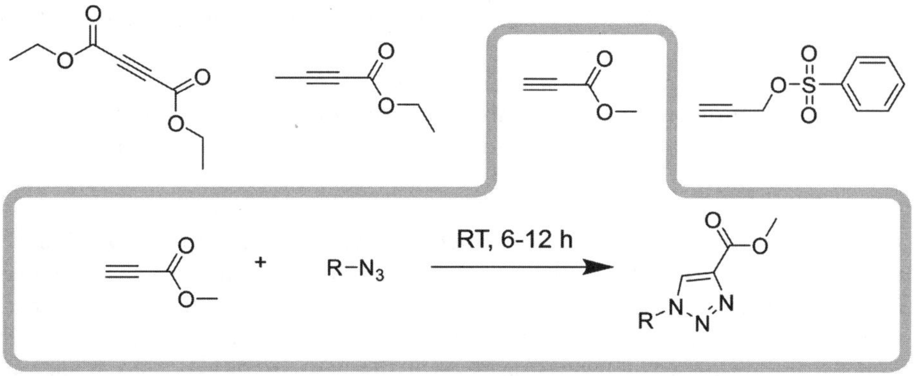

Scheme 2.2 Use of electron-deficient alkynes in catalyst-free cycloadditions with azides. (Reproduced with permission from A.J. Inglis and C. Barner-Kowollik, "Ultra rapid approaches to mild macromolecular conjugation," *Macromolecular Rapid Communications*, 2010, **31**, 14, 1247–1266. © Wiley-VCH Verlag GmbH & Co. KGaA.)

Scheme 2.3 (Gouin and Kovensky, 2009). After initial trials with azido-diethylene glycol at 100 °C with and without microwave irradiation, it was discovered that by quickly removing the reaction solvent under reduced pressure at 16 °C (<10 min), high conversions could be obtained within 2 h. The explanation given for such an observation is the high concentrations of reactants that are obtained upon solvent removal. Interestingly, the rapid evaporating process produced the corresponding 1,4-regioisomer with high selectivity (>90%).

Upon extending the process to include a variety of azides, the high rate of reaction and high regioselectivity was further confirmed in the vast majority of cases. Although the alkyne investigated is far simpler than those already discussed, the requirement of rapid solvent removal can be rather inconvenient. Thus, the reaction would not be suitable for many biological applications where dilute conditions are necessary. In addition, a series of experiments in which the *p*-toluene sulfonyl alkyne was reacted with unprotected amine-bearing

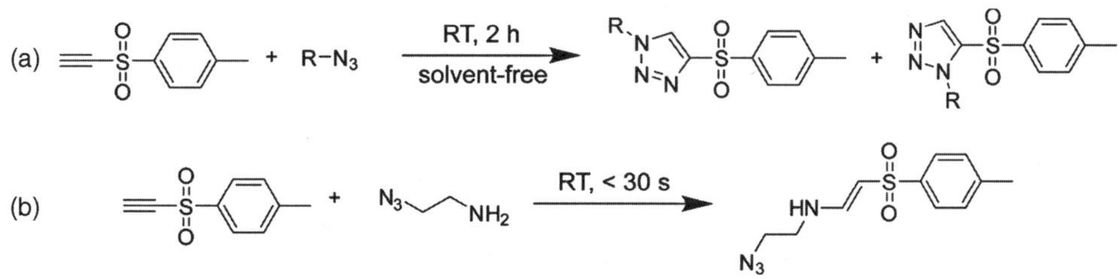

Scheme 2.3 (a) Solvent-free reaction between electron-deficient alkynes and azides; (b) ultrarapid side reaction of electron-deficient alkynes with primary amines. (Reproduced with permission from A. J. Inglis and C. Barner-Kowollik, "Ultra rapid approaches to mild macromolecular conjugation," *Macromolecular Rapid Communications*, 2010, **31**, 14, 1247–1266. © Wiley-VCH Verlag GmbH & Co. KGaA.)

azides resulted in an ultrarapid, selective and quantitative Michael-type addition of the amino group to the alkyne (Scheme 2.3). Although the authors use these results to highlight the lack of orthogonalilty of the reaction when it comes to amino functionality, the use of this "side reaction" should be further evaluated within the context of ultrafast and mild conjugations.

Thus far, the use of an azide as a dipolarophile in such cycloadditions had been discussed. However, very recently Pezacki and colleagues (McKay *et al.*, 2010) demonstrated the use of a series of nitrones in 1,3-dipolar cycloadditions with benzannulated cyclooctyne, some of the structures of which are presented in Figure 2.7a. For all 14 nitrones studied, very high (if not quantitative) conversions were achieved within reaction times of 2–110 min at ambient temperature, using a 1: 1 stoichiometry. Figure 2.7b compares measured second order rate coefficients (k) for selected nitrones with those observed for the reactions of benzyl azide with the benzannulated cyclooctyne and of benzyl azide with a DIFO compound. Upon correlating I, II, IV and V (as labeled in Figure 2.7a) with the values of k in Figure 2.7b, one notices that the reactivity of the nitrones are comparable to those of the reference reactions (IV and V). However, and rather impressively, the rate of the reaction of III is about 25-fold faster than that of IV and 20-fold faster than that of V. The authors partially attribute this enhanced reactivity to the ring strain in the cyclic nitrone. Although purely a small-molecule model study, such chemistry certainly holds promise for the field of macromolecular chemistry as a mild, ultrarapid conjugation technique.

Without a doubt, the CuAAC has led the way as far as *click* chemistry is concerned, but through its limitations it has also served as the inspiration of the development of the above-discussed catalyst-free variants. Although these technologies have only been applied in relatively low molecular weight chemistry, they certainly hold great potential to be brought to the same platform as the CuAAC. It is only the accessibility of the starting activated alkynes that prevents these methods from flourishing.

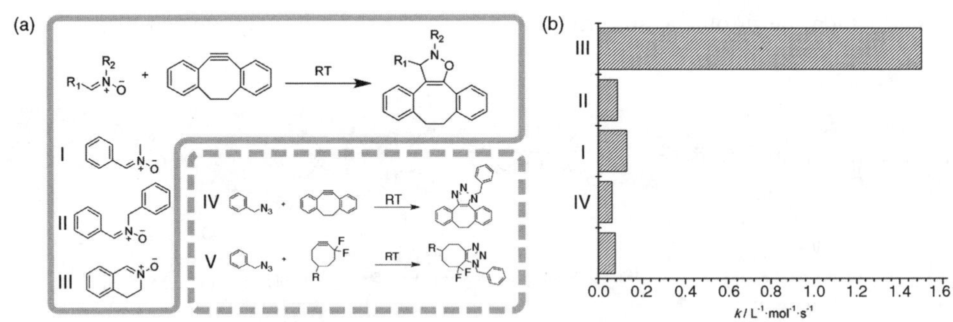

Figure 2.7 (a) 1,3-Dipolar cycloadditions of nitrones with a benzannualted cyclooctyne alongside the selected reference reactions of benzyl azide with the benzannualted cyclooctyne (IV) and of benzyl azide with a DIFO derivative (V); (b) comparison of the second-order rate coefficients of the reactions I–V presented in (a). (Reproduced with permission from A.J. Inglis and C. Barner-Kowollik, "Ultra rapid approaches to mild macromolecular conjugation," *Macromolecular Rapid Communications*, 2010, **31**, 14, 1247–1266. © Wiley-VCH Verlag GmbH & Co. KGaA.)

2.5 Thiol-Ene/Thiol-Yne Chemistry

Borne out of a rich history in the field of photopolymerization, the thiol-ene reaction has been widely investigated in recent years as a convenient conjugation technique in macromolecular science (Hoyle *et al.*, 2004; Lowe, 2010). However, it was the work of Schlaad and colleagues that made the initial efforts to extend such chemistry to the efficient modification of well-defined polymeric materials and surfaces (Gress *et al.*, 2007; ten Brummelhuis *et al.*, 2008; Bertin and Schlaad, 2009). Within this context, there are two main modes in which the thiol-ene reaction may be applied: the ionic mechanism or the radical mechanism. The ionic mechanism is taken to represent the various methods of performing Michael-type additions of the thiol moieties to an alkene, whereas the radical mechanism is taken to encompass the ways in which thiyl radicals are generated and add across the double bond of the alkene. As will be highlighted below, there are important differences between the two (each offering their own advantages and disadvantages) which should be taken into consideration when formulating an experimental design. Figure 2.8a presents the general thiol-ene reaction pathway.

In perhaps the most elegant example to date of the use of thiol-ene chemistry for macromolecular synthesis, Hawker and colleagues prepared up to fourth-generation dendrimers

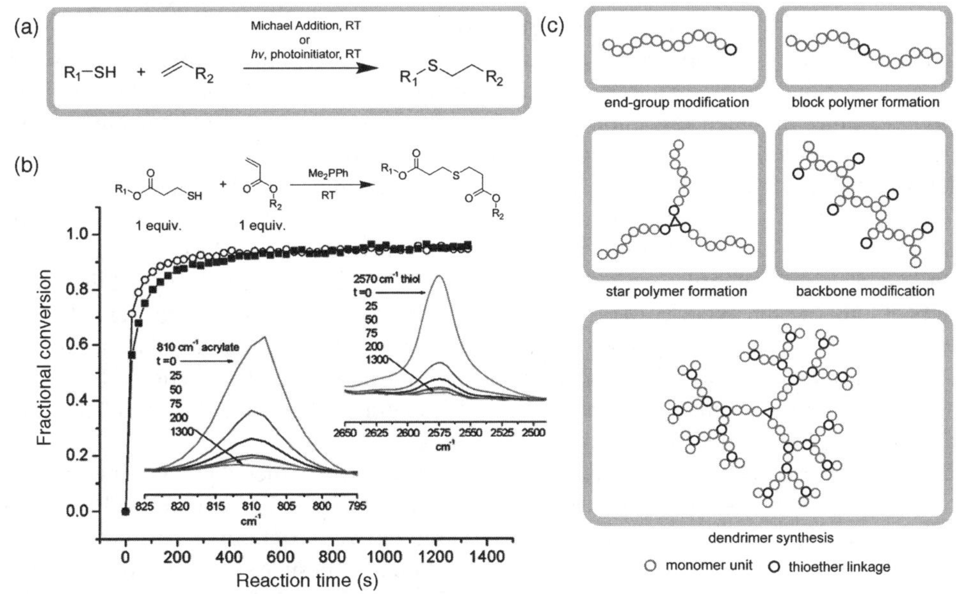

Figure 2.8 (a) UV-promoted radical addition and Michael-type addition of thiols with alkenes – the thiol-ene reaction; (b) kinetics of the thiol-ene reaction monitored via FT-IR spectroscopy (Chan *et al.*, 2009a); (c) various polymeric architectures achievable by thiol-ene chemistry. (Reprinted with permission from J.W. Chan, C.E. Hoyle and A.B. Lowe, "Sequential phosphine-catalyzed, nucleophilic thiol-ene/radical-mediated thiol-yne reactions and the facile orthogonal synthesis of polyfunctional materials," *Journal of the American Chemical Society*, **131**, 16, 5751–5753, 2009. © 2009 American Chemical Society.)

with unprecedented efficiency (Killops *et al.*, 2008). In a strategy involving sequential thiol-ene reactions and esterifications to build up the dendritic structure, the solvent-free thiol-ene reaction was performed under ambient conditions of atmosphere and temperature by simple irradiation with a hand-held UV-lamp ($\lambda_{ex} = 365$ nm). Trace amounts of the photoinitiator 2,2-dimethoxy-2-phenylacetophenone (DMPA) were required to kick-start the reaction, which took a total of 30 min.

Importantly, no perceptible reduction in efficiency was observed as the number of generations increased. However, when a 1: 1 stoichiometry between thiol and alkene groups was utilized, incomplete conversion was observed. In order to drive the reactions to completion, 1.5 equivalents of thiol to alkene was required. In this case, the excess was by no means a cause for concern as the reaction products could be easily isolated through simple precipitation procedures. This would generally be the case for all macromolecular thiol-ene reactions in which the species that must be used in excess is of small molecular weight, however, problems begin to arise when an excess of a macromolecular species is required.

In a further, and more comprehensive investigation into radical-based thiol-ene chemistry, Hawker and colleagues prepared a series of polymers with either single or multiple alkene functionality and *clicked* on a variety of commercially available mercaptans under both photoinitiation (with trace amounts of DMPA) and thermal initiation (in the presence of radical initiator 2,2′-azobis(2-methylpropionitrile)) (Campos *et al.*, 2008). In all cases studied, the photoinitiated thiol-ene reactions (all performed at ambient temperature) could be made to proceed to completion in times ranging from 30 min to 2 h.

These examples already prove the efficiency of thiol-ene reactions, however they have only dealt with couplings between alkene-functional macromolecules and small molecular weight thiols. Examples in which fast thiol-ene reactions have been used for truly macromolecular couplings have made use of the ionic mechanistic pathway.

Perhaps the most convincing example of Michael-type thiol-ene chemistry for macromolecular–macromolecular conjugations comes from Dove and colleagues (Pounder *et al.*, 2008). In the initial part of the investigation, maleimide-capped poly(lactic acid) (PLA) was prepared and the end groups modified via a triethylamine (TEA)-catalyzed Michael addition with a variety of thiols. In the majority of cases, a 1.05 equivalents of thiol and 1 equivalent of TEA to alkene resulted in quantitative conversions in times ranging from < 5 min to 2 h. Importantly, the structure of the thiol played an important role in the efficiency of the reactions. To point out just one example, the reaction with the sterically demanding ^{t}BuSH required 1 week to achieve quantitative conversion using 5 eq. thiol and 2 eq. TEA.

Extending these results to include macromolecular–macromolecular couplings, bi- and trifunctional thiol compounds were used to generate homocoupling between two PLA chains and 3-arm star-shaped PLA respectively. In both cases, an excess of thiol (1.1 eq.) was required to ensure complete conversion of all maleimide functionality. In such systems (where an excess of a linker is used) a statistical mixture of products is formed. In the case of the block homocoupling, a mixture of the required block, the product resulting from a single PLA chain adding to the thiol linker and bis-functional thiol linker is produced. It may be observed that only the target product contains no thiol functionality. A similar product distribution can also be expected for star formation. The thiol-containing side products are then cleverly (albeit relatively expensively) removed through the use of an iodoacetate resin. A subsequent precipitation produces very well-defined products, both of which were formed within 2 h at room temperature.

An alternative and convenient approach that is also highly versatile is the use of polymers prepared via RAFT polymerization as precursors to thiol-capped polymers. RAFT polymerization is the most versatile controlled radical polymerization method in terms of the wide variety of monomers that may be used. As such, the combination of this technology with thiol-ene chemistry offers, arguably, the greatest accessibility to synthetic functional materials.

Lowe, Hoyle and colleagues recently presented a one-pot strategy in which dithioester-capped polymer, alkene and a hexylamine (HexAM)/dimethylphenylphosphine (DMPP) catalyst system were used to perform a sequential aminolysis and Michael addition (Chan *et al.*, 2008; Chan *et al.*, 2009b). This was used to generate 3-arm star polymers quite rapidly at ambient temperature by employing a triacrylate as the star core. Although proceeding under ultrarapid and mild conditions, a considerable number of 2-arm and 1-arm star material was also formed, which the authors claim arose from a number of factors, including the use of impure starting materials and the required excess of thiol-capped polymer (1.5 eq. with respect to thiol functionality). Figures 2.8b and c show examples of the kinetics of the phosphine catalyzed thiol-ene reaction and the variety of macromolecular architectures that have been prepared (by either ionic or radical pathways) respectively. The above-described pathways to star polymers via thiol-ene chemistry is graphically depicted in Figure 2.9.

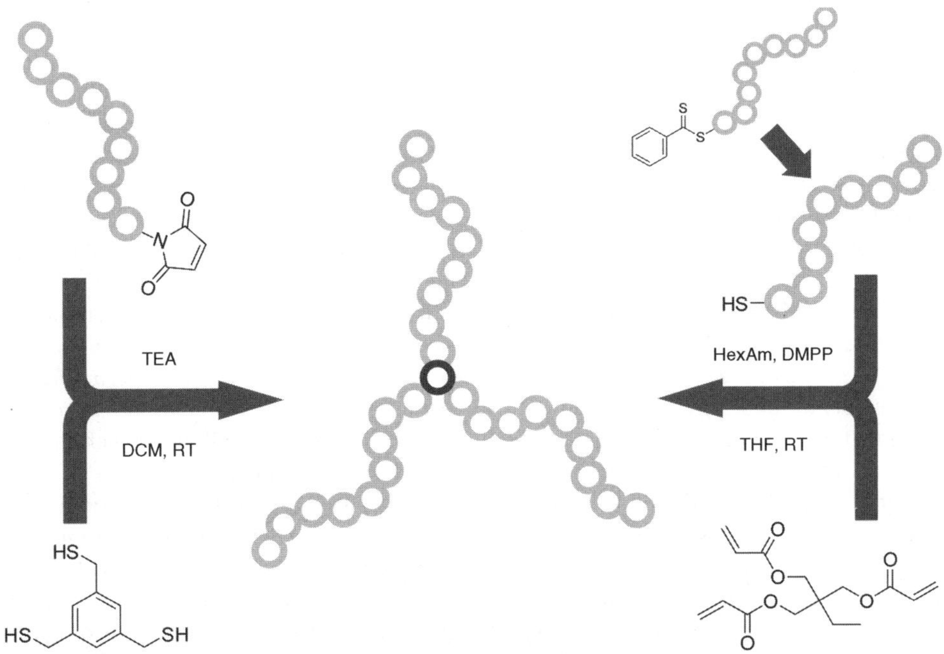

Figure 2.9 Strategies for polymeric star formation via the thiol-ene route as utilized by Dove (left) (Pounder *et al.*, 2008) and by Hoyle and Lowe (right) (Chan *et al.*, 2008; Chan *et al.*, 2009b). TEA: triethylamine; DCM: dichloromethane; HexAm: hexylamine; DMPP: dimethylphenylphosphine. (Reproduced with permission from A.J. Inglis and C. Barner-Kowollik, "Ultra rapid approaches to mild macromolecular conjugation," *Macromolecular Rapid Communications*, 2010, **31**, 14, 1247–1266. © Wiley-VCH Verlag GmbH & Co. KGaA.)

Making further use of the phosphine-catalyzed thiol-ene reaction, Lowe and Hoyle showed that it is orthogonal to the radical-based addition of thiols to alkynes, now referred to as thiol-yne chemistry (Chan *et al.*, 2009a). Operating under essentially the same mechanism as the radical-based thiol-ene chemistry, two thiols may add across the alkyne triple bond (in two steps) in a manner illustrated in Figure 2.10a. In the small-molecule investigation by Lowe and Hoyle, the thiol-ene reaction was observed to attain 90% conversion in 2.5 min, with 17 min required to reach quantitative conversion (increasing the phosphine content could speed up the reaction to <1 min). The thiol-yne reaction (which was performed under UV-irradiation at ambient temperature and in the presence of DMPA) was observed to attain >90% conversion in about 1.5 min, with full conversion requiring 10 min. The results from a FT-IR kinetic experiment are presented in Figure 2.10b. One limitation in this specific technique is that the thiol-ene reaction must be performed first as thiol-yne chemistry is not orthogonal to radical-based thiol-ene chemistry.

Taking advantage of the "two for one" mechansim of the thiol-yne reation, Stenzel and colleagues used this chemistry in the efficient construction of dendrimers in much the same way as Hawker reported using thiol-ene reactions (Chen *et al.*, 2009). In this case, however, the number of branching points that are generated in each generation of the synthesis are double that which can be obtained from the thiol-ene route. The thiol-yne reactions were carried out by UV-irradiation at ambient temperature in the presence of DMPA. As was similarly observed by Hawker, the efficiency of the reaction appeared to be "generation independent", requiring only 10 min to achieve quantitative conversions. Perrier and colleagues also recently made use of a similar approach to access functional, hyperbranched polymers starting from small molecules equipped with both thiol and alkyne functionality (Konkolewicz *et al.*, 2009). To give one final example, the use of direct sunlight as the radiation source was used to demonstrate the great practicality of thiol-yne chemistry in the modification of alkyne-functional polymeric brushes tethered to a surface (Hensarling *et al.*, 2009).

It is therefore apparent that thiol-ene and thiol-yne chemistry can be utilized to great advantage in macromolecular conjugation owing to their high-efficiency, rapid reaction kinetics and mild reaction conditions. However, it must be stressed that macromolecular–macromolecular conjugations have only been successfully reported through the use of the Michael-type variant of the reaction. Attempts to achieve efficient macromolecular–macromolecular couplings via the radical variant of thiol-ene chemistry has been met with profound difficulty and, although exceedingly long reaction times may allow for such a reaction to work as intended, it falls outside the realm of ultrarapid conjugation (Koo *et al.*, 2010). Although most of the small molecule kinetic studies have used a 1 : 1 stoichiometry of thiol to alkene (or 2 : 1 thiol to alkyne in the case of thiol-yne chemsitry); when multiple macromolecular species are involved, an excess of the thiol is typically required. For systems in which polymer chain-end modifications are performed, excesses of small molecules are not a problem as purification of the polymer is very straightforward. However, in macromolecular–macromolecular conjugations (such as block copolymer formation or convergent star synthesis), more elaborate purification methodologies are required, the most convenient (however relatively expensive) of which is the use of an iodoacetate resin to remove excesses of thiol.

2.6 Thiol-Isocyanate Chemistry

In yet another example of the chemical versatility of thiols, their base-catalyzed reaction with isocyanates is very fast and can be utilized with considerable ease. Such a reaction was used by

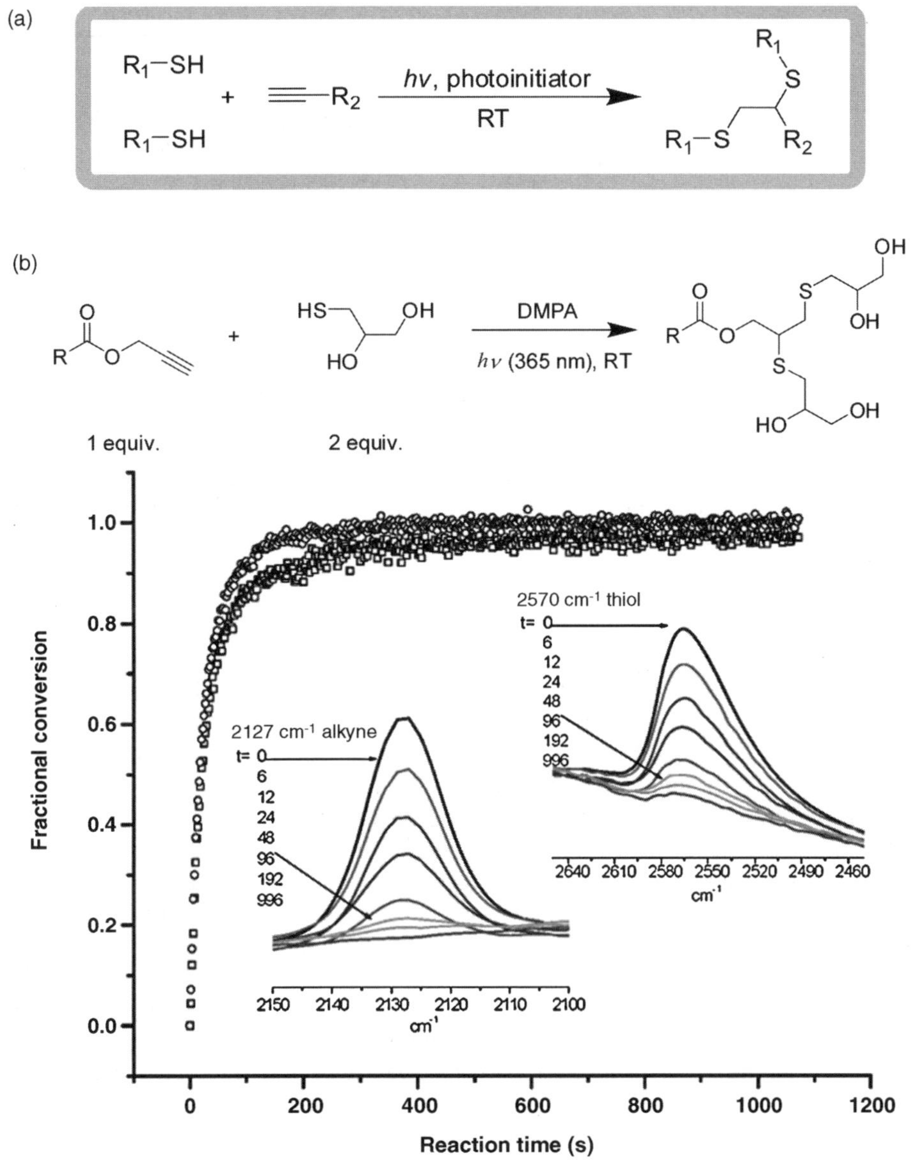

Figure 2.10 (a) The UV-promoted radical addition of thiols to alkynes – the thiol-yne reaction; (b) kinetics of the thiol-yne reaction monitored via FT-IR spectroscopy (Chan *et al.*, 2009a). DMPA: 2,2-dimethoxy-2-phenylacetophenone. (Reprinted with permission from J.W. Chan, C.E. Hoyle and A.B. Lowe, "Sequential phosphine-catalyzed, nucleophilic thiol-ene/radical-mediated thiol-yne reactions and the facile orthogonal synthesis of polyfunctional materials," *Journal of the American Chemical Society*, **131**, 16, 5751–5753, 2009. © 2009 American Chemical Society.)

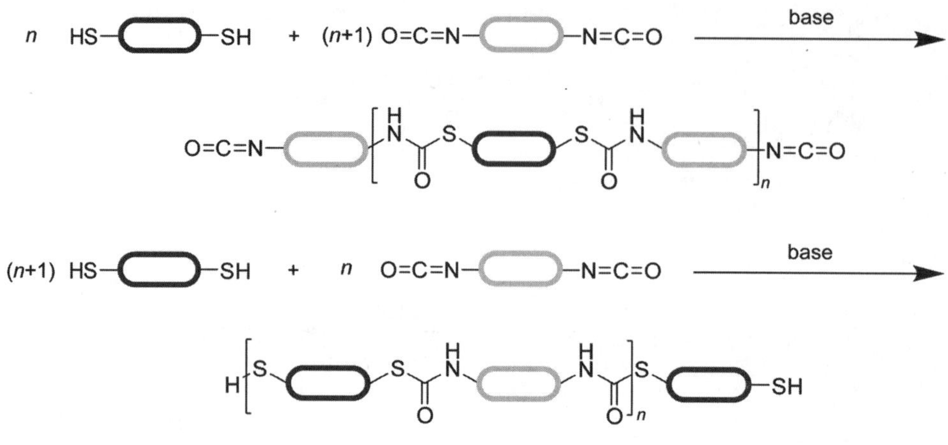

Scheme 2.4 Step-growth polymerization via the thiol-isocyanate reaction, yielding poly(thiourethane)s. (Reproduced with permission from A.J. Inglis and C. Barner-Kowollik, "Ultra rapid approaches to mild macromolecular conjugation," *Macromolecular Rapid Communications*, 2010, **31**, 14, 1247–1266. © Wiley-VCH Verlag GmbH & Co. KGaA.)

Klemm and Stöckle in the convenient preparation of thiol- and isocyanate-terminated thiocarbamate prepolymers, as depicted in Scheme 2.4 (Klemm and Stöckl, 1991). By varying the stoichiometry, either di-thiol or di-isocynate functional poly(thiourethane)s could be assembled within 1 h at ambient temperature.

Extrapolating from this work, Hoyle and colleagues combined this chemistry with thiol-ene reactions to achieve segmented polythiourethane elastomers (Shin *et al.*, 2009). A simple mixing of di-thiol prepolymer, diisocyanate linker and a catalytic amount of triethylamine (0.03–0.1 wt%) resulted in 95% conversion of the isocyanate within 10 min.

The use of the base-catalyzed thiol-isocyanate reaction in the quantitative preparation of ω-end functional polymers was also recently reported (Li *et al.*, 2009). Here, RAFT polymerization was used to prepare poly(*N, N*- diethylacrylamide), the dithioester end group of which was transformed into a thiol moiety by aminolysis. Subsequent end-group modification of the resulting thiol-terminated polymer was achieved in an overnight, ambient temperature and oxygen-free reaction with one of a collection of commercially available isocyanates in the presence of triethylamine (Figure 2.11a).

Although a twofold excess of isocyanate was used and the reactions were allowed to proceed overnight, a real-time reaction-kinetics experiment was performed. Hexyl isocyanate, thiol-terminated polymer and triethylamine were added in the ratio of 1 : 1 : 1.75 in THF and the reaction that followed was monitored at 20 °C. As can clearly be observed in Figure 2.11b, 94% conversion of the isocyanate is achieved within 15 min.

It is important to note that the thiol-isocyanate reaction is not orthogonal to the base-catalyzed Michael addition of thiols to alkenes, as was observed by Lowe when attempting to conjugate 2-isocyanoethyl methacrylate, although this system still showed significant selectivity towards the thiocarbamate adduct (Li *et al.*, 2009). Furthermore, it should be noted that isocyanates are very reactive (and toxic) compounds that can react with hydroxyl functionality,

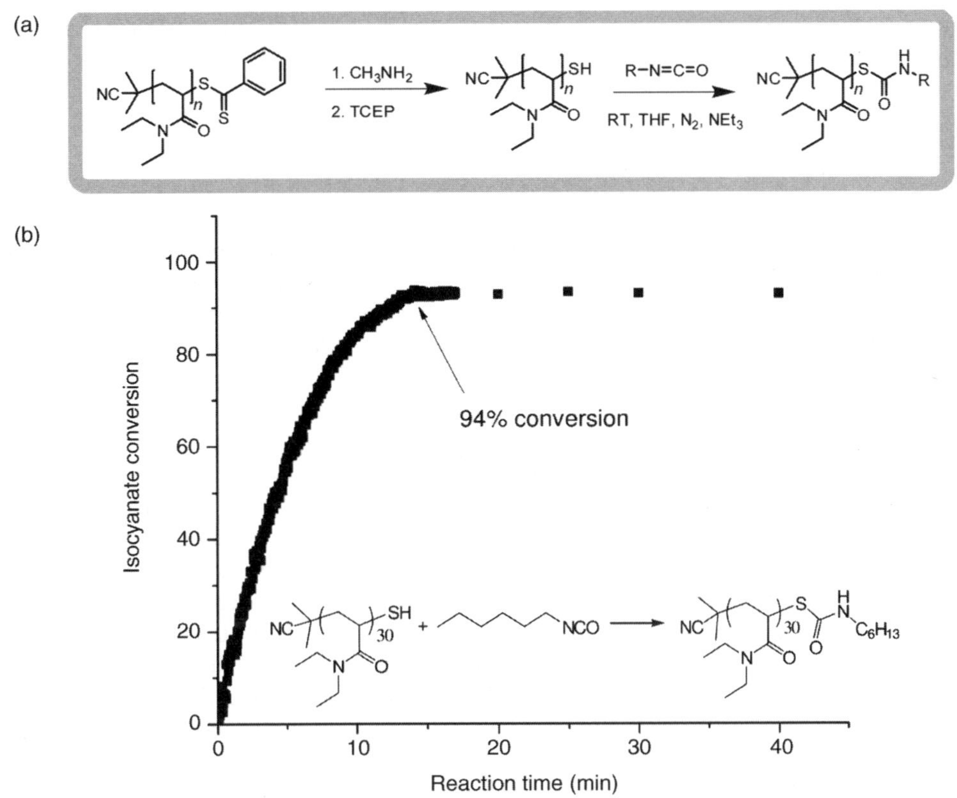

Figure 2.11 (a) Aminolysis of a polymer chain formed via RAFT polymerization and reaction of the resulting thiol with an isocyanate – the subsequent treatment with tris(2-carboxyethyl)phosphine (TCEP) is to regain thiol-capped polymers from possible disulfide species; (b) real-time kinetics of the thiol-isocyanate reaction, monitored by FT-IR spectroscopy (Li *et al.*, 2009). (Reprinted with permission from H. Li, B. Yu, H. Matsushima *et al.*, "The thiol-isocyanate click reaction: Facile and quantitative access to omega-end-functional poly(N,N-diethylacrylamide) synthesized by RAFT radical polymerization," *Macromolecules*, **42**, 17, 6537–6542, 2009. © 2009 American Chemical Society.)

amine functionality, water and even with themselves. Thus, this imposes obvious limitations to the applicability of such chemistry as a conjugation method in polymer science.

Therefore, the base-catalyzed thiol-isocyanate reaction qualifies as an ultrarapid and mild conjugation procedure that may be easily applied to the macromolecular sciences. As of yet, it has only been applied as a mechanism for step-growth polymerization and in the end-group modification of polymers with, by comparison, much smaller molecules. In order to be more widely adopted, macromolecular–macromolecular couplings must be attempted and in order for that to be achieved, a more universal strategy for equipping polymer chains with isocyanate groups is required. Unfortunately, the severe lack of orthogonality will surely prevent such chemistry from reaching the scope of other efficient conjugation technologies such as the CuAAC.

2.7 Thio-Bromo Chemistry

In this final example of the broad applicability the thiol moiety has within conjugative chemistry, the nucleophilic thio-bromo reaction is beginning to make an impact in macromolecular science. Typically, a base such as triethylamine is used to deprotonate the thiol, which can then efficiently react with α-bromoesters in a thio-etherification reaction.

The first noteworthy example comes from Percec and colleagues, in which the synthesis of up to G4 dendrimers was achieved through sequential thio-bromo and acylation reactions (Rosen *et al.*, 2009). The initial thio-bromo reaction of thioglycerol (1.2 equivalents) with a secondary α-bromoester in the presence of triethylamine (1.2 equivalents) in acetonitrile yielded the required G2 dendrimer in only 5 min at ambient temperature. However, unlike the observations of Hawker and colleagues (who made use of UV-initiated thiol-ene reactions) (Killops *et al.*, 2008), longer reaction times were required with increasing generation (20 min and 45 min required for G3 and G4 dendrimers respectively). Furthermore, silica gel chromatography was required to isolate the various dendrimers at each generation, something which was not required in the case of the UV-initiated thiol-ene reaction discussed earlier. Additionally, a potentially important note from the authors regarded the large exotherm that occurs at the beginning of the reaction. Therefore even though no external heating source is used, the reaction temperature could increase unless extra care is taken.

In a further example, dendritic macromolecules were synthesized via a combination of single-electron-transfer living radical polymerization (SET-LRP) and thio-bromo *click* reactions in a novel "branch and grow" strategy (Figure 2.12) (Rosen *et al.*, 2009). This involved SET-LRP of methyl acrylate from a multifunctional initiator, followed by conversion of the bromide end groups with thioglycerol via the thio-bromo reaction. Acylation of the resulting hydroxyl chain termini with 2-bromopropionyl bromide then yielded another multifunctional initiator for SET-LRP that contained twice as many initiating sites as the previous generation. Again, in each case an excess of triethylamine and thioglycerol were required (1.2 equivalents) and longer reaction times were required for high generations.

More recently, Davis and Lowe extended this concept to include polymers prepared via RAFT polymerization as macromolecular thiol precursors (Xu *et al.*, 2009). Dithiobenzoate-based RAFT agents were synthesized bearing double α-bromoester functionality in the R-group fragment. Two were synthesized, one bearing secondary bromide functionality; the other with tertiary bromide functionality. Use of these compounds to mediate controlled polymerization of various monomers yielded low molecular weight polymer chains bearing ω-dithiobenzoate end groups and α-bis bromoester end groups. The simple addition of a hexylamine/triethylamine (1: 2) catalyst system resulted in the *in situ* aminolysis of the dithioester end group and the subsequent reaction of the formed thiol with the bromide functionality of neighboring polymer chains. In consequence, multiblock and hyperbranched polymers were readily formed (Figure 2.13).

As part of an initial model study, poly(methyl acrylate) prepared via RAFT polymerization in the presence of a conventional RAFT agent (bearing no bromide functionality) was

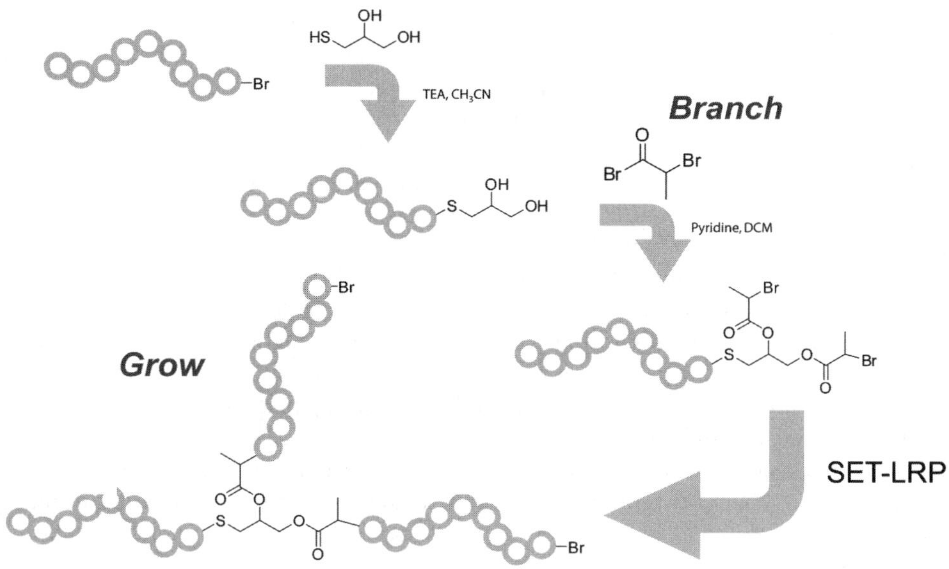

Figure 2.12 The "branch and grow" strategy of polymer design as utilized by Percec and colleagues (Rosen *et al.*, 2009). TEA: triethylamine; DCM: dichloromethane. (Reproduced with permission from A.J. Inglis and C. Barner-Kowollik, "Ultra rapid approaches to mild macromolecular conjugation," *Macromolecular Rapid Communications*, 2010, **31**, 14, 1247–1266. © Wiley-VCH Verlag GmbH & Co. KGaA.)

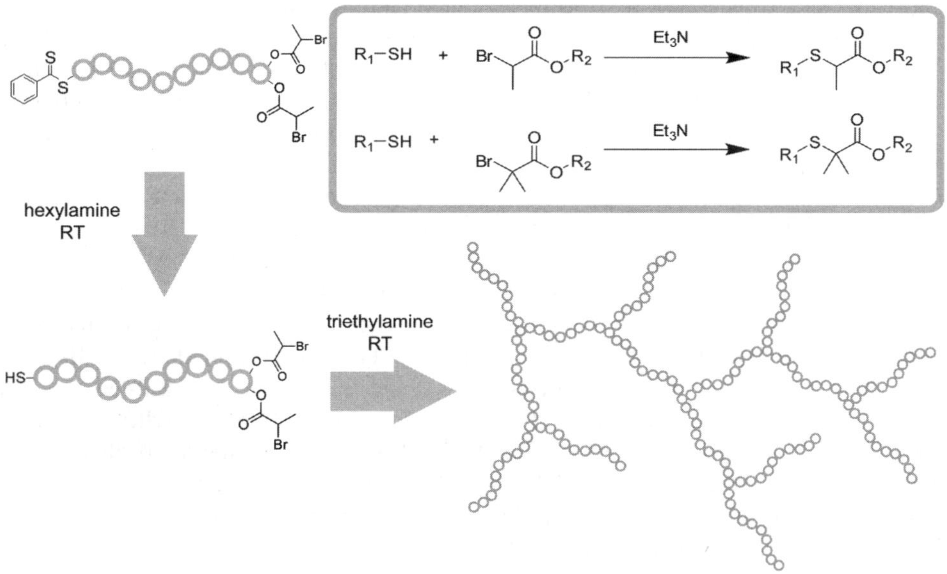

Figure 2.13 The use of the thio-bromo reaction to achieve highly branched polymers from monodisperse polymer precursors (Xu *et al.*, 2009). (Reproduced with permission from A.J. Inglis and C. Barner-Kowollik, "Ultra rapid approaches to mild macromolecular conjugation," *Macromolecular Rapid Communications*, 2010, **31**, 14, 1247–1266. © Wiley-VCH Verlag GmbH & Co. KGaA.)

separately reacted with methyl 2-bromopropionate (MBP) and ethyl 2-bromoisobutyrate (EBiB) (twofold excess), using a mixture of hexylamine and triethylamine (1 : 2). Based upon ESI-MS measurements, the reaction with MBP was completed within 30 min, however a significant quantity of side products (identified as cyclic thiolactone) were observed. The end-group functionality was subsequently estimated to be 73–80%. Similar experiments involving poly(*N*-isopropylacrylamide) (PNIPAAm) in place of PMA resulted in an estimated $\geq$95% ω-functionality. When MBP was replaced by EBiB, an expectedly longer reaction time (3 h) was required to obtain high yields of the corresponding thioether product. Translating these model studies to the generation of hyperbranched polymers, high yields were attains within 30 min when secondary bromides were utilized. When replacing secondary for tertiary bromides, a 20-h reaction time was required to achieve comparable yields.

An additional interesting observation is that the thio-bromo reaction is kinetically favored over the competing disulfide formation that must always be considered in such systems. The thio-bromo *click* reaction is obviously very convenient in that it makes use not only of readily available materials, but also some of the most common end groups encountered in modern polymer chemistry. It therefore has the potential to become a valuable addition to the *click* chemistry toolbox.

2.8 Inverse Electron Demand Diels–Alder

The reversibility of the conventional Diels–Alder reaction makes it a prime candidate for the synthesis of "smart materials" that exhibit a temperature-dependent transition between two physical properties (such as viscosity or color). However, many applications (such as biological labeling) require irreversible conjugation chemistry that also is very fast under mild and dilute reaction conditions. For such purposes, the inverse electron-demand Diels–Alder (IVED-DA) reaction between electron-deficient tetrazine derivatives and various alkenes is incredibly efficient.

The overall transformation (Figure 2.14) occurs through an initial catalyst-free [4 + 2] cycloaddition after which N_2 is released through a subsequent retro-[4 + 2] cycloaddition. To date, the most effective alkene for such reactions is *trans*-cyclooctene.

After publishing the facile synthesis of *trans*-cyclooctene derivatives from the readily accessible *cis*-cyclooctene (Royzen *et al.*, 2008), Fox and colleagues extended this work to develop a fast bioconjugation strategy based upon the IVED-DA cycloadditions (Blackman *et al.*, 2008). Maleimide functional *trans*-cyclooctene was coupled to thioredoxin (after reduction to produce the solvent exposed cysteine) via standard Michael addition chemistry. The now *trans*-cyclooctene functional thioredoxin (Trx-ene) could then be further modified extremely efficiently with an electron-deficient tetrazine (3,6-di-(2-pyidyl)-*s*-tetrazine) shown in the left side of Figure 2.15. The overall strategy is depicted in Scheme 2.5.

The reaction of 2 equivalents tetrazine (30 μM) with Trx-ene (15 μM) in aqueous media resulted in complete conversion into the expected Diels–Alder adduct within 5 min under ambient temperature. However, in a model reaction in which *trans*-cyclooctene and 3,6-di-(2-pyridyl)-*s*-tetrazine were reacted in a 1: 1 ratio and in low micromolar concentrations (5 μM); 40 min was required to attain quantitative conversion, thus further illustrating the extremely efficient and rapid nature of the transformation. Curiously, tetrazines bearing additional functionality were also synthesized, however not utilized in conjugation studies.

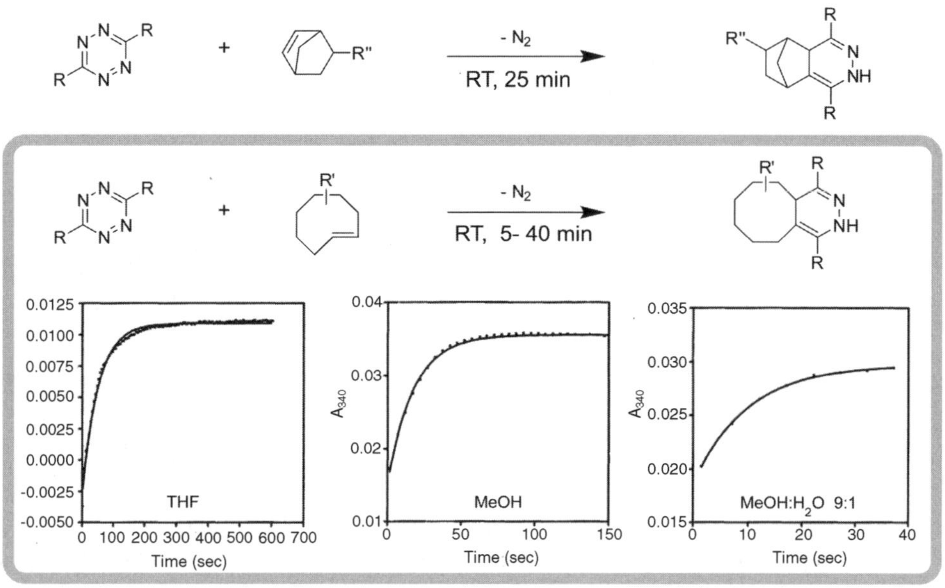

Figure 2.14 The inverse electron-demand Diels–Alder reaction between tetrazines and strained alkenes. The kinetics of the cycloaddition is greatly influenced by the solvent in which the reaction is performed (Blackman *et al.*, 2008). (Reprinted with permission from M.L. Blackman, M. Royzen and J.M. Fox, "Tetrazine ligation: Fast bioconjugation based on inverse-electron-demand Diels–Alder reactivity," *Journal of the American Chemical Society*, **130**, 41, 13518–13519, 2008. © 2008 American Chemical Society.)

Furthermore, experiments in which model reactions were performed in cell media and in cell lysate produced >80% yield of adduct (analyzed by ESI-MS against an internal standard), indicating that the reaction is tolerant to a broad range of biological functionality. In spite of these encouraging results, 3,6-di-(2-pyridyl)-*s*-tetrazine was also observed to undergo

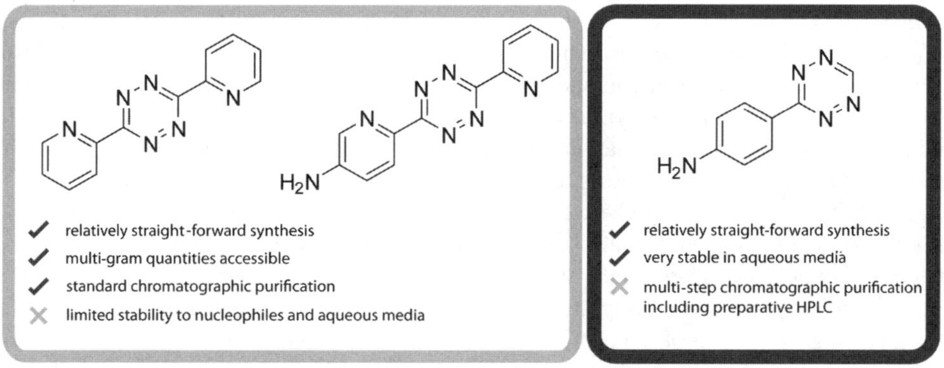

Figure 2.15 Comparison of the various tetrazines that have been successfully utilized in ultrarapid and mild conjugations. (Reproduced with permission from A.J. Inglis and C. Barner-Kowollik, "Ultra rapid approaches to mild macromolecular conjugation," *Macromolecular Rapid Communications*, 2010, **31**, 14, 1247–1266. © Wiley-VCH Verlag GmbH & Co. KGaA.)

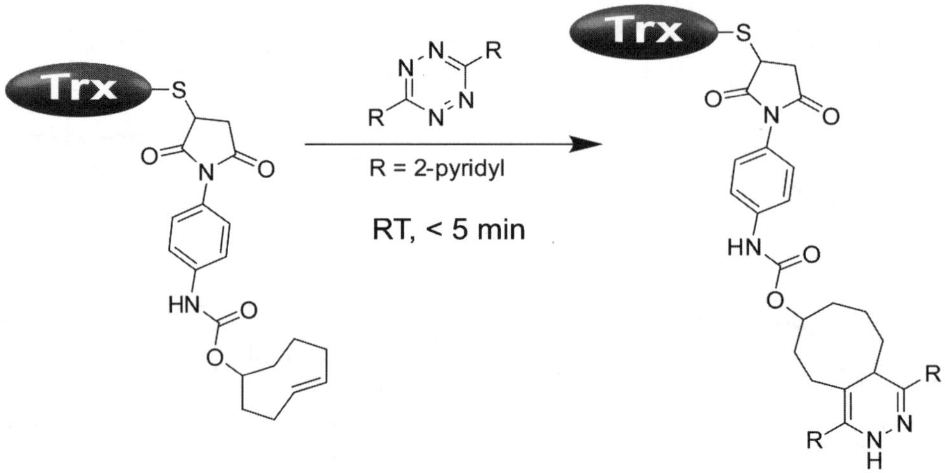

Scheme 2.5 Modification of thioredoxin (Trx) via the IVED-DA cycloaddition (Blackman, Royzen and Fox, 2008). (Reproduced with permission from A.J. Inglis and C. Barner-Kowollik, "Ultra rapid approaches to mild macromolecular conjugation," *Macromolecular Rapid Communications*, 2010, **31**, 14, 1247–1266. © Wiley-VCH Verlag GmbH & Co. KGaA.)

significant reaction with nucleophiles (50% conversion with $BuNH_2$ in 4 h, and 50% conversion with EtSH in 10 min) and also significant decomposition in pure water (50% after 2 h).

Hildebrand and colleagues used a similar strategy to achieve highly selective fluorescent labeling of live cells (Devaraj *et al.*, 2008). Unlike the various tetrazines described previously, the asymmetric, amine-functional tetrazine investigated here (3-(*p*-Benzylamino)-1,2,4,5-tetrazine, right side of Figure 2.15) had a significantly higher stability in aqueous and biological media. Furthermore, the tetrazine proved to also be highly reactive towards the IVED-DA reaction – initially with various norbornene derivatives (20–25 min required for quantitative conversion) (Devaraj *et al.*, 2008) and later with the highly active *trans*-cyclooctene (Devaraj *et al.*, 2009) (10 min required for quantitative conversion) – at ambient temperature. Despite this highly desirable improvement, one caveat lies in the generation of the 3-(*p*-Benzylamino)-1,2,4,5-tetrazine. Although the synthesis is relatively straightforward, purification on the other hand required a multistep chromatographic separation – including preparative HPLC. Thus, increasing the scale of the synthesis, at this stage, may prove problematic.

The IVED-DA cycloaddition between electron-deficient tetrazines and various alkenes (in particular *trans*-cyclooctene) – although nonregioselective – is an extremely efficient, rapid and mild chemical transformation. With respect to macromolecular science, however, the modification of a protein via this strategy by Fox and colleagues stands as the only example to date (Blackman *et al.*, 2008). Further extending this reaction to perform macromolecular conjugations on synthetic polymers is, arguably, limited to the relatively low accessibility of viable tetrazines. Although the tetrazines reported by Fox and colleagues proved to have limited compatibility with nucleophiles and aqueous environments, their ease of synthesis may be key in initial attempts at combining such a technology with CRP.

2.9 Cycloadditions Involving Nitrile Oxides

A reaction that is set to gain momentum within the field of macromolecular science over the next few years is the cycloaddition of nitrile oxides with alkynes and norbornenes to form isoxazoles and isoxazolines respectively (Scheme 2.6). Whilst strictly a catalyst-free reaction, the *in situ* generation of the requisite nitrile oxide from the respective oxime precursor typically makes use of a mixture of a dipole generating agent (such as Chloramine-T) and a weak base (such as aqueous $NaHCO_3$).

Heaney, Lutz and colleagues introduced the nitrile oxide-alkyne reaction as a mild means of modifying the end groups of alkyne-functional polymers formed by ATRP (Singh *et al.*, 2009a). Although proceeding to high yields and with great regioselectivity at room temperature, a 16-h reaction was required, utilizing a 50–70-fold excess of the precursor oxime. Interestingly, extension of this methodology to solid-phase oligonucleotide conjugation resulted in good conversions in just 10–30 min, an observation that was not readily explained (Singh *et al.*, 2009b).

Making use of norbornenes in place of alkynes, Carell and colleagues explored the modification of DNA with a variety of oximes via the *in-situ* nitrile oxide–norbornene cycloaddition (Gutsmiedl *et al.*, 2009). In this case, complete conversions were obtained at room temperature in 10–20 min, although a 10-fold excess of nitrile oxide with respect to the norbornene moieties was utilized.

Although the highly efficient copper(I)-catalyzed (Himo *et al.*, 2005) and ruthenium (II) -catalyzed (Grecian and Fokin, 2008) variants of the reaction have been previously reported, a definitive kinetic investigation of transition-metal-free reactions between nitrile oxides and alkynes (in macromolecular systems) is still lacking. Furthermore, such chemistry has also been effectively employed as a mechanism for step-growth polymerization, however, not under ambient conditions (Koyama *et al.*, 2008; Lee *et al.*, 2009). It is also worth noting that the isoxazole that is formed after such a cycloaddition may be partially reduced to β-aminoenone groups or fully reduced to β-aminoalcohol groups. Such generation of new functional groups (with relative ease) has already been proven to be successful in the formation of crosslinked polymer networks (Lee *et al.*, 2009). As a further note, nitrile oxides may also react

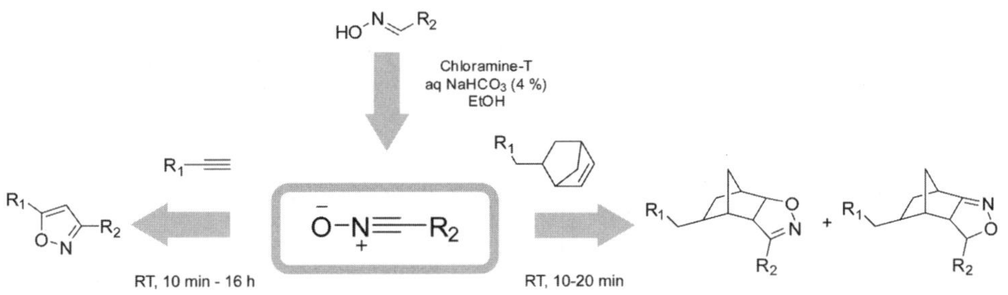

Scheme 2.6 Cycloaddition of nitrile oxides with alkynes/alkenes. (Reproduced with permission from A.J. Inglis and C. Barner-Kowollik, "Ultra rapid approaches to mild macromolecular conjugation," *Macromolecular Rapid Communications*, 2010, **31**, 14, 1247–1266. © Wiley-VCH Verlag GmbH & Co. KGaA.)

with a relatively wide range of nucleophiles (including thiols) (Grundmann and Frommeld, 1966). Thus, the orthogonality of such chemistry must be carefully taken into consideration on a case-by-case basis in macromolecular science.

Therefore, as it stands, the nitrile oxide–alkyne cycloaddition holds promise as a mild conjugation strategy for macromolecular systems; however, it still remains to be shown that efficient macromolecular–macromolecular conjugations are achievable through the use of equimolar stoichiometry of the respective reactive partners.

2.10 Oxime Formation

Amongst the list of highly efficient chemical transformations that were outlined in the definitive *click* chemistry paper by Sharpless and colleagues was carbonyl chemistry of the nonaldol type (Kolb *et al.*, 2001). One such reaction is oxime formation, resulting from the catalyst-free/ambient temperature reaction of aldehydes/ketones with aminooxy functionalities (Scheme 2.7).

The use of this strategy has rarely been encountered as a conjugation technique in macromolecular chemistry. Nevertheless, in a very nice example Maynard and colleagues prepared a variety of aminooxy functional polymers via ATRP and conjugated them to *N*-leuvinyl lysine modified bovine serum albumin (BSA) (Heredia *et al.*, 2007). The reaction was completed in only 30 min, producing the well-defined bioconjugate (Scheme 2.7).

In a further example, Kochendoefer and colleagues synthesized an aminooxy-functionalized CCL-5 protein for the site-specific attachment of a propionaldehyde-derivative of poly(ethylene glycol) monomethyl ether (Shao *et al.*, 2005). The oxime-forming reaction only required 1 h at ambient temperature to achieve almost quantitative conversion, utilizing a twofold excess of the

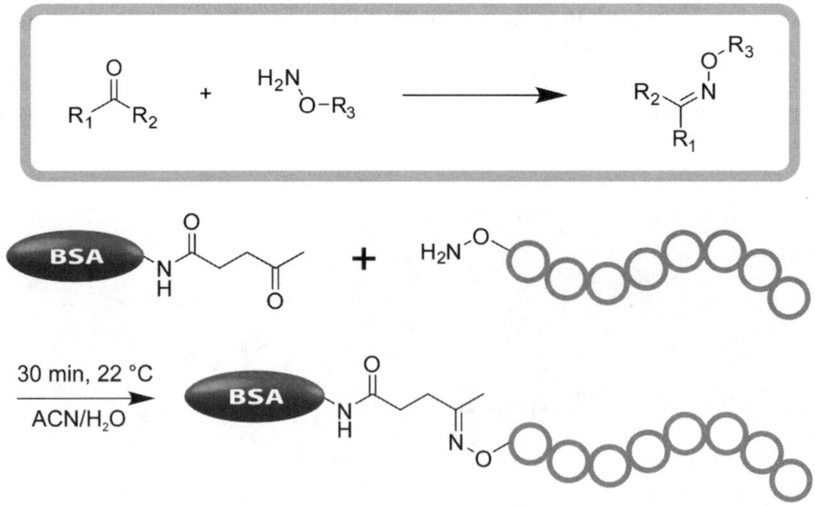

Scheme 2.7 The oxime-bond formation as a mild tool for bioconjugation. (Reproduced with permission from A.J. Inglis and C. Barner-Kowollik, "Ultra rapid approaches to mild macromolecular conjugation," *Macromolecular Rapid Communications*, 2010, **31**, 14, 1247–1266. © Wiley-VCH Verlag GmbH & Co. KGaA.)

synthetic polymer. Importantly, the conjugation reaction (in addition to the various synthetic steps required to attach the *clickable* functionality) was proven to have no observable impact on the correct folding of the protein, its tertiary structure or its antiviral activity.

Therefore, the oxime-forming reaction presents a mild and rapid conjugation approach that is biocompatible. However, the stability of the linkage that is generated must be further evaluated in a variety of macromolecular systems in order to determine in which applications it is suited, given that oximes are susceptible to further transformations such as hydrolysis, and rearrangements (Beckmann) that would result in an effective reversion of the conjugation.

2.11 Tetrazole–Ene Reaction

Ultrarapid conjugation reactions that may be triggered by some external stimulus (such as the addition of a catalyst or exposure to UV light) are highly desirable as it allows for exquisite reaction control in complex, multicomponent systems. One such reaction is the photoinducible 1,3-dipolar cycloaddition between 2,5-diaryl tetrazoles and alkenes (Figure 2.16).

Upon photoirradiation, 2,5-diaryl tetrazoles undergo an extremely fast cycloreversion to release N_2 and produce the corresponding nitrile-imine, which is then available to undergo a cycloaddition with a suitable dipolarophile such as an alkene. Using a hand-held UV lamp operating at a wavelength of 302 nm, Lin and colleagues successfully modified tetrazole-functionalized proteins with acrylamide within a few minutes at ambient temperature

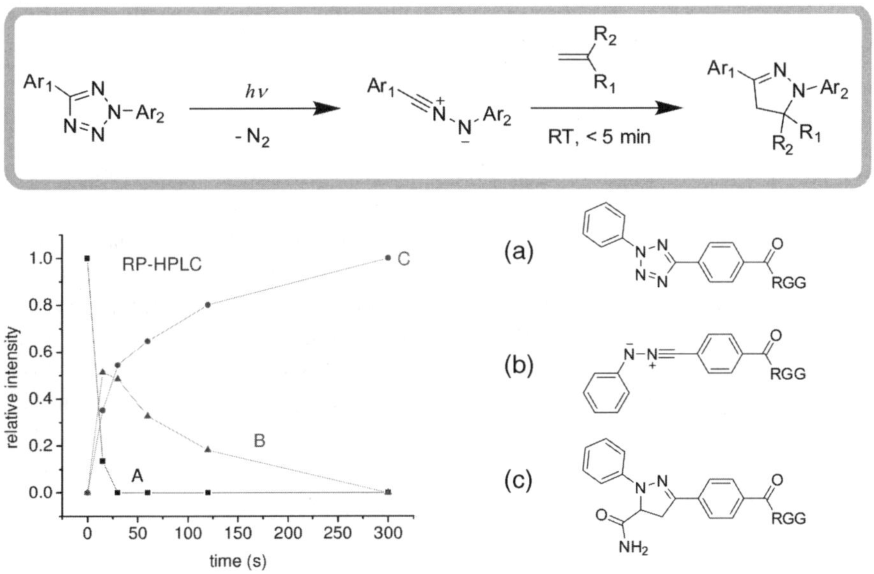

Figure 2.16 Kinetics of the photoinducible cycloaddition between 2,5-diaryl tetrazoles and alkenes (Song *et al.*, 2008). (Reproduced with permission from W. Song, Y. Wang, J. Qu *et al.*, "A photoinducible 1,3-dipolar cycloaddition reaction for rapid, selective modification of tetrazole-containing proteins," *Angewandte Chemie International Edition*, 2008, **47**, 15, 2832–2835. © Wiley-VCH Verlag GmbH & Co. KGaA.)

(Song *et al.*, 2008). It must be noted, however that a 50-fold excess of acrylamide was utilized and 90% conversion was achieved after 2 min, as determined by LC-MS anaylsis (Figure 2.16). Aside from the fast rate of reaction, the *click* conjugates are also fluorescent, which is particularly advantageous for monitoring the labeling of nonfluorescent proteins.

In a further example, a genetically encoded alkene-containing protein in *Escherichia coli* was modified with a library of 2,5-diaryl tetrazoles under similar conditions (Song *et al.*, 2008). Although the photoinduced formation of the nitrile-imine was very fast (2 min), the subsequent cycloaddition with the allyl phenyl ether functional protein was considerably slower (the reaction was allowed to proceed overnight at 4 °C). By comparison, the second-order rate constant of the cycloaddition with allyl phenyl ether was determined to be $0.00202 \pm 0.00007\,M^{-1}\,s^{-1}$, whereas that for acrylamide (as was used in the previous example) was determined to be $0.15\,M^{-1}\,s^{-1}$. A reasonable explanation for this observation is the lower LUMO energy of acrylamide than that of allyl phenyl ether, which offers better orbital overlap with the HOMO of the nitrile imine.

It is apparent from these studies that the photoinducible 1,3-cycloaddition between 2,5-diaryl tetrazoles and alkenes is bio-orthogonal, very rapid, highly selective and may be performed in biological media. Many of the reactions presented in this chapter have comparable reaction kinetics, many times at the expense of complicated synthetic strategies to arrive at the *clickable* functionality. In the present case, however, functional 2,5-diaryl tetrazoles are readily synthesized and may be purified by simple nonchromatographic techniques. The additional element of control that is offered by this reaction in terms of its UV trigger makes it a highly attractive inclusion in the design of other conjugates that may be desired in macromolecular science. As such, its implementation in the construction of functional materials would be of significant value.

2.12 Concluding Remarks

Synthetic polymers are finding more and more applications in a wide variety of fields of human endeavor. If one were to cite two recent milestone developments in polymer chemistry that have contributed to this, CRP and *click* chemistry would arguably be agreed to have provided the greatest impact. As such, the continued development of these methodologies to meet the unprecedented high demand for highly specialized materials lies at the foundation of modern polymer science. In an attempt to aid the successful continuation of such a development, the present chapter provides a critical overview of the very latest in highly efficient, ultrarapid and mild conjugation strategies for use in macromolecular chemistry. Furthermore, some strategies that have not yet been fully exploited within the context of macromolecular conjugations have been highlighted in the hope that future research steers in such a direction as to make use of these very latest inclusions in the *click* chemistry toolbox.

References

Agard, N.J., Prescher, J.A., and Bertozzi, C.R. (2004) A strain-promoted [3 + 2] azide-alkyne cycloaddition for covalent modification of biomolecules in living systems. *Journal of the American Chemical Society*, **126**, 15046.

Agard, N.J., Baskin, J.M., Prescher, J.A. *et al.* (2006) A comparative study of bioorthogonal reactions with azides. *ACS Chemical Biology*, **1**, 644.

Bailey, W.J. (1972) in *Step Growth Polymerizations* (ed. D.H. Solomon), Marcel Dekker, New York.

Barner-Kowollik, C. and Inglis, A.J. (2009) Has click chemistry lead to a paradigm shift in polymer materials design? *Macromolecular Chemistry and Physics*, **210**, 987.

Baskin, J.M., Prescher, J.A., Laughlin, S.T. *et al.* (2007) Copper-free click chemistry for dynamic in vivo imaging. *Proceedings of the National Academy of Sciences of the United States of America*, **104**, 16793.

Becer, C.R., Hoogenboom, R., and Schubert, U.S. (2009) Click chemistry beyond metal-catalyzed cycloaddition. *Angewandte Chemie-International Edition*, **48**, 4900.

Bertin, A. and Schlaad, H. (2009) Mild and versatile (bio-) functionalization of glass surfaces via thiol-ene photochemistry. *Chemistry of Materials*, **21**, 5698.

Binder, W.H. and Sachsenhofer, R. (2008) 'Click' chemistry in polymer and materials science: an update. *Macromolecular Rapid Communications*, **29**, 952.

Blackman, M.L., Royzen, M., and Fox, J.M. (2008) Tetrazine ligation: fast bioconjugation based on inverse-electron-demand Diels-Alder reactivity. *Journal of the American Chemical Society*, **130**, 13518.

Braunecker, W.A. and Matyjaszewski, K. (2007) Controlled/living radical polymerization: features, developments, and perspectives. *Progress in Polymer Science*, **32**, 93.

Campos, L.M., Killops, K.L., Sakai, R. *et al.* (2008) Development of thermal and photochemical strategies for thiol-ene click polymer functionalization. *Macromolecules*, **41**, 7063.

Chan, J.W., Yu, B., Hoyle, C.E., and Lowe, A.B. (2008) Convergent synthesis of 3-arm star polymers from RAFT-prepared poly(*N,N*-diethylacrylamide) via a thiol-ene click reaction. *Chemical Communications*, 4959.

Chan, J.W., Hoyle, C.E., and Lowe, A.B. (2009a) Sequential phosphine-catalyzed, nucleophilic thiol-ene/radical-mediated thiol-yne reactions and the facile orthogonal synthesis of polyfunctional materials. *Journal of the American Chemical Society*, **131**, 5751.

Chan, J.W., Yu, B., Hoyle, C.E., and Lowe, A.B. (2009b) The nucleophilic, phosphine-catalyzed thiol-ene click reaction and convergent star synthesis with RAFT-prepared homopolymers. *Polymer*, **50**, 3158.

Chen, G.J., Kumar, J., Gregory, A., and Stenzel, M.H. (2009) Efficient synthesis of dendrimers via a thiol-yne and esterification process and their potential application in the delivery of platinum anti-cancer drugs. *Chemical Communications*, 6291.

Codelli, J.A., Baskin, J.M., Agard, N.J., and Bertozzi, C.R. (2008) Second-generation difluorinated cyclooctynes for copper-free click chemistry. *Journal of the American Chemical Society*, **130**, 11486.

Dag, A., Durmaz, H., Demir, E. *et al.* (2008) Heterograft copolymers via double click reactions using one-pot technique. *Journal of Polymer Science Part A-Polymer Chemistry*, **46**, 6969.

Dag, A., Durmaz, H., Tunca, U., and Hizal, G. (2009) Multiarm star block copolymers via Diels-Alder click reaction. *Journal of Polymer Science Part A-Polymer Chemistry*, **47**, 178.

Debets, M.F., van Berkel, S.S., Schoffelen, S. *et al.* (2010) Aza-dibenzocyclooctynes for fast and efficient enzyme PEGylation via copper-free $(3+2)$ cycloaddition. *Chemical Communications*, **46**, 97.

Devaraj, N.K., Weissleder, R., and Hilderbrand, S.A. (2008) Tetrazine-based cycloadditions: application to pretargeted live cell imaging. *Bioconjugate Chemistry*, **19**, 2297.

Devaraj, N.K., Upadhyay, R., Hatin, J.B. *et al.* (2009) Fast and sensitive pretargeted labeling of cancer cells through a tetrazine/trans-cyclooctene cycloaddition. *Angewandte Chemie-International Edition*, **48**, 7013.

Durmaz, H., Colakoglu, B., Tunca, U., and Hizal, G. (2006a) Preparation of block copolymers via Diels-Alder reaction of maleimide- and anthracene-end functional polymers. *Journal of Polymer Science Part A-Polymer Chemistry*, **44**, 1667.

Durmaz, H., Karatas, F., Tunca, U., and Hizal, G. (2006b) Preparation of ABC miktoarm star terpolymer containing poly(ethylene glycol), polystyrene, and poly(tert-butylacrylate) arms by combining Diels-Alder reaction, atom transfer radical, and stable free radical polymerization routes. *Journal of Polymer Science Part A-Polymer Chemistry*, **44**, 499.

Durmaz, H., Dag, A., Altintas, O. *et al.* (2007) One-pot synthesis of ABC type triblock copolymers via in situ click $[3+2]$ and Diels-Alder $[4+2]$ reactions. *Macromolecules*, **40**, 191.

Fournier, D., Hoogenboom, R., and Schubert, U.S. (2007) Clicking polymers: a straightforward approach to novel macromolecular architectures. *Chemical Society Reviews*, **36**, 1369.

Gauthier, M.A., Gibson, M.I., and Klok, H.A. (2009) Synthesis of functional polymers by post-polymerization modification. *Angewandte Chemie-International Edition*, **48**, 48.

Gouin, S.G. and Kovensky, J. (2009) A procedure for fast and regioselective copper-free click chemistry at room temperature with p-toluenesulfonyl alkyne. *Synlett*, 1409.

Grecian, S. and Fokin, V.V. (2008) Ruthenium-catalyzed cycloaddition of nitrile oxides and alkynes: practical synthesis of isoxazoles. *Angewandte Chemie-International Edition*, **47**, 8285.
Gress, A., Volkel, A., and Schlaad, H. (2007) Thio-click modification of poly[2-(3-butenyl)-2-oxazoline]. *Macromolecules*, **40**, 7928.
Grundmann, C. and Frommeld, H.D. (1966) Nitrile oxides. 6. Some novel addition reactions of aromatic nitrile oxides. *The Journal of Organic Chemistry*, **31**, 157.
Gungor, E., Hizal, G., and Tunca, U. (2009) One-pot double click reactions for the preparation of H-shaped ABCDE-type quintopolymer. *Journal of Polymer Science Part A-Polymer Chemistry*, **47**, 3409.
Gutsmiedl, K., Wirges, C.T., Ehmke, V., and Carell, T. (2009) Copper-free “click” modification of DNA via nitrile oxide-norbornene 1,3-dipolar cycloaddition. *Organic Letters*, **11**, 2405.
Hawker, C.J. and Wooley, K.L. (2005) The convergence of synthetic organic and polymer chemistries. *Science*, **309**, 1200.
Hay, J.N., Boyle, J.D., Parker, S.F., and Wilson, D. (1989) Polymerization of N-phenylnadimide: a model for the cross-linking of PMR-15 polyimide. *Polymer*, **30**, 1032.
Hensarling, R.M., Doughty, V.A., Chan, J.W., and Patton, D.L. (2009) “Clicking” polymer brushes with thiol-yne chemistry: indoors and out. *Journal of the American Chemical Society*, **131**, 14673.
Heredia, K.L., Tolstyka, Z.P., and Maynard, H.D. (2007) Aminooxy end-functionalized polymers synthesized by ATRP for chemoselective conjugation to proteins. *Macromolecules*, **40**, 4772.
Himo, F., Lovell, T., Hilgraf, R. *et al.* (2005) Copper(I)-catalyzed synthesis of azoles. DFT study predicts unprecedented reactivity and intermediates. *Journal of the American Chemical Society*, **127**, 210.
Hoyle, C.E., Lee, T.Y., and Roper, T. (2004) Thiol-enes: chemistry of the past with promise for the future. *Journal of Polymer Science Part a-Polymer Chemistry*, **42**, 5301.
Inglis, A.J., Sinnwell, S., Davis, T.P. *et al.* (2008) Reversible addition fragmentation chain transfer (RAFT) and hetero-Diels-Alder chemistry as a convenient conjugation tool for access to complex macromolecular designs. *Macromolecules*, **41**, 4120.
Inglis, A.J., Sinnwell, S., Stenzel, M.H., and Barner-Kowollik, C. (2009a) Ultrafast click conjugation of macromolecular building blocks at ambient temperature. *Angewandte Chemie-International Edition*, **48**, 2411.
Inglis, A.J., Paulöhrl, T., and Barner-Kowollik, C. (2009b) Ambient temperature synthesis of a versatile macromolecular building block: cyclopentadienyl-capped polymers. *Macromolecules*. **43**, 33.
Inglis, A.J., Stenzel, M.H., and Barner-Kowollik, C. (2009c) Ultra-fast RAFT-HDA click conjugation: an efficient route to high molecular weight block copolymers. *Macromolecular Rapid Communications*, **30**, 1792.
Kamahori F K., Tada, S., Ito, K., and Itsuno, S. (1999) Optically active polymer synthesis by Diels-Alder polymerization with chirally modified Lewis acid catalyst. *Macromolecules*, **32**, 541.
Killops, K.L., Campos, L.M., and Hawker, C.J. (2008) Robust, efficient, and orthogonal synthesis of dendrimers via thiol-ene “click” chemistry. *Journal of the American Chemical Society*, **130**, 5062.
Klemm, E. and Stöckl, C. (1991) Synthesis of SH- and NCO- terminated thiocarbamate prepolymers. *Makromolekulare Chemie-Macromolecular Chemistry and Physics*, **192**, 153.
Köhn, M. and Breinbauer, R. (2004) The Staudinger ligation - a gift to chemical biology. *Angewandte Chemie-International Edition*, **43**, 3106.
Kolb, H.C., Finn, M.G., and Sharpless, K.B. (2001) Click chemistry: diverse chemical function from a few good reactions. *Angewandte Chemie-International Edition*, **40**, 2004.
Konkolewicz, D., Gray-Weale, A., and Perrier, S. (2009) Hyperbranched polymers by thiol-yne chemistry: from small molecules to functional polymers. *Journal of the American Chemical Society*, **131**, 18075.
Koo, S.P.S., Stamenovic, M.M., Prasath, A.R. *et al.* (2010) Limitations of radical thiol-ene reactions for polymer-polymer conjugation. *Journal of Polymer Science Part A-Polymer Chemistry*, **48**, 1699.
Koyama, Y., Yonekawa, M., and Takata, T. (2008) New click chemistry: click polymerization via 1,3-dipolar addition of homo-ditopic aromatic nitrile oxides formed in situ. *Chemistry Letters*, **37**, 918.
Lee, Y.G., Koyama, Y., Yonekawa, M., and Takata, T. (2009) New click chemistry: polymerization based on 1,3-dipolar cycloaddition of a homo ditopic nitrile N-oxide and transformation of the resulting polymers into reactive polymers. *Macromolecules*, **42**, 7709.
Li, Z., Seo, T.S., and Ju, J. (2004) 1,3-Dipolar cycloaddition of azides with electron-deficient alkynes under mild condition in water. *Tetrahedron Letters*, **45**, 3143.
Li, H.B., Yu, B., Matsushima, H. *et al.* (2009) The thiol-isocyanate click reaction: facile and quantitative access to omega-end-functional poly(N,N-diethylacrylamide) synthesized by RAFT radical polymerization. *Macromolecules*, **42**, 6537.

Lowe, A.B. (2010) Thiol-ene "click" reactions and recent applications in polymer and materials synthesis. *Polymer Chemistry*, **1**, 17.

Lutz, J.F. (2007) 1,3-dipolar cycloadditions of azides and alkynes: a universal ligation tool in polymer and materials science. *Angewandte Chemie-International Edition*, **46**, 1018.

Lutz, J.F. (2008) Copper-free azide-alkyne cycloadditions: new insights and perspectives. *Angewandte Chemie-International Edition*, **47**, 2182.

McKay, C.S., Moran, J., and Pezacki, J.P. (2010) Nitrones as dipoles for rapid strain-promoted 1,3-dipolar cycloadditions with cyclooctynes. *Chemical Communications*, **46**, 931.

Moad, G., Rizzardo, E., and Thang, S.H. (2008) Radical addition-fragmentation chemistry in polymer synthesis. *Polymer*, **49**, 1079.

Moad, G., Rizzardo, E., and Thang, S.H. (2009) Living radical polymerization by the RAFT process - a second update. *Australian Journal of Chemistry*, **62**, 1402.

Nebhani, L., Sinnwell, S., Inglis, A.J. *et al.* (2008) Efficient surface modification of divinylbenzene microspheres via a combination of RAFT and hetero Diels-Alder chemistry. *Macromolecular Rapid Communications*, **29**, 1431.

Nebhani, L., Sinnwell, S., Lin, C.Y. *et al.* (2009a) Strongly electron deficient sulfonyldithioformate based RAFT agents for hetero Diels-Alder conjugation: computational design and experimental evaluation. *Journal of Polymer Science Part A-Polymer Chemistry*, **47**, 6053.

Nebhani, L., Gerstel, P., Atanasova, P. *et al.* (2009b) Efficient and mild modification of Si surfaces via orthogonal hetero Diels-Alder chemistry *Journal of Polymer Science Part a-Polymer Chemistry*, **47**, 7090.

Nenov, N., Koch, M., Klapper, M., and Mullen, K. (2002) PEO-functionalized polystyrene as polymeric support in metallocene catalysed olefin polymerisation. *Polymer Bulletin*, **47**, 391.

Ning, X.H., Guo, J., Wolfert, M.A., and Boons, G.J. (2008) Visualizing metabolically labeled glycoconjugates of living cells by copper-free and fast huisgen cycloadditions. *Angewandte Chemie-International Edition*, **47**, 2253.

Ouchi, M., Terashima, T., and Sawamoto, M. (2009) Transition metal-catalyzed living radical polymerization: toward perfection in catalysis and precision polymer synthesis. *Chemical Reviews*, **109**, 4963.

Pounder, R.J., Stanford, M.J., Brooks, P. *et al.* (2008) Metal free thiol-maleimide 'Click' reaction as a mild functionalisation strategy for degradable polymers. *Chemical Communications*, 5158.

Quemener, D., Davis, T.P., Barner-Kowollik, C., and Stenzel, M.H. (2006) RAFT and click chemistry: A versatile approach to well-defined block copolymers. *Chemical Communications*, 5051.

Rosen, B.M., Lligadas, G., Hahn, C., and Percec, V. (2009a) Synthesis of dendrimers through divergent iterative thio-bromo "Click" chemistry. *Journal of Polymer Science Part A-Polymer Chemistry*, **47**, 3931.

Rosen, B.M., Lligadas, G., Hahn, C., and Percec, V. (2009b) Synthesis of dendritic macromolecules through divergent iterative thio-bromo "Click" chemistry and SET-LRP. *Journal of Polymer Science Part A-Polymer Chemistry*, **47**, 3940.

Royzen, M., Yap, G.P.A., and Fox, J.M. (2008) A photochemical synthesis of functionalized trans-cyclooctynes driven by metal complexation. *Journal of the American Chemical Society*, **130**, 3760.

Shao, H.Y., Crnogorac, M.M., Kong, T. *et al.* (2005) Site-specific polymer attachment to a CCL-5 (RANTES) analogue by oxime exchange. *Journal of the American Chemical Society*, **127**, 1350.

Shin, J., Matsushima, H., Chan, J.W., and Hoyle, C.E. (2009) Segmented polythiourethane elastomers through sequential thiol-ene and thiol-isocyanate reactions. *Macromolecules*, **42**, 3294.

Singh, I., Zarafshani, Z., Lutz, J.F., and Heaney, F. (2009a) Metal-free "click" chemistry: efficient polymer modification via 1,3-dipolar cycloaddition of nitrile oxides and alkynes. *Macromolecules*, **42**, 5411.

Singh, I., Vyle, J.S., and Heaney, F. (2009b) Fast, copper-free click chemistry: a convenient solid-phase approach to oligonucleotide conjugation *Chemical Communications*, 3276.

Sinnwell, S., Inglis, A.J., Davis, T.P. *et al.* (2008a) An atom-efficient conjugation approach to well-defined block copolymers using RAFT chemistry and hetero Diels-Alder cycloaddition. *Chemical Communications*, 2052.

Sinnwell, S., Inglis, A.J., Stenzel, M.H., and Barner-Kowollik, C. (2008b) Access to three-arm star block copolymers by a consecutive combination of the copper(I)-catalyzed azide-alkyne cycloaddition and the RAFT hetero Diels-Alder concept. *Macromolecular Rapid Communications*, **29**, 1090.

Song, W., Wang, Y., Qu, J. *et al.* (2008a) A photoinducible 1,3-dipolar cycloaddition reaction for rapid, selective modification of tetrazole-containing proteins. *Angewandte Chemie-International Edition*, **47**, 2832.

Song, W., Wang, Y., Qu, J., and Lin, Q. (2008b) Selective functionalization of a genetically encoded alkene-containing protein via "photoclick chemistry" in bacterial cells. *Journal of the American Chemical Society*, **130**, 9654.

Stork, M., Koch, M., Klapper, M. *et al.* (1999) Ethylene polymerization using crosslinked polystyrene as support for zirconocene dichloride/methylaluminoxane. *Macromolecular Rapid Communications*, **20**, 210.

Sumerlin, B.S. and Vogt, A.P. (2010) Macromolecular engineering through click chemistry and other efficient transformation. *Macromolecules*, **43**, 1.

ten Brummelhuis, N., Diehl, C., and Schlaad, H. (2008) Thiol-ene modification of 1,2-polybutadiene using UV light or sunlight. *Macromolecules*, **41**, 9946.

Watson, K.A. and Bass, R.G. (2000) Polymer synthesis via a Diels-Alder reaction between bis(isobenzofuran)s and bis (phenylacetylene)s containing preformed phthalimide rings. *High Performance Polymers*, **12**, 299.

Wittig, G. and Krebs, A. (1961) Zur existenz niedergliedriger cycloalkine, I. *Chemische Berichte*, **94**, 3260.

Xu, J., Tao, L., Boyer, C. *et al.* (2010) Combining thio-bromo "click" chemistry and RAFT polymerization: a powerful tool for preparing functionalized multiblock and hyperbranched polymers. *Macromolecules*. **43**, 20.

3

Synthesis and Self-Assembly of Hydrogen-Bonded Supramolecular Polymers

Wolfgang H. Binder, Claudia Enders, Florian Herbst and Katharina Hackethal

Martin-Luther University Halle-Wittenberg, Lehrstuhl Macromolecular Chemistry, Institute of Chemistry, Faculty of Natural Sciences II (Chemistry and Physics), Germany

3.1 Introduction

Besides conventional ("covalently linked") polymers, supramolecular polymers have gained in importance over the past two decades (Binder and Zirbs, 2007; Binder and Herbst, 2010; Bouteiller, 2007; Brunsveld *et al.*, 2001; ten Brinke *et al.*, 2007; Xu *et al.*, 2007). Supramolecular polymers are polymers linked by defined and often directed noncovalent chemical interactions as the main structural element, thus allowing the construction of polymeric architectures (i.e. homo, block, star, graft, dendritic) in a similar manner as is known from their covalently linked counterparts. Basically, the substitution of covalent bonds into supramolecular bonds serves to introduce an element of dynamic ordering and thus an additional element of structuring. Hydrogen bonds (Wilson, 2007) in particular fulfil many aspects of the "ideal" supramolecular interaction, as they are: (a) directed supramolecular bonds, (b) tunable in their strength over a wide range of binding constants, (c) highly controllable in their dynamics, (d) switchable by addition of small additives with large changes in the association constant and (e) specific in the sense of a lock–key interaction. This strongly discriminates the hydrogen-bonded supramolecular polymers from other supramolecular polymers (i.e. those linked by metal–metal bonds, pi–pi stacking, or dipole–dipole forces) and simple blend mixing, generated by nondirected interactions (Jiang *et al.*, 1999).

Complex Macromolecular Architectures: Synthesis, Characterization, and Self-Assembly, First Edition.
Edited by Nikos Hadjichristidis, Akira Hirao, Yasuyuki Tezuka and Filip Du Prez.

The basic principles of supramolecular polymer chemistry were introduced by Lehn *et al.* (Berl *et al.*, 2002) and by Stadler *et al.* (de Lucca Freitas and Stadler, 1988; Stadler, 1993a) either as a "small-molecule ordering" into polymeric molecules or "crosslinking" of polymers by statistically distributed hydrogen-bonding moieties. Stadler's concept relied on the statistical introduction of hydrogen bonds into the polybutadiene, thus creating defined noncovalent crosslinking sites between the polymer chains. It was found that a hydrogen-binding donor–acceptor unit for each 50 butadiene units (i.e. at approx. 2 mol%) is sufficient to significantly change the thermal and mechanical properties of the material, thus demonstrating that a couple of weak bonds on the side of a polymer chain can be very efficient in determining the final materials properties. By use of these concepts, polymers became a much larger group of materials, which now included even higher molecular weight aggregates with thermally reversible linkages at room temperature or below. Intermolecular hydrogen bonds at defined positions of these building blocks as well as their respective starting geometry and the initial size determine the mode of assembly into supramolecular polymers. Ordering can form network, rod-like, fibrous, disc-like, helical, lamellar and chain-like structures (Figure 3.1). The concept of "highly ordered" structures can be extended to planar surfaces such as Langmuir–Blodgett films, self-assembled monolayers or thin films. This includes the arrangement of nanosized objects (i.e. nanoparticles) on polymeric thin films to generate defined order by use of directed hydrogen bonds.

More recently (Cordier *et al.*, 2008; Kim *et al.*, 2009a; Montarnal *et al.*, 2008), the dynamics in supramolecular polymers have been recognized not only as a principle of structuring but also as a corrective force. Thus, elastomeric properties can be induced by hydrogen bonds, able to repair materials after scission of bonds – that is, generating self-healing materials. In this respect, the use of hydrogen bonds is outstanding, as the dynamics of the H bond can be directly transformed into the dynamics of the underlying polymeric material.

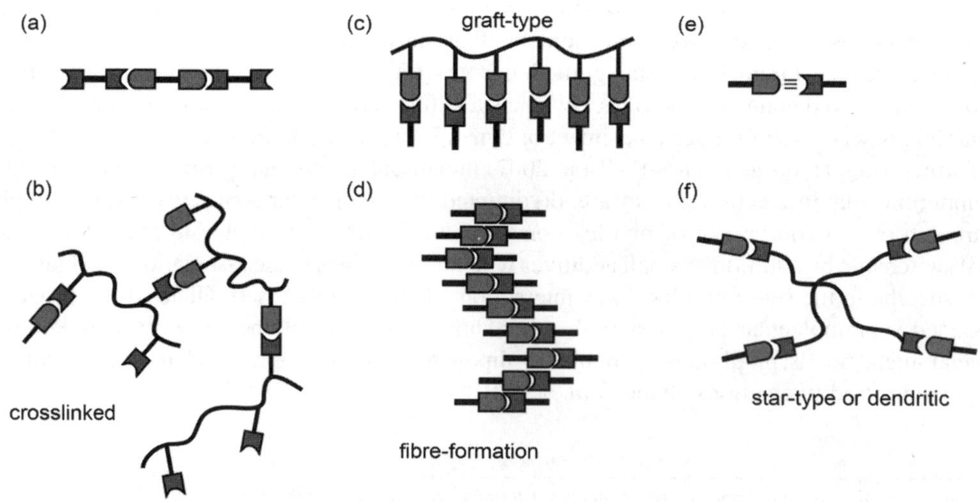

Figure 3.1 Architectures of hydrogen-bonded supramolecular polymers.

The present chapter deals with on synthetic and structural concepts to prepare hydrogen-bonded polymers, with a focus on the synthetic methodologies to affix hydrogen bonds efficiently to polymeric building blocks. Basically, hydrogen-bond-containing molecules are highly to moderately reactive compounds, which require efficient chemistry for their binding onto the respective polymer chain. The present review focuses on literature dated 2000–11/2009, many of them in junction with living/controlled polymerization methods, to position hydrogen-bonding moieties on macromolecules mostly based on our own work in this field. The synthetic methodology towards H-bonded polymers was reviewed by us in 2004 (Farnik *et al.*, 2004), therefore, this chapter focuses on the references published thereafter. Especially with the advent of highly efficient chemical methods ("click" chemistry) (Binder and Kluger, 2006; Binder and Sachsenhofer, 2007, 2008; Binder and Zirbs, 2009; Kolb *et al.*, 2001; Meldal and Tornoe, 2001; Tornoe *et al.*, 2002), the possibility to affix hydrogen bonds to polymers has been strongly improved in terms of yield, ease and overall efficiency. There are a number of recent reviews relating to the topic of hydrogen-bonded supramolecular polymers, mostly with a focus on specific hydrogen bonds (Jahnke *et al.*, 2008; Sessler *et al.*, 2007; Todd and Zimmerman, 2008), biomaterials (Dankers and Meijer, 2007) or liquid crystallinity (Hammond and Mezzenga, 2008). Section 3.2 will give an overview on the various hydrogen bonds used in supramolecular polymer science, as well as an account of their dynamics and the experimental methods used to elucidate their structure and dynamics. Section 3.3 will provide synthetic methods to prepare hydrogen-bonded polymers, mostly related to living polymerization methods and their combination with hydrogen-bonding systems. Section 3.4 will describe some newer aspects (with a focus on 2004–2009) of polymeric ordering by hydrogen bonds, in particular the generation of microphase-separated structures, the ordering of supramolecular polymers on surfaces, the ordering and dynamics of small molecules into polymeric aggregates (linear chains, tubes, rods and fibers) and finally applications of H-bonded polymers in science and technology, focusing on aspects of optoelectronics, bio- and self-healing materials.

3.1.1 Dynamics of Hydrogen Bonds

Three different classes of hydrogen-bonding systems can be discriminated (a) strong hydrogen bonds, (b) medium or weak hydrogen bond and (c) nonclassical hydrogen bonds. Strong two-center hydrogen bonds (such as $F\text{–}H\cdots F^-$; $O\text{–}H\cdots O^-$; $O^+\text{–}H\cdots O$ bonds) display short distances, a strongly directional nature and association energies higher than $\sim$40 kJ mol^{-1}. Medium and weak hydrogen bonds are classified by a $D\text{–}H\cdots A$ structure with bond energies between 20–40 kJ mol^{-1}. Usually, the acceptor-residue A is strongly electronegative, whereas the donor-residue D may be electronegative or even a carbon atom. Nonclassical (also termed unconventional) hydrogen bonds involve the interaction of D–H with $A = \pi$-systems as well as transition metals (interaction directly with the metal or via the metal hydride) or boron hydrides.

Basically, one hydrogen bond contributes about 7.4 kJ mol^{-1}/hydrogen bond, leading to stronger bonds if more hydrogen bonds are involved in the respective functional group. Additional structural contributions influencing the association strength include preorganization, cooperativity and secondary electrostatic effects (Wilson, 2007). More recently (Lafitte *et al.*, 2006; Park *et al.*, 2005), tautomerization has been included as a direct functional element able to control the strength and selectivities of H-bonding systems. H bonds often used in supramolecular polymers are displayed in Figure 3.2, depicting (a) single hydrogen bonds

(Takashi Kato, 2006) (b, c) double and triple hydrogen bonds (Binder and Zirbs, 2007; Binder *et al.*, 2004a; Bouteiller, 2007; Kunz *et al.*, 2004b; Lin *et al.*, 2008; Binder *et al.*, 2010; Sessler *et al.*, 2007; Sivakova and Rowan, 2005) and (d) multiple hydrogen bonds with often complex patterns of H-bonding interactions, often fixed into rigid (aromatic) systems (Binder *et al.*, 2005a; Binder *et al.*, 2006; Binder *et al.*, 2005b; Olobummo and Binder, 2011; Binder *et al.*, 2004a; Binder *et al.*, 2004a; Jahnke *et al.*, 2008; Kunz *et al.*, 2004a; Lafitte *et al.*, 2006; Park *et al.*, 2005; Sijbesma *et al.*, 1997; Sinkeldam *et al.*, 2006). Suffice it to say that especially multiple H bonds lead to stronger effects in aggregation, as higher-molecular weight aggregates can be formed with increasing association strength (Bosnian *et al.*, 2003).

More important than the individual binding strength are the underlying bonding dynamics of hydrogen bonds. Certainly, the strength of each individual hydrogen bond is strongly dependent on solvent effects, most of all polar and protic solvents as well as temperature effects, both influencing the dynamics of the H-bond. Therefore, the supramolecular chemistry of hydrogen-bonded polymers is mostly done in aprotic and nonpolar solvent such as linear and cyclic alkanes, toluene, dichloromethane and chloroform. Moreover, the often attached length of a polymer chain can influence the association behavior by formation of hydrophobic pockets, such as demonstrated recently by Bouteiller *et al.* (Obert *et al.*, 2007a). Thus, small ethylene-oxide units bound in close proximity to H-bonds can lead to a strong reduction

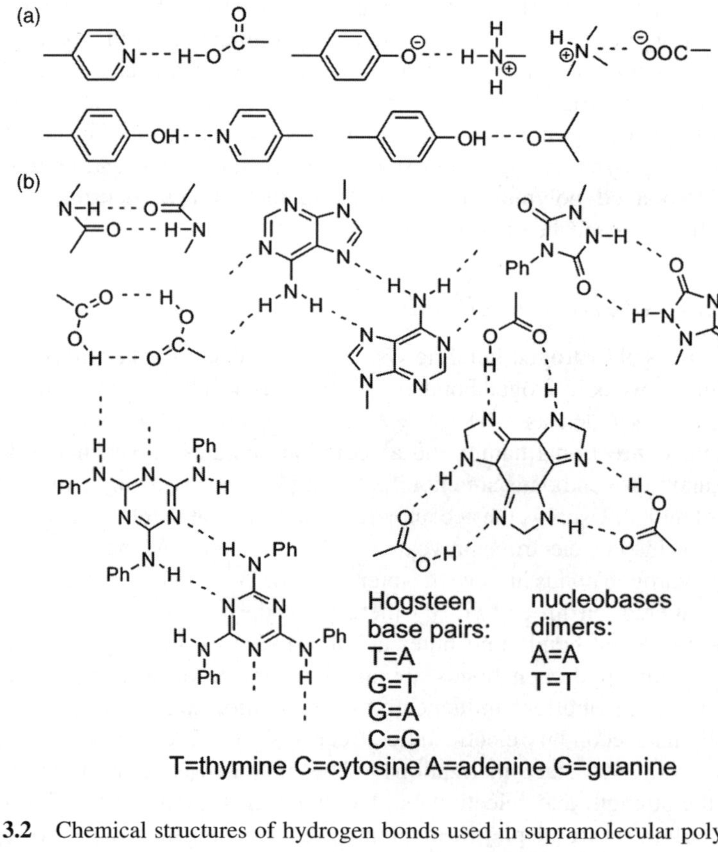

Figure 3.2 Chemical structures of hydrogen bonds used in supramolecular polymer science.

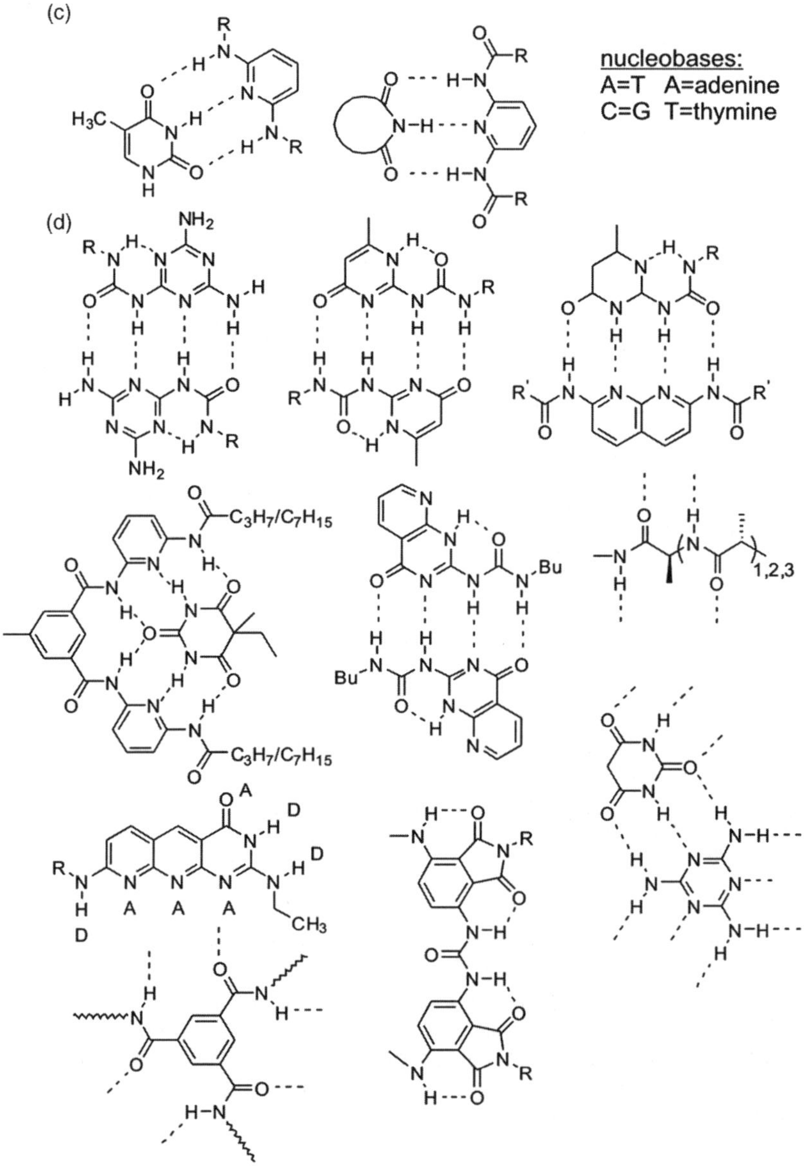

Figure 3.2 (*Continued*)

(up to 1000-fold) of the association constant due to hydrophilic effects between the EO units and the hydrogen-bonding system (Greef *et al.*, 2008).

The basic dynamics of hydrogen bonds can be described according to Figure 3.3a, represented by a simple A (acceptor) plus B (donor) equilibrium. Due to the temperature and concentration dependence of ($K_{assn.}$), changes in association can be followed by NMR-spectroscopy, as shown in Figure 3.3b (Kunz *et al.*, 2004b). Thus, association constants

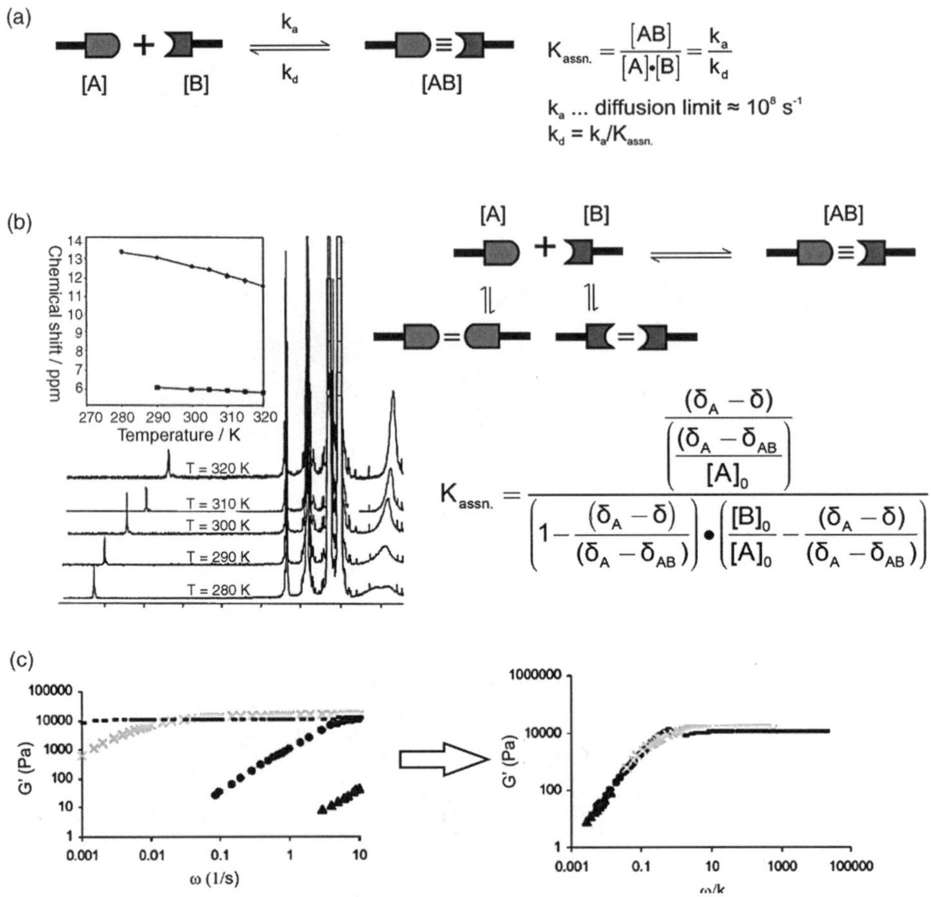

Figure 3.3 Dynamics of H bonds: (a) dynamic equilibrium characterized via the association constant ($K_{assn.}$). (b) Dynamic properties of supramolecular polymer as seen in temperature-dependent NMR-spectroscopy according to Kunz *et al.* (2004a). (Reprinted from M.J. Kunz, G. Hayn, R. Saf and W.H. Binder "Hydrogen-bonded supramolecular poly(ether ketone)s," *Journal of Polymer Science Part A: Polymer Chemistry*, **42**, 661–674. © 2004, with the permission of John Wiley & Sons, Inc.) ($K_{assn.}$) can be calculated by concentration-dependent measurements via the chemical shift (δ) of: the associating species (δ_A) with the concentration $[A]_o$; (δ_B) with the concentration $[B]_o$ and the aggregate (δ_{AB}). (c) (Right graph): Dependence of the storage modulus (G′) vs. the frequency (ω) (left graph) G' vs. ω/k_d – according to Yount, Loveless and Craig (2005a). (Reprinted with permission from W.C. Yount, D.M. Loveless and S.L. Craig, "Small-molecule dynamics and mechanisms underlying the macroscopic mechanical properties of coordinatively cross-linked polymer networks," *Journal of the American Chemical Society*, **127**, 41, 14488–14496, 2005. © 2005 American Chemical Society.)

($K_{assn.} = k_a/k_d$) can be calculated via determination of the concentration of each of the species [A], [B], or [AB], often by variation of the concentration of the fraction of [A]/[B] and monitoring the shift of the N-H resonances within the hydrogen bonds (Macomber, 1992). Therefore, ($K_{assn.}$) can be obtained by fitting the equation shown in Figure 3.3b, as demonstrated also on telechelic polymers (Binder *et al.*, 2004a). Association constants are usually comparable between model-type H bonds and those affixed to the respective longer

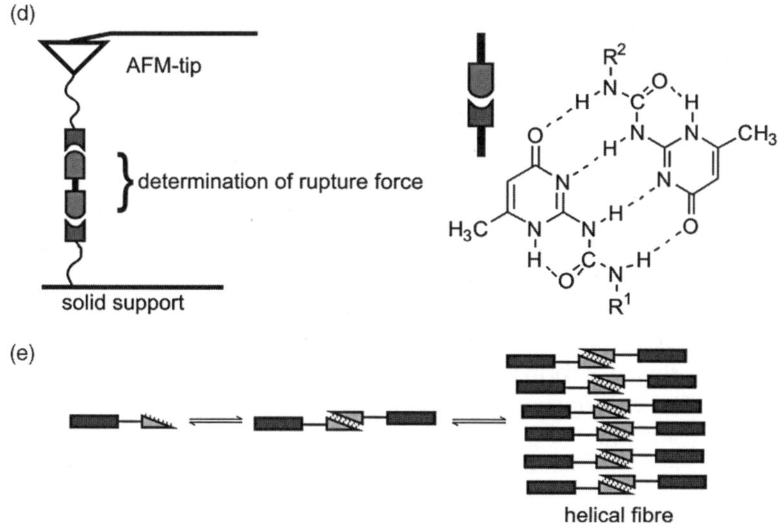

Figure 3.3 (*Continued*)

polymer chains. In solution, dynamics are often described by a diffusion-controlled association ($k_a \sim 10^8\,s^{-1}$), thus easily transforming the (thermodynamic) association constant ($K_{assn.}$) into the kinetic constant of dissociation of the H bond (k_d) (Hammes and Park, 1968; Sontjens *et al.*, 2000). The value $1/k_d$ is then a measure of the average lifetime of the respective H bond. Alternatively, isothermal titration calorimetry (ITC) (Arnaud *et al.*, 2004; Bouteiller *et al.*, 2005) is a highly efficient tool to determine the association energy and extracting free enthalpy, entropy and enthalpy of the binding process. Also, IR spectroscopy in solution and in the solid state as well as temperature-dependent SAXS methods have been found to be powerful tools to investigate nature and dynamics of the hydrogen bonds (Binder *et al.*, 2005a; Binder *et al.*, 2007b).

Another important method to determine dynamics in supramolecular systems is the measurement of viscoelastic properties, mostly in gelating systems. Craig *et al.* (Kersey *et al.*, 2007; Loveless *et al.*, 2006; Serpe and Craig, 2007; Yount *et al.*, 2005a; Yount *et al.*, 2005b) and Stuart *et al.* (Knoben *et al.*, 2005; Knoben *et al.*, 2007; Vermonden *et al.*, 2004) have determined the dynamics of crosslinked networks (as gels) via their storage-moduli (G'). Thus, crosslinked systems displayed significant falloffs of G' vs. the measuring frequency (ω) (Figure 3.3c, left). Scaling the storage modulus (G') with the rate of dissociation (k_d) leads to a superposition of the curves and allows to estimate the (k_d) values of the crosslinked materials (Figure 3.3c, right). An indirect measure of dynamics in the solid state can be achieved either by dielectric-relaxation spectroscopy (Herbst and Binder, 2010; Müller *et al.*, 1995; Muller *et al.*, 1995; Shikata *et al.*, 2008) or via DMA measurements (de Lucca Freitas and Stadler, 1988; Keizer *et al.*, 2003; Mather *et al.*, 2007b; Wisse *et al.*, 2006a; Wisse *et al.*, 2009) on solid samples. However, in both cases no direct values of (k_d) can be obtained, as often the inherent dynamics of the entangled polymer chains overrule the dynamics of the H bond itself.

A direct measurement of H-bonding interaction has been achieved by Vancso *et al.* (Embrechts *et al.*, 2008; Vancso, 2007; Zou *et al.*, 2005) by fixing two complementing

H bonds between a solid surface and an AFM-tip (Figure 3.3d). Addition of associating building blocks enabled measurement of the individual rupture force within the growing supramolecular system. This method is unique to determine binding forces in relatively strongly associating systems (such as the ureidopyrimidones). However, it is not useful for medium-to-weak H-binding systems.

Alternatively, solid-state NMR-spectroscopy (^{13}C-MAS- and ^{1}H-MAS-NMR, the latter usually under fast spinning conditions) can provide deeper insight into the dynamics of supramolecular polymers (Brown, 2009; Spiess, 2003). Thus, chain dynamics can be measured via the respective $T_{1\rho H}$ times in solid pseudoblock copolymers (Binder *et al.*, 2004a), whereas direct binding dynamics of the H bond can be monitored via ^{1}H-MAS-NMR while spinning at 35–50 kHz. ^{1}H–^{1}H DQ (double quantum) spectra (Schnell *et al.*, 2002; Wasserfallen *et al.*, 2005) can provide information about proton exchange and association dynamics, whereas ^{1}H-CRAMPS experiments lead to enhanced resolution of even strongly overlapping types of H bonds (Li *et al.*, 2007).

An important aspect of dynamics in solution concerns the generation of fibers and helical aggregates (Jonkheijm *et al.*, 2006b). Based on a model by Meijer, the aggregation of small molecules bearing self-complementary ureidopyrimidones (Figure 3.3e) or benzene-1,3,5-tricarboxamides (Stals *et al.*, 2009) were studied by CD spectroscopy. A solvent-assisted cooperative association pathway, initiated by nucleation was observed. This pathway is reminiscent of many biological systems.

3.1.2 Other Experimental Methods in Supramolecular Polymer Science

Besides the specific methods described in Section 3.1.1, which were directed towards the analysis of bonding strength and dynamics of the H bond, a large variety of nonspecific methods have been used to characterize supramolecular systems. In particular, rheological methods in melt and solution have been widely used to address the mechanical properties of associated supramolecular systems (de la Fuente *et al.*, 2006; Langela *et al.*, 2002; Oelschlaeger *et al.*, 2007). Thus, melt rheology, often in junction with IR or scattering methods can be used to elucidate the microphase behavior of block copolymers. As supramolecular polymers are often highly dynamic systems, there is an expected influence of the applied shear on the structure formation of the respective BCP (Bondzic *et al.*, 2007; Oelschlaeger *et al.*, 2007). In addition, the formation of transient networks can be elucidated via these rheological measurements (Stadler *et al.*, 2008). Rheology correlated to dynamic light scattering (DLS) revealed insight into the relaxation process and verified a model, which is in agreement with theoretical predictions based on the reptation model (Buhler *et al.*, 2007). Weakly interacting supramolecular systems have been characterized via a mathematical model, based on NMR-titration values (Odille *et al.*, 2007). Positron-annihilation spectroscopy has been used to study order–disorder transitions in self-assembled polymers (Ramani *et al.*, 2009).

3.2 Synthetic Strategies Towards Hydrogen-Bonded Supramolecular Polymers

The attachment of hydrogen-bonding moieties to polymers can be based on several strategies (Figure 3.4). As most of the attachment methods require activated reagents to achieve a complete addition of the hydrogen-bonding moiety to the polymeric chain, the inherent

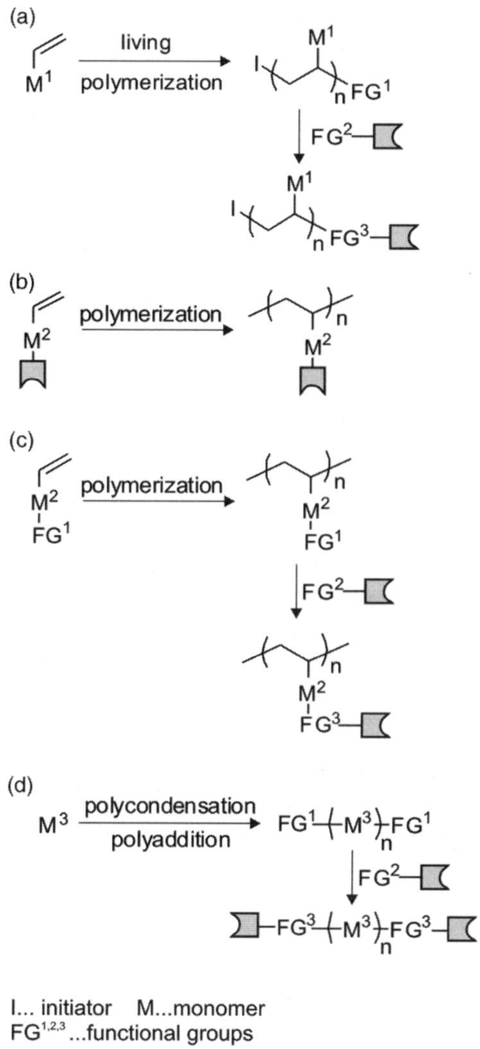

Figure 3.4 Synthetic strategies towards H-bonded supramolecular polymers.

reactivity of the H bond towards these reagents has to be taken into account. Most H bonds contain one or more nucleophilic reaction centers (i.e. N–H; O–C, O=C bonds and other nucleophilic O-, N-, S-heteroatoms), which often act as quenching agents in many living polymerization methods, in particular living anionic and living cationic polymerization. Therefore, these moieties either have to be blocked by protective groups, or reaction conditions must be optimized. Additionally, the functionalization on the polymer chain should be quantitative, as insufficiently functionalized polymer cannot be separated easily from the functionalized polymer structure.

Thus, (Figure 3.4a) a living polymerization can generate an end group-functionalized polymer bearing a reactive group FG^1, which subsequently can be reacted with a H-bonding molecule bearing the crossreactive functional group FG^2, finally generating the telechelic polymer with the H bond attached (Kunz *et al.*, 2004a; Rowan *et al.*, 2003). Alternatively, the H-bonding moiety can also be present in the initiator moiety (I) (Binder *et al.*, 2007a). Of course, competing reactivity between the H bond and either FG^1 or FG^2 must be strictly avoided.

Another strategy (Figure 3.4b) relies on the (co)polymerization using an already functionalized monomer, to generate "graft-type" H-bonded polymers (Binder and Herbst, 2010; Olobummo and Binder, 2011; Dalphond *et al.*, 2002; Ilhan *et al.*, 2001; Kluger and Binder, 2004; Li *et al.*, 2003; Mather *et al.*, 2007a; Pollino *et al.*, 2004; Rieth, Eaton and Coates, 2001; Yamauchi *et al.*, 2003). In this case, the usually living polymerization method must not be disturbed by the presence of the H bond itself. If this is the case the strategy shown in Figure 3.4c must be used, attaching a reactive group FG^1 to the monomer, polymerizing directly and subsequently reacting with a reactive group FG^2, bearing the H-bonding moiety.

If polycondensation reactions are used (Figure 3.4d), usually two reactive end groups (FG^1 = often −OH, −COOH, or $-NH_2$ moieties) are generated, which can be reacted with the H bond bearing the functional group FG^2 to yield the respective final bifunctional polymer (Kunz *et al.*, 2004a; Yamauchi *et al.*, 2004).

Table 3.1 displays some often used combinations of functional groups to achieve attachment of H bonds to polymers. Thus, simple acylation reactions or isocyanate additions have proven useful in some selected cases, especially the use of isocyanates for the attachment of the ureidopyrimidone moiety (Folmer *et al.*, 2000; Hirschberg *et al.*, 1999; Lange *et al.*, 1999;

Table 3.1 Selection of chemical reactivities of the functional groups FG^1 and FG^2, useful to efficiently attach H bonds onto telechelic polymers

FG^1	Type	FG^2
−OH; $-NH_2$	acylation	−COOH/DCC; 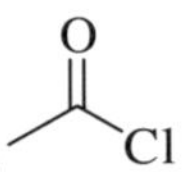
−OH; $-NH_2$	isocyanate addition	−NCO; 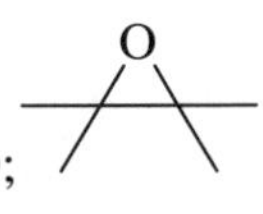
$-N_3$; —≡—H	"click reaction"	—≡—H; $-N_3 + Cu^I$
$-O-CH_2-Cl$	SN_2/activated	N,O nucleophile
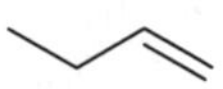	Karstedt reaction	H−Si− / Pt
$-CH_2X$	SN_2/nonactivated	N,O nucleophile
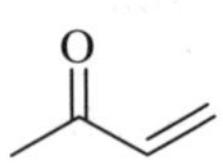	Micheal addition	N nucleophile

Viswanathan *et al.*, 2006; Yamauchi *et al.*, 2004). Another, highly successful example was introduced by our group applying the azide–alkine "click" reaction for the attachment of H bonds to polymers (Binder and Kluger, 2006; Binder and Sachenshofer, 2009; Binder and Sachsenhofer, 2007, 2008; Binder and Zirbs, 2009). This strategy is particularly useful, as the azide–alkine "click" reaction is not interfering with nucleophilic moieties, such as present on the case of H bonds (Binder and Kluger, 2004; Binder *et al.*, 2006; Binder and Herbst, 2010; Olobummo and Binder *et al.*, 2011; Binder *et al.*, 2005b; Binder *et al.*, 2004c; Binder *et al.*, 2007c; Kluger and Binder, 2004, 2005, 2007; Roth *et al.*, 2005). Therefore, this method is nowadays among the most simple and straightforward ones. Finally, SN_2-type reactions of activated (i.e. $O-CH_2-Cl$) (Binder *et al.*, 2004a) or nonactivated ($-CH_2-X$, which $X = Cl$, Br, J) (Kunz *et al.*, 2004a; Ma *et al.*, 2002) substrates as well as epoxide opening (Lee *et al.*, 2007b) can be used to attach the nucleophilic H-bonding moieties. A special case concerns Michael-type addition reactions and the use of the Karstedt-reaction (Hirschberg *et al.*, 1999) for the attachment of H bonds to preformed polymers.

It should be mentioned that living polymerization methods are preferred, as they allow the generation of polymeric architectures (block, star, graft, etc.) and control over block lengths in the final polymers. Therefore, primarily the living polymerization methods in conjunction with the generation of H-bonded supramolecular polymers will be discussed.

3.2.1 *Carbocationic Polymerization*

Living carbocationic polymerization (LCCP) (Faust and Kennedy, 1987; Iván and Kennedy, 1990; Joseph, 1999) relies on complex ion equilibria, which are normally maintained in the absence of polar solvents and monomers. Only specific polar additives (such as di-tert. butyl pyridine, or N,N-dimethyl acetamide) are compatible with a LCCP, so that the direct copolymerization of monomer containing H bonds is not possible. In the same manner, end-quenching reactions with quenchers containing H bonds is not possible, as often only chlorine-telechelic polymers would result.

Most strategies to attach hydrogen bonds to cabocationically polymerized monomers rely on postmodification reaction, using highly efficient attachment strategies to generate telechelic polymers. Thus, (Figure 3.5a), polyisobutylene (PIB) bearing nucelobases may only be prepared by use of bis-allyltelechelic PIB (**1**), which – after transformation into the respective bis-chloromethyl-telechelic PIB (**2**) yielded the final bis-thymine-telechelic PIB (**3**) with silylated nucleobases (**4**) via this strongly activated nucleophilic substitution reaction (Binder *et al.*, 2004a). It must be noted that a "conventional" nucleophilic substitution reaction with the silylated nucleobases on the bromine-telechelic PIB was not successful due to the low solubility of the silylated nucleobases (**4**) in apolar solvents. The corresponding attachment of more complex hydrogen-bonding systems (such as the Hamilton receptor, Figure 3.5b) was performed via an SN_2 reaction on bis-bromo-telechelic PIB (**5**), which led to quantitative results if crown-ethers were used, furnishing the intermediate compound (**6**), which subsequently was transformed into the final bis-(Hamilton-receptor)-telechelic PIB (**7**) in good yields. The corresponding MALDI-spectrum (Figure 3.5c) clearly shows the integrity and the high level of functionalization, also proven by ^{1}H-NMR and ^{13}C-NMR spectroscopy.

However, in the case of star polymers the above-mentioned strategies are not sufficiently active to achieve a full functionalization of the telechelic polymers. Thus, (Figure 3.6a) the use of the azide–alkine "click" reaction in combination with LCCP (Binder *et al.*, 2004c; Binder

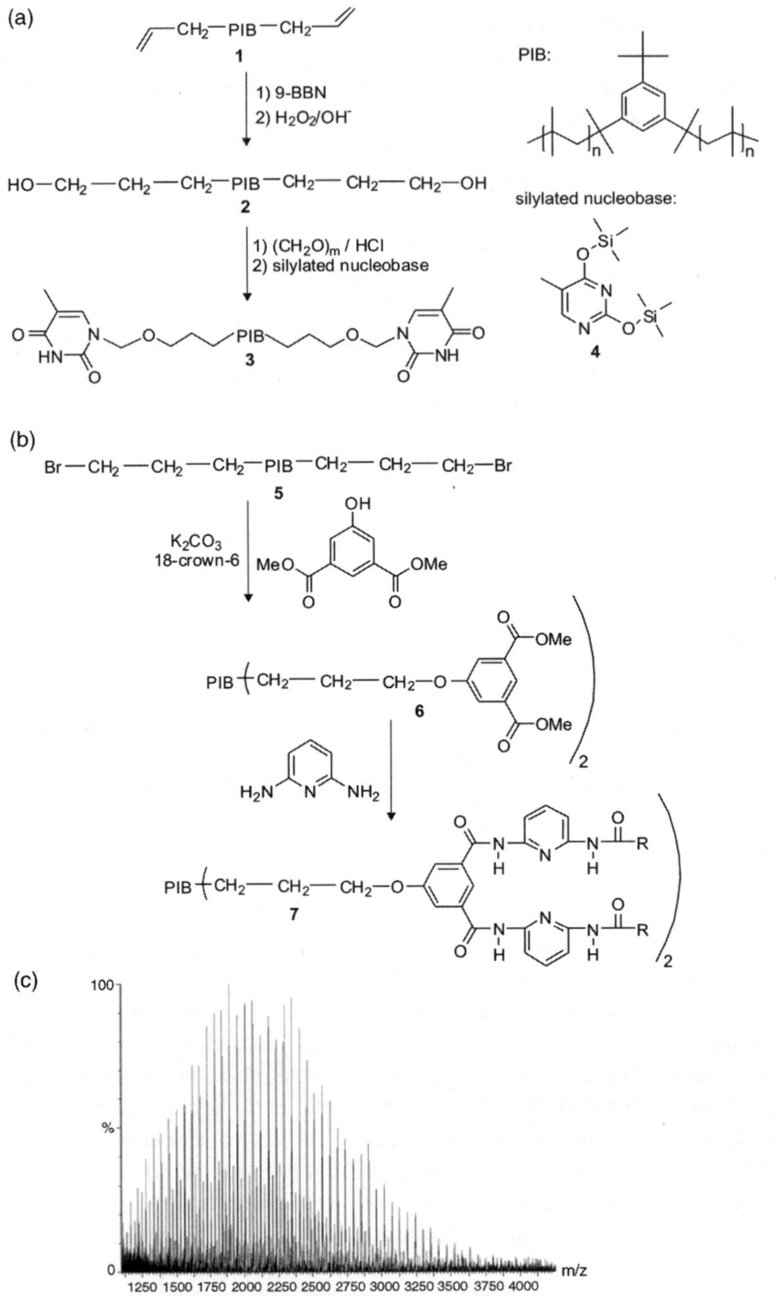

Figure 3.5 Attachment of complex hydrogen-bonding systems to bis-telechelic PIB prepared by LCCP. (Reprinted with permission from W.H. Binder, M.J. Kunz, C. Kluger *et al.*, "Synthesis and analysis of telechelic polyisobutylenes for hydrogen-bonded supramolecular pseudo-block copolymers," *Macromolecules*, **37**, 5, 1749–1759, 2004. © 2004 American Chemical Society.)

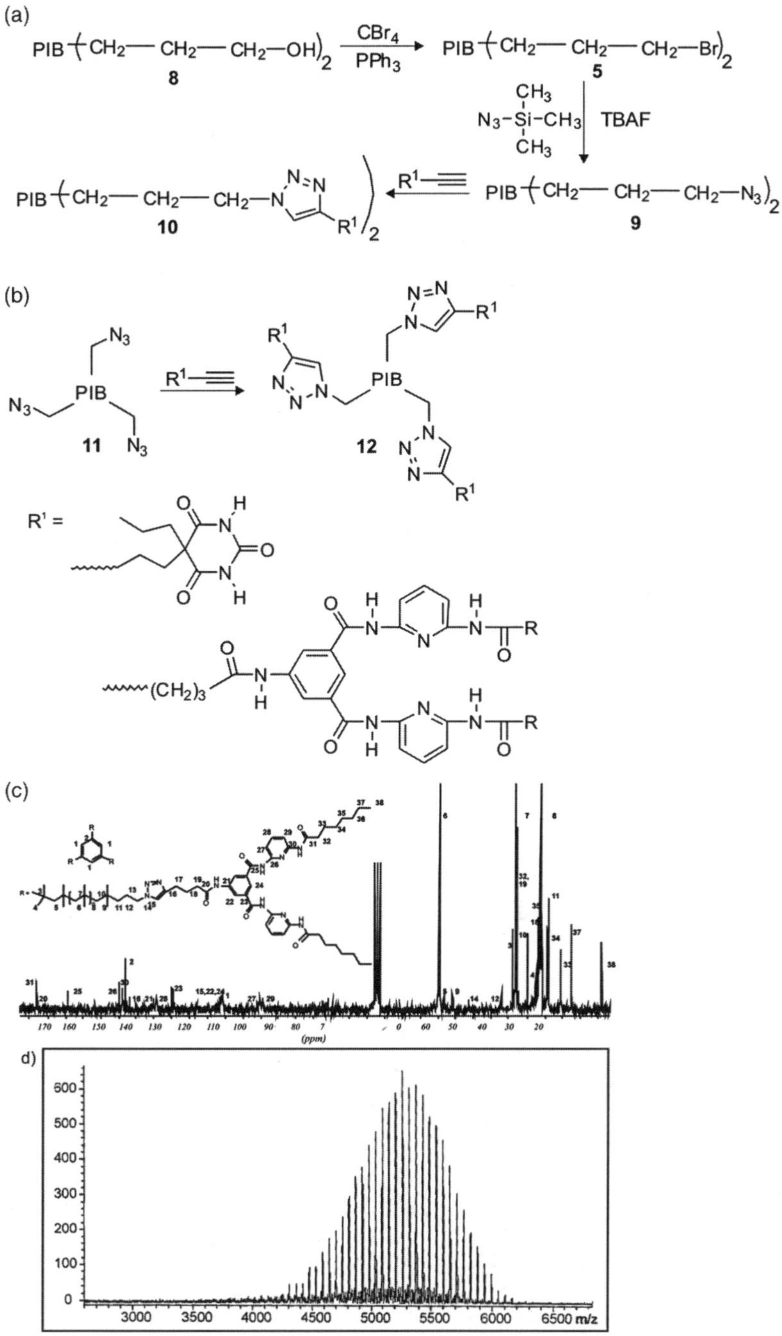

Figure 3.6 Azide–alkine "click" chemistry for the synthesis of three-arm-star functionalized PIBs. (Reproduced with permission from W. Binder, L. Petraru, T. Roth *et al.*, "Magnetic and temperature-sensitive release gels from supramolecular polymers," *Advanced Functional Materials*, 2007, **17**, 1317–1326. © Wiley-VCH Verlag GmbH & Co. KGaA.)

et al., 2007c; Roth *et al.*, 2005, Herbst and Binder *et al.*, 2010) is efficient, given that telechelic PIB-polymers with azide moieties can be prepared. Starting from a bis-hydroxy telechelic PIB (**8**), transformation into the bromide (**5**) can be achieved quantitatively via the use of CBr_4/PPh_3 and a subsequently activated SN_2-reaction into the bis-azide-telechelic PIB (**9**) based on the use of trimethyl silyl azide/TBAF. A click reaction then furnishes the final H-bond-modified PIB (**10**). This method can then be extended to the respective star polymer, starting from the three-arm star PIB (**11**), functionalized with three azide moieties. Azide–alkine "click" reaction then furnishes the three-arm-star-telechelic PIB (**12**), quantitatively bearing the H-bonding moieties on each arm. As shown in Figures 3.6c and d, the respective NMR and MALDI spectra demonstrate the pure three-arm-star structure of polymer (**12**) (Binder *et al.*, 2007c), which is required to generate responsive magnetic gels, using polymer (**12**) as crosslinker.

The first example of a copolymerization of polar monomers in LCCP was described in our group, copolymerizing the styrene monomer (**13**) with isobutylene in amounts of 1–5 mol% of comonomer (Figure 3.7) (Hackethal *et al.*, 2010). Incorporation of the polar monomer (**13**) can be achieved in amounts up to 2.5 mol%, as proven by NMR spectroscopy and MALDI methods. As followed by *in-situ* kinetic analysis, the polymerization follows a linear chain growth, together with a linear consumption of monomer when plotted as $\ln[M_o]/[M_t]$ vs. time (t).

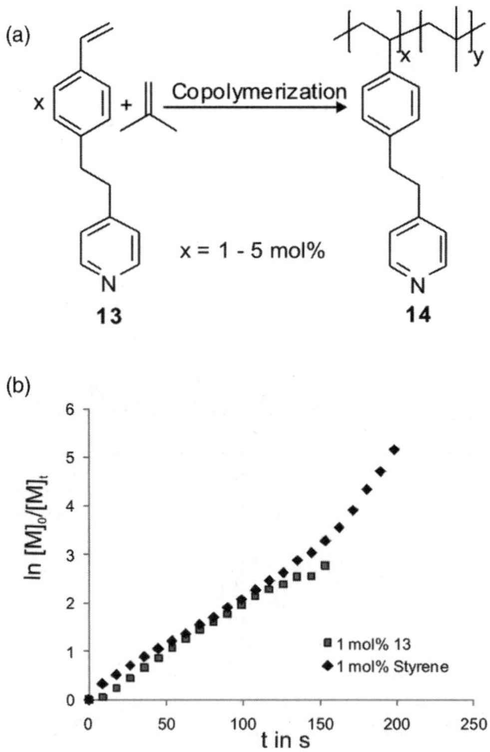

Figure 3.7 (a) LCCP of H-bonding monomer (**13**) by direct copolymerization with isobutylene. (b) *In-situ* kinetic plots of the copolymerization in comparison to the copolymerization of IB with pure styrene.

Indeed, this for the first time opens the possibility to introduce specific polar monomers during a LCCP.

3.2.2 *Anionic Polymerization*

Naturally, anionic polymerization is not compatible with H-bonding moieties, thus requiring efficient modification *after* the polymerization reaction (Ilhan *et al.*, 2001; Schadler *et al.*, 1998). An excellent example (Karatzas *et al.*, 2006) of an efficient postmodification method was presented starting from living PS-PI anions, which was quenched by only one unit of oxirane, furnishing the hydroxy-telechelic PS-PI BCP (**15**). This in turn was reacted with the isocyanate (**16**), attaching the ureidopyrimidone moiety to the final polymer (**17**).

Isoprene (Figure 3.8b) can be polymerized anionically followed by hydrogenation, furnishing polymer (**18**) functionalized with amino groups. This in turn was used for conventional acylation reactions to attached fmoc-protected alanine moieties, which could be added to the tetrapeptide, followed by addition of an bisacetylene moiety. In this manner, peptide fragments such as (**19**) can be attached to highly defined PI-polymers (Frauenrath *et al.*, 2009; Weiss *et al.*, 2008) for the study of further self-assembly processes in solution (Jahnke *et al.*, 2008). Sidechain-modified PBD polymers (Figure 3.8c) can be generated by

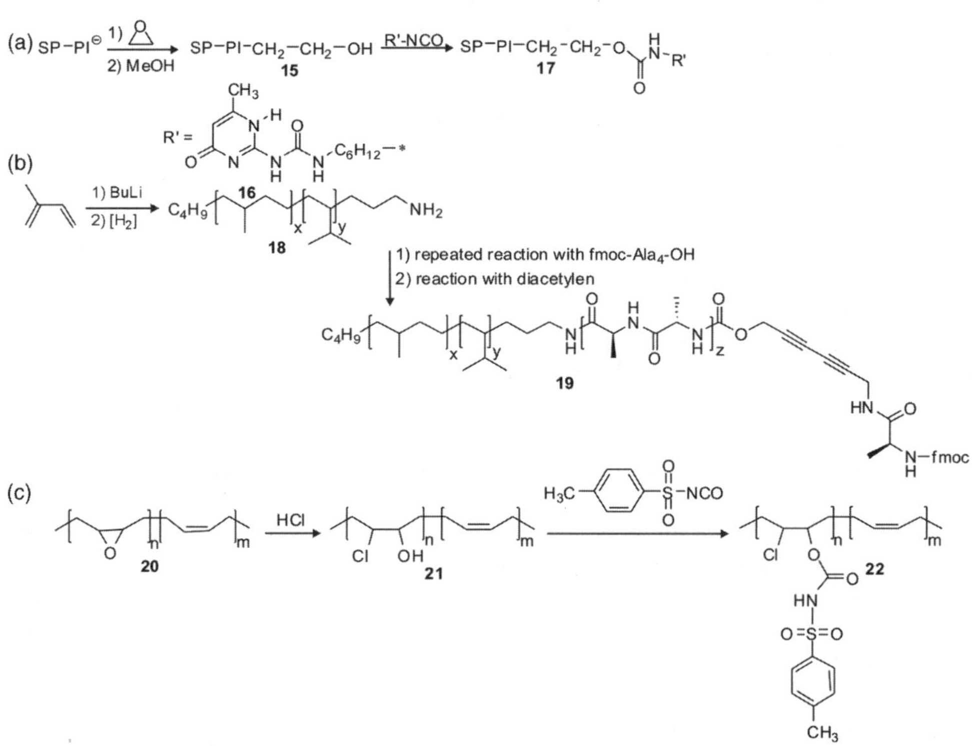

Figure 3.8 Living anionic polymerization and attachment of hydrogen bonding moieties via post-modification methods.

partial epoxidation of 1,4-PBD, yielding polymer (**20**). Hydrolysis of the epoxide moiety generates polymer (**21**), which in turn can be reacted with tolyl sulfonyl isocyanate, to generate the PBD polymer (**22**) with statistically distributed pendant hydrogen-bonding moieties (Peng and Abetz, 2005).

3.2.3 Ring-Opening Metathesis Polymerization (ROMP)

Polymerization via ROMP (Bielawski and Grubbs, 2007; Buchmeiser, 2000) offers an excellent access towards supramolecular polymers, as the functional group tolerance of Grubbs catalysts is enormously high. Thus, both strategies – the direct copolymerization as well as the postmodification can be used to incorporate monomers bearing H bonds into the chains under living polymerization conditions. However, broad molecular weight distributions may occur due to poor initiator/propagation rates, which may vary from H-bonding system to H-bonding system.

We have developed a simple method to attach H bonds to a ROMP-derived polymer after the polymerization reaction via azide–alkine "click" chemistry (Figures 3.9a–d) (Binder and Kluger, 2004; Binder *et al.*, 2006; Binder *et al.*, 2005b; Binder *et al.*, 2009; Kluger and Binder, 2004; Jurzhals and Binder, 2010). Thus, the phtalimide-derived oxy-norbornene polymer (**23**) can be substituted quantitatively with alkine moieties by reaction with propargylic bromide to yield the alkine-modified polymer (**24**). This in turn then serves as a scaffold for the azide/alkine-"click"-reaction, attaching a wide range of H-bonding structures to the polymers (**25**) (Binder *et al.*, 2004a). Thus, not only can the nucleobases (such as thymine) be attached in multiple copies to the polymer, but also the sterically highly demanding Hamilton receptor. The advantage of this process lies in the direct transformation of the previously achieved monomer distribution into the final H-bonded architecture. Alternatively, the strategy shown in Figure 3.9b can be used: in this case, a different monomer was polymerized into the polymer (**26**), in which the bromine end groups were transformed quantitatively into the azide moieties (**27**). Again, the azide–alkine "click" reaction can be used to quantitatively attach the H-bonding moieties onto the polymer backbone (polymer **28**) (Kluger and Binder, 2007). This strategy can be extended for the synthesis of BCPs (**29**) with perfluorinated moieties in one block, and H-bonding structures (such as thymine, barbiturate or Hamilton receptors) in the second block (Binder *et al.*, 2006; Binder *et al.*, 2005b; Binder *et al.*, 2009; Kluger and Binder, 2007; Kurzhals and Binder, 2010). As the polymerization method takes place before the modification with the H bonds, the polymer precursors serve as a simple scaffold for the attachment of a large multitude of H bonds in a library-type approach. Alternatively, N-hydroxysuccinimides can be used, although these reactions do not yield fully quantitative modifications of the monomers (Drechsler *et al.*, 2002).

The direct copolymerization of monomers containing H bonds has been described earlier, demonstrating the often good compatibility of Grubbs catalysts with polar moieties such as H bonds (Bazzi and Sleiman, 2002). However, in order to achieve good polymerization results with lower polydispersities, a "supramolecular" protecting group, matching the hydrogen-boding pattern of the H bonds on the norbornene monomer was used. We have demonstrated (Figure 3.9d) (Kluger and Binder, 2004) that monomer (**30**), containing a Hamilton-type receptor can be directly polymerized into the respective polymer (**31**) using Grubbs second-generation and Hoveyda-type catalysts.

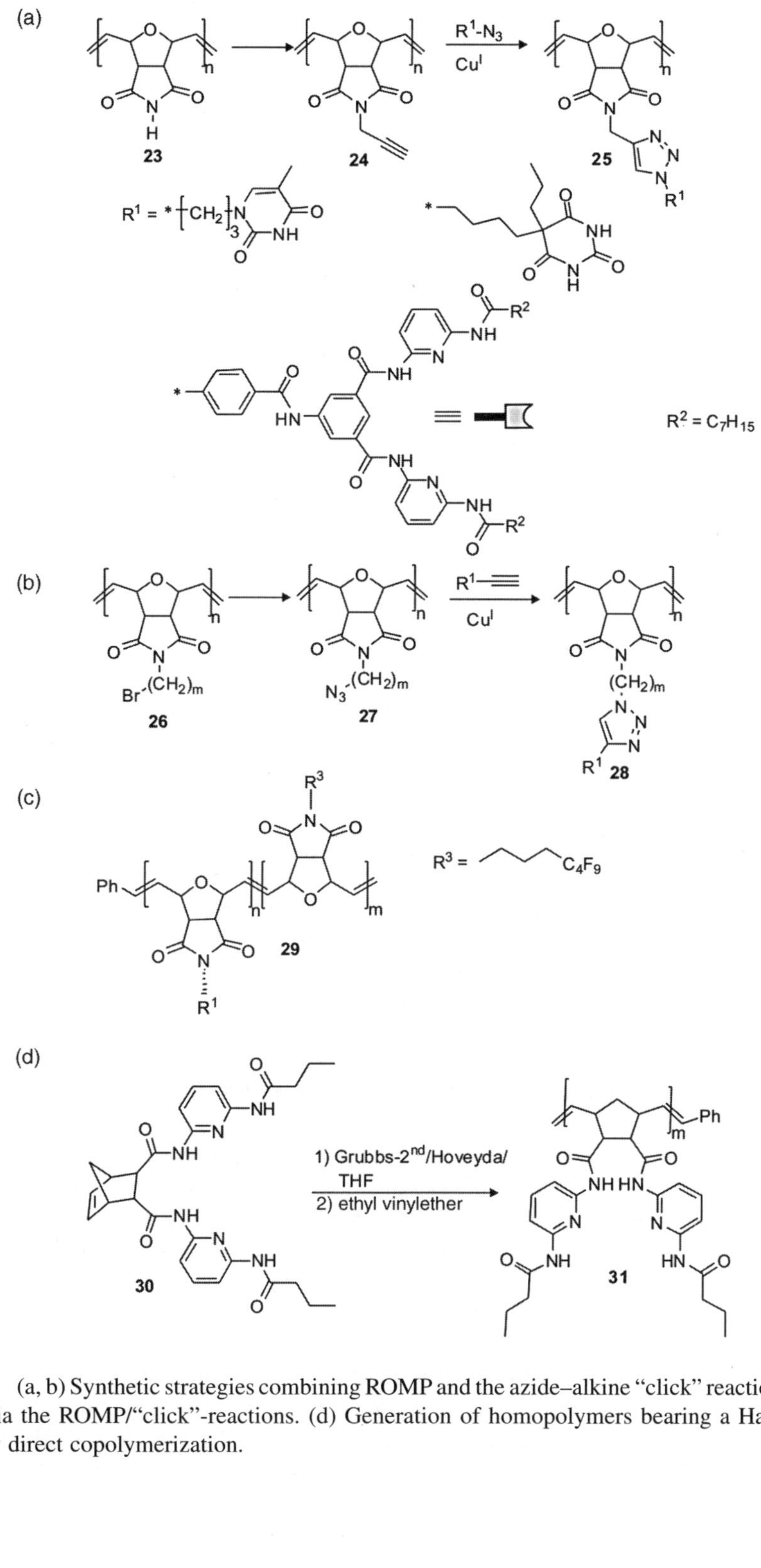

Figure 3.9 (a, b) Synthetic strategies combining ROMP and the azide–alkine "click" reaction. (c) BCPs prepared via the ROMP/"click"-reactions. (d) Generation of homopolymers bearing a Hamilton-type receptor by direct copolymerization.

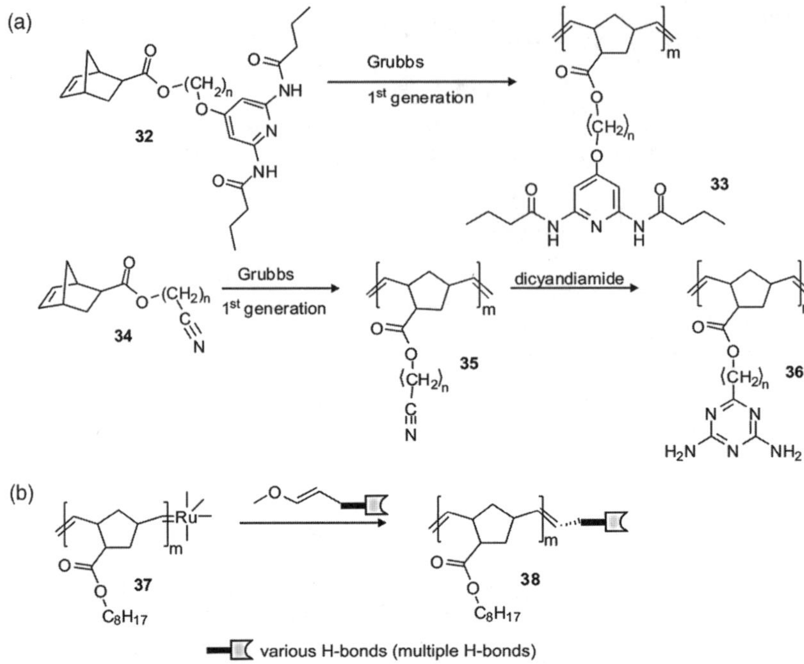

Figure 3.10 (a) Direct copolymerization of H bonds via ROMP and the alternative reaction pathway to yield a similar polymer. (b) End-group-modified telechelic polynorbornenes via quench with functionalized ethyl-vinyl ethers.

The direct copolymerization of hydrogen bonds can be easily achieved as demonstrated by us (Kluger and Binder, 2004) and others (Nair *et al.*, 2008; Nair *et al.*, 2006; Pollino *et al.*, 2004; South *et al.*, 2006) (Figures 3.9d and 3.10). If a spacer of appropriate length (i.e. $-C_{11}H_{22}$) is placed between the H-bonding unit and the reactive norbornene monomer (**32**) the polymerization proceeds smoothly. Alternatively, the nitrile-containing monomer (**34**) can be polymerized using Grubbs first-generation catalyst (polymer **35**), and subsequently reacted with dicyandiamide to build the 2,6-diaminopyridine unit on the final polymer (**36**) (Nair *et al.*, 2006).

Telechelic end-group-functionalized PNB polymers with H bonds on only one end have been generated by Weck *et al.* (Ambade *et al.*, 2009) relying on the quantitative and fast reaction of ethyl-vinyl ethers with the growing ruthenium-carbene on the chain end. Thus, the living polymer (**37**) can be directly reacted with ethyl-vinyl ethers containing the respective H bond resulting in the quantitative attachment to yield the fully functionalized polymer (**38**). A major drawback of this method, however, is the tedious synthetic approach to the respective substituted ethyl-vinyl-ethers, which require a multistep procedure.

An entirely different, but highly innovative approach to supramolecular polymers via metathesis methods has been developed by Meijer *et al.* (Ohkawa *et al.*, 2007; Scherman *et al.*, 2006a; Scherman *et al.*, 2006b) (Figure 3.11). The synthesis of biheaded telechelic PNB-polymers with H-bonds has been accomplished by Binder *et al.* (Kurzhals and Binder, 2010) via simple crossover reactions. Based on the two selectively interacting

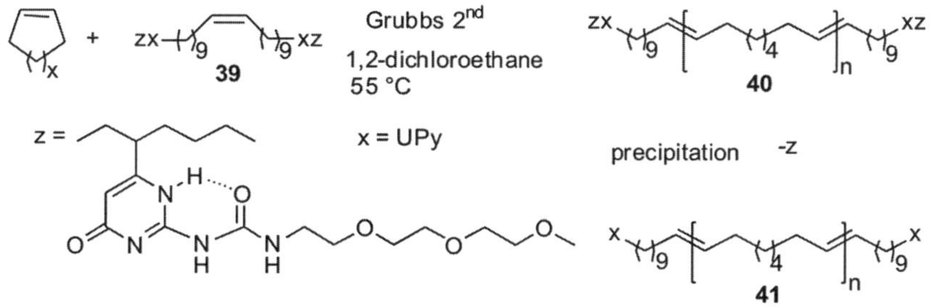

Figure 3.11 Generation of alternating copolymers via ADMET.

hydrogen-bonding monomers **39** and **40**, an ADMET-copolymerization reaction with cyclic olefins was performed using Grubbs second-generation catalyst. Due to the complementarity of the H bonds in the two different monomers, an A–B-type copolymer was obtained selectively. This definitely represents one of the most innovative dynamic polymerization systems during recent decades.

3.2.4 Controlled Radical Polymerization (CRP)

Controlled radical polymerization (CRP) (Braunecker and Matyjaszewski, 2007) reactions are mechanistically very well suited for the direct combination with monomers containing H bonds, as radicals are only rarely disturbed by the presence of hydrogen-bonding units. Thus, all three mainly used CRP-methods (NMP, ATRP and RAFT) can be combined with hydrogen-bonding units, enabling their direct incorporation.

The use of NMP for the synthesis of telechelic, monofunctionalized polymers bearing H bonds has been described as shown in Figure 3.12 (Binder *et al.*, 2007a). Thus, a modified nitroxide initiator (**42**) was synthesized, bearing a triazole-linked hydrogen bonding structure (barbituric acid, dihydroxyalkyl moieties), which can be easily prepared via azide–alkine "click" reaction from the respective nitroxide bearing the azido moiety. The nitroxide compound (**42**) can serve as initiator for the polymerization of n-butyl acrylate (n-BuA) or N-isopropyl acrylamide (NIPAM), furnishing monofunctionalized n-BuA or NIPAM polymers (**43**) with the incorporated hydrogen-bonding unit. The presence of the hydrogen-bonding units was proven by MALDI methods, the resulting polymer (**43**) can be used as surface-bound thermoresponsive polymer attached to iron-oxide nanoparticles. Another approach uses the modified NMP initiator (**44**) by copolymerization of the modified styrene-monomers (**45**) with n-butyl acrylate, furnishing the final polymer (**46**) (Mather *et al.*, 2007b).

ATRP has had an early use by generating adenine and thymine derivatives with a pendant bromo-isobutyrate as initiator (Khan *et al.*, 1999; Marsh *et al.*, 1999). A similar approach has been used recently for introducing 2H-ureido-4[1H]-pyrimidone or the 2,7-diamido-1,8-naphtyridine-unit via the respective attached bromo-isobutyrate moiety (**47**) (Feldman *et al.*, 2008). Conventional ATRP yielded the final polymer (**48**) in good yields and small polydispersities. Other concepts used acrylate monomers with amide-bound H bonds with good yields, often in copolymerization with other monomers (Kavitha and Singha, 2009; Li *et al.*, 2006). As shown in Figure 3.13b, block copolymers consisting of tBuA-PHEMA (**49**) (prepared by ATRP) were modified via a simple acylation procedure, using

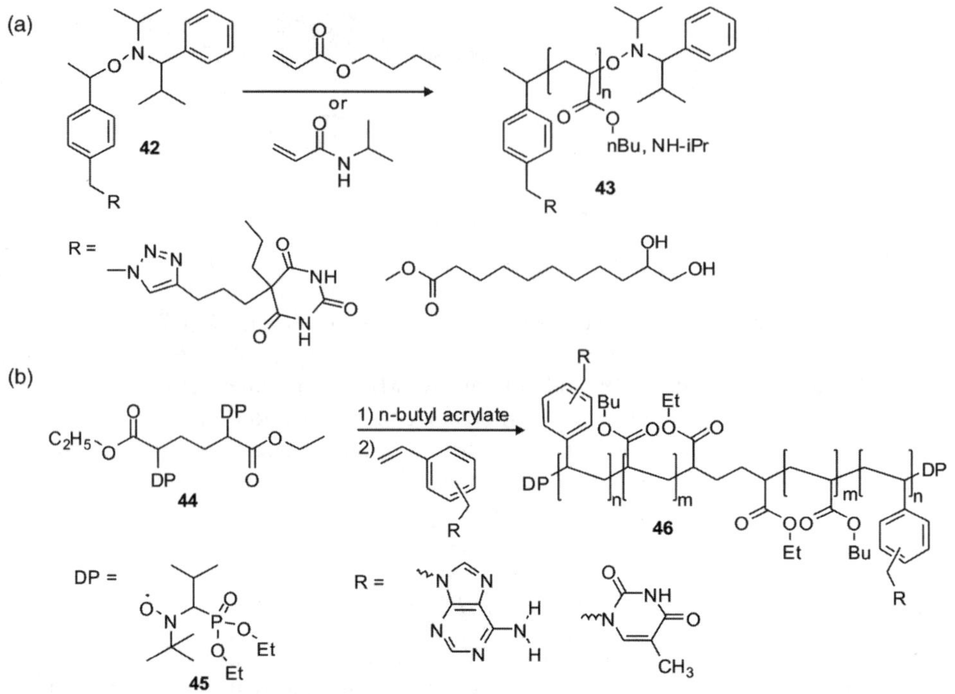

Figure 3.12 Using nitroxide mediated polymerization to introduce H bonds into telechelic polymers.

2-acrylamido-6-carboxybutylamidopyridine (ACAP) and DCC as reagent (Li *et al.*, 2006). The resulting copolymer (**50**) contains about 0.4 mol% of the attached hydrogen-bonding unit ACAP in the PHEMA-block. Classical copolymerization via ATRP enabled the generation of the copolymer (**51**) (Figure 3.13c), containing incorporated pyridine or phenol units. There polymers were subsequently used as blend for BCP-lithography (Tang *et al.*, 2008). RAFT (13d) has been used to incorporate 2,6-diamidopyridines or thymine-type H-bonds into poly (vinyl acetate) (PVAc) by use of chain-transfer agents such as compound (**52**) (Bernard *et al.*, 2009). These reaction generate PVAc polymers in molecular weights up to 22 000 Dalton and polydispersities of about 1.2–1.4. Additionally, the incorporation of 2-vinyl-pyridine monomers via RAFT has been reported (Noro *et al.*, 2008).

3.2.5 Polycondensation Methods

PEK (polyetherketone) and PSU (polysulfone) polymers can be easily prepared by the mentioned polycondensation methods, furnishing polymers with controllable end groups (either –OH, or fluorine (–F)) (Figure 3.14). These polymers can then be subjected to further polycondensation reactions to attach appropriate end-group moieties (such as 2,6-diamino-triazine (Figure 3.14a) or barbituric acid (such as in Figure 3.14b) (Kunz *et al.*, 2004a; Petraru *et al.*, 2004).

Basically, the main issue concerns either to overcome the often strong differences in solubility between the hydrogen-bonded moieties and the final product as well as the inherently present further condensation reactions of the H bonds after attachment to the main polymer chain.

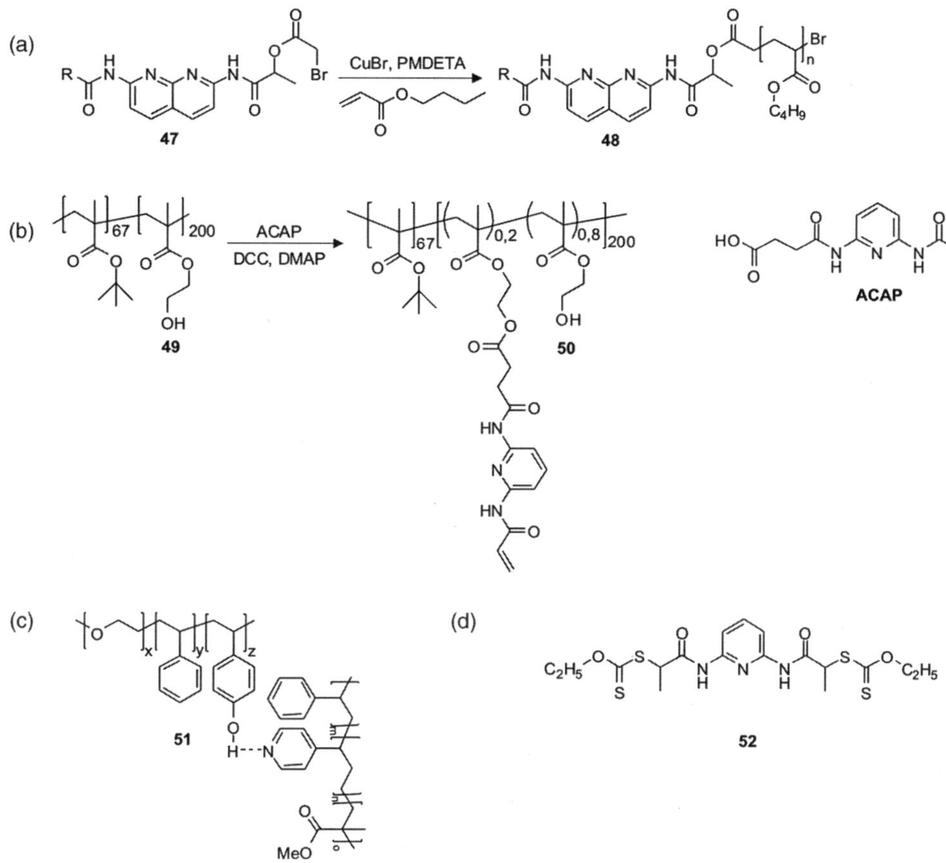

Figure 3.13 Introduction of H bonds into polymers via ATRP and RAFT.

Thus, (Figure 3.14a) reaction of the initial fluorine-telechelic PEK can either be conducted directly with the diaminotriazine building block (**53**), or by subsequent reaction with 3-cyanophenol (**54**) and subsequent formation of the diaminotriazine ring by condensation with dicyandiamide (Kunz *et al.*, 2004a). Both reaction pathways yield the desired product (**55**), with the first method definitely yielding a more complete functionalization, as proven by MALDI methods.

The attachment of the barbituric-acid moieties to PEK polymers proved more difficult, as the barbituric acid is more nucleophilic than the 2,6-diamino-pyridine, thus preventing a condensation of preformed building blocks due to the inherent self-reactivity of the H-bonding building block (Petraru *et al.*, 2004). Thus, starting from the hydroxy-telechelic PEK (**56**) an aldehyde moiety was condensed directly onto the PEK via reaction with 7-fluoro benzaldehyde, resulting in the bis-aldehyde telechelic PEK (**57**). This in turn can be reacted with barbituric acid to condense the barbiturate quantitatively onto the aldehyde moieties, only if the solvent mixtures contain a significant amount of methanol (which basically is a nonsolvent for PEK). Thus, the use of a specific solvent mixture containing a sufficient amount of methanol is required, to push the reaction to completion, yielding the bis-barbituric

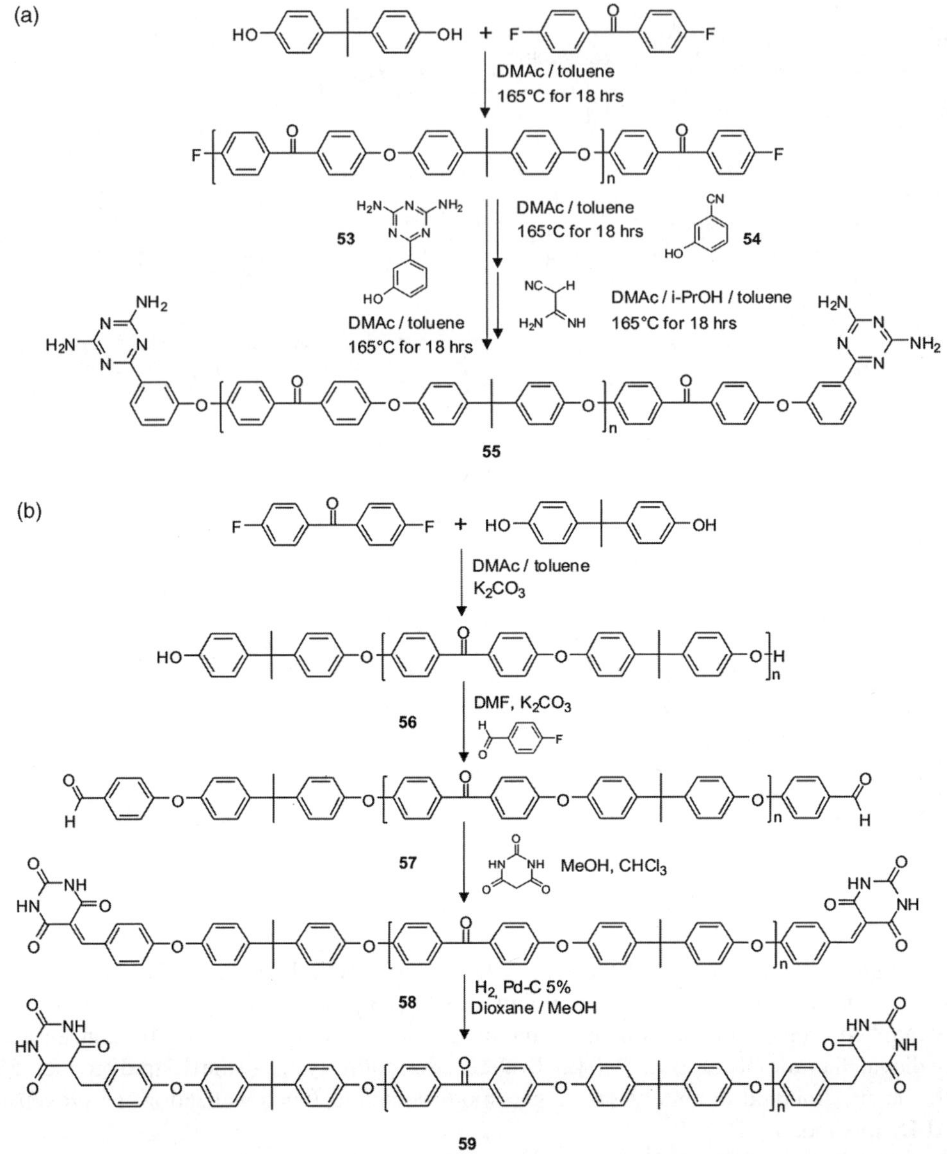

Figure 3.14 Telechelic PEK or PSU polymers bearing H bonds via postmodification reactions of the –OH or –F telechelic polymers. (a) Synthesis of PEK-polymers with 2,6-diaminotriazine end groups. (b) Synthesis of barbituric acid telechelic PEK polymers.

acid-telechelic PEK (**58**), which can be further hydrogenated to the final polymer (**59**). The MALDI shows the clear transformation to polymer (**57**) and subsequently to polymer (**58**).

Another strategy to prepare poly(caprolactone) polymers runs via the use of ROP-reactions and subsequent azide–alkine "click" reactions. Thus, starting from butenol (**59**) the ROP

(c)

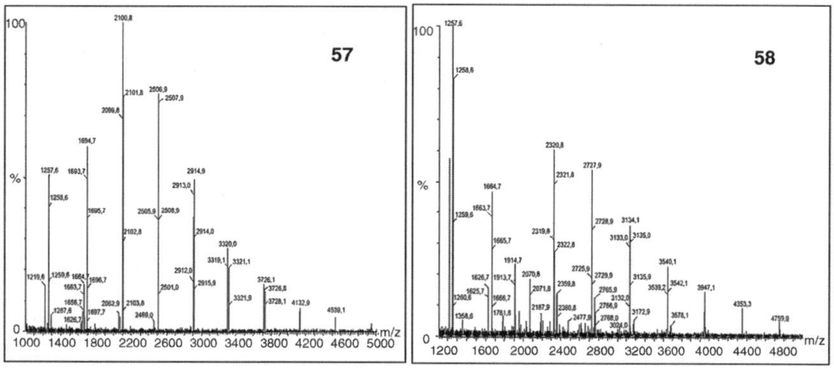

(d)

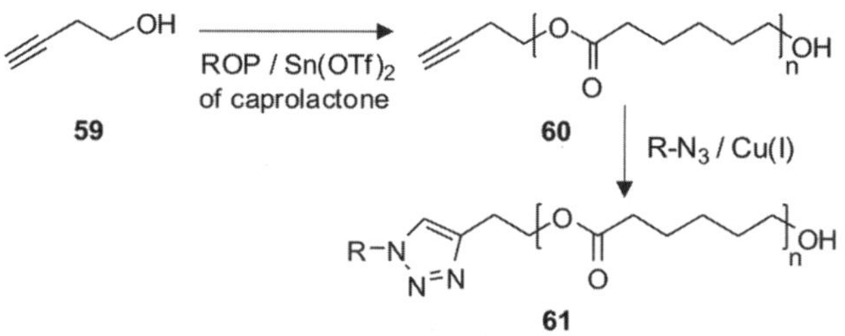

Figure 3.14 (*Continued*) (c) MALDI of the final PEK product bearing barbiturates at both chain ends according to Petraru *et al.* (2004). (Reprinted with permission from L. Petraru, D. Farnik, R. Saf, W.H. Binder, "Supramolecular, telechelic poly(etherketones) bearing barbituric acid as hydrogen bonding unit," *Polymer Preprints*, **45**, 2, 690–691, 2004. © 2004 Division of Polymer Chemistry, Inc. of the American Chemical Society.) (d) Synthesis of telechelic PCLs with hydrogen bonds via azide–alkine "click" reactions.

induced by $Sn(OTf)_2$ was used to generate the mono-alkine-telechelic PCL (**60**), and subsequently attaching a complex H-bonding unit via azide–alkine "click" reaction (Ostas and Binder,). The final PCL-polymer is obtained quantitatively, as proven by NMR spectroscopy and MALDI.

Recently, the modification of polythiophenes (PThs) with hydrogen-bonding units was reported (Enders and Binder, 2010). 1,5-Dibromothiophene was polymerized yielding the polythiophene (**62**) as proven by extensive ESI-TOF-measurements. However, polymer (**62**) consists of a mixture of bisbromotelechelic PTh (**62a**) and the monobromo-telechelic PTh (**62b**). Subsequent Sonogashira reaction of (**62**) with TMS-acetylene furnished the respective TMS-acetylene-functionalized PThs (**63**), again as a mixture between the mono- (**63b**), bi- (**63a**) and nonfunctionalized PTh's (**64b**). These polymers can be used to run the azide–alkine "click" reaction again, yielding the final polymer (**64a**), functionalized with the H-bonding moieties. The successfully conductance of this reaction can only be analyzed by LC-ESI-TOF methods, which enable a quantification of the individual species present in the reaction mixture. Thus, Figures 3.15b and c show the corresponding LC traces, together with the ESI-TOF-spectra of the now individually assignable species **62a, b** and **63a, b** and **64a**, **b**.

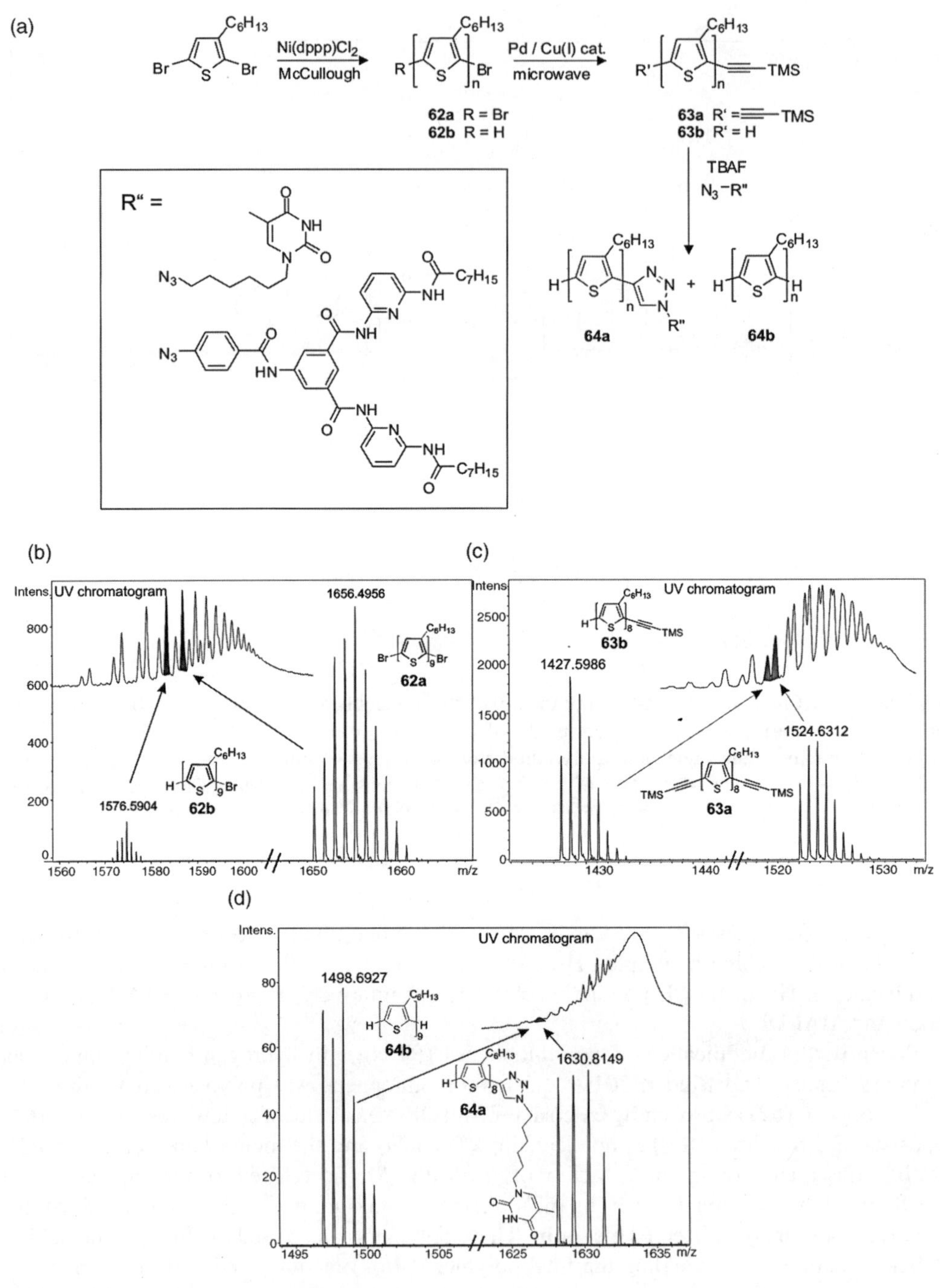

Figure 3.15 Preparation of polythiophenes with pendant H bonds via Sonogashira click reactions. (a) Reaction scheme. (b, c) LC-ESI-TOF spectra of the compound **62** and **63** and **64**.

3.2.6 Ring-Opening Polymerization (ROP)

Most polymer generated by ROP processes in junction with H bonds are poly(lactides), poly(carpolactones) and poly(ethylene oxide) polymers. As most of the nucleophilic reaction centers present in the H-bonding units are interferring with many ROP processes, usually efficient postmodification reactions after the ROP process is used to affix the H-bonding moiety to the central polymer. Again the azide–alkine "click" reaction is a highly valuable tool, that is, in the attachment of carbohydrate units to PEO polymers and oligomers, as shown in Figure 3.16a.

The conjugate compound (**65**) is obtained in excellent yield by reaction of a telechelic PEO-N_3 with the respective 1-O-alkinyl-glucoside. Another coupling method relies on the use of activated carbonyl-imidazoles with reactive OH– or $-NH_2$ moieties of a preformed PCL after introducing sufficiently reactive $-NH_2$ moieties at the chain end (Figure 3.16b). This method allows the generation of polymer (**66**) by coupling with the reactive imidazolium/H-bond (**67**) (van Beek *et al.*, 2007b). However, a two-step preparative procedure is required to generate the amino-telechelic PCL via DCC-mediated esterification with 6-N-(benzyloxycarbonylamino)-hexanoic acid, which is then deprotected by acidic cleavage to generate the amino-telechelic PCL (Wisse *et al.*, 2006b). Alternatively, the use

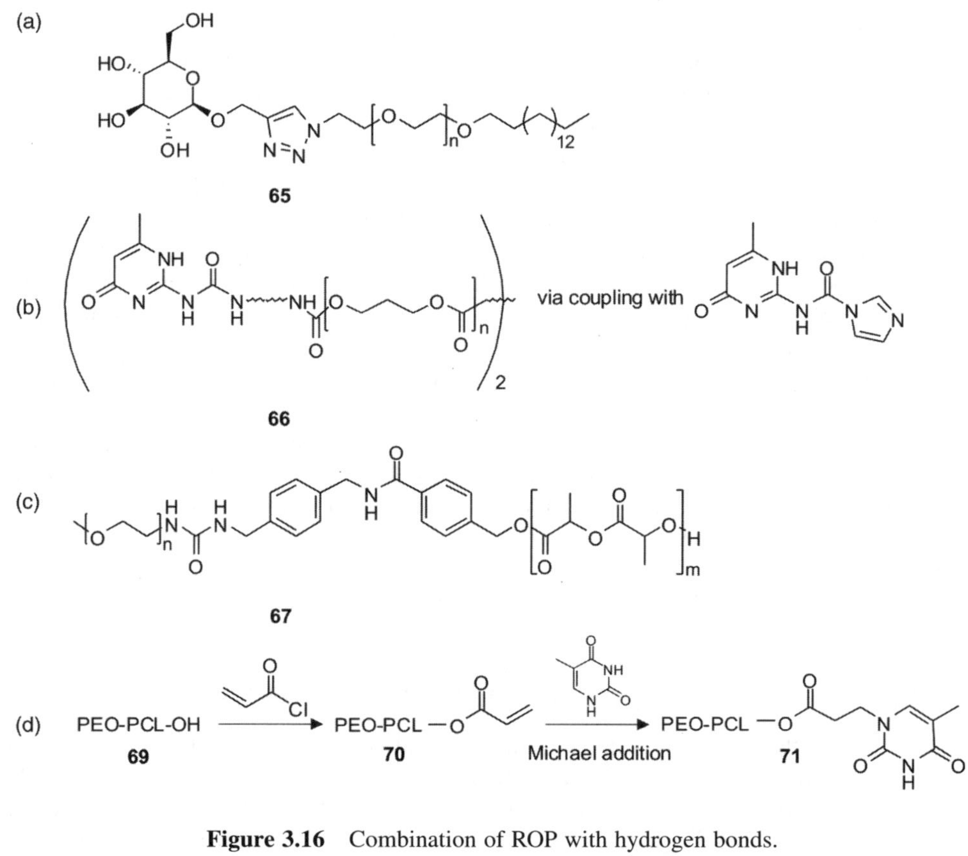

Figure 3.16 Combination of ROP with hydrogen bonds.

of highly reactive moieties is possible, that is, the well-known use of isocyanates to attach or generate hydrogen-bonding moieties onto a polymer chain (Wisse *et al.*, 2006a; Yamauchi *et al.*, 2004) (Figure 3.16c). Thus, urea moieties can be directly introduced by reaction with an isocyanate, or – alternatively – be introduced at the junction of two blocks (i.e. PEO-PLA-block (**68**)) (Kim *et al.*, 2009b). Less-reactive intermediates such as acrylic acid side chains can be used by reaction of the –OH end group in a PEO-PCL-diblockcopolymer (**69**) to yield the fully functionalized acryloyl-modified PEO-PCL polymer (**70**) (Karikari *et al.*, 2005; Karikari *et al.*, 2007). Subsequent Michael-type addition reaction yields the finally N^4-thymine-telechelic PEO-PCL polymer (**71**). It is, however, unclear to what extent alkylation at the N^6-position of the thymine moiety takes place as well. Alternative methods use reactive epoxide derivatives and the respective nucleophilic ROP process of PEO (Lee *et al.*, 2007a; Lee *et al.*, 2007b).

3.2.7 Other Postmodification Methods

Besides the above-discussed combinations of living polymerization methods with post-modification methods, other methods can be applied. This field was opened by applying ene-type reactions for the attachment of 4-(3,5-dioxo-1,2,4-triazolidin-4-yl) zoic acid onto poly (1,4-butadiene) (Stadler, 1993a). More simply, the attachment of the widely used (quadruple) uradiazole unit can be achieved simply by reaction of amino- or hydroxyl-telechelic polymers (**72**), an excess of a bivalent isocyanate (**73**) and the respective reactive H-bonding unit to finally yield the functionalized polymer (**74**) (Figure 3.17a) (Abbel *et al.*, 2009; Dankers *et al.*, 2006a; Dankers *et al.*, 2006b; Folmer *et al.*, 2000; Kautz *et al.*, 2006; Lange *et al.*, 1999; Wisse *et al.*, 2006a; Wisse *et al.*, 2006b; Woodward *et al.*, 2009a; Woodward *et al.*, 2009b).

Usually, technically used isocyanates such as TDI, MDI, HDI or IPDI are used to this purpose. However, this method requires reactivity on only one position of the H-bonding unit with the isocyanate, which often cannot be achieved in many multiple or more complex H-bonding systems. Due to its simplicity, the method has revolutionized the use of hydrogen-bonded supramolecular polymers, and is definitely the method for industrial applications of these materials. Recently, the synthesis of poly(ureidophtalimides) (**77**) has been accomplished by a simple reaction between diamino-phtalimides (**75**) and the bis-isocyanate (**76**) (Figure 3.17b). Another type of reaction to attach multiple H bonds uses epoxide opening (Lee *et al.*, 2007b). Thus, the bis-epoxy-telechelic PEO (**78**) (prepared by reaction of PEO with glycidyl chloride) can be reacted with methyl isocytosine (**79**) at a temperature of 120 °C to yield the fully functionalized polymer (**80**). Other SN_2-type reactions (Sivakova *et al.*, 2006) or hydrosilylation-reactions (Hirschberg *et al.*, 1999) provide additional attachment strategies towards telechelic polymer with attached H bonds.

3.3 Self-Assembly of Supramolecular Polymers via Hydrogen Bonds

Despite the fact that the structuring of supramolecular polymers is highly diverse with respect to the formed structures, three basic cases can be discriminated: (a) microphase separation in supramolecular polymers (b) ordering on surfaces – that is, under confinement on two dimensions (c) behavior of small molecules and their ordering effect into

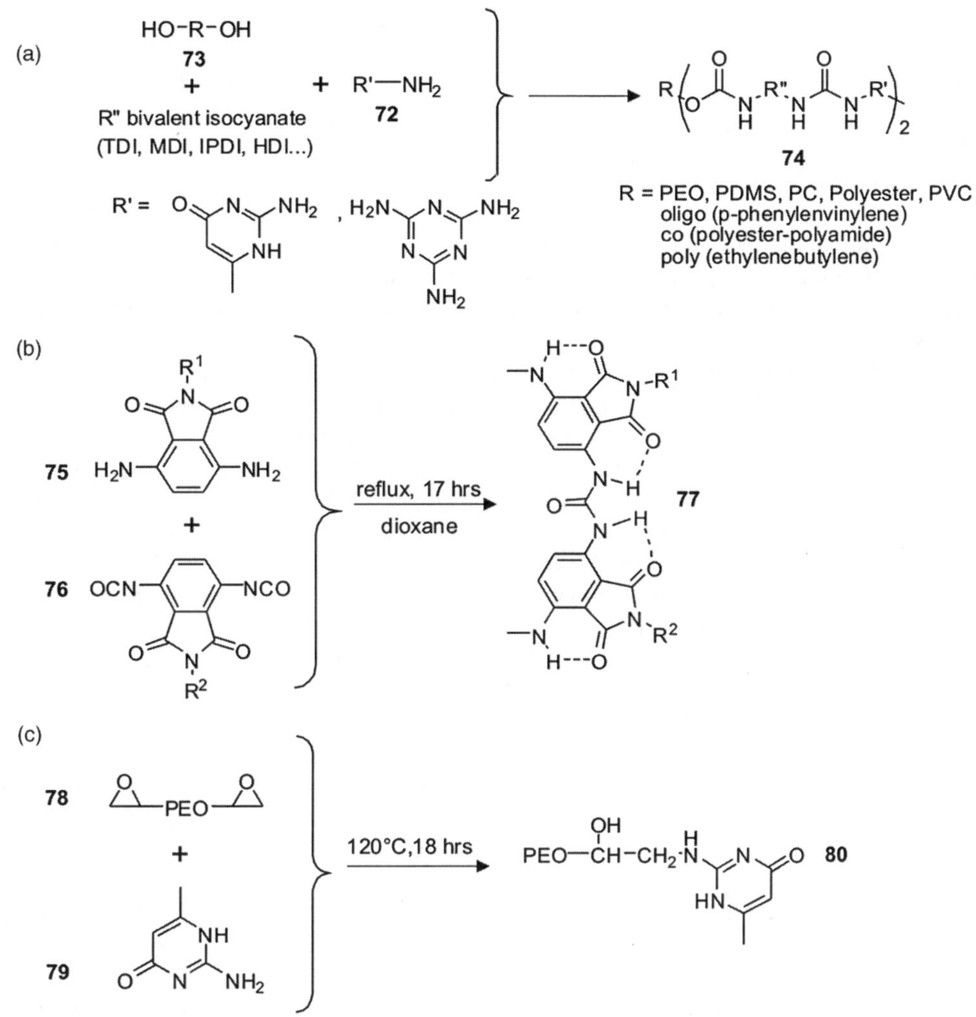

Figure 3.17 Functionalization of telechelic polymers by reactive isocyanates and epoxides.

supramolecular structures. In the following, the focus will be directed on recent aspects of these three issues.

3.3.1 *Microphase-Separated H-Bonded Polymers: Towards Pseudoblock Copolymers*

Microphase separation can only take place in polymeric structures, most of all in di-, tri- and multiblock copolymers – thus requiring polymeric structures already. In addition, graft-type polymers can show microphase separation. The hydrogen bond in supramolecular polymers introduces an additional attractive force acting together with the Flory–Huggins parameters in such immiscible polymeric systems. Thus, for diblock copolymers the structure and behavior of such supramolecular polymers displays an additional element of dynamics, which leads to a set of materials with advanced properties.

The synthesis and dynamics of "pseudo"-block copolymers was accomplished for the first time by use of bivalent telechelic building blocks, derived from PEK- and PIB-bistelechelic polymers as shown in Figure 3.18a (Binder *et al.*, 2005a; Binder *et al.*, 2004b; Binder *et al.*, 2004a; Kunz *et al.*, 2004a; Petraru *et al.*, 2004). Thus, 2,6-diamino-triazine/thymine-linked ($K_{assn.} \sim 10^3\,M^{-1}$) PEK and PIB telechelics were mixed, generating alternating PEK–PIB supramolecular polymers. As PEK and PIB poylmers are immiscible, individual PEK and PIB domains are generated. Moreover, the dynamics behavior of such polymers, now displaying different association constants, i.e. also those with the Hamilton receptor, barbituric acid, with ($K_{assn.} \sim 10^5\,M^{-1}$) were prepared, and their behavior studied by temperature-dependent SAXS (measuring the microphase behavior) and IR measurements (revealing the breaking of the H bond) (Binder *et al.*, 2005a). In contrast to the covalently linked PEK-PIB di- and triblock copoylmers (Binder and Machl, 2005), the polymers displayed a thermoreversible behavior, which persisted until cleavage of the corresponding H bond: a stronger bond (such as the Hamilton-receptor–barbituric-acid bond with ($K_{assn.} \sim 10^5\,M^{-1}$) allowed thermoreversibility far above the T_g of the PEK component (till 220 °C), whereas the weaker triazine–thymine bond ($K_{assn.} \sim 10^3\,M^{-1}$) allowed reversibility only up to ~135 °C, after which the system displayed macrophase separation. Similar examples with a large structural variety of different pseudo-BCPs have been prepared and a similar dynamic behavior observed (Ambade *et al.*, 2009; Asari *et al.*, 2006; Dobrosielska *et al.*, 2008; Feldman *et al.*, 2008; Huh *et al.*, 2006). Similar effects have been observed in dendritic systems (Chen *et al.*, 2005) and supramolecular magnetic gels (Binder *et al.*, 2007c).

Another intensely studied system concerns a "graft-type" polymer (Figure 3.18b), consisting of PS_n-$(PVP\text{-}PDP)_m$ block copolymers, where the PDP chains are bound via ionic hydrogen bonds towards the PS-PVP backbone (Brinke and Ikkala, 2004; Kosonen *et al.*, 2003; Polushkin *et al.*, 2005; Ramani *et al.*, 2009; Ruotsalainen *et al.*, 2007; ten Brinke *et al.*, 2007; Tung *et al.*, 2008; Vera *et al.*, 2008). As shown in Figure 3.18c, these polymers display a phase diagram similar to the one known for conventional BCPs, which, however, can be changed when the polymers are processed via electrospinning (Ruotsalainen *et al.*, 2007). A variety of physical effects and experiments (such as *in-situ* rheology (Laiho and Ikkala, 2007), shear alignment and orientation (van Ekenstein *et al.*, 2003), positron-annihilation spectroscopy (Ramani *et al.*, 2009), annealing effects by solvent vapors (Zoelen *et al.*, 2008)) have been studied using this versatile system. Polymers with thermoplastic elastomeric properties are generated if H bonds able to form extended H-bonding networks are affixed into polymers (Botterhuis *et al.*, 2008; van Beek *et al.*, 2007b; Versteegen *et al.*, 2006; Herbst and Binder, 2010). As shown in Figure 3.18d, cluster formation can take place, leading to stacks of associated hydrogen bonds, which can rearrange during shear and melt during heating.

In solution, strong effects are observed when multiple hydrogen bonds are connecting two H-bonding polymers, such as PMMA-b-P4VP–PS-b-PVP mixtures. A strong change of association is observed when changing the solvent from tetrahydrofurane (THF, as weakly hydrogen-binding solvent) to DMF (a solvent participating strongly in H bonding) (Kuo *et al.*, 2009; Kuo *et al.*, 2008). Competitive binding of the different solvents thus changes the association behavior of the two polymers, resulting in the formation of either block-copolymer micelles or polymersomes.

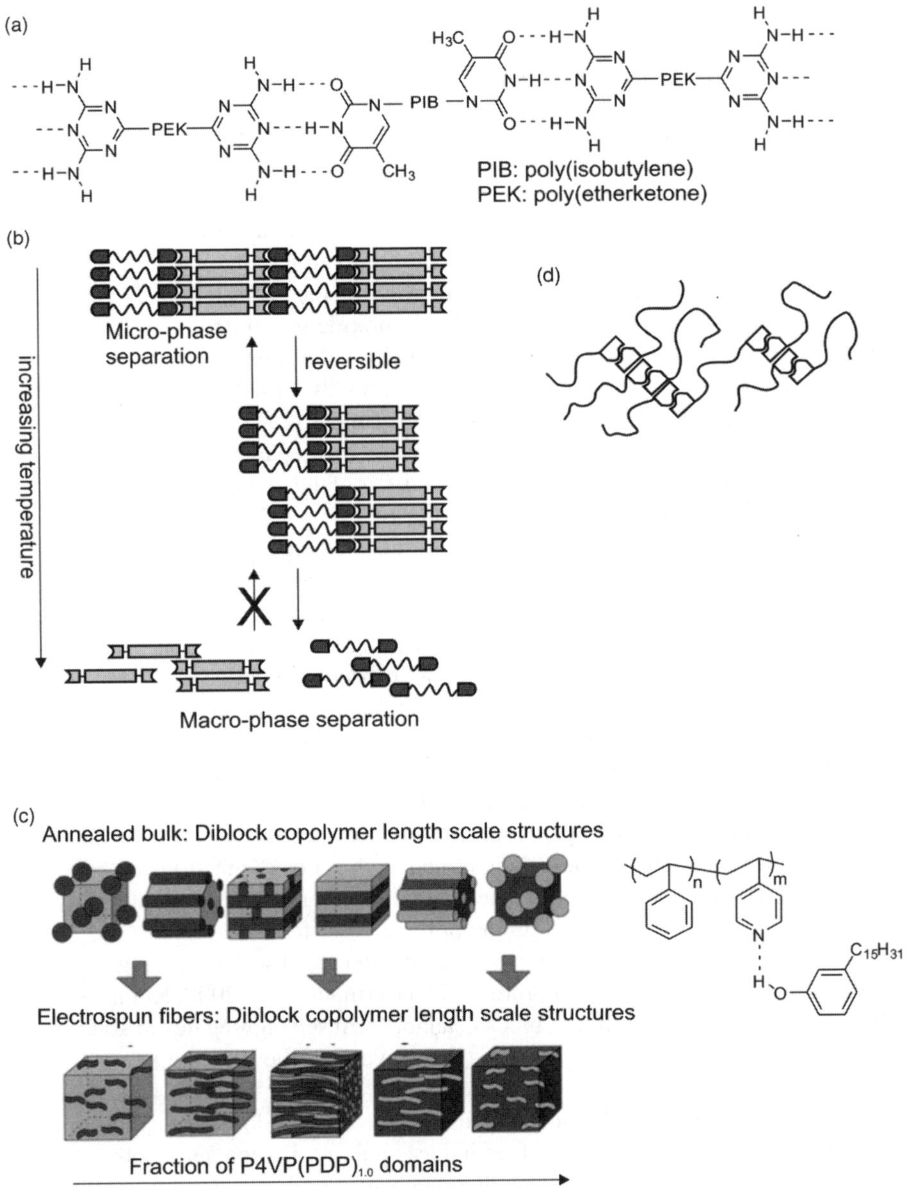

Figure 3.18 (a) Formation of pseudoblock copolymers by association of telechelic PEK and PIB polymers. (b) Thermoreversible behavior as studied by SAXS and IR-methods according to Binder *et al.* (2005a). (c) Microphase separation of "graft-type" supramolecular polymers according to Ruotsalainen *et al.* (2007). (Reproduced from T. Ruotsalainen, J. Turku, P. Hiekkataipale *et al.*, "Tailoring of the hierarchical structure with electrospun fibers due to supramolecular comb-coil block copolymers: polystyrene-block-poly(4-vinyl pyridine) plasticized by hydrogen bonded pentadecylphenol," *Soft Matter*, **3**, 978–985, 2007, by permission of the Royal Society of Chemistry.) (d) Formation of thermoplastic elastomeric networks by formation of clusters of H bonds.

3.3.2 Ordering on Surfaces

Similar to the ordering in bulk, microphase separation is an important factor for the ordering of supramolecular polymers on surfaces and interfaces. Due to the presence of (dynamic) H bonds in these polymers, rearrangement of chains in a thin film can happen much faster than in conventional films. An example to exploit this advantage was presented using stoichiometric blends of PEO-b-(PS-4HS)–P (S-rnd-4VP)-b-(PMMA) polymers, as shown in Figure 3.19a (Tang *et al.*, 2008). Thus, the formation of highly regular hexagonal arrays could be achieved by formation of a thin ($\sim$50 nm) film. Subsequent lithographic templating (i.e. irradiation with deep-UV light to degrade the PMMA matrix and crosslink the PS matrix) led to the transfer of this nanoscale patterns onto the underlying silicon dioxide substrate.

Another example demonstrating the versatility of H-bonded supramolecular polymers is the attachment of nanoparticles to surfaces via hydrogen bonds, based on Microphase-separated polymers (Binder *et al.*, 2006; Binder *et al.*, 2005b; Binder *et al.*, 2009; Zirbs *et al.*, 2005). Basically, homo-, block or random (co-) polymers were prepared, containing the multiple hydrogen bonds (Hamilton receptor) in specific positions of the polymer (either concentrated in one block, or randomly distributed. Spin coating of these polymers leads to a thin film, presenting the H bonds to the polymer/air interface. Using appropriately functionalized nanoparticles (Au (Zirbs *et al.*, 2005), Fe_2O_3 (Binder *et al.*, 2008; Binder and Weinstabl, 2007), CdSe (Binder *et al.*, 2007d)) bearing the matching barbiturate moieties, the nanoparticles are binding to specific regions of the surface, presenting the H bonds. Thus, the architecture of the block copolymer can be transferred to a respective (random or block) pattern of attached nanoparticles on the surface. This approach has recently been extended to a recognition–binding process on surfaces (Figure 3.19b) (Binder *et al.*, 2009). Polymer films presenting hydrogen bonds and azide moieties are spin coated to a surface. CdSe nanoparticles bearing (a) the corresponding barbituric acid moiety (for recognition) and an alkine moiety are then bound selectively to the hydrogen bonds on the polymer surface. Due to the then close proximity between the residual azide–alkine moieties, the "click" reaction can be induced by addition of copper(I)-salts. Subsequently, the nanoparticles are fixed covalently only after the respective supramolecular recognition on the polymeric surface.

Supramolecular polymerization reactions can be conducted at surfaces, using small building blocks with a defined amphiphilicity (Figure 3.19c) (Hermans *et al.*, 2008; Kumar *et al.*, 2007). Thus, the adenine-containing building blocks cannot form self-aggregates in solution, as this assembly is blocked by addition of DMSO, thus preventing the gelation via aggregation in solution (Kumar *et al.*, 2007). As the molecules adsorb to the surface as hydrophobic interactions between the hydrophobic segment and the HOPG dominate. Subsequently, the two-dimensional confinement on the surface induces polymerization, as adsorbate–adsorbate interactions are favored entropically. This then leads to the formation of a two-dimensional polymer on the HOPG surface.

3.3.3 Small-Molecule Ordering into Polymers via H Bonds

The arrangement of small molecules via hydrogen bonds into polymers is one of the most fruitful parts of supramolecular polymer science (Bouteiller, 2007), as developed by Lehn *et al.* (Berl *et al.*, 2002; Buhler *et al.*, 2007; Kolomiets *et al.*, 2006; Kolomiets and Lehn, 2005), Mejier *et al.* (Sijbesma *et al.*, 1997) and recently by Bouteiller *et al.* (Bellot and

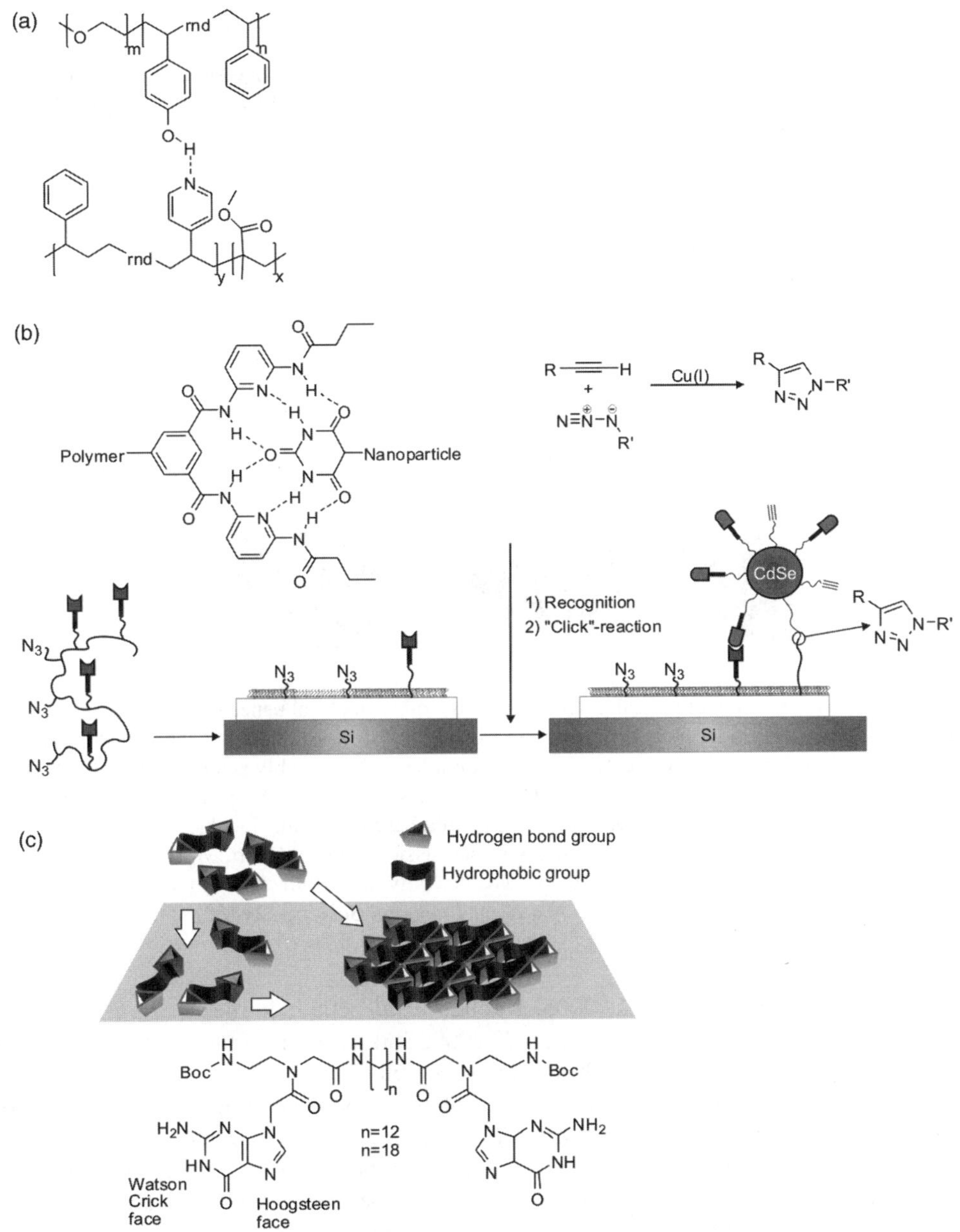

Figure 3.19 (a) Formation of microphase separated polymers by blending two block copolymers via the phenol–pyridyl interaction. (b) Selective attachment of nanoparticles to surfaces presenting a supramolecular interaction and subsequent covalent fixation after recognition. (c) Assembly and surface-induced polymerization of amphiphilic adenine-containing building blocks on highly oriented pyrolytic graphite (HOPG). (Adapted from A.M.S. Kumar, S. Sivakova, R.E. Marchant and S.J. Rowan, "Surface-aided supramolecular polymerization: a route to controlled nanoscale assemblies," *Small*, 2007, **3**, 783–787. © Wiley-VCH Verlag GmbH & Co. KGaA.)

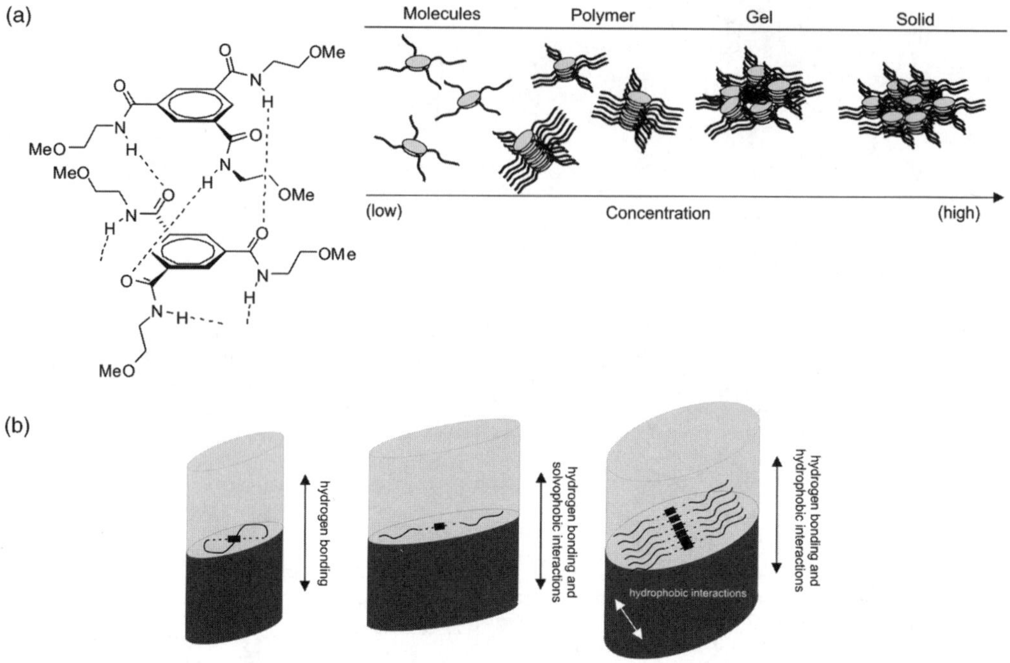

Figure 3.20 (a) Formation of helical aggregates by assembly of chiral benzene-1,3,5-tricarboxamides in apolar solvents. (Reproduced with permission from P. Stals, M. Smulders, R. Martín-Rapún *et al.*, "Asymmetrically substituted benzene-1,3,5-tricarboxamides: self-assembly and odd–even effects in the solid state and in dilute solution," *Chemistry - A European Journal*, 2009, **15**, 2071–2080. © Wiley-VCH Verlag GmbH & Co. KGaA.) (b) Ordering of bis-urea-based building blocks into supramolecular polymers in aqueous environment. (Reprinted with permission from E. Obert, M. Bellot, L. Bouteiller *et al.*, "Both water- and organo-soluble supramolecular polymer stabilized by hydrogen-bonding and hydrophobic interactions," *Journal of the American Chemical Society*, **129**, 50, 15601–15605, 2007. © 2007 American Chemical Society.)

Bouteiller, 2008; Bouteiller *et al.*, 2005; Knoben *et al.*, 2005; Lortie *et al.*, 2003; Lortie *et al.*, 2005; Obert *et al.*, 2007a; Pinault *et al.*, 2009; Shikata *et al.*, 2008; Simic *et al.*, 2003). A plethora of different structures has been observed (Bosman *et al.*, 2004; Brunsveld *et al.*, 2001), mostly showing the formation of fibers, rods, nanoparticles and the resulting entangled networks at higher concentrations (Foster *et al.*, 2009). Three main issues have been addressed in this context recently: (a) the mechanism of the formation of fibers in solution (b) the induction of helicity in the formed fibers in the case of pendant chiral sidechain moieties and (c) the formation of supramolecular aggregates in aqueous environments.

As shown in Figure 3.20a, the formation of supramolecular aggregates naturally follows a concentration-dependent process, with induction of aggregate formation at higher concentrations. This process follows a nucleation-driven aggregation process (Jonkheijm *et al.*, 2006b), as demonstrated for quadrupolar hydrogen bonds and benezene-1,3,5-tricarboxamides (Stals *et al.*, 2009). In the solid state, aggregation into columnar helical structures via intermolecular hydrogen bonding is observed.

More importantly, the induction of helicity and amplification of chirality during the growth of fibers has been studied (George *et al.*, 2007; Wolffs *et al.*, 2007). Thus, achiral OPV-building

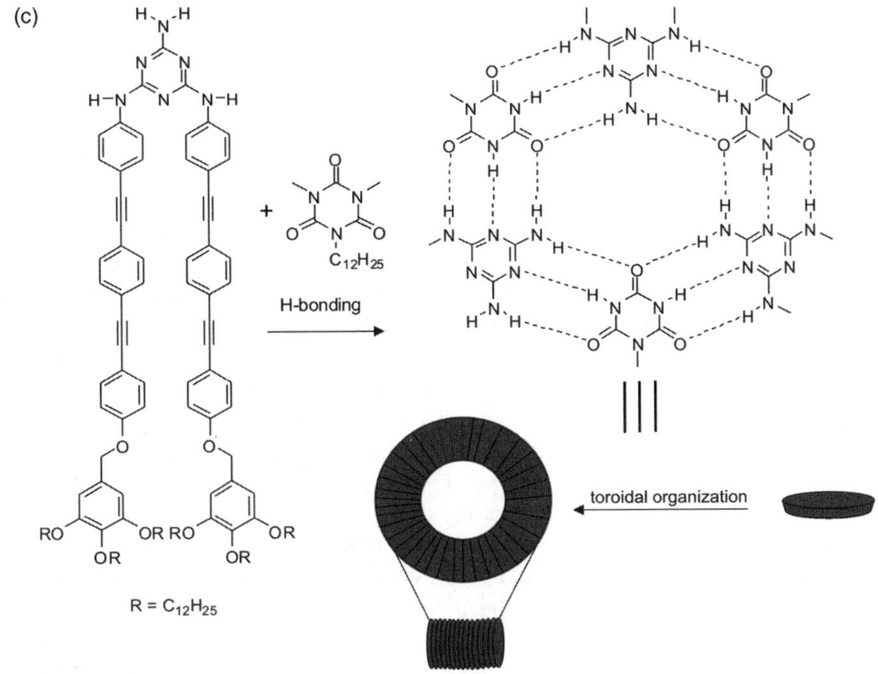

Figure 3.20 (*Continued*) (c) Formation of toroidal aggregates via assembly of barbituric acid/2,4,6-triamino-1,3,5-triazine building blocks. (Reproduced with permission from S. Yagai, S. Mahesh, Y. Kikkawa *et al.*, "Toroidal nanoobjects from rosette assemblies of melamine-linked oligo(p-phenyleneethynylene)s and cyanurates.," *Angewandte Chemie International Edition*, 2008, **47**, 4691–4694. © Wiley-VCH Verlag GmbH & Co. KGaA.)

blocks with pendant ureidotriazine moieties were studied under addition of added chiral acids as guest during the assembly process, observing significant induction of helical "handedness" in this system.

The formation of supramolecular aggregates in different solvent systems has been recently addressed by Bouteiller (Bellot and Bouteiller, 2008; Bouteiller *et al.*, 2005; Knoben *et al.*, 2005; Lortie *et al.*, 2003; Lortie *et al.*, 2005; Obert *et al.*, 2007a; Pinault *et al.*, 2009; Shikata *et al.*, 2008; Simic *et al.*, 2003), using a system of small molecules aggregating via bis-aryl-urea-compounds in water (20b). These systems are highly dynamic in contrast to the gel formation of many low molecular weight species, relying on crystallization phenomena within a supramolecular fiber. By careful design, the formation of such aggregates in water has been achieved (Obert *et al.*, 2007b). This is only possible by choosing small hydrophobic segments close to the H bond (thus preventing the access of water to the newly formed fibrous aggregate) and the attachment of short hydrophilic oligo(ethyleneoxide) chains to the outer part of the molecule, thus permitting sufficient water solubility and leading to an entropy-driven association behavior.

A final example of complex assembly of small molecules via H bonds is shown in Figure 3.20c (Yagai *et al.*, 2008). The formation of a toroidal aggregate is described via a combination of (a) hydrogen bonds of the barbituric acid/2,4,6-triamino-1,3,5-triazine system (Whitesides and Grzybowski, 2002) (b) pi–pi stacking interactions and (c) steric constraints of the attached alkyl chains. This example demonstrates that supramolecular design via assembly of small molecules is possible, given that rigidity is present in the underlying building blocks, thus enabling modeling of molecular parameters in a sufficient mode.

3.3.4 Applications of H-Bonded Supramolecular Polymers

After more than 10 years of H-bonded supramolecular polymers, applications and usefulness of these polymers can be clearly demonstrated. As the most primitive property of the H-bonded polymers, the strongly temperature-dependent aggregation–deaggregation process can be named. Based on the initial observation of Stadler *et al.* (Stadler, 1993b) that plasticity is induced via self-aggregation along 1,4-poly(butadiene) chains leading to a transformation of the PBD-polymer into a thermoplastic elastomer at room temperature, a plethora of thermoplastic structures formed via H bonds have been published (Binder *et al.*, 2005a; Binder *et al.*, 2004b; van Beek *et al.*, 2007a; Wisse *et al.*, 2009; Woodward *et al.*, 2009b; Yamauchi *et al.*, 2004). Additionally, the use of H-bonded polymers as rheological modifiers to achieve a stronger temperature-dependent melt rheology has been described and patented.

Among the recent applications, the field of self-healing materials, optoelectronic polymers and biomaterials has taken a large step to evolution in H-bonded polymers. Thus, an excellent approach to self-healing thermoplasts has been described, which strongly takes the lifetime and thus dynamics of H bonds and the aggregating systems into account. Basically, a detailed tuning between the dynamics of the H-bonded system and the respective chain dynamics has to be achieved, in order to generate a self-healing rubber (Cordier *et al.*, 2008; Montarnal *et al.*, 2008).

Application of H-bonded polymers in optoelectronics was recognized early, as the association can be used to arrange light-emitting/absorbing molecules into close proximity (Abbel *et al.*, 2009; Durkut *et al.*, 2006; Jonkheijm *et al.*, 2006a; Jonkheijm *et al.*, 2006c; Liang and Lin, 2009a; Liang and Lin, 2009b; Lin *et al.*, 2006; Precup-Blaga *et al.*, 2003; Priimagi *et al.*, 2008; Seki *et al.*, 2008; Würthner *et al.*, 2004; Würthner *et al.*, 2000). Thus, examples for application as electron-acceptor structures in solar cells (Kang *et al.*, 2007) or in OLEDs as emissive polymers (inducing, i.e. white light (Abbel *et al.*, 2009)) have been described. In many of these cases, the H-bonding systems allow more flexibility and easier processing (often into thin-film devices) due to the strong ordering of the H bond at lower temperature, but a defined higher dynamics at higher temperatures.

This also holds true for biomaterials, where the quadruple bonds associated to PCL polymers and different peptidic receptor are useful in the generation of a scaffold for cell growth, most of all stem cells, certainly, this is just the beginning of applicative features in supramolecular hydrogen-bonded polymers (Dankers *et al.*, 2005; Dankers and Meijer, 2007).

3.4 Conclusions and Outlook

Now being well developed, supramolecular polymer chemistry has definitely formed an important area in polymer science. With the advent of new synthetic methodologies, most of all the "click" reaction, the inherent drawback of designed attachment of supramolecular entities to polymers has now been solved efficiently. Moreover, the enormously high efficiency of these reactions enable generation of functionalized polymer molecules approaching the purity of organic chemistry. Thus, more complex polymer molecules can now be generated, putting H-bonded polymers in the limelight of supramolecular science as the high level of structural diversity of many hydrogen-bonding systems is unbeaten. A high directionality, a tunable strength over many orders of magnitudes, as well as the now thorough understanding of their dynamics enables the generation of this new set of dynamic materials, ranging from elastomeric to thermoplastic and self-healing materials. Thus, the field of hydrogen-bonded polymers already comprises the most active field in stimuli responsive and adaptable materials of the future.

Acknowledgment

The authors gratefully acknowledge grants from the DFG INST 271/249-1; INST 271/247-1 and INST 271/248-1.

References

Abbel, R., Grenier, C., Pouderoijen, M.J. *et al.* (2009) White-light emitting hydrogen-bonded supramolecular copolymers based on pi-conjugated oligomers. *Journal of the American Chemical Society*, **131**, 833–843.

Adekunle, O., Herbst, F., Hackethal, K., and Binder, W. H. (2011). Synthesis of Nonsymmetrically Chain End Functionalized Polyisobutylenes. *Journal of Polymer Science Part A: Polymer Chemistry*, submitted.

Arnaud, A., Belleney, J., Boué, F. *et al.* (2004) Aqueous supramolecular polymer formed from an amphiphilic perylene derivative. *Angewandte Chemie-International Edition*, **43**, 1718–1721.

Ambade, A.V., Yang, S.K., and Weck, M. (2009) Supramolecular ABC triblock copolymers. *Angewandte Chemie-International Edition*, **48**, 2894–2898.

Asari, T., Arai, S., Takano, A., and Matsushita, Y. (2006) Archimedean tiling structures from ABA/CD block copolymer blends having intermolecular association with hydrogen bonding. *Macromolecules*, **39**, 2232–2237.

Bazzi, H.S. and Sleiman, H.F. (2002) Adenine-containing block copolymers via ring-opening metathesis polymerization: Synthesis and self-assembly into rod morphologies. *Macromolecules*, **35**, 9617–9620.

Bellot, M. and Bouteiller, L. (2008) Thermodynamic description of bis-urea self-assembly: Competition between two supramolecular polymers. *Langmuir*, **24**, 14176–14182.

Berl, V., Schmutz, M., Krische, M.J. *et al.* (2002) Supramolecular polymers generated from heterocomplementary monomers linked through multiple hydrogen-bonding arrays-formation, characterization, and properties. *Chemistry - A European Journal*, **8**, 1227–1244.

Bernard, J., Lortie, F., and Fenet, B. (2009) Design of heterocomplementary H-bonding RAFT agents - towards the generation of supramolecular star polymers. *Macromolecular Rapid Communications*, **30**, 83–88.

Bielawski, C.W. and Grubbs, R.H. (2007) Living ring-opening metathesis polymerization. *Progress in Polymer Science*, **32**, 1–29.

Binder, W. and Zirbs, R. (2007) Supramolecular polymers and networks with hydrogen bonds in the main- and side-chain. *Advances in Polymer Science*, **207**, 1–78.

Binder, W. H, Herbst, F., Schultz, M., (2010). Ordentlich Dynamisch : Supramolekulare Polymere. *Nachrichten aus der Chemie.*, 734–740.

Binder, W.H., Bernstorff, S., Kluger, C. *et al.* (2005a) Tunable materials from hydrogen-bonded pseudo block copolymers. *Advanced Materials*, **17**, 2824–2828.

Binder, W.H., Gloger, D., Weinstabl, H. *et al.* (2007a) Telechelic poly(N-isopropylacrylamides) via nitroxide-mediated controlled polymerization and "click" chemistry: Livingness and "grafting-from" methodology. *Macromolecules*, **40**, 3097–3107.

Binder, W.H. and Kluger, C. (2004) Combining ring-opening metathesis polymerization (ROMP) with sharpless-type "click" reactions: An easy method for the preparation of side chain functionalized poly(oxynorbornenes). *Macromolecules*, **37**, 9321–9330.

Binder, W.H. and Kluger, C. (2006) Azide/alkyne-“click” reactions: applications in material science and organic synthesis. *Current Organic Chemistry*, **10**, 1791.

Binder, W.H., Kluger, C., Josipovic, M. *et al.* (2006) Directing supramolecular nanoparticle binding onto polymer films: Film formation and influence of receptor density on binding densities. *Macromolecules*, **39**, 8092–8101.

Binder, W.H., Kluger, C., Straif, C.J., and Friedbacher, G. (2005b) Directed nanoparticle binding onto microphase-separated block copolymer thin films. *Macromolecules*, **38**, 9405–9410.

Binder, W.H., Kunz, M.J., and Ingolic, E. (2004a) Supramolecular poly(ether ketone)-polyisobutylene pseudo-block copolymers. *Journal of Polymer Science Part A-Polymer Chemistry*, **42**, 162–172.

Binder, W.H., Kunz, M.J., Kluger, C. *et al.* (2004b) Synthesis and analysis of telechelic polyisobutylenes for hydrogen-bonded supramolecular pseudo-block copolymers. *Macromolecules*, **37**, 1749–1759.

Binder, W.H., Lomoschitz, M., Sachsenhofer, R., and Friedbacher, G. (2009) Reversible and irreversible binding of nanoparticles to polymeric surfaces. *Journal of Nanomaterials*. ID 613813. doi: 613810.611155/612009/613813

Binder, W.H. and Machl, D. (2005) Poly(ether ketone)-polyisobutylene block copolymers: Synthesis and phase behavior. *Journal of Polymer Science Part A-Polymer Chemistry*, **43**, 188–202.

Binder, W.H., Machl, D., and Kluger, C. (2004c) Connecting polymeric fragments by sharpless-type “click”-reactions. *Polymer Preprints*, **45**, 692.

Binder, W.H., Petraru, L., Roth, T. *et al.* (2007b) Magnetic and temperature-sensitive release gels from supramolecular polymers. *Advanced Functional Materials*, **17**, 1317–1326.

Binder, W.H., Petraru, L., Roth, T. *et al.* (2007c) Magnetic and temperature-sensitive release gels from supramolecular polymers. *Advanced Functional Materials*, **17**, 1317–1326.

Binder, W.H. and Sachenshofer, R. (2009) “Click”-chemistry on supramolecular materials, in *Click Chemistry for Biotechnology and Materials Science* (ed. J. Lahann), p. 119–175: Wiley-Blackwell.

Binder, W.H. and Sachsenhofer, R. (2007) “Click” Chemistry in polymer and materials science. *Macromolecular Rapid Communications*, **28**, 15–54.

Binder, W.H. and Sachsenhofer, R. (2008) “Click”-chemistry in polymer and material science: An update. *Macromolecular Rapid Communications*, **29**, 952–981.

Binder, W.H., Sachsenhofer, R., Straif, C.J., and Zirbs, R. (2007d) Surface-modified nanoparticles via thermal and Cu (i)-mediated “click” chemistry: Generation of luminescent CdSe nanoparticles with polar ligands guiding supramolecular recognition. *Journal of Materials Chemistry*, **17**, 2125–2132.

Binder, W.H., Weinstabl, H., and Sachsenhofer, R. (2008) Superparamagnetic iron oxide nanoparticles via ligand exchange reactions: organic 1,2-diols as versatile building blocks for surface engineering. *Journal of Nanomaterials*, Article ID 383020.

Binder, W.H. and Weinstabl, H.C. (2007) Surface-modified superparamagnetic iron-oxide nanoparticles. *Monatshefte fur Chemie/Chemical Monthly*, **138**, 315–320.

Binder, W.H. and Zirbs, R. (2009) “Click”-chemistry in macromolecular synthesis, in *Encyclopedia of Polymer Science and Technology*, John Wiley & Sons, Inc. doi: 10.1002/0471440264.pst0471440565

Bondzic, S., Polushkin, E., Schouten, A.J. *et al.* (2007) The influence of grain size on the alignment of hexagonally ordered cylinders of self-assembled diblock copolymer-based supramolecules. *Polymer*, **48**, 4723–4732.

Bosman, A.W., Sijbesma, R.P., and Meijer, E.W. (2004) Supramolecular polymers at work. *Materials Today*, **7**, 34–39.

Bosnian, A.W., Brunsveld, L., Folmer, B.J.B. *et al.* (2003) Supramolecular polymers: from scientific curiosity to technological reality. *Macromolecular Symposia*, **201**, 143–154.

Botterhuis, N.E., Beek, D.J.M.v., Gemert, G.M.L.v., Bosman, A.W., and Sijbesma, R.P. (2008) Self-assembly and morphology of polydimethylsiloxane supramolecular thermoplastic elastomers. *Journal of Polymer Science Part A-Polymer Chemistry*, **46**, 3877–3885.

Bouteiller, L. (2007) Assembly via hydrogen bonds of low molar mass compounds into supramolecular polymers. *Advances in Polymer Science*, **207**, 79–112.

Bouteiller, L., Colombani, O., Lortie, F., and Terech, P. (2005) Thickness transition of a rigid supramolecular polymer. *Journal of the American Chemical Society*, **127**, 8893–8898.

Braunecker, W.A. and Matyjaszewski, K. (2007) Controlled/living radical polymerization: Features, developments, and perspectives. *Progress in Polymer Science*, **32**, 93–146.

Brinke, G.T. and Ikkala, O. (2004) Smart materials based on self-assembled hydrogen-bonded comb-shaped supramolecules. *The Chemical Record*, **4**, 219–230.

Brown, S.P. (2009) Recent advances in solid-state MAS NMR methodology for probing structure and dynamics in polymeric and supramolecular systems. *Macromolecular Rapid Communications*, **30**, 688–716.

Brunsveld, L., Folmer, B.J.B., Meijer, E.W., and Sijbesma, R.P. (2001) Supramolecular polymers. *Chemical Reviews*, **101**, 4071–4098.

Buchmeiser, M.R. (2000) Homogeneous metathesis polymerization by well-defined group VI and group VIII transition-metal alkylidenes: Fundamentals and applications in the preparation of advanced materials. *Chemical Reviews*, **100**, 1565–1604.

Buhler, E., Candau, S.-J., Kolomiets, E., and Lehn, J.-M. (2007) Dynamical properties of semidilute solutions of hydrogen-bonded supramolecular polymers. *Physical Review E*, **76**, 064804-061804-064808.

Chen, C.-P., Dai, S.A., Chang, H.-L. *et al.* (2005) Polyurethane elastomers through multi-hydrogen-bonded association of dendritic structures. *Polymer*, **46**, 11849–11857.

Cordier, P., Tournilhac, F., Soulie-Ziakovic, C., and Leibler, L. (2008) Self-healing and thermoreversible rubber from supramolecular assembly. *Nature*, **451**, 977–980.

Dalphond, J., Bazzi, H.S., Kahrim, K., and Sleiman, H.F. (2002) Synthesis and self-assembly of polymers containing dicarboximide groups by living ring-opening metathesis polymerization. *Macromolecular Chemistry and Physics*, **203**, 1988–1994.

Dankers, P.Y.W., Harmsen, M.C., Brouwer, L.A. *et al.* (2005) A modular and supramolecular approach to bioactive scaffolds for tissue engineering. *Nature Materials*, **4**, 568–574.

Dankers, P.Y.W. and Meijer, E.W. (2007) Supramolecular biomaterials. A modular approach towards tissue engineering. *Bulletin of the Chemical Society of Japan*, **80**, 2047–2073.

Dankers, P.Y.W., van Leeuwen, E.N.M., van Gemert, G.M.L. *et al.* (2006a) Chemical and biological properties of supramolecular polymer systems based on oligocaprolactones. *Biomaterials*, **27**, 5490–5501.

Dankers, P.Y.W., Zhang, Z., Wisse, E. *et al.* (2006b) Oligo(trimethylene carbonate)-based supramolecular biomaterials. *Macromolecules*, **39**, 8763–8771.

de la Fuente, J.L., Fernández-García, M., Cerrada, M.L. *et al.* (2006) Small-angle X-ray scattering and linear melt rheology of poly(tert-butyl acrylate-g-styrene) graft copolymers. *Polymer*, **47**, 1487–1495.

de Lucca Freitas, L. and Stadler, R. (1988) Thermoplastic elastomers by hydrogen bonding 4. Influence of hydrogen bonding on the temperature dependence of the viscoelastic properties. *Colloid & Polymer Science*, **266**, 1095–1101.

Dobrosielska, K., Wakao, S., Takano, A., and Matsushita, Y. (2008) Nanophase-separated structures of AB block copolymer/C homopolymer blends with complementary hydrogen-bonding interactions. *Macromolecules*, **41**, 7695–7698.

Drechsler, U., Thibault, R.J., and Rotello, V.M. (2002) Formation of recognition-induced polymersomes using complementary rigid random copolymers. *Macromolecules*, **35**, 9621–9623.

Durkut, M., Mas-Torrent, M., and Hadley, P. (2006) Electrical transport measurements on self-assembled organic molecular wires. *Journal of Chemical Physics*, **124**, 154704.

Embrechts, A., Schönherr, H., and Vancso, G.J. (2008) Rupture force of single supramolecular bonds in associative polymers by AFM at fixed loading rates. *The Journal of Physical Chemistry. B*, **112**, 7359–7362.

Enders, C., Tanner, S., and Binder, W. H. (2010). End group telechelic Oligo- and Polythiophenes by "click"-Reactions: Synthesis and Analysis via LC-ESI-TOF MS. *Macromolecules* **43**, 8436–8446.

Farnik, D., Kluger, C., Kunz, M.J. *et al.* (2004) Synthesis and self assembly of hydrogen-bonded supramolecular polymers. *Macromolecular Symposia*, **217**, 247–266.

Faust, R. and Kennedy, J.P. (1987) Living carbocationic polymerization. IV. Living polymerization of isobutylene. *Journal of Polymer Science Part A-Polymer Chemistry*, **25**, 1847–1869.

Feldman, K.E., Kade, M.J., de Greef, T.F.A. *et al.* (2008) Polymers with multiple hydrogen-bonded end groups and their blends. *Macromolecules*, **41**, 4694–4700.

Folmer, B.J.B., Sijbesma, R.P., Versteegen, R.M., and Rijt, J.A.J.v.d., and Meijer, E.W. (2000) Supramolecular polymer materials: Chain extension of telechelic polymers using a reactive hydrogen-bonding synthon. *Advanced Materials*, **12**, 874–878.

Foster, E.J., Berda, E.B., and Meijer, E.W. (2009) Metastable supramolecular polymer nanoparticles via intramolecular collapse of single polymer chains. *Journal of the American Chemical Society*, **131**, 6964–6966.

Frauenrath, H., Jahnke, E., Weiss, J. *et al.* (2009) Synthesis of diacetylene-containing peptide building blocks and amphiphiles, their self-assembly and topochemical polymerization in organic solvents. *Chemistry - A European Journal*, **15**, 388–404.

George, S.J., Tomovic, Z., Smulders, M.M.J. *et al.* (2007) Helicity induction and amplification in an oligo(p-phenylenevinylene) assembly through hydrogen-bonded chiral acids. *Angewandte Chemie-International Edition*, **46**, 8206–8211.

Greef, T.F.A.d., Nieuwenhuizen, M.M.L., and Stals, P.J.M. *et al.* (2008) The influence of ethylene glycol chains on the thermodynamics of hydrogen-bonded supramolecular assemblies in apolar solvents. *Chemical Communications*, 4306–4308.

Hackethal, K., Döhler, D., Tanner, S., and Binder, W.H. (2010) Introducing polar monomers into polyisobutylene by living cationic polymerization: Structural and kinetic effects, *Macromolecules*, **43**, 1761–1770.

Hammes, G.G. and Park, A.C. (1968) Kinetic studies of hydrogen bonding. 1-Cyclohexyluracil and 9-ethyladenine. *Journal of the American Chemical Society*, **90**, 4151–4157.

Hammond, M.R. and Mezzenga, R. (2008) Supramolecular routes towards liquid crystalline side-chain polymers. *Soft Matter*, **4**, 952–961.

Hermans, K., Tomatsu, I., Matecki, M. *et al.* (2008) Highly efficient surface relief formation via photoembossing of a supramolecular polymer. *Macromolecular Chemistry and Physics*, **209**, 2094–2099.

Hirschberg, J.H.K.K., Beijer, F.H., van Aert, H.A. *et al.* (1999) Supramolecular polymers from linear telechelic siloxanes with quadruple-hydrogen-bonded units. *Macromolecules*, **32**, 2696–2705.

Herbst, F., Schultz, M., and Binder, W. H. (2010). Ordentlich Dynamisch : Supramolekulare Polymere. *Nachrichten Aus Chemie Technik Und Laboratorium*, 734–740.

Huh, J., Park, H.J., Kim, K.H. *et al.* (2006) Giant thermal tunability of the lamellar spacing in block-copolymer-like supramolecules formed from binary-end-functionalized polymer blends. *Advanced Materials*, **18**, 624–629.

Herbst, F., Schröter, K., Gunkel, I., *et al.* (2010) Aggregation and chain dynamics in supramolecular polymers by dynamic rheology: cluster formation and self aggregation. *Macromolecules* **43**, 10006–10016.

Ilhan, F., Gray, M., and Rotello, V.M. (2001) Reversible side chain modification through noncovalent interactions. "Plug and Play" polymers. *Macromolecules*, **34**, 2597–2601.

Iván, B. and Kennedy, J.P. (1990) Living carbocationic polymerization. XXX. One-pot synthesis of allyl-terminated linear and tri-arm star polyisobutylenes, and epoxy- and hydroxy-telechelics therefrom. *Journal of Polymer Science Part A: Polymer Chemistry*, **28**, 89–104.

Jahnke, E., Severin, N., Kreutzkamp, P. *et al.* (2008) Molecular level control over hierarchical structure formation and polymerization of oligopeptide-polymer conjugates. *Advanced Materials*, **20**, 409–414.

Jiang, M., Li, M., Xiang, M., and Zhou, H. (1999) Interpolymer complexation and miscibility enhancement by hydrogen bonding. *Advances Polymer Science*, 121–196.

Jonkheijm, P., Stutzmann, N., Chen, Z. *et al.* (2006a) Control of ambipolar thin film architectures by co-self-assembling oligo(p-phenylenevinylene)s and perylene bisimides. *Journal of the American Chemical Society*, **128**, 9535–9540.

Jonkheijm, P., van der Schoot, P., Schenning, A.P.H.J., and Meijer, E.W. (2006b) Probing the solvent-assisted nucleation pathway in chemical self-assembly. *Science*, **313**, 80–83.

Jonkheijm, P., vanDuren, J.K.J., Kemerink, M. *et al.* (2006c) Control of film morphology by folding hydrogen-bonded oligo(*p*-phenylenevinylene) polymers in solution. *Macromolecules*, **39**, 784–788.

Joseph, P.K. (1999) Living cationic polymerization of olefins. How did the discovery come about? *Journal of Polymer Science Part A: Polymer Chemistry*, **37**, 2285–2293.

Kang, M.-S., Kim, J.H., Won, J., and Kang, Y.S. (2007) Oligomer approaches for solid-state dye-sensitized solar cells employing polymer electrolytes. *The Journal of Physical Chemistry. C*, **111**, 5222–5228.

Karatzas, A., Talelli, M., Vasilakopoulos, T. *et al.* (2006) Micellization of omega-functionalized diblock copolymers in selective solvent. Study on the effect of hydrogen bonds. *Macromolecules*, **39**, 8456–8466.

Karikari, A.S., Edwards, W.F., Mecham, J.B., and Long, T.E. (2005) Influence of peripheral hydrogen bonding on the mechanical properties of photo-cross-linked star-shaped poly(D, L-lactide) networks. *Biomacromolecules*, **6**, 2866–2874.

Karikari, A.S., Mather, B.D., and Long, T.E. (2007) Association of star-shaped poly(D, L-lactide)s containing nucleobase multiple hydrogen bonding. *Biomacromolecules*, **8**, 302–308.

Kautz, H., van Beek, D.J.M., Sijbesma, R.P., and Meijer, E.W. (2006) Cooperative end-to-end and lateral hydrogen-bonding motifs in supramolecular thermoplastic elastomers. *Macromolecules*, **39**, 4265–4267.

Kavitha, A.A. and Singha, N.K. (2009) Tailor-made poly(methyl acrylate) bearing amantadine functionality (amino adamantyl) via atom transfer radical polymerization (ATRP). A precursor of a supramolecular cross-linked polymer. *Macromolecules*, **42**, 5499–5508.

Keizer, H.M., Sijbesma, R.P., Jansen, J.F.G.A. *et al.* (2003) Polymerization-induced phase separation using hydrogen-bonded supramolecular polymers. *Macromolecules*, **36**, 5602–5606.

Kersey, F.R., Loveless, D.M., and Craig, S.L. (2007) A hybrid polymer gel with controlled rates of cross-link rupture and self-repair. *Journal of The Royal Society Interface*, **4**, 373–380.

Khan, A., Haddleton, D.M., Hannon, M.J. *et al.* (1999) Hydrogen bond template-directed polymerization of protected 5′-acryloylnucleosides. *Macromolecules*, **32**, 6560–6564.

Kim, I., Shin, N.-H., Jo, J.-K. *et al.* (2009a) Self-healing elastomers as dream smart materials. *Elastomers and Composites*, **44**, 196–208.

Kim, S.H., Nederberg, F., Jakobs, R. *et al.* (2009b) A Supramolecularly assisted transformation of block-copolymer micelles into nanotubes. *Angewandte Chemie-International Edition*, **48**, 4508–4512.

Kluger, C. and Binder, W.H. (2004) Side chain-functionalized poly(norbornenes) bearing new hydrogen bonding motives via ROMP. *Polymer Preprints*, **45**, 694.

Kluger, C. and Binder, W.H. (2005) Connecting supramolecular receptors onto blockcopolymers: Combining ring opening metathesis polymerization with 1,3-Dipolar cycloadditions. *Polymer Preprints*, **46**, 723.

Kluger, C. and Binder, W.H. (2007) Functionalized poly(oxanorbornene)-block-copolymers: Preparation via ROMP/click-methodology. *Journal of Polymer Science Part A-Polymer Chemistry*, **45**, 485–499.

Knoben, W., Besseling, N.A.M., Bouteiller, L., and Stuart, M.A.C. (2005) Dynamics of reversible supramolecular polymers: Independent determination of the dependence of linear viscoelasticity on concentration and chain length by using chain stoppers. *Physical Chemistry Chemical Physics*, 2390–2398.

Knoben, W., Besseling, N.A.M., and Cohen Stuart, M.A. (2007) Rheology of a reversible supramolecular polymer studied by comparison of the effects of temperature and chain stoppers. *Journal of Chemical Physics*, **126**, 024907/024901-024907/024909.

Kolb, H.C., Finn, M.G., and Sharpless, K.B. (2001) Click chemistry: Diverse chemical function from a few good reactions. *Angewandte Chemie-International Edition*, **40**, 2004–2021.

Kolomiets, E., Buhler, E., Candau, S.J., and Lehn, J.M. (2006) Structure and properties of supramolecular polymers generated from heterocomplementary monomers linked through sextuple hydrogen-bonding arrays. *Macromolecules*, **39**, 1173–1181.

Kolomiets, E. and Lehn, J.-M. (2005) Double dynamers: molecular and supramolecular double dynamic polymers. *Chemical Communications*, 1519–1521.

Kosonen, H., Valkama, S., Ruokolainen, J. *et al.* (2003) Processible conducting nanoscale cylinders due to self-organized polyaniline supramolecules. *Synthetic Metals*, **137**, 881–882.

Kumar, A.M.S., Sivakova, S., Marchant, R.E., and Rowan, S.J. (2007) Surface-aided supramolecular polymerization: a route to controlled nanoscale assemblies. *Small*, **3**, 783–787.

Kunz, M.J., Hayn, G., Saf, R., and Binder, W.H. (2004a) Hydrogen-bonded supramolecular poly(ether ketone)s. *Journal of Polymer Science Part A-Polymer Chemistry*, **42**, 661–674.

Kuo, S.-W., Tung, P.-H., and Chang, F.-C. (2009) Hydrogen bond mediated supramolecular micellization of diblock copolymer mixture in common solvents. *European Polymer Journal*, **45**, 1924–1935.

Kuo, S.-W., Tung, P.-H., Lai, C.-L. *et al.* (2008) Supramolecular micellization of diblock copolymer mixtures mediated by hydrogen bonding for the observation of separated coil and chain aggregation in common solvents. *Macromolecular Rapid Communications*, **29**, 229–233.

Kurzhals, S., and Binder, W. H. (2010). Telechelic Polynorbornenes with Hydrogen Bonding Moieties by Direct End Capping of Living Chains. *Journal of Polymer Science Part A: Polymer Chemistry* **48**, 5522–5532.

Lafitte, V.G.H., Aliev, A.E., Horton, P.N. *et al.* (2006) Quadruply hydrogen bonded cytosine modules for supramolecular applications. *Journal of the American Chemical Society*, **128**, 6544–6545.

Laiho, A. and Ikkala, O. (2007) A rheo-optical apparatus for real time kinetic studies on shear-induced alignment of self-assembled soft matter with small sample volumes. *Review of Scientific Instruments*, **78**, 015109–015106.

Lange, R.F.M., Gurp, M.V., and Meijer, E.W. (1999) Hydrogen-bonded supramolecular polymer networks. *Journal of Polymer Science Part A-Polymer Chemistry*, **37**, 3657–3670.

Langela, M., Wiesner, U., Spiess, H.W., and Wilhelm, M. (2002) Microphase reorientation in block copolymer melts as detected via FT rheology and 2D SAXS. *Macromolecules*, **35**, 3198–3204.

Lee, K.J., Kim, Y.W., Koh, J.H., and Kim, J.H. (2007a) Supramolecular polymer/metal salt complexes containing quadruple hydrogen-bonding units. *Journal of Polymer Science Part B-Polymer Physics*, **45**, 3181–3188.

Lee, K.J., Lee, D.K., Kim, Y.W., and Kim, J.H. (2007b) One-pot synthesis of supramolecular polymer containing quadruple hydrogen-bonding units. *European Polymer Journal*, **43**, 4460–4465.

Li, B., Xu, L., Wu, Q. *et al.* (2007) Various types of hydrogen bonds, their temperature dependence and water-polymer interaction in hydrated poly(acrylic acid) as revealed by 1H solid-state NMR spectroscopy. *Macromolecules*, **40**, 5776–5786.

Li, B.S., Cheuk, K.K.L., Ling, L. *et al.* (2003) Synthesis and hierarchical structures of amphiphilic polyphenylacetylenes carrying L-valine pendants. *Macromolecules*, **36**, 77–85.

Li, Z., Ding, J., Day, M., and Tao, Y. (2006) Molecularly imprinted polymeric nanospheres by diblock copolymer self-assembly. *Macromolecules*, **39**, 2629–2636.

Liang, T.-C. and Lin, H.-C. (2009a) Study of supramolecular side-chain and cross-linking polymers by complexation of various H-donor acids with H-acceptor copolymers containing pendent carbazole and fluorescent pyridyl units. *Journal of Polymer Science Part A-Polymer Chemistry*, **47**, 2734–2753.

Liang, T.-C. and Lin, H.-C. (2009b) Supramolecular assembly of H-bonded copolymers/complexes/nanocomposites and fluorescence quenching effects of surface-modified gold nanoparticles on fluorescent copolymers containing pyridyl H-acceptors and acid H-donors. *Journal of Materials Chemistry*, **19**, 4753–4763.

Lin, H.C., Tsai, C.M., Huang, G.H., and Tao, Y.T. (2006) Synthesis and characterization of light-emitting H-bonded complexes and polymers containing Bis(pyridyl) emitting acceptors. *Macromolecules*, **39**, 557–568.

Lin, Z., Li, Y., Slawin, A.M.Z., and Morris, R.E. (2008) Hydrogen-bond-directing effect in the ionothermal synthesis of metal coordination polymers. *Dalton Transactions*, 3989–3994.

Lortie, F., Boileau, S., and Bouteiller, L. (2003) N,N-disubstituted ureas: Influence of substituents on the formation of supramolecular polymers. *Chemistry - A European Journal*, **9**, 3008–3014.

Lortie, F., Boileau, S., Bouteiller, L. *et al.* (2005) Chain stopper-assisted characterization of supramolecular polymers. *Macromolecules*, **38**, 5283–5287.

Loveless, D.M., Abu-Lail, N.I., Kaholek, M. *et al.* (2006) Reversibly cross-linked surface-grafted polymer brushes. *Angewandte Chemie-International Edition*, **45**, 7812–7814.

Ma, Y., Kolotuchin, S.V., and Zimmerman, S.C. (2002) Supramolecular polymer chemistry: Self-assembling dendrimers using the DDA;AAD (GC-like) hydrogen bonding motif. *Journal of the American Chemical Society*, **124**, 13757–13769.

Macomber, R.S. (1992) An introduction to NMR titration for studying rapid reversible complexation. *Journal Chemical Education*, **69**, 375–378.

Marsh, A., Khan, A., Haddleton, D.M., and Hannon, M.J. (1999) Atom transfer polymerization: Use of uridine and adenosine derivatized monomers and initiators. *Macromolecules*, **32**, 8725–8731.

Mather, B.D., Baker, M.B., Beyer, F.L. *et al.* (2007a) Supramolecular triblock copolymers containing complementary nucleobase molecular recognition. *Macromolecules*, **40**, 6834–6845.

Meldal, M. and Tornoe, C.W. (2001) Peptidotriazoles: Copper(I)-catalyzed 1,3-dipolar cycloadditions on solid-phase. Proceedings of the Second International and the Seventeenth American Peptide Symposium, pp. 263–264.

Montarnal, D., Cordier, P., Soulié-Ziakovic, C. *et al.* (2008) Synthesis of self-healing supramolecular rubbers from fatty acid derivatives, diethylene triamine, and urea. *Journal of Polymer Science Part A-Polymer Chemistry*, **46**, 7925–7936.

Müller, M., Fischer, E.W., Kremer, F. *et al.* (1995) The molecular dynamics of thermoreversible networks as studied by broadband dielectric spectroscopy. *Colloid & Polymer Science*, **273**, 38–46.

Muller, M., Seidel, U., and Stadler, R. (1995) Influence of hydrogen bonding on the viscoelastic properties of thermoreversible networks: analysis of the local complex dynamics. *Polymer*, **36**, 3143–3150.

Nair, K.P., Breedveld, V., and Weck, M. (2008) Complementary hydrogen-bonded thermoreversible polymer networks with tunable properties. *Macromolecules*, **41**, 3429–3438.

Nair, K.P., Pollino, J.M., and Weck, M. (2006) Noncovalently functionalized block copolymers possessing both hydrogen bonding and metal coordination centers. *Macromolecules*, **39**, 931–940.

Noro, A., Matsushita, Y., and Lodge, T.P. (2008) Thermoreversible supramacromolecular ion gels via hydrogen bonding. *Macromolecules*, **41**, 5839–5844.

Obert, E., Bellot, M., Bouteiller, L. *et al.* (2007a) Both water- and organo-soluble supramolecular polymer stabilized by hydrogen-bonding and hydrophobic interactions. *Journal of the American Chemical Society*, **129**, 15601–15605.

Odille, F.G.J., Jónsson, S., Stjernqvist, S. *et al.* (2007) On the characterization of dynamic supramolecular systems: A general mathematical association model for linear supramolecular copolymers and application on a complex two-component hydrogen-bonding system. *Chemistry - A European Journal*, **13**, 9617–9636.

Oelschlaeger, C., Gutmann, J.S., Wolkenhauer, M. *et al.* (2007) Kinetics of shear microphase orientation and reorientation in lamellar diblock and triblock copolymer melts as detected via FT-rheology and 2D-SAXS. *Macromolecular Chemistry and Physics*, **208**, 1719–1729.

Ohkawa, H., Ligthart, G.B.W.L., Sijbesma, R.P., and Meijer, E.W. (2007) Supramolecular graft copolymers based on 2,7-diamido-1,8-naphthyridines. *Macromolecules*, **40**, 1453–1459.

Ostas, E., Schröter, K., Thurn-Albrecht, T., *et al.* (2011) Poly(ε-caprolactone)-poly(isobutylene): a crystallizing, hydrogen bonded pseudo-blockcopolymer. *Journal of Polymer Science Part A: Polymer Chemistry*, submitted.

Park, T., Zimmerman, S.C., and Nakashima, S. (2005) A highly stable quadruply hydrogen-bonded heterocomplex useful for supramolecular polymer blends. *Journal of the American Chemical Society*, **127**, 6520–6521.

Peng, C.C. and Abetz, V. (2005) A simple pathway toward quantitative modification of polybutadiene: A new approach to thermoreversible cross-linking rubber comprising supramolecular hydrogen-bonding networks. *Macromolecules*, **38**, 5575–5580.

Petraru, L., Farnik, D., Saf, R., and Binder, W.H. (2004) Supramolecular, telechelic poly(Etherketones) bearing barbituric acid as hydrogen-bonding unit. *Polymer Preprints*, **45**, 690.

Pinault, T., Cannizzo, C., Andrioletti, B. *et al.* (2009) Anions as efficient chain stoppers for hydrogen-bonded supramolecular polymers. *Langmuir*, **25**, 8404–8407.

Pollino, J.M., Stubbs, L.P., and Weck, M. (2004) One-step multifunctionalization of random copolymers via self-assembly. *Journal of the American Chemical Society*, **126**, 563–567.

Polushkin, E., Bondzic, S., deWit, J. *et al.* (2005) In-situ SAXS study on the alignment of ordered systems of comb-shaped supramolecules: A shear-induced cylinder-to-cylinder transition. *Macromolecules*, **38**, 1804–1813.

Precup-Blaga, F.S., Garcia-Martinez, J.C., Schenning, A.P.H.J., and Meijer, E.W. (2003) Highly emissive supramolecular oligo(p-phenylene vinylene) dendrimers. *Journal of the American Chemical Society*, **125**, 12953–12960.

Priimagi, A., Vapaavuori, J., Rodriguez, F.J. *et al.* (2008) Hydrogen-bonded polymer–azobenzene complexes: enhanced photoinduced birefringence with high temporal stability through interplay of intermolecular interactions. *Chemistry of Materials*, **20**, 6358–6363.

Ramani, R., Valkama, S., Kilpeläinen, S. *et al.* (2009) Order-disorder transitions in self-assembled polymers: A positron annihilation study. *Physica Status Solidi C*, **6**, 2414–2416.

Rieth, L.R., Eaton, R.F., and Coates, G.W. (2001) Polymerization of ureidopyrimidinone-functionalized olefins by using late-transition metal Ziegler–Natta catalysts: synthesis of thermoplastic elastomeric polyolefins. *Angewandte Chemie-International Edition*, **40**, 2153–2156.

Roth, T., Groh, P.W., Pálfi, V. *et al.* (2005) Supramolecular three-arm star polyisobutylenes by sharpless-type "click" reactions. *Polymer Preprints*, **46**, 1166.

Rowan, S.J., Suwanmala, P., and Sivakova, S. (2003) Nucleobase-induced supramolecular polymerization in the solid state. *Journal of Polymer Science Part A-Polymer Chemistry*, **41**, 3589–3596.

Ruotsalainen, T., Turku, J., Hiekkataipale, P. *et al.* (2007) Tailoring of the hierarchical structure within electrospun fibers due to supramolecular comb-coil block copolymers: polystyrene-block-poly(4-vinyl pyridine) plasticized by hydrogen bonded pentadecylphenol. *Soft Matter*, **3**, 978–985.

Schadler, V., Kniese, V., Thurn-Albrecht, T. *et al.* (1998) Self-assembly of ionically end-capped diblock copolymers. *Macromolecules*, **31**, 4828–4837.

Scherman, O.A., Ligthart, G.B.W.L., Sijbesma, R.P., and Meijer, E.W. (2006a) A selectivity-driven supramolecular polymerization of an AB monomer. *Angewandte Chemie-International Edition*, **45**, 2072–2076.

Scherman, O.A., Ligthart, G.B.W.L., Ohkawa, H. *et al.* (2006b) Olefin metathesis and quadruple hydrogen bonding: A powerful combination in multistep supramolecular synthesis. *PNAS*, **103**, 11850–11855.

Schnell, I., Langer, B., Söntjens, S.H.M. *et al.* (2002) Quadruple hydrogen bonds of ureido-pyrimidinone moieties investigated in the solid state by 1H double-quantum MAS NMR-spectroscopy. *Physical Chemistry Chemical Physics*, **4**, 3750–3758.

Seki, T., Yagai, S., Karatsu, T., and Kitamura, A. (2008) Formation of supramolecular polymers and discrete dimers of perylene bisimide dyes based on melamine–cyanurates hydrogen-bonding interactions. *The Journal of Organic Chemistry*, **73**, 3328–3335.

Serpe, M.J. and Craig, S.L. (2007) Physical organic chemistry of supramolecular polymers. *Langmuir*, **23**, 1626–1634.

Sessler, J.L., Lawrence, C.M., and Jayawickramarajah, J. (2007) Molecular recognition via base-pairing. *Chemical Society Reviews*, **36**, 314–325.

Shikata, T., Nishida, T., Isare, B. *et al.* (2008) Structure and dynamics of a bisurea-based supramolecular polymer in n-dodecane. *The Journal of Physical Chemistry. B*, **112**, 8459–8465.

Sijbesma, R.P., Beijer, F.H., Brunsveld, L. *et al.* (1997) Reversible polymers formed from self-complementary monomers using quadruple hydrogen bonding. *Science*, **278**, 1601–1604.

Simic, V., Bouteiller, L., and Jalabert, M. (2003) Highly cooperative formation of bis-urea based supramolecular polymers. *Journal of the American Chemical Society*, **125**, 13148–13154.

Sinkeldam, R.W., Hoeben, F.J.M., Pouderoijen, M.J. *et al.* (2006) Chiral alignment of OPV chromophores: Exploitation of the ureidophthalimide-based foldamer. *Journal of the American Chemical Society*, **128**, 16113–16121.

Sivakova, S. and Rowan, S.J. (2005) Nucleobases as supramolecular motifs. *Chemical Society Reviews*, 9–21.

Sivakova, S., Wu, J., Campo, C.J. *et al.* (2006) Liquid-crystalline supramolecular polymers formed through complementary nucleobase-pair interactions. *Chemistry - A European Journal*, **12**, 446–456.

Sontjens, S.H.M., Sijbesma, R.P., van Genderen, M.H.P., and Meijer, E.W. (2000) Stability and lifetime of quadruply hydrogen bonded 2-ureido-4[1H]-pyrimidinone dimers. *Journal of the American Chemical Society*, **122**, 7487–7493.

South, C.R., Leung, K.C.F., Lanari, D. *et al.* (2006) Noncovalent side-chain functionalization of terpolymers. *Macromolecules*, **39**, 3738–3744.

Spiess, H.W. (2003) Supramolecular aspects of polymer science: a challenge for solid state NMR. *Macromolecular Symposia*, **201**, 85–88.

Stadler, F.J., Pyckhout-Hintzen, W., and Baillya, C. (2008) Linear and nonlinear rheological characterization of temporary networks of telechelic polybutadiene. *AIP Conference Proceedings*, **1027**, 552–554.

Stadler, R. (1993a) Thermoplastic elastomers via supramolecular self assembling in random copolymers. *Kautschuk Gummi Kunststoffe*, **46**, 619–628.

Stals, P.J.M., Smulders, M.M.J., Rafael, M.-R. *et al.* (2009) Asymmetrically substituted benzene-1,3,5-tricarboxamides: self-assembly and odd–even effects in the solid state and in dilute solution. *Chemistry - A European Journal*, **15**, 2071–2080.

Takashi Kato, N.M.K.K. (2006) Functional liquid-crystalline assemblies: Self-organized soft materials. *Angewandte Chemie-International Edition*, **45**, 38–68.

Tang, C., Lennon, E.M., Fredrickson, G.H. *et al.* (2008) Evolution of block copolymer lithography to highly ordered square arrays. *Science*, **322**, 429–432.

ten Brinke, G., Ruokolainen, J., and Ikkala, O. (2007) Supramolecular materials based on hydrogen-bonded polymers. *Advances in Polymer Science*, **207**, 113–177.

Todd, E.M. and Zimmerman, S.C. (2008) Bis-ureidodeazapterin (Bis-DeAP) as a general route to supramolecular star polymers. *Tetrahedron*, **64**, 8558–8570.

Tornoe, C.W., Christensen, C., and Meldal, M. (2002) Peptidotriazoles on solid phase: [1,2,3]-triazoles by regiospecific copper(I)-catalyzed 1,3-dipolar cycloadditions of terminal alkynes to azides. *The Journal of Organic Chemistry*, **67**, 3057–3064.

Tung, S.-H., Kalarickal, N.C., Mays, J.W., and Xu, T. (2008) Hierarchical assemblies of block-copolymer-based supramolecules in thin films. *Macromolecules*, **41**, 6453–6462.

van Beek, D.J.M., Gillissen, M.A.J., vanAs, B.A.C. *et al.* (2007a) Supramolecular copolyesters with tunable properties. *Macromolecules*, **40**, 6340–6348.

van Beek, D.J.M., Spiering, A.J.H., Peters, G.W.M. *et al.* (2007b) Unidirectional dimerization and stacking of ureidopyrimidinone end groups in polycaprolactone supramolecular polymers. *Macromolecules*, **40**, 8464–8475.

van Ekenstein, A.G., Polushkin, E., Nijland, H. *et al.* (2003) Shear alignment at two length scales: Comb-shaped supramolecules self-organized as cylinders-within-lamellar hierarchy. *Macromolecules*, **36**, 3684–3688.

Vancso, G.J. (2007) Feeling the force of supramolecular bonds in polymers. *Angewandte Chemie-International Edition*, **46**, 3794–3796.

Vera, F., Almuzara, C., Orera, I. *et al.* (2008) Side-chain supramolecular polymers with induced supramolecular chirality through H-bonding interactions. *Journal of Polymer Science Part A-Polymer Chemistry*, **46**, 5528–5541.

Vermonden, T., vanSteenbergen, M.J., Besseling, N.A.M. *et al.* (2004) Linear rheology of water-soluble reversible neodymium(III) coordination polymers. *Journal of the American Chemical Society*, **126**, 15802–15808.

Versteegen, R.M., Kleppinger, R., Sijbesma, R.P., and Meijer, E.W. (2006) Properties and morphology of segmented copoly(ether urea)s with uniform hard segments. *Macromolecules*, **39**, 772–783.

Viswanathan, K., Ozhalici, H., Elkins, C.L. *et al.* (2006) Multiple hydrogen bonding for reversible polymer surface adhesion. *Langmuir*, **22**, 1099–1105.

Wasserfallen, D., Fischbach, I., Chebotareva, N. *et al.* (2005) Influence of hydrogen bonds on the supramolecular order of hexa-peri-hexabenzocoronenes. *Advanced Functional Materials*, **15**, 1585–1594.

Weiss, J., Jahnke, E., and Frauenrath, H. (2008) Chiroptical properties of multiple-helical, oligopeptide-substituted poly(diacetylene)s in solution. *Macromolecular Rapid Communications*, **29**, 330–339.

Whitesides, G.M. and Grzybowski, B. (2002) Self-Assembly at all scales. *Science*, **295**, 2418–2421.

Wilson, A.J. (2007) Non-covalent polymer assembly using arrays of hydrogen-bonds. *Soft Matter*, **3**, 409–425.

Wisse, E., Govaert, L.E., Meijer, H.E.H., and Meijer, E.W. (2006a) Unusual tuning of mechanical properties of thermoplastic elastomers using supramolecular fillers. *Macromolecules*, **39**, 7425–7432.

Wisse, E., Spiering, A.J.H., Pfeifer, F. *et al.* (2009) Segmental orientation in well-defined thermoplastic elastomers containing supramolecular fillers. *Macromolecules*, **42**, 524–530.

Wisse, E., Spiering, A.J.H., vanLeeuwen, E.N.M. *et al.* (2006b) Molecular recognition in poly(epsilon-caprolactone)-based thermoplastic elastomers. *Biomacromolecules*, **7**, 3385–3395.

Wolffs, M., George, S.J., Tomović, Ž., Meskers, S.C.J., Schenning, A.P.H.J., and Meijer, E.W. (2007) Macroscopic origin of circular dichroism effects by alignment of self-assembled fibers in solution. *Angewandte Chemie-International Edition*, **46**, 8203–8205.

Woodward, P., Clarke, A., Greenland, B.W. *et al.* (2009a) Facile bisurethane supramolecular polymers containing flexible alicyclic receptor units. *Soft Matter*, **5**, 2000–2010.

Woodward, P., Merino, D.H., Hamley, I.W. *et al.* (2009b) Thermally responsive elastomeric supramolecular polymers featuring flexible aliphatic hydrogen-bonding end-groups. *Australian Journal of Chemistry*, **62**, 790–793.

Wurthner, F., Chen, Z., Hoeben, F.J.M. *et al.* (2004) Supramolecular p-n-heterojunctions by co-self-organization of oligo (p-phenylene vinylene) and perylene bisimide dyes. *Journal of the American Chemical Society*, **126**, 10611–10618.

Würthner, F., Thalacker, C., Sautter, A. *et al.* (2000) Hierarchical self-organization of perylene bisimide-melamine assemblies to fluorescent mesoscopic superstructures. *Chemistry - A European Journal*, **6**, 3871–3886.

Xu, H., Srivastava, S., and Rotello, V. (2007) Nanocomposites based on hydrogen bonds. *Advances in Polymer Science*, **207**, 179–198.

Yagai, S., Mahesh, S., Kikkawa, Y. *et al.* (2008) Toroidal nanoobjects from rosette assemblies of melamine-linked oligo-p-phenyleneethynylene)s and cyanurates. *Angewandte Chemie-International Edition*, **47**, 4691–4694.

Yamauchi, K., Kanomata, A., Inoue, T., and Long, T.E. (2004) Thermoreversible polyesters consisting of multiple hydrogen bonding (MHB). *Macromolecules*, **37**, 3519–3522.

Yamauchi, K., Lizotte, J.R., and Long, T.E. (2003) Thermoreversible poly(alkyl acrylates) consisting of self-complementary multiple hydrogen bonding. *Macromolecules*, **36**, 1083–1088.

Yount, W.C., Loveless, D.M., and Craig, S.L. (2005a) Small-molecule dynamics and mechanisms underlying the macroscopic mechanical properties of coordinatively cross-linked polymer networks. *Journal of the American Chemical Society*, **127**, 14488–14496.

Yount, W.C., Loveless, D.M., and Craig, S.L. (2005b) Strong means slow: Dynamic contributions to the bulk mechanical properties of supramolecular networks. *Angewandte Chemie-International Edition*, **44**, 2746–2748.

Zirbs, R., Kienberger, F., Hinterdorfer, P., and Binder, W.H. (2005) Directed assembly of Au-nanoparticles onto planar surfaces via multiple hydrogen bonds. *Langmuir*, **21**, 8414–8421.

Zoelen, W.v., Asumaa, T., Ruokolainen, J., Ikkala, O., and Brinke, G.t. (2008) Phase behavior of solvent vapor annealed thin films of PS-b-P4VP(PDP) supramolecules. *Macromolecules*, **41**, 3199–3208.

Zou, S., Schönherr, H., and Vancso, G.J. (2005) Stretching and rupturing individual supramolecular polymer chains by AFM. *Angewandte Chemie-International Edition*, **44**, 956–959.

4

Recent Synthetic Developments in Miktoarm Star Polymers with More than Three Different Arms

Akira Hirao[1], Mayumi Hayashi[1], Tomoya Higashihara[1] and Nikos Hadjichristidis[2]

[1] *Division of Soft Materials Chemistry, Polymeric and Organic Materials Department, Graduate School of Science and Engineering, Tokyo Institute of Technology, S1-6, 2-12-1, Ohokayama, Meguro-ku, Tokyo 152-8552, Japan*

[2] *Department of Chemistry, University of Athens, Panepistimiopolis, Zografou, 15771 Athens, Greece*

4.1 Introduction

Star-branched polymers have been widely investigated for a long time because of the synthetic challenges associated with preparing them and because they offer quite different properties and behavior from those of corresponding linear polymers (Bauer and Fetters, 1978; Bywater, 1979; Roovers, 1985; Rempp and Herz, 1989; Fetters and Thomas, 1993; Hsieh and Quirk, 1996; Grest *et al.*, 1996; Lutz and Rein, 1999). Among the star-branched polymers, asymmetric ones having chemically different arm segments, the so-called miktoarm (from the Greek word μικτόζ meaning *mixed*) star polymers (μ-star polymers), have recently received much attention because they exhibit unique and unusual morphologies due to their characteristic branched architectures in addition to multiphase structures (Hadjichristidis, 1999; Hadjichristidis *et al.*, 2001; Hadjichristidis *et al.*, 2005; Hadjichristidis *et al.*, 2006; Hadjichristidis *et al.*, 2007; Hirao *et al.*, 2002e; Hirao *et al.*, 2005; Hirao *et al.*, 2008b). For example, miktoarm

Complex Macromolecular Architectures: Synthesis, Characterization, and Self-Assembly, First Edition.
Edited by Nikos Hadjichristidis, Akira Hirao, Yasuyuki Tezuka and Filip Du Prez.

star copolymers of 3 μ-A_2B type (e.g., (polystyrene)$_2$(polyisoprene), PS_2PI) exhibit interesting morphological features either by moving the borders of the classic morphology map of linear block copolymers or by forming new morphological structures (Tselikas *et al.*, 1996). Addition of a third arm [3 μ-ABC miktoarm star terpolymer, e.g., (PS)(PI)(PMMA)] tremendously increases the variety of microdomain morphologies (Loshe and Hadjichristidis, 1997). Such materials have extended the use of linear block copolymers for potential applications in the fields of nanoscience and nanotechnology including electronic and optical nanodevices, nanowires, nanofunctional particles, nanoreactors, nanoscale filters, functional micelles for drug and genetic material delivery and so on.

In general, it is far more difficult to synthesize miktoarm star polymers than regular stars having identical arms because there are strict requirements in terms of multistep quantitative reactions that correspond to introducing different arms. In addition, the isolation of intermediate polymers is often required to obtain pure products. Although several successful examples of well-defined μ-star polymers have been reported, most of them are composed of less than three different arms. Only a few examples of miktoarm star-branched polymers with four different arms have been synthesized to date. For this difficult synthetic problem, general and versatile methodologies for the synthesis of multiarmed and multicomponent μ-star polymers have been vitally desired.

In order to synthesize well-defined μ-star polymers, the use of a living polymerization system is essential. Among the living polymerization systems so far developed, the living anionic polymerizations of styrene, isoprene, 1,3-butadiene, alkyl methacrylates (mainly methyl and *tert*-butyl methacrylates) and some of their derivatives with functional groups are still the best living polymerization systems from the following reasons (Nakahama and Hirao, 1990; Hirao and Nakahama, 1994; Hirao and Nakahama, 1998; Hirao *et al.*, 2002d; Hsieh and Quirk, 1996). First, molecular weight can be precisely controlled in a wide range from 10^3 to even 10^6 g/mol. Secondly, molecular-weight distribution is nearly monodisperse, the M_w/M_n value being 1.05 or even smaller. Furthermore, the resulting living polymers from these monomers have chain-end anions that are highly reactive, but stable under appropriate conditions. Such characteristics are ideally suited for the synthesis of μ-star polymers with well-defined structures.

For this reason, the focus of this chapter will be on the recent developments (since 2000) in μ-star polymers synthesized by the above living anionic polymerization systems, with emphasis on the control of synthetic factors necessary to achieve well-defined structures of μ-star polymers, that is, molecular weight, molecular-weight distribution, arm number, and composition. In the last 20 years, rapid progress in living/controlled radical polymerization systems as well as the application of "click" makes possible the synthesis of several new μ-star polymers. Therefore, representative examples will also be described. The syntheses of μ-star polymers before 2000 are beyond the scope of this chapter, although they will be briefly described in Section 4.2, since such subjects have been covered elsewhere by several excellent reviews (Hadjichristidis, 1999; Hadjichristidis *et al.*, 2001).

4.2 Miktoarm Star-Branched Polymers up to 2000

As mentioned in the introduction, μ-star polymers are much more difficult to synthesize than corresponding regular stars. The methodologies currently developed can cover the synthesis of two-component A_xB_y-type and several 3-arm three-component ABC-type μ-star polymers.

A few examples of three-component stars with more than 3 arms and only one 4-arm four-component ABCD star were reported prior to 2000, but there is no report on the synthesis of 5-arm ABCDE or higher-arm/component stars other than our recent reports, which will be described in Section 4.3.

The most general methodology for the synthesis of star-branched polymers has been developed based on a linking reaction of living anionic polymers with multifunctional chlorosilane compounds. With this method, a variety of many-armed regular star-branched polymers were successfully synthesized. This silyl chloride method is also effective for the synthesis of miktoarm star-branched polymers. In fact, the first successful well-defined miktoarm star-branched polymer was a 3-arm AB_2 star composed of PS (A) and PI (B) segments synthesized by this methodology in 1990 (Mays, 1990). The different reactivity of a living anionic polymer toward the Si–Cl reaction site of $MeSiCl_3$ is the key in this methodology. Since more sterically hindered poly(styryllithium) (PSLi) is less reactive with $MeSiCl_3$ than poly(isoprenyllithium) (PILi), PSLi first reacts with a large excess of $MeSiCl_3$ to afford a chain-end-$SiCl_2$-functionalized PS (PS-$SiCl_2$) and avoid the linking of two or more PSLi with $MeSiCl_3$. After removal of the excess $MeSiCl_3$ under vacuum, a more-reactive PILi reacts with PS-$SiCl_2$, resulting in the target 3-arm AB_2 star.

This methodology was extended to the synthesis of a 3-arm ABC μ-star polymer by Hadjichristidis *et al.* (Iatrou and Hadjichristidis, 1992). These authors also utilized the difference in anionic reactivity toward the Si–Cl reaction sites caused by the chain-end steric bulkiness of PSLi, PILi, and poly(1,3-butadienyllithium) (PBLi), which decreased in this order. The 3-arm ABC star could be synthesized by sequentially reacting PILi and PSLi, followed by PBLi, with $MeSiCl_3$, as illustrated in Scheme 4.1. The resulting polymer was well defined in architecture with a very narrow molecular-weight distribution ($M_n = 45$ 400 g/mol, $M_w/M_n = 1.03$). Similarly, another 3-arm ABC star composed of PS, PI, and poly(2-vinylpyridine) (P2VP) was also successfully synthesized ($M_w = 70$ 000 g/mol, $M_w/M_n = 1.09$).

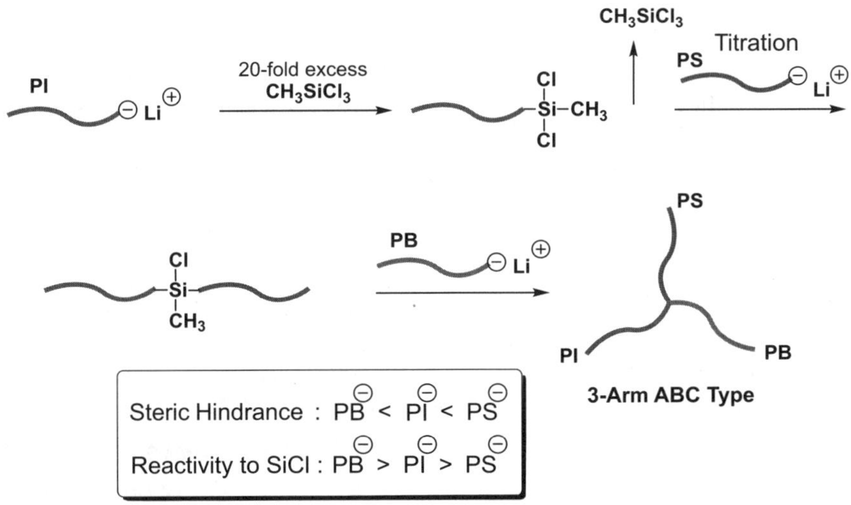

Scheme 4.1 Synthesis of 3-arm ABC miktoarm star polymer by silyl chloride methodology.

Soon after, a 4-arm ABCD star was prepared by developing the same methodology with the sequential addition of PSLi, poly(4-methylstyryl)lithium (PMSLi), PILi, and PBLi ($M_n = 62\,500$ g/mol, $M_w/M_n = 1.05$) (Iatrou and Hadjichristidis, 1993). Moreover, a variety of miktoarm stars of the 4-arm AB_3 (Tsoukatos and Hadjichristidis, 2002; Mavroudis *et al.*, 2003; Cho *et al.*, 2003), 4-arm A_2B_2 (Iatrou and Hadjichristidis, 1993; Allgaier *et al.*, 1996), 6-arm AB_5 (Velis and Hadjichristidis, 1999; Mavroudis *et al.*, 2006), and 16-arm A_8B_8 type (Avgeropoulos *et al.*, 1996) were synthesized, in which the A and B arms are corresponding to PS, PI, or PB segments.

A new ABC star was also synthesized by modifying the above-mentioned procedure. An in-chain-SiCl-functionalized AB diblock copolymer (A: PS and B: PI) was first prepared by a procedure similar to that described above, followed by reacting the silyl chloride function with a dilithium agent, $Li(CPh_2CH_2CH_2CPh_2)Li$, to change the reaction site to 1,1-diphenylalkyllithium (Sioula *et al.*, 1998). This new anionic reaction site was used to polymerize methyl methacrylate (MMA), yielding a well-defined ABC star composed of PS, PI, and PMMA ($M_n = 77\,000$ g/mol, $M_w/M_n = 1.06$).

In 1992, Fujimoto *et al.* were first successful in synthesizing a 3-arm ABC miktoarm star-branched polymer by developing a methodology using a nonhomopolymerizable 1,1-diphenylethylene (DPE) poly(dimethylsiloxane) (PDMS) macromonomer (Fujimoto *et al.*, 1992). As shown in Scheme 4.2, the synthesis involved an addition reaction of PSLi to the DPE macromonomer and the subsequent anionic polymerization of *tert*-butyl methacrylate (^{t}BMA) with the DPE anion produced by the addition reaction. In this methodology, DPE acts as a dually functionalized agent, that is, a linking site and an initiation site. Almost at the same time, Quirk *et al.* reported the synthesis of 3-arm AA′A″ (Quirk *et al.*, 1994) and 4-arm A_2B_2 stars composed of PS (A) and PB (B) segments (Quirk *et al.*, 1992) by utilizing similar dual functionalities of 1,4-bis(phenylethenyl)benzene and its 1,3-isomer for PSLi and PBLi. These methodologies were greatly developed for the synthesis of several 3-arm ABC and 4-arm A_2B_2 miktoarm star-branched polymers by Abetz *et al.* (Huckstaedt *et al.*, 1996; Huckstadt *et al.*, 2000; Walther *et al.*, 2009), Dumas *et al.* (Lambert *et al.*, 1997; Reutenauer *et al.*, 2001; Meyer *et al.*, 2002; Nasser-Eddine *et al.*, 2004), Young *et al.* (Fernyhough *et al.*, 1999), and recently Matsushita *et al.* (Takano *et al.*, 2004; Takano *et al.*, 2005; Hayashida *et al.*, 2006a; Hayashida *et al.*, 2006b; Hayashida *et al.*, 2007; Takano *et al.*, 2007; Matsushita, 2007; Hayashida *et al.*, 2008; Matsushita, 2008; Matsushita *et al.*, 2009). Interestingly, Faust *et al.* applied this methodology to living cationic polymerization to synthesize a well-defined 4-arm A_2B_2 star composed of poly (isobutylene) and poly(methyl vinyl ether) segments (Bae and Faust, 1998). Representative results are summarized in Table 4.1.

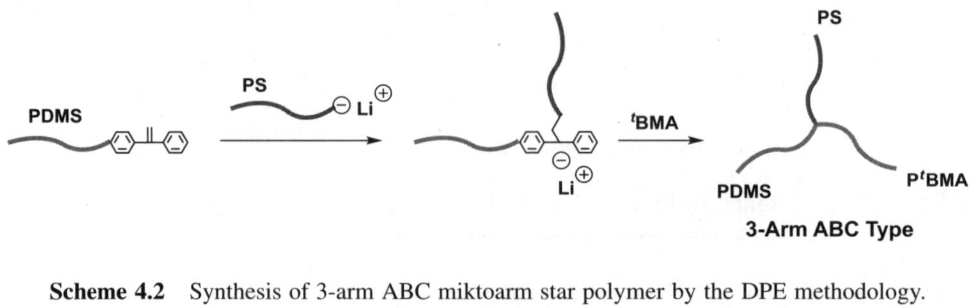

Scheme 4.2 Synthesis of 3-arm ABC miktoarm star polymer by the DPE methodology.

Table 4.1 Miktoarm star polymers obtained by DPE chemistry

Author	Type	A	B	C	M_n (kg/mol) Obsd	M_w/M_n SEC
Fujimoto	ABC	PDMS	PS	P^{t}BMA	59.0	1.06
Quirk	AA′A″	PS	—	—	58.3	1.02
	A_2B_2	PS	PB		23.4	1.05
Abetz	ABC	PS	PB	PMMA	250	1.13
	ABC	PB	PS	P2VP	288	1.04
Dumas	ABC	PS	PEO	PCL	112	1.29
Young	A_2B_2	PI	PMMA		58.9	1.03
Matsushita	ABC	PI	PS	P2VP	199	1.04
Faust	A_2B_2	PIB	PMeVE		25.3	1.12

Hirao *et al.* demonstrated that newly synthesized chain-end- and in-chain-functionalized polymers with a plural number of benzyl halide (BnX) moieties react readily and quantitatively with living anionic polymers, resulting in various miktoarm stars (Hayashi *et al.*, 1999a; Hayashi *et al.*, 1999b; Hirao *et al.*, 2002c; Hirao and Haraguchi, 2002; Hirao and Matsuo, 2003). For instance, a μ-star of the AB_n type was obtained by the reaction of a chain-end-$(BnX)_n$-functionalized polymer (A) with a living polymer (B). It was possible to synthesize ABC_n-type stars by the use of in-chain-$(BnX)_n$-functionalized AB diblock copolymers in a reaction with a living polymer (C). Moreover, a reaction of the same in-chain-$(BnX)_n$-functionalized AB diblock copolymer with either living polymer (A) or (B) allowed miktoarm stars of $A_{n+1}B$ as well as AB_{n+1} types to be synthesized. The results are summarized in Table 4.2. In order to avoid undesirable side reactions between a living anionic polymer and a BnX reaction site, living polymers were end capped with DPE prior to the linking reaction and the reaction was carried out under the conditions in THF at reaction temperatures lower than −40 °C (Ryu and Hirao, 2000; Ryu and Hirao, 2001; Ryu *et al.*, 2002; Hirao and Ryu, 2003; Ryu *et al.*, 2004; Tsoukatos *et al.*, 2000).

Table 4.2 Miktoarm star polymers by the reaction of chain-end- or in-chain-$(BnX)_n$-functionalized polymers with various living anionic polymers

Type	A	B	C	M_w (kg/mol)		M_w/M_n
				Calcd	SLS	SEC
AA'_3	PS			22.9	25.0	1.04
AB_4	PS	PI		27.8	27.0	1.03
A_9	PS			44.0	44.7	1.03
AB_8	PS	PI		40.5	41.2	1.02
AB_8	PS	PS-d_8		44.4	46.4	1.02
A_{17}	PS			88.1	87.6	1.05
A_{33}	PS			190	196	1.02
A_2B_4	PS	PI		28.4	30.5	1.05
A_2B_8	PS	PI		54.1	51.0	1.04
A_2B_{12}	PS	PI		88.4	83.2	1.04
$AA'A''_2$	PS			21.0	23.1	1.05
ABC_2	PS	PαMS	PI	22.7	23.0	1.04
ABC_4	PS	PαMS	PI	36.4	39.9	1.04

Most of the μ-star polymers synthesized by the three methodologies possess well-defined structures with precisely controlled arm segments and narrow molecular-weight distributions ($M_w/M_n < 1.1$ or smaller) as introduced above.

4.3 Novel and Versatile Methodology Based on an "Iterative Approach" for Miktoarm Star Polymer Syntheses

Although the generality and versatility of the methodologies mentioned in the preceding section can be evidenced by the variation of synthesized miktoarm star-branched polymers, the reaction site(s) always disappears after the synthesis of the target star-branched polymer (or the introduction of an arm segment). Accordingly, none of the methodologies can be used for the further synthesis of star-branched polymers with more arms and compositions. In general, the introduction of a number of different arm segments into a star-branched polymer requires the same number of selective reactions. For such a requirement very complicated molecular and reaction designs, are needed.

Hirao, Higashihara, and their coworkers have demonstrated a novel and versatile methodology since 2001, which is based on a new conceptual "iterative" approach, for the synthesis of miktoarm star-branched polymers. In this methodology, the reaction system is designed in such a way that the same reaction site is always regenerated after the introduction of an arm segment in each iterative process and this "arm introduction–regeneration of the reaction site" synthetic sequence is repeatable. Accordingly, a series of μ-star polymers can be successively and, in principle, limitlessly synthesized by repeating the synthetic sequence. This iterative methodology with the regeneration of reaction sites has been widely developed so far and allows access to a wide variety of well-defined μ-star polymers.

4.3.1 Iterative Methodology with Regeneration of DPE Function

As mentioned above, the key point of the iterative methodology is the regeneration of the same functional group that propagates the synthetic sequence. In the first methodology, the dual functionality of DPE is utilized for this purpose (Higashihara *et al.*, 2006). The methodology is illustrated in Scheme 4.3. The DPE function is capable of not only reacting with a living anionic polymer to link the polymer chain but also of offering the reaction site for the regeneration of the DPE function. It involves only two reaction steps for the iterative synthetic sequence: (1) a linking reaction of a living anionic polymer with a chain-DPE-functionalized polymer and (2) a regeneration of the DPE function by reacting a specially designed DPE-functionalized agent, 1-(4-(3-bromopropyl)phenyl)-1-phenylethylene (**1**), with the DPE anion produced by the linking reaction. The two reactions are repeated several times to successively synthesize a series of star polymers. In the iterative methodology, the star polymer synthesized in the preceding process always corresponds to the starting material of the next process.

The synthesis was initiated by the reaction of poly(4-trimethylsilylstyryl)lithium (PMSiSLi) with DPE used as the core, followed by reacting with **1** to introduce the DPE function at the chain end (the regeneration step). In the second iteration, poly(4-methoxystyryl)lithium (PMOSLi) reacted with the DPE terminus to link the two polymer chains with a change of the terminal styryl anion to a DPE anion. The agent, **1**, was again reacted *in situ* with the DPE anion to regenerate the DPE function, resulting in an in-chain-DPE-functionalized AB

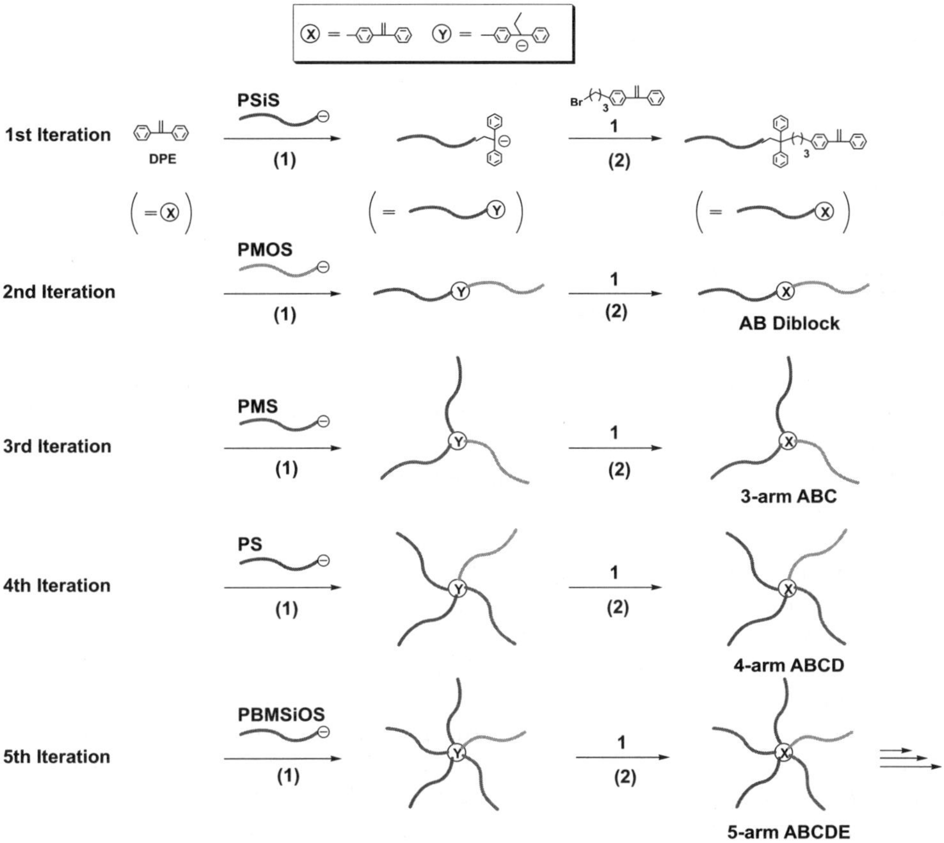

Scheme 4.3 Successive synthesis of miktoarm star polymers by the iterative methodology using **1**.

diblock copolymer. The repetition of the linking reaction with poly(4-methylstyryl)lithium (PMSLi) and the reaction *in situ* with **1** to regenerate the DPE function yielded a 3-arm ABC μ-star polymer with the DPE function at the core. The same synthetic sequence was repeated twice with the sequential use of PSLi, followed by poly(3-*tert*-butyldimethylsilyloxystyryl) lithium (PBMSiOSLi) in the linking step. Thus, a 4-arm four-component ABCD, followed by a 5-arm five-component ABCDE μ-star polymer, was synthesized.

All living anionic polymers herein used were adjusted to be around 10 000 g/mol in M_n value. The two reaction steps, (1) and (2), were carried out in THF at −78 °C for 1 h and 10 min, respectively. The linking reaction and the regeneration of the DPE function in each iterative process were confirmed by SEC and ^{1}H NMR analysis to be essentially quantitative. Therefore, the targeted polymers were always obtained in nearly 100% yields and isolated in >80% yields by fractional precipitation. The isolated polymers, that is, AB diblock, 3-arm ABC, 4-arm ABCD, and 5-arm ABCDE stars, exhibited sharp monomodal SEC peaks without shoulders and/or tailings.

As can be seen from the results summarized in Table 4.3, the molecular weights of the resulting polymers were in good agreement with those observed by SLS and ^{1}H NMR.

Table 4.3 Miktoarm star polymers by the iterative methodology using **1**

Type[a]	M_w (kg/mol)		M_w/M_n	Composition (wt%)	
	Calcd	SLS	SEC	Calcd	^{1}H NMR
A	10.5	10.7	1.03	100	100
AB Diblock	21.3	22.0	1.02	48/52	48/52
3-Arm ABC	32.0	34.4	1.02	32/35/33	31/35/34
4-Arm ABCD	42.8	43.0	1.02	24/29/23/24	23/28/23/26
5-Arm ABCDE	59.3	57.0	1.02	17/21/17/19/26	17/19/18/19/27

[a] A, B, C, D, and E were PMSiS, PMOS, PMS, PS, and PBMSiOS, respectively.

The compositions observed by ^{1}H NMR were consistent with those calculated from the feed ratios. Accordingly, these results clearly indicate that well-defined 3-arm ABC, 4-arm ABCD, and 5-arm ABCDE μ-star polymers are successfully synthesized by repeating the processes with five different living polymers. Thus, the first proposed iterative methodology with the regeneration of DPE reaction sites works very well.

The 5-arm ABCDE star synthesized by this method is the first successful μ-star polymer with five different arm segments. Furthermore, the E segment of the resulting star was readily and quantitatively converted to a poly(4-vinylphenol) segment by treatment with Bu_4NF, resulting in a new functional 5-arm ABCDE star with an acidic and ionic segment. Since the final 5-arm star still possesses a DPE function at the core, this procedure enables further synthetic sequence of stars with more arms and components.

The synthesis of different series of miktoarm star-branched polymers is also possible by using the same iterative methodology (Zhao *et al.*, 2005; Zhao *et al.*, 2007). As can be seen in Scheme 4.4, the intermediate polymers are the corresponding AB block copolymer anion and various star polymer anions. They can serve as the macroinitiators in the living anionic polymerization of appropriate monomers to grow one more arm segment. For example, the AB diblock polymer anion is capable of anionically polymerizing 4-methylphenyl vinyl sulfoxide (MPVS) in a living manner to afford a new 3-arm ABC μ-star polymer. Similarly, 4-arm ABCD and 5-arm ABCDE stars having PMPVS segments were obtained by the polymerization of MPVS with the corresponding star-branched polymer anions. Very interestingly, the PMPVS segment was readily and quantitatively converted to a conductive, π-conjugated rod-like poly (acetylene) (PA) simply by thermal treatment at 150 °C for a few hours. Thus, the procedure, "the iterative methodology and subsequent living polymerization", demonstrates another possibility to synthesize a series of quite new functional μ-star polymers bearing a conductive PA segment.

The advantage of this procedure is that, in essence, any monomer that undergoes living anionic polymerization from a DPE anion can be used. On the other hand, the successive synthesis of star-branched polymers can no longer be continued.

4.3.2 Iterative Methodology with Regeneration of Two or More DPE Functions

The number of arms introduced in the above-mentioned methodology is one segment in each iterative process. With the use of 1,3-bis(1-phenylethenyl)benzene (**2**), composed of two DPE

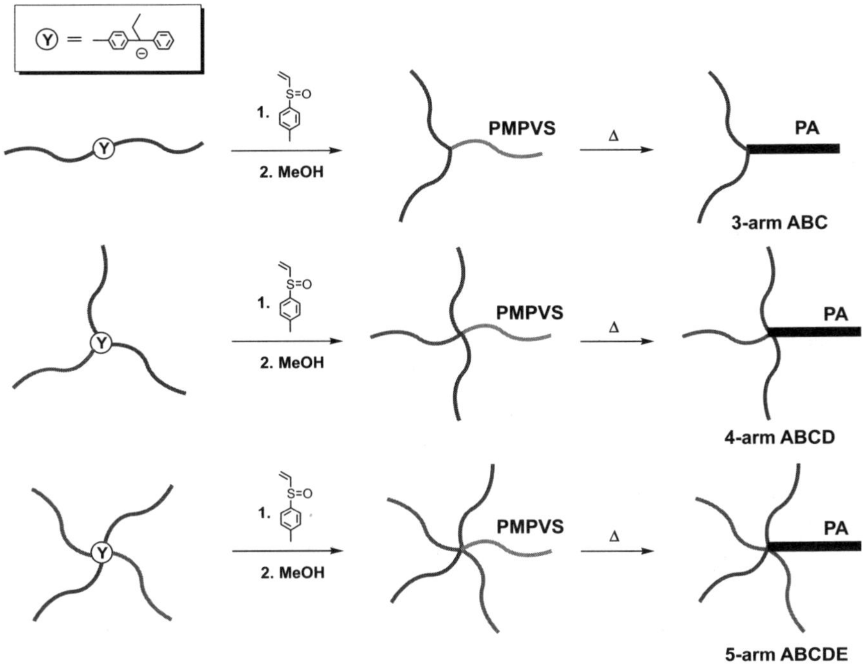

Scheme 4.4 Synthesis of 3-arm ABC, 4-arm ABCD, and 5-arm ABCDE miktoarm star polymers with PMPVS or PA segments using intermediate polymer anions.

skeletons as the core instead of DPE in the same iterative methodology, the arm number may increase by two in each process, as illustrated in Scheme 4.5 (Hirao *et al.*, 2001; Higashihara *et al.*, 2002). PSLi reacted with **2** to link two PS chains with the production of two DPE anions at the linking points. Then, two DPE functions were regenerated by reacting the DPE anions with **1**. In the second iteration, poly(α-methylstyryl)lithium (PαMSLi) reacted with the DPE functions between the two blocks, followed by reacting *in situ* with **1**, resulting in a 4-arm A_2B_2 star-branched polymer core functionalized with two DPE functions. The same reaction sequence with PMOSLi and **1** was repeated to afford a 6-arm core-$(DPE)_2$-functionalized $A_2B_2C_2$ miktoarm star-branched polymer. Similarly, a different 6-arm $A_2B_2C_2$ star could be obtained by using PILi instead of PMOSLi in the linking step in the third iteration. All of the analytical results by SEC, ^{1}H NMR, and SLS exhibited these polymers to be well defined and requisite in structure as listed in Table 4.4.

With the use of 1,1-bis(3-(1-phenylethenyl)phenyl)ethylene (**3**), composed of three DPE skeletons as the core, the number of arm segments could increase by three in each process (Hirao and Higashihara, 2002). A 3-arm A_3 and a 6-arm A_3B_3, followed by a quite new 9-arm $A_3B_3C_3$ μ-star polymer, whose A, B, and C segments are, respectively, PS, PMOS, and PI segments, were successively synthesized by repeating the synthetic sequence three times. Moreover, the methodology allowed for the successful synthesis of two more different 9-arm $A_3B_3C_3$ stars with PMSiS or PBMSiOS segments convertible to poly(4-vinylphenol) as the C arm segments. The results are summarized in Table 4.5. Thus, the starting core compounds, **2** and **3**, were effective agents capable of introducing two and three arm segments in each process.

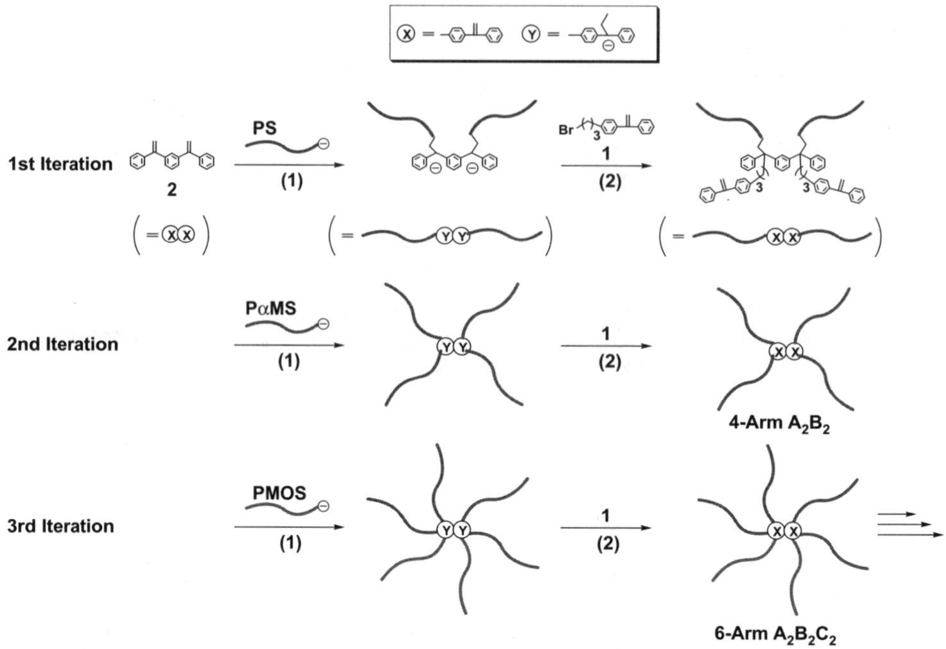

Scheme 4.5 Successive synthesis of 4-arm A_2B_2 and 6-arm $A_2B_2C_2$ miktoarm star polymers by the iterative methodology using **1** and **2**.

In a separate experiment, a 15-arm star PS was readily synthesized by repeating the synthetic sequence with PSLi five times in the methodology using **3** as the core. This strongly indicates that there is no steric hindrance at least up to the fifth iteration stage. Accordingly, a 12-arm $A_3B_3C_3D_3$, followed by a 15-arm $A_3B_3C_3D_3E_3$ μ-star and probably higher than 15-armed stars may possibly be synthesized. Furthermore, the same living polymer can also be used. For example, a 6-arm A_3B_2C star-branched polymer can be synthesized by reacting a living polymer (A) in the first, second, and third iterations, a living polymer (B) in the fourth and fifth iterations, and finally a living polymer (C) in the sixth iteration. This also indicates the possible synthesis of a wider variety of μ-star polymers by developing the methodology.

In a similar sense, it may also be possible to increase the number of arms in each process by replacing the DPE-functionalized agent **1** with 3,5-bis(3-(4-(1-phenylethenyl)pheny)propoxy)

Table 4.4 Miktoarm star polymers by the iterative methodology using **1** and **2**

Type	M_w (kg/mol)		M_w/M_n	Composition (wt%)	
	Calcd	SLS	SEC	Calcd	^{1}H NMR
4-Arm A_2B_2[a]	41.1	41.8	1.02	55/45	54/46
6-Arm $A_2B_2C_2$[a]	61.7	62.4	1.02	37/30/33	38/30/32
6-Arm $A_2B_2C_2$[b]	34.8	31.8	1.06	32/30/38	31/30/39

[a] A, B, and C were PS, PαMS, and PMOS, respectively.
[b] A, B, and C were PI, PαMS, and PS, respectively.

Table 4.5 Miktoarm star polymers by the iterative methodology using **1** and **3**

Type	M_w (kg/mol)		M_w/M_n	Composition (wt%)	
	Calcd	SLS	SEC	Calcd	^{1}H NMR
6-Arm A_3B_3[a]	67.8	66.7	1.03	50/50	52/48
9-Arm $A_3B_3C_3$[a]	97.0	97.1	1.02	34/35/31	35/34/31
9-Arm $A_3B_3C_3$[b]	101	103	1.04		
9-Arm $A_3B_3C_3$[c]	98.3	97.2	1.03	34/35/31	33/35/32

[a] A, B, and C were PS, PMOS, and PMSiS, respectively.
[b] A, B, and C were PS, PMOS, and PI, respectively.
[c] A, B, and C were PS, PMOS, and PBMSiOS, respectively.

benzyl bromide (**4**), consisting of two DPE moieties and one BnBr function (Hirao *et al.*, 2006). This agent, **4**, is designed in such a way that two DPE moieties can be introduced via one DPE anion produced by the introduction of an arm segment. The methodology is basically the same as those mentioned above except for the use of **4**. As illustrated in Scheme 4.6, the number of regenerated DPE moieties and hence the number of introduced arm segments increases exponentially in each process. The ability of **4** to increase the number of arm segments is thus remarkable. The number of arms increases from 3 to 7, 15, and 31 according to the equation, $f = 2^{(n+1)} - 1$ (f and n are the arm and repetition numbers, respectively), if the iterative process efficiently proceeds.

The starting chain-end-$(DPE)_2$-functionalized polymer was prepared by a reaction of DPE-end-capped PSLi with **4**. In the first iteration, two equivalents of PαMSLi reacted with the starting DPE-functionalized PS and then **4** reacted *in situ* with the two DPE anions produced by the linking reaction, regenerating four DPE functions. The two reaction steps efficiently proceeded to afford a 3-arm core-$(DPE)_4$-functionalized AB_2 star-branched polymer. Similarly, repetition of the same synthetic sequence twice with sequential uses of PMSLi and PMOSLi yielded 7-arm core-$(DPE)_8$-functionalized AB_2C_4 and 15-arm core-$(DPE)_{16}$-functionalized $AB_2C_4D_8$ μ-star polymers, respectively. Their yields were virtually quantitative in THF at −78 °C for a few hours. It was, however, observed that the next linking reaction of PMSiSLi with the 15-arm core-$(DPE)_{16}$-functionalized $AB_2C_4D_8$ star was not quantitative under the same conditions even for a longer reaction time of 48 h. On average 10 PMSiMS arms were linked, resulting in a 25-arm star (target: a 31-arm star). This may be attributed to the steric hindrance around the core. In addition, electrostatic repulsion among PMSiSLi and intermediate stars having several DPE anions may also cause incomplete linking.

The reaction conditions were changed to those using nonpolar *tert*-butylbenzene in order to reduce electrostatic repulsion. Raising the reaction temperature from −78 °C to 30 °C may significantly facilitate the reaction. Under the conditions in *tert*-butylbenzene at 30 °C for a longer reaction time of 168 h, the linking reaction was observed to go completion, resulting in the targeted 31-arm $AB_2C_4D_8E_{16}$ μ-star polymer. This strongly demonstrates the importance of setting the reaction conditions for the synthesis of multiarmed star polymers. The characterization results summarized in Table 4.6 clearly show that all polymers possess the requisite well-defined structures with a high degree of homogeneity in molecular weight and composition. Thus, the iterative methodology employing **4** also works effectively for the successive synthesis of a series of multiarmed and multicomponent μ-star polymers with less repetition

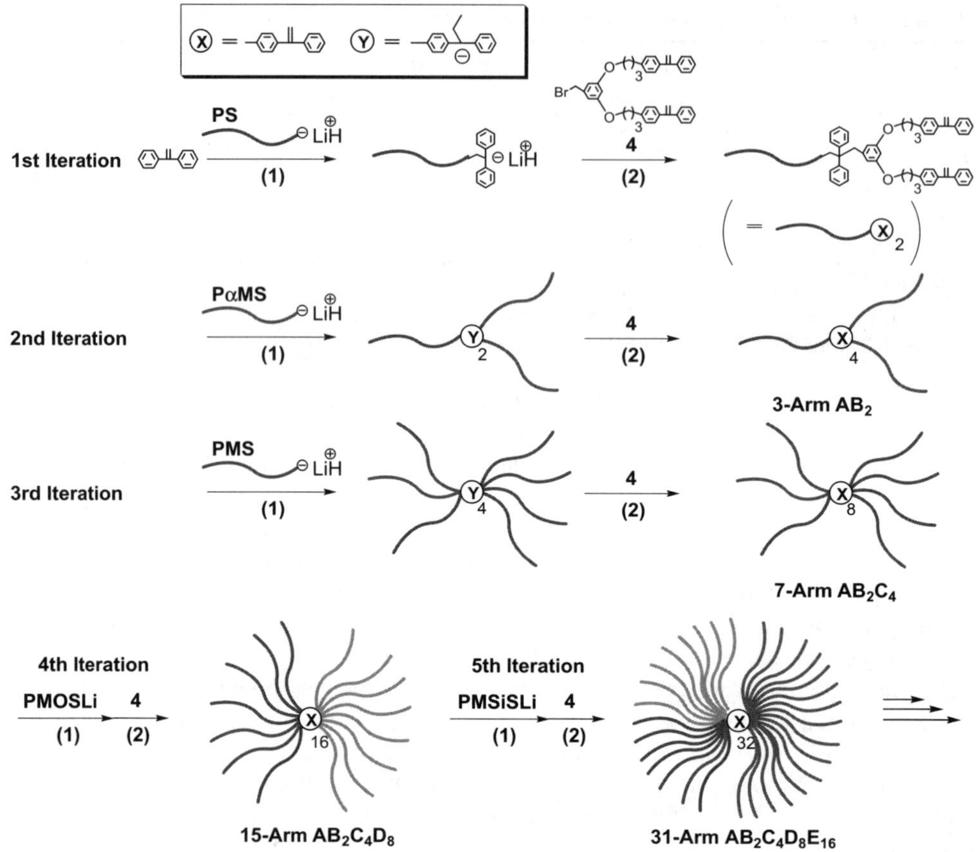

Scheme 4.6 Successive synthesis of 3-arm AB_2, 7-arm AB_2C_4, 15-arm $AB_2C_4D_8$, and 31-arm $AB_2C_4D_8E_{16}$ miktoarm star polymers by the iterative methodology using **4**.

of the synthetic sequence. Except for the 3-arm AB_2 star, 7-arm AB_2C_4, 15-arm $AB_2C_4D_8$, and 31-arm $AB_2C_4D_8E_{16}$ star polymers all are quite novel in terms of star-branched architecture.

The structural variations of multiarmed multicomponent miktoarm stars have been greatly extended by the iterative methodology using **4**, but it is still limited, especially to the synthesis

Table 4.6 Miktoarm star polymers by the iterative methodology using **4**

Type[a]	M_w (kg/mol)		M_w/M_n	Composition (wt%)	
	Calcd	SLS	SEC	Calcd	^{1}H NMR
3-Arm AB_2	32.1	35.7	1.02	36/64	36/64
7-Arm AB_2C_4	77.0	79.2	1.02	16/28/56	17/27/56
15-Arm $AB_2C_4D_8$	162	172	1.02	8/12/27/53	8/10/30/52
31-Arm $AB_2C_4D_8E_{16}$	339	363	1.02	4/6/13/27/50	4/7/15/28/46

[a] A, B, C, D, and E were PS, PαMS, PMS, PMOS, and PMSiS, respectively.

of stars with more than 10 arms (only two types of 15-arm $AB_2C_4D_8$ and 31-arm $AB_2C_4D_8E_{16}$ stars were synthesized). The limitation comes from the rapid divergence of the number of arms formulated as $f = 2^{n+1} - 1$ when only **4** is used in each process. To fill in the intervals of the f values and further pursue new star-branched polymers, a methodology with an alternative choice (**4** or **1**) in each process has recently been proposed (Higashihara *et al.*, 2009). The advantageous feature of the new methodology is that the number of arms and components are controlled to a larger extent simply by choosing **4** or **1**. As illustrated in Scheme 4.7, the same methodology using **4** is developed and the only difference is the DPE regeneration reaction step, in which either **4** or **1** reacts alternatively with the DPE anions. In the reaction with **1**, the number of regenerated DPE functions remains unchanged, while it doubles when using **4** in the reaction.

The synthesis was initiated to prepare a chain-end-$(DPE)_2$-functionalized polymer (A) by a reaction of a living anionic polymer (A) with **4**. In the first iteration, a linking reaction of the DPE-functionalized polymer (A) with a living polymer (B), followed by reacting either **1** or **4**, afforded two kinds of 3-arm AB_2 star-branched polymers. The number of core-functionalized DPE functions was the same (two) as that of the starting polymer in the former reaction, while it doubled to four in the latter reaction. The polymers abbreviated as 3-arm AB_2 $(DPE)_2$ and 3-arm AB_2 $(DPE)_4$, respectively. In the second iteration by repeating the same reaction

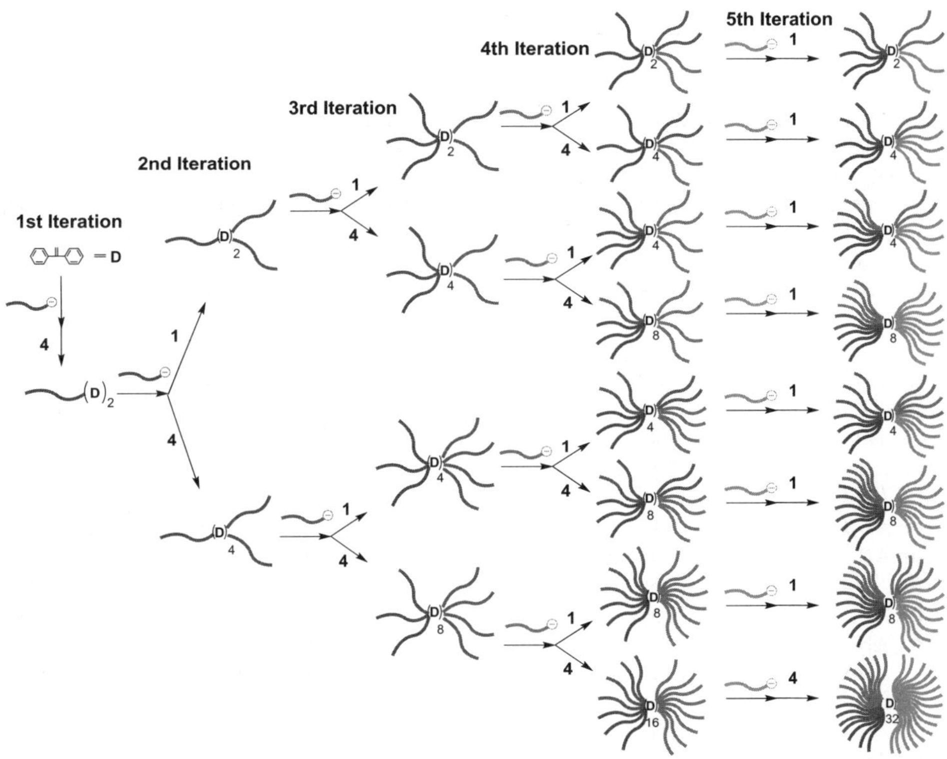

Scheme 4.7 Successive synthesis of miktoarm star polymers by the iterative methodology using either **1** or **4**.

sequence with a living polymer (C), the following four star-branched polymers were obtained: 5-arm AB_2C_2 $(DPE)_2$, AB_2C_2 $(DPE)_4$, 7-arm AB_2C_4 $(DPE)_4$, and 7-arm AB_2C_4 $(DPE)_8$. One additional repetition with a living polymer (D) gave the following eight star-branched polymers: 7-arm $AB_2C_2D_2$ $(DPE)_2$, 7-arm $AB_2C_2D_2$ $(DPE)_4$, 9-arm $AB_2C_2D_4$ $(DPE)_4$, 9-arm $AB_2C_2D_4$ $(DPE)_8$, 11-arm $AB_2C_4D_4$ $(DPE)_4$, 11-arm $AB_2C_4D_4$ $(DPE)_8$, 15-arm $AB_2C_4D_8$ $(DPE)_8$, and 15-arm $AB_2C_4D_8$ $(DPE)_{16}$.

Similarly, 16 μ-stars having five chemically different arms were obtained in the fourth iteration using a living polymer (E) as follows: 9-arm $AB_2C_2D_2E_2$ $(DPE)_2$, 9-arm $AB_2C_2D_2E_2$ $(DPE)_4$, 11-arm $AB_2C_2D_2E_4$ $(DPE)_4$, 11-arm $AB_2C_2D_2E_4$ $(DPE)_8$, 13-arm $AB_2C_2D_4E_4$ $(DPE)_4$, 13-arm $AB_2C_2D_4E_4$ $(DPE)_8$, 15-arm $AB_2C_4D_4E_4$ $(DPE)_4$, 15-arm $AB_2C_4D_4E_4$ $(DPE)_8$, 17-arm $AB_2C_2D_4E_8$ $(DPE)_8$, 17-arm $AB_2C_2D_4E_8$ $(DPE)_{16}$, 19-arm $AB_2C_4D_4E_8$ $(DPE)_8$, 19-arm $AB_2C_4D_4E_8$ $(DPE)_{16}$, 23-arm $AB_2C_4D_8E_8$ $(DPE)_8$, 23-arm $AB_2C_4D_8E_8$ $(DPE)_{16}$, 31-arm $AB_2C_4D_8E_{16}$ $(DPE)_{16}$. and 31-arm $AB_2C_4D_8E_{16}$ $(DPE)_{32}$. The yields of all polymers were about 100%, since both linking and DPE regeneration reactions proceeded essentially quantitatively. These polymers were well defined in structure, as evidenced by the results listed in Table 4.7. Thus, it is quite surprising that the alternative choice (**1** or **4**) in the DPE regeneration step enables the synthetic range of μ-star polymers to be broadened significantly.

Table 4.7 Miktoarm star polymers by the iterative methodology using either **1** or **4**

Type[a]	M_n (kg/mol)		M_w/M_n	Composition (wt%)	
	Calcd	RALLS	SEC	Calcd	^{1}H NMR
3-Arm AB_2(2)	33.7	32.6	1.04	31/69	30/70
3-Arm AB_2(4)	31.5	35.0	1.02	36/64	38/62
5-Arm AB_2C_2(2)	54.6	55.9	1.04	19/42/39	21/39/40
5-Arm AB_2C_2(4)	56.5	58.0	1.02	19/41/40	17/42/41
7-Arm AB_2C_4(4)	74.4	74.4	1.02	15/27/58	16/26/58
7-Arm AB_2C_4(8)	75.5	77.6	1.04	16/28/56	17/27/56
7-Arm $AB_2C_2D_2$(2)	72.3	78.4	1.05	14/32/30/24	20/29/29/22
7-Arm $AB_2C_2D_2$(4)	73.2	83.1	1.05	14/32/29/25	17/28/29/26
9-Arm $AB_2C_2D_4$(4)	98.4	100	1.02	10/24/23/43	10/25/23/42
9-Arm $AB_2C_2D_4$(8)	98.2	106	1.05	11/24/23/42	10/25/22/43
9-Arm $AB_2C_2D_2E_2$(2)	87.5	92.9	1.05	12/26/24/20/18	14/23/25/18/20
11-Arm $AB_2C_4D_4$(4)	116	119	1.02	10/17/37/36	10/17/37/36
11-Arm $AB_2C_4D_4$(8)	118	121	1.04	9/17/37/37	10/16/33/41
11-Arm $AB_2C_2D_2E_4$(4)	116	109	1.04	9/20/18/16/37	7/18/18/16/41
13-Arm $AB_2C_2D_4E_4$(4)	136	150	1.05	8/17/17/29/29	6/19/19/28/28
15-Arm $AB_2C_4D_8$(8)	159	163	1.04	7/9/27/57	6/9/28/57
15-Arm $AB_2C_4D_8$(16)	159	168	1.02	8/12/27/53	7/10/31/52
15-Arm $AB_2C_4D_4E_4$(4)	158	154	1.04	7/13/27/26/27	7/13/27/26/27
17-Arm $AB_2C_2D_4E_8$(8)	171	178	1.04	6/14/13/24/43	9/15/15/24/37
19-Arm $AB_2C_4D_4E_8$(8)	188	181	1.04	6/11/23/23/37	7/14/25/20/34
23-Arm $AB_2C_4D_8E_8$(8)	243	260	1.04	4/6/18/37/35	3/11/21/33/32
31-Arm $AB_2C_4D_8E_{16}$(32)	332	356	1.02	4/6/13/27/50	4/7/15/28/46

[a] A, B, C, D, and E were PS, PαMS, PMS, PMOS, and PMSiS, respectively. The number in brackets denotes that of DPE moieties introduced at the core.

Since all μ-star polymers still possess several DPE functions, the synthetic sequence may possibly further continue to synthesize higher armed and compositional star-branched polymers. It should be mentioned that styrene derivatives were generally used herein both for facile synthesis of highly reactive living anionic polymers and for easy characterization of the resulting stars by ^{1}H NMR, but living anionic polymers of 1,3-butadiene, isoprene, and certain functional styrene and 1,3-dienes can also be employed in the iterative methodology.

4.3.3 Iterative Methodology with Regeneration of 1,3-Butadiene Function

Hirao *et al.* previously observed that chain-end-1,3-butadiene (Bd)-functionalized PS reacted quantitatively with PSLi in a 1 : 1 addition manner to link the two PS chains with a change of the terminal styryl anion to a new anion (Bd anion) (Hirao, unpublished results). Neither the oligomerization nor the polymerization of the Bd terminus occurred under specific conditions in THF at −78 °C even with the use of a twofold excess of PSLi after 24 h. This reaction behavior was exactly the same as that of chain-end-DPE-functionalized PS in the reaction with PSLi. Furthermore, the Bd function was quantitatively regenerated by reacting 6-bromo-3-methylene-1-hexene (**5**) with the Bd anion. Such abilities of the Bd function to link living anionic polymers and to regenerate the same Bd function may access the iterative methodology, as is the case of the DPE function (Higashihara *et al.*, 2003; Hirao *et al.*, 2008a; Sugiyama *et al.*, 2009).

Similar to the already-mentioned methodologies, the two reaction steps were used in the synthetic sequence, that is, (1) a linking reaction of a living anionic polymer with a chain-Bd-functionalized polymer and (2) a regeneration of the Bd function by reacting **5** with the Bd anion produced by the linking reaction. The effectiveness of the Bd function in the methodology was first examined through the successive synthesis of star-branched (PS)s. Following the reaction of **5** with PSLi end capped with DPE, the starting chain-end-Bd-functionalized PS (PS-Bd) was prepared. Then, PSLi reacted with the resulting PS-Bd and subsequently **5** was reacted *in situ* with the produced Bd anion. Both reactions efficiently proceeded to link the two PS chains with the introduction (or the regeneration) of the Bd function between the two chains. As a result, the objective in-chain-Bd-functionalized PS was successfully synthesized. The same reaction sequence was repeated two more times, yielding a 3-arm, followed by a 4-arm star-branched PS in about 100% yields. The resulting three polymers, after isolation by fractional precipitation, exhibited narrow molecular-weight distributions and good agreement between the calculated and observed molecular weights. The 3- and 4-arm stars were 38 700 g/mol (1.02) and 48 600 g/mol (1.02) in M_n value (M_w/M_n), respectively. This result clearly indicates that the Bd function, similar to the DPE function, is effective in the iterative methodology. Accordingly, the iterative methodology with the regeneration of the Bd function was applied to the synthesis of μ-star polymers.

As shown in Scheme 4.8, the synthesis is started with the same chain-end-Bd-functionalized PS and the synthetic sequence is repeated using PαMSLi, yielding an in-chain-Bd-functionalized AB diblock copolymer. Subsequent repetitions including the linking and regeneration reaction steps enabled the synthesis of a 3-arm ABC, a 4-arm ABCD, and a 5-arm ABCDE, followed by a 6-arm ABCDEF star by the sequential addition of PMSLi, PMOSLi, PMSiSLi, and PBMSiOSLi in this order in the linking step. The results, listed in Table 4.8, show that all

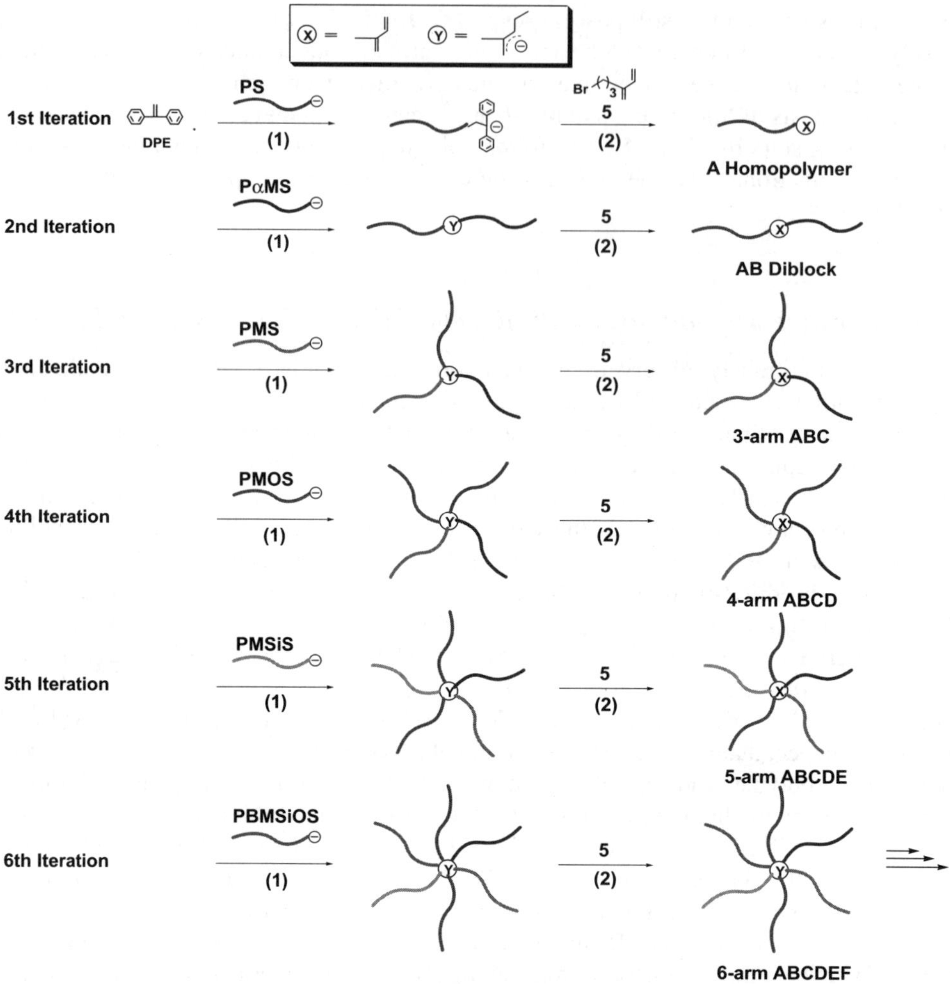

Scheme 4.8 Successive synthesis of 3-arm ABC, 4-arm ABCD, 5-arm ABCDE, and 6-arm ABCDEF type miktoarm star polymers by the iterative methodology using **5**.

polymers, after isolation (> 80%) by fractional precipitation, possess the requisite and well-defined structures. At the moment, the 6-arm ABCDEF polymer thus synthesized is the first μ-star polymer with six different arms successfully prepared.

This iterative methodology is the second successful procedure that can synthesize a variety of multiarmed and multicomponent μ-star polymers. Since the following core compounds as well as an alkyl bromide substituted with two Bd functions (Figure 4.1) were prepared, the same methodologies mentioned in Section 4.3.2 could be developed. Another advantage of the Bd function is the expectation that it is able to convert to highly reactive and synthetically useful anhydride and two epoxy functions by the Diels–Alder and oxidation reactions. Indeed, the core-functionalized Bd moieties of the star (PS)s were converted to the corresponding anhydrides and two epoxy functions in nearly quantitative yields by using such reactions as

Table 4.8 Miktoarm star polymers by the iterative methodology using **5**

Type[a]	M_w (kg/mol)		M_w/M_n	Composition (wt%)	
	Calcd	SLS	SEC	Calcd	^{1}H NMR
A	11.2	11.5	1.03	100	100
AB Diblock	24.2	25.3	1.04	47/53	49/51
3-Arm ABC	34.8	35.9	1.01	32/36/32	33/34/33
4-Arm ABCD	47.1	47.2	1.01	24/26/24/26	25/27/25/23
5-Arm ABCDE	58.8	58.5	1.02	19/22/19/21/19	20/21/20/21/18
6-Arm ABCDEF	71.8	71.5	1.03	16/18/16/17/16/17	17/18/17/17/15/16

[a] A, B, C, D, E, F, and G were PS, PαMS, PMS, PMOS, PMSiS, PDMSiOS, and P2VP, respectively.

illustrated in Scheme 4.9. Thus, the use of Bd functionality in the iterative methodology enabled further flexibility in the star-branched polymer synthesis. For instance, 3-arm core-(epoxide)$_2$-functionalized and 3-arm core-anhydride-functionalized star-branched (PS)s reacted with PSLi and chain-end-NH_2-functionalized PS, yielding the objective 5-arm and 4-arm star (PS)s (Scheme 4.9). Interestingly, their cores were functionalized by either two hydroxyl groups or one carboxyl group.

4.3.4 Iterative Methodology with Regeneration of Benzyl Bromide Function

The third iterative methodology is based on the regeneration of the benzyl bromide function, which is somewhat different in its reaction mechanism from that of the methodologies described in the preceding sections. It utilizes the transformability of the 3-*tert*-butyldimethylsilyloxymethylphenyl (SMP) group into a BnBr function capable of reacting with a living anionic polymer (Higashihara *et al.*, 2005). As illustrated in Scheme 4.10, 1-(3-*tert*-butyldimethylsilyloxmethylphenyl)-1-phenylethylene (**6**) is used as the key agent in this methodology. In the first iteration, PSLi reacted with **6** to introduce the SMP group at the chain-end, which was then transformed into the BnBr function by treatment with a 1 : 1 mixture of $(CH_3)_3SiCl$ and LiBr. Following the reaction of PαMSLi with **6**, the SMP was introduced at the chain end of PαMSLi and then *in-situ* reacted with the chain-end-BnBr-functionalized PS prepared in the first iteration to afford an in-chain-SMP-functionalized PS-*block*-PαMS. The SMP group was

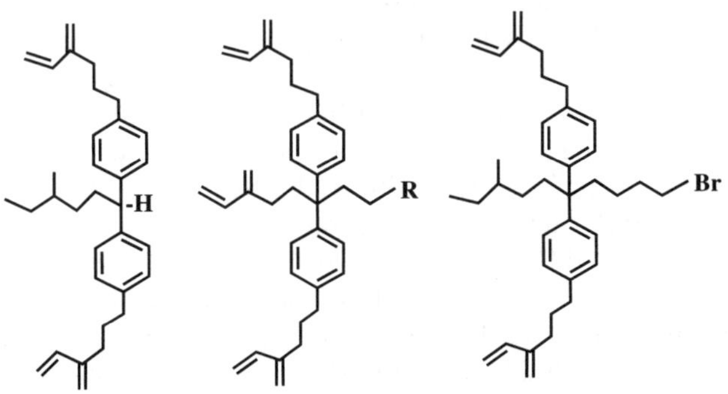

Figure 4.1 Multibutadiene-substituted compounds.

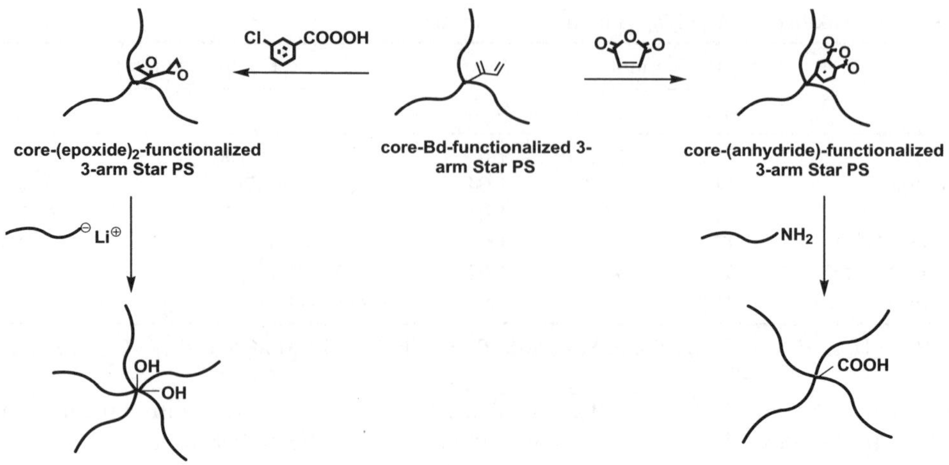

Scheme 4.9 Synthesis of core-(anhydride)- or core-(epoxyde)$_2$-functionalized star-branched (PS)s and synthesis of 5-arm core-(OH)$_2$-functionalized and 4-arm core-(COOH)-functionalized star (PS)s.

transformed into the BnBr function by the same treatment. Thus, the BnBr function was again regenerated. The same synthetic sequence with the use of PMSLi was repeated to result in a 3-arm ABC star-branched polymer core-functionalized with the BnBr function. In this methodology, the chain-end-SMP-functionalized living polymer is prepared by the reaction of the corresponding living polymer with **6** and then coupled with the benzyl brominated polymer obtained in the preceding transformation step to link the polymer chains with the simultaneous introduction of the SMP group. The BnBr function is regenerated from the SMP group and enables further coupling to the next star.

In the third iteration, living poly(methyl methacrylate) (PMMA) reacted with the 3-arm core-BnBr-functionalized ABC star, yielding a 4-arm ABCD star-branched polymer bearing PS, PαMS, PMS, and PMMA arms. The reaction steps were monitored by ^{1}H NMR and SEC and observed to be almost quantitative in all cases. As listed in Table 4.9, the resulting polymers are the target ABCD star-branched polymers confirmed by ^{1}H NMR, SEC, and SLS analyses. In the resulting star, the SMP group was not incorporated at the core because living PMMA could not react with **6** and the synthetic sequence was no longer repeated. The use of a living anionic polymer capable of reacting with **6**, for instance PILi and PBLi in place of living PMMA, can further continue the synthetic sequence.

With an in-chain-(BnBr)$_2$-functionalized PS as the starting polymer, two polymer chains could be introduced in each iteration (Higashihara *et al.*, 2005). At first, PSLi reacted with **6**, followed by coupling with α,α′-dichloro-*p*-xylene, to afford an in-chain-(SMP)$_2$-functionalized PS. Following treatment with $(CH_3)_3SiCl$-LiBr, the SMP groups could be transformed into two BnBr functions (regeneration step). The same synthetic sequence was repeated twice with PαMSLi and living poly(*tert*-butyl methacrylate), which were sequentially reacted in the second and third iterations. As a result, the target 4-arm A_2B_2 and 6-arm $A_2B_2C_2$ miktoarm star-branched polymers with well-defined structures were obtained in $\sim$ 100% yields (Table 4.9). In this synthesis, the number of arm segments increases by two in each iteration. Thus, the BnBr functionality can also be used in the iterative methodology.

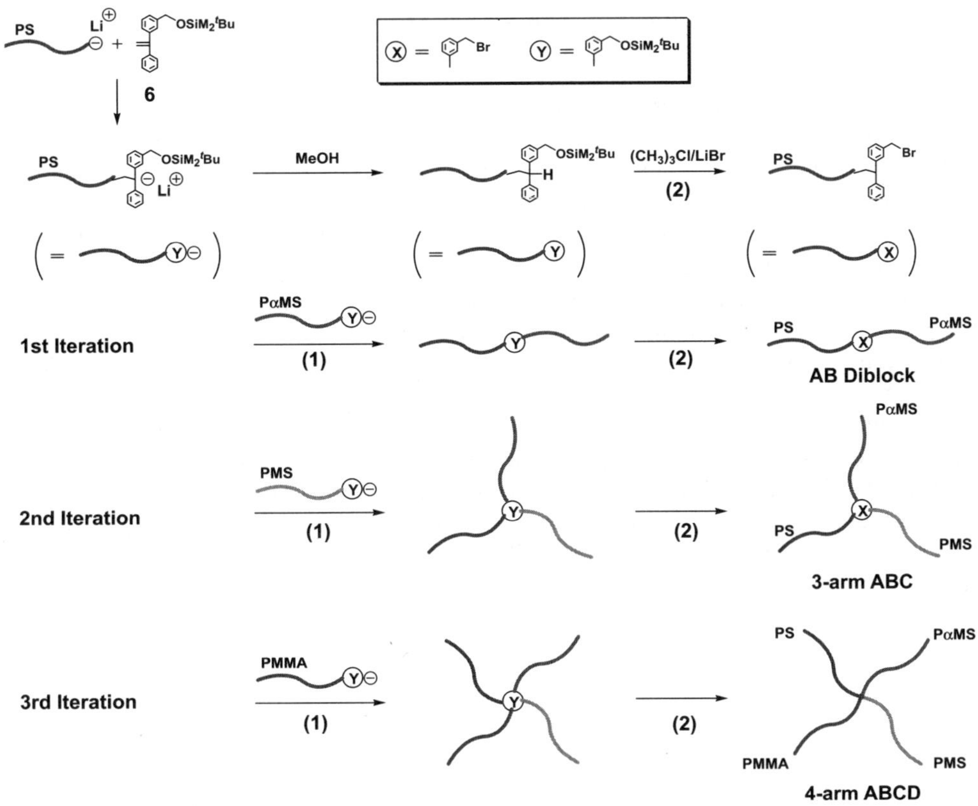

Scheme 4.10 Successive synthesis of 3-arm ABC and 4-arm ABCD miktoarm star polymer by the iterative methodology using **6**.

Table 4.9 Miktoarm star polymers by the iterative methodology using **6**

Type	M_w (kg/mol)		M_w/M_n	Composition (wt%)	
	Calcd	SLS	SEC	Calcd	^{1}H NMR
A[a]	10.6	11.2	1.02	100	100
AB Diblock[a]	20.2	21.2	1.02	50/50	48/52
3-Arm ABC[a]	32.3	35.7	1.02	32/34/34	34/32/34
4-Arm ABCD [a]	46.7	48.8	1.02	26/24/25/25	27/23/25/25
A_2[b]	24.1	25.2	1.02	100	100
4-Arm A_2B_2[b]	44.3	43.7	1.03	54/46	53/47
6-Arm $A_2B_2C_2$[b]	65.7	67.0	1.02	36/32/32	36/32/32

[a] A, B, C, and D correspond to PS, PαMS, PMS, and PMMA, respectively.
[b] A, B, and C correspond to PS, PαMS, and P^{t}BMA, respectively.

4.3.5 Convergent Synthesis of Miktoarm Star-Branched Polymers Using Polymer Anions

For the synthesis of dendrimers, two complementary stepwise iterative methodologies based on the divergent and convergent approaches are presented. In the latter convergent approach, dendritic fragments having a single reactive site, the so-called dendron, are first synthesized and react to link to each other to afford a higher-generation dendron. After sufficient repetition of the linking steps, these dendrons can be attached to a multifunctional core through the single reactive site to prepare a target dendrimer. The convergent approach using a polymer anion as the dendron has been recently proposed for the synthesis of μ-star polymers. The term "polymer anion" is herein defined as a polymer consisting of more than two of the same or different segments having a single anionic reactive site located at the junction of such segments.

The synthetic outline of the new convergent approach is illustrated in Scheme 4.11 (Hirao *et al.*, 2006a). The starting polymer anion was prepared by reacting PSLi with chain-end-DPE-functionalized PαMS. The resulting AB diblock copolymer bearing a DPE anion was coupled with 1,1-bis(3-chloromethylphenyl)ethylene (**7**), resulting in a 4-arm A_2B_2 star-branched polymer having the DPE moiety at the core. Thus, the introduction of arm segments and the regeneration of the DPE functionality were achieved. After the core

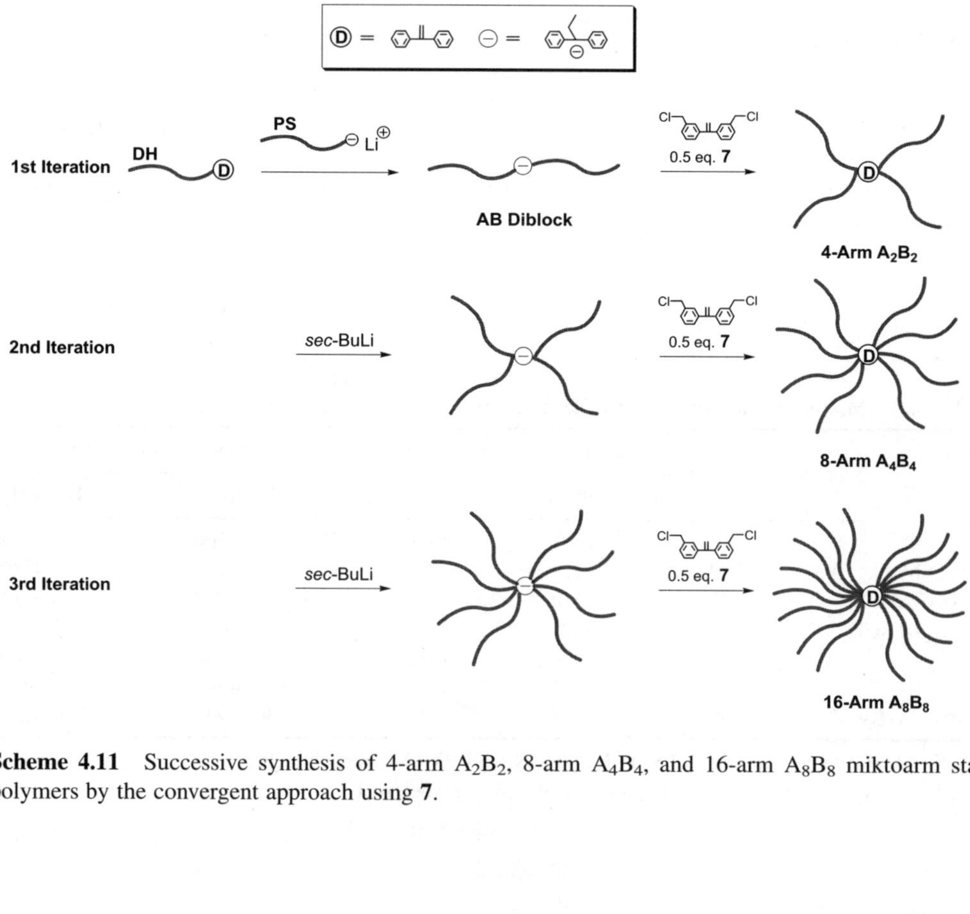

Scheme 4.11 Successive synthesis of 4-arm A_2B_2, 8-arm A_4B_4, and 16-arm A_8B_8 miktoarm star polymers by the convergent approach using **7**.

DPE moiety was converted to a reactive DPE anion by treatment with *sec*-BuLi, the coupling step was repeated between the star-branched anion and **7** in the second iteration. An 8-arm A_4B_4 star was obtained in 60% yield, but not quantitatively. The yield was not improved after several attempts under various conditions. A subsequent repetition of the same synthetic sequence resulted in a 16-arm A_8B_8 star. The reaction yield was significantly reduced to less than 20%.

The insufficient coupling reaction is undoubtedly due to steric hindrance around the reaction sites of intermediate polymers. In order to improve the coupling efficiency in the second iteration, a new methodology was proposed, as illustrated in Scheme 4.12. In this methodology, after the transformation of core DPE into a DPE anion, 1-(4-(4-bromobutyl) phenyl)-1-phenylethylene (**8**) reacts with the DPE anion to re-introduce the DPE function, which is designed to be separated from the core by four methylene chains to reduce steric hindrance. With this new 4-arm core-DPE-functionalized star-branched polymer, the same reaction steps were repeated. In this case, the target 8-arm A_4B_4 star was quantitatively synthesized. Thus, **8** is effective to improve the yield in the second process, although an additional step is needed. The same sequence was repeated once more with the use of the 8-arm core-DPE-functionalized A_4B_4 star. The coupling reaction proceeded, but the yield of a 16-arm A_8B_8 star was reduced to 70%.

Finally, an attempt to synthesize a 64-arm $A_{32}B_{32}$ star was made by a coupling reaction of the 16-arm A_8B_8 star-branched polymer anion with a core bearing four BnBr functions. The coupling reaction proceeded insufficiently and the target 64-arm $A_{32}B_{32}$ star was obtained in only a few per cent.

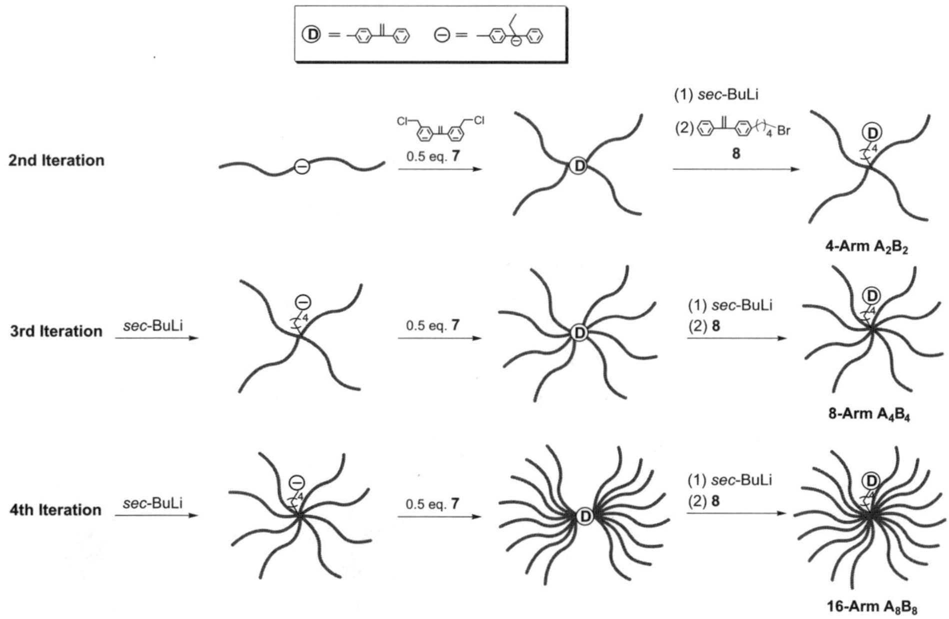

Scheme 4.12 Successive synthesis of 4-arm A_2B_2, 8-arm A_4B_4, and 16-arm A_8B_8 miktoarm star polymers by the convergent approach using **7** and **8**.

The convergent synthesis using a 3-arm ABC star-branched polymer anion was also carried out to synthesize multiple 3-component μ-star polymers (Higashihara *et al.*, submitted). In this approach, a new DPE-functionalized agent, 1-(4-(5,5-bis(3-bromomethylphenyl)-7-methylnonyl)phenyl)-1-phenylethylene (**9**) (Scheme 4.13) was used to save one reaction step, while the distance (four methylene chains) between the DPE molecule and the core was maintained. Preparation of the starting ABC star-branched polymer anion involved a linking reaction of PMSLi with an in-chain-DPE-functionalized AB diblock copolymer composed of PS and PαMS, prepared in a manner similar to that reported elsewhere. The resulting 3-arm star polymer anion was coupled *in situ* with **9**. The reaction efficiently proceeded to give the corresponding 6-arm $A_2B_2C_2$ star in 100% yield. However, a limitation of this convergent growth was observed when pursing the synthesis in the second, followed by the third iteration. The yield of a 12-arm $A_4B_4C_4$ star dropped to 70% and the next coupling reaction leading to a 24-arm $A_8B_8C_8$ star occurred only to a small extent. Furthermore, a coupling reaction between a 12-arm $A_4B_4C_4$ star-branched polymer anion and a core having four BnBr functions was not successful and the target polymer, a 48-arm $A_{16}B_{16}C_{16}$ star, could be obtained in less than 5% yield. Similar synthetic results were obtained when another 3-arm ABC star-branched polymer anion composed of PS, PαMS, and PI arms was used in the

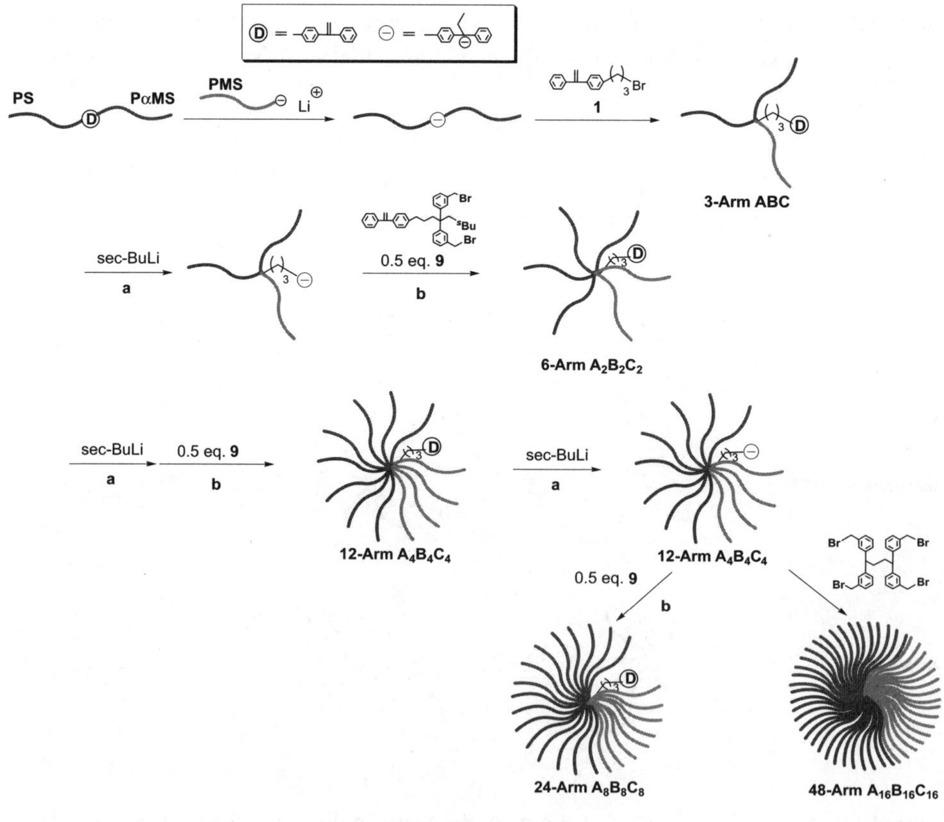

Scheme 4.13 Successive synthesis of miktoarm star polymers by the convergent approach using **9**.

Table 4.10 Miktoarm star polymers by a convergent approach based on the iterative methodology

Type	M_w (kg/mol)		M_w/M_n	Composition (wt%)	
	Calcd	SLS	SEC	Calcd	^{1}H NMR
AB Diblock	22.1	23.2	1.03	52/48	52/48
4-Arm A_2B_2	46.6	48.3	1.03	52/48	51/49
8-Arm A_4B_4	99.0	108	1.04	52/48	52/48
16-Arm A_8B_8	208	199	1.02	52/48	—
3-Arm ABC	32.6	34.1	1.02	35/32/33	37/32/31
6-Arm $A_2B_2C_2$	68.4	67.9	1.02	35/32/33	32/33/35
12-Arm $A_4B_4C_4$	137	137	1.02	35/32/33	35/32/33

[a] A, B, and C were PS, PαMS, and PMS, respectively.

convergent synthesis. As expected, the yield of a 6-arm $A_2B_2C_2$ star was 100%, but the next 12-arm $A_4B_4C_4$ star was obtained in 65% yield.

Compared to dendrimer synthesis, the use of high molecular weight polymer anions in the coupling reaction was required for star polymer synthesis and hence the synthesis suffered from enormous steric interactions around the reaction sites at the cores even at the second dendron stage. Moreover, the molecular weight of a polymer anion exactly doubles with each generation increment to significant reduce the concentration of the reaction site. Accordingly, the present convergent synthesis seems to be limited to the synthesis of stars having less than 20 arms, in which each arm is 10 000 g/mol or higher in M_n value. All of the results of the star-branched polymers obtained by convergent synthesis are summarized in Table 4.10. As expected, the resulting polymers are all well defined in structure with narrow molecular-weight distributions.

In conclusion, the synthesis of multiarmed star polymers was not successful by either the repetition of the synthetic sequence or the coupling reaction of the star-branched polymer anion with a multifunctional core. However, it should be mentioned that the methodology is facile and of great value to be able to successfully obtain some synthetically difficult miktoarm star-branched polymers, such as 8-arm A_4B_4, 16-arm A_8B_8, 6-arm $A_2B_2C_2$, and 12-arm $A_4B_4C_4$ stars.

4.4 Miktoarm Star Polymers by Other Methodologies Based on Living Anionic Polymerization

As mentioned in Section 4.2, several 3-arm ABC star-branched polymers were synthesized by utilizing the dual functionality of DPE toward living anionic polymers. This procedure was further extended by Hillmyer *et al.* (Li *et al.*, 2004; Saito *et al.*, 2008; Liu *et al.*, 2009). They synthesized a series of 3-arm ABC μ-star polymers using two successive living anionic polymerization steps and one polymer-polymer coupling reaction (Li *et al.*, 2004). In this procedure, the dual functionality of the ethylene oxide derivative was utilizing in place of DPE. As illustrated in Scheme 4.14, the synthesis is initiated by reacting PBLi with 2-methoxymethoxyoxirane to prepare a heterobifunctional PB with a hydroxyl group and a protected hydroxyl functionality at the chain end. Catalytic hydrogenation of this PB gave the corresponding poly(ethylethylene) (PEE). Transformation of the terminal hydroxyl group in the PEE to a potassium alkoxide, followed by the anionic ring-opening polymerization

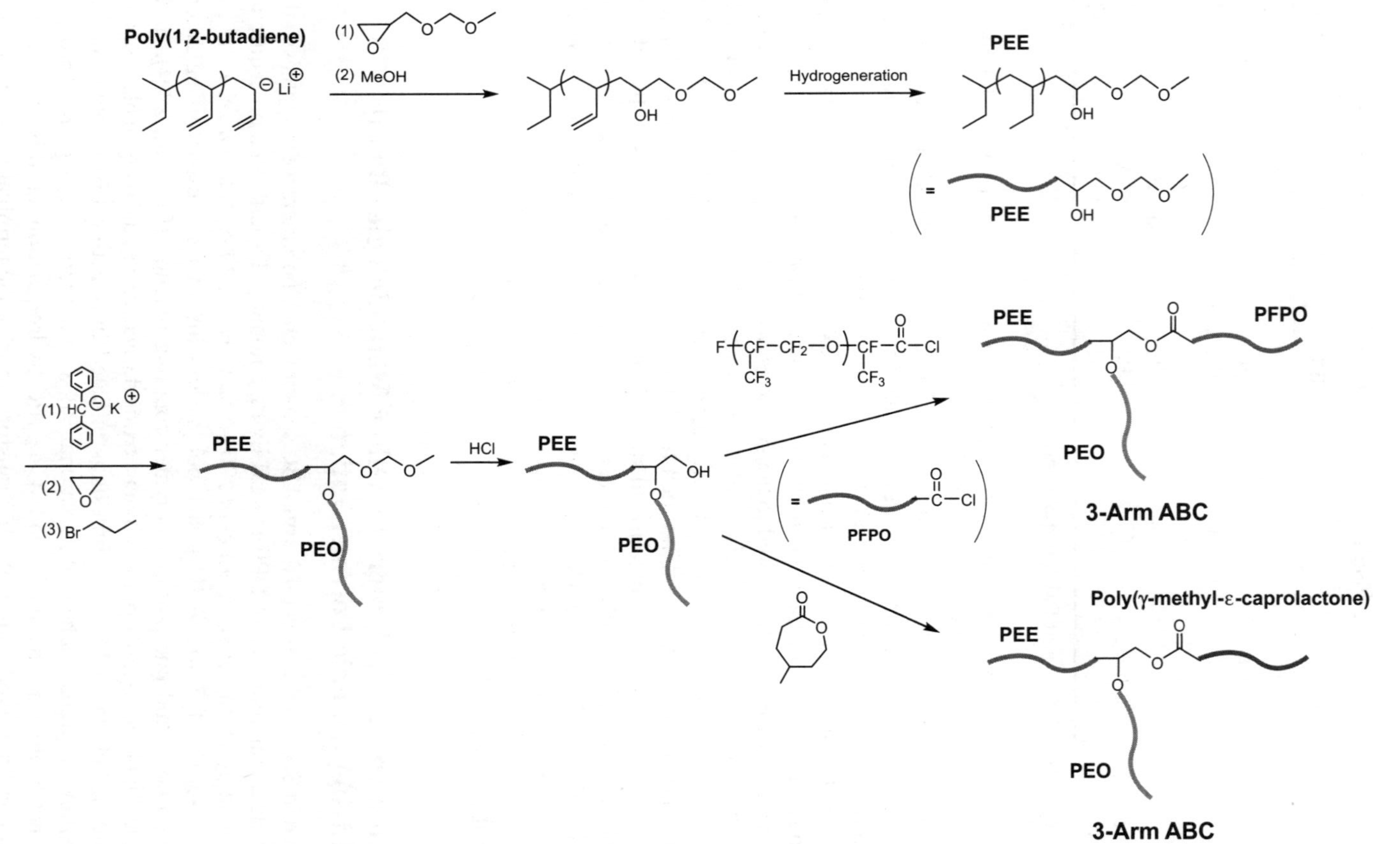

Scheme 4.14 Synthesis of 3-arm ABC miktoarm stars by using the dual functionality of the ethylene oxide derivative.

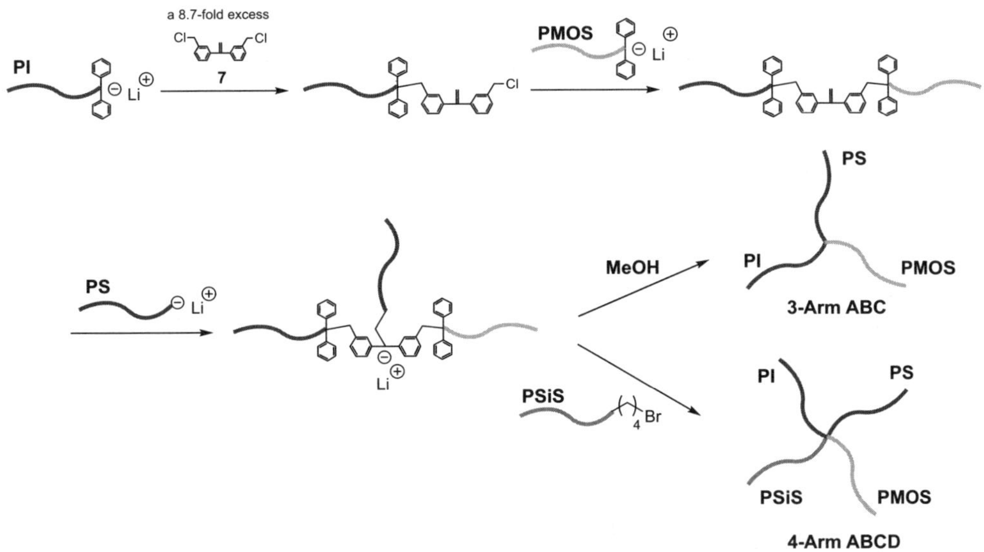

Scheme 4.15 Synthesis of 3-arm ABC and 4-arm ABCD miktoarm star polymers by using **7**.

of ethylene oxide (EO) and subsequent end capping with ethyl bromide, yielding a PEE-*block*-PEO with a protected hydroxyl group. Deprotection of the methoxymethyl group followed by coupling with acyl chloride end-capped poly(perfluoropropylene oxide), afforded the objective ABC star-branched polymer ($M_n = 17\,900$ g/mol, $M_w/M_n = 1.08$). This group also synthesized another ABC star composed of PEE, PEO, and poly(γ-methyl-ε-caprolactone) where the intermediate in-chain-OH-functionalized PEE-*block*-PEO served as a macroinitiator in the anionic ring-opening polymerization of γ-methyl-ε-caprolactone ($M_n = 18\,800$ g/mol, $M_w/M_n = 1.22$) (Saito *et al.*, 2008).

Hirao *et al.* successfully synthesized a 3-arm ABC, followed by a 4-arm ABCD μ-star polymer using the DPE functionalized agent, **7** (Higashihara and Hirao, 2004). As illustrated in Scheme 4.15, a chain-end-functionalized PI with both benzyl chloride and DPE moieties was prepared in 62% yield by the slow addition of PILi, after end capping with DPE, with a 8.7-fold excess of **7**. After the isolation of the chain-end-functionalized PI by fractional precipitation, DPE-end-capped PMOSLi was added to react selectively with the benzyl chloride functionality, followed by reacting PSLi with the residual DPE functionality, resulting in a 3-arm ABC star-branched polymer having an anion at the core ($M_n = 31\,800$ g/mol, $M_w/M_n = 1.03$).

The ABC star-branched polymer anion thus prepared was then reacted with a chain-end-butyl bromide-functionalized PSiS, prepared from 1,4-dibromobutane and PSiSLi end capped with DPE (yield = 95%). The linking reaction was observed to proceed nearly quantitatively to afford the objective 4-arm ABCD μ-star polymer ($M_n = 43\,000$ g/mol, $M_w/M_n = 1.04$). This is the first successful synthesis of a 4-arm ABCD μ-star polymer by the methodology using the dual functionality of the DPE derivative.

Wang *et al.* also synthesized a 4-arm ABCD μ-star polymer by ingeniously utilizing twice the dual functionality of 1,4-bis(1-phenylethenyl)benzene (**10**) in the living anionic polymerization (Wang *et al.*, 2007). As illustrated in Scheme 4.16, **10** undergoes a 1 : 1 addition reaction

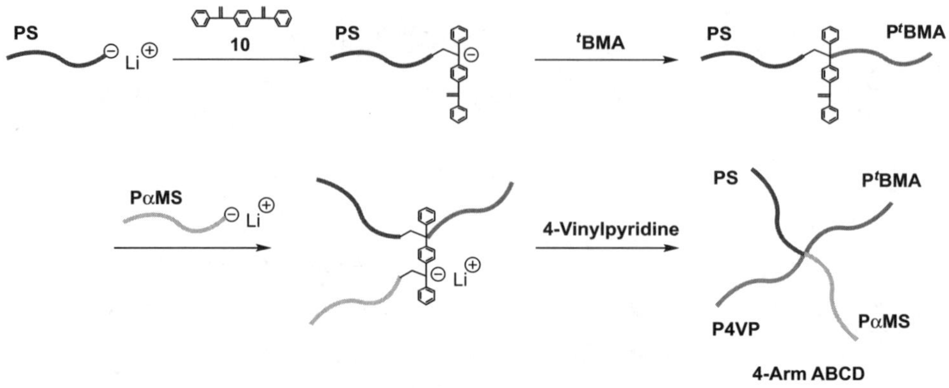

Scheme 4.16 Synthesis of 4-arm ABCD miktoarm star polymers by using **10**.

with PSLi, yielding a chain-end-DPE-functionalized PSLi end capped with DPE. This chain-end modified living polymer served as a macroinitiator of ᵗBMA to afford an in-chain-DPE-functionalized AB diblock copolymer, PS-*block*-PᵗBMA. Then, PαMSLi reacted with the DPE moiety incorporated into the diblock copolymer, followed by the living anionic polymerization of 4-vinylpyridine to grow the D arm segment. The well-defined objective ABCD star-branched polymer was obtained ($M_n = 30\ 000$ g/mol and $M_w/M_n = 1.05$).

Recently, four-miktoarm ABCD star quarterpolymers of styrene, isoprene, dimethylsiloxane (DMS) and 2-vinylpyridine (2VP) were synthesized using a different strategy by Hadjichristidis *et al.* (Mavroudis and Hadjichristidis, 2006). 4-(Dichloromethylsilyl)diphenylethylene (DCMSDPE), a dual-functionality compound with two SiCl groups and a nonhomopolymerizable double bond, was used as the nodule for the incorporation of four incompatible arms. As illustrated in Scheme 4.17, the synthetic approach involved the selective replacement of the two chlorines of DCMSDPE with PI and PDMS by titration with the

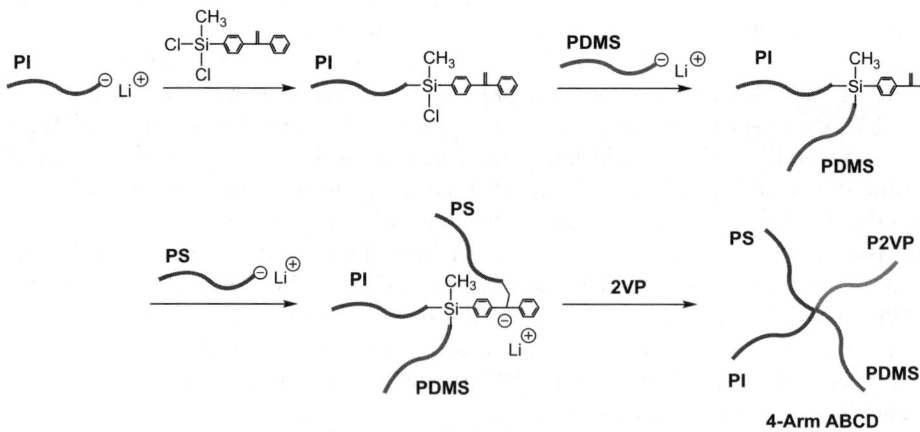

Scheme 4.17 Synthesis of 4-arm ABCD miktoarm star polymers using 4-(dichloromethylsilyl) diphenylethylene.

Table 4.11 Miktoarm star polymers obtained by the combination of silyl chloride and DPE methodologies

Type	M_w (kg/mol)	M_w/M_n	Composition (wt%)	
	RALLS	SEC	Calcd	^{1}H NMR
4-Arm ABCD[a]	75.9[b]	1.06	33/17/17/33	33/16/17/34
4-Arm A_2B_2[c]	88.7	1.05	52/48	57/43
4-Arm A_2BC[c]	96.4	1.02	58/18/24	64/17/19
5-Arm A_2B_2C[c]	109	1.02	41/37/22	37/39/23

[a] A, B, C, and D were PI, PDMS, PS, and P2VP, respectively.
[b] Determined by membrane osmometry.
[c] A, B, and C were PDMS, PS, and PI, respectively.

corresponding living chains, addition of PSLi to the in-chain double bond of the PI-PDMS intermediate and finally polymerization of 2VP. Extensive characterization by SEC/LALLS, membrane osmometry, ^{1}H NMR spectroscopy and DSC confirmed the high degree of molecular and compositional homogeneity and the incompatibility of the arms.

A series of 4-arm A_2B_2, 4-arm A_2B_2C, and 5-arm A_2B_2C stars, where A, B, C are PDMS, PS, and PI, were also successfully synthesized by the similar strategy (Fragouli *et al.*, 2006). Typical characterization results are summarized in Table 4.11. Thus, the variety of monomers has gradually increased since 2000, thanks to new methodologies applicable to multiarmed and multicomponent stars.

4.5 Miktoarm Star Polymers by Living/Controlled Radical Polymerization

There has been rapid growth in the area of living/controlled radical polymerization in the last decade. These polymerization systems involve atom-transfer radical polymerization (ATRP) (Matyjaszewski and Xia, 2001; Wang and Matyjaszewski, 1995; Patten *et al.*, 1996), transition-metal-mediated living radical polymerization (Kamigaito *et al.*, 2001; Ouchi *et al.*, 2009), nitroxide-mediated radical polymerization (NMRP) (Hawker *et al.*, 2001; Hawker, 2002), reversible addition-fragmentation transfer radical polymerization (RAFT) (Chiefari *et al.*, 1998; Chiefari and Rizzardo, 2002), and organotellurium (or stibine and bismuth) -mediated radical polymerization (Yamago, 2009). The development of such living/controlled polymerization systems has opened the way for the synthesis of regular and μ-star polymers. Indeed, several synthetic examples of miktoarm star-branched polymers by these systems have emerged since 2000. Among these stars, recent synthetic examples of 3-arm ABC and 4-arm ABCD μ-star polymers will be introduced in this section.

4.5.1 ABC Miktoarm Star Polymers

The first example was an ABC μ-star polymer composed of poly(tertahydrofuran) (PTHF), poly(1,3-dioxepane) (PDOP), and PS segments and was synthesized by Pan and coworkers in 2002 (Feng and Pan, 2002). The objective star-branched polymer was synthesized by the

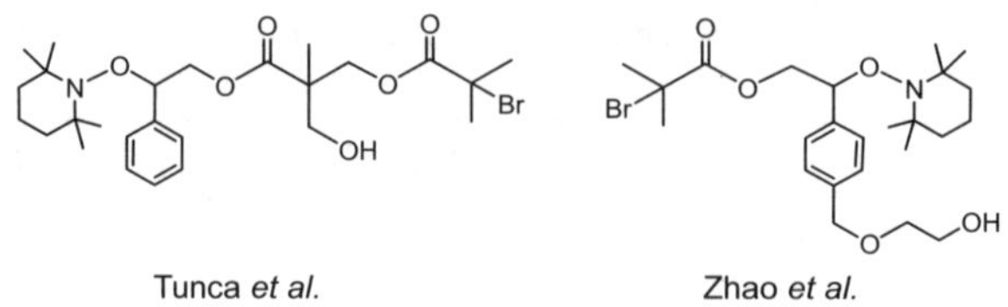

Figure 4.2 Trifunctional initiators having three initiation sites.

combination of cationic ring-opening polymerization and ATRP. The methodology involves (1) preparation of chain-end-functionalized PTHF with carboxylic acid and –OC(O)CHBr, (2) preparation of a block copolymer, PTHF-*block*-PDOP, by treatment of the chain-end-functionalized PTHF with $SOCl_2$ and then $AgClO_4$, followed by the cationic ring-opening polymerization of DOP, (3) ATRP of styrene with the in-chain-OC(O)CHBr-functionalized PTHF-*block*-PDOP to grow a PS arm. The resulting μ-star was 45 900 g/mol in M_n value and showed a broad molecular-weight distribution ($M_w/M_n = 1.52$).

Soon after, ABC miktoarm stars was also synthesized by combining three different polymerization systems with newly designed trifunctional initiators having three initiation sites, as shown in Figure 4.2 (Tunca *et al.*, 2004) and Zhao *et al.* (He *et al.*, 2004). The methodologies involved sequential ring-opening polymerization of ε-caprotactone (εCL), NMRP of styrene, ATRP of *tert*-butyl acrylate (^{t}BA), and ring-opening polymerization of εCL, ATRP of MMA, NMRP of styrene. Typically, stars with 23 600 g/mol in M_n value ($M_w/M_n = 1.35$) and with 44 200 g/mol in M_n value ($M_w/M_n = 1.37$) were synthesized. In 2004, Pan *et al.* also reported the synthesis of a series of ABC stars composed of poly(ethylene oxide) methyl ether, PS, and poly(L-lactide) arms by a similar sequential polymerization technique using RAFT and ring-opening polymerization with a purposely modified initiator (Shi, Li and Pan, 2004). The molecular weights of the resulting polymers were in the range of between 24 500 and 75 900 g/mol ($M_w/M_n = 1.05 - 1.14$).

In 2005, an ABC μ-star polymer composed of PEO, PS, and poly(*tert*-butyl acrylate) (P^{t}BA) was synthesized by combining the Diels–Alder reaction to make a macroinitiator with two initiation sites, followed by NTRP and ATRP, in a similar manner mentioned above ($M_n = 18\,500$ g/mol, $M_w/M_n = 1.27$) (Durmaz *et al.*, 2005). The same group first used "click" chemistry in conjunction with living/controlled polymerization systems to synthesize two ABC miktoarm stars (Altintas *et al.*, 2006). As shown in Scheme 4.18, a trifunctional core having two initiation sites for ATRP and NMRP, and a propargyl group available for the "click" reaction site was prepared. With this functionalized core, the NMRP of MMA was carried out, followed by the ATRP of styrene, resulting in an in-chain-propargyl-functionalized AB diblock copolymer. The ABC star polymers were prepared by the click reaction of the block copolymers with chain-end-N_3-functionalized PEO or P^{t}BA (M_n = about 11 000 g/mol, $M_w/M_n \sim 1.2$). Similarly, Xu *et al.* reported the synthesis of an ABC star with PEO, PS, and PεCL arms ($M_n = 93\,000$ g/mol, $M_w/M_n = 1.10$) (Deng, Ma and Xu, 2007).

Recently, two different water-soluble ABC μ-stars composed of PEO, P^{t}BMA, and poly(2-(diethylamino)ethyl methacrylate) arms plus PEO, P^{t}BMA, and poly(*N*-isopropylacrylamide) arms were synthesized by combining ATRP with "click" chemistry in a manner similar to that

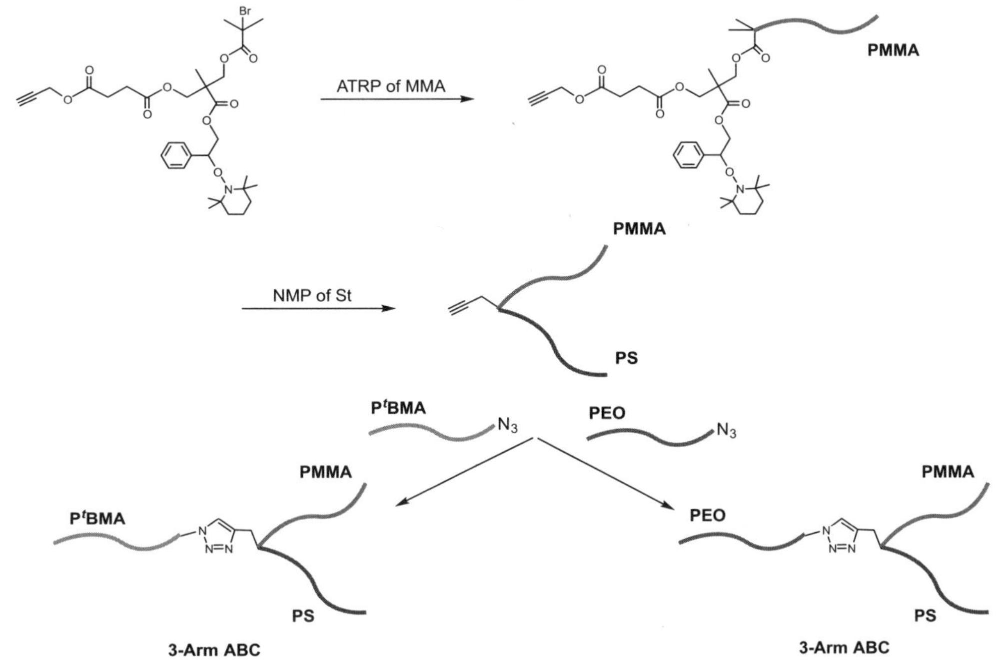

Scheme 4.18 Synthesis of 3-arm ABC miktoarm star polymers based on living/controlled radical polymerization and "click" reaction.

mentioned above ($M_n = 20\,000$–$30\,000$ g/mol, $M_w/M_n = 1.1$–1.2) (Liu *et al.*, 2009a). Their pH-responsive supramolecular self-assemblies were studied under various conditions.

4.5.2 ABCD Miktoarm Star Polymers

In addition to several examples of ABC stars synthesized by the living/controlled polymerization systems mentioned above, three different ABCD μ-star polymers were synthesized almost at the same time in 2008 using very similar methodologies (combining living/controlled polymerization systems with "click" chemistry) (Wang *et al.*, 2008; Altintas *et al.*, 2008), and (Yang *et al.*, 2008).

The synthetic outline of the ABCD star reported by Huang *et al.* is illustrated in Scheme 4.19. First, PSLi and PILi were reacted with ethoxyethyl glycidyl ether to prepare chain-end-functionalized polymers with hydroxyl and ethoxyethyl-protected hydroxyl groups. After these hydroxyl groups were sequentially converted to propargyl and the 2-bromoisobutyryl group for PS, chain-end-functionalized PS served as the macroinitiator of ATRP of ^{t}BA, resulting in an in-chain-propargyl-functionalized PS-*block*-P^{t}BA. In a separate experiment, an in-chain-(OH)-functionalized PI-*block*-PEO was prepared by the ring-opening polymerization of EO using chain-end-(OH)-functionalized PI as the macroinitiator, followed by termination with ethyl bromide, yielding a PI-*block*-PEO. The protected hydroxyl group remaining in the PI-*block*-PEO was deprotected and the resulting hydroxyl

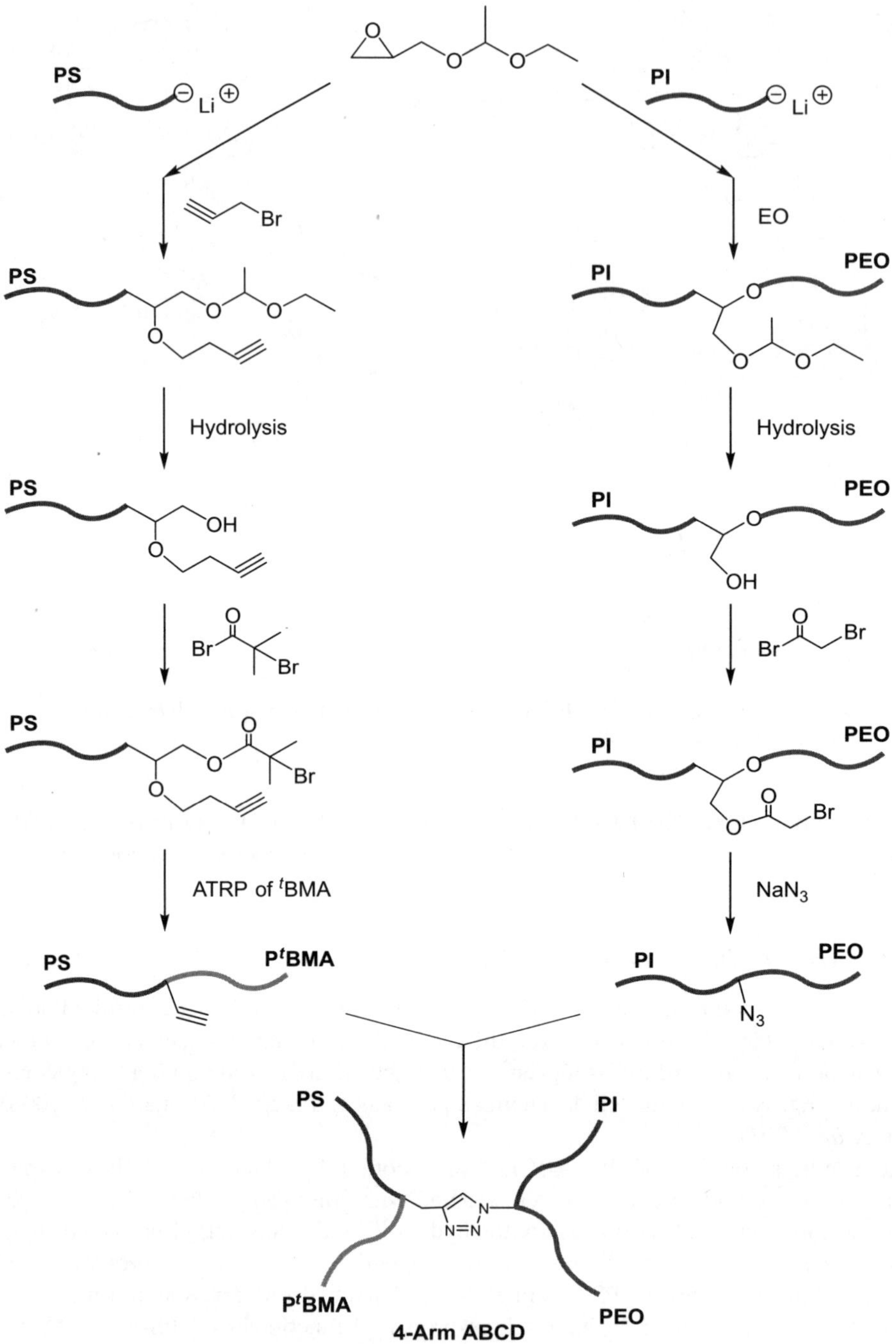

Scheme 4.19 Synthesis of 4-arm ABCD miktoarm star polymers by combination of living anionic polymerization, living/controlled radical polymerization, and "click" reaction.

group was converted to an azide function. Finally, the "click" reaction between both in-chain-functionalized block copolymers was performed in DMF at 80 °C for 48 h to afford the target ABCD star. A polymer with an M_n value of 27 900 g/mol ($M_w/M_n = 1.26$) was the best among a series of polymers synthesized.

Tunca *et al.* synthesized two different ABCD star-branched polymers with PS (A), PεCL (B), PMMA (C), and P^{t}BA or PEO as the D segment by linking between an in-chain-propargyl-functionalized AB block copolymer and an in-chain-N_3-functionalized CD block copolymer in the "click" manner. The resulting polymers were 13 000 g/mol and 16 000 g/mol in M_n value ($M_w/M_n \sim 1.2$). Pan *et al.* synthesized both ABC and ABCD stars with PS (A), PεCL (B), poly(methyl acrylate) (C), and PEO (D) arms in a similar manner based on a combination of RAFT and ring-opening polymerization with "click" chemistry ($M_n = 36\,600$ g/mol, $M_w/M_n = 1.16$).

For the synthesis of all 3-arm ABC and 4-arm ABCD μ-star polymers introduced in this section, combinations of living/controlled polymerization (including ATRP, NMRP, and RAFT), ring-opening polymerization, and/or a linking reaction based on "click" chemistry are employed. The polymerizations and the reactions were carried out in a stepwise manner in order to introduce different arm segments to the stars. Since the intermediate polymers, in addition to the final polymers, were characterized by SEC and ^{1}H NMR except for a few cases, the structures of the final polymers may be correct as indicated. There remains, however, a big problem in that no essential characterization of absolute molecular weight or molecular-weight distribution by light scattering was carried out in most cases. Considering the living/controlled radical polymerization systems used in the syntheses, higher molecular weights than 20 000 g/mol and narrow molecular-weight distributions of less than 1.1 in arm segments are very difficult to realize at the present time. On the other hand, facile experimental operation and the absence of rigorous purification of the reagents used for the synthesis are among the most attractive features of the living/controlled radical polymerization systems. "Click" chemistry also has advantages for the star-branched polymer synthesis, such as high yield, multifunctional group tolerance, and high selectivity. Although only moderate success has been achieved thus far in the synthesis of μ-star polymers by methodologies using living/controlled radical polymerization systems, various synthetic possibilities have been demonstrated. The next subject will be precise control of structural variables leading to "well-defined" architectures.

4.6 Concluding Remarks

Before 2000 the synthesis of miktoarm stars was mainly based on chlorosilane linking agents, characterized by a limit of four different arms possible. With the implementation of the "iterative strategy", this limit was surpassed and the synthesis of 5 μ-star quintopolymers was carried out and that of even more complex stars, for example, 6-arm ABCDEF, (even 7-arm ABCDEFG (Hirao, Higashihara and Inoue, 2008a)), 6-arm $A_2B_2C_2$, 9-arm $A_3B_3C_3$, 7-arm AB_2C_4, 15-arm $AB_2C_4D_8$, 31-arm $AB_2C_4D_8E_{16}$ and so on, become possible. The progress in polymer synthesis needs to be matched by the equivalent progress in polymer morphology in order to fully take advantage of the plethora of novel morphologies that could bring about an explosion of nanoapplications, including multifunctional sensors and multiselective catalysts for sequential or simultaneous chemical reactions. At the present time, however, the morphological studies of miktoarm stars with four or more different arms are very complicated and often impossible to carry out due to the difficulty of obtaining clearly defined microphase domains. More optimized conditions for domain formation and more improved analytical techniques are matters of urgency at the moment.

References

Allgaier, J., Young, R.N., Efstratiadis, V., and Hadjichristidis, N. (1996) Synthesis and characterization of polyisoprene/polybutadiene A_2B_2 star copolymers. *Macromolecules*, **29**, 1794–1797.

Altintas, O., Hizal, G., and Tunca, U. (2006) ABC-type hetero-arm star terpolymers through "click" chemistry. *Journal of Polymer Science Part A-Polymer Chemistry*, **44**, 5699–5707.

Altintas, O., Hizal, G., and Tunca, U. (2008) ABCD 4-miktoarm star quaterpolymers using click $[3+2]$ reaction strategy. *Journal of Polymer Science Part A-Polymer Chemistry*, **46**, 1218–1228.

Avgeropoulos, A., Poulos, Y., Hadjichristidis, N., and Roovers, J. (1996) Synthesis of model 16-miktoarm (vergina) star copolymers of the A_8B_8 type. *Macromolecules*, **29**, 6076–6078.

Bae, Y.C. and Faust, R. (1998) Living coupling reaction in living cationic polymerization. 2. Synthesis and characterization of amphiphilic A_2B_2 star-block copolymer: poly[bis(isobutylene)-star-bis(methyl vinyl ether)]. *Macromolecules*, **31**, 2480–2487.

Bauer, B.J. and Fetters, L.J. (1978) Synthesis and dilute-solution behavior of model star-branched polymers. *Rubber Chemistry and Technology*, **51**, 406–436.

Bywater, S. (1979) Preparation and properties of star-branched polymers. *Advances in Polymer Science*, **30**, 89–116.

Chiefari, J. and Rizzardo, E. (2002) *Handbook of Radical Polymerization* (eds. K. Matyjaszewskiand T.P. Davis), Wiley-Interscience, Hoboken, NJ, pp. 629–690.

Chiefari, F.J., Chong, Y.K., Ercole, F. *et al.* (1998) Living free-radical polymerization by reversible addition-fragmentation chain transfer: The RAFT process. *Macromolecules*, **31**, 5559–5562.

Cho, D., Park, S., Chang, T. *et al.* (2003) Characterization of a 4-miktoarm star copolymer of the (PS-*b*-PI)$_3$PS type by temperature gradient interaction chromatography. *European Polymer Journal*, **39**, 2155–2160.

Deng, G., Ma, D., and Xu, Z. (2007) Synthesis of ABC-type miktoarm star polymers by "click" chemistry, ATRP and ROP. *European Polymer Journal*, **43**, 1179–1187.

Durmaz, H., Karatas, F., Tunca, U., and Hizal, G. (2005) Preparation of ABC miktoarm star terpolymer containing poly (ethylene glycol), polystyrene, and poly(*tert*-butylacrylate) arms by combining Diels–Alder reaction, atom transfer radical, and stable free radical polymerization routes. *Journal of Polymer Science Part A-Polymer Chemistry*, **44**, 499–509.

Feng, X.S. and Pan, C.Y. (2002) Block and star block copolymers by mechanism transformation. 7. Synthesis of polytetrahydrofuran/poly(1,3-dioxepane)/polystyrene ABC miktoarm star copolymers by combination of CROP and ATRP. *Macromolecules*, **35**, 2084–2089.

Fernyhough, C.M., Young, R.N., and Tack, R.D. (1999) Synthesis and characterization of polyisoprene-poly(methyl methacrylate) AB diblock and A_2B_2 heteroarm star copolymers. *Macromolecules*, **32**, 5760–5764.

Fetters, L.J. and Thomas, E.L. (1993) *Material Science and Technology*, vol. **12**, VCH Verlangesellschaft, Weinheim, Germany, pp. 1–31.

Fragouli, P., Iatrou, H., Hadjichristidis, N. *et al.* (2006) Synthesis of well-defined miktoarm star polymers of poly (dimethylsiloxane) by the combination of chlorosilane and benzyl chloride linking chemistry. *Journal of Polymer Science Part A-Polymer Chemistry*, **44**, 6587–6599.

Fujimoto, T., Zhang, H., Kazama, T. *et al.* (1992) Preparation and characterization of novel star-shaped copolymers having three different branches. *Polymer*, **33**, 2208–2213.

Grest, G.S., Fetters, L.J., Huang, J.S., and Richter, D. (1996) Star polymers: experiment, theory, and simulation. *Advances in Chemical Physics*, **94**, 67–163.

Hadjichristidis, N. (1999) Synthesis of miktoarm star (μ-star) polymers. *Journal of Polymer Science Part A-Polymer Chemistry*, **37**, 857–871.

Hadjichristidis, N., Pitsikalis, M., Pispas, S., and Iatrou, H. (2001) Polymers with complex architecture by living anionic polymerization. *Chemical Reviews*, **101**, 3747–3792.

Hadjichristidis, N., Iatrou, H., Pitsikalis, M. *et al.* (2005) Linear and non-linear triblock terpolymers. Synthesis, self-assembly in selective solvents and in bulk. *Progress in Polymer Science*, **30**, 725–782.

Hadjichristidis, N., Iatrou, H., Pitsikalis, M., and Mays, J.W. (2006) Macromolecular architectures by living and controlled/living polymerizations. *Progress in Polymer Science*, **31**, 1068–1132.

Hadjichristidis, N., Pitsikalis, M., and Iatrou, H. (2007) Polymers with star-related structures. *Macromolecular Engineering*, **2**, 909–972.

Hawker, C.J. (2002) *Handbook of Radical Polymerization* (eds. K. Matyjaszewskiand T.P. Davis), Wiley-Interscience, Hoboken, NJ, pp. 463–521.

Hawker, C.J., Bosman, A.W., and Harth, E. (2001) New polymer synthesis by nitroxide mediated living radical polymerizations. *Chemical Reviews*, **101**, 3661–3688.

Hayashi, M., Kojima, K., and Hirao, A. (1999a) Synthesis of star-branched polymers by means of anionic living polymerization coupled with functional group transformation. *Macromolecules*, **32**, 2425–2433.

Hayashi, M., Negishi, Y., and Hirao, A. (1999b) Synthesis of heteroarm star-branched polymers by means of anionic living polymerization in conjunction with functional group transformation. *Proceedings of the Japan Academy, Series B*, **75**, 93–96.

Hayashida, K., Kawashima, W., Takano, A. *et al.* (2006a) Archimedean tiling patterns of ABC star-shaped terpolymers studied by microbeam small-angle X-ray scattering. *Macromolecules*, **39**, 4869–4872.

Hayashida, K., Takano, A., Arai, S. *et al.* (2006b) Systematic transitions of tiling patterns formed by ABC star-shaped terpolymers. *Macromolecules*, **39**, 9402–9408.

Hayashida, K., Saito, N., Arai, S. *et al.* (2007) Hierarchical morphologies formed by ABC star-shaped terpolymers. *Macromolecules*, **40**, 3695–3699.

Hayashida, K., Takano, A., Dotera, T., and Matsushita, Y. (2008) Giant zincblende structures formed by an ABC star-shaped terpolymer/homopolymer blend system. *Macromolecules*, **41**, 6269–6271.

He, T., Li, D., Sheng, X., and Zhao, B. (2004) Synthesis of ABC 3-miktoarm star terpolymers from a trifunctional initiator by combining ring-opening polymerization, atom transfer radical polymerization, and nitroxide-mediated radical polymerization. *Macromolecules*, **37**, 3128–3135.

Higashihara, T. and Hirao, A. (2004) Synthesis of asymmetric star-branched polymers consisting of three or four different segments in composition by means of living anionic polymerization with a new dual-functionalized 1,1-bis (3-chloromethylphenyl)ethylene. *Journal of Polymer Science Part A-Polymer Chemistry*, **42**, 4535–4547.

Higashihara, T., Hayashi, M., and Hirao, A. (2002) Synthesis of branched polymers by means of living anionic polymerization, 9: Radical coupling reaction of 1,1-diphenylethylene-functionalized polymers with potassium naphthalenide and its application to syntheses of in-chain-functionalized polymers and star-branched polymers. *Macromolecular Chemistry and Physics*, **203**, 166–175.

Higashihara, T., Kitamura, M., Haraguchi, N. *et al.* (2003) Synthesis of well-defined star-branched polymers by using chain-end-functionalized polystyrenes with a definite number of 1,3-butadienyl groups and its derivatized functions. *Macromolecules*, **36**, 6730–6738.

Higashihara, T., Nagura, M., Inoue, K. *et al.* (2005) Successive synthesis of well-defined star-branched polymers by a new iterative approach involving coupling and transformation reactions. *Macromolecules*, **38**, 4577–4587.

Higashihara, T., Inoue, K., Nagura, M., and Hirao, A. (2006) Successive synthesis of well-defined star-branched polymers by an iterative approach based on living anionic polymerization. *Macromolecular Research*, **14**, 287–299.

Higashihara, T., Sakurai, T., and Hirao, A. (2009) Successive synthesis of asymmetric star-branched polymers based on iterative methodology using 1,1-diphenylethylene derivatives of alternative choice at each iteration. *Macromolecules*, **42**, 6006–6014.

Higashihara, T., Inoue, K., and Hirao, A., Successive synthesis of well-defined star-branched polymers by "convergent" iterative methodology using core-functionalized 3-arm star-branched polymer and a specially designed 1,1-diphenylethylene derivative. *Macromolecular Symposia*, **296**, 53–62.

Hirao, A. and Haraguchi, N. (2002a) Synthesis of branched polymers by means of living anionic polymerization. 12. Anionic synthesis of well-defined star-branched polymers by using chain-end-functionalized polystyrenes with dendritic benzyl bromide moieties. *Macromolecules*, **35**, 7224–7231.

Hirao, A. and Higashihara, T. (2002b) Synthesis of branched polymers by means of living anionic polymerization. 13. Synthesis of well-defined star-branched polymers via an iterative approach using living anionic polymers. *Macromolecules*, **35**, 7238–7245.

Hirao, A. and Matsuo, A. (2003) Synthesis of chain-end-functionalized poly(methyl methacrylate)s with a definite number of benzyl bromide moieties and their application to star-branched polymers. *Macromolecules*, **36**, 9742–9751.

Hirao, A. and Nakahama, S. (1994) Recent developments in anionic living polymerization. *Trends in Polymer Science*, **2**, 267–271.

Hirao, A. and Nakahama, S. (1998) Anionic living polymerization of functionalized monomers. *Acta Polymerica*, **49**, 133–144.

Hirao, A. and Ryu, S.W. (2003) Synthesis of branched polymers by means of living anionic polymerization, 11. Anionic synthesis of densely branched polystyrenes carrying double branches in each repeating unit. *Macromolecular Symposia*, **192**, 31–42.

Hirao, A., Hayashi, M., and Higashihara, T. (2001) Synthesis of branched polymers by means of living anionic polymerization, 8. Synthesis of well-defined star-branched polymers by an iterative approach based on living anionic polymerization using 1,1-diphenylethylene derivatives. *Macromolecular Chemistry and Physics*, **202**, 3165–3173.

Hirao, A., Hayashi, M., and Matsuo, A. (2002c) Synthesis of branched polymers by means of living anionic polymerisation. 10. Synthesis of well-defined heteroarm star-branched polymers by coupling reaction of chain-functionalized polystyrenes with benzyl halide moieties with living anionic polymers of *tert*-butyl methacrylate. *Polymer*, **43**, 7125–7131.

Hirao, A., Loykulnant, S., and Ishizone, T. (2002d) Recent advance in living anionic polymerization of functionalized styrene derivatives. *Progress in Polymer Science*, **27**, 1399–1471.

Hirao, A., Hayashi, M., Tokuda, Y. *et al.* (2002e) Precise synthesis of regular and asymmetric star polymers and densely branched polymers with starlike structures by means of living anionic polymerization. *Polymer Journal*, **34**, 633–658.

Hirao, A., Hayashi, M., Loykulnant, S. *et al.* (2005) Precise syntheses of chain-multi-functionalized polymers, star-branched polymers, star-linear block polymers, densely branched polymers, and dendritic branched polymers based on iterative approach using functionalized 1,1-diphenylethylene derivatives. *Progress in Polymer Science*, **30**, 111–182.

Hirao, A., Inoue, K., and Higashihara, T. (2006a) Successive synthesis of regular and asymmetric star-branched polymers by iterative methodology based on living anionic polymerization using functionalized 1,1-diphenylethylene derivatives. *Macromolecular Symposia*, **240**, 31–40.

Hirao, A., Higashihara, T., Nagura, M., and Sakurai, T. (2006b) Successive synthesis of well-defined many arm star-branched polymers by an iterative methodology using a specially designed 1,1-diphenylethylene. *Macromolecules*, **39**, 6081–6091.

Hirao, A., Higashihara, T., and Inoue, K. (2008a) Successive synthesis of well-defined asymmetric star-branched polymers up to seven-arm, seven-component ABCDEFG type by an iterative methodology based on living anionic polymerization. *Macromolecules*, **41**, 3579–3587.

Hirao, A., Inoue, K., Higashihara, T., and Hayashi, M. (2008b) Successive synthesis of well-defined star-branched polymers by iterative methodology based on living anionic polymerization. *Polymer Journal*, **40**, 923–941.

Hsieh, H.L. and Quirk, R.P. (1996) *Anionic Polymerization: Principles and Applications*, Marcel Dekker, New York, pp. 333–368.

Huckstaedt, H., Abetz, V., and Stadler, R. (1996) Synthesis of a polystyrene-arm-polybutadiene-arm-poly(methyl methacrylate) triarm star copolymer. *Macromolecular Rapid Communications*, **17**, 599–606.

Huckstadt, H., Gopfert, A., and Abetz, V. (2000) Synthesis and morphology of ABC heteroarm star terpolymers of polystyrene, polybutadiene, and poly(2-vinylpyridine). *Macromolecular Chemistry and Physics*, **201**, 296–307.

Iatrou, H. and Hadjichristidis, N. (1992) Synthesis of a model 3-miktoarm star terpolymer. *Macromolecules*, **25**, 4649–4651.

Iatrou, H. and Hadjichristidis, N. (1993) Synthesis and characterization of model 4-miktoarm star co- and quaterpolymers. *Macromolecules*, **26**, 2479–2484.

Kamigaito, M., Ando, T., and Sawamoto, M. (2001) Metal-catalyzed living radical polymerization. *Chemical Reviews*, **101**, 3689–3745.

Lambert, O., Dumas, P., Hurtrez, G., and Riess, G. (1997) Synthesis of an amphiphilic triarm star copolymer based on polystyrene, poly(ethylene oxide) and poly(ε-caprolactone). *Macromolecular Rapid Communications*, **18**, 343–351.

Li, Z., Hillmyer, M.A., and Lodge, T.P. (2004) Synthesis and characterization of triptych μ-ABC star triblock copolymers. *Macromolecules*, **37**, 8933–8940.

Liu, C., Hillmyer, M.A., and Lodge, T.P. (2009a) Multicompartment micelles from pH-responsive miktoarm star block terpolymers. *Langmuir*, **25**, 13718–13725.

Liu, H., Li, C., Liu, H., and Liu, S. (2009b) pH-Responsive supramolecular self-assembly of well-defined zwitterionic ABC miktoarm star terpolymers. *Langmuir*, **25**, 4724–4734.

Loshe, D.J. and Hadjichristidis, N. (1997) Microphase separation in block copolymers. *Current Opinion in Colloid and Interface Science*, **2**, 171–176.

Lutz, P.J. and Rein, D. (1999) *Star and Hyperbranched Polymers* (eds. M.K. Mishraand S. Kobayashi), Marcel Dekker, New York, pp. 27–57.
Matsushita, Y. (2007) Creation of hierarchically ordered nanophase structures in block polymers having various competing interactions. *Macromolecules*, **40**, 771–776.
Matsushita, Y. (2008) Precise molecular design of complex polymers and morphology control of their hierarchical multiphase structures. *Polymer Journal*, **40**, 177–183.
Matsushita, Y., Takano, A., Hayashida, K. *et al.* (2009) Hierarchical nanophase-separated structures created by precisely-designed polymers with complexity. *Polymer*, **50**, 2191–2203.
Matyjaszewski, K. and Xia, J. (2001) Atom transfer radical polymerization. *Chemical Reviews*, **101**, 2921–2990.
Mavroudis, A. and Hadjichristidis, N. (2006) Synthesis of well-defined 4-miktoarm star quarterpolymers (4 μ-SIDV) with four incompatible arms: polystyrene (S), polyisoprene-1,4 (I), poly(dimethylsiloxane) (D), and poly(2-vinylpyridine) (V). *Macromolecules*, **39**, 535–540.
Mavroudis, A., Avgeropoulos, A., Hadjichristidis, N. *et al.* (2003) Synthesis and morphological behavior of model linear and miktoarm star copolymers of 2-methyl-1,3-pentadiene and styrene. *Chemistry of Materials*, **15**, 1976–1983.
Mavroudis, A., Avgeropoulos, A., Hadjichristidis, N. *et al.* (2006) Synthesis and morphological behavior of model 6-miktoarm star copolymers, $PS(P2MP)_5$, of styrene (S) and 2-methyl-1,3-pentadiene (P2MP). *Chemistry of Materials*, **18**, 2164–2168.
Mays, J.W. (1990) Synthesis of "simple graft" poly(isoprene-*g*-styrene) by anionic polymerization. *Polymer Bulletin*, **23**, 247–250.
Meyer, N., Delaite, C., Hurtrez, G., and Dumas, P. (2002) A common route to A_2B and A_3B multiarm star block copolymers. *Polymer*, **43**, 7133–7139.
Nakahama, S. and Hirao, A. (1990) Protection and polymerization of functional monomers: anionic living polymerization of protected monomers. *Progress in Polymer Science*, **15**, 299–335.
Nasser-Eddine, M., Reutenauer, S., Delaite, C. *et al.* (2004) Synthesis of polystyrene-poly(*tert*-butyl methacrylate)-poly(ethylene oxide) triarm star block copolymers. *Journal of Polymer Science Part A-Polymer Chemistry*, **42**, 1745–1751.
Ouchi, M., Terashima, T., and Sawamoto, M. (2009) Transition metal-catalyzed living radical polymerization: Toward perfection in catalysis and precision polymer synthesis. *Chemical Reviews*, **109**, 4963–5050.
Patten, T.E., Xia, J., Abernathy, T., and Matyjaszewski, K. (1996) Polymers with very low polydispersities from atom transfer radical polymerization. *Science*, **272**, 866–868.
Quirk, R.P., Lee, B., and Schock, L.E. (1992) Anionic synthesis of polystyrene and polybutadiene heteroarm star polymers. *Makromolekulare Chemie, Macromolecular Symposia*, **53**, 201–210.
Quirk, R.P., Yoo, T., and Lee, B. (1994) Anionic synthesis of heteroarm, star-branched polymers. Scope and limitations. *Journal of Macromolecular Science-Pure and Applied Chemistry*, **A31**, 911–926.
Rempp, P. and Herz, J.E. (1989) *Encyclopedia of Polymer Science and Engineering*, 2nd edn (ed. J.I. Kroschwitz), Wiley-Interscience, New York, Suppl. vol. 2, pp. 493–510.
Reutenauer F.S., Hurtrez, G., and Dumas, P. (2001) A new route to model (A_2B) and regular graft copolymers. *Macromolecules*, **34**, 755–760.
Roovers, J. (1985) *Encyclopedia of Polymer Science and Engineering*, Suppl. vol. **2**, 2nd edn (ed. J.I. Kroschwitz), Wiley-Interscience, New York, pp. 478–499.
Ryu, S.W. and Hirao, A. (2000) Anionic synthesis of well-defined poly(*m*-halomethylstyrene)s and branched polymers via graft-onto methodology. *Macromolecules*, **33**, 4765–4771.
Ryu, S.W. and Hirao, A. (2001) Synthesis of branched polymers by means of living anionic polymerization. 7. Synthesis of well-defined highly branched polymers and graft copolymers having one branch per repeating unit using poly(*m*-halomethyl)styrenes as backbone polymers. *Macromolecular Chemistry and Physics*, **202**, 1727–1736.
Ryu, S.W., Asada, H., and Hirao, A. (2002) Synthesis of novel well-defined substituted polystyrenes functionalized with two and four benzyl bromide moieties in each monomer unit and their application to densely branched polymer synthesis. *Macromolecules*, **35**, 7191–7199.
Ryu, S.W., Watanabe, T., Asada, H., and Hirao, A. (2004) Synthesis of well-defined high-density branched polymers carrying two branch chains in each repeating unit by coupling reaction of benzyl bromide-functionalized polystyrenes with polymer anions comprised of two polymer segments. *Macromolecules*, **37**, 6291–6298.
Saito, N., Liu, C., Lodge, T.P., and Hillmyer, M.A. (2008) Multicompartment micelles from polyester-containing ABC miktoarm star terpolymers. *Macromolecules*, **41**, 8815–8822.

Shi, P.-J., Li, Y.-G., and Pan, C.-Y. (2004) Block and star block copolymers by mechanism transformation X. Synthesis of poly(ethylene oxide) methyl ether/polystyrene/poly(L-lactide) ABC miktoarm star copolymers by combination of RAFT and ROP. *European Polymer Journal*, **40**, 1283–1290.

Sioula, S., Hadjichristidis, N., and Thomas, E.L. (1998) Novel 2-dimensionally periodic non-constant mean curvature morphologies of 3-miktoarm star terpolymers of styrene, isoprene, and methyl methacrylate. *Macromolecules*, **31**, 5272–5277.

Sugiyama, K., Inoue, K., Higashihara, T. *et al.* (2009) Synthesis of well-defined block copolymers and star-branched polymers by using terminal 1,3-butadiene functionalized polymers as reactive building blocks. *Reactive & Functional Polymers*, **69**, 480–492.

Takano, A., Wada, S., Sato, S. *et al.* (2004) Observation of cylinder-based microphase-separated structures from ABC star-shaped terpolymers investigated by electron computerized tomography. *Macromolecules*, **37**, 9941–9946.

Takano, A., Kawashima, W., Noro, A. *et al.* (2005) A mesoscopic Archimedean tiling having a new complexity in an ABC star polymer. *Journal of Polymer Science Part B-Polymer Physics*, **3**, 2427–2432.

Takano, A., Kawashima, W., Wada, S. *et al.* (2007) Composition dependence of nanophase-separated structures formed by star-shaped terpolymers of the $A_{1.0}B_{1.0}C_x$ type. *Journal of Polymer Science Part B-Polymer Physics*, **45**, 2277–2283.

Tselikas, Y., Iatrou, H., Hadjichristidis, N. *et al.* (1996) Morphology of miktoarm star copolymers of styrene and isoprene. *Journal of Chemical Physics*, **105**, 2456–6078.

Tsoukatos, T. and Hadjichristidis, N. (2002) Synthesis of model polycyclohexylene/polyethylene miktoarm star copolymers with three and four arms. *Journal of Polymer Science Part A-Polymer Chemistry*, **40**, 2575–2582.

Tsoukatos, T., Pispas, S., and Hadjichristidis, N. (2000) Complex macromolecular architectures by combining TEMPO living free radical and anionic polymerization. *Macromolecules*, **33**, 9504–9511.

Tunca, U., Ozyurek, Z., Erdogan, T., and Hizal, G. (2004) Novel miktofunctional initiator for the preparation of an ABC-type miktoarm star polymer via a combination of controlled polymerization techniques. *Journal of Polymer Science Part A-Polymer Chemistry*, **42**, 4228–4236.

Velis, G. and Hadjichristidis, N. (1999) Synthesis of model $PS(PI)_5$ and $(PI)_5PS(PI)_5$ nonlinear block copolymers of styrene (S) and isoprene (I). *Macromolecules*, **32**, 534–536.

Walther, A., Yuan, J., Abetz, V., and Mueller, A.H.E. (2009) Structure-tunable bidirectional hybrid nanowires via multicompartment cylinders. *Nano Letters*, **9**, 2026–2030.

Wang, G., Luo, X., Liu, C., and Huang, J. (2008) Synthesis of ABCD 4-miktoarm star-shaped quarterpolymers by combination of the "click" chemistry with multiple polymerization mechanism. *Journal of Polymer Science Part A-Polymer Chemistry*, **46**, 2154–2166.

Wang, J.S. and Matyjaszewski, K. (1995) Controlled/"living" radical polymerization. Atom transfer radical polymerization in the presence of transition-metal complexes. *Journal of the American Chemical Society*, **117**, 5614–5615.

Wang, X., He, J., and Yang, Y. (2007) Synthesis of ABCD-type miktoarm star copolymers and transformation into zwitterionic star copolymers. *Journal of Polymer Science Part A-Polymer Chemistry*, **45**, 4818–4828.

Yamago, S. (2009) Precision polymer synthesis by degenerative transfer controlled/living radical polymerization using organotellurium, organostibine, and organobismuthine chain-transfer agents. *Chemical Reviews*, **109**, 5051–5068.

Yang, L., Zhou, H., Shi, G. *et al.* (2008) Synthesis of ABCD 4-miktoarm star polymers by combination of RAFT, ROP, and "click chemistry". *Journal of Polymer Science Part A-Polymer Chemistry*, **46**, 6641–6653.

Zhao, Y., Higashihara, T., Sugiyama, K., and Hirao, A. (2005) Synthesis of functionalized asymmetric star polymers containing conductive polyacetylene segments by living anionic polymerization. *Journal of the American Chemical Society*, **127**, 14158–14159.

Zhao, Y., Higashihara, T., Sugiyama, K., and Hirao, A. (2007) Synthesis of asymmetric star-branched polymers having two polyacetylene arms by means of living anionic polymerization using 1,1-diphenylethylene derivatives. *Macromolecules*, **40**, 228–238.

5

Precise Synthesis of Dendrimer-Like Star-Branched Polymers, a New Class of Well-Defined Hyperbranched Polymers

Hee-Soo Yoo and Akira Hirao

Polymeric and Organic Materials Department, Graduate School of Science and Engineering, Tokyo Institute of Technology, S1-6, 2-12-1, Ohokayama, Meguro-ku, Tokyo 152-8552, Japan

5.1 Introduction

It has long been known that branching architecture strongly influences the properties of polymers. Accordingly, clarification and quantification of the relationships among the architectures and the properties are quite important from both the theoretical and industrial viewpoints. For this purpose, taking into account the recent rapid progress of living/controlled polymerization systems, a variety of branched polymers has been synthesized. Among these polymers, a series of branched polymers (Figure 5.1) has emerged as a novel class of structurally well-defined branched polymers since 1995 (Six and Gnanou, 1995). Although several names have been proposed for such polymers, they have recently been called "dendrimer-like star-branched polymers" based on their branched architectures, similar to those of well-established dendrimers and introduced in this chapter.

In order to clarify the structures of dendrimer-like star-branched polymers, the representative third-generation (3G) and fifth-generation (5G) polymers are shown in Figure 5.1. As can be seen, these polymers are structurally well defined and similar in branched architecture to dendrimers, but are composed of many polymer chains linked to each other between the junctions in all generations. Accordingly, they are much higher in molecular weight and much

Complex Macromolecular Architectures: Synthesis, Characterization, and Self-Assembly, First Edition.
Edited by Nikos Hadjichristidis, Akira Hirao, Yasuyuki Tezuka and Filip Du Prez.

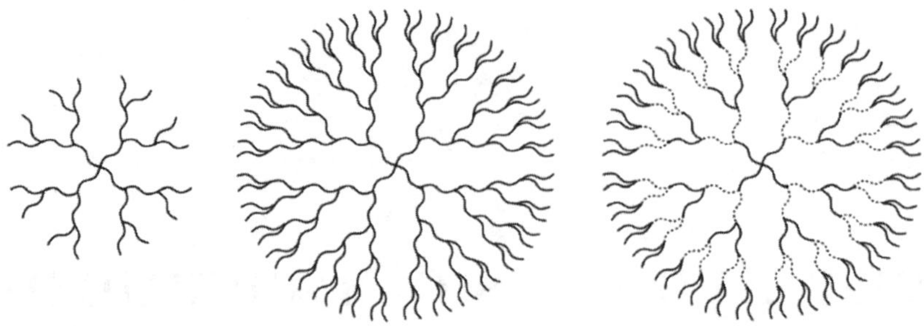

Figure 5.1 3G and 5G dendrimer-like star-branched polymers and block copolymer.

larger in molecular size than dendrimers of the same generation. Dendrimer-like star-branched polymers are recognized as nano-ordered three-dimensional globular macromolecules and have many characteristic features, such as specific topological hyperbranched architectures, hierarchic generation-based (or layered) structures, different branch densities between the core and outermost layer, plus many junctions and chain-end groups. Moreover, they are expected to behave as individual entities in solution like hyperbranched dendrigraft polymers.

Several synthetic examples of dendrimer-like star-branched block copolymers with different polymer segments introduced at each generation have also been reported. They are novel-type block copolymers possessing generation-based layered block structures. Such polymer segments may possibly be phase separated at the molecular level, followed by self-organizing, to generate quite new nanosize ordered suprastructures and supramolecular assemblies that are different from those formed by linear block copolymers and even asymmetric star-branched polymers having chemically different arm segments (so-called miktoarm star polymers) because of their hierarchic generation-based architectures. Indeed, some interesting morphologies and nanostructured materials have been reported. Furthermore, the formation of alternative or several separated layer structures is highly probable again due to the generation-based structures (Figure 5.1). Thus, dendrimer-like star-branched polymers and block copolymers are promising specialty functional materials with many potential applications in the fields of nanoscience and nanotechnology. For example, it may be expected that functionalized dendrimer-like star-branched polymers will be used as nanoreactors, enzymes and catalysts carrying layered reaction systems, multicompartment micelles, carriers of drugs and genetic materials, shape-persistent nanoscale electronic and optical devices, and so on.

For the synthesis of dendrimers composed of several generation-based layers (Grayson and Fréchet, 2001), stepwise iterative methodologies based on two complementary approaches, the divergent and the convergent, have been employed. Likewise, methodologies related to these approaches have now been developed in order to synthesize dendrimer-like star-branched polymers, which will be discussed in the following sections. However, there is a distinct difference between dendrimers and dendrimer-like star-branched polymers. Since the latter polymers have a branched structure in which there is a polymer chain connected between junctions in all generations, one more reaction step for introducing the polymer chain is required in the synthesis. This additional step often makes the synthesis of dendrimer-like star-branched polymers more difficult or limited.

This chapter reviews the synthesis of dendrimer-like star-branched polymers and block copolymers by stepwise iterative methodologies based on the divergent or convergent approach, especially focusing on the control of synthetic factors necessary for well-defined structures of dendrimer-like star-branched polymers, that is, molecular weight, molecular-weight distribution, arm number, and composition. Some characterization studies on molecular sizes, molecular shapes, and viscosity behavior of dendrimer-like star-branched polymers are also described. Various hyperbranched polymers called dendrigraft, comb-burst, arborescent, hyperMacs, and so on, structurally similar to dendrimer-like star-branched polymers, have also been synthesized, but these polymers exhibit an irregular architecture with incompletely reacted branched points throughout the structures. Dendrimer-like star-branched polymers, on the other hand, are highly ordered and regularly branched polymers and differ from such hyperbranched polymers in their structural perfection. For this reason, these hyperbranched polymers have not been dealt with in this review, although they have provided valuable and important contributions to the field of polymer chemistry and physics. Several reviews on dendrimer-like star-branched polymers have recently been reported and more information on this subject is available (Hirao *et al.*, 2005b; Hirao *et al.*, 2006; Hirao *et al.*, 2008; Hirao *et al.*, 2009a; Taton *et al.*, 2006; Taton *et al.*, 2007; Hutchings, 2008).

5.2 Synthetic Approach

As shown in Figure 5.1, dendrimer-like star-branched polymers are built by accumulating a number of generation-based layers of branched repeating units emanating from a core. Accordingly, a stepwise iterative methodology based on either a divergent or convergent approach used for dendrimer synthesis has been proposed and employed for the synthesis of dendrimer-like star-branched polymers. However, that synthesis requires one more reaction step for the introduction of the constituent polymer chains connected between the junctions in all generations. Since precise control of such polymer chains in terms of molecular weight and molecular-weight distribution is essential for the synthesis of well-defined dendrimer-like star-branched polymers, living polymerization systems are usually used. For the introduction of well-defined polymer chains, either an initiation reaction with multifunctional reaction sites (so-called "core-first") or a termination reaction with a premade living polymer (so-called "arm-first") is adopted.

5.2.1 "Core-First" Divergent Iterative Methodology

The first methodology herein introduced is a stepwise iterative methodology based on the divergent approach. In general, the divergent approach initiates growth at a multifunctionalized core and continues outward by iterating the same or similar synthetic sequence several times. Since one more reaction step using living polymerization is required to synthesize dendrimer-like star-branched polymers as mentioned above, the following stepwise iterative methodology was newly proposed. Basically, there are two reaction steps in the iterative synthetic sequence: (1) The living polymerization of an appropriate monomer with a defined multifunctional initiator used as the core in the first iteration and then with multifunctional initiation sites at the polymer chain end in the second or further iterations and (2) a chain-end modification reaction to introduce two or more initiation sites for preparing the branch points and for the formation of the next generation. The two reaction steps are repeated several times to synthesize a series of dendrimer-like star-branched polymers. Since the polymer chains are always prepared by initiation followed by propagation reaction throughout the synthesis, this procedure corresponds to an example of a "core-first" process.

The first dendrimer-like star-branched polymer synthesized by the methodology based on the "core-first" divergent approach was the second-generation (2G) dendrimer-like star-branched poly(ethylene oxide) (PEO) reported by Gnanou and coworkers in 1995 (Six and Gnanou, 1995). Although the dendrimer-like star-branched architecture of the synthesized polymer was shown, the detailed synthetic procedure and characterization results were not available in this report. In the meantime, the successful synthesis of a series of dendrimer-like star-branched poly(ε-caprolactone)s (PεCL)s up to the 3G stage was reported by Hedrick and coworkers (Trollsås and Hedrick, 1998). That synthetic route is illustrated in Scheme 5.1. In this synthesis, the following two reaction steps were used in the iterative synthetic sequence: (1) a stannous 2-ethylhexanoate mediated living ring-opening polymerization (ROP) of ε-caprolactone (εCL) with a hexa-functional 2,2-bis(hydroxymethyl)propionic acid derivative (**1**) as the core and (2) a modification reaction of the hydroxyl terminus with hydroxyl-protected propionic acid to introduce two hydroxyl groups. Thus, there are two hydroxyl initiation sites at

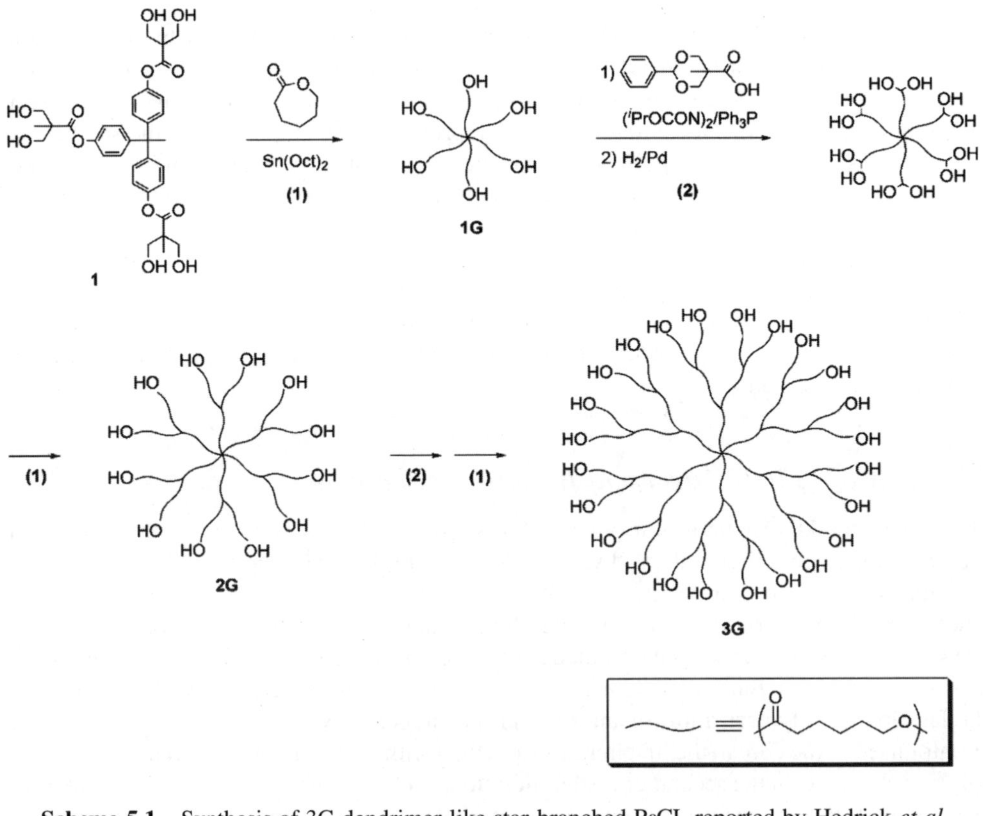

Scheme 5.1 Synthesis of 3G dendrimer-like star-branched PεCL reported by Hedrick *et al.*

each chain end. In the next iteration, the ROP of εCL is carried out with a resulting chain-end-$(OH)_2$-functionalized PεCL attached to the core by further iteration.

As seen in Scheme 5.1, preparation of the 1G polymer in this synthesis involved the ROP of εCL and subsequent chain-end modification. Conversion of the terminal hydroxyl group to two hydroxyl functions enabled further ROP to the 2G polymer. The 1G polymer synthesized by this procedure was a 6-arm star-branched PεCL. The target dendrimer-like star-branched polymer was obtained as a 2G polymer by the second iteration and possessed a minimum architectural unit. One more repetition of the synthetic sequence involving the two reaction steps resulted in a 3G dendrimer-like star-branched PεCL. The 3G polymer possessed six branches at the core and two branches at the junctions in both the 2G- and 3G-based layers, composed of 42 arm segments (6 (1G) + 12 (2G) + 24 (3G) = 42). The observed M_n value was 96 000 g/mol, close to the theoretical value, and the molecular-weight distribution was not narrow, but an acceptable value of 1.14.

The methodology further demonstrated that it was possible to synthesize a series of dendrimer-like star-branched block copolymers with additional functionalities by combining the living ROP of εCL with the atom-transfer radical polymerization (ATRP) of methacrylate monomers. A 6-arm star-branched PεCL synthesized in the same manner as that mentioned above was followed by chain-end modification to increase the hydroxyl terminus twofold or more and subsequently reacting the terminal hydroxyl groups with 2-bromo-2-methylpropionyl bromide, introducing 12 or 24 ATRP initiation sites at the chain ends. In the second iteration, the chain-end-modified polymers served as macroinitiators for the ATRP of MMA, a mixture of MMA and 2-hydroxyethyl methacrylate, or oligo(ethylene oxide)-substituted methacrylate, resulting in 2G and 3G dendrimer-like star-branched block copolymers (Hedrick *et al.*, 1998; Stancik *et al.*, 2003). An amphiphilic character could be introduced by designing the different generations to be either hydrophobic or hydrophilic. Thus, the switching of the polymerization system from the ROP to ATRP was readily achieved.

Various 2G and 3G dendrimer-like star-branched P(εCL)s, poly(L-lactide)s, and poly(alkyl or phenyl-substituted ε-CL)s with different branched architectures were also synthesized by using the same methodology with an intentional change in the number of initiation sites in the chain-end modification reaction step (Trollsås and Hedrick, 1998; Trollsås *et al.*, 1998a; Trollsås *et al.*, 1998b; Hedrick *et al.*, 1998; Trollsås *et al.*, 2000a; Trollsås *et al.*, 2000b; Würsch *et al.*, 2001; Trollsås *et al.*, 2004). Most of the polymers synthesized by this methodology were approximately 2.0×10^5 g/mol in maximum M_n value, the M_w/M_n values being between 1.1 and 1.5. The representative results are summarized in Table 5.1.

Almost at the same time, Gnanou *et al.* reported the synthesis of amphiphilic 2G dendrimer-like star-branched block copolymers composed of poly(ethylene oxide) (PEO) and polystyrene (PS) segments by the same divergent approach combining the living anionic ring-opening polymerization of EO with the ATRP of styrene (Angot *et al.*, 2000; Gnanou and Taton, 2001). As shown in Scheme 5.2, the process is initiated by the ROP of EO with an anionic tetra-functional initiator from pentaerythritol and Ph_2CHK to yield a 4-arm star PEO. Treatment of the terminal hydroxyl groups of the 4-arm star PEO with 2,2-bis(2-(bromomethyl)propionato) propionyl chloride (**2**) to introduce two ATRP initiation sites at each of the chain ends enabled the ATRP of styrene, resulting in a 2G dendrimer-like star-branched block copolymer composed of PEO (1G) and PS (2G) segments. It was also possible to first polymerize styrene with a tetra-functional ATRP initiator, followed by chain-end modification to introduce twice the number of hydroxyl groups that served to grow PEO, affording a 2G dendrimer-like

Table 5.1 Representative dendrimer-like star-branched polymers synthesized by the "core-first" divergent iterative methodology

Polymer Segment	Generation	Number of Branch		Number of Total Segment	M_n or M_w (kg/mol)	M_w/M_n SEC
		Core	Layer			
PEO	2	3	2	9	28.0	1.30
	8	3	2,2,2,2,2,2,2	765	650	1.18
PS	3	8	2,2	56	141	1.27
	7	4	2,2,2,2,2,2	508	1920	1.04
PεCL	2	6	8	54	221	1.12
	3	6	2,2	42	158	1.14
Poly(*L*-lactide)	3	6	2,2	42	101	1.26
PMMA	4	3	2,2,2	45	456	1.23
	4	2	2,2,2	30	19.8	1.26
PEO-*b*-PS	2	3	2	9	70.0	—
	2	4	2	12	117	—
PEO-*b*-P^{t}BA	2	3	2	9	17.5	1.28
PS-*b*-PEO	2	3	2	9	63.7	1.20
PS-*b*-P^{t}BA	3	6	2,2	42	128	1.25
PεCL-*b*-PMMA	2	6	4	30	220	1.14
PεCL-*b*-(PMMA-*co*-PHEMA)	2	6	4	30	76.0	1.19

star-branched block copolymer placing PS segments in 1G and PEO segments outside 2G (Francis *et al.*, 2003). Moreover, a water-soluble 2G polymer composed of PEO and poly (acrylic acid) could be synthesized by the ROP of EO, followed by the ATRP of *tert*-butyl acrylate and subsequent deprotection of the poly(*tert*-butyl acrylate) segment (Hou *et al.*, 2003).

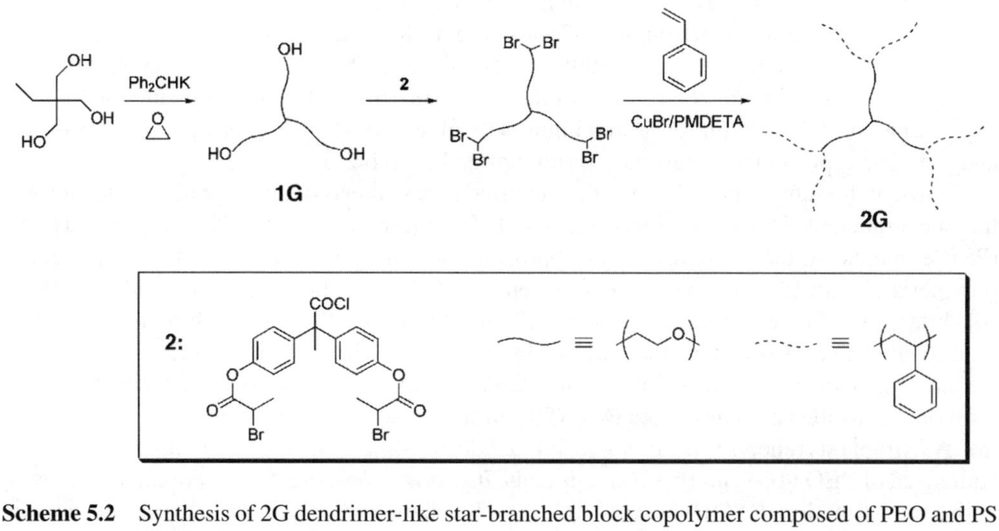

Scheme 5.2 Synthesis of 2G dendrimer-like star-branched block copolymer composed of PEO and PS segments reported by Gnanou *et al.*

The methodology using living ROP of EO and a chain-end modification reaction to increase the number of the hydroxyl reaction sites twofold was developed to allow the synthesis of a series of dendrimer-like star-branched (PEO)s up to the 8G stage (Feng *et al.*, 2005). The 8G polymer possessed three branches at the core and two branches at the junctions in all generations having an M_n value of 6.50×10^5 g/mol ($M_w/M_n = 1.18$) with 384 external hydroxyl termini. Thus, the developed methodology works quite well. Recently, an improved methodology for fast access to dendrimer-like star-branched (PEO)s was reported using nonprotected glycidol as the branching agent (Feng *et al.*, 2009).

Percec *et al.* reported the divergent synthesis of dendrimer-like star-branched poly(methyl methacrylate)s (PMMA)s using a new agent, **3**, named TERMINI, designed to serve as a terminator as well as a latent difunctional initiator (Percec *et al.*, 2003; Percec *et al.*, 2004; Percec *et al.*, 2005). As illustrated in Scheme 5.3, the TERMINI worked in such a way that the aryl *tert*-butyldimethylsilyl enol ether moiety reacted with the growing chain end and the *N,N′*-diethylthiocarbamate function was readily converted to aryl sulfonyl chloride capable of initiating the Cu-mediated living/controlled radical polymerization of MMA. A series of dendrimer-like star-branched (PMMA)s up to the 4G stage was successfully synthesized

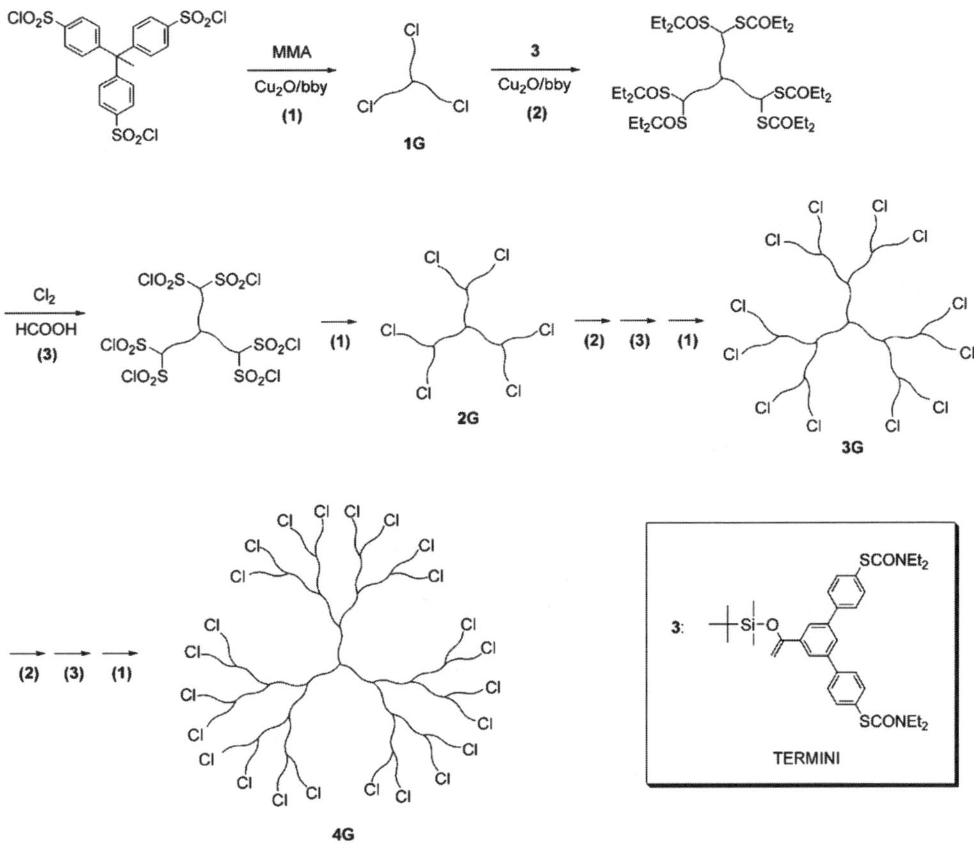

Scheme 5.3 Synthesis of 4G dendrimer-like star-branched PMMA reported by Percec *et al.*

by this methodology. Typically, the resulting 4G polymer possessed a predictable M_n value of 4.56×10^5 g/mol ($M_w/M_n = 1.23$).

Recently, Gnanou *et al.* reported that 1,1-bis(4-bromophenyl)ethylene (**4**) can be used in a way similar to that for the above-mentioned TERMINI in the synthesis of dendrimer-like star-branched (PS)s and poly(1,3-butadiene)s (PB)s (Matmour and Gnanou, 2008). The synthetic outline is illustrated in Scheme 5.4. In each reaction step, **4** reacted with the terminal potassium alkoxide in a 1:1 addition manner to introduce two phenyl bromide moieties at the chain end, which was subsequently transformed into two phenyl lithiums by the Li–Br exchange reaction with *sec*-BuLi. Following the living anionic polymerization of styrene or 1,3-butadiene,

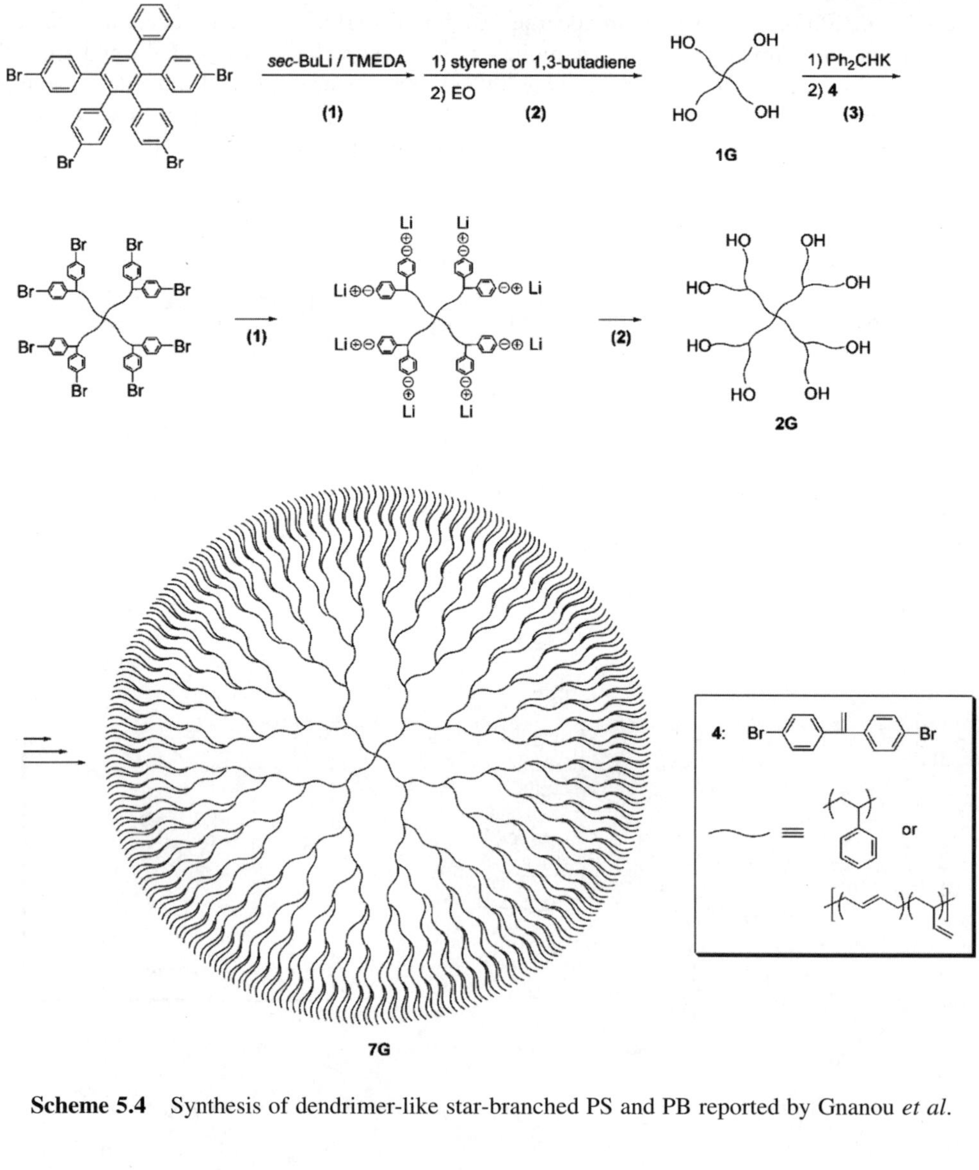

Scheme 5.4 Synthesis of dendrimer-like star-branched PS and PB reported by Gnanou *et al.*

initiated with each of the phenyllithium termini, well-controlled PS (or PB) chains were introduced to prepare the generation-based layer. The transformation reaction, followed by living polymerization, were then repeated several times, yielding the 7G and 3G dendrimer-like star-branched (PS)s and (PB)s, respectively. Thus, the agent, **4**, plays a satisfactory role as the TERMINI. In this methodology, a perfect 1:1 stoichiometry between the terminal phenyl bromide moiety and highly reactive *sec*-BuLi is always required in each of all transformation reaction steps, and not even a small deviation is permitted. Otherwise, significant amounts of undesirable polymers and structural imperfections are formed. Since perfect stoichiometry is experimentally difficult in practical operation, this is not a convenient synthetic procedure, even though the synthetic potential seems attractive.

Overall, the stepwise iterative methodology based on the "core-first" divergent approach so far developed can be summarized as follows. There is a repeatable synthetic sequence that involves the following two reaction steps: (1) the living/controlled polymerization of an appropriate monomer with a multifunctional core in the first iteration and with two or more initiation sites at the polymer chain ends at subsequent iterative processes and (2) a chain-end modification reaction to introduce two or more initiation sites. By repeating the synthetic sequence involving the two reaction steps, dendrimer-like star-branched polymers are progressively synthesized from the core in an outward direction. A variety of dendrimer-like star-branched polymers and their structural analogous amphiphilic block copolymers have been synthesized by this methodology. Representative examples are also listed in Table 5.1 showing generation, molecular weight, molecular-weight distribution, and the number of chain branches at the core and junctions. The molecular-weight distributions of the resulting polymers were not narrow, but were acceptable because living/controlled radical polymerization and ROP systems were employed in most cases. On the other hand, narrow molecular-weight distributions ($M_w/M_n < 1.1$) were achieved in a series of dendrimer-like star-branched (PS)s and (Bd)s by using the anionic living polymerization of styrene and 1,3-butadiene.

Although the defined structures of the resulting polymers so far synthesized have been generally suggested, there are some inherent problems in the methodology based on the "core-first" divergent approach. During the growth of dendrimer-like star-branched polymers, the number of initiation sites dramatically increases with generation. Any incomplete reactions in steps (1) and (2) lead to imperfections in structure that may significantly increase and accumulate generation by generation. In addition, there still remains a serious problem with the uniformity of all polymer chains propagated from the multifunctional initiation sites of all generations. Furthermore, as the number of initiation sites increases dramatically with each subsequent generation, it is not possible to accurately determine how many reaction sites actually work, especially at higher generations. For these reasons, the architecture of the resulting polymer synthesized by this methodology is not guaranteed even by agreement between the molecular weights, calculated (or predicted) and observed. Gnanou *et al.* have demonstrated that MALDI-TOF mass spectroscopy is a powerful technique for monitoring structural imperfection (Feng *et al.*, 2005). Detailed analysis of several low molecular weight PEO samples clearly shows that the hydroxyl termini can be quantitatively modified and the uniform growth of PEO was achieved, strongly indicating that the polymerization of EO in combination with the chain-end-modification procedure worked as expected. However, some problems still remain in the synthesis of high-generation and high molecular weight polymers.

Given the appropriate choice of living polymerization and chain-end modification in divergent synthesis, isolation of the polymer in each iterative process is not necessary and steric hindrance may possibly be reduced as the iteration proceeds. Therefore, the iterative methodology based on the "core-first" divergent approach is suited for large-scale preparation and high-generation polymer synthesis. However, it should be kept in mind that there are inherent problems, as mentioned above.

5.2.2 "Arm-First" Convergent Iterative Methodology

The synthesis of dendrimer-like star-branched polymers by the methodology based on the divergent approach was described in Section 5.2.1. While the developed methodology worked well, the inherent problems in the synthesis also became obvious. As mentioned before in the synthesis of dendrimers, a stepwise iterative methodology based on the convergent approach has been proposed as a complementary alternative. In this approach, the exterior subunit of a dendrimer is first prepared by repeating a linking reaction between dendritic fragments, so-called "dendrons" and then attaching it to a multifunctional core to form a dendrimer. Hadjichristidis *et al.* first proposed a new convergent approach in conjunction with a living anionic polymerization system for the synthesis of dendrimer-like star-branched polymers (Chalari and Hadjichristidis, 2002).

The synthetic outline is illustrated in Scheme 5.5. The process was initiated by an addition reaction of poly(isoprenyllithium) (PILi) to an ω-styryl poly(isoprene) (PI) macromonomer that was prepared by reacting 4-(chlorodimethylsilyl)styrene with PILi, to yield an in-chain-functionalized PI with a styrene-derived anion at the linking point. In both the preparation of the macromonomer and the addition reaction, an exact stoichiometry is required. Isoprene was then added to polymerize, affording an off-center living PI composed of three PI chains. Finally, the off-center living PI was linked with $MeSiCl_3$ to synthesize a dendrimer-like star-branched PI. The resulting polymer possesses three branches at the core and two branches at the junction in the 2G layer and is exactly the same in structure as the 2G dendrimer-like star-branched polymer synthesized by the above-mentioned divergent methodology. A 2G polymer with four branches at the core was also synthesized by linking the same off-center living PI with $SiCl_4$. It was also possible to synthesize two 2G dendrimer-like star-branched block copolymers composed of PS and PI segments by reacting the off-center living block copolymer composed of two PS segments and one PI segment with either $MeSiCl_3$ or $SiCl_4$. Thus, the off-center living polymer corresponds to the exterior subunit of each 2G dendrimer-like star-branched polymer.

The methodology was extended to the synthesis of 3G dendrimer-like star-branched PB (Orfanou *et al.*, 2006). As can be seen in Scheme 5.6, the synthetic procedure is identical to that employed for the 2G polymer except for the use of a multiple intermediate polymer anion called a "2G living dendron" prepared by performing a series of iterative reactions. In this synthesis, the necessity of preserving the stoichiometry of each reaction is much more critical than the synthesis of the 2G polymer. An off-center living PB (also called a "1G living dendron") was prepared in a manner similar to that mentioned above and subsequently reacted with the two Si-Cl reaction sites of 1-(4-dichloromethylsilylphenyl)-1-phenylethylene (**5**) to link the two 3-arm star (PB)s and to introduce the DPE moiety between both stars. The synthesis continued with the addition of a stoichiometric amount of *sec*-BuLi, followed by the living anionic polymerization of 1,3-butadeine to afford a 2G living dendron. This living dendron reacted *in situ* with $MeSiCl_3$, resulting in the 3G dendrimer-like star-branched PB.

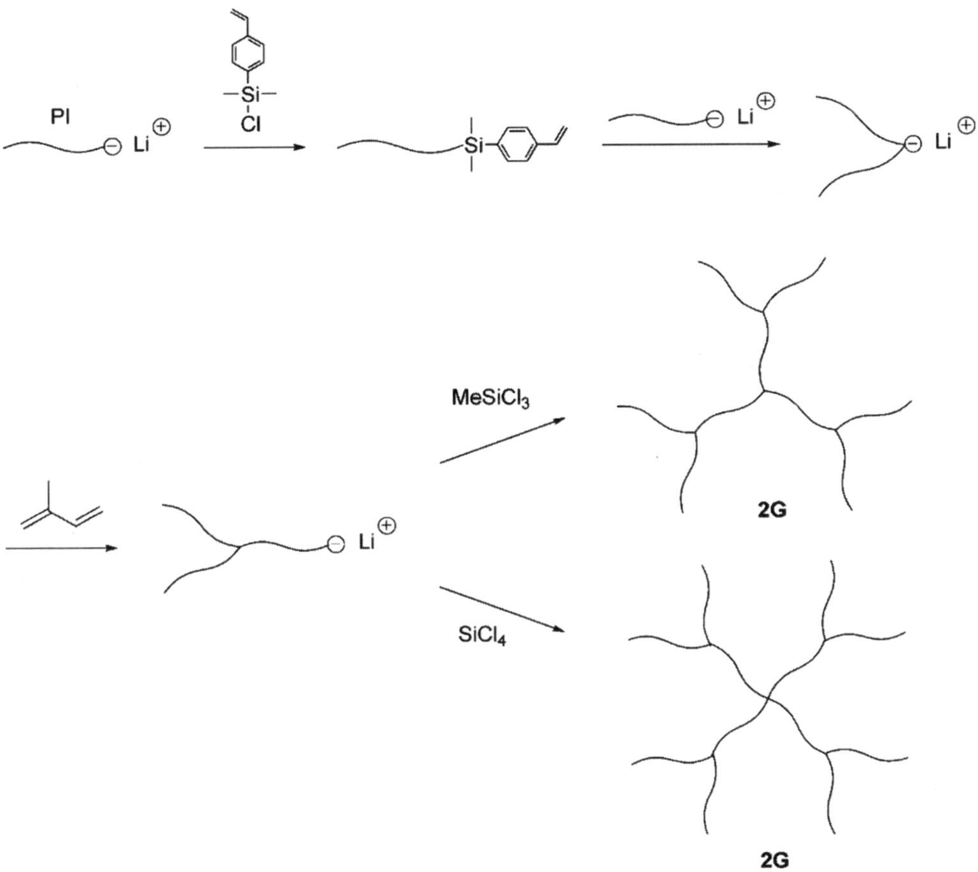

Scheme 5.5 Synthesis of 2G dendrimer-like star-branched (PI)s reported by Hadjichristidis *et al.*

The results of the 2G and 3G dendrimer-like star-branched polymers and block copolymers, after fractional precipitation, are summarized in Table 5.2. The resulting polymer all possessed the observed M_n values in good agreement with those calculated, and narrow molecular-weight distributions ($M_w/M_n \sim 1.1$). Since the off-center living polymers and the 2G living dendrons used as subunits are sampled during the synthesis and well characterized prior to the synthesis and then reacted with a multifunctional core to synthesize the 2G and 3G polymers, this procedure corresponds to an example of an "arm-first" process.

Hutchings *et al.* also developed a similar convergent synthesis for 2G and 3G dendrimer-like star-branched (PS)s (Hutchings and Roberts-Bleming, 2006). In this synthesis, an asymmetric α,ω-XY_2 PS macromonomer (X and Y: different functionalities) is used as the key building block leading to the 1G and 2G dendrons. As illustrated in Scheme 5.7, the living anionic polymerization of styrene, initiated with 3-*tert*-butyldimethylsilyloxy-1-propyllithium (SiO-PLi) and end capped with 1,1-bis(4-*tert*-butyldimethylsilyloxyphenyl)ethylene is followed by deprotection to afford the α,ω-XY_2 PS macromonomer, which is functionalized with two phenol groups and one primary alcohol group. In a separate experiment, a chain-end propyl

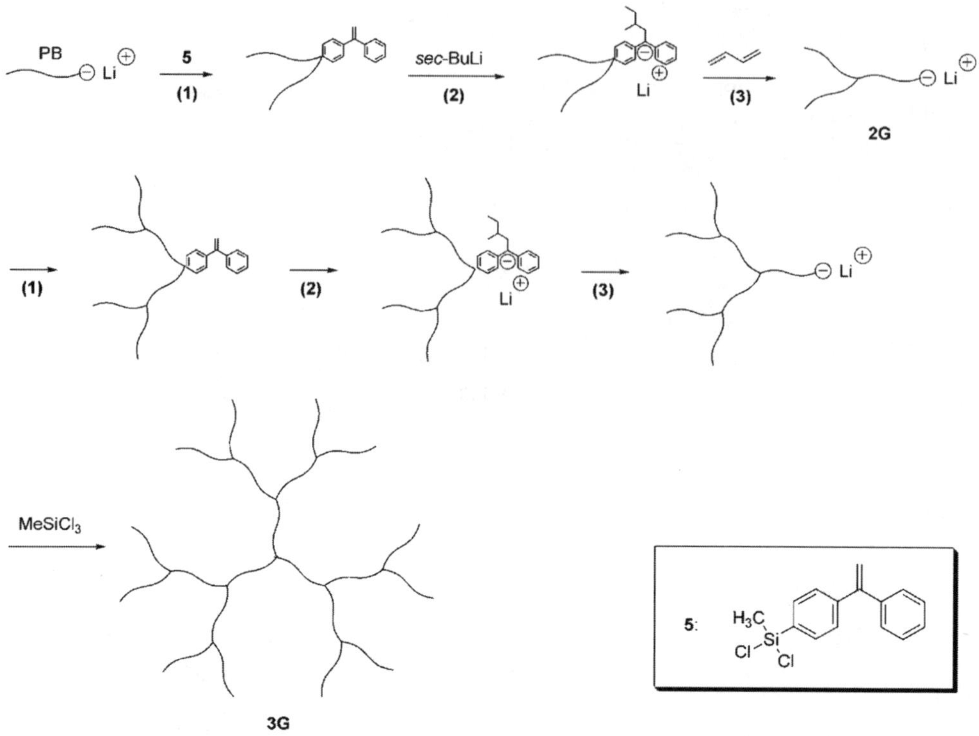

Scheme 5.6 Synthesis of 3G dendrimer-like star-branched PB reported by Hadjichristidis *et al.*

chloride-functionalized PS is prepared by the living anionic polymerization of styrene initiated with the same SiOPLi and deprotection, followed by chlorination with $SOCl_2$. Under basic conditions, the two phenol moieties of the macromonomer could be coupled efficiently with the terminal propyl chloride of the chain-end propyl chloride-functionalized PS. Following chlorination with $SOCl_2$, the alcohol moiety could be transformed to the corresponding propyl chloride, yielding a 1G dendron with a terminal propyl chloride reaction site. The synthesis

Table 5.2 Representative dendrimer-like star-branched polymers synthesized by the "arm-first" convergent iterative methodology

Polymer Segment	Generation	Number of Branch		Number of Total Segment	M_n or M_w (kg/mol)	M_w/M_n SEC
		Core	Layer			
PI	2	4	2	12	245	1.07
PB	3	3	2,2	21	380	1.09
PS	3	3	2,2	21	299	1.05
	2	5	4	25	288	1.15
PS-*b*-PI	2	3	2	9	167	1.07
PS-*b*-P^{t}BA	2	3	2	6	57.0	1.19

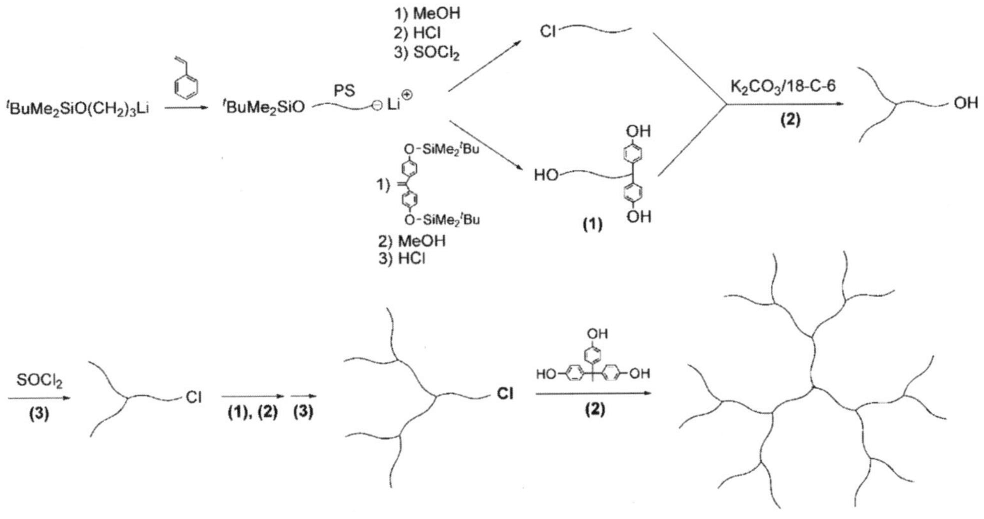

Scheme 5.7 Synthesis of 3G dendrimer-like star-branched PS reported by Hutchings *et al.*

utilized the Williamson reaction under basic conditions, where the propyl chloride selectively reacted with the phenol functionalities of the macromonomer, while the primary alcohol function remained intact. The Williamson reaction and subsequent chlorination was then repeated for the 2G dendron with the same propyl chloride reaction site.

In the final reaction step, the propyl chloride reaction site of the 1G or 2G dendron was linked with a multifunctional core, 1,1,1-tris(4-hydroxyphenyl)ethane, by the same Williamson reaction, resulting in 2G and 3G dendrimer-like star-branched (PS)s. The resulting polymers had branched structure, three branches at the cores and two branches at the junctions in each generation layer. After fractionation, both purified polymers showed that the molecular weights are in excellent agreement with predicted values, and narrow molecular-weight distributions were attained ($M_w/M_n \leq 1.05$), as listed in Table 5.2.

In the two stepwise iterative methodologies based on the "arm-first" convergent approach herein developed, the dendrons of known molecular weights are prepared prior to the synthesis and reacted with appropriate multifunctional core(s) to synthesize dendrimer-like star-branched polymers. Accordingly, this approach offers a distinct advantage in that the resulting polymer structure is readily estimated by comparing the molecular weight of the final product with that calculated from the molecular weight of the dendron and agreement between both values gives a complete guarantee for the requisite structure. Indeed, Table 5.2 shows good agreement between the calculated and determined values of M_n for all 2G and 3G polymers, strongly demonstrating the requisite structures. Furthermore, the narrow molecular-weight distributions in addition to the above agreement also indicate a high degree of molecular and structural homogeneity. There is an additional advantage in that structural imperfections can be minimized by performing a series of iterative reactions between dendrons having a single reaction site, in which the reaction site is always kept single with increasing iteration. The unreacted dendron and imperfectly branched materials can easily be detected by SEC analysis

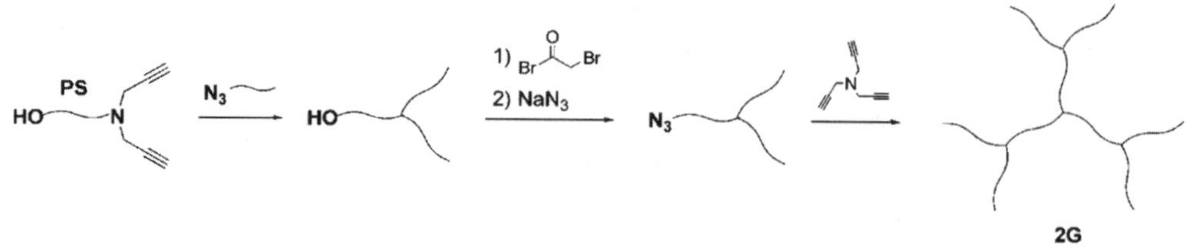

Scheme 5.8 Synthesis of 2G dendrimer-like star-branched PS reported by Monteiro *et al.*

of the reaction mixture because of the molecular-weight difference in the target product and other undesirable polymers.

The linking reaction in sterically bulky and high molecular weight dendrons is essential and unavoidable in the final step of the convergent process. It becomes increasingly difficult with each generation increment because the molecular weight of the dendron essentially multiples. Accordingly, the convergent approach is not suited to synthesizing high-generation and high molecular weight dendrimer-like star-branched polymers, and growth seems to be not practical beyond a generation higher than 3G in the present systems. Furthermore, fractionation as an additional step is usually needed to isolate the final product, since the dendron is used in excess in the reaction in order to go to completion.

Monteiro *et al.* recently reported the convergent synthesis of 2G dendrimer-like star-branched polymers by combining "click chemistry" with ATRP, as shown in Scheme 5.8 (Urbani *et al.*, 2008). This success indicates the possibility of click chemistry for the synthesis. However, as the molecular weights of the resulting polymers were not high (~50 000 g/mol, $M_w/M_n < 1.2$) and the final click reactions proceeded incompletely, more optimization is needed for general use.

Knauss and Huang demonstrated the successful synthesis of radially linked star-*block*-linear (PS)s with star-shaped branching, structurally analogous to 2G dendrimer-like star-branched (PS)s, by the convergent living anionic polymerization method (Knauss and Huang, 2003). As illustrated in Scheme 5.9, these polymers are synthesized by sequential addition of a dually functionalized 4-(chlorodimethylsilyl)styrene (**6**), styrene, and **6** again to a solution of living PS. The methodology involves four reaction steps. In the first step, styrene is polymerized with *sec*-BuLi to give living PS. In the second step, living PS is linked by reaction with

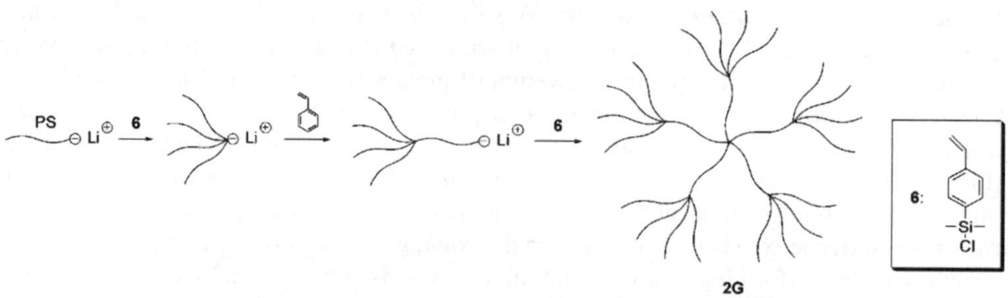

Scheme 5.9 Synthesis of 2G dendrimer-like star-branched PS reported by Knauss *et al.*

slowly added **6** to form a star PS. The addition was stopped prior to a stoichiometric amount of **6** relative to the living PS to maintain a living chain end at the focal point. In the third step, the living star PS is used as a macroinitiator for styrene to grow a PS chain. The final step involves the slow addition of **6** to link the star-*block*-linear diblock living chain. A stoichiometric amount of **6** relative to the living chain is added in this step to link the chains and terminate the polymer with the formation of a vinyl function at the core. A typical sample was 2.88×10^5 g/mol in M_n value and 1.15 in M_w/M_n value. Although the structure of the resulting polymer is not perfectly controlled in terms of precise synthesis and, while a further synthesis seems to be difficult, the advantage of this methodology is a facile and straightforward synthesis of a 2G-like dendrimer-like star-branched polymer. In addition, it is advantageous that polymers can be produced sequentially by a one-pot process.

5.2.3 "Arm-First" Divergent Iterative Methodology

As mentioned above, some difficulties arose in the synthesis of dendrimer-like star-branched polymers by the two aforementioned methodologies based on the "core-first" divergent and "arm-first" convergent approaches described in Sections 5.2.1 and 5.2.2. In fact, most of the polymers are limited to 4G stages and a few 10^5 g/mol orders in molecular weight except for two cases reported by Gnanou *et al.* (Tables 5.1 and 5.2). Moreover, the structures of the resulting polymers could not be well characterized and structural imperfections were indicated in several cases.

Recently, a new stepwise iterative methodology based on an "arm-first" divergent approach having the advantages of both of the above two methodologies was developed by Hirao *et al.*, in order to synthesize high-generation and high molecular weight dendrimer-like star-branched polymers with well-characterized structures (Hirao *et al.*, 2005b; Hirao *et al.*, 2006; Hirao *et al.*, 2008; Hirao *et al.*, 2009a; Matsuo, Watanabe and Hirao, 2004). In this methodology, a termination reaction using premade living anionic polymers is employed to introduce polymer segments like the case of the "arm-first" convergent approach. Accordingly, the structure of the resulting polymer can be readily examined by comparing the molecular weight of the final polymer with that calculated from the molecular weight of a premade living polymer, and it is completely guaranteed to be the requisite structure by agreement between both molecular weights. Furthermore, steric hindrance may possibly be reduced as the iteration proceeds, because the synthesis initiates growth at the core and continues outward by performing the iterative synthetic sequence in a divergent manner.

The synthetic outline of this methodology (Matsuo *et al.*, 2004) as given in Scheme 5.10. Similar to the two methodologies introduced above, two reaction steps are used in the iterative synthetic sequence: (1) a linking reaction of an α-chain-end two 3-*tert*-butyldimethylsilyloxymethylphenyl $(SiOMP)_2$-functionalized living anionic polymer with either a multifunctional core substituted with benzyl bromide (BnBr) moieties (only for the first iteration) or a chain-end-$(BnBr)_2$-functionalized polymer connected to the core and (2) a transformation reaction of the introduced SiOMP group to the BnBr function. By repeating the synthetic sequence involving the two reaction steps (1) and (2), a series of dendrimer-like star-branched polymers are synthesized. Thus, in addition to the above-mentioned advantages, both heterogeneous initiation and propagation reactions with multiple initiation sites in the methodology based on the "core-first" divergent approach (Section 5.2.1) and the linking reaction among high

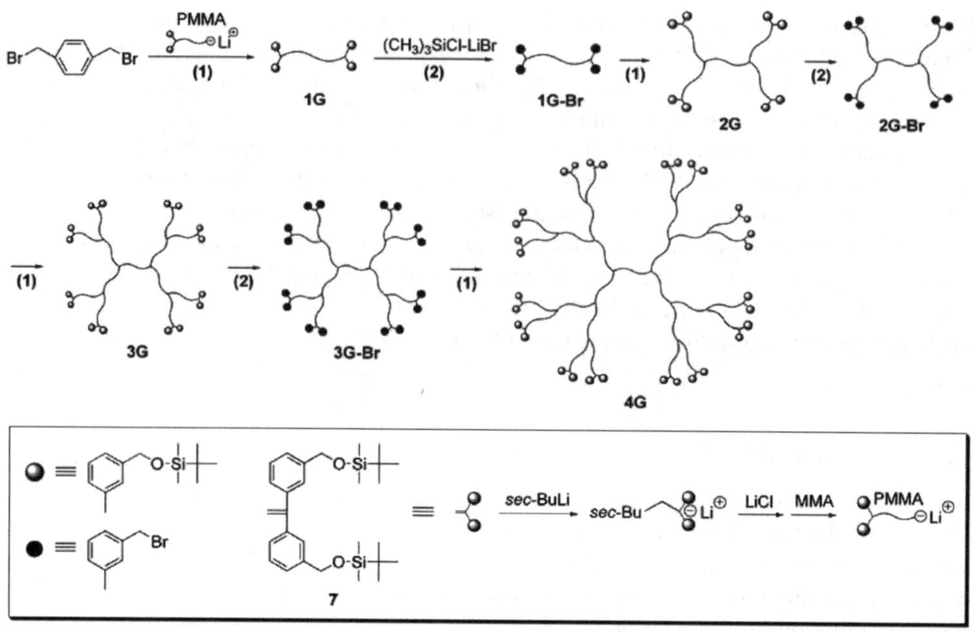

Scheme 5.10 Synthesis of 4G dendrimer-like star-branched PMMA with two branches at core reported by Hirao *et al.*

molecular weight dendrons in the methodology based on the "arm-first" convergent approach (Section 5.2.2) can be completely avoided.

As shown in Scheme 5.10, preparation of an α-chain-end $(SiOMP)_2$-functionalized living anionic PMMA used as the key building block in this synthesis involves the anionic polymerization of MMA with a specially designed functional initiator from *sec*-BuLi and 1,1-bis(3-*tert*-butyldimethylsilyloxymethylphenyl)ethylene (**7**) in THF at −78 °C. A threefold excess of LiCl was added prior to the polymerization to narrow the molecular-weight distribution. The molecular weight of the living PMMA was adjusted to be around 10 000 g/mol throughout the synthesis. The α-chain-end $(SiOMP)_2$-functionalized living PMMA thus prepared was coupled efficiently and quantitatively with α,α′-dibromoxylene in THF at −40 °C for 24 h, affording an α-$(SiOMP)_2$-ω-$(SiOMP)_2$-functionalized PMMA as the 1G polymer in this synthesis. Following treatment of the resulting polymer with a 1:1 mixture of $(CH_3)_3SiCl$ and LiBr, the four SiOMP groups could be transformed to four BnBr functions. Conversion of silyl ether to the BnBr functionality enables further coupling to the 2G polymer.

In the second iteration, preparation of the α-chain-end $(SiOMP)_2$-functionalized living PMMA in a manner similar to that as above is followed by linking with a 1G PMMA bearing four BnBr termini under the same conditions. The linking reaction proceeded nearly quantitatively to form a chain-end $(SiOMP)_2$-functionalized H-shaped PMMA as the 2G polymer. After bromination of the 2G polymer with $(CH_3)_3SiCl$/LiBr, the linking of the brominated polymer with α-chain-end $(SiOMP)_2$-functionalized living PMMA resulted in a 3G polymer, which is the first polymer possessing a dendrimer-like star-branched architecture in this synthesis. A 4G polymer could also be synthesized by one more repetition of the synthetic sequence involving

Table 5.3 Synthesis of dendrimer-like star-branched (PMMA)s by the "arm-first" divergent iterative methodology

		M_n (kg/mol)			M_w (kg/mol)		M_w/M_n
Generation	Type	Calcd	SEC	^{1}H NMR	Calcd	SLS	
1G	(A_2)	12.0	12.5	12.6	—	—	1.01
2G	$(A\text{-}A_2)_2$	55.6	48.6	56.1	56.7	57.9	1.02
3G	$(A\text{-}A_2\text{-}A_4)_2$	143	105	141	147	146	1.03
4G	$(A\text{-}A_2\text{-}A_4\text{-}A_8)_2$	321	195	320	331	330	1.03
1G	$(A_4)^a$	14.6	13.8	14.2	14.9	14.8	1.02
2G	$(A\text{-}A_2)_4{}^a$	43.0	36.9	42.9	43.9	44.2	1.02
3G	$(A\text{-}A_2\text{-}A_4)_4{}^a$	97.3	66.1	98.4	99.2	105	1.02
4G	$(A\text{-}A_2\text{-}A_4\text{-}A_8)_4{}^a$	219	115	219	223	230	1.02
5G	$(A\text{-}A_2\text{-}A_4\text{-}A_8\text{-}A_{16})_4{}^a$	452	178	449	462	472	1.03
6G	$(A\text{-}A_2\text{-}A_4\text{-}A_8\text{-}A_{16}\text{-}A_{32})_4{}^a$	980	282	974	1000	1060	1.02
7G	$(A\text{-}A_2\text{-}A_4\text{-}A_8\text{-}A_{16}\text{-}A_{32}\text{-}A_{64})_4{}^a$	1960	414	1940	1980	1970	1.02
1G	$(A_4)^b$	44.4	36.2	44.9	45.3	45.3	1.02
2G	$(A\text{-}A_2)_4{}^b$	131	93.1	131	134	134	1.02
3G	$(A\text{-}A_2\text{-}A_4)_4{}^b$	298	171	300	304	305	1.02
4G	$(A\text{-}A_2\text{-}A_4\text{-}A_8)_4{}^b$	639	301	649	658	661	1.03
5G	$(A\text{-}A_2\text{-}A_4\text{-}A_8\text{-}A_{16})_4{}^b$	1390	512	1430	1430	1500	1.03

[a] Molecular weight of arm polymer, $M_n \sim 4000$ g/mol.
[b] Molecular weight of arm polymer, $M_n \sim 10\,000$ g/mol.

the bromination and subsequent linking reactions. All of the reactions were followed by ^{1}H NMR, SEC, and SLS and confirmed to be virtually quantitative. In each linking reaction step, α-chain-end $(SiOMP)_2$-functionalized living PMMA was always used in excess to complete the reaction and, therefore, fractionation was needed to isolate the target polymers.

SEC profiles of the isolated polymers all exhibited narrow monomodal distributions, the M_w/M_n values being less than 1.03. The results are summarized in Table 5.3. The good agreement between the calculated molecular weights and those determined by ^{1}H NMR and SLS is observed in all polymers, clearly demonstrating that the 3G and 4G polymers are the target dendrimer-like star-branched (PMMA)s with the requisite structures. These polymers possess two branches at the cores as well as at the junctions in each generation, exactly as illustrated in Scheme 5.10. The 4G polymer ($M_n = 3.20 \times 10^5$ g/mol and $M_w/M_n = 1.03$) was made up of 30 PMMA segments (2(1G) + 4(2G) + 8(3G) + 16(4G)) and possessed 28 highly reactive BnBr termini.

Based on such successful results, the same methodology was applied to the synthesis of high-generation high molecular weight dendrimer-like star-branched (PMMA)s up to the 7G stage, as illustrated in Scheme 5.11 (Hirao *et al.*, 2005a). In this synthesis, α-chain-end $(SiOMP)_2$-functionalized living PMMA was adjusted to be about 4000 g/mol in M_n value. Since a core compound substituted with four BnBr functions was used, the expected dendrimer-like star-branched polymers should have four branches emanating from the core and two branches at the junctions in all generations. All of the reactions were monitored by ^{1}H NMR, SEC, and SLS and observed to be nearly quantitative. Accordingly, there were no steric effects even for the largest 7G polymer, although the reaction conditions were not the same throughout the

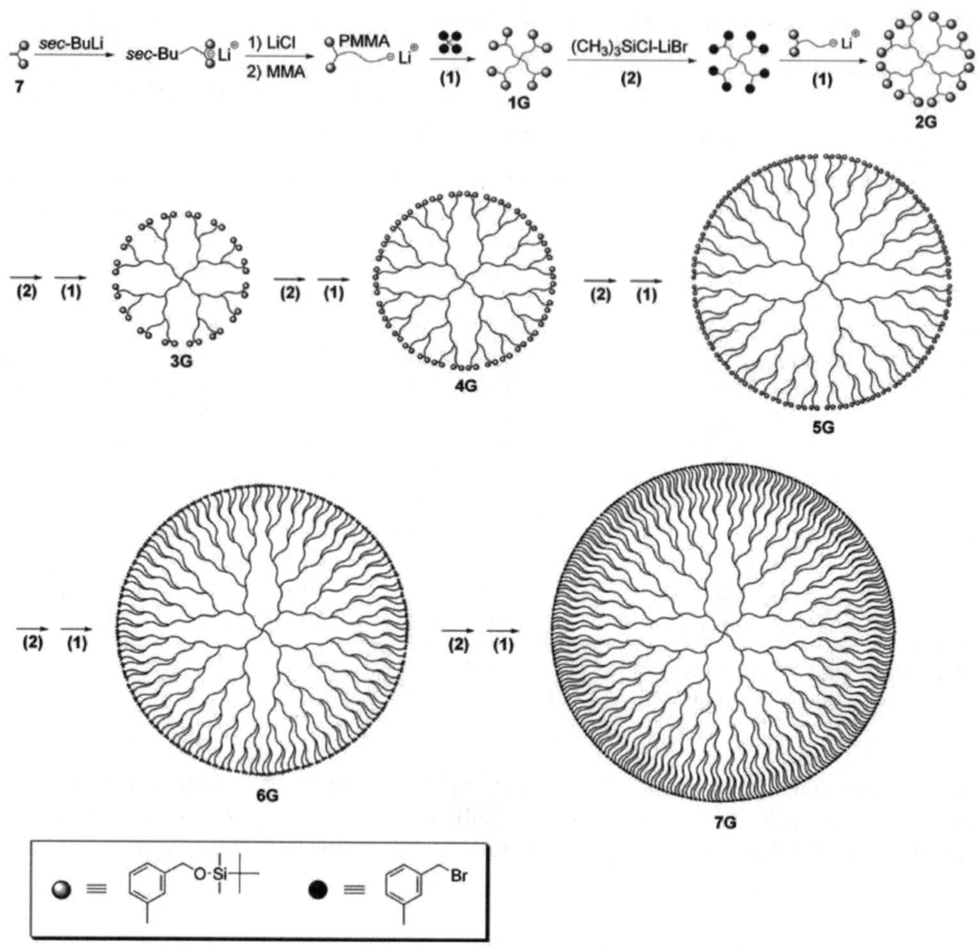

Scheme 5.11 Synthesis of dendrimer-like star-branched (PMMA)s up to the 7G stage reported by Hirao *et al.*

synthesis, but optimized as the iteration proceeds. For instance, a 5-fold excess of α-chain-end $(SiOMP)_2$-functionalized living PMMA and longer reaction times up to 72 h were used in the seventh iteration.

As seen in Table 5.3, the observed molecular weights of the polymers obtained at all stages were in excellent agreement with those calculated. Their SEC profiles shown in Figure 5.2 exhibited narrow monomodal distributions ($M_w/M_n \leq 1.03$). These results clearly demonstrated the successful synthesis of a series of well-defined dendrimer-like star-branched (PMMA)s up to the 7G stage. The final 7G polymer was a huge macromolecule having a well-controlled M_n value of 1.94×10^6 g/mol and a narrowly distributed M_w/M_n value of 1.02. It consisted of 508 PMMA segments and possessed 512 BnBr termini capable of continuing the iteration to form the next generation.

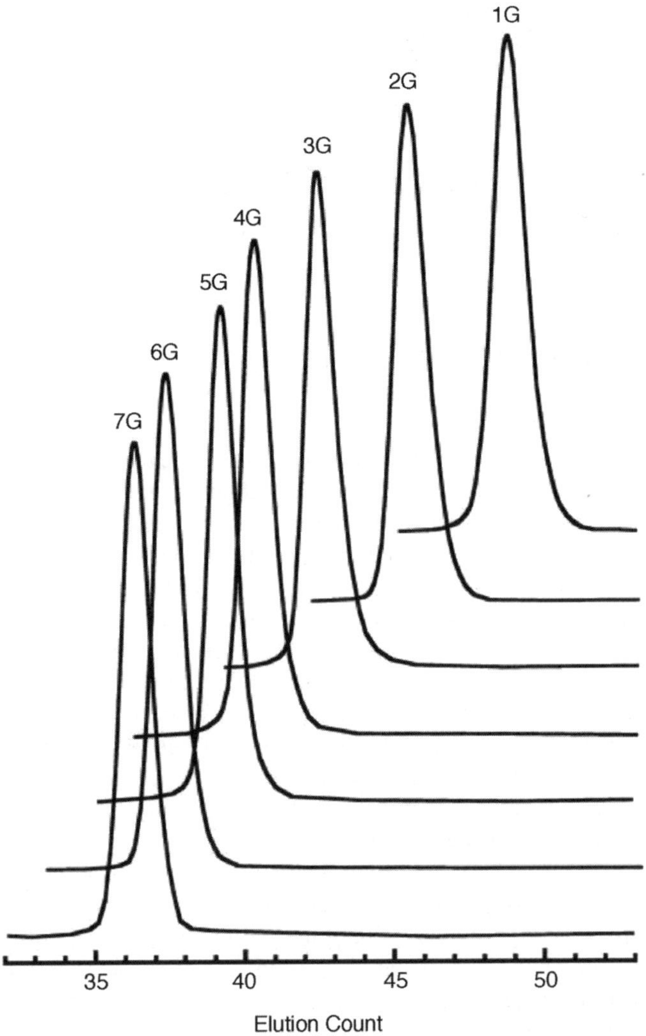

Figure 5.2 SEC profiles of dendrimer-like star-branched (PMMA)s from 1G to 7G.

A different series of dendrimer-like star-branched (PMMA)s up to the 5G stage were also successfully synthesized by the same methodology using a higher molecular weight α-chain-end $(SiOMP)_2$-functionalized living PMMA (M_n = about 10 000 g/mol) as a building block at each iteration (Hirao, unpublished results). In this series, a molecular weight of the polymer was reached 1.43×10^6 g/mol even in the 5G polymer (Table 5.3). Thus, the developed methodology works very satisfactorily for the synthesis of high-generation high molecular weight dendrimer-like star-branched (PMMA)s with well-defined structures.

The versatility of the present iterative methodology based on the "arm-first" divergent approach can be witnessed by the number of structural variations in the resulting polymers. For instance, a series of highly branched dendrimer-like star-branched (PMMA)s could be

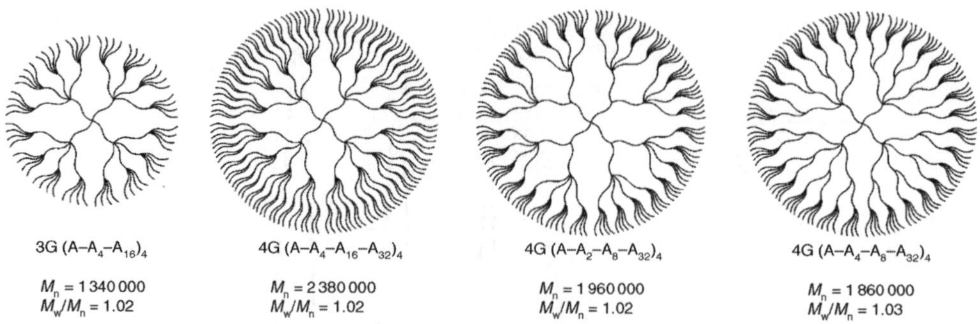

Figure 5.3 3G and 4G dendrimer-like star-branched (PMMA)s.

synthesized by developing the same methodology with the use of a living PMMA α-functionalized with four SiOMP groups (Watanabe *et al.*, 2006). The resulting polymers possess four polymer segments branched at the cores as well as at all junctions and a loading capacity of 256 BnBr termini, which are exactly equal to the number of BnBr termini of the 6G polymer mentioned above. Furthermore, with the appropriate choice of α-chain-end-functionalized living PMMA having either two or four SiOMP groups in each iterative process, the structural variations lead to significant changes in the branched architectures of the 3G and 4G dendrimer-like star-branched (PMMA)s, as shown in Figure 5.3 (Hirao *et al.*, 2009b).

Similar to living PMMA, another living anionic polymer should react with a brominated polymer to afford a next-generation dendrimer-like star-block copolymer with an additional functionality. For example, a 3G dendrimer-like star-branched block copolymer was obtained by the reaction of a 2G brominated polymer with a living anionic poly(2-hydroxyethyl methacrylate) whose hydroxyl group was masked with *tert*-butyldimethylsilyl group (Matsuo *et al.*, 2004). After removal of the silyl group, a 3G block copolymer bearing one thousand hydroxyl groups at the periphery could be obtained. It was also possible to synthesize 4G block copolymers with carboxylic acid, 1,2-diol, *tert*-amine, and ammonium ion functions by reacting a 3G brominated polymer with either a living anionic polymer of *tert*-butyl methacrylate, (2,2-dimethyl-1,3-dioxolan-4-yl)methyl methacrylate, or 2-vinylpyridine, followed by deprotection or quaternarization (Scheme 5.12) (Hirao *et al.*, 2009b). Thus, the possible introduction of functional groups has proven to be of particular interest in broadening the synthetic range of functional dendrimer-like star-branched block copolymers with many potential applications. The introduced different segments cause phase separation at the molecular level, followed by self-organization, and the formation of quite new nano-ordered suprastructures and molecular assemblies is expected due to their characteristic branched architectures. Moreover, highly reactive BnBr termini present in the periphery parts could readily be transformed into a variety of useful functional groups, as shown in Scheme 5.13 (Hirao *et al.*, 2006).

Very recently, a series of well-defined dendrimer-like star-branched (PS)s was synthesized by a new modified methodology (Yoo *et al.*, 2009). As illustrated in Scheme 5.14, the following three reaction steps are employed in the iterative synthetic sequence: (1) a linking reaction of α-chain-end 3-(*tert*-butyldimethylsilyloxy)propyl (SiOP)-functionalized PSLi end capped with 1,1-diphenylethylene (DPE) with either a core substituted with four BnBr functions or a chain-end $(BnBr)_2$-functionalized PS connected to the core, (2) a transformation reaction

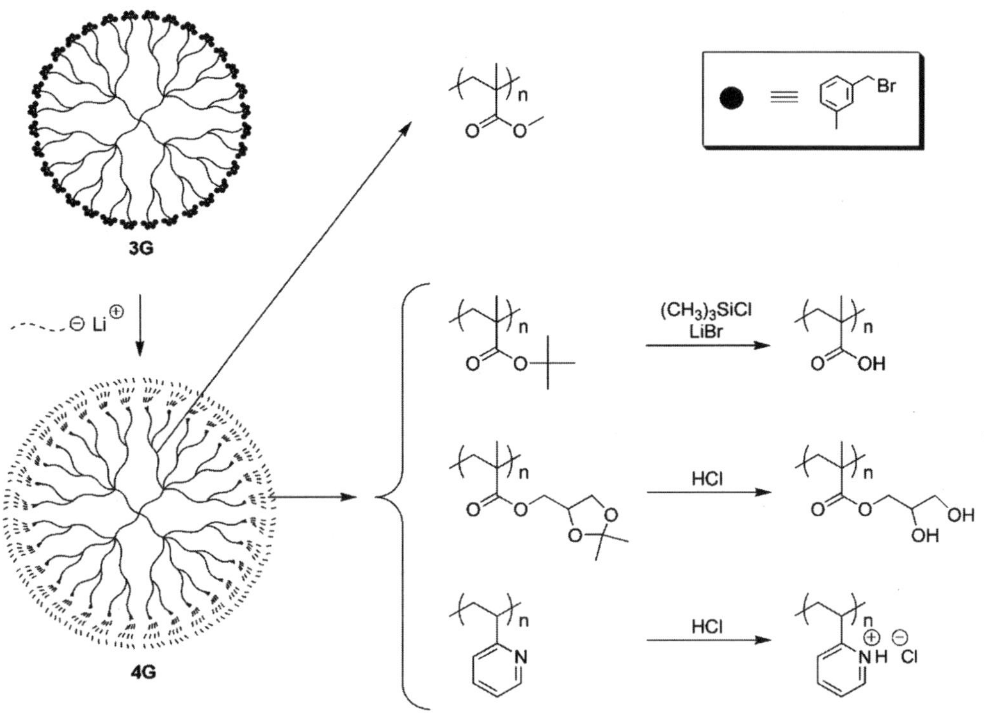

Scheme 5.12 Synthesis of 4G dendrimer-like star-branched block copolymers having functionalized segment at the fourth generation.

of the SiOP group to the 3-bromopropyl terminus, and (3) the introduction of two BnBr functions via the 3-bromopropyl group. The synthetic sequence involving the three reaction steps was repeated five times to synthesize dendrimer-like star-branched (PS)s up to the 5G stage. The final 5G polymer was 1.80×10^6 g/mol in M_n value ($M_w/M_n = 1.05$) consisting of 124 PS segments (M_n = about 15 000 g/mol) with 128 BnBr termini (Table 5.4). Two 3G

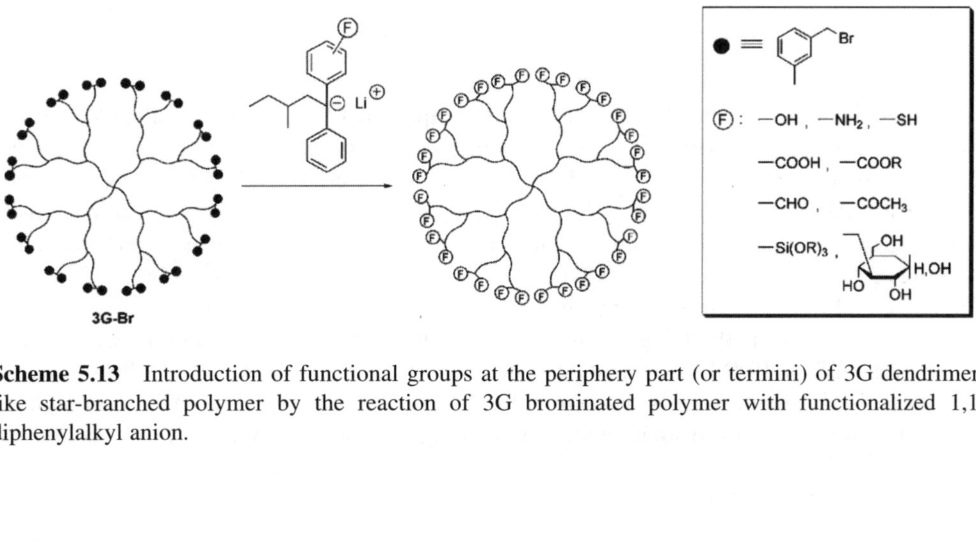

Scheme 5.13 Introduction of functional groups at the periphery part (or termini) of 3G dendrimer-like star-branched polymer by the reaction of 3G brominated polymer with functionalized 1,1-diphenylalkyl anion.

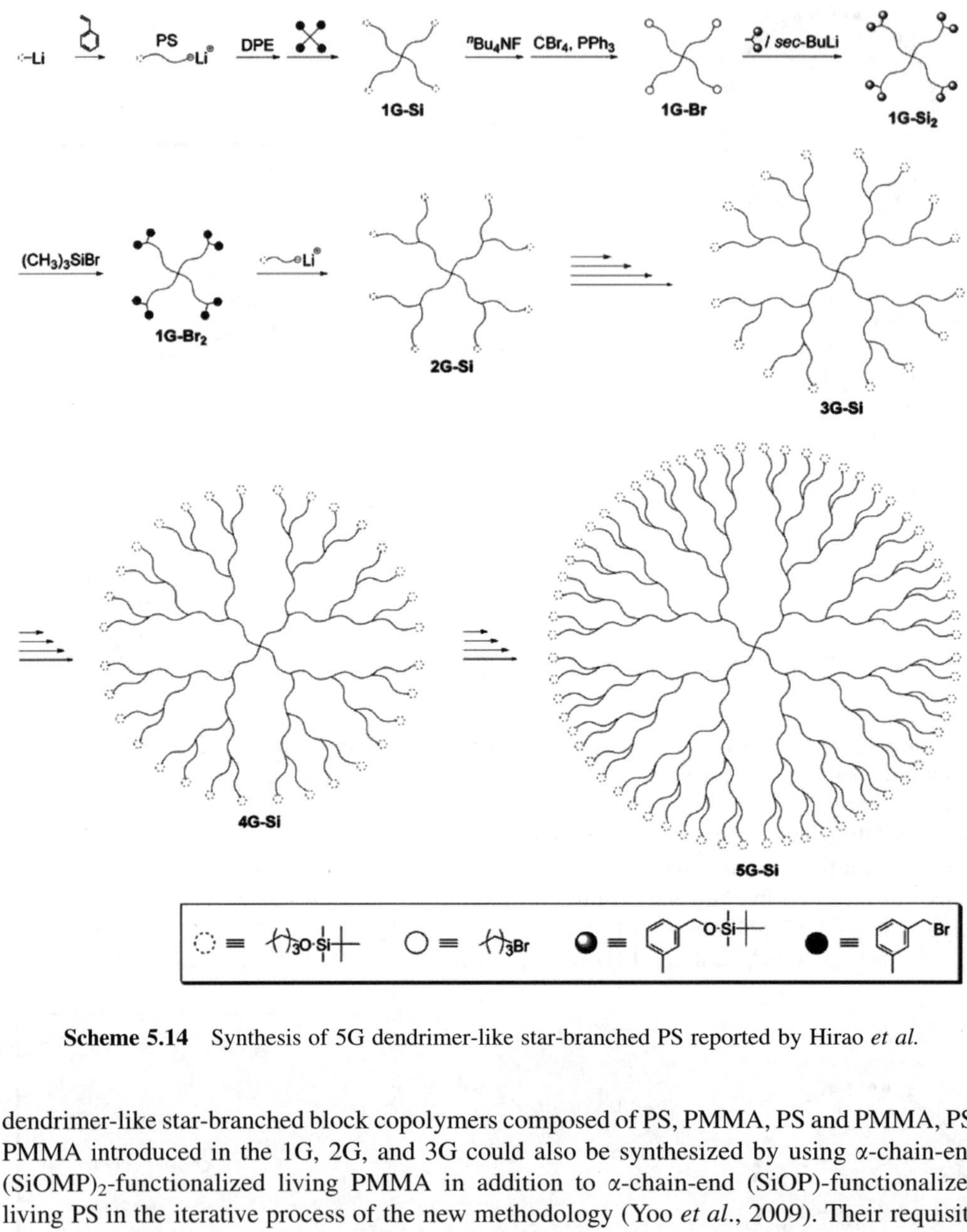

Scheme 5.14 Synthesis of 5G dendrimer-like star-branched PS reported by Hirao *et al.*

dendrimer-like star-branched block copolymers composed of PS, PMMA, PS and PMMA, PS, PMMA introduced in the 1G, 2G, and 3G could also be synthesized by using α-chain-end $(SiOMP)_2$-functionalized living PMMA in addition to α-chain-end (SiOP)-functionalized living PS in the iterative process of the new methodology (Yoo *et al.*, 2009). Their requisite structures and low degrees of structural and compositional homogeneity were clearly indicated by good agreement between both the observed and calculated molecular weights of the final polymers, and by the narrow molecular-weight distributions (Table 5.4).

Thus, the present methodology provides a general and versatile synthetic procedure with which either PS or PMMA segments can be introduced at any generation. The successful synthesis of such block copolymers strongly indicates the possible use of living polymers with different reactivities, from highly reactive living polymers of styrene, 1,3-butadeine, and

Table 5.4 Synthesis of dendrimer-like star-branched (PS)s and block copolymers composed of PS and PMMA segments by the "arm-first" divergent iterative methodology

		M_n (kg/mol)			M_w (kg/mol)		M_w/M_n
Generation	Type	Calcd	SEC	^{1}H NMR	Calcd	SLS	
1G	(A_4)	39.7	34.7	39.5	40.5	39.4	1.02
2G	$(A\text{-}A_2)_4$	113	102	123	116	128	1.03
3G	$(A\text{-}A_2\text{-}A_4)_4$	305	209	300	320	311	1.05
4G	$(A\text{-}A_2\text{-}A_4\text{-}A_8)_4$	688	314	667	722	704	1.05
5G	$(A\text{-}A_2\text{-}A_4\text{-}A_8\text{-}A_{16})_4$	1730	542	1800	1820	1880	1.05
1G	PS_4	39.7	34.7	39.5	40.5	39.4	1.02
2G	P(S-*b*-$MMA_2)_4$	130	87.7	122	134	145	1.03
3G	P(S-*b*-MMA_2-*b*-$S_4)_4$	284	176	284	295	278	1.04
1G	$PMMA_4$	40.8	35.2	40.0	41.6	43.6	1.02
2G	P(MMA-*b*-$S_2)_4$	112	84.7	121	115	123	1.03
3G	P(MMA-*b*-S_2-*b*-$MMA_4)_4$	281	168	295	289	291	1.03

isoprene to less reactive living polymers of 2-vinylpyridine, alkyl (meth)acrylate monomers, and *N,N*-(dialkyl)acrylamides. The versatility of the methodology is also demonstrated by the synthesis of a number of dendrimer-like star-branched polymers with several structural variations. The success of the present methodology is undoubtedly attributed to the choice of the reaction steps that proceed cleanly and nearly quantitatively. It is surprising that the precise synthesis of huge macromolecules with M_n values of over several million has been achieved simply by repeating the synthetic sequence involving two or three reaction steps.

In the present methodology, there is a rapid increase in the number of reaction sites as the number of generations increases similar to the case of the standard divergent approach. Accordingly, if there are any incomplete reactions at the reaction sites leading to structural imperfections in the subsequent generation, such imperfections significantly increase with the generation. As listed in Tables 5.3 and 5.4, however, agreement between the observed and calculated molecular weights is excellent within analytical error in each high-generation high molecular weight polymer. This result strongly indicates that the polymers are precisely controlled in structure and structural imperfections are very small even if present.

5.3 Hydrodynamic Radii and Radii of Gyration

Hirao *et al.* reported the values of the hydrodynamic radius, R_h, of a series of dendrimer-like star-branched (PMMA)s from 1G to 5G determined by RALLS in THF (Table 5.5) (Hirao, unpublished results). Each polymer possesses four branches at the core and two branches at each junction and can be abbreviated as 1G (A_4), 2G $(A\text{-}A_2)_4$, 3G $(A\text{-}A_2\text{-}A_4)_4$, 4G $(A\text{-}A_2\text{-}A_4\text{-}A_8)_4$, and 5G $(A\text{-}A_2\text{-}A_4\text{-}A_8\text{-}A_{16})_4$, respectively. The PMMA chain between the junctions is adjusted to be around 10 000 g/mol in M_n value. It was observed that the R_h value increased by increasing the generation and therefore the molecular weight. A good linear relationship is obtained by plotting log R_h as a function of log M_w, as shown in Figure 5.4, demonstrating that the molecular size demonstrating that the molecular size estimated from the R_h value strongly correlates with the M_w value. A similar linear relationship was also observed in a series of dendrimer-like star-branched (PS)s from 1G to 7G with the same branched architectures,

Table 5.5 Hydrodynamic radii and radii of gyration of dendrimer-like star-branched (PMMA)s

Generation	Type	M_w (kg/mol)[a]	R_h (nm)[a]	R_g (nm)[b]	R_g/R_h
1G	(A_4)	47.0	4.72		
2G	$(A\text{-}A_2)_4$	136	7.84		
3G	$(A\text{-}A_2\text{-}A_4)_4$	295	10.2	11.9	1.16
4G	$(A\text{-}A_2\text{-}A_4\text{-}A_8)_4$	661	13.9	13.6	0.978
5G	$(A\text{-}A_2\text{-}A_4\text{-}A_8\text{-}A_{16})_4$	1500	19.4		
3G	$(A\text{-}A_2\text{-}A_4)_4$	295	10.2	11.9	1.16
	$(A\text{-}A_2\text{-}A_8)_4$	514	14.7	12.8	0.871
	$(A\text{-}A_4\text{-}A_8)_4$	568	14.9	12.5	0.839
	$(A\text{-}A_4\text{-}A_{16})_4$	939	17.2	17.3	1.01
4G	$(A\text{-}A_2\text{-}A_4\text{-}A_8)_4$	661	13.9	13.6	0.978
	$(A\text{-}A_2\text{-}A_8\text{-}A_{32})_4$	1960	18.1	18.8	1.04
	$(A\text{-}A_4\text{-}A_8\text{-}A_{32})_4$	1920	20.1	19.3	0.960
	$(A\text{-}A_4\text{-}A_{16}\text{-}A_{32})_4$	2360	24.6	20.4	0.829
	$(A\text{-}A_4\text{-}A_{16}\text{-}A_{46})_4$	3960	30.2		

[a] Measured by RALLS in THF at 30 °C.
[b] Measured by SAX in THF at 25 °C.

reported by Gnanou and coworkers (Matmour and Gnanou, 2008). The diameter of the 5G PMMA ($M_w = 1.50 \times 10^6$ g/mol) was 38.8 (19.4 × 2) nm, comparable to the 31 nm of the 7G dendrimer-like star-branched PS ($M_n = 1.92 \times 10^6$ g/mol).

Similarly, the R_h values of the 3G and 4G dendrimer-like star-branched (PMMA)s with different branched architectures and molecular weights were measured and listed in the same table. These polymers include four 3G polymers, $(A\text{-}A_2\text{-}A_4)_4$, $(A\text{-}A_2\text{-}A_8)_4$, $(A\text{-}A_4\text{-}A_8)_4$, and $(A\text{-}A_4\text{-}A_{16})_4$, plus five 4G polymers, $(A\text{-}A_2\text{-}A_4\text{-}A_8)_4$, $(A\text{-}A_2\text{-}A_8\text{-}A_{32})_4$, $(A\text{-}A_4\text{-}A_8\text{-}A_{32})_4$, $(A\text{-}A_4\text{-}A_{16}\text{-}A_{32})_4$, and $(A\text{-}A_4\text{-}A_{16}\text{-}A_{46})_4$. The arm segments of all polymer samples were between 10 000 and 15 000 g/mol in M_n value. In all 3G and 4G polymers, log R_h is also observed to increase linearly with log M_w, as shown in Figure 5.5. More interestingly, the data

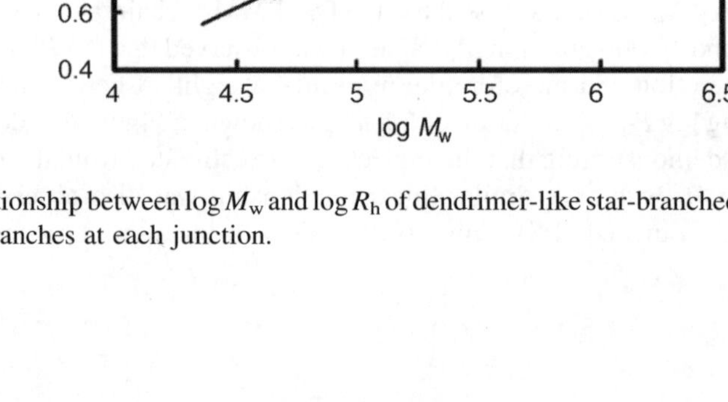

Figure 5.4 Relationship between log M_w and log R_h of dendrimer-like star-branched (PMMA)s from 1G to 5G with two branches at each junction.

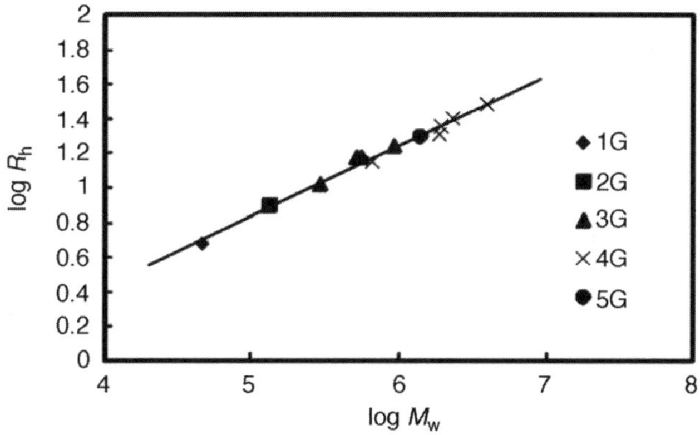

Figure 5.5 Relationship between log M_w and log R_h of dendrimer-like star-branched (PMMA)s with various branched architectures.

obtained by all polymers were almost on the same line, regardless of different generations and branched architectures. Thus, the dendrimer-like star-branched polymer has a larger molecular size, estimated from the R_h value as the M_w value increases. Although the M_w value changes via the branched architecture and/or generation, the R_h value is strongly correlated to the M_w value, but not directly reflected in the branched architecture and generation. Hedrick *et al.* also reported that similar R_h values were observed among the 2G and 3G dendrimer-like star-branched (PεCL)s with almost the same molecular weights, and the difference in both generations had a minimal effect on the R_h value (Trollsås *et al.*, 2000a).

The values of the radius of gyration, R_g, of the 3G and 4G polymers were also measured in THF by SAXS (Hirao *et al.*, 2009b). Similar to the relationship between R_h and M_w values, the R_g value of each 3G and 4G polymer increases by increasing the M_w value (Table 5.5). The ratios of R_g/R_h were between 0.83 and 1.16. Although the 3G and 4G polymers are considered to possess the most shell-weighted density distribution estimated from their branched architectures, all data are comparable to those of multiarmed star-branched polymers with high segment densities near the cores (Roovers *et al.*, 1983; Khasat *et al.*, 1988). The reason is not understood at the present time.

As mentioned in the introduction, dendrimer-like star-branched polymers are believed to be nanosize globular molecules, estimated by their generation-based branched architectures and some analytical results. The direct visualization of high molecular weight architectural polymers such as combs, dendrigrafts, comb-stars, and comb-cyclic polymers by AFM imaging has been reported with great success by Matyjaszewski *et al.* (Qin *et al.*, 2003; Matyjaszewski *et al.*, 2003; Sheiko *et al.*, 2003; Boyce *et al.*, 2004) and Deffieux *et al.* (Schappacher *et al.*, 2003; Viville *et al.*, 2004; Schappacher and Deffieux, 2004; Schappacher and Deffieux, 2005a; Schappacher and Deffieux, 2005b; Schappacher *et al.*, 2005a; Schappacher *et al.*, 2005b; Bernard *et al.*, 2006; Schappacher and Deffieux, 2008) Dendrimers have also been successfully visualized as special dome-shaped molecules by AFM and STM (Li *et al.*, 2000) (Zhang *et al.*, 2000). Therefore, Deffieux *et al.* first attempted to visualize the 4G dendrimer-like star-branched PMMA, $(A\text{-}A_4\text{-}A_{16}\text{-}A_{46})_4$ ($M_n = 3.96 \times 10^6$ g/mol, $M_w/M_n = 1.08$) by the AFM tapping mode on highly ordered graft (HOPG) (Hirao *et al.*, 2009b). Unfortunately, several isolated polymer molecules appeared on the AFM image as completely flattened objects with an average square size closer to 100 nm, and the height remained very small at 1.6 nm. Since the diameter of this sample is estimated to be 60 nm or smaller in THF from the R_h value, the average size observed by AFM is larger and the height is too small. Accordingly, the polymer shape that would be considered/expected to be globular is quite different from the resulting AFM image. The polymer molecule may be spread out on HOPG and stick flatly on the graphite surface due to the low polymer compactness and strong interaction between the polymer and the surface. Very recently, similar flattened square-shaped objects have also been observed by AFM using a 3G dendrimer-like star-branched PS with a higher M_n value of 1.43×10^7 g/mol (Deffieux *et al.*, 2008). Thus, these dendrimer-like star-branched polymers are still considered to be too soft for AFM analysis. It may be necessary to synthesize tougher polymers with crosslinking or the introduction of rigid polymer segments.

Ree *et al.* recently reported synchrotron SAXS studies of 3G (A-A_2-A_4)$_4$ and 4G (A-A_2-A_8-A_{32})$_4$ polymers showing that these polymers reveal ellipsoidal shapes composed of a core part and a shell part where the core part has a relatively higher density than the shell part (Jin *et al.*, 2008; Jin *et al.*, submitted). This indication is also quite different from the molecular shapes that would be imagined. Detailed SAXS study is expected to characterize the shapes of dendrimer-like star-branched polymers.

5.4 Viscosity Behavior

In order to study the solution behavior of dendrimer-like star-branched polymers, the intrinsic viscosity values, $[\eta]_{\text{dendrimer-like}}$, of two series of dendrimer-like star-branched (PMMA)s up to 7G were measured in THF at 25 °C (Hirao *et al.*, 2005a). (Hirao, unpublished results). Their branched structures have four branches at the cores and two branches at the junctions and are referred to as 1G (A_4), 2G (A-A_2)$_4$, 3G (A-A_2-A_4)$_4$, 4G (A-A_2-A_4-A_8)$_4$, 5G (A-A_2-A_4-A_8-A_{16})$_4$, 6G (A-A_2-A_4-A_8-A_{16}-A_{32})$_4$, and 7G (A-A_2-A_4-A_8-A_{16}-A_{32}-A_{64})$_4$. The PMMA arm segment between the junctions was adjusted to be either about 4000 g/mol or about 10 000 g/mol in M_n value. Their $[\eta]_{\text{dendrimer-like}}$, values are summarized in Table 5.6. For comparison, the intrinsic viscosity values of linear (PMMA)s, $[\eta]_{\text{linear}}$, with the same molecular weights, were calculated from the equation previously reported and added in the same table (Hirao and Matsuo, 2003):

$$[\eta]_{\text{linear}} = 1.11 \times 10^{-2} M_w^{0.695}.$$

In these polymers, the molecular weight rapidly increased and nearly doubled by increasing the generation, while the intrinsic viscosity was increased gradually. As expected from the branched architectures, the $[\eta]_{\text{dendrimer-like}}$ values were always smaller than those of linear polymers with the same molecular weights. For a better comparison to understand the relationship between viscosity and generation or molecular weight, log $[\eta]_{\text{dendrimer-like}}$ as a function of generation number or log M_w are plotted in Figure 5.6. As can be seen, log $[\eta]_{\text{dendrimer-like}}$ gradually increases with generation or the log M_w value even at the 7G generation or 1.97×10^7 g/mol, although it seems to be almost saturated. These relationships

Table 5.6 Intrinsic viscosities of dendrimer-like star-branched (PMMA)s with four branches at core and two branches at each junction

Generation	Type	M_w (kg/mol)[a]	[η] (mL/g)	
			$[\eta]_{dendrimer\text{-}like}$[b]	$[\eta]_{linear}$[c]
1G[d]	(A_4)	14.8	6.76	8.78
2G[d]	$(A\text{-}A_2)_4$	44.2	12.5	18.8
3G[d]	$(A\text{-}A_2\text{-}A_4)_4$	105	16.3	34.3
4G[d]	$(A\text{-}A_2\text{-}A_4\text{-}A_8)_4$	230	18.9	59.1
5G[d]	$(A\text{-}A_2\text{-}A_4\text{-}A_8\text{-}A_{16})_4$	472	22.5	97.4
6G[d]	$(A\text{-}A_2\text{-}A_4\text{-}A_8\text{-}A_{16}\text{-}A_{32})_4$	1060	24.0	171
7G[d]	$(A\text{-}A_2\text{-}A_4\text{-}A_8\text{-}A_{16}\text{-}A_{32}\text{-}A_{64})_4$	1970	26.3	263
1G[e]	(A_4)	47.0	16.7	19.6
2G[e]	$(A\text{-}A_2)_4$	136	26.1	41.0
3G[e]	$(A\text{-}A_2\text{-}A_4)_4$	295	35.2	70.3
4G[e]	$(A\text{-}A_2\text{-}A_4\text{-}A_8)_4$	661	42.2	123
5G[e]	$(A\text{-}A_2\text{-}A_4\text{-}A_8\text{-}A_{16})_4$	1500	48.6	218

[a] Measured by RALLS in THF at 30 °C.
[b] Measured in THF at 25 °C.
[c] Calculated from $[\eta]_{linear} = 1.11 \times M_w^{0.695}$.
[d] Molecular weight of arm polymer, $M_n \sim 4000$ g/mol.
[e] Molecular weight of arm polymer, $M_n \sim 10\,000$ g/mol.

are in contrast to those previously reported for dendrimers and the dendrimer-like star-branched (PS)s up to 7G, exhibiting a well-known bell-shaped curve (Matmour and Gnanou, 2008).

Since the molecular weight of dendrimer-like star-branched polymers can be made to vary widely by a change in the branched architecture, it is possible to compare the viscosity values between themselves, each in the same generation (Hirao *et al.*, 2009b). The $[\eta]_{dendrimer\text{-}like}$

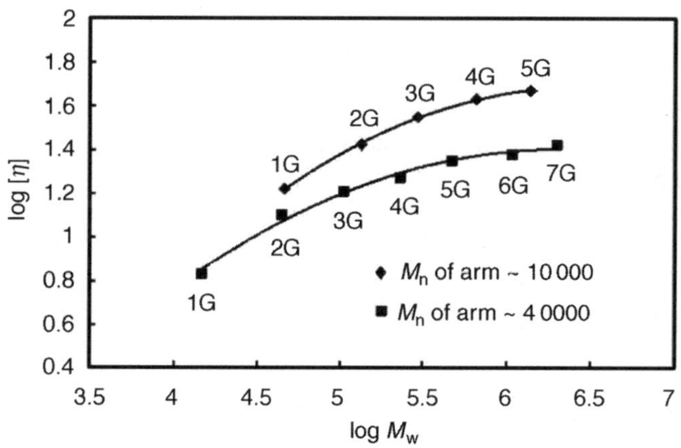

Figure 5.6 Relationship between log M_w and log $[\eta]$ of dendrimer-like star-branched (PMMA)s with two branches at each junction.

Table 5.7 Intrinsic viscosities of 3G and 4G dendrimer-like star-branched (PMMA)s with various branched architectures

Generation	Type	M_w (kg/mol)[a]	[η] (mL/g)	
			$[\eta]_{dendrimer\text{-}like}$[b]	$[\eta]_{linear}$[c]
3G	$(A\text{-}A_2\text{-}A_4)_4$	295	35.2	70.3
	$(A\text{-}A_2\text{-}A_8)_4$	514	33.2	103
	$(A\text{-}A_4\text{-}A_8)_4$	568	32.3	111
	$(A\text{-}A_4\text{-}A_{16})_4$	939	35.1	157
4G	$(A\text{-}A_2\text{-}A_4\text{-}A_8)_4$	661	42.2	123
	$(A\text{-}A_2\text{-}A_8\text{-}A_{32})_4$	1960	39.1	262
	$(A\text{-}A_4\text{-}A_8\text{-}A_{32})_4$	1920	43.3	258
	$(A\text{-}A_4\text{-}A_{16}\text{-}A_{32})_4$	2360	38.1	298
	$(A\text{-}A_4\text{-}A_{16}\text{-}A_{46})_4$	3960	42.1	427

[a] Measured by RALLS in THF at 30 °C.
[b] Measured in THF at 25 °C.
[c] Calculated from $[\eta]_{linear} = 1.11 \times 10^{-2} M_w^{0.695}$.

values of four 3G and five 4G polymers with different branched architectures (and therefore different M_w values) are also provided in Table 5.7.

Very surprisingly, the $[\eta]_{dendrimer\text{-}like}$ values of the four 3G polymers were close to each other (between 32.2 and 35.2 mL/g), although their M_w values varied considerably from 2.95×10^5 to 9.39×10^5 g/mol. For instance, the $(A\text{-}A_4\text{-}A_{16})_4$ polymer was almost equal in [η] value to the $(A\text{-}A_2\text{-}A_4)_4$ polymer, while the former was three times higher in M_w value than the latter. A similar trend was also observed among the 4G polymers. The [η] value of the $(A\text{-}A_4\text{-}A_{16}\text{-}A_{46})_4$ polymer was the same as that of the $(A\text{-}A_2\text{-}A_4\text{-}A_8)_4$ polymer, having only one-sixth the M_w value of the former sample. Thus, the insensitivity of molecular weight to viscosity seems to be apparent among the same-generation polymers. This means that the M_w value and branched architecture do not depend on the [η] value in the same generation. However, the dependence of molecular weight as well as generation on viscosity was clearly observed in the two polymer series, as mentioned above. Furthermore, the intrinsic viscosity shows a discernible increase from the 3G to the 4G polymers and the 4G polymers are always higher than the 3G polymers in [η] value. Thus obviously, the dependence of molecular weight on viscosity appears to be present among the generations. Such a contradiction may be attributed to various structural differences in the dendrimer-like star-branched polymers used. In fact, the molecular weight and branched architecture were different within the same-generation polymers. Even in the same polymer series, the branched architecture was changed by increasing the generation. The relationships between intrinsic viscosity and molecular weight are more complicated and should at first be discussed among polymers with the same or very similar branched architectures.

5.5 Branching Factor (g′ Value)

In regular-type star-branched polymers, the branching factor, g' value, defined as $[\eta]_{dendrimer\text{-}like}/[\eta]_{linear}$, is an important factor in examining the volume extended in solution by the number of arm segments. The g' value correlates with the number of arm segments and decreases rapidly by increasing the arm number according to the following well-established equation: (Douglas *et al.*, 1990).

$$g' = [(3f-2)/f^2]^{0.58}[0.724-0.015(f-1)]/0.724.(f = \text{arm number}(3<f<18).$$

In the case of star-branched polymers with more than 18 arms, the empirical equation, log $g' = 0.36–0.80 \log f$, is used (Roovers, 1999).

Similarly, the g' value was determined in each dendrimer-like star-branched polymer. The results are summarized in Table 5.8. As mentioned above, the $[\eta]_{\text{dendrimer-like}}$ values were

Table 5.8 g′ Values of dendrimer-like star-branched polymers

Type	M_w (kg/mol)[a]	Number of Outermost Arm Segments	g′	
			Dendrimer-Like[b]	Star
1G[c]	14.8	4	0.77	0.71[e]
2G[c]	44.2	8	0.66	0.46[e]
3G[c]	105	16	0.48	0.25[e]
4G[c]	230	32	0.32	0.14[f]
5G[c]	472	64	0.23	0.083[f]
6G[c]	1060	128	0.14	0.048[f]
7G[c]	1970	256	0.10	0.027[f]
1G[d]	47.0	4	0.85	
2G[d]	136	8	0.64	
3G[d]	295	16	0.50	
4G[d]	661	32	0.34	
5G[d]	1500	64	0.24	
$(A\text{-}A_2)_2$	57.9	4	0.80	
$(A\text{-}A_2)_4$	136	8	0.64	
$(A\text{-}A_4)_4$	272	16	0.37	
$(A\text{-}A_2\text{-}A_4)_2$	164	8	0.59	
$(A\text{-}A_2\text{-}A_4)_4$	295	16	0.50	
$(A\text{-}A_2\text{-}A_8)_4$	514	32	0.32	
$(A\text{-}A_4\text{-}A_8)_4$	568	32	0.29	
$(A\text{-}A_4\text{-}A_{16})_4$	939	64	0.22	
$(A\text{-}A_2\text{-}A_4\text{-}A_8)_2$	330	16	0.46	
$(A\text{-}A_2\text{-}A_4\text{-}A_8)_4$	661	32	0.34	
$(A\text{-}A_2\text{-}A_8\text{-}A_{32})_4$	1960	128	0.15	
$(A\text{-}A_4\text{-}A_8\text{-}A_{32})_4$	1920	128	0.17	
$(A\text{-}A_4\text{-}A_{16}\text{-}A_{32})_4$	2360	128	0.13	
$(A\text{-}A_4\text{-}A_{16}\text{-}A_{46})_4$	3960	184	0.099	

[a] Measured by RALLS in THF at 30 °C.
[b] Calculated from $g'_{\text{dendrimer-like}} = [\eta]_{\text{dendrimer-like}}/[\eta]_{\text{linear}}$.
[c] Molecular weight of arm polymer, $M_n \sim 4000$ g/mol.
[d] Molecular weight of arm polymer, $M_n \sim 10\,000$ g/mol.
[e] Calculated from $g'_{\text{star}} = [(3f-2)/f]^{0.58} \times [0.724 - 0.015 \times (f-1)]/0.724; f =$ number of arms $(3<f<18)$ (Douglas, Roovers and Freed, 1990).
[f] Calculated from $\log g'_{\text{star}} = 0.36 - 0.80 \times \log f$; $f =$ number of arms $(6<f)$ (Roovers, 1999).

always smaller than those of the corresponding linear polymers due to their branched architectures and, therefore, the g' values were less than unity for all polymer samples. In the same polymer series from 1G to 7G, the g' value decreased on increasing the generation from 1G to 7G, indicating that the dendrimer-like star-branched polymers are smaller in volume in solution than the corresponding linear polymers and the trend is more emphasized by increasing the generation (Hirao *et al.*, 2005a). With the same polymer series composed of higher molecular weight PMMA segments (M_n = about 10 000 g/mol), the same trend in g' value versus generation was observed (Hirao, unpublished results). Thus, the g' value is not influenced by the absolute molecular weight and can be used as a general factor in order to discuss the volume in solution.

Interestingly, a star-branched polymer is always somewhat smaller in g' value than that of a dendrimer-like star-branched polymer having the same number of outermost arm segments (Table 5.8). This seems reasonable when the branched architectures between star-branched polymers with a single branch point and dendrimer-like star-branched polymers with a generation-based hyperbranched architecture are considered. The g' values of dendrimer-like star-branched (PS)s with the same branched architectures reported by Gnanou *et al.* (Matmour and Gnanou, 2008) exhibited behavior similar to that of the above-mentioned dendrimer-like star-branched (PMMA)s between 1G and 3G. In the higher-generation PS samples of 4G–7G, however, the g' values were close to those of the corresponding star-branched polymers.

The relationship between g' value and branched architecture (and/or therefore molecular weight) is also interesting. In general, the g' value strongly depends on the branched architecture even in the same generation and decreases with increasing branch density (Matsuo *et al.*, 2004; Hirao *et al.*, 2009b), (Hirao, unpublished results). For instance, the g' value of (A-A_4)$_4$ was found to be the smallest among the 2G polymers, (A-A_4)$_4$, (A-A_2)$_4$, and (A-A_2)$_2$. Similarly, the g' value of the 3G polymers decreased by increasing the branch density in the order of (A-A_2-A_4)$_2$ to (A-A_2-A_4)$_4$ to (A-A_4-A_{16})$_4$. On the other hand, both 3G polymers, (A-A_2-A_8)$_4$ and (A-A_4-A_8)$_4$, with similar branch densities showed similar g' values. The same trend was observed among the 4G polymers. More interestingly, the g' value evidently appears to strongly correlate with the number of branch segments in the outermost layer even in different generations. For example, two 3G polymers and one 4G polymer with 32 outermost branch segments, (A-A_2-A_8)$_4$, (A-A_4-A_8)$_4$, and (A-A_2-A_4-A_8)$_4$, were very close to each other in g' value (0.29–0.34). By increasing the outermost branch segments to 64 in number, the g' value of the 3G polymer, (A-A_4-A_{16})$_4$, was reduced to 0.22 and was quite consistent with that (g' = 0.23) of a 5G polymer, (A-A_2-A_4-A_8-A_{16})$_4$. Moreover, the g' values of the (A-A_2-A_8-A_{32})$_4$, (A-A_4-A_8-A_{32})$_4$, and (A-A_4-A_{16}-A_{32})$_4$ polymers with 128 outermost branches were further reduced to those of polymers with 64 branches, but close to each other (0.13–0.17). Thus, the g' values decreased significantly by increasing the number of branch segments in the outermost layer, but were comparable to each other among samples with the same branch numbers even in different generations.

The g' value was plotted as a function of the logarithmic outermost branch number in Figure 5.7. Surprisingly, most of the plots fit well on one curve and, more interestingly, the data of the dendrimer-like star-branched (PS)s synthesized by Hutchings *et al.* (Hutchings and Roberts-Bleming, 2006) and Gnanou *et al.* (Matmour and Gnanou, 2008) are also on this curve, although the relationship is not currently theoretically understood. The g' value can be readily predicted only by the outermost branch number of dendrimer-like star-branched polymers.

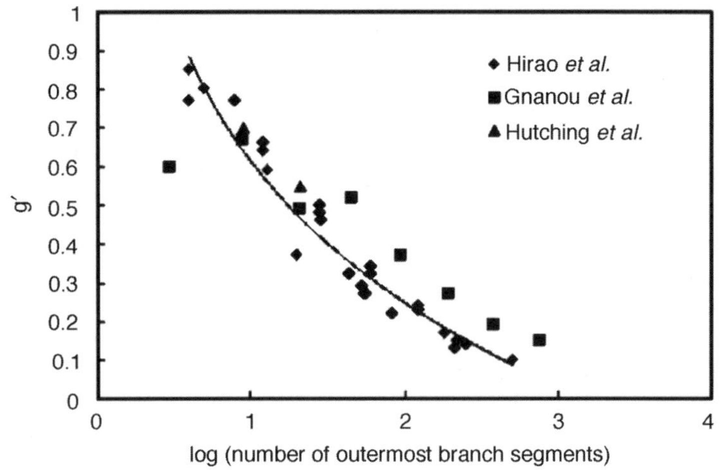

Figure 5.7 Relationship between log (number of outermost branch segments) and g' of dendrimer-like star-branched polymers.

This suggests that the g' value, that is, the degree of volume extent, depends strongly on the branch density at the periphery more than the inside branch density.

5.6 Concluding Remarks

This chapter reviews the precise synthesis of dendrimer-like star-branched polymers and block copolymers by stepwise iterative methodologies, focusing on the control of synthetic factors necessary for well-defined and precisely controlled structures of the polymers, that is, molecular weight, molecular-weight distribution, arm segment, and composition. The methodologies herein introduced are those based on either the "core-first" or "arm-first" divergent approach and the "arm-first" convergent approach in conjunction with living/controlled polymerization systems. Although each of these methodologies involves several advantageous points as well as inherent problems, well-defined and well-characterized dendrimer-like star-branched polymers up to the 7G stage with M_n values of a few million and with M_w/M_n values of less than 1.05 have been synthesized. Furthermore, their branched architectures can be varied to a certain extent and a variety of functional polymer segments can be introduced into any generation to add desirable functionalities. Moreover, intermediate and/or final polymers usually have numerous active termini at their peripheries capable of continuing the next iterative reaction step and, therefore, it is also possible to introduce many useful functional groups and molecules via such active termini.

As mentioned in the introduction, dendrimer-like star-branched polymers are well-defined huge macromolecules with regulated hyperbranched architectures, specific nanosize shapes, and many controllable structural variables. For such structures, a variety of not only molecular but also functional designs is possible, providing access to hyperbranched and specific nanosize globular macromolecules highly and appropriately functionalized with useful functions in any generation, at any internal/external position, or at the periphery. Furthermore, different segments and functional groups may be phase separated at the molecular level, followed by self-organizing, to form quite new nanosize ordered supramolecules and molecular assemblies. Thus, such functionalized dendrimer-like star-branched polymers and block copolymers are expected with many possible applications, especially in nanoscience and nanotechnology. Several preliminary applications have appeared in papers by Gnanou *et al.*, Taton *et al.* (2007), Feng *et al.* (2008) and Hedrick *et al.* (2002).

On the other hand, the basic properties and behavior of dendrimer-like star-branched polymers have not been characterized enough and related morphological and rheological studies have just started at the present time (Lee *et al.*, 2008). As mentioned in Section 5.3, direct and even indirect evidence for the real shapes of dendrimer-like star-branched polymers has not been obtained yet. Thus, it may be time to clarify the whole image of dendrimer-like star-branched polymers by developing not only synthetic, but also analytical and physical studies.

References

Angot, S., Taton, D., and Gnanou, Y. (2000) Amphiphilic stars and dendrimer-like architectures based on poly(ethylene oxide) and polystyrene. *Macromolecules*, **33**, 5418–5426.

Bernard, J.M., Schappacher, M., Deffieux, A. *et al.* (2006) Water-soluble dendrigrafts bearing saccharidic moieties: Elaboration and application to enzyme linked oligosorbent assay (ELOSA) diagnostic tests. *Bioconjugate Chemistry*, **17**, 6–14.

Boyce, J.R., Shirvayants, D., Sheiko, S.S. *et al.* (2004) multiarm molecular brushes: Effect of the number of arms on the molecular weight polydispersity and surface ordering. *Langmuir*, **20**, 6005–6011.

Chalari, I. and Hadjichristidis, N. (2002) Synthesis of well-defined second-generation dendritic polymers of isoprene (I) and styrene (S): $(S_2I)_3$, $(SI'I)_3$, $(I''I'I)_3$, and $(I'_2I)_4$. *Journal of Polymer Science Part A-Polymer Chemistry*, **40**, 1519–1526.

Deffieux, A., Schappacher, M., Hirao, A., and Watanabe, T. (2008) Synthesis and AFM structural imaging of dendrimer-like star-branched polystyrenes. *Journal of the American Chemical Society*, **130**, 5670–5672.

Douglas, J.F., Roovers, J., and Freed, K.F. (1990) Characterization of branching architecture through "universal" ratios of polymer solution properties. *Macromolecules*, **23**, 4168–4180.

Francis, R., Taton, D., Logan, J.L. *et al.* (2003) Synthesis and surface properties of amphiphilic star-shaped and dendrimer-like copolymers based on polystyrene core and poly(ethylene oxide) corona. *Macromolecules*, **36**, 8253–8259.

Feng, X., Taton, D., Chaikof, E.L., and Gnanou, Y. (2005) Toward an easy access to dendrimer-like poly(ethylene oxide)s. *Journal of the American Chemical Society*, **127**, 10956–10966.

Feng, X., Taton, D., Ibarboure, E. *et al.* (2008) Janus-type dendrimer-like poly(ethylene oxide)s. *Journal of the American Chemical Society*, **130**, 11662–11676.

Feng, X., Taton, D., Chaikof, E.L., and Gnanou, Y. (2009) Fast access to dendrimer-like poly(ethylene oxide)s through anionic ring-opening polymerization of ethylene oxide and use of nonprotected glycidol as branching agent. *Macromolecules*, **42**, 7292–7298.

Gnanou, Y. and Taton, D. (2001) Stars and dendrimer-like architectures by the divergent method using controlled radical polymerization. *Macromolecular Symposia*, **174**, 333–341.

Grayson, S.M. and Fréchet, J.M.J. (2001) Convergent dendrons and dendrimers: From synthesis to applications. *Chemical Reviews*, **101**, 3819–3868.

Hedrick, J.L., Trollsås, M., Hawker, C.J. *et al.* (1998) Dendrimer-like star block and amphiphilic copolymers by combination of ring opening and atom transfer radical polymerization. *Macromolecules*, **31**, 8691–8705.

Hedrick, J.L., Magbitang, T., Connor, E.F. *et al.* (2002) Application of complex macromolecular architectures for advanced microelectronic materials. *Chemistry - A European Journal*, **8**, 3308–3319.

Hirao, A. and Matsuo, A. (2003) Synthesis of chain-end-functionalized poly(methyl methacrylate)s with a definite number of benzyl bromide moieties and their application to star-branched polymers. *Macromolecules*, **36**, 9742–9751.

Hirao, A., Matsuo, A., and Watanabe, T. (2005a) Precise synthesis of dendrimer-like star-branched poly(methyl methacrylate)s up to seventh generation by an iterative divergent approach involving coupling and transformation reactions. *Macromolecules*, **38**, 8701–8711.

Hirao, A., Hayashi, M., Loykulnant, S. *et al.* (2005b) Precise syntheses of chain-multi-functionalized polymers, star-branched polymers, star-linear block polymers, densely branched polymers, and dendritic branched polymers based on iterative approach using functionalized 1,1-diphenylethylene derivatives. *Progress in Polymer Science*, **30**, 111–182.

Hirao, A., Sugiyama, K., Tsunoda, Y. *et al.* (2006) Precise synthesis of well-defined dendrimer-like star-branched polymers by iterative methodology based on living anionic polymerization. *Journal of Polymer Science Part A-Polymer Chemistry*, **44**, 6659–6687.

Hirao, A., Sugiyama, K., Matsuo, A. *et al.* (2008) Synthesis of well-defined dendritic hyperbranched polymers by iterative methodologies using living/controlled polymerizations. *Polymer International*, **57**, 554–570.

Hirao, A., Watanabe, T., and Yoo, H.-S. (2009a) Precise synthesis of dendrimer-like hyperbranched polymers. *Journal of Synthetic Organic Chemistry, Japan*, **67**, 1033–1043.

Hirao, A., Watanabe, T., Ishizu, K. *et al.* (2009b) Precise synthesis and characterization of fourth-generation dendrimer-like star-branched poly(methyl methacrylate)s and block copolymers by iterative methodology based on living anionic polymerization. *Macromolecules*, **42**, 682–693.

Hou, S., Chaikof, E.L., Taton, D., and Gnanou, Y. (2003) Synthesis of water-soluble star-block and dendrimer-like copolymers based on poly(ethylene oxide) and poly(acrylic acid). *Macromolecules*, **36**, 3874–3881.

Hutchings, L.R., and Roberts-Bleming, S.J. (2006) DendriMacs. well-defined dendritically branched polymers synthesized by an iterative convergent strategy involving the coupling reaction of AB_2 macromonomers. *Macromolecules*, **39**, 2144–2152.

Hutchings, L.R. (2008) DendriMacs and HyperMacs – emerging as more than just model branched polymers. *Soft Matter*, **4**, 2150–2159.

Jin, S., Jin, K.S., Yoon, J. *et al.* (2008) X-ray scattering studies on molecular structures of star and dendritic polymers. *Macromolecular Research*, **16**, 686–694.

Jin, S., Higashihara, T., Jin, K.S. *et al.* Synchrotron X-ray scattering characterization of the molecular structures of star polystyrenes with varying numbers of arms. *The Journal of Physical Chemistry B*, **114**, 6247–6257.

Khasat, N., Pennisi, R.W., Hadjichristidis, N., and Fetters, L.J. (1988) Dilute solution behavior of 3-arm asymmetric and regular 3- and 12-arm polystyrene stars. *Macromolecules*, **21**, 1100–1106.

Knauss, D.M. and Huang, T. (2003) $((PS)_nPS)_m$ star-shaped polystyrene with star-shaped branches at the terminal chain ends by convergent living anionic polymerization. *Macromolecules*, **36**, 6036–6042.

Lee, J.H., Orfanou, K., Driva, P. *et al.* (2008) Linear and nonlinear rheology of dendritic star polymers: experiment. *Macromolecules*, **41**, 9165–9178.

Li, J., Piehler, L.T., Qin, D.T. *et al.* (2000) Visualization and characterization of poly(amidoamine) dendrimers by atomic force microscopy. *Langmuir*, **16**, 5613–5616.

Matmour, R. and Gnanou, Y. (2008) Combination of an anionic terminator multifunctional initiator and divergent carbanionic polymerization: application to the synthesis of dendrimer-like polymers and of asymmetric and miktoarm stars. *Journal of the American Chemical Society*, **130**, 1350–1361.

Matsuo, A., Watanabe, T., and Hirao, A. (2004) Synthesis of well-defined dendrimer-like branched polymers and block copolymer by the iterative approach involving coupling reaction of living anionic polymer and functionalization. *Macromolecules*, **37**, 6283–6290.

Matyjaszewski, K., Qin, S., Boyce, J.R. *et al.* (2003) Effect of initiation conditions on the uniformity of three-arm star molecular brushes. *Macromolecules*, **36**, 1843–1849.

Orfanou, K., Iatrou, H., Lohse, D.J., and Hadjichristidis, N. (2006) Synthesis of well-defined second (G-2) and Third (G-3) generation dendritic polybutadienes. *Macromolecules*, **39**, 4361–4365.

Percec, V., Barboiu, B., Grigoras, C., and Bera, T.K. (2003) Universal iterative strategy for the divergent synthesis of dendritic macromolecules from conventional monomers by a combination of living radical polymerization and irreversible TERminator multifunctional INItiator (TERMINI). *Journal of the American Chemical Society*, **125**, 6503–6516.

Percec, V., Grigoras, C., and Kim, H.J. (2004) Toward self-assembling dendritic macromolecules from conventional monomers by a combination of living radical polymerization and irreversible terminator multifunctional initiator. *Journal of Polymer Science Part A-Polymer Chemistry*, **42**, 505–513.

Percec, V., Grigoras, C., Bera, T.K. *et al.* (2005) Accelerated iterative strategy for the divergent synthesis of dendritic macromolecules using a combination of living radical polymerization and an irreversible terminator multifunctional initiator. *Journal of Polymer Science Part A-Polymer Chemistry*, **43**, 4894–4906.

Qin, S., Matyjaszewski, K., Xu, H., and Sheiko, S.S. (2003) Synthesis and visualization of densely grafted molecular brushes with crystallizable poly(octadecyl methacrylate) block segments. *Macromolecules*, **36**, 605–612.

Roovers, J. (1999) Dilute solution properties of regular star polymers, in *Star and Hyperbranched Polymers* (eds. M.K. Mishra, and S. Kobayashi), Marcel Dekker Inc., New York, pp. 285–341.

Roovers, J., Hadjichristidis, N., and Fetters, L.J. (1983) Analysis and dilute solution properties of 12- and 18-arm-star polystyrenes. *Macromolecules*, **16**, 214–220.

Schappacher, M. and Deffieux, A. (2004) Encapsulation of metallo-porphyrins into water-soluble amphipatic dendrigrafts. *Polymer*, **45**, 4633–4639.

Schappacher, M. and Deffieux, A. (2005a) AFM Image analysis applied to the investigation of elementary reactions in the synthesis of comb star copolymers. *Macromolecules*, **38**, 4942–4946.

Schappacher, M. and Deffieux, A. (2005b) From combs to comb-*g*-comb centipedes. *Macromolecules*, **38**, 7209–7213.

Schappacher, M. and Deffieux, A. (2008) Synthesis of macrocyclic copolymer brushes and their self-assembly into supramolecular tubes. *Science*, **319**, 1512–1515.

Schappacher, M., Deffieux, A., Putaux, J.-L. *et al.* (2003) Synthesis and characterization of water-soluble amphipatic polystyrene-based dendrigrafts. *Macromolecules*, **36**, 5776–5783.

Schappacher, M., Deffieux, A., Putaux, J.-L. *et al.* (2005a) Synthesis and properties of PS–PEO core–shell amphiphilic dendrigrafts. *Polymer*, **46**, 6767–6776.

Schappacher, M., Putaux, J.L., Lefebvre, C., and Deffieux, A. (2005b) Molecular containers based on amphiphilic PS-*b*-PMVE dendrigraft copolymers: Topology, organization, and aqueous solution properties. *Journal of the American Chemical Society*, **127**, 2990–2998.

Sheiko, S.S., Da Silva, M., Shirvayants, D. *et al.* (2003) Measuring molecular weight by atomic force microscopy. *Journal of the American Chemical Society*, **125**, 6725–6728.

Six, J.L. and Gnanou, Y. (1995) From star-shaped to dendritic poly(ethylene oxide)s: toward increasingly branched architectures by anionic polymerization. *Macromolecular Symposia*, **95**, 137–150.

Stancik, C.M., Pople, J.A., Trollsås, M. *et al.* (2003) Impact of core architecture on solution properties of dendrimer-like star copolymers. *Macromolecules*, **36**, 5765–5775.

Taton, D., Gnanou, Y., Matmour, R. *et al.* (2006) Controlled polymerizations as tools for the design of star-like and dendrimer-like polymers. *Polymer International*, **55**, 1138–1145.

Taton, D., Feng, X., and Gnanou, Y. (2007) Dendrimer-like polymers: a new class of structurally precise dendrimers with macromolecular generations. *New Journal of Chemistry*, **31**, 1097–1110.

Trollsås, M. and Hedrick, J.L. (1998) Dendrimer-like star polymers. *Journal of the American Chemical Society*, **120**, 4644–4651.

Trollsås, M., Claesson, H., Atthoff, B., and Hedrick, J.L. (1998a) Layered dendritic block copolymers. *Angewandte Chemie (International Edition in English)*, **37**, 3132–3136.

Trollsås, M., Hawker, C.J., Remenar, J.F. *et al.* (1998b) Highly branched radial block copolymers via dendritic initiation of aliphatic polyesters. *Journal of Polymer Science Part A-Polymer Chemistry*, **36**, 2793–2798.

Trollsås, M., Atthoff, B., Würsch, A. *et al.* (2000a) Constitutional isomers of dendrimer-like star polymers: Design, synthesis, and conformational and structural properties. *Macromolecules*, **33**, 6423–6438.

Trollsås, M., Claesson, H., Atthoff, B. *et al.* (2000b) Conformational and structural properties of high functionality dendrimer-like star polymers synthesized from living polymerization techniques. *Macromolecular Symposia*, **153**, 87–108.

Trollsås, M., Atthoff, B., Claesson, H., and Hedrick, J.L. (2004) Dendritic Homopolymers and block copolymers: Tuning the morphology and properties. *Journal of Polymer Science Part A-Polymer Chemistry*, **42**, 1174–1188.

Urbani, C.N., Bell, C.A., Whittaker, M.R., and Monteiro, M.J. (2008) Convergent synthesis of second generation AB-type miktoarm dendrimers using "click" chemistry catalyzed by copper wire. *Macromolecules*, **41**, 1057–1060.

Viville, P., Leclère, P., Deffieux, A. *et al.* (2004) Atomic force microscopy study of comb-like vs. arborescent graft copolymers in thin films. *Polymer*, **45**, 1833–1843.

Watanabe, T., Tsunoda, Y., Matsuo, A. *et al.* (2006) Synthesis of dendrimer-like star-branched poly(methyl methacrylate)s of generations consisting of four branched polymer chains at each junction by iterative methodology involving coupling and transformation reactions. *Macromolecular Symposia*, **240**, 23–30.

Würsch, A., Möller, M., Glauser, T. *et al.* (2001) Dendritic–linear A_xB_x block copolymers prepared via controlled ring-opening polymerization of lactones from orthogonally protected multifunctional initiators. *Macromolecules*, **34**, 6601–6615.

Yoo, H.-S., Watanabe, T., and Hirao, A. (2009) Precise synthesis of dendrimer-like star-branched polystyrenes and block copolymers composed of polystyrene and poly(methyl methacrylate) segments by an iterative methodology using living anionic polymerization. *Macromolecules*, **42**, 4558–4570.

Zhang, H., Grim, P.C.M., Foubert, P. *et al.* (2000) Properties of single dendrimer molecules studied by atomic force microscopy. *Langmuir*, **16**, 9009–9014.

6

Arborescent Polymers with a Mesoscopic Scale

Toufic Nabil Aridi and Mario Gauthier
Department of Chemistry, Institute for Polymer Research, University of Waterloo, 200 University Avenue West, Waterloo, ON N2L 3G1, Canada

6.1 Introduction

Branched macromolecules fall into three main classes: star-branched polymers, characterized by multiple chains linked at one central point (Roovers, 1985), comb-branched polymers, having one linear backbone and side chains randomly distributed along it (Rempp *et al.*, 1988), and dendritic polymers, with a multilevel branched architecture (Tomalia and Fréchet, 2001). The cascade-branched structure of dendritic polymers is typically derived from polyfunctional monomers under more or less strictly controlled polymerization conditions. This class of macromolecules has a unique combination of features and, as a result, a broad spectrum of applications is being developed for these materials in areas including microencapsulation, drug delivery, nanotechnology, polymer processing additives, and catalysis.

Dendritic polymers may be further subdivided into three categories on the basis of their architecture (Tomalia and Fréchet, 2001), namely dendrimers, hyperbranched, and dendrigraft polymers (Figure 6.1). Dendrimers, the first dendritic polymers reported in the literature, are obtained from a generation-based scheme using small-molecule monomers as building blocks. Their synthesis is performed either in a core-first (divergent) or core-last (convergent) manner, using cycles of protection, condensation, and deprotection of AB_n-type monomers. Since strict control is attained over molecular architecture in this approach, dendrimers can have extremely narrow molecular-weight distributions (MWD), with polydispersity indices $M_w/M_n < 1.01$, and exactly predictable molecular weights. However, since small-molecule building blocks are used, many reaction cycles are necessary to synthesize high molecular weight dendrimers.

Complex Macromolecular Architectures: Synthesis, Characterization, and Self-Assembly, First Edition.
Edited by Nikos Hadjichristidis, Akira Hirao, Yasuyuki Tezuka and Filip Du Prez.

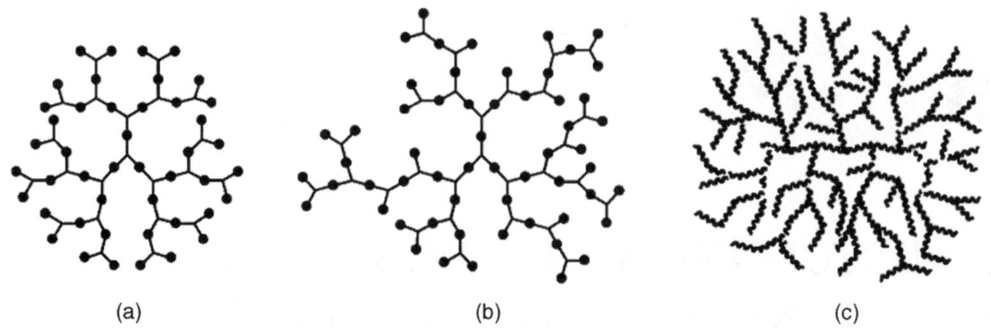

Figure 6.1 Schematic representation of (a) dendrimers, (b) hyperbranched, and (c) dendrigraft polymers. (Reprinted from *Progress in Polymer Science*, **29**, S.J. Teertstra and M. Gauthier, "Dendrigraft polymers: Macromolecular engineering on a mesoscopic scale," 277–327, 2004, with permission from Elsevier.)

These complex reaction sequences are avoided in the synthesis of hyperbranched polymers, the second family of dendritic polymers. Hyperbranched polymers with a high molecular weight can be obtained in one-pot self-condensation reactions of AB_n monomers without protecting groups, with the tradeoff that control over the molecular weight and branching is much more limited. The random nature of the condensation reaction leads to polymers with many structural flaws and a broad MWD ($M_w/M_n > 2$) in most cases.

The third family of dendritic polymers, the dendrigraft systems, was introduced simultaneously in 1991 as the comb-burst polymers by Tomalia *et al.*, (1991), and as the arborescent polymers by Gauthier and Möller (1991). Dendrigraft (or arborescent) polymers are synthesized in a generation-based scheme analogous to dendrimers, but using cycles of ionic polymerization and grafting reactions, and polymeric chains as building blocks rather than small molecules (Scheme 6.1). A linear polymer substrate functionalized with coupling sites is first reacted with "living" polymer chains to yield a comb-branched structure, also called a generation-zero (G0) arborescent polymer. A further cycle of functionalization to introduce coupling sites on the G0 substrate and grafting with side chains leads to the first-generation (G1) arborescent polymer, and subsequently to higher generations (G2, G3, etc.). It should be noted that the G1 (twice-grafted) macromolecules represent the first polymer generation with a true dendritic (multilevel) branched architecture. Because the building blocks in arborescent polymer syntheses are polymeric chain segments and a large number of coupling sites are introduced in each cycle (branching multiplicity $f = 10–15$ coupling sites per side chain), rapid molecular weight growth is observed over successive generations. Since the functionalization reaction used to introduce coupling sites on the substrates takes place randomly, the branched structure obtained also bears similarities to hyperbranched polymers. While this architecture is not as strictly controlled as for dendrimers, the MWD achieved for these materials still remains narrow ($M_w/M_n \approx 1.1$ typically).

This chapter focuses on the latest efforts in our laboratory aimed at the synthesis and the physical characterization of arborescent polymers. The synthesis of arborescent styrene and isoprene homopolymers is discussed, together with copolymers incorporating a polystyrene substrate grafted with side chains of different compositions such as poly(2-vinylpyridine),

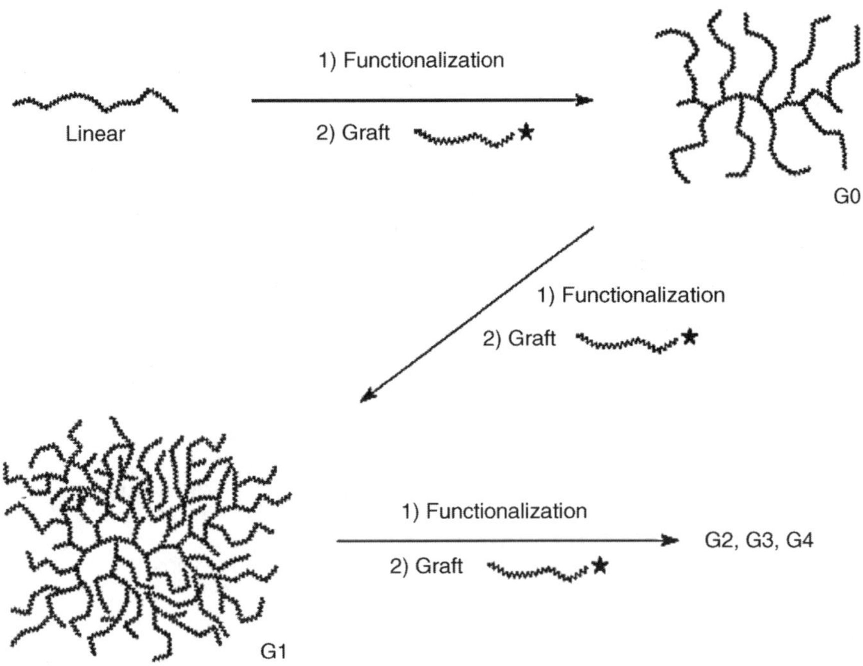

Scheme 6.1 General scheme for the synthesis of arborescent polymers. (Reprinted from *Progress in Polymer Science*, **29**, S. J. Teertstra and M. Gauthier, "Dendrigraft polymers: Macromolecular engineering on a mesoscopic scale," 277–327, 2004, with permission from Elsevier.)

polyisoprene, poly(*tert*-butyl methacrylate), and poly(ethylene oxide). Some of the physical properties of these materials are also examined. The peer-reviewed literature considered covers mainly the period from year 2000 on, since earlier developments in the field of arborescent polymers have been the topic of other reviews (Gauthier, 2000; Kee *et al.*, 2001; Teertstra and Gauthier, 2004).

6.2 Arborescent Polystyrene

Arborescent homopolymers are interesting, among others, as model compounds to study the properties of branched macromolecules beyond other systems such as star-branched polymers and microgels, due to their high branching functionalities and the extensive control achieved over the architecture (branching density, side-chain molecular weight) of the molecules. The synthesis of arborescent styrene homopolymers and some of their physical properties will be discussed first.

Polystyrene is a convenient starting material for the synthesis of arborescent polymers, since its aromatic pendent groups allow the facile introduction of coupling sites suitable for the synthesis of successive generations. The preparation of arborescent polystyrene has been achieved by three different methods. The first two are based on the "grafting-onto" procedure described in Scheme 6.1 using chloromethyl or acetyl coupling sites; the third approach uses a one-pot "grafting-from" technique.

6.2.1 Arborescent Polystyrene from Chloromethyl Coupling Sites

The first synthetic path, relying on the partial chloromethylation of linear polystyrene (Scheme 6.2), was introduced in 1991 (Gauthier and Möller, 1991) and reviewed on a few occasions. Consequently, only the main features of these reactions will be considered here for comparison with the methods developed more recently for copolymers. Chloromethylation poses two major obstacles to the synthesis of arborescent polymers. It is sensitive to cross-linking, since further reaction of the chloromethylated ring with another styrene unit leads to the formation of a methylene bridge either intra- or intermolecularly. These side reactions were minimized by working in dilute solution with a large excess of chloromethyl methyl ether, using Lewis acids such as $SnCl_4$ (Gauthier and Möller, 1991) or $AlCl_3$/1-nitropropane (Gauthier *et al.*, 1996). Another important issue is the occurrence of metal–halogen exchange competing with the coupling reaction and lowering the *grafting yield* (defined as the fraction

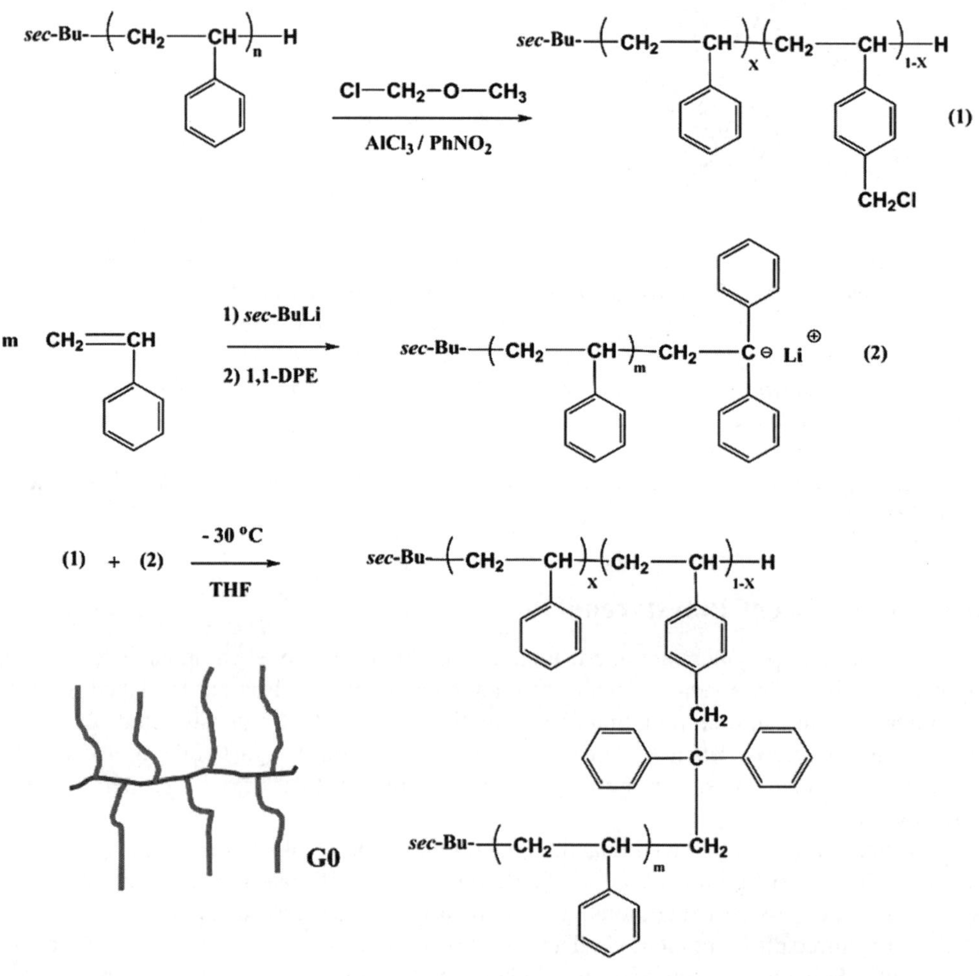

Scheme 6.2 Arborescent polystyrene synthesis by grafting onto chloromethylated polystyrene.

of living chains generated in the reaction becoming attached on the substrate). This problem was overcome by capping the polystyryllithium chains with a single 1,1-diphenylethylene (DPE) unit. Another interesting feature of DPE capping is that the resulting deep red coloration of the DPE macroanions is convenient for visual monitoring of the coupling step, as a solution of chloromethylated polystyrene substrate is slowly added to the living chains until the color fades. Repetition of cycles of chloromethylation and grafting reactions according to these procedures led to arborescent polystyrene of generations G0–G3 in up to 96% yield.

Anionic polymerization produces side chains with a controllable molecular weight and a narrow MWD. The number of coupling sites on the grafting substrate can also be varied, providing a handle over the branching density for each generation. The side-chain molecular weight and branching density may thus be controlled independently for each generation.

6.2.2 *Arborescent Polystyrene from Acetyl Coupling Sites*

The introduction of coupling sites on polystyrene via chloromethylation, besides being complicated by side reactions, also relies on hazardous and expensive reagents including chloromethyl methyl ether and carbon tetrachloride. An alternate grafting method was developed to eliminate these problems with acetyl coupling sites (Li and Gauthier, 2001) as depicted in Scheme 6.3. Acetyl chloride, apart from being less harmful than chloromethyl methyl ether, reacts quantitatively in the reaction and is not susceptible to crosslinking.

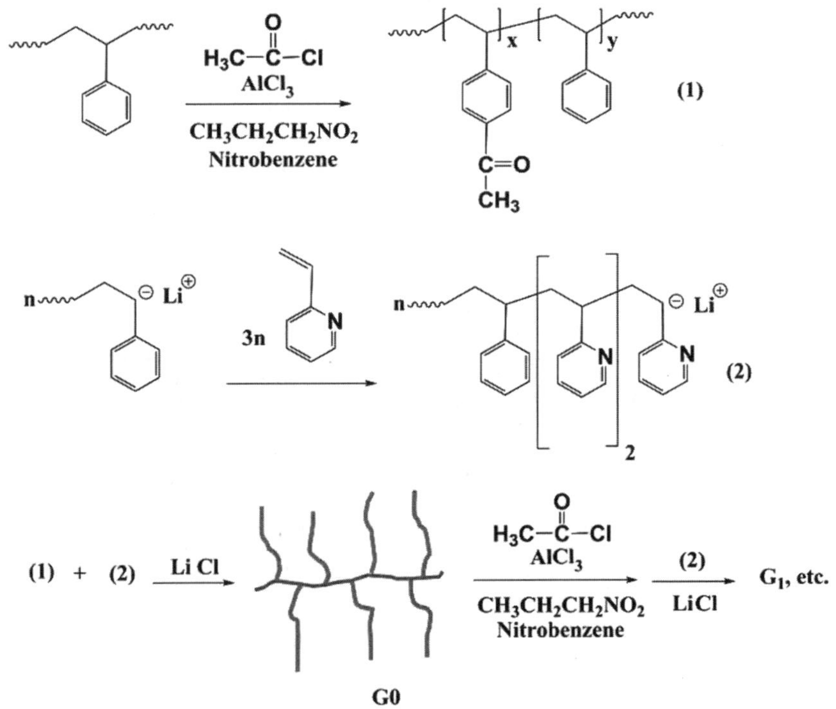

Scheme 6.3 Arborescent polystyrene synthesis by grafting onto acetylated polystyrene substrates.

The synthesis begins with the random acetylation of polystyrene with acetyl chloride and $AlCl_3$ in nitrobenzene, followed by coupling with polystyryllithium. In this approach, capping of the polystyryllithium chains with a few units of 2-vinylpyridine (2VP) and the addition of LiCl were necessary to avoid deactivation of the polystyryllithium species by proton abstraction from the acetyl groups. This side reaction was indeed found to limit the grafting yield to about 65% for direct coupling of polystyryllithium with acetylated polystyrene in the absence of reactivity modifiers, independently of the solvent type and temperature used. After capping the chains with 2VP, repetition of the acetylation and grafting reaction cycle yielded arborescent polystyrene of generations G0–G3 in 43–95% yield.

Characterization data for two series of arborescent polystyrene samples based on the acetylation path are provided in Table 6.1 as an example (Li and Gauthier, 2001). The polymers were synthesized from either $M_w \approx 5000$ (PS5) or $M_w \approx 30\,000$ (PS30) polystyrene side chains. The sample nomenclature used identifies the generation number of the substrate as well as the molecular weight of the side chains grafted in the last reaction. For example, G1PS-PS5 refers to a G1 polystyrene substrate grafted with $M_w \approx 5000$ side chains, while G1PS-PS30 refers to the same substrate grafted with $M_w \approx 30\,000$ side chains. It should be noticed that sample PS-PS5 became substrate G0PS after acetylation, and so on. The weight-average branching functionality of the polymers, defined as the number of chains added in the last grafting reaction, was calculated according to the equation

$$f_w = \frac{M_w(G) - M_w(G-1),}{M_w^{br}}$$

where $M_w(G)$, $M_w(G-1)$, and M_w^{br} are the absolute weight-average molecular weight of graft polymers of generation G, of the preceding generation and of the side chains, respectively. A wide range of molecular weights and branching functionalities were attained by the procedure described in Scheme 6.3 while maintaining a narrow MWD ($M_w/M_n \leq 1.1$). The absolute molecular weight (from light-scattering measurements) and branching functionality

Table 6.1 Characteristics of successive generations of two series of arborescent polystyrene (Adapted with permission from J. Li and M. Gauthier, "A novel synthetic path to arborescent graft polystyrenes," *Macromolecules,* **34**, 8918–8924, 2001. © 2001 American Chemical Society.)

	side chains			graft polymers			
Sample	$M_w{}^a$	$M_w/M_n{}^a$	yield (%)^b	$M_w{}^a$	$M_w/M_n{}^a$	$M_w{}^c$	f_w
PS-PS5	4400	1.09	95	5.3×10^4	1.08	3.3×10^4	11
G0PS-PS5	4500	1.07	89	4.3×10^5	1.08	1.3×10^5	84
G1PS-PS5	5000	1.08	84	3.9×10^6	1.09	4.5×10^5	690
G2PS-PS5	5500	1.09	75	2.5×10^7			3800
PS-PS30	28 500	1.09	87	3.2×10^5	1.08	1.9×10^5	11
G0PS-PS30	28 700	1.07	75	2.1×10^6	1.07	4.7×10^5	71
G1PS-PS30	27 800	1.06	55	1.1×10^7	1.09	7.1×10^5	380
G2PS-PS30	28 500	1.08	43	6.1×10^7			2000

^{a}Absolute values determined from SEC-MALLS or laser light-scattering measurements.
^{b}Fraction of side chains generated attached to the substrate.
^{c}Apparent values determined by SEC analysis with linear polystyrene standards calibration.

increase in an approximately geometric fashion over successive generations, however, the grafting yield decreases due to steric hindrance within the highly branched substrates limiting the accessibility of the coupling sites. Other factors lowering the grafting yield are residual protic impurities and side reactions with the solvent (THF) deactivating the living chains. Comparison of the apparent molecular weights (M_w, column 7 of Table 6.1), from size-exclusion chromatography (SEC) analysis using a linear polystyrene standards calibration curve, with the absolute M_w determined from light scattering measurements (column 5) shows that the linear standards calibration method strongly underestimates the molecular weight of arborescent polymers, due to their very compact structure.

The availability of arborescent polymers on a large scale (100 g and over) is important for their detailed physical characterization and in the development of applications. Munam and Gauthier (2008) recently demonstrated the large (100-g)-scale synthesis of arborescent polystyrene according to the acetylation path described in Scheme 6.3. When the large-scale reactions were carried out by the procedure developed for the small (10–15 g) scale synthesis, lower grafting yields were observed due in part to dimerization of the side chains prior to grafting. The origin of the problem was traced to decreased stirring efficiency in the large-scale reactions: The very quick consumption of 2VP as it was added to the reactor left a portion of the macroanions noncapped, simply as a result of mixing inhomogeneities. Under these conditions, the highly reactive residual polystyryl anions can attack the 2VP ring of another chain already capped and cause dimerization of the side chains, which greatly hinders the coupling reaction. This problem was solved by adding DPE as a capping agent before 2VP. Since DPE does not homopolymerize, using 1.2 equiv. of DPE per living end ensures that all the chains are capped with one DPE unit. The rate of the crossover reaction from DPE to 2VP being lower than from styrene to 2VP, the 2VP capping reaction becomes less sensitive to mixing inhomogeneities and dimer formation is suppressed under these conditions. It should be noted that DPE alone is not suitable as a capping agent for the polystyryllithium chains, because the congested DPE macroanions favor proton abstraction from the acetyl coupling sites at the expense of nucleophilic addition. Grafting yields reaching 95% were achieved after capping with DPE and 3 equiv. of 2VP for short ($M_w \approx 5000$) polystyrene side chains, while longer side chains ($M_w \approx 30\,000$) made it necessary to increase the amount of 2VP in the capping step to 6 equiv. per living end to maintain a high grafting yield.

6.2.3 One-Pot Synthesis of Arborescent Polystyrene

The step-wise synthesis of arborescent polymers, while able to generate large amounts of materials in good yield, requires isolation and purification of the crude product prior to substrate functionalization. For less-demanding applications not requiring stringent removal of the linear polymer contaminant, a one-pot method was recently developed by Yuan and Gauthier (2006) for the synthesis of arborescent polystyrene without isolation of the reaction intermediates.

The synthesis, represented in Scheme 6.4, starts with the preparation of a linear copolymer of styrene and 1,3-diisopropenylbenzene (DIB) in a semibatch copolymerization process. To this end, it was necessary to optimize the reaction conditions in terms of the initiation method, the reaction temperature, and the monomer addition protocol used. The copolymerization reaction

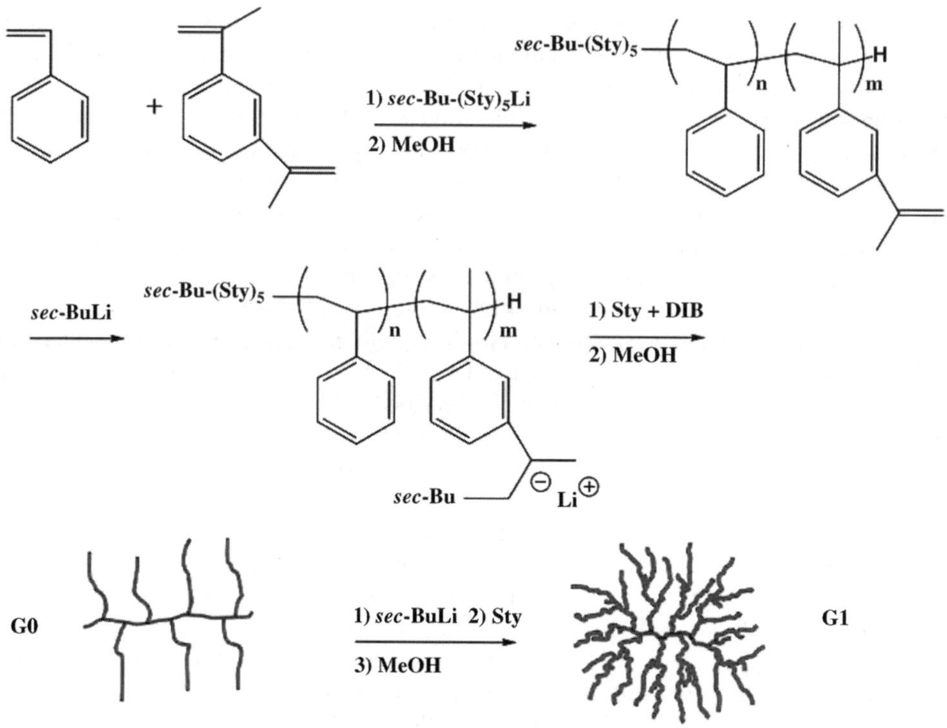

Scheme 6.4 One-pot synthesis of a G1 arborescent polystyrene.

was initiated with oligostyryllithium rather than *sec*-butyllithium (*sec*-BuLi), because of its greater selectivity toward monoaddition on DIB. A temperature of −78 °C was selected for the copolymerization with stepwise addition of the monomer, to ensure a uniform distribution of DIB units within the copolymer chains. The semibatch addition protocol developed corresponds to successive additions of a monomer mixture containing 3 equiv. of styrene and 1 equiv. of DIB per living end, followed by a delay of 80–120 s to allow incorporation of the DIB unit at the chain end. This protocol, apart from achieving a more uniform distribution of branching points along the substrate, also yielded superior results in terms of polydispersity index and reproducibility in comparison with the dropwise continuous addition of a styrene-DIB monomer mixture. The average spacing between branching points (DIB units) is thus expected to be 3 styrene units under these conditions (for a 3 : 1 styrene : DIB ratio). After terminating the linear copolymer chains by titration with methanol, the pendant isopropenyl moieties of the DIB units were activated *in situ* with 0.95 equiv. of *sec*-BuLi for 4 h at −20 °C to generate a polyfunctional anionic macroinitiator. Further additions of styrene-DIB monomer mixture to the activated linear substrate yielded a comb-branched (generation G0) copolymer, also suitable as a polyfunctional anionic initiator after activation with *sec*-BuLi. The addition of styrene to the linear and G0 macroinitiators yielded G0 and G1 arborescent polystyrene structures, respectively.

The synthesis of the G1 polymers with substoichiometric activation of the DIB units was sensitive to crosslinking or gelation in the side-chain growth step, due to the attack of residual (nonactivated) isopropenyl moieties on the substrate by the polystyryllithium propagating centers. This problem was minimized by the addition of 6 equiv. of LiCl per living end as a common-ion salt to shift the dissociation equilibrium to the ion-pair form. While the addition of styrene monomer to the activated G0 substrate in the absence of LiCl led to gelation in less than 1–2 min, no gelation was observed in the presence of LiCl over a few minutes; the reaction was nonetheless terminated 2 min after monomer addition as a precaution. Another strategy to avoid crosslinking was the addition of excess *sec*-BuLi in the activation step, to ensure the full consumption of the pendant isopropenyl moieties on the substrate, with the drawback of increased linear polymer contaminant formation.

The molecular weight of the side chains in the G1 polymers was conveniently controlled by varying the amount of monomer added to the G0 macroinitiator. Polymers with relatively low polydispersity indices ($M_w/M_n = 1.1$–1.3) and absolute M_w reaching 7×10^6 were thus obtained.

6.2.4 *Physical Characterization of Arborescent Polystyrene*

The influence of branching functionality and side-chain molecular weight on the physical properties of arborescent polystyrene has been the topic of a number of investigations already reviewed (Gauthier, 2000; Kee *et al.*, 2001; Teertstra and Gauthier, 2004). The main conclusion of these studies is that arborescent polymers behave increasingly like rigid spheres as their branching functionality is increased and/or the molecular weight of the side chains is decreased. Many of these investigations hinted at a *dual-phase* morphology for arborescent polystyrene: Even the upper-generation molecules retained residual interpenetrability, while their central (core) portion was much more hindered and less accessible. The validity of the hard-core–soft-shell – or perturbed hard sphere – model for arborescent polystyrene was verified more recently in a number of investigations including the small-angle neutron scattering (SANS) experiments outlined below.

6.2.4.1 Neutron Scattering

The radius of gyration (R_g) of two series of arborescent polystyrene with different side-chain molecular weights ($M_w \approx 5000$, PS5; $M_w \approx 30\,000$, PS30) was determined in SANS experiments as a function of molecular weight over successive generations (Il Yun *et al.*, 2008b). The scaling relation $R_g \propto M_w^{\nu}$ was determined with scaling exponents $\nu = 0.26$ and 0.32 for the PS5 and PS30 series, respectively. These scaling factors correspond to an increase in average density over successive generations for molecules with short side chains (PS5), while the average segmental density of the polymers with long side chains (PS30) remains approximately constant. The SANS data collected in cyclohexane-*d* displayed a pronounced maximum for the G1-G3 polymers in Kratky plots (scattering intensity $I(q)\,q^2$ vs. q, where $q = \sin(\theta)4\pi/\lambda$ is the scattering vector for a scattering angle θ and a wavelength λ), also observed for star and hyperbranched polymers, and predicted for spherical objects. Interestingly, no maximum was detected for the G0 polymer, corresponding to a comb-branched structure with twelve PS5 side chains grafted onto a PS5 backbone and expected to have a more ellipsoidal shape. The scattering peak became more pronounced for upper generations, the G3 polymer having multiple peaks characteristic for hard spheres of uniform size.

The morphology of the arborescent polystyrene molecules was also investigated by the SANS contrast matching method (Il Yun *et al.*, 2008b). To this end, G3 and G4 "copolymers" were synthesized by grafting deuterated polystyrene (PS-*d*) side chains onto G2 and G3 protonated arborescent polystyrene substrates in the last reaction cycle. The SANS measurements were carried out in both PS core-matching and PS-*d* shell-matching solvents. The R_g of the core should be obtained in a shell-matching solvent, since the polymer chains in the shell are invisible. When the scattering length density of the solvent matches the core, the R_g determined from the SANS data corresponds to a hollow sphere. The R_g for the G3 (G2PS-*graft*-PS-*d*) and G4 (G3PS-*graft*-PS-*d*) polymers was larger in THF-*d* is than in cyclohexane-*d*, indicating excluded volume effects (core swelling) in the good solvent. SANS contrast matching also allowed detailed profiling of the radial segment density within the core and shell phases of arborescent graft polymers. It was thus shown that the higher-generation G3PS-*graft*-PS-*d* polymer had better defined core–shell morphology (sharper core/shell interface) than the lower generation G2PS-*graft*-PS-*d* sample. Enhanced phase separation between the core and the shell is an expected consequence of the preferential reaction of coupling sites located at the periphery of the grafting substrates, favored by the inaccessibility of coupling sites buried deeper within their structure.

SANS also served to probe interpenetration in blends of arborescent polystyrene with linear deuterated poly(vinyl methyl ether) (*d*-PVME) and polystyrene (*d*-PS; Choi *et al.*, 2000). The interactions between the blend components were monitored as a function of the molar mass of the two components and temperature. The radius of gyration of the arborescent polymers decreased in all polymer blends in comparison to the values measured in solution. For G0, G1, and G2 polymers the R_g was comparable in blends with *d*-PVME and *d*-PS while for the G3 sample, the R_g in *d*-PS was smaller than in *d*-PVME and close to the value calculated for a collapsed sphere having the bulk density of PS. This indicates that the G3 molecules behave mainly like noninterpenetrating spheres, the linear *d*-PS matrix chains being largely excluded from the interior of the arborescent molecules. Fitting the single particle form factor for the arborescent polymers in the blends with a power-law model for the density profile, corresponding to a diffuse arborescent polymer/matrix interface in the blends, was found to be more appropriate than fitting with a hard-sphere model representing a sharp interface.

6.2.4.2 Solution Thermodynamics

The influence of polymer architecture on intermolecular interactions in dilute solutions was investigated by membrane osmometry in toluene (good solvent for polystyrene), cyclohexane (theta or θ solvent), and methylcyclohexane (poor solvent; Striolo *et al.*, 2001). The osmotic second virial coefficient (B_{22}) measured for arborescent polystyrene in toluene was lower than for homologous linear polymers, as expected due to their smaller R_g. In a θ solvent (cyclohexane), branching lowered the θ temperature from 34.5 °C (linear homolog) to 32.2 °C (G0 polymer). The θ temperature for the G0 polystyrene sample in methylcyclohexane was likewise lowered to 36 °C, as compared to values estimated between 60 and 70 °C for linear polystyrene samples. The experimental osmotic pressure data were successfully fitted with a molecular-thermodynamic equation suitable for colloids, indicating that the behavior of arborescent polystyrene molecules in dilute solution corresponds to a perturbed (weakly interacting or interpenetrable) hard sphere.

Vapor–liquid equilibrium (VLE) measurements for binary solutions of arborescent and linear polystyrene samples in chloroform, toluene, and cyclohexane were performed by a classic gravimetric sorption method (Lieu *et al.*, 2000). In cyclohexane, the VLE data indicated that solvent absorption depends on the generation number (branching functionality) of the polymers. No dependence on generation number was observed in the VLE data for toluene or chloroform, however. The solvent–polymer interaction parameter (Flory parameter χ_{12}) calculated from the VLE data was compared with that calculated from the second virial coefficient derived from light-scattering measurements (Gauthier *et al.*, 1998). While the interaction parameter increased with the generation number for all solvents based on the second virial coefficients method, the trend was not as clear for the VLE results. The poor agreement between the two methods was attributed to the concentrated solution regime used for the VLE measurements, causing the solvent–polymer interactions to become primarily a function of enthalpic contributions. For example, chloroform and toluene are good solvents for arborescent polystyrene and favorable polymer–solvent interactions may result from favorable enthalpic contributions with relatively small entropic contributions. In contrast, cyclohexane is a poor solvent for arborescent polystyrene, perhaps because the enthalpic contribution to the solvation process is low. As the entropic contribution to solvation may be significant in cyclohexane, the VLE results for polymers in cyclohexane may become dependent on the generation number.

The systematic manner in which the characteristics of dendrigraft homopolymers can be varied is useful to establish structure–property relationships for branched polymers. When considering potential applications, however, materials with a wider range of physical and chemical properties are of interest. An easy way to broaden the range of properties observed for arborescent macromolecules is by extending the scope of the synthetic techniques developed for arborescent polystyrene to allow the preparation of copolymers. The generation-based grafting scheme developed for arborescent polystyrene indeed relies on the presence of aromatic units enabling further functionalization with coupling sites, but this does not preclude the introduction of side chains with a different chemical composition in the last grafting reaction. The synthesis and properties of copolymers incorporating polymer segments of different compositions grafted on arborescent polystyrene substrates will be reviewed in the following sections.

6.3 Arborescent Polystyrene-*graft*-Poly(2-vinylpyridine) Copolymers

The synthesis of arborescent copolymers from polystyrene substrates and poly(2-vinylpyridine) side chains (PS-*graft*-P2VP), useful as precursors for highly branched cationic polyelectrolytes, was achieved using both chloromethyl (Kee and Gauthier, 2002) and acetyl (Gauthier *et al.*, 2003) coupling sites. The synthesis of the copolymers follows the paths illustrated in Schemes 6.2 and 6.3 with minor changes, by substituting polystyryllithium with poly(2-vinylpyridinyllithium) in the final grafting step. Two series of arborescent PS-*graft*-P2VP copolymers were synthesized by grafting either short ($M_w \approx 5000$, P2VP5) or long ($M_w \approx 30\,000$, P2VP30) side chains on the polystyrene substrates. Similar copolymer architectures and compositions were obtained using both chloromethyl and acetyl coupling sites (Gauthier *et al.*, 2003). Due to the relatively low reactivity of the P2VP anions, no capping was necessary prior to grafting on chloromethylated polystyrene to suppress metal–halogen exchange reactions, but the addition of LiCl was still beneficial to avoid proton abstraction from the acetylated substrates. Since the synthesis of P2VP copolymers from chloromethylated substrates was already reviewed, it will not be discussed in detail here. The general characteristics of the arborescent P2VP copolymers obtained (molecular weight, branching functionality, polydispersity index) are comparable with those quoted for the styrene homopolymers (Table 6.1), the P2VP content varying from about 90 to 98% depending on the molecular weight of the side chains and the characteristics of the substrate.

A one-pot synthesis of arborescent PS-*graft*-P2VP copolymers incorporating P2VP chains according to the method illustrated in Scheme 6.4, by substituting styrene with 2-vinylpyridine in the final side-chain growth step, was described more recently (Yuan and Gauthier, 2007). Monomer addition to the linear and G0 styrene–diisopropenylbenzene copolymers activated with *sec*-BuLi led to the formation of the G0 and G1 copolymers, respectively, the side-chain length being controlled by the amount of monomer added in the reaction. Interestingly, the MWD of the copolymers became narrower as the length of the side chains was increased. This effect was attributed to averaging of the propagating site reactivity in the larger molecules. The chemical compositions estimated by NMR analysis and calculated from the absolute molecular weight determined from light-scattering measurements were in good agreement. This divergent "grafting-from" strategy is clearly a powerful method to synthesize arborescent copolymers, providing control over both copolymer composition and structure.

6.3.1 Physical Characterization of Arborescent Polystyrene-graft-Poly(2-vinylpyridine)

6.3.1.1 Neutron and Light Scattering

The morphology of G0–G3 arborescent PS-*graft*-P2VP copolymers was characterized by SANS measurements in deuterated methanol (CD_3OD; Il Yun *et al.*, 2003). Both polydispersed core–shell and (monodispersed) power-law radial density models worked reasonably well to fit the SANS data but the power-law model, corresponding to a somewhat diffuse P2VP shell surrounding a dense polystyrene core, provided the best fit. The density profiles obtained for all generations of arborescent PS-*graft*-P2VP copolymers in deuterated methanol otherwise provided no clear indication for the presence of a collapsed polystyrene core in the copolymers. This was explained by the low weight fraction (10–20%) of the polystyrene component in the copolymers, and a relatively diffuse core/shell interface due to the random distribution of grafting sites on the substrate. The scaling exponent for $R_g \propto M^{\nu}$ was determined as $\nu = 0.24$, hinting that the overall density of the molecules increased over successive generations.

The aggregation of weakly charged PS-*graft*-P2VP copolymers in methanol-d_4 and in D_2O was investigated by a combination of the SANS and dynamic light-scattering (DLS) techniques in dilute solutions (Il Yun *et al.*, 2006). The most notable effect upon addition of a substoichiometric amount of acid (<1 equiv. HCl/2VP unit) was the appearance of a scattering peak characteristic for polyelectrolyte solutions, and in agreement with previous observations for dendritic polyelectrolytes (Briber *et al.*, 1992). In analogy to the latter, the peak was interpreted as a correlation peak defined by the interparticle distance in arborescent polymer aggregates resulting from long-range Coulombic interactions. The interparticle distance calculated from the scattering peak for a uniform particle size distribution was consistent with liquid-like ordering of the charged arborescent graft copolymer molecules within the aggregates. The lower dielectric constant of methanol-d_4 led to long-range electrostatic interactions persisting to lower polymer concentrations than in D_2O. The DLS data displayed two diffusive relaxation processes upon addition of HCl to a G0PS-*graft*-P2VP copolymer solution. The slow-mode dynamics were attributed to local structural inhomogeneities (aggregation) in the arborescent polyelectrolyte solutions: Both the SANS scattering peak and the slow diffusion mode disappeared upon addition of excess HCl to the solutions, which screened the electrostatic interactions and induced breakup of the aggregates. Molecular expansion was more significant for copolymers with long (P2VP30K) vs. short polyelectrolyte side chains (P2VP5K), which led to the P2VP30K molecules forming a gel upon addition of a sub-stoichiometric amount of HCl at low concentrations.

The extensive association of PS-*graft*-P2VP copolymers observed at low HCl/2VP ratios is in stark contrast with the situation encountered when excess HCl is present in solution: Dynamic light-scattering measurements have shown that P2VP copolymers derived from polystyrene substrates of generations G0 and above exist as isolated (unimolecular) species under these conditions (Gauthier *et al.*, 2003). DLS measurements served to monitor changes in molecular dimensions upon ionization of the P2VP side chains (Kee and Gauthier, 2002), by comparing the hydrodynamic radius of the arborescent copolymer molecules measured in methanol and in methanol with 0.1 N HCl. The structure dependence of hydrodynamic volume expansion, expressed as the ratio of the hydrodynamic volume in acidic methanol (ionized) to

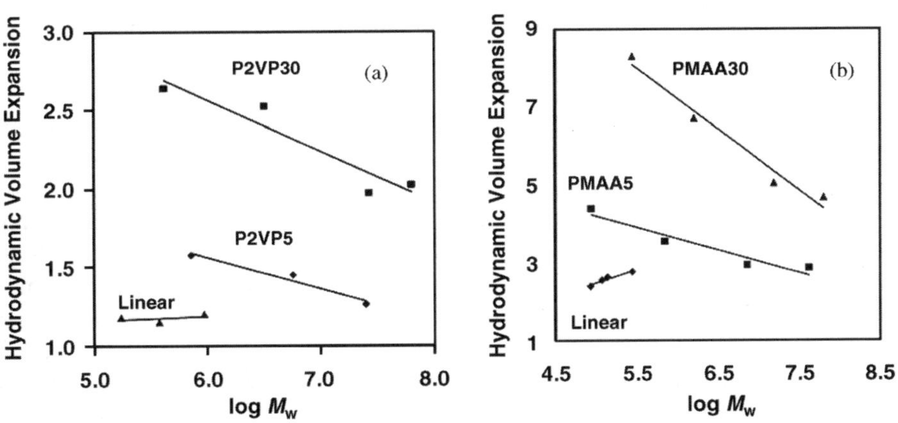

Figure 6.2 (a) Molecular weight dependence of hydrodynamic volume expansion for linear and arborescent P2VP graft copolymers dissolved in MeOH/H_2O 95/5 with HCl. (b) Molecular weight dependence of hydrodynamic volume expansion for linear and arborescent PMAA graft copolymers dissolved in MeOH/H_2O 95/5 with 0.05N NaCl and NaOH. Volume expansion is expressed as the ratio of hydrodynamic volumes in the ionized and neutral forms. (Adapted with permission from (a) R.A. Kee and M. Gauthier, "Arborescent polystyrene-*graft*-poly(2-vinylpyridine) copolymers: Synthesis and enhanced polyelectrolyte effect in solution," *Macromolecules*, **35**, 6526–6532, 2002. © 2002 American Chemical Society; (b) R.A. Kee and M. Gauthier, "Arborescent polystyrene-*graft*-poly(*tert*-butyl methacrylate) copolymers: Synthesis and enhanced polyelectrolyte effect in solution," *Journal of Polymer Science, Part A: Polymer Chemistry*, **46**, 2335–2346 (Figure 5), © 2008 John Wiley & Sons, Inc.)

that in pure methanol (neutral), is compared for linear and arborescent PS-*graft*-P2VP samples in Figure 6.2a. The DLS results show that the arborescent P2VP copolymers expand much more in solution than the linear homologous polymers when protonated with HCl. This effect is attributed to the higher charge density attained in the branched, compact molecules. Copolymers with long, flexible P2VP30 side chains displayed the largest volume increases due to their enhanced flexibility relative to short side chain materials (P2VP5). Interestingly, volume expansion decreased within each copolymer series as the generation number increased, most likely as a result of structural stiffening.

6.3.1.2 Micellar Properties

The amphiphilic character and the covalently bonded, stable structure of the PS-*graft*-P2VP copolymers is particularly interesting for applications as unimolecular micelles in the solubilization of hydrophobic compounds in aqueous environments, and the sustained release of bioactive compounds.

The solubilization of various model hydrophobic probes (polycyclic aromatic hydrocarbons, PAH) by arborescent PS-*graft*-P2VP micelles of different generations and compositions in aqueous solutions was monitored by UV and fluorescence spectroscopy analysis (Njikang *et al.*, 2008c). The partitioning behavior of hydrophobic probes including naphthalene and pyrene derivatives between the arborescent micelles and water was found to depend on both the characteristics of the copolymers and the nature of the probes. Not surprisingly, arborescent copolymers with a higher polystyrene content had a greater solubilization capacity, but the solubilization rate and capacity also decreased for the upper-generation copolymers, presumably due to their stiffer structure. Partition coefficients reaching around 10^6 were observed for very hydrophobic probes (e.g. pyrene), but the less-hydrophobic PAH (e.g. naphthalene derivatives) gave higher micelle loadings due nonspecific sorption in within the polystyrene core and the P2VP shell of the micelles.

The release kinetics of two drugs (indomethacin and lidocaine) from the dendritic PS-*graft*-P2VP micelles in dilute HCl solutions were also investigated by fluorescence and UV spectroscopy (Njikang *et al.*, 2008d). The copolymers were loaded with the drugs by dissolution in a common solvent (CH_2Cl_2 or CH_3Cl) before precipitation and washing with hexanes. Release from the micelles was monitored by dialysis in 0.05 M HCl. The release profiles obtained for indomethacin- and lidocaine-loaded G1 arborescent PS-*graft*-P2VP copolymers are compared in Figure 6.3. In contrast to the rapid equilibration observed for the free probes from dialysis tubing (100% release in less than 5 h) the release from the dendritic micelles was slow, corresponding to sustained characteristics. For both indomethacin and lidocaine, an initial burst in release over the first 5 h was followed by more gradual release until equilibration was attained over 1–2 days. Furthermore, while roughly 80% of the loaded lidocaine was released from the micelles over the duration of the experiment, only about 40% of the indomethacin was released under the same conditions. Lidocaine being less hydrophobic than indomethacin, a large fraction of it was presumably located within the P2VP shell after encapsulation. This represents the fraction completely released within the time scale of the experiment. Incomplete release of the drugs from the micelles is more likely related to physical entrapment within the hydrophobic core and/or the palisade (interfacial) regions of the micelles. Another possibility, in the case of indomethacin, could be hydrogen-bonding

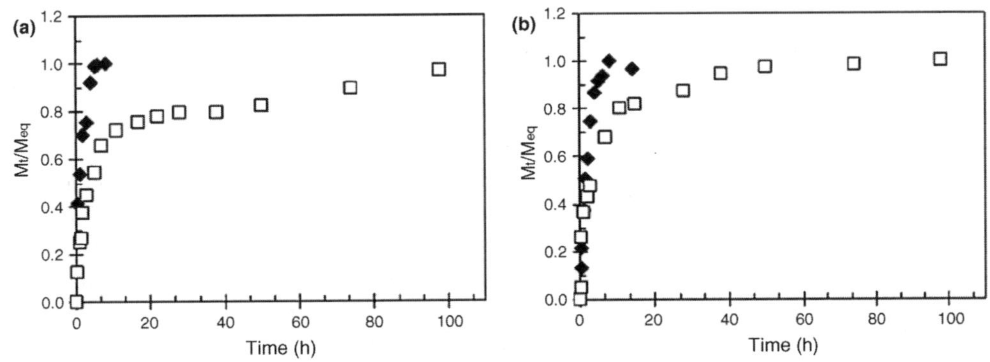

Figure 6.3 Release profiles for (a) indomethacin and (b) lidocaine in 0.05M HCl in the free state (◆) and from G1PS-*graft*-P2VP copolymer (□). M_t is the cumulative mass released at time t, and M_{eq} is the mass released at equilibrium. (Reprinted from *Polymer*, **49**, G.N. Njikang, M. Gauthier and J. Li, "Sustained release properties of arborescent polystyrene-*graft*-poly (2-vinylpyridine) copolymers," 5474–5481, 2008, with permission from Elsevier.)

interactions between the pyridine ring of the P2VP chains and the carboxylate group of the drug. There are no such interactions between lidocaine and the arborescent micelles, which could also explain why a larger fraction of the encapsulated drug was released.

The results discussed above show that arborescent PS-*graft*-P2VP copolymers can be designed to solubilize or release hydrophobic compounds of different polarities at controllable rates.

6.3.1.3 Self-Assembly in Solution

In contrast to micelles formed by the self-assembly of linear block copolymer chains, rigid arborescent PS-*graft*-P2VP copolymer molecules are unable to reverse their core and shell components when exposed to a solvent selective for the polystyrene core, thus forcing the P2VP shell to directly contact the poor solvent. This unusual micellar morphology with a solvophilic core and a solvophobic shell is interesting with respect to the core–shell structure of unimolecular micelles and their self-assembly. The self-assembly of arborescent PS-*graft*-P2VP copolymers derived from a G1PS substrate was investigated as a function of temperature in toluene (Il Yun *et al.*, 2008a), a solvent selective for the polystyrene core (Figure 6.4a). DLS measurements were used to monitor the hydrodynamic diameter (D_h) of G1PS-*graft*-P2VP5 in methanol and in toluene as a function of temperature. The D_h of the copolymer in methanol, a solvent selective for the P2VP shell, was essentially independent of temperature and no aggregation was observed, in agreement with previous results (Il Yun *et al.*, 2003). The copolymer also did not dissolve in toluene at low temperatures, as expected for this copolymer having a composition with 87% of insoluble P2VP branches within its shell. However, the solution became transparent and the copolymer was molecularly dissolved above 26 °C, defined as the critical self-assembly temperature (cst). The size of the molecularly dissolved G1PS-*graft*-P2VP5 copolymer in toluene at these temperatures was about 10% lower than in methanol, indicating shrinkage of the P2VP shell in toluene as a consequence of the poor solvent quality. Upon decreasing the temperature to 25 °C the D_h increased abruptly to a few micrometers, due to the formation of large aggregates having a somewhat broad size distribution but nonetheless a spherical topology. The DLS measurements confirmed that the supramolecular structures (Figure 6.4b) persisted at low temperatures, albeit some very large aggregates precipitated at temperatures below 20 °C. The influence of the P2VP chain length on the temperature-sensitive self-assembly process was analyzed by comparing copolymers with P2VP side chains having a number-average molecular weight (M_n) of either 5000 (P2VP5) or 30 000 (P2VP30). Similar thermoreversible self-assembly was observed for the copolymer with the P2VP30 side chains, but the cst was shifted to a higher temperature (41 °C). This shows that the critical temperature for the thermal responsiveness is tunable by varying the length of the P2VP segments within the shell. For comparison, the copolymer grafted with shorter (P2VP5) side chains at the same concentration displayed no aggregation whatsoever down to 27 °C. Clearly, aggregation becomes more favorable as the size of the insoluble branches increases. The self-assembly of arborescent copolymers in solution provides a new path for the

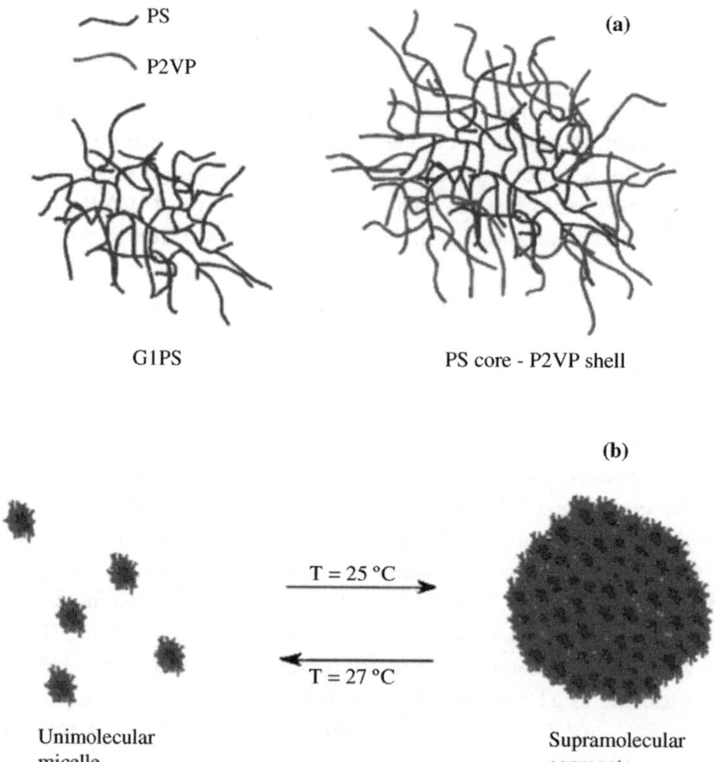

Figure 6.4 Schematic representation of (a) core–shell morphology and (b) temperature-responsive supramolecular self-assembly of arborescent PS-*graft*-P2VP5 copolymers in toluene. (Reprinted with permission from S.I. Yun, G.E. Gadd, V. Lo *et al.*, "Temperature-responsive supramolecular assembly of dendrigraft micelles with a solvophilic core – solvophobic shell structure," *Macromolecules*, **41**, 7166–7172. © 2008 American Chemical Society.) (Colour version of this figure is available on the book companion web site.)

creation of well-defined supramolecular structures of large dimensions. Most interestingly, the temperature-responsive self-assembly of the arborescent copolymers offers the possibility to produce new types of "smart" supramolecular structures.

6.4 Arborescent Polystyrene-*graft*-Polystyrene-*block*-Poly(2-vinylpyridine)

Beyond the core–shell morphologies already described, the grafting-onto technique may also be applied to the generation of more complex molecular architectures such as layered morphologies. This was recently demonstrated with the synthesis of arborescent polystyrene-*graft*-[poly(2-vinylpyridine)-*block*-polystyrene] copolymers and their application as templates for the preparation of gold nanoparticles (Dockendorff *et al.*, 2008). The synthesis of these templates, represented schematically in Figure 6.5, was performed by grafting living polystyrene-*block*-poly(2-vinylpyridine) macroanions onto acetylated linear and branched (G1) polystyrene. The resulting dendritic species have a covalently bonded, layered structure with an inner P2VP shell that can be loaded with metallic compounds such as tetrachloroauric acid ($HAuCl_4$). The copolymer templates loaded at various levels were subsequently reduced in solution with hydrazine to produce gold nanoparticles. The characteristics of the arborescent copolymer templates governed the distribution of gold salt within the molecules: Templates containing the large G1PS hydrophobic core displayed lower salt loading in the center of the molecules, while smaller templates such as PS-*graft*-(P2VP-*block*-PS) (derived from linear polystyrene) had a uniform distribution of gold throughout. The solution-reduction studies revealed that the size and uniformity of the gold nanoparticles obtained were independent of the template generation and rather reliant on the amount of reducing agent used. Excess hydrazine minimized the coalescence and led to a more uniform size distribution of gold nanoparticles within each template molecule.

6.5 Arborescent Polystyrene-*graft*-Polyisoprene

The synthesis of polystyrene-*graft*-polyisoprene (PS-*graft*-PIP) copolymers was achieved from both chloromethylated and acetylated polystyrene substrates as illustrated in Schemes 6.2

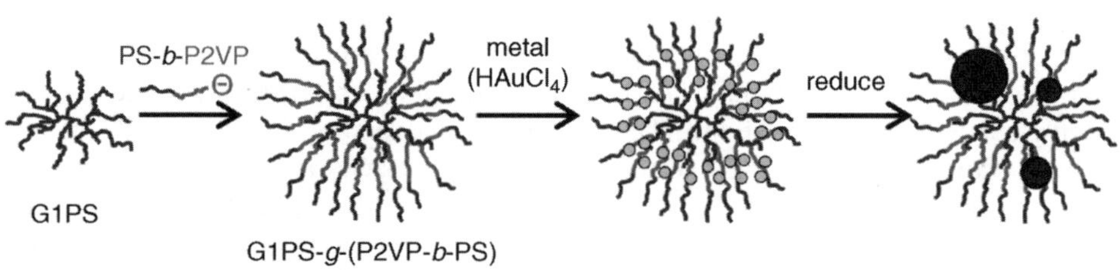

Figure 6.5 Schematic representation of an arborescent G2 copolymer template synthesis, metallic salt loading, and reduction. (Reprinted with permission from J. Dockendorff, M. Gauthier, A. Mourran and M. Möller, "Arborescent amphiphilic copolymers as templates for the preparation of gold nanoparticles," *Macromolecules*, **41**, 6621–6623. © 2008 American Chemical Society.)

and 6.3, respectively, using polyisoprenyllithium rather than polystyryllithium in the final grafting step.

The reaction with the acetylated substrates involved the polymerization of isoprene with *sec*-BuLi either in THF to yield polyisoprenyllithium with a mixed microstructure (approximately 1 : 1 : 1 ratio of 1,2-, 3,4-, and 1,4-units; Li *et al.*, 2004), or in cyclohexane to obtain a high ($\geq$70%) *cis*-1,4-microstructure content (Teertstra and Gauthier, 2007). The living polymer was then titrated with an acetylated substrate to generate the copolymer. The grafting yield was maximized at 25 °C in the presence of 5 equiv. of LiCl per living end, to attenuate the reactivity of polyisoprenyllithium. Capping with a few 2VP units was not critical as in the synthesis of arborescent styrene homopolymers, but still convenient to take advantage of the darker coloration of the polyvinylpyridinyl anions in the colorimetric titration process used to monitor the stoichiometry of the coupling reaction.

Arborescent copolymers were synthesized by grafting PIP side chains with a M_w of either 5000 (PIP5) or 30 000 (PIP30) onto linear, comb-branched (G0), G1, and G2 acetylated polystyrene. The polyisoprene content attained varied from 84 to 91% by weight for copolymers with short (PIP5) side chains, and from 92 to over 97% by weight for long (PIP30) side chains. The graft copolymers also exhibited the expected geometric increases in branching functionality and molecular weight over successive generations, as well as decreased grafting yields for longer polyisoprene side chains and higher-generation substrates, due to steric congestion and increased sensitivity of the living ends to impurities.

A core–shell (heterogeneous) morphology was clearly observed by atomic force microscopy imaging in the phase-contrast mode, even for copolymers with PIP30 side chains (Figure 6.6), due to differences in viscoelastic response for the glassy polystyrene-rich core and the rubbery polyisoprene-rich shell. The type of solvent serving to prepare the monolayer films by spin casting also influenced the extent of phase separation between the core and the shell: Phase contrast was enhanced when using heptane, a solvent selective for the polyisoprene component, but decreased for nonselective solvents such as toluene and chloroform.

6.5.1 Rheological Properties

The viscoelastic properties of arborescent PS-*graft*-PIP copolymers of generations G0–G3 having *cis*-1,4-polyisoprene side chains of different lengths were investigated (Teertstra and Gauthier, 2007) with dynamic mechanical measurements. The zero-shear viscosity (η_o) of the copolymers was 1–3 orders of magnitude lower than for linear polyisoprene samples of comparable molecular weight, but its scaling ($\eta_o \propto M_w^{\nu}, n = 3.5-4$) was nonetheless consistent with entanglement formation. The zero-shear recoverable compliance J_e^o increased linearly with the molecular weight of the copolymers to reach values up to 10 times larger than the limiting value observed for the linear PIP analogues. The modulus–frequency master curves obtained for the G0 and G1 copolymers were comparable to other branched polymers such as star and comb-branched systems. The G2 and G3 copolymers, in contrast, displayed features analogous to microgels and filled polymer systems. The change in behavior was attributed to a transition from a flexible branched structure (for generations G0 and G1) to a spherical rigid structure for generations G2 and above.

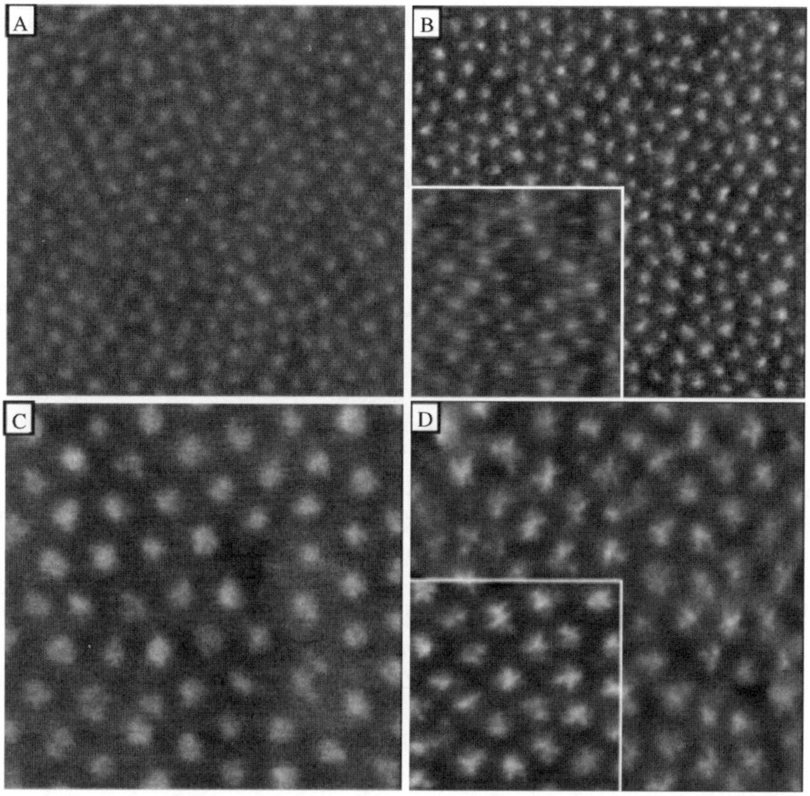

Figure 6.6 AFM images for samples prepared by spin-casting from solutions (concentration 1mg/mL): (A) G0PS-PIP30 in toluene, (B) G0PS-PIP30 in heptane, (C) G1PS-PIP30 in toluene, (D) G1PS-PIP30 in heptane. The width of each picture is 500 nm. All images are shown in the phase mode to allow visualization of the molecular morphology due to differences in viscoelastic behaviors of the polystyrene-rich and polyisoprene-rich phases. The insets of (B) and (D) are the height images, showing the topology of the monolayers. (Adapted with permission from J. Li, S.J. Teertstra, M. Gauthier *et al.*, "Synthesis of arborescent polystyrene-*graft*-polyisoprene copolymers using acetylated substrates," *Macromolecules*, **37**, 795–802. © 2004 American Chemical Society.)

6.6 Arborescent Polystyrene-*graft*-Poly(*tert*-Butyl Methacrylate)

The synthesis of arborescent copolymers incorporating poly(*tert*-butyl methacrylate) (PtBMA) side chains, useful as precursors for highly branched anionic polyelectrolytes, was achieved from bromomethylated polystyrene substrates (Kee and Gauthier, 2008). Two series of arborescent copolymers with either short ($M_w \approx 5000$) or long ($M_w \approx 30\,000$) PtBMA side chains grafted onto linear, G0, G1, and G2 substrates were obtained by chloromethylation of the corresponding arborescent polystyrene and conversion to its bromomethylated analog with NaBr in DMF/dibromomethane. The side chains were generated in THF from *tert*-butyl methacrylate and 1,1-diphenyl-2-methylpentyllithium in the presence of LiCl, and the graft copolymers by addition of a solution of the bromomethylated polystyrene substrate at 0 °C.

Branching functionalities f_w ranging from 9–4500 and molecular weights ranging from 8.8×10^4–6.3×10^7 with narrow MWD ($M_w/M_n \approx 1.05$–1.10) were thus obtained.

A one-pot method for the synthesis of arborescent polystyrene-*graft*-poly(*tert*-butyl methacrylate) copolymers was also reported (Yuan and Gauthier, 2007) following the reaction path illustrated in Scheme 6.4, the purified *tert*-butyl methacrylate monomer being added to the anionic macroinitiator in the last side-chain growth step. The synthesis of G0 and G1 arborescent copolymers was thus achieved by initiating the polymerization of the methacrylate monomer with the linear and G0 styrene-diisopropenylbenzene copolymer substrates activated with *sec*-BuLi, respectively. Copolymers with PtBMA side chains having a target $M_n \approx 2500$, 5000, 10 000, and 20 000 were obtained through successive monomer additions to the polymerization reactor.

6.6.1 Solution Properties

The highly branched structure of the arborescent polystyrene-*graft*-poly(methacrylic acid) polyelectrolytes (PS-*graft*-PMAA), generated by hydrolysis of the *tert*-butyl methacrylate copolymer precursors, should lead to unusual properties in solution. Indeed, it was shown (Kee and Gauthier, 2008) using dynamic light-scattering measurements that the arborescent PMAA copolymers expand more than linear PMAA upon neutralization with NaOH (Figure 6.2b). Interestingly, volume expansion was also more pronounced in this case than for arborescent polyelectrolytes derived from the analogous P2VP copolymers (Figure 6.2a), possibly due to the closer proximity of the charge to the polymer backbone in the PMAA system.

6.7 Arborescent Polystyrene-*graft*-Poly(ethylene Oxide)

The synthesis of an amphiphilic arborescent polystyrene-*graft*-poly(ethylene oxide) (PS-*graft*-PEO) copolymer starting from a G1PS core is represented in Scheme 6.5 as an example (Gauthier *et al.*, 1996). A comb-branched (G0) styrene homopolymer was first synthesized by grafting polystyrene side chains with $M_w \approx 5000$ or 30 000 onto chloromethylated linear polystyrene ($M_w \approx 5000$). The G1 substrate was synthesized by chloromethylation and grafting of the G0 substrate with polystyrene side chains having a molecular weight similar to the G0 substrate but carrying a terminal acetal functionality, obtained by using (6-lithiohexyl) acetaldehyde acetal (LHAA) as the polymerization initiator. The PEO copolymers were obtained by acid-catalyzed hydrolysis of the terminal acetal functionalities, deprotonation of the hydroxyl chain termini with a base (potassium naphthalide), and addition of ethylene oxide to the polyfunctional anionic macroinitiator thus obtained. Before adding the base, a metal–halogen exchange reaction (not shown in Scheme 6.5) served to deactivate residual chloromethyl groups on the substrates and avoid crosslinking in the deprotonation step.

6.7.1 Self-Assembly at the Air/Water Interface

The self-assembly of arborescent PS-*graft*-PEO copolymers spread as monolayers at the air/water interface was recently investigated (Njikang *et al.*, 2008a). AFM imaging was used to examine the influence of molecular structure and composition on the characteristics of the

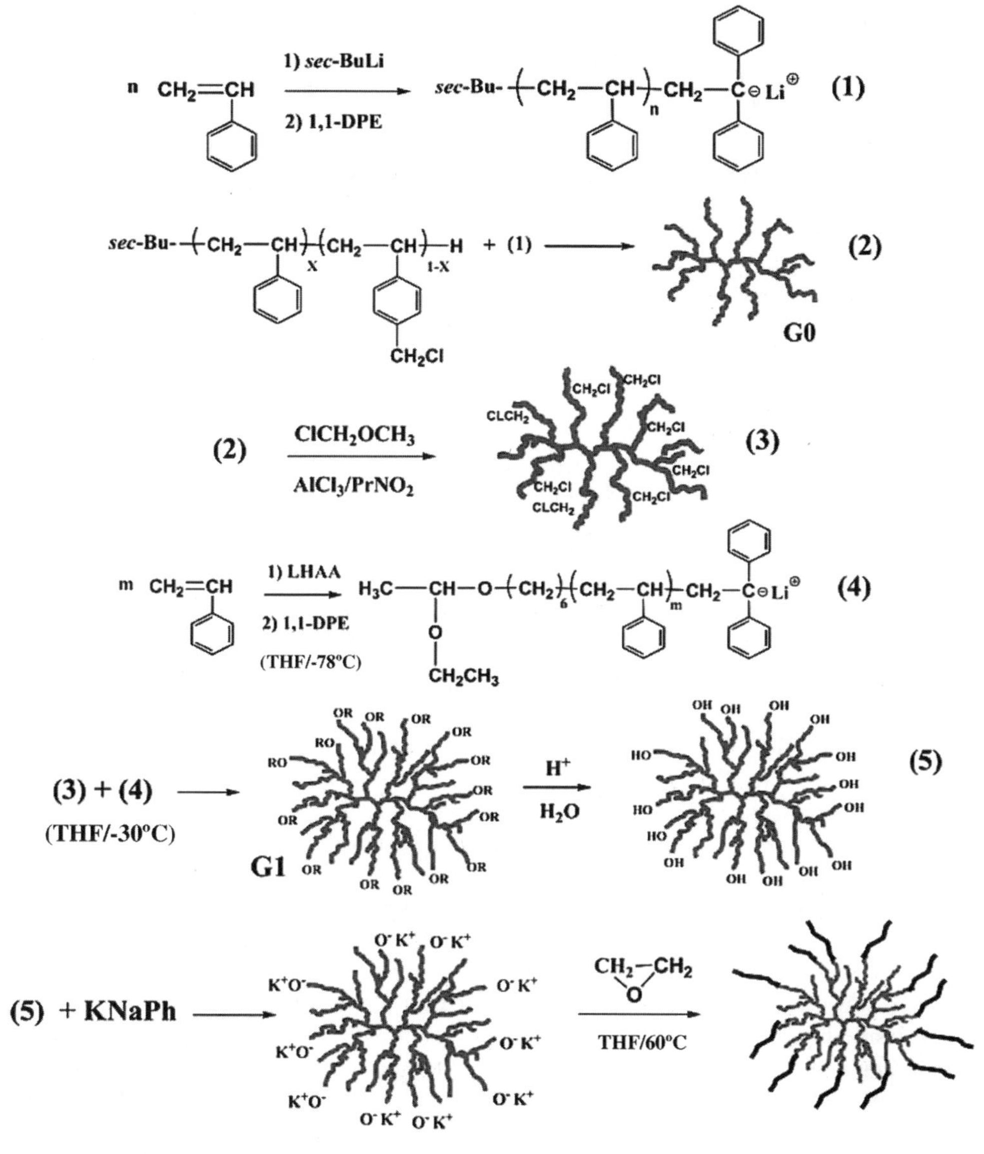

Scheme 6.5 Reaction steps used to synthesize an arborescent copolymer incorporating a G1 polystyrene core and a poly(ethylene oxide) shell.

monolayers formed by the amphiphilic copolymers under these conditions. The copolymer solutions were initially spread on the water subphase without compression. Variations in the amount of solution spread had no influence on the film topologies observed, apart from increasing the number of molecules present on the subphase. The amphiphilic molecules assembled spontaneously at the air/water interface to form stable monolayers that were easily transferred onto solid substrates as Langmuir–Blodgett films. Molecules with PEO weight

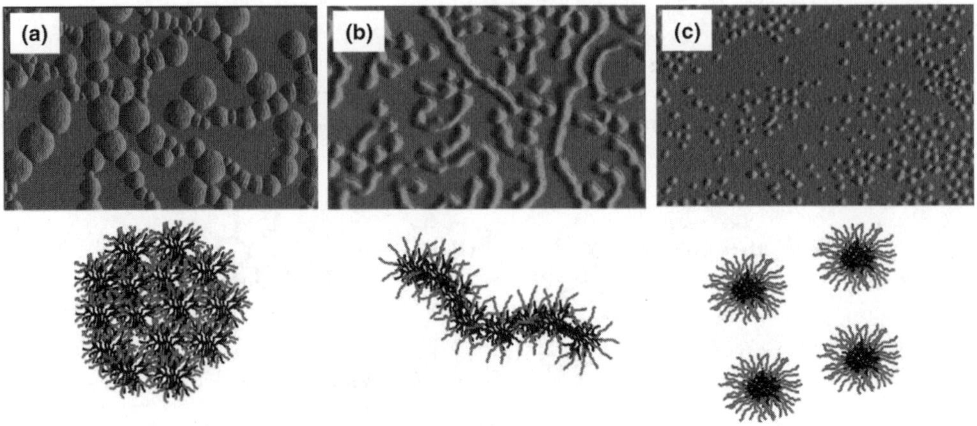

Figure 6.7 Self-assembly of arborescent G1 polystyrene-*graft*-poly(ethylene oxide) copolymers at the air/water interface: Copolymers with PEO contents of (a) 15%, (b) 31%, and (c) 74% by weight. The width of each picture is 1.5 μm. (Adapted with permission from G.N. Njikang, L. Cao and M. Gauthier, "Self-assembly of arborescent polystyrene-*graft*-poly(ethylene oxide) copolymers at the air/water interface," *Macromolecular Chemistry and Physics,* 2008, **209**, 907–918. © Wiley-VCH Verlag GmbH & Co. KGaA.) (Colour version of this figure is available on the book companion web site.)

fractions in the 19–31% range formed ribbon-like structures irrespective of architecture, whereas island-like clusters were obtained for low PEO contents, and nonassociated species dominated for high PEO contents (Figure 6.7). The predominance of van der Waals attractive interactions between the hydrophobic polystyrene cores for low PEO content samples explains the observed coalescence of the molecules into large clusters (Figure 6.7a). For higher PEO contents, the degree of deformation of the chains in the corona should become an important factor controlling the association process. While repulsive forces arising from the elastic deformation of the PEO chains disfavor coalescence, deformation of the PEO corona as the molecules move closer on the water subphase may give rise to significant van der Waals attraction forces between adjacent cores, inducing the formation of the ribbon-like superstructures shown below (Figure 6.7b). According to the proposed model, molecular assembly is limited to side-by-side aggregation due to the PEO chains within the corona being pushed aside and building a thicker stabilizing layer alongside the ribbon superstructures. Along the same line, the predominance of isolated molecules for copolymers with a high PEO content (Figure 6.7c) can be explained by weakened van der Waals core–core interactions vs. strong repulsive forces associated with the deformation of their dense PEO corona. Another factor influencing the assembly process is the structure of the arborescent polystyrene cores: Copolymer micelles incorporating a very rigid core had a more pronounced tendency to form island-like clusters, even for high PEO contents.

Changes in monolayer topology resulting from surface pressure and subphase temperature variations were also reported for selected arborescent PS-*graft*-PEO copolymers (Njikang *et al.*, 2008b). Samples with different branching densities were selected to study the influence of conformational constraints on the assembly process in the presence of these external stimuli. The study was limited to samples with long polystyrene side chains ($M_w \approx 30\,000$) because of their interesting association behavior noticed earlier (Njikang *et al.*, 2008a). The formation of superstructures by the dendritic micelles was enhanced by compression and, to a lesser extent, at higher temperatures. The influence of temperature on film topology was examined for copolymers with high PEO contents transferred at a low surface pressure (4 mN/m), to ensure a uniform distribution of molecules on the water. Molecules with minimal aggregation were observed at 12 °C, whereas a significant number of short ribbon-like features could be identified at 37 °C due to decreased solvation of the PEO chains in the corona. AFM micrographs obtained from monolayers transferred at surface pressures of 0–8 mN/m for low branching functionality ($f_w = 62$) samples with PEO contents ≥22% by weight showed that some of the molecules associated to form short ribbon structures even before compression, albeit low-level aggregates and isolated molecules were also present. Under the influence of compression, however, the number of isolated molecules clearly decreased, while the average length of the ribbons and their packing density both increased. Enhanced ribbon formation upon compression was also noticed for samples with a high branching density ($f_w = 270$), but the ribbon-like structures were significantly shorter. The height and width measured for the ribbons roughly corresponded to the diameter and thickness of the individual molecules for both the low and high branching density samples, indicating that ribbon formation was taking place mainly through side-by-side assembly of the molecules as indicated in Figure 6.7b. AFM images obtained for some of the samples before and after a compression–decompression cycle did not show a significant increase in the extent of association, demonstrating that the compression-induced formation of ribbon-like superstructures was reversible, at least in some cases.

6.8 Arborescent Polyisoprene

The synthesis of arborescent isoprene homopolymers by coupling polyisoprenyllithium with epoxidized polyisoprene substrates as outlined in Scheme 6.6 was recently demonstrated (Yuan

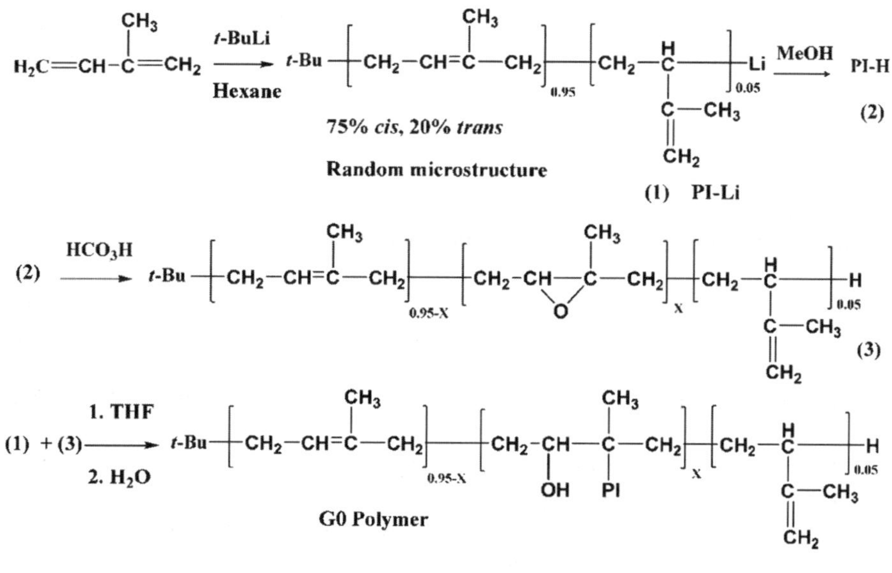

Scheme 6.6 Synthesis of arborescent polyisoprene by grafting onto epoxidized polyisoprene.

and Gauthier, 2005). The reaction path was developed, in analogy to arborescent styrene homopolymers, on the basis of successive functionalization and grafting reaction cycles. Linear polyisoprene was synthesized in hexane with *tert*-butyllithium to afford a large proportion of *cis*-1,4-isoprene units, and modified by epoxidation with performic acid, generated *in situ* from formic acid and hydrogen peroxide. The epoxidized polyisoprene substrate was then reacted with living *cis*-1,4-polyisoprenyllithium in the presence of THF to yield a comb-branched (G0) polymer. In the absence of coupling promoters, the grafting yield attained ranged from 31–78%. The low grafting yield was linked to the slow polymerization of isoprene in hexane, leading to a small amount of residual monomer reacting in the presence of THF to produce 1,2- and 3,4-isoprene units at the chain ends and modifying the reactivity of the macroanions. The 1,2-units are expected to be less reactive than 3,4- and 1,4-units because they are more sterically hindered, which may explain the limiting grafting yields observed in the absence of promoters.

Three different types of promoters were examined to overcome this hurdle: (1) A Lewis base (*N,N,N′,N′*-tetramethylethylenediamine, TMEDA) to complex the lithium counterion and increase the nucleophilicity of the polyisoprenyl anions; (2) Lewis acids to increase the reactivity of the epoxide ring via coordination; and (3) lithium salts to stabilize (decrease the reactivity of) the polyisoprenyl anions by the common-ion effect, while also increasing the reactivity of the epoxide ring, since they are weak Lewis acids. The additives TMEDA, BF_3, and $AlMe_3$ all improved the grafting yield, but LiCl and LiBr were most efficient as promoters. This is presumably due to their ability to minimize the termination of polyisoprenyllithium as well as to promote the coupling reaction with the epoxide functionalities. Using a 25% excess of epoxide functionalities versus living ends and lithium salt promoters, grafting yields reaching 92% were attained in the G0 polymer synthesis. Repetition of the epoxidation and grafting reactions also produced G1 and G2 arborescent polyisoprene with $M_w \approx 5000$ side chains in 83 and 75% yields, respectively, and polydispersity indices $M_w/M_n \leq 1.05$. The attempted synthesis of a G3 polymer failed, possibly due to the inaccessibility of coupling sites within the highly branched substrates and/or limited solubility of the G2 epoxidized substrate.

6.9 Conclusions

In this chapter we presented various synthetic methods developed for the preparation of dendrigraft (arborescent) homo- and copolymers achieved in our research group over the past ten years. These techniques yield polymers differing in terms of composition, distribution and density of branching sites, ultimately leading to distinct molecular architectures and morphologies. Some of the interesting physical properties observed for arborescent polymers have been outlined; these open the door to a wide range of applications in areas including microencapsulation (solubilization and release of hydrophobic compounds), smart materials (temperature- and pressure-sensitive self-assembly), and catalysis (metallic nanoparticle preparation). Our current synthetic and physical characterization efforts aim at further extending the scope of arborescent polymer chemistry to develop novel grafting strategies and new materials derived from fully biocompatible components, while also examining other potential applications including polymer processing additives and surface modification. Reports on these new developments are forthcoming.

References

Briber, R.M., Bauer, B.J., Hammouda, B., and Tomalia, D. (1992) Small-angle neutron scattering from solutions of dendrimer molecules: Intermolecular interactions. *Polymeric Materials Science and Engineering*, **67**, 430–431.

Choi, S., Briber, R.M., Bauer, B.J. *et al.* (2000) Small-angle neutron scattering of blends of arborescent polystyrenes. *Macromolecules*, **33**, 6495–6501.

Dockendorff, J., Gauthier, M., Mourran, A., and Möller, M. (2008) Arborescent amphiphilic copolymers as templates for the preparation of gold nanoparticles. *Macromolecules*, **41**, 6621–6623.

Gauthier, M. (2000) Arborescent polymers: Designed macromolecules with a dendritic structure, in *Tailored Polymers and Applications* (eds. Y. Yagci, M.K. Mishra, O. Nuyken *et al.*) VSP, Utrecht, pp. 43–61.

Gauthier, M. and Möller, M. (1991) Uniform highly branched polymers by anionic grafting: Arborescent graft polymers. *Macromolecules*, **24**, 4548–4553.

Gauthier, M., Tichagwa, L., Downey, J.S., and Gao, S. (1996) Arborescent graft copolymers: Highly branched macromolecules with a core-shell morphology. *Macromolecules*, **29**, 519–527.

Gauthier, M., Chung, J., Choi, L., and Nguyen, T.T. (1998) Second virial coefficient of arborescent polystyrenes and its temperature dependence. *Journal of Physical Chemistry, Part B*, **102**, 3138–3142.

Gauthier, M., Li, J., and Dockendorff, J. (2003) Arborescent polystyrene-*graft*-poly(2-vinylpyridine) copolymers as unimolecular micelles. Synthesis from acetylated substrates. *Macromolecules*, **36**, 2642–2648.

Il Yun, S., Briber, R.M., Kee, R.A., and Gauthier, M. (2003) Small-angle neutron scattering of arborescent polystyrene-*graft*-poly(2-vinylpyridine) copolymers. *Polymer*, **44**, 6579–6787.

Il Yun, S., Briber, R.M., Kee, R.A., and Gauthier, M. (2006) Dilute-solution structure of charged arborescent graft polymer. *Polymer*, **47**, 2750–2759.

Il Yun, S., Gadd, G.E., Lo, V. *et al.* (2008a) Temperature-responsive supramolecular assembly of dendrigraft micelles with a solvophilic core – solvophobic shell structure. *Macromolecules*, **41**, 7166–7172.

Il Yun, S., Lai, K.-C., Briber, R.M. *et al.* (2008b) Conformation of arborescent polymers in solution by small-angle neutron scattering: Segment density and core-shell morphology. *Macromolecules*, **41**, 175–183.

Kee, R.A. and Gauthier, M. (2002) Arborescent polystyrene-*graft*-poly(2-vinylpyridine) copolymers: Synthesis and enhanced polyelectrolyte effect in solution. *Macromolecules*, **35**, 6526–6532.

Kee, R.A. and Gauthier, M. (2008) Arborescent polystyrene-*graft*-poly(*tert*-butyl methacrylate) copolymers: Synthesis and enhanced polyelectrolyte effect in solution. *Journal of Polymer Science, Part A: Polymer Chemistry*, **46**, 2335–2346.

Kee, R.A., Gauthier, M., and Tomalia, D.A. (2001) Semi-controlled dendritic structure synthesis, in *Dendrimers and Other Dendritic Polymers* (eds. J.M.J. Fréchet and D.A. Tomalia) Wiley, New York, pp. 209–236.

Li, J. and Gauthier, M. (2001) A novel synthetic path to arborescent graft polystyrenes. *Macromolecules*, **34**, 8918–8924.

Li, J., Teertstra, S.J., Gauthier, M. *et al.* (2004) Synthesis of arborescent polystyrene-*graft*-polyisoprene copolymers using acetylated substrates. *Macromolecules*, **37**, 795–802.

Lieu, J.G., Prausnitz, J.M., and Gauthier, M. (2000) Vapor-liquid equilibria for binary solutions of arborescent and linear polystyrenes. *Polymer*, **41**, 219–224.

Munam, A. and Gauthier, M. (2008) Large-scale synthesis of arborescent polystyrenes. *Journal of Polymer Science, Part A: Polymer Chemistry*, **46**, 5742–5751.

Njikang, G.N., Cao, L., and Gauthier, M. (2008a) Self-assembly of arborescent polystyrene-*graft*-poly(ethylene oxide) copolymers at the air-water interface. *Macromolecular Chemistry and Physics*, **209**, 907–918.

Njikang, G.N., Cao, L., and Gauthier, M. (2008b) Pressure- and temperature-induced association of arborescent polystyrene-*graft*-poly(ethylene oxide) copolymers at the air-water interface. *Langmuir*, **24**, 12919–12927.

Njikang, G., Gauthier, M., and Li, J. (2008c) Arborescent polystyrene-*graft*-poly(2-vinylpyridine) copolymers as unimolecular micelles. Solubilization studies. *Polymer*, **49**, 1276–1284.

Njikang, G.N., Gauthier, M., and Li, J. (2008d) Sustained release properties of arborescent polystyrene-*graft*-poly (2-vinylpyridine) copolymers. *Polymer*, **49**, 5474–5481.

Rempp, P., Franta, E., and Herz, J.-E. (1988) Macromolecular engineering by anionic methods. *Advances in Polymer Science*, **86**, 145–173.

Roovers, J. (1985) Branched polymers, in *Encyclopedia of Polymer Science and Engineering*, 2nd edn, vol. **2** (ed. J.I. Kroschwitz) Wiley, New York, pp. 478–499.

Striolo, A., Prausnitz, J.M., Bertucco, A. *et al.* (2001) Dilute-solution properties of arborescent polystyrenes: Further evidence for perturbed-hard-sphere behavior. *Polymer*, **42**, 2579–2584.

Teertstra, S.J. and Gauthier, M. (2004) Dendrigraft polymers: Macromolecular engineering on a mesoscopic scale. *Progress in Polymer Science*, **29**, 277–327.
Teertstra, S.J. and Gauthier, M. (2007) Viscoelastic properties of arborescent polystyrene-*graft*-polyisoprene copolymers. *Macromolecules*, **40**, 1657–1666.
Tomalia, D.A. and Fréchet, J.M.J. (2001) Introduction to the dendritic state, in *Dendrimers and Other Dendritic Polymers* (eds. J.M.J. Fréchet and D.A. Tomalia) Wiley, New York, pp. 3–44.
Tomalia, D.A., Hedstrand, D.M., and Ferritto, M.S. (1991) Comb-burst dendrimer topology: New macromolecular architecture derived from dendritic grafting. *Macromolecules*, **24**, 1435–1438.
Yuan, Z. and Gauthier, M. (2005) Arborescent isoprene homopolymers. *Macromolecules*, **38**, 4124–4132.
Yuan, Z. and Gauthier, M. (2006) One-pot synthesis of arborescent polystyrenes. *Macromolecules*, **39**, 2063–2071.
Yuan, Z. and Gauthier, M. (2007) Synthesis of arborescent copolymers by a one-pot method. *Macromolecular Chemistry and Physics*, **208**, 1615–1624.

7

Hyperbranched Glyco-Conjugated Polymers

Toshifumi Satoh and Toyoji Kakuchi
Division of Biotechnology and Macromolecular Chemistry, Faculty of Engineering, Hokkaido University, Sapporo, 060-8628, Japan

7.1 Introduction

The spherical architectures of highly branched macromolecules, such as dendrimers, star-shaped polymers, and hyperbranched polymers, have attracted much attention from the viewpoint of nanotechnology, because their numerous terminal units can be converted into various functional groups leading to novel nanomaterials (Zeng and Zimmerman, 1997; Hirao *et al.*, 2007; Satoh, 2009). Thus, various types of dendrimers, star polymers, and hyperbranched polymers have been synthesized and their properties were compared to the linear analogues (Stiriba *et al.*, 2002; Hirao *et al.*, 2005).

Spherical macromolecular architectures with their surfaces covered with sugars are of increasing interest due to their unique structures along with specific properties, such as the glyco-cluster or multivalent effect on carbohydrate–protein interactions. For a spherical polymer with many terminal sugar residues, there are a few types of glycodendrimers, for example, dendrimers prepared from the reaction of the dendrimer surface with mono- and disaccharides (sugar ball) (Lundquist and Toone, 2002), and dendrimers prepared through the convergent method using semidendritic branched oligosaccharides (Jayaraman *et al.*, 1997). Although the synthesis of such glycodendrimer types was accomplished, dendrimers essentially consisting of sugar units, that is, dendritic carbohydrate polymers, have not been prepared until now due to the complicated and time-consuming synthetic process involving numerous protection, deprotection, and purification procedures. Star-shaped glyco-conjugated polymers were also prepared using the "core-first" method based on the RAFT polymerization (Bernard *et al.*, 2005) and the "arm-first" method based on the TEMPO-mediated radical polymerization

Complex Macromolecular Architectures: Synthesis, Characterization, and Self-Assembly, First Edition.
Edited by Nikos Hadjichristidis, Akira Hirao, Yasuyuki Tezuka and Filip Du Prez.

(Narumi and Kakuchi, 2008). Hyperbranched glyco-conjugated polymers from the polymerization of AB_m-type monomers can be prepared more easily than glycodendrimers and star-shaped glyco-conjugated polymers (Kadokawa *et al.*, 1998). Recently, we proposed that the ring-opening multibranching polymerization of anhydro and dianhydro sugars as latent AB_m-type monomers should be a convenient method of synthesizing hyperbranched glyco-conjugated polymers as a novel spherical macromolecular architecture (Satoh and Kakuchi, 2007). For our synthetic strategy of hyperbranched glyco-conjugated polymers, it is important to elucidate the effect of the structure of the sugar monomers on the ring-opening multibranching polymerization tendency and to clarify the relationship between the structure and solution properties of the hyperbranched glyco-conjugated polymers.

In this chapter, we introduce our comprehensive study of synthesizing novel hyperbranched glyco-conjugated polymers from the ring-opening multibranching polymerizations of 1,6-anhydro-β-D-hexopyranose, 1,4-anhydrotetritol, 2,3-anhydrotetritol, 1,2:5,6-dianhydro-D-mannitol, and 5,6-anhydro-1,2-*O*-isopropylidene-α-D-glucofuranose. In addition, we also introduce the synthesis and property of novel unimolecular reversed micelles consisting of a hyperbranched glyco-conjugated polymer core and a trityl (Kitajyo *et al.*, 2007a), polylactide (Satoh *et al.*, 2006) or L-leucine ethyl ester (Kitajyo *et al.*, 2007b) shell, which had encapsulation and slow-release abilities for hydrophilic molecules.

7.2 Synthesis of Hyperbranched Glyco-Conjugated Polymers

7.2.1 Hyperbranched Polysaccharides by Polymerization of 1,6-Anhydro-β-D-hexopyranose

Schuerch *et al.* and other groups have reported the synthesis of highly branched polysaccharides prepared by the ring-opening polymerization of 1,6-anhydro-β-D-hexopyranose using monochloroacetic acid (Carvalho *et al.*, 1959). This solid-phase polymerization, however, has the disadvantage of a heterogeneous initiation reaction, which may lead to a polymer with a broad polydispersity. In order to alleviate this problem, we chose the solution polymerizations of 1,6-anhydro-β-D-hexopyranose (**1a–c**) using a thermally induced cationic initiator, for example, (*S*-2-butenyl) tetramethylenesulfonium hexafluoroantimonate (**2**) in propylene carbonate, as shown in Figure 7.1 (Satoh *et al.*, 2003; Satoh *et al.*, 2004; Satoh *et al.*, 2005). Table 7.1 summarizes the polymerization results of 1,6-anhydro-β-D-glucopyranose (**1a**), 1,6-anhydro-β-D-mannopyranose (**1b**), and 1,6-anhydro-β-D-galactopyranose (**1c**).

When propylene carbonate was used as the solvent, the resulting polysaccharide was a gel-free white solid that was completely soluble in water and dimethyl sulfoxide, slightly soluble in pyridine, and insoluble in other organic solvents such as toluene, chloroform, acetone and methanol. The weight-average molecular weights ($M_{w,SLS}$) measured by static light scattering (SLS) of the resulting polysaccharide (**Poly1a–c**) gradually increased with the increasing polymerization temperature and monomer concentration, and varied in the range from 2.2×10^4 to 6.4×10^5. The values were significantly higher than the weight-average molecular weight values ($M_{w,SEC}$: 2.1×10^3 to 1.1×10^4) measured by size-exclusion chromatography (SEC). In general, highly branched polymers, such as dendrimers and hyperbranched polymers, are known to have spherical conformations in solution, and the $M_{w,SEC}$ values were claimed to be too low because the hydrodynamic volumes of these polymers are smaller than those of the corresponding linear polymers used for calibration. The results of the molecular

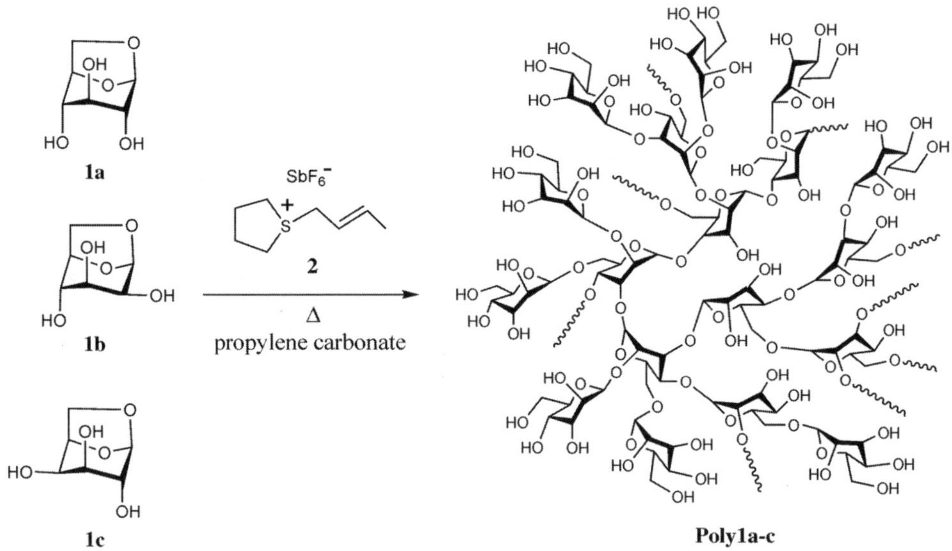

Figure 7.1 Synthesis of hyperbranched polysaccharide (**Poly1a–c**) by cationic ring-opening polymerization of 1,6-anhydro-β-D-hexopyranose (**1a–c**).

weight suggest that **Poly1a–c** have a more compact form in solution compared to the linear polymer because of their branched structure. The polydispersities ($M_{w,SEC}/M_{n,SEC}$) for **Poly1a–c** were found to be relatively narrow values in the range of 1.13–1.43 and increased with increasing monomer conversion. The melt polymerization of **1b** using **2** at 210 °C immediately occurred to produce a brown cake, and the precipitated polysaccharide was a light

Table 7.1 Thermally induced cationic polymerization of 1,6-anhydro-β-D-hexopyranose (**1a–c**) using (S-2-butenyl)tetramethylenesulfonium hexafluoroantimonate (**2**)[a]

M	[M] mol L^{-1}	[M]/[**2**]	Time min.	Temp. °C	Yield %[b]	$M_{w,SLS}$[c] $\times 10^{-3}$	$[\alpha]_D$[d]	DB[e]
1a	4.0	700	40	150	60.0	33.0	+99.5	0.38
	6.4	350	20	130	35.9	24.5	+90.6	0.37
	6.4	700	40	150	73.8	70.6	+101.2	0.38
1b	6.4	350	20	130	61.6	80.5	+64.7	0.43
	6.4	700	40	150	63.0	637	+64.4	0.44
	Melt	350	<1	210	23.8	80.9	+51.9	0.44
1c	6.4	350	20	130	30.7	22.3	+95.8	0.50
	6.4	700	40	150	63.1	26.9	+86.2	0.49

[a] Solvent, propylene carbonate.
[b] Water-soluble but methanol-insoluble parts.
[c] Determined by static light scattering (SLS) in 0.2 mol L^{-1} aqueous $NaNO_3$ solution.
[d] Measured in H_2O at 23 °C (c 1.0).
[e] Degree of branching.

brown solid with an $M_{w,SLS}$ of 8.1×10^4 in 24% yield. The $M_{w,SEC}/M_{n,SEC}$ (1.84) was broader than that obtained from the solution polymerizations in propylene carbonate.

In the viscosity measurements of **Poly1a–c** in aqueous $NaNO_3$ solution (0.2 mol L^{-1}, 23 °C), the reduced viscosities (η_{sp}/c) linearly increased with the increase of the polymer concentration (0–10 g L^{-1}), meaning that no aggregation of **Poly1a–c** occurred over the range of the measured polymer concentration. The solution viscosities of **Poly1a–c** were also very low, that is, the intrinsic viscosities ([η]) were in the range of 0.023–0.047 dL g^{-1}, while the [η] values for the linear polymers of (1 → 6)-α-D-glucan (M_n of 4.3×10^4) and (1 → 6)-α-D-mannan (M_v of 8.1×10^4) were 0.37 (Uryu and Schuerch, 1971) and 0.41 dL g^{-1} (Tkacz *et al.*, 1972) in aqueous solution. In the steady shear flow measurement in the aqueous **Poly1a** solutions (30–60 wt% at 25 °C), all the aqueous **Poly1a** solutions ranging in shear rate from 10 to 382 s^{-1} exhibited Newtonian behavior with steady shear viscosities independent of the shear rate, even at the high concentrations, as shown in Figure 7.2. These experimental results from the viscosity and steady shear flow measurements of the hyperbranched polysaccharides indicated that the very low viscosities and the Newtonian behavior were attributed to the spherical structure of the hyperbranched polysaccharides without any physical entanglements.

In order to confirm the branching and manner of linkage of the repeating units, a methylation analysis of **Poly1a–c** was carried out according to a reported method (Satoh *et al.*, 2003; Satoh *et al.*, 2005). The identified repeating units were classified into five categories of terminal units (T), linear units (L), two semidendritic units with three and four linkages (sD_1 and sD_2, respectively), and perfect dendritic units (D) with five linkages, as shown in Figure 7.3. The main components were D-hexopyranosyl terminal units (T-*p*, 25–35 mol%) and

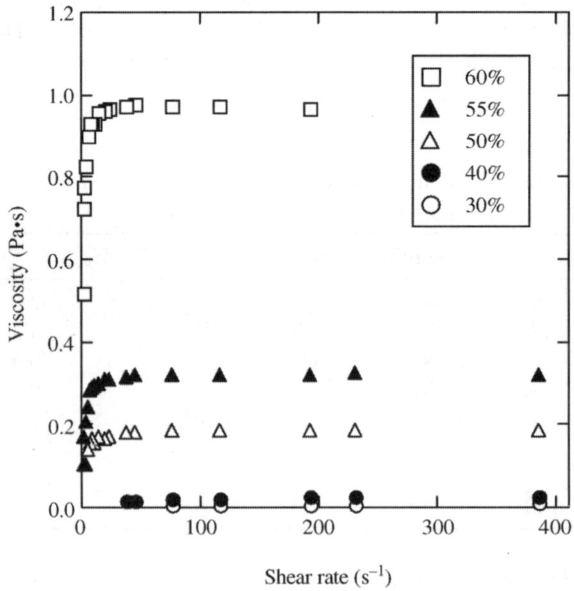

Figure 7.2 Steady shear viscosity as a function of shear rate for **Poly1a** ($M_{w,\ SLS}$ of 52 400) in aqueous solution at 25 °C.

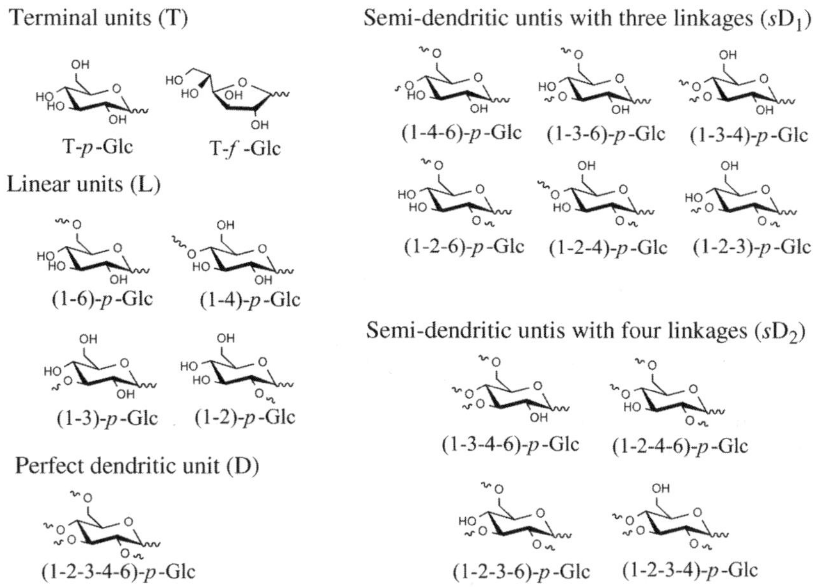

Figure 7.3 Repeating units in hyperbranched D-glucan (**Poly1a**). T-*p* and T-*f* mean a nonreducing terminal D-glucopyranosyl and a D-glucofuranosyl residue, respectively. (1–4)-*p* and (1–6)-*p* mean a (1 → 4)-linked D-glucopyranosyl and a (1 → 6)-linked glucopyranosyl residue, respectively, and so on.

(1 → 6)-linked D-hexopyranosyl units ((1–6)-*p*, 20–30 mol%). The results of the methylation analysis provided information about the reactivity of the three hydroxyl groups in **1a–c**. For the solution polymerization of **1a**, the C-3 hydroxyl groups were less reactive than the C-2 and C-4 hydroxyl groups, that is, the reactivity increased in the order of C-2 = C-4 > C-3, but C-2 > C-4 > C-3 for **1b**, and C-3 > C-2 > C-4 for **1c**.

The formation of the hyperbranched polysaccharides proceeded through a mechanism with the ring-opening reaction of the 1,6-anhydro moieties and the proton-transfer reaction, as shown in Figure 7.4. In the polymerization system, **1** was rapidly protonated at the 1,6-ether oxygen atom and the protonated **1** underwent the ring opening to afford the carbonium-oxonium ion (**3**). When **3** reacted with the 1,6-ether oxygen atom of a second monomer, two kinds of α- and β-D-linked disaccharides (**4**) were formed. The process produced a terminal unit and a (1 → 6)-linkage. On the contrary, when **3** reacted with the hydroxyl group of another monomer, a disaccharide monomer (**5**) with a terminal residue and a (1 → 2)-, (1 → 3)-, or (1 → 4)-linkage was obtained along with the regeneration of a proton. A branched structure was obtained by the reaction of **3** (or **4**) with a hydroxyl group of the resulting oligo- and polysaccharides. These reactions mentioned above simultaneously occurred in the polymerization systems and consequently produced the hyperbranched polysaccharides **Poly1a-c**.

To determine the degree of branching (DB) for the hyperbranched polysaccharide, **1a–c** were classified as latent AB_4-type monomers, which have one anhydro moiety capable of ring opening and three hydroxyl groups in a molecule, even though the reactivity of their hydroxyl groups was expected to be different in each case. The DBs of **Poly1a–c**, therefore, can be

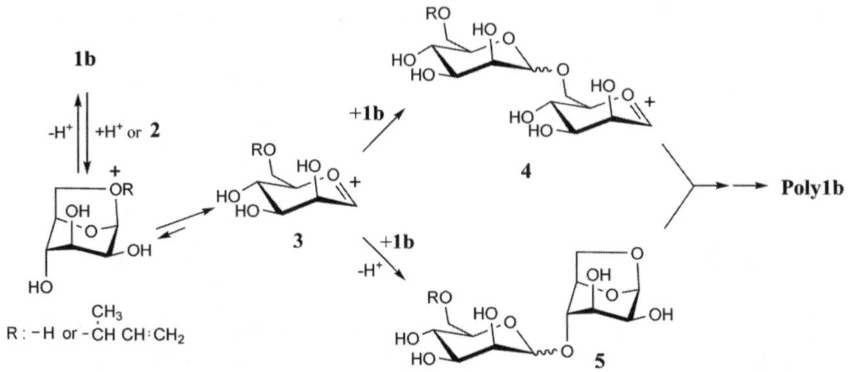

Figure 7.4 Proposed mechanism for the thermally induced cationic polymerization of 1,6-anhydro-β-D-mannopyranose (**1b**).

calculated from the ratios of T, L, sD_1, sD_2, and D units using Frey's equation (Hölter and Frey, 1997). Assuming that the B groups have the same reactivity, the statistical DB value for the polymerization of the AB_4-type monomers is 0.42 (Hölter and Frey, 1997). For the thermally induced cationic polymerization of **1a–c**, the DB values of **Poly1a-c** were in the range of 0.37–0.50. As a result, the thermally induced cationic polymerization of **1a–c** with **2** is a facile method leading to a hyperbranched polysaccharide with a high DB value.

7.2.2 Hyperbranched Polytetritol by Polymerization of 1,4-Anhydrotetritol

To utilize hyperbranched glyco-conjugated polymers as functional raw materials, it is important to expand the limit and scope of the ring-opening multibranching polymerization method in connection with the structure of the applicable monomer and the resulting polymers. Thus, there are various kinds of anhydro sugars, so that it is of great interest to elucidate the characteristics of the ring-opening multibranching polymerization based on monomer structures. In this section, we introduce the ring-opening multibranching polymerization of 1,4-anhydroerythritol (**6a**) and 1,4-anhydro-L-threitol (**6b**) using trifluoromethanesulfonic acid (CF_3SO_3H) or fluorosulfonic acid (FSO_3H), as shown in Figure 7.5 (Imai *et al.*, 2004).

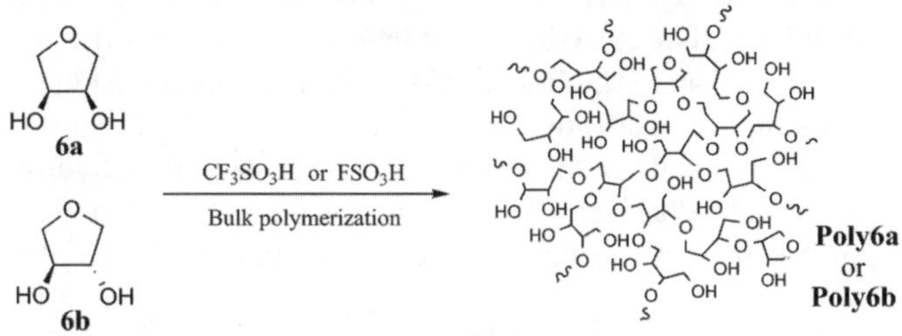

Figure 7.5 Synthesis of hyperbranched glyco-conjugated polymer (**Poly6a** and **Poly6b**) by ring-opening multibranching polymerization of 1,4-anhydroerythritol (**6a**) and 1,4-anhydro-L-threitol (**6b**) using trifluoromethanesulfonic acid (CF_3SO_3H) or fluorosulfonic acid (FSO_3H).

Table 7.2 Bulk polymerization of 1,4-anhydroerythritol (**6a**) and 1,4-anhydro-L-threitol (**6b**) using CF_3SO_3H or FSO_3H

M	Cat.	Temp. °C	[M]/[Cat.]	Time h	Yield%[a]	$M_{w,SLS}$[b] $\times 10^{-4}$	DB[c]
6a	CF_3SO_3H	60	10	240	43.7	3.2	0.45
6a	CF_3SO_3H	60	20	240	46.3	5.9	0.46
6a	CF_3SO_3H	60	40	240	37.4	11.9	0.47
6a	CF_3SO_3H	80	20	120	43.5	51.0	0.33
6a	FSO_3H	60	10	240	47.7	1.0	0.28
6b	CF_3SO_3H	80	20	120	41.0	22.6	0.37

[a] THF insoluble parts.
[b] Determined by static light scattering (SLS) in 0.2 mol L^{-1} aqueous $NaNO_3$ solution.
[c] Degree of branching.

The bulk polymerizations of **6a** and **6b** were carried out using CF_3SO_3H or F_3SO_3H at 60–80 °C. After the initiator was added to the monomer, the polymerization proceeded homogeneously, and the viscosity of the polymerization systems immediately increased with the polymerization progress. The resulting product was a white powdery polymer, which was soluble in water, sparingly soluble in methanol, and insoluble in other organic solvents. Table 7.2 summarized the polymerization results. For the polymerization of **6a** using CF_3SO_3H, the $M_{w,SLS}$s of the resulting polymer (**Poly6a**) were in the range from 3.2×10^4 to 5.1×10^5. On the other hand, in the polymerization of **6a** using FSO_3H as a catalyst, the $M_{w,SLS}$ was as low as 1.0×10^4, which should be related to the lower acidity of FSO_3H ($pK_a = -12.6$) in contrast to CF_3SO_3H ($pK_a = -19.1$) (Thiem *et al.*, 1989).

The $M_{w,SLS}$ values of the polymers (**Poly6b**) prepared from **6b** were lower than those of **Poly6a**, that is, for the [**6**]/[CF_3SO_3H] molar ratio of 20 at 80 °C for 120 h, the $M_{w,SLS}$ value of **Poly6b** was 2.26×10^5, while that of **Poly6a** was 5.10×10^5. The lower $M_{w,SLS}$ value of **Poly6b** compared to that of **Poly6a** should be caused by the difference in the ring-strain effect between **6a** and **6b**, that is, the ring-opening reaction of the tertrahydrofuran ring in **6a** more easily proceeded than that of **6b**.

Based on the NMR measurements, **Poly6a** consisted of numerous erythritol terminal units, which were characteristic of a hyperbranched polymer, that is, **Poly6a** is a hyperbranched glyco-conjugated polymer. Monomers **6a** and **6b** are tetrahydrofurans having two hydroxy groups in the molecule; therefore, they are classified as latent AB_3-type monomers. The DBs of **Poly6a** and **Poly6b** prepared from the AB_3-type monomers are determined from the ratio of the terminal unit using Frey's equation (Hölter and Frey, 1997). The DBs of **Poly6a** were determined to be 0.28–0.47, while the DB of **Poly6b** was determined to be about 0.37, as summarized in Table 7.2.

Hyperbranched glyco-conjugated polymer **Poly6a** mainly consisted of linear and branching erythritol units. However, **Poly6a** also contained the tetrahydrofuranosyl units of 2-bonded 1,4-anhydroerythritol (**8a**) as the terminal unit and 2,3-bonded 1,4-anhydroerythritol (**9a**) as the linear unit, which were estimated to be 6.8–14.7% in **Poly6a**. On the other hand, polymer **Poly6b** consisted of about 67% of the L-threitol unit and about 33% of the 1,4-anhydro-L-threitol unit. Figure 7.6 represents the proposed mechanism for the polymerization of **6a**. In the polymerization system, the 1,4-ether oxygen atom of **6a** was protonated, and the protonated **6a** was reacted with the 1,4-ether oxygen atom of another monomer, thus leading to a 1,4-bonded

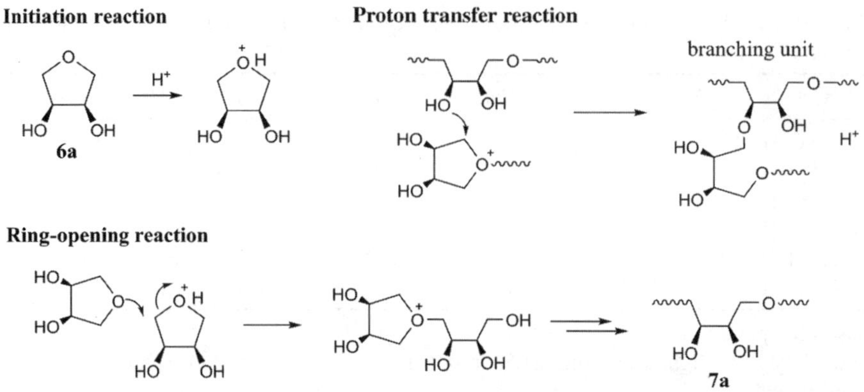

Figure 7.6 Proposed mechanism for cationic polymerization of 1,4-anhydroerythritol (**6a**).

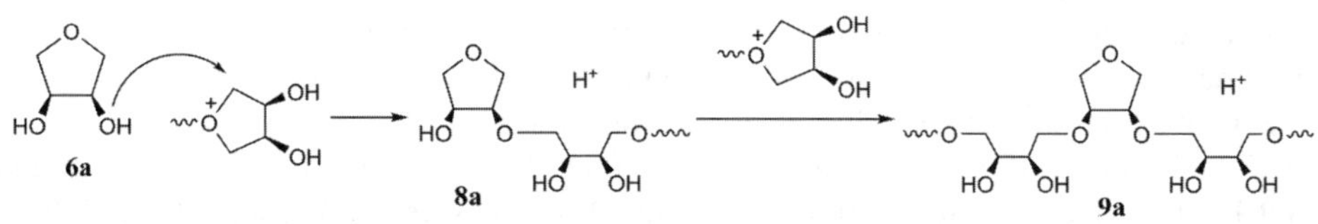

Figure 7.7 Proposed mechanism for the formation of 1,4-anhydroerythritol units, **8a** and **9a**.

erythritol unit (**7a**). The branched unit for **Poly6a** should be produced by the proton-transfer reaction of the hydroxyl groups in the erythritol unit with the oxonium cation of the protonated **6a**. The 1,4-anhydroerythritol units **8a** and **9a** should be formed by the proton-transfer reaction of the hydroxyl groups in **6a** with an oxonium cation in the polymer without the ring-opening reaction, as shown in Figure 7.7. For the polymerization mechanism of **6a**, Kubisa *et al.* also reported that the elimination of water occurs in the polymerization process, so the formed polymer contains considerably fewer hydroxyl groups than could be expected based on the monomer structure (Bednarek and Kubisa, 2006).

7.2.3 Hyperbranched Polytetritol by Polymerization of 2,3-Anhydrotetritol

For the synthesis of hyperbranched polytetritol by the ring-opening multibranching polymerization of 1,4-anhydroerythritol (**6a**) and 1,4-anhydro-L-threitol (**6b**) using super acids such as CF_3SO_3H and FSO_3H, the resulting polymer contained the 1,4-anhydrotetritol unit as an unexpected unit (Imai *et al.*, 2004), as mentioned in the previous section. Therefore, it is important to synthesize and characterize hyperbranched polytetritol without any anhydrotetritol unit. In this section, we introduce the synthesis of a hyperbranched polytetritol essentially consisting of tetritol units by the ring-opening multibranching polymerization of 2,3-anhydroerythritol (**10a**) and 2,3-anhydro-DL-threitol (**10b**) (Imai *et al.*, 2005), which are the structural isomers of 1,4-anhydrotetritol (**6a** and **6b**), as shown in Figure 7.8.

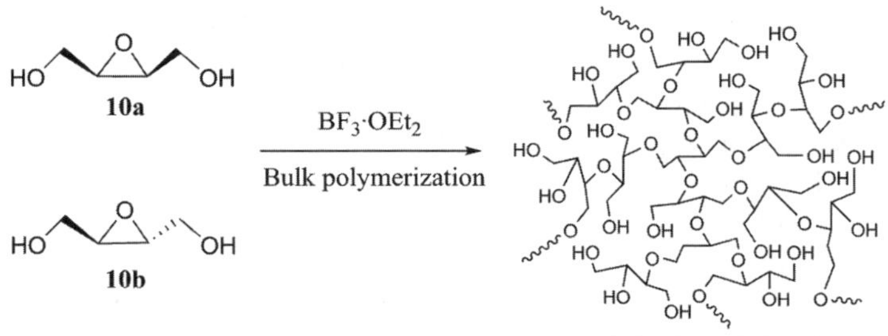

Figure 7.8 Synthesis of hyperbranched polytetritol (**Poly10a** and **Poly10b**) by ring-opening multibranching polymerization of 2,3-anhydroerythritol (**10a**) and 2,3-anhydro-DL-threitol (**10b**), respectively.

The bulk polymerization of 2,3-anhydroerythritol (**10a**) and 2,3-anhydro-DL-threitol (**10b**) was carried out using boron trifluoride diethyl etherate ($BF_3 \cdot OEt_2$) as the initiator. The resulting polymers were white solids, which were soluble in water, sparingly soluble in methanol, and insoluble in other organic solvents. Table 7.3 summarized the polymerization results. The $M_{w,SLS}$ value and the yield of the resulting polymer increased with the rising polymerization temperature; that is, when the [**10a**]/[$BF_3 \cdot OEt_2$] molar ratio of 100 was used for 120 h, then the $M_{w,SLS}$ value was 2.67×10^5 in 44.8% yield at 80 °C and 1.61×10^6 in 57.3% yield at 100 °C. The $M_{w,SLS}$ value and the yield also increased with the extension of the polymerization time. The $M_{w,SLS}$ values of **Poly10a** and **Poly10b** were higher than those of the hyperbranched glyco-conjugated polymers prepared by the ring-opening multibranching polymerizations of **6a** and **6b**.

In the ^1H NMR spectra of **Poly10a** and **Poly 10b**, the characteristic absorption due to the epoxy proton disappeared, indicating that the epoxy group was consumed and that the resulting polymers essentially consisted of tetritol repeating units. Because the ring strains of the oxirane-type monomers of **10a** and **10b** were larger than those of the tetrahydrofuran-type

Table 7.3 Bulk polymerization of 2,3-anhydroerythritol (**10a**) and 2,3-anhydro-DL-threitol (**10b**) using $BF_3 \cdot OEt_2$

M	Temp. °C	Time h	[M]/[$BF_3 \cdot OEt_2$]	Yield %[a]	$M_{w,SLS}$[b] $\times 10^{-5}$	DB[c]
10a	80	120	100	44.8	2.67	0.47
	100	60	100	41.2	3.38	0.47
	100	100	100	57.3	16.1	0.47
	100	120	100	60.8	31.2	0.47
	100	120	200	51.8	32.0	0.47
10b	100	120	100	61.4	30.8	0.45

[a] THF insoluble parts.
[b] Determined by static light scattering (SLS) in 0.2 mol L^{-1} aqueous $NaNO_3$ solution.
[c] Degree of branching.

monomers of **6a** and **6b**, the polymerizations of **10a** and **10b** led to hyperbranched polytetritols without the 2,3-anhydroerythritol and 2,3-anhydro-L-threitol units as the anhydrotetritol units.

Monomers **10a** and **10b** have an epoxy and two hydroxy groups in the molecule, so that they are classified as latent AB_3-type monomers. The DBs estimated by the ^{13}C NMR spectra of **Poly10a** and **Poly 10b** were 0.47 and 0.45, respectively. Although the DBs of **Poly 6a** and **Poly 6b** varied in the range of 0.28–0.47, the DBs of **Poly 10a** and **Poly 10b** did not change according to the polymerization condition.

The solution viscosities of **Poly10a** and **Poly 10b** were extremely low, and the $[\eta]$s were in the range of 0.019–0.025 $dL\,g^{-1}$. The α value in the Mark–Houwink–Sakurada equation ($[\eta] = KM^{\alpha}$) was 0.105. This small value was consistent with a highly branched compact structure.

Because hyperbranched glyco-conjugated polymers **Poly10a** and **Poly 10b** mainly consisted of DL-threitol and erythritol repeating units, respectively, the polymerizations of **10a** and **10b** proceeded through a mechanism similar to the ring-opening multibranching polymerization of glycidol, which was reported by Kubisa and Penczek (Tokar *et al.*, 1994). Figure 7.9 represents the proposed mechanism for the cationic polymerization of **10a**. In the polymerization system, the oxygen atom of the epoxy group of **10a** was protonated, and the protonated **10a** reacted with the oxygen atom of the epoxy group from another monomer, thus leading to the linear unit (**11a**). The branched unit (**12a**) should be produced by the proton-transfer reaction of the hydroxy groups in the polymer chain with the oxonium cation end of the another polymer chain end. The polymerization of **10b** also proceeded through a mechanism similar to that for **10a** to produce the hyperbranched glyco-conjugated polymer, **poly10b**.

The resulting polytetritols were the hyperbranched glyco-conjugated polymers **Poly10a** and **Poly10b** essentially consisting of DL-threitol and erythritol units, respectively, which differed from the hyperbranched polytetritols prepared from 1,4-anhydroerythritol (**6a**) and 1,4-anhydro-L-threitol (**6b**).

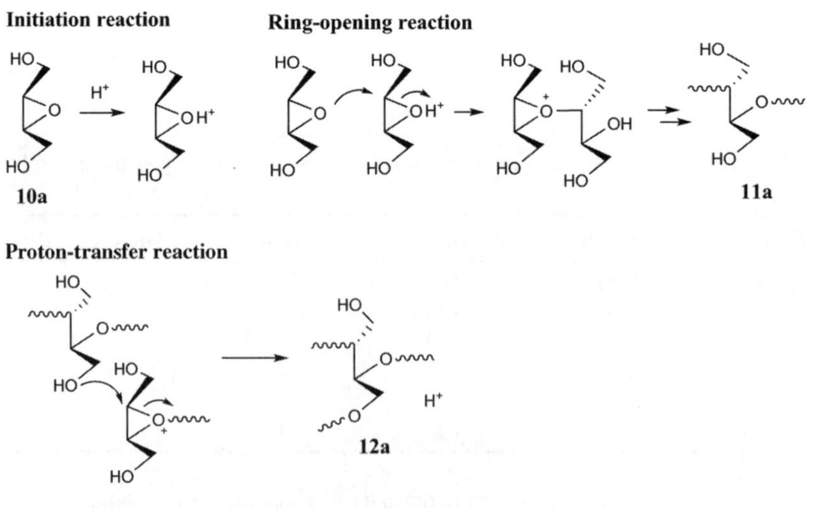

Figure 7.9 Proposed mechanism for cationic polymerization of 2,3-anhydroerythritol (**10a**).

7.2.4 *Hyperbranched Poly(2,5-anhydro-D-glucitol) by Polymerization of 1,2:5,6-Dianhydro-D-mannitol*

In the section, we focused on the ring-opening polymerization of diepoxy diol through a proton-transfer cyclopolymerization mechanism, which should be a new class of cyclopolymerization together with a novel type of hyperbranched glyco-conjugated polymer, that is, the ring-opening and ring-forming multibranching polymerization of 1,2:5,6-dianhydro-D-mannitol (**13**) was carried out using a cationic initiator, as shown in Figure 7.10 (Imai *et al.*, 2003). The proton-transfer ring-opening and ring-forming mechanism was examined by determining the structure of the resulting polymers, such as the cyclic constitutional units and the degree of branching.

The anionic and cationic polymerizations of **13** were carried out using potassium *tert*-butoxide (*t*-BuOK) in tetrahydrofuran and borontrifluoride etherate ($BF_3{\cdot}OEt_2$) in dichloromethane. For the anionic polymerization, the polymer yield was a relatively high value. However, the obtained polymer was mostly insoluble along with a small amount of a water-soluble polymer, which was quite different from the anionic cyclopolymerization of 3,4-di-*O*-alkyl-1,2:5,6-dianhydro-D-mannitol leading to gel-free stereoregular polymers (Satoh and Kakuchi, 2004). In the ^{1}H NMR spectrum of the polymer prepared by the anionic polymerization, the signal was broad and a characteristic absorption due to the epoxy protons was observed. Although the extent of the cyclization (f_c) of the water-soluble polymer was about 0.9, the polymer gradually became insoluble after storage for a few days. On the other hand, polymers obtained from the cationic polymerization were soluble in methanol and water. The characteristic absorption due to the epoxy proton of **13** completely disappeared in the ^{1}H NMR spectrum of the polymer prepared, so that the polymerization proceeded according to a cyclopolymerization mechanism leading to polymers consisting essentially of cyclic repeating units, that is, the f_c was 1.0 for every polymer.

Table 7.4 summarized the results of cationic polymerization. The $M_{w,SLS}$ values of the resulting polymer were in the range of 2.08×10^5 to 9.84×10^5, depending on the polymerization conditions. The $M_{w,SLS}$ values were greater than the $M_{w,SEC}$ values for all of the polymers prepared by cationic polymerization, indicating that the resulting polymers were highly branched spherical molecules, that is, a hyperbranched glyco-conjugated polymer (**Poly13**).

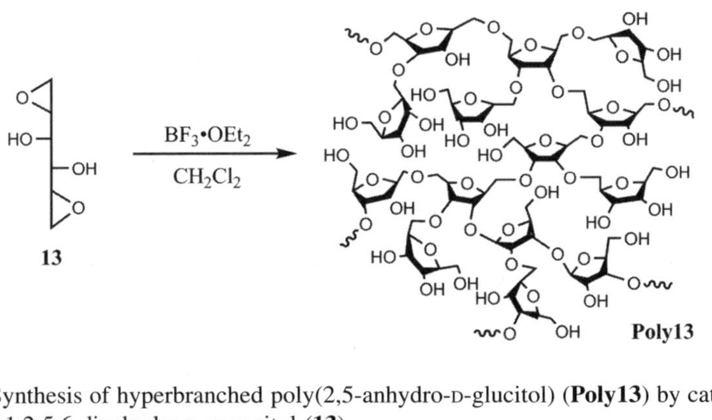

Figure 7.10 Synthesis of hyperbranched poly(2,5-anhydro-D-glucitol) (**Poly13**) by cationic cyclopolymerization of 1,2:5,6-dianhydro-D-mannitol (**13**).

Table 7.4 Cationic polymerization of 1,2:5,6-dianhydro-D-mannitol (**13**) using $BF_3{\cdot}OEt_2$[a]

[**13**]	Temp. °C	Time h	Yield%[b]	$M_{w,SLS}$[c] $\times 10^{-5}$	DB[d]
0.5	0	100	30.5	2.08	0.46
0.5	0	200	41.8	3.08	0.46
0.5	0	240	50.4	3.90	0.46
0.5	0	300	72.7	9.84	0.46

[a]Cationic polymerization was carried out at [**13**]/[$BF_3{\cdot}OEt_2$] of 100 in CH_2Cl_2.
[b]Water-soluble part.
[c]Determined by static light scattering (SLS) in 0.2 mol L^{-1} aqueous $NaNO_3$ solution.
[d]Degree of branching.

In the ^{13}C NMR spectrum of **Poly13**, the major signals were very close to those for 2,5-anhydro-D-glucitol (**14**), which was the model compound for the terminal units of **Poly13**. In addition, six small signals were consistent with those for a linear polymer, (1 → 6)-2,5-anhydro-D-glucitol (**15**) (Umeda *et al.*, 1998). The rest of the signals were close to those of 2,5-anhydro-3,4,6-tri-*O*-methyl-D-glucitol (**16**) (Satoh *et al.*, 1995), 2,5-anhydro-3,4-di-*O*-methyl-D-glucitol (**17**) (Kakuchi *et al.*, 1995), 2,5-anhydro-1,3,4,6-tetra-*O*-methyl-D-glucitol (**18**) (Satoh *et al.*, 1995), 2,5-anhydro-1,6-di-*O*-methyl-D-glucitol (**19**) (Umeda *et al.*, 1998), and (1 → 6)-bonded 3,4-*O*-methyl-2,5-anhydro-D-glucitol (**20**) (Satoh, Yokota, and Kakuchi, 1995), which were the model compounds for linear units, semidendritic units, and dendritic units of 2,5-anhydro-D-glucitol. Thus, the signals should be due to the multibonded 2,5-anhydro-D-gluitol units (Figure 7.11). These results indicated that the cationic polymer should be a branched structure having many terminal units, that is, a hyperbranched glyco-conjugated polymer consisting mainly of 2,5-anhydro-D-gluitol units.

Although monomer **13** has two epoxy and two hydroxy groups in the molecule, one of the two epoxy groups was consumed during the intramolecular cyclization in the cyclopolymerization, so that **13** was virtually a monomer possessing one epoxy and two hydroxy groups. Thus, **13** can be classified as a latent AB_3-type monomer, and the DBs of **Poly13** were calculated to be about 0.46, which was approximately consistent with the statistical value of 0.45 for the polymerization of an AB_3-type monomer (Hölter and Frey, 1997).

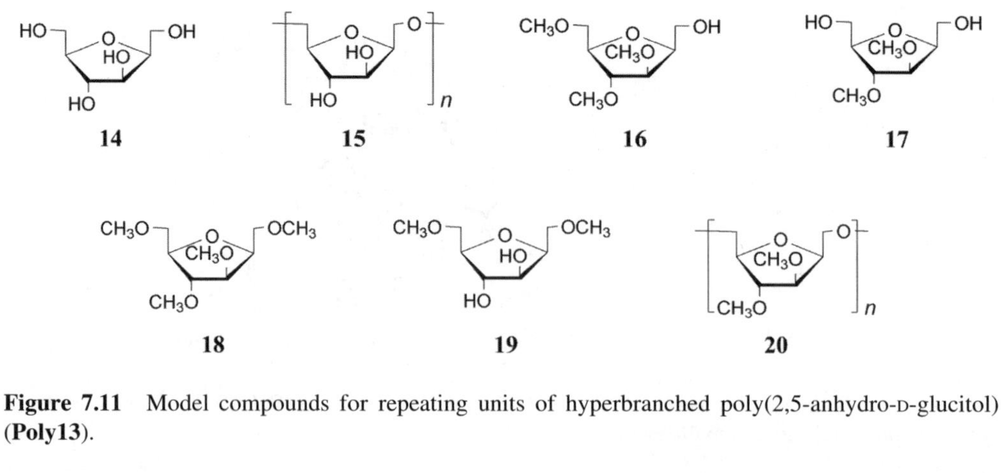

14 **15** **16** **17** **18** **19** **20**

Figure 7.11 Model compounds for repeating units of hyperbranched poly(2,5-anhydro-D-glucitol) (**Poly13**).

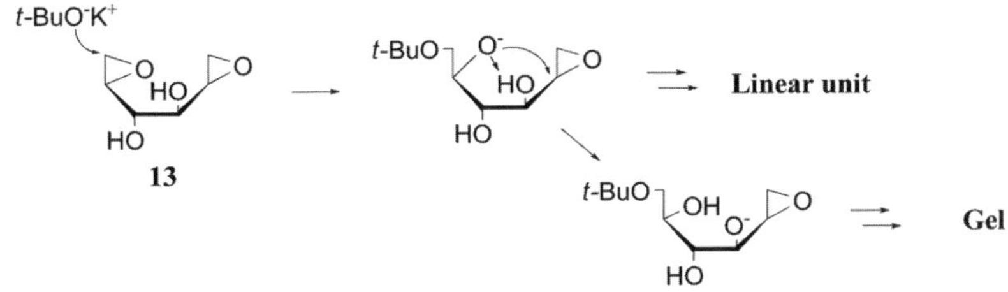

Figure 7.12 Proposed mechanism for anionic polymerization of 1,2:5,6-dianhydro-D-mannitol (**13**).

Figure 7.12 shows an anionic polymerization mechanism. During the polymerization, the intramolecular proton transfer occurred from the hydroxy group to the alkoxide, which was generated by the ring opening of the epoxide group, to afford an insoluble gel. On the other hand, the cationic polymerization of **13** proceeded through a cyclopolymerization mechanism to give the polymers consisting of 2,5-anhydro-D-glucitol units. Figure 7.13 represents a proposed mechanism for the cationic cyclopolymerization. In the propagation step, the repeating units of 2,5-anhydro-D-glucitol should be formed by the alternative reaction of intermolecular propagation and intramolecular cyclization through the β- and α-scissions of the two epoxy groups in **13**, respectively. The branches of the polymer should be produced by proton transfer from the hydroxyl groups in the polymer chains to the oxonium cations.

The three-dimensional property was characterized based on a viscosity study. The $[\eta]$ value increased with the increasing $M_{w,SLS}$ value and the α value of the Mark–Houwink–Sakurada equation was found to be 0.29, which was obtained from the slope of the logarithmic plots of $[\eta]$ versus $M_{w,SLS}$. It is well known that the α value is less than 0.5 for the various hyperbranched polymers, so that **Poly13** had a spherical shape in solution (Jikei and Kakimoto, 2001).

Although the anionic polymerization of **13** tended to form gels, the cationic polymerization proceeded through a proton-transfer reaction mechanism to produce hyperbranched glyco-conjugated polymer **Poly13** mainly consisting of 2,5-anhydro-D-glucitol units. This was the first report of the synthesis of a hyperbranched polymer via a cyclopolymerization mechanism.

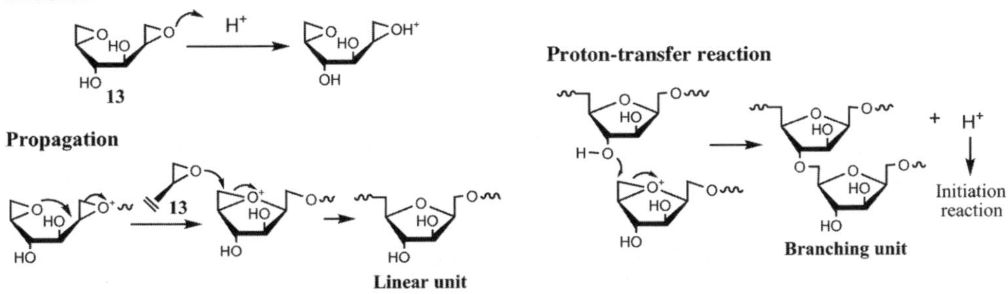

Figure 7.13 Proposed mechanism for cationic cyclopolymerization of 1,2:5,6-dianhydro-D-mannitol (**13**).

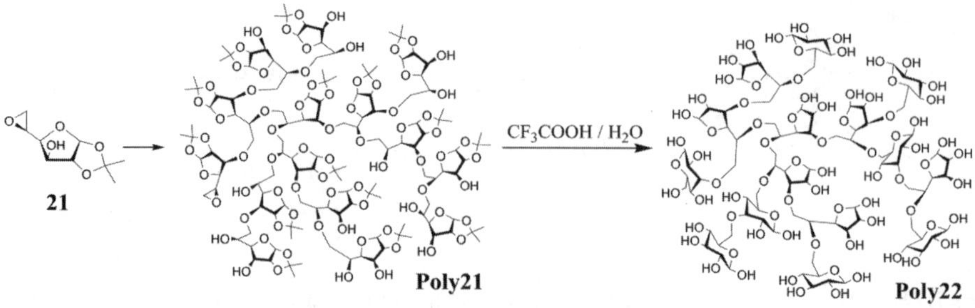

Figure 7.14 Polymerization of 5,6-anhydro-1,2-*O*-isopropylidene-α-D-glucofuranose (**21**) and synthesis of hyperbranched 5,6-glucan (**Poly22**) by hydrolysis of **Poly21**.

7.2.5 Hyperbranched 5,6-Glucan by Polymerization of 5,6-Anhydro-α-D-glucofuranose Derivative

In Section 7.2.1, we introduced the view that the ring-opening multibranching polymerization of 1,6-anhydro-*β*-D-hexopyranose is a convenient method of synthesizing hyperbranched glyco-conjugated polymer with numerous nonreducing sugar ends on the peripheries of the polymer (Satoh *et al.*, 2003; Satoh *et al.*, 2005). In contrast, in this section, we introduce the ring-opening multibranching polymerization of 5,6-anhydro-1,2-*O*-isopropylidene-α-D-glucofuranose (**21**) leading to the hyperbranched glyco-conjugated polymer (**Poly21**) with numerous reducing sugar ends, as shown in Figure 7.14. Since a hyperbranched 5,6-glucan (**Poly22**) obtained by hydrolysis of **Poly21** is a novel water-soluble hyperbranched glyco-conjugated polymer arranged with numerous reducing D-glucose units on the peripheries of the polymer, the reduction property of **Poly22** is also discussed from the viewpoint of the physical property as a "reducing sugar ball" (Tamaki *et al.*, 2010).

The ring-opening polymerizations of monomer **21** were carried out using potassium *tert*-butoxide (*t*-BuOK) as the anionic initiator or boron trifluoride diethyletherate ($BF_3 \cdot OEt_2$) as the cationic initiator. The typical results are summarized in Table 7.5. When *t*-BuOK was used

Table 7.5 The polymerization results of 5,6-anhydro-1,2-*O*-isopropylidene-α-D-glucofuranose (**21**)

Entry	Cat.	Temp °C	Time h	Addition time h	Yield%	$M_{w,SEC\text{-}MALLS}$ $(M_w/M_n)^c$	α^c	$[\eta]^c$ dL g^{-1}
1	*t*-BuOK[a]	60	160	—	52.9	7700 (1.22)	0.27	0.046
2	*t*-BuOK[a]	50	600	—	68.4	7400 (1.37)	0.25	0.040
3	*t*-BuOK[a]	50	770	—	69.5	8900 (1.33)	0.27	0.043
4	$BF_3 \cdot OEt_2$[b]	10	24	—	68.5	15 000 (1.45)	0.11	0.034
5	$BF_3 \cdot OEt_2$[b]	10	24	4.4	73.4	37 000 (1.19)	0.09	0.033
6	$BF_3 \cdot OEt_2$[b]	10	24	13.3	75.5	81 300 (1.23)	0.09	0.034
7	$BF_3 \cdot OEt_2$[b]	10	24	17.8	66.1	122 400 (1.70)	0.08	0.038

[a] Conditions: Ar atmosphere; [**21**]/[*t*-BuOK] = 20; solvent, THF; [**21**] = 2.0 mol L^{-1}.
[b] Condition: Ar atmosphere; [**21**]/[$BF_3 \cdot OEt_2$] = 40; solvent, CH_2Cl_2; [**21**] = 0.5 mol L^{-1}.
[c] The weight average molecular weight $M_{w,SEC\text{-}MALLS}$, polydispersity $M_{w,SEC\text{-}MALLS}/M_{n,SEC\text{-}MALLS}$, intrinsic viscosity $[\eta]$, and Mark–Houwink–Sakurada constant α ($[\eta] = KM_{w,SEC\text{-}MALLS}{}^{\alpha}$), were determined by SEC in THF equipped with MALLS and viscosity detectors.

in dry tetrahydrofuran as the solvent, the reaction mixtures immediately changed from colorless to dark brown as the reaction progressed, while the cationic polymerization using $BF_3{\cdot}OEt_2$ in dry methylene chloride homogeneously proceeded without any color change. The polymers obtained using *t*-BuOK or $BF_3{\cdot}OEt_2$ were a light-yellow solid or a white solid, respectively, which were soluble in chloroform, THF, DMF and methanol, but insoluble in water. The polymer yields for the anionic polymerization were in the range of 52.9–69.5% and the increase in conversion was very slow. The weight-average molecular weights ($M_{w,SEC\text{-}MALLS}$) measured by a multiangle laser light-scattering (MALLS) instrument of the resulting polymers ranged from approximately 7400 to 8900. In contrast, the polymer yield for the cationic polymerization (entry 4 in Table 7.5) was 68.5% for 24 h and the $M_{w,SEC\text{-}MALLS}$ was 15 000, indicating that the cationic polymerization is much better than the anionic one.

As the efficient preparation method of hyperbranched polymer with both a high molecular weight and a high branching value degree, the slow monomer-addition strategy (Sunder *et al.*, 1999) was used for the cationic polymerization of **21**. According to the established time of monomer addition, the monomer solution was slowly added to the mixture of monomer solution and $BF_3{\cdot}OEt_2$ using a dosing pump. These results are also summarized in Table 7.5. The slow monomer-addition strategy results in a very low monomer concentration actually present in the reaction mixture. Thus, the exclusive reaction of the monomer with the growing polyfunctional macromolecules occurs, resulting in a high molecular weight and a high degree of branching. Based on a longer addition time, that is, a decrease in the drop rate, the $M_{w,SEC\text{-}MALLS}$ values dramatically increase up to 122 400.

The $[\eta]$ values of the polymers obtained using *t*-BuOK were very low in the range from 0.040 to 0.046 dL g^{-1}. The $M_{w,SEC\text{-}MALLS}$ dependence of $[\eta]$ for the polymer is shown in Figure 7.15. On the basis of these results, the exponent α of the Mark–Houwink–Sakurada equations of the polymers obtained using *t*-BuOK were determined to be 0.25–0.27. Our experimental α values

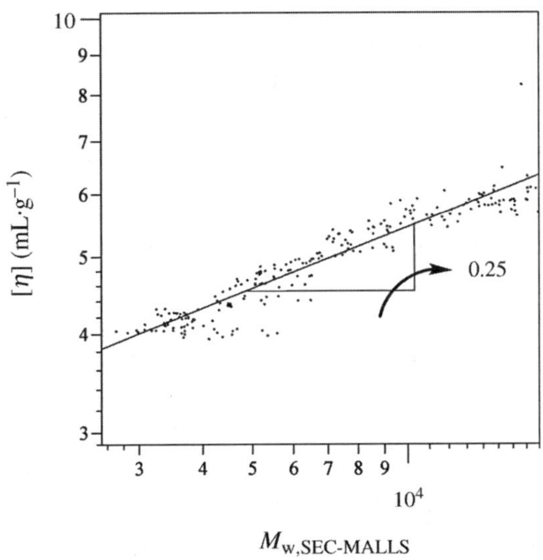

Figure 7.15 Double logarithmic plots of $[\eta]$ vs. $M_{w,SEC\text{-}MALLS}$ for **Poly21** in THF.

were noticeably low, which are ascribed to the compact spherical structure in solution. Therefore, the measurement results suggested that the polymer obtained using *t*-BuOK was a branched spherical molecule, that is, a hyperbranched poly(5,6-anhydro-1,2-*O*-isopropylidene-α-D-glucofuranose) (**Poly21**).

The [η]s of the polymers obtained using $BF_3 \cdot OEt_2$ were 0.034–0.038 dL g^{-1}, which were lower than those of the polymers obtained using *t*-BuOK, regardless of the higher molecular weight polymer. In addition, the α values were 0.08–0.11, which were less than those of the polymers obtained using *t*-BuOK. These results indicate that the polymer obtained using $BF_3 \cdot OEt_2$ should have a more compact spherical structure in solution than the polymer obtained using *t*-BuOK, that is, the hyperbranched glyco-conjugated polymer obtained using $BF_3 \cdot OEt_2$ should have a higher degree of branching value than the polymer obtained using *t*-BuOK.

The hydrolysis of **Poly21** was carried out using a mixture of trifluoroacetic acid and water (85/15, v/v) to produce the hyperbranched poly(5,6-anhydro-α-D-glucofuranose) (**Poly22**), that is, the hyperbranched 5,6-glucan. The physical data of **Poly22** are listed in Table 7.6. The obtained polymers **Poly22** were white powders that were soluble in water, and insoluble in chloroform, THF, DMF and methanol, which differed remarkably from the solubility of **Poly21**.

Poly22 is a novel hyperbranched glyco-conjugated polymer arranged with numerous reducing D-glucose units on the peripheries of the hyperbranched polymer, whereas the glyco-conjugated polymers obtained from 1,6-anhydro-β-D-hexopyranose consist only of nonreducing sugar ends. Therefore, **Poly22** is a particularly interesting material, which should be called a "reducing sugar ball." **Poly22** probably has a higher reducing ability than D-glucose because of the glyco-cluster effect or the multivalent effect of the reducing D-glucose units. Such a glyco-conjugated polymer with numerous reducing sugar units is also of considerable interest in a wide variety of fields such as biochemical and medicinal applications.

To estimate the reducing ability of **Poly22**, the formation of gold particles, which is derived from the reduction of $HAuCl_4$, is demonstrated in the aqueous solution of **Poly22**. After mixing of the **Poly22** and $HAuCl_4$ aqueous solutions, the color of the system changed from colorless to light red with time. The UV-Vis spectra of the **Poly22**-$HAuCl_4$ system have a λ_{max} that ranged from 512 nm to 514 nm, which is due to the formation of the gold particles. From previous studies (Nakamura, Kawabata, and Mori, 2003), the average gold particle diameter can be estimated from the λ_{max}, that is, gold particles of about 8–12 nm were obtained by the mixing of **Poly22** and the $HAuCl_4$ aqueous solutions. Thus, the average hydrodynamic diameter of the resulting Au particles, which was measured by a dynamic light-scattering (DLS) technique, was about 11.3 nm.

Table 7.6 The results of hydrolysis of **Poly21** to prepare **Poly22**

Entry	Poly21	$M_{w,SEC\text{-}MALLS}$ of **Poly22**[a] ($M_{w,SEC\text{-}MALLS}/M_{n,SEC\text{-}MALLS}$)	α[a]	[η][a] dL g^{-1}
1	7400	3740 (1.26)	0.36	0.044
2	122 400	74 400 (2.74)	0.26	0.050

[a]The weight-average molecular weight $M_{w,SEC\text{-}MALLS}$, polydispersity $M_{w,SEC\text{-}MALLS}/M_{n,SEC\text{-}MALLS}$, intrinsic viscosity [$\eta$], and Mark–Houwink–Sakurada constant α ([η] = $KM_{w,SEC\text{-}MALLS}{}^{\alpha}$), were determined by SEC in 0.2 mol L^{-1} $NaNO_3$ aqueous solution equipped with MALLS and viscosity detectors.

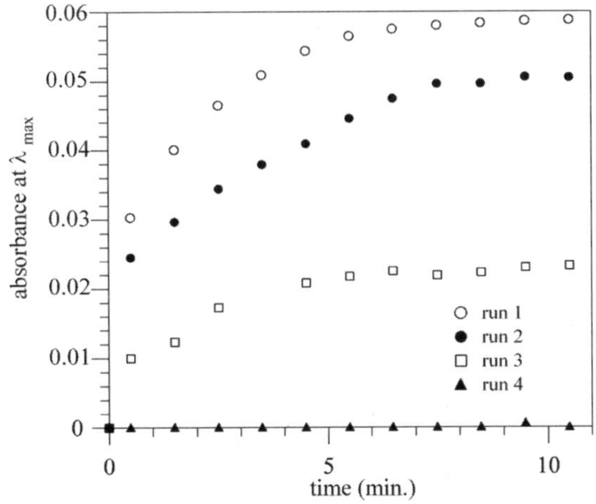

Figure 7.16 Change in the UV absorption at λ_{max} of mixtures of **Poly22** (or D-glucose) and $HAuCl_4$ as a function of reaction time at 70 °C. Run 1 (open circle); **Poly22** (46 unit-mmol L^{-1} of entry 1 in Table 7.6, 1.8 mL) and $HAuCl_4$ (0.01 mol L^{-1}, 6.67 mL) aqueous solutions at 512 nm, run 2 (closed circle); **Poly22** (25 unit-mmol L^{-1} of entry 1 in Table 7.6, 1.8 mL) and $HAuCl_4$ (0.01 mol L^{-1}, 6.67 mL) aqueous solutions at 514 nm, run 3 (open square); **Poly22** (46 unit-mmol L^{-1} of entry 2 in Table 7.6, 1.8 mL) and $HAuCl_4$ (0.01 mol L^{-1}, 6.67 mL) aqueous solutions at 514 nm, and run 4 (closed triangle); D-glucose (46 unit-mmol L^{-1}, 1.8 mL) and $HAuCl_4$ (0.01 mol L^{-1}, 6.67 mL) aqueous solutions at 526 nm.

To obtain the relationship between the formation of the gold particles and the reaction time, the increasing trend in the UV-Vis absorbance at λ_{max} is presented in Figure 7.16. For all the **Poly22**-$HAuCl_4$ systems (runs 1–3), the UV-Vis absorbance sharply increases without an induction time and finally became a constant value within 10 min. On the other hand, the D-glucose-$HAuCl_4$ system (run 4) shows a long induction time, then the absorbance sharply increases as an autocatalytic system (Tamaki *et al.*, 2010). A similar autocatalytic tendency with a long induction time was also reported using the sugar-persubstituted poly(amidoamine) dendrimer, that is, a sugar ball with numerous nonreducing sugar units (Esumi *et al.*, 2000). The results indicate that **Poly22** has a high reducing ability that originated from the high density of the reducing sugars in **Poly22**, that is, a glyco-cluster effect or multivalent effect of the reducing sugars. Thus, the reaction rates within 10 min for the **Poly22**-$HAuCl_4$ systems were much faster than that for the D-glucose-$HAuCl_4$ system, as shown in Figure 7.16. This indicates that "reducing sugar ball" **Poly22** effectively act as a sugar-based reductant.

7.3 Unimolecular Reversed Micelle Based on Hyperbranched Glyco-Conjugated Polymer Core

7.3.1 Structural Transition of Unimolecular Reversed Micelle

Hyperbranched polymers and dendrimers have been used as the key starting materials for the preparation of a globular amphiphile with a unimolecular micellar property. However, recent research has indicated that some of the amphiphilic dendritic polymers do not become a unimolecular micelle and form a large micellar structure in solution (Garamus *et al.*, 2004). The structural transition from a unimolecular form to an aggregate form should have an obvious effect on its molecular encapsulation–release property. The structural transition should be derived from the content of the shell part in a molecule, that is, the shell density. Therefore, it is very important to elucidate the effect of the shell density on the encapsulation–release properties of amphiphilic dendritic polymers. In order to investigate the relation between the shell density of the amphiphilic hyperbranched polymer and its encapsulation–release property for a guest molecule, the preparation and micellar property of a unimolecular reversed micelle (**23**) consisting of a hydrophilic hyperbranched polythreitol core and a hydrophobic trityl shell were introduced in the section (Kitajyo *et al.*, 2007a; Satoh, 2009).

To prepare the unimolecular reversed micelle **23**, two **Poly10a** with different molecular weights ($M_{w,SLS}$: 1.18×10^4 (**Poly10a-1**, $M_{w,SEC}/M_{n,SEC}$: 1.66) and 4.79×10^4 (**Poly10a-2**, $M_{w,SEC}/M_{n,SEC}$: 1.58)) (Imai *et al.*, 2005) were partially reacted with trityl chloride, as shown in Figure 7.17. The reaction homogeneously proceeded to produce a brown solid after purification by dialysis and preparative size-exclusion chromatography. The degree of substituents (DS) could be controlled by the feed molar ratio of the trityl chloride and hydroxyl groups in **Poly10a** ([Tr]/[OH]); that is, the increasing [Tr]/[OH] values lead to the increasing DS values. The DS values were 15.7 mol% at a condition of [Tr]/[OH] of 0.25 at 23 °C for 24 h (**23a**), 23.8 mol% at a [Tr]/[OH] of 0.50 (**23b**), and 38.7 mol% at a [Tr]/[OH] of 1.50 (**23c**) using **Poly10a-1**, and 23.2 mol% at a [Tr]/[OH] of 0.50 (**23d**), 43.2 mol% at a [Tr]/[OH] of 1.00 (**23e**), and 61.0 mol% at a [Tr]/[OH] of 2.00 (**23f**) using **Poly10a-2**. The disagreement between the [Tr]/[OH] and DS values, that is, the DS values were always lower than the values estimated by the [Tr]/[OH], could be caused by the lower reactivity of the sterically hindered inner hydroxyl groups in **Poly10a** than that of the outer one.

The solubility characteristics of **23a–f** with certain DS values were examined using various solvents and are summarized along with that of **Poly10a** in Table 7.7. **Poly10a** was soluble in water and methanol, but insoluble in other organic solvents, while **23a–f** were soluble in chloroform and acetone, but insoluble in water. The solubilities of **23a–f** were also changed by the DS values, that is, **23c**, **23e**, and **23f** with high DS values were soluble in toluene and ethyl acetate, while **23a**, **23b**, and **23d** with low DS values were insoluble or partially soluble in toluene and ethyl acetate.

To investigate the encapsulation properties of **23a–f**, the encapsulation characteristics toward the water-soluble dye, rose bengal (RB), were examined using the solid/liquid

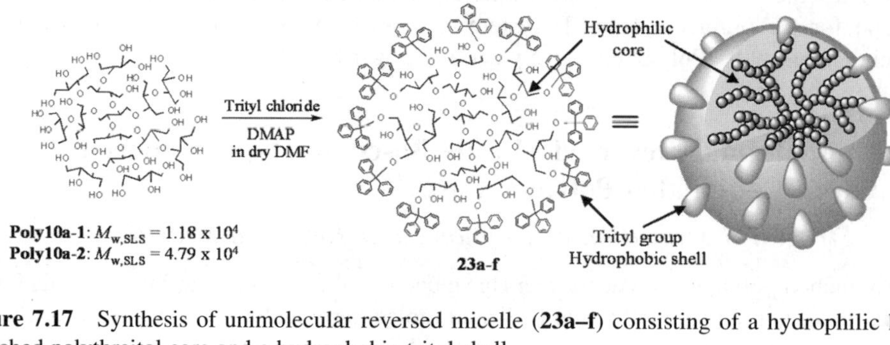

Figure 7.17 Synthesis of unimolecular reversed micelle (**23a–f**) consisting of a hydrophilic hyperbranched polythreitol core and a hydrophobic trityl shell.

Table 7.7 Solubilities of hyperbranched polythreitol (**Poly10a**) and tritylated hyperbranched polythreitol (**23a–f**)[a]

Polymer	DS mol%	H_2O	MeOH	Acetone	$CHCl_3$	AcOEt	Toluene
Poly10a	0	+ +	+ +	−	−	−	−
23a	15.7	−	+	+	+ +	−	−
23b	23.8	−	−	+ +	+ +	+	+
23c	38.7	−	−	+ +	+ +	+ +	+ +
23d	23.2	−	−	+ +	+ +	+	+
23e	43.2	−	−	+ +	+ +	+ +	+ +
23f	61.0	−	−	+ +	+ +	+ +	+ +

[a]Examined by using 5 mg of polymer in 0.5 mL of solvent at room temperature (+ +: soluble; +: swelling; −: insoluble).

phase-transfer method, as shown in Figure 7.18 (Kitajyo *et al.*, 2007a). For the RB-loaded **23a–f** (RB/**23a–f**) systems, the characteristic UV-Vis absorption due to RB appeared in the visible region from 470 to 600 nm even though the RB absorption did not appear for the control experiment, as shown in Figure 7.19, indicating that **23a–f** possessed an encapsulation property toward RB. In addition, the results also suggested that RB existed under a more compatible environment than in chloroform, that is, the hydrophilic hyperbranched polythreitol core in **23a–f**, as shown in Figure 7.18.

The average numbers of RB molecules per **23a–f** molecule (N_{RB}) are 0.5 for **23a**, 1.4 for **23b**, 1.3 for **23c**, 4.9 for **23d**, 4.4 for **23e**, and 4.3 for **23f**. Two tendencies between N_{RB} and the polymer structure were recognized from the results. One is that N_{RB} increased with the increasing hydrophilic core size, that is, 0.5–1.4 for **23a–c** with a small core and 4.3–4.9 for **23d–f** with a large core. The other tendency is that N_{RB} depends on the DS value, that is, the shell density of the trityl groups. Based on the results, it seems that polymers **23b** and **23d** with an about 23% DS value showed the maximum efficiency. The decrease in the N_{RB} value with the increasing DS value (e.g. **23b** > **23c**, and **23d** > **23e** > **23f**) should be due to a decrease in the hydrophilic core space, while the decrease in the N_{RB} value with the decreasing DS value (e.g. **23b** > **23a**) may be due to the defective encapsulation characteristics of **23a**, that is, the solubility difference between the core and shell is not quite enough to lead to an efficient unimolecular reversed micelle for **23a**.

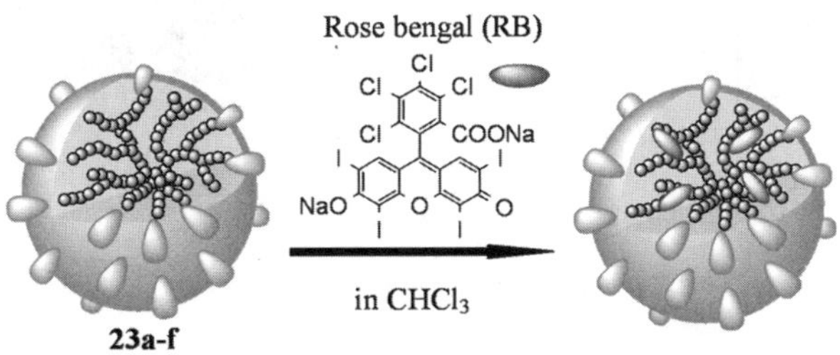

Figure 7.18 Encapsulation of rose bengal (RB) into **23a–f**.

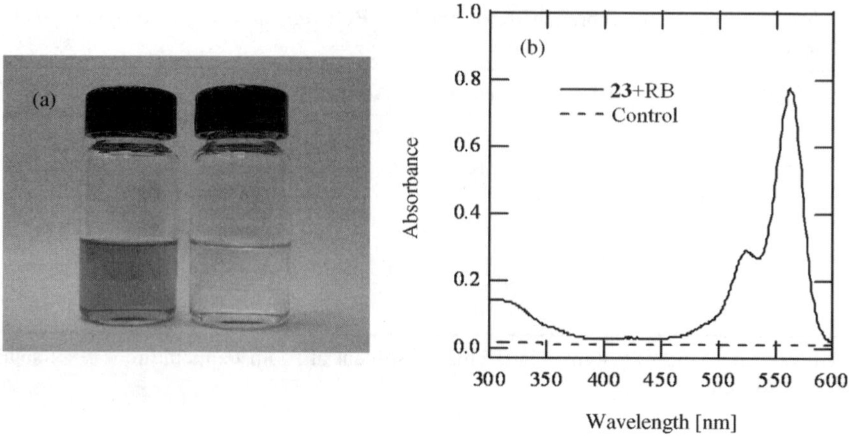

Figure 7.19 (a) Demonstration of the encapsulation property of **23c** toward RB in chloroform (left vial: RB-loaded **23c** (RB/**23c**); right vial: control). (b) UV-Vis spectra of RB/**23c** and control in chloroform.

Figure 7.20 shows the plot of the absorbance at the λ_{max} (562 nm) of RB versus the polymer concentration in chloroform. In general, the encapsulation capacities of the unimolecular micelles are invariable at any polymer concentration. Thus, if these polymers act as unimolecular reversed micelles, the absorbance values are expected to proportionally increase with the polymer concentration. RB/**23d** with a DS of 23.2% shows a proportional and linear relation between the absorbance value and polymer concentration even at extremely low concentrations, indicating that **23d** existed as a unimolecular reversed micelle in chloroform. However, a nonlinear relationship with a deflection point at about $1.2 \times 10^{-5}\,\text{mol}\,\text{L}^{-1}$ is obtained for RB/**23a** with the DS of 15.7%. This result suggested that the RB-encapsulation state of **23a** was quite different from that of the unimolecular reversed micelle.

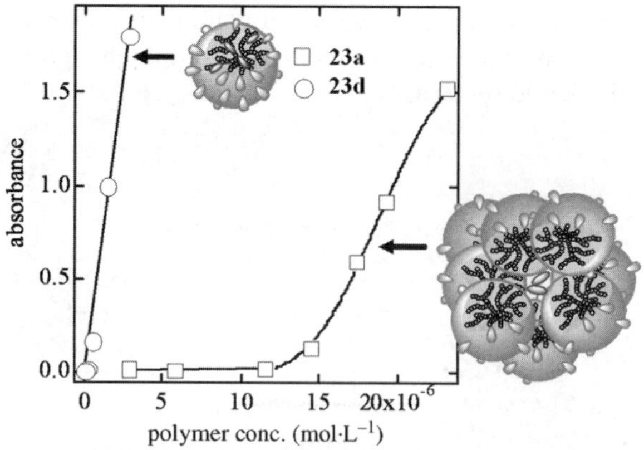

Figure 7.20 Absorbance at λ_{max} (562 nm) of RB/**23a** and RB/**23d** in chloroform versus polymer concentration.

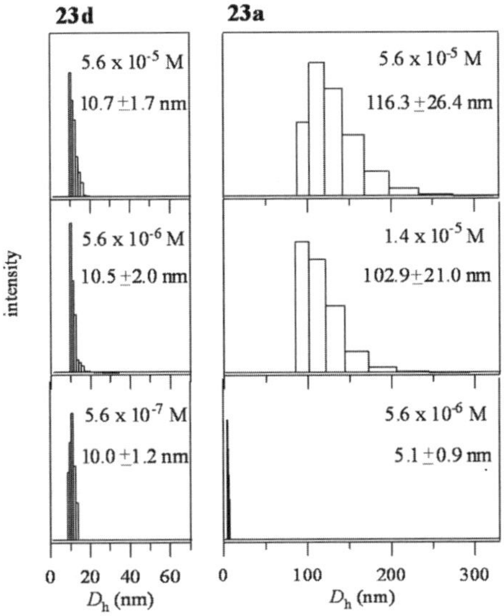

Figure 7.21 Particle-size distribution and number average hydrodynamic diameter ($<D_h>$) estimated by DLS measurements of **23d** (left side) and **23a** (right side) in chloroform.

To compare the solution structure of **23a** with that of **23d**, dynamic light-scattering (DLS) measurements were performed at various concentrations in chloroform (Figure 7.21). The number average hydrodynamic diameter ($<D_h>$) of **23d** was about 10 nm, which has no influence on the polymer concentration in the range from 5.6×10^{-7} to 5.6×10^{-5} mol L^{-1}. This result reveals that no self-aggregation occurred at these concentrations. On the other hand, the $<D_h>$ value of **23a** drastically changed with the changing polymer concentration, indicating that **23a** formed a large aggregate above 1.4×10^{-5} mol L^{-1} and its aggregate dissociated into a unimolecular form at 5.6×10^{-6} mol L^{-1}. In Figure 7.20, no absorbance of RB was observed at 5.6×10^{-6} mol L^{-1} of **23a**, thus, it is clear that **23a** was incapable of encapsulating RB as a unimolecular form. Because **23a** with a low DS value has many unreacted hydroxyl groups on the polymer surface, **23a** could not form a stable unimolecular reversed micelle and became an aggregate at a high polymer concentration, as shown in Figure 7.20.

Because **23a–f** are able to encapsulate a hydrophilic compound, release tests were performed. Figure 7.22 shows the RB-release results. The particulates of the RB-encapsulating polymers (RB/**23a–c**) were suspended in a 0.067 mol L^{-1} phosphate buffer solution (pH 7.4), and the resulting solutions were shaken at 37 °C. For the RB/**23a** aggregate, the encapsulated RB molecules were rapidly released from the polymer and the amount of the released RB remained constant at about 55% after 20 h. In contract, a slight release of less than 10% was observed for the RB/**23b–c** system even after more than 120 h. These results indicated that the release rate of the encapsulated RB molecules mostly depended on the type of the molecular form of **23** when RB was entrapped. A slight release was also observed for RB/**23d–f**, meaning that the unimolecular reversed micelles **23b–f** have a high RB-holding ability in water.

The unimolecular reversed micellar properties of **23** were dramatically changed with the tuning of the hydrophobic shell density of **23**. Hence, the study possibly provides a strategy for

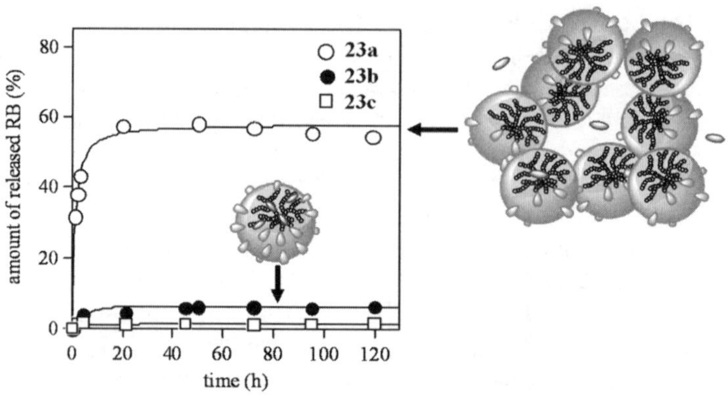

Figure 7.22 Release test of RB from RB/**23a–c** in phosphate buffer solution (pH 7.4).

the development of novel unimolecular micelles, and the further molecular design of amphiphilic hyperbranched polymers will be used for medical applications, such as drug delivery.

7.3.2 Biodegradable Unimolecular Reversed Micelle

Although the encapsulation–release property of guest molecules using a dendritic architectures have been studied by several research groups, relatively little is known about the active release of the encapsulated guest molecules by biodegradable cleavage of the shell. Therefore, of interest is the synthesis and encapsulation–release property of unimolecular reversed micelle with biodegradability. In this section, the synthesis of a novel unimolecular reversed micelle (**24**) consisting of a poly(L-lactide) (PLA) shell and hyperbranched D-mannan core was introduced, as shown in Figure 7.23 (Satoh *et al.*, 2006). In addition, the encapsulation–release

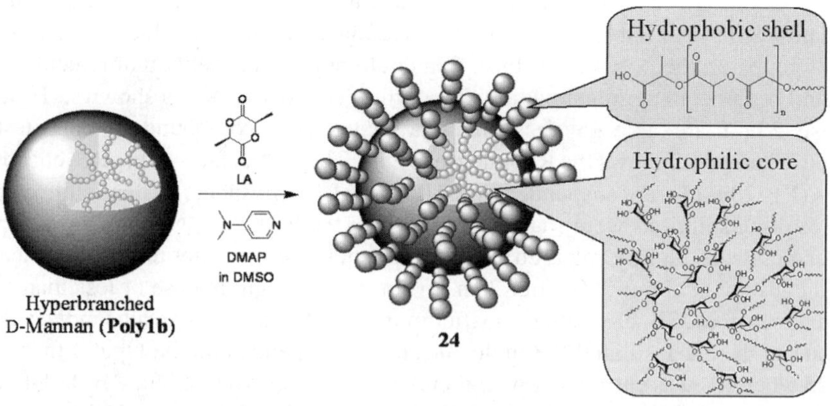

Figure 7.23 Synthesis of unimolecular reversed micelle (**24**) consisting of poly L-lactide shell and hyperbranched D-mannan core.

property of **24** toward RB is discussed from the viewpoint of the unimolecular reversed micellar property with biodegradability.

The synthesis of **24** consisting of the PLA shell and hyperbranched D-mannan core was carried out by the polymerization of L-lactide (LA) using hyperbranched D-mannan **Poly1b** as the initiator, as shown in Figure 7.23. For the study, **Poly1b** with an $M_{w,SLS}$ of 9.63×10^4 ($M_{w,SEC}$; 4.30×10^3, $M_{w,SEC}/M_{n,SEC}$; 1.48) as a multifunctional initiator was used (Satoh *et al.*, 2003). **Poly1b** consists of highly branched globular macromolecules having about 30–40 mol% of nonreducing D-mannopyranosyl terminal units. The outer hydroxyl groups in the globular polymer, therefore, act as initiators leading to the PLA chains. Thus, the polymerization of LA using **Poly1b** is a facile method for preparing the well-defined "chestnut-shaped" polymer **24**. The obtained copolymers were white solids that were soluble in DMSO, THF, and chloroform, but insoluble in H_2O.

The molecular weights of the PLA chain (M_{PLA}) on **24** were in the range from 700 to 2160 and tended to increase with the increasing polymerization time, as shown in Figure 7.24. The number of PLA chains (N_{PLA}) on **24** ranged from 380 to 790, which could be controlled by the ratio of DMAP to the sugar unit in **Poly1b** ([DMAP]/[sugar unit]).

In order to study the unimolecular reversed micellar property of **24**, the encapsulation characteristics for RB were examined using the solid/liquid phase-transfer method, as shown in Figure 7.25 (Satoh *et al.*, 2006). Figure 7.26 shows the UV-Vis spectra of the chloroform solution for the RB/**24** and RB/linear PLA systems. For the RB/linear PLA system, no absorption was observed, indicating that the linear PLA does not have an encapsulation ability for RB. In contrast, for the RB/**24** system, the characteristic absorption due to RB appeared in the visible region from 480 to 600 nm, indicating that RB existed under hydrophilic circumstances, that is, the hyperbranched D-mannan core in **24**, as shown in Figure 7.25. The number of RBs per unimolecular reversed nanocapsule was estimated to be from 18.4 to 53.7, which increased with the increasing hydrophobicity of **24**.

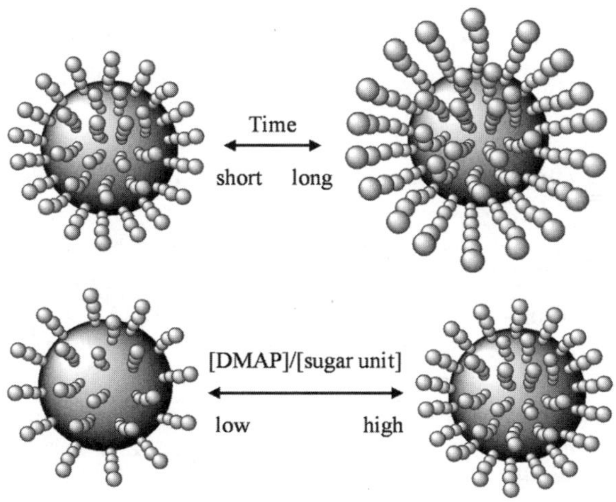

Figure 7.24 Effect of M_{PLA} and N_{PLA} for polymerization time and [DMAP]/[sugar unit].

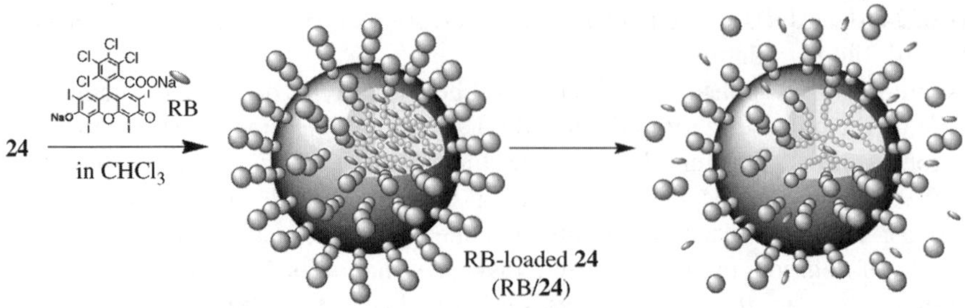

Figure 7.25 Encapsulation-release properties of **24** toward RB.

Since **24** is able to encapsulate hydrophilic compounds, the release tests *in vitro* were performed to evaluate the possible application of **24** for controlled-release systems. The RB/**24** was suspended in a 0.01 M sodium phosphate buffer (pH 7.4), and the resultant solutions were shaken at 37 °C. Figure 7.27 shows the RB-release results. For every polymer, the entrapped RB molecules were slowly released from the core of **24** and showed the sustained characteristics. The release rate increased with the decreasing hydrophobicity of **24**, suggesting that the diffusion of RB from the hydrophilic core in **24** depended on the hydrophobicity of the shell. After 21 days, proteinase K as the hydrolase of PLA was added to the systems. The RB-release rates were gradually accelerated by breaking the PLA chain of the shell. This fact suggested that the hydrophilicity of **24** increased by the hydrolysis of PLA chain on **24**, that is, the destruction of the hydrophobic shell on **24**. Hence, the unimolecular reversed micelle, **24**, is a good candidate for biodegradable controlled-release systems.

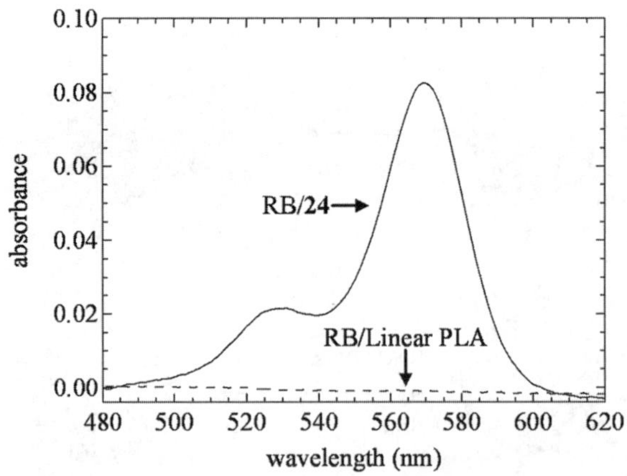

Figure 7.26 UV-Vis spectra of RB-loaded **24** (RB/**24**) and RB-loaded linear poly L-lactide (PLA) in chloroform.

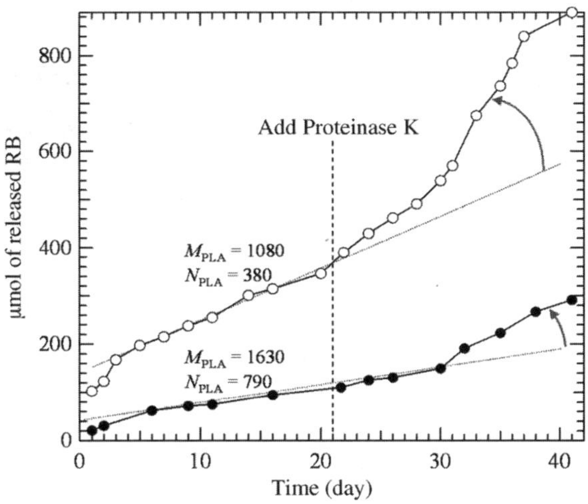

Figure 7.27 Cumulative molar amount of RB released from RB/**24**.

7.3.3 pH-Sensitive Unimolecular Reversed Micelle with Size-Selective Encapsulation Ability

As mentioned in the previous sections, the amphiphilic hyperbranched polymers, that is, unimolecular reversed micelles, have potential applications as a unimolecular nanocapsule. Thus, of great interest is the design, synthesis and functional extension of a unimolecular reversed micelle with a hyperbranched polymer core, which is capable of selective encapsulation and/or active release by other stimuli, such as pH change, of hydrophilic guest molecules. In this section, the synthesis of a novel unimolecular reversed micelle (**25**) consisting of a hyperbranched D-glucan core and an L-leucine ethyl ester shell were introduced, as shown in Figure 7.28 (Kitajyo *et al.*, 2007b; Satoh *et al.*, 2009). In addition, the size-selective encapsulations and pH-sensitive release properties of **25** toward water-soluble dyes, such as RB, thymol blue (TB as shown in Figure 7.29), alizarin yellow (AY), methyl blue (MB), and

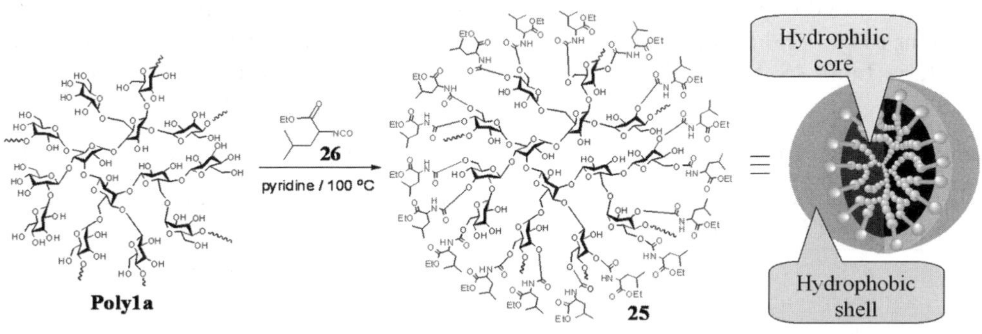

Figure 7.28 Synthesis of unimolecular reversed micelle (**25**) consisting of hydrophilic hyperbranched D-glucan core and hydrophobic L-leucine ethyl ester shell.

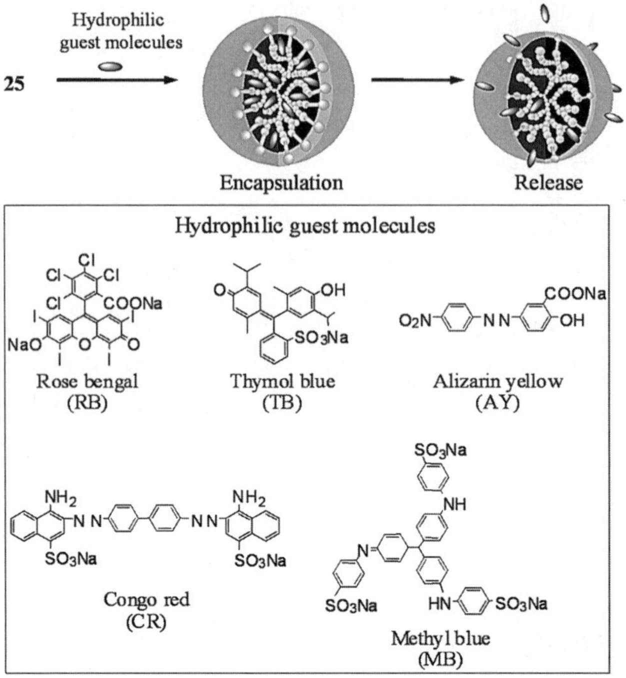

Figure 7.29 Encapsulation-release properties of **25** toward hydrophilic guest molecules.

congo red (CR), are discussed from the viewpoint of the unimolecular reversed micelle property of **25**.

In order to prepare the unimolecular reversed micelle with the hyperbranched glyco-conjugated polymer core, the reaction of the hyperbranched D-glucan (**Poly1a**, $M_{w,SLS}$, 70 600) (Satoh *et al.*, 2005) with the *N*-carbonyl L-leucine ethyl ester (**26**) was carried out in pyridine at 100 °C, as shown in Figure 7.28. The degrees of substitution (DS) determined by elemental analysis could be easily controllable by the feed rate of **26**, and these values gradually increased with the increasing molar ratio of **26** and glucose units in **Poly1a** ([**26**]/[glu]); the DS values were 46.0% (**25a**) for 2.3 of [**26**]/[glu], 68.7% (**25b**) for 3.6 of [**26**]/[glu], and 93.7% (**25c**) for 10.0 of [**26**]/[glu]. The solubilities of the resulting polymers **25** were compared with **Poly1a** in Table 7.8. **Poly1a** is soluble in water and insoluble in organic solvent. In contrast,

Table 7.8 Solubilities of hyperbranched D-glucan (**Poly1a**) and the hyperbranched D-glucan with L-leucine moiety (**25**)[a]

	DS mol%	Water	Methanol	Acetone	$CHCl_3$	CH_2Cl_2	AcOEt	Ether	Toluene
Poly1a	0	+	−	−	−	−	−	−	−
25a	46.0	−	+	+	+	+	−	−	−
25b	68.7	−	+	+	+	+	+	+	+
25c	93.7	−	+	+	+	+	+	+	+

[a] +: Soluble at *r.t.*, −: Insoluble at *r.t.* Ether: Diethyl ether.

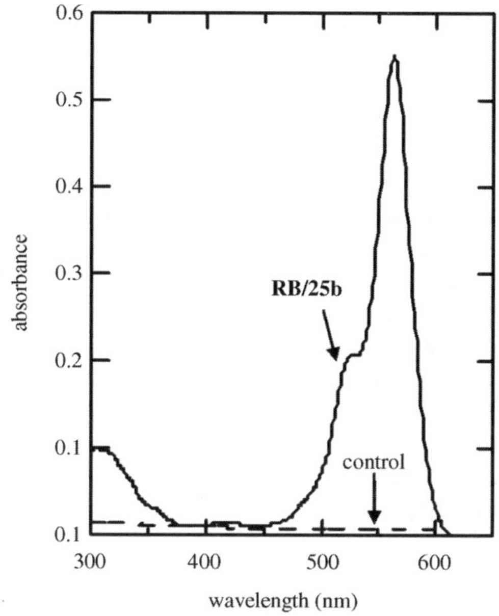

Figure 7.30 UV-Vis spectra of RB with **25b** and control after transfer of RB in $CHCl_3$.

the obtained **25** are soluble in a large variety of organic solvents, such as methanol, acetone, chloroform, and ethyl acetate, and insoluble in water, which differed remarkably from the solubility of **Poly1a**. The solubilities of **25** were also changed by the substitution degrees (DS) of the L-leucine moiety. **25b** and **25c** with a DS of 68.7% and 93.7%, respectively, were soluble in ethyl acetate, diethyl ether, and toluene, while **25a** with a DS of 46.0% was insoluble in these solvents.

In order to study the unimolecular reversed micellar property of **25**, the encapsulation-release characteristics for RB are examined using the solid/liquid phase-transfer method (Kitajyo *et al.*, 2007b). Figures 7.30 and 7.31 show the UV-Vis spectra and the pictures of the

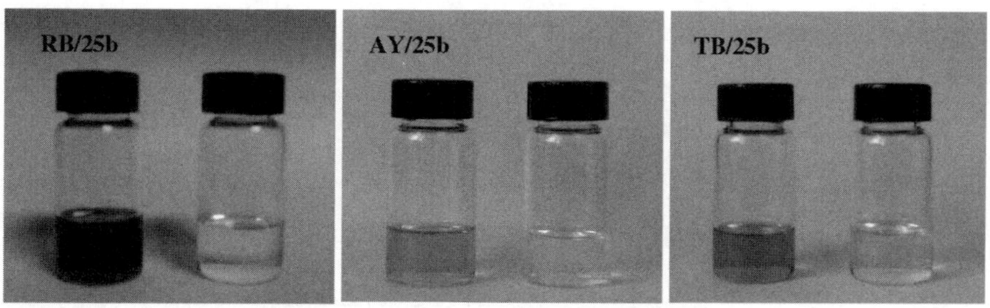

Figure 7.31 Demonstration of the encapsulation property of **25b** in $CHCl_3$ toward RB (left), AY (center), and TB (right). The left vials in each picture include **25b**, while the right vials in each picture are the control experiment in the absence of **25b**.

chloroform solutions for the RB/**25b** system, respectively. For the RB/**25b** system, the chloroform phases were apparently colorized, and the characteristic absorption due to RB appeared in the visible region from 470 to 600 nm. In contrast, the control experiment in the absence of **25b** showed no coloration.

The encapsulation amounts of RB per **25a–c** ([RB]/[**25**]) were 27.9 for **25a**, 30.0 for **25b**, and 21.6 for **25c**. **25b** more efficiently encapsulated RB than did **25a** and **25c**. The DS value of **25a** was 46.0%, which was lower than the expected value when all the external hydroxyl groups of **Poly1a** were reacted (about 51%), indicating that a somewhat denser shell was required to encapsulate many dyes. In contrast, the lower encapsulation values of **25c** were ascribed to the fact that the higher DS value of **25c** led to a narrower hydrophilic interior than in **25b**.

Additionally, the chiroptical property of the RB/**25b** system was estimated using circular dichroism (CD) spectroscopy (Figure 7.32). In the CD spectrum, a negative Cotton effect was observed in the range from 540 to 580 nm corresponding to the absorption of RB, indicating that RB existed under chiral hydrophilic circumstances, that is, the hyperbranched glycoconjugated polymer core in **25b**. Furthermore, in the ^1H NMR spectra of the RB/**25b** system, the signals due to all protons appeared in $CDCl_3$/dimethyl sulfoxide-d_6 (DMSO-d_6) (3/2, v/v), whereas the signal due to RB did not appear in $CDCl_3$. These results showed that the intermolecular mobility of the RB protons was significantly limited because the RB encapsulation in $CDCl_3$ occurred inside the hydrophilic core of **25b**.

The encapsulation ability of **25** was also examined using other water-soluble dyes, such as TB, AY, MB, and CR. Although TB and AY were readily encapsulated by **25** as well as RB, MB and CR were not encapsulated because these dyes had greater extended structures than the other dyes. The results indicated that the encapsulation ability of **25** toward these dye molecules

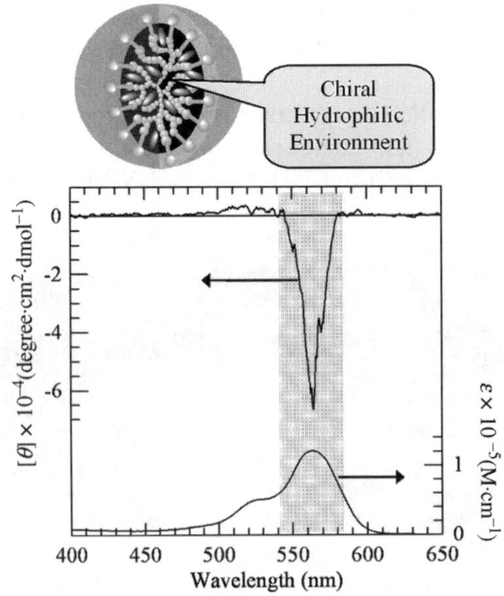

Figure 7.32 CD (upper) and UV (bottom) spectra of RB/**25b** in chloroform.

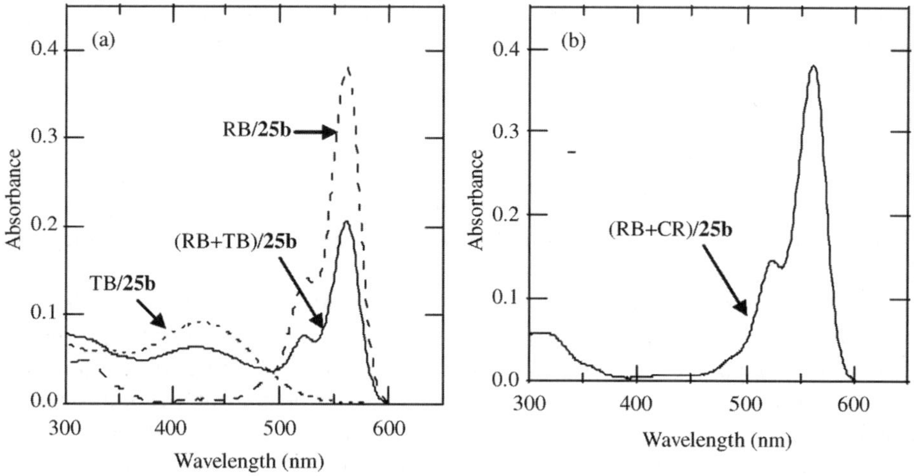

Figure 7.33 UV-Vis spectra of (a) RB/**25b**, TB/**25b**, and (RB + TB)/**25b** systems and (b) (RB + CR)/**25b** system. Conditions of (RB + TB)/**25b** system; **25b**:11 mg, RB: about 10 mmol, TB: about 10 mmol, 2 mL $CHCl_3$. After shaking for 24 h at 37 °C, any insoluble dyes were removed using a 0.45 μm membrane filter.

should relate to the molecular size of the dye, that is, **25** do not have enough space needed for encapsulating MB and CR.

In order to study the size selectivity for the encapsulation ability of **25**, the encapsulation characteristics toward the mixed dyes was examined using the solid/liquid phase-transfer method (Satoh *et al.*, 2009). Figure 7.33a shows the UV-Vis spectra of the chloroform solutions for the RB/**25b**, TB/**25b**, and (RB + TB)/**25b** systems. For the (RB + TB)/**25b** mixed system, the characteristic absorptions due to RB and TB appeared in the visible region of 500–600 nm and 360–500 nm, respectively. The encapsulation amounts of hydrophilic dyes per **25b** ([Dye]/[**25b**]) were determined from the UV-Vis spectra of the (RB + TB)/**25b** system and the experimental results are summarized in Table 7.9. In the experiment, TB was more efficiently encapsulated in **25b** than RB, for example, the encapsulation amounts of hydrophilic dyes per **25b** were a [RB]/[**25b**] of 8.7 and a [TB]/[**25b**] of 15.5. In contrast, the encapsulation

Table 7.9 Encapsulation amounts of hydrophilic dyes per **25b**[a]

	[RB]/[**25b**]	[TB]/[**25b**]	[AY]/[**25b**]
RB and TB	8.7	15.5	—
RB and AY	17.6	—	11.5

[a]The encapsulation amounts of the hydrophilic dyes per **25**b ([Dye]/[**25**b]) were determined by quantitative analysis of the UV-Vis spectra of the Dye/**25**b system diluted with DMSO (rose bengal (RB), $CHCl_3$/DMSO = 1/7 (v/v), $\lambda_{max} = 565$ nm, $\varepsilon_{max} = 1.47 \times 10^5$ $mol^{-1}\,L\,cm^{-1}$; thymol blue (TB), $CHCl_3$/DMSO = 1/7 (v/v), $\lambda_{max} = 391$ nm, $\varepsilon_{max} = 1.41 \times 10^4$ $mol^{-1}\,L\,cm^{-1}$; alizarin yellow (AY), $CHCl_3$/DMSO = 1/1 (v/v), $\lambda_{max} = 369$ nm, $\varepsilon_{max} = 1.67 \times 10^4$ $mol^{-1}\,L\,cm^{-1}$).

experiment using the (RB + AY)/**25b** mixed system showed that RB was more efficiently encapsulated than AY, for example, a [RB]/[**25b**] of 17.6 and a [AY]/[**25b**] of 11.5. These results indicated that **25b** has a molecular size-selective encapsulation property, and the order of the selectivity is TB > RB > AY, which is the reverse order of the major axis of the dyes, that is, TB (12.5 Å) < RB (13.2 Å) < AY (14.1 Å).

Figure 7.33b shows the UV-Vis spectrum of the chloroform solution for the (RB + CR)/**25b** mixed system. The spectrum showed that only RB was encapsulated in **25b**, indicating that **25b** has a size-selective encapsulation ability, that is, **25b** can preferentially encapsulate small molecules from the mixture.

The RB-release property for the RB/**25a–c** systems in water was examined using UV-Vis spectroscopy. Figure 7.34 shows the RB-release results. In order to clarify the effect of the unimolecular reversed micelle on the RB-release, the permeation of RB (free RB) through a cellulose membrane was carried out in the absence of **25** as the control. In contrast to the fast permeation of the free RB through the cellulose membrane, the release of RB/**25** was very slow and no initial release burst was observed. The release rate decreased with the increasing DS value, suggesting that the hydrophobic shells in **25** prevented the diffusion of RB from the hydrophilic core in **25**.

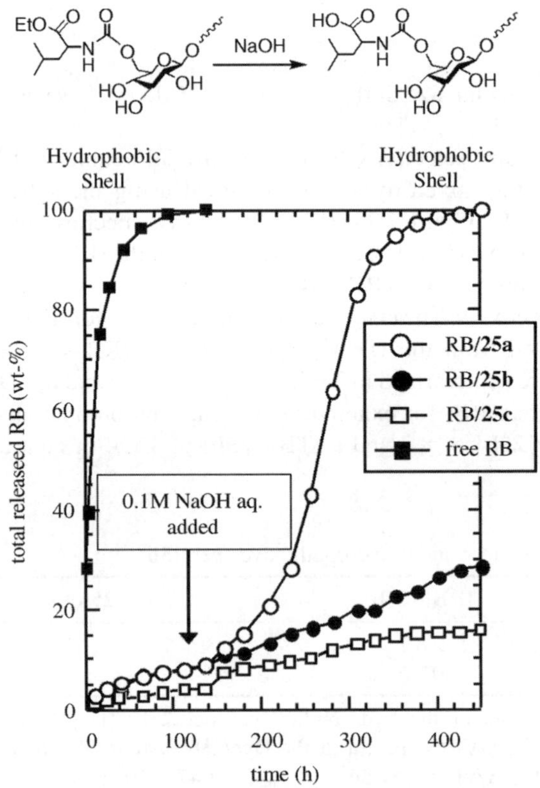

Figure 7.34 Release of RB from the RB/**25a–c** systems in water at 37 °C.

To clarify the influence of the hydrophobic shell on the release rate of RB, an aqueous NaOH solution (0.1 mol L^{-1}) was added until the release solution reached pH 12.0 after the beginning of the release experiment for 120 h. The release rates for RB/**25b** and RB/**25c** only slightly changed due to the dense shell that could not be adequately destroyed under these conditions, whereas the release rate of RB/**25a** gradually accelerated and the released RB for RB/**25a** reached 100% at 450 h. After the release experiment, the release solution of RB/**25a** became homogeneous, and the hydrolysis of the ethyl ester group of **25a** was confirmed by the ^{1}H NMR spectrum of the system. These results clearly indicated that the release rate of the encapsulated RB could be accelerated by breaking the ester groups linked with the hydrophobic shell in **25a** and that the pH sensitivity for the release rate depended on the shell density.

The encapsulation ability of **25** toward water-soluble dyes indicated that **25** was a unimolecular reversed micelle with an encapsulation ability toward hydrophilic dye molecules. In addition, **25** showed a molecular size-selective encapsulation ability and a pH-sensitive controlled-release property. Hence, **25** was a good candidate for mixture-separation systems and pH-sensitive controlled-release systems.

7.4 Conclusions

The ring-opening polymerizations of 1,6-anhydro-β-D-hexopyranose, 1,4-anhydrotetritol, 2,3-anhydrotetritol, 1,2:5,6-dianhydro-D-mannitol, and 5,6-anhydro-α-D-glucofuanose proceeded without gelation to produce a hyperbranched glyco-conjugated polymer with a high degree of branching. The weight-average molecular weights of the resulting polymers can be controlled by the polymerization conditions, such as the ratio of monomer to initiator, the monomer concentration, the polymerization time, the temperature, and the addition time of monomer. The results of the NMR, light scattering, and viscosity measurements indicated that the resulting polymers were highly branched spherical macromolecules with very low intrinsic viscosities and a lack of entanglements. Therefore, the ring-opening multibranching polymerization of an anhydro sugar is a useful method for preparing a hyperbranched glyco-conjugated polymer. The obtained hyperbranched glyco-conjugated polymer also leads a relatively new class of materials, a unimolecular reversed micelle, with the encapsulation ability for guest molecules and the controlled release ability of the encapsulated guest molecules. These unimolecular micellar abilities are dramatically changed by the combination of core and shell and the tuning of the shell density. This review chapter also describes a strategy for the development of novel unimolecular reversed micelles, and the further molecular design of amphiphilic hyperbranched polymers will be used for industrial and medical applications such as a mixture separator, molecular container, and carrier for a controlled drug-delivery system.

Acknowledgments

This study was supported by the Industrial Technology Research Grant Program in 2005–2008 from the New Energy and Industrial Technology Development Organization (NEDO) of Japan. The authors thank the OPEN FACILITY (Hokkaido University Sousei Hall) for the polymer characterization.

References

Bednarek, M.and Kubisa, P. (2006) Cationic polymerization of 2-hydroxymethyltetrahydrofuran. *Journal of Polymer Science, Part A: Polymer Chemistry*, **44**, 6484–6493.

Bernard, J., Favier, A., Zhang, L. *et al.* (2005) poly(vinyl ester) star polymers via xanthate-mediated living radical polymerization: from poly(vinyl alcohol) to glycopolymer stars. *Macromolecules*, **38**, 5475–5484.

Carvalho, J.D.S., Prins, W., and Schuerch, C. (1959) Addition polymerization of anhydrosugar derivatives. I. A polyanhydroglucose. *Journal of the American Chemical Society*, **81**, 4054–4058.

Esumi, K., Hosoya, T., Suzuki, A., and Torigoe, K. (2000) Formation of gold and silver nanoparticles in aqueous solution of sugar-persubstituted poly(amidoamine) dendrimers. *Journal of Colloid and Interface Science*, **226**, 346–352.

Garamus, V.M., Maksimova, T.V., Kautz, H. *et al.* (2004) Hyperbranched polymers: Structure of hyperbranched polyglycerol and amphiphilic poly(glycerol ester)s in dilute aqueous and nonaqueous solution. *Macromolecules*, **37**, 8394–8399.

Hirao, A., Sugiyama, K., and Yokoyama, H. (2007) Precise synthesis and surface structures of architectural per- and semifluorinated polymers with well-defined structures. *Progress in Polymer Science*, **32**, 1393–1438.

Hirao, A., Hayashi, M., Loykulnant, S. *et al.* (2005) Precise syntheses of chain-multi-functionalized polymers, star-branched polymers, star-linear block polymers, densely branched polymers, and dendritic branched polymers based on iterative approach using functionalized 1,1-diphenylethylene derivatives. *Progress in Polymer Science*, **30**, 111–183.

Hölter, D.and Frey, H. (1997) Degree of branching in hyperbranched polymers. *Acta Polymerica*, **48**, 30–35.

Imai, T., Satoh, T., Kaga, H. *et al.* (2003) Synthesis of hyperbranched 2,5-anhydro-D-glucitol by proton-transfer cyclopolymerization of 1,2:5,6-dianhydro-D-mannitol. *Macromolecules*, **36**, 6359–6363.

Imai, T., Satoh, T., Kaga, H. *et al.* (2004) Synthesis of hyperbranched carbohydrate polymer by ring-opening multibranching polymerization of 1,4-anhydroerythritol and 1,4-anhydro-L-threitol. *Macromolecules*, **37**, 3113–3119.

Imai, T., Nawa, Y., Kitajyo, Y. *et al.* (2005) Synthesis of hyperbranched polytetritol by ring-opening multibranching polymerizations of 2,3-anhydroerythritol and 2,3-anhydro-Dl-threitol. *Macromolecules*, **38**, 1648–1654.

Jayaraman, N., Nepogodiev, S.A., and Stoddart, J.F. (1997) Synthetic carbohydrate-containing dendrimers. *Chemistry: A European Journal*, **3**, 1193–1199.

Jikei, M.and Kakimoto, M.-A. (2001) Hyperbranched polymers: a promising new class of materials. *Progress in Polymer Science*, **26**, 1233–1285.

Kakuchi, T., Satoh, T., Umeda, S. *et al.* (1995) Regio- and stereoselectivity in cationic cyclopolymerizations of 1,2:5,6-dianhydro-3,4-di-*O*-methyl-D-mannitol and -L-iditol and the synthesis of Poly[(1 → 6)-2,5-anhydro-3,4-di-O-methyl-D-glucitol]. *Macromolecules*, **28**, 5643–5648.

Kadokawa, J.-I., Sato, M., Karasu, M. *et al.* (1998) Synthesis of hyperbranched aminopolysaccharides. *Angewandte Chemie International Edition*, **37**, 2373–2376.

Kitajyo, Y., Nawa, Y., Tamaki, M. *et al.* (2007a) A unimolecular nanocapsule: Encapsulation property of amphiphilic polymer based on hyperbranched polythreitol. *Polymer*, **48**, 4683–4690.

Kitajyo, Y., Imai, T., Sakai, Y. *et al.* (2007b) Encapsulation-release property of amphiphilic hyperbranched D-glucan as a unimolecular reverse micelle. *Polymer*, **48**, 1237–1244.

Lundquist, J.J.and Toone, E.J. (2002) The cluster glycoside effect. *Chemical Reviews*, **102**, 555–578.

Nakamura, K., Kawabata, T., and Mori, Y. (2003) Size distribution analysis of colloidal gold by small angle X-ray scattering and light absorbance. *Powder Technology*, **131**, 120–128.

Narumi, A.and Kakuchi, T. (2008) Synthesis of glycoconjugated branched macromolecular architectures. *Polymer Journal*, **40**, 383–397.

Satoh, T., Yokota, K., and Kakuchi, T. (1995) Synthesis of (1 → 6)-2,5-anhydro-3,4-di-*O*-methyl-D-glucitol via highly regio- and stereospecific cyclopolymerization of 1,2:5,6-dianhydro-3,4-di-O-methyl-D-mannitol with potassium tert-butoxide. *Macromolecules*, **28**, 4762–4764.

Satoh, T., Imai, T., Ishihara, H. *et al.* (2003) Synthesis of hyperbranched polysaccharide by thermally induced cationic polymerization of 1,6-anhydro-β-D-mannopyranose. *Macromolecules*, **36**, 6364–6370.

Satoh, T., Imai, T., Kitajyo, Y. *et al.* (2004) Synthesis of hyperbranched polysaccharide by thermally induced cationic polymerization of 1,6-anhydrohexopyranose. *Macromolecular Symposia*, **217**, 39–46.

Satoh, T.and Kakuchi, T. (2004) Cyclopolymerization of dianhydro sugar leading to novel carbohydrate polymers as macromolecular ionophores. *Progress in Polymer Science*, **29**, 13–43.

Satoh, T., Imai, T., Ishihara, H. *et al.* (2005) Synthesis, branched structure, and solution property of hyperbranched D-glucan and D-galactan. *Macromolecules*, **38**, 4202–4210.

Satoh, T., Tamaki, M., Kitajyo, Y. *et al.* (2006) Synthesis of unimolecular reversed micelle consisting of a poly(L-lactide) shell and hyperbranched D-mannan core. *Journal of Polymer Science, Part A: Polymer Chemistry*, **44**, 406–413.

Satoh, T.and Kakuchi, T. (2007) Synthesis of hyperbranched carbohydrate polymers by ring-opening multibranching polymerization of anhydro sugar. *Macromolecular Bioscience*, **7**, 999–1009.

Satoh, T. (2009) Unimolecular micelles based on hyperbranched polycarbohydrate cores. *Soft Matter*, **5**, 1972–1982.

Satoh, T., Kitajyo, Y., Sakai, R. *et al.* (2009) Size-selective encapsulation property of unimolecular reverse micelle consisting of hyperbranched D-glucan core and L-leucine ethyl ether shell. *Macromolecular Symposia*, **279**, 145–150.

Stiriba, S.-E., Kautz, H., and Frey, H. (2002) Hyperbranched molecular nanocapsules: comparison of the hyperbranched architecture with the perfect linear analogue. *Journal of the American Chemical Society*, **124**, 9698–9699.

Sunder, A., Hanselmann, R., and Frey, H. (1999) Controlled synthesis of hyperbranched polyglycerols by ring-opening multibranching polymerization. *Macromolecules*, **32**, 4240–4246.

Tamaki, M., Taguchi, T., Nakabayashi, S. *et al.* (2010) Hyperbranched 5,6-glucan as reducing sugar ball. *Polymer Chemistry*, **1**, 82–92.

Thiem, J., Stieholt, W.A., and Häring, T. (1989) Ring-opening polymerization of carbohydrate-derived dialkoxyoxolane. *Makromolekulare Chemie*, **190**, 1737–1753.

Tkacz, J.S., Lampen, J.O., and Schuerch, C. (1972) Stereoregularity of chemically synthesized (1 → 6)-**α**-D-mannopyranan as revealed by enzymic degradation with arthrobacter-**α**-D-mannanase. *Carbohydrate Research*, **21**, 465–472.

Tokar, R., Kubisa, P., Penczek, S., and Dworak, A. (1994) Cationic polymerization of glycidol: coexistence of the activated monomer and active chain end mechanism. *Macromolecules*, **27**, 320–322.

Umeda, S., Satoh, T., Saitoh, K. *et al.* (1998) Synthesis of (1 → 6)-2,5-anhydro-D-glucitol through cyclopolymerization of 3,4-di-*O*-allyl-1,2:5,6-dianhydro-D-mannitol and optical resolution ability of its derivative in HPLC. *Journal of Polymer Science, Part A: Polymer Chemistry*, **36**, 901–909.

Uryu, T.and Schuerch, C. (1971) Preparation of high molecular weight 2,3,4-tri-O-benzyl-[1 → 6]-**α**-D-gluco- and -galactopyranan and [1 → 6]-**α**-D-glucopyranan. *Macromolecules*, **4**, 342–345.

Zeng, F.and Zimmerman, S.C. (1997) Dendrimers in supramolecular chemistry: from molecular recognition to self-assembly. *Chemical Reviews*, **97**, 1681–1712.

8

Highly Branched Functional Polymer Architectures by Click-Chemistry Strategies

Mieke Lammens and Filip Du Prez
Polymer Chemistry Research Group (PCR), Ghent University, Krijgslaan 281 S4-bis, B-9000 Ghent, Belgium

8.1 Introduction

Highly branched structures such as dendritic polymers and star-shaped polymers have received significant interest for a long time as a direct consequence of their compact structure and high end-group functionality (Mishra and Kobayashi, 1999; Newkome *et al.*, 2001; Fréchet and Tomalia, 2001; Bosman *et al.*, 1999). Due to their unusual bulk and solution properties, they provide a perfect platform for numerous applications, for example in pharmaceutics as drug and gene carriers, as catalytic containers and for coating applications. However, despite the increase in demand of these complex structures, the conventional synthetic methods are often not applicable for industrial use as they are often tedious and time consuming. On the other hand, by combining recent developments in click chemistry with traditional strategies, complex structures have recently become accessible via robust and versatile synthetic pathways, making them available to the broader materials community. While the first reports on the use of click chemistry were in the area of dendrimer synthesis (Morgenroth *et al.*, 1998; Muller *et al.*, 1995; Morgenroth *et al.*, 1997), this knowledge was quickly transferred to the preparation of other types of highly functionalized polymer structures such as hyperbranched polymers (Stumpe *et al.*, 2006; Kim and Webster, 1992).

Click chemistry comprises a number of organic heteroatom coupling procedures that comply with the stringent criteria as defined by Sharpless (see also Chapter 2) (Kolb *et al.*, 2001). Among these click reactions, the copper-catalyzed azide–alkyne cycloaddition reaction

Complex Macromolecular Architectures: Synthesis, Characterization, and Self-Assembly, First Edition.
Edited by Nikos Hadjichristidis, Akira Hirao, Yasuyuki Tezuka and Filip Du Prez.

(CuAAC) (Johnson *et al.*, 2008), historically the first one that has been given this name, is up to now the most applied method. However, despite the popularity of this reaction on a lab scale, its widespread use is limited by the presence of a cytotoxic transition metal and the need for potentially explosive azides. For that reason, a range of copper-free click approaches are now intensively studied, including thiol-ene/yne addition reactions and (ultrafast) hetero-Diels–Alder cycloaddition (HDA) reactions, which are believed to have great potential to become as popular as the CuAAC in the click chemistry toolbox (Lutz, 2008; Becer *et al.*, 2009).

In this book chapter, we intend to show how these click chemistry developments have allowed for more straightforward synthetic strategies with regard to the preparation of highly branched polymer architectures such as dendrimers, hyperbranched polymers, unsymmetrical dendrimers, dendronized polymers and star-shaped (co)polymers. Where other reviews provide a succinct overview of the current state of art (Sumerlin and Vogt, 2010), or focus on one specific topic, for example, azide-alkyne click chemistry (Fournier *et al.*, 2007; Binder and Sachsenhofer, 2007; Binder and Sachsenhofer, 2008) or thiol-ene click chemistry (Kade *et al.*, 2010) and its use in the area of dendrimers and hyperbranched polymers (Voit, 2007; Voit and Lederer, 2009; Gao and Matyjaszewski, 2009), the intention of this chapter is to highlight in detail the latest click chemistry developments for the preparation of all kinds of highly branched polymer structures, including star-shaped polymer structures, and to compare the methods in terms of efficiency, orthogonality and purification simplicity. Moreover, the limitations of the click methods for those complex structures are highlighted.

In this way, our aim is that this chapter may serve to any polymer chemist, physicist or material engineer to make the most appropriate choice of the design methodology to prepare the desired branched polymer structures.

Before describing the various click strategies for the different types of polymer architectures, the definition of click chemistry and the various approaches known until now, in particular the ones used for the construction of the highly branched polymers, are briefly explained. For a more extensive overview, we refer to Chapter 2 by Barner-Kowollik *et al.*

8.2 What's Available in the Click Chemistry Toolbox?

The introduction of the click chemistry philosophy in the area of polymer chemistry is probably the one that had the largest impact on the area of polymer chemistry in the twenty-first century, mostly because conjugation of any building blocks (low and high molecular weight), obtained by independent mechanisms, became easily available through universal approaches.

Click chemistry enclosures a group of powerful linking reactions that are easy to perform, have high yields, require no or minimal purification and can work in a broad range of solvents and in the presence of other functional groups (Kolb *et al.*, 2001). Four major classifications of click reactions have been distinguished:

- cycloaddition reactions of unsaturated species such 1,3-dipolar cycloaddition and hetero-Diels–Alder cycloaddition reactions;
- additions to carbon–carbon multiple bonds including thiol-ene chemistry;
- nucleophilic substitution reactions such as ring-opening reactions of strained heterocyclic electrophiles and thio-bromo click chemistry;
- "non-aldol"-type carbonyl chemistry such as the formation of oxime ethers.

The first years after Sharpless introduced this definition, the term click chemistry was almost exclusively denoted to the 1,3-dipolar cycloaddition between azides and alkynes. The original reaction, known as the huisgen [3 + 2] cycloaddition, involves a coupling reaction between azides and alkynes in the absence of a catalyst. As azides are generally poor 1,3-dipolar acceptors, azide–alkyne coupling reactions are relatively slow. Also, they require high temperature to reach acceptable yields and are not regioselective. These major drawbacks were successfully overcome in 2002 by two independent groups led by Sharpless (Rostovtsev *et al.*, 2002) and Meldal (Tornoe *et al.*, 2002), who discovered that this reaction can be efficiently catalyzed with copper(I), leading to the exclusive formation of the 1,4-substituted-1,2,3-triazole ring (A, Scheme 8.1) (Fournier *et al.*, 2007; Binder and Sachsenhofer, 2007; Binder and Sachsenhofer, 2008; Lutz, 2007; Bock *et al.*, 2005; Binder, 2008). Moreover, the use of copper(I) accelerates the reaction rate tremendously and allows copper-catalyzed azide–alkyne cycloaddition (CuAAC) reactions at room temperature. In practice, the copper(I) catalyst can be used as a copper(I) salt stabilized by a nitrogen-based ligand or can be generated *in situ* by reduction of copper(II) salts, usually in aqueous medium. However, concerns about the cytotoxicity led to the exploration of other click reactions that do not require any metal catalyst.

Indeed, great interest was recently attributed to metal-free click reaction of azides and alkynes, comprising the development of different approaches to increase the reactivity of the alkyne compound (Lutz, 2008; Baskin and Bertozzi, 2007). One elegant approach is the reaction of azides with strained cycloalkynes, as developed by Bertozzi *et al.* (Laughlin *et al.*, 2008). However, the rate of the reaction is relatively low in comparison to the CuAAC reaction. Therefore electron-withdrawing substituents were added on the α-position of the triple bond (Ess *et al.*, 2008). Another approach makes use of substituted dibenzocyclooctynes, where the aromatic rings impose additional ring strain and conjugate with the alkyne, thereby increasing the reactivity of the alkyne in metal-free [3 + 2] cycloaddition reactions with azides (Ning *et al.*, 2008). However, as the synthetic procedures to obtain the starting alkynes (and alkyne-terminated polymer chains) are quite cumbersome, these strategies are for the moment limited to specific biorelated applications and are not expected to have a large breakthrough in the development of polymer materials (Lutz, 2008).

Among the cycloaddition reactions, the Diels–Alder [4 + 2] cycloaddition (DA) between anthracene derivatives and maleimides has also been successfully used in the synthesis of polymeric structures (B, Scheme 8.1) (Durmaz *et al.*, 2006; Durmaz *et al.*, 2007; Gacal *et al.*, 2006). However, this strategy bears characteristics that may prove to be problematic in certain applications. For example, high temperatures are required for this type of coupling reaction, which makes it unsuitable for the formation of polymeric structures from thermally unstable compounds. Recently, the research groups of Barner-Kowollik and Stenzel reported an alternative hetero-Diels–Alder (HDA) strategy for synthesizing polymeric conjugates based on the coupling reaction between a terminal electron-deficient thiocarbonylthio group of RAFT-generated polymers and an appropriate diene (C, Scheme 8.1) (Inglis *et al.*, 2008). The thiocarbonylthio moieties bearing electron-withdrawing Z-groups were designed in such a way that they could be sequentially used as a RAFT agent for reversible addition fragmentation chain-transfer reaction (RAFT) and as a reactive heterodienophile in the HDA reaction with a reactive diene. To facilitate the HDA reaction, the dienophilicity of the dithioester end group was further increased by using a catalyst ($ZnCl_2$ or TFA) that enhances the electron-withdrawing effect of the Z-group, which lowers the LUMO energy of the thiocarbonyl group (Inglis *et al.*, 2008; Sinnwell *et al.*, 2008a). The reactions proceed, without the

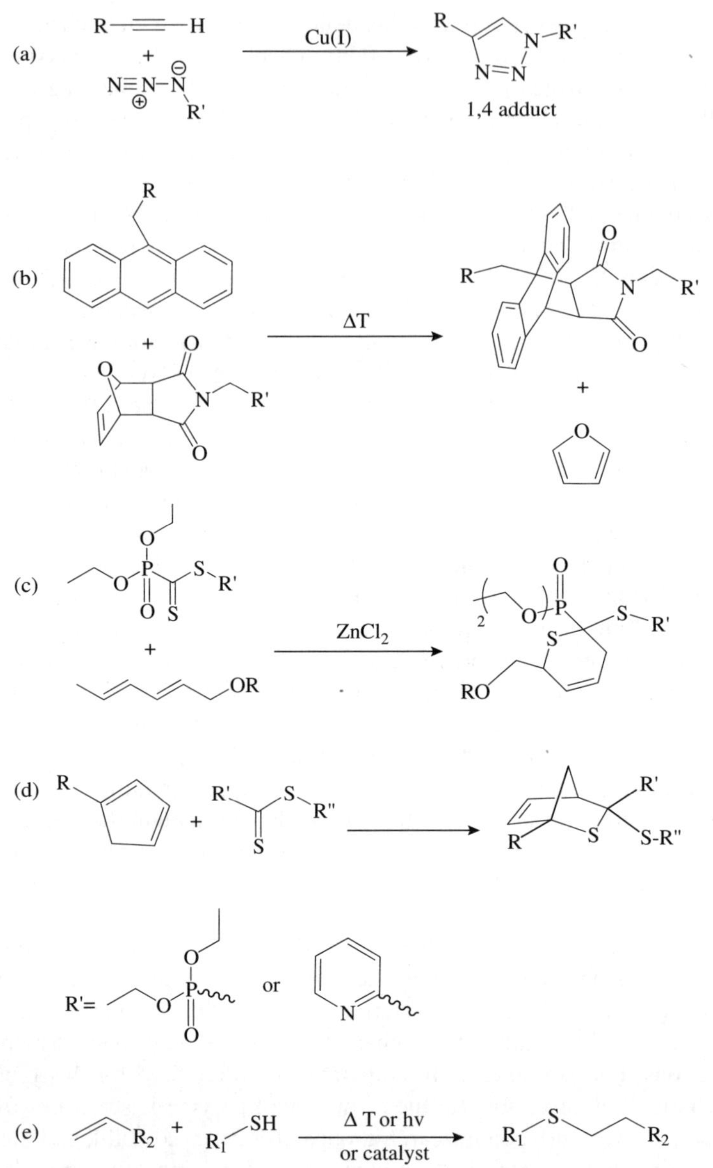

Scheme 8.1 Examples of click reactions commonly employed in polymer synthesis and functionalization reactions. (a) Copper(I)-catalyzed azide alkyne cycloaddition reaction. (b) Diels–Alder [4 + 2] cycloaddition between anthracene derivatives and maleimides. (c) Hetero-Diels–Alder between a terminal electron-deficient thiocarbonylthio group of RAFT-generated polymers and an appropriate diene. (d) Ultrafast hetero-Diels–Alder reaction between a terminal electron-deficient thiocarbonylthio group of RAFT-generated polymers and a reactive cyclopentadiene. (e) Thiol-ene click chemistry.

need of additional modification steps at the chain ends, to high conversions at 50 °C in reaction times between 2 and 24 h (Sinnwell *et al.*, 2009).

Very recently, the same authors reported on a quite efficient ultrafast hetero–Diels–Alder reaction by making use of a more reactive cyclopentadiene compared to the linear dienes (D, Scheme 8.1) (Inglis *et al.*, 2009). High yields were obtained in just a few minutes at ambient temperature, without any catalyst being added to the reaction mixture. For these reasons, this methodology is expected to become an important one in the click chemistry toolbox.

Also, a series of thiol-based reactions, already known for a long time in the area of photochemistry, have recently been highlighted as powerful linking chemistries and include the thiol-ene/yne (Cook *et al.*, 2007; Carioscia *et al.*, 2007) reactions (E, Scheme 8.1). Thiol-ene/yne is used to denote the addition of a thiol to an ene/yne bond (Hoyle *et al.*, 2004). Depending on the reaction conditions used and the chemical nature of the unsaturated bond, the thiol-ene reaction can proceed either by a free radical (UV-light or thermally initiated thiol-ene coupling reaction) or an ionic mechanism (Michael-type additions), while thiol-yne click reactions only proceeds using radical initiated reaction. Thiol-click reactions have a bright future for the synthesis of complex architectures on the larger scale due to the simplicity, high reactivity and broad variety of available compounds, short reaction times and ambient reaction conditions (Campos *et al.*, 2008).

8.3 Click Approaches for the Synthesis of Dendrimers

Dendritic structures comprise subclasses such as perfectly branched dendrimers, hyperbranched polymers and all kinds of polymer structures in between both extremes, such as dendrigrafts and dendronized polymers (Scheme 8.2) (Voit, 2007). Due to their highly branched, spherical structure with a high end-group functionality, they exhibit special reactivity, solubility and rheology features, leading to applications in catalysis (Mery and Astruc, 2006), magnetic resonance imaging (Stiriba *et al.*, 2002), encapsulation of dyes (Froehling, 2001) and drug delivery (Esfand and Tomalia, 2001; Gillies and Frechet, 2005).

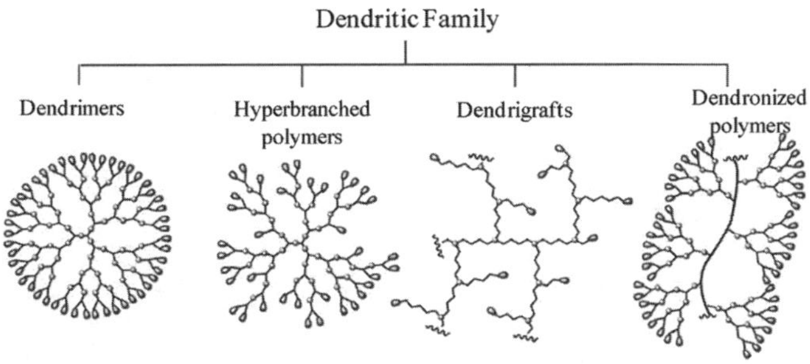

Scheme 8.2 General scheme of the dendritic family (Carlmark *et al.*, 2009). (Reproduced from A. Carlmark, C. Hawker, A. Hult and M. Malkoch, "New methodologies in the construction of dendritic materials," *Chemical Society Reviews*, **38**, 2, 352–362, 2009, by permission of the Royal Society of Chemistry.)

Below, first a brief summary of the state-of-the-art for more traditional design of dendrimers is given, after which we focus on the use of the CuAAC, Diels–Alder cycloaddition reactions and thiol-ene chemistry for the preparation of these dendritic structures.

8.3.1 Traditional Synthetic Strategies for Dendrimers

Since the first reports by Vogtle (Buhleier *et al.*, 1978) and Newkome *et al.* (1985), the field of dendrimers has been intensively explored. Dendrimers are defined as highly ordered tree-like structures that comprise a single core that is capped by layers of repeating units that are radially branched (Carlmark *et al.*, 2009; Ma *et al.*, 2009; Flomenbom *et al.*, 2005). In dendrimer synthesis many chemical pathways can be followed, but there are two main strategies: the divergent and convergent approaches. Using the divergent route, the dendrimers are built up from a central core (Scheme 8.3, upper part). In a repeated reaction cycle, the building blocks are added layer by layer, whereby each additional step leads to a higher generation dendrimer. In the convergent synthesis, complete dendrons are synthesized and subsequently coupled onto a central core (Scheme 8.3, lower part).

Both approaches involve repeated reactions between functional groups A and B in multifunctional monomers AB_X or $A_X + B_Y$, whereby A denotes one type of functional group, B denotes another type of functional group, and X ($\geq$2) and Y ($\geq$1) are the number of reactive functional groups. As A can only react with B, and not with itself, it leads to an ($A_X + B_Y$) or (AB_X)-type branched polymer. When using the ($A_X + B_Y$)-type of monomers, a large excess of the reactant must be used in each step to minimize crosslinking, resulting in a large excess of unreacted monomer that has to be removed to isolate the dendrimer. When using the AB_X-type of monomer, the B groups are generally protected to make the A group react solely with the B groups of the prior generation of dendrimer. Deprotection is needed to activate the B group for the subsequent coupling reaction. Due to the activation steps/deprotection steps, these AB_X type dendrimers need, for example for a third-generation dendrimer,

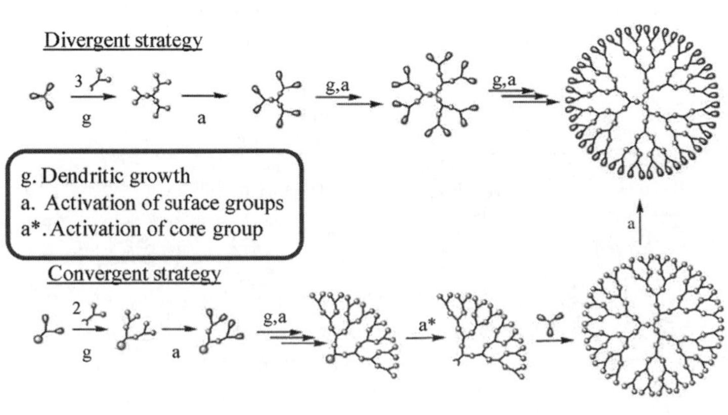

Scheme 8.3 Two synthetic routes towards dendrimers: the divergent synthesis (upper part) and the convergent synthesis (lower part) (Carlmark *et al.*, 2009). (Reproduced from A. Carlmark, C. Hawker, A. Hult and M. Malkoch, "New methodologies in the construction of dendritic materials," *Chemical Society Reviews*, **38**, 2, 352–362, 2009, by permission of the Royal Society of Chemistry.)

a minimum of six reaction steps, which makes the process time consuming as chromatographic separations are often needed, while at the same time significant waste streams are generated. Moreover, these deprotection steps are sometimes incomplete, yielding defect dendrimers.

8.3.2 CuAAC for the Synthesis of Dendrimers

As mentioned in the previous paragraph, traditional dendrimer synthesis requires high monomer concentrations and tedious workup. The application of CuAAC reactions for the synthesis of dendrimers has been reported to be successful as the click reactions are highly efficient, requiring only stoichiometric amounts of reagents and leading to simple purification steps (Voit and Lederer, 2009; Carlmark *et al.*, 2009). In 2004, Hawker, Sharpless and Fokin were the first to explore the convergent synthesis of 1,2,3-triazole-based dendrimers (Wu *et al.*, 2004). Up to fourth-generation dendrons were synthesized by successive CuAAC reactions and activation steps, and were subsequently coupled to a range of different core molecules (Figure 8.1). AB_2 monomers were synthesized as building blocks for those dendrons, in which the A functionality is a chloromethyl group and the B_2-functionalities are acetylene end groups. After each CuAAC reaction, the chloride was easily replaced with an azido functionality via an S_N2 reaction, activating it for the next CuAAC reaction. The obtained dendrons could be easily isolated as pure white solids by simple filtration or aqueous workup with yields exceeding 90%. Finally, third- and fourth-generation 1,4-disubstituted 1,2,3-triazole dendrimers were constructed by anchoring these dendrons to various polyacetylene cores. This work was the first example of an easy and efficient synthesis of dendrimers by

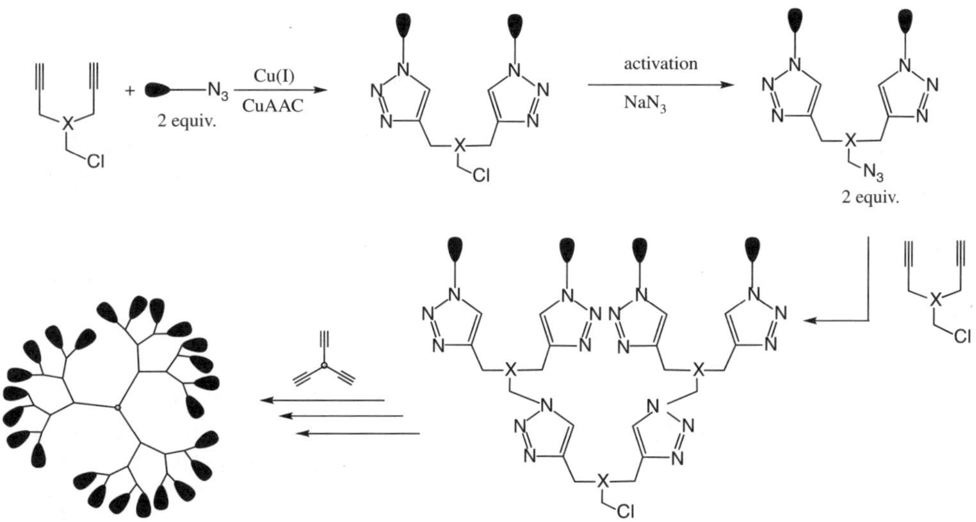

Figure 8.1 General method for convergent synthesis of 1,2,3-triazole based dendrimers via CuAAC, as initially reported by Hawker, Sharpless and Fokin (Wu *et al.*, 2004). (Reproduced with permission from P. Wu, A.K. Feldman, A.K. Nugent *et al.*, "Efficiency and fidelity in a click-chemistry route to triazole dendrimers by the copper(I)-catalyzed ligation of azides and alkynes", *Angewandte Chemie International Edition*, 2004, **43**, 3928–3932. © Wiley-VCH Verlag GmbH & Co. KGaA.)

CuAAC reactions. However, activation/deprotection steps and long reaction times (up to 30 h per coupling reaction) were still needed.

Hawker and Wooley later extended this CuAAC approach to prepare azide- or alkyne-terminated Fréchet-type dendrimers via a divergent strategy (Joralemon *et al.*, 2005).

More recently, Malkoch developed a divergent strategy without activation/deprotection steps whereby CuAAC is combined with traditional esterification/etherification reactions steps, for the synthesis of 2,2-bis(methylol)propionic acid (bis-MPA) and Fréchet-type of dendrimers (Antoni *et al.*, 2007). The bis-MPA-type dendrimers were synthesized by sequential esterification reactions and CuAAC (Figure 8.2). First, bis-MPA based AB_2 monomers containing an acyl chloride and two azide functionalities were coupled to a triphenol core via an esterification reaction, yielding a first-generation dendrimer. Subsequently, CD_2 monomers, which are propargyl esters of bis-MPA, are clicked onto the first-generation dendrimer, yielding a second-generation dendrimer. Repeating of this sequence of reactions resulted in a well-defined fourth-generation dendrimer in only four steps with excellent yield and in multigram quantities. As a comparison, a traditional synthesis would require eight steps and multiple purification steps to obtain a similar reaction product. Moreover, the dendrimer containing hydroxyl or alkyne end groups can be further postfunctionalized, so that they can for example be used in catalysis and drug-delivery applications.

The versatility of the concept was also demonstrated for the synthesis of Fréchet-type dendrimers by sequential etherification and CuAAC reactions starting from a triphenol core, using AB_2 and CD_2 types of monomers (Antoni *et al.*, 2007). Because of the simplicity of this strategy, a wide variety of other macromolecular architectures, including star-shaped polymers and hyperbranched polymers, could be easily synthesized with unique characteristics that are tuned by the nature of the monomers.

In the literature, a few examples are found on the convergent synthesis of dendrimers that use CuAAC in the final coupling step. Jin *et al.* (Lee *et al.*, 2006a) reported on the convergent synthesis of dendrimers by coupling Fréchet-type polyaromatic ether dendrons modified with an acetylene focal group, to trivalent azide cores. The coupling reaction was carried out at 70 °C for several days to achieve complete conversion. For higher-generation dendrons (third generation), incomplete coupling reactions occurred due to steric hindrance. The same group also reported on the convergent CuAAC synthesis of poly(amido amine) (PAMAM) dendrimers by coupling azido-functionalized dendritic wedges to bifunctional and tetrafunctional acetylene core molecules (Lee *et al.*, 2006b). Via a similar approach, acetylene-functionalized wedges were prepared and subsequently coupled onto azide-containing cores (Lee *et al.*, 2006c; Van Wuytswinkel *et al.*, 2000).

Although CuAAC allows for the efficient synthesis of dendritic structures, it should be noted that click chemistry has demonstrated its highest impact for the derivatization or functionalization of well-known dendrimers with bioactive compounds, UV-active compounds or ferrocenyl units. For the detailed description of these topics, we refer to literature (Malkoch *et al.*, 2005; Vestberg *et al.*, 2007).

8.3.3 Thiol-Ene Click Chemistry for the Synthesis of Dendrimers

More recently, thiol-ene click chemistry has been applied in a very effective way for the synthesis of dendrimers. Hawker *et al.* (Killops *et al.*, 2008) were the first to describe the

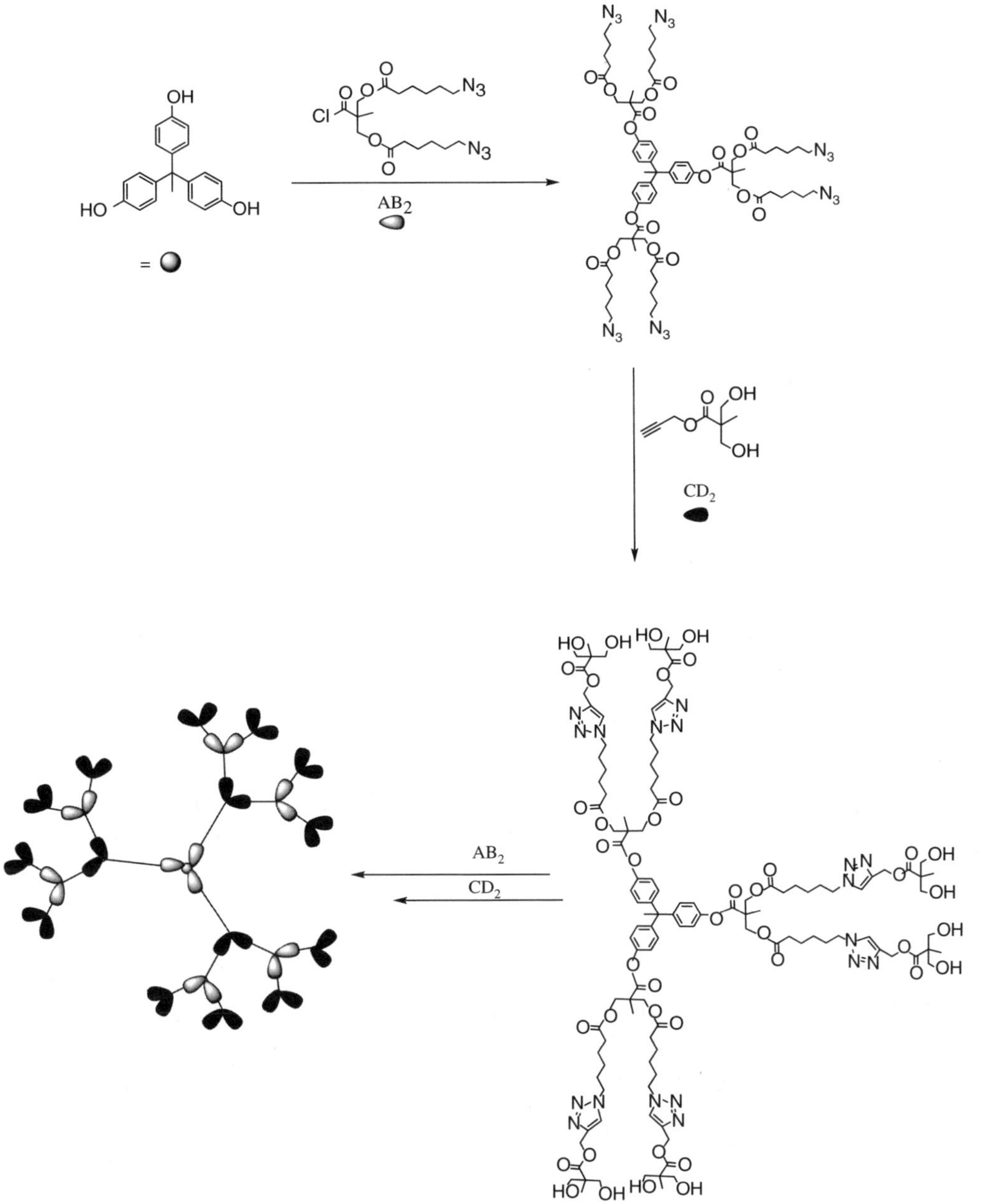

Figure 8.2 General scheme for the divergent synthesis of bis(MPA)-type dendrimers using a combination of CuAAC and traditional esterification reactions, as synthesized by Malkoch *et al.* (Antoni *et al.*, 2007). (Reproduced from P. Antoni, D. Nystrom, C.J. Hawker *et al.*, "A chemoselective approach for the accelerated synthesis of well-defined dendritic architectures," *Chemical Communications*, **22**, 2249–2251, 2007, by permission of the Royal Society of Chemistry.)

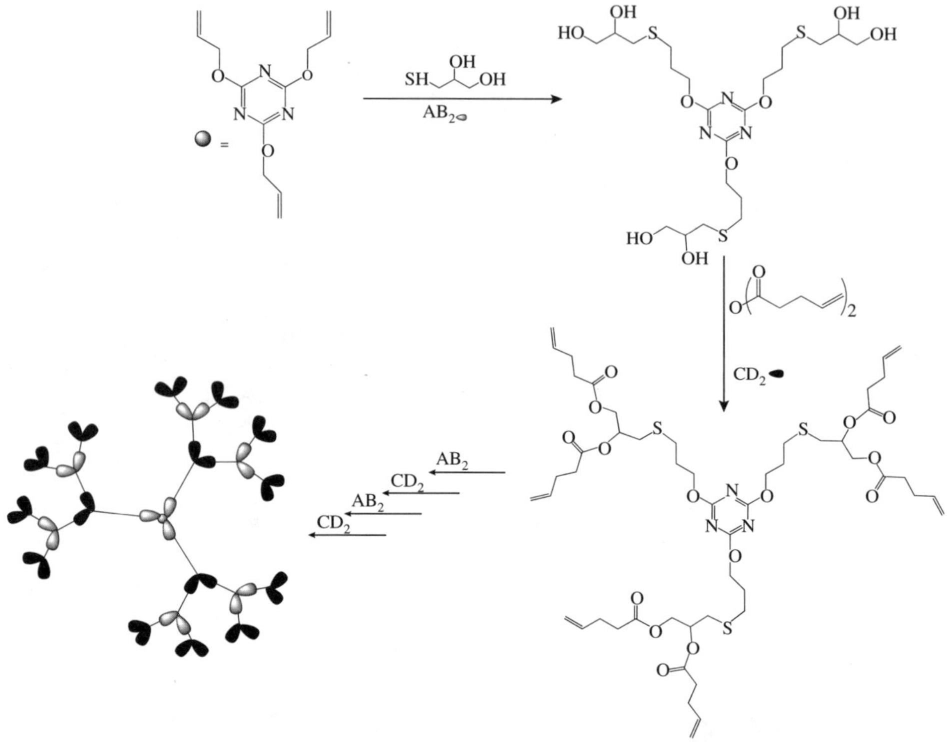

Figure 8.3 General scheme for the divergent synthesis of dendrimers using a photoinitiated thiol-ene click chemistry, as developed by Hawker *et al.* (Reprinted with permission from K.L. Killops, L.M. Campos and C.J. Hawker, "Robust, efficient, and orthogonal synthesis of dendrimers via thiol-ene 'click' chemistry," *Journal of the American Chemical Society*, **130**, 15, 5062–5064, 2008. © 2010 American Chemical Society.)

divergent synthesis of fourth-generation poly(thioether) dendrimers by using photoinitiated thiol-ene click chemistry. The overall strategy is highlighted in Figure 8.3. The AB_2 monomer, 1-thioglycerol, was reacted in a solvent-free way with a trisalkene core in the presence of a trace amount of photoinitiator by irradiation with a hand-held UV lamp for 30 min. Subsequent esterification of the hydroxyl end groups with 4-pentenoic anhydride resulted in the ene-functional dendrimer. After repetition of this stepwise procedure, the fourth-generation dendrimers were obtained.

In contrast to the CuAAC reaction, the combination of thiol-ene chemistry with a traditional esterification reaction allowed for an efficient synthesis of dendrimers without the use of a metal catalyst. Moreover, as thiol-ene click can occur in the absence of solvent under benign reaction conditions, this reaction seems to be quite a promising alternative.

Following this work, Rissing and Son (2009) recently reported on the synthesis of carbosilane–thioether dendrimers via a synthetic method based on an alternating thiol-ene/Grignard substitution reaction. While the thiol-ene based steps proceeded smoothly, the vinylation steps with vinylmagnesium bromide (ViMgBr) (Grignard substitution) required slightly more care. Moreover, Grignard substitution reactions left for long reaction times or employing a high excess of ViMgBr, resulted in the formation of high molecular weight impurities. However, keeping this in mind, fifth-generation dendrimers were synthesized.

In a similar way, Stenzel and coworkers (Chen *et al.*, 2009) recently synthesized dendrimers by successive thiol-yne chemistry and esterification reactions. As for the thiol-ene case, it is an efficient process and does not require a metal catalyst. For the thiol-yne system, the reaction proceeds by reacting two equivalents of thiol with the alkyne, via a two-step process. Therefore, after the reaction, each alkyne functional group will be combined with two thiols, which is very attractive and advantageous in the synthesis of dendrimers.

All thiol-containing systems mentioned above involve intermediate reactions, particularly activation steps, which can introduce defects into the dendrimers. Therefore, Radosz and coworkers (Ma *et al.*, 2009) developed a sequential clicking concept for the synthesis of polyester dendrimers using kinetically and mechanically chemoselective monomers. First, the reactive acrylate functionality of a kinetically asymmetric 2-[(methacryloyl)oxy]ethyl acrylate monomer (MAEA) (A, Figure 8.4) containing both an acrylate and methacrylate end group, was selectively reacted with a diamine via a Michael addition to produce a tetrafunctional core. Although the Michael addition to the primary amine is quick and quantitative, successive reaction of the formed secondary amine with the second acrylate function proceeds slower due to steric hindrance. However, heating the reaction to 40–70 °C pushes the reaction to completion. The pendant methacrylate groups of the first-generation dendrimer were then selectively reacted with the thiol group of a second monomer cysteamine (B, Figure 8.4). An alternative addition of the two monomers produced the

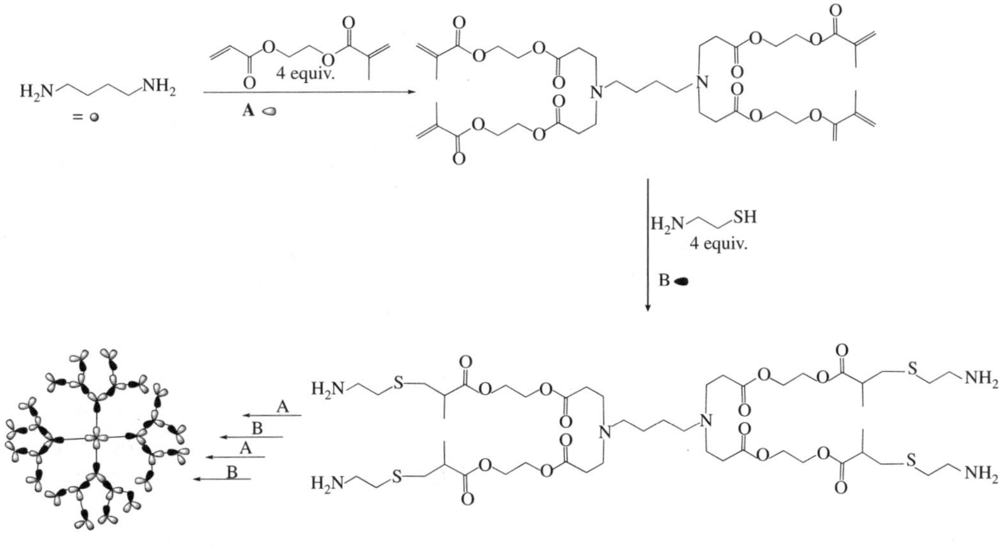

Figure 8.4 Divergent synthesis of poly(ester) dendrimers by sequential click reactions. (Reprinted with permission from X.P. Ma, J.B. Tang, Y.Q. Shen *et al.*, "Facile synthesis of polyester dendrimers from sequential click coupling of asymmetrical monomers," *Journal of the American Chemical Society*, **131**, 41, 14795–14803, 2009. © 2010 American Chemical Society.)

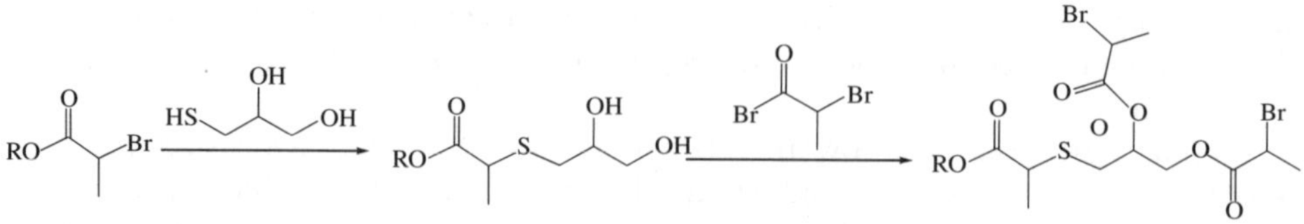

Figure 8.5 Iterative two-step divergent approach by thio-bromo reaction and subsequent acylation (Brad *et al.*, 2009). (Reprinted from M.R. Brad, L. Gerard, H. Christian and P. Virgil, "Synthesis of dendrimers through divergent iterative thio-bromo 'click' chemistry," *Journal of Polymer Science, Part A: Polymer Chemistry*, 2009, **47**, 15, 3931–3939 (Scheme 1), © 2009, with the permission of John Wiley & Sons, Inc.)

third and fourth generations of dendrimers at high yields. Thanks to its simplicity and easy workup, this click process is particularly suitable for the synthesis of custom-made dendrimers.

Very recently, a so-called "thio-bromo" strategy has been used for the synthesis of a new class of poly(thioglycerol-2-propionate) (PTP) dendrimers. The combination of thio-bromo (S_N2) reaction with subsequent acylation with 2-bromopropionyl bromide onto the divergent thioglycerol as a branching point, provides an iterative two-step divergent growth approach to the synthesis of fourth-generation dendrimers with high structural fidelity (Figure 8.5) (Brad *et al.*, 2009).

8.3.4 [4 + 2] Cycloaddition (Diels–Alder) Reaction for the Synthesis of Dendrimers

As already briefly mentioned, also the [4 + 2] Diels–Alder reactions show characteristics of a click approach. Diels–Alder reactions involve the simultaneous formation and destruction of carbon–carbon bonds, mostly requiring very little energy that leads to successful coupling, even at lower reaction temperatures. Among the various Diels–Alder reactions, only the reaction between cyclopentadienones and acetylene units was well described for the synthesis of dendrimers, already in the 1990s. Mullen *et al.* (Morgenroth *et al.*, 1998; Muller *et al.*, 1995; Morgenroth *et al.*, 1997) were the first to synthesize dendrimers via a divergent method starting from AB_2 compounds, in which the A functionality is a cyclopentadienone and the B functionalities are protected acetylenes, which were added to a multifunctional alkyne core molecule (Figure 8.6). After deprotection of the acetylene units of the AB_2 blocks, the reaction steps are repeated to yield higher-generation dendrimers. When using the convergent strategy, first 2,3,4,5-tetraphenyl-cyclopenta-2,4-dienone was reacted with 4,4′-diethynylbenzil to form the first-generation dendritic precursor, which was further reacted with 1,3-diphenylacetone leading to the first-generation dendron. However, this strategy is, due to steric constraints, limited to the second generation. Thus, for obtaining higher-generation dendrimers, the divergent strategy has to be employed. For a more extensive literature overview, we refer to the reviews of Voit *et al.* (Voit, 2007; Voit and Lederer, 2009).

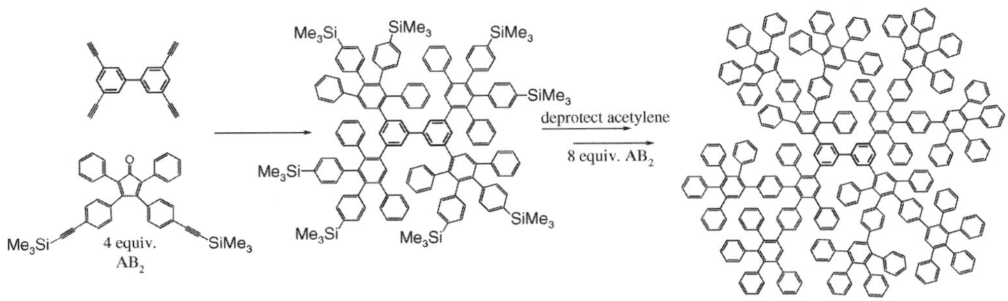

Figure 8.6 Divergent synthesis of polyphenylene dendrimers through Diels–Alder reaction (Morgenroth *et al.*, 1998). (Reproduced from F. Morgenroth, A.J. Berresheim, M. Wagner and K. Mullen, "Spherical polyphenylene dendrimers via Diels–Alder reactions: The first example of an A(4)B building block in dendrimer chemistry," *Chemical Communications*, 1139–1140, 1998, by permission of the Royal Society of Chemistry.

8.4 Click Approaches for Hyperbranched Polymers, Dendronized Polymers and Unsymmetrical Dendrimers

8.4.1 Overview of Definitions and Traditional Synthetic Strategies

8.4.1.1 Hyperbranched Polymers

Hyperbranched polymers are randomly branched structures, similar to dendrimers, but containing imperfections in the branching points due to their traditional one-pot synthesis (Voit and Lederer, 2009; Carlmark *et al.*, 2009; Feng *et al.*, 2005; Kitajyo *et al.*, 2007). These hyperbranched polymers, which can be synthesized by condensation reactions, cationic procedures, free radical polymerizations and ring opening polymerization of AB_X or $(A_X + B_Y)$-type monomers, are the least challenging from a synthetic point of view as they do not require any protection/deprotection procedures or tedious workups (Scheme 8.2).

When using AB_X-type monomers, hyperbranched polymers with polydispersity indices of 2 or higher are typically obtained. However, by adding a functional core with a B_X-functionality, the polydispersity of the final polymer is lowered and a better control of the molecular weight is obtained (for example Boltorn H40$^{(R)}$). Still, it is rather difficult to synthesize hyperbranched polymers via a traditional one-pot AB_X approach, as a result of the high reactivity of the monomers leading to premature polymerizations. Therefore, the $A_X + B_Y$ approach has better perspectives for commercial applications, particularly as the monomers are readily available. The obtained hyperbranched polymers have many of the properties of dendrimers, as they are highly branched, have a multitude of end groups, possess a globular structure and show a similar viscosity behavior, attributed to the reduced entanglements.

8.4.1.2 Dendronized Polymers

Dendronized polymers, originally termed "rod-shaped polymers", are polymers having a linear backbone with dendritic side chains (Frauenrath, 2005; Tonga *et al.*, in press; Schluter and Rabe, 2000; Malkoch *et al.*, 2004). This structure may be synthesized via three conceptually

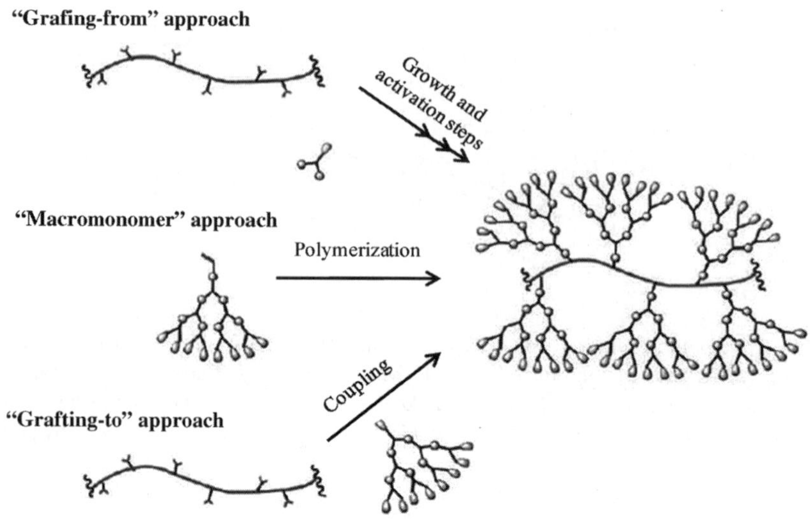

Scheme 8.4 The three strategies to synthesize dendronized polymers (Carlmark *et al.*, 2009). (Reproduced from A. Carlmark, C. Hawker, A. Hult and M. Malkoch, "New methodologies in the construction of dendritic materials," *Chemical Society Reviews*, **38**, 2, 352–362, 2009, by permission of the Royal Society of Chemistry.)

different strategies: the grafting-to, the grafting-from and the macromonomer approach (Scheme 8.4). In the grafting-to approach, premade dendrons are coupled onto a functionalized linear polymer. Using the grafting-from approach, dendrons are divergently grown onto the functional linear polymer. The third technique to synthesize dendronized polymers consists of the macromonomer approach, involving the synthesis of dendrons bearing polymerizable focal points, which are subsequently polymerized with small monomers. However, these macromonomers are tedious to make and their synthesis becomes increasingly difficult with higher dendron generation. Moreover, the actual polymerization of the dendrons becomes increasingly difficult to perform with relatively higher generation of dendronized side groups. This limitation can be overcome by adding a spacer in between the polymerizable end group and the dendron. For more detailed information, we refer to the review of Frauenrath (2005).

8.4.1.3 Unsymmetrical Dendrimers

Unsymmetrical dendrimers, also referred to as bifunctional dendrimers, are dendrimers in which the attached dendrons have different chemical structures (Scheme 8.2) (Carlmark *et al.*, 2009). These dendrons can differ in fundamental AB_X or $A_X + B_Y$ repeating units or functional end groups, leading to interesting dendrimers containing multiple functionalities.

Such dendritic structures are traditionally synthesized via the convergent strategy. First, two types of dendrons containing complementary focal groups are synthesized, after which they are coupled together. Although this strategy looks straightforward and easy, it is generally complicated due to undesired deprotection steps, which leads to tedious purification procedures (Carlmark *et al.*, 2009).

8.4.2 CuAAC for the Synthesis of Hyperbranched Polymers, Dendronized Polymers and Unsymmetrical Dendrimers

8.4.2.1 Hyperbranched Polymers

Since it is still rather difficult to produce high molecular weight (> 3000 g/mol) hyperbranched polymers via the highly reactive AB_X approach, the CuAAC reaction opens perspectives for the synthesis of branched polymers as the compounds can react at room temperature in the presence of a copper catalyst. Voit *et al.* (Scheel *et al.*, 2004) were among the first to synthesize hyperbranched poly(triazoles) based on the one-pot synthesis of an AB_2 monomer having an azide group as A-unit and two terminal acetylene bounds as B-unit, using thermal polymerization or copper(I)-catalyzed click reactions. In the thermal polymerization approach, soluble hyperbranched polymers of high molecular weight were obtained but stereocontrol of the formed triazoles was lost. The 1,4- and 1,5-disubstituted triazole rings were found in a 64: 36 ratio (Scheel *et al.*, 2004). The copper(I)-catalyzed approach yielded insoluble products, which was probably due to very high molecular weight or possible side reactions. However, full stereocontrol of the formed 1,4-substituted triazoles was obtained.

The use of AB_2 monomers containing an internal triple bond as A-unit and two azides as B-units for the synthesis of hyperbranched polymers was developed by the same group (Figure 8.7) (Scheel *et al.*, 2004). Hyperbranched polymers containing a mixture of 1,4- and 1,5-substituted triazoles were obtained, as the internal alkynes can only react via the classical thermal induced 1,3-dipolar cycloaddition reaction. However, it was possible to synthesize fully soluble products by low-temperature (45 °C) autopolymerization in bulk. The end product contains a large number of reactive azide functionalities that can be further postfunctionalized by CuAAC reaction with the desired alkyne-containing compound.

Recently, Mikhailov and coworkers reported another suitable AB_2 monomer, 2-azido-((bis) propynyloxy)-1,3,5-triazine, which was also successfully used in the one-pot synthesis of hyperbranched polymers through CuAAC (Malkov *et al.*, 2007).

When using the $A_2 + B_3$ strategy, azobenzene-containing hyperbranched polymers were synthesized via a one-pot CuAAC reaction of 4-(bis(3-azidopropyl)amino-4′-nitroazobenzene (A_2) and 1,3,5-tris(alkynyloxy)benzene (B_3) (Xie *et al.*, 2008). Soluble polymers with molecular weights of about 15 000 g/mol have been obtained.

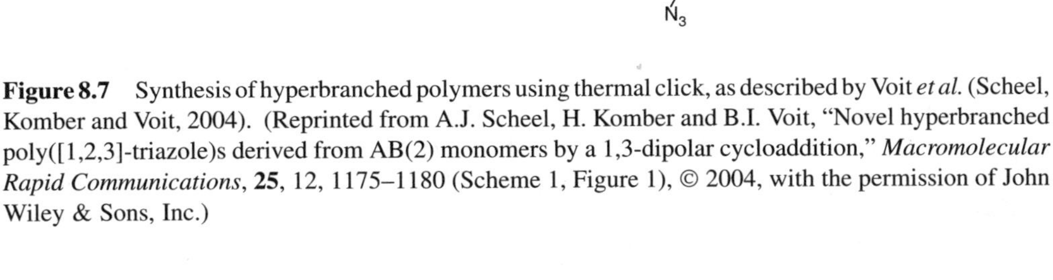

Figure 8.7 Synthesis of hyperbranched polymers using thermal click, as described by Voit *et al.* (Scheel, Komber and Voit, 2004). (Reprinted from A.J. Scheel, H. Komber and B.I. Voit, "Novel hyperbranched poly([1,2,3]-triazole)s derived from AB(2) monomers by a 1,3-dipolar cycloaddition," *Macromolecular Rapid Communications*, **25**, 12, 1175–1180 (Scheme 1, Figure 1), © 2004, with the permission of John Wiley & Sons, Inc.)

8.4.2.2 Dendronized Polymers

Analogous to the successful application to dendrimer and hyperbranched polymer synthesis, the CuAAC reaction appears to be an excellent coupling method for the attachment of dendrons to a polymer backbone. In 2004, Fréchet and Hawker used CuAAC to fix first- to fourth-generation Fréchet-type dendrons with focal azide groups onto a poly(vinyl acetylene) backbone to yield a dendronized polymer. While the coupling efficiency was quantitative for the first- and second-generation dendrons, and 98% in the case of the third-generation dendrons, no reaction was observed for the fourth generation (Helms *et al.*, 2004).

In order to overcome the problems of steric hindrance close to the backbone of the linear polymer chain, a different approach was used. Indeed, dendrons were coupled onto smaller dendrons that were already attached to the polymer backbone, leading to doubly dendronized linear polymers (Mynar *et al.*, 2005). For this, linear poly(hydroxystyrene) was synthesized, that was divergently dendronized up to third generation with a dendritic poly(aliphatic ester) based on bishydroxy propionic acid (Figure 8.8). Then, the hydroxyl end groups of these dendronized polymers were postmodified with pentynoic acid, providing terminal alkyne groups. In the final step, CuAAC was applied to link third-generation dendrons containing azido focal points to the corona of these dendronized polymers, yielding sixth-generation

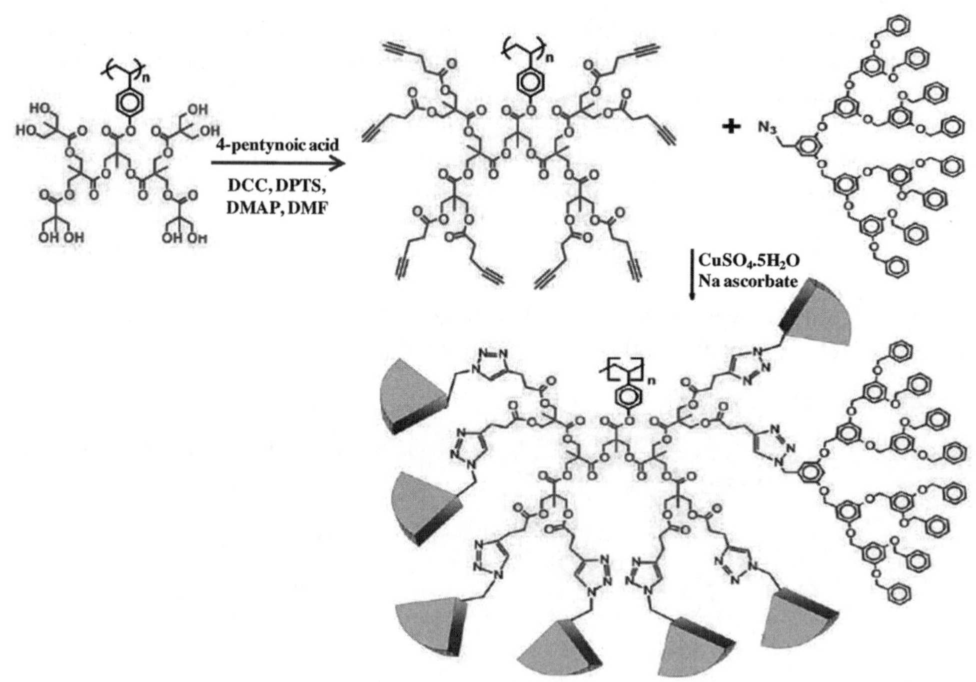

Figure 8.8 Synthesis of doubly dendronized linear polymers as developed by Fréchet (Mynar *et al.*, 2005). (Reproduced from J.L. Mynar, T.L. Choi, M. Yoshida *et al.*, "Doubly-dendronized linear polymers," *Chemical Communications*, **41**, 5169–5171, 2005, by permission of the Royal Society of Chemistry.)

dendronized polymers. A nearly quantitative CuAAC reaction was assumed from SEC and spectroscopic analysis.

8.4.2.3 Unsymmetrical Dendrimers

Despite the high potential of unsymmetrical dendrimers, the synthetic protocols utilized to prepare these structures are tedious. By taking advantage of the CuAAC reaction, their synthesis has been simplified tremendously. In 2005, Hawker *et al.* (Wu *et al.*, 2005a) reported the synthesis of bifunctional dendrimers using CuAAC (Scheme 8.5). First, hydrophobic and hydrophilic dendrons with either acetylene or azide groups at the focal point were grown using anhydride chemistry. Subsequently, these dendrons were coupled together via CuAAC, yielding unsymmetrical dendrimer up to fourth generations. To further elevate the structural complexity, CuAAC was used for functionalization of the outer shell. Dyes and sugars were added onto the periphery of individual blocks, allowing preparation of agents with dual function, namely recognition and detection, which may be useful in treatment of pathological conditions via multivalent interactions.

8.4.3 Thiol-Yne and Thio-Bromo Chemistry for the Synthesis of Hyperbranched Polymers

In a recent paper of Perrier *et al.* (Konkolewicz *et al.*, 2009), thiol-yne click chemistry has been used for the synthesis of functional hyperbranched polymers. First, low and high molecular weight AB_2-type molecules containing a thiol (**A**) and one bifunctional alkyne

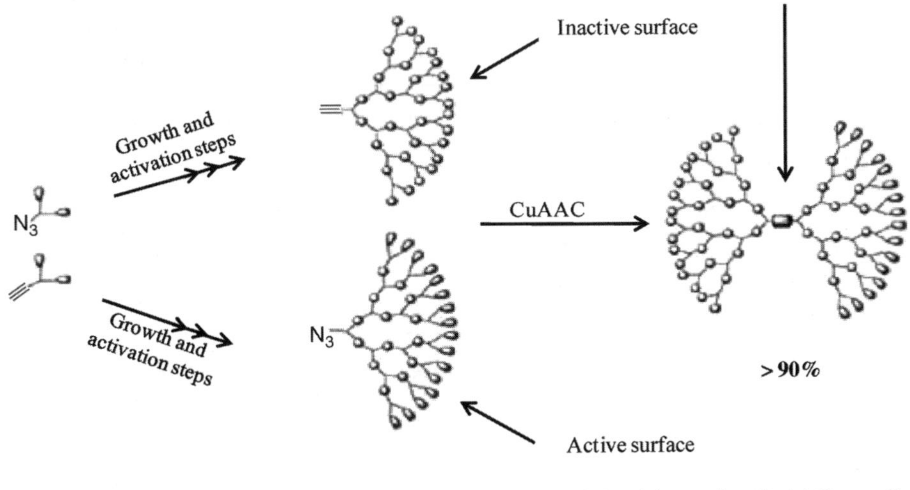

Scheme 8.5 General strategy of synthesizing unsymmetrical dendrimers by CuAAC coupling of different dendrons (Carlmark *et al.*, 2009). (Reproduced from A. Carlmark, C. Hawker, A. Hult and M. Malkoch, "New methodologies in the construction of dendritic materials," *Chemical Society Reviews*, **38**, 2, 352–362, 2009, by permission of the Royal Society of Chemistry.)

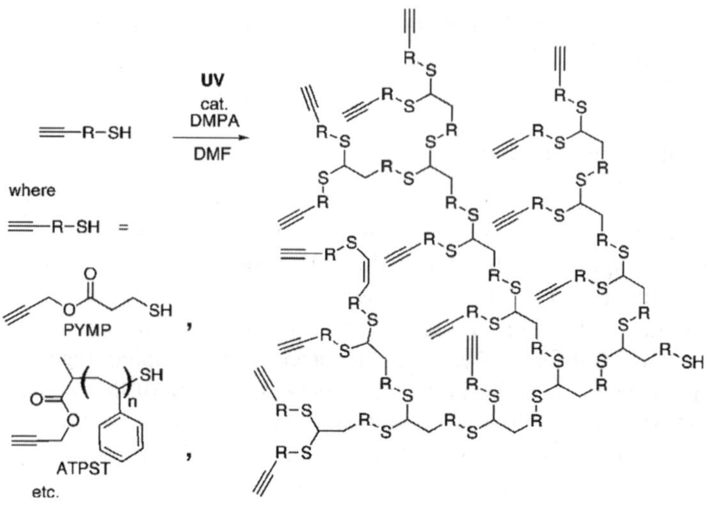

Figure 8.9 General strategy for the synthesis of hyperbranched polymers by thiol-yne click chemistry. (Reprinted with permission from D. Konkolewicz, A. Gray-Weale and S. Perrier, "Hyperbranched polymers by thiol-yne chemistry: From small molecules to functional polymers," *Journal of the American Chemical Society*, **131**, 50, 18075–18077, 2009. © 2010 American Chemical Society.)

(**B$_2$**) were synthesized. Subsequently, hyperbranched polymers were formed by a thiol-yne click reaction under UV radiation at room temperature, whereby yields over 95% were reached within less than 2 h (Figure 8.9). Furthermore, the reaction was applied to RAFT polymerized linear polymers bearing an alkyne and a thiol group at the α and ω end of the polymer, allowing subsequently formed hyperbranched polymer to incorporate additional functionalities.

Very recently, Davis and coworkers synthesized high molecular weight hyperbranched polymers by using a combination of thio-bromo chemistry and RAFT polymerization (Figure 8.10) (Xu *et al.*, 2009). First, RAFT agents bearing two tertiary bromoesters in the R group fragment were synthesized. Then, homopolymerization of methyl acrylate (MA) using this RAFT agent yielded novel AB$_2$ macromonomers bearing an α-bis bromoester and

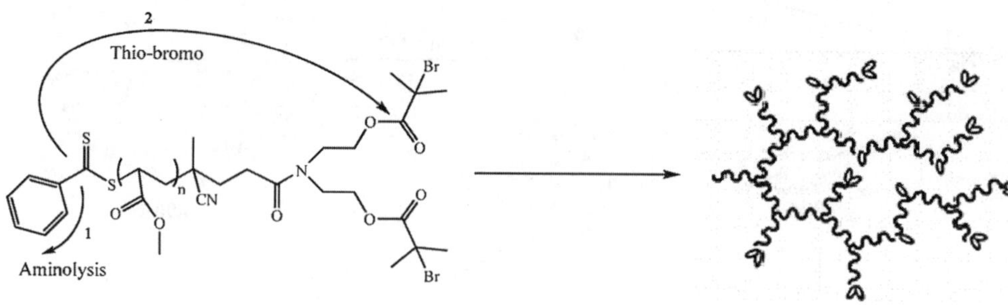

Figure 8.10 One-pot preparation of hyperbranched polymers using thio-bromo reaction. (Reprinted with permission from J. Xu, L. Tao, C. Boyer *et al.*, "Combining thio-bromo 'click' chemistry and RAFT polymerization: A powerful tool for preparing functionalized multiblock and hyperbranched polymers," *Macromolecules*, **43**, 1, 20–24, 2010. © 2010 American Chemical Society.)

ω-dithiobenzoate end groups. The ω-dithiobenzoate end groups were cleaved via aminolysis to form the corresponding thiol, which reacted intermolecularly with α-bromoester terminal groups of other AB_2 monomers, yielding a hyperbranched structure. The aminolysis and subsequent click reaction of the AB_2 macromonomers were monitored by size-exclusion chromatography (SEC), showing a shift toward high molecular weight with increasing progress of the aminolysis reaction. Although the reaction proceeded more than 20 h to reach high yields due to the steric hindered tertiary α-bromoester, hyperbranched structures with high molecular weights (64 000 g/mol) could be obtained, which has not been reported using traditional strategies. The excess of α-bromoester functionalities onto the periphery of the hyperbranched polymer can be easily modified to other functionalities and provide potential anchoring points for graft copolymerization via ATRP and other living radical processes.

Until now, thio-bromo click chemistry has only been used for the synthesis of dendrimers and hyperbranched polymers. Without any doubt, further research will lead to other complex architectures, such as dendronized polymers, dendrigrafts and unsymmetrical dendrimers.

8.4.4 [4 + 2] Cycloaddition Diels–Alder Reaction for the Synthesis of Hyperbranched Polymers and Dendronized Polymers

8.4.4.1 Hyperbranched Polymers

As already mentioned in Section 4.1, hyperbranched polymers can be synthesized using AB_X-type monomers or via the $A_2 + B_3$ approach. Synthesis of hyperbranched polymers by Diels–Alder reactions using both strategies has already been described in various papers and reviews (Voit, 2007; Voit and Lederer, 2009). The first hyperbranched polyphenylenes using the Diels–Alder strategy were synthesized by Kim and Webster through C–C coupling starting from AB_2-type monomers, for example, 3,5-dibromophenylboronic acid and 3,5-dihalophenyl Grignard reagents, which were polymerized by Pd(0) and Ni(II)-catalyzed aryl–aryl coupling reactions, respectively (Kim and Webster, 1992).

In 1997, Müllen *et al.* (Morgenroth and Mullen, 1997) reported the Diels–Alder reaction of AB_2 monomers 3,4-bis-(4-(tri-isopropylsilylethynyl)phenyl)-2,5-diphenylcyclopentadienone or 3,4-bis(4-(phenylethynyl)-2,5-diphenylcyclopentadienone containing an ethynyl A unit and a cyclopentadienone B unit by thermal treatment at 180 °C and reaction times up to 45 h. The obtained hyperbranched products had molar masses up to 100 000 g/mol and broad polydispersity indices (see also Figure 8.6 for comparable dendrimers).

To provide some control over the degree of branching in the branched polyphenylenes, this approach has recently been extended by combining AB_2 monomers (3,4-bis(4-(phenylethynyl)-2,5-diphenylcyclopentadienone) with a corresponding AB monomer (Stumpe *et al.*, 2006). On the other hand, it has to be mentioned that the addition of the AB monomer limited the achievable molar masses: example values of up to 75 000 g/mol (for AB_2/AB ratio of 3/1) were obtained compared to 600 000 g/mol for the pure AB_2 monomer, under identical reaction conditions.

8.4.4.2 Dendronized Polymers

Recently, a new strategy toward the synthesis of dendronized polymers via Diels–Alder click reaction was designed by Sanyal and coworkers (Tonga *et al.*, 2010). First, a styrene-based polymer with anthracene groups as reactive side chains was synthesized via RAFT and, third-generation polyester dendrons containing furan-protected maleimide groups at their focal point were prepared. *In-situ* thermal deprotection of the focal points of these dendrons exposes the reactive maleimide dienophile that undergoes a reagent-free thermal Diels–Alder cycloaddition with the anthracyl moieties on the polymer side chain. High efficiency of the formation of the DA adduct on the copolymer was clearly evidenced by ^{1}H NMR, UV spectroscopy and SEC analysis.

Dendronizing of polymers is an effective manner to modulate the physical and chemical characteristics of the parent polymer. In this particular case, the hydroxyl groups on the appended third-generation dendrons render the polymer more hydrophilic, making it soluble in methanol. As expected, this also changes the surface characteristics of coated films of these polymers. Moreover, the hydroxyl groups of these films should be amenable to further functionalization with molecules of interest to fabricate functional surfaces.

8.5 Click Approaches for the Synthesis of Star-Shaped Polymers

The star-shaped polymers are the simplest case of branched species where all chains of a given macromolecule are connected to a single molecule, referred to as the core (Mishra and Kobayashi, 1999; Gao and Matyjaszewski, 2009; Inoue, 2000; Ouchi *et al.*, 2009). Because of their unique structure, star-shaped polymers have several distinct features, including a core–shell microstructure with multiple radiating arms and chain-end functionalities. Due to these characteristics, star-shaped polymers are used in many fields including catalysis (Terashima *et al.*, 2006), drug delivery (Wiltshire and Qiao, 2007), host–guest chemistry (Kul *et al.*, 2008; Meier *et al.*, 2004) and coatings (Groll *et al.*, 2005). Based on the chemical composition of the arms, star-shaped polymers can be classified into two categories: (1) regular (homoarm) arm star-shaped polymers and (2) miktoarm (heteroarm) star copolymers. The regular star-shaped polymers (Scheme 8.6, left) consist of a symmetrical structure, containing arms with similar molecular weight and identical chemical composition, and are particularly

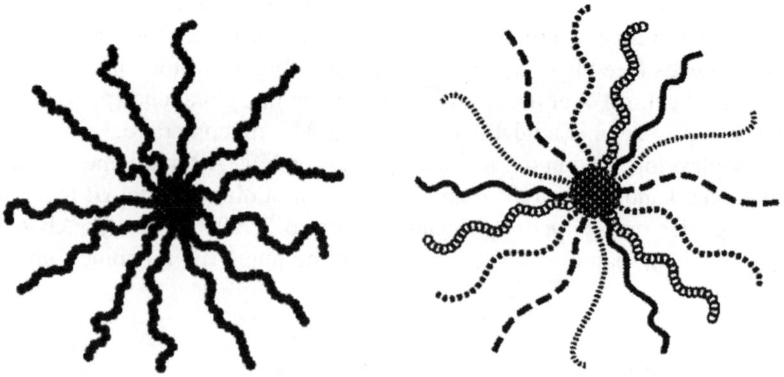

Scheme 8.6 Graphical representation of a regular star-shaped polymer (left), and a miktoarm star-shaped polymer (right).

interesting as they can easily be synthesized via living (anionic) or controlled radical polymerization (CRP) techniques (Gao and Matyjaszewski, 2009).

Miktoarm star-copolymers (Scheme 8.6, right), consisting of arms with different molecular weight and/or composition, were mostly synthesized by combining different polymerization techniques including anionic polymerization, RAFT, ring-opening polymerization (ROP) and ring-opening metathesis polymerization (ROMP).

Three different pathways are used to synthesize star-shaped polymer structures: (1) "arm-first", (2) "coupling-onto" and (3) "core-first" approach. Using the "arm-first" method, monofunctional living linear macromolecules are initially synthesized. Star formation then occurs by adding a difunctional comonomer, which provides crosslinking through propagation. However, by using this strategy, star-shaped polymers with a large, often heterogeneous number of arms are produced (Kanaoka *et al.*, 1991; Wu *et al.*, 2005b; Xia, Zhang and Matyjaszewski, 1999). In the "coupling-onto" approach near polymer chains are coupled onto a multifunctional coupling agent (Van Renterghem *et al.*, 2006). Using this method, coupling efficiency between the reactive chain ends and the coupling agent is low due to the slow coupling reaction. Moreover, separation techniques are required to isolate stars from residual uncoupled linear polymers. In the "core-first" method, the arms grow from multifunctional initiators or precursor molecules (Huang *et al.*, 2004; Angot *et al.*, 1998; Kreutzer *et al.*, 2006; Heise *et al.*, 2000; Heise *et al.*, 1999; Ohno *et al.*, 2001; Stenzel and Davis, 2002). Because the polymer chains start to grow at the same time, star-shaped polymers with controlled molecular weight and low polydispersity indices can be obtained. By variation of the composition, size, and number of arms, the physical properties of these star-shaped polymers can be adjusted.

8.5.1 CuAAC for the Synthesis of Star-Shaped Polymers

8.5.1.1 Regular Star-Shaped Polymers

The combination of CRP with high yield chemical transformations, like the CuAAC reaction, has been shown to be a powerful strategy in the preparation of star-shaped polymers. Matyjaszewski and coworkers (Gao and Matyjaszewski, 2006) were the first to describe the synthesis of star-shaped polymers by using a combination of atom-transfer radical polymerization (ATRP) and CuAAC (Table 8.1, entry 1). Azido-terminated PS was coupled onto alkyne-containing multifunctional compounds. The coupling agents included di-, tri-, and tetra-alkynes, to which the azide-terminated polystyrene (PS) was attached with a click efficiency of 95, 90 and 83%, respectively. All coupling reactions were completed within 3 h. The decrease in click efficiency with increasing number of arms could be attributed to the difficulties in diffusion of the azido polymer chain ends to the alkyne molecules, especially if one or more polymers are already attached to the alkyne core. The effect of several parameters, such as molecular weight of the azido-containing polymer, the presence of Cu(0) as reducing agent, and the molar ratio of azide/alkyne groups on the final yield of coupling was studied. It was found that the efficiency of the click reaction decreased with increasing molecular weight of the arm precursors, due to higher sterical constraints and less precisely balanced stoichiometry between the azide and alkyne groups. The presence of a small amount of Cu(0) enhanced the coupling efficiency, and the highest yield of coupling product was obtained when the molar ratio of azido/alkyne groups was 1. Almost simultaneously, Tunca and coworkers synthesized

Table 8.1 Overview CuAAC reactions for the synthesis of regular star-shaped (block co-) polymers and heteroarm star-shaped polymers

Entry	Polymer	R-group	Catalyst	Ref.
1	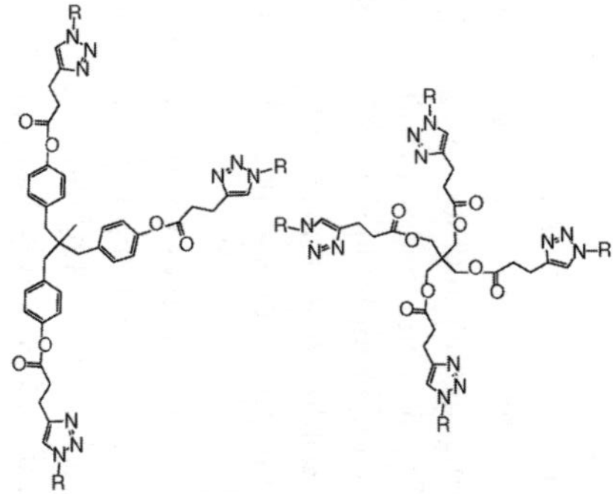	R = PS	Cu^{I}/PMDETA/DMF/RT	Gao and Matyjaszewski, 2006
2	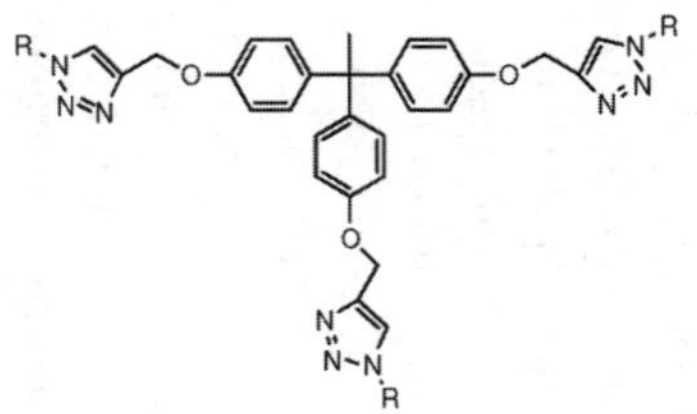	R = PS, PtBuA or PEG	Cu^{I}/PMDETA/RT	Altintas *et al.*, 2006a
3	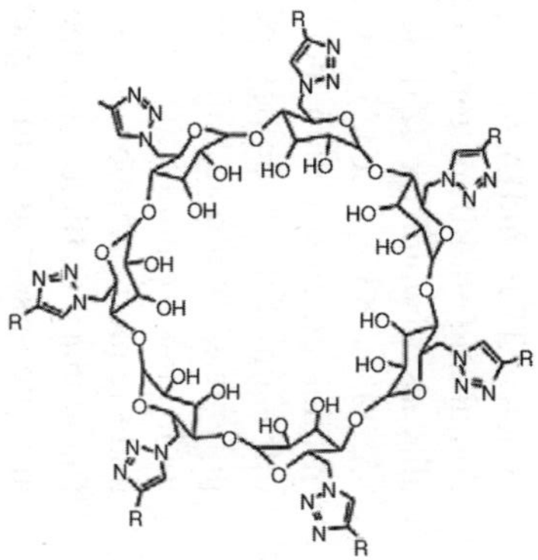	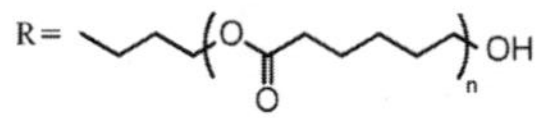	$CuSO_4$/sodium ascorbate/100 C/mW-irradiation	Hoogenboom *et al.*, 2006

4	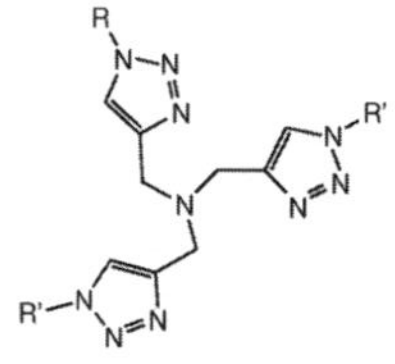	R = PS, PtBuA, PMA	Cu^I/PMDETA/80 C	Whittaker *et al.*, 2006
5	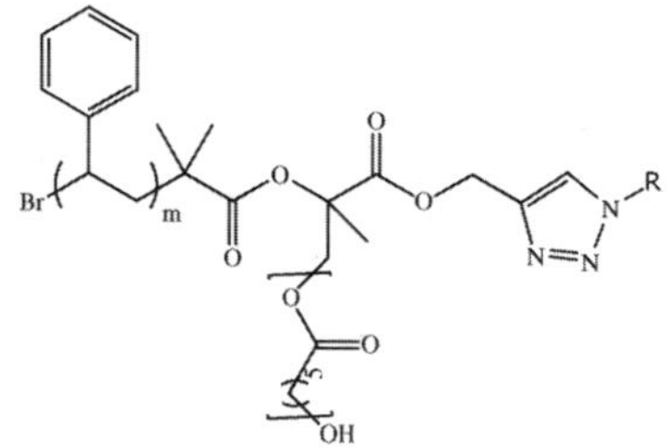	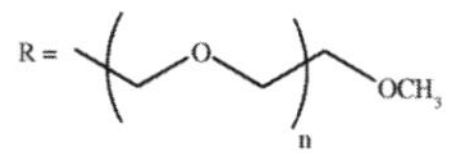	Cu^I/PMDETA/RT	Deng *et al.*, 2007
6	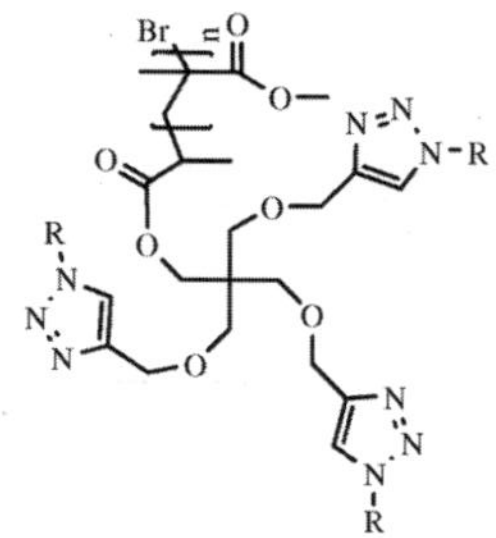	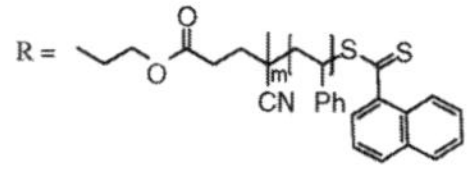	$CuSO_4 \cdot 5H_2O$/sodium ascorbate/70 C	Zhu *et al.*, 2007
7	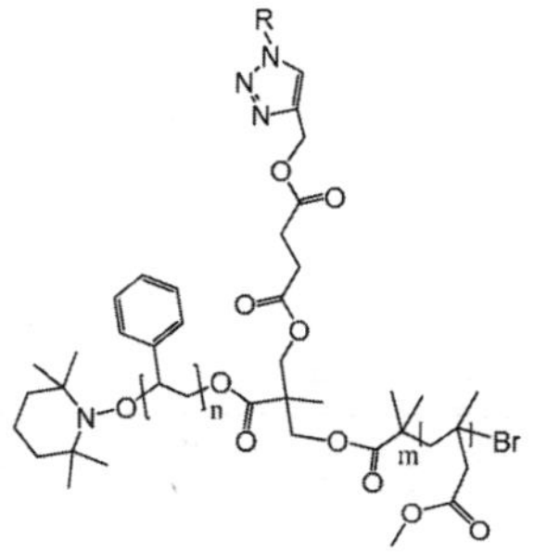	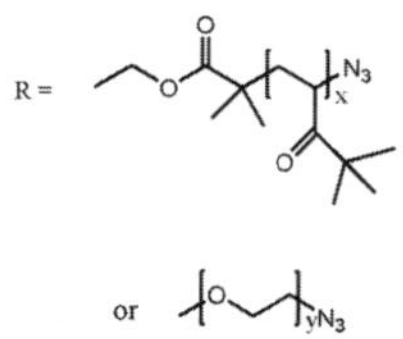	Cu^I/PMDETA/RT	Altintas *et al.*, 2006

(continued)

Table 8.1 *(continued)*

Entry	Polymer	R-group	Catalyst	Ref.
8	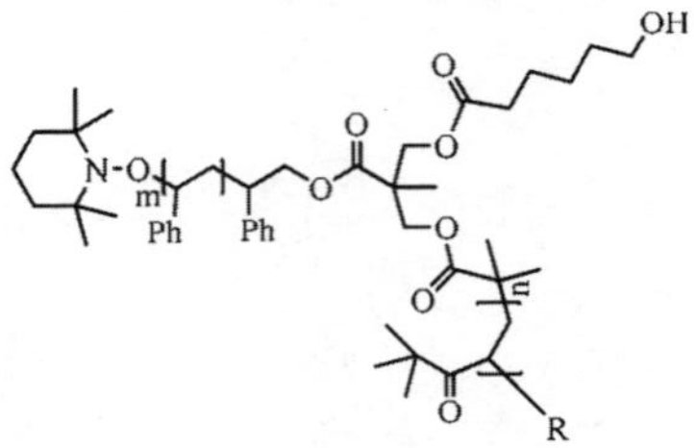	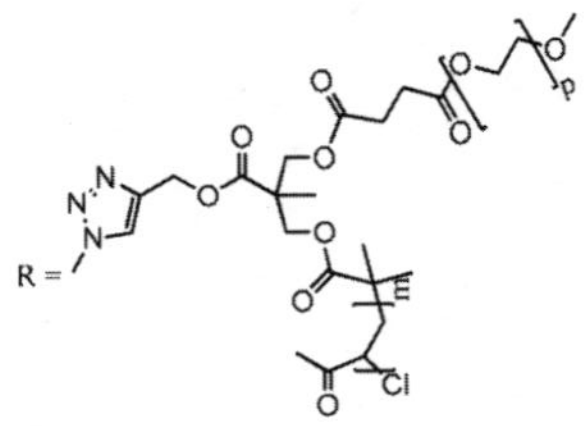	Cu^I/PMDETA/RT	Gungor *et al.*, 2008

3-arm PS, poly(*tert*-butyl acrylate) (PtBuA) or poly(ethylene glycol) (PEG) star-shaped polymer using the same strategy, with a 87% click efficiency (Table 8.1, Entry 2) (Altintas *et al.*, 2006a).

Following the same arm-first approach, Schubert *et al.* synthesized acetylene functionalized poly(ε-caprolactones) (PCL), which were subsequently coupled onto a heptakis-azido-β-cyclodextrin resulting in a 7-arm star-shaped heptakis-PCL-β-cyclodextrin (Table 8.1, Entry 3) (Hoogenboom *et al.*, 2006). However, harsh reaction conditions were required to drive the reaction to completion (microwave-assisted CuAAC at 100 °C). ^{1}H NMR, SEC analysis (UV and RI) and MALDI-TOF MS spectra confirmed the quantitative character.

Also, star block copolymers were synthesized using the CuAAC strategy. Matyjaszewski *et al.* (Gao *et al.*, 2007) synthesized 3-arm star block copolymers via the CuAAC of 3-arm azido-containing PS star with alkyne-terminated PEO precursors. First, 3-arm star-shaped PS with bromine chain-end functionality was synthesized by ATRP using the "core-first" method. After modifying the bromine groups into azide groups, these polymers were used for CuAAC reaction with alkyne-terminated poly(ethylene oxide) (PEO), in order to form amphiphilic structures. This strategy led to well-defined structures but with a limited number of arms.

Similar methodologies for the synthesis of star-shaped copolymers with a higher number of arms were utilized shortly thereafter. Recently, the group of Du Prez synthesized 12-arm star block copolymers based on poly(isobornyl acrylate) (PiBA) and poly(2-ethyloxazoline)$_{20}$ (P(EtOx)$_{20}$ or poly(ε-caprolactone) (PCL) (Lammens *et al.*, 2009) First, bromine end-functionalized PiBA star-shaped polymers with well-defined structures were synthesized using ATRP. The optimal ATRP conditions were determined to minimize star–star coupling and to preserve high end-group functionality (Dervaux *et al.*, 2008; Van Renterghem *et al.*, 2008). The bromine end groups were then converted into reactive azido end groups, and subsequently coupled with alkyne-end functionalized P (EtOx)$_{20}$, PCL and PiBA (Figure 8.11) (Van Camp *et al.*, 2007).

When CuAAC was applied using standard conditions (Cu(I)Br/PMDETA, room temperature in THF), a maximum click efficiency of 64% was obtained. By changing the catalytic system, catalytic concentration or temperature, no positive trend was observed. This inefficient CuAAC of the highly branched PiBA star-shaped polymers with the linear polymers is ascribed to the entropic penalty, as also observed by Matyjaszewski (Gao and Matyjaszewski, 2006). In an attempt to improve this click efficiency, the effect of microwave heating on the Cu (I)-mediated click reaction of these bulky star-shaped polymers with other polymer segments

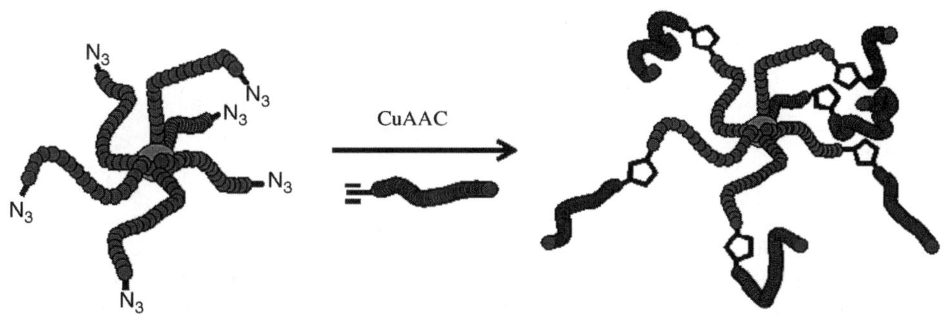

Figure 8.11 Synthesis of star block copolymers by the combination of ATRP and CuAAC.

was investigated. Using 7 equivalents of copper in a microwave for 30 min at 100 °C, quantitative coupling of the alkyne end-functionalized polymers onto the PiBA star-shaped polymers, starting from the dodecafunctional initiator, was obtained. However, to guarantee a full conversion of the azide functionalities, a twofold excess of the alkyne polymer compared to the azide end groups was used. The resulting star block copolymers were isolated by removing the excess of linear polymer by preparative SEC after which the separated peaks were investigated via SEC using a triple detection system. Figure 8.12 shows the SEC chromatograms of the PiBA star-shaped polymer, $P(EtOx)_{20}$ and the coupling product, before and after preparative SEC. A clear shift of the complete distribution to higher masses was observed, indicating the successful formation of the star block copolymer structure under these conditions.

Another approach to synthesize multiarm star block copolymers is based on a combination of CuAAC and the "arm-first" method (Durmaz *et al.*, 2010). Protected alkyne PS polymers were prepared via ATRP and subsequently crosslinked by a divinyl containing compound. The formed 27-arm star-shaped polymers containing a protected alkyne periphery, was deprotected and subsequently coupled with azide-end-functionalized PEG and PtBuA to form star block or mixed block copolymers. The CuAAC reactions occurred at room temperature for 24 h, surprisingly leading to a full click efficiency. The quantitative character of the latter click reaction at ambient temperatures for such dense polymer structures is in contrast to those obtained by other research groups, as mentioned in previous paragraphs.

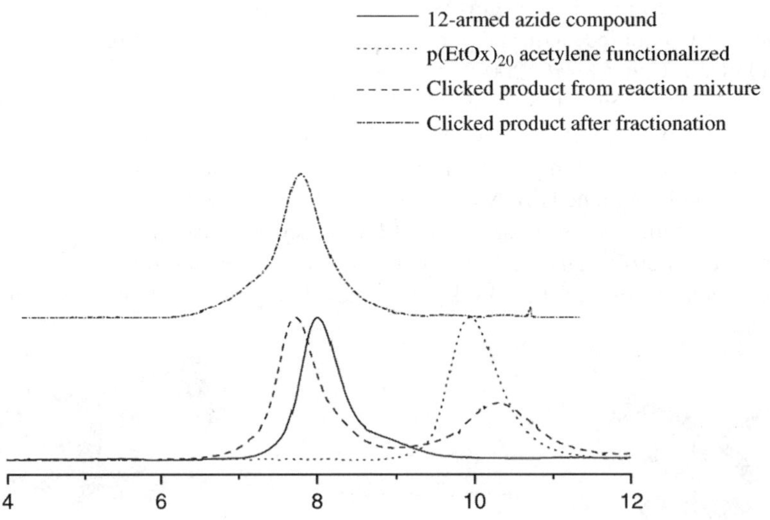

Figure 8.12 The SEC analysis of the PiBA star-shaped polymer, $P(EtOx)_{20}$ and the coupling product before (bottom) and after preparative SEC (top) (Lammens *et al.*, 2009). (Reprinted from M. Lammens, D. Fournier, M.W.M. Fijten *et al.*, "Star-shaped polyacrylates: Highly functionalized architectures via CuAAC click conjugation," *Macromolecular Rapid Communications*, **30**, 23, 2049–2055 (Table 2, Figure 3), © 2009, with the permission of John Wiley & Sons, Inc.)

To synthesize ABA star block copolymers, a combination of ROP, ATRP and CuAAC was used (Gou *et al.*, 2009). First, a resorcinarene-centered 8-arm PCL star-shaped polymer was synthesized via ROP. After esterification of the hydroxyl end groups into 2-bromoesters, the star-shaped polymer is used as macroinitiator for the ATRP of tBuA. Then, the bromine end groups of the PCL-PtBuA-Br were converted into the azide end groups and subsequently coupled onto alkyne-containing PCL at 40 °C for 48 h, in an almost quantitative yield.

8.5.1.2 Miktoarm Star-Shaped Polymers

Various methods have been used to synthesize miktoarm star-shaped polymers, ranging from combinations of ring-opening, ATRP, nitroxide-mediated, anionic polymerization, and the "in-out" ATRP method (Gao and Matyjaszewski, 2009). A combination of CRP and click chemistry facilitates the synthesis of these miktoarm star-shaped polymers.

Monteiro reported on the synthesis of 3-arm miktoarm star-shaped polymers and first-generation polymeric mikto-dendrimers consisting of PS, (P*t*BA), poly(acrylic acid) (PAA) and poly(methyl acrylate) (PMA) (Table 8.1, Entry 4) (Whittaker *et al.*, 2006). Well-defined 3-arm miktoarm stars were synthesized by first coupling a large excess of tripropargylamine to azido-terminated polymer to form the dialkyne containing analog. After purification, this polymer is added to a solution of a different azido-terminated polymer. The resulting yield of the 3-arm miktoarm star-shaped polymer went from 75 up to 95% depending on the composition.

Other groups have also combined different CRP techniques with CuAAC and synthesized various miktoarm star copolymers containing 3–5 arms with different molecular weights and chemical compositions (Table 8.1, Entries 5–7) (Deng, Ma and Xu, 2007; Zhu *et al.*, 2007; Altintas *et al.*, 2006b). For instance, Tunca *et al.* prepared ABC hetero-arm terpolymers by a combination of ATRP, nitroxide-mediated polymerization (NMP) and CuAAC (Table 8.1, Entry 7) (Altintas *et al.*, 2006b). A trifunctional initiator containing an ATRP-initiating site, an alkyne group and an alkoxyamine were sequentially used for the ATRP of methyl methacrylate (MMA) and NMP of styrene, followed by CuAAC with alkyne-terminated PEO or PtBuA. In addition to this work, Tunca synthesized an H-shaped (ABCDE type) polymer consisting of a PtBA block as main chain and PCL, PS, PEG and poly(methyl methacrylate) (PMMA) blocks as side chains (Table 8.1, Entry 8) (Gungor *et al.*, 2008). First, azide-end-functionalized miktoarm star terpolymer (PCL-PS-PtBuA-N_3) was obtained by sequential ROP of ε-caprolactone (ε-CL), NMP of styrene and ATRP of tBuA starting from a trifunctional initiator. After ATRP of tBuA, the bromine end group was converted into an azide functionality. In addition to the above-mentioned miktoarm star terpolymer, a PEG-PMMA-alkyne block copolymer with the alkyne at the junction point was prepared as a second precursor. In a last step, both polymers were coupled together with the formation of a H-shaped polymer. From ^{1}H NMR it could be derived that the click efficiency was about 60%, which is in agreement with the majority of the literature examples.

8.5.2 Thiol-Ene Click Chemistry for the Synthesis of Star-Shaped Polymers

Hoyle, Lowe and coworkers developed a new strategy in synthesizing 3-arm star-shaped polymers by combining RAFT and thiol-ene click chemistry (Chan *et al.*, 2008;

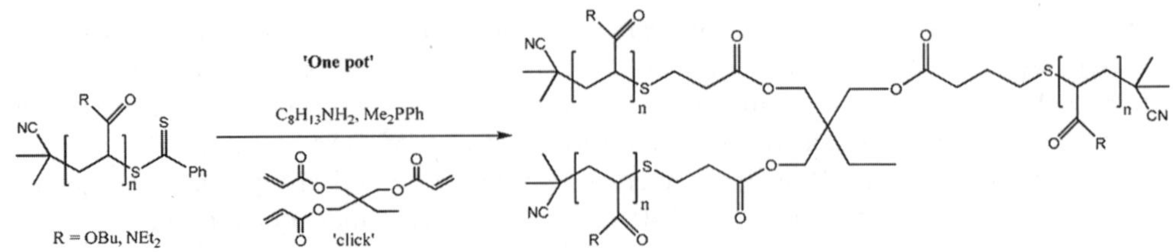

Figure 8.13 Convergent approach for the synthesis of 3-arm star-shaped polymers under nucleophilic phosphine-catalyzed conditions using RAFT-prepared precursor homopolymers. (Reprinted from *Polymer*, **50**, 14, J.W. Chan, B. Yu, C.E. Hoyle and A.B. Lowe, "The nucleophilic, phosphine-catalyzed thiol-ene click reaction and convergent star synthesis with RAFT-prepared homopolymers," 3158–3168, 2009, with permission from Elsevier.)

Chan *et al.*, 2009a). The thiocarbonylthio end groups of a RAFT-prepared poly(N,N-diethylacrylamide) (PDEAm) and polybutylacrylate (PBuA) were reduced into thiol end functionalities, and subsequently coupled onto trimethylolpropane triacrylate via the Michael-type addition click chemistry to form 3-arm star-shaped polymers (Figure 8.13). Mixing the PDEAm and PBuA with trimethylolpropane triacrylate in the presence of a primary or secondary amine ($C_6H_{13}NH_2$) as reductants and a catalyst dimethylphenylphosphine (DMPP), results in a sequential reduction of the thiocarbonylthio end group of PDEAm and PBuA to a thiol end, followed by Michael-type click chemistry in the presence of DMPP to yield the targeting star. This one-pot synthesis was finished within minutes, as indicated from FT-IR and ^{1}H NMR results. The major advantage of using the combination of RAFT and Michael-type addition chemistry in comparison to CuAAC, is that readily available starting materials and catalysts can be used and that reaction times are much lowered. This is in contrast with the extended reaction times, extreme conditions and elevated temperatures that are mostly required in CuAAC to make complex architectures.

On the other hand, when aiming at the synthesis of star-shaped polymers via thermally and photochemically induced radical thiol-ene click chemistry, serious limitations of side reactions and incomplete conversions have been clearly demonstrated by the groups of Du Prez and Barner-Kowollik. From this study, it was concluded that this type of thiol-ene reactions cannot be applied for the well-defined synthesis of complex polymer architectures in which polymer–polymer conjugation is needed (Koo *et al.*, 2010).

Son *et al.* (Rim and Son, 2009) demonstrated that 3-arm oligomeric star-shaped polymers could be prepared in high yields from 1,3,5-triacryloylhexahydro-1,3,5-triazine (TAT) as a core by alternating amine-catalyzed thiol-ene and acrylate esterification reactions (Figure 8.14). The first-generation star-shaped oligomer was synthesized in 96% by the amine-catalyzed thiol-ene reaction of the TAT and stoichiometric amounts of 2-mercaptoethanol in the presence of a catalytic amount of propylamine (A, Figure 8.14). Subsequently, the hydroxyl end groups were esterified with acryloyl chloride in the presence of K_2CO_3 under efficient and rapid stirring (B, Figure 8.14). Acryloyl chloride was taken in excess as it prevented the formation of side products. After reaction, the latter can easily be removed from the reaction product by simple filtration through a silica-gel column. A pure reaction product was obtained in a 98% yield. Larger star oligomers were synthesized by repeating the previous two steps. The newly synthesized star-shaped oligomers can be used as metal-chelating ligands or as precursors for more complex molecular architectures.

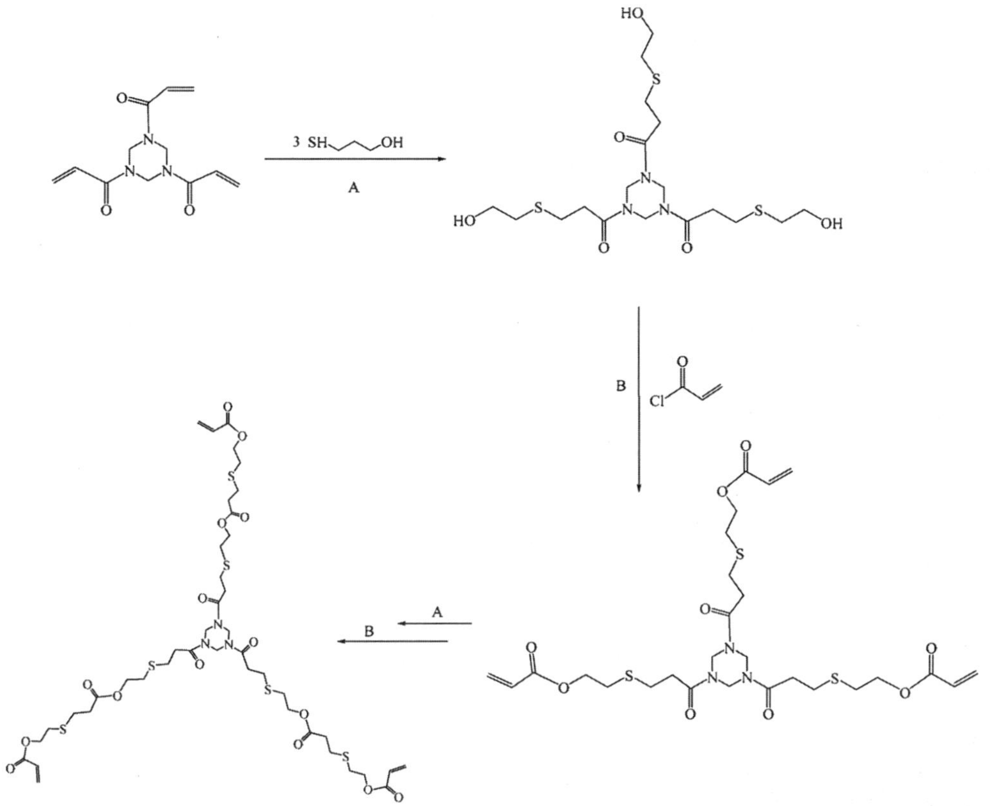

Figure 8.14 Divergent synthesis of star-shaped thioether oligomers by using successive thiol-ene click reaction and esterification reaction. (Reprinted from *Tetrahedron Letters*, **50**, 28, C. Rim, D.Y. Son, "Facile and efficient synthesis of star-shaped oligomers from a triazine core," 4161–4163, 2009, with permission from Elsevier.)

Besides thiol-ene click chemistry, the reaction of a thiol and an yne-compound has been widely studied in organic and organometallic chemistry. Only limited research on its applicability in the synthesis of star-shaped structures is reported. Lowe *et al.* (Chan *et al.*, 2009b) were the first to synthesize star-shaped oligomers by taking advantage of the extreme rapid and quantitative reaction of a thiol with an activated ene under nucleophilic-initiated conditions (Michael-type click reaction) in conjunction with the radical-mediated thiol-yne reaction, yielding exclusively a 1,4-addition product. First, a Michael-type click reaction between the thiol functionalities of the core and the acrylate function of propargyl acrylate by using DMPP as a catalyst occurred in a rapid and selective way (17 min), in 100% yield. Then, the propargyl end groups of the star-shaped oligomer were reacted with thioglycerol in the presence of a UV-initiator, α-dimethoxy-α-phenylacetophenone at 365 nm at ambient temperature under air, yielding a dendrimer-like structure. In all instances, the resulting multifunctional materials required no clean-up, or were easily purified by simple washing/extraction steps.

8.5.3 Diels–Alder Click Reaction for the Synthesis of Star-Shaped Polymers

In addition to the broad application of the CuAAC reaction, other types of highly efficient click reactions were used for the synthesis of star-shaped polymers. Tunca and coworkers used the Diels–Alder click reaction for the preparation of 3-arm star-shaped polymers using furan protected maleimide end-functionalized PEO, PMMA or PtBuA and trianthracene functional linking agents (Dag *et al.*, 2008). However, the resulting yields are comprised between 82 and 93% depending on the coupled polymer, the reaction conditions are more drastic than those used in CuAAC.

The same group also reported on the synthesis of multiarm star block copolymers by using Diels–Alder cycloaddition reactions (Dag *et al.*, 2009). First, an α-anthracene-end functionalized PS (PS-anthr) and furan-protected maleimide-end-functionalized polymers, including PMMA and PtBuA, were prepared via ATRP. The maleimide functionalities were protected as they can contribute to the copolymerization with MMA or tBuA. Moreover, the polymerization temperature was kept below 60 °C to prevent deprotection during the polymerization. In the next step, a 33-arm anthracene-end functionalized (PS) star polymer was obtained using PS-anthr as macroinitiator and divinyl benzene as crosslinker. These star polymers were then reacted with the unprotected maleimide end-functionalized PMMA or PtBuA to give multiarm star block copolymers via Diels–Alder click reaction. The efficiencies were found to be 96 and 88%, respectively.

This result proves that well-defined structures with low degree of heterogeneity of the multiarm star-shaped polymers can be synthesized. Moreover, the method reported herein can also provide a synthetic pathway for the introduction of block copolymers synthesized via different polymerization routes (RAFT, ROP, etc.) onto the anthracene-end-functionalized multiarm star-shaped polymers. Although the Diels–Alder cycloaddition between anthracene and maleimide derivatives has proven to provide good results in the formation of complex architectures, the major drawback of this method remains the requirement of high temperature and relatively long reaction times.

Recently, Barner-Kowollik reported an alternative strategy for synthesizing star-shaped polymers up to 4 arms by a combination of RAFT chemistry and hetero-Diels–Alder (HDA) cycloaddition reactions (Inglis *et al.*, 2008). Styrene was polymerized via RAFT onto an electron-deficient dithioester and subsequently coupled onto 2-arm, 3-arm and 4-arm diene-containing coupling agents to form star-shaped polymers with 2, 3 and 4 arms, respectively. The coupling reactions were performed in chloroform at 50 °C in the presence of $ZnCl_2$ for 24 h. The effectiveness of this click reaction was monitored via UV/Vis spectroscopy, SEC and ^{1}H NMR analysis. Moreover, this click reaction was also visualized as after a reaction time of 24 h, the pink reaction mixture (coming from the RAFT-agent in the presence of $ZnCl_2$) turned colourless.

It is well known that thioacetals and thioketals are prone to fragmentation under radical conditions and are susceptible to retro-Diels–Alder reactions. Therefore, the authors investigated the conditions that may induce the cleavage of these star-shaped structures. The 4-arm PS star-shaped polymers were subjected to a high-temperature environment for 24 h, and subsequently measured via SEC. These results showed quantitative cleavage after 24 h at 160 °C. Because of its reversible character, HDA can be used in the design of macromolecular architectures with predetermined or targeted fragmentation patterns.

When combining the hetero-Diels–Alder strategy with the CuAAC reaction, star block copolymers have been obtained (Sinnwell *et al.*, 2008b). First, α-diene-ω-alkyne functionalized PCL, PS equipped with an electron-deficient dithioester end group, and a triazide coupling agent were synthesized as building blocks. The diene-functionalized PCL was obtained after ROP of ε-caprolactone using sorbic alcohol as the initiator, and subsequently functionalized with an alkyne by esterification of the terminal hydroxyl group with 4-pentynoic acid. Electron-deficient dithioesters-functionalized PS was synthesized via RAFT polymerization in the presence of benzyl pyridine-2-yldithioformate (BPDF) as controlling agent. BPDF was chosen as RAFT agent because of its good control over the polymerization and high HDA activity. The trifunctional coupling agent was obtained by esterification of 1,3,5-tris (bromomethyl)-benzene with azido-1-propanol. To synthesize these star block copolymers, two different methods are used: (1) the "arm-first" method, involving the coupling of PCL and PS polymers via the HDA cycloaddition, and subsequent CuAAC reaction to the coupling agent or (2) the "core-first" method, for which the PCL is first coupled onto the coupling agent resulting in a 3-arm star PCL and further converted with PS in a HDA cycloaddition. Both methods led to the desired product, however, for the "core first" method, the basic triazole ring in the 3-arm PCL is suspected to reduce the reactivity of the pyridinyl dithioester in PS, leading to lower overall yields.

Very recently, the same group, in collaboration with the group of Du Prez, reported on the synthesis of multiarm star block copolymers by the combination of RAFT-hetero-Diels–Alder and ATRP (Figure 8.15). Poly(isobornyl acrylate) star-shaped polymers containing 11 bromine end groups were synthesized via ATRP from a multifunctional initiator containing 12 bromine end groups (Lammens *et al.*, 2009). To use the star-shaped PiBA as a precursor in the HDA approach, diene end groups were introduced onto the star-shaped structure. Subsequently, RAFT-polymerized styrene containing an electron-deficient dithioester, was coupled onto the diene-end functionalized PiBA star-shaped polymer in a quantitative yield. The major advantage of this technique in comparison to the well-known and most commonly used CuAAC, is that polymer chains can be quantitatively coupled onto complex architectures without using harsh conditions (Lammens *et al.*, 2009).

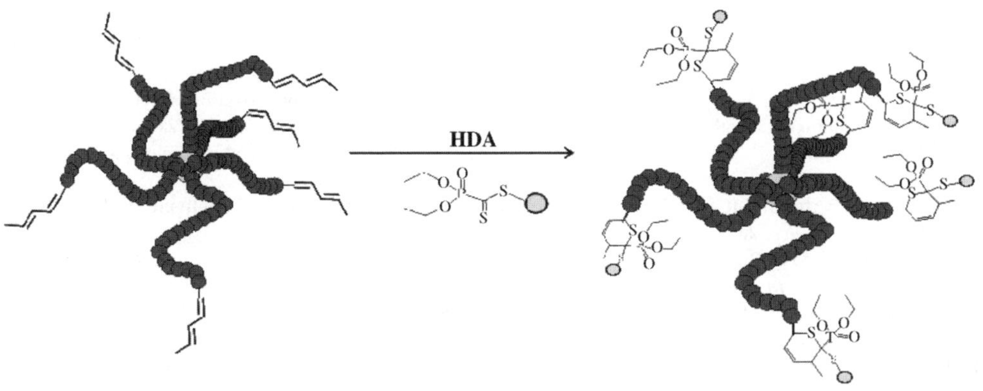

Figure 8.15 Synthesis of a PiBA-PS multiarm star block copolymers by the combination of RAFT-hetero-Diels–Alder and ATRP. (Reprinted from S. Sinnwell, M. Lammens, M.H. Stenzel *et al.*, "Efficient access to multi-arm star block copolymers by a combination of ATRP and RAFT-HDA click chemistry," *Journal of Polymer Science, Part A: Polymer Chemistry*, **47**, 8, 2207–2213 (Figure 3), © 2009, with the permission of John Wiley & Sons, Inc.)

8.6 Conclusion

Several types of dendritic structures and star-shaped polymers possess unique qualities, making them excellent candidates as materials in a variety of high value applications. However, until recently the synthesis of those structures was known on the one hand to be tedious and costly but well controlled, or on the other hand to be fast but with a lack of control. With the introduction of all types of click chemistries in polymer science since 2002, many novel, efficient synthetic strategies have been developed to make dendrimers, hyperbranched polymers, dendronized polymers, star-shaped polymers and unsymmetrical dendrimers, often with physical properties that were not known earlier. Indeed, utilizing click chemistry allows the synthesis to be accelerated, lowering the number of reaction steps and the need for tedious purification procedures, while at the same time diminishing the necessity for large excess of reagents. In the meantime, different click approaches are known. Among them, CuAAC and Diels–Alder cycloaddition reactions have been used intensively, while thiol-ene/yne click chemistry is actually enjoying much research interest. Because of the metal-free character and efficiency of these thio-click reactions, many more developments to obtain branched macromolecular structures are expected in the upcoming years.

Acknowledgments

Mieke Lammens thanks the Institute for the Promotion of Innovation through Science and Technology in Flanders (IWT-Vlaanderen) for a PhD scholarship. The Belgian Program on Interuniversity Attraction Poles initiated by the Belgian State, Prime Minister's office (Program P6/27) is acknowledged for financial support.

References

Altintas, O., Yankul, B., Hizal, G., and Tunca, U. (2006a) A_3-type star polymers via click chemistry. *Journal of Polymer Science Part A – Polymer Chemistry*, **44**, 6458.

Altintas, O., Hizal, G., and Tunca, U. (2006b) ABC-type hetero-arm star terpolymers through "click" chemistry. *Journal of Polymer Science Part A – Polymer Chemistry*, **44**, 5699.

Angot, S., Murthy, K.S., Taton, D., and Gnanou, Y. (1998) Atom transfer radical polymerization of styrene using a novel octafunctional initiator: Synthesis of well-defined polystyrene stars. *Macromolecules*, **31**, 7218.

Antoni, P., Nystrom, D., Hawker, C.J. *et al.* (2007) A chemoselective approach for the accelerated synthesis of well-defined dendritic architectures. *Chemical Communications*, 2249.

Baskin, J.M. and Bertozzi, C.R. (2007) Bioorthogonal click chemistry: Covalent labeling in living systems. *QSAR & Combinatorial Science*, **26**, 1211.

Becer, R., Hoogenboom, R., and Schubert, U.S. (2009) Click chemistry beyond metal-catalyzed cycloaddition. *Angewandte Chemie-International Edition*, **48**, 4900.

Binder, W.H. and Sachsenhofer, R. (2007) 'Click' chemistry in polymer and materials science. *Macromolecular Rapid Communications*, **28**, 15.

Binder, W.H. and Sachsenhofer, R. (2008) 'Click' chemistry in polymer and material science: An update. *Macromolecular Rapid Communications*, **29**, 952.

Binder, W.H. (2008) "Click" chemistry in polymer and material science: the update. *Macromolecular Rapid Communications*, **29**, 951.

Bock, V.D., Hiemstra, H., and van Maarseveen, J.H. (2005) Cu-I-catalyzed alkyne-azide "click" cycloadditions from a mechanistic and synthetic perspective. *European Journal of Organic Chemistry*, 51.

Bosman, A.W., Janssen, H.M., and Meijer, E.W. (1999) About dendrimers: Structure, physical properties, and applications. *Chemical Reviews*, **99**, 1665.

Brad, M.R., Gerard, L., Christian, H., and Virgil, P. (2009) Synthesis of dendrimers through divergent iterative thio-bromo "click" chemistry. *Journal of Polymer Science Part A-Polymer Chemistry*, **47**, 3931.

Buhleier, E., Wehner, W., and Vogtle, F. (1978) Cascade-chain-like and nonskid-chain-like syntheses of molecular cavity topologies. *Synthesis*, 155.

Campos, L.M., Killops, K.L., Sakai, R. *et al.* (2008) Development of thermal and photochemical strategies for thiol-ene click polymer functionalization. *Macromolecules*, **41**, 7063.

Carioscia, J.A., Stansbury, J.W., and Bowman, C.N. (2007) Evaluation and control of thiol-ene/thiol-epoxy hybrid networks. *Polymer*, **48**, 1526.

Carlmark, A., Hawker, C., Hult, A., and Malkoch, M. (2009) New methodologies in the construction of dendritic materials. *Chemical Society Reviews*, **38**, 352.

Chan, J.W., Yu, B., Hoyle, C.E., and Lowe, A.B. (2008) Convergent synthesis of 3-arm star polymers from RAFT-prepared poly(N,N-diethylacrylamide) via a thiol-ene click reaction. *Chemical Communications*, 4959.

Chan, J.W., Yu, B., Hoyle, C.E., and Lowe, A.B. (2009a) The nucleophilic, phosphine-catalyzed thiol-ene click reaction and convergent star synthesis with RAFT-prepared homopolymers. *Polymer*, **50**, 3158.

Chan, J.W., Hoyle, C.E., and Lowe, A.B. (2009b) Sequential phosphine-catalyzed, nucleophilic thiol-ene/radical-mediated thiol-yne reactions and the facile orthogonal synthesis of polyfunctional materials. *Journal of the American Chemical Society*, **131**, 5751.

Chen, G.J., Kumar, J., Gregory, A., and Stenzel, M.H. (2009) Efficient synthesis of dendrimers via a thiol-yne and esterification process and their potential application in the delivery of platinum anti-cancer drugs. *Chemical Communications*, 6291.

Cook, W.D., Chen, F., Pattison, D.W. *et al.* (2007) Thermal polymerization of thiol-ene network-forming systems. *Polymer International*, **56**, 1572.

Dag, A., Durmaz, H., Hizal, G., and Tunca, U. (2008) Preparation of 3-arm star polymers (A_3) via Diels-Alder click reaction. *Journal of Polymer Science Part A-Polymer Chemistry*, **46**, 302.

Dag, A., Durmaz, H., Tunca, U., and Hizal, G. (2009) Multiarm star block copolymers via Diels-Alder click reaction. *Journal of Polymer Science Part A-Polymer Chemistry*, **47**, 178.

Deng, G., Ma, D., and Xu, Z. (2007) Synthesis of ABC-type miktoarm star polymers by "click" chemistry, ATRP and ROP. *European Polymer Journal*, **43**, 1179.

Dervaux, B., Van Camp, W., Van Renterghem, L., and Du Prez, F.E. (2008) Synthesis of poly(isobornyl acrylate) containing copolymers by atom transfer radical polymerization. *Journal of Polymer Science Part A-Polymer Chemistry*, **46**, 1649.

Durmaz, H., Colakoclu, B., Tunca, U., and Hizal, G. (2006) Preparation of block copolymers via Diels–Alder reaction of maleimide- and anthracene-end functionalized polymers. *Journal of Polymer Science Part A-Polymer Chemistry*, **44**, 1667.

Durmaz, H., Dag, A., Altintas, O. *et al.* (2007) One-pot synthesis of ABC type triblock copolymers via in situ click [3 + 2] and Diels-Alder [4 + 2] reactions. *Macromolecules*, **40**, 191.

Durmaz, H., Dag, A., Erdogan, E. *et al.* (2010) Multiarm star block and multiarm star mixed-block copolymers via azide-alkyne click reaction. *Journal of Polymer Science Part A-Polymer Chemistry*, **48**, 99.

Esfand, R. and Tomalia, D.A. (2001) Poly(amidoamine) (PAMAM) dendrimers: from biomimicry to drug delivery and biomedical applications. *Drug Discovery Today*, **6**, 427.

Ess, D.H., Jones, G.O., and Houk, K.N. (2008) Transition states of strain-promoted metal-free click chemistry: 1,3-dipolar cycloadditions of phenyl azide and cyclooctynes. *Organic Letters*, **10**, 1633.

Feng, X., Taton, D., Elliot, L., and Gnanou, Y. (2005) Toward an easy access to dendrimer-like poly(ethylene oxide)s. *Journal of the American Chemical Society*, 11.

Flomenbom, O., Amir, R.J., Shabat, D., and Klafter, J. (2005) Some new aspects of dendrimer applications. *Journal of Luminescence*, **111**, 315.

Fournier, D., Hoogenboom, R., and Schubert, U.S. (2007) Clicking polymers: a straightforward approach to novel macromolecular architectures. *Chemical Society Reviews*, **36**, 1369.

Frauenrath, H. (2005) Dendronized polymers--building a new bridge from molecules to nanoscopic objects. *Progress in Polymer Science*, **30**, 325.

Fréchet, J.M. and Tomalia, D.A. (2001) *Dendrimers and Other Dendritic Polymers*. John Wiley and Sons, Chichester.

Froehling, P.E. (2001) Dendrimers and dyes - a review. *Dyes and Pigments*, **48**, 187.

Gacal, B., Durmaz, H., Tasdelen, M.A. *et al.* (2006) Anthracene-maleimide-based Diels-Alder "click chemistry" as a novel route to graft copolymers. *Macromolecules*, **39**, 5330.

Gao, H. and Matyjaszewski, K., (2006) Synthesis of star polymers by a combination of ATRP and the "click" coupling method. *Macromolecules*, **39**, 4960.

Gao, H. and Matyjaszewski, K. (2009) Synthesis of functional polymers with controlled architecture by CRP of monomers in the presence of cross-linkers: From stars to gels. *Progress in Polymer Science*, **34**, 317.

Gao, H.F., Min, K., and Matyjaszewski, K. (2007) Synthesis of 3-arm star block copolymers by combination of "core-first" and "coupling-onto" methods using ATRP and click reactions. *Macromolecular Chemistry and Physics*, **208**, 1370.

Gillies, E.R. and Frechet, J.M.J. (2005) Dendrimers and dendritic polymers in drug delivery. *Drug Discovery Today*, **10**, 35.

Gou, P.F., Zhu, W.P., Zhu, N., and Shen, Z.Q., (2009) Synthesis and characterization of novel resorcinarene-centered amphiphilic star-block copolymers consisting of eight ABA triblock arms by combination of ROP, ATRP, and click chemistry. *Journal of Polymer Science Part A-Polymer Chemistry*, **47**, 2905.

Groll, J., Ademovic, Z., Ameringer, T. *et al.* (2005) Comparison of coatings from reactive star shaped PEG-stat-PPG prepolymers and grafted linear PEG for biological and medical applications. *Biomacromolecules*, **6**, 956.

Gungor, E., Durmaz, H., Hizal, G., and Tunca, U. (2008) H-shaped (ABCDE type) quintopolymer via click reaction [3 + 2] strategy. *Journal of Polymer Science Part A-Polymer Chemistry*, **46**, 4459.

Heise, A., Hedrick, J.L., Trollsas, M. *et al.* (1999) Novel starlike poly(methyl methacrylate)s by controlled dendritic free radical initiation. *Macromolecules*, **32**, 231.

Heise, A., Nguyen, C., Malek, R. *et al.* (2000) Starlike polymeric architectures by atom transfer radical polymerization: Templates for the production of low dielectric constant thin films. *Macromolecules*, **33**, 2346.

Helms, B., Mynar, J.L., Hawker, C.J., and Frechet, J.M.J. (2004) Dendronized linear polymers via "click chemistry". *Journal of the American Chemical Society*, **126**, 15020.

Hoogenboom, R., Moore, B.C., and Schubert, U.S. (2006) Synthesis of star-shaped poly(epsilon-caprolactone) via 'click' chemistry and 'supramolecular click' chemistry. *Chemical Communications*, 4010.

Hoyle, C.E., Lee, T.Y., and Roper, T. (2004) Thiol-enes: Chemistry of the past with promise for the future. *Journal of Polymer Science Part A-Polymer Chemistry*, **42**, 5301.

Huang, C.F., Lee, H.F., Kuo, S.W. *et al.* (2004) Star polymers via atom transfer radical polymerization from adamantane-based cores. *Polymer*, **45**, 2261.

Inglis, A.J., Sinnwell, S., Davis, T.P. *et al.* (2008) Reversible addition fragmentation chain transfer (RAFT) and hetero-Diels-Alder chemistry as a convenient conjugation tool for access to complex macromolecular designs. *Macromolecules*, **41**, 4120.

Inglis, A.J., Sinnwell, S., Stenzel, M.H., and Barner-Kowollik, C. (2009) Ultrafast click conjugation of macromolecular building blocks at ambient temperature. *Angewandte Chemie-International Edition*, **48**, 2411.

Inoue, K. (2000) Functional dendrimers, hyperbranched and star polymers. *Progress in Polymer Science*, **25**, 453.

Johnson, J.A., Finn, M.G., Koberstein, J.T., and Turro, N.J. (2008) Construction of linear polymers, dendrimers, networks, and other polymeric architectures by copper-catalyzed azide-alkyne cycloaddition "click" chemistry. *Macromolecular Rapid Communications*, **29**, 1052.

Joralemon, M.J., O'Reilly, R.K., Matson, J.B. *et al.* (2005) Dendrimers clicked together divergently. *Macromolecules*, **38**, 5436.

Kade, M.J., Burke, D.J., and Hawker, C.J. (2010) The power of thiol-ene chemistry. *Journal of Polymer Science Part A-Polymer Chemistry*, **48**, 743.

Kanaoka, S., Sawamoto, M., and Higashimura, T. (1991) Star-shaped polymers by living cationic polymerization. 1. Synthesis of star-shaped polymers of alkyl and vinyl ethers. *Macromolecules*, **24**, 2309.

Killops, K.L., Campos, L.M., and Hawker, C.J. (2008) Robust, efficient, and orthogonal synthesis of dendrimers via thiol-ene "click" chemistry. *Journal of the American Chemical Society*, **130**, 5062.

Kim, Y.H. and Webster, O.W. (1992) Hyperbranched polyphenylenes. *Macromolecules*, **25**, 5561.

Kitajyo, Y., Kinugawa, Y., Tamaki, M. *et al.* (2007) Synthesis, characterization, and functionalization of hyperbranched poly(3,4-epoxycyclohexanemethanol). *Macromolecules*, **40**, 9313.

Kolb, H.C., Finn, M.G., and Sharpless, K.B. (2001) Click chemistry: Diverse chemical function from a few good reactions. *Angewandte Chemie-International Edition*, **40**, 2004.

Konkolewicz, D., Gray-Weale, A., and Perrier, S. (2009) Hyperbranched polymers by thiol-yne chemistry: From small molecules to functional polymers. *Journal of the American Chemical Society*, **131**, 18075.

Koo, S., Stamenovic, M., Prasath, A. *et al.* (2010) Limitations of radical thiol-ene reactions for polymer-polymer conjugation. *Journal of Polymer Science Part A-Polymer Chemistry*, **48**, 1699.

Kreutzer, G., Ternat, C., Nguyen, T.Q. *et al.* (2006) Water-soluble, unimolecular containers based on amphiphilic multiarm star block copolymers. *Macromolecules*, **39**, 4507.

Kul, D., Van Renterghem, L.M., Meier, M.A.R. *et al.* (2008) Encapsulation and release by star-shaped block copolymers as unimolecular nanocontainers. *Journal of Polymer Science Part A-Polymer Chemistry*, **46**, 650.

Lammens, M., Fournier, D., Fijten, M.W.M., Hoogenboom, R., and Du Prez, F. (2009) Star-shaped polyacrylates: Highly functionalized architectures via CuAAC click conjugation. *Macromolecular Rapid Communications*, **30**, 2049.

Laughlin, S.T., Baskin, J.M., Amacher, S.L., and Bertozzi, C.R. (2008) *In vivo* imaging of membrane-associated glycans in developing zebrafish. *Science*, **320**, 664.

Lee, J.W., Kim, J.H., Kim, B.-K. *et al.* (2006a) Synthesis of Fréchet type dendritic benzyl propargyl ether and Fréchet type triazole dendrimer. *Tetrahedron*, **62**, 894.

Lee, J.W., Kim, J.H., and Kim, B.K. (2006b) Synthesis of azide-functionalized PAMAM dendrons at the focal point and their application for synthesis of PAMAM-like dendrimers. *Tetrahedron Letters*, **47**, 2683.

Lee, J.W., Kim, B.-K., Kim, H.J. *et al.* (2006c) Convergent synthesis of symmetrical and unsymmetrical PAMAM dendrimers. *Macromolecules*, **39**, 2418.

Lutz, J.F. (2007) 1,3-dipolar cycloadditions of azides and alkynes: A universal ligation tool in polymer and materials science. *Angewandte Chemie-International Edition*, **46**, 1018.

Lutz, J.F. (2008) Copper-free azide-alkyne cycloadditions: New insights and perspectives. *Angewandte Chemie-International Edition*, **47**, 2182.

Ma, X.P., Tang, J.B., Shen, Y.Q. *et al.* (2009) Facile synthesis of polyester dendrimers from sequential click coupling of asymmetrical monomers. *Journal of the American Chemical Society*, **131**, 14795.

Malkoch, M., Carlmark, A., Wodegiorgis, A. *et al.* (2004) Dendronized aliphatic polymers by a combination of ATRP and divergent growth. *Macromolecules*, **37**, 322.

Malkoch, M., Schleicher, K., Drockenmuller, E. *et al.* (2005) Structurally diverse dendritic libraries: A highly efficient functionalization approach using Click chemistry. *Macromolecules*, **38**, 3663.

Malkov, G.V., Lodygina, V.P., Shastin, A.V. *et al.* (2007) New hyperbranched poly([1,2,3]-triazole-[1,3,5]-triazines). *Polymer Science, Series B. Polymer Chemistry*, **49**, 301.

Meier, M., Gohy, J.F., Fustin, C.A., and Schubert, U.S. (2004) Combinatorial synthesis of star-shaped block copolymers: Host-guest chemistry of unimolecular reversed micelles. *Journal of the American Chemical Society*, **126**, 11517.

Mery, D. and Astruc, D. (2006) Dendritic catalysis: Major concepts and recent progress. *Coordination Chemistry Reviews*, **250**, 1965.

Mishra, M. and Kobayashi, S. (1999) *Star and Hyperbranched Polymers*. Marcel Dekker Inc., New York.

Morgenroth, F. and Mullen, K. (1997) Dendritic and hyperbranched polyphenylenes via a simple Diels-Alder route. *Tetrahedron*, **53**, 15349.

Morgenroth, F., Reuther, E., and Mullen, K. (1997) Polyphenylene dendrimers: From three-dimensional to two-dimensional structures. *Angewandte Chemie – International Edition in English*, **36**, 631.

Morgenroth, F., Berresheim, A.J., Wagner, M., and Mullen, K. (1998) Spherical polyphenylene dendrimers via Diels-Alder reactions: the first example of an A(4)B building block in dendrimer chemistry. *Chemical Communications*, 1139.

Muller, M., Mauermanndull, H., Wagner, M. *et al.* (1995) A cycloaddition cyclodehydrogenation route from stilbenoids to extended aromatic-hydrocarbons. *Angewandte Chemie – International Edition in English*, **34**, 1583.

Mynar, J.L., Choi, T.L., Yoshida, M. *et al.* (2005) Doubly-dendronized linear polymers. *Chemical Communications*, 5169.

Newkome, G.R., Yao, Z.Q., Baker, G.R., and Gupta, V.K. (1985) Micelles. 1. Cascade molecules - A new approach to micelles - A [27]-Arborol. *The Journal of Organic Chemistry*, **50**, 2003.

Newkome, G.R., Moorefield, C.N., and Vögtle, F. (2001) *Dendrimers And Dendrons: Concepts, Syntheses, Applications*. Wiley-VCH Verlag Gmbh Weinheim, Germany.

Ning, X.H., Guo, J., Wolfert, M.A., and Boons, G.J. (2008) Visualizing metabolically labeled glycoconjugates of living cells by copper-free and fast huisgen cycloadditions. *Angewandte Chemie-International Edition*, **47**, 2253.

Ohno, K., Wong, B., and Haddleton, D.M. (2001) Synthesis of well-defined cyclodextrin-core star polymers. *Journal of Polymer Science Part A-Polymer Chemistry*, **39**, 2206.

Ouchi, M., Terashima, T., and Sawamoto, M. (2009) Transition metal-catalyzed living radical polymerization: Toward perfection in catalysis and precision polymer synthesis. *Chemical Reviews*, **109**, 4963.

Rim, C. and Son, D.Y. (2009) Facile and efficient synthesis of star-shaped oligomers from a triazine core. *Tetrahedron Letters*, **50**, 4161.

Rissing, C. and Son, D.Y. (2009) Application of thiol-ene chemistry to the preparation of carbosilane-thioether dendrimers. *Organometallics*, **28**, 3167.

Rostovtsev, V.V., Green, L.G., Fokin, V.V., and Sharpless, K.B. (2002) A stepwise Huisgen cycloaddition process: Copper(I)-catalyzed regioselective "ligation" of azides and terminal alkynes. *Angewandte Chemie-International Edition*, **41**, 2596.

Scheel, A.J., Komber, H., and Voit, B.I. (2004) Novel hyperbranched poly([1,2,3]-triazole)s derived from AB(2) monomers by a 1,3-dipolar cycloaddition. *Macromolecular Rapid Communications*, **25**, 1175.

Schluter, A.D. and Rabe, J.P. (2000) Dendronized polymers: Synthesis, characterization, assembly at interfaces, and manipulation. *Angewandte Chemie-International Edition*, **39**, 864.

Sinnwell, S., Synatschke, C.V., Junkers, T. *et al.* (2008a) A study into the stability of 3,6-dihydro-2H-thiopyran rings: key linkages in the RAFT hetero-Diels-Alder click concept. *Macromolecules*, **41**, 7904.

Sinnwell, S., Inglis, A.J., Stenzel, M.H., and Barner-Kowollik, C. (2008b) Access to three-arm star block copolymers by a consecutive combination of the copper(I)-catalyzed azide-alkyne cycloaddition and the RAFT hetero Diels-Alder concept. *Macromolecular Rapid Communications*, **29**, 1090.

Sinnwell, S., Lammens, M., Stenzel, M.H. *et al.* (2009) Efficient access to multi-arm star block copolymers by a combination of ATRP and RAFT-HDA *click* chemistry. *Journal of Polymer Science Part A-Polymer Chemistry*, **47**, 2207.

Stenzel, M.H. and Davis, T.P. (2002) Star polymer synthesis using trithiocarbonate functional beta-cyclodextrin cores (Reversible addition-fragmentation chain-transfer polymerization). *Journal of Polymer Science Part A-Polymer Chemistry*, **40**, 4498.

Stiriba, S.E., Frey, H., and Haag, R. (2002) Dendritic polymers in biomedical applications: From potential to clinical use in diagnostics and therapy. *Angewandte Chemie-International Edition*, **41**, 1329.

Stumpe, K., Komber, H., and Voit, B.L. (2006) Novel branched polyphenylenes based on A(2)/B-3 and AB(2)/AB monomers via Diels-Alder cycloaddition. *Macromolecular Chemistry and Physics*, **207**, 1825.

Sumerlin, B.S. and Vogt, A.P. (2010) Macromolecular engineering through click chemistry and other efficient transformations. *Macromolecules*, **43**, 13.

Terashima, T., Ouchi, M., Ando, T. *et al.* (2006) Metal-complex-bearing star polymers by metal-catalyzed living radical polymerization: Synthesis and characterization of poly(methyl methacrylate) star polymers with Ru(II)-embedded microgel cores. *Journal of Polymer Science Part A-Polymer Chemistry*, **44**, 4966.

Tonga, M., Cengiz, N., Kose, M.M. *et al.* Dendronized polymers via Diels-Alder "click" reaction. (2010) *Journal of Polymer Science Part A-Polymer Chemistry*, **48**, 410.

Tornoe, C.W., Christensen, C., and Meldal, M. (2002) Peptidotriazoles on solid phase: [1,2,3]-triazoles by regiospecific copper(I)-catalyzed 1,3-dipolar cycloadditions of terminal alkynes to azides. *The Journal of Organic Chemistry*, **67**, 3057.

Van Camp, W., Germonpré, V., Mespouille, L. *et al.* (2007) New poly(acrylic acid) containing segmented copolymer structures by combination of "click" chemistry and atom transfer radical polymerization. *Reactive & Functional Polymers*, **67**, 1168.

Van Renterghem, L.M., Lammens, M., Dervaux, B. *et al.* (2008) Design and use of organic nanoparticles prepared from star-shaped polymers with reactive end groups. *Journal of the American Chemical Society*, **130**, 10802.

Van Renterghem, L.M., Goethals, E.J., and DuPrez, F.E. (2006) Star-shaped poly(tetrahydrofuran) with reactive end groups: Design, MALDI-TOF study, and solution behavior. *Macromolecules*, **39**, 528.

Van Wuytswinkel, G., Verheyde, B., Compernolle, F. *et al.* (2000) A convergent synthesis of heterocyclic dendrimers using the 1,3-dipolar cycloaddition reaction of organic azides and acetylenedicarboxylate esters. *Journal of the Chemical Society-Perkin Transactions 1*, **9**, 1337.

Vestberg, R., Malkoch, M., Kade, M. *et al.* (2007) Role of architecture and molecular weight in the formation of tailor-made ultrathin multilayers using dendritic macromolecules and click chemistry. *Journal of Polymer Science Part A-Polymer Chemistry*, **45**, 2835.

Voit, B. (2007) The potential of cycloaddition reactions in the synthesis of dendritic polymers. *New Journal of Chemistry*, **31**, 1139.

Voit, B.I. and Lederer, A. (2009) Hyperbranched and highly branched polymer architectures-synthetic strategies and major characterization aspects. *Chemical Reviews*, **109**, 5924.

Whittaker, M.R., Urbani, C.N., and Monteiro, M.J. (2006) Synthesis of 3-miktoarm stars and 1st generation mikto dendritic copolymers by "living" radical polymerization and "click" chemistry. *Journal of the American Chemical Society*, **128**, 11360.

Wiltshire, J. and Qiao, A. (2007) Recent advances in star polymer design: Degradability and the potential for drug delivery. *Australian Journal of Chemistry*, **60**, 7.

Wu, P., Feldman, A.K., Nugent, A.K. *et al.* (2004) Efficiency and fidelity in a click-chemistry route to triazole dendrimers by the copper(I)-catalyzed ligation of azides and alkynes. *Angewandte Chemie-International Edition*, **43**, 3928.

Wu, P., Malkoch, M., Hunt, J.N. *et al.* (2005a) Multivalent, bifunctional dendrimers prepared by click chemistry. *Chemical Communications*, 5775.

Wu, Y., Shi, Y., and Fu, Z.F. (2005b) Synthesis of heteroarm star-shaped (poly styrene)(n)-[poly(ethyl acrylate)](m) via atom transfer radical polymerization. *Polymer*, **46**, 12722.

Xia, J., Zhang, X., and Matyjaszewski, K. (1999) Synthesis of star-shaped polystyrene by atom transfer radical polymerization using an "arm first" approach. *Macromolecules*, **32**, 4482.

Xie, J.D., Hu, L.H., Shi, W.F. *et al.* (2008) Synthesis and characterization of hyperbranched polytriazole via an 'A(2) + B-3' approach based on click chemistry. *Polymer International*, **57**, 965.

Xu, J., Tao, L., Boyer, C. *et al.* (2009) Combining thio-bromo "click" chemistry and RAFT polymerization: A powerful tool for preparing functionalized multiblock and hyperbranched polymers. *Macromolecules*, **43**, 22.

Zhu, J., Zhu, X., Kang, E.T., and Neoh, K.G. (2007) Design and synthesis of star polymers with hetero-arms by the combination of controlled radical polymerizations and click chemistry. *Polymer*, **48**, 6992.

9

Living Alkene Polymerization for Polyolefin Architectures

Amelia M. Anderson-Wile, Joseph B. Edson and Geoffrey W. Coates
Department of Chemistry and Chemical Biology, Baker Lab, Cornell University, Ithaca, NY 14850

9.1 Introduction

One of the ultimate challenges in polymer chemistry is the development of new synthetic methods for the polymerization of a wide range of monomers to obtain polymers with well-defined stereochemistry (Resconi *et al.*, 2000) while controlling molecular weight and molecular-weight distribution (Coates *et al.*, 2002; Domski *et al.*, 2007b; Edson *et al.*, 2009a). A primary goal of synthetic polymer chemistry over the last half-century has been the development of chain-growth polymerization methods that enable consecutive enchainment of monomer units without termination. These living polymerization systems (Szwarc, 1998) allow precise molecular weight control as well as the synthesis of a wide array of polymer architectures (Webster, 1991). Additionally, living polymerization methods allow the synthesis of end-functionalized polymers in addition to the creation of virtually limitless types of new materials from a basic set of monomers.

Currently, polyolefins are by far the largest volume class and most important commercial synthetic polymers (Vasile, 2000). Since the initial discoveries of Ziegler (Ziegler *et al.*, 1955) and Natta (Natta *et al.*, 1955), remarkable advances in activity and comonomer incorporation have been reported. Homogeneous olefin polymerization catalysts now exist that are unparalleled in all of polymer chemistry concerning the detailed control of macromolecular stereochemistry (Coates, 2000). However, olefin polymerization catalysts have traditionally been inferior to their other chain-growth counterparts in one respect. While extraordinary advances in living/controlled polymerization have been discovered using anionic (Hsieh and Quirk, 1996), cationic (Matyjaszewski, 1996; Aoshima and Kanaoka, 2009), and radical-based

Complex Macromolecular Architectures: Synthesis, Characterization, and Self-Assembly, First Edition.
Edited by Nikos Hadjichristidis, Akira Hirao, Yasuyuki Tezuka and Filip Du Prez.

polymerization (Hawker *et al.*, 2001; Kamigaito *et al.*, 2001; Matyjaszewski and Xia, 2001; Braunecker and Matyjaszewski, 2007; Ouchi *et al.*, 2009), until very recently there existed a comparative lack of living olefin polymerization systems. Alkene polymerization catalysts often undergo irreversible chain transfer to metal alkyls and β-elimination reactions that result in chain termination and the initiation of new polymer chains by the catalyst. However, truly living polymerization systems are now available that have acceptable rates of propagation with negligible rates of termination for alkenes such that block copolymer synthesis is now possible.

Perhaps the most important application of living olefin polymerization is the production of block copolymers, which is typically achieved via sequential monomer addition. Physical blends or random copolymers often give rise to materials whose properties are intermediate between those of the respective homopolymers. Block copolymers, on the other hand, often furnish materials whose mechanical properties are superior to the sum of their parts. This unique behavior is often due to microphase separation of the different segments of the block copolymer into discrete domains that give rise to otherwise unattainable morphologies (Bates, 1991; Ruzette and Leibler, 2005; Bates and Fredrickson, 1999). One of the most highly sought after goals in the field of olefin polymerization is the synthesis of block copolymers containing isotactic polypropylene (*i*PP) domains, which may serve as compatibilizers in blends containing *i*PP homopolymers (Ittel *et al.*, 2000). Additionally, one of the most actively pursued block copolymer structures are those with "hard" or semicrystalline end blocks (e.g. PE, *i*PP, *s*PP) and amorphous midblocks (e.g. *a*PP, poly(E-*co*-P)); triblock copolymers of this type have been shown to behave as thermoplastic elastomers (Mallin *et al.*, 1990; Coates and Waymouth, 1995; Chien *et al.*, 1997; Lieber and Brintzinger, 2000; Harney *et al.*, 2006a; Hotta *et al.*, 2006; Edson *et al.*, 2008; Rose *et al.*, 2008a).

This review will primarily focus on living polymerization of terminal-alkenes with some coverage of nonconjugated dienes and cyclic olefins. Special attention will be paid to systems capable of producing complex architectures such as block, star and graft copolymers. Living polymerization catalysts that have not been utilized in the production of advanced polymer architectures (Fukui *et al.*, 1999; Nomura and Fudo, 2004; Sassmannshausen *et al.*, 1997; Baumann *et al.*, 1997; Schrock *et al.*, 1999; Mehrkhodavandi *et al.*, 2003; Mehrkhodavandi and Schrock, 2001; Mehrkhodavandi *et al.*, 2000; Baumann and Schrock, 1998; Schrock *et al.*, 2003; Liang *et al.*, 1999; Schrock, *et al.* 2001; Schrock *et al.*, 2005; Manivannan and Sundararajan, 2002; Sudhakar and Sundararajan, 2005; Sudhakar, 2007; Kretschmer *et al.*, 2007; Lee *et al.*, 2008; Nienkemper *et al.*, 2008; Annunziata *et al.*, 2009; Ward *et al.*, 2005; Jeske *et al.*, 1985; Wang and Nomura, 2005; Bambirra *et al.*, 2003; Mashima *et al.*, 1993; Kotzabasakis *et al.*, 2009; Starzewski *et al.*, 2008; Di Lena and Chen, 2009; Jansen *et al.*, 2003; Jansen *et al.*, 2001; Mashima *et al.*, 1995; Mashima *et al.*, 1994; Macadams *et al.*, 2005; Ravasio *et al.*, 2008; Azoulay *et al.*, 2009; Azoulay *et al.*, 2008; Hicks *et al.*, 2003; Yang *et al.*, 2009) will not be covered in this review, however, many of these systems are included in our previous work (Domski *et al.*, 2007b; Edson *et al.*, 2009a; Coates *et al.*, 2002).

9.2 Living Olefin Polymerization

9.2.1 Poly(1-hexene)

1-Hexene is one of the most commonly employed monomers for detailed studies of living olefin polymerization due to the fact that it is an easily handled liquid, and molecular-weight determination of its polymers is accomplished at or slightly above room temperature (Figure 9.1). However, due to their poor mechanical properties, poly(1-hexene) (PH) and homopolymers derived from higher α-olefins (with the exception of poly(4-methyl-1-pentene)) are of little commercial significance for applications requiring rigid materials. One application

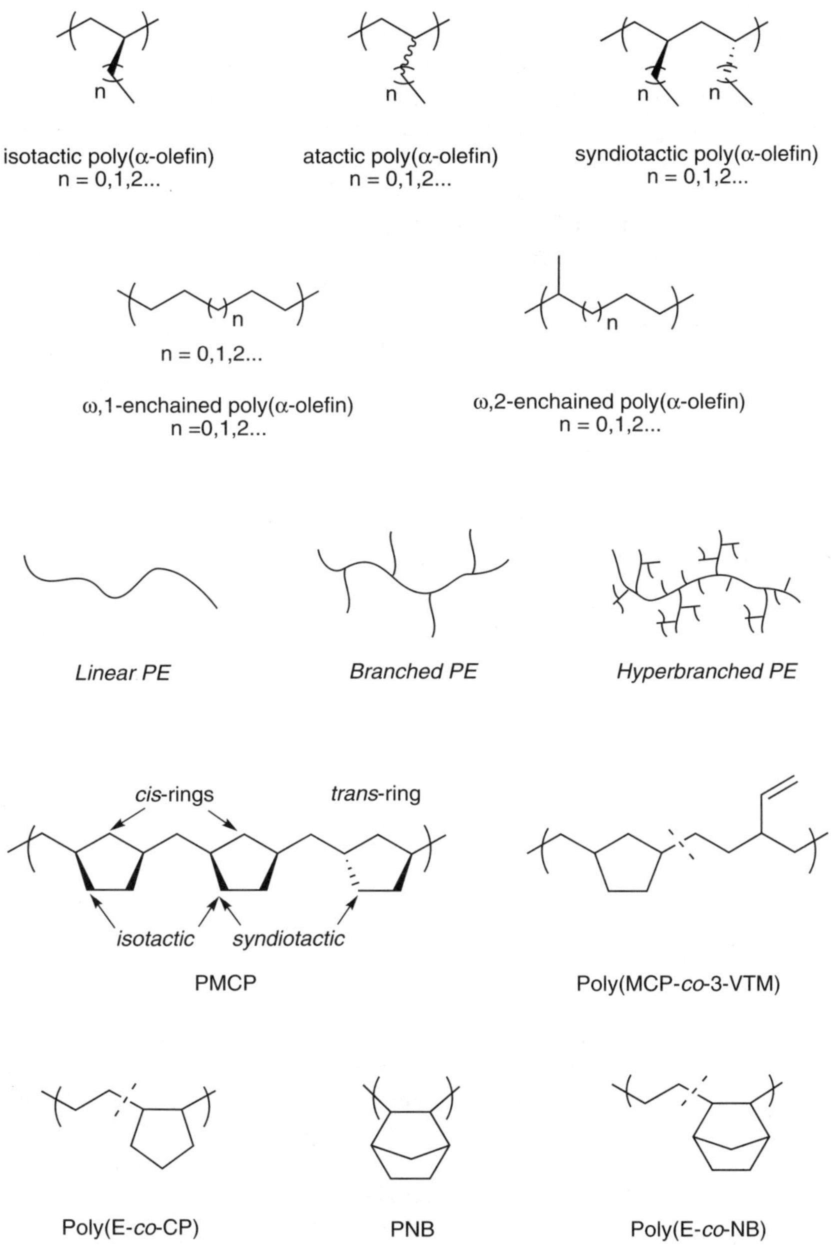

Figure 9.1 Olefin homopolymers and copolymers.

of amorphous poly(α-olefin)s are as impact strength modifiers when blended with polypropylenes (Vasile, 2000).

9.2.2 Polypropylene

While poly(1-hexene) is an amorphous material regardless of the level of tacticity, polypropylene (Figure 9.1) can range from amorphous to semicrystalline due to the variability in the level of tacticity. The bulk properties of the polymer are intimately related to its tacticity. For example, atactic polypropylene (*a*PP) is an amorphous material with limited industrial uses (e.g. adhesives, sealants, and caulks) whereas syndiotactic polypropylene (*s*PP) and isotactic polypropylene (*i*PP) are semicrystalline materials with relatively high melting temperature (T_m) values of ~150 °C and ~165 °C, respectively (Vasile, 2000). The lower crystallinity and the fact that only a very few syndiospecific propylene polymerization catalysts have been discovered to date have limited the commercial impact of *s*PP. On the other hand, numerous catalysts, both heterogeneous and homogeneous, are capable of isospecific propylene polymerization. When combined with *i*PP's highly desirable mechanical properties (durability, chemical resistance, and stiffness), it is obvious why the vast majority of industrially produced polypropylenes are of the isotactic variety (Karian, 2003).

9.2.3 Polyethylene

In addition to *i*PP, the polymer obtained from the polymerization of ethylene represents another commercially important material. The annual production of polyethylene (PE) world wide is in excess of 80 billion pounds (Peacock, 2000). Considering the simplicity of ethylene, the range of different polymer architectures derived from it is truly impressive (Figure 9.1). The properties of PEs vary greatly depending on the polymer's microstructure (i.e. branched or linear) from high-density plastics with relatively high melting points (linear PE: $T_m \sim 135$ °C), to low-density, branched material with T_m as low as 105 °C. While most early transition-metal olefin polymerization catalysts furnish linear PE, late-metal catalysts based on nickel or palladium typically give rise to branched structures (Ittel *et al.*, 2000).

9.2.4 Polyolefins from Conjugated Dienes, Cyclic Olefins and Polar Monomers

Beyond simple α-olefins, the polymers obtained from both nonconjugated dienes and cyclic olefins provide materials with desirable properties. Nonconjugated dienes are versatile monomers, which can furnish polymers with a variety of microstructures depending on the mechanism by which they are polymerized. For example, 1,5-hexadiene can be cyclopolymerized to furnish a polymer with methylene-1,3-cyclopentane (MCP) units (Figure 9.1). Depending on the selectivity of the ring-closing reaction, *cis* or *trans* rings may be formed. Polymers containing mostly *cis* rings exhibit higher T_m values than those with mostly *trans* rings. For example, poly(methylene-1,3-cyclopentane) (PMCP) having >90% *cis* ring content displays a T_m of 189 °C, whereas PMCP containing 82% *trans* rings exhibits a T_m of 102 °C (Coates and Waymouth, 1991; Kumudini *et al.*, 2000). Polymerization of 1,5-hexadiene can also lead to vinyl-tetramethylene (VTM) units, which may serve as a synthetic handle by which polymer functionalization can be achieved (Mathers and Coates, 2004).

Homopolymers of cyclic olefins (e.g. polycyclopentene, polynorbornene (PNB), Figure 9.1) are characterized by extremely high melting points and low solubility in most organic solvents. Taken together, these properties of cyclic olefin homopolymers make them difficult to process and therefore commercially insignificant. However, materials with desirable properties can be obtained upon incorporation into copolymers with α-olefins. The copolymers typically exhibit high chemical resistance, good optical properties, and facile processability (Tritto *et al.*, 2006). In addition to simple hydrocarbon systems, copolymers of polar monomers with olefins are attractive due to enhanced physical properties such as biocompatibility and ease of processing (Boffa and Novak, 2000).

9.2.5 Criteria for Living Polymerization

The homopolymerization of the aforementioned alkenes will be discussed with an order of early metal to late-metal catalyzed polymerizations with special emphasis paid to block copolymers and new polymer architectures. Note that in this review, we refer to living species for alkene polymerization as catalysts, not initiators, to emphasize the fundamental catalytic event of monomer enchainment, not polymer-chain formation. There are seven generally accepted criteria for a living polymerization: (1) polymerization proceeds to complete monomer conversion, and chain growth continues upon further monomer addition; (2) number-average molecular weight (M_n) of the polymer increases linearly as a function of conversion; (3) the number of active centers remains constant for the duration of the polymerization; (4) molecular weight can be precisely controlled through stoichiometry; (5) polymers display narrow molecular-weight distributions, described quantitatively by the ratio of the weight-average molecular weight to the number-average molecular weight ($M_w/M_n \sim 1$); (6) block copolymers can be prepared by sequential monomer addition; and (7) end-functionalized polymers can be synthesized (Quirk and Lee, 1992). Whether ionic, radical, or metal mediated, few polymerization systems that are claimed to proceed by a living mechanism have been shown to meet all of these criteria. This review will therefore include all systems that claim living alkene polymerization and have been used in the production of complex polymer architectures such as block, star and graft polymers.

9.3 Early Metal Olefin Polymerization Catalysts

9.3.1 Vanadium Acetylacetonate Catalysts

In the 1960s, Natta and coworker discovered that activation of VCl_4 with Et_2AlCl at −78 °C in the presence of propylene furnished syndio-enriched polypropylene. Linear growth of molecular weight over time for a period of 25 h was observed (Zambelli *et al.*, 1967). In a subsequent report, the polypropylenes produced by VCl_4/Et_2AlCl were shown to possess monomodal molecular-weight distributions ($M_w/M_n = 1.4$–1.9) (Doi *et al.*, 1979a).

The first example of a truly living alkene polymerization catalyst was reported by Doi and coworkers in 1979 (Doi *et al.*, 1979b; Doi *et al.*, 1979c). In the presence of propylene at temperatures ≤ −65 °C, **1**/Et_2AlCl (Figure 9.2) furnished syndio-enriched PP ($[r] = 0.81$)

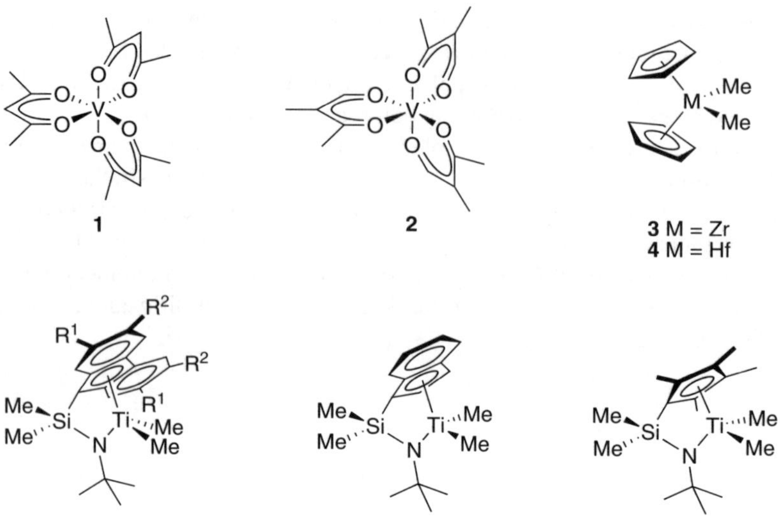

1 2 3 M = Zr
4 M = Hf

5 R¹ = R² = H
6 R¹ = *t*Bu, R² = H
7 R¹ = H, R² = *t*Bu
8 R¹,R² = -$CMe_2CH_2CH_2CMe_2$-

9 10

Figure 9.2 Early-metal catalyst precursors for living polymerization.

exhibiting narrow molecular-weight distributions ($M_w/M_n = 1.07$–1.18) and M_n as high as 100 000 g/mol. Over 15 h, a linear increase in M_n with time was shown. Initially, only about 4% of vanadium centers were shown to be active, however, addition of anisole to the polymerization led to a threefold increase in the number of active vanadium centers (Ueki *et al.*, 1981). Upon replacing the acetylacetonate ligands of **1** with 2-methyl-1,3-butanedionato ligands (**2**, Figure 9.2), Doi and coworkers found that nearly all of the vanadium centers were active for polymerization with essentially the same degree of syndioselectivity as **1**/Et_2AlCl (Doi *et al.*, 1985b). Additionally, the living character of propylene polymerization by **2**/Et_2AlCl was maintained up to −40 °C (M_w/M_n as low as 1.4) (Doi *et al.*, 1986; Doi *et al.*, 1988). Copolymerization of propylene and ethylene by **2**/Et_2AlCl was also shown to be living (Doi *et al.*, 1987b).

Utilizing the living nature of **1**/Et_2AlCl, Doi and coworkers were able to synthesize block copolymers of propylene and ethylene (Doi *et al.*, 1982). Specifically, a *s*PP-*block*-poly(E-*co*-P)-*block*-*s*PP triblock copolymer was synthesized via sequential monomer addition and exhibited a narrow molecular-weight distribution ($M_w/M_n = 1.24$) with $M_n = 94\,000$ g/mol and a propylene content of 70 mol%.

Beyond block copolymer synthesis, one of the important applications of living olefin polymerization is in the synthesis of end-functionalized polymers, which is typically achieved by reaction of the living chain end with an electrophile. The vanadium-based living olefin polymerization catalysts discovered by Doi and coworkers proved to be particularly amenable to this application (Doi *et al.*, 1983; Doi *et al.*, 1984; Doi and Keii, 1986; Doi *et al.*, 1987a; Ueki *et al.*, 1995). In addition to end-functionalized polymers, one challenging goal in polymer synthesis is the incorporation of polar monomers because of the limited ability of many catalysts to tolerate polar functionalities. In 1983, Doi reported the use of **1**/Et_2AlCl for the synthesis of a series of PP-*block*-poly(tetrahydrofuran) AB-type diblock copolymers by quenching a living propylene polymerization with iodine and using the iodide-terminated PP to initiate cationic polymerization of THF (Doi *et al.*, 1983). Catalyst **1**/Et_2AlCl has also been used to synthesize PP-*block*-PMMA (PMMA = poly(methylmethacrylate) (Doi *et al.*, 1985a). At −78 °C, **1**/Et_2AlCl was used to polymerize propylene to which methyl methacrylate (MMA) was then added at 25 °C to form the diblock copolymer by a proposed radical polymerization of the MMA.

In addition to living propylene polymerization, vanadium acetylacetonate complexes have also been shown to be living for 1,5-hexadiene polymerization and 1,5-hexadiene/propylene copolymerization (Doi *et al.*, 1989). At −78 °C, **1**/Et_2AlCl polymerized 1,5-hexadiene to produce a low molecular weight polymer (M_n = 6600 g/mol, M_w/M_n = 1.4) that contained a mixture of MCP and VTM units in a 54 : 46 ratio. The distribution of these two units varied in 1,5-hexadiene-propylene random copolymers as a function of 1,5-hexadiene incorporation.

9.3.2 Metallocene Catalysts

Since the discovery of their catalytic activity, Group 4 metallocene complexes have found extensive use as olefin polymerization catalysts (Resconi *et al.*, 2000). There have only been few examples of living metallocene-based catalysts due to their high propensity toward termination via chain transfer (e.g. β-H elimination/transfer, transfer to alkylaluminum species). However, several groups have shown that by employing well-defined boron-based activators (Chen and Marks, 2000) at low reaction temperatures these termination pathways can be suppressed. For example, Fukui and coworkers have reported that Cp_2ZrMe_2 (**3**, Figure 9.2) activated with $B(C_6F_5)_3$ at −78 °C in the presence of $Al(nOct)_3$ produces PP with $M_w/M_n \leq 1.15$ (M_n = 9400–27 300 g/mol) (Fukui *et al.*, 1999). The polymerization shows a linear increase in M_n with time and it was later reported that quenching the polymerization with CO resulted in aldehyde-functionalized polymer chains (Fukui and Murata, 2001; Fukui and Murata, 2002). At −50 °C, the hafnium analog (**4**, Figure 9.2) was also shown to be living.

The synthesis of diblock copolymers of ethylene and propylene was achieved by Hlatky and Turner using a hafnocene catalyst in 1991 (Turner and Hlatky, 1991). Utilizing **4**/$[PhNMe_2H][B(C_6F_5)_4]$ at 0 °C, the rate of termination via β-H transfer was slow enough to allow for the synthesis of *a*PP-*block*-PE via sequential monomer addition. Both orders of monomer addition (ethylene followed by propylene and propylene followed by ethylene) were successful in furnishing a polymeric product that contained a majority (50–60%) of diblock material isolated by hexanes extraction. In a subsequent report by Fukui and coworkers, the synthesis of *a*PP-*block*-poly(E-*co*-P) diblock copolymers using **3**, **4**, and $Cp^*_2HfMe_2$ activated with $B(C_6F_5)_3$ and employing $Al(nOct)_3$ as a scavenger was achieved through sequential monomer addition at low temperatures (T_{rxn} = − 78 °C for **3** and $Cp^*_2HfMe_2$; T_{rxn} = − 50 °C for **4**) (Fukui and Murata, 2002). The resultant polymers exhibited narrow molecular-weight distributions (M_w/M_n = 1.07–1.30) and M_n = 71 000–155 000 g/mol with propylene contents between 65 and 75 mol%.

9.3.3 *Catalysts Bearing Monocyclopentadienyl-Amido Ligands*

9.3.3.1 1-Hexene Polymerization

In spite of the large number of metallocene-based polymerization catalysts, there remain relatively few examples of living systems. Titanium complexes bearing a linked monocyclopentadienyl-amido ligand, such as **5** (Figure 9.2), have been shown to polymerize 1-hexene in a living fashion when activated with $B(C_6F_5)_3$ in the presence of Al(*n*Oct)$_3$ at −50 °C (Hagihara *et al.*, 1998). The poly(1-hexene) formed was syndio-enriched ([*rr*] = 0.49) with M_n up to 26 000 g/mol and M_w/M_n = 1.07–1.12. A linear relationship between M_n and polymer yield was also demonstrated. In a subsequent report, **5**/($B(C_6F_5)_3$/Al(*n*Oct)$_3$ was also shown to polymerize 1-octene and 1-butene in a living fashion (Nishii *et al.*, 2002).

9.3.3.2 Propylene Polymerization

In addition to living 1-hexene polymerization, **5**/$B(C_6F_5)_3$ at −50 °C in the presence of Al(*n*Oct)$_3$ can also produce syndio-enriched PP ([*rr*] ~ 0.49) in living fashion (Hagihara *et al.*, 1998). Subsequent studies showed that when activated with "dried" MAO (dMAO) (free of trimethylaluminum), **5** could polymerize propylene at 0 °C producing PP with a higher degree of syndiotacticity ([*rr*] ~ 0.63) and with a relatively narrow molecular-weight distribution (M_w/M_n = 1.22, M_n = 157 000 g/mol) (Hasan *et al.*, 2001). Shiono and coworkers also demonstrated significant solvent effects on the tacticity of the resulting PP (Nishii *et al.*, 2004a). For example, polymerization of propylene with **5**/dMAO in heptane at 0 °C results in polymer with higher tacticity than when the reaction is carried out in toluene ([*rr*] = 0.73 vs. 0.60) or chlorobenzene ([*rr*] = 0.42).

Utilizing these solvent effects, Shiono and coworkers cleverly prepared stereoblock copolymers of propylene containing *s*PP and *a*PP segment by initial polymerization in heptane followed by addition of more propylene and chlorobenzene (Nishii *et al.*, 2004b). The resultant *s*PP-*block*-*a*PP had M_n = 94 700 g/mol with a narrow PDI (M_w/M_n = 1.27). Shiono and coworkers have also described similar effects of the tacticity of PP generated from **5**/dMMAO under varying propylene pressures (Cai *et al.*, 2008b). At low pressure (0.2 atm) atactic PP is furnished, but at higher pressure (1 atm) syndiotactic PP is generated. A *s*PP-*block*-*a*PP copolymer and *s*PP-*block*-*a*PP-*block*-*s*PP copolymer were synthesized by varying propylene pressure over the course of the polymerization.

Shiono and coworkers examined structural variants of **5** (**6** and **7**, Figure 9.2) by introducing *tert*-butyl substituents into the fluorenyl ligand framework (Cai *et al.*, 2005). At 0 °C, **6**/dMMAO produced *s*PP ([*rr*] ~ 0.83) with a broadened molecular-weight distribution (M_w/M_n = 1.68, M_n = 202 000 g/mol). However, a two-stage sequential polymerization of 0.63 g of propylene revealed a nearly doubling of molecular weight compared with a single-stage polymerization. Catalyst **7**/dMMAO furnished polymer with even higher tacticity ([*rr*] ~ 0.93) and lower molecular-weight distribution (M_w/M_n = 1.45). Upon further expanding the sterics about the fluorenyl, **8**/dMMAO was utilized in the production of polypropylene with decreased syndioselectivity ([*rr*] ~ 0.45) compared to the polymer obtained from **7**/dMMAO (Shiono *et al.*, 2009). While molecular weights (M_n = 44 000–150 000 g/mol) increased with polymer yield, molecular-weight distributions were broadened (M_w/M_n = 2.91–4.61). Although some characteristics of a living polymerization were exhibited, a two-stage sequential polymerization resulted in polypropylene that was nearly identical to that obtained from the single-stage polymerization. Employing an indenyl-based ligand, **9**/dMAO at 0 °C affords iso-enriched PP ([*mm*] = 0.40) with quasiliving behavior (Nishii *et al.*, 2005). Dare and coworkers reported that a similar complex **10**/MAO furnished PP at 0 °C that was syndio-enriched ([*rr*] = 0.56) and exhibited a somewhat narrow PDI ($M_w/M_n = 1.37$, $M_n = 108\,000$ g/mol) (Dare *et al.*, 2004).

9.3.3.3 Copolymerization of Ethylene/Norbornene and Propylene/Norbornene

In addition to linear α-olefins, Shino and coworkers reported that **5**/MAO catalyzed the living copolymerization of ethylene and norbornene. At 0 °C, **5**/MAO can furnish poly(E-*co*-NB) with 53 mol% norbornene and $M_n = 78\,000$ g/mol with $M_w/M_n = 1.16$ (Hasan *et al.*, 2004b). In addition, a linear increase in M_n with reaction time was observed for this system. At 40 °C, a similar compound, **10**/MAO, also provided ethylene-norbornene copolymers with fairly narrow PDIs ($M_w/M_n = 1.21$–1.27) (Hasan *et al.*, 2004a).

Shiono and coworkers also reported the living copolymerization of propylene and norbornene with **5**/dMAO to produce copolymers with very high T_g values (249 °C) and narrow molecular-weight distributions ($M_w/M_n = 1.16$) (Hasan *et al.*, 2005). In a later report, the copolymerization of higher α-olefins (1-hexene, 1-octene, and 1-decene) with norbornene by **5**/MAO was reported, however, molecular-weight distributions were somewhat broadened ($M_w/M_n = 1.36$–1.72) (Shiono *et al.*, 2008).

At 20 °C, activation of **7** with dMMAO containing 0.4 mol% of triisobutylaluminum (TIBA) in the presence of norbornene resulted in living polymerization (Cai *et al.*, 2008a). The molecular-weight distributions obtained were narrow ($M_w/M_n = 1.07$–1.08). A two-stage reaction was shown to increase the molecular weight of the second step by double that of the first step when each was carried to quantitative conversion. Furthermore, Shiono and coworkers were able to show that **7**/MAO could also copolymerize propylene and norbornene to form random and block copolymers (Cai *et al.*, 2006). For example, three *s*PP-*block*-poly(P-*co*-NB) diblock copolymers were synthesized through sequential monomer addition that had similar molecular weights ($M_n \sim 20\,000$ g/mol, $M_w/M_n = 1.21$–1.32). Shiono and coworkers were also able to show that **7**/dMMAO containing 1.8 mol% TIBA could furnish PNB-*block*-poly(P-*co*-NB)-*block*-PP triblock copolymers in a catalytic fashion (Cai *et al.*, 2008a). The successive addition of norbornene and propylene before complete consumption of norbornene gives PNB-*block*-poly(P-*co*-NB)-*block*-PP terminated with a Ti-PP bond, which can be exchanged with TIBA. Repeated addition of norbornene and propylene gives a catalytic synthesis of the triblock copolymers in this system (Scheme 9.1).

9.3.4 Monocyclopentadienylzirconium Amidinate Catalysts

9.3.4.1 1-Hexene Polymerization

In 2000, Sita and Jayaratne reported monocyclopentadienyl acetamidinate zirconium dimethyl compounds that exhibited living polymerization behavior at temperatures between −10 and 0 °C (Jayaratne and Sita, 2000). At 0 °C, **11a**/[PhNMe$_2$H][B(C$_6$F$_5$)$_4$] (Figure 9.3) formed atactic poly(1-hexene) with a narrow polydispersity ($M_w/M_n = 1.10$) that lacked olefinic

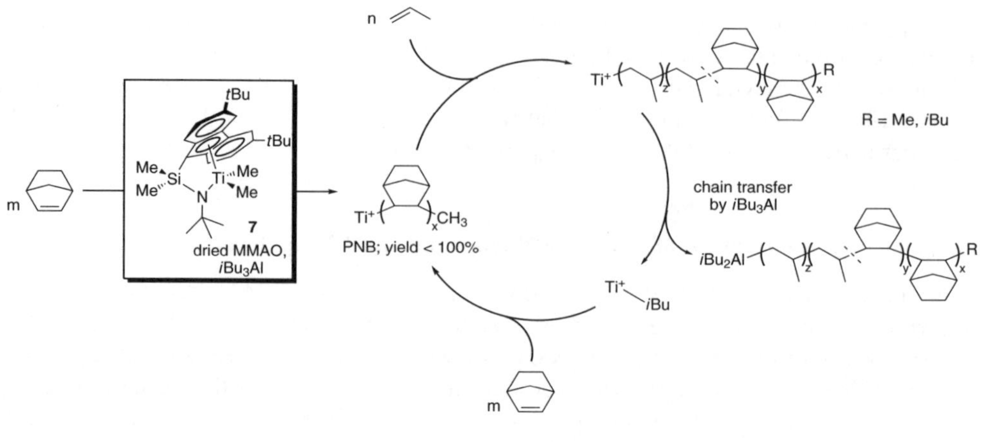

Scheme 9.1 Catalytic synthesis of block copolymers from norbornene and propylene using **7**/dMMAO-TIBA.

resonances in ^{13}C and ^{1}H NMR spectra. The C_1-symmetric complex, **11b**, when activated in an identical manner at $-10\,°C$ furnished highly isotactic poly(1-hexene) ([*mmmm*] > 0.95) with a narrow molecular-weight distribution ($M_w/M_n = 1.03$–1.13). The molecular weight was shown to increase linearly with conversion. This was the first report of a Ziegler–Natta polymerization catalyst that was both living and highly isospecific for α-olefin polymerization. Living and isoselective 1-hexene polymerization was also observed upon covalently

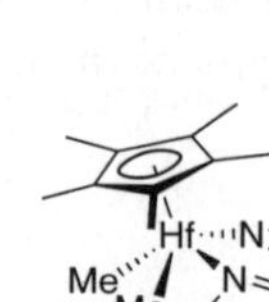

	M	X	R[1]	R[2]	R[3]
11a	Zr	Me	Cy	Cy	Me
11b	Zr	Me	*t*Bu	Et	Me
11c	Zr	Me	*t*Bu	Cy	Me
13a	Hf	Me	*t*Bu	Et	Me
13b	Hf	*i*Bu	*t*Bu	Et	Me
15a	Zr	Me	*t*Bu	Et	Ph
15b	Zr	Me	*t*Bu	Et	H
15c	Zr	Me	*t*Bu	Et	*t*Bu
17	Zr	Me	*t*Bu	$CH_2C(CH_3)_3$	Me
19	Hf	Me	Et	Et	Me

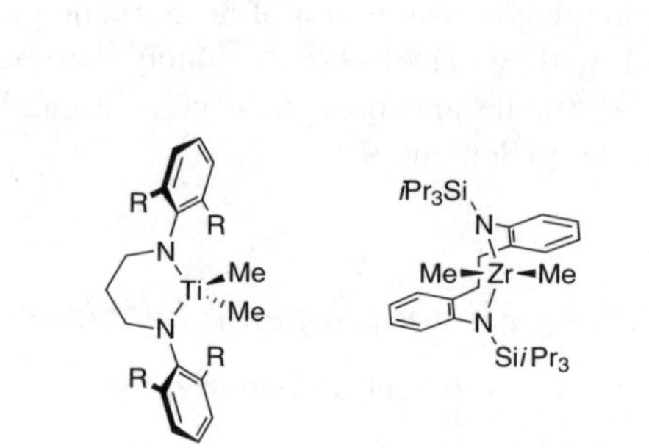

12

14a R[1] = R[2] = Cy
14b R[1] = R[2] = *i*Pr
14c R[1] = *t*Bu, R[2] = Cy

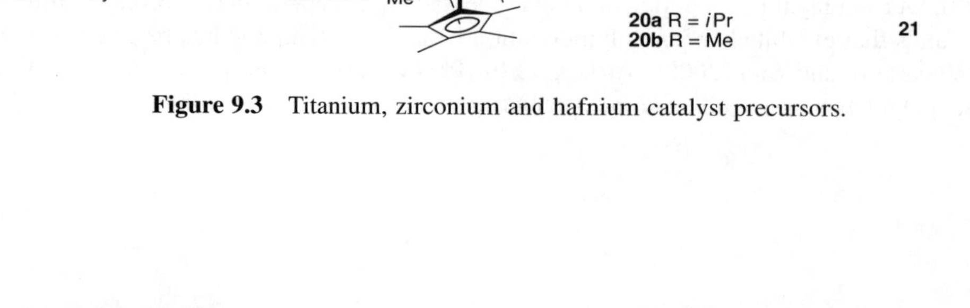

16a R = 4-MeOC$_6$H$_4$
16b R = 4-MeC$_6$F$_4$
16c R = Cy

18a n = 2
18b n = 4
18c n = 6

20a R = *i*Pr
20b R = Me

21

Figure 9.3 Titanium, zirconium and hafnium catalyst precursors.

attaching **11b** to a crosslinked PS support (**12**, Figure 9.3) (Zhang and Sita, 2003). The hafnium congener of **11b** (**13a**, Figure 9.3) and its diisobutyl analog (**13b**) were also shown to be living and isospecific 1-hexene polymerization catalysts albeit with a rate about 60 times slower than **11b** (Kissounko *et al.*, 2005).

As discussed above, **11b**/$[PhNMe_2H][B(C_6F_5)_4]$ produces highly isotactic poly(1-hexene) in a living fashion. However, when a substoichiometric amount of the borate is used (e.g. $[PhNMe_2H][B(C_6F_5)_4]$: [**11b**] = 0.5), the resulting polymer is considerably less isotactic ([*mm*]=0.45–0.50) resulting from degenerative-transfer polymerization (Scheme 9.2), which will be discussed in detail in the following section. This system has been employed to make a poly(α-olefin) diblock copolymer (Zhang *et al.*, 2003). Initially, **11b** was activated with 0.5 equivalents of $[PhNMe_2H][B(C_6F_5)_4]$ and used to produce an atactic poly(1-hexene) block. After 2 hours, 1-octene was added along with an additional 0.5 equivalents of the borate leading to the growth of an isotactic poly(1-octene) block. GPC analysis revealed clean formation of the *a*PH-*block*-*i*-poly(1-octene) diblock copolymer with $M_n = 12\,400$ g/mol and $M_w/M_n = 1.04$.

Sita and coworkers were able to greatly increase the 1-hexene polymerization activity for this class of catalysts upon replacing the Cp* moiety with the less sterically demanding Cp ligand (Keaton *et al.*, 2001). At −10 °C, compounds **14a–c**/$[PhNMe_2H][B(C_6F_5)_4]$ (Figure 9.3) furnished atactic poly(1-hexene)s with narrow molecular-weight distributions ($M_w/M_n = 1.03$–1.10). Although a decrease in enantiofacial selectivity was observed, the more

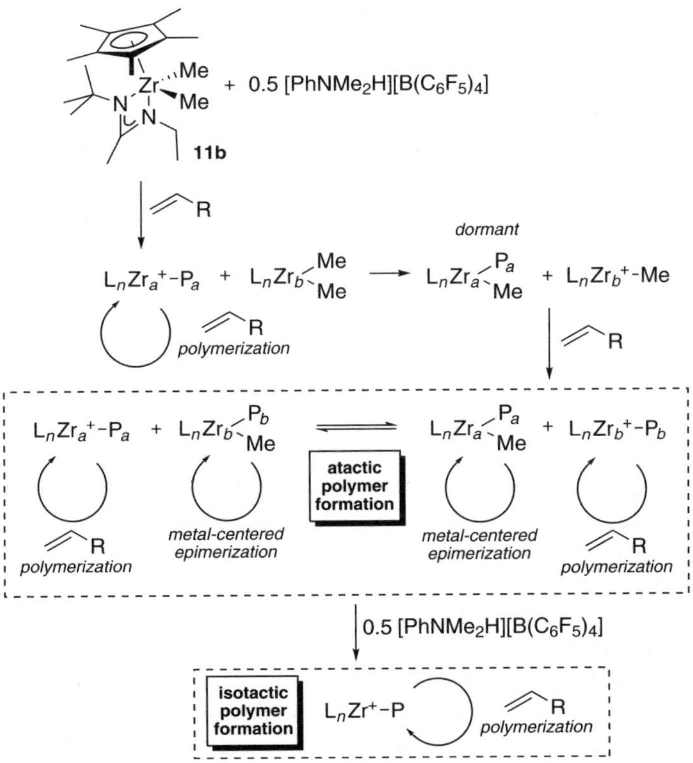

Scheme 9.2 Degenerative group transfer polymerization employing **11b**.

open environment of the active site did impart the ability to polymerize the more challenging vinylcyclohexane. Upon activation with $[PhNMe_2H][B(C_6F_5)_4]$ at $-10\,°C$, **14a,b** furnished highly isotactic poly(vinylcyclohexane)s ([*mmmm*] = 0.95) with narrow polydispersities ($M_w/M_n = 1.04$–1.10). The authors postulate that the high degree of isoselectivity displayed is likely the result of chain-end control. Sita and coworkers were able to exploit the living nature of **14c**/$[Ph_3C][B(C_6F_5)_4]$ for 1-hexene and vinylcyclohexane (VCH) polymerization to prepare a triblock copolymer of isotactic poly(vinylcyclohexane)-*block*-atactic-poly(1-hexene)-*block*-isotactic-poly(vinylcyclohexane) by sequential monomer addition (Keaton *et al.*, 2001). The triblock copolymer exhibited a narrow polydispersity ($M_w/M_n = 1.08$) and $M_n = 24\,400$ g/mol with a vinylcyclohexane content of 33 mol%.

In 2004, the effect of further structural elaboration of the amidinate ligand framework on polymerization behavior was reported (Zhang *et al.*, 2004). Altering the identity of the distal R^3 substituent (Figure 9.3) led to dramatic effects on both the living character and stereospecificty of 1-hexene polymerization. At $-10\,°C$, polymerization of 1-hexene by **15a** or **15b** (R^3 = Ph or H)/$[PhNMe_2H][B(C_6F_5)_4]$ furnished polymer with a significantly lower degree of isotacticity than the poly(1-hexene) produced by **11b**/$[PhNMe_2H][B(C_6F_5)_4]$, and in the case of **15b**, the polymerization is no longer living. Furthermore, **15c** (R^3 = *t*Bu)/$[PhNMe_2H][B(C_6F_5)_4]$ was found to be completely inactive for polymerization. The loss in stereocontrol of **15a** was attributed to a "buttressing effect" by which the *t*Bu and Et groups were "pushed" forward toward the active site leading to a lack of steric discrimination at the metal center for olefin coordination. The decrease in stereoselectivity of **15b** was attributed to a low barrier to metal-centered epimerization relative to **11b.**

Interested in varying the amidinate portion of the catalyst, Mashima and coworkers recently reported the synthesis and 1-hexene polymerization behavior of Cp* hafnium dimethyl complexes bearing an iminopyrrolyl ligand (Yasumoto *et al.*, 2005). At 0 °C or below, compounds **16a–c**/$[Ph_3C][B(C_6F_5)_4]$ (Figure 9.3) polymerized 1-hexene to furnish polymers with narrow molecular-weight distributions ($M_w/M_n = 1.07$–1.12) and M_n = 9000–36 100 g/mol. The poly(1-hexene)s were all significantly isoenriched ([*mmmm*] = 0.90 for **16b**) at $-20\,°C$. Furthermore, the polymerization of 1-hexene with **16a** exhibited a linear dependence of M_n versus time at -20 and 0 °C.

9.3.4.2 Propylene Polymerization

In addition to living 1-hexene polymerization, Sita and coworkers have shown that **11b**/$[PhNMe_2H][B(C_6F_5)_4]$ furnished highly isotactic PP ([*mmmm*] = 0.71) in a living fashion ($M_w/M_n \leq 1.20$) (Harney *et al.*, 2006a). Interestingly, activation with 0.5 equivalents of $[PhNMe_2H][B(C_6F_5)_4]$ resulted in the production of atactic PP where the M_n was also shown to increase linearly with time ($M_w/M_n \sim 1.05$). A degenerative-transfer (DT) polymerization that proceeds through a rapid and reversible methyl-group transfer between the cationic (active) and neutral (dormant) zirconium centers was considered as the mechanism for this living system (Scheme 9.2) (Zhang *et al.*, 2003). Additionally, the methyl-polymeryl dormant species can undergo epimerization that is faster than propagation. Thus, through a combination of methyl-group transfer and epimerization at dormant sites, each occurring faster than propagation, stereocontrol is greatly diminished, leading to an atactic microstructure. However, upon addition of a second 0.5 equivalents of $[PhNMe_2H][B(C_6F_5)_4]$, all Zr species become active for polymerization thereby "turning off" DT and initiating isospecific polymerization, which can advantageously be used for the production of block copolymers. Sita and coworkers synthesized stereogradient PP by initial polymerization under DT polymerization conditions followed by slow introduction of $[PhNMe_2H][B(C_6F_5)_4]$ to 100% activation (Harney *et al.*, 2006b).

Formation of a PP diblock copolymer with **11b**/$[PhNMe_2H][B(C_6F_5)_4]$ was accomplished by initial activation with 0.5 equivalents of $[PhNMe_2H][B(C_6F_5)_4]$ furnishing an atactic PP segment followed by complete activation with another 0.5 equivalents of $[PhNMe_2H][B(C_6F_5)_4]$ to furnish the *a*PP-*block*-*i*PP. It was found that related complex **17** could effectively reinvoke degenerative-transfer by *irreversibly* transferring a methyl group to the active polymerization species. In addition to synthesizing an *a*PP-*block*-*i*PP diblock copolymer, an *a*PP-*block*-*i*PP-*block*-*a*PP triblock and an *a*PP-*block*-*i*PP-*block*-*a*PP-*block*-*i*PP tetrablock sample were formed (Scheme 9.3). The polymers had very similar molecular weights ($M_n = 164\,200$–$172\,400$ g/mol, $M_w/M_n = 1.19$). Testing of the tensile properties of the block copolymers showed good elastomeric behavior. For example, the triblock copolymer displayed an elongation to break of 1530%, the highest of the three samples.

In a later report, Sita and coworkers prepared bimetallic analogs of **11b** to investigate further the effects of DT polymerization (**18a–c**, Figure 9.3) (Zhang and Sita, 2008b). At $-10\,°C$, compounds **18a–c**/$[PhNMe_2H][B(C_6F_5)_4]$ were all found to be living and isoselective for propylene polymerization (M_n up to 50 000 g/mol, $M_w/M_n = 1.1$–1.2), with the degree of stereoselectivity decreasing as the two metal centers are brought closer together. Activation under substoichiometric conditions led to living DT polymerization. Under these conditions, the frequency of [*mr*] stereoerrors in the PP decrease as the two metal centers are brought closer together resulting from an increased barrier to metal-centered epimerization of the dormant site.

Sita and coworkers have also reported a modified amidinate hafnium catalyst (**19**, Figure 9.3) that furnished atactic PP of high molecular weight ($M_n = 137\,000$ g/mol, $M_w/M_n = 1.12$) upon activation with one equivalent of $[PhNMe_2H][B(C_6F_5)_4]$ at $-10\,°C$ (Zhang and Sita, 2008a) M_n of up to 830 000 g/mol could be obtained with this system, however, significant broadening of the PDI was observed ($M_w/M_n = 2.43$). Furthermore, this system demonstrated the first

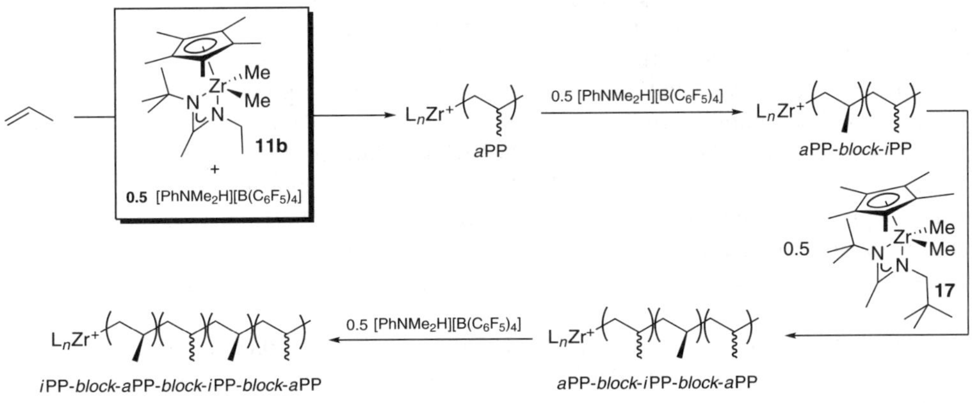

Scheme 9.3 Synthesis of propylene-based block copolymers using **11b**/MAO.

example of living coordinative chain-transfer polymerization (CCTP) (Kempe, 2007) of propylene with diethyl zinc. It was used for the living CCTP of ethylene, 1-hexene, 1-octene, and 1,5-hexadiene in addition to living CCTP copolymerization of ethylene with the aforementioned higher α-olefins (Zhang *et al.*, 2008b).

9.3.4.3 Polymerization of 1,5-Hexadiene

In addition to simple α-olefins such propylene, Sita and coworkers showed that **11a–c**/[$PhNMe_2H$][$B(C_6F_5)_4$] were active for the cyclopolymerization of 1,5-hexadiene at −10 °C (Jayaratne *et al.*, 2000). The polymers produced possessed ≥98% MCP units, and exhibited narrow polydispersities ($M_w/M_n = 1.03–1.09$). The selectivity of ring closure was ubiquitously *trans*; the stereoselectivity increasing with increasing steric bulk of the amidinate ligand (**11a**:% *trans* = 64; **11c**:% *trans* = 82).

Utilizing the living behavior of **11b**/[$PhNMe_2H$][$B(C_6F_5)_4$], Sita and coworkers were able to prepare diblock and triblock copolymers of 1-hexene and 1,5-hexadiene (Jayaratne *et al.*, 2000). At −10 °C, isotactic poly(1-hexene)-*block*-PMCP and isotactic poly(1-hexene)-*block*-PMCP-*block*-poly(1-hexene) were obtained from **11b**/[$PhNMe_2H$][$B(C_6F_5)_4$] through sequential monomer addition. The diblock copolymer possessed $M_w/M_n = 1.05$ and $M_n = 22\,800$ g/mol with $T_m = 91$ °C. The triblock copolymer had $M_w/M_n = 1.10$, $M_n = 30\,900$ g/mol, and $T_m = 79$ °C. Atomic force microscopy (AFM) revealed that microphase separation of the crystalline PMCP and amorphous poly(1-hexene) had occurred.

9.3.5 Catalysts Bearing Diamido Ligands

While Group 4 metallocene-based olefin polymerization catalysts have dominated the field of homogenous olefin polymerization catalysis since the late 1950s (Resconi *et al.*, 2000), the development of complexes bearing non-Cp ligands as potential olefin polymerization catalysts has become a rapidly expanding area over the last 15 years (Britovsek *et al.*, 1999; Gibson and Spitzmesser, 2003). McConville and Scollard reported that titanium complexes bearing diamide ligands, compounds **20a,b** (Figure 9.3), polymerized 1-hexene, 1-octene, and 1-decene to high molecular weight ($M_n = 121\,500–164\,200$ g/mol) and narrow polydispersity index (PDI) ($M_w/M_n = 1.07$) upon activation with $B(C_6F_5)_3$ at room temperature (Scollard and Mcconville, 1996). Polymerization of 1-hexene by **20b**/$B(C_6F_5)_3$ exhibited a linear increase of M_n with time. Later, Kim and coworkers reported the living polymerization of 1-hexene catalyzed by a structurally similar zirconium diamide (**21**, Figure 9.3) with an ethylene bridging unit (Jeon *et al.*, 1998). At −10 °C, **21**/$B(C_6F_5)_3$ furnished poly(1-hexene)s with $M_n = 30\,000–175\,000$ g/mol and $M_w/M_n = 1.18–1.27$. It was also shown that the M_n increased linearly with increasing monomer loading. Utilizing sequential monomer addition, Kim and coworkers prepared a block copolymer of 1-hexene and 1-octene (Jeon *et al.*, 1998). The polymer produced at 0 °C possessed a narrow polydispersity ($M_w/M_n = 1.21$) and a $M_n = 109\,000$ g/mol.

In 2002, Shiono and coworkers reported that activation of **20a**/MMAO at 0 °C in the presence of propylene furnished PP in a living manner (Hagimoto *et al.*, 2002b). The resultant polymers were atactic and displayed narrow molecular-weight distributions (M_w/M_n = about 1.16–1.3). The M_n was shown to increase linearly with polymerization time from 10 to 25 min

(M_n up to about 30 000 g/mol). In later reports, Shiono and coworkers discussed the effects of supported MMAOs on the propylene polymerization behavior of **20a** (Hagimoto *et al.*, 2002a; Hagimoto *et al.*, 2003; Shiono, 2003; Hagimoto *et al.*, 2004). Regardless of the support (SiO_2, Al_2O_3, and MgO), polymerization of propylene by **20a**/supported MMAO at 0 °C exhibited a linear increase of M_n with time.

9.3.6 Amine-Phenolate Titanium and Zirconium Catalysts

9.3.6.1 Polymerization of α-Olefins

In 2000, Kol and coworkers reported the synthesis and olefin polymerization behavior of a titanium complex bearing an amine bis(phenolate) ligand that incorporated an additional amino side-arm donor (**22a**, Figure 9.4) (Tshuva *et al.*, 2000b). At room temperature,

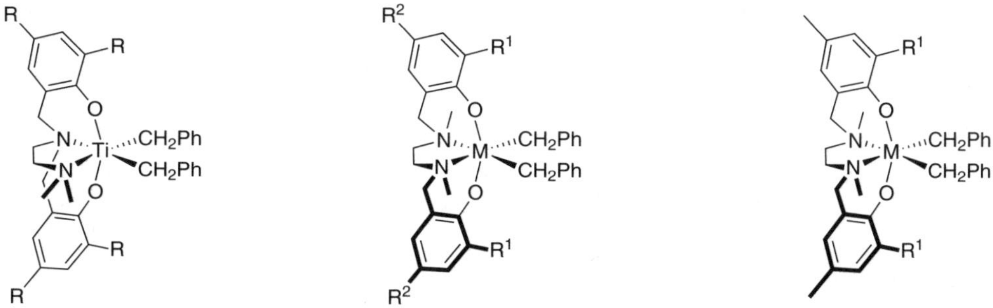

22a R = *t*Bu
22b R = Cl

	M	R¹	R²
24a	Zr	*t*Bu	*t*Bu
24b	Zr	Me	Me
24c	Zr	Cl	Cl
24d	Ti	Cl	Cl
24e	Ti	Br	Br

	M	R¹
28a	Zr	1-adamantyl
28b	Zr	cumyl
29a	Hf	1-adamantyl
29b	Hf	cumyl

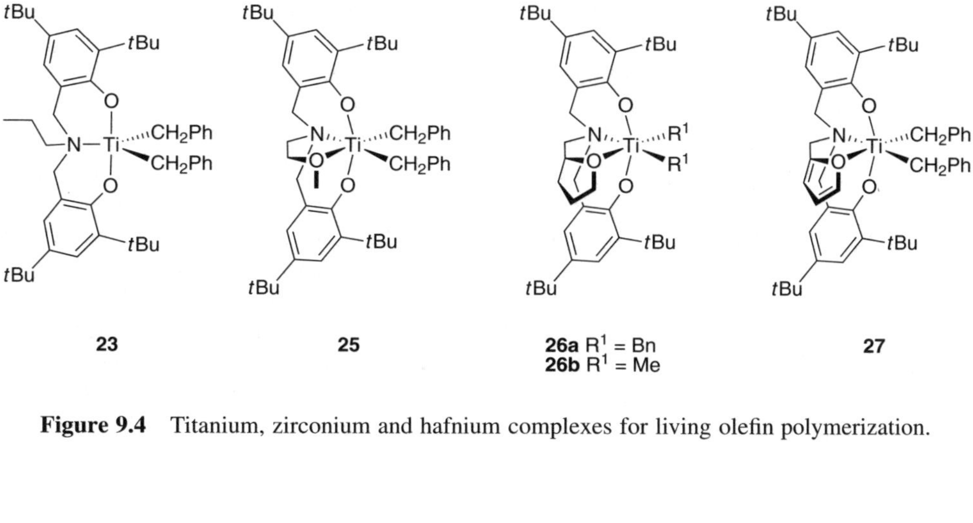

23 **25** **26a** R¹ = Bn **26b** R¹ = Me **27**

Figure 9.4 Titanium, zirconium and hafnium complexes for living olefin polymerization.

22a/$B(C_6F_5)_3$ furnished atactic poly(1-hexene)s with narrow molecular distributions ($M_w/M_n = 1.09–1.18$) and the M_n was shown to increase linearly with time. Upon omission of the amino side-arm donor (**23**), only low molecular weight poly(1-hexene) ($M_n = 2000$ g/mol) with $M_w/M_n = 1.92–2.43$ was obtained. However, replacing the bulky *t*Bu groups with sterically less-demanding chlorides (**22b**) allowed the living polymerization of 4-methyl-1-pentene to furnish atactic poly(4-methyl-1-pentene) (Gendler *et al.*, 2006).

Kol and coworkers also reported in 2000 the synthesis and polymerization behavior of the C_2-symmetric, ethylene-bridged zirconium analog (**24a**, Figure 9.4) of **22a** (Tshuva *et al.*, 2000a). At room temperature, **24a**/$B(C_6F_5)_3$ furnished highly isotactic poly(1-hexene) and poly(1-octene). The poly(1-hexene)s exhibited narrow polydispersities ($M_w/M_n = 1.11–1.15$) with $M_n = 4000–12\,000$ g/mol; the M_n was shown to increase linearly with monomer consumption. Reducing the substituent size on the phenoxide moiety (**24b**) resulted in atactic poly(1-hexene) with a broadened molecular-weight distribution ($M_w/M_n = 1.57$). In a subsequent report, it was shown that replacing the *ortho*- and *para-tert*-butyl substituents of **24a** with chlorides (**24c**) resulted in greatly diminished living behavior (Segal *et al.*, 2005). Importantly, the titanium congener (**24d**) and the analogous 2,4-dibromophenol-bearing complex (**24e**) polymerized 1-hexene in a living manner for a period of 40 to 75 min when activated with $B(C_6F_5)_3$. The poly(1-hexene)s exhibited extremely high molecular weights (M_n up to 1 750 000 g/mol, $M_w/M_n \leq 1.2$) and moderate degrees of isotacticity (**24d**: [*mm*] = 0.60, **24e**: [*mm*] = 0.80).

A third class of novel olefin polymerization catalysts featuring [ONOO] Group 4 metal complexes bearing a methoxy side-arm donor (**25**) was introduced by Kol and coworkers in 2001 (Tshuva *et al.*, 2001). At room temperature, **25**/$B(C_6F_5)_3$ (Figure 9.4) furnished poly(1-hexene)s with narrow polydispersities ($M_w/M_n = 1.07–1.12$) and high molecular weights (M_n up to 445 000 g/mol). A linear increase in M_n with increasing reaction time was observed for up to 31 h. The living character of the polymerization was maintained upon heating to 65 °C for one hour, as evidenced by the narrow polydispersity of the resultant polymer ($M_w/M_n = 1.30$). Kol and coworkers were able to apply this catalyst system to the synthesis of block copolymers of 1-hexene and 1-octene (Tshuva *et al.*, 2001; Groysman *et al.*, 2003b). Using **25**/$B(C_6F_5)_3$, a block copolymer of 1-hexene and 1-octene was prepared via sequential monomer addition where each domain had an atactic microstructure. The polymer possessed a narrow molecular-weight distribution ($M_w/M_n = 1.2$) and a $M_n = 11\,600$ g/mol. Further modification of the catalyst system through introduction of the 2,4-dimethyl- or 2,4-dichlorophenoxide moiety led to a loss of living character (Groysman *et al.*, 2004). Additionally, the zirconium and hafnium analogs of **25** were also shown to deviated from living behavior ($M_w/M_n = 1.4–3.0$) (Tshuva *et al.*, 2002).

The effect of the neutral oxygen donor's identity on the polymerization behavior of the [ONOO] titanium complexes has also been investigated. Substituting the methoxy donor of **25** with a THF-moiety (**26a**, Figure 9.4) led to similar polymerization results, however, replacing the benzyl ligands with methyl ligands (**26b**) results in a dramatic increase in the duration of the living period for up to 6 days at room temperature upon activation with $B(C_6F_5)_3$. The resultant atactic poly(1-hexene) had a M_n up to 816 000 g/mol and $M_w/M_n = 1.04–1.12$ (Groysman *et al.*, 2003b). The extremely long-lived catalyst generated from **26a** (Figure 9.4) was used to prepare a block copolymer of 1-hexene and 1-octene through sequential monomer addition to furnish poly(1-hexene)-*block*-poly(1-octene). The block copolymer had $M_n = 34\,000$ g/mol, while maintaining the low $M_w/M_n = 1.16$.

Further modification of the system through introduction of a furan donor (**27**, Figure 9.4) into the ligand framework led to a 10-fold increase in polymerization activity relative to **26a**/$B(C_6F_5)_3$ furnishing poly(1-hexene)s of high molecular weight (M_n up to 500 000 g/mol, $M_w/M_n \leq 1.37$) (Groysman *et al.*, 2003a). The increase in activity of **27**/$B(C_6F_5)_3$ resulted in diminished living character of the 1-hexene polymerization exhibiting a linear increase in M_n over the course of only two hours.

9.3.6.2 Propylene Polymerization

Busico and coworkers have investigated the propylene polymerization behavior of Kol's octahedral [ONNO] zirconium complexes **24a,b** (Figure 9.4) (Busico *et al.*, 2001). In contrast to the living 1-hexene polymerization observed for **24a**/$B(C_6F_5)_3$, the polypropylenes produced by **24a** and **24b**/$[PhNMe_2H][B(C_6F_5)_4]$/Al(iBu)$_3$ showed evidence of termination via chain transfer to Al and β-H transfer to monomer. Utilizing **24a**, Busico and coworkers reported the preparation of a diblock copolymer of *i*PP and PE under "quasiliving" conditions in 2003 (Busico *et al.*, 2003). Using **24a**/$[PhNMe_2H][B(C_6F_5)_4]$ with 2,6-di-*tert*-butylphenol modified Al(iBu)$_3$ as scavenger, a diblock copolymer of ethylene and propylene was prepared by sequential monomer addition of ethylene (1.5 min) and propylene (20 min). The resultant copolymer possessed a narrow polydispersity (M_w/M_n as low as 1.2 when $M_n = 6500$ g/mol). These results represented the first synthesis of an *i*PP-*block*-PE copolymer via sequential monomer addition at polymerization durations greater than one minute.

In a subsequent report, Busico and coworkers reported that modification of the ligand framework resulted in the controlled polymerization of propylene with this class of catalysts (Busico *et al.*, 2004). By installing bulky 1-adamantyl (**28a**) or cumyl (**28b**) substituents at the *ortho*-position of the phenol moiety (Figure 9.4), PPs with narrow molecular-weight distributions were obtained ($M_w/M_n = 1.2$–1.6) under the same activation procedure. For **28a**/$[PhNMe_2H][B(C_6F_5)_4]$/Al(iBu)$_3$, a linear increase of M_n with time is observed over the course of 3 h, however, after three hours resonances consistent with terminal vinylidene groups were apparent in the ^{13}C NMR spectrum. The resultant PP was highly isotactic ([*mmmm*] = 0.985, $T_m = 151$ °C). Chain transfer to aluminum was suppressed by the addition of 2,6-di-*tert*-butylphenol. Utilizing sequential monomer addition of ethylene and propylene, **28a**/$[PhNMe_2H][B(C_6F_5)_3]$/Al(iBu)$_3$ with 2,6-di-*tert*-butylphenol were employed in the synthesis of a diblock copolymer. The resultant *i*PP-*block*-PE displayed higher molecular weight ($M_n = 22\,000$ g/mol, $M_w/M_n = 1.3$) and T_m of the *i*PP block (152 °C) than the analogous block copolymer obtained from **24a** (Busico *et al.*, 2004).

In 2009, Busico and coworkers reported the polymerization behavior of **29a,b** (Figure 9.4), the hafnium analogs of **28a,b**, which were both found to be living for the polymerization of ethylene and propylene (Cipullo *et al.*, 2009). In the presence of propylene, **29b**/MAO/2,6-di-*tert*-butylphenol produced isotactic polypropylene ([*mmmm*] = 0.970) with increasing molecular weight ($M_n = 6200$–$13\,900$ g/mol) over the course of 9 h while maintaining narrow molecular-weight distributions ($M_w/M_n = 1.3$–1.5). In addition to propylene, **29b**/MAO/2,6-di-*tert*-butylphenol was found to be living for the polymerization of ethylene over the course of 4 h, a significant improvement relative to **28a**. Utilizing this improved living behavior, **29b**/MAO/2,6-di-*tert*-butylphenol was used to prepare a triblock copolymer. The resultant *i*PP-*block*-poly(ethylene-*co*-propylene)-*block*-*i*PP possessed narrow polydispersity ($M_w/M_n = 1.2$, $M_n = 22\,000$ g/mol) while maintaining a high T_m (143 °C).

9.3.7 *Bis(phenoxyimine)titanium Catalysts*

9.3.7.1 Propylene Polymerization

In 1999, Fujita and coworkers reported on a class of Group IV complexes bearing chelating phenoxyimine ligands, including **30a** (Figure 9.5). When activated with MAO, these complexes showed extremely high activity for ethylene polymerization (Matsui *et al.*, 1999a; Matsui *et al.*, 1999b; Matsui *et al.*, 2000). Interested in the development of catalysts that could produce stereoregular polymers, Coates and coworkers used a pooled combinatorial approach to screen bis(phenoxyimine)titanium complexes for propylene polymerization behavior. Despite the C_2-symmetry of the catalyst precursor, **30b**/MAO furnished syndiotactic PP ($[r] = 0.94$) resulting from a chain-end control mechanism (Tian and Coates, 2000). Later, several studies revealed an unusual 2,1-insertion mechanism (Saito *et al.*, 2001c; Hustad *et al.*, 2002; Lamberti *et al.*, 2002). In addition, calculations on the system have supported a ligand isomerization event that interconverts the Λ and Δ isomers of the active species between consecutive insertions, causing an alternation between *si* and *re* coordination of propylene that leads to syndiotactic polymer formation (Milano *et al.*, 2002; Talarico *et al.*, 2003; Corradini *et al.*, 2004).

It was later found that the incorporation of fluorinated *N*-aryl moieties into the bis (phenoxyimine) ligand framework could provide catalyst precursors for the syndioselective and living polymerization of propylene. At 0 °C, **31**/MAO (Figure 9.5) produced highly syndiotactic PP ($[rrrr] = 0.96$) that had a peak melting temperature of 148 °C (Tian *et al.*, 2001). The polymerization exhibited a linear increase in M_n with PP yield while polydispersities remained narrow ($M_w/M_n \leq 1.11$) for M_n up to 100 000 g/mol. Additionally, **31**/MAO has been shown to produce *s*PP-*block*-poly(E-*co*-P) diblock copolymers of high molecular weight

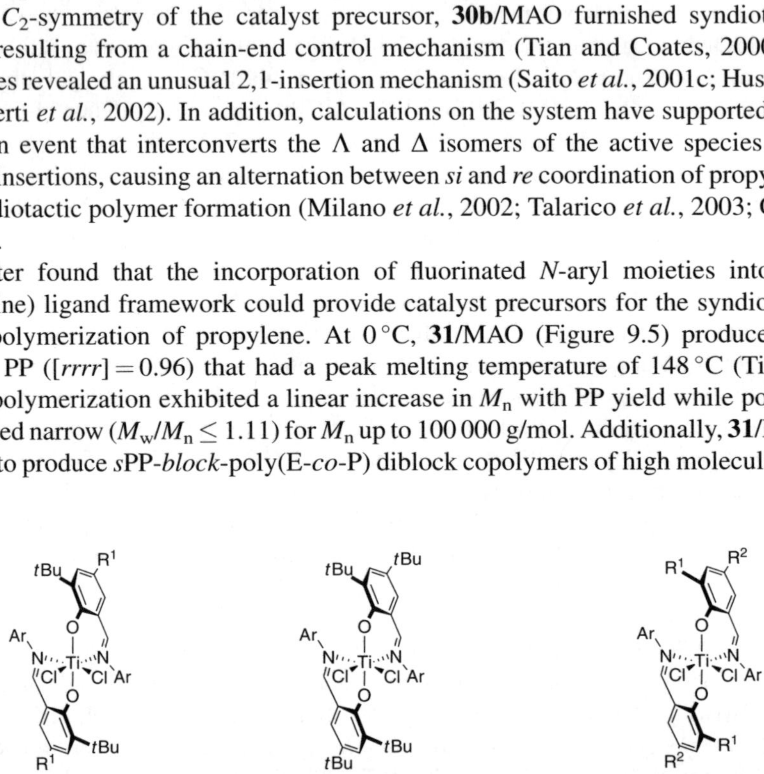

	R¹	Ar
30a	H	C_6H_5
32	H	C_6F_5
39a	Me	C_6H_5
39b	Me	Cy
41a	H	2-FC_6H_4
41b	H	2,6-$F_2C_6H_3$
41c	H	2,4,6-$F_3C_6H_2$
42b	H	2,4,6-$F_3C_6H_2$

	Ar
30b	C_6H_5
31	C_6F_5
33a	2-FC_6H_4
33b	2,6-$F_2C_6H_3$
33c	2,4,6-$F_3C_6H_2$
33d	2,3,5,6-F_4C_6H
33e	2,3,5,6-F_4-4-CF_3-C_6
38	C_6H_5,C_6F_5

	R¹	R²	Ar
34	$SiMe_3$	H	C_6F_5
35a	$SiEt_3$	H	C_6F_5
35b	Me	H	C_6F_5
35c	*i*Pr	H	C_6F_5
36	I	I	2,6-$F_2C_6H_3$
37a	Me	Me	3,5-$F_2C_6H_3$
37b	*i*Pr	H	3,5-$F_2C_5H_3$
40a	Cy	H	C_6F_5
40b	H	H	C_6F_5
42a	CMe_2Ph	Me	2,4,6-$F_3C_6H_2$

Figure 9.5 Bis(phenoxyimine) titanium complexes.

($M_n = 145\,100$ g/mol, $M_w/M_n = 1.12$) through sequential monomer addition (Tian *et al.*, 2001). Several studies on the physical properties of sPP-*block*-poly(E-*co*-P) diblock copolymers made using **31**/MAO have been conducted, including those involving the morphology (Ruokolainen *et al.*, 2005), thermodynamic behavior, and self-assembly (Radulescu *et al.*, 2004) of the materials. In subsequent work, the addition of a third block was employed in the formation of a *s*PP-*block*-poly(E-*co*-P)-*block*-*s*PP triblock copolymer (Hotta *et al.*, 2006). TEM revealed that the polymer exhibited a microphase-separated morphology with *s*PP cylinders in a poly(E-*co*-P) matrix. Tensile testing revealed a strain to break of about 550%.

Fujita and coworkers independently reported that **32**/MAO (Figure 9.5) was also living and syndioselective ([*rr*] = 87%) for propylene polymerization at room temperature, producing polymer with $M_n = 28\,500$–$108\,000$ g/mol and $M_w/M_n = 1.07$–1.14 (Saito *et al.*, 2001a). Furthermore, employing a supported cocatalyst ($MgCl_2/i\text{-}Bu_nAl(OCH_2CH(Et)(CH_2)_3CH_3)_{3-n}$) with **32** has also shown characteristics of living behavior with the polypropylene formed exhibiting narrow PDIs ($M_w/M_n = 1.09$–1.17, $M_n = 53\,000$–$132\,000$ g/mol) (Nakayama *et al.*, 2005). Living ethylene/propylene copolymerization and block copolymer formation has also been demonstrated with **32**/MAO (Kojoh *et al.*, 2001). Exploiting this living behavior, *s*PP-*block*-poly(E-*co*-P), PE-*block*-*s*PP, and PE-*block*-poly(E-*co*-P)-*block*-*s*PP have been prepared through sequential monomer addition (Mitani *et al.*, 2002b; Sakuma *et al.*, 2007).

Studies on the effect of the fluorination pattern of the *N*-aryl ring revealed that complexes bearing the 2,4-di-*tert*-butyl phenoxide moiety require at least one *ortho* fluorine on the *N*-aryl ring to exhibit living propylene polymerization behavior (Mason *et al.*, 2002; Mason and Coates, 2004b). As the amount of fluorination of the *N*-aryl moiety is decreased from the perfluoro complex **31** to the monofluoro complex **33a** (Figure 9.5), activities and tacticities for propylene polymerization also decreased, while the polydispersities remain consistently low ($M_w/M_n \leq 1.11$, M_n up to 28 900 g/mol) upon MAO activation. Installing a trifluoromethyl group at the *para*-position of the *N*-aryl moiety (**33e**, Figure 9.5) led to an increase in activity of approximately 1.5 times that of **31**/MAO with similar tacticity ([*rrrr*] = 0.91) (Mason, 2005). Interestingly, complexes related to **33a–c** where the *para*-substituent of the phenoxide moiety is H, produced amorphous PP upon MAO activation. These samples gave bimodal GPC traces each composed of a narrow peak ($M_w/M_n \leq 1.10$) and a broad peak ($M_w/M_n = 4.19$–14.9). (Saito *et al.*, 2003).

Modification of the *ortho* substituents on the phenolate ring has yielded a number of new complexes. Of note, a complex bearing a *ortho*-phenolate trimethylsilyl group, **34** (Figure 9.5), has been shown to produce *s*PP with very high melting temperatures (T_m up to 156 °C) in a living fashion (Mitani *et al.*, 2002a). Changing the aforementioned *ortho* position to a larger triethylsilyl group (**35a**) gave similar results as **34** with lower activity (Mitani *et al.*, 2003a). Decreasing sterics through employment of a methyl (**35b**) or isopropyl (**35c**) group in the *ortho* position resulted in a substantial loss of stereocontrol producing amorphous PP that exhibited fairly narrow polydispersities ($M_w/M_n \sim 1.2$).

In further modifications, **36** (Figure 9.5) containing a 2,6-$F_2C_6H_3$*N*-aryl moiety and iodine substituents on the phenolate ring was synthesized. At 25 °C, **36**/MAO produced amorphous PP that exhibited a narrow molecular weight distribution ($M_w/M_n = 1.17$, $M_n = 200\,000$ g/mol) (Weiser and Mulhaupt, 2006; Weiser *et al.*, 2006b). A PE-*block*-poly(E-*co*-P) diblock copolymer was synthesized using **36**/MAO with $M_n = 2\,000\,000$ g/mol and a broad molecular-weight distribution ($M_w/M_n = 1.60$) (Weiser and Mulhaupt, 2006; Weiser *et al.*, 2006a). Additionally, complexes **37a,b** employing 3,5-difluorophenyl *N*-aryl groups and substituents smaller than *tert*-butyl in the *ortho* position of the phenoxide moiety furnished amorphous PP ([*rrrr*] $\leq$ 0.48) with narrow molecular-weight distributions ($M_w/M_n = 1.13$–1.16, M_n up to 240 000 g/mol) upon MAO activation (Cherian *et al.*, 2005a). This finding was surprising in that both complexes lack *ortho* fluorines on the *N*-aryl moiety.

In a final variation, two different phenoxyimine ligands were coordinated to one titanium center. Using gel-permeation chromatography as a combinatorial screening method, a number of heteroligated phenoxyimine complexes bearing one nonliving (*ortho*-nonfluorinated ligand) and one living ligand (*ortho*-fluorinated ligand) displayed superior activities over their homoligated counterparts (Mason and Coates, 2004b). For example, PP produced with **30b**/MAO exhibited a broad PDI ($M_w/M_n = 1.41$) and a turn-over frequency (TOF) of 42 h^{-1}, while **31**/MAO exhibited a narrow PDI ($M_w/M_n = 1.06$) and a TOF of 221 h^{-1}. Alternatively, the heteroligated catalyst **38**/MAO (Figure 9.5) produced syndiotactic PP ([*rrrr*] = 0.91]) with $M_n = 70\,200$ g/mol and $M_w/M_n = 1.16$ and exhibited a TOF of 760 h^{-1}.

9.3.7.2 Ethylene Polymerization

Many of the same titanium bis(phenoxyimine) catalysts used for living propylene polymerization have also been reported for the living polymerization of ethylene. In 2001, Fujita and coworkers found that **32**/MAO at 25 °C furnished linear PE with high molecular weight and narrow molecular-weight distribution ($M_n = 412\,000$ g/mol, $M_w/M_n = 1.13$) (Saito *et al.*, 2001b). Furthermore, polymerizations at 25 and 50 °C exhibited a linear increase in M_n with reaction time. Using **32**/MAO (Figure 9.5), a PE-*block*-poly(E-*co*-P) diblock and a PE-*block*-poly(E-*co*-P)-*block*-PE triblock copolymer have been synthesized through sequential monomer addition (Mitani *et al.*, 2002b). It was also found that addition of an equimolar amount of functionalized α-olefin, $H_2C{=}CH(CH_2)_n$-Y (Y = $OAlMe_2$, $n = 4$ and Y = $OSiMe_3$, $n = 9$), to **32**/MAO and subsequent living ethylene polymerization furnished hydroxyl terminated PEs upon acidic workup (Makio and Fujita, 2007). This strategy was also successful for the production of hydroxyl-terminated syndiotactic PP from **32**/MAO. Futhermore, addition of the aforementioned functionalized α-olefins as a chain-end-capping agent furnished telechelic syndiotactic PPs bearing hydroxyl groups at both chain ends upon acidic workup.

Utilizing a related complex, Coates and coworkers reported that **30b**/MAO at 50 °C polymerized ethylene to produce polyethylene with $M_w/M_n = 1.10$ and $M_n = 44\,500$ g/mol (Reinartz *et al.*, 2003). Additionally, Ivanchev *et al.* reported a near-linear increase in M_v with time for the polymerization of ethylene with **30a**/MAO at 30 °C (Ivanchev *et al.*, 2004). In later work on the same system, Fujita and coworkers found that molecular-weight distributions were low at a reaction time of one minute ($M_w/M_n = 1.12$, $M_n = 52\,000$ g/mol) while the PDI broadened significantly at reaction times of just 5 min ($M_w/M_n = 1.61$, $M_n = 170\,000$ g/mol) (Furuyama *et al.*, 2005b). Related complexes **39a,b**/MAO (Figure 9.5) in the presence of ethylene also showed a near-linear increase in M_v with times up to about 20 min (Ivanchev *et al.*, 2005).

The effect of the substitution at the *ortho* position of the phenoxide moiety was investigated utilizing complexes **35b,c** and **40a,b** (Figure 9.5) (Furuyama *et al.*, 2005a). When activated with MAO at 25 °C, each complex produced PE with a narrow molecular-weight distribution ($M_w/M_n = 1.05$–1.16, M_n up to 75 000 g/mol), however, reaction times were kept to one minute. While all the catalysts were living, activities were about an order of magnitude less than **32**/MAO.

Having shown that **32**, **35b**, **40a,b**/MAO (Figure 9.5) were living for ethylene polymerization, Fujita and coworkers investigated the ability of these catalysts to produce ethylene/α-olefin copolymers in a living fashion (Furuyama *et al.*, 2005a). Copolymerizations with ethylene and either 1-hexene, 1-octene, or 1-decene were carried out with each catalyst at 25 °C. In all cases, polymers with narrow molecular-weight distributions were obtained ($M_w/M_n \leq 1.22$). As the steric bulk of the *ortho*-substituent decreased, increased α-olefin incorporation was observed. A series of PE-*block*-poly(E-*co*-1-hexene) samples were produced using **35b**/MAO via sequential monomer addition. Molecular-weight distributions for the block copolymers were generally low ($M_w/M_n = 1.11$–1.31 for M_n up to 121 000 g/mol) and 1-hexene contents of up to 28.9 mol% were estimated.

The role of *N*-aryl fluorination on ethylene polymerization behavior has also been explored. Each of the catalysts **41a–c**/MAO (Figure 9.5) has been shown to be well behaved for ethylene polymerization at 50 °C and **41a**/MAO and **41b**/MAO produced polymer with narrow molecular-weight distributions at reaction times between 1 and 5 min ($M_w/M_n \sim 1.05$, $M_n = 13\,000$–$64\,000$ g/mol) (Makio *et al.*, 2002; Mitani *et al.*, 2002b).

Finally, Fujita and coworkers reported that $ZnEt_2$ could be used as a chain-transfer agent in the living ethylene polymerization employing **42a**/MAO (Figure 9.5) leading to zinc end-functionalized chains and a titanium species that reinitiates living ethylene polymerization upon the addition of monomer (Mitani *et al.*, 2003b). Despite being living for ethylene polymerization, **42b** was no longer living in the presence of $ZnEt_2$.

9.3.7.3 Polymerization of Cyclopentene, Norbornene and 1,5-Hexadiene

Bis(phenoxyimine) titanium complexes have also been employed in the living copolymerization of ethylene/cyclic olefins and ethylene/1,5-hexadiene. Utilizing **31**/MAO and varying ethylene pressure, a series of poly(E-*co*-CP)s with different cyclopentene contents were prepared (Fujita and Coates, 2002). When ethylene pressure was low (<1 psi), an almost perfectly alternating copolymer was formed ($M_n = 21\,000$ g/mol, $M_w/M_n = 1.34$, $T_g = 10.1$ °C). However, the use of higher ethylene pressures (3 psi) resulted in the formation of a random copolymer containing 36 mol% cyclopentene ($M_n = 133\,000$ g/mol, $M_w/M_n = 1.24$, $T_g = -4.5$ °C). Tri- and multi-block copolymers were synthesized in which the constituent blocks differed in their cyclopentene content.

In addition to cyclopentene, copolymers from ethylene and norbornene have also been made using **31**/MAO (Yoon *et al.*, 2006). With this catalyst, a high molecular weight, low PDI poly(E-*co*-NB) sample was prepared ($M_n = 238\,000$ g/mol, $M_w/M_n = 1.05$) containing 62 mol% ethylene and a T_g of 86.5 °C. In addition, **31**/MAO was also used to synthesize a high molecular weight poly(E-*co*-P)-*block*-poly(E-*co*-NB) sample ($M_n = 576\,000$ g/mol, $M_w/M_n = 1.13$).

Finally, Coates and coworkers found that bis(phenoxyimine) titanium complex **31** was also capable of living 1,5-hexadiene polymerization and 1,5-hexadiene/propylene copolymerization (Hustad and Coates, 2002). Homopolymerization of 1,5-hexadiene with **31**/MAO at 0 °C produced a high molecular weight polymer with a narrow PDI ($M_n = 268\,000$ g/mol, $M_w/M_n = 1.27$). The polymer showed the presence of two distinct units – the expected MCP units as well as 3-vinyl tetramethylene (3-VTM) units. As shown in Scheme 9.4, the MCP units are proposed to arise from 1,2-insertion of 1,5-hexadiene followed by a 1,2-cyclization. However, an initial 2,1-insertion of 1,5-hexadiene followed by a 1,2-cyclization forms a

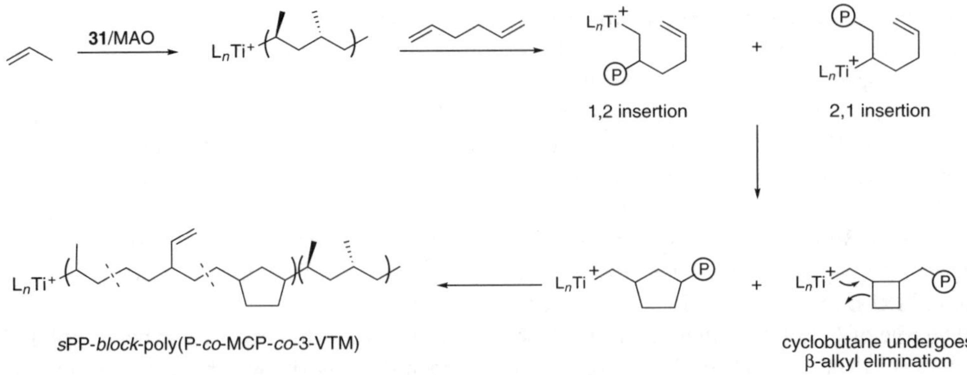

Scheme 9.4 Synthesis of propylene/1,5-hexadiene block copolymers using **31**/MAO.

strained cyclobutane species. After a β-alkyl elimination, the 3-VTM unit is generated. Additionally, propylene/1,5-hexadiene copolymers with high molecular weights were also produced ($M_n = 119\,000$–$145\,000$ g/mol, $M_w/M_n = 1.09$–1.16). Through sequential monomer addition, a *s*PP-*block*-poly(P-*co*-MCP-*co*-3-VTM) diblock copolymer was synthesized with **31**/MAO (Scheme 9.4). The molecular-weight distribution of the block copolymer was low ($M_w/M_n = 1.11$, $M_n = 93\,300$ g/mol) and contained 4.3 mol% MCP units and 2.7 mol% 3-VTM units. A poly(E-*co*-P)-*block*-poly(MCP-*co*-3-VTM) was also synthesized ($M_n = 524\,700$ g/mol, $M_w/M_n = 1.13$). Lastly, **31**/MAO has been used to produce a high molecular weight poly(MCP-*co*-3-VTM)-*block*-poly(E-*co*-NB) sample with $M_n = 451\,000$ g/mol and $M_w/M_n = 1.41$ (Yoon *et al.*, 2006).

9.3.8 Bis(phenoxyketimine)titanium Catalysts

While bis(phenoxyimine) titanium complexes furnish sPP, it had been proposed that placing a substituent at the imine carbon of the ligand could prevent the isomerization responsible for the production of sPP and lead to the formation of *i*PP (Mason and Coates, 2004a). Ketimine complexes **43a–c** (Figure 9.6) were synthesized and found to be sparingly active for propylene polymerization, despite the ability to polymerize ethylene in a living fashion upon activation (Reinartz *et al.*, 2003; Mason and Coates, 2004a). Therefore, complexes bearing smaller *ortho* substituents (**44a–d**) were obtained and screened for propylene polymerization (Mason and Coates, 2004a). Upon activation with MAO at 0 °C, each complex produced PP with a narrow molecular-weight distribution ($M_w/M_n = 1.12$–1.17, $M_n = 2700$–$35\,400$ g/mol) and **44c**/MAO was shown to exhibit a linear increase in M_n as a function of yield. The resulting polymers displayed a variety of tacticities with **44c**/MAO furnishing PP with the highest tacticity ([*mmmm*] = 0.53, $T_m = 69$ °C). As a comparison, the analogous aldimine of **44c** (**44e**) in which $R^3 = H$, furnishes atactic PP with $M_n = 123\,100$ g/mol and $M_w/M_n = 1.13$. Further exemplifying the living nature of **44c**/MAO, an *i*PP-*block*-poly(E-*co*-P) sample was produced. After polymerization of propylene to form an *i*PP block of $M_n = 28\,100$ g/mol ($M_w/M_n = 1.10$), ethylene was added to the reaction yielding a diblock copolymer with $M_n = 62\,000$ g/mol and $M_w/M_n = 1.10$.

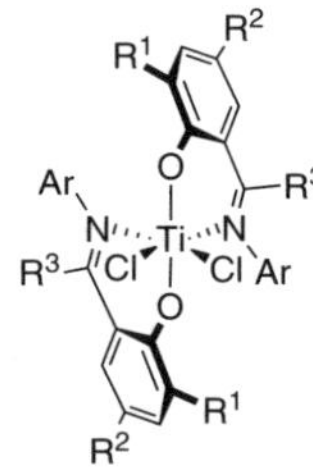

	R^1	R^2	R^3	Ar
43a	*t*Bu	*t*Bu	Me	C_6H_5
43b	*t*Bu	*t*Bu	Et	C_6H_5
43c	*t*Bu	*t*Bu	Ph	C_6H_5
44a	*i*Pr	*i*Pr	C_6H_5	C_6F_5
44b	Et	Me	C_6H_5	C_6F_5
44c	Me	Me	C_6H_5	C_6F_5
44d	H	Me	C_6H_5	C_6F_5
44e	Me	Me	H	C_6F_5
45a	Cl	Me	C_6H_5	C_6F_5
45b	Br	Me	C_6H_5	C_6F_5
45c	Ph	Me	C_6H_5	C_6F_5
45d	Bn	Me	C_6H_5	C_6F_5
46	*t*Bu	H	Et	C_6H_5

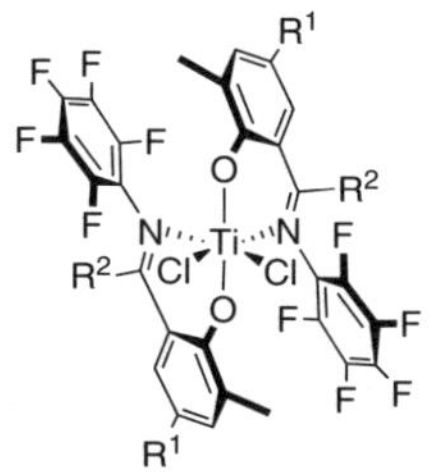

	R^1	R^2
45e	Cl	C_6H_5
45f	F	C_6H_5
45g	OMe	C_6H_5
45h	Me	Me
45i	Me	*i*Pr
45j	Me	Cy
45k	Me	Cycloheptyl
45l	Me	CF_3
45m	Me	2-Napthyl
45n	Me	1-Napthyl
45o	Me	4-Methoxyphenyl
45p	Me	Mesityl
45q	Me	C_6F_5

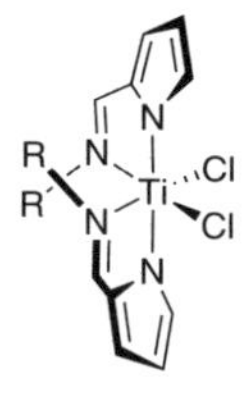

47a R = Ph
47b R = Cy
47c R = Cyclooctyl
47d R = 4-*t*Bu-Cy

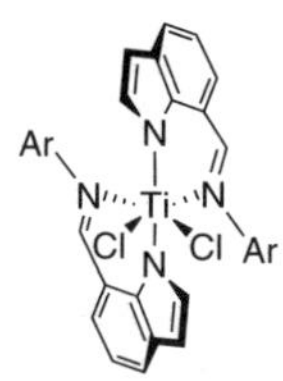

48a Ar = Ph
48b Ar = 2,6-F_2-C_6H_3
48c Ar = 2,4,6-F_3-C_6H_2
48d Ar = C_6F_5

Figure 9.6 Bis-ligated titanium catalyst precursors.

To obtain higher isoselectivity, Coates and coworkers systematically varied *ortho-*, *meta-*, and *para*-substituents on the phenoxide moiety in addition to the ketimine substituent in complexes **45a–q** (Figure 9.6) (Edson *et al.*, 2008). All the complexes produced PP with a narrow molecular-weight distribution ($M_w/M_n = 1.07$–1.33, $M_n = 3000$–$364\,000$ g/mol) upon activation with MAO except those bearing ancillary methoxy groups in the ligand framework (**45g** and **45o**). The tacticities of the resulting polymers differed with **45j**/MAO producing PP with the highest tacticity ([*mmmm*] = 0.73, $T_m = 116.8\,°C$) in addition to exhibiting a linear increase in M_n as a function of polymer yield. Exploiting the living nature of the catalyst, block copolymers were synthesized with **45j**/MAO via sequential monomer addition (Edson *et al.*, 2008). Specifically, an *i*PP-*block*-poly(E-*co*-P)-*block*-*i*PP, *i*PP-*block*-poly(E-*co*-P)-*block*-*i*PP-*block*-poly(E-*co*-P)-*block*-*i*PP, and *i*PP-*block*-poly(E-*co*-P)-*block*-*i*PP-*block*-poly(E-*co*-P)-*block*-*i*PP-*block*-poly(E-*co*-P)-*block*-*i*PP copolymer were prepared. The polymers had narrow molecular-weight distributions ($M_w/M_n = 1.13$–1.30) and high molecular weights ($M_n = 102\,000$–$235\,000$ g/mol). Testing of the tensile properties of the block copolymers showed good elastomeric behavior with the triblock copolymer displaying an elongation to break of 1000%.

In addition to propylene, Coates and coworkers reported that **43a–c** when activated with MAO at 0 and 20 °C all produced PE that exhibited a narrow molecular-weight distribution ($M_w/M_n \leq 1.08$) and had number-average molecular weights ($M_n = 15\,000$–$47\,000$ g/mol) that coincided with M_n^{theo} (Reinartz *et al.*, 2003). A linear increase in M_n with polymer yield for the polymerization catalyzed by **43c**/MAO at 0 °C and for **43b**/MAO at 50 °C was demonstrated.

A related complex (**46**, Figure 9.6), when activated with MAO at 50 °C, produced PE with $M_w/M_n = 1.08$ ($M_n = 9000$ g/mol) (Furuyama *et al.*, 2005b).

9.3.9 *Bis(pyrrolide-imine)titanium Catalysts*

In 2000, Fujita and coworkers reported bis(pyrrolide-imine)titanium complexes for ethylene polymerization, however, living behavior was not observed (Yoshida *et al.*, 2000). As a result, Fujita and coworkers turned their attention to the copolymerization of ethylene and norbornene (Yoshida *et al.*, 2002; Yoshida *et al.*, 2003; Yoshida *et al.*, 2004). At 25 °C, **47a–d**/MAO (Figure 9.6) furnished poly(E-*alt*-NB) with narrow molecular-weight distributions and high molecular weights ($M_w/M_n = 1.10$–1.24, $M_n = 127\,000$–$600\,000$ g/mol). A linear increase of M_n with time over the course of 20 min was observed. The copolymers were found to contain 95.4% perfectly alternating units. The polymer chain-end structures were consistent with chain initiation by insertion of norbornene into the Ti—Me bond, and a last-inserted norbornene unit after termination by protonolysis. This suggests norbornene plays a stabilizing role for the active species against termination processes.

Utilizing **47b**/MAO (Figure 9.6), Fujita and coworkers were able to prepare block copolymers of the type poly(E-*co*-NB)$_x$-*block*-poly(E-*co*-NB)$_y$ with 7.6 mol% norbornene incorporation in the first block and 27.4 mol% norbornene overall (Yoshida *et al.*, 2004). Additionally, PE-*block*-poly(E-*co*-NB) was prepared through sequential monomer addition. The diblock copolymer exhibited a narrow polydispersity ($M_w/M_n = 1.56$) and $M_n = 414\,000$ g/mol with a norbornene content of 31 mol%.

9.3.10 *Bis(indolide-imine)titanium Catalysts*

Expanding on imine-ligated catalysts, Fujita and coworkers synthesized bis(indolide-imine) titanium complexes and evaluated their potential as ethylene polymerization catalysts (Matsugi *et al.*, 2001; Matsugi *et al.*, 2002a; Matsugi *et al.*, 2002b; Matsugi *et al.*, 2003; Yoshida *et al.*, 2005). At room temperature, compounds **48a–c**/MAO (Figure 9.6) furnished PE with narrow molecular-weight distributions ($M_w/M_n = 1.11$–1.23) and $M_n = 11\,000$ – $90\,000$ g/mol. At 25 °C, a linear increase of M_n with increasing polymer yield was observed for **48a–c**/MAO. Exhaustive fluorination of the *N*-aryl moiety (**48d**) resulted in living behavior at −10 °C ($M_w/M_n = 1.12$–1.15), while polymerization at 25 °C led to a broadened molecular-weight distribution ($M_w/M_n = 1.93$) (Matsugi *et al.*, 2002a; Matsugi *et al.*, 2002b). Using **48c**/MAO, PE-*block*-poly(E-*co*-P) copolymers were prepared containing 8.0 mol% propylene via sequential monomer addition with $M_w/M_n = 1.17$ and $M_n = 31\,400$ g/mol (Matsugi *et al.*, 2002b; Matsugi *et al.*, 2003). TEM visualization of the block copolymer revealed microphase separation of the poly(E-*co*-P) and PE domains, which were evenly dispersed throughout the sample.

9.3.11 *Bis(enaminoketonato)titanium Catalysts*

The synthesis and ethylene polymerization activity of bis(enaminoketonato)titanium complexes was reported by Li and coworkers (Li *et al.*, 2004). At 25 °C, **49a,b**/MMAO (Figure 9.7) furnish linear PEs with narrow molecular-weight distributions ($M_w/M_n = 1.25–1.45$) and $M_n = 51\,000$ g/mol–129 000 g/mol. At 25 °C, a linear increase in M_n with reaction time is observed with **49a**/MMAO. Mecking and coworkers reported *ortho*-fluorination on the *N*-aryl moiety (**50a**, Figure 9.7) furnished living and thermally robust ethylene polymerization catalysts upon MAO activation (Yu and Mecking, 2008). A linear increase in M_n over time was observed at 25, 50, and up to 75 °C. Nonliving behavior observed for **50b**/MAO supports the fact that the living behavior of **50a**/MAO is not steric in nature illustrating another example in which *ortho*-fluorination appears beneficial for living polymerization. Utilizing sequential monomer addition, a PE-*block*-*a*PP block copolymer with $M_n = 190\,000$ g/mol and $M_w/M_n = 1.12$ was produced with **50a**/MAO.

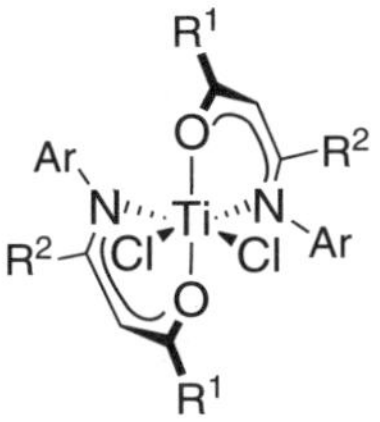

	R1	R2	Ar
49a	C_6H_5	CF_3	C_6H_5
49b	2-thiophenyl	CF_3	C_6H_5
49c	2-furanyl	CF_3	C_6H_5
49d	CF_3	CH_3	C_6H_5
49e	Ph	CH_3	C_6H_5
50a	CF_3	CH_3	2,6-$F_2C_6H_3$
50b	CF_3	CH_3	2,6-$(CH_3)_2C_6H_3$
50c	C_6H_5	CF_3	2-FC_6H_4
50d	C_6H_5	CF_3	2,4-$F_2C_6H_3$
50e	C_6H_5	CF_3	2,6-$F_2C_6H_3$
51	C_6H_5	CF_3	4-MeC_6H_4

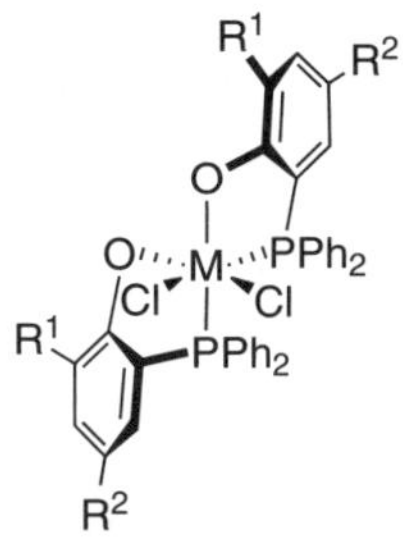

	M	R1	R2
52a	Ti	H	H
52b	Ti	Ph	H
52c	Ti	*t*Bu	H
52d	Zr	*t*Bu	H
52e	Hf	*t*Bu	H
52f	Ti	*t*Bu	*t*Bu
52g	Ti	$SiMe_3$	H

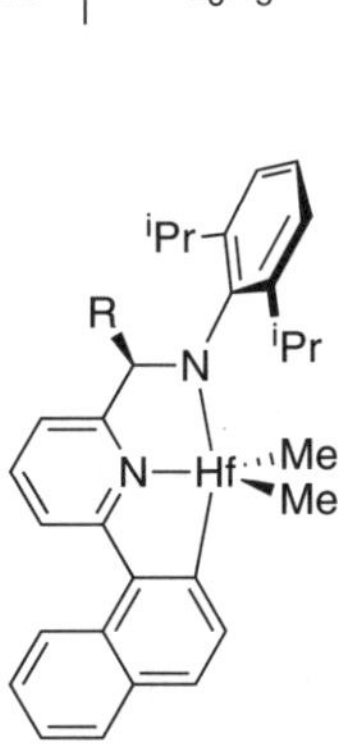

53 R = 2-*i*PrC_6H_4
54 R = H

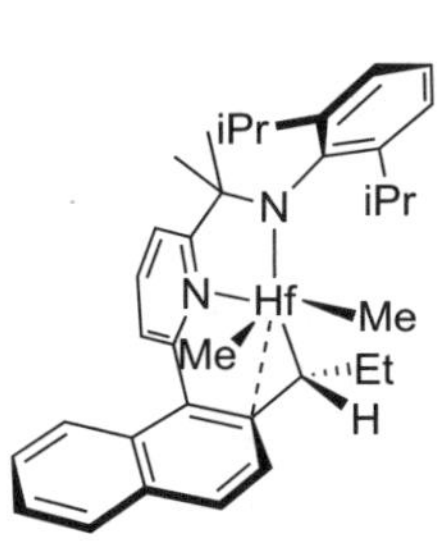

rac-**55**

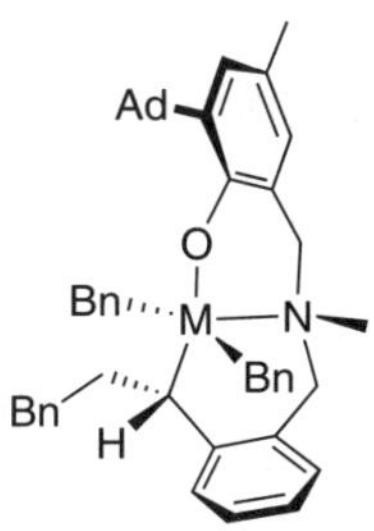

rac-**56a** M = Zr
rac-**56b** M = Hf

Figure 9.7 Early-metal precatalysts for olefin polymerization.

In addition to ethylene homopolymerization, the copolymerization of ethylene and norbornene by **49a–d**/MMAO (Figure 9.7) was also shown to possess some characteristics of a living polymerization (Li *et al.*, 2004). The polymers obtained from polymerizations conducted at 25 °C displayed narrow molecular-weight distributions ($M_w/M_n = 1.07–1.54$) with $M_n = 150\,000–580\,000$ g/mol. The norbornene content ranged from 35 to 55 mol%. At 25 °C, a linear increase in M_n with reaction time was demonstrated for **49a**/MMAO over the course of 20 min. Li and coworkers discussed the effects of further ligand modifications on the copolymerization behavior of bis(enaminoketonato)titanium catalysts in a subsequent report (Tang *et al.*, 2005b). At 25 °C, **51**/MMAO (Figure 9.7) produced poly(E-*co*-NB) with narrow polydispersities ($M_w/M_n = 1.18–1.31$, $M_n = 200\,000–570\,000$ g/mol) and norbornene content ranging from 45 to 48 mol%. The polymerization displayed a linear increase of M_n with time from $t = 5–20$ minutes. Utilizing the living nature **49a**/MAO, Li and coworkers prepared a PE-*block*-poly(E-*co*-NB) block copolymer with a narrow polydispersity ($M_w/M_n = 1.38$, $M_n = 143\,000$ g/mol) and a norbornene content of 11.1 mol% (Li *et al.*, 2004). Furthermore, the copolymerization of ethylene and cyclopentene (CP) by **49a**/MMAO and **49d**/MMAO was also shown to possess some characteristics of living behavior.(Tang *et al.*, 2005a) At temperatures from −10 to 30 °C, poly(E-*co*-CP)s with narrow molecular-weight distributions ($M_w/M_n = 1.23–1.82$) were produced. Additionally, a PE-*block*-poly(E-*co*-CP) was prepared in a nearly identical manner to the PE-*block*-poly(E-*co*-NB).

In addition to norbornene and cyclopentene, Li and coworkers found **49a**/MAO and **50a,c,d**/MAO (Figure 9.7) were active for the polymerization of ethylene and dicyclopentadiene at 25 °C (Pan *et al.*, 2009). Depending on the monomer feed, the resultant polymers contained up to 48 mol% dicyclopentadiene with a nearly alternating structure. Over 20 min, **49a**/MAO produced a copolymer with increasing molecular weights ($M_n = 47\,000$ g/mol) while maintaining narrow polydispersities ($M_w/M_n = 1.06–1.14$). Analysis of the ^{13}C NMR spectrum revealed a copolymer containing only unreacted cyclopentene units, suggesting the copolymerization proceeds through enchainment of the norbornene portion of the dicyclopentadiene exclusively. Utilizing sequential monomer addition, a PE-*block*-poly(ethylene-*alt*-dicyclopentadiene) was produced. Characterization using atomic force microscopy (AFM) and transmission electron micrography (TEM) revealed a microphase-separated material. Additionally, functionalized diblock copolymers were produced through reaction of the remaining alkene.

9.3.12 Bis(phosphanylphenoxide)titanium Catalysts

In 2006, Gibson and coworkers reported the synthesis of group 4 metal complexes (**52a–f**, Figure 9.7) featuring two bidentate ligands equipped with phenoxide and phosphine donors (Long *et al.*, 2006). At 25 °C, **52a–f**/MAO were found to be active for both the polymerization of ethylene and propylene although polydispersities were broadened ($M_w/M_n > 1.49$). In subsequent report, Li and coworkers synthesized **52g,h**. In the presence of ethylene, **52g**/MAO exhibited a nearly linear increase M_n over the first 10 min of the polymerization and the resultant polymers had relatively narrow molecular-weight distributions ($M_w/M_n = 1.31–1.33$) (He *et al.*, 2009).

To enhance the living nature of the system, Li and coworkers explored the copolymerization of norbornene and ethylene. They found that the copolymerization of ethylene and norbornene by **52g**/MAO resulted in polymers with narrow molecular-weight distributions ($M_w/M_n < 1.2$)

for norbornene incorporation of more than 10 mol% (He *et al.*, 2009). Additionally, the molecular weight was found to increase linearly with polymer yield. A poly(E-*co*-NB)$_x$-*block*-poly(E-*co*-NB)$_y$ containing 25 mol% norbornene in the first block and 44 mol% norbornene in the second block was produced. The resultant block copolymer exhibited an $M_n = 32\,000$ g/mol and a narrow molecular-weight distribution ($M_w/M_n = 1.17$).

9.3.13 Catalysts Supported by sp^2 and sp^3 Carbon Donors

One of the more recent classes of catalysts to emerge are the C_1-symmetric pyridylamidohafnium complexes (**53**, Figure 9.7) developed by Dow and Symyx that furnish high molecular weight and highly isotactic poly(α-olefin)s at high reaction temperatures upon activation (Boussie *et al.*, 2006; Froese *et al.*, 2007; Zuccaccia *et al.*, 2008). A C_s-symmetric pyridylamidohafnium complex (**54**, Figure 9.7) developed by Coates and coworkers produces isotactic poly(1-hexene) in a living fashion when activated with $B(C_6F_5)_3$ (Domski *et al.*, 2007a). The resultant poly(1-hexene)s exhibited narrow polydispersities ($M_w/M_n \leq 1.20$, M_n up to 152 000 g/mol) and the M_n was shown to increase linearly with monomer conversion. At 50 °C, the molecular-weight distribution of the polymer produced by **54**/$B(C_6F_5)_3$ remains narrow suggesting that living behavior is maintained at elevated temperatures.

Additionally, Coates and coworkers have shown that **54**/$B(C_6F_5)_3$ furnished isotactic PP ([*mmmm*] = 0.56) with a narrow molecular-weight distribution ($M_n = 68\,600$ g/mol, $M_w/M_n = 1.05$) at 20 °C (Domski *et al.*, 2007a). The mechanism of stereocontrol proceeded by an enantiomorphic site-control mechanism, which is quite unusual for a C_s-symmetric catalyst. Detailed mechanistic studies with **53** (Figure 9.7) (Froese *et al.*, 2007; Zuccaccia *et al.*, 2009) have shown the 1,2-insertion of an olefin into the Hf–C_{Aryl} bond generates an sp^3-hybridized carbon donor atom that supports the active metal center; it is likely that the isoselectivity observed with **54**/$B(C_6F_5)_3$ results from a similar activation mechanism. With this in mind, Coates and coworkers prepared a new pyridylamidohafnium complex (*rac*-**55**, Figure 9.7) supported by an sp^3-carbon donor that was generated as a mixture of diastereomers (61 : 39 ratio) (Domski *et al.*, 2008). Upon activation with $B(C_6F_5)_3$, *rac*-**55** furnished isotactic PP ([*mmmm*] = 0.80) with a narrow molecular-weight distribution ($M_w/M_n \leq 1.05$, M_n up to 124 400 g/mol) and TOF of 2800 h^{-1}. The M_n was shown to increase linearly with polymer yield over the course of 45 min.

Coates and coworkers were able to further exploit the area of olefin polymerization catalysts supported by sp^3-carbon donors through the synthesis and subsequent metallation of vinyl-appended phenoxyamine ligands (Edson *et al.*, 2009b). The resultant zirconium and hafnium complexes (*rac*-**56a** and *rac*-**56b**, Figure 9.7) bearing six-membered metallacycles were obtained as diastereo–isomeric mixtures. Upon activation, *rac*-**56a,b** formed highly active polymerization catalysts (TOF > 930 h^{-1}) that isoselectively polymerized both 1-hexene and propylene. At 0 °C, *rac*-**56b**/$B(C_6F_5)_3$ produced isotactic poly(1-hexene) in a living fashion. Molecular weights (M_n = 70 000–240 000 g/mol) increased linearly as a function of conversion, while polydispersities remained narrow (M_w/M_n = 1.10–1.15). Living 1-hexene polymerization behavior was not observed for the zirconium analog (*rac*-**56a**).

Alternatively, living propylene polymerization behavior was not observed for the hafnium analog (*rac*-**56b**), however, living isoselective polymerization of propylene was observed for the zirconium analog (*rac*-**56a**). At 0 °C, *rac*-**56a**/$B(C_6F_5)_3$ produced polypropylene with narrow polydispersities ($M_w/M_n = 1.17$–1.19) while molecular weights ($M_n = 60\,000$–$220\,000$ g/mol) were observed to increase linearly with propylene yield. Utilizing the living nature of *rac*-**56a**, *i*PP-*block*-poly(E-*co*-P) was synthesized by sequential monomer addition. The resultant polymer had a $M_n = 122\,000$ g/mol and a narrow molecular-weight distribution ($M_w/M_n = 1.20$).

9.3.14 Rare-Earth Metal Catalysts

In addition to Group 4 transition metals, there are examples of rare-earth metal complexes capable of producing block copolymers. Yasuda and coworkers described the synthesis of block copolymers containing polyethylene (insertion mechanism) with several polar monomers (noninsertion mechanism) such as MMA, methyl acrylate (MA), ethyl acrylate (EA), d-valerolactone (VL), and e-caprolactone (CL) using **57** and **58** (Figure 9.8) (Yasuda *et al.*, 1992). First, ethylene was polymerized ($M_w/M_n = 1.39$–2.01, $M_n = 6600$–$27\,000$ g/mol) followed by addition of the respective polar monomer to form a diblock copolymer, however, reversal of monomer addition led to no block copolymer formation. Thus, a PE-*block*-PMMA, PE-*block*-PMA, PE-*block*-PEA, PE-*block*-PVL, and PE-*block*-PCL were synthesized and showed good material properties such as deep coloration with dyes. In a later report, the structurally related **59** (Figure 9.8) was also shown to be a viable block copolymerization catalyst for ethylene, MMA, and CL (Desurmont *et al.*, 2000a). Furthermore, Yasuda and coworkers reported on the synthesis of diblock copolymers of poly(1-pentene)/poly(1-hexene) and PMMA/PCL using the structurally similar bridged Cp-bearing yttrium and scandium catalysts **60** and **61** (Figure 9.8) (Desurmont *et al.*, 2000c).

It had been previously shown that Cp^*_2Sm is active for ethylene homopolymerization involving coordination of ethylene by two Sm^{II} centers followed by electron transfer to form a telechelic initiator (**62**) (Evans *et al.*, 1995). Yasuda and coworkers cleverly applied this observation to the synthesis of PMMA-*block*-PE-*block*-PMMA, PCL-*block*-PE-*block*-PCL,

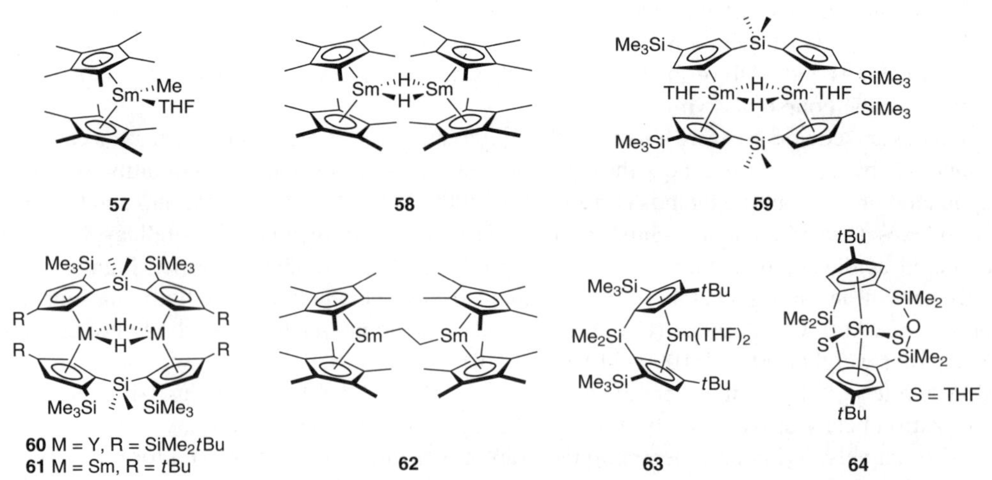

Figure 9.8 Rare-earth metal olefin polymerization precatalysts.

and PDTC-*block*-PE-*block*-PDTC (PDTC = poly(2,2-dimethyltrimethylene carbonate)) utilizing sequential monomer addition with **63** and **64** (Figure 9.8) (Desurmont *et al.*, 2000b).

9.4 Late-Metal Olefin Polymerization Catalysts

9.4.1 *Nickel and Palladium α-Diimine Catalysts*

9.4.1.1 Polymerization of α-Olefins

In the mid-1990s, Brookhart and coworkers reported the synthesis and olefin polymerization activity of nickel and palladium α-diimine complexes (Johnson *et al.*, 1995). These systems were unique among late-metal catalysts in their ability to produce high molar mass materials, rather than oligomers, from both ethylene and higher α-olefins. Furthermore, the metal centers were shown to migrate along polymer chains ("chain walking") (Guan *et al.*, 1999), allowing access to polyolefins with a wide variety of microstructures simply by varying ligand substitution patterns, temperature, or pressure. Shortly after the initial reports, **65** and **66** (Figure 9.9) were reported to polymerize 1-hexene and 1-octadecene in a living fashion (Killian *et al.*, 1996). Upon activation with MAO or MMAO at −10 °C and low monomer concentrations, **65** and **66** resulted in living systems furnishing polymers of M_n = 19 000 to 91 000 g/mol and narrow molecular weight distributions (M_w/M_n as low as 1.09). The systems were also shown to exhibit a linear increase in M_n with time. Branching density was less than that calculated for perfect sequential 1,2-insertions as a result of ω,1-enchainment, or "chain straightening" (Mohring and Fink, 1985). Poly(1-hexene) with as few as 118 branches/1000 carbons (vs. 167 expected for perfect 1,2-insertions) and poly(1-octadecene) with as few as 39 branches/1000 carbons (vs. 56 expected) were produced at −10 °C. Branching density was controlled by reaction conditions and catalyst structure, with **66** producing more linear polymers than **65**. Studying several palladium and nickel complexes, Merna *et al.* (Merna *et al.*, 2005; Merna *et al.*, 2009) reported living 1-hexene polymerization behavior with **65**/MAO and **67**/MAO. These values are considerably narrower than those reported by Brookhart for similar reactions at 23 °C, although the reason for the improvement is not clear.

A siloxy-substituted analog of **65** (**68**, Figure 9.9) was synthesized by Marques, Gomes and coworkers (Yuan *et al.*, 2005). At 16 °C, the polydispersity increases somewhat (M_w/M_n = 1.12–1.21), and the increase in molecular weight is only linear for the first 40 min. Upon changing the temperature, the polymer has a different microstructure, with branching greatly decreased at 16 °C relative to −10 °C (83 vs. 132 branches/1000 carbons, respectively).

Utilizing a cyclophane diimine ligand (**69**, Figure 9.9), Camacho and Guan reported the first living polymerization at elevated temperatures with nickel (Camacho and Guan, 2005). Up to 75 °C, **69**/MMAO furnished poly(1-hexene)s with narrow molecular weight distributions (M_w/M_n = 1.13–1.22) and branching densities (52–58 branches/1000 carbons) approximately one-half of those reported for **65**. The authors attribute the improved behavior to the cyclophane framework that very effectively blocks the axial sites of the nickel center preventing chain transfer.

Coates and coworkers employed the chiral, C_2-symmetric nickel diimine complex **70** (Figure 9.9) to control polymer microstructure for living polymerization of α-olefins (Rose *et al.*, 2006). The ability of nickel diimine catalysts to undergo successive β-hydride eliminations/reinsertions may lead to regioirregular polymers with less branching than expected (Ittel *et al.*, 2000). Careful tailoring of reaction conditions (low temperatures and

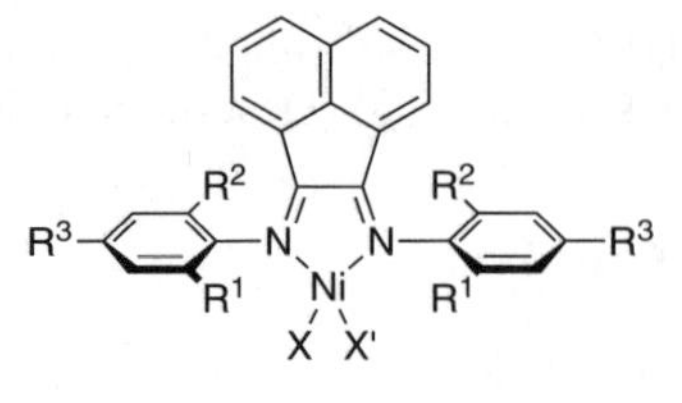

	M	R¹	R²	R³	X	X'
65	Ni	*i*Pr	*i*Pr	H	Br	Br
66	Ni	*t*Bu	H	H	Br	Br
67	Ni	Et	Et	H	Br	Br
68	Ni	H	H	$OSiPh_2$*t*Bu	Br	Br
81	Pd	*i*Pr	iPr	H	Me	Cl

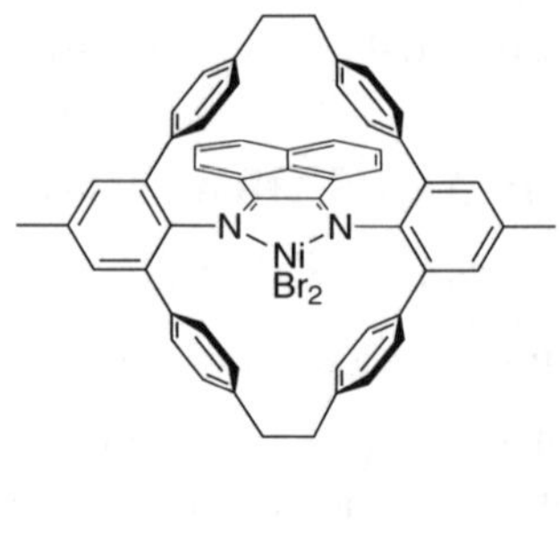

69

	M	R¹, R¹	R²	X	X'
71a	Ni	Me, Me	*i*Pr	Br	Br
71b	Ni	Me, Me	Me	Br	Br
71c	Ni	1,8-napthyl	*i*Pr	Br	Br
71d	Pd	Me, Me	*i*Pr	Cl	Cl
71e	Pd	Me, Me	*i*Pr	Cl	Me
74	Ni	Me, Me	4-*t*BuPh	Br	Br

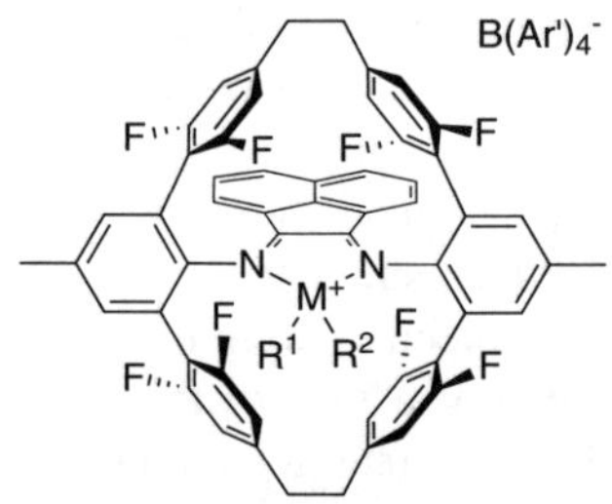

75 M = Ni, $R^1 = R^2$ = $-OC(CH_3)CHC(CH_3)O-$
76 M = Pd, R^1 = Me, R^2 = NCAr

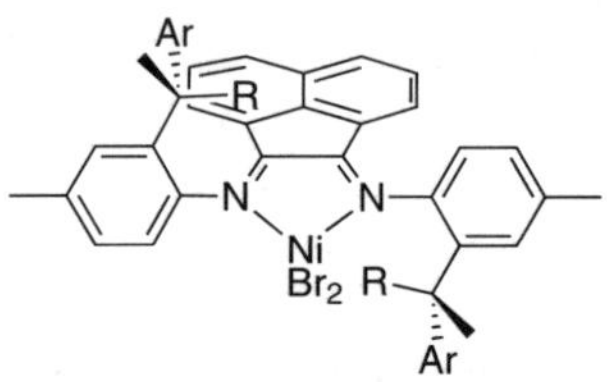

	70	**73a**	**73b**	**73c**	**73d**	**73e**	**73f**
R	H	Me	Me	Me	Me	Me	Me
Ar							

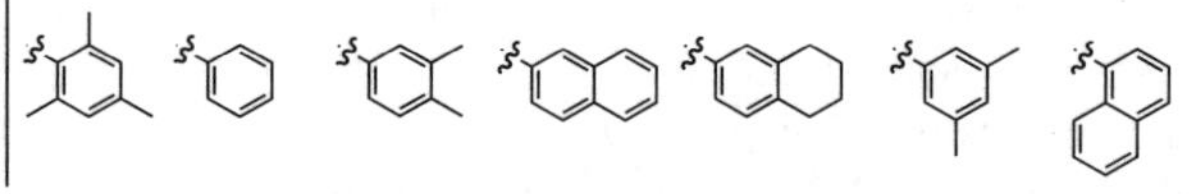

Figure 9.9 Nickel and palladium α-diimine catalyst for living olefin polymerization.

high monomer concentration) with catalyst **70** generates high selectivity for ω,2-enchainment (illustrated for 1-hexene in Scheme 9.5). Selectivity for ω,2-enchainment was shown to decrease with increasing chain length (96 mol% ω,2-enchainment for 1-butene vs. 70% for 1-octene).

In addition to nickel (**71a–c**), Suzuki *et al.* have explored α-diimine complexes of palladium (**71d,e**, Figure 9.9) for the polymerization of 1-hexene at very high pressures (up to 750 MPa) (Suzuki *et al.*, 2003). While nickel catalysts displayed non-living behavior, the palladium catalysts **71d,e** were living for 1-hexene polymerization and polydispersities decreased at higher pressures (M_w/M_n = 1.27–1.29 at 0.1 MPa vs. 1.11–1.17 at 500 MPa with **71e**).

Also using palladium, Gottfried and Brookhart have demonstrated conditions that allow for living polymerizations of 1-hexene and 1-octadecene at 0 °C where quenching with Et_3SiH is

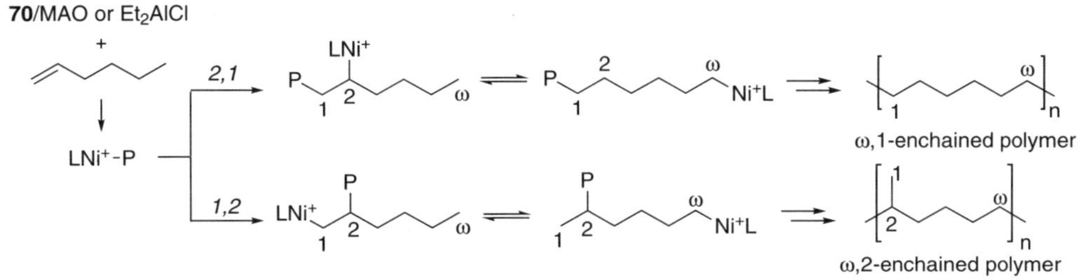

Scheme 9.5 ω,1-and ω,2-enchainment of 1-hexene.

required to prevent chain coupling (Gottfried and Brookhart, 2003). Relative to **72b,** catalyst **72a** (Figure 9.10) exhibited improved living behavior for 1-hexene polymerization. This was attributed to the ability of the nitrile donor to compete with 1-hexene for the open coordination site in **72b**. Over the course of 3 h, both **72a** and **72b** showed a linear increase in M_n with time and furnished poly(1-hexene)s with narrow molecular-weight distributions ($M_w/M_n = 1.10$–1.15) and branching densities of 75–85 branches/1000 carbons. Catalyst **72b** was likewise applied to the living polymerization of 1-octadecene. The M_n was observed to increase linearly over the first 3 h at 0 °C, however, the molecular-weight distribution increased as well ($M_w/M_n = 1.34$ after 3 h). Sen and coworkers reported a structurally similar complex bearing a five-membered ester chelate arising from 1,2-insertion of methyl methacrylate into the cationic palladium precursor (**72c**, Figure 9.10) that displayed "quasiliving" behavior for 1-hexene polymerization over 22 h (Borkar *et al.*, 2007).

It was discovered that addition of carbon monoxide (CO) in the presence of 1-hexene and **72c** afforded an alternating copolymer (Borkar *et al.*, 2007). Taking advantage of this, a poly(1-hexene)-*block*-poly(1-hexene-*alt*-CO) diblock copolymer was made through sequential monomer addition. Similarly, a PE-*block*-poly(ethylene-*alt*-CO) diblock copolymer was also synthesized through sequential monomer addition.

9.4.1.2 Propylene Polymerization

α-Diimine complexes of nickel were the first late transition-metal catalysts reported for the living polymerization of propylene (Johnson *et al.*, 1995). At −10 °C, catalyst **65**/MMAO afforded PP with a narrow molecular-weight distribution ($M_w/M_n = 1.13$; $M_n = 160\,000$ g/mol) and M_n was shown to increase linearly with conversion. The material obtained had 159 branches/1000 carbons, far less than the theoretical value of 333 for sequential 1,2-insertions, which was attributed to chain straightening. As branching decreased, T_g values decreased as well (as low as −55 °C), illustrating the dramatic effects of enchainment mechanism on physical properties.

Brookhart and coworkers demonstrated the living nature of nickel catalysts **65** and **66** with the synthesis of well-defined di- and tri-block copolymers of α-olefins (Killian *et al.*, 1996). Using **66**/MAO at −15 °C, a PP-*block*-poly(1-hexene) was produced with monomodal, narrow molecular-weight distributions ($M_w/M_n = 1.11$–1.13), which exhibited less branching than predicted, due to partial chain straightening. Additionally, a poly(1-octadecene)-*co*-PP-*co*-poly(1-octadecene) was prepared by activation of **65** or **66** with MMAO at −10 °C

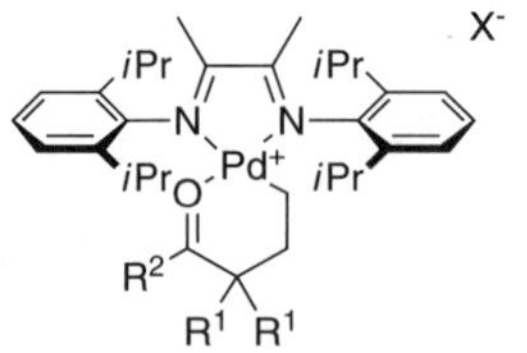

	R^1	R^2	X
72a	H	OMe	$B(Ar)_4^-$
72c	Me	OMe	$B(Ar)_4^-$
77	H	OCH_2CH_2POSS	SbF_6^-
79	H	$CH_2CH_2CH_2OCO(CH_2)CH_3$	$B(Ar)_4^-$
80	H	$OCH_2CH_2OCO(CH_3)_2Br$	SbF_6^-

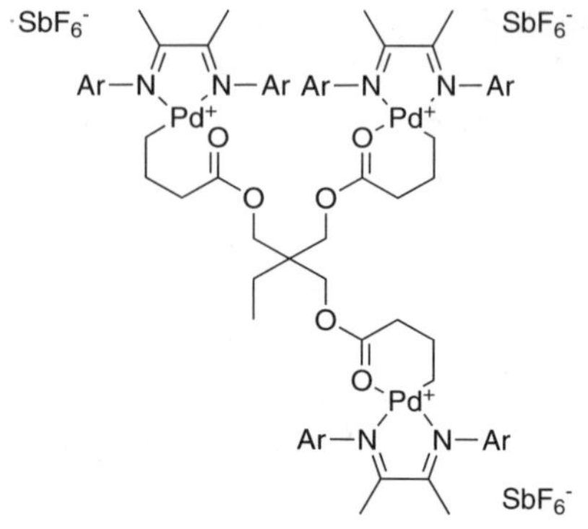

78 Ar = 2,6-iPrC_6H_3

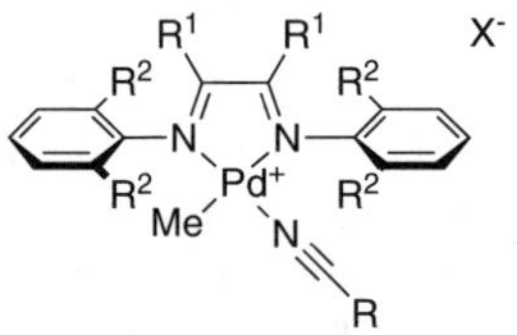

	R^1	R^2	R^2	X
72b	Me	iPr	3,5-$(CF_3)_2C_6H_3$	$B(Ar)_4^-$
82	Me	iPr	Me	SbF_6^-
83a	H	Me	Me	$B(Ar)_4^-$
83b	Me	iPr	Me	$B(Ar)_4^-$

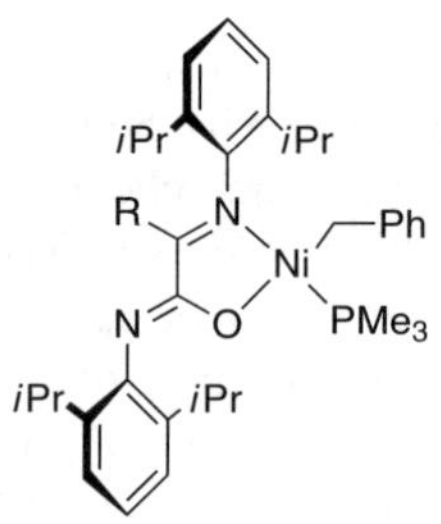

84a R = Me
84b R = Et
84c R = CH_2CHMe_2
84d R = $CHMe_2$
84e R = Ph
84f R = p-CF_3-Ph
84g R = p-OCH_3-Ph

Figure 9.10 Nickel and palladium precatalysts for living olefin polymerization.

(Scheme 9.6). Polymerization of 1-octadecene afforded a semicrystalline, chain-straightened block, as observed in homopolymerizations with that monomer. The resultant materials exhibited elastomeric properties as expected based on the "hard–soft–hard" triblock structure of the polymers.

Living polymerization of propylene was also achieved by Marques and Gomes with **68** (Yuan *et al.*, 2005). At −11 °C, **68**/MAO produced PP with a narrow molecular weight distribution ($M_w/M_n = 1.17$–1.19) and a nearly linear increase in molecular weight over 2 h. Minimal chain straightening was observed (316 branches/1000 carbons) and the polymer was moderately syndioenriched ($[rr] = 0.54$). The living nature of **68** was exploited for the production of di- and triblock copolymers (Yuan *et al.*, 2005). At −15 °C, **68**/MAO polymerized propylene and following removal of excess monomer under vacuum, addition of 1-hexene at the same temperature generated a diblock copolymer with $M_w/M_n = 1.18$ and $M_n = 46\,400$ g/mol. Triblocks were produced with this catalyst in a similar fashion. Polymerization of 1-hexene first, followed by introduction of propylene (with 1-hexene still present), followed by venting and then allowing the residual 1-hexene to react allowed for formation of a shorter

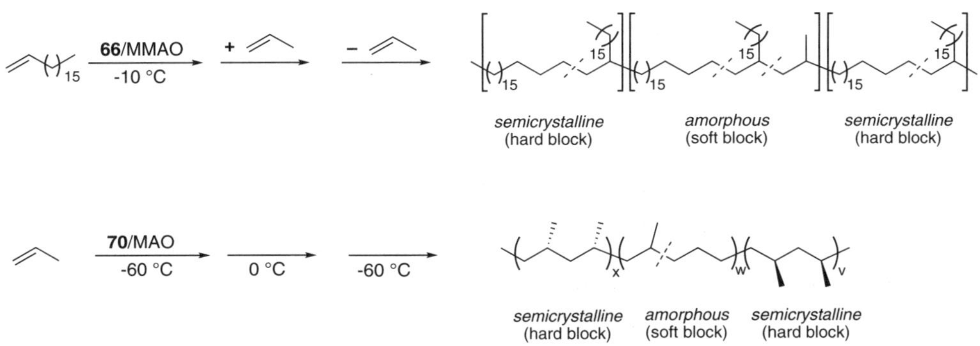

Scheme 9.6 Synthesis of 1-octadecene/propylene and propylene-based block copolymers using **66**/MAO and **70**/MAO.

($M_n = 31\,100$ g/mol) ABA triblock copolymer, with the middle block composed of poly (propylene-*ran*-(1-hexene)).

Guan and coworkers reported a dramatic demonstration of chain straightening of propylene with the cyclophane Ni catalyst **69** (Camacho and Guan, 2005). At temperatures up to 50 °C, **69**/MAO showed good activity for propylene polymerization, with narrow molecular-weight distributions ($M_w/M_n = 1.06$–1.16). In addition, M_n was shown to increase linearly with time at 50 °C. The PPs contain 104–113 branches/1000 carbons, indicative of extensive chain straightening. This implies that the cyclophane-ligand geometry favors a 2,1-insertion mechanism.

Utilizing **70**, Coates and coworkers polymerized propylene in a living fashion at temperatures up to 22 °C in the presence of MAO, to obtain polymers with narrow polydispersities ($M_w/M_n \leq 1.11$) (Cherian *et al.*, 2005b). Both the regio- and stereocontrol of enchainment are temperature dependent, allowing access to a wide variety of polymer microstructures from a single monomer. At low temperatures (−78 °C), no chain straightening is observed, furnishing highly isotactic PP, but the percentage of 3,1-enchainment increases up to 56.2% at 22 °C, producing an amorphous and regioirregular PP. Utilizing this temperature dependence, **70**/MAO was used to synthesize regioblock PPs having good elastomeric properties (Hotta *et al.*, 2006). Both a triblock and pentablock copolymer were synthesized by simply varying the reaction temperature during the course of the polymerization (Scheme 9.6). For example, an *i*PP-*block*-*rir*PP-*block*-*i*PP ($M_n = 109\,000$ g/mol, $M_w/M_n = 1.14$) and *i*PP-*block*-*rir*PP-*block*-*i*PP-*block*-*rir*PP-*block*-*i*PP ($M_n = 159\,000$ g/mol, $M_w/M_n = 1.39$) pentablock copolymer were prepared by toggling the reaction temperature between −60 and 0 °C. Transmission electron microscopy (TEM) revealed no microphase separation. However, the pentablock copolymer exhibited an exceptional strain to break of 2400% and good elastomeric recovery out to strains of 1000%.

In an attempt to obtain higher regio- and isoselectivity at low reaction temperatures, new chiral, C_2-symmetric nickel diimine complexes featuring cumyl-derived ligands were prepared (Rose *et al.*, 2008a). Each of the complexes (**73a–f**, Figure 9.9) exhibited higher regioselectivity than **70** at −60 °C in the presence of MAO. Of the complexes studied at −78 °C, **73f** furnished isotactic PP with the highest melting temperature ($T_m = 149$ °C). In addition, a linear increase in M_n with polymer yield is observed at 0 °C. Regioblock polypropylenes were also produced with **73f**/MAO that exhibited improved elastomeric performance at elevated temperatures (e.g. 65 °C) over block copolymers synthesized with **70**/MAO (Rose *et al.*, 2008a).

Using Pd complexes **72b** at 0 °C, Gottfried and Brookhart observed a relatively linear increase in polypropylene M_n with time up to approximately 40 000 g/mol (Gottfried and Brookhart, 2003). However, living systems were not obtained when the palladium ester chelate catalyst **72a** was employed due to slow initiation relative to propagation. This initiation problem could be circumvented utilizing **72b** because the weakly bound nitrile group is more easily displaced by propylene. Consistent with previous results, PPs generated by this catalyst are chain straightened, containing approximately 253 branches per 1000 carbons.

9.4.1.3 Ethylene Polymerization

First-generation nickel α-diimine catalysts such as **65** do not polymerize ethylene in a living fashion due to relatively facile chain transfer. Rieger and coworkers have investigated modifications of this framework to prevent chain transfer by enhancing the steric bulk about the metal center (Schmid *et al.*, 2001). When **74** (Figure 9.9) was activated with MAO at ambient temperature, molecular-weight distributions were decreased markedly (M_w/M_n as low as 1.3) for short reaction times and ultrahigh molecular weight (>4 500 000 g/mol), highly linear PE was produced.

Guan and coworkers have extended the study of hindered diimine catalysts even further with cyclophane complex **69** (Camacho *et al.*, 2004). Upon activation with MMAO, **69** is highly active for production of branched PEs (66–97 branches/1000 carbons) with relatively narrow polydispersities (M_w/M_n as low as 1.23 at 50 °C). Most significantly, these catalysts exhibit impressive thermal stability, with good activities even up to 90 °C. However, the polydispersity increases, and the activities decrease somewhat at higher temperatures. Interestingly, a related alkyl cyclophane Ni complex demonstrated almost no activity for ethylene polymerization (Camacho *et al.*, 2005).

In a subsequent report, Guan and coworkers found a significant effect on polymer properties and reactivity upon fluorination of the cyclophane ligand (**75**, Figure 9.9) (Popeney *et al.*, 2009). Complex **75** showed improved thermal stability relative to its nonfluorination counterpart **69**. At 105 °C, **75**/TIBA produced high molecular weight PE (M_n = 190 000 g/mol) with a unimodal molecular-weight distribution (M_w/M_n = 2.7). The related palladium complex (**76**, Figure 9.9) produced higher molecular weight polymer than variations of the catalyst containing nonfluorinated ligands. At 35 °C, **76**/TIBA yielded PE with a narrow polydispersity (M_w/M_n = 1.3) and high molecular weight (M_n = 264 000 g/mol). Both the nickel (**75**) and palladium (**76**) complexes exhibited a significant decrease in chain branching compared to their nonfluorinated counterparts. The authors postulate that the fluorine atom stabilizes the electrophilic metal center through electron donation. This type of stabilization is well known in early-metal polymerization systems, but this report represents the first example of positive fluorine interactions in late-metal olefin polymerization.

To achieve living ethylene polymerization with palladium catalysts **72a,b**, Brookhart and coworkers demonstrated that specific reaction conditions, particularly quenching reactions with Et_3SiH to prevent chain coupling, were crucial (Gottfried and Brookhart, 2001). At 5 °C, highly branched (~100 branches/1000 carbons), amorphous PEs with very narrow molecular-weight distributions ($M_w/M_n < 1.1$) were produced, and M_n was shown to increase linearly over at least 6 h. At 27 °C, M_n of 237 000 g/mol could be obtained in 2 h with broadened molecular-weight distributions ($M_w/M_n = 1.19$). To ensure rapid initiation of **72a**, high pressures of ethylene (400 psig) were required to displace the chelated carbonyl group, which is retained in the highly branched PE product. A telechelic polymer could be produced with **72a** by addition of alkyl acrylates before the silane quench (Gottfried and Brookhart, 2003). Acrylates undergo one insertion into the growing chain, forming stable chelates, but do not insert further, allowing for clean end functionalization to generate polymers with two distinct ester end groups. Additionally, aldehyde end groups could be generated by quenching with 4-penten-1-ol. The vinyl group inserts, followed by Pd migration down the chain, and finally β-hydride elimination to generate the difunctional polymers.

Compound **72b** exhibited similar activity to **72a** at high pressures, while also yielding polymers with relatively narrow polydispersities ($M_w/M_n = 1.15$ at 5 °C) at 1 atm ethylene. Given that palladium diimine catalysts produce highly branched, amorphous materials with ethylene, and produce semicrystalline polymers from long-chain α-olefins via a "chain-straightening" mechanism, copolymers containing these two segments are an attractive goal. Using **72b**, PE-*block*-poly(1-octadecene) and poly(1-octadecene)-*block*-PE were produced with narrow molecular-weight distributions ($M_w/M_n = 1.06$–1.22) (Gottfried and Brookhart, 2003).

Ye and coworkers modified the palladium ester chelate, **72a**, by immobilization on a polyhedral oligomeric silsesquioxane (POSS) support (**77**, Figure 9.10) which furnished POSS end-functionalized PEs (Zhang and Ye, 2008a). The M_n was shown to increase linearly with time. In a subsequent report, Ye and coworkers immobilized the palladium ester chelate on silica nanoparticles as a versatile surface-initiated living ethylene polymerization technique for grafting from silica nanoparticles (Zhang and Ye, 2008b). After cleavage of the PE brushes from the silica nanoparticles, the polymers were found to possess narrow PDIs ($M_w/M_n \sim 1.18$).

Utilizing a triacrylate, Subramanian and coworkers synthesized a trinuclear palladium α-diimine catalyst (**78**, Figure 9.10) for the production of star PE (Zhang *et al.*, 2009). At 5 °C in the presence of ethylene, **78** produced three-arm star polymers with narrow molecular-weight distributions ($M_w/M_n = 1.05$–1.12) and molecular weights ($M_n = 32\,000$–$135\,000$) that increased over the course of five hours. Cleavage of the star polymer revealed arm molecular weights ($M_n = 11\,000$–$44\,000$ g/mol) that were approximately three times smaller than the corresponding star polymer.

To synthesize graft copolymers, Brookhart and Matyjaszewski combined living insertion polymerization with living ATRP (atom-transfer radical polymerization) techniques. Palladium diimine chelate complexes have been previously used for living polymerization of ethylene, affording end-functionalized, branched polyethylenes (Hong *et al.*, 2002). Catalyst **79** (Figure 9.10) builds on this strategy by appending an acrylate ester forming a PE macromonomer that was incorporated into a living ATRP polymerization of *n*-butyl acrylate to generate poly(*n*-butyl acrylate)-*graft*-PE. Graft copolymers of moderate molecular weight (M_n = up to 115 000 g/mol) were obtained with approximately 4–5 grafts per chain and relatively narrow molecular-weight distribution (M_w/M_n as low as 1.4).

Combining living insertion polymerization with living ATRP techniques, Ye and coworkers were also able to generate block copolymers with a functionalized palladium diimine catalyst (Zhang *et al.*, 2008a). Catalyst **80** (Figure 9.10) appended with bromo-functionality forms a PE macroinitiator that was incorporated into a living ATRP polymerization of *n*-butyl acrylate or styrene to generate poly(*n*-butyl acrylate)-*block*-PE and PS-*block*-PE diblock copolymers.

9.4.1.4 Other Monomers

In addition to simple α-olefins, Coates and coworkers have polymerized poly(E-*co*-P) macromonomers to generate a series of poly(E-*co*-P) comb polymers (Rose *et al.*, 2008b). Using the 3,5-difluoro isomer of the nonliving bis(phenoxyimine)titanium catalyst **33b** (Figure 9.5), poly(E-*co*-P) macromonomers featuring one unsaturated chain end were synthesized. The macromonomers were then homopolymerized using a living nickel diimine catalyst, **65**/MAO, to generate poly(E-*co*-P) comb polymers featuring approximately 7 to 14 arms/molecule after fractionation that corresponded well to the theoretical values based on reaction stoichiometry. The molecular-weight distributions remained relatively low ($M_w/M_n = 1.51$–1.90, $M_n = 74\,000$–$209\,000$ g/mol).

Beyond unfunctionalized olefins, **65** is also capable of polymerizing olefin monomers that incorporate polar functional groups. Coates and coworkers synthesized polyolefin elastomers by addition of small amounts of ureidopyrimidone (UP) functionalized hexene to polymerizations of 1-hexene (Rieth, Eaton, and Coates, 2001). Polymers incorporating ~2% UP-functionalized monomer were obtained with narrow molecular-weight distributions ($M_w/M_n = 1.2$–1.4). The resultant polymers exhibited reversible, noncovalent crosslinking through hydrogen-bonding interactions, and thus have elastomeric properties at room temperature.

Utilizing 1,6-dienes, Osakada and coworkers reported the synthesis of polymers with *trans*-1,2-disubstituted five-membered rings (Park, Takeuchi, and Osakada, 2006). In a subsequent report, **81**/$NaBArF_4$ (Figure 9.9) was shown to polymerize 5-allyl-5-crotyl-2,2-dimethyl-1,3-dioxane in a living fashion (Okada *et al.*, 2007). A linear increase in M_n with monomer conversion was demonstrated with molecular-weight distributions remaining relatively narrow ($M_w/M_n = 1.20$–1.24). Later, Osakada and coworkers reported that **65**/MMAO was capable of polymerizing 9,9-diallylfluorene yielding a polymer with $M_n = 6100$ g/mol and $M_w/M_n = 1.36$ after three hours (Takeuchi *et al.*, 2009). The resultant polymer contained predominately six-membered rings with *cis* geometry, as evidenced by ^{13}C NMR spectroscopy.

In another application of 1,6-dienes, Liu and coworkers investigated the copolymerization of ethylene and diethyl diallylmalonate using **82** (Figure 9.10) (Xiang *et al.*, 2009). Due to extensive chain walking, ethylene homopolymerization produces a polymer with a hyperbranched structure. Incorporation of the diethyl diallylmalonate reduces the amount of branching in the resultant polymer because the catalyst is unable to chain walk through the resultant five-membered ring. At 35 °C, **82** produced PE containing 97 branches per 1000 carbons with $M_w = 78\,000$ g/mol and a narrow molecular-weight distribution ($M_w/M_n = 1.44$). Upon incorporation of diethyl diallylmalonate (0.40 M), molecular weight ($M_w = 25\,000$ g/mol) and polydispersity ($M_w/M_n = 1.20$) decreased along with branch density, which was reduced to 81 per 1000 carbons. Therefore, polymer topology was tuned from hyperbranched to linear as a function of diethyl diallylmalonate incorporation.

Finally, Kaminsky and Kiesewetter applied a combinatorial screening approach to identify catalysts for the copolymerization of norbornene with ethylene (Kiesewetter and Kaminsky, 2003). Catalysts **83a** and **83b** (Figure 9.10) were identified and furnished poly(E-*co*-NB) with 9–62 mol% norbornene incorporation and relatively narrow molecular-weight distributions (M_w/M_n as low as 1.4) indicating “quasiliving” behavior.

9.4.2 Other Nickel Catalysts

Bazan and coworkers have investigated nickel diimine variants **84a–g** (Figure 9.10) (Diamanti *et al.*, 2003; Azoulay *et al.*, 2008). At 20 °C, **84a**/Ni(COD)$_2$ produced PE with low branching (12–19 methyl branches/1000 carbons) and M_n was observed to increase linearly with time up to 30 min. Additionally, the molecular-weight distributions remained narrow for all the catalysts studied ($M_w/M_n = 1.1–1.4$). Investigation into the role of the ligand revealed that as the steric bulk increased an increase in polymerization activity was observed.

Utilizing **84**, Bazan and coworkers achieved the first quasiliving copolymerization of ethylene with a polar monomer, 5-norbornen-2-yl acetate (Diamanti *et al.*, 2003; Azoulay *et al.*, 2008). Upon activation with Ni(COD)$_2$, **84a–g** incorporated 1–17 mol% 5-norbornen-2-yl acetate into a polyethylene backbone. Molecular-weight distributions were relatively narrow ($M_w/M_n = 1.2–1.6$), and the M_n exhibited a nearly linear increase with conversion. To synthesize block copolymers with different ratios of ethylene, Bazan and coworkers adopted a strategy based on a pressure-jump technique (Diamanti *et al.*, 2004). Polymerization of 5-norbornen-2-yl acetate and 50 psi ethylene generates an amorphous copolymer with approximately 25% polar monomer incorporation. After a given reaction time (8–45 min), the ethylene pressure was increased to 1100 psi leading to formation of an essentially PE block with relatively narrow polydispersities ($M_w/M_n = 1.3–1.6$). Analysis of GPC and DSC data is consistent with diblock formation. TEM analysis demonstrated that the materials are microphase separated, consistent with blocks of distinct compositions. Tapered copolymers (TCP) have also been prepared with this system that allows depletion of 5-norbornen-2-yl acetate under a constant ethylene pressure (Diamanti *et al.*, 2006). As 5-norbornen-2-yl acetate concentration depletes semicrystalline properties are obtained, consistent with the fact that ethylene-rich segments are formed, indicating pseudodiblock copolymer formation. This strategy was used to make pseudotetrablock copolymers by addition of a second batch of 5-norbornen-2-yl acetate after a prescribed time (Coffin *et al.*, 2007). Tensile testing revealed a strain to break of about 1000% at 65 °C with 80% elastic recovery.

In a later report, Bazan and coworkers combined living insertion polymerization with living ATRP techniques to synthesize graft copolymers (Schneider *et al.*, 2008). Polymerization of ethylene followed by copolymerization with 5-norbornen-2-yl-2′-bromo-2′-methyl propanoate using **84c** activated with Ni(COD)$_2$ furnished a PE macroinitiator. Subsequent polymerization with MMA by living ATRP methods furnished PE-*graft*-PMMA copolymers.

9.5 Outlook and Summary

The last decade has seen significant new advances achieved in the field of living olefin polymerization. Many efficient and selective catalysts are now available for the living polymerization of ethylene in addition to living and stereoselective polymerization of α-olefins, resulting in the creation of a variety of new polymer architectures, such as block copolymers and end-functionalized macromolecules. The ability to synthesize such polymers will allow the detailed study of the effect of polymer microstructure on the mechanical and physical properties of this new class of materials.

As we wrote in our previous reviews (Coates *et al.*, 2002; Domski *et al.*, 2007b; Edson *et al.*, 2009a), the main challenge facing this field is that these expensive metal complexes only form one polymer chain during the polymerization reaction, resulting in economically nonviable materials for commodity applications. Significant research in developing catalytic systems that can produce multiple chains per metal center must be conducted. One strategy to accomplish this goal is to add excess amounts of an inexpensive metal complex that will rapidly transmetallate the active living catalyst, producing many chains per metal center (Kaneyoshi *et al.*, 2005; Van Meurs *et al.*, 2005; Alfano *et al.*, 2007; Kempe, 2007; Hustad *et al.*, 2008; Zhang and Sita, 2008a; Zhang *et al.*, 2008b; Sita, 2009). This strategy can then be used to create multiple block copolymers per metal center by varying polymerization conditions (Hustad *et al.*, 2007; Cai *et al.*, 2008a). A second strategy is to add an external agent at specific intervals during the living polymerization to terminate a chain and begin a new one (Mitani *et al.*, 2003b). Third, metal complexes that transmetallate at rates slower than monomer enchainment but faster than chain formation have the potential to produce block copolymers when two different polymerization catalysts are used (Chien *et al.*, 1997; Chien, *et al.* 1999; Lieber and Brintzinger, 2000; Jayaratne and Sita, 2001; Arriola *et al.*, 2006). Finally, nonliving catalysts that can be induced to introduce blocks on a timescale faster than that of chain formation can be used to make block copolymers (Coates and Waymouth, 1995).

Future research will continue to uncover new living systems capable of making unique polyolefin structures, and these advances will greatly expand the range of available materials. New strategies for developing catalyst systems capable of furnishing multiple chains per metal center will allow commodity polyolefin production from living catalysts. Undoubtedly, the future for specialty materials is a bright one in light of continued new developments in the field of living olefin polymerization.

References

Alfano, F., Boone, H.W., Busico, V. *et al.* (2007) Polypropylene "chain shuttling" at enantiomorphous and enantiopure catalytic species: Direct and quantitative evidence from polymer microstructure. *Macromolecules*, **40**, 7736–7738.

Annunziata, L., Pappalardo, D., Tedesco, C., and Pellecchia, C. (2009) Isotactic-specific polymerization of propene by a C-s-symmetric zirconium(IV) complex bearing a dianionic tridentate [-NNN-] amidomethylpyrrolidepyridine ligand. *Macromolecules*, **42**, 5572–5578.

Aoshima, S. and Kanaoka, S. (2009) A renaissance in living cationic polymerization. *Chemical Reviews*, **109**, 5245–5287.

Arriola, D.J., Carnahan, E.M., Hustad, P.D. *et al.* (2006) Catalytic production of olefin block copolymers via chain shuttling polymerization. *Science*, **312**, 714–719.

Azoulay, J.D., Itigaki, K., Wu, G., and Bazan, G.C. (2008) Influence of steric and electronic perturbations on the polymerization activities of α-iminocarboxamide nickel complexes. *Organometallics*, **27**, 2273–2280.

Azoulay, J.D., Schneider, Y., Galland, G.B., and Bazan, G.C. (2009) Living polymerization of ethylene and alpha-olefins using a nickel α-keto-β–diimine initiator. *Chemical Communications*, 6177–6179.

Bambirra, S., Van Leusen, D., Meetsma, A. *et al.* (2003) Yttrium alkyl complexes with a sterically demanding benzamidinate ligand: synthesis, structure and catalytic ethene polymerisation. *Chemical Communications (Cambridge, United Kingdom)*, 522–523.

Bates, F.S. (1991) Polymer-polymer phase-behavior. *Science*, **251**, 898–905.

Bates, F.S. and Fredrickson, G.H. (1999) Block copolymers-designer soft materials. *Physics Today*, **52**, 32–38.

Baumann, R., Davis, W.M., and Schrock, R.R. (1997) Synthesis of titanium and zirconium complexes that contain the tridentate diamido ligand, $[((t\text{-Bu-}d_6\text{-})N\text{-}o\text{-}C_6H_4)_2O]^{2-}$ ($[NON]^{2-}$) and the living polymerization of 1-hexene by the activated $[NON]ZrMe_2$. *Journal of the American Chemical Society*, **119**, 3830–3831.
Baumann, R. and Schrock, R.R. (1998) NMR detection of living intermediates prepared from activated $[NON]ZrMe_2$ ($[NON]^{2-}=[(t\text{-Bu-}d_6\text{-}N\text{-}o\text{-}C_6H_4)_2O]^{2-}$) and olefins. *Journal of Organometallic Chemistry*, **557**, 69–75.
Boffa, L.S. and Novak, B.M. (2000) Copolymerization of polar monomers with olefins using transition-metal complexes. *Chemical Reviews*, **100**, 1479–1493.
Borkar, S., Yennawar, H., and Sen, A. (2007) Methacrylate Insertion into cationic diimine palladium(II)-alkyl complexes and the synthesis of poly(alkene-*block*-alkene/carbon monoxide) copolymers. *Organometallics*, **26**, 4711–4714.
Boussie, T.R., Diamond, G.M., Goh, C. *et al.* (2006) Nonconventional catalysts for isotactic propene polymerization in solution developed by using high-throughput-screening technologies. *Angewandte Chemie-International Edition*, **45**, 3278–3283.
Braunecker, W.A. and Matyjaszewski, K. (2007) Controlled/living radical polymerization: Features, developments, and perspectives. *Progress in Polymer Science*, **32**, 93–146.
Britovsek, G.J.P., Gibson, V.C., and Wass, D.F. (1999) The search for new-generation olefin polymerization catalysts: Life beyond metallocenes. *Angewandte Chemie, International Edition*, **38**, 428–447.
Busico, V., Cipullo, R., Ronca, S., and Budzelaar, P.H.M. (2001) Mimicking Ziegler-Natta catalysts in homogeneous phase, 1 – C_2-symmetric octahedral Zr(IV) complexes with tetradentate [ONNO]-type ligands. *Macromolecular Rapid Communications*, **22**, 1405–1410.
Busico, V., Cipullo, R., Friederichs, N. *et al.* (2003) The first molecularly characterized isotactic polypropylene-block-polyethylene obtained via "quasi-living" insertion polymerization. *Macromolecules*, **36**, 3806–3808.
Busico, V., Cipullo, R., Friederichs, N. *et al.* (2004) Block copolymers of highly isotactic polypropylene via controlled Ziegler-Natta polymerization. *Macromolecules*, **37**, 8201–8203.
Cai, Z., Nakayama, Y., and Shiono, T. (2008a) Catalytic synthesis of a monodisperse olefin block copolymer using a living polymerization system. *Macromolecular Rapid Communications*, **29**, 525–529.
Cai, Z.G., Ikeda, T., Akita, M., and Shiono, T. (2005) Substituent effects of *tert*-butyl groups on fluorenyl ligand in syndiospecific living polymerization of propylene with ansa-fluorenylamidodimethyltitanium complex. *Macromolecules*, **38**, 8135–8139.
Cai, Z.G., Nakayama, Y., and Shiono, T. (2006) Living random copolymerization of propylene and norbornene with *ansa*-fluorenylamidodimethyltitanium complex: Synthesis of novel syndiotactic polypropylene-*block*-poly(propylene-*ran*-norbornene). *Macromolecules*, **39**, 2031–2033.
Cai, Z.G., Nakayama, Y., and Shiono, T. (2008b) Facile synthesis of tailor-made stereoblock polypropylenes via successive variation of monomer pressure. *Macromolecules*, **41**, 6596–6598.
Camacho, D.H., Salo, E.V., Ziller, J.W., and Guan, Z.B. (2004) Cyclophane-based highly active late-transition-metal catalysts for ethylene polymerization. *Angewandte Chemie-International Edition*, **43**, 1821–1825.
Camacho, D.H. and Guan, Z.B. (2005) Living polymerization of α-olefins at elevated temperatures catalyzed by a highly active and robust cyclophane-based nickel catalyst. *Macromolecules*, **38**, 2544–2546.
Camacho, D.H., Salo, E.V., Guan, Z.B., and Ziller, J.W. (2005) Nickel(II) and palladium(II) complexes with an alkane-bridged macrocyclic ligand: Synthesis, characterization, and polymerization tests. *Organometallics*, **24**, 4933–4939.
Chen, E.Y.X. and Marks, T.J. (2000) Cocatalysts for metal-catalyzed olefin polymerization: Activators, activation processes, and structure-activity relationships. *Chemical Reviews*, **100**, 1391–1434.
Cherian, A.E., Lobkovsky, E.B., and Coates, G.W. (2005a) Synthesis of allyl-terminated syndiotactic polypropylene: macromonomers for the synthesis of branched polyolefins. *Macromolecules*, **38**, 6259–6268.
Cherian, A.E., Rose, J.M., Lobkovsky, E.B., and Coates, G.W. (2005b) A C_2-symmetric, living α-diimine Ni(II) catalyst: Regioblock copolymers from propylene. *Journal of the American Chemical Society*, **127**, 13770–13771.
Chien, J.C.W., Iwamoto, Y., Rausch, M.D. *et al.* (1997) Homogeneous binary zirconocenium catalyst systems for propylene polymerization.1. Isotactic/atactic interfacial compatibilized polymers having thermoplastic elastomeric properties. *Macromolecules*, **30**, 3447–3458.
Chien, J.C.W., Iwamoto, Y., and Rausch, M.D. (1999) Homogeneous binary zirconocenium catalysts for propylene polymerization. II. Mixtures of isospecific and syndiospecific zirconocene systems. *Journal of Polymer Science, Part A: Polymer Chemistry*, **37**, 2439–2445.
Cipullo, R., Busico, V., Fraldi, N. *et al.* (2009) Improving the behavior of bis(phenoxyamine) group 4 metal catalysts for controlled alkene polymerization. *Macromolecules*, **42**, 3869–3872.

Coates, G.W. and Waymouth, R.M. (1991) Enantioselective cyclopolymerization: optically active poly(methylene-1,3-cyclopentane). *Journal of the American Chemical Society*, **113**, 6270–6271.

Coates, G.W. and Waymouth, R.M. (1995) Oscillating stereocontrol: A strategy for the synthesis of thermoplastic elastomeric polypropylene. *Science*, **267**, 217–219.

Coates, G.W. (2000) Precise control of polyolefin stereochemistry using single-site metal catalysts. *Chemical Reviews*, **100**, 1223–1252.

Coates, G.W., Hustad, P.D., and Reinartz, S. (2002) Catalysts for the living insertion polymerization of alkenes: Access to new polyolefin architectures using Ziegler-Natta chemistry. *Angewandte Chemie, International Edition*, **41**, 2236–2257.

Coffin, R.C., Diamanti, S.J., Hotta, A. *et al.* (2007) Pseudo-tetrablock copolymers with ethylene and a functionalized comonomer. *Chemical Communications*, 3550–3552.

Corradini, P., Guerra, G., and Cavallo, L. (2004) Do new century catalysts unravel the mechanism of stereocontrol of old Ziegler-Natta catalysts? *Accounts of Chemical Research*, **37**, 231–241.

Dare, E.O., Ogunniyi, D.S., Olatunji, G.A., and Chattopadhyay, P. (2004) Polymerization of propene with (BuNSi-$Me_2C_5Me_4TiMe_2$)-Bu-t: Effects of trialkylaluminiums as additives. *Bulletin of the Chemical Society of Ethiopia*, **18**, 131–141.

Desurmont, G., Li, Y., Yasuda, H. *et al.* (2000a) Reaction pathway for the formation of binuclear samarocene hydride from monomeric alkyl samarocene derivative and the effective catalysis of samarocene hydride for the block copolymerization of ethylene with polar monomers. *Organometallics*, **19**, 1811–1813.

Desurmont, G., Tanaka, M., Li, Y. *et al.* (2000b) New approach to block copolymerization of ethylene with polar monomers by the unique catalytic function of organolanthanide complexes. *Journal of Polymer Science, Part A: Polymer Chemistry*, **38**, 4095–4109.

Desurmont, G., Tokimitsu, T., and Yasuda, H. (2000c) First controlled block copolymerizations of higher 1-olefins with polar monomers using metallocene type single component lanthanide initiators. *Macromolecules*, **33**, 7679–7681.

Di Lena, F. and Chen, P. (2009) Unexpected kinetics in the polymerization of ethene by $Cp_2^*ZrCl_2$/MAO. *Helvetica Chimica Acta*, **92**, 890–896.

Diamanti, S.J., Ghosh, P., Shimizu, F., and Bazan, G.C. (2003) Ethylene homopolymerization and copolymerization with functionalized 5-norbornen-2-yl monomers by a novel nickel catalyst system. *Macromolecules*, **36**, 9731–9735.

Diamanti, S.J., Khanna, V., Hotta, A. *et al.* (2004) Synthesis of block copolymer segments containing different ratios of ethylene and 5-norbornen-2-yl acetate. *Journal of the American Chemical Society*, **126**, 10528–10529.

Diamanti, S.J., Khanna, V., Hotta, A. *et al.* (2006) Tapered block copolymers containing ethylene and a functionalized comonomer. *Macromolecules*, **39**, 3270–3274.

Doi, Y., Takada, M., and Keii, T. (1979a) Polymerization of propylene with VCl_4/Et_2AlCl. *Bulletin of the Chemical Society of Japan*, **52**, 1802.

Doi, Y., Ueki, S., and Keii, T. (1979b) Living coordination polymerization of propene initiated by the soluble $V(acac)_3$-$Al(C_2H_5)_2Cl$ System. *Macromolecules*, **12**, 814–819.

Doi, Y., Ueki, S., and Keii, T. (1979c) Preparation of living polypropylenes by a soluble vanadium-based Ziegler Catalyst. *Makromolekulare Chemie-Macromolecular Chemistry and Physics*, **180**, 1359–1361.

Doi, Y., Ueki, S., and Keii, T. (1982) Block copolymerization of propylene and ethylene with the living coordination catalyst $V(acac)_3/Al(C_2H_5)_2Cl$/Anisole. *Makromolekulare Chemie, Rapid Communications*, **3**, 225–229.

Doi, Y., Watanabe, Y., Ueki, S., and Soga, K. (1983) Synthesis of a propylene-tetrahydrofuran block co-polymer via living coordination polymerization. *Makromolekulare Chemie, Rapid Communications*, **4**, 533–537.

Doi, Y., Murata, M., and Soga, K. (1984) Reaction of carbon-monoxide with living polypropylene prepared with a vanadium-based catalyst. *Makromolekulare Chemie, Rapid Communications*, **5**, 811–814.

Doi, Y., Koyama, T., and Soga, K. (1985a) Synthesis of a propene – methyl-methacrylate diblock copolymer via living coordination polymerization. *Makromolekulare Chemie-Macromolecular Chemistry and Physics*, **186**, 11–15.

Doi, Y., Suzuki, S., and Soga, K. (1985b) A perfect initiator for living coordination polymerization of propene - tris(2-methyl-1,3-butanedionato)vanadium diethylaluminum chloride system. *Makromolekulare Chemie, Rapid Communications*, **6**, 639–642.

Doi, Y. and Keii, T. (1986) Functional polypropylenes. *Advances in Polymer Science*, **73/74**, 201.

Doi, Y., Suzuki, S., and Soga, K. (1986) Living coordination polymerization of propene with a highly-active vanadium-based catalyst. *Macromolecules*, **19**, 2896–2900.

Doi, Y., Hizal, G., and Soga, K. (1987a) Syntheses and characterization of terminally functionalized polypropylenes. *Makromolekulare Chemie*, **188**, 1273–1279.

Doi, Y., Tokuhiro, N., Suzuki, S., and Soga, K. (1987b) Synthesis and characterization of living copolymer of ethylene and propene. *Makromolekulare Chemie, Rapid Communications*, **8**, 285–290.

Doi, Y., Suzuki, S., Hizal, G., and Soga, K. (1988) Living coordination polymerization of propylene and synthesis of tailor-made polymers, in *Transition Metal Catalyzed Polymerization* (eds R.P. Quirk and R.E. Hoff), University of Akron, pp. 182–194.

Doi, Y., Tokuhiro, N., and Soga, K. (1989) Synthesis and structure of a living copolymer of propylene and 1,5-hexadiene. *Makromolekulare Chemie-Macromolecular Chemistry and Physics*, **190**, 643–651.

Domski, G.J., Lobkovsky, E.B., and Coates, G.W. (2007a) Polymerization of α-olefins with pyridylamidohafnium catalysts: Living behavior and unexpected isoselectivity from a Cs-symmetric catalyst precursor. *Macromolecules*, **40**, 3510–3513.

Domski, G.J., Rose, J.M., Coates, G.W. *et al.* (2007b) Living alkene polymerization: New methods for the precision synthesis of polyolefins. *Progress in Polymer Science*, **32**, 30–92.

Domski, G.J., Edson, J.B., Keresztes, I. *et al.* (2008) Synthesis of a new olefin polymerization catalyst supported by an sp^3-C donor via insertion of a ligand-appended alkene into the Hf-C bond of a neutral pyridylamidohafnium complex. *Chemical Communications*, **46**, 6137–6139.

Edson, J.B., Wang, Z.G., Kramer, E.J., and Coates, G.W. (2008) Fluorinated bis(phenoxyketimine)titanium complexes for the living, isoselective polymerization of propylene: Multiblock isotactic polypropylene copolymers via sequential monomer addition. *Journal of the American Chemical Society*, **130**, 4968–4977.

Edson, J.B., Domski, G.J., Rose, J.M. *et al.* (2009a) *Controlled and Living Polymerizations: From Mechanisms to Applications*, Wiley-VCH, Weinheim, Germany, pp. 169–237.

Edson, J.B., Keresztes, I., Lobkovsky, E.B., and Coates, G.W. (2009b) Alkene polymerization catalysts bearing tridentate phenoxyamine ligands with sp^3 C donors. *ChemCatChem*, **1**, 122–130.

Evans, W.J., Decoster, D.M., and Greaves, J. (1995) Field Desorption Mass-Spectrometry studies of the samarium-catalyzed polymerization of ethylene under hydrogen. *Macromolecules*, **28**, 7929–7936.

Froese, R.D.J., Hustad, P.D., Kuhlman, R.L., and Wenzel, T.T. (2007) Mechanism of activation of a hafnium pyridyl-amide olefin polymerization catalyst: Ligand modification by monomer. *Journal of the American Chemical Society*, **129**, 7831–7840.

Fujita, M. and Coates, G.W. (2002) Synthesis and characterization of alternating and multiblock copolymers from ethylene and cyclopentene. *Macromolecules*, **35**, 9640–9647.

Fukui, Y., Murata, M., and Soga, K. (1999) Living polymerization of propylene and 1-hexene using bis-Cp type metallocene catalysts. *Macromolecular Rapid Communications*, **20**, 637–640.

Fukui, Y. and Murata, M. (2001) Reaction of carbon monoxide with living PP prepared with a $Cp_2ZrMe_2/B(C_6F_5)_3$ catalyst system. *Macromolecular Chemistry and Physics*, **202**, 1430–1434.

Fukui, Y. and Murata, M. (2002) Living polymerizations of propylene and syntheses of atactic polypropylene-block-poly(ethylene-*co*-propylene) using metallocene catalyst systems. *Applied Catalysis, A: General*, **237**, 1–10.

Furuyama, R., Mitani, M., Mohri, J. *et al.* (2005a) Ethylene/higher α-olefin copolymerization behavior of fluorinated bis(phenoxy-imine)titanium complexes with methylalumoxane: Synthesis of new polyethylene-based block copolymers. *Macromolecules*, **38**, 1546–1552.

Furuyama, R., Saito, J., Ishii, S. *et al.* (2005b) Fluorinated bis(phenoxy-imine) Ti complexes with MAO: Remarkable catalysts for living ethylene and syndioselective living propylene polymerization. *Journal of Organometallic Chemistry*, **690**, 4398–4413.

Gendler, S., Groysman, S., Goldschmidt, Z. *et al.* (2006) Polymerization of 4-methylpentene and vinylcyclohexane by amine bis(phenolate) titanium and zirconium complexes. *Journal of Polymer Science, Part A: Polymer Chemistry*, **44**, 1136–1146.

Gibson, V.C. and Spitzmesser, S.K. (2003) Advances in non-metallocene olefin polymerization catalysis. *Chemical Reviews*, **103**, 283–315.

Gottfried, A.C. and Brookhart, M. (2001) Living polymerization of ethylene using Pd(II) α-diimine catalysts. *Macromolecules*, **34**, 1140–1142.

Gottfried, A.C. and Brookhart, M. (2003) Living and block copolymerization of ethylene and α-olefins using palladium (II)-α–diimine catalysts. *Macromolecules*, **36**, 3085–3100.

Groysman, S., Goldberg, I., Kol, M. *et al.* (2003a) From THF to furan: Activity tuning and mechanistic insight via sidearm donor replacement in group IV amine bis(phenolate) polymerization catalysts. *Organometallics*, **22**, 3013–3015.

Groysman, S., Goldberg, I., Kol, M. *et al.* (2003b) Group IV complexes of an amine bis(phenolate) ligand featuring a THF sidearm donor: from highly active to living polymerization catalysts of 1-hexene. *Inorganica Chimica Acta*, **345**, 137–144.

Groysman, S., Tshuva, E.Y., Goldberg, I. *et al.* (2004) Diverse structure-activity trends in amine bis(phenolate) titanium polymerization catalysts. *Organometallics*, **23**, 5291–5299.

Guan, Z., Cotts, P.M., Mccord, E.F., and Mclain, S.J. (1999) Chain walking: a new strategy to control polymer topology. *Science*, **283**, 2059–2062.

Hagihara, H., Shiono, T., and Ikeda, T. (1998) Living polymerization of propene and 1-hexene with the [*t*-BuNSiMe$_2$Flu]TiMe$_2$/B(C$_6$F$_5$)$_3$ catalyst. *Macromolecules*, **31**, 3184–3188.

Hagimoto, H., Shiono, T., and Ikeda, T. (2002a) Living polymerization of propene with a chelating (diamide) dimethyltitanium complex using silica-supported methylaluminoxane. *Macromolecules*, **35**, 5744–5745.

Hagimoto, H., Shiono, T., and Ikeda, T. (2002b) Living polymerization of propene with a chelating diamide complex of titanium using dried methylaluminoxane. *Macromolecular Rapid Communications*, **23**, 73–76.

Hagimoto, H., Shiono, T., and Ikeda, T. (2003) Supporting effects of methylaluminoxane on propene polymerization with an alkyltitanium complex. *Science and Technology in Catalysis, 2002*, **145**, 129–132.

Hagimoto, H., Shiono, T., and Ikeda, T. (2004) Supporting effects of methylaluminoxane on the living polymerization of propylene with a chelating (diamide)dimethyltitanium complex. *Macromolecular Chemistry and Physics*, **205**, 19–26.

Harney, M.B., Zhang, Y.H., and Sita, L.R. (2006a) Discrete, multiblock isotactic-atactic stereoblock polypropene microstructures of differing block architectures through programmable stereomodulated living Ziegler-Natta polymerization. *Angewandte Chemie, International Edition*, **45**, 2400–2404.

Harney, M.B., Zhang, Y.H., and Sita, L.R. (2006b) Bimolecular control over polypropene stereochemical microstructure in a well-defined two-state system and a new fundamental form: Stereogradient polypropene. *Angewandte Chemie-International Edition*, **45**, 6140–6144.

Hasan, T., Ioku, A., Nishii, K. *et al.* (2001) Syndiospecific living polymerization of propene with [*t*-BuNSiMe$_2$Flu] TiMe$_2$ using MAO as cocatalyst. *Macromolecules*, **34**, 3142–3145.

Hasan, T., Ikeda, T., and Shiono, T. (2004a) Ethene-norbornene copolymer with high norbornene content produced by *ansa*-fluorenylamidodimethyltitanium complex using a suitable activator. *Macromolecules*, **37**, 8503–8509.

Hasan, T., Shiono, T., and Ikeda, T. (2004b) Living random copolymerization of ethene and norbornene using *ansa*-fluorenylamidodimethyltitanium complex. *Macromolecular Symposia*, **213**, 123–129.

Hasan, T., Ikeda, T., and Shiono, T. (2005) Random copolymerization of propene and norbornene with *ansa*-fluorenylamidodimethyltitanium-based catalysts. *Macromolecules*, **38**, 1071–1074.

Hawker, C.J., Bosman, A.W., and Harth, E. (2001) New polymer synthesis by nitroxide mediated living radical polymerizations. *Chemical Reviews*, **101**, 3661–3688.

He, L.P., Liu, J.Y., Li, Y.G. *et al.* (2009) High-temperature living copolymerization of ethylene with norbornene by titanium complexes bearing bidentate [O,P] ligands. *Macromolecules*, **42**, 8566–8570.

Hicks, F.A., Jenkins, J.C., and Brookhart, M. (2003) Synthesis and ethylene polymerization activity of a series of 2-anilinotropone-based neutral nickel(II) catalysts. *Organometallics*, **22**, 3533–3545.

Hong, S.C., Jia, S., Teodorescu, M. *et al.* (2002) Polyolefin graft copolymers via living polymerization techniques: Preparation of poly(*n*-butyl acrylate)-*graft*-polyethylene through the combination of Pd-mediated living olefin polymerization and atom transfer radical polymerization. *Journal of Polymer Science, Part A: Polymer Chemistry*, **40**, 2736–2749.

Hotta, A., Cochran, E., Ruokolainen, J. *et al.* (2006) Semi-crystalline thermoplastic elastomeric polyolefins: Advances through catalyst development and macromolecular design. *Proceedings of the National Academy of Sciences of the United States of America*, **42**, 15327–15332.

Hsieh, H.L. and Quirk, R.P. (1996) *Anionic Polymerization: Principles and Practical Applications*, Marcel Dekker Inc., New York.

Hustad, P.D. and Coates, G.W. (2002) Insertion/isomerization polymerization of 1,5-hexadiene: Synthesis of functional propylene copolymers and block copolymers. *Journal of the American Chemical Society*, **124**, 11578–11579.

Hustad, P.D., Tian, J., and Coates, G.W. (2002) Mechanism of propylene insertion using bis(phenoxyimine)-based titanium catalysts: An unusual secondary insertion of propylene in a group IV catalyst system. *Journal of the American Chemical Society*, **124**, 3614–3621.

Hustad, P.D., Kuhlman, R.L., Arriola, D.J. *et al.* (2007) Continuous production of ethylene-based diblock copolymers using coordinative chain transfer polymerization. *Macromolecules*, **40**, 7061–7064.

Hustad, P.D., Kuhlman, R.L., Carnahan, E.M. *et al.* (2008) An exploration of the effects of reversibility in chain transfer to metal in olefin polymerization. *Macromolecules*, **41**, 4081–4089.

Ittel, S.D., Johnson, L.K., and Brookhart, M. (2000) Late-metal catalysts for ethylene homo- and copolymerization. *Chemical Reviews*, **100**, 1169–1203.

Ivanchev, S.S., Badaev, V.K., Ivancheva, N.I., and Khaikin, S.Y. (2004) Living polymerization of ethylene on the bis [*N*-(3-*tert*-butylsalicylidene)anilinato]titanium dichloride-methylalumoxane catalyst system. *Doklady Physical Chemistry*, **394**, 46–49.

Ivanchev, S.S., Trunov, V.A., Rybakov, V.B. *et al.* (2005) Structure and catalytic activity of titanium and zirconium phenoxyimine complexes. *Doklady Physical Chemistry*, **404**, 165–168.

Jansen, J.C., Mendichi, R., Locatelli, P., and Tritto, I. (2001) Evidence for the quasi-living character of the *ansa*-zirconocene/MAO-catalyzed copolymerization of ethylene and norbornene. *Macromolecular Rapid Communications*, **22**, 1394–1398.

Jansen, J.C., Mendichi, R., Sacchi, M.C., and Tritto, I. (2003) Kinetic studies of the copolymerization of ethylene with norbornene by ansa-zirconocene/methylaluminoxane catalysts: Evidence of a long-lasting "quasi-living" initial period. *Macromolecular Chemistry and Physics*, **204**, 522–530.

Jayaratne, K.C., Keaton, R.J., Henningsen, D.A., and Sita, L.R. (2000) Living Ziegler-Natta cyclopolymerization of nonconjugated dienes: New classes of microphase-separated polyolefin block copolymers via a tandem polymerization/cyclopolymerization strategy. *Journal of the American Chemical Society*, **122**, 10490–10491.

Jayaratne, K.C. and Sita, L.R. (2000) Stereospecific living Ziegler-Natta polymerization of 1-hexene. *Journal of the American Chemical Society*, **122**, 958–959.

Jayaratne, K.C. and Sita, L.R. (2001) Direct methyl group exchange between cationic zirconium Ziegler-Natta initiators and their living polymers: Ramifications for the production of stereoblock polyolefins. *Journal of the American Chemical Society*, **123**, 10754–10755.

Jeon, Y.M., Park, S.J., Heo, J., and Kim, K. (1998) Zirconium complexes with the new ancillary diamido ligand 2,2′-ethylenebis(*N,N*′-(triisopropylsilyl)anilinido)$^{2-}$: Syntheses, structures, and living α-olefin polymerization activities. *Organometallics*, **17**, 3161–3163.

Jeske, G., Lauke, H., Mauermann, H. *et al.* (1985) Highly reactive organolanthanides. Systematic routes to and olefin chemistry of early and late bis(pentamethylcyclopentadienyl) 4f hydrocarbyl and hydride complexes. *Journal of the American Chemical Society*, **107**, 8091–8103.

Johnson, L.K., Killian, C.K., and Brookhart, M. (1995) New Pd(II)- and Ni(II)-based catalysts for polymerization of ethylene and α-olefins. *Journal of the American Chemical Society*, **117**, 6414–6415.

Kamigaito, M., Ando, T., and Sawamoto, M. (2001) Metal-catalyzed living radical polymerization. *Chemical Reviews*, **101**, 3689–3746.

Kaneyoshi, H., Inoue, Y., and Matyjaszewski, K. (2005) Synthesis of block and graft copolymers with linear polyethylene segments by combination of degenerative transfer coordination polymerization and atom transfer radical polymerization. *Macromolecules*, **38**, 5425–5435.

Karian, H.G. (2003) *Handbook of Polypropylene and Polypropylene Composites*, Marcel Dekker Inc., New York.

Keaton, R.J., Jayaratne, K.C., Henningsen, D.A. *et al.* (2001) Dramatic enhancement of activities for living Ziegler-Natta polymerizations mediated by "exposed" zirconium acetamidinate initiators: The isospecific living polymerization of vinylcyclohexane. *Journal of the American Chemical Society*, **123**, 6197–6198.

Kempe, R. (2007) How to polymerize ethylene in a highly controlled fashion? *Chemistry – A European Journal*, **13**, 2764–2773.

Kiesewetter, J. and Kaminsky, W. (2003) Ethene/norbornene copolymerization with palladium(II) α–diimine catalysts: From ligand screening to discrete catalyst species. *Chemistry – A European Journal*, **9**, 1750–1758.

Killian, C.M., Tempel, D.J., Johnson, L.K., and Brookhart, M. (1996) Living polymerization of α-olefins using Ni(II)-α-diimine catalysts. Synthesis of new block polymers based on α-olefins. *Journal of the American Chemical Society*, **118**, 11664–11665.

Kissounko, D.A., Zhang, Y.H., Harney, M.B., and Sita, L.R. (2005) Evaluation of (η^5-C_5Me_5)HfR_2 N(Et)C(Me)N(*t*-Bu) (R=Me and *i*-Bu) for the stereospecific living and degenerative transfer living Ziegler-Natta polymerization of a-olefins. *Advanced Synthesis & Catalysis*, **347**, 426–432.

Kojoh, S., Matsugi, T., Saito, J. *et al.* (2001) New monodisperse ethylene-propylene copolymers and a block copolymer created by a titanium complex having fluorine- containing phenoxy-imine chelate ligands. *Chemistry Letters*, 822–823.

Kotzabasakis, V., Kostakis, K., Pitsikalis, M. *et al.* (2009) Polymerization of higher α-olefins using a Cs-symmetry hafnium metallocene catalyst. Kinetics of the polymerization and microstructural analysis. *Journal of Polymer Science Part a-Polymer Chemistry*, **47**, 4314–4325.

Kretschmer, W.P., Hessen, B., Noor, A. *et al.* (2007) Highly active/selective and adjustable zirconium polymerization catalysts stabilized by aminopyridinato ligands. *Journal of Organometallic Chemistry*, **692**, 4569–4579.

Kumudini, C.J., Keaton, R.J., Henningsen, D.A., and Sita, L.R. (2000) Living Zieger-Natta cyclopolymerization of nonconjugated dienes: New classes of microphase-separated polyolefin block copolymers via tandem polymerization/cyclopolymerization strategy. *Journal of the American Chemical Society*, **122**, 10490–10491.

Lamberti, M., Pappalardo, D., Zambelli, A., and Pellecchia, C. (2002) Syndiospecific polymerization of propene promoted by bis(salicylaldiminato)titanium catalysts: Regiochemistry of monomer insertion and polymerization mechanism. *Macromolecules*, **35**, 658–663.

Lee, H., Nienkemper, K., and Jordan, R.F. (2008) Synthesis and reactivity of a sterically crowded tris(pyrazolyl)borate hafnium benzyl complex. *Organometallics*, **27**, 5075–5081.

Li, X.F., Dai, K., Ye, W.P. *et al.* (2004) New titanium complexes with two β-enaminoketonato chelate ligands: Syntheses, structures, and olefin polymerization activities. *Organometallics*, **23**, 1223–1230.

Liang, L.C., Schrock, R.R., Davis, W.M., and Mcconville, D.H. (1999) Synthesis of group 4 complexes that contain the diamidoamine ligands, $[(2,4,6\text{-}Me_3C_6H_2NCH_2CH_2)_2NR]^{2-}$ ($[Mes_2N_2NR]^{2-}$; R=H or CH_3), and polymerization of 1-hexene by activated $[Mes_2N_2NR]ZrMe_2$ complexes. *Journal of the American Chemical Society*, **121**, 5797–5798.

Lieber, S. and Brintzinger, H.H. (2000) Propene polymerization with catalyst mixtures containing different ansa-zirconocenes: Chain transfer to alkylaluminum cocatalysts and formation of stereoblock polymers. *Macromolecules*, **33**, 9192–9199.

Long, R.J., Gibson, V.C., White, A.J.P., and Williams, D.J. (2006) Combining hard and soft donors in early-transition-metal olefin polymerization catalysts. *Inorganic Chemistry*, **45**, 511–513.

Macadams, L.A., Buffone, G.P., Incarvito, C.D. *et al.* (2005) A chromium catalyst for the polymerization of ethylene as a homogeneous model for the Phillips catalyst. *Journal of the American Chemical Society*, **127**, 1082–1083.

Makio, H., Kashiwa, N., and Fujita, T. (2002) FI catalysts: A new family of high performance catalysts for olefin polymerization. *Advanced Synthesis & Catalysis*, **344**, 477–493.

Makio, H. and Fujita, T. (2007) Synthesis of chain-end functionalized polyolefins with a bis(phenoxy imine) titanium catalysts. *Macromolecular Rapid Communications*, **28**, 698–703.

Mallin, D.T., Rausch, M.D., Lin, Y.G. *et al.* (1990) Rac-[Ethylidene(1-η^5-tetramethylcyclopentadienyl)(1-η^5-indenyl)]dichlorotitanium and its homopolymerization of propylene to crystalline amorphous block thermoplastic elastomers. *Journal of the American Chemical Society*, **112**, 2030–2031.

Manivannan, R. and Sundararajan, G. (2002) Latent bimodal polymerization of 1-hexene by a titanium-based diastereomeric catalyst containing a rac/meso-aminodiol ligand. *Macromolecules*, **35**, 7883–7890.

Mashima, K., Fujikawa, S., Urata, H. *et al.* (1993) Polymerization of ethylene catalyzed by the system Ta(η^5-C_5Me_5)(η^4-Diene)(CH_3)- 2-MAO: An isoelectronic analogue for a group 4 metallocene catalyst. *Journal of the American Chemical Society*, **115**, 10990–10991.

Mashima, K., Fujikawa, S., Urata, H. *et al.* (1994) Polyethylene with extremely narrow polydispersity obtained from the new catalyst systems Nb(η^5-C_5Me_5)(η^4-Diene)Cl- 2-MAO and Nb(η^5-C_5Me_5)(η^4-Diene)$_2$-MAO. *Journal of the Chemical Society, Chemical Communications*, 1623–1624.

Mashima, K., Fujikawa, S., Tanaka, Y. *et al.* (1995) Living polymerization of ethylene catalyzed by diene complexes of niobium and tantalum, M(η^5-C_5Me_5)(η^4-Diene)X_2 and M(η^5-C_5Me_5)(η^4-Diene)$_2$(M= Nb and Ta), in the presence of methylaluminoxane. *Organometallics*, **14**, 2633–2640.

Mason, A.F., Tian, J., Hustad, P.D. *et al.* (2002) Syndiospecific living catalysts for propylene polymerization: Effect of fluorination on activity, stereoselectivity, and termination. *Israel Journal of Chemistry*, **42**, 301–306.

Mason, A.F. and Coates, G.W. (2004a) New phenoxyketimine titanium complexes: Combining isotacticity and living behavior in propylene polymerization. *Journal of the American Chemical Society*, **126**, 16326–16327.

Mason, A.F. and Coates, G.W. (2004b) Gel permeation chromatography as a combinatorial screening method: Identification of highly active heteroligated phenoxyimine polymerization catalysts. *Journal of the American Chemical Society*, **126**, 10798–10799.

Mason, A.F. (2005) Ph. D. Thesis. Ithaca, NY, Cornell University.

Mathers, R.T. and Coates, G.W. (2004) Cross metathesis functionalization of polyolefins. *Chemical Communications*, 422–423.

Matsugi, T., Matsui, S., Kojoh, S. *et al.* (2001) New titanium complexes having two indolide-imine chelate ligands for living ethylene polymerization. *Chemistry Letters*, 566–567.
Matsugi, T., Kojoh, S., Fujita, T., and Kashiwa, N. (2002a) Living polymerization of ethylene with titanium complexes having two indolide-imine chelate ligands. *Kobunshi Ronbunshu*, **59**, 410–414.
Matsugi, T., Matsui, S., Kojoh, S. *et al.* (2002b) New titanium complexes bearing two indolide-imine chelate ligands for the polymerization of ethylene. *Macromolecules*, **35**, 4880–4887.
Matsugi, T., Matsui, S., Kojoh, S. *et al.* (2003) II Catalysts: New titanium complexes having two indolide-imine chelate ligands for ethylene polymerization. *Science and Technology in Catalysis, 2002*, **145**, 523–524.
Matsui, S., Mitani, M., Saito, J. *et al.* (1999a) Post-metallocenes: A new bis(salicylaldiminato) zirconium complex for ethylene polymerization. *Chemistry Letters*, 1263–1264.
Matsui, S., Tohi, Y., Mitani, M. *et al.* (1999b) New bis(salicylaldiminato) titanium complexes for ethylene polymerization. *Chemistry Letters*, 1065–1066.
Matsui, S., Mitani, M., Saito, J. *et al.* (2000) Post-metallocenes: Catalytic performance of new bis(salicylaldiminato) zirconium complexes for ethylene polymerization. *Chemistry Letters*, 554–555.
Matyjaszewski, K. (1996) *Cationic Polymerizations: Mechanisms, Synthesis, and Applications*, Marcel Dekker Inc., New York.
Matyjaszewski, K. and Xia, J.H. (2001) Atom transfer radical polymerization. *Chemical Reviews*, **101**, 2921–2990.
Mehrkhodavandi, P., Bonitatebus, P.J., and Schrock, R.R. (2000) A comparison of cationic zirconium methyl and isobutyl initiators that contain an arylated diamido-pyridine ligand for polymerization of 1-hexene. Elucidation of a dramatic "initiator effect". *Journal of the American Chemical Society*, **122**, 7841–7842.
Mehrkhodavandi, P. and Schrock, R.R. (2001) Cationic hafnium alkyl complexes that are stable toward β-hydride elimination below 10 °C and active as initiators for the living polymerization of 1-hexene. *Journal of the American Chemical Society*, **123**, 10746–10747.
Mehrkhodavandi, P., Schrock, R.R., and Pryor, L.L. (2003) Living polymerization of 1-hexene by cationic zirconium and hafnium complexes that contain a diamido/donor ligand of the type $H_3CC(2\text{-}C_5H_4N)(CH_2Nmesityl)_2^{2-}$. A comparison of methyl and isobutyl initiators. *Organometallics*, **22**, 4569–4583.
Merna, J., Cihlar, J., Kucera, M. *et al.* (2005) Polymerization of hex-1-ene initiated by diimine complexes of nickel and palladium. *European Polymer Journal*, **41**, 303–312.
Merna, J., Host'alek, Z., Peleska, J., and Roda, J. (2009) Living/controlled olefin polymerization initiated by nickel diimine complexes: The effect of ligand ortho substituent bulkiness. *Polymer*, **50**, 5016–5023.
Milano, G., Cavallo, L., and Guerra, G. (2002) Site chirality as a messenger in chain-end stereocontrolled propene polymerization. *Journal of the American Chemical Society*, **124**, 13368–13369.
Mitani, M., Furuyama, R., Mohri, J. *et al.* (2002a) Fluorine- and trimethylsilyl-containing phenoxy-imine Ti complex for highly syndiotactic living polypropylenes with extremely high melting temperatures. *Journal of the American Chemical Society*, **124**, 7888–7889.
Mitani, M., Mohri, J., Yoshida, Y. *et al.* (2002b) Living polymerization of ethylene catalyzed by titanium complexes having fluorine-containing phenoxy-imine chelate ligands. *Journal of the American Chemical Society*, **124**, 3327–3336.
Mitani, M., Furuyama, R., Mohri, J. *et al.* (2003a) Syndiospecific living propylene polymerization catalyzed by titanium complexes having fluorine-containing phenoxy-imine chelate ligands. *Journal of the American Chemical Society*, **125**, 4293–4305.
Mitani, M., Mohri, J., Furuyama, R. *et al.* (2003b) Combination system of fluorine-containing phenoxy-imine Ti complex and chain transfer agent: A new methodology for multiple production of monodisperse polymers. *Chemistry Letters*, **32**, 238–239.
Mohring, V.M. and Fink, G. (1985) Novel polymerization of α-olefins with the catalyst system nickel/aminobis(imino) phosphorane. *Angewandte Chemie, International Edition*, **24**, 1001–1003.
Nakayama, Y., Saito, J., Bando, H., and Fujita, T. (2005) Propylene polymerization behavior of fluorinated bis (phenoxy-imine) Ti complexes with an $MgCl_2$-based compound($MgCl_2$-supported Ti-based catalysts). *Macromolecular Chemistry and Physics*, **206**, 1847–1852.
Natta, G., Pino, P., Corradini, P. *et al.* (1955) Crystalline high polymers of α-olefins. *Journal of the American Chemical Society*, **77**, 1708–1710.
Nienkemper, K., Lee, H., Jordan, R.F. *et al.* (2008) Synthesis of double-end-capped polyethylene by a cationic tris (pyrazolyl)borate zirconium benzyl complex. *Organometallics*, **27**, 5867–5875.

Nishii, K., Shiono, T., and Ikeda, T. (2002) Propagation mechanism for living polymerization of olefins using η^1,η^3-tert-Butyl(dimethylfluorenylsilyl)amido dimethyltitanium-based catalyst. *Kobunshi Ronbunshu*, **59**, 371–376.

Nishii, K., Matsumae, T., Dare, E.O. *et al.* (2004a) Effect of solvents on living polymerization of propylene with *t*-BuNSiMe_2FluTiMe_2-MMAO catalyst system. *Macromolecular Chemistry and Physics*, **205**, 363–369.

Nishii, K., Shiono, T., and Ikeda, T. (2004b) A novel synthetic procedure for stereoblock poly(propylene) with a living polymerization system. *Macromolecular Rapid Communications*, **25**, 1029–1032.

Nishii, K., Ikeda, T., Akita, M., and Shiono, T. (2005) Polymerization of propylene with [*t*-BuNSiMe_2Ind]TiMe_2-MAO catalyst systems. *Journal of Molecular Catalysis A: Chemical*, **231**, 241–246.

Nomura, K. and Fudo, A. (2004) Efficient living polymerization of 1-hexene by Cp*TiMe_2(O-2,6-*i*$Pr_2C_6H_3$)-borate catalyst systems at low temperature. *Journal of Molecular Catalysis A: Chemical*, **209**, 9–17.

Okada, T., Park, S., Takeuchi, D., and Osakada, K. (2007) Pd-catalyzed polymerization of dienes that involves chain-walking isomerization of the growing polymer end: Synthesis of polymers composed of polymethylene and five-membered-ring units. *Angewandte Chemie-International Edition*, **46**, 6141–6143.

Ouchi, M., Terashima, T., and Sawamoto, M. (2009) Transition metal-catalyzed living radical polymerization: toward perfection in catalysis and precision polymer synthesis. *Chemical Reviews*, **109**, 4963–5050.

Pan, L., Hong, M., Liu, J.Y. *et al.* (2009) Living copolymerization of ethylene with dicyclopentadiene using uitanium catalyst: Formation of well-defined polyethylene-block-poly(ethylene-co-dicyclopentadiene)s and their transformation into novel polyolefin-block-(functional polyolefin)s. *Macromolecules*, **42**, 4391–4393.

Park, S., Takeuchi, D., and Osakada, K. (2006) Pd complex-promoted cyclopolymerization of functionalized α, ω-dienes and copolymerization with ethylene to afford polymers with cyclic repeating units. *Journal of the American Chemical Society*, **128**, 3510–3511.

Peacock, A.J. (2000) *Handbook of Polyethylene*, Marcel Dekker Inc., New York.

Popeney, C.S., Rheingold, A.L., and Guan, Z.B. (2009) Nickel(II) and palladium(II) polymerization catalysts bearing a fluorinated cyclophane ligand: stabilization of the reactive intermediate. *Organometallics*, **28**, 4452–4463.

Quirk, R.P. and Lee, B. (1992) Experimental Criteria for Living Polymerization. *Polymer International*, **27**, 359–367.

Radulescu, A., Mathers, R.T., Coates, G.W. *et al.* (2004) A SANS study of the self-assembly in solution of syndiotactic polypropylene homopolymers, syndiotactic polypropylene-block-poly(ethylene-co-propylene) diblock copolymers, and an alternating atactic-isotactic multisegment polypropylene. *Macromolecules*, **37**, 6962–6971.

Ravasio, A., Zampa, C., Boggioni, L. *et al.* (2008) Copolymerization of ethylene with norbornene catalyzed by cationic rare-earth metal half-sandwhich complexes. *Macromolecules*, **41**, 9565–9569.

Reinartz, S., Mason, A.F., Lobkovsky, E.B., and Coates, G.W. (2003) Titanium catalysts with ancillary phenoxyketimine ligands for living ethylene polymerization. *Organometallics*, **22**, 2542–2544.

Resconi, L., Cavallo, L., Fait, A., and Piemontesi, F. (2000) Selectivity in propene polymerization with metallocene catalysts. *Chemical Reviews*, **100**, 1253–1345.

Rieth, L.R., Eaton, R.F., and Coates, G.W. (2001) Polymerization of ureidopyrimidinone-functionalized olefins by using late-transition metal Ziegler-Natta catalysts: synthesis of thermoplastic elastomeric polyolefins. *Angewandte Chemie-International Edition*, **40**, 2153–2156.

Rose, J.M., Cherian, A.E., and Coates, G.W. (2006) Living polymerization of α-olefins with an α–diimine Ni(II) catalyst: Formation of well-defined ethylene-propylene copolymers through controlled chain-walking. *Journal of the American Chemical Society*, **128**, 4186–4187.

Rose, J.M., Deplace, F., Lynd, N.A. *et al.* (2008a) C_2-Symmetric Ni(II) α–diimines featuring cumyl-derived ligands: Synthesis of improved elastomeric regioblock polypropylenes. *Macromolecules*, **41**, 9548–9555.

Rose, J.M., Mourey, T.H., Slater, L.A. *et al.* (2008b) Poly(ethylene-co-propylene macromonomer)s: Synthesis and evidence for starlike conformations in dilute solution. *Macromolecules*, **41**, 559–567.

Ruokolainen, J., Mezzenga, R., Fredrickson, G.H. *et al.* (2005) Morphology and thermodynamic behavior of syndiotactic polypropylene-poly(ethylene-*co*-propylene) block polymers prepared by living olefin polymerization. *Macromolecules*, **38**, 851–860.

Ruzette, A.V. and Leibler, L. (2005) Block copolymers in tomorrow's plastics. *Nature Materials*, **4**, 19–31.

Saito, J., Mitani, M., Mohri, J. *et al.* (2001a) Highly syndiospecific living polymerization of propylene using a titanium complex having two phenoxy-imine chelate ligands. *Chemistry Letters*, 576–577.

Saito, J., Mitani, M., Mohri, J. *et al.* (2001b) Living polymerization of ethylene with a titanium complex containing two phenoxy-imine chelate ligands. *Angewandte Chemie-International Edition*, **40**, 2918–2920.

Saito, J., Mitani, M., Onda, M. *et al.* (2001c) Microstructure of highly syndiotactic “living” poly(propylene)s produced from a titanium complex with chelating fluorine- containing phenoxyimine ligands (an FI catalyst). *Macromolecular Rapid Communications*, **22**, 1072–1075.

Saito, J., Mitani, M., Onda, M. *et al.* (2003) Syndiospecific living polymerization of propylene with fluorine-containing titanium FI catalyst. *Science and Technology in Catalysis*, **145**, 515–516.

Sakuma, A., Weiser, M.S., and Fujita, T. (2007) Living olefin polymerization and block copolymer formation with FI catalysts. *Polymer Journal*, **39**, 193–207.

Sassmannshausen, J., Bochmann, M.E., Rosch, J., and Lilge, D. (1997) Titanium-catalysed formation of high molecular weight elastomeric polypropene: evidence for living propene polymerisation. *Journal of Organometallic Chemistry*, **548**, 23–28.

Schmid, M., Eberhardt, R., Klinga, M. *et al.* (2001) New C_{2v}- and chiral C_2-symmetric olefin polymerization catalysts based on nickel(II) and palladium(II) diimine complexes bearing 2,6-diphenyl aniline moieties: Synthesis, structural characterization, and first insight into polymerization properties. *Organometallics*, **20**, 2321–2330.

Schneider, Y., Azoulay, J.D., Coffin, R.C., and Bazan, G.C. (2008) New polyethylene macroinitiators and their subsequent grafting by atom transfer radical polymerization. *Journal of the American Chemical Society*, **130**, 10464–10465.

Schrock, R.R., Baumann, R., Reid, S.M. *et al.* (1999) Synthesis of titanium, zirconium, and hafnium complexes that contain diamido donor ligands of the type $[(t\text{-BuN-}o\text{-}C_6H_4)_2O]^{2-}$ and an evaluation of activated versions for the polymerization of 1-hexene. *Organometallics*, **18**, 3649–3670.

Schrock, R.R., Bonitatebus, P.J., and Schrodi, Y. (2001) C-H bond activation in cations of the type $\{[(2,4,6\text{-}Me_3C_6H_2NCH_2CH_2)_2NMe]ZrR\}^+$ and a simple solution that yields a catalyst for the living polymerization of 1-hexene. *Organometallics*, **20**, 1056–1058.

Schrock, R.R., Adamchuk, J., Ruhland, K., and Lopez, L.P.H. (2003) Zirconium and hafnium complexes that contain the electron-withdrawing diamido/donor ligands $(2,6\text{-}X_2C_6H_3NCH_2)_2C(2\text{-}C_5H_4N)(CH_3)^{2-}$ (X=Cl or F). An evaluation of the role of ortho halides in 1-hexene polymerization. *Organometallics*, **22**, 5079–5091.

Schrock, R.R., Adamchuk, J., Ruhland, K., and Lopez, L.P.H. (2005) Evaluation of zirconium and hafnium complexes that contain the electron-withdrawing diamido/donor ligand $(2,6\text{-}Cl_2C_6H_3NCH_2CH_2)^2NCH_3)^{2-}$ for the polymerization of 1-hexene. *Organometallics*, **24**, 857–866.

Scollard, J.D. and Mcconville, D.H. (1996) Living polymerization of α-olefins by chelating diamide complexes of titanium. *Journal of the American Chemical Society*, **118**, 10008–10009.

Segal, S., Goldberg, I., and Kol, M. (2005) Zirconium and titanium diamine bis(phenolate) catalysts for α-olefin polymerization: From atactic oligo(1-hexene) to ultrahigh-molecular-weight isotactic poly(1-hexene). *Organometallics*, **24**, 200–202.

Shiono, T. (2003) Living polymerization of propene with alkyltitanium-based catalysts. *Catalysis Surveys from Asia*, **7**, 47–62.

Shiono, T., Sugimoto, M., Hasan, T. *et al.* (2008) Random copolymerization of norbornene with higher 1-alkene with ansa-fluorenylamidodimethyltitanium catalyst. *Macromolecules*, **41**, 8292–8294.

Shiono, T., Harada, R., Cai, Z.G., and Nakayama, Y. (2009) A highly active catalyst composed of ansa-fluorenylamidodimethyltitanium derivative for propene polymerization. *Topics in Catalysis*, **52**, 675–680.

Sita, L.R. (2009) Ex uno plures (“out of one, many”). *Angewandte Chemie-International Edition*, **48**, 2464–2472.

Starzewski, A.O., Steinhauser, N., and Xin, B.S. (2008) Decisive progress in metallocene-catalyzed elastomer synthesis. *Macromolecules*, **41**, 4095–4101.

Sudhakar, P. and Sundararajan, G. (2005) Tuning the reactivity of N,O,O,O-non-metallocene catalysts for α-olefin polymerization: Issues related to ligand symmetry and derivatization. *Macromolecular Rapid Communications*, **26**, 1854–1859.

Sudhakar, P. (2007) Titanium and zirconium complexes containing modified TREN ligands for the polymerization of 1-alkenes – A comparative study. *Journal of Polymer Science Part A:Polymer Chemistry*, **45**, 5470–5479.

Suzuki, N., Yu, J., Masubuchi, Y. *et al.* (2003) Nickel- and palladium-catalyzed olefin polymerization under high pressures. *Journal of Polymer Science, Part A: Polymer Chemistry*, **41**, 293–302.

Szwarc, M. (1998) Living polymers. Their discovery, characterization, and properties. *Journal of Polymer Science, Part A: Polymer Chemistry*, **36**, IX–XV.

Takeuchi, D., Fukuda, Y., Park, S., and Osakada, K. (2009) Cyclopolymerization of 9,9-diallylfluorene promoted by Ni complexes. Stereoselective formation of six- and five-membered rings during the polymer growth. *Macromolecules*, **42**, 5909–5912.

Talarico, G., Busico, V., and Cavallo, L. (2003) Origin of the regiochemistry of propene insertion at octahedral column 4 polymerization catalysts: Design or serendipity? *Journal of the American Chemical Society*, **125**, 7172–7173.

Tang, L.M., Duan, Y.Q., Pan, L., and Li, Y.S. (2005a) Copolymerization of ethylene and cyclopentene with bis (β-enaminoketonato) titanium complexes. *Journal of Polymer Science, Part A: Polymer Chemistry*, **43**, 1681–1689.

Tang, L.M., Hu, T., Bo, Y.J. *et al.* (2005b) Titanium complexes bearing aromatic-substituted β-enaminoketonato ligands: Syntheses, structure and olefin polymerization behavior. *Journal of Organometallic Chemistry*, **690**, 3125–3133.

Tian, J. and Coates, G.W. (2000) Development of a diversity-based approach for the discovery of stereoselective polymerization catalysts: Identification of a catalyst for the synthesis of syndiotactic polypropylene. *Angewandte Chemie-International Edition*, **39**, 3626–3629.

Tian, J., Hustad, P.D., and Coates, G.W. (2001) A new catalyst for highly syndiospecific living olefin polymerization: Homopolymers and block copolymers from ethylene and propylene. *Journal of the American Chemical Society*, **123**, 5134–5135.

Tritto, I., Boggioni, L., and Ferro, D.R. (2006) Metallocene catalyzed ethene- and propene co-norbornene polymerization: Mechanisms from a detailed microstructural analysis. *Coordination Chemistry Reviews*, **250**, 212–241.

Tshuva, E.Y., Goldberg, I., and Kol, M. (2000a) Isospecific living polymerization of 1-hexene by a readily available nonmetallocene C_2-symmetrical zirconium catalyst. *Journal of the American Chemical Society*, **122**, 10706–10707.

Tshuva, E.Y., Goldberg, I., Kol, M., and Goldschmidt, Z. (2000b) Living polymerization of 1-hexene due to an extra donor arm on a novel amine bis(phenolate) titanium catalyst. *Inorganic Chemistry Communications*, **3**, 611–614.

Tshuva, E.Y., Goldberg, I., Kol, M., and Goldschmidt, Z. (2001) Living polymerization and block copolymerization of a-olefins by an amine (bis)phenolate titanium catalyst. *Chemical Communications*, 2120–2121.

Tshuva, E.Y., Groysman, S., Goldberg, I. *et al.* (2002) [ONXO]-type amine bis(phenolate) zirconium and hafnium complexes as extremely active 1-hexene polymerization catalysts. *Organometallics*, **21**, 662–670.

Turner, H.W. and Hlatky, G.G. (1991) Block copolymers from ionic catalysts. *PCT Int. Appl.* 9112285. Exxon.

Ueki, S., Doi, Y., and Keii, T. (1981) Activation effect of anisole on the living polymerization of propene with the soluble $V(acac)_3$-$Al(C_2H_5)_2Cl$ System. *Makromolekulare Chemie, Rapid Communications*, **2**, 403–406.

Ueki, S., Furuhashi, H., Murakami, N. *et al.* (1995) Synthesis of propylene macromonomer with terminal methacryl group by vanadium-based catalyst. *Science and Technology in Catalysis*, **92**, 359–362.

Van Meurs, M., Britovsek, G.J.P., Gibson, V.C., and Cohen, S.A. (2005) Polyethylene chain growth on zinc catalyzed by olefin polymerization catalysts: A comparative investigation of highly active catalyst systems across the transition series. *Journal of the American Chemical Society*, **127**, 9913–9923.

Vasile, C. (2000) *Handbook of Polyolefins*, Marcel Dekker Inc., New York.

Wang, W. and Nomura, K. (2005) Remarkable effects of aluminum cocatalyst and comonomer in ethylene copolymerizations catalyzed by (arylimido)(aryloxo)vanadium complexes: Efficient synthesis of high molecular weight ethylene/norbornene copolymer. *Macromolecules*, **38**, 5905–5913.

Ward, B.D., Bellemin-Laponnaz, S., and Gade, L.H. (2005) C_3- chirality in polymerization catalysis: A highly active dicationic scandium(III) catalyst for the isoselective polymerization of 1-hexene. *Angewandte Chemie, International Edition*, **44**, 1668–1671.

Webster, O.W. (1991) Living polymerization methods. *Science*, **251**, 887–893.

Weiser, M.S. and Mulhaupt, R. (2006) Living olefin polymerisation and block copolymer synthesis in the presence of a single-site catalyst containing a phenoxyimine ligand. *Macromolecular Symposia*, **236**, 111–116.

Weiser, M.S., Thomann, Y., Heinz, L.C. *et al.* (2006a) Living random and block copolymerization of ethene and propene. *Polymer*, **47**, 4505–4512.

Weiser, M.S., Wesolek, M., and Mulhaupt, R. (2006b) The synthesis and X-ray structure of a phenoxyimine catalyst tailored for living olefin polymerization and the synthesis of ultra-high molecular weight polyethylene and atactic polypropylene. *Journal of Organometallic Chemistry*, **691**, 2945–2952.

Xiang, P., Ye, Z.B., Morgan, S. *et al.* (2009) Tuning polyethylene chain topology via ring incorporation in chain walking ethylene polymerization. *Macromolecules*, **42**, 4946–4949.

Yang, F.-Z., Chen, Y.-C., Lin, Y.-F. *et al.* (2009) Nickel catalysts bearing bidentate α-aminoaldimines for ethylene polymerization--independent and cooperative structure/reactivity relationship resulting from unsymmetric square planar coordination. *Dalton Transactions*, 1243–1250.

Yasuda, H., Furo, M., Yamamoto, H. *et al.* (1992) New approach to block copolymerizations of ethylene with alkyl methacrylates and lactones by unique catalysis with organolanthanide complexes. *Macromolecules*, **25**, 5115–5116.

Yasumoto, T., Yamagata, T., and Mashima, K. (2005) Isoselective living polymerization of 1-hexene catalyzed by half-metallocene dimethyl complexes of hafnium with bidentate N-substituted (iminomethyl)pyrrolyl ligands. *Organometallics*, **24**, 3375–3377.

Yoon, J., Mathers, R.T., Coates, G.W., and Thomas, E.L. (2006) Optically transparent and high molecular weight polyolefin block copolymers toward self-assembled photonic band gap materials. *Macromolecules*, **39**, 1913–1919.

Yoshida, Y., Matsui, S., Takagi, Y. *et al.* (2000) Post-metallocenes: New bis(pyrrolyl-2-aldiminato) titanium complexes for ethylene polymerization. *Chemistry Letters*, 1270–1271.

Yoshida, Y., Saito, J., Mitani, M. *et al.* (2002) Living ethylene/norbornene copolymerisation catalyzed by titanium complexes having two pyrrolide-imine chelate ligands. *Chemical Communications*, 1298–1299.

Yoshida, Y., Matsui, S., Takagi, Y. *et al.* (2003) PI catalysts: New titanium complexes having two pyrrolide-imine chelate ligands for olefin polymerization. *Science and Technology in Catalysis*, **145**, 521–522.

Yoshida, Y., Mohri, J., Ishii, S. *et al.* (2004) Living copolymerization of ethylene with norbornene catalyzed by bis(pyrrolide-imine) titanium complexes with MAO. *Journal of the American Chemical Society*, **126**, 12023–12032.

Yoshida, Y., Matsui, S., and Fujita, T. (2005) Bis(pyrrolide-imine) Ti complexes with MAO: A new family of high performance catalysts for olefin polymerization. *Journal of Organometallic Chemistry*, **690**, 4382–4397.

Yu, S.M. and Mecking, S. (2008) Extremely narrow-dispersed high molecular weight polyethylene from living polymerization at elevated temperatures with o-F substituted Ti enolatoimines. *Journal of the American Chemical Society*, **130**, 13204–13205.

Yuan, J.C., Silva, L.C., Gomes, P.T. *et al.* (2005) Living and block polymerization of α-olefins using a Ni(II)-α-diimine catalyst containing ($OSiPh_2Bu$)-Bu-*t* groups. *Polymer*, **46**, 2122–2132.

Zambelli, A., Natta, G., Pasquon, I., and Signorin, R. (1967) Polymerization of propylene to syndiotactic polymer. 2. Behavior of catalytic system VCl_4-$Al(C_2H_5)_2Cl$. *Journal of Polymer Science Part C-Polymer Symposium*, **16**, 2485.

Zhang, K., Ye, Z., and Subramanian, R. (2008a) Synthesis of block copolymers of ethylene with styrene and n-butyl acrylate via a tandem strategy combining ethylene "living" polymerization catalyzed by a functionalized Pd-diimine catalyst with atom transfer radical polymerization. *Macromolecules*, **41**, 640–649.

Zhang, K.J., Ye, Z.B., and Subramanian, R. (2009) A trinuclear Pd-diimine catalyst for "core-first" synthesis of three-arm star polyethylenes via ethylene "living" polymerization. *Macromolecules*, **42**, 2313–2316.

Zhang, W. and Sita, L.R. (2008a) Highly efficient, living coordinative chain-transfer polymerization of propene with $ZnEt_2$: Practical production of ultrahigh to very low molecular weight amorphous atactic polypropenes of extremely narrow polydispersity. *Journal of the American Chemical Society*, **130**, 442–443.

Zhang, W. and Sita, L.R. (2008b) Investigation of dynamic intra- and intermolecular processes within a tether-length dependent series of group 4 bimetallic initiators for stereomodulated degenerative transfer living Ziegler-Natta propene polymerization. *Advanced Synthesis & Catalysis*, **350**, 439–447.

Zhang, W., Wei, J., and Sita, L.R. (2008b) Living coordinative chain-transfer polymerization and copolymerization of ethene, α-olefins, and α, ω-nonconjugated dienes using dialkylzinc as "surrogate" chain-growth sites. *Macromolecules*, **41**, 7829–7833.

Zhang, Y.H., Keaton, R.J., and Sita, L.R. (2003) Degenerative transfer living Ziegler-Natta polymerization: Application to the synthesis of monomodal stereoblock polyolefins of narrow polydispersity and tunable block length. *Journal of the American Chemical Society*, **125**, 9062–9069.

Zhang, Y.H. and Sita, L.R. (2003) Solid-supported stereospecific living Ziegler-Natta polymerization of α-olefins. *Chemical Communications*, 2358–2359.

Zhang, Y.H., Reeder, E.K., Keaton, R.J., and Sita, L.R. (2004) Goldilocks effect of a distal substituent on living Ziegler-Natta polymerization activity and stereoselectivity within a class of zirconium amidinate-based initiators. *Organometallics*, **23**, 3512–3520.

Zhang, Y.W. and Ye, Z.B. (2008a) Homogeneous polyhedral oligomeric silsesquioxane (POSS)-supported Pd- diimine complex and synthesis of polyethylenes end-tethered with a POSS nanoparticle via ethylene "living" polymerization. *Chemical Communications*, 1178–1180.

Zhang, Y.W. and Ye, Z.B. (2008b) Covalent surface grafting of branched polyethylenes on silica nanoparticles by surface-initiated ethylene "living" polymerization with immobilized Pd-diimine catalysts. *Macromolecules*, **41**, 6331–6338.

Ziegler, K., Holzkamp, E., Breil, H., and Martin, H. (1955) Polymerisation von athylen und anderen olefinen. *Angewandte Chemie*, **67**, 426.

Zuccaccia, C., Macchioni, A., Busico, V. *et al.* (2008) Intra- and intermolecular NMR studies on the activation of arylcyclometallated hafnium pyridyl-amido olefin polymerization precatalysts. *Journal of the American Chemical Society*, **130**, 10354–10368.

Zuccaccia, C., Busico, V., Cipullo, R. *et al.* (2009) On the first insertion of α-olefins in hafnium pyridyl-amido polymerization catalysts. *Organometallics*, **28**, 5445–5458.

10

Precision Polyolefins

Erik B. Berda[1] and Kenneth B. Wagener[2]
[1] *Department of Chemistry and Materials Science Program, University of New Hampshire, Durham, NH 03824*
[2] *The George and Josephine Butler Polymer Research Laboratory and Center for Macromolecular Science and Engineering, Department of Chemistry, University of Florida, Gainesville, FL 32611-7200*

10.1 Introduction

10.1.1 ADMET Polycondensation Chemistry

Acyclic diene metathesis (ADMET) polymerization is the step-growth polycondensation reaction of a linear diolefin (Figure 10.1), which provides an efficient route to polyolefins with precise distribution of an exactly defined functional group, either within or pendant to the polymer backbone (Baughman and Wagener, 2005; Berda *et al.*, 2006; Lehman and Wagener, 2003; Rojas *et al.*, 2008). The mechanism of this polymerization (Figure 10.1) begins with a single turnover of the olefin metathesis catalyst, which eliminates a small molecule condensate to yield an active metal alkylidene and begin the catalytic cycle. The catalytic cycle continues with the addition of a second diene monomer to form a dimer and release an active metal methylidene. This species then reacts with another diene of any size (monomer, dimer, trimer, etc.) releasing ethylene to reform the active metal alkylidene and restart the catalytic cycle. Like all step-growth polymerizations, ADMET is an equilibrium that is driven to completion by the removal of the condensate (ethylene in this case) and requires conversions greater than 99% for the formation of a high molecular weight polymer. Being a gas at room temperature and atmospheric pressure, ethylene is easily removed using mild conditions (i.e. gentle heating and reduced pressure). This reaction has been used to create a vast library of functionalized materials, including most recently amino acid functionalized polymers (Leonard *et al.*, 2008; Terada *et al.*, 2008), silicon-containing elastomers (Matloka *et al.*, 2008; Hyun *et al.*, 2006; Matloka and Wagener, 2006), supramolecular graft copolymers

Complex Macromolecular Architectures: Synthesis, Characterization, and Self-Assembly, First Edition.
Edited by Nikos Hadjichristidis, Akira Hirao, Yasuyuki Tezuka and Filip Du Prez.

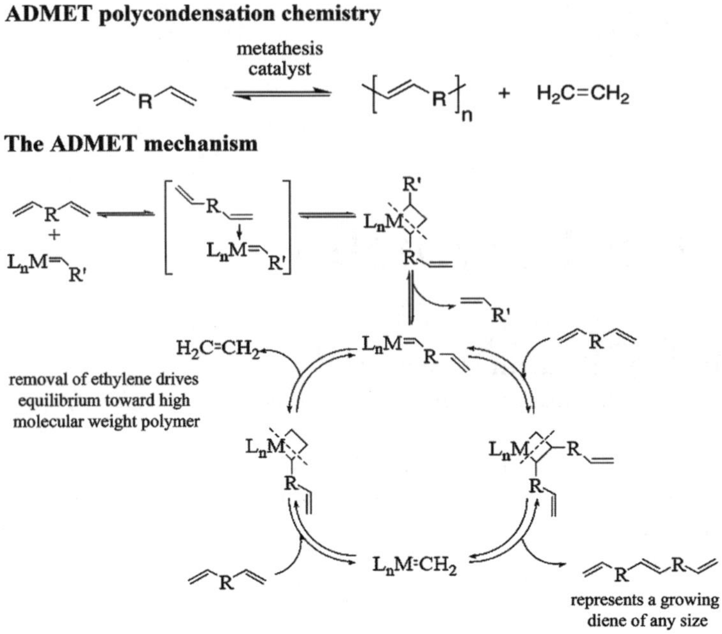

Figure 10.1 Acyclic diene metathesis.

(Ohkawa *et al.*, 2007), metal-containing polymers (Shultz *et al.*, 2008a; Shultz *et al.*, 2009; Karabulut *et al.*, 2007a; Karabulut *et al.*, 2006; Karabulut *et al.*, 2007b; Shultz and Tyler, 2009; Shultz *et al.*, 2008b; Solmaz *et al.*, 2006), sequenced copolymers of ethylene and vinyl comonomers (Baughman *et al.*, 2007; Baughman *et al.*, 2006a; Baughman *et al.*, 2005; Baughman *et al.*, 2006b; Berda *et al.*, 2007; Berda and Wagener, 2008a; Berda and Wagener, 2008b; Boz *et al.*, 2007a; Boz *et al.*, 2007b; Boz *et al.*, 2008; Boz *et al.*, 2006; Lehman *et al.*, 2007; Opper *et al.*, 2009; Opper and Wagener, 2009; Rojas *et al.*, 2009; Rojas and Wagener, 2009; Sworen and Wagener, 2007; Zuluaga *et al.*, 2009), conjugated polymers (Qin and Hillmyer, 2009; Yamamoto *et al.*, 2009; Mukherjee and Peetz, 2008; Weychardt and Plenio, 2008), hyperbranched polymers (Gorodetskaya *et al.*, 2009), liquid-crystalline polymers (Galli *et al.*, 2005), nanoparticles (Pecher and Mecking, 2007), mechanically interlocked daisy chains (Guidry *et al.*, 2007), and even in templated macrocycle synthesis (Hou *et al.*, 2006).

10.1.2 The Evolution of ADMET

By the mid-1980s considerable advancements had been made in the field of olefin metathesis chemistry (Grubbs, 2004). This mild carbon-bond-forming reaction was discovered by accident in the late 1960s when researchers at Goodyear exposed a mixture of α-olefins to a combination of tungsten hexachloride and a Lewis acid with the intent to find a new catalyst for the polymerization of vinyl olefins (Calderon *et al.*, 1968; Calderon *et al.*, 1967). Instead of high polymer the research team observed a complex mixture of scrambled olefin products. The mechanism of this reaction, first proposed by Yves Chauvin (Herrison and Chauvin, 1971) in

1971 and later confirmed by Thomas Katz (Katz and McGinnis, 1975) in 1975, involves the 2 + 2 cycloaddition of an olefin to a metal carbene to form a metallocyclobutane, followed by a 2 + 2 cycloreversion to yield a new olefin and metal carbene. A majority of the research in the years to follow involved the development of stable metal carbenes that could facilitate this useful transformation, which has become an indispensable tool to the synthetic organic chemist (Grubbs, 2004; Grubbs, 2003).

This reaction was used from its inception in the synthesis of polymers. Despite the lack of well-defined catalyst systems, a large volume of work had been produced by the mid-1980s on the ring-opening polymerization of strained cyclic olefins (Grubbs, 2003). This reaction, coined ROMP (ring-opening metathesis polymerization) has become one of the most useful methods in the fabrication of functionalized polymers for use in every imaginable application. Although it was proposed early on that this reaction could also be used in the polycondensation of a linear diolefin, it was not until 1987 when the first study proving the feasibility of this concept appeared (Lindmark-Hamberg and Wagener, 1987). In this report the Wagener Group at the University of Florida attempted the polymerization of 1,9-decadiene using an ill-defined tungsten hexachloride/Lewis acid mixture, resulting in viscous oil and an intractable solid. Monitoring the disappearance of terminal olefin signal via ^{13}C NMR and IR confirmed that the polymerization had occurred. Since any polycondensation requires that the reaction involved proceeds quantitatively, the self-metathesis of styrene to stilbene was attempted as a test reaction. The isolated product in this case was polystyrene, rather than stilbene, due to the cationic polymerization of styrene initiated by the Lewis acid present in the ill-defined catalyst system. The authors were quick to realize that while the concept was feasible, a well-defined metal carbene catalyst would be necessary to prevent this detrimental side reaction (Lindmark-Hamberg and Wagener, 1987).

Fortunately, excellent progress was being made in catalyst design at this time. By utilizing well-defined metal-carbene catalysts developed independently by the Grubbs and Schrock groups, researchers in the Wagener group successfully synthesized stilbene from styrene in quantitative yields and reported the first viable ADMET of 1,9-decadiene to polyoctenamer (Wagener *et al.*, 1990).

It was obvious even in the initial ADMET report that this reaction could have utility in the fabrication of unique structures (Lindmark-Hamberg and Wagener, 1987; Wagener *et al.*, 1990). Virtually any compound that can be functionalized with two olefins could potentially become an ADMET monomer. Not surprisingly, early efforts in this area involved the synthesis of a wide variety of functionalized polymers using this chemistry. As metathesis catalysts became increasingly robust and tolerant of functional groups the number of moieties that could be used in this reaction increased and the library of ADMET polymers grew considerably. Several recent review articles cover this topic (Baughman and Wagener, 2005; Berda *et al.*, 2006; Lehman and Wagener, 2003; Rojas *et al.*, 2008).

10.2 Precision Polyolefins

Polyolefins, and in particular polyethylene (PE) represent the largest production volume material in the world (Gedde and Mattozzi, 2004). This is due in large part to its wide range of material responses, structural simplicity, and ease of production. Because of this PE, its copolymers, and its many end-use products have made their way into every facet of daily life.

The range of properties available in polyolefins is staggering. Simply altering comonomer identity, concentration, and distribution can change the morphology of a material from completely amorphous to highly crystalline and thus change the properties from a low-melting rubber to a high impact strength engineering material (Berda *et al.*, 2006).

Chain-propagation chemistry is generally used to produce these materials, during which undesirable chain transfer or chain walking side reactions incorporate small fractions of structural imperfections, like undesired branches, into the polymer microstructure. These irrepressible side reactions lead to a random distribution of defects throughout almost every chain polymerization leading to imperfect microstructure for many industrial materials. Furthermore, mismatches in reactivity ratios between comonomers can result in poor levels of incorporation coupled with poor control over comonomer distribution, which makes characterization of primary structure tedious and complicates the interpretation of structure–property relationships. Modeling unwanted branching and the effects of comonomer distribution is a topic of great interest since branch and comonomer content, distribution, and identity ultimately determine the material properties of the final polymer. The variability of these features results in the myriad of properties polyolefins exhibit in many end-use products. While many polymer-modeling studies have been conducted on materials made by chain-propagation chemistries, to accurately study the effects of these structural attributes model polymers of known chemical microstructure and no unwanted branching are necessary (Gedde and Mattozzi, 2004).

Step-growth polycondensation chemistry is not subject to the inherent side reactions and random comonomer incorporation associated with chain-propagation mechanisms; these materials are perfectly linear and possess pristine microstructures that mirror the structure of the monomers used during polymerization. ADMET offers a unique method to create perfectly linear, defect-free model polyolefins in precisely this fashion. The "comonomer" content and distribution is perfectly controlled *a priori* by controlling the monomer structure. Polymerization of a symmetric diene containing a pendant functional group (in essence mimicking an oligomer of ethylene and a vinyl comonomer), followed by exhaustive saturation of the resulting unsaturated backbone, results in polyolefins with some exactly known functionality at perfectly regular intervals along a linear polymer backbone that contains no structural defects or irregularities (Figure 10.2). This is the meaning of "precision polyolefins."

Precisely knowing the identity, content, and distribution of a pendant functional group allows for specific structural features to be isolated in polyolefins and their effects on material behavior and property to be precisely understood. This chapter describes this intensive investigation of structure and property by examining several separate studies on precision polyolefins and presenting the results in a systematic fashion. Beginning with perfectly linear defect-free PE we show that even though these materials are not made in the same way as the commercial materials they model, their behaviors are nonetheless identical. We then discuss the effects of introducing regularly spaced defects, systematically increasing in size and bulk, to fully understand the role of pendant functional group size and placement on the behavior of these materials. We delineate the requirements for a pendant-group's inclusion or exclusion from the polymer unit cell. We address the role interactions between these pendant groups play and the properties these materials possess, and demonstrate how this knowledge can be used to impart specific properties in a highly controllable fashion using a simple concept: precise structure manifests in precise properties.

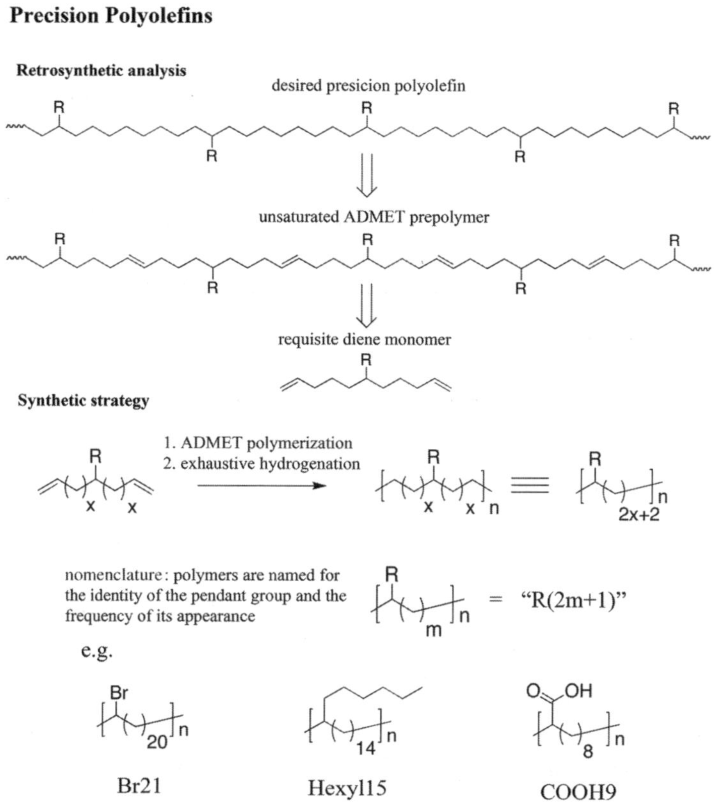

Figure 10.2 Precision polyolefins via ADMET polymerization followed by exhaustive hydrogenation.

For simplicity of discussion, the polymers described in this chapter will follow a systematic nomenclature based on the identity and distribution of the pendant functional group they contain. The name (or abbreviation) of this pendant group is given first, followed by a number corresponding to how often it regularly appears along the polymer backbone. For example Br21 would correspond to PE with a bromine atom appearing on every 21st backbone carbon, COOH9 indicates a carboxylic acid group on every 9th backbone carbon and so one.

10.3 Linear ADMET Polyethylene: Meeting the Benchmark

Unbranched, or perfectly linear polyethylene is of considerable interest, particularly for studying the behavior of this homopolymer during crystallization. Much of this work has been conducted on large *n*-paraffins (monodisperse PE oligomers) up to 390 carbons long (Tracz and Ungar, 2005), since defect-free high molecular weight PE is an elusive goal. While small-molecule paraffin compounds allow indepth study of structurally perfect model materials (Tracz and Ungar, 2005; Ungar *et al.*, 1998), end-group concentration becomes a problem and leads to irregularities when trying to extrapolate morphological behavior to a macromolecular system. The methyl end groups present in *n*-paraffins are regarded as defects that

Defect-free ADMET polyethylene

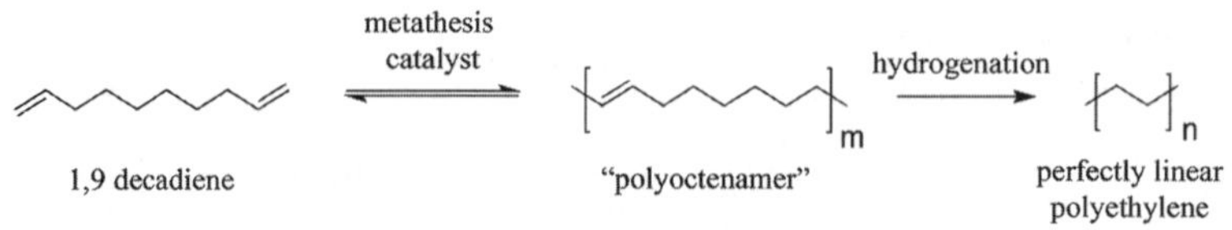

DSC data for linear ADMET polyethylenes

M_n	T_m (°C) (peak)	Δh_m (J/g)
2400	130.7	252
7600	131.3	213
11000	132.0	221
15000	133.9*	204

* melting temperature of commercial PE = 134 °C

Figure 10.3 Linear ADMET polyethylene. (Reprinted from E.B. Berda, T.W. Baughman and K.B. Wagener, "Precision branching in ethylene copolymers: Synthesis and thermal behavior," *Journal of Polymer Science Part A: Polymer Chemistry*, **44**, 17, 4981–4989, © 2006, with the permission of John Wiley & Sons, Inc.)

hinder crystallization. Polymer chains up to millions of grams per mol, on the other hand, possess an infinitesimal number of methyl end groups relative to backbone carbons, thus comparisons between these two systems can lead to ambiguous results (Gedde and Mattozzi, 2004). The synthesis of perfectly linear, defect-free PE via ADMET is therefore a choice model system for studying the behavior of HDPE (O'Gara *et al.*, 1993). One possible drawback in this reasoning is the staggering differences in molecular weight between these step-growth models and the commercial materials. Commercial PE samples can have molecular weights in the millions, orders of magnitude higher than polymers made using metathesis polycondensation chemistry. To prove that these ADMET polymers can truly measure up, a family of linear PE with various molecular weights was synthesized to find the molecular weight required in these materials to meet the benchmark melting temperature possessed by HDPE: 134 °C.

Perfectly linear ADMET PE (Figure 10.3) is easily prepared via the polymerization of 1,9-decadiene, followed by saturation of the product polyoctenamer with hydrogen. Molecular weights of products can be controlled by regulating reaction time, temperature and the monomer/catalyst ratio, affording a family of ADMET PE polymers with varied thermal responses. The samples all exhibit thermal behavior similar to that of high-density polyethylene; despite the relatively low molecular weights (between 2000 and 15 000 g/mol), these polymers all display sharp DSC melting transitions above 130 °C. While the peak melting points increase slightly with increasing molecular weight, the polymer of $M_n = 15\,000$ shows a peak melt of 134 °C, exactly that of high-density polyethylene. It is evident based on this work that beyond the molecular-weight threshold of roughly $M_n = 15\,000$ ADMET polymers can effectively model analogous materials of higher molecular weight made via chain-propagation chemistry (O'Gara *et al.*, 1993).

10.4 Precision Halogenated Polyolefins

ADMET polyolefins with precisely placed halogen atoms provide an excellent system for studying crystallization and melting behavior of precision polyolefins. By synthesizing polymers decorated with fluorine, chlorine, and bromine the effect of this systematic increase in pendant defect size can be probed. Four separate studies (Boz *et al.*, 2007a; Boz *et al.*, 2007b; Boz *et al.*, 2008; Boz *et al.*, 2006) were conducted on this class of precision polyolefins. In the first, the static methylene sequence length between defects was held constant (18 backbone carbons) and the defect identity altered (Boz *et al.*, 2006). In the other three the defect identity was held constant (fluorine, chlorine, or bromine) and the distance between defects altered (Boz *et al.*, 2007a; Boz *et al.*, 2007b; Boz *et al.*, 2008). In the first study it was found, not surprisingly, that with the increased defect size came a decrease in melting temperature and enthalpy (Boz *et al.*, 2006). When the defect identity was held constant the melting temperature and enthalpy decreased with decreasing distance between defects (Boz *et al.*, 2007a; Boz *et al.*, 2007b), the notable exception being the fluorine-containing polymers that all have approximately the same melting temperature despite differences in defect distribution (Boz *et al.*, 2008).

10.4.1 Synthesis of Precision Halogenated Polyolefins

The synthesis of the precision halogenated polyolefins is relatively straightforward (Boz *et al.*, 2006). Regardless of the identity of the halogen atom being incorporated, synthesis begins with the preparation of a secondary diene alcohol via a double Grignard reaction using an alkenyl bromide and ethyl formate. The halogen distribution in the final polymer is controlled by the size of the alkenyl bromide used in the Grignard reaction. This alcohol is converted to the desired halogen atom following simple transformations. In the case of the fluorine- and bromine-containing polymers a single reaction is required: DAST and catalytic pyridine for the fluorine-containing polymers or carbon tetrabromide and triphenylphosphine for the bromine-containing polymers. The chlorine-containing monomers are synthesized in two steps from the diene alcohol; tosylation of the alcohol followed by nucleophilic substitution using lithium chloride. Polymerization of these monomers via 1st-generation Grubb's catalyst, followed by exhaustive hydrogenation using toluene sulfonylhydrazide affords the final precision halogenated polyolefins. Careful NMR characterization confirmed the precise nature of the structures. The precise composition of these polymers in terms of the halogen incorporation was confirmed via thermogravimetric analysis by monitoring the loss of HF, HCl, or HBr during the decomposition of the sample. In all cases the experimental results perfectly match the theoretically predicted values. Figure 10.4 outlines the synthetic route to precision halogenated polyolefins and their thermal behavior (discussed below).

10.4.2 Behavior of Precision Halogenated Polyolefins

10.4.2.1 Precise Fluorine Placement

Regardless of branch distribution, the melting temperatures of ADMET polymers containing precisely placed fluorine atoms are consistent with what is witnessed for linear ADMET PE, evidence that the orthorhombic crystal structure of HDPE is not significantly affected by the addition of the fluorine atom (Boz *et al.*, 2008). WAXS measurements confirm this; only small shifts in the reflections for the orthorhombic unit cell are witnessed, indicating only a minor lattice expansion due to fluorine incorporation. This is not surprising; considering the similarity in size between hydrogen and fluorine the steric requirements for housing this defect in the

Precise halogen placement

Synthesis of precise halogen family

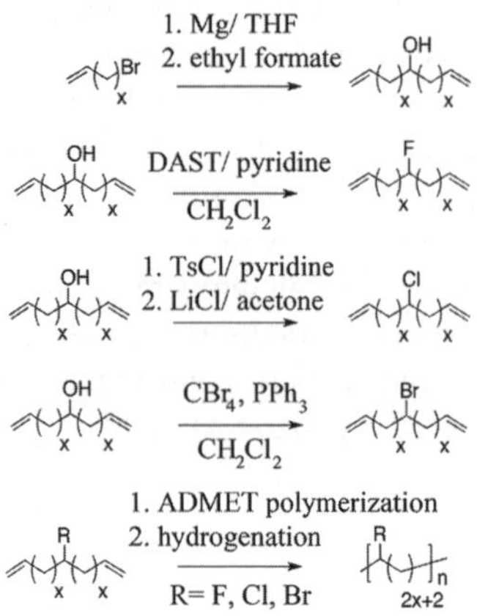

DSC data for precise halogen family

Polymer	T_m (°C) (peak)	Δh_m (J/g)
F9	124	137
F15	124	174
F19	127	207
F21	124	205
Cl 9	41	27
Cl 15	63	87
Cl 19	72	105
Cl 21	81	11
Br9	-14	21
Br15	49	35
Br19	63	55
Br21	70	48

Figure 10.4 Precision halogenated polyolefins.

crystal should not be significantly different. The melting enthalpy of this family of copolymers does, however, decrease with increasing fluorine content. This is much more significant in the cases of F15 and F9 than in the cases of F19 and F21. The decrease in enthalpy for the higher defect concentrations is evidence that electronic repulsions and bond polarity, as well as sterics, play an important role in the crystallization behavior for precision polyolefins (Boz *et al.*, 2008; Boz *et al.*, 2006). Polymers containing the same fluorine content, however with a random rather than precise distribution are not expected to show behavior significantly different from the precision versions (Boz *et al.*, 2008). However, a slightly higher melting temperature was witnessed for these random copolymers, hinting at a different crystallization mechanism in which long methylene sequences without fluorine atoms are able to crystallize first, creating populations of crystallites less affected by the fluorine incorporation.

10.4.2.2 Precise Chlorine Placement

The consequences for the incorporation of chlorine atoms at precise intervals along polyethylene's backbone are much more severe (Boz *et al.*, 2007b). There is a marked decrease in melting temperature compared with both ADMET PE and the precise fluorine family. There is also a change in crystal structure, from orthorhombic to triclinic. Solid-state ^{13}C data show that the distribution of chlorine atoms in the crystalline and amorphous regions is uniform. Further, the lamellar thickness (estimated using atomic force microscopy) far exceeds the distance between defects on the backbone. This confirms that the chlorine atoms are without doubt included within the crystal. The steric requirements, however, are too severe for the orthorhombic crystal structure to remain intact. There is also a clear relationship between the distribution of chlorine atoms and melting behavior. As the chlorine content increases, a decrease in melting temperature and enthalpy is witnessed. This is not surprising; although the chlorine atom is small enough to reside in a triclinic crystal the ability of the polymer crystallize becomes strained as higher amounts of defects become incorporated. In the case of C19, which has the highest amount of chlorine incorporation, a complex melting and recrystallization process is observed, highlighting the stress increased chlorine content imparts on the crystallization of this polymer. This behavior is solely a result of the uniform defect distribution, as random analogs with similar branch content possess slightly broadened melting profiles with higher peak melting temperatures (Boz *et al.*, 2008). This difference in behavior is easily explained by the difference in defect distribution between the random and precise samples. Long, chlorine-free ethylene sequences in the random samples crystallize first. This results in populations of lamella with various thicknesses, which manifest in the broadened melting behavior. This is quite unlike the sharp melting endotherm and homopolymer-like crystallization witnessed in the precise samples, which possess uniform lamella thickness.

10.4.2.3 Precise Bromine Placement

ADMET models with precise bromine placement represent the upper size limit for the study of halogen-containing polymers (iodine-containing samples proved too unstable for use in these studies). The increase in size on moving from a precisely spaced chlorine atom to a precisely spaced bromine atom results in a decrease in both melting temperature and melting enthalpy (Boz *et al.*, 2007a). Like the precision chlorine polymers a triclinic crystal structure was assigned for the bromine-containing polymers. Again, solid-state ^{13}C data confirmed the distribution of the bromine atoms was uniform in both crystalline and amorphous regions like the chlorine polymers. Based on AFM measurements, as is the case for chlorine, the bromine atom must reside within the crystal. In light of this it is clear that the detrimental effects on thermal behavior here are a direct result of the increased steric requirements of the bromine atom, which more severely perturbs the PE crystal than do the smaller chlorine atoms.

The difference in thermal behavior between the precision brominated polyolefins and their random counterparts is much more pronounced than in the previous chlorinated examples. Here, no sharp melting peak is witnessed for the random samples, rather a very broad ill-defined endotherm is observed (Boz *et al.*, 2008). As is the case with the random chlorinated samples a sequence-selection type crystallization mechanism is in operation, resulting in a broad distribution of lamellar thicknesses. The precise versions, similar to the precise chlorine family, all possess sharp, well-defined melting peaks attributed again to uniform lamellar thickness. Perhaps the most interesting feature of the precision bromine polymers is their similarity in thermal response to ADMET polymers with precise methyl branch placement (discussed in the following section). This similarity is based on the similarity in the size between a methyl group and a bromine atom. In conjunction with the data for the precision fluorinated polymers, that is negligible relationship between fluorine content and behavior, it appears that defect steric requirements are perhaps the most important factor in dictating the crystallization and melting behavior of the resulting ADMET polymer crystal. Or, more simply put: a defect is a defect, regardless of its identity (Boz *et al.*, 2007a).

10.5 Precision Alkyl-Branched Polyolefins

For more than half a century studies on structure and morphology have been central to polyolefin research (Gedde and Mattozzi, 2004). Discoveries in this area have allowed for the synthesis of materials with a wide range of properties and behaviors. This is exemplified particularly in the case of linear low-density PE (LLDPE), which are synthesized industrially by the copolymerization of ethylene and α-olefins. Here, the behavior of the polymer can be greatly altered simply by varying the content and identity of the comonomer chosen. In typical industrial polymerizations both interchain and intrachain structural heterogeneities exist, which in turn leads to a broad lamellar thickness distribution and results in decreased mechanical performance. The pristine polymers available via ADMET, free from intrachain and interchain heterogeneities, can serve the dual purpose of elucidating the effects of specific structural features and providing precise control over properties.

10.5.1 Synthesis of Precision Alkyl-Branched Polyolefins

Although relatively straightforward, the syntheses of the polymers used in modeling studies described here are rather tedious. This is in large part because the synthesis of symmetric alkyl branched dienes requires several steps, often with dismal overall yields. In addition, a different synthetic scheme was necessary for each type of branch studied. The synthesis of these materials has been discussed in detail in a number of recent reviews (Baughman and Wagener, 2005; Berda *et al.*, 2006; Lehman and Wagener, 2003; Rojas *et al.*, 2008).

Recently, a new synthetic route (Figure 10.5) to symmetric alkyl branched dienes has been presented and utilized in the synthesis of a whole family of precision alkyl-branched polyolefins (Rojas *et al.*, 2007). This general route, which can be used to create dienes with virtually any branch, involves only two steps with quantitative yields. The first step is the dialkylation of a primary nitrile using freshly prepared lithium diisopropylamide (LDA). Alkylation in this fashion affords a symmetric diene premonomer with the desired alkyl branch and a nitrile group on the central carbon. In the next step, this nitrile is removed by reductive elimination in the presence of potassium metal, HMPA, and *t*-butanol.

The mechanism of this reaction (Rojas and Wagener, 2008) (Figure 10.5) proceeds first with the insertion of an electron from the metal surface into the nitrile forming a radical anion. The

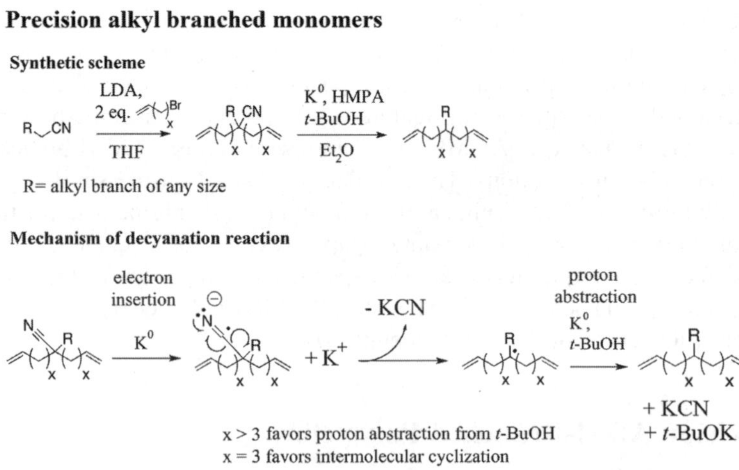

Figure 10.5 Synthesis of precision alkyl-branched monomers.

nitrile then cleaves, liberating a cyanide anion and leaving a tertiary radical on the diene monomer. The diene radical then quickly abstracts a proton from *t*-butanol, resulting in the final ADMET monomer. HMPA is present simply to help solvate the radical species. The only drawback associated with this method, from a synthetic standpoint, is the inability to create monomers that space branches every 9th backbone carbon in the resulting polymer. This is because the rapid radical cyclization of the diene is favored at this particular spacing (5- and 6-membered rings are formed). For dienes with longer spacing between the branch and terminal olefin this cyclization is not favored, however olefin isomerization was observed when other decyanation conditions were employed. Although this method provides a high yielding, efficient route to a variety of ADMET monomers it does involve the use of highly toxic and carcinogenic reagents with an extremely reactive alkali metal in highly flammable solvents. Extreme caution is necessary when conducting such reactions, and small reaction scales should be employed to minimize personal risk.

10.5.2 Behavior of Precision Alkyl-Branched Polyolefins

10.5.2.1 Precise Methyl Placement

Besides being commercially significant, ethylene–propylene (EP) copolymers can provide a general insight into the structure–property relationship when viewed as model systems. Consider polypropylene: highly crystalline when the orientation of the pendant methyl group is highly regular (syndiotactic or isotactic), however, completely amorphous when the backbone methyl groups are randomly oriented (atactic), a simple structural difference resulting in significantly altered behavior. Linear defect-free polyethylene, the other extreme, is highly crystalline. However, this crystallinity can be disrupted by the incorporation of defects, clearly evident in the aforementioned halogen work. Between the extremes of amorphous attactic polypropylene and highly crystalline, defect-free polyethylene lie EP copolymers. Simply varying the amount and placement of the incorporated methyl defect allows the response of the final material to be significantly altered and ultimately controlled (Berda, Baughman, and Wagener, 2006).

Although numerous methods are available for producing model polymeric systems, ADMET modeling controls comonomer content and distribution, therefore leading to fewer ambiguities relative to other model systems when relating structure on the molecular level to macroscopic properties (Berda *et al.*, 2006; Baughman *et al.*, 2006a; Sworen *et al.*, 2003; Smith *et al.*, 2000). Polymerization of symmetrical methyl branched terminal dienes, followed by exhaustive saturation of the resultant polymer, afford these precise EP models (Smith *et al.*, 2000). To date Me5, Me7, Me9, Me11, Me15, Me19, and Me21 have been investigated. The syntheses of Me9 through Me21 are rather straightforward using this simple yet elegant polymerization/hydrogenation approach (Smith *et al.*, 2000). Placing branches every 5th or 7th backbone carbon requires the synthesis of a symmetrical diene dimer, as the corresponding diene monomers undergo ring closing metathesis rather than ADMET (Baughman *et al.*, 2006a). Attempts to place methyl groups every third backbone carbon, using a diene trimer, were unsuccessful due to the placement of the methyl group in the allylic position. This allylic methyl decreases the yield of the crossmetathesis reaction allowing for only the partial oligermerization of the diene, rather than polymerization (Courchay *et al.*, 2006).

The effects of branch distribution are clear when examining the thermal behavior of the precision EP copolymer family (Smith *et al.*, 2000). As defect content increases melting temperature and enthalpy decrease. These precise models are semicrystalline even at branch contents high enough to render random EP copolymers completely amorphous. Not until methyl groups are placed on every 5th carbon do these precise ADMET EP copolymers lose the ability to crystallize (Baughman *et al.*, 2006a). By precisely controlling the sequence of EP copolymers in this fashion it is possible to control the melting temperature of PE over a range of ~200 °C.

Copolymerization of ADMET EP monomers with 1,9-decadiene, thereby forming linear EP copolymers with random branch distribution, has also been accomplished (Sworen *et al.*, 2003). In this study it was again found that as the branch content increased, overall crystallinity as well as the melting temperatures and enthalpies decreased. In the cases of the highest amount of branch incorporation the random materials exhibited a broad, ill-defined melting behavior in contrast to the sharp melting endotherm observed for the precise models with similar branch content. This drastic difference in the behavior between precise and random models punctuates the effect of precise branch placement (Sworen *et al.*, 2003; Smith *et al.*, 2000).

Me21 and Me15 have been further characterized by using X-ray diffraction, TEM, and Raman spectroscopy to further understand their structure and morphology (Lieser *et al.*, 2004). TEM results indicate a lamellar thickness far exceeding the interbranch distance along the backbone, proving that like chlorine and bromine atoms the methyl group must be included within the crystal. The diffraction work elucidated crystal structure: this data showed that the chains pack into a triclinic lattice, which allows inclusion of methyl branches as lattice defects. Further, it was found that methylene sequences between defects participate in a hexagonal sublattice. In order for the chains to pack in this way the defects must be contained within planes oblique to the chain stems. This is more prevalent in the case of Me15 than in Me21 due to the greater defect content, a result confirmed by Raman spectroscopy. The melting-point depression witnessed as methyl content increases is no surprise given these observations. Further scattering experiments and exhaustive DSC experiments performed on Me21 lead to the same conclusion involving defects concentrated in planes between stacks of hexagonally packed methylene sequences; however, a monoclinic lattice rather than triclinic was used by the authors to describe the main unit cell within which the defects planes and hexagonal sublattice resides (Qiu *et al.*, 2006a; Qiu *et al.*, 2006b). The effects of precise methyl branching are summarized in Figure 10.6.

10.5.2.2 Precise Geminal Dimethyl Placement

Precision geminal dimethyl ADMET models display the effect increasing steric bulk has on the polymer's thermal behavior (Schwendeman and Wagener, 2005). The addition of the second methyl group when moving from Me9 to 2Me9 disrupts the polymers ability to pack into crystals resulting in an amorphous material for 2Me9. Extending the interdefect sequence length to 14 or 20 carbons renders the polymer semicrystalline with a depressed melting temperature when compared to the analogous EP models. Interestingly, 2Me15 shows much less melting-point depression from Me15 than does 2Me21 from Me21. Further, 2Me21 shows thermal behavior unlike either 2Me15 or any of the EP family. Exhaustive DSC studies on this material reveal that much of this behavior is dependent on thermal history. WAXS studies show

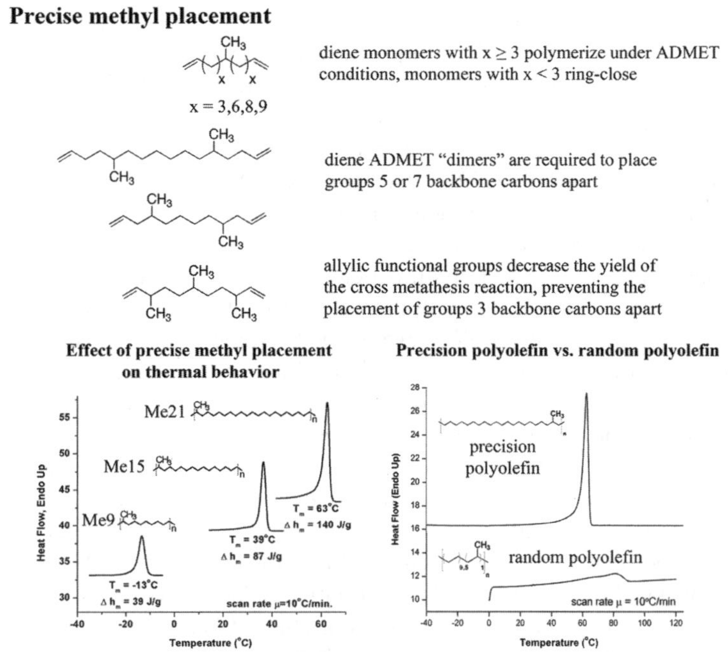

DSC Data for precise methyl family

Polymer	T_m (°C)	Δh_m (J/g)
Me5	Amorphous	–
Me7	−60	19
Me9	−14	28
Me11	11	66
Me15	39	82
Me19	57	96
Me21	63	103

Figure 10.6 Effects of precise methyl-branch placement. (Reprinted from E.B. Berda, T.W. Baughman and K.B. Wagener, "Precision branching in ethylene copolymers: Synthesis and thermal behavior," *Journal of Polymer Science Part A: Polymer Chemistry*, **44**, 17, 4981–4989, © 2006, with the permission of John Wiley & Sons, Inc.)

reflections associated with hexagonal, monoclinic, and triclinic packing, pointing towards polymorphism as a possible cause of this complex behavior. The melting temperature in this instance is coincident with that of eicosane (a 20-carbon *n*-paraffin), implying that crystallization behavior of 2Me21 is strongly related to the branch-to-branch distance. It has been suggested that placing two methyl branches on the chain in this fashion creates a favorable place to induce a fold in the chain stem, resulting in the similarity between the melting point of this polymer and that of eicosane (Qiu *et al.*, 2006b).

10.5.2.3 Precise Ethyl-Branch Placement

Ethylene–butene (EB) copolymers featuring precisely placed ethyl branches were the next logical step in this study; moving from two single carbon defects to a single two-carbon defect

(Sworen *et al.*, 2004). These are of particular interest as EB copolymers are important materials commercially. Like 2Me9, Et9 is fully amorphous. Again, extending the space between defects allows for crystallization in both Et15 and Et21. Like the geminal dimethyl models the EB models show greatly depressed melting temperatures when compared to the EP models. Another point of interest is the difference in the observed thermal behavior from both the geminal dimethyl family and the EP family. Et21 and Et15 exhibit bimodal melting profiles very much unlike the sharp, uniform melt exhibited by the EP family (Sworen *et al.*, 2004).

WAXS investigations (Sworen *et al.*, 2004), as well as exhaustive DSC analysis (Qiu *et al.*, 2006b) have helped in explaining this behavior. Like 2Me21, the melting behavior of Et21 can be correlated with that of eicosane and is therefore very much dependent on the branch-to-branch distance. The WAXS results shows some lattice expansion implying the partial inclusion of ethyl groups into the crystal, however to a lesser extent than in EP21. Comparing these results for 2Me21 and Et21 implies that much of the melting behavior is attributed to crystallization of methylene sequences between defects. The bimodal melt could therefore be a result of polymorphism involving the inclusion and exclusion of these defects, or a melting and simultaneous crystallization mechanism.

A second study on the morphology of Et21 using WAXS, SAXS, DSC, and TEM has further helped in understanding the behavior of this system (Hosoda *et al.*, 2009). The data collected using this variety of techniques indicated the formation of lamella of uniform thickness for this polymer. The measured thickness of the lamella, about 55 Å, is approximately double the distance between branches along the backbone, indicating on average that about one ethyl branch will be included in the lamella stem. In this case, scattering experiments point towards a psuedohexagonal unit cell for this polymer. The unusual melting behavior witnessed, then, is likely due to crystal forms that both partially include and exclude this branch. Samples of EB copolymers made using chain-propagation techniques with similar branch content displayed a much broader distribution of lamellar thicknesses, again highlighting the ability of precise defect placement to create materials with precise morphology.

10.5.2.4 Precise Butyl-Branch Placement

ADMET ethylene/hexene (EH) copolymers provide an interesting system to model commercial LLDPE materials containing butyl branches (synthesized by the copolymerization of ethylene with 1-hexene) and industrial LDPE synthesized via radical chemistry (where the short-chain branch identity is primarily butyl due to back-biting during polymerization). These industrial materials typically show broad ill-defined melting behavior due to selective crystallization of long defect-free ethylene run lengths. The precision butyl polymers prepared by ADMET, however, show the sharp, well-defined melting and crystallization transition typical of other ADMET polymers. The melting temperature for Butyl21 is 14 °C, significantly depressed from that of Et21. This is not surprising considering the increased steric requirements of a 4-carbon branch compared with a 2-carbon branch. The low melting temperature likely indicates very small or very distorted crystals. Even Butyl15 is semicrystalline (albeit with an extremely low melting temperature, −33 °C). This is far above the branch concentration required to render similar polymers with random branch placement completely amorphous. When comparing the precise butyl polymer to a random sample with similar branch content, the effects of precise placement once again become perfectly clear: the precise material displays a sharp well-defined melt in contrast to a broad, ill-defined melting behavior of the random versions.

10.5.2.5 Precise Hexyl-Branch Placement

ADMET Ethylene–octene (EO) copolymers serve as LLDPE models with precise hexyl branch placement (Sworen and Wagener, 2007). These precise EO models follow a similar trend in behavior as the previously discussed families; that is with increasing branch content decreasing melting temperature and enthalpy are observed. It is no surprise that Hexyl9 is totally amorphous, as the much smaller ethyl branch is able to completely disrupt crystallinity at this branch concentration. A semicrystalline morphology is observed for Hexyl15; an interesting result considering all other known EO copolymers with similar branch content are completely amorphous (Sworen and Wagener, 2007). Hexyl21 is also semicrystalline. The very low melting temperature (16 °C) is indicative of very small crystallites. Interestingly, the melting enthalpy of Hexyl21 is similar to that of Et21, which is likewise surprising considering the notable decreases in enthalpy from Me21 to 2Me21 to Et21. The melting profile of the Hexyl21 closely mimics that of Me21 with a single sharp melting endotherm, rather than the complex endotherms displayed by 2Me21 and Et21.

Like Et21, Hexyl21 was thoroughly characterized in a separate study via WAXS, SAXS, DSC, and TEM (Hosoda *et al.*, 2009). The result in this case showed again that the precision sample contained uniform lamella thickness. The scattering data support monoclinic packing, although with much less lattice expansion than either the ethyl or methyl branched ADMET polymers. In fact, these data agreed quite well with monoclinic homopolyethylene reflections, implying that the hexyl branch doesn't significantly affect the lattice dimensions. The measured lamellar thickness in this case was 26 Å, which correlates precisely to the distance between hexyl branches along the backbone. These results prove, beyond any doubt, that the hexyl branch must be completely excluded from the crystal. Figure 10.7 compares the TEM images of commercial PE, Et21, and Hexyl21.

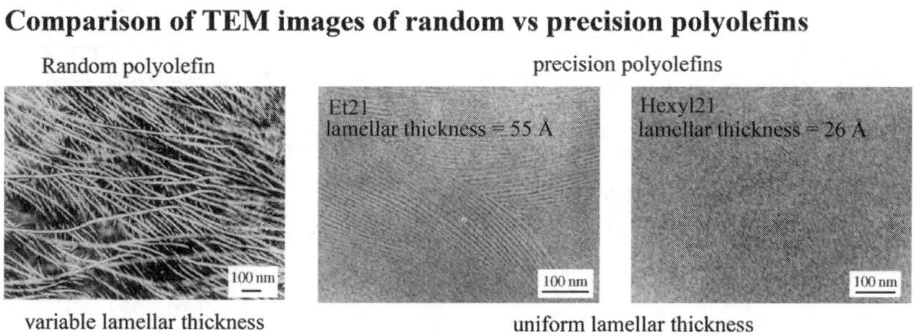

Figure 10.7 TEM images clearly show the effects of precise branch placement: random polyolefins show a broad distribution of lamellar thicknesses; precision polyolefins feature lamella of uniform thickness. (Reprinted from S. Hosoda, Y. Nozue, Y. Kawashima *et al.*, "Perfectly controlled lamella thickness and thickness distribution: A morphological study on ADMET polyolefins," *Macromolecular Symposia*, **282**, 50–64 © 2009, with the permission of John Wiley & Sons, Inc.)

10.5.2.6 Probing the Limit for Branch Inclusion vs. Exclusion

There is an interesting correlation between the melting behavior of Butyl21, and Hexyl21: the thermal profiles are nearly identical. Considering that the hexyl branch is completely excluded from the crystalline lamella in Hexyl21, it is likely that the same is true for Butyl21. This prompts an interesting question: if an ethyl group is partially included but butyl or hexyl groups are excluded what is the limit on branch length for inclusion in or exclusion from the crystal? Also, is crystallization with branch exclusion limited to *n*-alkane branches or can polyolefins retain the ability to crystallize even with bulkier pendant groups such as isopropyl or *t*-butyl? To probe this behavior an entire family of precision alkyl-branched polyolefins with increasing branch length (methyl, ethyl, propyl, butyl, pentyl, and hexyl), and increasing steric bulk (isopropyl, sec-butyl, t-butyl, cyclohexyl, and adamantyl), were examined (or re-examined) by DSC and WAXS in an attempt to find these limits (Rojas *et al.*, 2009).

The thermal behavior (Figure 10.8) exhibited a clear result: all branches longer than ethyl display nearly identical behavior. When increasing steric bulk, even the polymer decorated with an enormous adamantyl group is able to crystallize. Although there is a slight depression in peak melting temperature, the behavior of the bulky substitutes closely resembles that of

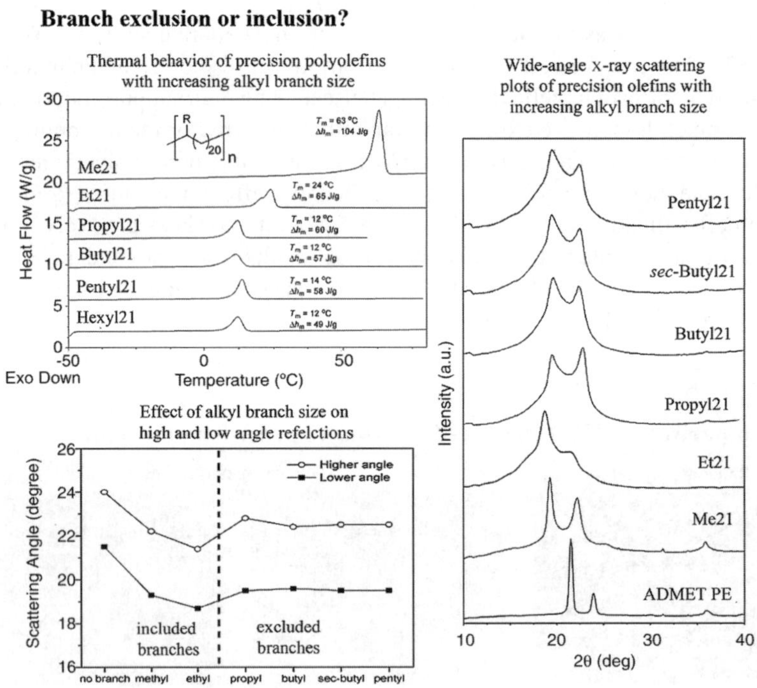

Figure 10.8 Probing the limits of branch inclusion: alkyl branches larger than an ethyl group will be completely excluded to amorphous regions. (Reprinted with permission from G. Rojas, B. Inci, Y. Wei and K.B. Wagener, "Precision polyethylene: Changes in morphology as a function of alkyl branch size," *Journal of the Americal Chemical Society*, **131**, 17376–17386, 2009. © 2009 American Chemical Society.)

the linear branched analogs. The scattering data shows clear lattice expansion when going from Me21 to Et21, as discussed above. Beyond Et21, however, the reflections show a contraction in the lattice dimensions that remain constant even as branches get larger and bulkier. This is clear proof that the ethyl branch is the size limit for alkyl branch inclusion in PE; anything larger in terms of branch length or steric bulk will be excluded to amorphous regions (Rojas *et al.*, 2009).

10.5.3 Increasing the Spacing Between Alkyl Branches in Precision Polyolefins

The results in this section bring up an extremely important point about the structure–property relationship in PE: by perfectly controlling branch placement and ensuring that this branch possesses sufficient steric bulk to be excluded from crystalline lamella (i.e. the branch dictates where a chain stem will fold) it is possible to precisely control the secondary structure and thus the physical properties of these polymers. Although this is intellectually interesting, it is of little practical value if these materials melt below room temperature. For this reason, a synthetic route was developed to reduce the branching frequency and ultimately increase the melting point (Zuluaga *et al.*, 2009). As a proof of concept study, Butyl39 was synthesized using the quantitative nitrile alkylation/decyanation route. The alkenyl bromides used in this case are not commercially available, but can be easily made by coupling 11-bromo-undecene via crossmetathesis, hydrogenating the olefin, and performing a monoelimination on the resulting terminal dibromide. This method, although demonstrated for spacing branches every 39th backbone carbon, could be adopted to space them even further. DSC and WAXS measurements on the polymer prove that this concept works perfectly: scattering reflections indicate that the orthorhombic crystal structure of polyethylene remains intact at this branch spacing. The melting behavior, similar to other polymers in this family, is very sharp and well defined, consistent with uniform lamella thickness. The melting temperature, 75 °C, is coincident with that of a 40-carbon *n*-paraffin molecule suggesting that the thickness of the lamella is consistent with the branch-to-branch distance. This method shows great promise for the synthesis of high melting precision polyolefins with mechanical properties superior to any current commercial material.

10.6 Precison Ether-Branched Polyolefins

Precision polyolefins with the regular placement of ether groups represent the first examples of linear ethylene-*co*-vinyl ether (EVE) copolymers ever made (Baughman *et al.*, 2005; Baughman, vanderAa, and Wagener, 2006b); the difference in reactivity ratios between ethylene and vinyl ethers prevented the synthesis of this type of material via chain-propagation chemistry until very recently (Luo *et al.*, 2007). Although they do not model a commercially important industrial copolymer, this class of precision polyolefins is closely related to the alkyl-branched precision polyolefin family, the only difference being the exchange of a methylene group for an oxygen atom directly off the PE backbone. While this slight change in sterics and bond polarity seems subtle, comparisons between the two provide a great deal of information concerning the consequences of this type of structural manipulation and its effect on the behavior of precision polyolefins.

10.6.1 Synthesis of Precise Ether-Branched Polyolefins

The diene monomer required to make polyolefins with precisely placed ether groups is synthesized in one step from a symmetric secondary diene alcohol (the synthesis of which is discussed in Section 10.4.1) using the classic Williamson ether synthesis. Polymerization followed by hydrogenation results in a defect-free, linear-sequenced copolymer of ethylene and either methyl vinyl ether or ethyl vinyl ether, as confirmed by careful NMR characterization.

10.6.2 Behavior of Precise Ether-Branched Polyolefins

Similar to other ADMET precision polyolefin families the precision EVE family shows a clear relationship between branch size and placement and the observed thermal behavior. As branch content increases, melting temperature and enthalpy decrease. Likewise, increasing strerics from a pendant methyl ether group to a pendant ethyl ether group generally results in a decrease in peak melting temperature. There is a notable exception in this case: OMe11 and OEt11, where the polymer with the larger pendant group has a higher peak melting temperature and exhibits a very complex thermal profile.

The most useful information provided by this family of precision polyolefins comes when comparing their behavior to that of the corresponding alkyl-branched polymers, that is OMe21 with Et21 and OEt21 with Propyl21. The substitution of an oxygen atom for a methylene group results in only a slight increase in the peak melting temperature and melting enthalpy for OMe21 when compared to Et21. As is the case in several of the previous examples, this melting point is coincident with that of eicosane, suggesting a strong relationship to the branch-to-branch distance. The FTIR data on this polymer indicates a significant number of kink defects, similar to Et21. As the steric requirements are roughly the same for these two polymers it is reasonable to assume that this group could be included, at least partially, within the unit cell.

The effect of this slight structural change is much more pronounced when the size of the ether branch is increased by one methylene unit. When compared to Propyl21, OEt21 has an increase of 12 °C in melting temperature and 19 J/g in melting enthalpy. Based on the alkyl-branching data it seems unlikely that a defect of this size is included within the unit cell, as the smaller ethyl group delineates this boundary when placing defects every 21st backbone carbon. In this case it can be reasonably concluded that the polarity of the bond and the electronics of the oxygen create a favorable place to initiate a fold in the polymer backbone, allowing for a more complete crystallization of the interdefect paraffic unit cell, resulting in the increased melting temperature and enthalpy.

Comparing the behaviors of OMe15, OEt15, and Et15 leads to another rather interesting observation. The thermal profile of Et15 shows two distinct melting endotherms, likely due to two populations of lamella that partially include or completely exclude the ethyl-branch defect at this branch content. OMe15 and OEt15 each show a single melting endotherm, closely matching the two melting points of Et15. It appears that at this branch concentration the addition of the oxygen atom allows the material to prefer a single crystal form at each of these defect sizes. Figure 10.9 summarizes the synthesis and thermal properties of these polymers.

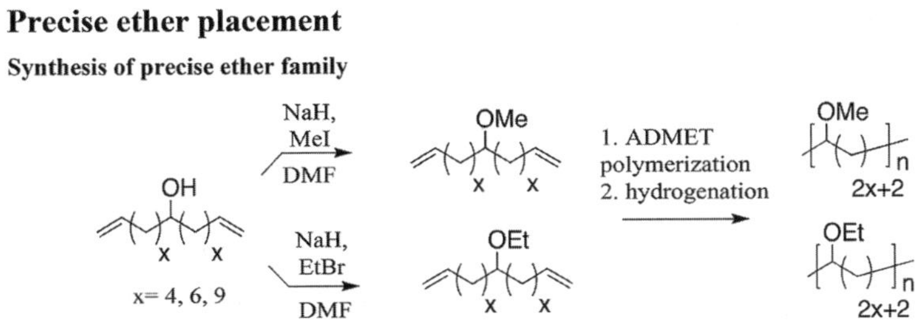

DSC comparison of Et21 and OMe 21

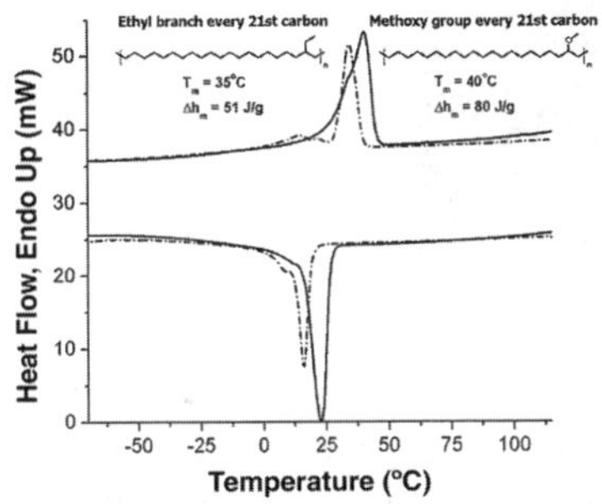

DSC data for precise ether family

Polymer	T_m (°C) (peak)	Δh_m (J/g)
OMe11	−41	35
OMe15	−10	62
OMe21	40	78
OEt11	−4	66
OEt15	−33	82
OEt21	28	96

Figure 10.9 Synthesis and thermal behavior of polyolefins with precise ether-branch placement. (Reprinted from E.B. Berda, T.W. Baughman and K.B. Wagener, "Precision branching in ethylene copolymers: Synthesis and thermal behavior," *Journal of Polymer Science Part A: Polymer Chemistry*, **44,** 17, 4981–4989, © 2006, with the permission of John Wiley & Sons, Inc.)

10.7 Precision Acid-Functionalized Polyolefins

10.7.1 Precise Carboxylic Acid Placement

Ethylene/acrylic acid copolymers are a commercially important class of materials due to ease and low cost of production and multiple end use applications (Baughman *et al.*, 2007). By simply varying the acid content in the polymer the properties of the resultant material can be tuned over a wide range of responses. The combination of polyethylene's crystallinity and hydrogen bonding between carboxylic acid groups provides these materials with improved mechanical performance over homopolyethylene. The synthesis of these materials industrially requires chain-propagation techniques that again result in materials subject to intrachain and interchain heterogeneities. By utilizing ADMET chemistry to synthesize precision carboxylic-acid-containing polyolefins, these irregularities can be minimized and the effects of specific structural features probed. This study also compared the behavior of the precision carboxylic acid containing polymers to that of linear, defect-free polymers prepared using ROMP that feature random acid placement.

10.7.1.1 Synthesis of Precision Carboxylic Acid-Functionalized Polyolefins

Synthesis of the symmetrical diene acid (Figure 10.10) is easily prepared in 3 steps beginning with diethyl malonate and the appropriate alkenyl bromides. To ensure that the acid moiety

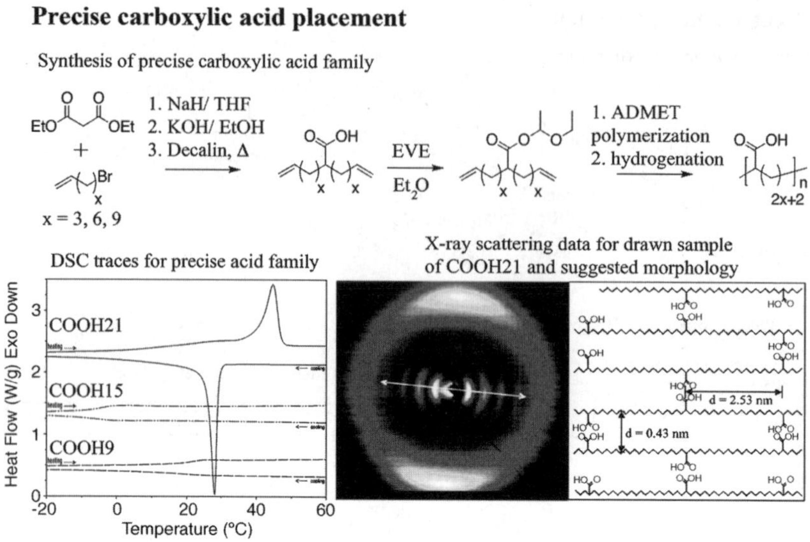

Figure 10.10 Synthesis, thermal behavior, and scattering data of precision carboxylic acid functionalized polyolefins suggesting a layered morphology. (Reprinted with permission from T.W. Baughman, C.D. Chan, K.I. Winey and K.B. Wagener, "Synthesis and morphology of well-defined poly(ethylene-co-acrylic acid) copolymers," *Macromolecules*, **40**, 18, 6564–6571, 2007. © 2007 American Chemical Society.)

does not poison the ruthenium catalyst during polymerization hemiacetal ester protection chemistry was employed. This protecting group is easily cleaved during the hydrogenation step or thermally to afford the final material. Careful ^{1}H and ^{13}C NMR experiments confirm the precise structure. IR spectroscopy revealed, interestingly, that all carboxylic acids are dimerized in the final material.

10.7.1.2 Behavior of Precision Carboxylic Acid-Functionalized Polyolefins

The effect of incorporating a functional group capable of supramolecular interactions is evident in behavior of these materials relative to the precision alkyl-branched polyolefins (Baughman *et al.*, 2007) (Figure 10.10). In the case of COOH21 the melting temperature is ~20 °C higher than Et21. COOH15, however, is completely amorphous. This is clearly an affect of the carboxylic acid dimerization. In the case of the lower acid content this interaction between acid groups has a positive effect on the polymer's crystallization, serving to produce more stable, higher-melting crystals than analogous precision polyolefins with pendant groups of similar steric requirements. However, as acid content increases these interactions overwhelm the system preventing the interacid methylene sequences from organizing into lamella. COOH9 has a higher T_g than COOH15, attributed again to dimerized carboxylic acids, essential physical crosslinks, restricting mobility to a greater extent at higher concentrations. The materials with random acid placement made by ROMP possess much broader melting and crystallization profiles with higher peak melting temperatures compared to the precision materials, again attributed to the selection of long ethylene run lengths during crystallization.

At the highest amount of acid incorporation studied, equivalent to that of COOH9, the random materials become amorphous as well, although with a significantly depressed T_g compared to the precision material; displaying again the effect precise placement has on these materials.

Despite the fact that no crystallinity is observed in DSC experiments on COOH9 and COOH15, there is still evidence that these systems possess order. Scattering data show reflections corresponding exactly to the interacid distance along the backbone for all three polymers in this family. In COOH21 this is very pronounced and multiple orders of diffractions corresponding to 2.53 nm are present, indicating a high degree of ordering based on the interacid distance along the backbone.

To better understand this phenomenon thermal and scattering experiments were conducted on drawn samples, a well-known method for inducing anisotropy in polymer samples. For COOH9 and COOH15 no anisotropy was observed upon drawing. Drawing COOH21, however, produced noticeable changes in both the thermal and scattering behavior. The melting profile of the drawn sample is bimodal, suggesting two populations of crystallites. The scattering data (Figure 10.10) showed several reflections perpendicular to the draw axis that again perfectly correspond to the acid–acid distance along the backbone and multiples thereof. The authors speculate that in the case of the drawn sample dimerized carboxylic acids are able to lie within the orthorhombic unit cell perpendicular to the chain stems. This unique behavior is likely responsible for this polymer's excellent mechanical strength. This system is a superb example demonstrating the ability of using precise primary structure to control and improve the properties of traditional ill-defined chain-made materials.

10.7.2 Precise Phosphonic Acid and Sulfonic Acid Ester Placement

Polyethylene containing sulfonic and phosphonic acids is attractive for a wide range of applications, particularly in the realm of biomaterials and membrane technologies (Opper *et al.*, 2009; Opper and Wagener, 2009). These materials are difficult to make industrially: reactivity ratios prevent efficient comonomer incorporation. Postfunctionalization, for example via phosphorolation of PE or sulfonation of PS results in materials with no microstructural control. Homopolymers of vinyl sulfonic, styrene sulfonic, and vinyl phosphonic acids are well studied, but again these materials are made using chain-propagation techniques that are again subject to interchain and intrachain heterogeneities as well as structural defects. Studying precision polyolefins containing these groups can simultaneously probe the effects of efficient incorporation of these groups into PE and the effects of regular spacing and controlled composition.

10.7.2.1 Synthesis of Precision Phosphonic Acid and Sulfonic Acid Ester-Functionalized Polyolefins

The synthesis of symmetric dienes utilized in the fabrication of these polymers begins with the sequential dialkylation of ethyl methane sulfonate, ethyl toluene sulfonate (Opper and Wagener, 2009), or ethyl methane phosphonate (Opper *et al.*, 2009) using freshly prepared LDA and an alkenyl bromide. Standard ADMET polymerization/hydrogenation techniques are used to convert the protected acid-containing dienes to the corresponding protected acid-functionalized precision polyolefins. Deprotection of the posphonic acid esters proceeds quantitatively upon treatment with trimethylsilyl bromide in methanol; the deprotection of the sulfonic acid ester-containing precision polyolefins has not been reported as of yet. Thorough ^{1}H, ^{13}C, and ^{31}P

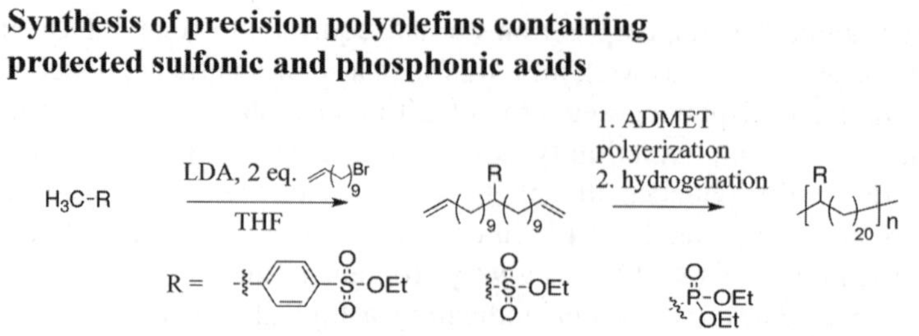

Figure 10.11 Synthetic route to precision polyolefins functionalized with phosphonic acids and sulfonic acid esters.

(in the case of the phosphonic acid polymers) NMR characterization confirms the pristine structure of these materials. Figure 10.11 shows the synthetic route to these polymers.

10.7.2.2 Behavior of Precision Phosphonic Acid and Sulfonic Acid Ester-Functionalized Polyolefins

Thermal characterization using DSC reveals very different behavior for the two precision sulfonic acid ester polyolefins (Opper and Wagener, 2009). SO_3Et21 is semicrystalline, exhibiting a sharp melting endotherm and crystallization exotherm typical of ADMET precision polyolefins. The melting temperature in this case is 29 °C, very similar to the behavior of OEt21, albeit with a decreased melting enthalpy (not surprising considering the increased steric requirements of this bulkier defect). This data suggests, as is the case for the precise ether family, that this group is likely excluded from the crystal. Here again, the polarity of this group creates favorable conditions for the crystallization of the inter defect ethylene run length. Scattering data is not yet available for this polymer, however, so drawing absolute conclusions about the morphology of this system becomes difficult.

The placement of a phenyl group spacer between the polyolefin backbone and the sulfonic acid ester moiety induces a significant change in behavior (Opper *et al.*, 2009). PhSO_3Et21 is completely amorphous with a glass transition at 24 °C. Apparently, the sterics of the phenyl group are too severe in this case to allow crystallization of the backbone. It is also possible that interactions between pendant functional groups (e.g. the relatively electron deficient phenyl ring on one pendant pairing with the relatively electron rich SO_3Et group on another) in essence create supramolecular physical crosslinks, similar to the precise carboxylic acid family, which impedes the ability of the backbone to organize into lamella.

Although no thermal data was reported on the precision phosphonic acid functionalized polymer solid-state NMR experiments can provide insight into the morphology and behavior of this system. The ^{13}C-CP MAS spectrum for this polymer shows peaks consistent with all-*trans* crystalline PE, indicating some degree of order for this polymer. Temperature-dependent NMR experiments reveal that the motion of the phosphonic acids groups are minimal even at elevated temperatures. This suggests strong hydrogen-bonding interactions between acid groups; the authors attribute the toughness of this polymer and its poor solubility to these supramolecular interactions. Chemical crosslinking (via anhydride formation) can be ruled out as the cause of this behavior based on ^{31}P NMR data.

10.8 Precision Amphiphilic Copolymers

We have discussed in the previous sections how increasing the size of a pendant branch and its concentration lead to perturbations in the preferred crystal structure of polyethylene, until, at a certain point, the branch is entirely excluded from the crystal or the material itself is rendered completely amorphous. We have likewise discussed how properties can further be affected by noncovalent interactions between these pendant branches. The studies presented here on precision amphiphilic copolymers combine these principles to show how such information could be used to design materials with a specific behavior or application in mind. The concept is relatively straightforward: placing an oligo(polyethylene glycol) (PEG) branch at regular intervals along a PE backbone takes advantage of the branch's size (large enough to be excluded) and the incompatibility of the branch with the backbone to afford materials with two distinct phases, one phase of crystalline PE and one phase of amorphous PEG (Berda *et al.*, 2007; Berda and Wagener, 2008b). Then, by altering the identity of the PEG branch end group the interaction of the PEG phase can be adjusted to separately tune its properties without significantly affecting the PE segments (Berda and Wagener, 2008a). Results indicate this is indeed feasible, and that it is even possible to "trick" the backbone into crystallizing with the excluded pendant group.

10.8.1 Synthesis of Precision Amphiphilic Copolymers

Synthesizing these intricate materials involves only a few simple organic transformations (Berda *et al.*, 2007; Berda and Wagener, 2008a; Berda and Wagener, 2008b) (Figure 10.12). Reduction of the symmetric carboxylic acid diene described in Section 10.7.1 using lithium aluminum hydride affords a symmetric diene primary alcohol, which serves as the basis for all

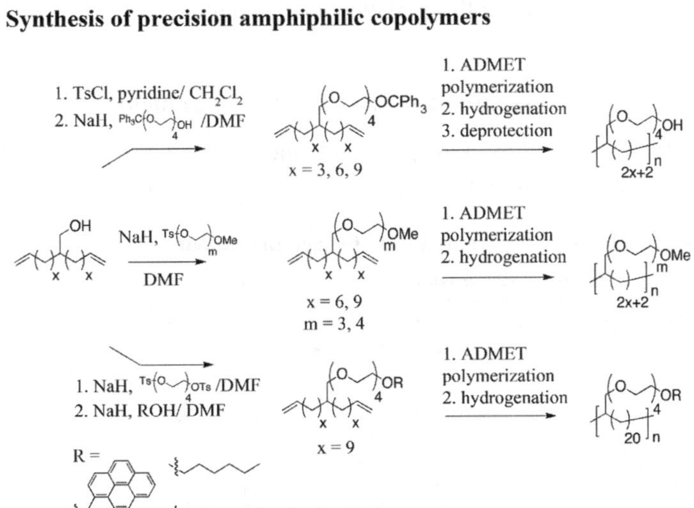

Figure 10.12 Synthetic route a variety of precision amphiphilic copolymers. (Reprinted from E.B. Berda and K.B. Wagener, "ADMET polycondensation of diketopiperazine-based dienes. Polymerization behavior and effect of diketopiperazine on the properties of the formed polymers," *Macromolecular Chemistry and Physics*, **209**, 1601–1611 © 2008, with the permission of John Wiley & Sons, Inc.)

of the monomers described here. This alcohol can be deprotonated and used as the nucleophile in a Williamson ether synthesis to attach the desired branch (Berda and Wagener, 2008a; Berda and Wagener, 2008b), or it can be tosylated and used as an electrophile (Berda *et al.*, 2007). Standard ADMET polymerization/hydrogenation techniques are employed to yield the final precision amphiphilic copolymers.

10.8.2 Behavior of Precision Amphiphilic Copolymers

10.8.2.1 Precise Methoxy-Terminated PEG Branch Placement

To ensure the concept works and the branches are indeed excluded from the crystal a set of polymers with methoxy-terminated PEG branches of two different sizes (3- or 4-oxyethylene repeating units) were synthesized and their behaviors compared (Berda and Wagener, 2008b). Evaluating the thermal profiles for 3EGOMe branched and 4EGOMe branched materials shows that indeed the size of the PEG branch has no effect on the behavior of the material. These polymers all display a T_g at about −75 °C regardless of pendant branch size or distribution, which corresponds to a fully amorphous, primarily polyether phase. These polymers also exhibit the sharp crystalline melting endotherms characteristic of ADMET polyolefins. The melting temperatures, as is the case for other precision polyolefins, decrease with increasing branch content. The temperature of the melt in the cases of 3EGOMe21 and 4EGOMe21 are very similar to that of the Hexyl21 (and its relatives), indicating that this endotherm is surely an effect of the PE backbone crystallizing to exclude the PEG branch. This is punctuated when examining the thermal profile of the corresponding polymers with unsaturated backbone: this behavior is severely disrupted (for branches every 21st backbone carbon) or completely destroyed (for branches every 15th backbone carbon). Figure 10.13 compares the thermal behavior of the polymers in this family.

10.8.2.2 Precise Hydroxy-Terminated PEG Branch Placement

To evaluate the affect interactions of the pendant branches could have on the behavior of these precision amphiphilic copolymers a family of similar polymers with hydroxy-terminated PEG

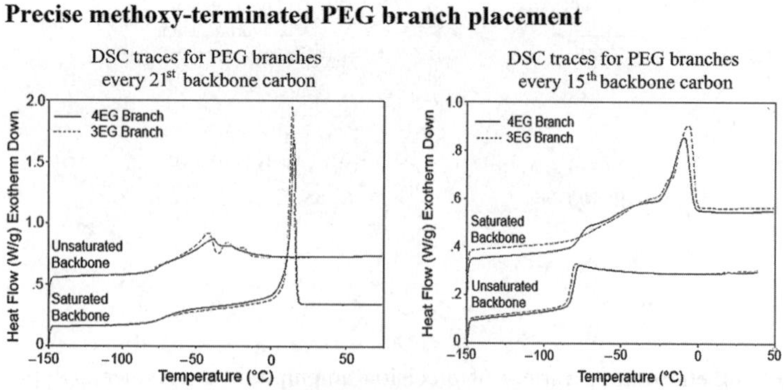

Figure 10.13 Thermal behavior of precision amphiphilic copolymers featuring precise placement of methoxy terminated PEG branches.

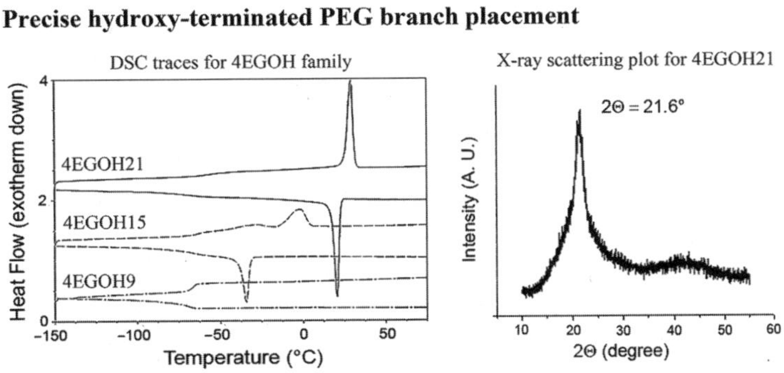

Figure 10.14 Thermal behavior and scattering data of precision amphiphilic copolymers featuring precisely placed hydroxy terminated PEG branches. (Reprinted with permission from E.B. Berda, R.E. Lande and K.B. Wagener, "Precisely defined amphiphilic graft copolymers," *Macromolecules*, **40**, 8547–8552, 2007. © 2007 American Chemical Society.)

branches was synthesized (Berda *et al.*, 2007). In this study the typical full ADMET series with branches placed every 9th, 15th, and 21st backbone carbon atom was examined. The DSC traces for these polymers are shown in Figure 10.14. 4EGOH9 is completely amorphous, unsurprisingly, with a T_g at −65 °C. On increasing the branch spacing, 4EGOH15 is semicrystalline with complex melting behavior (peak melting temperature of −3 °C) that is very much dependent on thermal history. The T_g for the polyether phase is −63 °C in this case. Again increasing the distance between branches to 4EGOH21 doesn't affect the T_g of the excluded polyether phase, but as expected results in an increase in melting temperature to 29 °C. Scattering experiments conducted on this sample show reflections indicative of a slightly contracted hexagonal unit cell for this polymer and absolutely confirms that the polyether branch is excluded from the crystal in these materials (Figure 10.14).

Comparing the behavior of 4EGOMe21 and 4EGOH21 shows the roll branch end-group interactions play in the thermal behavior of precision amphiphilic copolymers. The T_m of 4EGOH21 is 13 °C higher than 4EGOMe21. IR spectroscopy confirms that hydrogen bonding between the free −OH groups occurs, and the increase in T_g for the 4EGOH family compared to the 4EGOMe family reflects this. The more restricted motion of amorphous PEG phase, due to the interaction between branch end groups, apparently imparts a bit more order to this system. This ultimately allows for a more complete crystallization of the methylene sequences between PEG branches, resulting in the increase in melting temperature.

10.8.2.3 Precise Amphiphilic Branch Placement

Given the noticeable effect of this structural manipulation in the previous example, three polymers with different PEG branch end groups were synthesized to probe the limits of this behavior (Berda and Wagener, 2008a). In this study, the distance between the branches was held constant (branches every 21st carbon) to maximize the propensity of the backbone to crystallize in order to isolate the effects of the branches. By the same reasoning the behavior of the unsaturated version of these polymers was investigated as well (a lowercase "u" denotes an unsaturated backbone in the discussion below). The three polymers exhibit remarkably different thermal behavior (Figure 10.15), a clear indication that changing the graft end-group moiety has significant affect on the morphology of these systems. The difference in behavior between the saturated and unsaturated polymers is also significant, emphasizing the role of backbone crystallization in these materials.

The heating and cooling curves for 4EGOPy21 and 4EGOPy21u show that the saturation of the backbone allows for crystallization to occur, while the unsaturated polymer remains completely amorphous. This suggests that the pendant defect does not contribute to the crystallization as expected. The T_g of the saturated polymer is also slightly increased compared to the unsaturated analog, an indication that segmental motion of the grafts is restricted by the crystallinity of

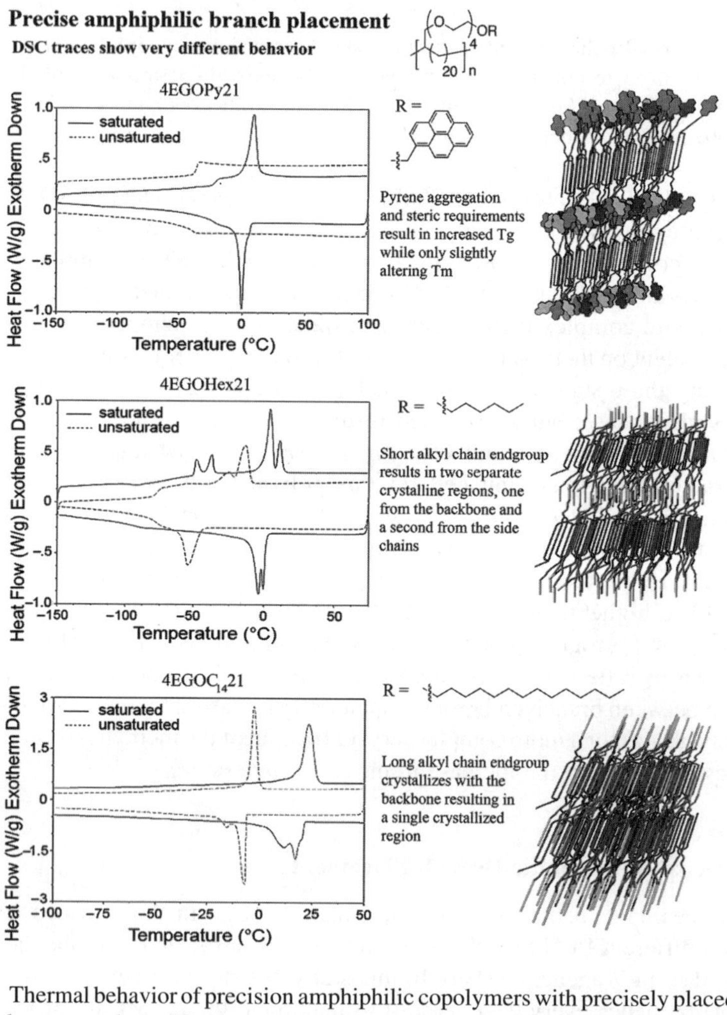

Figure 10.15 Thermal behavior of precision amphiphilic copolymers with precisely placed amphiphilic branches: simple structural modifications at the PEG branch terminus result in significant changes in behavior and morphology. (Reprinted with permission from E.B. Berda and K.B. Wagener, "Inducing pendant group interactions in precision polyolefins: Synthesis and thermal behavior," *Macromolecules*, **41**, 5116–5122, 2008. © 2008 American Chemical Society.)

the PE backbone. The completely amorphous behavior of 4EGOPy21u is an interesting result, considering that previous reported unsaturated ADMET polymers with the same distribution of pendant functionality are semicrystalline (Baughman and Wagener, 2005; Berda *et al.*, 2006; Lehman and Wagener, 2003). The lack of crystallinity in 4EGOPy21u, as well as the depressed melting point for 4EGOPy21 compared to other polymers in this family is most likely a result of pyrene aggregation (confirmed by fluorescence spectroscopy) that impedes backbone crystallization, similar to the effect or carboxylic acid dimers in the precise COOH family.

The difference between 4EGOHex21 and 4EGOHex21u in terms of thermal response are especially interesting. The unsaturated polymer is semicrystalline with a melting endotherm at −13 °C. There is significant amorphous content to this material as well, indicated by the distinct T_g at −76 °C. The thermal behavior of the saturated analog is completely different, however. A single, bimodal crystallization at −4 °C is witnessed on cooling. Upon heating, a small exotherm is barely observed at −91 °C, followed by a bimodal melting endotherm with peaks at −48 °C and −37 °C. A second bimodal melting endotherm occurs with peaks at 4 °C and 11 °C. This complex behavior for 4EGOHex21 was further investigated by MDSC, which provides increased sensitivity compared to traditional DSC and allows for resolution of overlapping transitions (Shieh and Liu, 2007; Wunderlich, 2006; Schawe, 1997; Jones *et al.*, 1997). The MDSC data indicates a melting and annealing process is at work in these two endotherms. Because the melting enthalpy of this higher temperature peak matches the enthalpy of crystallization, and also because its melting temperature and enthalpy (4 °C and 27 J/g) are in good agreement with those of TEGOPy21 (9 °C and 24 J/g), it can be concluded that this behavior is a result of crystallization of the backbone excluding the pendant branch during cooling. During subsequent heating, the excluded branches gain sufficient mobility to self-crystallize, noted by the exotherm at −100 °C. The pendant crystals then melt at about −50 °C, followed by the melting of the backbone crystals.

The thermal behavior of $TEGOC_{14}21u$ and $TEGOC_{14}21$ differs significantly from the previous two examples. Both polymers exhibit extremely sharp melting transitions at temperatures much higher than the other polymers in this family. This deviates from the well-known trend discussed in this chapter for ADMET polymers, which typically show a decrease in melting temperature and enthalpy as the pendant group size increases. For $TEGOC_{14}21u$ and $TEGOC_{14}21$, the increase in pendant group length results in marked increases in both melting temperature and enthalpy. Further, no T_g is observed in either the saturated or unsaturated analogs, indicating the absence of appreciable amorphous content for these polymers. Thus, the long alkyl chains at the termini of the PEG branches must also be crystallizing. Since there is only one melting peak, the C_{14} chain must be long enough to extend back into the crystallizing PE backbone, forming a single crystallized region. By cleverly playing with structure the authors in this case were able to completely exclude the branch from the crystal at the point of attachment, but "trick" the backbone into reincluding the terminus of the excluded branch, effectively bringing this study of structure and property full circle.

10.9 Summary and Outlook

This chapter has discussed the behavior of precision polyolefins and their synthesis via metathesis polycondensation chemistry. The term "precision polyolefin" used here refers to a class of linear, defect-free polymers that mimic perfectly sequenced copolymers of ethylene and a vinyl comonomer. These materials, which are only available though ADMET polymerization/hydrogentation chemistry, allow for the systematic study of structure and its relationship to property in ethylene-based materials. The effects of both the size and identity of the "comonomer" (i.e. pendant functional group) have been evaluated, as well as the effects of changing the frequency of its appearance.

Perfectly linear polyethylene synthesized by ADMET, although of lower molecular weight, behaves similarly to the commercial materials it mimics, indicating the effectiveness of this method in creating reliable model systems. Introducing a regularly spaced fluorine atom results in negligible changes in the materials behavior, showing that despite having different bond polarity fluorine can still be included within the orthorhombic crystal structure of PE. Increasing the size of this regularly placed "defect" to chlorine or bromine perturbs the crystal, resulting in a decrease in melting point, melting enthalpy, and a change in crystal structure.

When going from a precisely placed bromine atom to a precisely placed methyl group the character of this "defect" is altered but the size remains similar. The behaviors of the polymers are very similar, indicating that the sterics of the pendant group are the most important factor in dictating the properties of these precision materials. Changing the frequency of this regularly occurring methyl group imparts complete control of polyethylene's melting temperature over a range of 200 °C.

Increasing the size of this pendant alkyl branch results in a decrease in melting temperature when going from a methyl branch to an ethyl branch and again when moving from an ethyl branch to a propyl branch. Beyond a propyl branch, however, the properties are similar regardless of the size of the branch. Thermal analysis, electron microscopy, and X-ray scattering experiments confirm that these larger branches are completely excluded from the crystal. Incorporating one of these larger branches and decreasing the frequency of its appearance results in precise control over the lamellar thickness and melting temperature of the corresponding polymer.

The incorporation of highly polar functional groups such as carboxylic, sulfonic and phosphonic acids produce materials that prove difficult to make industrially. These precision polyolefins are not only useful models in understanding structure and property, but also attractive in advanced applications such as precision ionomers, biomedical materials, and membrane technologies.

Finally, by incorporating a large, chemically incompatible branch and altering its end group it is possible to exclude a pendant group and control its interactions to separately tune the behavior of the crystalline and amorphous regions of the material. The backbone can even be "tricked" into reincluding the excluded pendant branch with the right structural manipulations.

It is clearly evident through this volume of work how important structural control is when attempting to fine tune behavior in ethylene based materials. The model polymers available through ADMET have helped isolate the effects of very specific structural features. Applying this knowledge to the design of new materials will allow for the synthesis of polymers with exactly predictable behavior. Creatively tailoring materials with such controllable behavior for useful applicability represents a promising topic for future work.

Acknowledgments

The authors thank both the National Science Foundation and the Army Research Office for their generous and continued support.

References

Baughman, T.W. and Wagener, K.B. (2005) Recent advances in ADMET polymerization. *Advances in Polymer Science*, **176**, 1–42.

Baughman, T.W., van der Aa, E., Lehman, S.E., and Wagener, K.B. (2005) Circumventing the reactivity ratio dilemma: synthesis of ethylene-*Co*-Methyl vinyl ether copolymer. *Macromolecules*, **38**, 2550–2551.

Baughman, T.W., Sworen, J.C., and Wagener, K.B. (2006a) Sequenced ethylene-propylene copolymers: effects of short ethylene run lengths. *Macromolecules*, **39**, 5028–5036.

Baughman, T.W., vander Aa, E., and Wagener, K.B. (2006b) Linear ethylene-vinyl ether copolymers: synthesis and thermal characterization. *Macromolecules*, **39**, 7015–7021.

Baughman, T.W., Chan, C.D., Winey, K.I., and Wagener, K.B. (2007) Synthesis and morphology of well-defined poly (ethylene-Co-acrylic acid) copolymers. *Macromolecules*, **40**, 6564–6571.

Berda, E.B.and Wagener, K.B. (2008a) Inducing pendant group interactions in precision polyolefins: synthesis and thermal behavior. *Macromolecules*, **41**, 5116–5122.

Berda, E.B.and Wagener, K.B. (2008b) Probing the effects of hydrophilic branch size, distribution, and connectivity in amphiphilic polyethylene. *Macromolecular Chemistry and Physics*, **209**, 1601–1611.

Berda, E.B., Baughman, T.W., and Wagener, K.B. (2006) Precision branching in ethylene copolymers: synthesis and thermal behavior. *Journal of Polymer Science Part A: Polymer Chemistry*, **44**, 4981–4989.

Berda, E.B., Lande, R.E., and Wagener, K.B. (2007) Precisely defined amphiphilic graft copolymers. *Macromolecules*, **40**, 8547–8552.

Boz, E., Wagener, K.B., Ghosal, A. *et al.* (2006) Synthesis and crystallization of precision ADMET polyolefins containing halogens. *Macromolecules*, **39**, 4437–4447.

Boz, E., Nemeth, Alexander J., Alamo, Rufina G., and Wagener, Kenneth B. (2007a) Precision ethylene/vinyl bromide polymers. *Advanced Synthesis & Catalysis*, **349**, 137–141.

Boz, E., Nemeth, A.J., Ghiviriga, I. *et al.* (2007b) Precision ethylene/vinyl chloride polymers via condensation polymerization. *Macromolecules*, **40**, 6545–6551.

Boz, E., Ghiviriga, I., Nemeth, A.J. *et al.* (2008a) Random, defect-free ethylene/vinyl halide model copolymers via condensation polymerization. *Macromolecules*, **41**, 25–30.

Boz, E., Nemeth, A.J., Wagener, K.B. *et al.* (2008b) Well-defined precision ethylene/vinyl fluoride polymers: synthesis and crystalline properties. *Macromolecules*, **41**, 1647–1653.

Calderon, N., Chen, H.Y., and Scott, K.W. (1967) Olefin metathesis - a novel reaction for skeletal transformations of unsaturated hydrocarbons. *Tetrahedron Letters*, **8**, 3327–3329.

Calderon, N., Ofstead, E.A., Ward, J.P. *et al.* (1968) Olefin metathesis. I. Acyclic vinylenic hydrocarbons. *Journal of the American Chemical Society*, **90**, 4133–4140.

Courchay, F.C., Baughman, T.W., and Wagener, K.B. (2006) Understanding the effect of allylic methyls in olefin cross-metathesis. *Journal of Organometallic Chemistry*, **691**, 585–594.

Galli, G., Demel, S., Slugovc, C. *et al.* (2005) Novel liquid crystal ADMET polymers with banana main-chain mesogens. *Molecular Crystals and Liquid Crystals*, **439**, 43–53.

Gedde, U.W.and Mattozzi, A. (2004) *Long Term Properties of Polyolefins*, Springer-Verlag, Berlin Heidelberg New York, pp. 29–74.

Gorodetskaya, I.A., Gorodetsky, A.A., Vinogradova, E.V., and Grubbs, R.H. (2009) Functionalized hyperbranched polymers via olefin metathesis. *Macromolecules*, **42**, 2895–2898.

Grubbs, R.H. (ed.) (2003) *Handbook of Metathesis*, Wiley-VHC, New York.

Grubbs, R.H. (2004) Olefin metathesis. *Tetrahedron*, **60**, 7117–7140.

Guidry, E.N., Li, J., Stoddart, J.F., and Grubbs, R.H. (2007) Bifunctional [C2]daisy-chains and their incorporation into mechanically interlocked polymers. *Journal of the American Chemical Society*, **129**, 8944–8945.

Herrison, J.-L. and Chauvin, Y. (1971) Catalyse de transformation des oléfines par les complexes du tungstène. II. Télomérisation des oléfines cycliques en présence d'oléfines acycliques. *Makromolekulare Chemie-Macromolecular Chemistry and Physics*, **141**, 161–176.

Hosoda, S., Nozue, Y., Kawashima, Y. *et al.* (2009) Perfectly controlled lamella thickness and thickness distribution: a morphological study on ADMET polyolefins. *Macromolecular Symposia*, **282**, 50–64.

Hou, H., Leung, K.C.F., Lanari, D. *et al.* (2006) Template-directed one-step synthesis of cyclic trimers by ADMET. *Journal of the American Chemical Society*, **128**, 15358–15359.

Hyun, J., Han, J., Ryu, C.Y., and Interrante, L.V. (2006) Synthesis of cyclolinear poly(carbosilane)-g-poly(methyl methacrylate or styrene) random copolymers. *Macromolecules*, **39**, 8684–8691.

Jones, K.J., Kinshott, I., Reading, M. *et al.* (1997) The origin and interpretation of the signals of Mtdsc. *Thermochimica Acta*, **304–305**, 187–199.

Karabulut, S., Aydogdu, C., Düz, B., and İmamoglu, Y. (2006) Metal-containing polymers synthesized via acyclic diene metathesis (ADMET) polymerization using electrochemically reduced tungsten-based catalyst: polycarbogermanes. *Journal of Inorganic and Organometallic Polymers and Materials*, **16**, 115–122.

Karabulut, S., Aydogdu, C., Düz, B., and İmamoglu, Y. (2007a) Metal-containing polymers synthesized via acyclic diene metathesis (ADMET) polymerization using electrochemically reduced tungsten-based catalyst: polycarbogermanes *Journal of Inorganic and Organometallic Polymers and Materials*, **17**, 313–313.

Karabulut, S., Aydogdu, C., Düz, B., and İmamoglu, Y. (2007b) Metal-containing polymers synthesized via acyclic diene metathesis (ADMET) polymerization using electrochemically-reduced tungsten-based catalyst: polycarbosilanes *Journal of Inorganic and Organometallic Polymers and Materials*, **17**, 517–523.

Katz, T.J.and McGinnis, J. (1975) Mechanism of the olefin metathesis reaction. *Journal of the American Chemical Society*, **97**, 1592–1594.

Lehman, S.E.and Wagener, K.B. (2003) *The Handbook of Metathesis*, (ed R.H. Grubbs) Wiley, New York.

Lehman, S.E., Wagener, K.B., Baugh, L.S. *et al.* (2007) Linear copolymers of ethylene and polar vinyl monomers via olefin metathesis-hydrogenation: synthesis, characterization, and comparison to branched analogues. *Macromolecules*, **40**, 2643–2656.

Leonard, J.K., Hopkins, T.E., Chaffin, K., and Wagener, K.B. (2008) Semicrystalline lysine functionalized precision polyolefins. *Macromolecular Chemistry and Physics*, **209**, 1485–1494.

Lieser, G., Wegner, G., Smith, J.A., and Wagener, K.B. (2004) Morphology and packing behavior of model ethylene/propylene copolymers with precise methyl branch placement. *Colloid and Polymer Science*, **282**, 773–781.

Lindmark-Hamberg, M.and Wagener, K.B. (1987) Acyclic metathesis polymerization: the olefin metathesis reaction of 1,5-hexadiene and 1,9-decadiene. *Macromolecules*, **20**, 2949–2951.

Luo, S., Vela, J., Lief, G.R., and Jordan, R.F. (2007) Copolymerization of ethylene and alkyl vinyl ethers by a (phosphine- sulfonate)Pdme catalyst. *Journal of the American Chemical Society*, **129**, 8946–8947.

Matloka, P.P.and Wagener, K.B. (2006) The acyclic diene metathesis (ADMET) polymerization approach to silicon containing materials. *Journal of Molecular Catalysis A: Chemical*, **257**, 89–98.

Matloka, P.P., Kean, Z., Greenfield, M., and Wagener, K.B. (2008) Synthesis and characterization of oligo(oxyethylene)/carbosilane copolymers for thermoset elastomers via ADMET. *Journal of Polymer Science Part A: Polymer Chemistry*, **46**, 3992–4011.

Mukherjee, N.and Peetz, R.M. (2008) Acyclic diene metathesis synthesis of silylene- and siloxane-containing conjugated polymers and macrocycles. *Macromolecules*, **41**, 6677–6685.

O'Gara, J.E., Wagener, K.B., and Hahn, S.F. (1993) Acyclic diene metathesis (ADMET) polymerization. Synthesis of perfectly linear polyethylene. *Die Makromolekulare Chemie, Rapid Communications*, **14**, 657–662.

Ohkawa, H., Ligthart, G.B.W.L., Sijbesma, R.P., and Meijer, E.W. (2007) Supramolecular graft copolymers based on 2,7-diamido-1,8-naphthyridines. *Macromolecules*, **40**, 1453–1459.

Opper, K.L.and Wagener, K.B. (2009) Precision sulfonic acid ester copolymers. *Macromolecular Rapid Communications*, **30**, 915–919.

Opper, K.L., Fassbender, B., Brunklaus, G. *et al.* (2009) Polyethylene functionalized with precisely spaced phosphonic acid groups. *Macromolecules*, **42**, 4407–4409.

Pecher, J.and Mecking, S. (2007) Nanoparticles from step-growth coordination polymerization. *Macromolecules*, **40**, 7733–7735.

Qin, Y.and Hillmyer, M.A. (2009) Poly(3-hexyl-2,5-thienylene vinylene) by ADMET polymerization of a dipropenyl monomer. *Macromolecules*, **42**, 6429–6432.

Qiu, W., Pyda, M., Nowak-Pyda, E. *et al.* (2006a) Crystallization and melting of a branched polyethylene with precisely controlled chemical structure. *Journal of Polymer Science Part B: Polymer Physics*, **44**, 3461–3474.

Qiu, W., Sworen, J., Pyda, M. *et al.* (2006b) Effect of the precise branching of polyethylene at each 21st CH_2 group on its phase transitions, crystal structure, and morphology. *Macromolecules*, **39**, 204–217.

Rojas, G.and Wagener, K.B. (2008) Avoiding olefin isomerization during decyanation of alkylcyano α,ω-dienes: a deuterium labeling and structural study of mechanism. *The Journal of Organic Chemistry*, **73**, 4962–4970.

Rojas, G.and Wagener, K.B. (2009) Precisely and irregularly sequenced ethylene/1-hexene copolymers: a synthesis and thermal study. *Macromolecules*, **42**, 1934–1947.

Rojas, G., Baughman, T.W., and Wagener, K.B. (2007) Quantitative α-alkylation of primary nitriles. *Synthetic Communications: An International Journal for Rapid Communication of Synthetic Organic Chemistry*, **37**, 3923–3931.

Rojas, G., Berda, E.B., and Wagener, K.B. (2008) Precision polyolefin structure: modeling polyethylene containing alkyl branches. *Polymer*, **49**, 2985–2995.

Rojas, G., Inci, B., Wei, Y., and Wagener, K.B. (2009) Precision polyethylene: changes in morphology as a function of alkyl branch size. *Journal of the American Chemical Society*, **131**, 17376–17386.

Schawe, J.E.K. (1997) Principles for the interpretation of temperature-modulated dsc measurements. Part 2: a thermodynamic approach *Thermochimica Acta*, **304–305**, 111–119.

Schwendeman, J.E.and Wagener, K.B. (2005) *Gem*-dimethyl effects in the thermal behavior of polyethylene. *Macromolecular Chemistry and Physics*, **206**, 1461–1471.

Shieh, Y.-T.and Liu, G.-L. (2007) Temperature-modulated differential scanning calorimetry studies on the origin of double melting peaks in isothermally melt-crystallized poly(L-lactic acid), *Journal of Polymer Science Part B: Polymer Physics*, **45**, 466–474.

Shultz, G.and Tyler, D. (2009) Preparation of functionalized organometallic metal–metal bonded dimers used in the synthesis of photodegradable polymers *Journal of Inorganic and Organometallic Polymers and Materials*, **19**, 423–435.

Shultz, G.V., Zakharov, L.N., and Tyler, D.R. (2008a) Transition-metal-containing polymers by ADMET: Polymerization of cis-Mo(Co)4(Ph2p(Ch2)3ch=ch2)2. *Macromolecules*, **41**, 5555–5558.

Shultz, G., Berryman, O., Zakharov, L., and Tyler, D. (2008b) Preparation of photodegradable oligomers containing metal–metal bonds using ADMET. *Journal of Inorganic and Organometallic Polymers and Materials*, **18**, 149–154.

Shultz, G.V., Zemke, J.M., and Tyler, D.R. (2009) Preparation of Photoreactive Oligomers by ADMET Polymerization of [(C5h4(Ch2)8ch=ch2)Mo(Co)3]2. *Macromolecules*, **42**, 7644–7649.

Smith, J.A., Brzezinska, K.R., Valenti, D.J., and Wagener, K.B. (2000) Precisely controlled methyl branching in polyethylene via acyclic diene metathesis (ADMET) polymerization. *Macromolecules*, **33**, 3781–3794.

Solmaz, K., Cemil, A., Bülent, D.z., and Imamoglu, Y. (2006) Synthesis and characterization of Ge- and Sn-containing ADMET polymers. *Journal of Molecular Catalysis A: Chemical*, **254**, 186–191.

Sworen, J.C.and Wagener, K.B. (2007) Linear low-density polyethylene containing precisely placed hexyl branches. *Macromolecules*, **40**, 4414–4423.

Sworen, J.C., Smith, J.A., Wagener, K.B. *et al.* (2003) Modeling random methyl branching in ethylene/propylene copolymers using metathesis chemistry: synthesis and thermal behavior. *Journal of the American Chemical Society*, **125**, 2228–2240.

Sworen, J.C., Smith, J.A., Berg, J.M., and Wagener, K.B. (2004) Modeling branched polyethylene: copolymers possessing precisely placed ethyl branches. *Journal of the American Chemical Society*, **126**, 11238–11246.

Terada, K., Berda, E.B., Wagener, K.B. *et al.* (2008) ADMET polycondensation of diketopiperazine-based dienes. Polymerization behavior and effect of diketopiperazine on the properties of the formed polymers *Macromolecules*, **41**, 6041–6046.

Tracz, A.and Ungar, G. (2005) AFM study of lamellar structure of melt-crystallized n-alkane C390h782. *Macromolecules*, **38**, 4962–4965.

Ungar, G., Zeng, X., Brooke, G.M., and Mohammed, S. (1998) Structure and formation of noninteger and integer folded-chain crystals of linear and branched monodisperse ethylene oligomers. *Macromolecules*, **31**, 1875–1879.

Wagener, K.B., Boncella, J.M., Nel, J.G. *et al.* (1990) The key to successful acyclic diene metathesis polymerization chemistry. *Die Makromolekulare Chemie*, **191**, 365–374.

Weychardt, H.and Plenio, H. (2008) Acyclic diene metathesis polymerization of divinylarenes and divinylferrocenes with grubbs-type olefin metathesis catalysts. *Organometallics*, **27**, 1479–1485.

Wunderlich, B. (2006) The contributions of MDSC to the understanding of the thermodynamics of polymers. *Journal of Thermal Analysis & Calorimetry*, **85**, 179–187.

Yamamoto, N., Ito, R., Geerts, Y., and Nomura, K. (2009) Synthesis of all-trans high molecular weight poly (N-alkylcarbazole-2,7-vinylene)S and poly(9,9-dialkylfluorene-2,7-vinylene)S by acyclic diene metathesis (ADMET) polymerization using ruthenium-carbene complex catalysts. *Macromolecules*, **42**, 5104–5111.

Zuluaga, F., Inci, B., Nozue, Y. *et al.* (2009) Reducing branch frequency in precision polyethylene. *Macromolecules*, **42**, 4953–4955.

11

Polyhomologation: The Living Polymerization of Ylides

Jun Luo and Kenneth J. Shea
Department of Chemistry, University of California, Irvine, Irvine, CA 92697-2025

11.1 Introduction

The physical properties of synthetic macromolecules are strongly coupled to their molecular weight (MW), topology and polydispersity index (PDI). Other factors that contribute to their utility include the control of functionality at the macromolecule termini and copolymer composition. Conventional polymerization reactions that produce carbon backbone polymers (ionic, free radical and coordination) provide little opportunity for controlling these variables. Living polymerizations, sometimes referred to as controlled polymerizations, have provided the control for achieving these goals. Not surprisingly, these reactions have had a profound impact on polymer and materials science (Odian, 2004; Moad and Solomon, 2006).

Simple nonfunctional hydrocarbon polymers such as polyethylene (PE), polypropylene, poly-α-olefins and their copolymers are synthesized by uncontrolled high-pressure-high-temperature radical, metathesis or transition-metal-catalyzed coordination polymerization (Natta, 1956; Ziegler *et al.*, 1955; Wu and Grubbs, 1994; Chanda and Roy, 1993). Although catalysts of exceptional efficiency that produce polymers on a huge scale are in common use, control that approaches a "living polymerization" for these methods has not been realized.

There are now a number of alternative approaches for synthesizing linear "polyethylene" (Arthur and Charles, 1959; O'Gara *et al.*, 1993; Sita, 2009; Wagener *et al.*, 1991; Wu and Grubbs, 1994; Wu *et al.*, 1992; Zhang *et al.*, 2008; Morita *et al.*, 2000). Although at present these reactions are not competitive with existing commercial processes for bulk polymer production, they can provide quantities of specialized materials for study.

We initiated a search for novel polymerization reactions that would produce simple hydrocarbon polymers with the potential for molecular-weight and topological control.

Complex Macromolecular Architectures: Synthesis, Characterization, and Self-Assembly, First Edition.
Edited by Nikos Hadjichristidis, Akira Hirao, Yasuyuki Tezuka and Filip Du Prez.

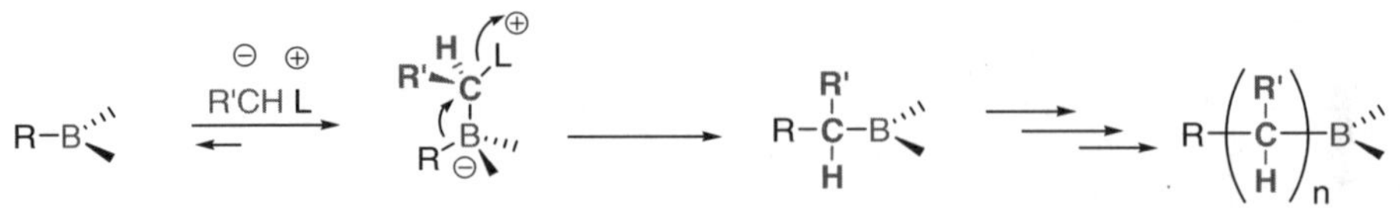

Scheme 11.1 General illustration for organoborane-catalyzed/initiated C1 polymerization.

Our research has focused on polymerization reactions that employ nonolefin monomers, more specifically the polymerization of ylides and diazoalkanes. The carbon backbone is built one carbon at a time (C1 polymerization). These studies draw upon earlier investigations of the Lewis-acid-catalyzed polymerization of diazoalkanes (Imoto and Nakaya, 1972; Maria and Bernhard, 1974; Bawn *et al.*, 1959) and build upon our discovery of the trialkylborane-initiated living polymerization of dimethylsulfoxonium methylide **1** (Shea *et al.*, 1997).

The general reaction involves formation of an organoboron "ate" complex between an ylide and a Lewis-acidic borane (Scheme 11.1). The "ate" complex breaks down by an intramolecular 1,2-migration. This step results in formation of a new sp^3–sp^3 carbon–carbon bond and extends by one carbon the alkyl group. This step also regenerates the Lewis-acid site. Repetition of this process results in formation of a perfectly linear hydrocarbon chain. The living polymerization of ylides results in a controlled synthesis of linear polymethylene (PM), a surrogate for polyethylene, the world's largest commodity synthetic polymer.

11.2 Motivation for Developing a Polyethylene Surrogate

PE is the largest volume commodity polymer with an annual world-wide production of over 79 million tons (Piringer and Baner, 2008). Polyethylene's range of physical properties, along with its low cost, processability and recyclability offers applications that range from plastic shopping bags to artificial joint implants (Chung, 2002). PE is ubiquitous in contemporary society.

The low cost and desirable physical properties of PE motivate efforts to extend its use to products of higher value. Ironically, it is PEs physical properties and methods of production that contribute to limiting realization of this goal (Brintzinger *et al.*, 1995; Hadjichristidis *et al.*, 2000). For example, to extend the potential applications of PE, it is important to develop structure–property relationships of this polymer, as well as overcome some of its drawbacks, such as incompatibility with other materials. The lack of model polyethylene with tailored structure, controlled molecular weight and low polydispersity has limited structure–property studies of this polymer (Bielawski *et al.*, 2002). The absence of functionality or the ability to incorporate PE segments in block copolymers has limited the compatibility of commercial polyethylene products with other materials (Dong and Hu, 2006).

The majority of commercial polyethylene is prepared by high-pressure–high-temperature free radical polymerization, Ziegler–Natta or metallocene catalysts (Natta, 1956; Ziegler *et al.*, 1955; Chanda and Roy, 1993; Brintzinger *et al.*, 1995). These methods result in polyethylene with a spectrum of physical properties including: low-density polyethylene (LDPE), linear low-density polyethylene (LLDPE), and high-density polyethylene (HDPE). As a result of chain-transfer and termination reactions during polymerization, these polymers have a broad molecular-weight distribution and differ in branching along the main polymer chain (Hadjichristidis *et al.*, 2000). The industrial ethylene polymerization methods are not well suited for fine control of polymer architecture or molecular weight.

In this review, we describe the polyhomologation reaction, a living polymerization that results in the synthesis of linear polymethylene with no branches, controlled molecular weight, narrow molecular-weight distribution, well-defined topology and functionality. Polymethylene is a unique member of the polyethylene family. The carbon backbone is built up one carbon at a time, in contrast to the two carbons in ethylene polymerization (Jellema *et al.*, 2010). The polyhomologation reaction can provide polymethylene samples for model studies of structure–property relationships and functionalized polymethylene for copolymers that can function as polyethylene compatibilizers. In addition, since many vinyl compounds do not readily polymerize, the polyhomologation reaction can serve as an entry to completely new substances.

11.3 A Living Polymerization of Ylides

Controlled synthesis of linear hydrocarbon polymers has been achieved by the Lewis-acid-catalyzed/initiated polymerization of ylides and ylide-like monomers. The hydrocarbon backbone is extended one carbon at a time. Dimethylsulfoxonium methylide **1** has been the most common monomer used for these polymerizations (Shea *et al.*, 1997).

Dimethylsulfoxonium methylide **1**, an ylide developed by Corey and coworker (Corey and Chaykovsky, 1965), is synthesized from trimethylsulfoxonium iodide **2** which is converted to trimethylsulfoxonium chloride **3** via ion exchange with benzyltributylammonium chloride. The ylide **1** is produced by refluxing a heterogeneous mixture of trimethylsulfoxonium chloride **3** and sodium hydride in toluene (Scheme 11.2). Homogeneous solutions of ylide **1** of approximately 0.6 M can be stored at −20 °C for >3 weeks (Busch *et al.*, 2002).

In a representative polyhomologation reaction, a solution of tris(4-methoxyphenylethyl) borane **4** (1 equiv.), prepared by hydroboration of 4-vinylanisole with $BH_3{\cdot}THF$ (Brown and Sharp, 1966), was rapidly injected into a solution of ylide **1** [300 equiv., theoretical average degree of polymerization ($\overline{DP}_{theo.}$) = 100] in toluene preheated to 70 °C (Figure 11.1a). Ylide **1** was consumed within 10 min. The cooled solution was treated with trimethylamine *N*-oxide dihydrate ($TAO{\cdot}2H_2O$) (Kabalka and Hedgecock, 1975) and the resulting polymer was isolated by precipitation to afford α-hydroxy-ω-(4-methoxyphenyl)polymethylene **5** in yields ranging from 85 to 95% (Busch *et al.*, 2002).

α-Hydroxy-ω-(4-methoxyphenyl)polymethylene **5** was synthesized with a wide range of $\overline{DP}$. Molecular-weight analyses of polymethylene **5** were performed using ^1H NMR spectroscopy, field desorption mass spectrometry (FDMS), and gel-permeation chromatography (GPC) (Figure 11.1) (Busch *et al.*, 2002). The results, summarized in Table 11.1, establish an excellent correlation between the theoretical and observed molecular weights,

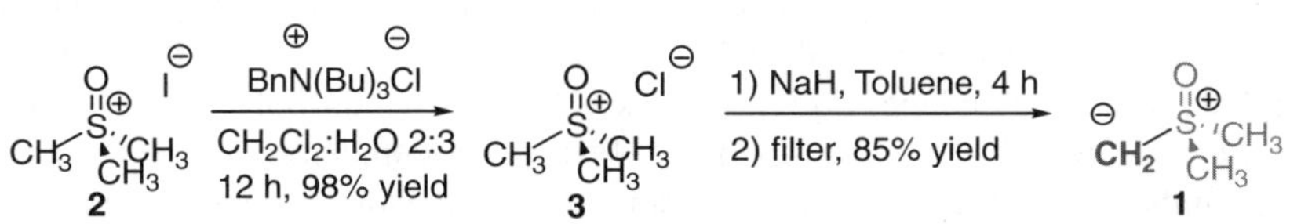

Scheme 11.2 Synthesis of dimethylsulfoxonium methylide **1**.

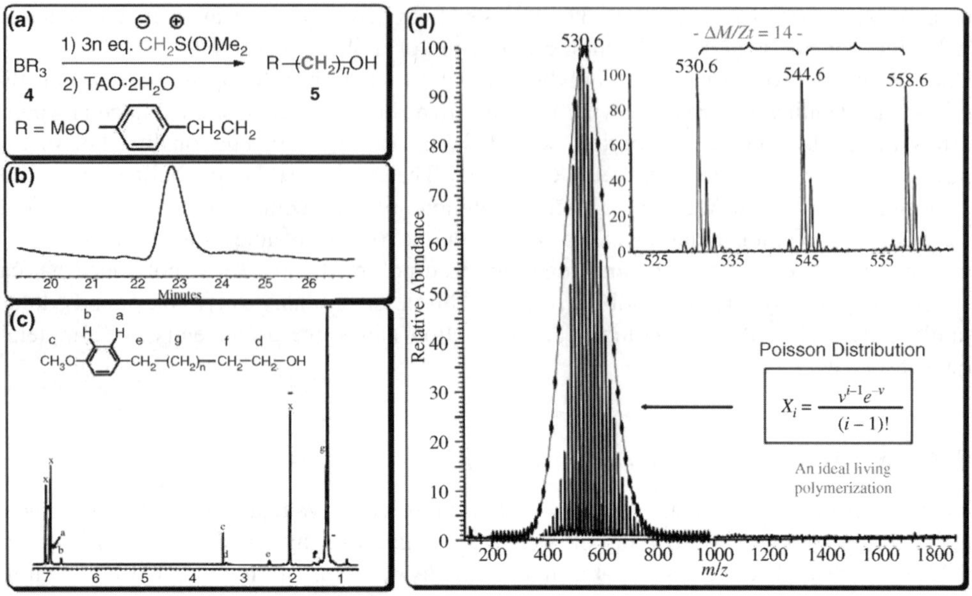

Figure 11.1 Synthesis and characterization of α-hydroxy-ω-(4-methoxyphenyl)polymethylene **5**. (a) Synthesis of α-hydroxy-ω-(4-methoxyphenyl)polymethylene **5**. (b) GPC of α-hydroxy-ω-(4-methoxyphenyl)polymethylene **5** ($\overline{DP}_{expt.} = 106$, $M_n = 1639$, PDI = 1.01) prepared from 300 molar equiv. of dimethylsulfoxonium methylide **1** and tris(4-methoxyphenylethyl)borane **4**. (c) 500 MHz ^1H NMR (x: toluene-d_8) of α-hydroxy-ω-(4-methoxyphenyl)polymethylene **5** ($\overline{DP}_{expt.} = 397$). (d) FDMS spectrum of α-hydroxy-ω-(4-methoxyphenyl)polymethylene **5** ($M_n = 552$, $\overline{DP}_{expt.} = 31$, PDI = 1.01). Inset shows an expanded scale revealing isotope patterns with each main peak separated by 14 *m/e*. The curve is a plot of the simulated distribution of a theoretical polymer of comparable molecular weight produced by an ideal living polymerization. The plot is generated by a Poisson distribution function.

consistent with a living polymerization. Low polydispersities further support the living character of polyhomologation. Representative values for PDI range from 1.01 to 1.04, approaching the Poisson distribution of a living polymerization (Mueller *et al.*, 1995).

A typical ^1H NMR spectrum of polymethylene **5** confirms the linear structure with a hydroxyl and 4-methoxyphenyl group at each termini of the polymer chain (Figure 11.1c) (Busch *et al.*, 2002). The spectrum includes the methoxy signal (s, 3.79 ppm), the hydroxyl bearing methylene (t, 3.64 ppm), and the benzylic methylene (t, 2.54 ppm) in the ratio of 3: 2: 2. The measured $\overline{DP}$ of 397, determined by the integration of chain methylenes (s, 1.25 ppm and m, 1.57 ppm), is in close agreement with the theoretical $\overline{DP}$ of 400.

FDMS analysis of polymethylene **5** shows a distribution of peaks separated by 14 amu (Figure 11.1d) (Busch *et al.*, 2002). In contrast, the mass spectrum of polyethylene has peak separations of 28 amu (Evans *et al.*, 1996). Mass spectrometry highlights the distinction between polymethylene and polyethylene. Polymethylene samples contain polymer chains of integral carbon numbers ($n, n+1, n+2, \ldots$), while polyethylene samples contain polymer chains of alternating numbers ($n, n+2, n+4, \ldots$). Although we are not aware of the consequences of this difference on the physical properties of these materials, the polymethylene "signature" confirms that the reaction builds the polymethylene chain one carbon at a time with $-CH_2-$ repeat units.

Table 11.1 Experimental degrees of polymerization $(\overline{DP}_{expt.})^a$ and polydispersities^b of α-hydroxy-ω-(4-methoxyphenyl)polymethylene **5** (Reprinted with permission from B.B. Busch, M.M. Paz, K.J. Shea *et al.*, "The boron-catalyzed polymerization of dimethylsulfoxonium methylide. A living polymethylene synthesis," *Journal of the American Chemical Society*, **124**, 14, 3636–3646, 2002. © 2002 American Chemical Society.)

Entry	$\overline{DP}_{theo.}$	- $\overline{DP}_{expt.}$			- PDI	
		NMR	FDMS	GPC	FDMS	GPC
1	23	—	25	—	1.02	—
2	30	35	31	32	1.01	1.02
3	30	36	35	—	1.03	—
4	40	45	43	39	1.03	1.01
5	40	47	44	43	1.02	1.03
6	70	—	—	75	—	1.01
7	140	130	—	149	—	1.03
8	228	—	—	226	—	1.04

$^a\overline{DP}_{expt.}$ is the experimental average number of repeat units based on the M_n.
^{b}Polydispersity (PDI) was calculated from M_w/M_n.

The curve in Figure 11.1d is a plot of a simulated distribution of a theoretical polymer produced by an ideal living polymerization (Zhou, 2002). The plot is generated by a Poisson distribution function (Flory, 1940). The polymer distribution (FDMS) produced by the polyhomologation reaction closely approximates the Poisson distribution. This is strong support for the living nature of the polyhomologation reaction.

A representative GPC of polymethylene **5** shows a narrow, monodisperse polymer distribution with $\overline{DP}_{expt.}$ in agreement with the calculated molar ratio of [ylide]/3[BR_3] (Figure 11.1b) (Busch *et al.*, 2002). The stoichiometry of the reaction between ylide **1** (3*n* equiv.) and tris(4-methoxyphenylethyl)borane **4** (1 equiv.) is consistent with the formation of an intermediate three-arm star polymethylene **6** ($\overline{DP} = 3n$) (Scheme 11.3). To account for the very low PDI, all three alkyl groups on boron must propagate with equal probability.

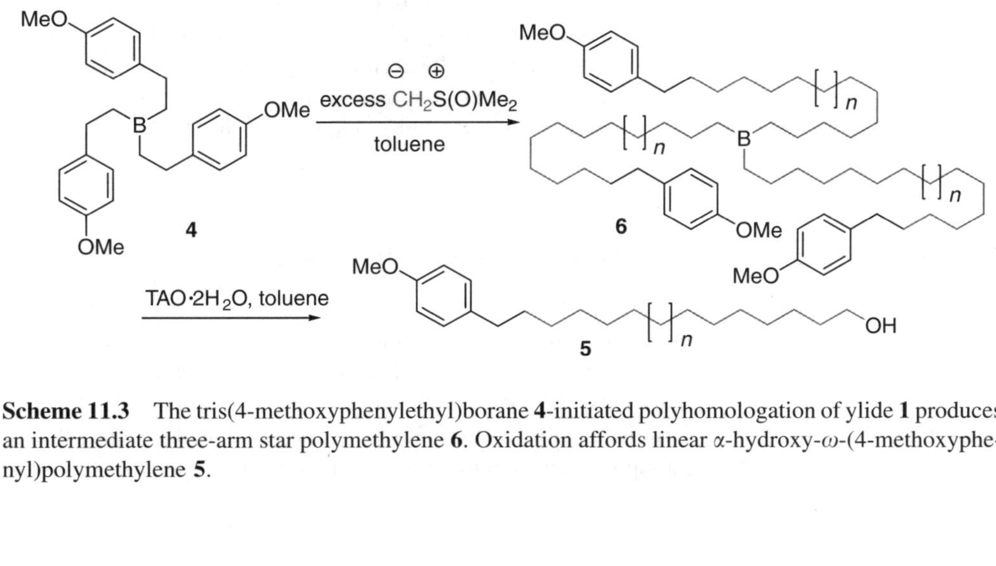

Scheme 11.3 The tris(4-methoxyphenylethyl)borane **4**-initiated polyhomologation of ylide **1** produces an intermediate three-arm star polymethylene **6**. Oxidation affords linear α-hydroxy-ω-(4-methoxyphenyl)polymethylene **5**.

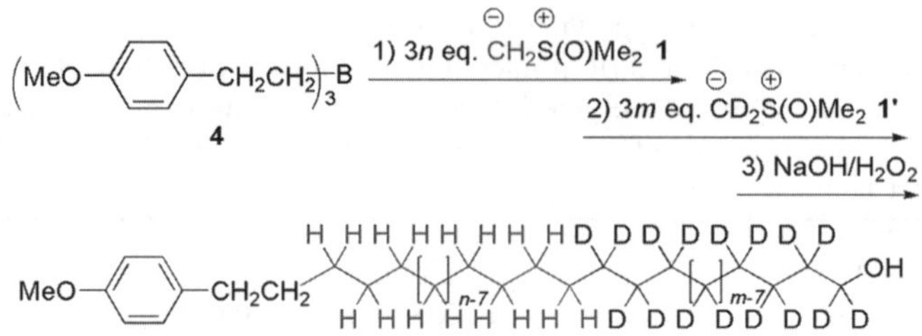

Scheme 11.4 Synthesis of α-hydroxy-ω-(4-methoxyphenyl)polymethylene-*co*-polyperdeuteriomethylene.

Oxidation of star polymethylene **6** affords 3 equiv. of α-hydroxy-ω-(4-methoxyphenyl) polymethylene **5** with $\overline{DP} = n$. The strict correspondence between stoichiometry and $\overline{DP}$ in these examples is consistent with a rapid initiation and the absence of termination or chain transfer reactions in the polymerization of ylide **1**. The absence of a termination step is reasonable, since the "living end" of this polymer is a trialkylborane, which is stable under the reaction conditions. Dimethyl sulfoxide (DMSO), a byproduct that accumulates during the reaction does not inhibit the initiator/catalyst.

The absence of a termination step during the polyhomologation reaction was demonstrated by the synthesis of α-hydroxy-ω-(4-methoxyphenyl)polymethylene-*co*-polyperdeuterio-methylene (Scheme 11.4) (Shea *et al.*, 1997). The copolymer was prepared by the addition of tris(4-methoxyphenylethyl)borane **4** to a toluene solution of ylide **1**. After ylide **1** consumption, a solution of *perdeuterio* ylide **1′** was added. Following consumption of this second batch of ylide **1′**, the tris-polyhomologated organoborane was oxidized to the linear-terminal alcohol. FDMS analysis reveals a distribution of molecular ions consistent with a copolymer of α-hydroxy-ω-(4-methoxyphenyl)polymethylene containing $(CH_2)_n(CD_2)_m$ units where $3n$ and $3m$ represent the equivalents of *protio* ylide **1** and *perdeuterio* ylide **1′** added, respectively.

The polyhomologation reaction has been used to prepare polymethylene with a M_n as high as 354 KDa corresponding to an intermediate tris(polymethylene)borane with $M_n > 1.0 \times 10^6$ (Busch *et al.*, 2002). A M_n of 1.06 million for the star tris(polymethylene)borane represents approximately 7.6×10^4 turnovers per boron atom. The reaction is complete in <10 min, from which one estimates a lower limit turnover frequency of 6.4×10^6 g of polymethylene (mol boron)$^{-1}$ h^{-1} at 120 °C. The turnover frequency for the boron catalytic center is comparable to some of the most efficient homogeneous ethylene polymerization catalysts, such as neutral anilinotropone-Ni(II) (8.8×10^6 g of polyethylene (mole catalyst)$^{-1}$ h^{-1}) (80 °C, 200 psi ethylene) (Hicks and Brookhart, 2001).

11.4 Mechanism of the Polyhomologation Reaction

A proposed mechanism for the formation of polymethylene is shown in Scheme 11.5 (Shea, 2000). The reaction is initiated by attack of the nucleophilic ylide **1** on the Lewis-acidic trialkylborane to form a zwitterionic complex **7**, which subsequently undergoes a 1,2-migration of one of the three alkyl substituents on boron to the methylene group. This results in a homologated borane **8** and a molecule of DMSO. Since the homologated product **8** is a trialkylborane, after each migration the Lewis-acidic boron is regenerated so it can react with

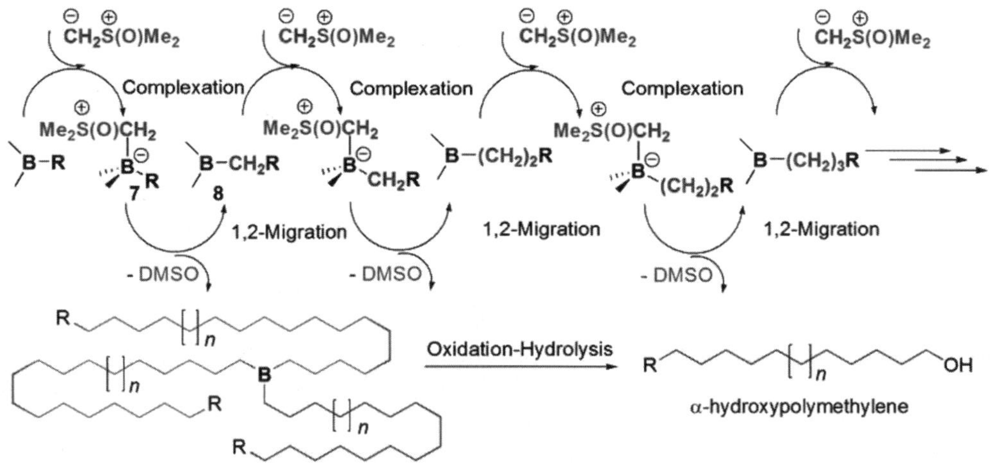

Scheme 11.5 Proposed mechanism for the polyhomologation reaction.

another ylide **1** monomer. Repetition of this affords a polymethylene chain. The low PDI of these polymerizations implies that all three alkyl groups on boron undergo migration with equal probability resulting in the three polymethylene chains with the same average chain length.

The proposed mechanism was supported by an ^{11}B NMR study (Busch *et al.*, 2002). In noncoordinating solvents, there is a significant chemical-shift difference between trialkylboranes (80–90 ppm) and tetraalkylborate complexes (−20 to −10 ppm) (Eaton and Lipscomb, 1969). The ^{11}B NMR spectrum of tri-*n*-hexylborane in toluene at 40 °C exhibits a single resonance at 87.4 ppm (Figure 11.2a). When an equimolar amount of ylide **1** was added at −78 °C followed by warming to −15 °C, the peak at 87.4 ppm disappeared and a

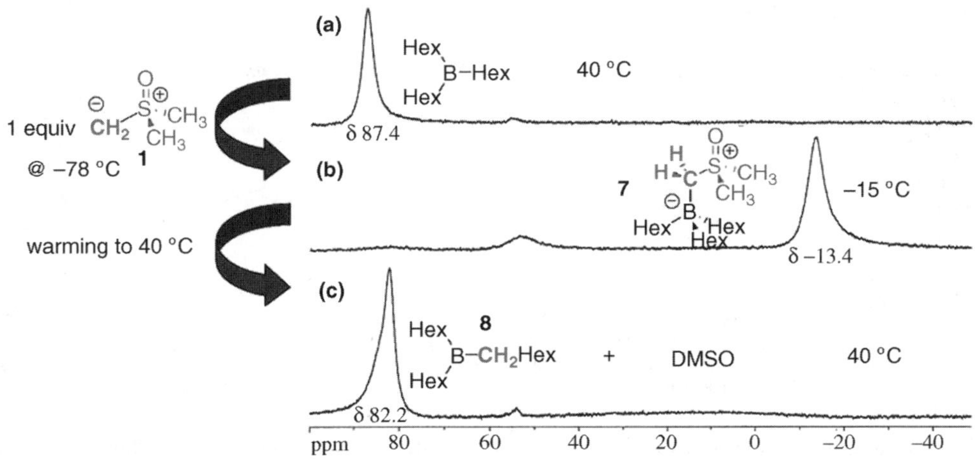

Figure 11.2 ^{11}B NMR spectra of (a) tri-*n*-hexylborane (40 °C), (b) tri-*n*-hexylborane·ylide complex **7** (−15 °C), and (c) trialkylboranes **8** and DMSO product mixture (40 °C) in toluene. The small peak at 58 ppm is a borinic ester impurity (<2%).

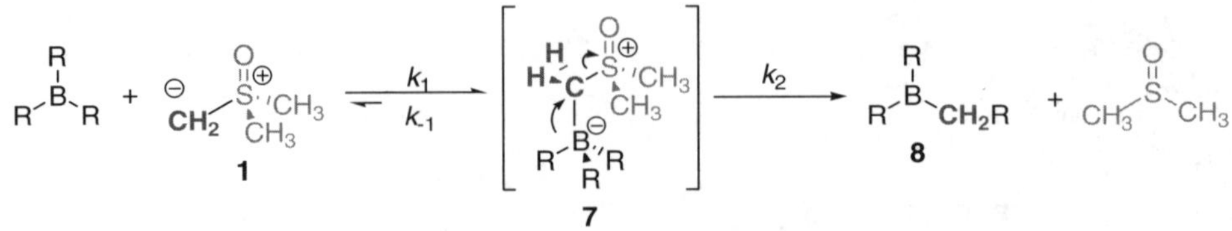

Scheme 11.6 Pre-equilibrium followed by a *rate-limiting 1,2-migration*.

new resonance was observed at −13.4 ppm (Figure 11.2b). The observation is consistent with quantitative formation of the zwitterionic complex **7** (Scheme 11.6). Warming the mixture to 40 °C results in conversion to a single resonance at 82.2 ppm (Figure 11.2c). The new signal is attributed to the homologated trialkylborane **8**, formed after 1,2-migration. (Note – the rate of the polyhomologation reaction at −15 °C is very slow).

Based on the preceding ^{11}B NMR spectra, a pre-equilibrium exists between the borane initiator and ylide that strongly favors formation of the zwitterionic complex **7**. The pre-equilibrium is established immediately and complex **7** is stable at −15 °C. These results are consistent with the proposed mechanism of a fast pre-equilibrium followed by a *rate-limiting 1,2-migration step* since homologation was observed only after warming above 0 °C (Busch *et al.*, 2002).

Additional support for a *rate-limiting 1,2-migration step* was obtained from kinetic studies (Busch *et al.*, 2002). The rate of polyhomologation can be monitored by the consumption of ylide **1**. Triethylborane was reacted with excess ylide **1** (ylide:borane >21). Plots of the consumption of ylide **1** versus time are linear, indicating that accumulated DMSO does not inhibit the polyhomologation reaction (Figure 11.3a). When ylide **1** is in large excess, doubling the concentration of the ylide does not affect the rate of consumption. However, doubling the concentration of triethylborane doubled the rate of consumption of ylide **1**. Under these pseudofirst-order conditions, the reaction is first order in organoborane and zero order in ylide **1**, a result consistent with the proposed two-step mechanism, with a *rate-limiting*

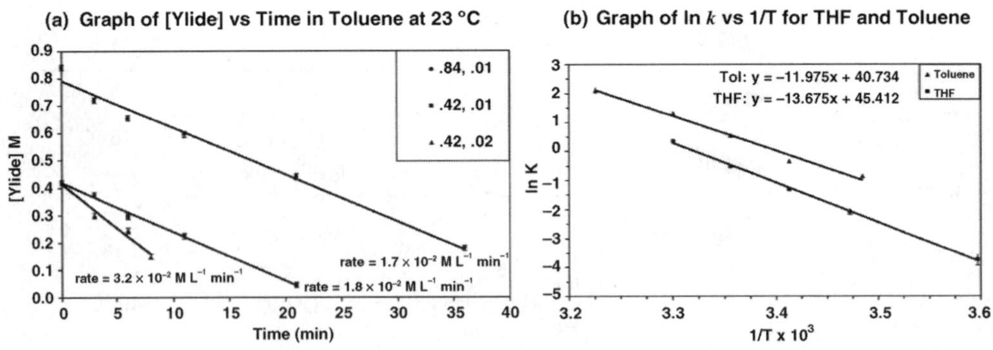

Figure 11.3 Kinetic study of the reaction of Et_3B with ylide **1** in toluene. (a) Graph of [ylide] vs. time in toluene at 23 °C. (b) Graph of ln *k* vs. 1/*T* for toluene and THF. (Reprinted with permission from B.B. Busch, M.M. Paz, K.J. Shea *et al.*, "The boron-catalyzed polymerization of dimethylsulfoxonium methylide. A living polymethylene synthesis," *Journal of the American Chemical Society*, **124**, 14, 3636–3646, 2002. © 2002 American Chemical Society.)

1,2-migration of the *ate* complex **7** (Scheme 11.6). The rate expression is d[product]/dt = rate = k_2[R_3B·ylide]. When ylide **1** is in large excess, the concentration of complex **7** will be approximated the initial trialkylborane concentration. The rate expression can be written as rate = $k_2[R_3B]_0$, consistent with the kinetic results.

The Arrhenius activation parameters were determined from the temperature dependence of the rate constants k over the temperature range of 15–35 °C in toluene and in THF (Figure 11.3b) (Busch *et al.*, 2002). A summary of the kinetic parameters are given in Table 11.2. The free energies of activation in toluene and THF are the same within experimental error. However, the enthalpy and entropy of activation in the two solvents differ significantly. This difference is consistent with related solvent effects for reactions involving charge annihilation in the rate-determining step (Reichardt, 1990).

The rate-determining step of the polyhomologation reaction is a 1,2-migration of an alkyl group from boron to carbon in the zwitterionic complex **7** (Busch *et al.*, 2002). Structural analysis of zwitterionic complexes (X-ray crystallography and computational study) provides insight into the rearrangement.

The structure of ylide·BF_3 complex **9**, a stable surrogate for complex **7**, the proposed intermediate in the polyhomologation reaction, was obtained by single-crystal X-ray analysis (Figure 11.4a). The complex, prepared from ylide **1** and BF_3 (Stoddard and Shea, 2003), has a staggered conformation about the B1–C1 bond with an alignment of one of the fluorine atoms *anti* to the sulfoxonium group (F1–B1–C1–S1 = – 161.21°). The *anti* fluorine has a slightly shorter B–F distance (1.394 Å) compared to the two remaining fluorines (1.400 Å). In addition, the F1–B1–C1 angle (107.97°) is less than the two remaining F3–B1–C1 (112.45°) and F2–B1–C1 (110.38°) angles. The S1–O1 bond of complex **9** is aligned *anti* to the B1–C1 bond (B1–C1–S1–O1 = 166.42°), which can be attributed in part to the opposing dipole moments of the BF_3 and S–O groups. We suggest that this ground-state conformation can be understood in terms of a stabilizing orbital interaction that aligns the migrating and leaving groups in zwitterionic complex **7**.

A computational study of the ground-state conformation of ylide·BMe_3 complex **10**, a surrogate for **7**, was obtained from a DFT/B3LYP/6-311G(d,p) calculation (Figure 11.4b) (Luo and Shea, 2010). Comparison of the structures of ylide·BMe_3 complex **10** and ylide·BF_3 complex **9** reveals many similarities. The calculation positions one of the alkyl groups *anti* to the sulfoxonium group (C–B–C–S = 165.4°). And the S–O bond of complex **10** is also aligned *anti* to the B–C bond (B–C–S–O = 165.8°).

Table 11.2 Kinetic parameters from Arrhenius plot (Reprinted with permission from B.B. Busch, M.M. Paz, K.J. Shea *et al.*, "The boron-catalyzed polymerization of dimethylsulfoxonium methylide. A living polymethylene synthesis," *Journal of the American Chemical Society*, **124**, 14, 3636–3646, 2002. © 2002 American Chemical Society.)

Parameter	THF	Toluene
E_a kcal mol^{-1}	27.1 ± 1.5	23.8 ± 1.4
$\Delta H^{\neq}$ kcal mol^{-1}	26.5 ± 1.5	23.2 ± 1.4
$\Delta S^{\neq}$ cal deg^{-1} mol^{-1}	21.5 ± 5.2	12.6 ± 4.7
$\Delta G^{\neq}$ kcal mol^{-1} (23 °C)	20.1 ± 3.0	19.5 ± 2.8

$\Delta H^{\neq} = E_a - RT$. $\Delta S^{\neq} = R\ \ln(Ah/ekT)$. $\Delta G^{\neq} = \Delta H^{\neq} - T\,\Delta S^{\neq}$.

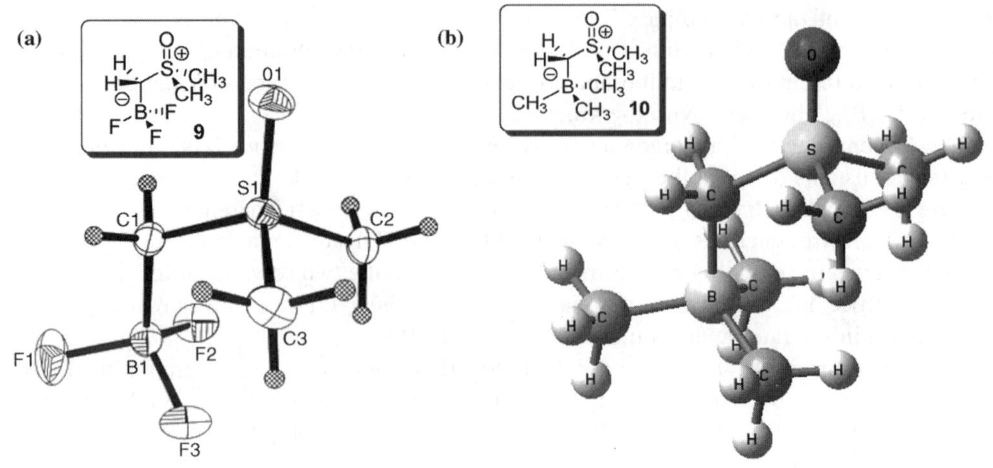

Figure 11.4 Molecular structure of ylide·BF_3 complex **9** and ylide·BMe_3 complex **10**. (a) Molecular structure of ylide·BF_3 complex **9** obtained from a single-crystal X-ray analysis. (b) Molecular structure of ylide·BMe_3 complex **10** obtained from a DFT/B3LYP/6-311G(d,p) calculation.

Both the X-ray diffraction structure and computational result reveal a *trans anti*-periplanar relationship between the migrating R–B bond and the C–S leaving group. Similar conformations have been observed in related zwitterionic species (Bottoni *et al.*, 2003; Aggarwal *et al.*, 2006; Stoddard and Shea, 2003; Stoddard and Shea, 2004; Bai and Shea, 2006). Factors that contribute to the ground-state structures of **9** and **10** may be viewed as foreshadowing the transition states for the following 1,2-migration. The low-energy conformation arises from a stabilizing interaction between the filled B–C σ orbital and the empty S–O σ^* orbital (Figure 11.5a). This would be optimal when the migrating group R is oriented *anti*-periplanar to the C–S leaving group. This orientation allows the filled R–B σ orbital (HOMO) to interact with the unfilled C–S σ^* orbital (LUMO) during 1,2-migration (Figure 11.5b). The transition state for the 1,2-migration would involve stretching and bending of the R–B bond in the direction of the migration terminus.

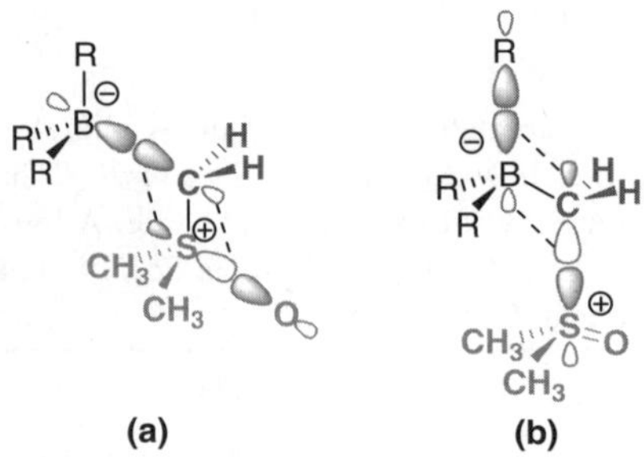

Figure 11.5 Proposed molecular orbital interactions of zwitterionic complex **7** deduced from X-ray and calculated ground-state structures of **9** and **10**. (a) The filled B–C σ orbital aligns *anti*-periplanar to the empty S–O σ^* orbital for stabilization. (b) The migrating group (HOMO: R–B σ) aligns *anti*-periplanar to the sulfoxonium leaving group (LUMO: C–S σ^*) during 1,2-migration.

11.5 Topological Control of Polymethylene

For polymers with identical chemical composition and similar molecular weight, topology can be used to alter the physical properties including density, crystallinity, viscosity, melt rheology, diffusion rate and mechanical strength (Bielawski *et al.*, 2002; Everaers *et al.*, 2004; McLeish, 2002; Lohse *et al.*, 2002; Bick and McLeish, 1996). Considerable effort has been invested in the development of methods to control polymer architecture and tailor polymer properties for specific applications (Laurent and Grayson, 2006). For example, the use of star and comb polymers allows better control over pore size and regularity compared to the corresponding linear polymers in honeycomb-structured porous film fabrication (Stenzel *et al.*, 2006). In biological applications, molecular architecture can also affect the polymer–drug pharmacokinetics profile and biodistribution when soluble polymers are used as drug carriers (Fox *et al.*, 2009). However, despite the potential utility of novel polymer topologies, literature examples have been restricted to a limited number of polymers since polymerization methods for many monomers, including ethylene, are not suitable for precise control of architecture.

The lack of polyethylene samples with tailored architectures, controlled molecular weight, low PDI, and absence of branching defects has limited a systematic study of the relationships between molecular structure and physical properties of polyethylene. Several notable exceptions include cyclic polyethylene that has been synthesized by ring-opening metathesis polymerization (ROMP) (PDI of 2) (Bielawski *et al.*, 2002; Bielawski *et al.*, 2003) and star, comb, H-shape and pom–pom polyethylenes, containing 17–25 ethyl branching defects per 1000 carbons, which were produced by anionic polymerization (Hadjichristidis *et al.*, 2000).

Postpolymerization modification of PM and the use of structurally novel borane initiators provide control of polymer topology and synthesis of novel PM architectures. Examples including three-arm star polymers, giant tube-like structures and macrocycles are illustrated below.

11.5.1 Postpolymerization Topological Control of Polymethylene

Known standard protocols allow trialkylboranes to be converted to a diverse number of architectures (Scheme 11.7) (Brown, 1975; Pelter *et al.*, 1988). For example, trialkylboranes can undergo a "three-arm stitching" reaction to join the three alkyl groups on boron to a common carbon atom. This transformation produces a trialkyl-substituted carbinol. Alternatively, a "two-arm stitching" reaction can connect two alkyl groups on boron to form a linear ketone. These stitching transformations can be used following polyhomologation to build up polymethylene with novel structures incorporating functionality that can be used for further modification.

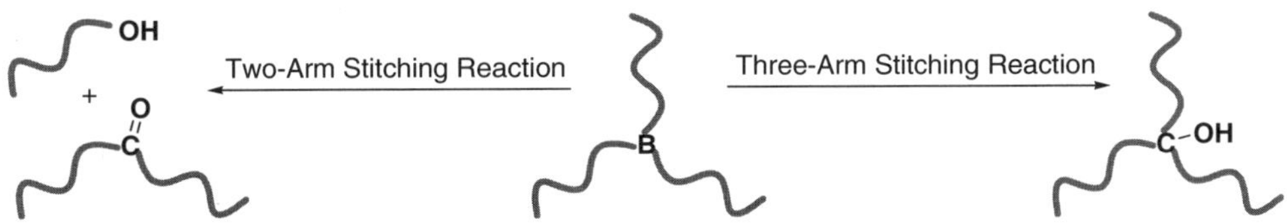

Scheme 11.7 An illustration of two-arm and three-arm stitching reactions.

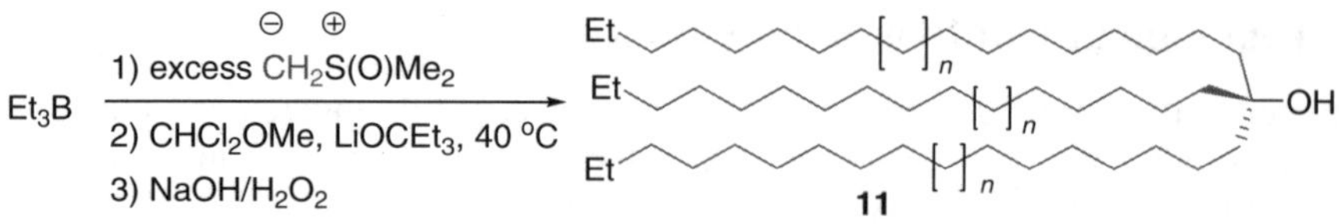

Scheme 11.8 Synthesis of tris(polymethylene)carbinol **11** from Et_3B.

For example, a three-arm star polymethylene structure was synthesized from a polyhomologation–"stitching"–oxidation sequence. An example is given by the synthesis of tris(polymethylene)carbinol **11** (Scheme 11.8) (Shea *et al.*, 1998a). Following polyhomologation of a simple trialkylborane, the resulting tris(polymethylene)borane was treated with α,α-dichloromethyl methyl ether (DCME) followed by lithium 3-ethylpentan-3-olate ($LiOCEt_3$) then oxidized to afford three-arm star polymer, tris(polymethylene)carbinol **11** (Hine *et al.*, 1960). The experimental molecular weights of product **11** were in close agreement with the calculated molecular weights and the resultant PDI ranged from 1.02 to 1.13. In this example, with a postpolymerization modification of a polyhomologated organoborane, initiator/catalyst BEt_3 produces three-arm polymethylene architectures with well-controlled molecular weight and low PDI. These materials are not readily available by conventional ethylene polymerization.

11.5.2 Initiator-Based Topological Control

Structurally novel borane initiators can also provide access to well-defined polymethylene architectures. For example, polycyclic organoborane initiators offer an opportunity for the construction of "tube-like" macrotricyclic organoboranes. These structures may be elaborated to hydrocarbon star polymer structures following the oxidative cleavage of the boron–carbon bond.

For example, the polyhomologation reaction of 1-boraadamantane·THF **12** with ylide **1** produced a "tube-like" macropolycylcic organoborane **13**, which upon oxidation gave a regular A_3 three-arm star polymethylene **14** (Scheme 11.9) (Wagner *et al.*, 2003; Wagner and Shea, 2001). Both ^1H NMR spectra and GPC (triple detector) analyses of product **14** are consistent with the proposed regular A_3 star architecture. ^1H NMR spectrum of the product reveals the clear threefold pseudosymmetry of the equatorial and axial protons, as well as the multiplicity of the terminal hydroxymethylene groups (Figure 11.6). This is a consequence of all three branches of the propagating species participating in migration to produce a regular

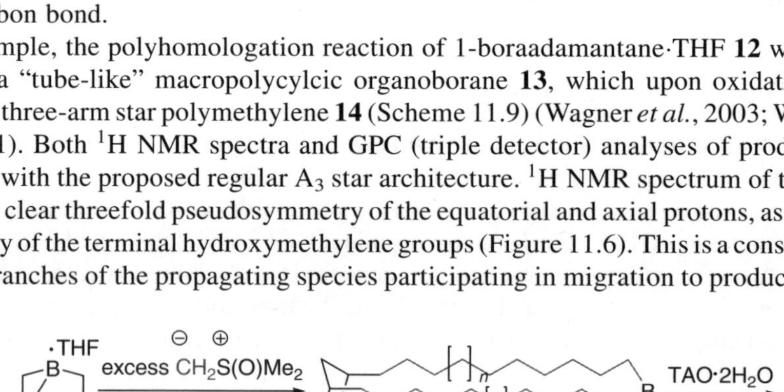

Scheme 11.9 Synthesis of three-arm star polymethylene **14** incorporating a *cis*,*cis*-1,3,5-trisubstituted cyclohexane core from 1-boraadamantane·THF **12**.

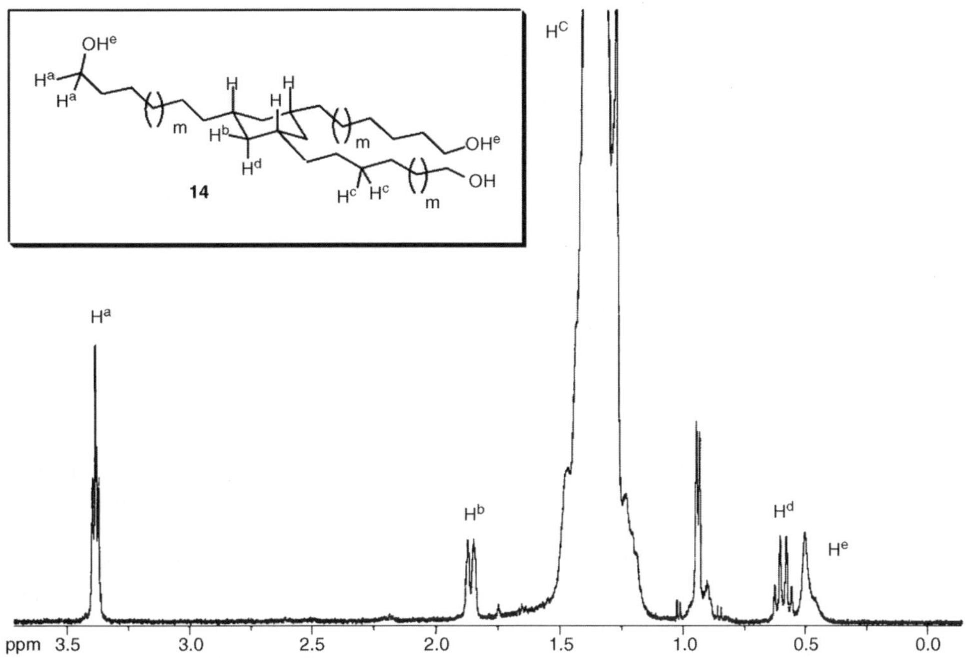

Figure 11.6 Representative ^1H NMR spectrum of polymeric triol **14** (C_6D_6, 78 °C, 500 MHz). The triplet at 3.38 ppm ($J = 6.4$ Hz) corresponds to the terminal hydroxymethylene protons of **14**. The apparent doublet at 1.85 ppm ($J = 12.2$ Hz) and a quartet at 0.59 ppm ($J = 11.8$ Hz) arise from the equatorial and axial methylene protons of the cyclohexane ring of **14**. The broad singlet at 0.5 ppm has been assigned to the hydroxy protons of the end groups.

star polymethylene. The branching of a star polymer can also be characterized by the ratio of the square of the radius of gyration of the star polymer compared to that of a linear polymer of the same molecular weight (g). For a perfectly regular A_3 star polymer, the theoretical prediction of g is 0.778 (Zimm and Stockmayer, 1949; Berry and Orofino, 1964). The g values of polymer **14** (0.790–0.755) were obtained by GPC (Table 11.3) (Drott and Mendelso, 1970a,1970b; Zimm and Kilb, 1959; Wood-Adams *et al.*, 2000). The correspondence between the experimental and the theoretical g value supports the assigned structure of **14** as regular A_3 star polymers.

Interestingly, $\overline{DP}_{expt.}$ of star polymer **14**, established by GPC and ^1H NMR end-group analyses, is consistently three times greater than $\overline{DP}_{theo.}$ (Wagner *et al.*, 2003; Wagner and Shea, 2001). Based on the fact that the PDIs of star polymers **14** are <1.12, it is proposed that approximately two-thirds of the propagating species died in the early stages of the polyhomologation reaction and did not proceed to form polymer.

Examination of the products and product distributions at the *initial* stages of polymerization, were helpful in accounting for the discrepancy between the calculated and experimental molecular weight. The results are summarized in Figure 11.7 (Wagner *et al.*, 2003). The bold arrows show the main pathway for polyhomologation of 1-boraadamantane·THF **12**. The first step of the reaction with ylide **1** is the formation of monohomologated products **15**. Following the first insertion, the reaction proceeds by a path-selective series of insertions

Table 11.3 Light scattering, viscosity, and GPC data for three-arm star polymethylene **14** (Reprinted with permission from C.E. Wagner, J.-S. Kim and K.J. Shea, "The polyhomologation of 1-boraadamantane: Mapping the migration pathways of a propagating macrotricyclic trialkylborane," *Journal of the American Chemical Society*, **125**, 40, 12179–12195, 2003. © 2003 American Chemical Society.)

Trial	$\overline{DP}_{theo}$ [a]	M_w [b] (g/mol)	$\overline{DP}_{expt.}$ [b]	$[\eta]$ [c] (dL/g)	g' [c]	g^d ($\in = 0.5$)	PDI [e]
1	296	24 270	573	0.567	0.889	0.790	1.12
2	593	39 650	938	0.805	0.869	0.755	1.12

[a] $\overline{DP}$ refers to *n* (Scheme 11.9).
[b] Data refer to light scattering according to the technique of deGroot and coworkers.
[c] Data refer to viscometry according to the technique of deGroot and coworkers.
[d] Calculated from the experimental value of g'. $g' = g^{\in}$.
[e] Data obtained by GPC in *o*-xylene at 100 °C.

governed by the relative activation energies of the migration step of the individual branches of the tricyclic organoboranes. Homologated product **16** is formed over species **17** in the second homologation. The trihomologated product **18** is produced from **16**, and gives rise to species **19** and **20** in the fourth generation, followed by **21** and **22** in the fifth step. Computational analysis of the early steps of the reaction confirms that there are significant energetic barriers that arise at the fifth and sixth homologation stages (Wagner *et al.*, 2003).

Following the polyhomologation–oxidation sequence with a large excess of ylide **1**, analysis of the low molecular weight species shows that the residual (not polyhomologated) organoboranes **21**, **22**, **23** account for 63% of the initiator used in the polyhomologation (Wagner *et al.*, 2003). These low molecular weight intermediate organoboranes explain the higher molecular weight of the polyhomologated product. The result confirms that two-thirds of the organoborane catalysts (**21**, **22**, **23**), terminate at an early stage of the polymerization, only the remaining one-third of catalysts escape to continue propagating and form polymethylene with 3 × higher molecular weight.

We have proposed that the active propagating species, which comprise only one-third of the original initiator, arise predominantly from the organoborane species **21**. The majority of species that are diverted to pathways **22** and **23** terminate. This is supported by the following observation (Wagner *et al.*, 2003). Resubjecting the terminated species (**21**, **22**, **23**) with excess ylide **1** followed by oxidation produced polymethylene. Analysis of the low molecular weight residue from this experiment revealed that intermediate **21** was largely depleted, while the majority of **22** and **23** remained in the residue. This result suggests that **21** is a more active initiator/catalyst than **22** and **23** and is the primary contributor to the high molecular weight polymer.

The origin of differences in reactivity of specific polycyclic organoboron intermediates arises from formation of caged tricyclic structures with collapsed, *inverted* pyramidal boron centers that have a low reactivity toward ylide **1** (Figure 11.8) (Wagner *et al.*, 2003). For example, after five methylene insertions, organoborane species **22** forms a caged tricyclic structure. To reduce transannular interactions and angle strain, the boron center is distorted from trigonal planarity. Formation of the zwitterionic complex **24**, results in a distorted tetrahedron. The high-energy of complex **24** arises from transannular interactions in the congested tricyclic core and angle strain. Transannular strain energy can be partially relieved at the transition state **25** during 1,2-migration. There are several organoboron intermediates that are generated during the *initial* stages of this reaction, each with different reactivity toward

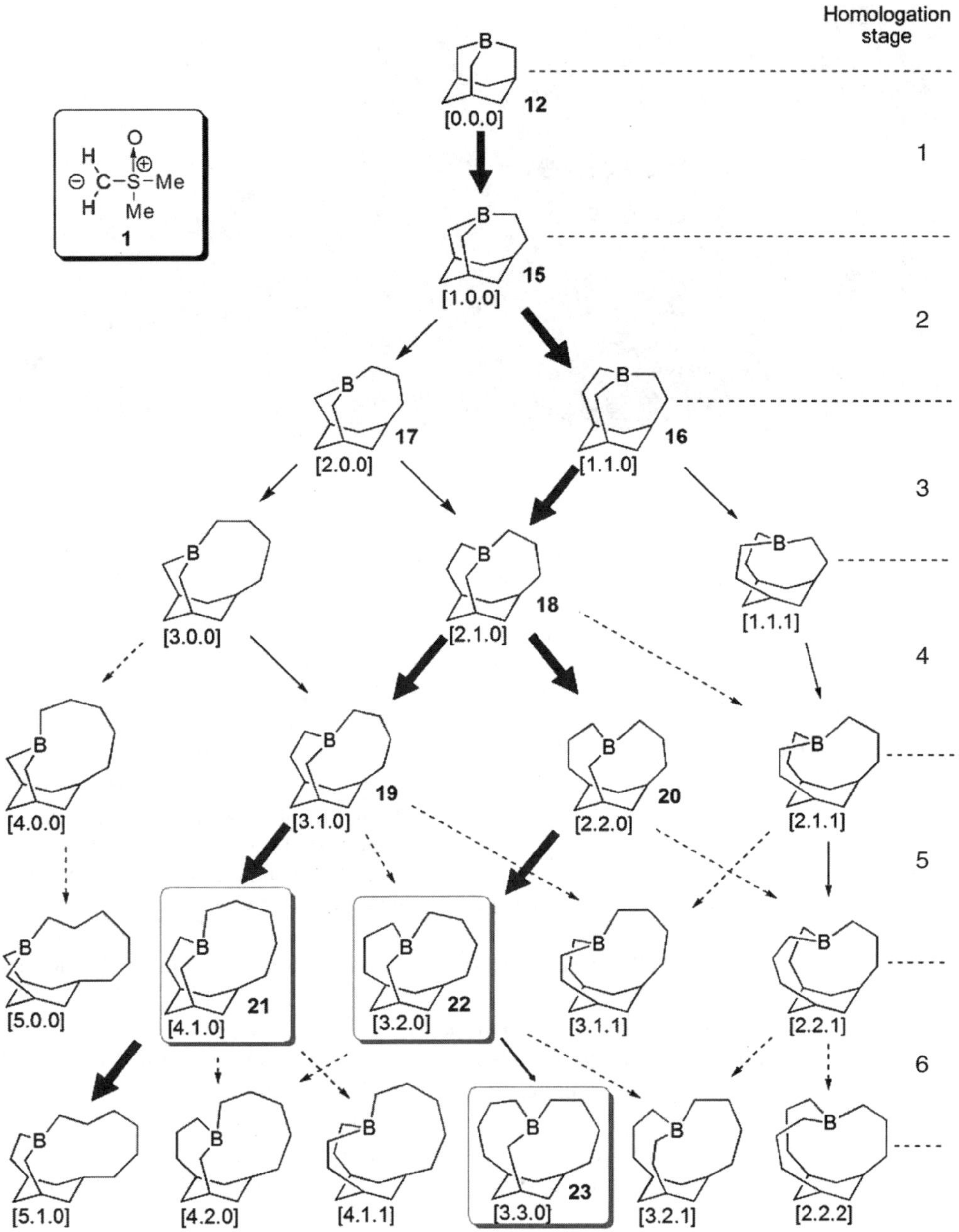

Figure 11.7 Proposed reaction pathways at the *initial* stage of the polyhomologation of 1-boraadamantane·THF **12**. The bold arrows indicate the main pathway determined by a combination of calculated activation energies and experimental findings. The compounds in the highlighted boxes (**21**, **22**, **23**) are the terminated organoborane intermediates recovered from the polymerization reaction.

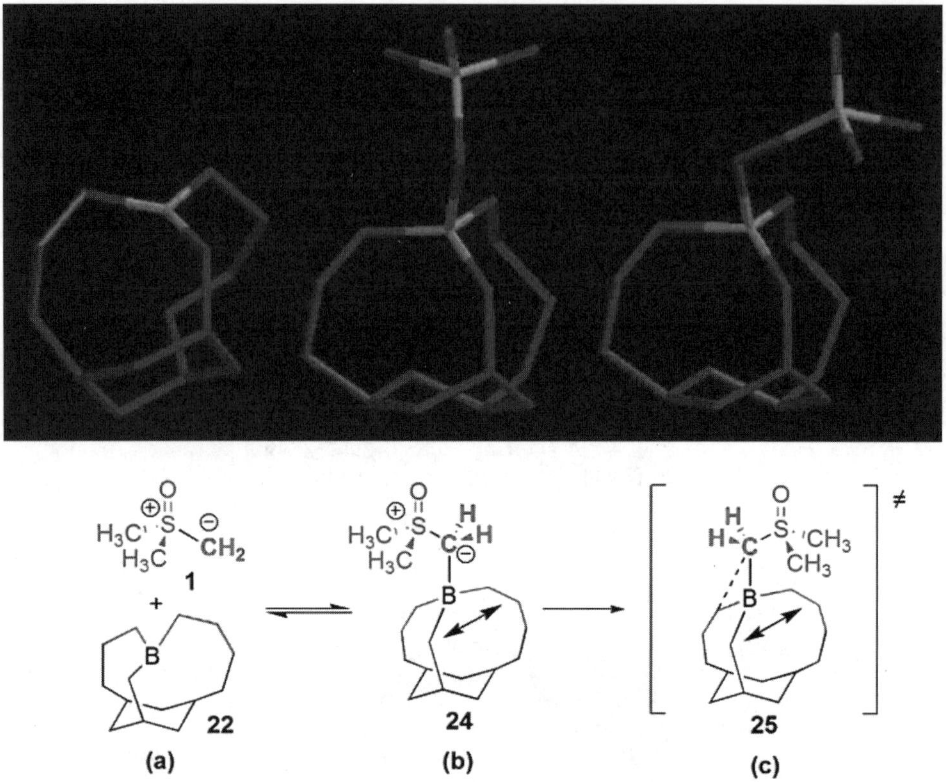

Figure 11.8 Computational results of the [3.2.0] system **22**, its complex **24**, and transition state **25**. (a) Tricyclic boranes at the fourth through sixth homologated stages contain inverted pyramidalized boron centers arising to reduce transannular interactions and angle strain. (b) Complex formation with the ylide **1** is endothermic. (c) Transannular strain energy is partially relieved at the transition state.

ylide **1**. It is this differentiation that results in a selective main pathway for the polyhomologation reaction.

The active organoborane species on the main pathway produce a regular A_3 three-arm star polymer **14**. After overcoming the *initial* stages of the polyhomologation, all the three arms of active initiators will propagate with equal probability. Because the polycyclic boranes are eventually expanded to very large rings, they behave like acyclic alkylboranes with the same average length of each arm. Despite the mechanistic complication at the *initial* stages, the reaction still produces regular A_3 three-arm star polymers **14** with controlled molecular weight and low PDI.

11.5.3 Topological Control from a Combination of Initiator and Postpolymerization Modification

Other novel polymethylene architectures can be obtained by related "stitching" reactions of polyhomologated organoboranes produced from structurally novel initiators. For example, the

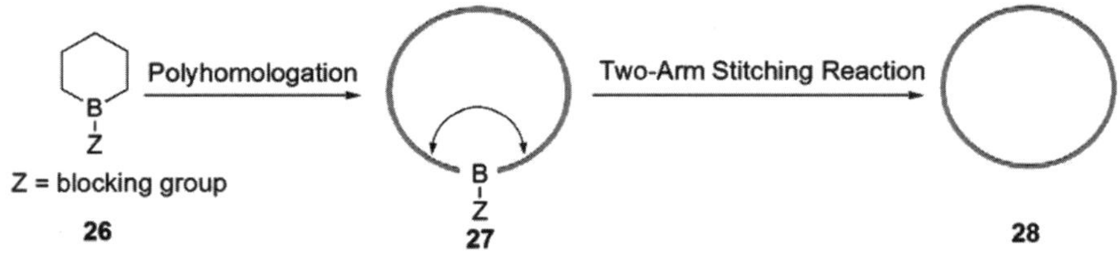

Scheme 11.10 An illustration of the synthesis of macrocyclic polymethylene **28** from a side-chain-blocked boracyclane initiator **26** followed by "two-arm stitching" reaction.

construction of macrocyclic polymethylene is achieved by "two-arm stitching" and a side-chain-blocked boracyclane initiator (Scheme 11.10). If a blocking group occupies one of the positions on a boracyclane, that is, **26**, polyhomologation results in expansion to produce a macrocyclic ring **27**. The subsequent "two-arm stitching" joins the two branches to construct the macrocyclic polymer **28**. Stitching the two arms transfers the macrocyclic organoborane intermediate to a carbocyclic ring. Design of the organoborane initiator followed by chemical "stitching" can result in novel polymethylene architectures.

B-Thexylborocane **29** was used for construction of macrocyclic polymethylene (Scheme 11.11) (Shea *et al.*, 1998a). In the event, a thexyl group, a substituent with known low migratory aptitude (Aggarwal *et al.*, 2006; Slayden, 1981; Brown, 1969), was employed as a blocking group. *B*-Thexylborocane **29** was synthesized by the hydroboration of 1,5-hexadiene with thexylborane in THF at −10 °C (Brown and Negishi, 1967). *B*-Thexylborocane was added to a solution of ylide **1** in toluene at 5–10 °C. Following ylide consumption, the expanded boracycle **30** was treated with sodium cyanide and benzyl chloride, a known "stitching" protocol (Pelter *et al.*, 1970; Pelter *et al.*, 1975). A subsequent oxidation produced cyclic ketone **31**.

The spectral properties of the product were consistent with the assigned macrocyclic structure (Shea *et al.*, 1998b). ^1H NMR analysis permitted the assignment of $\overline{DP}$ as 20 based on the ratio of integrated intensities of methylenes α to the carbonyl to those of the ring methylenes. The distribution of ketone macrocycles **31** is shown in the gas chromatograph of the reaction mixture (Figure 11.9). The identity of the individual peaks was established by high-resolution GC mass spectrometry. The most abundant molecular ion (M^+ 322.3222) corresponds to a 21-membered cyclic ketone (322.3235). The molecular ions of the surrounding peaks were separated by 14 amu. The molecular weights, calculated from the mass spectral

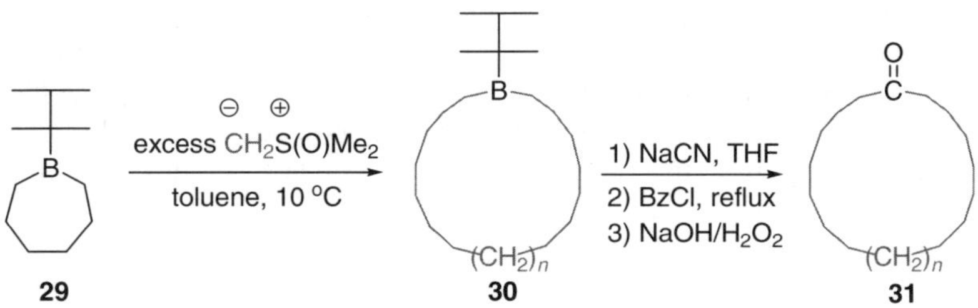

Scheme 11.11 Synthesis of oligomeric cyclic ketone **31** from *B*-thexylborocane **29**.

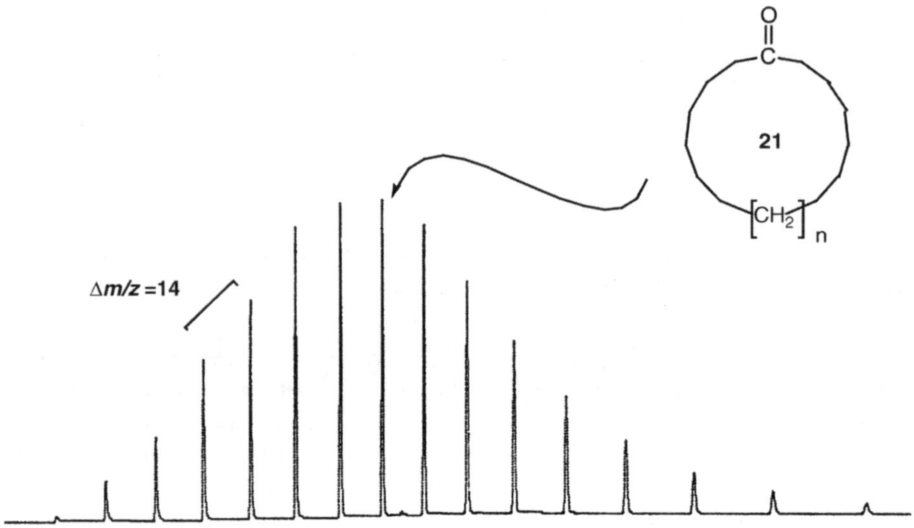

Figure 11.9 Gas chromatograph of the distribution of macrocyclic ketones **31**. HRMS found the most abundant ion at $t_R = 19.4$ to be 322.3222, the calculated M^+ for $n = 21$ is 322.3235. (Reprinted with permission from K.J. Shea, S.Y. Lee and B.B. Busch, "A new strategy for the synthesis of macrocycles. The polyhomologation of boracyclanes," *The Journal of Organic Chemistry*, **63**, 17, 5746–5747, 1998. © 1998 American Chemical Society.)

data, are $M_n = 290$ and $M_w = 295$ (PDI = 1.02). Both ^{1}H NMR analysis and GC–MS characterization indicate that the size of the macrocyclic ketone can be controlled by the living nature of the polyhomologation reaction.

11.6 Copolymers of Polymethylene

Despite the widespread use of polyethylene, some inherent limitations of polyethylene prevent its wider use in areas where other more expensive and less environmentally friendly materials are currently in use (Godoy-Lopez *et al.*, 2007). For example, nonpolarity and a high degree of crystallinity impart polyethylene with poor adhesion properties and incompatibility with many materials, including pigments, glass fibers, clays, carbon black and most polar polymers (Chung, 2002).

To extend the use of polyethylene, it is desirable to enhance polyethylene's polarity, toughness, adhesion and compatibility with other materials. One approach is by incorporating polyethylene in block copolymer structures (Hong *et al.*, 2002). Polyethylene block copolymers can maintain some of the superior properties of polyethylene while introducing the desired new properties from the other copolymer segments. In this way, the utility of polyethylene can be expanded to higher value areas, especially in polymer blends or composites, the preparation of micelles and the fabrication of nanoporous membranes (Wang and Hillmyer, 2001; Chen *et al.*, 2009; Uehara *et al.*, 2006; Uehara *et al.*, 2009).

The properties of polyethylene block copolymers depend on the structural and compositional parameters including the distribution and relative amounts of each segment, as well as molecular weight and molecular-weight distribution (Chung, 1995). To control these parameters, it is desirable to introduce each block in a controlled manner (Dong and Hu, 2006). Since the polyhomologation reaction provides accurate control of the polymethylene segments (which are expected to have the same properties as polyethylene), it can be combined with other polymerization methods to synthesize polymethylene copolymers. With efficient hydroboration reactions, organoborane initiators can be readily installed on other polymer segments from alkene functional groups then used for the polyhomologation reaction to construct polymethylene segments.

The strategy has been successfully used to prepare well-defined poly(methylene-*b*-styrene) (PM-*b*-PS) **36** (Scheme 11.12) (Zhou and Shea, 2001). Two living polymerizations were utilized to control the molecular weight of both blocks. Polystyrene block **33** was first installed from initiator **32** via TEMPO-mediated free-radical polymerization of styrene. Following polymerization, the hydroxyl group of the polystyrene block **33** was allylated to give allyl-terminated polystyrene **34**, which was subsequently converted to organoborane macroinitiator **35** by hydroboration. During the hydroboration reaction, borane was used as the limiting reagent, and the concentration of **35** was calculated on the assumption of quantitative hydroboration. The polyhomologation reaction with ylide **1** constructed the polymethylene block of copolymer **36**. Matrix-assisted laser desorption ionization time-of-flight (MALDI-TOF) mass spectrometry analysis of PS block **33** and GPC analysis of poly(methylene-*b*-styrene) **36** revealed that the molecular weight of both PS block and PM block were well controlled by nitroxide-mediated polymerization (NMP), a living free-radical polymerization and the polyhomologation reaction, respectively. In this example, polyhomologation can be combined with other living polymerizations to prepare polymethylene copolymer with well-defined composition of both blocks.

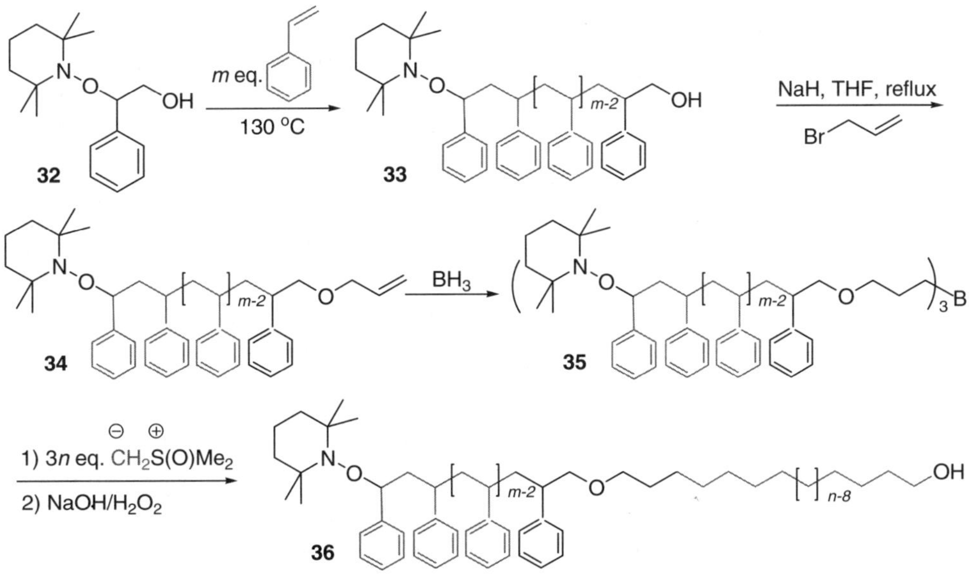

Scheme 11.12 Procedure for the synthesis of poly(methylene-*b*-styrene) **36**.

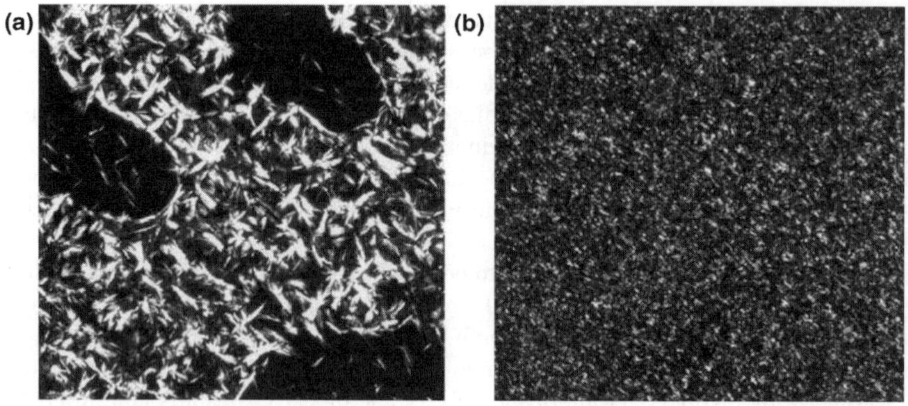

Figure 11.10 Polarized optical micrographs of polymer blends. (a) Two homopolymers with PE/PS = 50/50. (b) Two homopolymers and PM-*b*-PS copolymer with PS/PS-*b*-PM/PE = 50/8/50.

The PS-*b*-PM copolymer **36** was evaluated as compatibilizer for blends of immiscible homopolymers such as PS and PE (Zhou, 2002). PE shows poor miscibility with PS due to differences in solubility and crystallinity of each polymer. The relative block size was varied and optimized for compatibility effectiveness. Two blends PS/PE (50/50) and PS/PS-*b*-PM/PE (50/8/50) were prepared by solution mixing in *o*-xylene followed by precipitation with hexane. The phase separation of the two blends was observed using polarized optical microscopy (Figure 11.10). Two distinct phases, the crystalline PE phase (light area) and an amorphous PS phase (dark area), are visible in the PS/PE (50/50) blend (Figure 11.10a). The PE domains are segregated from the PS matrix due to the weak interfacial adhesion between PS and PE phases. On the other hand, for the ternary blends PS/PS-*b*-PM/PE (50/8/50) with PM-*b*-PS as the compatibilizer, the size of PE domains decreases dramatically, approaching a continuous phase (Figure 11.10b). The PM segments of the copolymer can disperse into the PE domain and act as surfactants to improve the miscibility between the two separate phases. The optimal block size for each segment can be found by systematically varying each block size with the two living polymerizations.

Copolymers containing very disparate blocks have also been prepared. Poly(ethylene glycol-*b*-methylene) was prepared by polyhomologation from commercially available methoxy terminated poly(ethylene glycol) (mPEG) **37** (Scheme 11.13) (Shea *et al.*, 1999).

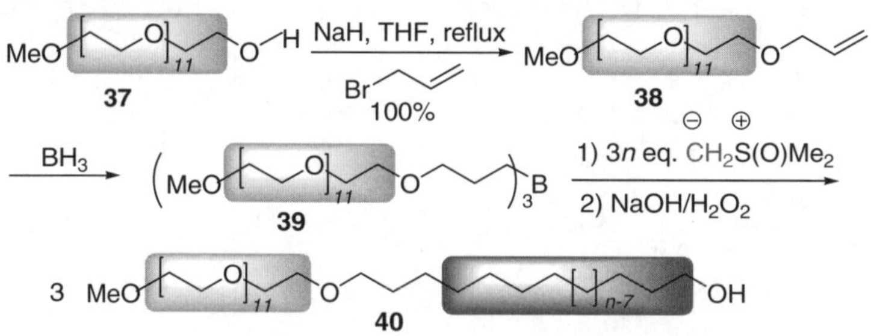

Scheme 11.13 Procedure for the synthesis of poly(ethylene glycol-*b*-methylene) **40**.

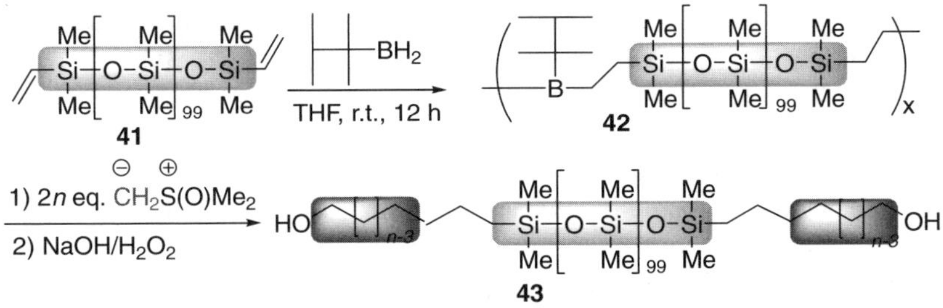

Scheme 11.14 Procedure for the synthesis of ABA triblock copolymer poly(methylene-b-dimethylsiloxane-*b*-methylene) **43**.

The terminal hydroxyl groups of mPEG **37** were quantitatively allylated by reaction with NaH and allyl bromide in refluxing THF. Hydroboration of intermediate **38** was carried out with borane as the limiting reagent to prepare macroinitiator **39**. Macroinitiator **39** was used immediately for the polyhomologation reaction to install the polymethylene block. The molecular weight of polymethylene block was controlled by varying the mole ratio of ylide **1** to borane **39**. Block copolymers **40** were isolated in yields >80%.

The synthesis of ABA triblock polymethylene copolymer requires a modified strategy. Hydroboration of long-chain diolefins with borane leads to an insoluble gel that is not suitable for use as a polyhomologation initiator (Shea *et al.*, 1999). Network formation can be suppressed by reducing the functionality of the hydroborating agent. Since the thexyl group has a low migratory aptitude under polyhomologation conditions (25 °C), thexylborane was used to hydroborate diolefins resulting in viscous solutions of organoborane initiators instead of gels.

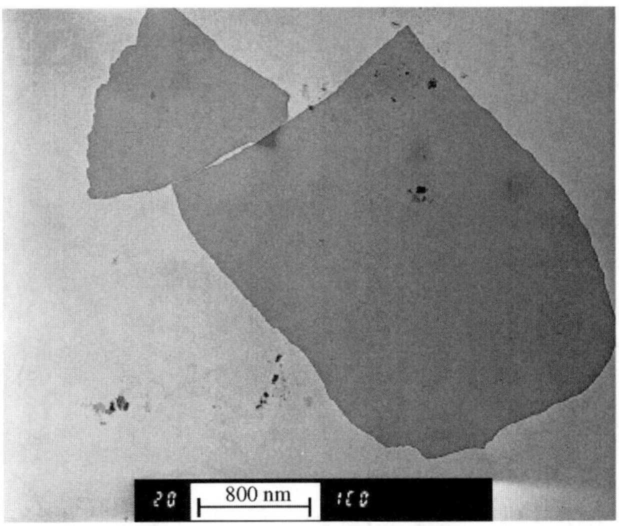

Figure 11.11 TEM image of PM-PDMS-PM crystal plates. (Reprinted from Polymer, **48**, 14, J. Wang, J.H. Horton, G. Liu *et al.*, "Polymethylene-*block*-poly(dimethyl siloxane)-*block*-polymethylene nanoaggregates in toluene at room temperature," 4123–4129, 2007, with permission from Elsevier.)

α,ω-Divinylpoly(dimethylsiloxane) **41** (PDMS, $\overline{DP} = 100$) was hydroborated with thexylborane to produce soluble macroinitiator **42**. The macroinitiator was added to a solution of an excess of ylide **1** in toluene. Following ylide consumption, the polyhomologated borane was oxidized to form triblock copolymer **43** (Scheme 11.14) (Shea *et al.*, 1999). GPC analysis of the triblock copolymer **43** shows a unimodal distribution, indicating the siloxane block retained its integrity throughout the polyhomologation and oxidation. The $\overline{DP}$ of polymethylene block was controlled by the organoborane–ylide stoichiometry.

Aggregates of the PM-PDMS-PM triblock copolymer sample can be prepared in toluene by slow cooling of a toluene solution of the sample from 105 °C to room temperature (Wang *et al.*, 2007). At room temperature, the ^{1}H NMR spectra of PM-PDMS-PM triblock

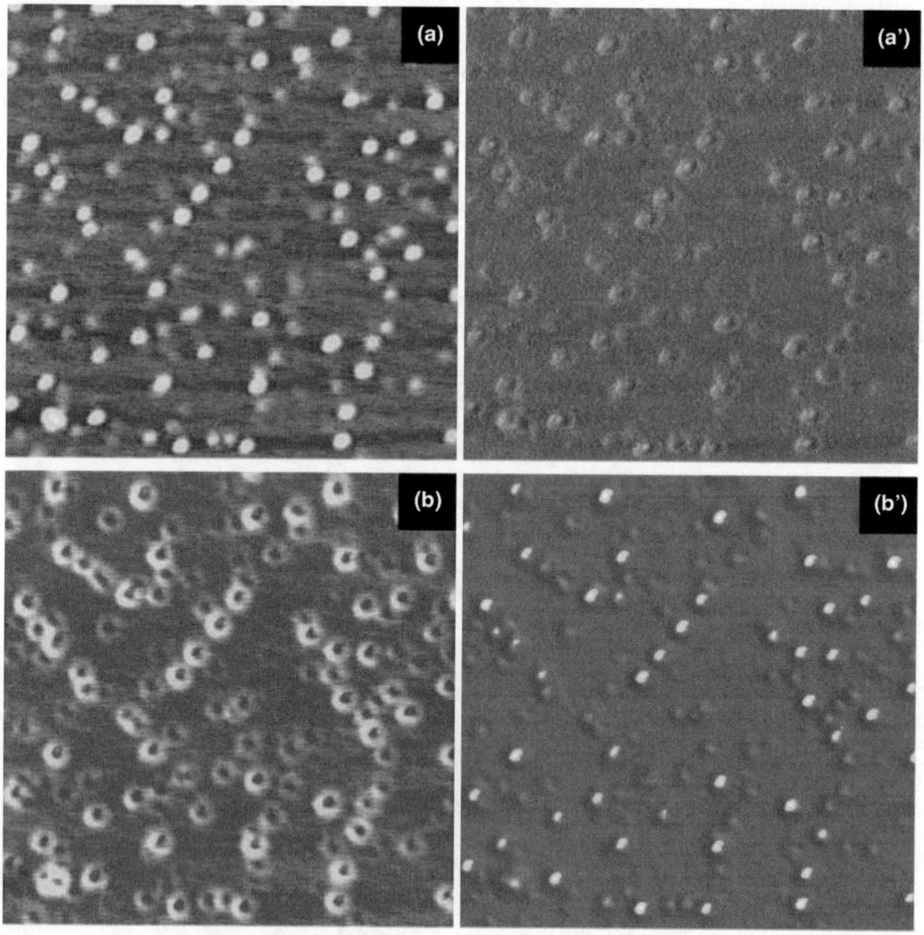

Figure 11.12 Topographic (left) and phase (right) AFM images of PM-PDMS-PM nanoaggregates aspirated on mica from toluene. The A_0 value used for AFM imaging was always ~45 nm. The r_{sp} values used were 0.95 (top), 0.75 (middle), and 0.60 (bottom). (Reprinted from Polymer, **48**, 14, J. Wang, J.H. Horton, G. Liu *et al.*, "Polymethylene-*block*-poly(dimethyl siloxane)-*block*-polymethylene nanoaggregates in toluene at room temperature," 4123–4129, 2007, with permission from Elsevier.)

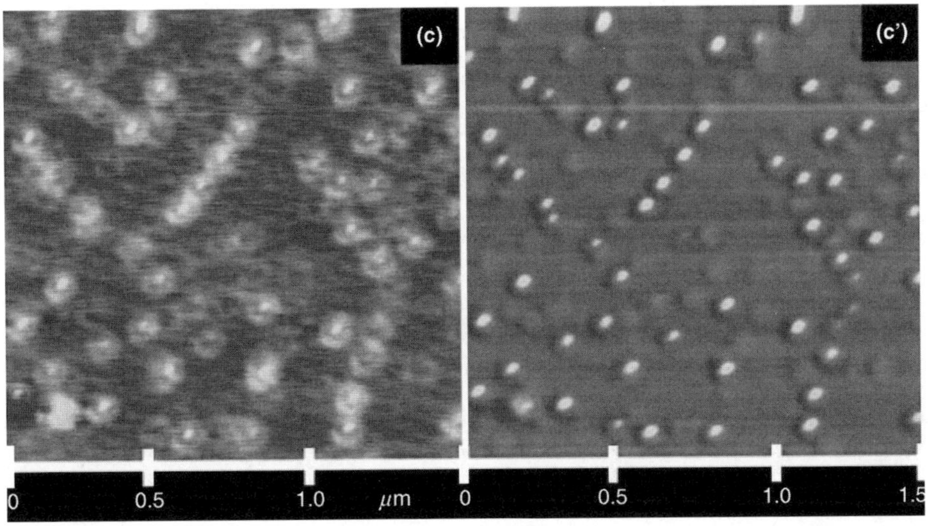

Figure 11.12 (*Continued*)

copolymer in toluene-d_8 showed *only* the siloxane methyl resonance at 0.07 ppm (Shea *et al.*, 1999). The characteristic polymethylene resonance at 1.25 ppm was absent. Only upon heating to 85 °C in toluene-d_8 was the polymethylene resonance observed. These results suggest that above 85 °C, both polydimethylsiloxane and polymethylene segments are soluble in toluene. Upon cooling to room temperature, only the polydimethylsiloxane segments are solvated and mobile in toluene, the polymethylene segments are aggregated, perhaps as microcrystallites, and cannot undergo rapid rotation. Due to the different solubility of each segment in toluene, aggregates of the triblock copolymer can be prepared.

The morphology of PM-PDMS-PM aggregates was studied by TEM and AFM (Wang *et al.*, 2007). Crystal plates of PM-PDMS-PM block copolymers with sizes up to tens of micrometers were observed by TEM (Figure 11.11). Polyethylene is insoluble in most solvents at room temperature and forms square crystalline plates by slow evaporation from dilute solution (Lin and Gast, 1996). The plate structures of the polymethylene-containing block copolymers must be driven by the polymethylene blocks. Many nanoparticles are also present in the sample. Because these nanoparticles are unstable and "melt" immediately in the TEM electron beam, AFM was used to examine the structure of these nanoaggregates (Figure 11.12). It is interesting that the topographic images of the nanoaggregates obtained by tapping-mode AFM changed from nanodiscs at light tapping to nanodonuts at moderate tapping to filled nanodonuts at heavy tapping. These different forms can be explained by a combination of phase separation in the nanoparticles and the difference in hardness between PM and PDMS components of the nanoparticles. The AFM tip can penetrate to different depths within the PM and PDMS phases. In the same domain, the AFM tip reaches different depths due to the different forces under different tapping conditions.

Based on observations at heavier tapping modes of AFM, and the fact that at room temperature PM is insoluble in toluene and PDMS is soluble, a structure of the

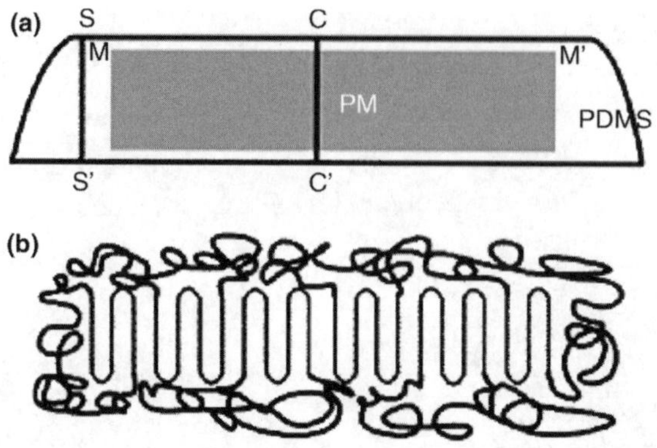

Figure 11.13 (a) Proposed domain structures in PM-PDMS-PM nanodiscs deposited on mica surfaces. (b) Possible chain packing in such nanodiscs in toluene. (Reprinted from Polymer, **48**, 14, J. Wang, J.H. Horton, G. Liu *et al.*, "Polymethylene-*block*-poly(dimethyl siloxane)-*block*-polymethylene nanoaggregates in toluene at room temperature," 4123–4129, 2007, with permission from Elsevier.)

nanoaggregate is proposed. The core of the nanoaggregate is the insoluble PM block and the PDMS serves as its shell (Figure 11.13a) (Wang *et al.*, 2007). Since PM crystallizes and favors parallel chain packing, it is possible that these nanodiscs have well-ordered chain packing inside the PM core and are covered with PDMS loops (Figure 11.13b) (Sperling, 2006).

The formation of PM-PDMS-PM triblock copolymer aggregates is a kinetic rather than thermodynamic process (Wang *et al.*, 2007). Micrometer-sized crystal plates are produced if the sample solution is cooled slowly, while nanodiscs are formed when the hot polymer solution is quenched with water. Given sufficient time, polymethylene chains may be reorganized to form the larger crystal plates.

11.7 Conclusion

A new approach for the controlled synthesis of polymethylene has been described. The polyhomologation reaction uses an organoborane as the initiator/catalyst and dimethylsulfoxonium methylide **1** as the monomer to produce perfectly linear polymethylene. The reaction involves association of ylide **1** and trialkylborane to form a zwitterionic complex, which subsequently undergoes a *rate-limiting 1,2-migration step* to insert a methylene unit into a boron-terminated polymer chain. The polymer backbone is built up one carbon at a time. The polyhomologation reaction is a living polymerization, providing accurate control over molecular weight and polydispersity. Cyclic and polycyclic organoborane initiators coupled with postpolymerization modification permit construction of novel polymethylene topologies including three-arm star polymers, giant tube-like structures and macrocycles. The polyhomologation can be combined with other living polymerizations to synthesize AB and ABA polymethylene block copolymers. An example includes PS-PM diblock copolymers that were shown to be an effective compatibilizing agent for PS/PE blends. By combining polymethylene with disparate polymer blocks such as PM-PDMS-PM triblock copolymers, complex architectures ranging from micrometer-sized crystal plates to nanodiscs were prepared.

Acknowledgment

We thank the Chemistry Division of the National Science Foundation for financial support of this work.

References

Aggarwal, V.K., Fang, G.Y., Ginesta, X. *et al.* (2006) Toward an understanding of the factors responsible for the 1,2-migration of alkyl groups in borate complexes. *Pure and Applied Chemistry*, **78**, 215–229.

Arthur, V.T. and Charles, E.R. (1959) Isoprene polymerization by organometallic compounds. II. *Journal of Polymer Science*, **40**, 73–89.

Bai, J. and Shea, K.J. (2006) Reaction of boranes with TMS diazomethane and dimethylsulfoxonium methylide. Synthesis of poly(methylidene-*co*-TMSmethylidene) random copolymers. *Macromolecular Rapid Communications*, **27**, 1223–1228.

Bawn, C.E.H., Ledwith, A., and Matthies, P. (1959) The mechanism of the polymerization of diazoalkanes catalyzed by boron compounds. *Journal of Polymer Science*, **34**, 93–108.

Berry, G.C. and Orofino, T.A. (1964) Branched polymers. III. Dimensions of chains with small excluded volume. *The Journal of Chemical Physics*, **40**, 1614–1621.

Bick, D.K. and McLeish, T.C.B. (1996) Topological contributions to nonlinear elasticity in branched polymers. *Physical Review Letters*, **76**, 2587–2590.

Bielawski, C.W., Benitez, D., and Grubbs, R.H. (2002) An “endless” route to cyclic polymers. *Science*, **297**, 2041–2044.

Bielawski, C.W., Benitez, D., and Grubbs, R.H. (2003) Synthesis of cyclic polybutadiene via ring-opening metathesis polymerization: The importance of removing trace linear contaminants. *Journal of the American Chemical Society*, **125**, 8424–8425.

Bottoni, A., Lombardo, M., Neri, A., and Trombini, C. (2003) Migratory aptitudes of simple alkyl groups in the anionotropic rearrangement of quaternary chloromethyl borate species: A combined experimental and theoretical investigation. *The Journal of Organic Chemistry*, **68**, 3397–3405.

Brintzinger, H.H., Fischer, D., Mulhaupt, R. *et al.* (1995) Stereospecific olefin polymerization with chiral metallocene catalysts. *Angewandte Chemie-International Edition*, **34**, 1143–1170.

Brown, H.C. (1969) Organoborane–carbon monoxide reactions. Synthesis of carbon structures. *Accounts of Chemical Research*, **2**, 65–72.

Brown, H.C. (1975) *Organic Syntheses* via *Boranes*, Wiley Inc., New York.

Brown, H.C. and Negishi, E. (1967) Carbonylation of the organoboranes from the cyclic hydroboration of dienes with thexylborane. Simple procedure for the conversion of dienes into cyclic ketones. *Journal of the American Chemical Society*, **89**, 5477–5478.

Brown, H.C. and Sharp, R.L. (1966) Hydroboration. XXIV. Directive effects in the hydroboration of some substituted styrenes. *Journal of the American Chemical Society*, **88**, 5851–5854.

Busch, B.B., Paz, M.M., Shea, K.J. *et al.* (2002) The boron-catalyzed polymerization of dimethylsulfoxonium methylide. A living polymethylene synthesis. *Journal of the American Chemical Society*, **124**, 3636–3646.

Chanda, M. and Roy, S.K. (1993) *Plastics Technology Handbook*, 2nd edn, (Revised and Expanded), Marcel Dekker, Inc., New York.

Chen, J.Z., Zhao, Q.L., Shi, L.P. *et al.* (2009) Synthesis, structure, and property of well-defined polymethylene-based diblock copolymers by a combination of living polymerization of ylides and atom transfer radical polymerization. *Journal of Polymer Science Part A: Polymer Chemistry*, **47**, 5671–5681.

Chung, T.C. (1995) Recent developments in functionalization of polyolefins by borane approach. *Trends in Polymer Science*, **3**, 191–198.

Chung, T.C. (2002) Synthesis of functional polyolefin copolymers with graft and block structures. *Progress in Polymer Science*, **27**, 39–85.

Corey, E.J. and Chaykovsky, M. (1965) Dimethyloxosulfonium methylide ($(CH_3)_2SOCH_2$) and dimethylsulfonium methylide ($(CH_3)_2SCH_2$). Formation and application to organic synthesis. *Journal of the American Chemical Society*, **87**, 1353–1364.

Dong, J.-Y. and Hu, Y. (2006) Design and synthesis of structurally well-defined functional polyolefins via transition metal-mediated olefin polymerization chemistry. *Coordination Chemistry Reviews*, **250**, 47–65.

Drott, E.E. and Mendelso, R.A. (1970a) Determination of polymer branching with gel-permeation chromatography. 1. Theory. *Journal of Polymer Science Part A:2 Polymer Physics*, **8**, 1361–1371.

Drott, E.E. and Mendelso, R.A. (1970b) Determination of polymer branching with gel-permeation chromatography. 2. Experimental results for polyethylene. *Journal of Polymer Science Part A:2 Polymer Physics*, **8**, 1373–1385.

Eaton, G.R. and Lipscomb, W.N. (1969) *NMR Studies of Boron Hydrides and Related Compounds*, Benjamin, New York.

Evans, W.J., Decoster, D.M., and Greaves, J. (1996) Evaluation of field desorption mass spectrometry for the analysis of polyethylene. *Journal of the American Society for Mass Spectrometry*, **7**, 1070–1074.

Everaers, R., Sukumaran, S.K., Grest, G.S. *et al.* (2004) Rheology and microscopic topology of entangled polymeric liquids. *Science*, **303**, 823–826.

Flory, P.J. (1940) Molecular size distribution in ethylene oxide polymers. *Journal of the American Chemical Society*, **62**, 1561–1565.

Fox, M.E., Szoka, F.C., and Fréchet, J.M.J. (2009) Soluble polymer carriers for the treatment of cancer: The importance of molecular architecture. *Accounts of Chemical Research*, **42**, 1141–1151.

Godoy-Lopez, R., D'Agosto, F., and Boisson, C. (2007) Synthesis of well-defined polymer architectures by successive catalytic olefin polymerization and living/controlled polymerization reactions. *Progress in Polymer Science*, **32**, 419–454.

Hadjichristidis, N., Xenidou, M., Iatrou, H. *et al.* (2000) Well-defined, model long chain branched polyethylene. 1. Synthesis and characterization. *Macromolecules*, **33**, 2424–2436.

Hicks, F.A. and Brookhart, M. (2001) A highly active anilinotropone-based neutral nickel(II) catalyst for ethylene polymerization. *Organometallics*, **20**, 3217–3219.

Hine, J., Rosscup, R.J., and Duffey, D.C. (1960) Formation of the intermediate methoxychloromethylene in the reaction of dichloromethyl methyl ether with base. *Journal of the American Chemical Society*, **82**, 6120–6123.

Hong, S.C., Jia, S., Teodorescu, M. *et al.* (2002) Polyolefin graft copolymers via living polymerization techniques: Preparation of poly(n-butyl acrylate)-graft-polyethylene through the combination of Pd-mediated living olefin polymerization and atom transfer radical polymerization. *Journal of Polymer Science Part A: Polymer Chemistry*, **40**, 2736–2749.

Imoto, M. and Nakaya, T. (1972) Polymerization by carbenoids, carbenes, and nitrenes. *Polymer Reviews*, **7**, 1–48.

Jellema, E., Jongerius, A.L., Reek, J.N.H., and Debruin, B. (2010) C1 polymerization and related C–C bond forming carbene insertion reactions. *Chemical Society Reviews*, **39**, 1706–1723.

Kabalka, G.W. and Hedgecock, H.C. (1975) Mild and convenient oxidation procedure for the conversion of organoboranes to the corresponding alcohols. *The Journal of Organic Chemistry*, **40**, 1776–1779.

Laurent, B.A. and Grayson, S.M. (2006) An efficient route to well-defined macrocyclic polymers via "click" cyclization. *Journal of the American Chemical Society*, **128**, 4238–4239.

Lin, E.K. and Gast, A.P. (1996) Semicrystalline diblock copolymer platelets in dilute solution. *Macromolecules*, **29**, 4432–4441.

Lohse, D.J., Milner, S.T., Fetters, L.J. *et al.* (2002) Well-defined, model long chain branched polyethylene. 2. Melt rheological behavior. *Macromolecules*, **35**, 3066–3075.

Luo, J. and Shea, K.J. (2010) Polyhomologation. A living C1 polymerization, *Accounts of Chemical Research*, **43**, 1420–1433.

Maria, M. and Bernhard, W. (1974) Crystallization during polymerization of diazomethane. I. The boron trifluoride catalyzed reaction. *Journal of Polymer Science: Polymer Physics Edition*, **12**, 1993–2018.

McLeish, T. (2002) Chemistry: Polymers without beginning or end. *Science*, **297**, 2005–2006.

Moad, G. and Solomon, D.H. (2006) *The Chemistry of Radical Polymerization*, Elsevier Science, New York.

Morita, T., Maughon, B.R., Bielawski, C.W., and Grubbs, R.H. (2000) A ring-opening metathesis polymerization (ROMP) approach to carboxyl- and amino-terminated telechelic poly(butadiene)s. *Macromolecules*, **33**, 6621–6623.

Mueller, A.H.E., Yan, D., Litvinenko, G. *et al.* (1995) Kinetic analysis of "living" polymerization processes exhibiting slow equilibria. 2. Molecular weight distribution for degenerative transfer (direct activity exchange between active and "dormant" species) at constant monomer concentration. *Macromolecules*, **28**, 7335–7338.

Natta, G. (1956) Stereospezifische katalysen und isotaktische polymere. *Angewandte Chemie*, **68**, 393–403.

Odian, G. (2004) *Principles of Polymerization*, Wiley Inc., New York.

O'Gara, J.E., Wagener, K.B., and Hahn, S.F. (1993) Acyclic diene metathesis (ADMET) polymerization. Synthesis of perfectly linear polyethylene. *Die Makromolekulare Chemie, Rapid Communications*, **14**, 657–662.

Pelter, A., Hutchings, M.G., Rowe, K., and Smith, K. (1975) The chemistry of organoborates. Part II. High yield syntheses of trialkylmethanols by the cyanoborate process. *Journal of the Chemical Society, Perkin Transactions 1*, 138–142.

Pelter, A., Hutchings, M.G., and Smith, K. (1970) The reactions of electrophilic reagents with trialkylcyanoborates: Some new efficient ketone syntheses. *Journal of the Chemical Society D: Chemical Communications*, 1529–1530.

Pelter, A., Smith, K., and Brown, H.C. (1988) *Borane Reagents*, Academic Press, London.

Piringer, O.G. and Baner, A.L. (2008) *Plastic Packaging: Interactions with Food and Pharmaceuticals*, VCH Verlagsgessellschaft, Weinheim, Germany, New York.

Reichardt, C. (1990) *Solvents and Solvent Effects in Organic Chemistry*, VCH Verlagsgessellschaft, Weinheim, Germany, New York.

Shea, K.J. (2000) Polyhomologation: The living polymerization of ylides. *Chemistry-A European Journal*, **6**, 1113–1119.

Shea, K.J., Busch, B.B., and Paz, M.M. (1998a) Polyhomologation: Synthesis of novel polymethylene architectures by a living polymerization of dimethylsulfoxonium methylide. *Angewandte Chemie-International Edition*, **37**, 1391–1393.

Shea, K.J., Lee, S.Y., and Busch, B.B. (1998b) A new strategy for the synthesis of macrocycles. The polyhomologation of boracyclanes. *The Journal of Organic Chemistry*, **63**, 5746–5747.

Shea, K.J., Staiger, C.L., and Lee, S.Y. (1999) Synthesis of polymethylene block copolymers by the polyhomologation of organoboranes. *Macromolecules*, **32**, 3157–3158.

Shea, K.J., Walker, J.W., Zhu, H. *et al.* (1997) Polyhomologation. A living polymethylene synthesis. *Journal of the American Chemical Society*, **119**, 9049–9050.

Sita, L.R. (2009) Ex uno plures ("out of one, many"): New paradigms for expanding the range of polyolefins through reversible group transfers. *Angewandte Chemie-International Edition*, **48**, 2464–2472.

Slayden, S.W. (1981) Relative migratory aptitudes of alkyl groups in the iodination of lithium ethynyltrialkylborates. *The Journal of Organic Chemistry*, **46**, 2311–2314.

Sperling, L.H. (2006) *Introduction to Physical Polymer Science*, Wiley Inc., New York.

Stenzel, M.H., Barner-Kowollik, C., and Davis, T.P. (2006) Formation of honeycomb-structured, porous films via breath figures with different polymer architectures. *Journal of Polymer Science Part A: Polymer Chemistry*, **44**, 2363–2375.

Stoddard, J.M. and Shea, K.J. (2003) Isolation and characterization of borane complexes of dimethylsulfoxonium methylide. *Organometallics*, **22**, 1124–1131.

Stoddard, J.M. and Shea, K.J. (2004) Thermal stability and kinetics of ylide-borane complexes. *Thermochimica Acta*, **424**, 149–155.

Uehara, H., Kakiage, M., Sekiya, M. *et al.* (2009) Size-selective diffusion in nanoporous but flexible membranes for glucose sensors. *ACS Nano*, **3**, 924–932.

Uehara, H., Yoshida, T., Kakiage, M. *et al.* (2006) Nanoporous polyethylene film prepared from bicontinuous crystalline/amorphous structure of block copolymer precursor. *Macromolecules*, **39**, 3971–3974.

Wagener, K.B., Boncella, J.M., and Nel, J.G. (1991) Acyclic diene metathesis (ADMET) polymerization. *Macromolecules*, **24**, 2649–2657.

Wagner, C.E., Kim, J.-S., and Shea, K.J. (2003) The polyhomologation of 1-boraadamantane: Mapping the migration pathways of a propagating macrotricyclic trialkylborane. *Journal of the American Chemical Society*, **125**, 12179–12195.

Wagner, C.E. and Shea, K.J. (2001) 1-Boraadamantane blows its top, sometimes. The mono- and polyhomologation of 1-boraadamantane. *Organic Letters*, **3**, 3063–3066.

Wang, J., Horton, J.H., Liu, G. *et al.* (2007) Polymethylene-*block*-poly(dimethyl siloxane)-*block*-polymethylene nanoaggregates in toluene at room temperature. *Polymer*, **48**, 4123–4129.

Wang, Y.B. and Hillmyer, M.A. (2001) Polyethylene-poly(L-lactide) diblock copolymers: Synthesis and compatibilization of poly(L-lactide)/polyethylene blends. *Journal of Polymer Science Part A: Polymer Chemistry*, **39**, 2755–2766.

Wood-Adams, P.M., Dealy, J.M., deGroot, A.W., and Redwine, O.D. (2000) Effect of molecular structure on the linear viscoelastic behavior of polyethylene. *Macromolecules*, **33**, 7489–7499.

Wu, Z. and Grubbs, R.H. (1994) Synthesis of narrow dispersed linear polyethylene and block copolymers from polycyclobutene. *Macromolecules*, **27**, 6700–6703.

Wu, Z., Wheeler, D.R., and Grubbs, R.H. (1992) Living ring-opening metathesis polymerization of cyclobutene: The thermodynamic effect of a reversibly binding ligand. *Journal of the American Chemical Society*, **114**, 146–151.

Zhang, W., Wei, J., and Sita, L.R. (2008) Living coordinative chain-transfer polymerization and copolymerization of ethene, α-olefins, and α,ω-nonconjugated dienes using dialkylzinc as "surrogate" chain-growth sites. *Macromolecules*, **41**, 7829–7833.

Zhou, X.Z. (2002) Polyhomologation, in *Application for the Synthesis of Polymethylene Block Copolymers and Discovery of New Polymerizable Ylides*, Department of Chemistry. Irvine, University of California, Irvine.

Zhou, X.-Z. and Shea, K.J. (2001) Synthesis of poly(methylene-b-styrene) by sequential living polymerization. *Macromolecules*, **34**, 3111–3114.

Ziegler, K., Holzkamp, E., Breil, H., and Martin, H. (1955) Das mülheimer normaldruck-polyäthylen-verfahren. *Angewandte Chemie*, **67**, 541–547.

Zimm, B.H. and Kilb, R.W. (1959) Dynamics of branched polymer molecules in dilute solution. *Journal of Polymer Science*, **37**, 19–42.

Zimm, B.H. and Stockmayer, W.H. (1949) The dimensions of chain molecules containing branches and rings. *The Journal of Chemical Physics*, **17**, 1301–1314.

12

Phenylenevinylene Homopolymers and Block Copolymers via Ring-Opening Metathesis Polymerization

Chin-Yang Yu and Michael L. Turner
Organic Materials Innovation Centre, School of Chemistry, University of Manchester, UK

12.1 Introduction

Conjugated polymers have been widely studied for application in devices such as light-emitting diodes (Kugler *et al.*, 1999; Friend *et al.*, 1999), field effect transistors (Zaumseil and Sirringhaus, 2007; Naber *et al.*, 2005) and solar cells (Gunes *et al.*, 2007; Schmidt-Mende *et al.*, 2001). Polymers with a backbone of repeating phenylenevinylene units have attracted the most interest (Burroughes *et al.*, 1990) as these polymers were employed as the active semiconducting layer in the first polymeric light-emitting diode. A number of synthetic routes to phenylenevinylene polymers have been reported in the literature for example, pyrolysis of sulfonium polymer precursors (Wessling route) (Wessling, 1985), dehydrohalogenation of xylylene dihalides (Gilch route) (Gilch and Wheelwright, 1966), polycondensation of xylyene diphosphonates with phthaldehydes (Wittig–Horner type condensation reaction) (Rost *et al.*, 1997), palladium-catalyzed polycondensation of divinylbenzene with dihaloarenes (Heck type condensation reaction) (Bao *et al.*, 1993; Yu and Bao, 1994). In the synthetic routes reported to date there is little control over the polydispersity, molecular weight or the end groups of the polymers. The optoelectronic properties of conjugated polymers are profoundly affected by small variations in the chemical structure or the conformation of the delocalized backbone (Miao and Bazan, 1994a).

Complex Macromolecular Architectures: Synthesis, Characterization, and Self-Assembly, First Edition.
Edited by Nikos Hadjichristidis, Akira Hirao, Yasuyuki Tezuka and Filip Du Prez.

Ring-opening metathesis polymerization (ROMP) has been extensively investigated for the polymerization of strained cyclic monomers (e.g. norbornene) and leads to precise control of polymer molecular weight, end groups and low polydispersities (Ahmed *et al.*, 2003). ROMP has been previously applied to the synthesis of polyacetylene **4** by the ring opening of cyclobutenes **1** or cyclooctatetraenes **3** by Feast (Edwards and Feast, 1980) and Grubbs, respectively (Scheme 12.1) (Gorman *et al.*, 1991; Jozefiak *et al.*, 1993).

It has also been exploited in the synthesis of phenylenevinylene polymers and copolymers. (Scheme 12.2). In general, these studies focused on the preparation of precursors that could be converted to insoluble phenylenevinylene homopolymer **8** (Thorn-Csányi *et al.*, 1993; Thorn-Csányi and Pleug, 1993; Thorn-Csányi *et al.*, 1994), for example ROMP of (Friend *et al.*, 1999) paracyclophane-1,9-diene **9** gives PPV as an insoluble, yellow fluorescent powder. Soluble phenylenevinylene copolymers have been prepared by ROMP of (Friend *et al.*, 1999) paracyclophane-1,9-diene with cyclopentene, cyclooctene and cyclocta-1,5-diene comonomers (Thorn-Csányi *et al.*, 1993; Thorn-Csányi and Pleug, 1993; Thorn-Csányi *et al.*, 1994). Unfortunately, the incorporation of more than 5% of phenylenevinylene units gives insoluble polymers, limiting the potential application of these materials.

An alternative approach was reported by Grubbs and coworkers (Conticello *et al.*, 1992) who demonstrated that PPV precursors can be obtained by ROMP of functionalized bicyclo[2.2.2] octanedienes **11** with excellent control of the polymer backbone structure. Final conversion to the conjugated PPV requires a high-temperature treatment ($> 200\,°C$) and the presence of a base catalyst. PPV has also been prepared by ROMP of 9-[(*tert*-butyldimethylsilyl)oxy][2.2] paracyclophane-1-ene **5** followed by thermal and chemical treatment (Miao and Bazan, 1994a, 1994b). In general, solubilization of conjugated polymers such as polyphenylenevinylenes requires long alkyl chains to be appended to the delocalized backbone or incorporated into the main chain. The latter approach leads to segmented polymers and breaks the delocalization of the backbone. The appending of "bound solvent" reduces the interactions between adjacent chains and leads to an increase in entropy when dissolving or melting the polymer.

This chapter discusses the use of ROMP to prepare soluble phenylenevinylene polymers and fully conjugated block copolymers of well-defined microstructures with control of the molecular weight.

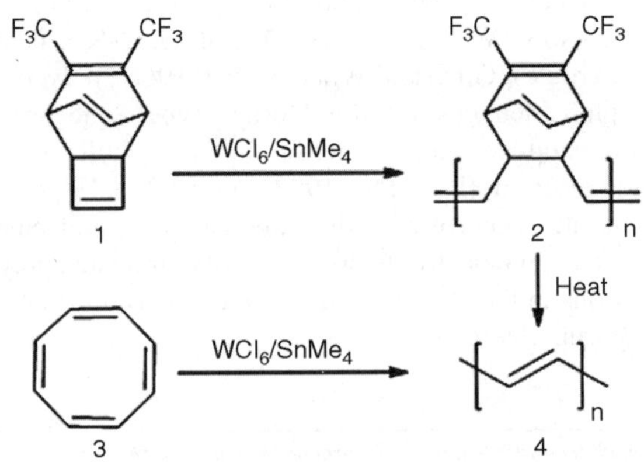

Scheme 12.1 The synthesis of polyacetylene **4** by ROMP.

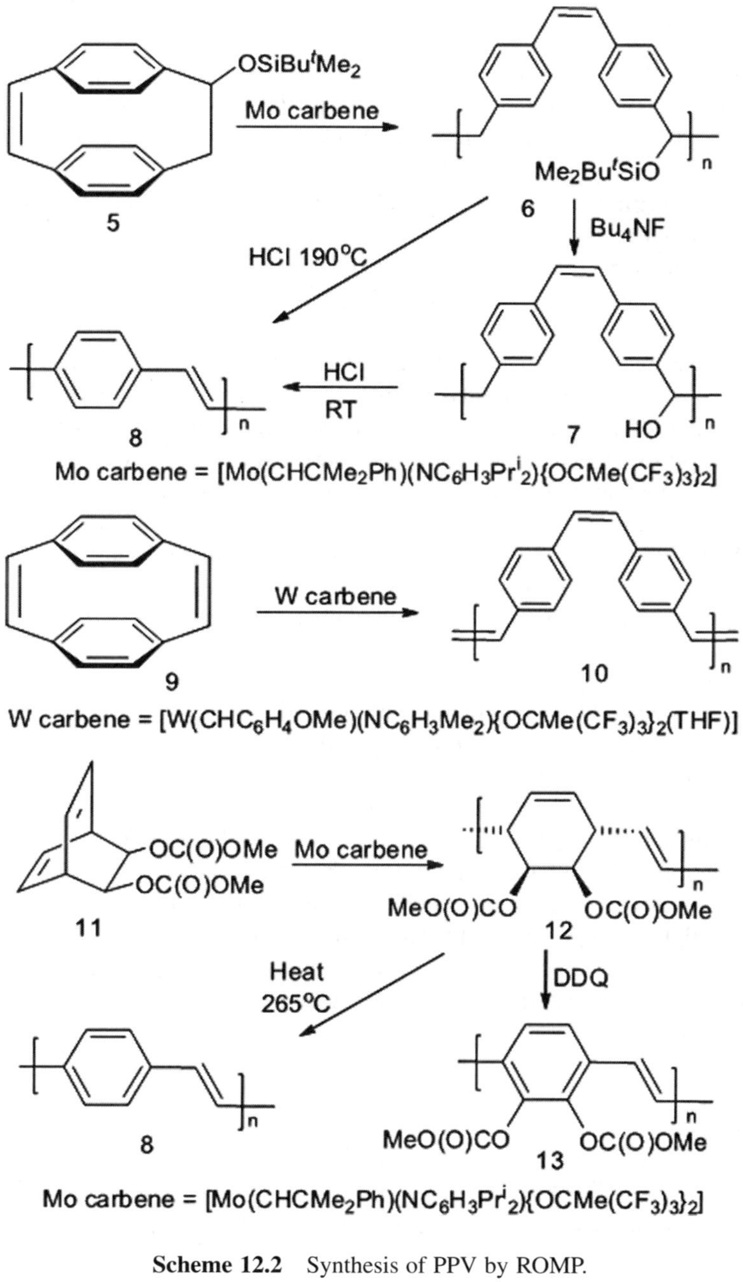

Scheme 12.2 Synthesis of PPV by ROMP.

12.2 Phenylenevinylene Homopolymers by Ring-Opening Metathesis Polymerization

The ROMP of the 4,7,12,15-tetraoctyloxy-[2.2]paracyclophane-1,9-diene **14** (Yu and Turner, 2006) dissolved in THF, was initiated by addition of a THF solution of the second-generation Grubbs catalyst (Bielawski and Grubbs, 2000) (Scheme 12.3). The reaction mixture was heated at reflux and then quenched by addition of ethyl vinyl ether. The resulting polymers **15** were purified by filtering through silica gel and isolated by evaporation of the solvent. A range of polymer molecular weights was prepared by varying the initial monomer to catalyst ratio. An extended reaction time (36 h) was necessary to achieve complete conversion of monomer, indicating a slow rate of polymerization (Wagaman and Grubbs, 1997; Thorn-Csányi *et al.*, 1998).

The molecular weights of the polymers **15** were determined by gel permeation chromatography (GPC) in THF solution, calibrated against narrow polydispersity polystyrene standards, using refractive index (RI) detection. The M_n increased monotonically with initial monomer to catalyst ratio (Figure 12.1) with a correlation coefficient of 0.999 and the polydispersities were all close to 1.2. The apparent discrepancy between the expected and the observed molecular weights is due to the difference in the hydrodynamic radii of PPV and polystyrene (Cotts *et al.*, 1996; Stalmach *et al.*, 2001). From the linear relationship between monomer to catalyst ratio and degree of polymerization, the low polydispersities and the complete consumption of monomer it is clear that this is a well-controlled chain-growth process (Ivin and Mol, 1997; Matyjaszewski, 1993).

The living nature of this polymerization was confirmed by an experiment in which a second charge of monomer was added to a polymerization reaction. The initial reaction gave a polymer with a M_n of 7400 (expected $M_n = 7264$) and a polydispersity of 1.20. A second addition of one third amount of monomer increased the M_n to 9600 (expected $M_n = 9412$) with no change in the polydispersity, consistent with a living polymerization and demonstrating the potential of this chemistry to prepare conjugated block copolymers. The MALDI-TOF mass spectrum of polymer **15** (Figure 12.2), isolated by a filtration through silica gel followed by washing with methanol, shows a series of peaks separated by an interval of 716 mass units, corresponding to the molecular weight of the monomer **14**. This series is consistent with PPVs capped with vinyl and phenyl end groups, as expected and is accompanied by peaks associated with the addition of one sodium per polymer chain. There was no evidence for the defect structures commonly seen in other synthetic routes (Kraft *et al.*, 1998).

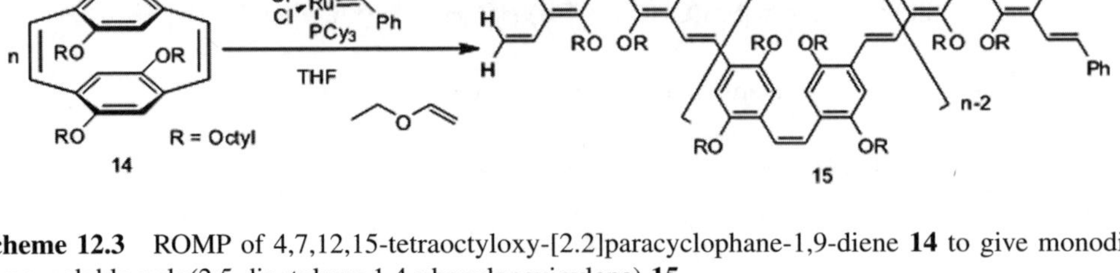

Scheme 12.3 ROMP of 4,7,12,15-tetraoctyloxy-[2.2]paracyclophane-1,9-diene **14** to give monodisperse, soluble poly(2,5-dioctyloxy-1,4-phenylenevinylene) **15**.

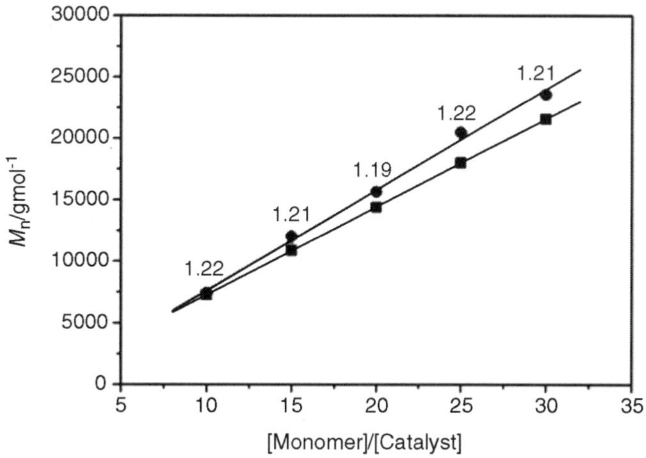

Figure 12.1 Dependence of molecular weight (M_n) of polymer **15** on the monomer **14** to catalyst ratio. The polydispersities are given for expected $n = 10, 15, 20, 25, 30$. ● RI detector, ■ calculated (including end groups). (Reproduced with permission from C.-Y. Yu and M.L. Turner, "Soluble poly(p-phenylenevinylene)s through ring-opening metathesis polymerization," *Angewandte Chemie International Edition*, 2006, **45**, 7797. © Wiley-VCH Verlag GmbH & Co. KGaA.).

The ^{1}H NMR spectrum showed the high level of control achieved in this polymerization (Figure 12.3). Relief of significant ring strain within monomer **14** is achieved by opening of only one vinylene linkage and it is therefore expected that polymer **15** has a backbone of regularly alternating *cis* and *trans* vinylene linkages. Peaks at $\delta = 4.00$ and 3.61 ppm can be assigned to the methylene groups attached to the oxygen for the *trans* and *cis* vinylene links of

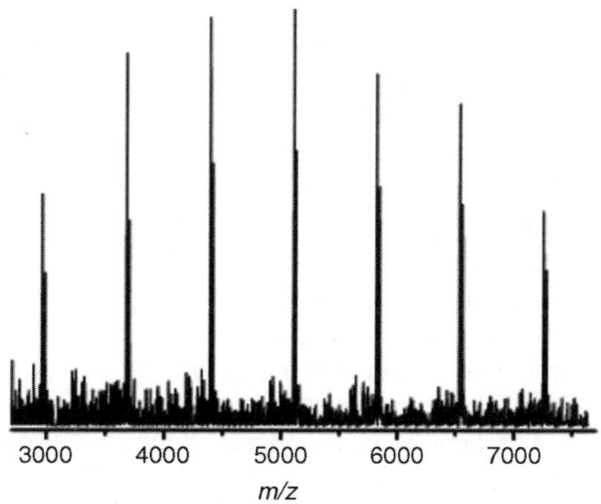

Figure 12.2 MALDI-TOF mass spectrum of polymer **15** with an expected degree of polymerization of 10 repeat units. (Reproduced with permission from C.-Y. Yu and M.L. Turner, "Soluble poly(p-phenylenevinylene)s through ring-opening metathesis polymerization," *Angewandte Chemie International Edition*, 2006, **45**, 7797. © Wiley-VCH Verlag GmbH & Co. KGaA.).

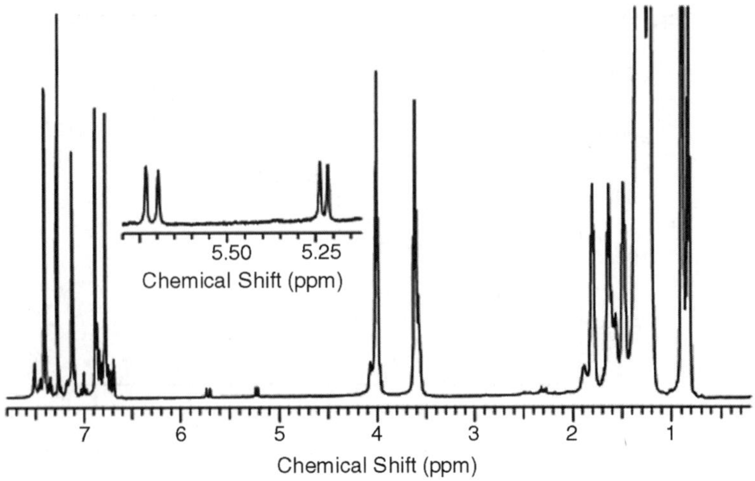

Figure 12.3 ^{1}H NMR spectrum of polymer **15** in $CDCl_3$ solution. Inset shows an expansion of the signals associated with the vinyl end groups.

the polymer backbone, respectively (Liao *et al.*, 2001; Jacobs *et al.*, 1993; Becker *et al.*, 1999). Integration of these signals gave the expected 1: 1 ratio. The precise control of the polymer microstructure was confirmed by the simplicity of the signals associated with the aromatic and vinylene groups (i.e., the peaks between $\delta = 6.77$ and 7.41) in the ^{1}H NMR spectra. The signals at 6.77 ppm and 6.88 ppm are associated with the aromatic and *cis* vinylene units, respectively; whereas those at 7.41 ppm and 7.12 ppm are assigned to hydrogens of the *trans* vinylene and associated aromatic groups. Again, the integration of these signals is consistent with a 1: 1 ratio of *cis*:*trans* linkages. Absolute values for the degree of polymerization were determined by integration of the signals for the vinyl end groups observed at 5.23 and 5.72 ppm against those for the methylene groups attached to oxygen. In all cases the values obtained are consistent with the expected number of repeat units, derived from the monomer to catalyst ratio.

It is clearly desirable to extend this polymerization route to the preparation of one of the most widely studied conjugated polymers, namely poly[2-methoxy-5-(2′-ethylhexyloxy)-1,4-phenylenevinylene], which is normally abbreviated to MEH-PPV. It is also important to reduce the rather long reaction times required to achieve complete monomer conversion. One approach to reduce reaction times is to use microwave heating, as this technique has been extensively applied in synthetic organic chemistry (Kappe, 2004) and is now being exploited in polymerization chemistry (Larhed *et al.*, 2002). The main advantage of microwave irradiation over conventional heating is that it results in rapid, homogeneous heating of the reaction mixture and can lead to a reduction in potential side reactions and shorter reaction times. 4,12-Di-2′-ethylhexyloxy-7,15-dimethoxy-[2.2]paracyclophane-1,9-diene **16** was polymerized by the third-generation Grubbs catalyst (Love *et al.*, 2002) when heated by microwave irradiation in anhydrous 1,2-dichloroethane (Scheme 12.4) (Spring *et al.*, 2009). The optimum reaction condition for monomer conversion, determined by sampling the reaction mixture during the polymerization, was found to be heating at 80 °C for 1 h. This compares with reaction times of up to 36 h required for complete monomer consumption using conventional heating in THF solution.

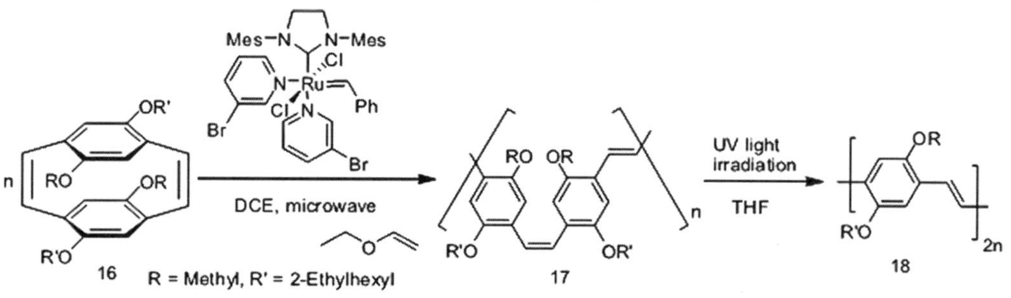

Scheme 12.4 ROMP of 4,12-di-2′-ethylhexyloxy-7,15-dimethoxy-[2.2]paracyclophane-1,9-diene **16** to give monodisperse, soluble PPV.

The polymerization was quenched by addition of an excess of ethyl vinyl ether and the resulting polymers were purified by reprecipitation in methanol. Polymers **17** with a range of molecular weights were prepared by varying the initial monomer to catalyst ratio. For MEH-PPVs the M_n was close to that expected from the initial monomer to catalyst ratio and the polydispersities were in the range of 1.18–1.28. Isomerization of the as-formed *cis,trans*-microstructure to a polymer with an all *trans*-vinylene backbone was expected to increase the conjugation length of the polymer and allow a direct comparison with those polymers prepared by more conventional methods such as the Gilch process. Heating polymers in the solid state for extended periods of time at up to 80 °C proved ineffective and the polymer was recovered essentially unchanged. The as-formed *cis,trans* polymer **17** can be isomerized to the *trans*-vinylene polymer **18** by prolonged irradiation at 365 nm in THF solution (20 mg in 10 ml) (Katayama *et al.*, 2005; Katayama *et al.*, 2006). The reaction was followed by ^{1}H NMR spectroscopy by measuring the ratio of the signal at 3.91 ppm to that at 3.41 ppm, and 95% conversion was achieved by irradiation for 36 h. The disappearance of the signals between 6.60 and 6.90 ppm associated with the *cis*-vinylene and the phenyl groups neighboring to *cis*-vinylene groups confirmed the isomerization to an all *trans*-vinylene microstructure. The polydispersities of all *trans* polymers **18** were very similar to those of the *cis,trans*-polymers **17**, however, in all cases the molecular weight determined by GPC against polystyrene standards was approximately 2.5 times that of the precursor. A plausible explanation is that the *trans* polymer **18** has a larger hydrodynamic radius than that of the isomeric polymer, resulting in a larger polystyrene equivalent molecular weight when determined by GPC against polystyrene standards.

The optical properties of all polymers in solution and in the solid state are summarized in Table 12.1, which also includes the properties of MEH-PPV prepared by the Gilch route. The maximum absorption observed for *trans* polymer **18**, in both solution and thin film, shifts to a longer wavelength on increasing the molecular weight (M_n). Interestingly, the absorption maximum of the highest molecular weight polymer ($M_n = 90\,600$) is higher in solution and thin film than that of MEH-PPV ($M_n = 106\,800$) prepared via the Gilch method. This suggests that MEH-PPV prepared via ROMP has fewer chain defects and a more conjugated structure than that prepared by the Gilch method despite being of a lower molecular weight.

The energy levels of the highest occupied molecular orbital (HOMO) and the lowest unoccupied molecular orbital (LUMO) for the polymers were estimated from the onset of

Table 12.1 Physical, optical and electronic properties of MEH-PPV **17** and **18**

Polymer	M_n[a]	M_n[b]	M_w/M_n[b]	λ_{max}/nm		λ_{PL}/nm		$E_{g\text{-}opt}$/nm[c]		HOMO/eV[d]	LUMO/eV[d]	$E_{g\text{-}ec}$/eV[e]
				solution	film	solution	film	solution	film			
17d	26 205	34 100	1.28	456	458	552	592	2.24	2.06	−5.08	−2.66	2.42
17c	20 985	23 800	1.21	467	462	552	592	2.24	2.07	−5.09	−2.64	2.44
17b	15 765	16 000	1.22	455	458	552	592	2.25	2.08	−5.08	−2.62	2.46
17a	10 545	11 200	1.18	444	444	550	592	2.25	2.08	−5.08	−2.62	2.46
18d	26 205	90 600	1.32	510	537	555	596	2.23	2.05	−5.03	−2.71	2.31
18c	20 985	57 000	1.34	508	537	554	596	2.23	2.05	−5.02	−2.64	2.38
18b	15 765	38 800	1.27	502	530	554	596	2.23	2.05	−5.05	−2.63	2.42
18a	10 545	28 800	1.30	502	528	554	595	2.23	2.05	−5.04	−2.64	2.40
Gilch	—	106 800	6.96	498	530	554	592	2.24	2.03	−4.99	−2.63	2.36

[a] Expected molecular weight from monomer to catalyst ratio.
[b] Calculated from GPC measurements carried out using THF as the solvent and calibrated by polystyrene standards.
[c] $E_{g\text{-}opt}$ = optical band gap.
[d] CV of polymer films on Pt plate were measured in MeCN solution containing 0.10 M n-Bu_4NPF_6. HOMO = − (4.8 + $E_{pa\text{-}onset}$ − E_{Fc}). Half-wave potential of ferrocene, E_{Fc} (= 0.105 V vs. $Ag/AgNO_3$), was measured in MeCN solution. LUMO = − (4.8 + $E_{pc\text{-}onset}$ − E_{Fc}).
[e] Electrochemical band gap was calculated from $E_{g\text{-}ec}$ = LUMO − HOMO.

the oxidation and reduction peaks, respectively (Table 12.1). The small difference between the electrochemical band gap (about 0.06 eV) of the *trans*-vinylene polymers **18** and that of the *cis,trans* analogues **17** is consistent with an increase in the conjugation length of all *trans* polymer **18**. The electrochemical band gap of *cis,trans*-**17** and all-*trans* polymers **18** decrease with increasing molecular weight and the highest molecular weight *trans*-vinylene polymer ($M_n = 90\ 600$) is smaller than MEH-PPV prepared via the Gilch method ($M_n = 106\ 800$).

ROMP of tetra-alkoxy substituted cyclophanedienes is only possible for the meso-isomer in which the ruthenium catalyst is able to coordinate to the alkene bridges. The racemic isomer is completely inert under the reaction conditions investigated to date and can be recovered unreacted from the final polymer. This is a significant problem for the overall yield of the polymerization. It can be overcome by the use of the asymmetric 4,7-diethylhexyloxy-[2.2]paracyclophane-1,9-dienes **19** and 4,7-diethylhexyloxy-[2.2]metaparacyclophane-1,9-diene **21** as these monomers have no isomers and all of the monomer can be converted to PPV. The ROMP of these monomers can be initiated by addition of a THF solution of the second-generation Grubbs catalyst (Scheme 12.5). The reaction mixture was heated at 40 °C and the reactions were quenched by the addition of an excess of ethyl vinyl ether. The resulting polymers **20**, **22** were purified by filtering through silica gel and isolated by evaporation of the solvent. A range of polymer molecular weights was prepared by varying the initial monomer to initiator (Grubbs catalyst) ratio. Extended reaction times (20 h) were necessary to achieve complete conversion of monomer (as measured by TLC), indicating a slow rate of polymerization.

The absorption and emission spectra of the longest polymers containing only *p*-phenylenevinylenes **20** showed an absorption maximum at 467 nm, an emission peak at 527 nm and a shoulder at 566 nm. The absorption maximum of the longest chain was slightly red-shifted by about 10 nm over that of the related polymer, poly(*p*-phenylenevinylene-*alt*-2,5-dioctyloxy-*p*-phenylenevinylene)s reported previously and prepared by a simple

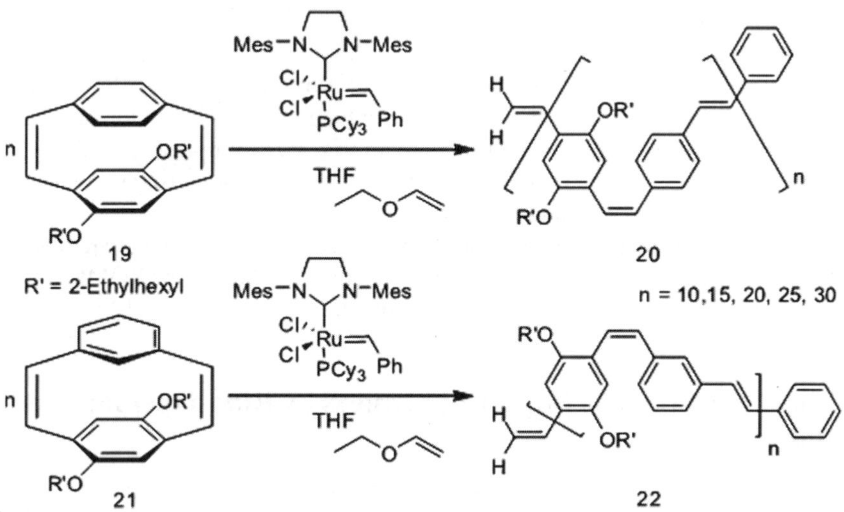

Scheme 12.5 ROMP of dialkoxy-substituted cyclophanedienes **19**, **21** using the second-generation Grubbs catalyst.

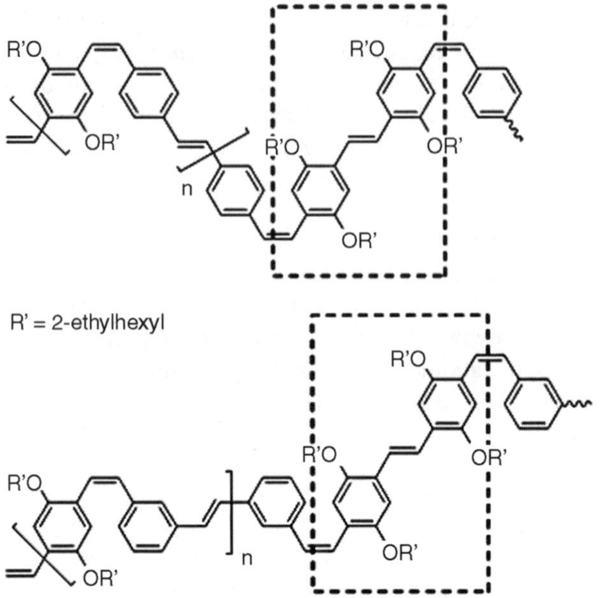

Figure 12.4 Backbone microstructure of PPV prepared by ROMP of asymmetric cyclophanedienes.

condensation polymerization (Barashkov *et al.*, 1995). The emission maximum of this polymer at 527 nm was also red-shifted by 10 nm when compared to that of polymers reported by Ferraris *et al.* (Barashkov *et al.*, 1995). This was unexpected given the *cis,trans*-microstructure of these polymers as the previously reported poly(*p*-phenylenevinylene-*alt*-2,5-dioctyloxy-*p*-phenylenevinylene)s were synthesized by Wittig–Horner polycondensation and had an entirely *trans*-vinylene microstructure. Polymers with *trans*-vinylene linkages are known to be more conjugated than the corresponding *cis*-vinylene polymers (Hu and Karasz, 1998). The more extended conjugation of the polymers prepared by ROMP appears to be due to the random nature of the ROMP that results in polymers in which the microstructure contains some segments with two dialkoxy phenylene rings bound to a *trans*-vinylene unit. Emission from these segments is expected to be red-shifted over the regularly alternating polymers prepared by Wittig–Horner polycondensation (Barashkov *et al.*, 1995).

It is important to note that the regularly alternating microstructures depicted for the ROMP of asymmetric cyclophanedienes are an idealized representation and that the ROMP process gives regiorandom polymers with the backbone segments identified in Figure 12.4.

12.3 Phenylenevinylene Block Copolymers by Ring-Opening Metathesis Polymerization

Conjugated block copolymers are of considerable interest as these polymers can self-assemble into a variety of morphologies (Matsen and Bates, 1996; Matsen and Thompson, 1999; Klok and Lecommandoux, 2001). In conventional block copolymers the particular morphology adopted depends primarily on the volume fraction of the two blocks, the interaction between the blocks and the molecular weight of each of the blocks (Krevelen, 1990). A number of groups have examined the supramolecular assembly of A-B- or A-B-A-type rod–coil block copolymers containing one conjugated block. However, the commonly used coil blocks, such as polystyrene or poly(ethylene oxide), only support the self-assembly and do not carry any electronic function (Asawapirom *et al.*, 2005; Schmitt *et al.*, 2001; Scherf and List, 2002; Tu *et al.*, 2006; Scherf *et al.*, 2008; Wang *et al.*, 2002; Liang *et al.*, 2007).

Block copolymers of poly(*p*-phenylenevinylene-2,5-diethylhexyloxy-*p*-phenylenevinylene) **20** and poly(*m*-phenylenevinylene-2,5-diethylhexyloxy-*p*-phenylenevinylene) **22** with different volume fractions were made by ROMP of 4,7-di-2′-ethylhexyloxy-[2.2]paracyclophane-1,9-diene **19** followed by addition of the second monomer, 4,7-di-2′-ethylhexyloxy-[2.2]metaparacyclophane-1,9-diene **21** (Scheme 12.6) (Yu *et al.*, 2010). In order to maximize the chances of forming desirable morphologies such as cylindrical and lamella it was necessary to estimate the volume fractions and the length of the polymer. This was achieved by using estimates of the Van der Waals volume of each monomer, the Van der Waals volume of the repeat unit of 4,7-di-2′-ethylhexyloxy-[2.2]paracyclophane-1,9-diene **19** and 4,7-di-2′-ethylhexyloxy-[2.2]metaparacyclophane-1,9-diene **21** were both calculated to be 300.98 cm^3/mol. Polymers **24** were designed with volume fractions of 3: 1, 1: 1 and 1: 3 by using the calculated repeat unit molar volumes. These volume fractions equate to degrees of polymerization of the constituent monomers of 15: 5, 15: 15 and 15: 45. Scheme 12.6 shows the synthetic route employed to prepare block copolymers.

The first monomer **19** was dissolved in THF and ROMP was initiated by addition of a THF solution of the second-generation Grubbs catalyst. The reaction mixture was heated at 40 °C and the conversion of monomer monitored by TLC. After complete consumption of

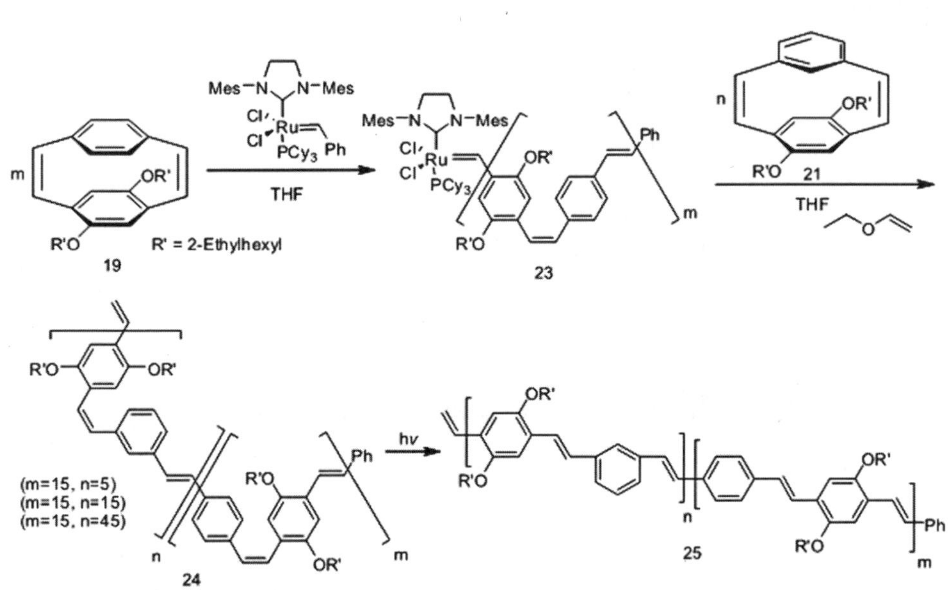

Scheme 12.6 Preparation of PPV block copolymers **24** and **25** by ROMP.

the first monomer **19** a THF solution of the second monomer **21** was added to a polymerization reaction at the same temperature. The reaction was finally quenched by addition of an excess of ethyl vinyl ether at room temperature. The resulting block copolymers **24** were purified by filtering through a short plug of silica gel and isolated by evaporation of the solvent. For block copolymers, a M_n of 13 200 (expected $M_n = 9304$) with a polydispersity of 1.76, a M_n of 17 600 (expected $M_n = 13\,904$) with a polydispersity of 1.67 and a M_n of 32 800 (expected $M_n = 27\,704$) with a polydispersity of 1.46 were obtained. The polydispersities of the block copolymers are comparable with or higher than the polydispersity (1.46) of the starting homopolymer. This is presumably due to intramolecular chain-transfer reactions that take place during the extended reaction time needed to prepare the block copolymers. The polydispersity decreases with increasing chain length as expected. Absolute values for the degree of polymerization in block copolymers **24** were determined using the ^{1}H NMR spectrum by integration of the signals for the vinyl end groups observed at around 5.23 and 5.75 ppm against those for the methylene groups attached to oxygen. The values obtained are higher than expected for the monomer:catalyst ratio employed. This suggests that secondary metathesis reactions may occur during the polymerization. The end groups of the polymer were also analyzed by MALDI-TOF mass spectrometry. The results of MALDI-TOF mass spectrometry depend strongly on the sample and the matrix preparation and it was difficult to obtain high-quality spectra for the higher molecular weight polymers (Anders *et al.*, 2002; Klarner *et al.*, 1998). The MALDI-TOF mass spectrum of block copolymer **24** ($m = 15$, $n = 5$) (Figure 12.5) shows a main series of peaks separated by an interval of 460 mass units, corresponding to the molecular weight of monomers. This series (▲) is consistent with polymers capped with vinyl and phenyl end groups. Further series of peaks are also observed, the series (•) at 102 mass units (C_8H_6) less than the major series can be attributed to the formation of linear polymers with one less phenylenevinylene unit and the series of

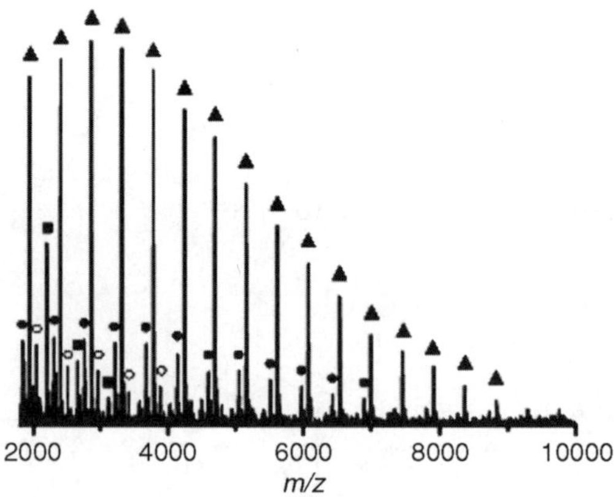

Figure 12.5 MALDI-TOF mass spectrum of block copolymer **24** ($m = 15$, $n = 5$). (Reprinted with permission from C.-Y. Yu, M. Horie, A.M. Spring *et al.*, "Homopolymers and block copolymers of p-phenylenevinylene-2,5-diethylhexyloxy-p-phenylenevinylene and m-phenylenevinylene-2,5-diethylhexyloxy-p-phenylenevinylene by ring-opening metathesis polymerization," *Macromolecules*, **43**, 222, 2010. © 2010 American Chemical Society.).

peaks (○) at 358 less than the primary series can be assigned to those polymers with one less diethylhexyloxy substituted phenylenevinylene unit. These sequences are consistent with intramolecular secondary metathesis before termination. Low molar mass cyclic oligomers (■) generated by this reaction are observed with one extra diethylhexyloxy substituted phenylenevinylene unit ($C_{24}H_{38}O_2$, mass = 358). There is no evidence for intermolecular chain-transfer reactions as no peaks are detected for polymers with two phenyl or two vinyl end groups. The detection of only one series of linear polymers that are integral repeats of the monomer units (460) associated with intramolecular chain transfer suggests that chain transfer is restricted to the more active *cis*-vinylene units of the polymer backbone. Intramolecular chain transfer to the *trans*-vinylene linkages should lead to a second series of peaks that are integral repeats of the monomer units, containing even numbers of phenylenevinylene units; this series of peaks was not observed. Initial studies of the solid-state morphology by AFM and X-ray diffraction have been inconclusive due to the limited contrast between two blocks that carry ethylhexyloxy substituents. To enhance the phase separation between the blocks, polymers with linear alkoxy side chains in one of the blocks were synthesized.

The sequential ROMP of cyclophanediene monomers **14** and **21** generated poly[(2,5-dioctyloxy-1,4-phenylenevinylene)-*block*-poly(1,3-phenylenevinylene-2,5-diethylhexyloxy-1,4-phenylenevinylene)] **27** with a range of block volume fractions (Scheme 12.7) (Yu *et al.*, 2009). Polymers were designed with volume fractions of 3: 1, 1: 1 and 1: 3 by using the calculated repeat unit molar volumes. These volume fractions equate to degrees of polymerization of the constituent monomers of 10: 5, 10: 16 and 10: 48. The initial block derived from the first monomer **14** was designed to have 10 repeat units (20 phenylenevinylene units) and an expected M_n of 7264. GPC against polystyrene standards gave a M_n of 7400 and a polydispersity of 1.22. Addition of a second block derived from the second monomer **21** gave copolymers with a M_n of 10 000 (expected $M_n = 9564$) and a polydispersity of 1.34, a M_n of

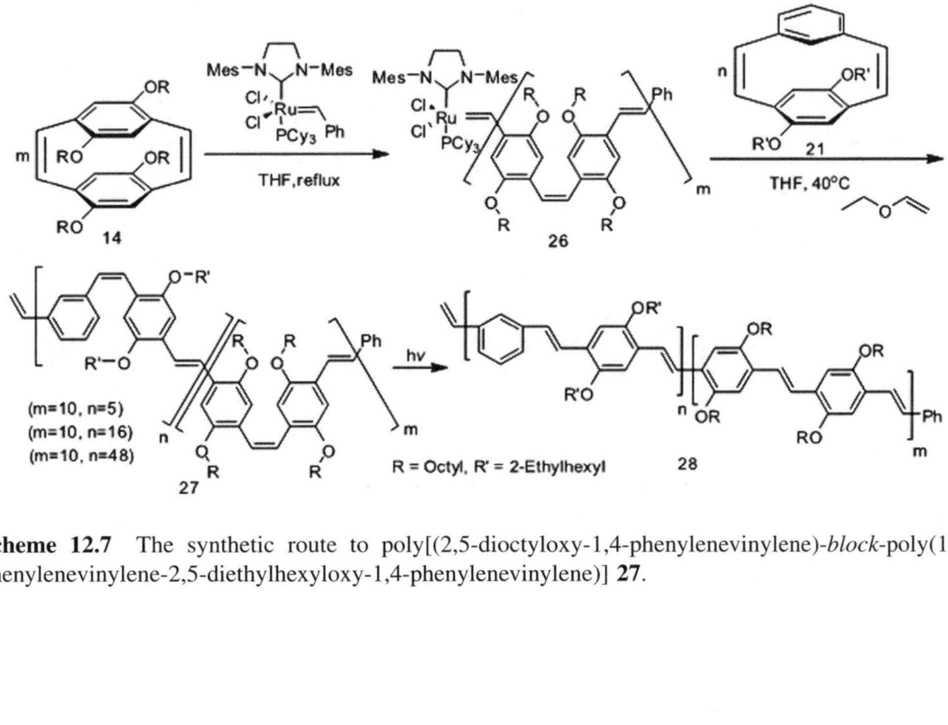

Scheme 12.7 The synthetic route to poly[(2,5-dioctyloxy-1,4-phenylenevinylene)-*block*-poly(1,3-phenylenevinylene-2,5-diethylhexyloxy-1,4-phenylenevinylene)] **27**.

15 200 (expected $M_n = 14\,624$) and a polydispersity of 1.37 and a M_n of 28 800 (expected $M_n = 29\,344$) and a polydispersity of 1.39. The increase in polydispersity of the block copolymers over that of the initial block may be related to slow initiation relative to propagation at the second stage that leads to a broadened molecular-weight distribution.

The DSC traces for these block copolymers exhibited a small transition at $-5\,°C$ corresponding to the glass-transition temperature of poly(1,3-phenylenevinylene-2,5-diethylhexyloxy-1,4-phenylenevinylene) **22**. An endothermic peak at $110\,°C$ for block copolymer can be assigned to the melting of the octyloxy side chains of the first blocks. On cooling exothermic peaks for side-chain recrystallization were observed. The observation of transitions associated with both homopolymers **15**, **22** in the DSC traces of block copolymers suggests that phase separation of the block copolymer is extensive.

Thin films of poly(2,5-dioctyloxy-1,4-phenylenevinylene) **15**, poly(1,3-phenylenevinylene-2,5-diethylhexyloxy-1,4-phenylenevinylene) **22** and block copolymer **28** were prepared by spin coating a chlorobenzene solution on glass substrates at room temperature. After annealing at $220\,°C$ for 24 h, AFM was used to examine the morphology of polymers in these thin films (Figure 12.6). Thin films of poly(1,3-phenylenevinylene-2,5-diethylhexyloxy-1,4-phenylenevinylene) **22** were amorphous (Figure 12.6a), whereas the phase image for films of poly(2,5-dioctyloxy-1,4-phenylenevinylene) **15** (20 phenylenevinylene units, Figure 12.6b) showed spherical particles with an average diameter of around 30 nm. This observation is consistent with double the calculated fully extended molecular length of poly(2,5-dioctyloxy-1,4-phenylenevinylene) **15** of about 13 nm and previous studies of the aggregation of PPVs which show truncated beads with a diameter of 10–20 nm (Chen *et al.*, 2004). The phase image of block copolymer **28** (Figure 12.6c) showed partly aligned cylindrical structures with a cylinder width of approximately 30 nm and a length of several hundred nanometers. The width is consistent with the morphology observed for the parent homopolymer and the formation of these nanostructures can be attributed to the phase separation of the crystalline, poly(2,5-dioctyloxy-1,4-phenylenevinylene) **15** and amorphous, poly(1,3-phenylenevinylene-2,5-diethylhexyloxy-1,4-phenylenevinylene) **22** blocks when the volume fraction was 1:3.

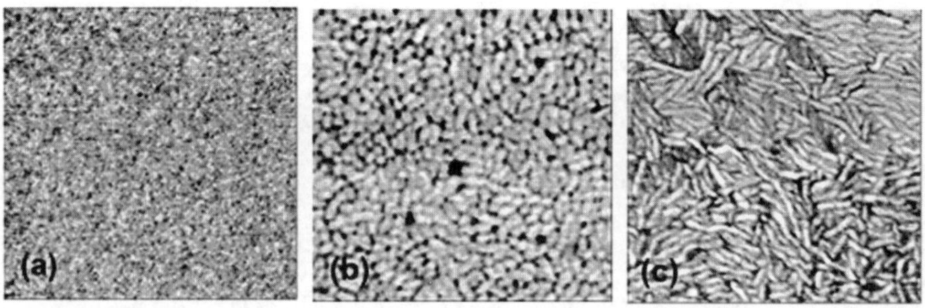

Figure 12.6 AFM phase image (1 μm × 1 μm) of (a) poly(1,3-phenylenevinylene-2,5-diethylhexyloxy-1,4-phenylenevinylene) **22**, (b) poly(2,5-dioctyloxy-1,4-phenylenevinylene) **15**, and (c) poly[(2,5-dioctyloxy-1,4-phenylenevinylene)-*block*-poly(1,3-phenylenevinylene-2,5-diethylhexyloxy-1,4-phenylenevinylene)] **28**. (Reproduced with permission from C.-Y. Yu, J.W. Kingsley, D.G. Lidzey and M.L. Turner, "Phenylenevinylene block copolymers via ring-opening metathesis polymerization," *Macromolecular Rapid Communications*, 2009, **30**, 1889. © Wiley-VCH Verlag GmbH & Co. KGaA.).

12.4 Conclusions

Ring-opening metathesis polymerization of strained cyclophanedienes using ruthenium carbene complexes gives phenylenevinylene homopolymers and block copolymers by a chain-growth process with control of the polymer molecular weight and the block volume fraction. Metathesis of only one vinylene group of the cyclophanediene leads to polymers with a backbone of alternating *cis-* and *trans-*vinylene linkages with little or no control of the regiochemistry of the ring opening for asymmetric cyclophanediene monomers. The as-formed polymers can be photochemically isomerized to the all *trans-*vinylene derivatives with a decrease in the electrochemical band gap of the polymers consistent with an increase in the conjugation length.

References

Ahmed, S.R., Bullock, S.E., Cresce, A.V., and Kofinas, P. (2003) Polydispersity control in ring opening metathesis polymerization of amphiphilic norbornene diblock copolymers. *Polymer*, **44**, 4943.

Anders, U., Nuyken, O., Buchmeiser, M.R., and Wurst, K. (2002) Fine-tuning of molybdenum imido alkylidene complexes for the cyclopolymerization of 1,6-heptadiynes to give polyenes containing exclusively five-membered rings. *Macromolecules*, **35**, 9029.

Asawapirom, U., Guntner, R., Forster, M., and Scherf, U. (2005) Semiconducting block copolymers—synthesis and nanostructure formation. *Thin Solid Films*, **477**, 48.

Bao, Z.N., Chen, Y.M., Cai, R.B., and Yu, L.P. (1993) Conjugated liquid-crystalline polymers - soluble and fusible poly (phenylenevinylene) by the Heck coupling reaction. *Macromolecules*, **26**, 5281.

Barashkov, N.N., Guerrero, D.J., Olivos, H.J., and Ferraris, J.P. (1995) Synthesis and optical properties of oligo- and poly(2,5-dialkoxy-1,4-*para*-phenylenevinylene)s. *Synthetic Metals*, **75**, 153.

Becker, H., Spreitzer, H., Ibrom, K., and Kreuder, W. (1999) New insights into the microstructure of GILCH-polymerized PPVs. *Macromolecules*, **32**, 4925.

Bielawski, C.W. and Grubbs, R.H. (2000) Highly efficient ring-opening metathesis polymerization (ROMP) using new ruthenium catalysts containing N-heterocyclic carbene ligands. *Angewandte Chemie-International Edition*, **39**, 2903.

Burroughes, J.H., Bradley, D.D.C., Brown, A.R. *et al.* (1990) Light-emitting diodes based on conjugated polymers. *Nature*, **347**, 539.

Chen, S.H., Su, A.C., Han, S.R. *et al.* (2004) Molecular aggregation and luminescence properties of bulk poly(2,5-di-*n*-octyloxy-1,4-phenylenevinylene). *Macromolecules*, **37**, 181.

Conticello, V.P., Gin, D.L., and Grubbs, R.H. (1992) Ring-opening metathesis polymerization of substituted bicyclo [2.2.2]octadienes: a new precursor route to poly(1,4-phenylenevinylene). *Journal of the American Chemical Society*, **114**, 9708.

Cotts, P.M., Swager, T.M., and Zhou, Q. (1996) Equilibrium flexibility of a rigid linear conjugated polymer. *Macromolecules*, **29**, 7323.

Edwards, J.H. and Feast, W.J. (1980) A new synthesis of poly(acetylene). *Polymer*, **21**, 595.

Friend, R.H., Gymer, R.W., Holmes, A.B. *et al.* (1999) Electroluminescence in conjugated polymers. *Nature*, **397**, 121.

Gilch, H.G. and Wheelwright, W.L. (1966) Polymerization of **α**-halogenated *p*-xylenes with base. *Journal of Polymer Science Part A-Polymer Chemistry*, **4**, 1337.

Gorman, C.B., Ginsburg, E.J., Sailor, M.J. *et al.* (1991) Substituted polyacetylenes through the ring-opening metathesis polymerization (ROMP) of substituted cyclooctatetraenes: A route into soluble polyacetylene. *Synthetic Metals*, **41**, 1033.

Gunes, S., Neugebauer, H., and Sariciftci, N.S. (2007) Conjugated polymer-based organic solar cells. *Chemical Reviews*, **107**, 1324.

Hu, B. and Karasz, F.E. (1998) Comparison of electroluminescent quantum efficiencies of PPV and a PPV-related alternating block copolymer. *Synthetic Metals*, **92**, 157.

Ivin, K.J. and Mol, J.C. (1997) *Olefin Metathesis and Metathesis Polymerization*, Academic Press, San Diego, CA, 6–19, 41–43, 233.

Jacobs, S., Eevers, W., Verreyet, G. *et al.* (1993) Spectroscopy and conductivity of the separate cis and trans isomers of a PPV oligomer. *Synthetic Metals*, **61**, 189.

Jozefiak, T.H., Ginsburg, E.J., Gorman, C.B. *et al.* (1993) Voltammetric characterization of soluble polyacetylene derivatives obtained from the ring-opening metathesis polymerization (ROMP) of substituted cyclooctatetraenes. *Journal of the American Chemical Society*, **115**, 4705.

Kappe, C.O. (2004) Controlled microwave heating in modern organic synthesis. *Angewandte Chemie-International Edition*, **43**, 6250.

Katayama, H., Nagao, M., Nishimura, T. *et al.* (2005) Stereocontrolled synthesis and optical properties of all-cis poly (phenylene vinylenes) (PPVs): A method for direct patterning of PPVs. *Journal of the American Chemical Society*, **127**, 4350.

Katayama, H., Nagao, M., Ozawa, F. *et al.* (2006) Stereoselective synthesis of *cis-* and *trans-*Oligo(phenylenevinylene)s via palladium-catalyzed cross-coupling reactions *The Journal of Organic Chemistry*, **71**, 2699.

Klarner, G., Former, C., Martin, K. *et al.* (1998) Connective CC double bond formation for the synthesis of donor- and acceptor-substituted poly(*p*-phenylenevinylene)s. *Macromolecules*, **31**, 3571.

Klok, H.A. and Lecommandoux, S. (2001) Supramolecular materials via block copolymer self-assembly. *Advanced Materials*, **13**, 1217.

Kraft, A., Grimsdale, A.C., and Holmes, A.B. (1998) Electroluminescent conjugated polymers – seeing polymers in a new light. *Angewandte Chemie-International Edition*, **37**, 402.

Krevelen, D.W.V. (1990) *Properties of Polymers- their Correlation with Chemical Structure; their Numerical Estimation and Prediction from Additive Group Contributions*, Elsevier.

Kugler, T., Logdlund, M., and Salaneck, W.R. (1999) Photoelectron spectroscopy and quantum chemical modeling applied to polymer surfaces and interfaces in light-emitting devices. *Accounts of Chemical Research*, **32**, 225.

Larhed, M., Moberg, C., and Hallberg, A. (2002) Microwave-accelerated homogeneous catalysis in organic chemistry. *Accounts of Chemical Research*, **35**, 717.

Liang, Y.Y., Wang, H.B., Yuan, S.W. *et al.* (2007) Conjugated block copolymers and co-oligomers: from supramolecular assembly to molecular electronics. *Journal of Materials Chemistry*, **17**, 2183.

Liao, L., Pang, Y., Ding, L., and Karasz, F.E. (2001) Synthesis, characterization and luminescence of poly[(*m*-phenylenevinylene)-*alt*-(1,4-dibutoxy-2,5-phenylenevinylene)] with different content of *cis-* and *trans-*olefins. *Macromolecules*, **34**, 6756.

Love, J.A., Morgan, J.P., Trnka, T.M., and Grubbs, R.H. (2002) A practical and highly active ruthenium-based catalyst that effects the cross metathesis of acrylonitrile *Angewandte Chemie-International Edition*, **41**, 4035.

Matsen, M.W. and Bates, F.S. (1996) Unifying weak- and strong-segregation block copolymer theories. *Macromolecules*, **29**, 1091.

Matsen, M.W. and Thompson, R.B. (1999) Equilibrium behavior of symmetric ABA triblock copolymer melts. *Journal of Chemical Physics*, **111**, 7139.

Matyjaszewski, K. (1993) Ranking living systems. *Macromolecules*, **26**, 1787.

Miao, Y.J. and Bazan, G.C. (1994a) Paracyclophene route to poly(p-phenylenevinylene). *Journal of the American Chemical Society*, **116**, 9379.

Miao, Y.J. and Bazan, G.C. (1994b) Stereoselective polymerization of [2.2]paracyclophan-1-ene. *Macromolecules*, **27**, 1063.

Naber, R.C.G., Tanase, C., Blom, P.W.M. *et al.* (2005) High-performance solution-processed polymer ferroelectric field-effect transistors. *Nature Materials*, **4**, 243.

Rost, H., Teuschel, A., Pfeiffer, S., and Horhold, H.H. (1997) Novel light emitting and photoconducting polyarylenevinylene derivatives containing phenylene arylamine and phenylene oxide units in the main chain. *Synthetic Metals*, **84**, 269.

Scherf, U. and List, E.J.W. (2002) Semiconducting polyfluorenes—towards reliable structure–property relationships. *Advanced Materials*, **14**, 477.

Scherf, U., Gutacker, A., and Koenen, N. (2008) All-conjugated block copolymers. *Accounts of Chemical Research*, **41**, 1086.

Schmidt-Mende, L., Fechtenkotter, A., Mullen, K. *et al.* (2001) Self-organized discotic liquid crystals for high-efficiency organic photovoltaics. *Science*, **293**, 1119.

Schmitt, C., Nothofer, H.G., Falcou, A., and Scherf, U. (2001) Conjugated polyfluorene/polyaniline block copolymers. *Macromolecular Rapid Communications*, **22**, 624.

Spring, A.M., Yu, C.-Y., Horie, M., and Turner, M.L. (2009) MEH-PPV by microwave assisted ring-opening metathesis polymerisation. *Chemical Communications*, 2676.

Stalmach, U., de Boer, B., Post, A.D. *et al.* (2001) Synthesis of a conjugated macromolecular initiator for nitroxide-mediated free radical polymerization. *Angewandte Chemie International Edition*, **40**, 428.

Thorn-Csányi, E. and Pleug, K.P. (1993) Copolymers of 1,5-cyclooctadiene and [2.2]paracyclophane-1,9-diene: metathetic preparation and structure analysis. *Macromolecular Chemistry*, **194**, 2287.

Thorn-Csányi, E., Hohnk, H.D., and Pflug, K.P. (1993) Ring-opening metathesis copolymerization of cyclopentene and [2.2] paracyclophane-1,9-diene: UV-Vis-spectroscopic determination of sequence-length distribution of *p*-phenylenevinylene units. *Journal of Molecular Catalysis*, **84**, 253.

Thorn-Csányi, E., Kraxner, P., and Hammer, J. (1994) Ring-opening copolymerization of [2.2] paracyclophane-1,9-diene (PCPDE) and cyclooctene (COE); a route to soluble products with *p*-phenylenevinylene sequences. *Journal of Molecular Catalysis*, **90**, 15.

Thorn-Csányi, E., Kraxner, P., and Strachota, A. (1998) Synthesis of soluble all-trans oligomers of 2,5-diheptyloxy-p-phenylenevinylene via olefin metathesis. *Macromolecular Rapid Communications*, **19**, 223.

Tu, G.L., Li, H.B., Forster, M. *et al.* (2006) Conjugated triblock copolymers containing both electron-donor and electron-acceptor blocks. *Macromolecules*, **39**, 4327.

Wagaman, M.W. and Grubbs, R.H. (1997) Synthesis of organic and water soluble poly(1,4-phenylenevinylenes) containing carboxyl groups: Living ring-opening metathesis polymerization (ROMP) of 2,3-dicarboxybarrelenes. *Macromolecules*, **30**, 3978.

Wang, H.B., Ng, M.K., Wang, L.M. *et al.* (2002) Synthesis and characterization of diblock copolymers. *Chemistry – A European Journal*, **8**, 3246.

Wessling, R.A. (1985) The polymerization of xylylene bisdialkyl sulfonium salts. *Journal of Polymer Science Part C-Polymer Symposium*, **72**, 55.

Yu, L.P. and Bao, Z.N. (1994) Conjugated polymers exhibiting liquid crystallinity. *Advanced Materials*, **6**, 156.

Yu, C.-Y. and Turner, M.L. (2006) Soluble poly(p-phenylenevinylene)s through ring-opening metathesis polymerization. *Angewandte Chemie-International Edition*, **45**, 7797.

Yu, C.-Y., Kingsley, J.W., Lidzey, D.G., and Turner, M.L. (2009) Phenylenevinylene block copolymers via ring-opening metathesis polymerization. *Macromolecular Rapid Communications*, **30**, 1889.

Yu, C.-Y., Horie, M., Spring, A.M. *et al.* (2010) Homopolymers and block copolymers of *p*-phenylenevinylene-2,5-diethylhexyloxy-*p*-phenylenevinylene and *m*-phenylenevinylene-2,5-diethylhexyloxy-*p*-phenylenevinylene by ring-opening metathesis polymerization. *Macromolecules*, **43**, 222.

Zaumseil, J. and Sirringhaus, H. (2007) Electron and ambipolar transport in organic field-effect transistors. *Chemical Reviews*, **107**, 1296.

13

Block Copolymers Containing Rod Segments

Tomoya Higashihara and Mitsuru Ueda

Department of Organic and Polymeric Materials, Graduate School of Science and Engineering, Tokyo Institute of Technology, 2-12-1-H-120, Ookayama, Meguro-ku, Tokyo 152-8552, Japan

13.1 Introduction

Block copolymers (BCPs) have been of great interest for more than 40 years in many areas, including self-assembly studies, micelles, processing, material sciences, and functional device applications (Krausch and Magerle, 2002; Ishizu *et al.*, 2002; Mori and Mueller, 2003; Park *et al.*, 2003; Hillmyer 2005a; Olson *et al.*, 2008a; Hoeben *et al.*, 2005; Tseng *et al.*, 2009; Ho *et al.*, 2009a). Many researchers have pursued well-defined shapes and functions derived from such BCPs, making fascinating supramolecular architectures accessible in advanced nanotechnology. Indeed, well-defined BCPs with predictable molecular weights and low polydispersities (PDIs) have been synthesized based on the feasibility of living/controlled polymerization systems such as living anionic (Hsieh and Quirk, 1996; Hadjichristidis *et al.*, 2001; Hirao and Nakahama, 1992), living cationic (Sawamoto and Higashimura, 1990; Sawamoto, 1993; Percec and Tomazos, 1992; Kennedy, 1999; Kwon and Faust, 2004; Aoshima *et al.*, 2007; Aoshima and Kanaoka, 2009), and living radical polymerization (Matyjaszewski *et al.*, 1999; Matyjaszewski and Xia, 2001; Heise *et al.*, 1999a; Heise *et al.*, 1999b; Cloutet *et al.*, 1999; Angot *et al.*, 2000; Stenzel-Rosenbaum *et al.*, 2001; Barner-Kowollik *et al.*, 2003). Other new living/controlled systems have recently been developed, including ring-opening polymerization (ROP) (Bazan and Schrock, 1991), coordination polymerization (Mason and Coates, 2007), metathesis polymerization (Bielawski and Grubbs, 2007; Frenzel and Nuyken, 2002; Baughman and Wagener, 2005; Berda *et al.*, 2006), group-transfer polymerization (Webster, 2000; Webster, 2004), and condensative

Complex Macromolecular Architectures: Synthesis, Characterization, and Self-Assembly, First Edition.
Edited by Nikos Hadjichristidis, Akira Hirao, Yasuyuki Tezuka and Filip Du Prez.

chain-growth polymerization (Yokozawa and Yokoyama, 2004; Yokozawa and Yokoyama, 2007; Yokoyama and Yokozawa, 2007; Miyakoshi *et al.*, 2008a; Yokozawa *et al.*, 2008; Yokozawa and Yokoyama, 2009; Shibasaki and Ueda, 2003), all of which allow for the creation of novel BCPs. Typically, for synthesizing well-defined BCPs, after completely consuming the first monomer, the second monomer is *in situ* charged to realize an efficient crossover, followed by polymerization in a living manner. Under the current situation whereby the requirements for polymer properties and performances have become stricter and more versatile, not only monomer varieties but also combinations of polymer segments should be broadened.

BCPs containing rod segments have emerged as intriguing and unique materials due to their stiff rod-like structure in solution, melt, and/or solid states, as well as heterophase structures (Klok and Lecommandoux, 2001; Lee *et al.*, 2001; Lim *et al.*, 2008). These BCPs are categorized into several types by different rod materials, that is, nonconjugated rods such as polypeptides, polyisocyanates, aromatic polyamides, and other π-conjugated rods such as polyacetylenes, polyphenylenes, polyfluorenes, poly(phenylene vinylene)s, and polythiophenes, as seen in Figure 13.1. There is also an alternative categorization, sorted by topological architecture, for example, rod–coil AB, coil–rod–coil ABA, rod–coil–rod ABA, rod–rod AB, and star type (e.g. the 3-arm A_2B type). Another important categorization is also possible, sorted by synthetic methodology for constructing rod segments. The first, second, and third groups of BCPs are comprised of monodisperse oligomer rods prepared by stepwise reactions, oligomers or polymers with broad PDIs via a conventional polymerization system, and well-defined polymers with predicted molecular weights and low PDIs prepared via living/controlled systems, respectively.

In recent years, novel living/controlled systems in the polymerization of α-amino acid *N*-carboxyanhydrides (NCAs) and isocyanates that form well-defined helical rods, polypeptides and polyisocyanates have been developed, and their copolymerization with other coil blocks have been synthesized, while the breakthrough for condensative chain-growth polymerization also permits synthesizing well-defined aromatic polyamides and π-conjugated polymers via living/controlled systems and BCPs containing low polydisperse rod segments (Yokozawa and Yokoyama, 2004; Yokozawa and Yokoyama, 2007; Yokoyama and Yokozawa, 2007; Miyakoshi *et al.*, 2008a; Yokozawa *et al.*, 2008; Yokozawa and Yokoyama, 2009; Shibasaki and Ueda, 2003). These efforts have contributed to the accelerated evolution of many application areas such as polymeric catalysts, tissue engineering, biomedication, sensors, engineering plastics, photonics, electronics and so on.

Herein, we review recent progress in the synthesis of well-defined BCPs containing rod segments, especially those based on living/controlled polymerization systems. It should be noted that BCPs with mesogenic rods have already been established and well reviewed by Myongsoo Lee and coworkers (Lee *et al.*, 2001; Lim *et al.*, 2008), therefore they are beyond the scope of this review.

13.2 Block Copolymers Containing Nonconjugated Rod Segments

13.2.1 Block Copolymers Containing Polypeptide Segments

Polypeptides have received much attention because they exhibit complex and versatile structures, and reversible transitions depending on external environments (pH, *T*, solvent, etc.) (Alberts *et al.*, 1989). Much effort has gone into the synthesis of homo- and

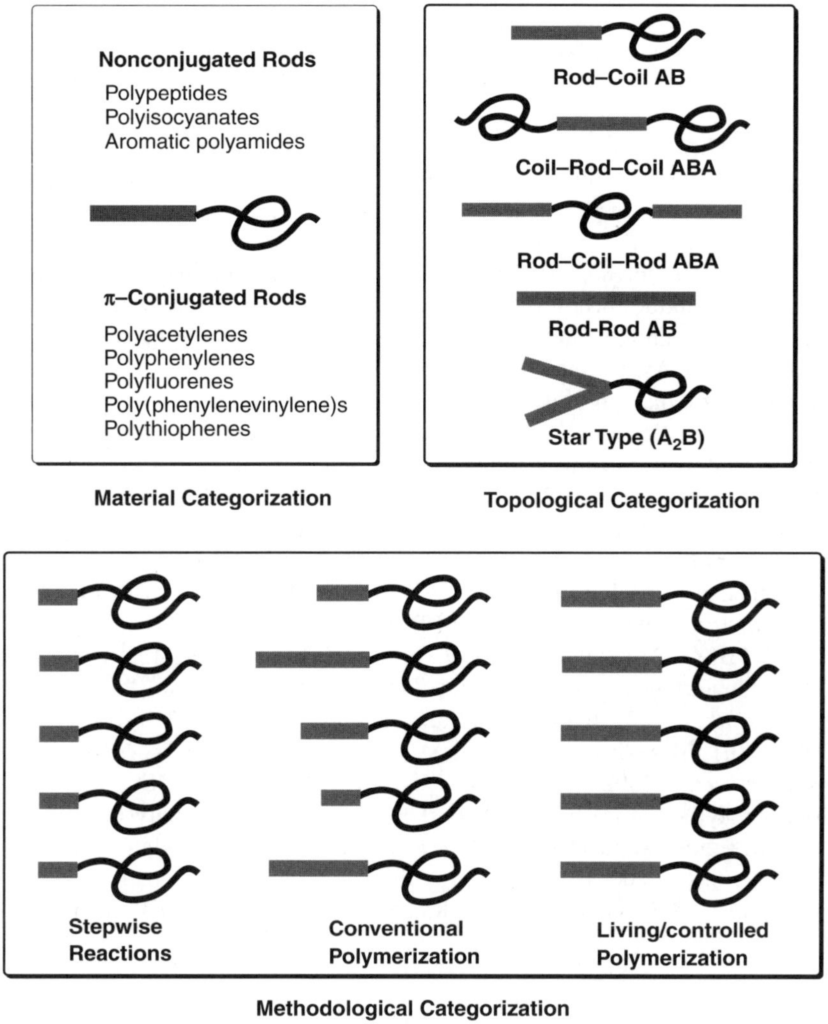

Figure 13.1 Categorization of block copolymers with rod segments.

copolypeptides using ROP of α-amino acid-NCAs (Kricheldorf, 1987). However, controlled polymerization based on conventional methods could not be accomplished due to concurrent side reactions, such as chain-transfer and termination reactions. In 1997, Deming and coworkers succeeded in a living polymerization of NCAs initiated with a Ni(0) complex of NCAs as well as block copolymerization (Deming, 1997). These initiators provide propagating active species, which are less susceptible to side reactions due to steric and electronic reasons. Schlaad *et al.* employed the HCl salt of amino-initiators to control the polymerization of NCAs and eliminate the activated monomer mechanism that mostly leads to high molecular weight and branched polypeptides (Dimitrov and Schlaad, 2003). Hadjichristidis and coworkers reported living polymerization of NCAs and the synthesis of an array of block copolypeptides based on high-vacuum techniques. They reported that the complete removal of impurities suppressed side reactions (Aliferis *et al.*, 2004). Surprisingly, the molecular weight of poly (γ-benzyl-L-glutamate) (PBLG) could be controlled up to 166 K ($DP_n = 743$) by this method. Recently, Cheng and coworkers reported a hexamethyldisilane-mediated controlled polymerization of NCAs where a trimethylsilyl group was transferred from the terminal trimethylsilyl carbamate group to the incoming monomer to form a new trimethylsilyl-propagating group, resulting in the reduction of side reactions (Lu and Cheng, 2007).

These developments in the living/controlled polymerization of NCAs encouraged synthetic, morphological, and biomedical studies of BCPs containing polypeptides (Hamley, 2003). Indeed, many polypeptide-*b*-(nonpeptide polymer) BCPs were synthesized by the methodology in which primary amino-functionalized nonpeptide macroinitiators were utilized for the polymerization of NCAs. For example, Klok *et al.* reported the synthesis of peptide-based diblock oligomers by the combination of NCA polymerization and living anionic polymerization. As shown in Scheme 13.1, primary amino chain-end-functionalized polystyrene (PS-NH_2) was first synthesized by living anionic polymerization of styrene, followed by a reaction of 1-(3-chloropropyl)-2,2,5,5-tetramethyl-1-aza-2,5-disilacyclopentane and spontaneous deprotection. PS-NH_2 was then utilized as a macroinitiator for the polymerization of an NCA monomer of γ-benzyl-L-glutamate, giving PS-*b*-PBLG (Klok *et al.*, 2000). As seen in Figure 13.2, other successful examples include PBLG-*b*-polybutadiene-*b*-PBLG (**1**) (Nakajima *et al.*, 1979), poly(ethylene glycol)(PEG)-*b*-PBLG (**2**) (Brzezinska *et al.*, 2002), PBLG-*b*-poly (octenamer)-*b*-PBLG (**3**) (Brzezinska *et al.*, 2002), PBLG-*b*-PEG-*b*-PBLG (**4**) (Brzezinska *et al.*, 2002; Floudas *et al.*, 2003), PBLG-*b*-poly(dimethylsiloxane)(PDMS)-*b*-PBLG (**5**) (Brzezinska *et al.*, 2002; Ibarboure *et al.*, 2006), PEG-*b*-PBLG-*b*-poly(octenamer)-*b*-PBLG-*b*-PEG (**6**) (Brzezinska *et al.*, 2002), PEG-*b*-PBLG-*b*-PEG-*b*-PBLG-*b*-PEG (**7**) (Brzezinska *et al.*, 2002), PEG-*b*-PBLG-*b*-PDMS-*b*-PBLG-*b*-PEG (**8**) (Brzezinska *et al.*, 2002), poly(L-lactide)-*b*-PBLG (**9**) (Caillol *et al.*, 2003), poly(methyl acrylate)-*b*-PBLG (**10**) (Brzezinska and Deming, 2004), poly(methyl methacrylate)(PMMA)-*b*-PBLG (**11**) (Agut *et al.*, 2007), poly[2-(*N*,*N*-dimethylamino)ethyl methacrylate]-*b*-PBLG (**12**) (Agut, *et al.*, 2007), poly(ferrocenyldimethylsilane)-*b*-PBLG (**13**) (Kim *et al.*, 2005),

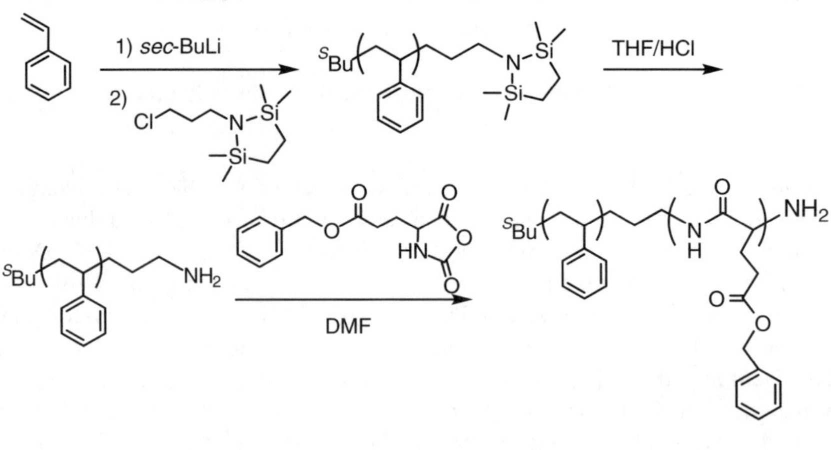

Scheme 13.1 Synthesis of PS-*b*-PBLG.

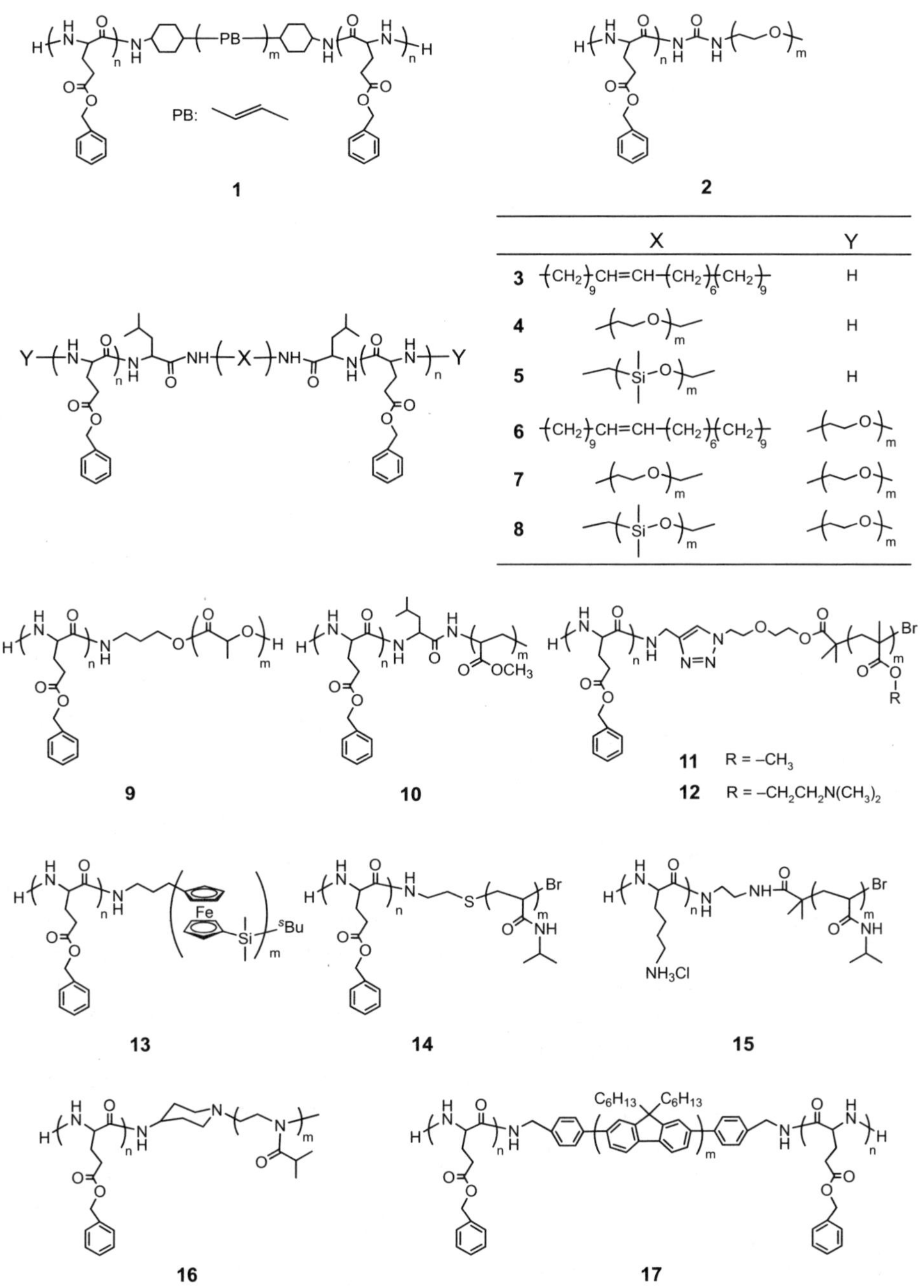

Figure 13.2 Chemical structures of polypeptides-based BCPs.

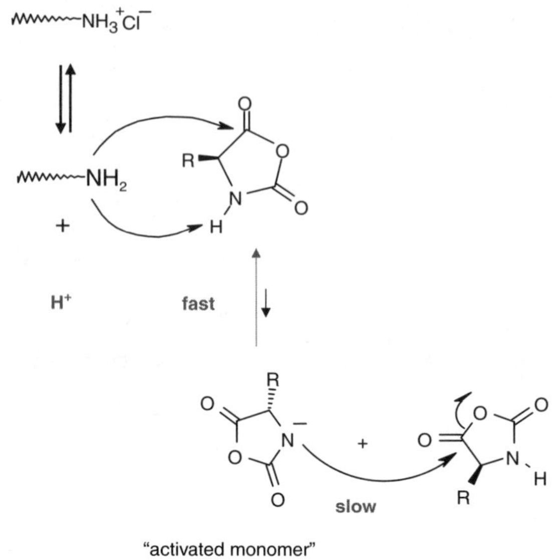

Scheme 13.2 Plausible mechanism for ammonium-mediated polymerization.

poly(*N*-isopropylacrylamide)-*b*-PBLG (**14**) (Rao *et al.*, 2007), and poly(*N*-isopropylacrylamide)-*b*-polylysine (**15**) (Huang and Chang, 2008).

The combination of living cationic polymerization and NCA polymerization has also been reported. Schlaad synthesized poly(2-isopropyl-2-oxazoline)-*b*-PBLG (**16**) based on an "ammonium-mediated polymerization" system. By adding an acid in the NCA polymerization, a primary amino (NH_2)-propagating chain end could be transformed into an ammonium ($NH_3{}^+$) chain end via equilibrium and this system suppresses the generation of the aminyl anion of an NCA monomer that causes byproducts via activated monomer mechanism (Scheme 13.2) (Dimitrov and Schlaad, 2003).

Higashihara and Faust synthesized new elastomers of PBLG-*b*-poly(isobutylene)(PIB)-*b*-PBLG and (PBLG-*b*-PIB-*b*-PBLG)$_n$ multiblock copolymers by the combination of living carbocationic and NCA polymerization via the ammonium-mediated system (Scheme 13.3) (Higashihara and Faust, 2009). NCA polymerization from the macroinitiator α,ω-amino-functionalized PIB, in the presence of acetic acid, afforded a well-defined PBLG-*b*-PIB-*b*-PBLG building block. Furthermore, the polyaddition between the resulting polymer and the chain extender, 4,4′-methylene-bis(phenyldiisocyanate), produced a (PBLG-*b*-PIB-*b*-PBLG)$_n$ multiblock copolymer.

Jenekhe and coworkers reported the synthesis of ABA triblock copolymers containing poly (9,9-dihexylfluorene-2,7-diyl)(PHF) and PBLG segments (**17**) by NCA polymerization from benzylamine end-functionalized PHF (Rubatat *et al.*, 2008). The copolymers exhibited two different conformations (rod–rod–rod and coil–rod–coil) as a result of a controllable rod-to-coil transition of the polypeptide helices. X-ray scattering data showed that the coil–rod–coil form of the copolymer leads to lamellar ordering, whereas the rod–rod-rod form exhibits a clusterlike structure. These bioactive and fluorescent materials can be explored as

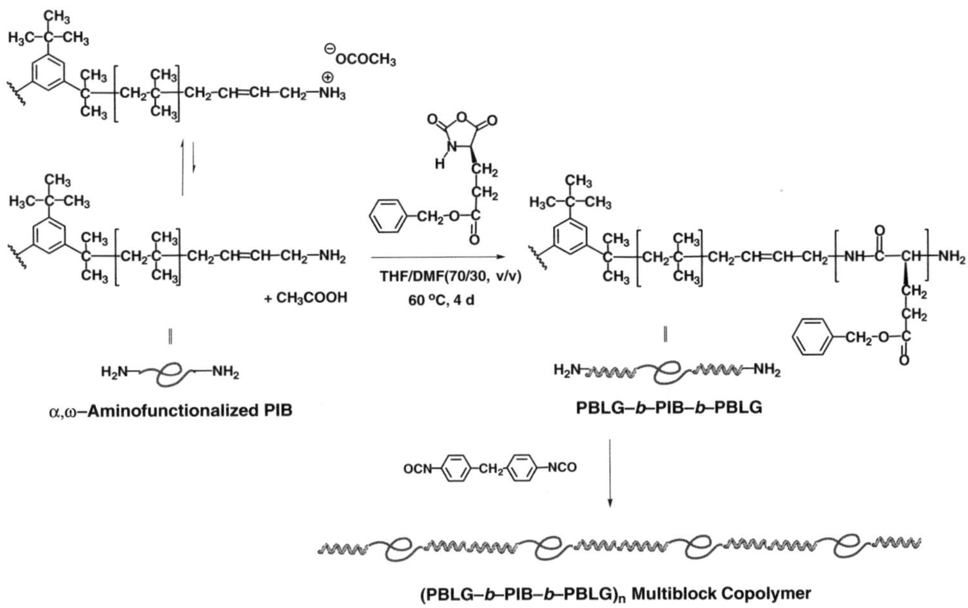

Scheme 13.3 Synthesis of PBLG-*b*-PIB-*b*-PBLG and (PBLG-*b*-PIB-*b*-PBLG)$_n$ multiblock copolymer based on ammonium-mediated polymerization.

biocompatible surfaces or wires for delivering electrical stimuli that influence cell growth, migration, and morphology.

Recently, Hadjichristidis, Hirao, and coworkers synthesized "complex macromolecular chimeras" (a Greek term), which include novel multicomponent linear and miktoarm star polymers containing polypeptide blocks based on a high-vacuum technique (Karatzas *et al.*, 2008). Indeed, 3-arm AB_2, 3-arm ABC, 4-arm A_2B_2, and 4-arm AB_2C were synthesized, where A, B, and C were polypeptides, PS, and PI or poly(α-methylstyrene) (PαMS), respectively (Figure 13.3).

Basically, intermediate polymeric anions were reacted with 1-(3-bromopropyl)-2,2,5,5-tetramethyl-aza-2,5-disilacyclopentane, followed by spontaneous deprotection during workup to prepare well-defined in-chain- or core-functionalized star-branched polymers with primary amine group(s). These functionalized polymers could be used as macroinitiators for the living polymerization of NCAs. The success of those star-polymer syntheses may be derived from the 100% introduction of primary amino groups at the precise position and extremely clean media for NCA polymerization in completely sealed reactors based on the high-vacuum technique.

Taton and Lecommandoux reported the self-assembly behavior of AB diblock copolymers and miktoarm AB_2 star polymers comprised of PS (=A) and poly(glutamic acid) (=B) segments (Babin *et al.*, 2005). AFM image and X-ray scattering data indicated that those polymers displayed hexagonally packed cylinders of helical rods within lamellar phase.

In the course of these studies, many other hierarchical structures induced by the self-assembly of BCPs containing polypeptide rods have been discovered. However, there is a limitation to the synthesis of polypeptide-based BCPs due to less commercial availability of high-grade NCA monomers. In particular, the conventional phosgene-based synthetic

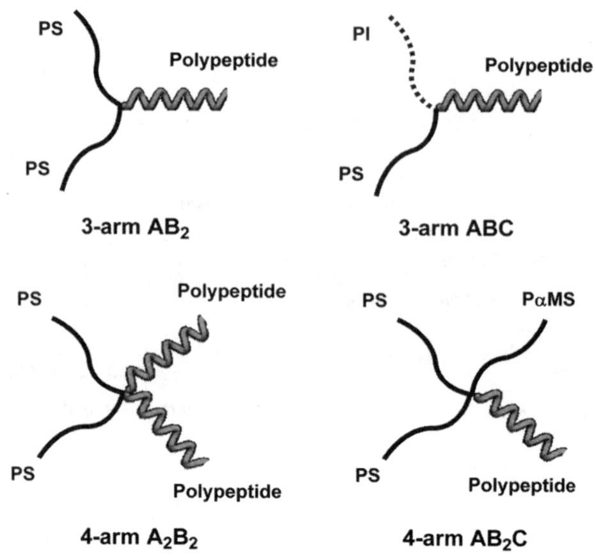

Figure 13.3 Synthesis of "macromolecular chimeras."

approach seems to prevent the further development of this area. Endo and coworkers proposed a new, phosgene-free, and universal strategy for the synthesis of NCA monomers utilizing diphenyl carbonate derivatives (Fujita *et al.*, 2007). If this system can be established in the near future, the second breakthrough in polypeptide-based science and engineering will be realized.

13.2.2 Block Copolymers Containing Polyisocyanate Segments

Polyisocyanates are comprised of only amide bonding in the main chain and maintain their helical rod structures, so they normally show liquid crystallinity or high crystallinity. For instance, the helix conformation (12/5) of poly(hexylisocyanate) (PHIC) and its high crystallinity in the solid state are well observed by X-ray scattering (Wu *et al.*, 1992). PHIC displays rod structures with a very long persistence length of 50–60 nm in the solution state (Bur and Fetters, 1976). It also shows lyotropic liquid-crystalline behavior at high concentrations (Aharomi, 1979; Aharoni and Walsh, 1979).

Polyisocyanates have been synthesized by anionic polymerization of isocyanate but the system has suffered from a serious side reaction of depolymerization, which generates cyclic trimers. Novak *et al.* reported a useful strategy for controlled coordination polymerization of isocyanates by employing organic titanium compounds (Patten and Novak, 1991; Patten and Novak, 1996). Facile synthesis of initiators by reacting alcohols with titanium chlorides and gentle polymerization conditions at room temperature brought about a breakthrough in this area. On the other hand, Lee and coworkers recently developed a living system in the anionic polymerization of isocyanates that achieved low polydisperse polyisocyanates by blocking active chain ends with bulky additives (Shin *et al.*, 2001).

In contrast to homopolymers, interesting new supermolecular structures are expected when a flexible coil segment is connected to a rigid polyisocyanate rod, because the molecules impart

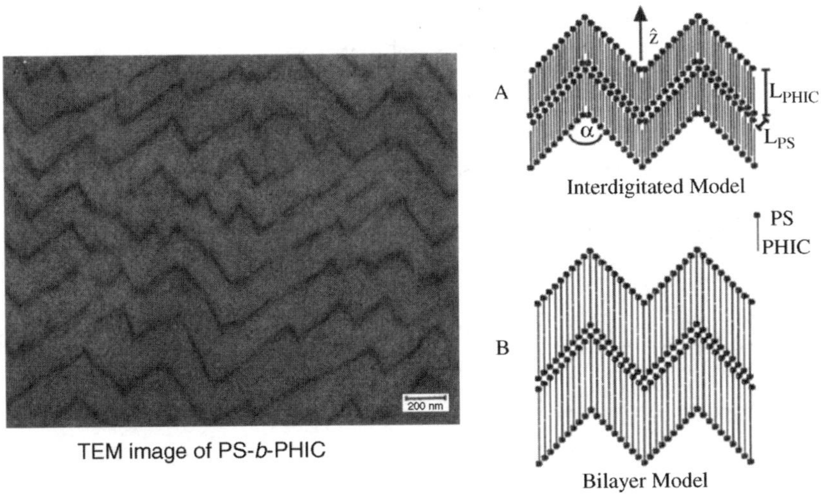

Figure 13.4 TEM image of a PS-*b*-PHIC thin film and speculated models for molecule orientations. (Reprinted with permission from J.T. Chen, E.L. Thomas, C.K. Ober and S.S. Hwang, "Zigzag morphology of a poly(styrene-b-hexyl isocyanate) rod-coil block copolymer," *Macromolecules*, **28**, 5, 1688–1697, 1995. © 1995 American Chemical Society.)

both microphase separation characteristics of each block and a tendency for the rod segments to form anisotropic order. Ober and Thomas *et al.* reported that AB diblock copolymers consisting of PHIC and PS segments displayed liquid-crystalline behavior in concentrated solutions, suggesting an anisotropical order of rod segments (Chen *et al.*, 1995). Transmission electron microscopy (TEM) of thin-film samples showed a novel zigzag morphology (Figure 13.4) with a high degree of a smectic-like long-range order. In addition, electron diffraction patterns confirmed that the rod segments are tilted along the interface separating the rod and coil domains. Based on these results, the authors speculated on the molecular orientations of the zigzag morphology, as shown in Figure 13.4.

Lee and coworkers have recently contributed a lot of work on well-defined BCPs containing PHIC segments based on living anionic polymerization using Na^+ as the countercation in the presence of sodium tetraphenylborate. These BCPs include PHIC-*b*-PS-*b*-PHIC (**18**) (Ahn and Lee, 2003), PHIC-*b*-polyisoprene(PI)-*b*-PHIC (**19**) (Ahn *et al.*, 2003), PHIC-*b*-poly(2-vinyl-pyridiene)(P2VP)-*b*-PHIC (**20**) (Rahman, Samal, and Lee, 2006), P2VP-*b*-PHIC-*b*-P2VP (**21**) (Rahman *et al.*, 2007), and PHIC-*b*-P2VP (**22**) (Rahman *et al.*, 2007) (Figure 13.5). They reported unique self-assembled micelle and vesicle structures using PHIC-*b*-P2VP in a rod-selective solvent (Koh *et al.*, 2009). The rod-like PHIC blocks tangentially wrap around the surface of the aggregate hydrophilic P2VP cores to form a surface layer a few nanometers thick. The micelle and vesicle structures could be reversibly interchanged by varying the polymer concentration. Ishizu, Uchida, and coworkers first synthesized a novel rod–brush BCP containing PHIC segments by titanium-based coordination polymerization combined with atom-transfer radical polymerization (ATRP) (Ishizu *et al.*, 2008). On the basis of dynamic scattering data, they stated that such rod–brush BCPs exhibited a tadpole-shaped structure in which a linear rod was connected to the brush head in solution.

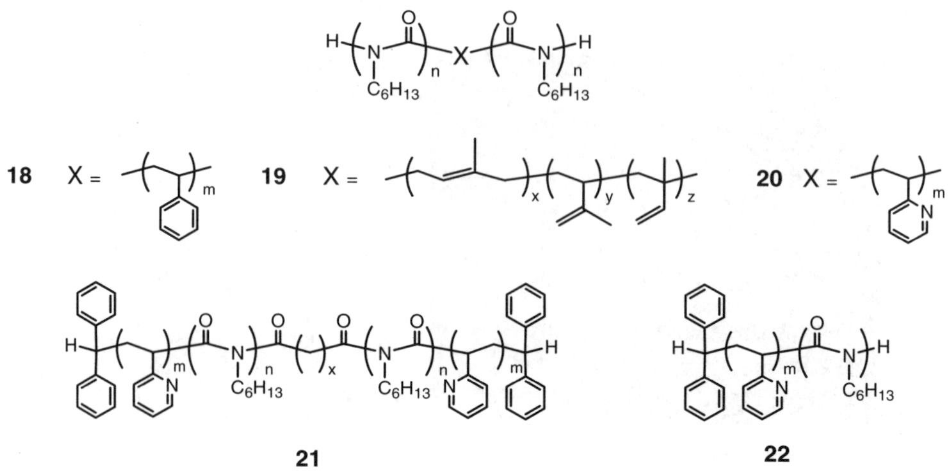

Figure 13.5 Chemical structures of PHIC-based BCPs.

13.2.3 Block Copolymers Containing Aromatic Polyamide Segments

N-Alkyl-substituted aromatic polyamides are well known to exhibit, in general, helical rod structures. Yokozawa and coworkers reported the chain-growth condensation polymerization of AB monomers, utilizing different substituent effects between monomers and polymers to synthesize well-defined aromatic polyamides with controlled molecular weights and low PDIs (<1.1) (Yokozawa and Yokoyama, 2004; Yokozawa and Yokoyama, 2007; Yokoyama and Yokozawa, 2007; Miyakoshi *et al.*, 2008a; Yokozawa *et al.*, 2008; Yokozawa and Yokoyama, 2009). As shown in Scheme 13.4, the base, *N*-octyl-*N*-triethylsilylamide, along with CsF and 18-crown-6, abstracts the proton of the amino group of **23** to generate aminyl anion monomer **24**. It deactivates the phenyl ester group of **24** by its strong electron-donating resonance effect, preventing oligomerization between the monomers (Yokozawa *et al.*, 2000). Monomer **24** reacts with initiator **25**, in which the phenyl ester group is activated by an electron-withdrawing nitro group. The obtained amide possesses a phenyl group with a weak electron-donating amide linkage, and is still more active toward the aminyl anion group of **24** than that of **24** itself. Thus, only the polymer chain end participates in the consumption of the monomer, without oligomerization between the monomers, as seen in conventional step-growth polycondensation. Consequently, the molecular weights increase by increasing the feed ratio and reach low PDIs if the rate constant for initiation is reasonably faster than or at least similar to that of propagation. Soon after, Shibasaki, Ueda and coworkers modified the control system by employing an acryl-chloride-type initiator and an activated amide monomer containing benzoxazoline-2-thione with the use of a Grignard reagent as the base (Scheme 13.4) (Shibasaki *et al.*, 2002). Yokozawa *et al.* further investigated this and found a useful base, lithium hexamethyldisilazide, for the chain-growth condensation polymerization of general ester monomers (Yokozawa *et al.*, 2005).

Based on these successful results, novel rod–coil BCPs have been synthesized by the combination of chain-growth condensation polymerization with other living polymerization systems. For instance, Yokozawa synthesized rod–coil AB diblock (**26**) and rod–coil–rod ABA

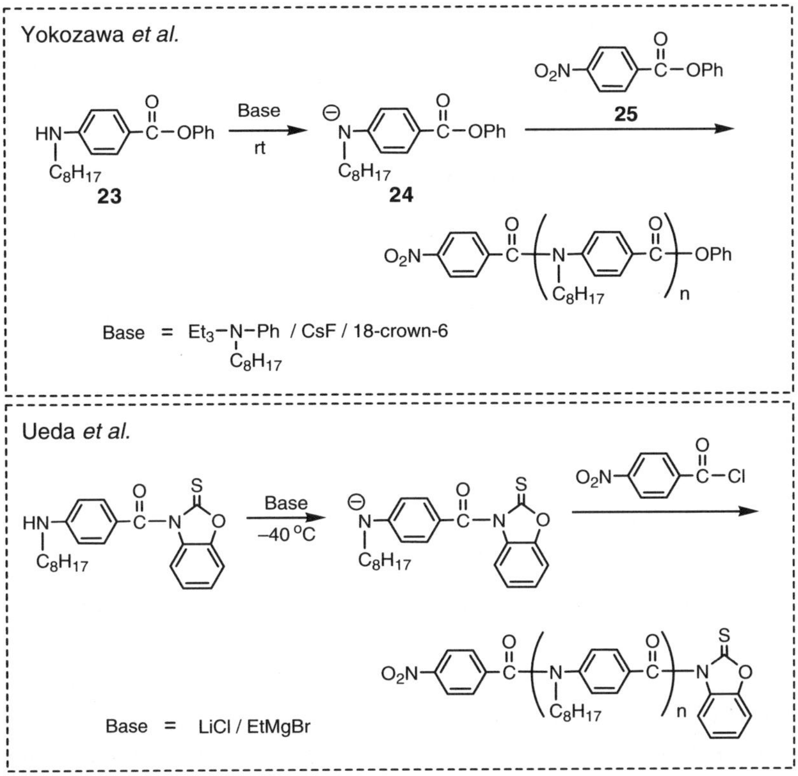

Scheme 13.4 Synthesis of well-defined aromatic polyamides by chain-growth condensation polymerization.

triblock copolymers (**27**) of aromatic polyamides and poly(tetrahydrofuran) (PTHF) by a reaction of the terminal amino group of a polyamide with the cationic propagating group of PTHF (Figure 13.6) (Sugi *et al.*, 2003a). Another type of BCP containing aromatic polyamides and PEO segments (**28**) were synthesized (Sugi *et al.*, 2003b). They also used a living anionic polymerization system to synthesize well-defined rod–coil BCPs with aromatic

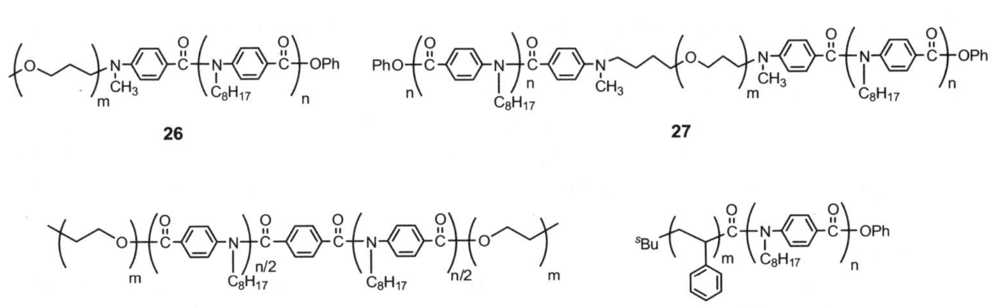

Figure 13.6 Chemical structures of aromatic-polyamide-based BCPs.

polyamide and PS segments (**29**) (Kim *et al.*, 2007). The micellation behavior of this polymer in 2-propanol was investigated, but there have not been any reports yet on detailed morphological studies of such rod–coil BCPs in bulk.

13.3 Block Copolymers Containing π-Conjugated Rod Segments

Much attention has recently been paid to BCPs containing π-conjugated stiff-rod segments, because of their many potential optical and electronic applications, including batteries, capacitors, light-emitting diodes, transistors, photovoltaics, and so on, by exploiting their semiconductivity (Grimsdale *et al.*, 2009). The introduction of flexible coil segments directly connected to the π-conjugated segments should improve the solubility and processability of BCPs. In addition, their specific morphologies and self-organizing suprastructures, induced by microphase separation and/or (liquid) crystallinity, may affect and occasionally assist high-performance device application. In this chapter, we focus on BCPs containing the important and most frequently studied rod materials, such as polyacetylene, polyphenylene, polyfluorene, poly(phenylene vinylene), and polythiophene segments.

13.3.1 Block Copolymers Containing Polyacetylene Segments

Polyacetylene (PA) is a classical polymeric material that serves as the prototype for other conducting polymers (Shirakawa *et al.*, 1977; Simionescu and Percec, 1982; Akagi, 2009). Soluble PI-*b*-PA was first synthesized by a combination of living anionic polymerization of isoprene, converted to a transition-metal alkyl end, which was a suitable macroinitiator for acetylene, followed by coordination polymerization of acetylene (Aldissi, 1984). Schrock and coworkers synthesized BCPs containing PA and norbornene based on ring-opening metathesis polymerization (ROMP) and reported the self-assembled lamellar phases of BCP thin film observed by TEM and SAX (Krouse and Schrock, 1988). Recently, Grubbs and coworkers synthesized BCPs consisting of PA rod segments, and PS, PMMA, or PEG coil segments based on ROMP in the presence of vinyl-terminated coil polymers (Scherman *et al.*, 2003). Hogen-Esch reported an alternative approach to the synthesis of well-defined PA-*b*-PS and PA-*b*-PS-*b*-PA by sequential living anionic polymerization of styrene and phenyl vinyl sulfoxide (PVS), followed by thermal elimination of sulfenic acid from poly(phenyl vinyl sulfoxide) (PPVS), as shown in Scheme 13.5 (Kanga *et al.*, 1990).

Hirao and coworkers applied this living system to prepare either covalently bonded or ion-bonded BCPs and miktoarm star polymers containing PA rod segments (Hirao *et al.*, 2004; Zhao *et al.*, 2005; Sugiyama *et al.*, 2006; Zhao *et al.*, 2007). In a typical experiment, 4-methylphenyl vinyl sulfoxide (MPVS) was anionically polymerized from intermediate polymeric anions, followed by thermal treatment. As outlined in Scheme 13.6, an AB diblock copolymer anion, as well as 3-arm ABC and 4-arm ABCD star polymer anions were first prepared and MPVS was then added *in situ* to each polymeric anion to polymerize them in a living manner (Zhao *et al.*, 2005). As a result, the targeted 3-arm ABC, 4-arm ABCD, and 5-arm ABCDE stars could be successfully synthesized. As coil blocks, polystyrene derivatives such as PS, PαMS, poly(4-methoxystyrene) (PMOS), poly(4-trimethylsilylstyrene) (PMSiS), and poly(4-methylstyrene) (PMS) were employed. Finally, the resulting polymers were thermally treated to quantitatively convert them to new stars consisting of one conductive

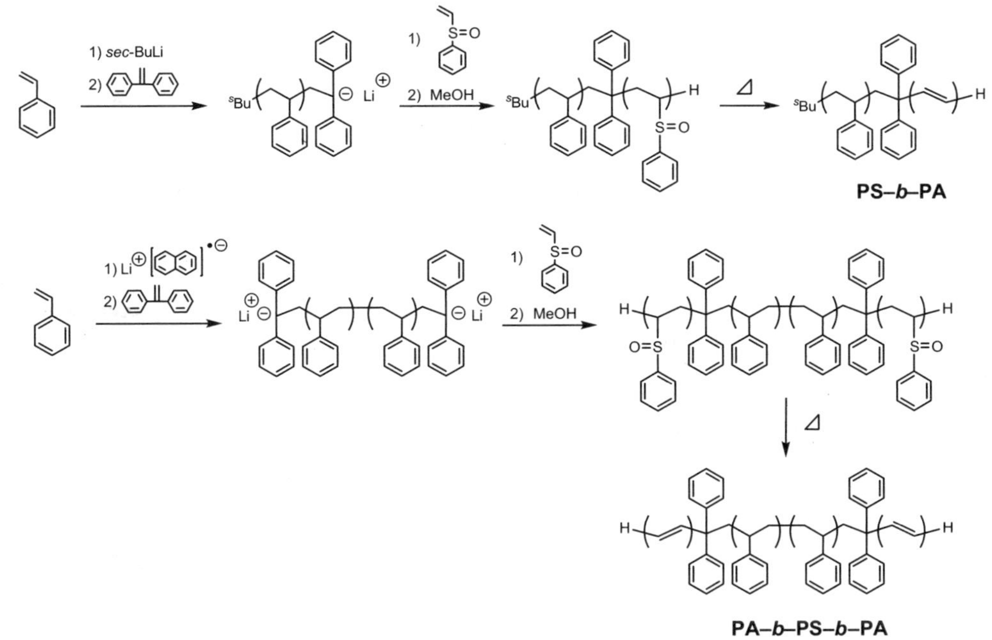

Scheme 13.5 Synthesis of PA-*b*-PS and PA-*b*-PS-*b*-PA.

rod, PA, and other coil segments. Two PA arm segments could also be introduced by the living anionic polymerization of MPVS with the following polymers having two anions at the chain-end position, in-chain position, or at the core, resulting in the formation of miktoarm stars (Zhao *et al.*, 2007). They involve two series of 3-arm AE_2, 4-arm ABE_2, 5-arm $ABCE_2$, and 5-arm AD_2E_2, 6-arm ABD_2E_2, 7-arm $ABCD_2E_2$ stars where the A, B, C, D, and E segments are the PS, PαMS, PMOS, PMSiS, and PMPVS chains, respectively. After thermal treatment at 150 °C for a few hours, these star polymers were quantitatively converted to new functional asymmetric stars having two conductive PA segments. In these stars, it is expected that phase separation would occur at the molecular level to create new and unique nano-ordered suprastructures involving conductive PA domains. At present, a simple lamellar morphology with PA nanolayers was observed in the cast films of AB_2 type star polymers, where A and B were PA and PS, respectively (Hirao *et al.*, 2004).

13.3.2 Block Copolymers Containing Polyphenylene Segments

The history of well-defined BCPs with polyphenylene segments is quite long, similar to polyacetylenes (Section 13.3.1). In 1988, Francois *et al.* synthesized poly(*p*-phenylene)(PPP)-*b*-PS rod–coil BCPs based on living anionic polymerization (Zhong and Francois, 1988). After the sequential polymerization of styrene and cyclohexadiene, the poly(cyclohexa-1,3-diene)-*b*-PS was synthesized and subsequently aromatized with *p*-chloranil to afford the desired PPP-*b*-PS. Although the aromatization was not complete, they discovered special nonequilibrium honeycomb structures in which monodisperse pores are arranged in a hexagonal array by

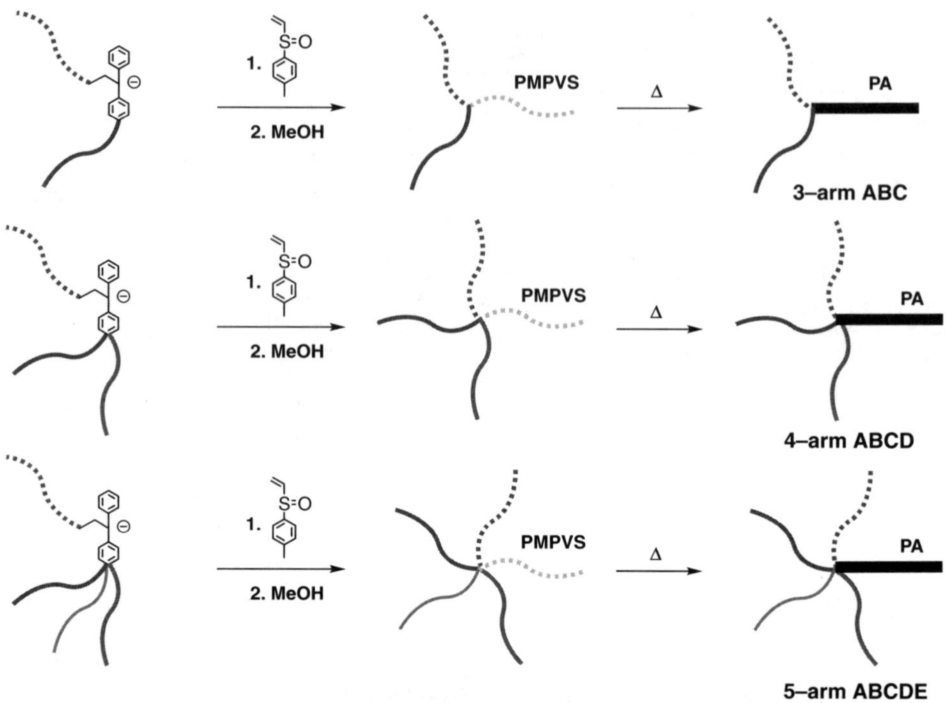

Scheme 13.6 Synthesis of 3-arm ABC, 4-arm ABCD, and 5-arm ABCDE miktoarm star polymers having PA segments.

evaporating from a CS_2 solution in moist air (Widawski, Rawiso, and Francois, 1994). This novel morphology was proposed to be due to micelle formation. Wang *et al.* recently synthesized PPP-*b*-PS and PPP-*b*-PEO based on TEMPO-mediated living radical polymerization of 3,5-cyclohexadiene-1,2-diol-derived monomers, followed by thermal aromatization of the polymer precursor (Wang *et al.*, 2007).

13.3.3 Block Copolymers Containing Polyfluorene Segments

Polyfluorene is one of the widely studied π-conjugated polymers for optoelectronic applications due to its high-fluorescence quantum yields, plus good thermal and chemical stability in solution and the solid state. However, stiff-rod polyfluorene chains generally induce stacking cofacially with each other due to the favorable interchain π–π interactions between flat aromatic units, which causes problems, for instance in light-emitting devices, because it may enhance nonradiative decay. In order to suppress such interactions and to increase the solubility for processing, many rod–coil BCPs have been synthesized, including PHF-*b*-PS (**30**) (Tsolakis and Kallitsis, 2003), PHF-*b*-P^{t}BMA (**31**) (Tsolakis and Kallitsis, 2003), PHF-*b*-PMMA (**32**) (Kuo *et al.*, 2008), PHF-*b*-poly(acrylic acid) (**33**) (Tung *et al.*, 2006), PHF-*b*-poly (*N*-isopropylacrylamide)(PNIPAM) (**34**) (Tian *et al.*, 2009), PHF-*b*-poly[3-(trimethoxysilyl) propyl methacrylate] (**35**) (Tung and Chen, 2009), poly[2-(9-carbozoyl)ethyl methacrylate]-*b*-PHF-*b*-poly[2-(9-carbozoyl)ethyl methacrylate] (**36**) (Lu *et al.*, 2005), poly[2-(dimethylamino)ethyl methacrylate]-*b*-PHF-*b*-poly[2-(dimethylamino)ethyl methacrylate] (**37**) (Fang *et al.*, 2007), and poly[penta(ethylene glycol) methyl ether methacrylate]-*b*-PHF-*b*-poly[penta(ethylene glycol) methyl ether methacrylate] (**38**) (Kwon *et al.*, 2009) (Figure 13.7). Recently, two new type rod–coil–coil ABC triblock copolymers, PHF-*b*-poly[poly(ethylene glycol) methyl ester methacrylate]-*b*-poly[3-(tripropoxysilyl)propyl methacrylate] (**39**) (Wu *et al.*, 2007) and PHF-*b*-PNIPAM-*b*-poly(*N*-hydroxyethyl acrylamide) (**40**) (Lin *et al.*, 2009), were synthesized for micelle and morphological studies. It should be mentioned that all PHF segments basically possess broad PDI due to the uncontrollable polymerization technique, and coil segments are typically prepared by living/controlled radical polymerization systems. Exceptionally, Jenekhe reported novel rod–rod–rod/coil–rod–coil BCPs of

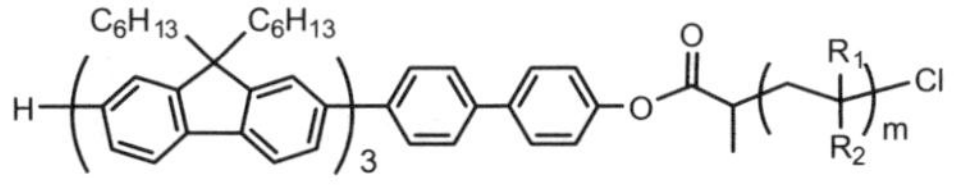

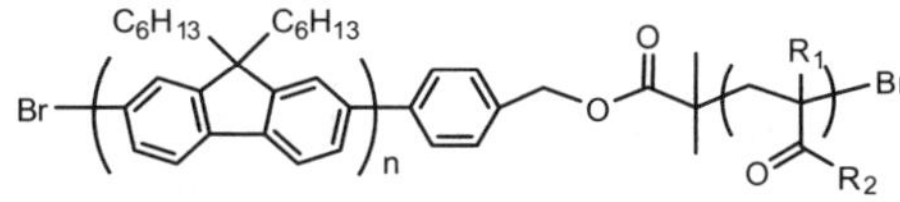

	R_1	R_2
30	–H	–Ph
31	$–CH_3$	$–COOC(CH_3)_3$

	R_1	R_2
32	$–CH_3$	$–OCH_3$
33	–H	–OH
34	–H	$–NHCH(CH_3)_2$
35	$–CH_3$	$–OCH_2CH_2CH_2Si(OCH_3)_3$

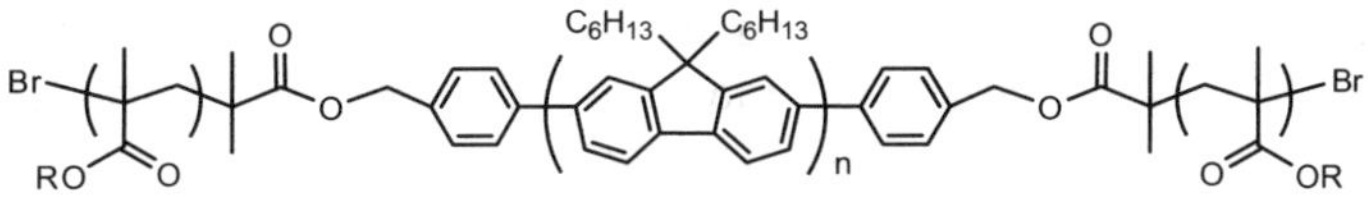

	R		R
36	$–CH_2CH_2–N$ 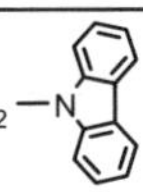	**37**	$–CH_2CH_2N(CH_3)^+I^–$
		38	$–(CH_2CH_2O)_5CH_3$

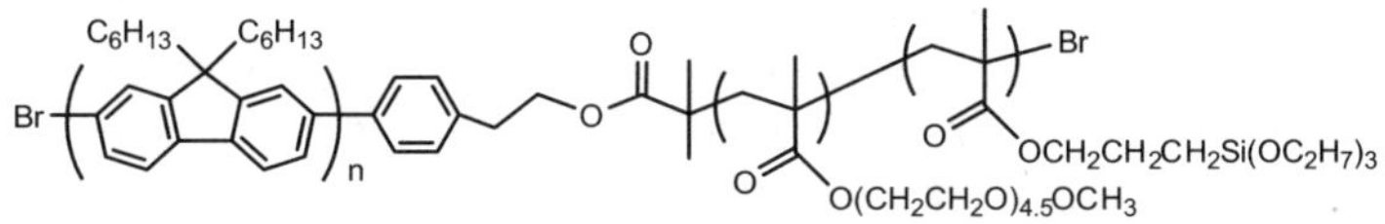

39

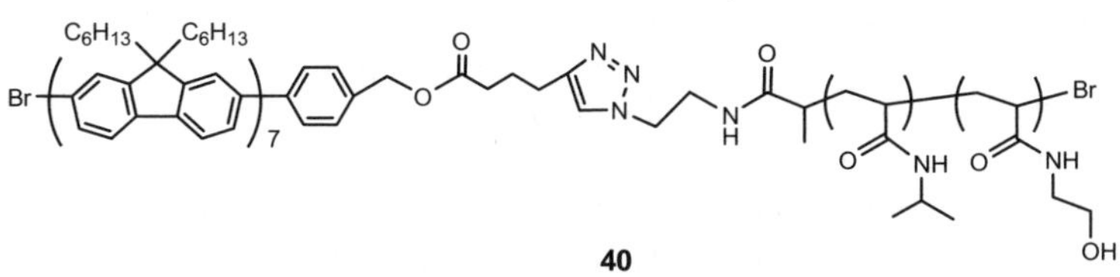

40

Figure 13.7 Chemical structures of PHF-based BCPs.

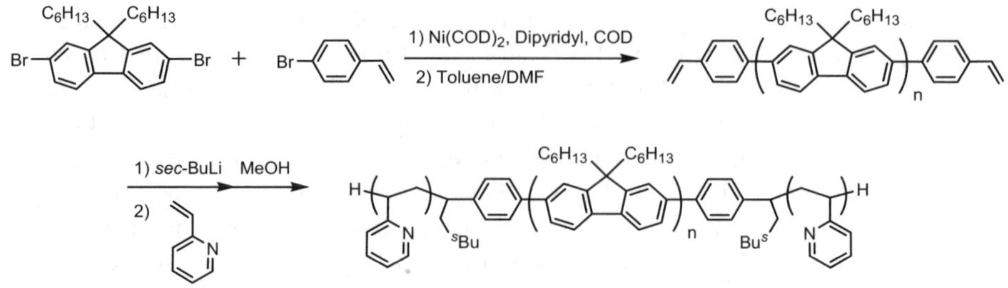

Scheme 13.7 Synthesis of P2VP-*b*-PHF-*b*-P2VP.

PBLG-*b*-PHF-*b*-PBLG based on NCA polymerization from amino-terminated PHF (Rubatat *et al.*, 2008), as described in Section 13.2.1.

Chen and coworkers successfully synthesized rod–coil and coil–rod–coil BCPs comprised of PHF and P2VP by combining the coupling reaction for rod segments and living anionic polymerization (Lin *et al.*, 2008). First, ω-styryl-terminated PHF was prepared by the Suzuki coupling protocol quenched with 4-bromostyrene. After replacing the α-bromo chain end with a phenyl group, a living anionic polymer of 2VP was then coupled with the ω-styryl chain end of PHF to afford the expected rod–coil diblocks. To synthesize coil–rod–coil triblocks, they employed α,ω-styryl-difunctionalized PHF as the precursor (Scheme 13.7). Interestingly, the diblock and triblock copolymers showed different morphologies of micellar aggregates in THF/MeOH mixtures where the diblock copolymers maintained spherical micellar aggregates, but the triblocks readily aggregated in elongated cylinders. They also revealed the effect of polymer chain architecture and aggregated structures on photophysical properties. Because triblock copolymer chains have more favorable stacking in cylindrical arrays, they can provide stronger π–π interchain stacking compared with the spherical aggregates observed for the diblock, leading to a higher absorption maximum.

13.3.4 Block Copolymers Containing Poly(phenylene vinylene) Segments

Poly(phenylene vinylene)s (PPV)s are quite important π-conjugated polymers possessing stiff-rod-like structures, although the molecular weight and molecular-weight distribution of a PPV segment itself cannot be well controlled yet even at the present time. Segalman *et al.* established synthetic and morphological studies of PPV-*b*-PI based on polycondensation and living anionic polymerization (Olsen and Segalman, 2005). They recently reported universalized phase diagrams of PPV-*b*-PI in terms of rod–rod interaction (μ) and rod–coil interaction (χ) in a weakly segregated system (Olsen *et al.*, 2008). They also demonstrated the magnetic-field alignment of rod–coil BCPs that proceed through coupling to the diamagnetic moment of individual rod blocks. The self-assembly of BCPs led to the alignment of the interfaces perpendicular to the field lines and long-range order on a 10-nm length scale (Tao *et al.*, 2007).

To pursue a segregation system stronger than PPV-*b*-PI, PPV-*b*-PMMA (Qu *et al.*, 2007; Ho *et al.*, 2009b) and PPV-*b*-poly(4-vinylpyridine) (P4VP) (Sary *et al.*, 2007), which have larger χ values, were subsequently synthesized based on living anionic polymerization, and their self-assembly behavior was investigated. For example, Sary reported the phase diagrams of

PPV-*b*-P4VP, including the sphere, hexagonal cylinder, and lamellar, as well as the smectic-in-lamellar phases (Sary *et al.*, 2007).

Quite interestingly, Hadziioannou and coworkers described novel donor–acceptor rod–coil BCPs containing PPV and fullerene-substituted coil segments PPV-*b*-(PS-C_{60}) (Stalmach *et al.*, 2000) and PPV-*b*-[poly(*n*-butyl methacrylate)(P^{n}BMA)-*stat*-(PS-C_{60})] (Barrau *et al.*, 2008) based on nitroxy-mediated controlled radical polymerization (NMP). In a typical synthetic method, a macroinitiator, nitroxy-chain-end-functionalized PPV, was synthesized by a reaction between aldehyde-terminated PPV and a corresponding Grignard reagent. The macroinitiator was then used for controlled radical polymerization of a mixture of ^{t}BMA and 4-chloromethylstyrene, followed by the azidation of chloromethyl units. Finally, replacement of azide by C_{60} groups afforded the expected donor–acceptor rod–coil BCP, PPV-*b*-[P^{n}BMA-*stat*-(PS-C_{60})] (Scheme 13.8). The thin film PL spectra of the BCPs were measured and revealed significant PL quenching. This suggests that an efficient exciton dissociation takes place at the PPV/C_{60} interface, although the existence of a nonradiative recombination cannot

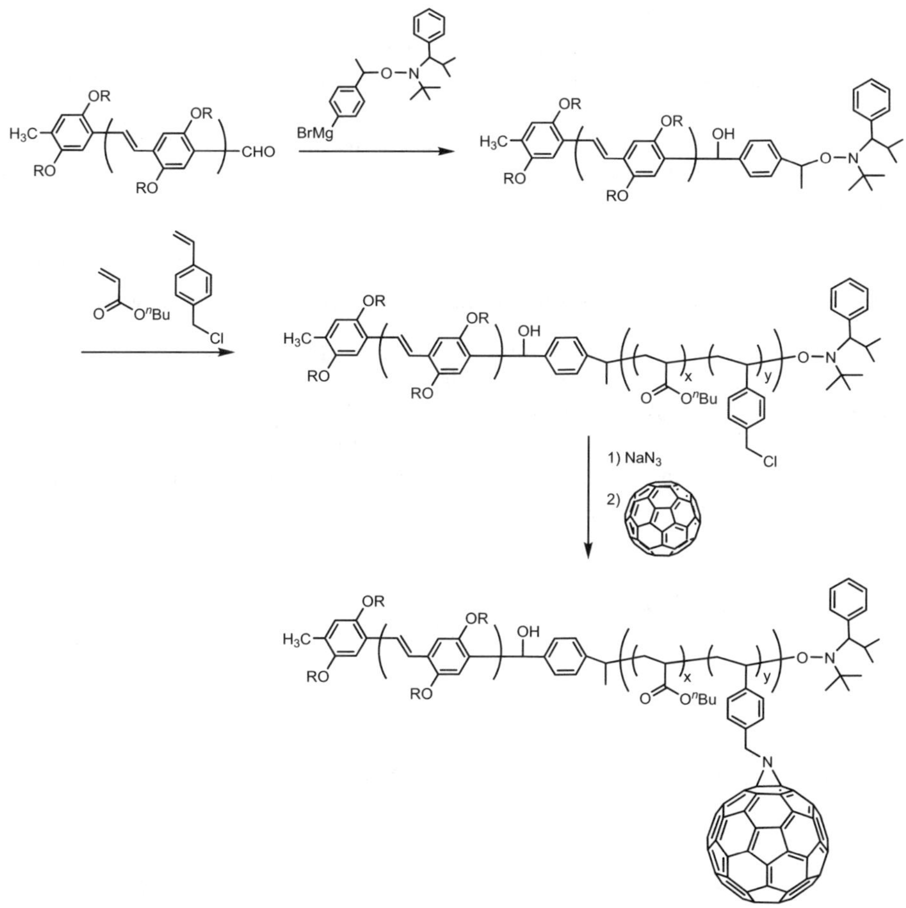

Scheme 13.8 Synthesis of donor–acceptor-type PPV-*b*-[P^{n}BMA-*stat*-(PS-C_{60})].

be excluded. Unfortunately, this material did not show good solar cell performance, probably due to the limitation of using PPV as a *p*-type semiconductor.

13.3.5 Block Copolymers Containing Polythiophene Segments

Polythiophene (PT) has so far been in the best class of balanced high-performance materials as a *p*-type semiconductor among all π-conjugated polymers in terms of solubility, chemical stability, charge mobility, and commercial availability (Roncali, 1992; Chan and Ng, 1998; Perepichka and Perepichka, 2009). Therefore, much attention has been paid to these materials for prompt application to polymer electronic devices, such as organic field-effect transistors and photovoltaic cells. Conventionally, well-defined BCPs with PT rod segments and other coil segments were prepared by a stepwise procedure, as described in the excellent work on PS-*b*-oligothiophene(11-mer)-*b*-PS by Hempenius *et al.* (Hempenius *et al.*, 1998). In 2002, McCullough reported the synthesis of AB diblock and ABA triblock copolymers by ATRP using the macroinitiators of α-chain-end- and α,ω-chain-ends-functionalized head-to-tail regioregular poly(3-hexylthiophene) (P3HT)s (Liu *et al.*, 2002).

Since McCullough *et al.* discovered the quasiliving Grignard metathesis (GRIM) polymerization system for regioregular poly(3-hexylthiophene) (P3HT) (Iovu *et al.*, 2005a), followed by the improvement by Yokozawa *et al.* (Miyakoshi *et al.*, 2005), the control synthesis of poly (alkylthiophene)s with predictable molecular weights and narrow molecular-weight distributions ($\sim$1.1) has become very practical and broke through the synthetic facility for BCPs containing poly(alkylthiophene)s.

13.3.5.1 Combination of GRIM and Living/Controlled Radical Polymerization

To synthesize rod–coil BCPs containing P3HT and poly(acrylate) (PMA) segments, McCullough combined GRIM polymerization with an ATRP system involving several site-transformation steps, as shown in Scheme 13.9 (Iovu *et al.*, 2005b). First, ω-chain-end-vinyl-functionalized P3HT was synthesized by terminating quasiliving P3HT having an ω-Ni-Br chain end with an excess of vinylmagnesium bromide. Then, the introduced vinyl

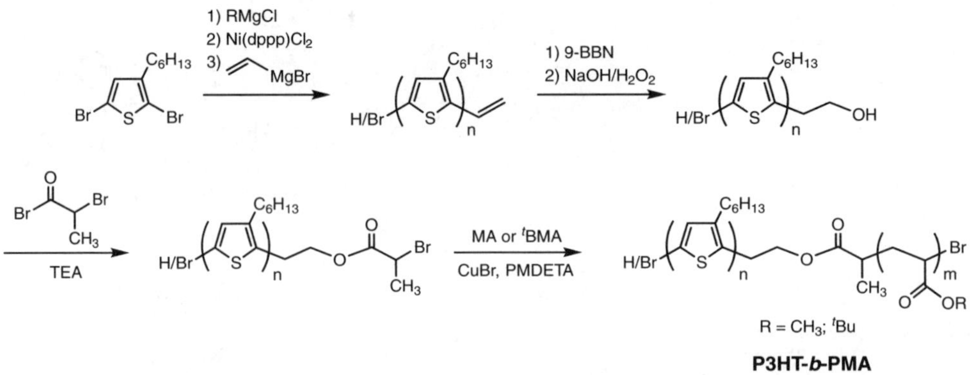

Scheme 13.9 Synthesis of P3HT-*b*-PMAs by the combination of GRIM polymerization and ATRP.

group was transformed into a hydroxyl one by a hydroboration/oxidation process. After transforming the hydroxyl group into a 2-bromopropionate one by an esterfication reaction, ATRP of the acrylates was performed to afford P3HT-*b*-PMAs. They also succeeded in synthesizing P3HT-*b*-PI (**41**) and P3HT-*b*-PS (**42**) (Figure 13.8) using NMP and addition fragmentation chain-transfer polymerization (RAFT), respectively, and found that the nanofibril surface structures of those polymer films could be observed by atomic force microscopy (AFM) (Iovu *et al.*, 2007). Furthermore, new conducting amphiphilic BCPs of P3HT-*b*-poly (acrylic acid) (PAA) (**43**) were synthesized by hydrolysis of P3HT-*b*-P^{t}BMA, which was prepared by a similar method (Craley *et al.*, 2009). Zou *et al.* synthesized a series of P3HT

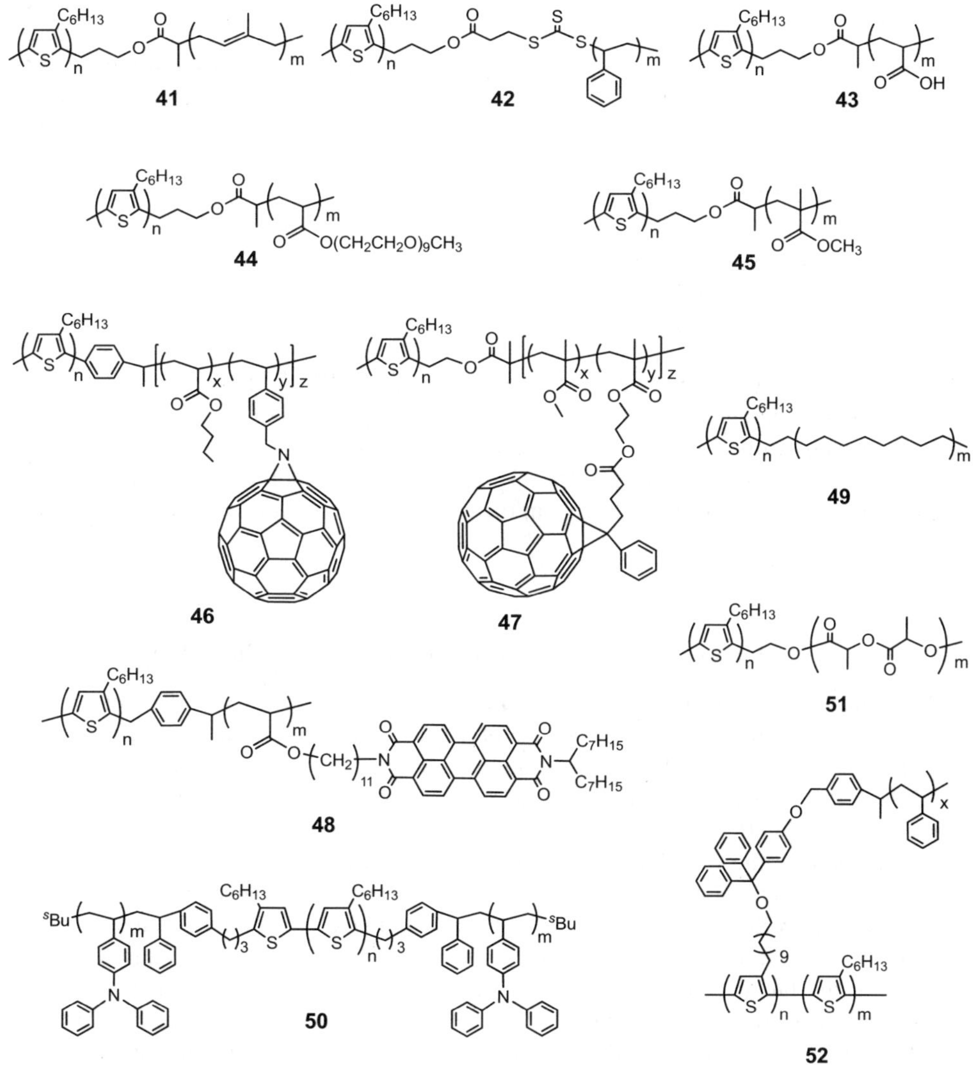

Figure 13.8 Chemical structures of P3HT-based BCPs.

containing BCPs with PAA, poly(PEG acrylate) (**44**), and PMMA (**45**) and used it as an efficient dispersant for carbon nanotubes (Zou *et al.*, 2009). Urien and coworkers reported an alternative synthetic approach via a copper (I)-catalyzed Huisgen [3 + 2] dipolar cycloaddition reaction based on "click" chemistry, combined with GRIM polymerization and ATRP, to afford AB and ABA BCPs containing P3HT and PS segments (Urien *et al.*, 2008). Very recently, a series of interesting donor–acceptor BCPs was synthesized based on living/ controlled radical polymerization cosystems. Richard *et al.* synthesized P3HT-based rod–coil BCPs with fullerene-grafted PS segments (**46**) using the NMP system (Richard *et al.*, 2008). Soon thereafter, Lee *et al.* synthesized P3HT-based rod–coil BCPs with fullerene-grafted PMMA segments (**47**) (Lee *et al.*, 2009). Yang *et al.* reported photovoltaic device application using similar donor–acceptor BCPs as the processing surfactant in the bulk heterojunction (BHJ) system (P3HT/PCBM) (Yang *et al.*, 2009). They achieved higher power conversion efficiency (PCE = 3.5%) than that in the BHJ system of P3HT/PCBM without the surfactant (PCE = 2.6%), probably due to the construction of stable p/n phase domains. Thelakkat and coworkers synthesized another type of donor-acceptor BCPs containing P3HT and perylene-bisimide-grafted polyacrylate segments based on NMP (**48**) (Sommer *et al.*, 2008). Subsequently, Ruseell, Emrick and coworkers synthesized similar BCPs and applied for photovoltaic cells and achieved PCE = 0.49% which was higher than the blend system of P3HT and perylene-bisimide-based compound (PCE = 0.4%) (Zhang *et al.*, 2009). Segalman *et al.* also reported the synthesis of similar BCPs by the combination of GRIM and ATRP polymerization together with click chemistry. Unfortunately, their photovoltaic cell performances were not high (PCE < 0.1%) (Tao *et al.*, 2009).

Although many examples have been developed based on GRIM polymerization in conjunction with the living/controlled radical polymerization system, tedious site-transformation reactions as well as the limited control nature of radical polymerization frequently cause broad molecular-weight distributions (> 1.3). Furthermore, a thorough purification process to remove the residual copper catalyst from the products is sometimes essential for optoelectronic application.

13.3.5.2 Combination of GRIM Polymerization and Living Anionic Polymerization

Besides radical polymerization cosystems, Jannssen and coworkers reported the synthesis of crystalline–crystalline BCPs comprised of P3HT and polyethylene segments (**49**) by a combination of GRIM and ROP (Radano *et al.*, 2005). Recently, Chi-An Dai synthesized new P3HT-*b*-P2VP by a combination of GRIM polymerization and living anionic polymerization (Dai *et al.*, 2007). As shown in Scheme 13.10, ω-chain-end-vinyl-functionalized P3HT was synthesized by the established method. Then, the α-Br chain end was converted into a phenyl group so that the unwanted side reactions such as Li–Br exchange or single-electron transfer reactions could be eliminated. After lithiation of the terminal vinyl groups with ^{s}BuLi, living anionic polymerization of 2VP was carried out to synthesize the desired BCPs. Simply by changing the block compositional ratio, spherical, cylindrical, lamellar and nanofibril structures could clearly be observed by TEM and SAX. The molecular-weight distributions of all BCP samples were relatively narrow due to the use of the living anionic polymerization system (1.18–1.28). However, there are still multistep reactions, especially for the modification reaction of the α-Br chain end of P3HT. Furthermore, in the lithiation reaction, a possible

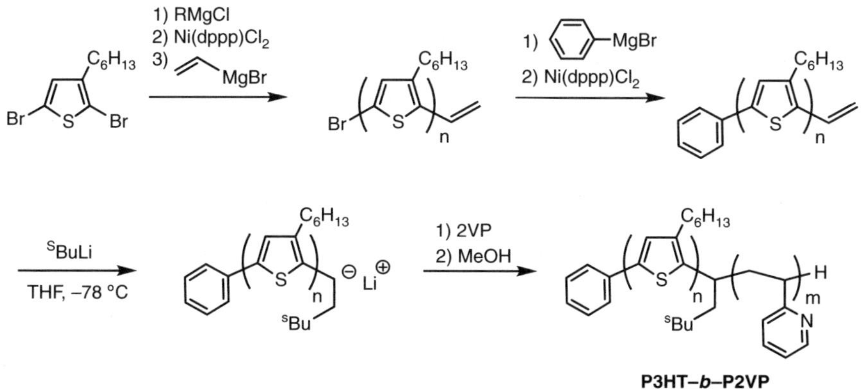

Scheme 13.10 Synthesis of P3HT-*b*-P2VP by the combination of GRIM and living anionic polymerization.

homopolymerization of the terminal thiophenylvinyl units might generate higher coupling byproducts.

Our group has developed a facile synthesis of PS-*b*-P3HT-*b*-PS based on a novel approach involving only two steps, an *in-situ* introduction of nonhomopolymerizable 1,1-diphenylethylene (DPE) moieties at the α,ω-ends of P3HT, and a linking reaction with a living anionic polymer of styrene (Higashihara *et al.*, 2008). The synthetic routes are shown in Scheme 13.11. The key step is an *in-situ* introduction of DPE moieties simultaneously at the α,ω-ends of P3HT, which readily and quantitatively link with living PS without homopolymerization of the DPE units due to its steric hindrance and resonance effect. The disubstitution feature was experimentally supported by MALDI-TOF mass spectroscopy (>90%) and a previous paper

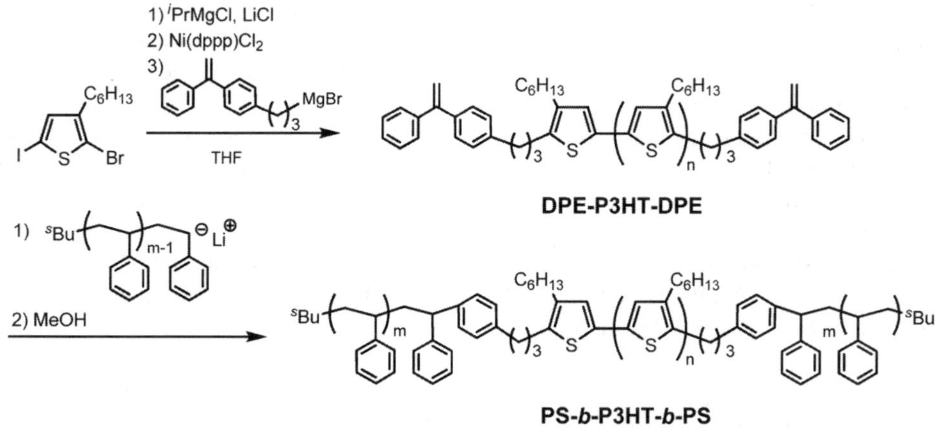

Scheme 13.11 Synthesis of PS-*b*-P3HT-*b*-PS by the combination of GRIM and living anionic polymerization.

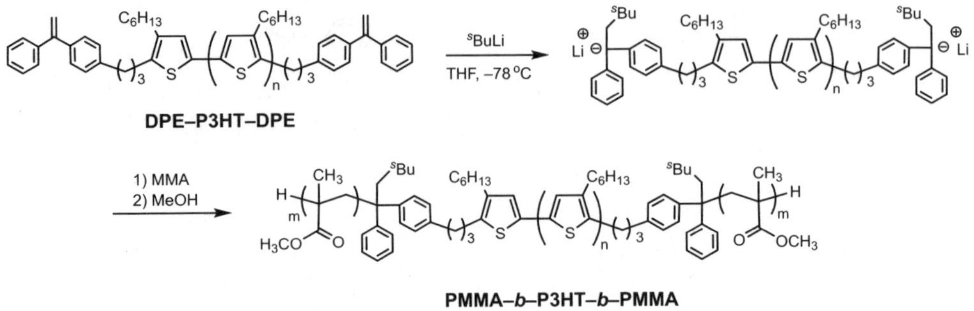

Scheme 13.12 Synthesis of PMMA-*b*-P3HT-*b*-PMMA by the combination of GRIM and living anionic polymerization.

describing the end functionalization of P3HT with an alkyl-type terminating Grignard reagent (Jeffries-El *et al.*, 2004; Jeffries-El *et al.*, 2005). This spontaneous disubstitution permits skipping the extra steps of the transformation reactions at the chain end. In addition, the developed polymer–polymer linking method provides more reliable molecular weights and compositions of BCPs than by conventional macroinitiator methods because both segments can be premade and separately characterized. The purified PS-*b*-P3HT-*b*-PS possesses a molecular weight of 18 900 and a PDI of 1.15, as determined by GPC. The surface morphology of the BCP thin films was observed by AFM and continuous nanofibril and nanowire structures were observed. The formation of the nanowire structures may be dictated by the immiscibility of PS and P3HT segments, as reported in the literature describing the nanowire structures of the P3HT-*b*-PS diblock copolymer (Iovu *et al.*, 2007). The precursor α,ω-ends-DPE-functionalized P3HT was also used for synthesizing well-defined PMMA-*b*-P3HT-*b*-PMMA with low PDI (1.10) according to Scheme 13.12 (Higashihara and Ueda, 2009a).

Unfortunately, these BCPs contain insulating segments, so their electronic performance is probably limited. Therefore, the introduction of optoelectronically active coil segments into P3HT-based BCPs is highly desired. We further synthesized novel BCPs containing P3HT and optoelectronically active poly(4-vinyltriphenylamine) (PVTPA) by the similar approach (Higashihara and Ueda, 2009b). PVTPA-*b*-P3HT-*b*-PVTPA (**50**) may be one of the candidates for better hole-transporting material than PS-*b*-P3HT-*b*-PS or PMMA-*b*-P3HT-*b*-PMMA. The surface morphology of cast film **50** was observed by AFM. When the cast film was treated by solvent annealing under exposure of a toluene vapor, a neat, uniform, and periodic morphology emerged over a large area, in which each phase-separated domain with 12–15-nm widths was well connected with some micrometer length (Figure 13.9). These periodic self-assemble structures may possibly be derived from the well-defined structure of PVTPA-*b*-P3HT-*b*-PVTPA with low PDI (1.15). A TEM image of the sample film also supported such microphase separation.

13.3.5.3 Preparation of Nanoporous P3HT Films

The thin films of BCPs can self-assemble into many types of highly ordered nanostructures on a wide variety of substrates after being cast from a number of solvents. Selective etching of one of

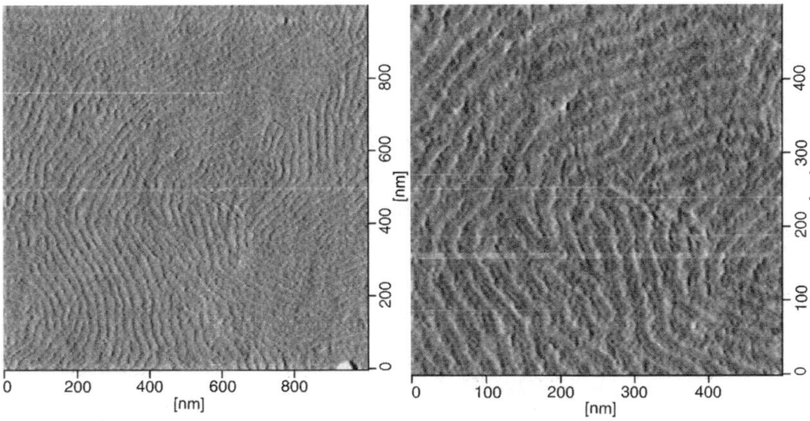

Figure 13.9 Tapping mode AFM phase images of PVTPA-*b*-P3HT-*b*-PVTPA after solvent annealing.

the domains can lead to nanoporous materials that are useful for separations and templates of nanotechnology (Lee *et al.*, 1988; Lee *et al.*, 1989; Hillmyer, 2005b; Olson *et al.*, 2008b). Nanoporous P3HT films derived from BCP templates are quite interesting and are expected to be used for such applications as voltaic chemical sensors and, especially, BHJ organic photovoltaic cells, after a back-fill of *n*-type semiconductors into the porous films. Due to the large interface areas and well-connected charge-transporting channels having widths comparable to exciton diffusion lengths ($\sim$10 nm), it is expected that efficient charge separation with less recombination and fast charge transport will be realized, resulting in improved PCE.

Hillmyer and coworkers reported nanoporous P3HT films for the first time by using a template of P3HT-*b*-poly(D,L-lactide) (**51**) (Boudouris *et al.*, 2008). Starting from ω-chain-end-vinyl-functionalized P3HT, the chain end is transformed into a hydroxy group, followed by controlled ROP of D,L-lactide from an aluminum alkoxide macroinitiator to synthesize **51**. The back-fill of water-soluble fullerene derivatives into a similar nanoporous P3HT, derived from **51**, was recently performed to create an ideal p/n BHJ active layer, however, the characterization of photovoltaic cells was not shown (Botiz and Darling, 2009). Hawker and coworkers recently reported the creation of nanoporous P3HT films by the selective solvent etching of PS segments after being cleaved off from the template of P3HT grafted with PS segments via the trityl ether linkage (**52**) (Sivanandan *et al.*, 2009).

Our group developed an alternative approach for nanoporous P3HT by employing P3HT-based BCPs with ionic interaction. Although covalently bonded P3HT-based BCPs are presently very competitive, a blend system of well-defined chain-end-functionalized polymers to achieve similar nanostructures has scarcely been focused on yet. The advantageous points of using such a blend system are as follows: (1) more facile synthetic approaches are available for chain-end-functionalized polymers than for BCPs, thereby being preferred for industry; (2) the combination of block segments is versatile; (3) the facile preparation of nanoporous films by the removal of one of the segments with a selective solvent after cleaving the ionic bonds under acidic or alkaline conditions. The first example of P3HT-based BCPs via ionic

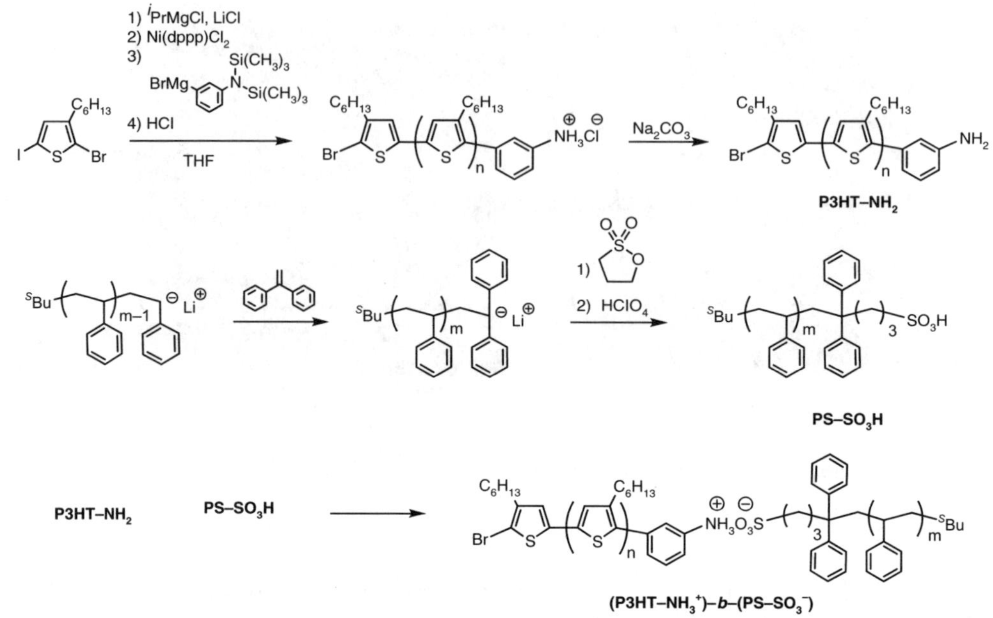

Scheme 13.13 Synthesis of (P3HT-NH_3^+)-*b*-(PS-SO_3^-).

interaction was prepared by blending ω-chain-end-aniline-functionalized P3HT (P3HT-NH_2) and ω-chain-end-carboxy-functionalized PS (PS-COOH) (Higashihara *et al.*, 2010) or sulfonic-acid-chain-end-functionalized PS (PS-SO_3H) (Takahashi *et al.*, 2010). The representative synthetic routes for the BCPs of (P3HT-NH_3^+)-*b*-(PS-SO_3^-) are depicted in Scheme 13.13.

The termination of P3HT with an ω-Ni-Br chain end with an excess of a Grignard reagent carrying a protected aniline moiety, followed by quenching the system with 5N-HCl aq. afforded an HCl salt of P3HT-NH_2. After washing with an alkaline solution, P3HT-NH_2 could be regenerated. The PDIs were very low, in the range of from 1.08 to 1.12, and NH_2-end functionalities were in the range of 85–92%. On the other hand, PS-SO_3H with 100% end functionality was synthesized based on living anionic polymerization according to Scheme 13.13.

AFM images of the surface morphology of a (P3HT-NH_3^+)-*b*-(PS-SO_3^-) thin film showed nanofibril structures having 15–20-nm widths and a few hundred nanometer lengths are observed (Figure 13.10a). In order to create the desired nanoporous P3HT films, selective etching of PS domains with ethyl acetate/triethylamine (100/1, v/v) was performed by exploiting the pH stimulating ionic interaction between the P3HT and PS segments. Ethyl acetate and triethylamine play important roles in the selective solvent for PS and the cleavage of aniline–sulfonic acid interaction, respectively. Nanoporous textures are indeed observed in some areas in AFM (Figure 13.10b) and SEM images (Figure 13.10c), but P3HT domains tend to collapse, probably due to the wet process and high aspect ratio of more than 5. Further improvements of this system are now under investigation.

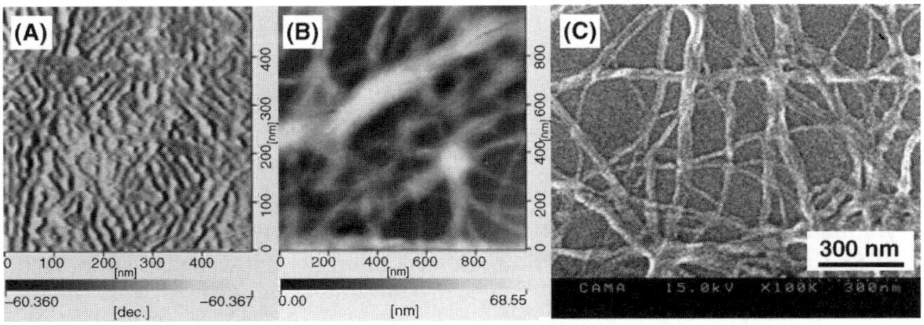

Figure 13.10 (a) Tapping-mode AFM phase image of $(P3HT\text{-}NH_3^+)$-*b*-$(PS\text{-}SO_3^-)$ thin film, (b) tapping-mode AFM height image of nanoporous P3HT thin film, and (c) SEM image of nanoporous P3HT thin film.

13.4 Rod–Rod Block Copolymers

As mentioned in the preceding section, a wide variety of rod–coil BCPs has been synthesized and well investigated in regard to self-assembly behavior, morphologies, and micelles, as well as possible application to many fields, such as biomaterial, electronics, energy technology and so on. On the other hand, binary-phase rod–rod BCPs, which are entirely comprised of rod segments via covalent bonding, are quite novel and advanced materials from the viewpoints of both fundamental and application research. Scherf and coworkers initially contributed all-conjugated rod–rod BCPs, including polyfluorene segments (Schmitt *et al.*, 2001; Güntner *et al.*, 2002; Asawapirom *et al.*, 2005; Tu *et al.*, 2007; Park *et al.*, 2008), reviewed elsewhere (Scherf *et al.*, 2008). In Section 13.2.3, a new concept of chain-growth condensation polymerization of AB monomers was introduced for precisely controlling the molecular weights and PDIs of rod blocks. Indeed, an array of well-defined rod–coil BCPs was synthesized by combining different polymerization systems. On the other hand, rod–rod aromatic copolyamides can be synthesized simply by the sequential polymerization of different monomers in a one-pot process. Indeed, rod–rod aromatic copolyamide **53** (Figure 13.11) was synthesized by the sequential polymerization of a 4-(*N*-octylamino)benzoic

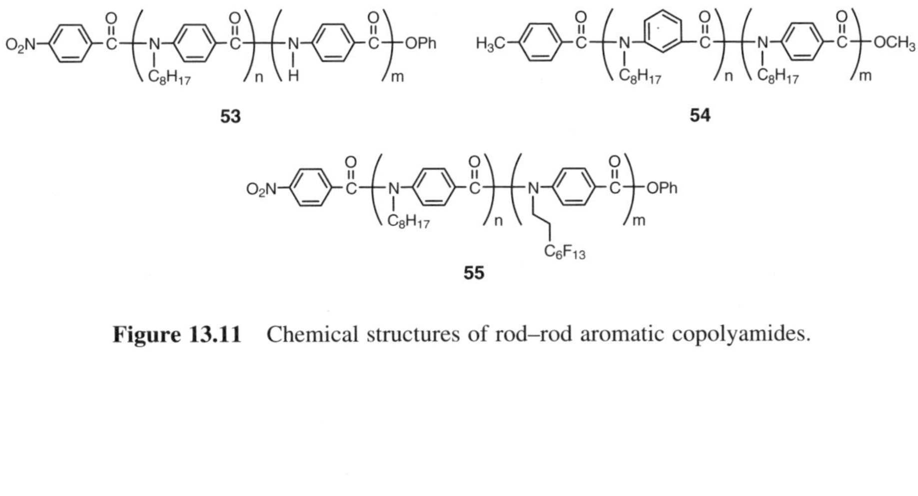

Figure 13.11 Chemical structures of rod–rod aromatic copolyamides.

acid ester and 4-(*N*-(4-octyloxybenzyl)amino)benzoic acid ester, followed by deprotection of a 4-octyloxybenzyl group with trifluoroacetic acid, resulting in the generation of a second block of *N*-H polyamide (Yokozawa *et al.*, 2002). Similarly, *m*- and *p*-substituted aromatic regioblock copolyamide **54** was synthesized (Sugi *et al.*, 2005). Our group also succeeded in synthesizing rod–rod aromatic copolyamide **55** by sequential polymerization of 3-[4-(*N*-octylamino)benzoyl]benzoxazoline-2-thione and 3-[4-(*N*-1,1,1,2,2,3,3,4,4,5,5,6,6-trideca-fluorooctylamino)benzoyl]benzoxazoline-2-thione. The tapping mode phase images of the thin film of **55** showed phase-separated domain structures, which was also supported by DSC analysis (Ohshimizu *et al.*, 2007).

Since the development of quasiliving GRIM polymerization using $Ni(dppp)Cl_2$ as the initiator, new rod–rod block copolythiophenes have been synthesized. First, McCullough and coworkers reported an all-conjugated block copolythiophene **56** containing P3HT and poly(3-dodecylthiophene) (Figure 13.12) (Iovu *et al.*, 2005a). Then, Yokozawa and coworkers synthesized amphiphilic P3HT-*b*-poly(3-methoxyethoxyethoxymethylthiophene) (**57**) (Yokozawa *et al.*, 2007). Soon thereafter, crystalline-amorphous block copolythiophenes, P3HT-*b*-poly(3-phenoxymethylthiophene)(P3PT) (**58**) (Ohshimizu and Ueda, 2008) and P3HT-*b*-poly[3-(2-ethylhexyl)thiophene] (**59**) (Zhang *et al.*, 2008) were independently synthesized almost at the same time by our group and Tajima *et al.*, respectively. Many other examples emerged in the literature subsequently, **60–64** (Benanti *et al.*, 2008; Ouhib *et al.*, 2008a; Van den Bergh *et al.*, 2008; Ouhib *et al.*, 2008b; Wu *et al.*, 2009). New all-conjugated BCPs of P3HT-*b*-poly(2,5-dihexyloxy-*p*-phenylene) (**65**) (Miyakoshi, Yokoyama, and Yokozawa, 2008b) and poly(2,5-dihexyloxy-*p*-phenylene)-*b*-poly(*N*-hexylpyrrole) (**66**)

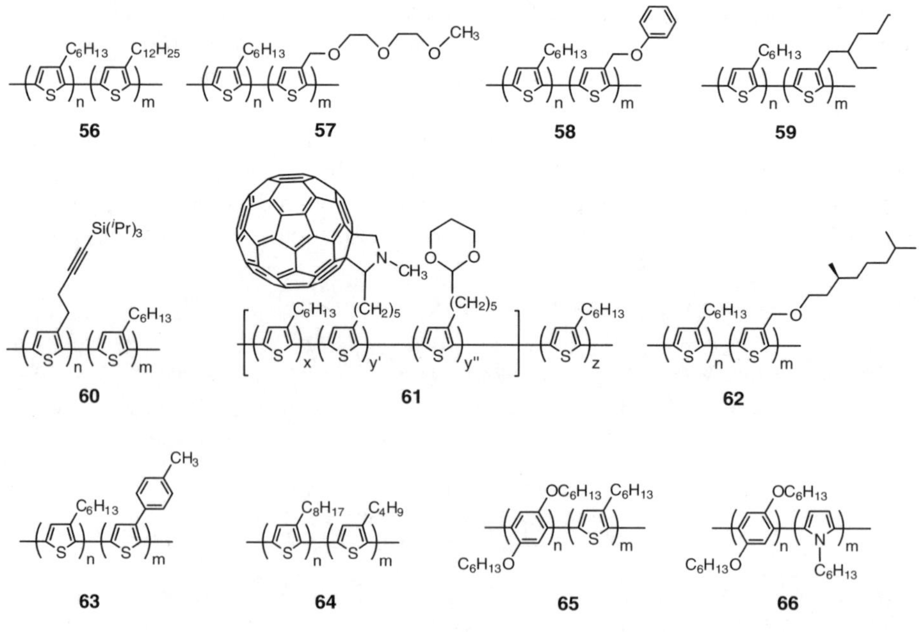

Figure 13.12 Chemical structures of all-conjugated rod–rod BCPs.

(Yokoyama *et al.*, 2008), could also be synthesized by sequential GRIM polymerization using Ni(dppe)Cl_2 as the initiator.

The phase-separated domains of P3HT have been thought to be extremely important for effective charge separation and smooth charge transportation in photovoltaic cell application, especially when such phase-separated domains are arranged out-of-plane to the film on the order of 10 nm, corresponding to the exciton diffusion length. Block copolythiophene has the possibility to meet these requirements. Very recently, Chen, Ueda and coworkers first demonstrated the manipulation of optoelectronic device uses of block copolythiophene **58** in thin-film transistor (TFT) and photovoltaic-cell applications (Chueh *et al.*, 2009). The optical absorption spectra showed the enhanced intensity of the crystalline absorption peak at around 605 nm from the film of **58**, compared to a pristine P3HT film. The amorphous and relatively flexible P3PT domains might afford the chance to force the crystal alignment of P3HT domains. The AFM images also suggested an orderly fiber-like structure of P3HT-*b*-P3PT. The TFT hole mobility of **58** was as high as $2.0 \times 10^{-2}\,cm^2\,V^{-1}\,s^{-1}$, which is comparable to pristine P3HT. The PCE of **58**/PCBM (1: 1, w/w)-based photovoltaic cells was 2.60%, processed from the 1,2-dichlorobenzene solution, was higher than 1.88% using $CHCl_3$. The PCE of **58**/PCBM was further improved to 2.80% with a 1: 0.7 blend ratio (w:w) of **58** to PCBM. In order to gain insight into the morphology–performance relationship, further detailed investigation is ongoing.

13.5 Concluding Remarks

BCPs containing rod segments have undoubtedly received much attention in many fields of bio-, nano-, and energy technologies. In practice, there have been vital evolutions based on such BCPs in the following applications: polymeric catalysts, tissue engineering, biomedication, sensors, engineering plastics, photonics, electronics and so on. Due to the chemical bonding between dissimilar segments of BCPs, microphase separation on a nanometer scale spontaneously happens so as to minimize the contact energy. At the same time, (liquid) crystallinity of rod segments also affords the chance to self-organize suprastructual nano-objects with a wide variety of morphological shapes. This direct, prompt, and economical nanofabrication approach seems to be suitable for future industrial needs. We have reviewed herein recent advances in BCPs, roughly categorized by nonconjugated and π-conjugated rod segments, followed by subdivision into individual materials. Rapid growth in living/controlled polymerization studies not only for coil segments but also for rod segments could open new windows for well-defined BCPs. These prototype samples are essential for investigation of structure–morphology–performance relationships and may be strong candidates for virtual materials. Therefore, throughout this review, controlled synthesis of BCPs in regard to molecular weight, molecular-weight distribution, and composition, on the basis of living/controlled polymerization systems, has been focused on. In particular, the discovery of the chain-growth condensative polymerization system using AB monomers has stimulated this research area. Our current motivation is directed to the synthesis, characterization, self-assembly behavior, and optoelectronic device application of new BCPs containing polythiophenes and their related materials based on chain-growth condensative polymerization. Also, we expect a lot of further research to broaden the currently limited acceptable monomers for such a controlled polymerization system in the near future.

References

Agut, W., Taton, D., and Lecommandoux, S. (2007) A versatile synthetic approach to polypeptide based rod–coil block copolymers by click chemistry. *Macromolecules*, **40**, 5653–5661.

Aharomi, S.M. (1979) Rigid backbone polymers. 2. Polyisocyanates and their liquid-crystal behavior. *Macromolecules*, **12**, 94–103.

Aharoni, S.M. and Walsh, E.K., (1979) Rigid backbone polymers. 4. Solution properties of two lyotropic mesomorphic poly(isocyanates). *Macromolecules*, **12**, 271–276.

Ahn, J.-H. and Lee, J.-S. (2003) Synthesis of well-defined rod–coil-rod polyhexylisocyanate-block-polystyrene-block-polyhexylisocyanate via one-pot anionic polymerization. *Macromolecular Rapid Communications*, **24**, 571–575.

Ahn, J.-H., Shin, Y.-D., Kim, S.-Y., and Lee, J.-S. (2003) Synthesis of well-defined block copolymers of n-hexyl isocyanate with isoprene by living anionic polymerization. *Polymer*, **44**, 3847–3854.

Akagi, K. (2009) Helical polyacetylene: Asymmetric polymerization in a chiral liquid-crystal field. *Chemical Reviews*, **109**, 5354–5401.

Alberts, B., Bray, D., Lewis, J. *et al.* (1989) *Molecular Biology of the Cell*, 2nd edn, Garland Publishing, New York, pp. 87–134, 820–830.

Aldissi, M. (1984) Soluble conducting poly(acetylene) block copolymers. *Journal of the Chemical Society. Chemical Communications*, **20**, 1347–1348.

Aliferis, T., Iatrou, H., and Hadjichristidis, N. (2004) Living polypeptides. *Biomacromolecules*, **5**, 1653–1656.

Angot, S., Murthy, K.S., Taton, D., and Gnanou, Y. (2000) Scope of the copper halide/bipyridyl system associated with calixarene-based multihalides for the synthesis of well-defined polystyrene and poly(meth)acrylate stars. *Macromolecules*, **33**, 7261–7274.

Aoshima, S. and Kanaoka, S. (2009) A renaissance in living cationic polymerization. *Chemical Reviews*, **109**, 5245–5287.

Aoshima, S., Yoshida, T., Kanazawa, A., and Kanaoka, S. (2007) New stage in living cationic polymerization: an array of effective Lewis acid catalysts and fast living polymerization in seconds. *Journal of Polymer Science Part A-Polymer Chemistry*, **45**, 1801–1813.

Asawapirom, U., Güntner, R., Forster, M., and Scherf, U. (2005) Semiconducting block copolymers-synthesis and nanostructure formation. *Thin Solid Films*, **477**, 48–52.

Babin, J., Leroy, C., Lecommandoux, S. *et al.* (2005) Towards an easy access to amphiphilic rod–coil miktoarm star copolymers. *Chemical Communications*, **25**, 1993–1995.

Barner-Kowollik, C., Davis, T.P., Heuts, J.P.A. *et al.* (2003) RAFTing down under: tales of missing radicals, fancy architectures, and mysterious holes. *Journal of Polymer Science Part A-Polymer Chemistry*, **41**, 365–375.

Barrau, S., Heiser, T., Richard, F. *et al.* (2008) Self-assembling of novel fullerene-grafted donor-acceptor rod–coil block copolymers. *Macromolecules*, **41**, 2701–2710.

Baughman, T.W. and Wagener, K.B. (2005) Recent advances in ADMET polymerization. *Advances in Polymer Science*, **176**, 1–42.

Bazan, G.C. and Schrock, R.R. (1991) Synthesis of star block copolymers by controlled ring-opening metathesis polymerization. *Macromolecules*, **24**, 817–823.

Benanti, T.L., Kalaydjian, A., and Venkataraman, D. (2008) Protocols for efficient postpolymerization functionalization of regioregular polythiophenes. *Macromolecules*, **41**, 8312–8315.

Berda, E.B., Baughman, T.W., and Wagener, K.B. (2006) Precision branching in ethylene copolymers: synthesis and thermal behavior. *Journal of Polymer Science Part A-Polymer Chemistry*, **44**, 4981–4989.

Bielawski, C.W. and Grubbs, R.H. (2007) Living ring-opening metathesis polymerization. *Progress in Polymer Science*, **32**, 1–29.

Botiz, I. and Darling, S.B. (2009) Self-assembly of poly(3-hexylthiophene)-*block*-polylactide block copolymer and subsequent incorporation of electron acceptor material. *Macromolecules*, **42**, 8211–8217.

Boudouris, B.W., Frisbie, C.D., and Hillmyer, M.A. (2008) Nanoporous poly(3-alkylthiophene) thin films generated from block copolymer templates. *Macromolecules*, **41**, 67–75.

Brzezinska, K.R. and Deming, T.J. (2004) Synthesis of AB diblock copolymers by atom-transfer radical polymerization (ATRP) and living polymerization of **α**-amino acid-*N*-carboxyanhydrides. *Macromolecular Bioscience*, **4**, 566–569.

Brzezinska, K.R., Curtin, S.A., and Deming, T.J. (2002) Polypeptide end-capping using functionalized isocyanates: preparation of pentablock copolymers. *Macromolecules*, **35**, 2970–2976.

Bur, A.J. and Fetters, L.J. (1976) The chain structure, polymerization, and conformation of polyisocyanates. *Chemical Reviews*, **76**, 727–746.

Caillol, S., Lecommandoux, S., Mingotaud, A.-F. *et al.* (2003) Synthesis and self-assembly properties of peptide-polylactide block copolymers. *Macromolecules*, **36**, 1118–1124.

Chan, H.S.O. and Ng, S.C. (1998) Synthesis, characterization and applications of thiophene-based functional polymers. *Progress in Polymer Science*, **23**, 1167–1231.

Chen, J.T., Thomas, E.L., Ober, C.K., and Hwang, S.S. (1995) Zigzag morphology of a poly(styrene-*b*-hexyl isocyanate) rod–coil block copolymer. *Macromolecules*, **28**, 1688–1697.

Chueh, C.-C., Higashihara, T., Tsai, J.-H. *et al.* (2009) All-conjugated diblock copolymer of poly(3-hexylthiophene)-*block*-poly(3-phenoxymethylthiophene) for field-effect transistor and photovoltaic applications. *Organic Electronics*, **10**, 1541–1548.

Cloutet, E., Fillaut, J.-L., Astruc, D., and Gnanou, Y. (1999) Star block copolymers and hexafullerene stars via derivatization of star-shaped polystyrenes. *Macromolecules*, **32**, 1043–1054.

Craley, C.R., Zhang, R., Kowalewski, T. *et al.* (2009) Regioregular poly(3-hexylthiophene) in a novel conducting amphiphilic block copolymer. *Macromolecular Rapid Communications*, **30**, 11–16.

Dai, C.-A., Yen, W.-C., Lee, Y.-H. *et al.* (2007) Facile synthesis of well-defined block copolymers containing regioregular poly(3-hexyl thiophene) via anionic macroinitiation method and their self-assembly behavior. *Journal of the American Chemical Society*, **129**, 11036–11038.

Deming, T. (1997) Facile synthesis of block copolypeptides of defined architecture. *Nature*, **390**, 386–389.

Dimitrov, I. and Schlaad, H. (2003) Synthesis of nearly monodisperse polystyrene-polypeptide block copolymers via polymerisation of *N*-carboxyanhydrides. *Chemical Communications*, **23**, 2944–2945.

Fang, C., Qi, X.-Y., Fan, Q.-L. *et al.* (2007) A facile route to semiconductor nanocrystal-semiconducting polymer complex using amine-functionalized rod–coil triblock copolymer as multidentate ligand. *Nanotechnology*, **18**, 35704.

Floudas, G., Papadopoulos, P., Klok, H.-A. *et al.* (2003) Hierarchical self-assembly of poly(γ-benzyl-L-glutamate)-poly(ethyleneglycol)-poly(γ-benzyl-L-glutamate) rod–coil-rod triblock copolymers. *Macromolecules*, **36**, 3673–3683.

Frenzel, U. and Nuyken, Oskar (2002) Ruthenium-based metathesis initiators: development and use in ring-opening metathesis polymerization. *Journal of Polymer Science Part A-Polymer Chemistry*, **40**, 2895–2916.

Fujita, Y., Koga, K., Kim, H.-K. *et al.* (2007) Phosgene-free synthesis of *N*-carboxyanhydrides of α-amino acids based on bisarylcarbonates as starting compounds. *Journal of Polymer Science Part A-Polymer Chemistry*, **45**, 5365–5369.

Grimsdale, A.C., Leok Chan, K., Martin, R.E. *et al.* (2009) Synthesis of light-emitting conjugated polymers for applications in electroluminescent devices. *Chemical Reviews*, **109**, 897–1091.

Güntner, R., Asawapirom, U., Forster, M. *et al.* (2002) Conjugated polyfluorene/polyaniline block copolymers-improved synthesis and nanostructure formation. *Thin Solid Films*, **417**, 1–6.

Hadjichristidis, N., Pitsikalis, M., Pispas, S., and Iatrou, H. (2001) Polymers with complex architecture by living anionic polymerization. *Chemical Reviews*, **101**, 3747–3792.

Hamley, I.W. (2003) Nanotechnology with soft materials. *Angewandte Chemie-International Edition*, **42**, 1692–1712. and references therein.

Heise, A., Hedrick, J.L., Frank, C.W., and Miller, R.D. (1999a) Starlike block copolymers with amphiphilic arms as models for unimolecular micelles. *Journal of the American Chemical Society*, **121**, 8647–8648.

Heise, A., Hedrick, J.L., Trollss, M. *et al.* (1999b) Novel starlike poly(methyl methacrylate)s by controlled dendritic free radical initiation. *Macromolecules*, **32**, 231–235.

Hempenius, M.A., Langeveld-Voss, B.M.W., van Haare, J.A.E.H. *et al.* (1998) A polystyrene-oligothiophene-polystyrene triblock copolymer. *Journal of the American Chemical Society*, **120**, 2798–2804.

Higashihara, T. and Faust, R. (2009) Synthesis of novel ABA triblock and (ABA)n multiblock copolymers comprised of polyisobutylene and poly(γ-benzyl-L-glutamate) segments. *Reactive & Functional Polymers*, **69**, 429–434.

Higashihara, T. and Ueda, M. (2009a) Synthesis and characterization of a novel coil-rod–coil triblock copolymers comprised of regioregular poly(3-hexylthiophene) and poly(methyl methacrylate) segments. *Reactive & Functional Polymers*, **69**, 457–462.

Higashihara, T. and Ueda, M. (2009b) Living anionic polymerization of 4-vinyltriphenylamine for synthesis of novel block copolymers containing low-polydisperse poly(4-vinyltriphenylamine) and regioregular poly(3-hexylthiophene) segments. *Macromolecules*, **42**, 8794–8800.

Higashihara, T., Ohshimizu, K., Hirao, A., and Ueda, M. (2008) Facile synthesis of ABA triblock copolymer containing regioregular poly(3-hexylthiophene) and polystyrene segments via linking reaction of poly(styryl)lithium. *Macromolecules*, **41**, 9505–9507.

Higashihara, T., Takahashi, A., Tajima, S. *et al.* (2010) Synthesis of block copolymers consisting of poly(3-hexylthiophene) and polystyrene segments via ionic interaction and their self-assembly behavior. *Polymer Journal*, **42**, 43–50.

Hillmyer, M.A. (2005a) Nanoporous materials from block copolymer precursors. *Advances in Polymer Science*, **190**, 137–181.

Hillmyer, M.A. (2005b) Nanoporous materials from block copolymer precursors. *Advances in Polymer Science*, **190**, 137–181.

Hirao, A. and Nakahama, S. (1992) Anionic living polymerization of monomers with functional silyl groups. *Progress in Polymer Science*, **17**, 283–317.

Hirao, A., Karasawa, Y., Higashihara, T. *et al.* (2004) Synthesis of block co-polymers and star-branched polymers consisting of conducting polyacetylene segments via ionic interaction to form ionic bonds. *Designed Monomers and Polymers*, **7**, 647–660.

Ho, R.M., Chen, C.K., and Chiang, Y.W. (2009a) Novel nanostructures from self-assembly of chiral block copolymers. *Macromolecular Rapid Communications*, **30**, 1439–1456.

Ho, C.-C., Lee, Y.-H., Dai, C.-A. *et al.* (2009b) Synthesis and self-assembly of poly(diethylhexyloxy-*p*-phenylene-vinylene)-*b*-poly(methyl methacrylate) rod–coil block copolymers. *Macromolecules*, **42**, 4208–4219.

Hoeben, F.J.M., Jonkheijm, P., Meijer, E.W., and Schenning, A.P.H.J. (2005) About supramolecular assemblies of π-conjugated systems. *Chemical Reviews*, **105**, 1491–1546.

Hsieh, H.L. and Quirk, R.P. (1996) *Anionic Polymerization: Principles and Applications*, Marcel Dekker, Inc., New York, Basel, Hong Kong.

Huang, C.-J. and Chang, F.-C. (2008) Polypeptide diblock copolymers: syntheses and properties of poly(*N*-isopropylacrylamide)-*b*-polylysine. *Macromolecules*, **41**, 7041–7052.

Ibarboure, E., Rodriguez-Hernandez, J., and Papon, E. (2006) Thermotropic liquid crystal behavior on PBLG-PDMS-PBLG triblock copolymers. *Journal of Polymer Science Part A-Polymer Chemistry*, **44**, 4668–4679.

Iovu, M.C., Sheina, E.E., Gil, R.R., and McCullough, R.D. (2005a) Experimental evidence for the quasi-"living" nature of the Grignard metathesis method for the synthesis of regioregular poly(3-alkylthiophenes). *Macromolecules*, **38**, 8649–8656.

Iovu, M.C., Jeffries-EL, M., Sheina, E.E. *et al.* (2005b) Regioregular poly(3-alkylthiophene) conducting block copolymers. *Polymer*, **46**, 8582–8586.

Iovu, M.C., Craley, C.R., Jeffries-EL, M. *et al.* (2007) Conducting regioregular polythiophene block copolymer nanofibrils synthesized by reversible addition fragmentation chain transfer polymerization (RAFT) and nitroxide mediated polymerization (NMP). *Macromolecules*, **40**, 4733–4735.

Ishizu, K., Tsubaki, K., Mori, A., and Uchida, S. (2002) Architecture of nanostructured polymers. *Progress in Polymer Science*, **28**, 27–54.

Ishizu, K., Hatoyama, N., and Uchida, S. (2008) Architecture of rod–brush block copolymers synthesized by a combination of coordination polymerization and atom transfer radical polymerization. *Journal of Applied Polymer Science*, **108**, 3346–3352.

Jeffries-El, M., Sauvé, G., and McCullough, R.D. (2004) *In-situ* end-group functionalization of regioregular poly(3-alkylthiophene) using the Grignard metathesis polymerization method. *Advanced Materials*, **16**, 1017–1019.

Jeffries-El, M., Sauvé, G., and McCullough, R.D. (2005) Facile synthesis of end-functionalized regioregular poly(3-alkylthiophene)s via modified Grignard metathesis reaction. *Macromolecules*, **38**, 10346–10352.

Kanga, R.S., Hogen-Esch, T.E., Randrianalimanana, E. *et al.* (1990) Studies of the anionic polymerization of phenylvinylsulfoxide and its copolymer with styrene. *Macromolecules*, **23**, 4235–4240.

Karatzas, A., Iatrou, H., Hadjichristidis, N. *et al.* (2008) Complex macromolecular chimeras. *Biomacromolecules*, **9**, 2072–2080.

Kennedy, J.P. (1999) Living cationic polymerization of olefins. How did the discovery come about? *Journal of Polymer Science Part A-Polymer Chemistry*, **37**, 2285–2293.

Kim, K.T., Vandermeulen, G.W.M., Winnik, M.A., and Manners, I. (2005) Organometallic-polypeptide block copolymers: synthesis and properties of poly(ferrocenyldimethylsilane)-*b*-poly-(γ-benzyl-L-glutamate). *Macromolecules*, **38**, 4958–4961.

Kim, S., Kakuda, Y., Yokoyama, A., and Yokozawa, T. (2007) Synthesis of controlled rod–coil block copolymers by a macroinitiator method: chain-growth polycondensation for an aromatic polyamide from a polystyrene macroinitiator. *Journal of Polymer Science Part A-Polymer Chemistry*, **45**, 3129–3133.

Klok, H.A. and Lecommandoux, S. (2001) Supramolecular materials via block copolymer self-assembly. *Advanced Materials*, **13**, 1217–1229.

Klok, H.-A., Langenwalter, J.F., and Lecommandoux, S. (2000) Self-assembly of peptide-based diblock oligomers. *Macromolecules*, **33**, 7819–7826.

Koh, H.-D., Changez, M., Rahman, M.S., and Lee, J.-S. (2009) Formation of intermicellar-chained and cylindrical micellar networks from an amphiphilic rod–coil block copolymer: poly(n-hexyl isocyanate)-*block*-poly(2-vinylpyridine). *Langmuir*, **25**, 7188–7192.

Krausch, G. and Magerle, R. (2002) Nanostructured thin films via self-assembly of block copolymers. *Advanced Materials*, **14**, 1579–1583.

Kricheldorf, H. (1987) *α-Amino acid-N-Carboxyanhydrides and Related Materials*, Springer-Verlag, New York, pp. 1–280.

Krouse, S.A. and Schrock, R.R. (1988) Preparation of polyacetylene chains in low-polydispersity diblock and triblock copolymers. *Macromolecules*, **21**, 1885–1888.

Kuo, C.-C., Tung, Y.-C., Lin, C.-H., and Chen, W.-C. (2008) Novel luminescent electrospun fibers prepared from conjugated rod–coil block copolymer of poly[2,7-(9,9-dihexylfluorene)]-block-poly(methyl methacrylate). *Macromolecular Rapid Communications*, **29**, 1711–1715.

Kwon, Y. and Faust, R. (2004) Synthesis of polyisobutylene-based block copolymers with precisely controlled architecture by living cationic polymerization. *Advances in Polymer Science*, **167**, 107–135.

Kwon, Y.K., Kim, H.S., Kim, H.J. *et al.* (2009) Reduced excimer formation in polyfluorenes by introducing coil-like poly[penta(ethylene glycol) methyl ether methacrylate] block segments. *Macromolecules*, **42**, 887–891.

Lee, J.S., Hirao, A., and Nakahama, S. (1988) Polymerization of monomers containing functional silyl groups. 5. Synthesis of new porous membranes with functional groups. *Macromolecules*, **21**, 274–276.

Lee, J.S., Hirao, A., and Nakahama, S. (1989) Polymerization of monomers containing functional silyl groups. 7. Porous membranes with controlled microstructures. *Macromolecules*, **22**, 2602–2606.

Lee, M., Cho, B.-K., and Zin, W.-C. (2001) Supramolecular structures from rod–coil block copolymers. *Chemical Reviews*, **101**, 3869–3892, and references therein.

Lee, J.U., Cirpan, A., Emrick, T. *et al.* (2009) Synthesis and photophysical property of well-defined donor-acceptor diblock copolymer based on regioregular poly(3-hexylthiophene) and fullerene. *Journal of Materials Chemistry*, **19**, 1483–1489.

Lim, Y., Moon, K.-S., and Lee, M. (2008) Rod-coil block molecules: their aqueous self-assembly and biomaterials applications. *Journal of Materials Chemistry*, **18**, 2909–2918.

Lin, C.-H., Tung, Y.-C., Ruokolainen, J. *et al.* (2008) Poly[2,7-(9,9-dihexylfluorene)]-*block*-poly(2-vinylpyridine) rod–coil and coil-rod–coil block copolymers: Synthesis, morphology and photophysical properties in methanol/THF mixed solvents. *Macromolecules*, **41**, 8759–8769.

Lin, S.-T., Fuchise, K., Chen, Y. *et al.* (2009) Synthesis, thermomorphic characteristics, and fluorescent properties of poly[2,7-(9,9-dihexylfluorene)]-block-poly(N-isopropylacrylamide)-block-poly(N-hydroxyethylacrylamide) rod–coil-coil triblock copolymers. *Soft Matter*, **5**, 3761–3770.

Liu, J., Sheina, E., Kowalewski, T., and McCullough, R.D. (2002) Tuning the electrical conductivity and self-assembly of regioregular polythiophene by block copolymerization: nanowire morphologies in new di- and triblock copolymers. *Angewandte Chemie-International Edition*, **41**, 329–332.

Lu, H. and Cheng, J. (2007) Hexamethyldisilazane-mediated controlled polymerization of α-amino acid *N*-carboxyanhydrides. *Journal of the American Chemical Society*, **129**, 14114–14115.

Lu, S., Liu, T., Ke, L. *et al.* (2005) Polyfluorene-based light-emitting rod–coil block copolymers. *Macromolecules*, **38**, 8494–8502.

Mason, A.F., and Coates, G.W. (2007) Coordination polymerization: synthesis of new homo- and copolymer architectures from ethylene and propylene using homogeneous Ziegler–Natta polymerization catalysts. *Macromolecular Engineering*, **1**, 217–249.

Matyjaszewski, K. and Xia, J. (2001) Atom transfer radical polymerization. *Chemical Reviews*, **101**, 2921–2990.

Matyjaszewski, K., Miller, P.J., Pyun, J. *et al.* (1999) Synthesis and characterization of star polymers with varying arm number, length, and composition from organic and hybrid inorganic/organic multifunctional initiators. *Macromolecules*, **32**, 6526–6535.

Miyakoshi, R., Yokoyama, A., and Yokozawa, T. (2005) Catalyst-transfer polycondensation. Mechanism of Ni-catalyzed chain-growth polymerization leading to well-defined poly(3-hexylthiophene). *Journal of the American Chemical Society*, **127**, 17542–17547.

Miyakoshi, R., Yokoyama, A., and Yokozawa, T. (2008a) Development of catalyst-transfer condensation polymerization. Synthesis of π-conjugated polymers with controlled molecular weight and low polydispersity. *Journal of Polymer Science Part A-Polymer Chemistry*, **46**, 753–765.

Miyakoshi, R., Yokoyama, A., and Yokozawa, T. (2008b) Importance of the order of successive catalyst-transfer condensation polymerization in the synthesis of block Copolymers of polythiophene and poly(*p*-phenylene). *Chemistry Letters*, **37**, 1022–1023.

Mori, H. and Mueller, A.H.E. (2003) New polymeric architectures with (meth)acrylic acid segments. *Progress in Polymer Science*, **28**, 1403–1439.

Nakajima, A., Hayashi, T., Kugo, K., and Shinoda, K. (1979) Synthesis and structural study of the A-B-A type block copolymer consisting of poly(γ-benzyl$_L$-glutamate) as the A component and polybutadiene as the B component. *Macromolecules*, **12**, 840–843.

Ohshimizu, K. and Ueda, M. (2008) Well-controlled synthesis of block copolythiophenes. *Macromolecules*, **41**, 5289–5294.

Ohshimizu, K., Shibasaki, Y., Komura, M. *et al.* (2007) Synthesis and characterization of aromatic block copolyamides by condensative chain polymerization. *Chemistry Letters*, **36**, 742–743.

Olsen, B.D. and Segalman, R.A. (2005) Structure and thermodynamics of weakly segregated rod–coil block copolymers. *Macromolecules*, **38**, 10127–10137.

Olsen, B.D., Shah, M., Ganesan, V., and Segalman, R.A. (2008) Universalization of the phase diagram for a model rod–coil diblock copolymer. *Macromolecules*, **41**, 6809–6817.

Olson, D.A., Chen, L., and Hillmyer, M.A. (2008a) Templating nanoporous polymers with ordered block copolymers. *Chemistry of Materials*, **20**, 869–890.

Olson, D.A., Chen, L., and Hillmyer, M.A. (2008b) Templating nanoporous polymers with ordered block copolymers. *Chemistry of Materials*, **20**, 869–890.

Ouhib, F., Khoukh, A., Ledeuil, J.-B. *et al.* (2008a) Diblock and random donor/acceptor "double cable" polythiophene copolymers via the GRIM method. *Macromolecules*, **41**, 9736–9743.

Ouhib, F., Hiorns, R.C., de Bettignies, R. *et al.* (2008b) Photovoltaic cells based on polythiophenes carrying lateral phenyl groups. *Thin Solid Films*, **516**, 7199–7204.

Park, C., Yoonand, J., and Thomas, E.L. (2003) Enabling nanotechnology with self assembled block copolymer patterns. *Polymer*, **44**, 6725–6760.

Park, J.Y., Koenen, N., Forster, M. *et al.* (2008) Interplay of vesicle and lamellae formation in an amphiphilic polyfluorene-*b*-polythiophene all-conjugated diblock copolymer at the air–water interface. *Macromolecules*, **41**, 6169–6175.

Patten, T.E. and Novak, B.M. (1991) "Living" titanium(IV) catalyzed coordination polymerizations of isocyanates. *Journal of the American Chemical Society*, **113**, 5065–5066.

Patten, T.E. and Novak, B.M. (1996) Living organotitanium(IV)-catalyzed polymerizations of isocyanates. *Journal of the American Chemical Society*, **118**, 1906–1916.

Percec, V. and Tomazos, D. (1992) Molecular engineering of side-chain liquid-crystalline polymers by living cationic polymerization. *Advanced Materials*, **4**, 548–561.

Perepichka, I.F. and Perepichka, D.F. (eds) (2009) *Handbook of Thiophene-Based Materials*, John Wiley-VCH and Sons Ltd, Chichester, UK.

Qu, G., Jiang, F., Zhang, S., and Usuda, S. (2007) A novel poly(*p*-phenylene vinylene)-*b*-poly(methyl methacrylate) rod–coil diblock copolymer. *Materials Letters*, **61**, 3421–3424.

Radano, C.P., Scherman, O.A., Stingelin-Stutzmann, N. *et al.* (2005) Crystalline-crystalline block copolymers of regioregular poly(3-hexylthiophene) and polyethylene by ring-opening metathesis polymerization. *Journal of the American Chemical Society*, **127**, 12502–12503.

Rahman, M.S., Samal, S., and Lee, J.-S. (2006) Synthesis and self-assembly studies of amphiphilic poly(n-hexyl isocyanate)-*block*-poly(2-vinylpyridine)-*block*-poly(n-hexyl isocyanate) rod–coil-rod triblock copolymer. *Macromolecules*, **39**, 5009–5014.

Rahman, M.S., Samal, S., and Lee, J.-S. (2007) Quantitative in situ coupling of living diblock copolymers for the preparation of amphiphilic coil-rod–coil triblock copolymer poly(2-vinylpyridine)-*b*-poly(n-hexyl isocyanate)-*b*-poly(2-vinylpyridine). *Macromolecules*, **40**, 9279–9283.

Rao, J., Luo, Z., Ge, Z. *et al.* (2007) "Schizophrenic" micellization associated with coil-to-helix transitions based on polypeptide hybrid double hydrophilic rod–coil diblock copolymer. *Biomacromolecules*, **8**, 3871–3878.

Richard, F., Brochon, C., Leclerc, N. *et al.* (2008) Design of a linear poly(3-hexylthiophene)/fullerene-based donor-acceptor rod–coil block copolymer. *Macromolecular Rapid Communications*, **29**, 885–891.

Roncali, J. (1992) Conjugated poly(thiophenes): synthesis, functionalization, and applications. *Chemical Reviews*, **92**, 711–738.

Rubatat, L., Kong, X., Jenekhe, S.A. *et al.* (2008) Self-assembly of polypeptide/π-conjugated polymer/polypeptide triblock copolymers in rod-rod-rod and coil-rod–coil conformations. *Macromolecules*, **41**, 1846–1852.

Sary, N., Rubatat, L., Brochon, C. *et al.* (2007) Self-assembly of poly(diethylhexyloxy-*p*-phenylenevinylene)-*b*-poly (4-vinylpyridine) rod–coil block copolymer systems. *Macromolecules*, **40**, 6990–6997.

Sawamoto, M. (1993) The synthesis of well-defined polymers by living cationic polymerization of vinyl monomers. *Trends in Polymer Science*, **1**, 111–115.

Sawamoto, M. and Higashimura, T. (1990) Living polymer systems, living cationic polymerization, alkenyl. *Encyclopedia of Polymer Science and Engineering*, (Suppl), **399–412**.

Scherf, U., Gutacker, A., and Koenen, N. (2008) All-conjugated block copolymers. *Accounts of Chemical Research*, **41**, 1086–1097.

Scherman, O.A., Rutenberg, I.M., and Grubbs, R.H. (2003) Direct synthesis of soluble, end-functionalized polyenes and polyacetylene block copolymers. *Journal of the American Chemical Society*, **125**, 8515–8522.

Schmitt, C., Nothofer, H.-G., Falcou, A., and Scherf, U. (2001) Conjugated polyfluorene/polyaniline block copolymers. *Macromolecular Rapid Communications*, **22**, 624–628.

Shibasaki, Y. and Ueda, M. (2003) Precise control of molecular weight and its distribution in polycondensation. *Kino Zairyo*, **23**, 22–29.

Shibasaki, Y., Araki, T., Okazaki, M., and Ueda, M. (2002) Control of molecular-weight distribution in polycondensation polymers. Polyamide synthesis. *Polymer Journal*, **34**, 261–266.

Shin, Y.-D., Kim, S.-Y., Ahn, J.-H., and Lee, J.-S. (2001) Synthesis of poly(n-hexyl isocyanate) by controlled anionic polymerization in the presence of $NaBPh_4$. *Macromolecules*, **34**, 2408–2410.

Shirakawa, H., Louis, E.J., MacDiarmid, A.G. *et al.* (1977) Synthesis of electrically conducting organic polymers: halogen derivatives of polyacetylene, (CH)x. *Journal of the Chemical Society. Chemical Communications*, **16**, 578–580.

Simionescu, C.I. and Percec, V. (1982) Progress in polyacetylene chemistry. *Progress in Polymer Science*, **8**, 133–214.

Sivanandan, K., Chatterjee, T., Treat, N. *et al.* (2009) High surface area poly(3-hexylthiophenes) thin films from cleavable graft copolymers. *Macromolecules*, **43**, 233–241.

Sommer, M., Lang, A.S., and Thelakkat, M. (2008) Crystalline-crystalline donor-acceptor block copolymers. *Angewandte Chemie-International Edition*, **47**, 7901–7904.

Stalmach, U., de Boer, B., Videlot, C. *et al.* (2000) Semiconducting diblock copolymers synthesized by means of controlled radical polymerization techniques. *Journal of the American Chemical Society*, **122**, 5464–5472.

Stenzel-Rosenbaum, M., Davis, T.P., Chen, V., and Fane, A.G. (2001) Star-polymer synthesis via radical reversible addition-fragmentation chain-transfer polymerization. *Journal of Polymer Science Part A-Polymer Chemistry*, **39**, 2777–2783.

Sugi, R., Yokoyama, A., and Yokozawa, T. (2003a) Synthesis of well-defined telechelic aromatic polyamides by chain-growth polycondensation: Application to the synthesis of block copolymers of polyamide and poly(tetrahydrofuran). *Macromolecular Rapid Communications*, **24**, 1085–1090.

Sugi, R., Hitaka, Y., Sekino, A. *et al.* (2003b) Bidirectional propagation of chain-growth polycondensation: Its application to poly(ethylene glycol)-aromatic polyamide-poly(ethylene glycol) triblock copolymer with low polydispersity. *Journal of Polymer Science Part A-Polymer Chemistry*, **41**, 1341–1346.

Sugi, R., Yokoyama, A., Furuyama, T. *et al.* (2005) Inductive effect-assisted chain-growth polycondensation. Synthetic development from para- to meta-substituted aromatic polyamides with low polydispersities. *Journal of the American Chemical Society*, **127**, 10172–10173.

Sugiyama, K., Karasawa, Y., Higashihara, T. *et al.* (2006) Synthesis of block copolymers and asymmetric star-branched polymers comprised of polyacetylene and polystyrene segments via ionic bond formation. *Monatshefte fur Chemie*, **137**, 869–880.

Takahashi, A., Ryo, Y., Higashihara, T., Ahn, B., Ree, M., Ueda, M. (2010) Preparation of nanoporous poly(3-hexylthiophene) films based on a template system of block copolymers via ionic interaction. *Macromolecules*, **43**, 4843–4852.

Tao, Y., McCulloch, B., Kim, S., and Segalman, R.A. (2009) The relationship between morphology and performance of donor-acceptor rod–coil block copolymer solar cells. *Soft Matter*, **5**, 4219–4230.

Tian, Y., Chen, C.-Y., Yip, H.-L. *et al.* (2009) Synthesis, nanostructure, functionality, and application of polyfluorene-*block*-poly(*N*-isopropylacrylamide)s. *Macromolecules*, **43**, 282–291.

Tseng, W.H., Chen, C.K., Chiang, Y.W. *et al.* (2009) Helical nanocomposites from chiral block copolymer templates. *Journal of the American Chemical Society*, **131**, 1356–1357.

Tsolakis, P.K. and Kallitsis, J.K. (2003) Synthesis and characterization of luminescent rod–coil block copolymers by atom transfer radical polymerization: Utilization of novel end-functionalized terfluorenes as macroinitiators. *Chemistry – A European Journal*, **9**, 936–943.

Tu, G., Li, H., Forster, M. *et al.* (2007) Amphiphilic conjugated block copolymers: Synthesis and solvent-selective photoluminescence quenching. *Small*, **3**, 1001–1006.

Tung, Y.-C. and Chen, W.-C. (2009) Poly[2,7-(9,9-dihexylfluorene)]-*block*-poly[3-(trimethoxysilyl)propyl methacrylate] (PF-*b*-PTMSPMA) rod–coil block copolymers: Synthesis, morphology and photophysical properties in mixed solvents. *Reactive & Functional Polymers*, **69**, 507–518.

Tung, Y.-C., Wu, W.-C., and Chen, W.-C. (2006) Morphological transformation and photophysical properties of rod–coil poly[2,7-(9,9- dihexylfluorene)]-block-poly(acrylic acid) in solution. *Macromolecular Rapid Communications*, **27**, 1838–1844.

Urien, M., Erothu, H., Cloutet, E. *et al.* (2008) Poly(3-hexylthiophene) based block copolymers prepared by Click chemistry. *Macromolecules*, **41**, 7033–7040.

Van den Bergh, K., Huybrechts, J., Verbiest, T., and Koeckelberghs, G. (2008) Transfer of supramolecular chirality in block copoly(thiophene)s. *Chemistry – A European Journal*, **14**, 9122–9125.

Wang, H., Jost, R., and Wudl, F. (2007) Facile synthesis of conjugated rod–coil block copolymers. *Journal of Polymer Science Part A-Polymer Chemistry*, **45**, 800–808.

Webster, O.W. (2000) The discovery and commercialization of group-transfer polymerization. *Journal of Polymer Science Part A-Polymer Chemistry*, **38**, 2855–2860.

Webster, O.W. (2004) Group transfer polymerization: Mechanism and comparison with other methods for controlled polymerization of acrylic monomers. *Advances in Polymer Science*, **167**, 1–34.

Widawski, G., Rawiso, M., and Francois, B. (1994) Self-organized honeycomb morphology of star-polymer polystyrene films. *Nature*, **369**, 387–389.

Clough, S.B. (1975) *Characterization of Materials in Research Ceramics and Polymers*, Burke, J.J., Weiss, V. (Eds.), Syracuse University Press, Syracuse, New York, pp. 417–436.

Wu, W.-C., Tian, Y., Chen, C.-Y. *et al.* (2007) Theoretical and experimental studies on the surface structures of conjugated rod–coil block copolymer brushes. *Langmuir*, **23**, 2805–2814.

Wu, P.-T., Ren, G., Li, C. *et al.* (2009) Crystalline diblock conjugated copolymers: Synthesis, self-assembly, and microphase separation of poly(3-butylthiophene)-*b*-poly(3-octylthiophene). *Macromolecules*, **42**, 2317–2320.

Yang, C., Lee, J.K., Heeger, A.J., and Wudl, F. (2009) Well-defined donor-acceptor rod–coil diblock copolymers based on P3HT containing C60: the morphology and role as a surfactant in bulk-heterojunction solar cells. *Journal of Materials Chemistry*, **19**, 5416–5423.

Yokoyama, A. and Yokozawa, T. (2007) Converting step-growth to chain-growth condensation polymerization. *Macromolecules*, **40**, 4093–4101.

Yokoyama, A., Kato, A., Miyakoshi, R., and Yokozawa, T. (2008) Precision synthesis of poly(*N*-hexylpyrrole) and its diblock copolymer with poly(*p*-phenylene) via catalyst-transfer polycondensation. *Macromolecules*, **41**, 7271–7273.

Yokozawa, T. and Yokoyama, A. (2004) Chain-growth polycondensation: Living polymerization nature in polycondensation and approach to condensation polymer architecture. *Polymer Journal*, **36**, 65–83.

Yokozawa, T. and Yokoyama, A. (2007) Chain-growth polycondensation: The living polymerization process in polycondensation. *Progress in Polymer Science*, **32**, 147–172.

Yokozawa, T. and Yokoyama, A. (2009) Chain-growth condensation polymerization for the synthesis of well-defined condensation polymers and π-conjugated polymers. *Chemical Reviews*, **109**, 5595–5619.

Yokozawa, T., Asai, T., Sugi, R. *et al.* (2000) Chain-growth polycondensation for nonbiological polyamides of defined architecture. *Journal of the American Chemical Society*, **122**, 8313–8314.

Yokozawa, T., Ogawa, M., Sekino, A. *et al.* (2002) Chain-growth polycondensation for well-defined aramide. Synthesis of unprecedented block copolymer containing aramide with low polydispersity. *Journal of the American Chemical Society*, **124**, 15158–15159.

Yokozawa, T., Muroya, D., Sugi, R., and Yokoyama, A. (2005) Convenient method of chain-growth polycondensation for well-defined aromatic polyamides. *Macromolecular Rapid Communications*, **26**, 979–981.
Yokozawa, T., Adachi, I., Miyakoshi, R., and Yokoyama, A. (2007) Catalyst-transfer condensation polymerization for the synthesis of well-defined polythiophene with hydrophilic side chain and of diblock copolythiophene with hydrophilic and hydrophobic side chains. *High Performance Polymers*, **19**, 684–699.
Yokozawa, T., Ajioka, N., and Yokoyama, A. (2008) Reaction control in condensation polymerization. *Advances in Polymer Science*, **217**, 1–77.
Zhang, Y., Tajima, K., Hirota, K., and Hashimoto, K. (2008) Synthesis of all-conjugated diblock copolymers by quasi-living polymerization and observation of their microphase separation. *Journal of the American Chemical Society*, **130**, 7812–7813.
Zhang, Q., Cirpan, A., Russell, T.P., and Emrick, T. (2009) Donor-acceptor poly(thiophene-*block*-perylene diimide) copolymers: Synthesis and solar cell fabrication. *Macromolecules*, **42**, 1079–1082.
Zhao, Y., Higashihara, T., Sugiyama, K., and Hirao, A. (2005) Synthesis of functionalized asymmetric star polymers containing conductive polyacetylene segments by living anionic polymerization. *Journal of the American Chemical Society*, **127**, 14158–14159.
Zhao, Y., Higashihara, T., Sugiyama, K., and Hirao, A. (2007) Synthesis of asymmetric star-branched polymers having two polyacetylene arms by means of living anionic polymerization using 1,1-diphenylethylene derivatives. *Macromolecules*, **40**, 228–238.
Zhong, X.F. and Francois, B. (1988) Synthesis of soluble polystyrene (PS)-poly(p-phenylene) (PPP) block copolymers: a new way towards pure PPP films. *Die Makromolekulare Chemie, Rapid Communications*, **9**, 411–416.
Zou, J., Khondaker, S.I., Huo, Q., and Zhai, L. (2009) A general strategy to disperse and functionalize carbon nanotubes using conjugated block copolymers. *Advanced Functional Materials*, **19**, 479–483.

14

Synthesis of Well-Defined Poly(meth)acrylamides with Varied Stereoregularity by Living Anionic Polymerization

Takashi Ishizone
Department of Organic and Polymeric Materials, Tokyo Institute of Technology, 2-12-1-S1-13, Ookayama, Meguro-ku, Tokyo, 152-8552, Japan

14.1 Introduction

Alkyl acrylates and methacrylates, α,β-unsaturated carbonyl compounds, possess ester groups showing an electron-withdrawing character and exhibit high affinity toward nucleophilic reagents, including various anionic initiators (Davis *et al.*, 1994; Hatada and Kitayama, 2000; Webster, 2000; Baskaran, 2003). Although these polar (meth)acrylate monomers show high anionic polymerizability, anionic polymerization has often encountered serious problems with yield and molecular characteristics, such as molecular weight and the distribution due to inherent ester carbonyl attack and/or α-proton abstraction of the anionic initiators and the propagating chain ends. It is also considered that exothermic instantaneous propagation and partially aggregated propagating species cause a broadening of the molecular-weight distribution (MWD) of the polymer for kinetic reasons. The intramolecular self-condensation of the propagating enolates and the carbonyl group at the antepenultimate monomer unit, the so-called back-biting termination, is, in particular, the most serious side reaction observed in the final stage of polymerization. It has been already demonstrated that suitable selection of initiator, counterion, solvent, and polymerization temperature is essential for quantitative and controlled polymerization of alkyl (meth)acrylates.

Complex Macromolecular Architectures: Synthesis, Characterization, and Self-Assembly, First Edition.
Edited by Nikos Hadjichristidis, Akira Hirao, Yasuyuki Tezuka and Filip Du Prez.

A number of new living anionic polymerization systems for (meth)acrylates have been intensively developed in order to overcome the above-mentioned drawbacks (Anderson *et al.*, 1981; Hatada *et al.*, 1986; Fayt *et al.*, 1987; Kitayama *et al.*, 1988; Reetz, 1988; Varshney *et al.*, 1990; Ballard *et al.*, 1992; Kunkel *et al.*, 1992; Ozaki, Hirao, and Nakahama, 1995; Zagala and Hogen-Esch, 1996; Baskaran and Sivaram, 1997; Ishizone *et al.*, 1998a). In these polymerization systems, control of the primary structures of the resulting polymers, such as molecular weight, MWD and, in some cases, the stereoregularity of the repeating units, is realized. One of the most important examples is the combination of sterically hindered π-conjugated organolithium initiators, such as 1,1-diphenylhexyllithium (DPHLi, Chart 14.1), and LiCl in a polar solvent of THF, which produces well-defined poly(*tert*-butyl acrylate) and polymethacrylates having extremely narrow MWD (Fayt *et al.*, 1987; Varshney *et al.*, 1990). Also, we have found that the simple addition of diethylzinc (Et_2Zn) to a bulky organopotassium initiator, such as diphenylmethylpotassium (Ph_2CHK) in THF, induces living anionic polymerization of alkyl (meth)acrylates to produce a stabilized propagating species (Ozaki *et al.*, 1995; Ishizone *et al.*, 1998a). It is believed that the added Et_2Zn acts as a weak Lewis acid to dissociate the aggregated propagating enolate anions and to form a single nonaggregated active species after shifting the equilibrium state. The undesirable back-biting side reaction is also eliminated to form stabilized less-nucleophilic active chain ends by the association. A lowering of the propagation rate is apparently observed after the addition of Et_2Zn, even in a polar solvent of THF showing a Lewis-basic character. A similar additive effect of Lewis-acidic compounds, such as triethylborane (Et_3B) and triphenylborane (Ph_3B), is also found in the polymerization of *tert*-butyl acrylate (Ishizone *et al.*, 1999) and methacrylonitrile (Ishizone *et al.*, 2001). In these cases, a more drastic retardation of the polymerization is observed compared with Et_2Zn, and no apparent polymerization virtually occurs at −78 °C in THF. By contrast, polymerization certainly proceeds with Ph_2CHK/R_3B to

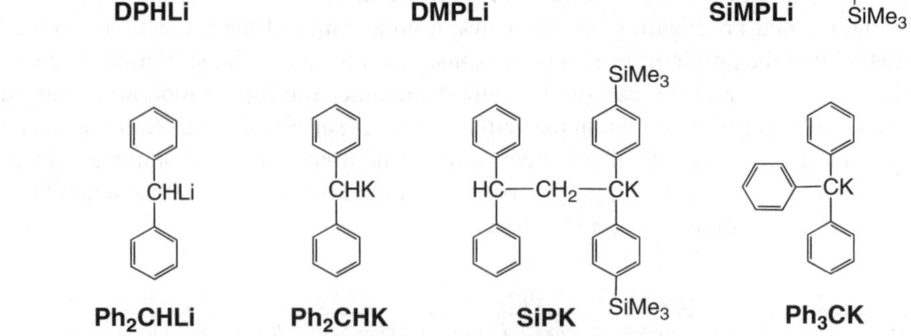

Chart 14.1 Anionic initiators for *N,N*-dialkyl(meth)acrylamides.

afford well-defined polymers at elevated temperatures of 0 or 30 °C. This contrastive polymerization behavior clearly indicates that a dormant species between the propagating carbanions and the Et_3B- or Ph_3B-added forms in the polymerization systems of *tert*-butyl acrylate and methacrylonitrile upon the addition of Lewis-acidic boranes at −78 °C.

Interestingly, the two binary initiator systems of organolithium/LiCl and organopotassium/Et_2Zn are widely applicable to living anionic polymerization of a variety of methacrylate derivatives carrying functional groups (Hild *et al.*, 1993; Mori *et al.*, 1994a; Mori *et al.*, 1994b; Hoogeveen *et al.*, 1996; Ishizone *et al.*, 1998b; Zhang and Ruckenstein, 1998a; Zhang and Ruckenstein, 1998b; Ishizone *et al.*, 2002; Sugiyama *et al.*, 2003; Ishizone *et al.*, 2003; Han *et al.*, 2003; Ishizone *et al.*, 2008; Gallei *et al.*, 2009). In several cases, serious side reactions toward reactive functional groups are effectively suppressed in the presence of the additives. However, no apparent additive effect on the stereoregularity of the resulting poly(meth)acrylates has been observed so far, whereas the association of the added LiCl or Et_2Zn toward the propagating species is strongly suggested by the retardation of the polymerization and the stabilization of the propagating enolates.

Similar to the ester monomers, including acrylates and methacrylates, it is well known that *N,N*-dialkylacrylamides, α,β-unsaturated amides, readily undergo anionic polymerization with high efficiency, since the electron-withdrawing amide moiety markedly reduces the electron density of the C=C double bond, resulting in high anionic polymerizability. The purpose of this review is to discuss the anionic polymerization behavior of *N,N*-dialkylacrylamides and their methacryloyl counterparts, *N,N*-dialkylmethacrylamides. In particular, stereospecific living polymerization of several amide monomers is achieved by the new initiator systems, such as organolithium/LiCl, organolithium/Et_2Zn, and organopotassium/Et_2Zn.

14.2 Anionic Polymerization of *N,N*-Dialkylacrylamides

It is reported that the anionic polymerizability of vinyl monomers can be roughly predicted by the ^{13}C NMR chemical shifts of the vinyl β-carbons of monomers, since the β-carbon chemical shifts usually reflect the electron densities of polymerizable C=C bonds (Ishizone *et al.*, 1993). The β-carbon chemical shifts of *N,N*-dimethyl- (DMA) and *N,N*-diethylacrylamide (DEA) are observed at 127.1 and 127.2 ppm, which are located between acrylates (129–131 ppm) and methacrylates (124–125 ppm), indicating the lowered π-electron density due to the electron-withdrawing amide substituents. In fact, the observed high anionic polymerizability of various *N,N*-dialkylacrylamides is in good accordance with the β-carbon chemical shifts observed in the downfield region.

In 1960, anionic polymerization of *N,N*-dialkylacrylamides such as DMA, *N,N*-dipropylacrylamide, *N,N*-dibutylacrylamide, *N,N*-diphenylacrylamide, and *N*-acryloylmorpholine (AMO) was attempted with alkyllithium in toluene (Butler *et al.*, 1960), but the primary structures of the polymers were not well characterized in early studies. The high crystallinity of the resulting poly(DMA) was suggested from the results of sharp X-ray scattering patterns and a high softening temperature around 300 °C. Two decades later, McGrath reported from the results of ^1H and ^{13}C NMR measurements that highly isotactic poly(DMA) was produced from the polymerization system of DMA with *sec*-BuLi in benzene or THF (Gia and McGrath, 1980). Although anionic polymerization of *N,N*-dialkylacrylamides has been performed since the 1960s, living polymerization was rather recently achieved. In 1996, Xie and Hogen-Esch reported that bulky and stabilized organocesium initiators, such as triphenylmethylcesium and (1,1,4,4-tetraphenylbutyl)dicesium, and the radical anion, cesium naphthalenide, quantitatively gave poly(DMA) of predicted molecular weight and narrow MWD in THF at −78 °C (Xie and Hogen-Esch, 1996). The resulting poly(DMA) showed water solubility, indicating the high polarity of the amide function. This polymerization system was also effective for controlled polymerization of *N*-acryloyl-*N*′-methylpiperazine (AMP). The resulting poly (AMP) was a novel well-defined functional polymer, since it possessed a reactive tertiary amine moiety in the side chain as well as tailored chain structures. Anionic polymerization of *N,N*-diphenylacrylamide with *n*-BuLi in toluene afforded a polymer rich in *meso* diad ($m = 0.95$) and with broad MWD (Shiohara *et al.*, 1996).

N,N-Dialkylacrylamides also undergo group-transfer polymerization to provide polymers in good to quantitative yields. Sogah and Webster have reported that the polymerization of DMA can be initiated with dimethylketene methyl trimethylsilyl acetal in THF (Sogah *et al.*, 1987). Poly(DMA) was quantitatively obtained, while the number-average molecular weight was apparently lower than the theoretical one and the MWD was fairly broad ($M_w/M_n = 1.62$). On the other hand, group-transfer polymerization of DEA was also performed with dimethylketene methyl trimethylsilyl acetal in THF at 0 °C (Eggert and Freitag, 1994; Freitag *et al.*, 1994; Baltes *et al.*, 1999). Polymerization proceeded to afford tailored poly(DEA) in 50–70% yields. The polydispersity indexes, M_w/M_n, of the resulting poly(DEA)s were around 1.2, indicating relatively narrow MWD. However, in these reports on group-transfer polymerization, the living character of the polymerization was not demonstrated clearly.

We have attempted to polymerize a series of *N,N*-dialkylacrylamides, such as DMA, DEA, *N*-ethyl-*N*-methylacrylamide (EMA), *N*-acryloyl-2-methylaziridine (**A3**), *N*-acryloyl-azetidine (**A4**), *N*-acryloylpyrrolidine (**A5**), *N*-acryloylpiperidine (**A6**), and AMO, under anionic conditions (Nakahama *et al.*, 1997; Kobayashi *et al.*, 1999; Suzuki *et al.*, 2008; Suzuki *et al.*, 2010). The alkyl substituents on the amide nitrogen of the monomers were purposefully varied to investigate the effect of alkyl substituents on the polymerizability of monomers and the properties of polymers (Chart 14.2). Polymerization of monomers was initiated with bulky and π-stabilized organolithium initiators such as diphenylmethyllithium (Ph_2CHLi), 1,1-diphenyl-3-methylpentyllithium (DMPLi), and 1,1-bis[(4′-trimethylsilyl)phenyl]-3-methylpentyllithium (SiMPLi) in THF at −78 °C, as shown in Table 14.1. All the monomers were found to give the polymers in quantitative yields, indicating the high anionic polymerizability

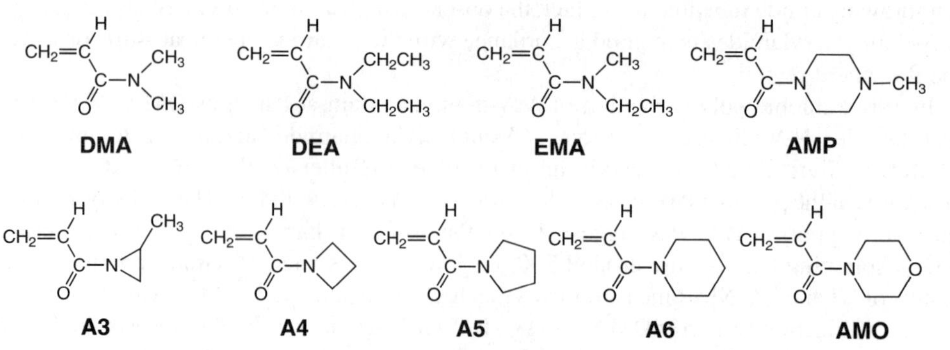

Chart 14.2 *N,N*-Dialkylacrylamides.

Table 14.1 Anionic polymerization of *N,N*-dialkylacrylamides with organolithium initiators in THF at −78 °C[a]

Initiator	Monomer	LiCl/Li$^+$	Et_2Zn/Li^+	Time h	$M_n \times 10^{-3}$		M_w/M_n[d]
					calcd.[b]	obsd.[c]	
SiMPLi	DMA			1	9.4	11	2.58
SiMPLi	DEA			1	9.9	11	2.83
SiMPLi	EMA			0.5	11	16	4.75
DMPLi	**A3**			5 min	14	21	2.30
Ph_2CHLi	**A4**			2	6.3	6.8	1.45
SiMPLi	**A5**			0.5	8.2	14	2.26
SiMPLi	**A6**			0.5	9.3	11	2.42
SiMPLi	AMO			1	14	13	2.73
SiMPLi/LiCl	DMA	4.4		1	11	13	6.17
SiMPLi/LiCl	DEA	2.8		1	10	11	3.00
SiMPLi/LiCl	EMA	6.2		0.5	11	16	8.18
DMPLi/LiCl	**A3**	6.8		2	12	15	1.12
Ph_2CHLi/LiCl	**A4**	4.8		2	6.2	10	1.65
SiMPLi/LiCl	**A6**	6.0		0.5	10	11	10.3
SiMPLi/Et_2Zn	DMA		12	1	9.2	9.3	1.14
SiMPLi/Et_2Zn	DEA		18	1	9.6	11	1.12
SiMPLi/Et_2Zn	EMA		12	15 min	8.3	8.2	1.35
DMPLi/Et_2Zn	**A3**		11	1	7.0	9.3	2.03
Ph_2CHLi/Et_2Zn	**A4**		35	2	14	18	1.06
SiMPLi/Et_2Zn	**A5**		17	15 min	9.5	10	1.11
SiMPLi/Et_2Zn	**A6**		19	1	5.1	6.7	1.08
SiMPLi/Et_2Zn	AMO		14	1	10	10	1.23

[a] Yield~100%.
[b] M_n(calcd) = (MW of monomer) × [monomer]/[initiator] + (MW of initiator).
[c] M_n(obsd) was determined by the ^{1}H NMR.
[d] M_w/M_n was determined by the SEC calibration using PMMA standards.

of *N,N*-dialkylacrylamides, as described above. Polymerization of *N,N*-dialkylacrylamides in the absence of additives usually proceeded in a heterogeneous state, and the resulting polymer was precipitated in THF during the course of polymerization. In most cases, the resulting polymers possessed predictable molecular weights based on the molar ratios between the monomers and initiators, but the MWD was very broad. In fact, the M_w/M_n values were between 1.5 and 4.8, when polymerization was simply performed with organolithiums. Unexpectedly, the addition of LiCl to the polymerization system of *N,N*-dialkylacrylamides caused further broadening of the MWD of the polymers in sharp contrast to the polymerization behavior of (meth)acrylates. This is probably due to the heterogeneous polymerization system. Only the **A3** monomer, possessing a small and highly strained aziridine ring, gave a polymer of narrow MWD with DMPLi in the presence of 6.8-fold LiCl. In fact, the M_w/M_n value of poly (**A3**) was markedly reduced from 2.30 to 1.12 upon the addition of LiCl. On the other hand, the drastic additive effect of Et_2Zn was ascertained in the narrowing of the MWD of poly(*N,N*-dialkylacrylamide)s. In most cases, polymerization homogeneously proceeded in the presence of 11–35-fold Et_2Zn, and the polydispersity indexes were lowered to around 1.1. As an exception, only poly(**A3**), obtained with DMPLi/Et_2Zn, possessed very broad MWD ($M_w/M_n = 2.03$).

Table 14.2 shows the polymerization results using organopotassium initiators such as Ph_2CHK and 1,1-bis[(4′-trimethylsilyl)phenyl]-3,3-diphenylpropylpotassium (SiPK) (Chart 14.1). These organopotassium initiators induced quantitative polymerization of *N,N*-dialkylacrylamides in the homogeneous state to afford polymers in THF at −78 °C in contrast to the heterogeneous system of organolithiums. The observed molecular weights agreed with the calculated values, indicating the quantitative initiation efficiency of the employed organopotassium initiators. The MWD of the polymers was rather broad ($M_w/M_n = 1.4\sim3.0$), as well as the results with organolithium initiators. In fact, polymerization of DEA with SiPK proceeded very rapidly, and all the monomers were consumed within 15 s even at −78 °C. On the other hand, the addition of Lewis-acidic Et_2Zn (11–19 equivalents) to the polymerization system significantly lowered the polymerization rate of the monomers to narrow the MWD of the polymers, as shown in Table 14.2. The yields of poly(DMA) and poly(DEA) were around 60% after 1 h of polymerization at −78 °C, suggesting that the retardation of polymerization certainly occurred in the presence of Et_2Zn. By contrast, the rapid propagation of DMA and DEA was possible at the elevated temperature of 0 °C, and the complete consumption of the

Table 14.2 Anionic polymerization of *N,N*-dialkylacrylamides with organopotassium initiators in THF

Initiator	Monomer	Et_2Zn/K^+	Temp. °C	Time h	Yield %	$M_n \times 10^{-3}$		M_w/M_n[c]
						calcd.[a]	obsd.[b]	
SiPK	DMA		−78	1	100	11	14	2.85
SiPK	DEA		−78	1	100	10	15	2.37
SiPK	DEA		−78	15 sec	97	21	23	1.63
SiPK	EMA		−78	0.5	100	9.6	8.1	2.18
Ph_2CHK	**A3**		−78	1	100	6.7	8.2	1.40
Ph_2CHK	**A4**		−78	2	100	5.5	7.2	1.79
SiPK	**A5**		−78	15 min	100	7.9	8.0	3.03
SiPK	**A6**		−78	0.5	100	9.9	11	1.60
SiPK	AMO		−78	1	100	9.4	9.0	2.73
SiPK/Et_2Zn	DMA	15	−78	1	60	8.5	5.6	1.07
SiPK/Et_2Zn	DMA	18	0	0.5	100	11	9.8	1.07
SiPK/Et_2Zn	DEA	15	−78	1	57	10	7.8	1.12
SiPK/Et_2Zn	DEA	18	−78	6	93	6.6	6.8	1.16
SiPK/Et_2Zn	DEA	15	−40	1	100	10	8.6	1.09
SiPK/Et_2Zn	DEA	17	0	20 min	100	41	45	1.05
SiPK/Et_2Zn	DEA	16	40	1	100	9.2	8.7	1.15
SiPK/Et_2Zn	EMA	17	0	0.5	100	10	8.0	1.12
Ph_2CHK/Et_2Zn	**A3**	11	−78	1	85	6.0	6.9	1.18
Ph_2CHK/Et_2Zn	**A4**	14	−78	2	100	5.4	7.0	1.08
SiPK/Et_2Zn	**A5**	16	0	15 min	100	9.6	8.2	1.10
SiPK/Et_2Zn	**A6**	19	0	0.5	100	10	10	1.07
SiPK/Et_2Zn	AMO	16	0	1	100	10	10	1.09

[a] M_n(calcd) = (MW of monomer) × yield/100 × [monomer]/[initiator] + (MW of initiator).
[b] M_n(obsd) was determined by the ^{1}H NMR.
[c] M_w/M_n was determined by the SEC calibration using PMMA standards.

monomers was always attained within 30 min. Very interestingly, the polymers derived from DMA, DEA, EMA, **A5**, **A6**, and AMO formed at 0 °C also possessed narrow MWD ($M_w/M_n < 1.12$), as well as the predicted molecular weights. Thus, precise control of the molecular weight and the distribution of poly(*N,N*-dialkylacrylamide)s is achieved by the addition of Et_2Zn to the organolithium and organopotassium initiators in THF. The added Lewis-acidic Et_2Zn might play an important role in providing a single propagating species through a plausible coordination with enolate anions. The Et_2Zn-coordinated propagating species cause the slow and uniform propagation reaction and narrow the MWD of polymers. Furthermore, we consider that the coordinated Et_2Zn also prevents side reactions such as α-proton abstraction and carbonyl attack, and stabilizes the propagating amide enolate anions, as previously reported in the living anionic polymerization of various (meth)acrylates. In fact, the persistency of the propagating carbanion derived from DEA was confirmed by the quantitative efficiency in the postpolymerization (monomer-addition experiment) initiated with Ph_2CHK/Et_2Zn. The propagating species maintained the living character for at least 1 min even at 0 °C, strongly indicating the tolerance of amide carbonyl groups.

A similar but more drastic additive effect of Et_3B was ascertained in the polymerization systems of DMA and DEA (Kobayashi *et al.*, 2000a), as can be seen in Table 14.3. No apparent polymerization of monomers was observed at −78 °C after the addition of Et_3B (2–6 equivalents) to the anionic initiators associated with lithium or potassium counterions. In the case of Et_2Zn, polymerization certainly proceeded even at −78 °C to afford polymeric products, while apparent polymerization retardations were also detected. At higher temperatures of 0 or 30 °C, polymerization of DMA and DEA with $DMPLi/Et_3B$ or Ph_2CHK/Et_3B could be accelerated, similar to previous reports on *tert*-butyl acrylate (Ishizone *et al.*, 1999) and methacrylonitrile (Ishizone *et al.*, 2001). However, the rate of polymerization for DEA seemed very slow even at 0 °C, since only 56% of the monomer was consumed with Ph_2CHK/Et_3B after 24 h. A prolonged polymerization time of 96 h was required for the completion of the polymerization of DEA at 0 °C, while quantitative propagation was achieved within 6 h at 30 °C. In fact, the k_p values of the anionic polymerization of DMA at 30 °C in THF were estimated to be higher than 10^2 (with Ph_2CHK) and $8.2 \times 10^{-2}\ L\ mol^{-1}\ s^{-1}$ (with Ph_2CHK in

Table 14.3 Anionic polymerization of *N,N*-dialkylacrylamides in THF in the presence of Et_3B

Initiator	Monomer	Et_3B/M^+	Temp. °C	Time h	Yield %	$M_n \times 10^{-3}$		M_w/M_n[c]
						calcd.[a]	obsd.[b]	
DMPLi	DMA		−78	0.02	100	5.3	6.5	3.41
$DMPLi/Et_3B$	DMA	5.9	−78	1	0	—	—	—
$DMPLi/Et_3B$	DMA	2.1	0	4	100	16	16	1.13
Ph_2CHK	DEA		−78	0.02	100	12	11	1.51
Ph_2CHK/Et_3B	DEA	2.4	−78	1	0	—	—	—
Ph_2CHK/Et_3B	DEA	2.3	0	24	56	6.5	4.7	1.07
Ph_2CHK/Et_3B	DEA	2.3	0	96	100	11	8.9	1.08
Ph_2CHK/Et_3B	DEA	3.4	30	6	100	28	30	1.06

[a] M_n(calcd) = (MW of monomer) × yield/100 × [monomer]/[initiator] + (MW of initiator).
[b] M_n(obsd) was determined by the ^1H NMR.
[c] M_w/M_n was determined by the SEC calibration using PMMA standards.

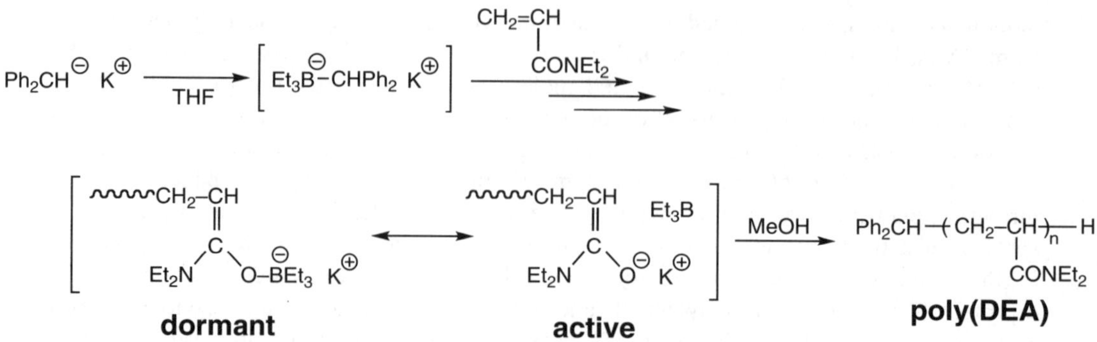

Scheme 14.1 Plausible polymerization mechanism of DEA in the presence of Et_3B.

the presence of 2.3-fold Et_3B). Interestingly, poly(DMA) and poly(DEA), produced with DMPLi/Et_3B or Ph_2CHK/Et_3B, possessed controlled molecular weights based on the molar ratios of the monomers and initiators and narrow MWD ($M_w/M_n < 1.13$). This clearly shows the effectiveness of the newly developed binary initiator systems for the synthesis of well-defined poly(*N,N*-dialkylacrylamide)s. It is demonstarted from the results of postpolymerization that the active chain end of poly(DEA) was completely stable at least at 30 °C for 1 h in THF. These polymerization results indicate that a dormant species like the borate complex forms at −78 °C in THF by the strong coordination of Lewis-acidic Et_3B and the amide enolate anion, as shown in Scheme 14.1. By increasing the temperature to 0 or 30 °C, the equilibrium shifts from the dormant species toward the active enolate form to propagate the polymerization rather slowly.

In addition to the control of the molecular weight and its distribution, the stereoregularity of the resulting poly(*N,N*-dialkylacrylamide)s is of great interest and is largely influenced by the initiator system. In particular, the stereospecific anionic polymerization of DEA has been achieved by careful tuning of the polymerization system, as shown in Figure 14.1. The addition of LiCl to an organolithium initiator such as Ph_2CHLi further increased the isotacticity of poly (DEA) compared to polymerization with Ph_2CHLi. It should be emphasized again that no additive effect of LiCl on the stereoregularity was observed in the anionic polymerization of (meth)acrylates. Isotactic poly(DEA) was also produced with *tert*-BuMgBr/Et_2Zn in THF (Kobayashi *et al.*, 2000b), although the molecular characteristics such as molecular weight and its distribution were ill-regulated. The initiator system of DPMLi/Et_2Zn induced syndiotactic-specific polymerization of DEA along with excellent molecular weight control. The poly (DEA) obtained with Ph_2CHK/Et_2Zn or Ph_2CHK/Et_3B at 0 °C was highly heterotactic (rich in *mr* triads) and possessed the narrow MWD and the predicted molecular weight, as described above. Since the coordination of Et_2Zn with an amide enolate anion is stronger than an ester enolate, the coordinated active chain ends allow the stereospecific propagation for *N,N*-dialkylacrylamides but not for (meth)acrylates.

It is well known that the tacticity of a polymer often strongly affects various properties, such as solubility, crystallinity, and glass-transition temperature. The observed stereoregularity of the resulting poly(DEA)s largely altered its solubility in water, although the poly(DEA) was recognized as a typical thermosensitive water-soluble polymer. In fact, the heterotactic poly (DEA)s produced with Ph_2CHK/Et_2Zn or Ph_2CHK/Et_3B were readily soluble in water, and

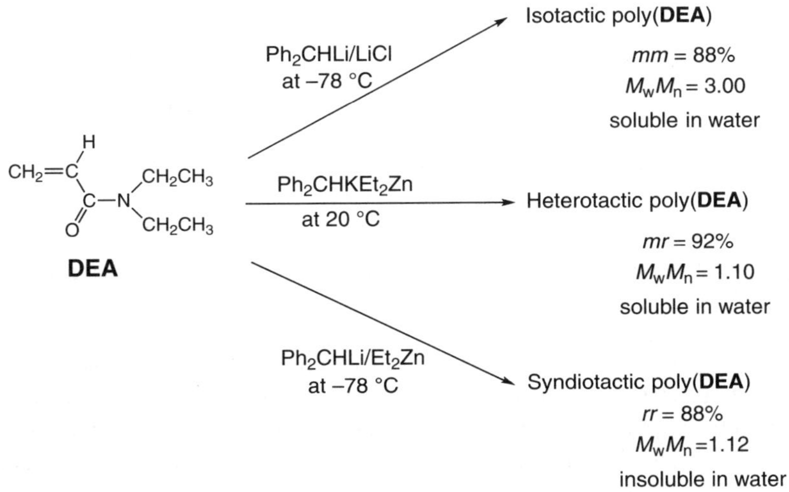

Figure 14.1 Stereospecific anionic polymerization of DEA.

their aqueous solutions exhibited reversible phase transitions at around 27–30 °C with very small hysteresis. On the other hand, isotactic poly(DEA)s showed water solubility and presented different cloud points for the heating and cooling processes at 37 and 29 °C, respectively. Similar large hysteresis of the coil–globule transition in water has been reported for the isotactic poly(DEA) obtained with *n*-BuLi in THF at −78 °C (Eggert and Freitag, 1994; Freitag *et al.*, 1994). In sharp contrast to the observed water solubility for isotactic and heterotactic polymers, the highly syndiotactic poly(DEA) obtained with DPMLi/Et_2Zn was not soluble in water. Thus, the solubility and thermosensitivity of poly(DEA) in water are significantly affected by the stereoregularity. On the other hand, all poly(DMA)s with varied stereoregularity were readily soluble in water, showing no cloud points in aqueous solutions.

14.3 Enolates of *N,N*-Dialkylamides as Novel Anionic Initiators

In organic chemistry, enolates derived from ketones, esters, amides and related carbonyl compounds are versatile synthetic intermediates and are utilized in regioselective and asymmetric syntheses, since these anionic species are readily available via the deprotonation of the above carbonyl compounds and react with a wide variety of electrophiles. Among the enolates, ester-derived enolates can be structurally regarded as the propagating terminal units in the anionic polymerization of (meth)acrylates, as shown in Figure 14.2. In fact, the various ester enolates have been widely employed as anionic initiators of (meth)acrylates. Lithium ester enolates readily initiated the polymerization of methyl methacrylate (MMA) in quantitative efficiencies without side reactions in THF (Tsvetanov *et al.*, 1985; Janata *et al.*, 1990; Janata *et al.*, 1992; Wang *et al.*, 1993; Dvořánek and Vlček, 1994). The enolate complex of samarium could induce living anionic polymerization of MMA to afford a highly syndiotactic poly(MMA) with high molecular weight and narrow MWD (Yasuda *et al.*, 1992; Yasuda *et al.*, 1993). On the other hand, the initiation of MMA with zirconium enolates gave isotactic

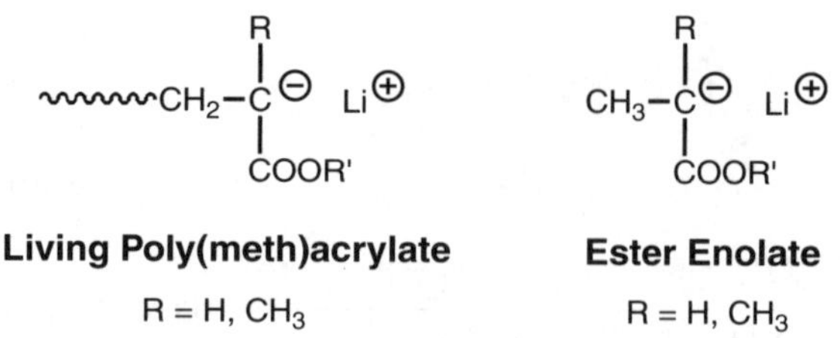

Figure 14.2 Propagating chain end of living poly(meth)acrylates and ester enolate.

poly(MMA) (Collins *et al.*, 1994). Group-transfer polymerization of (meth)acrylates was initiatied with trialkylsilyl ester enolates in the presence of a nucleophilic catalyst (Webster *et al.*, 1983). Butyl acrylate could be polymerized with tetrabutylammonium diethyl malonate (a metal-free enolate) to give a polymer having a tailored molecular weight and a narrow MWD (Reetz *et al.*, 1988). Thus, the polymerization feature is strongly affected by the associated counterions of the ester enolates.

Various amide-derived enolates have also been prepared by the reactions of *N,N*-dialkylamides with strong bases, and used as building blocks in the multistep syntheses of complex target molecules (Gassman and Fox, 1966; Woodbury and Rathke, 1977; Oare *et al.*, 1990; Krom and Streitwieser, 1996). The amide-derived enolates generally have greater stability than the enolates of esters, indicating the tolerance of the amide moiety toward nucleophilic attack compared with the labile ester function. Contrary to a number of publications concerning polymerization with ester enolates, amide-derived enolates are very limited as the anionic initiators. As only one example, the polymerization of styrene and butadiene could be initiated with a potassium enolate of *N,N*-dimethylpropionamide, but the initiation efficiency was less than 30% (Nakhmanovich *et al.*, 2001).

We prepared a series of novel α-potassio-enolates, (DEAcK, DEPrK, and DEIBK) from *N,N*-diethylacetamide (DEAc), *N,N*-diethylpropionamide (DEPr), and *N,N*-diethylisobutyramide (DEIB), respectively, as shown in Scheme 14.2 (Ishizone *et al.*, 2007). Since these new amide enolates are recognized as models for the active chain ends of poly(DEA) (Figure 14.3), the polymerization results with these enolates offer useful information on the reactivity and stability of the propagating chain ends of poly(DEA). The α-protons of DEAc could be abstracted with Ph_2CHK to form DEAcK, while a stronger base, such as potassium

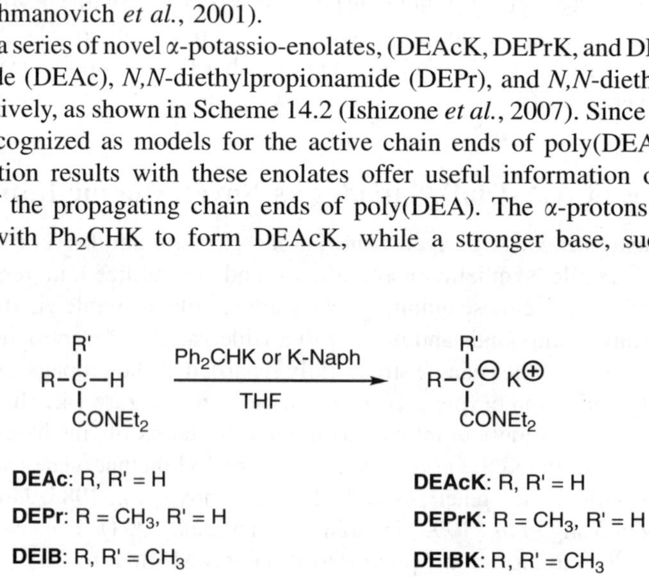

Scheme 14.2 Synthesis of amide enolates.

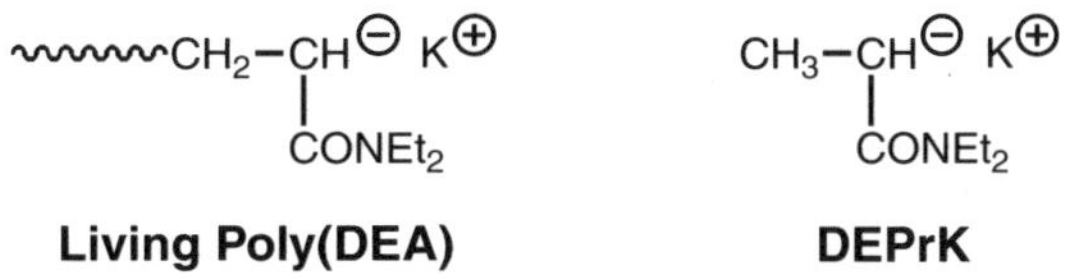

Figure 14.3 Propagating chain end of living poly(DEA) and amide enolate DEPrK.

naphthalenide (K-Naph), was required to abstract the α-protons of DEPr and DEIB. Thus, the relative acidity of the related *N,N*-diethylamides was estimated in the following order:

Acidity: DEIB < DEPr < DEAc

Under preferred conditions, the nucleophilic alkylation of DEAcK, DEPrK, and DEIBK with methyl iodide quantitatively proceeded in THF to provide the corresponding products, such as DEPr, DEIB, and *N,N*-diethylpivalamide (DEPv), respectively. For example, DEPrK and DEIBK underwent quantitative alkylations even at −78 °C within 1 min, while only 5% of DEAcK reacted with methyl iodide at −78 °C after 1 h. The complete alkylation of DEPrK rapidly occurred at −78 °C within 1 min in the presence of Et_2Zn, whereas the conversion of DEIBK was less than 13% after a 1 h reaction at −78 °C. These results indicate the relative nucleophilicity of α-potassio-amide enolates as follows:

Nucleophilicity: DEAcK < DEIBK < DEPrK

Similarly, the relative basicity of amide enolates was estimated from the results of the proton-transfer reactions:

Basicity: DEAcK < DEPrK ∼ DEIBK

Thus, DEPrK is considered as the best initiator for anionic polymerization from the viewpoints of nucleophilicity, basicity, and ease of preparation. Therefore, the anionic polymerization of a series of anionically polymerizable monomers, such as styrene, 2-vinylpyridine (2VP), and MMA, was performed with DEPrK in THF to confirm the initiation ability of the potassium amide enolate, as shown in Table 14.4. Polymerization was also investigated with Ph_2CHK under identical conditions to compare the reactivity with DEPrK. Very interestingly, polymerization of all monomers could be initiated with DEPrK to give polymers in 100% yields in THF at −78 °C for 1 h. For example, during the course of the polymerization of styrene, the reaction mixture was pale red, indicating the formation of the propagating carbanion of polystyrene. The unexpectedly high nucleophilicity of DEPrK was thus demonstrated, since the initiation of styrene showing low anionic polymerizability was possible regardless of the low efficiency (∼2%). The observed initiation efficiency of DEPrK was slightly lower than that of Ph_2CHK but rather comparable from a synthetic viewpoint. On the other hand, the initiation efficiency (nucleophilicity) of DEPrK apparently decreased in the presence of Et_2Zn. No polymerization of styrene occurred with DEPrK/Et_2Zn in THF at −78 °C. The polymerization of 2VP still proceeded quantitatively with DEPrK/Et_2Zn at −78 °C within 1 h, whereas the initiation efficiency decreased from 67% to 27% in the presence of Et_2Zn. By contrast, this novel binary initiator system afforded poly(MMA)s with predicted molecular weights and narrow MWD via the rapid and quantitative initiation reaction. Thus, the ability of the DEPrK/Et_2Zn system as an anionic initiator is almost comparable to that of

Table 14.4 Anionic polymerization of styrene, 2VP, and MMA with DEPrK or Ph_2CHK in THF at −78 °C for 1 h

Initiator	Additive[a]	Monomer	Yield %	$10^{-3}M_n$		M_w/M_n[c]	Initiator efficiency[d]
				calcd[b]	obsd[c]		%
DEPrK		styrene	100	7.7	490	1.31	2
DEPrK		2VP	100	9.4	14	1.27	67
DEPrK		MMA	100	3.5	3.3	1.67	quantitative
DEPrK	Et_2Zn	styrene	~0	—	—	—	—
DEPrK	Et_2Zn	2VP	100	9.3	34	1.48	27
DEPrK	Et_2Zn	MMA	100	3.5	3.3	1.09	quantitative
DEPrK	Et_2Zn	MMA	100	13	14	1.08	quantitative
Ph_2CHK		styrene	100	11	68	1.36	16
Ph_2CHK		2VP	100	10	9.6	1.06	quantitative
Ph_2CHK		MMA	100	10	11	1.99	quantitative
Ph_2CHK	Et_2Zn	styrene	<10	—	oligomers	—	—
Ph_2CHK	Et_2Zn	2VP	100	10	11	1.09	quantitative
Ph_2CHK	Et_2Zn	MMA	100	4.2	4.4	1.07	quantitative

[a] $[Et_2Zn]/[\text{initiator}] = 12\sim16$.
[b] $M_n(\text{calcd}) = (\text{MW of monomer}) \times \text{yield}/100 \times [\text{monomer}]/[\text{initiator}] + (\text{MW of initiator})$.
[c] M_n(obsd) and M_w/M_n were determined by the SEC calibration using standard polymer samples.
[d] $M_n(\text{calcd})/M_n(\text{obsd}) \times 100$.

the Ph_2CHK/Et_2Zn system, while a small difference was observed in the polymerization result of 2VP.

Table 14.5 shows the results of the polymerization of DEA with DEPrK in THF at 0 or −78 °C. In the absence of Et_2Zn, the polymerization of DEA with DEPrK proceeded smoothly at either 0 or −78 °C within 1 h to give polymers having broad MWD ($M_w/M_n = 1.7$–1.9) in quantitative yields. On the addition of Et_2Zn, the rate of polymerization apparently decreased at −78 °C, and a longer reaction (for 20 h) was required for polymerization completion. As expected from the additive effects of Lewis-acidic Et_2Zn, significant narrowing of MWD ($M_w/M_n < 1.2$) was achieved in the polymer samples produced with $DEPrK/Et_2Zn$, in addition to the molecular-weight control. It should be noted that the MWD was further narrowed for the polymers obtained at 0 °C. The polymerization results obtained with DEPrK and $DEPrK/Et_2Zn$ were in accordance with those of Ph_2CHK and Ph_2CHK/Et_2Zn, respectively, indicating the effectiveness of the novel amide enolate initiator. Similarly, controlled anionic polymerization of DEA could be initiated with DEAcK and DEIBK at 0 °C in the presence of Et_2Zn to form well-defined poly(DEA)s quantitatively.

The stereoregularity of the poly(DEA)s obtained with DEPrK was estimated from the ^{13}C NMR spectra of amide carbonyl carbons. Figure 14.4 shows the carbonyl carbon signals in the ^{13}C NMR spectroscopy of poly(DEA)s formed with DEPrK. The stereoregularity is greatly varied under reaction conditions including additive and polymerization temperature, as described above. In particular, the poly(DEA)s produced with DEAcK, DEPrK, and DEIBK

Table 14.5 Anionic polymerization of DEA with DEAcK, DEPrK, and DEIBK in THF

Initiator	Additive[a]	Temp. °C	Time h	Yield %	$10^{-3}M_n$		M_w/M_n[c]	Tacticity (%)[d]		
					calcd[b]	obsd[c]		*mm*	*mr*	*rr*
DEPrK		−78	1	100	8.6	12	1.92	40	54	6
DEPrK		0	1	100	5.2	6.6	1.73	22	71	7
DEPrK	Et_2Zn	−78	1	64	4.8	5.5	1.12	28	65	7
DEPrK	Et_2Zn	−78	20	100	8.4	8.2	1.23	39	45	16
DEPrK	Et_2Zn	0	1	100	9.8	11	1.10	11	83	6
DEPrK	Et_2Zn	0	1	100	32	28	1.04	6	89	5
DEAcK	Et_2Zn	0	1	100	10	7.5	1.20	18	76	6
DEIBK	Et_2Zn	0	1	100	8.0	9.5	1.15	14	85	1
Ph_2CHK		−78	1	100	10	14[e]	2.07	52	32	16
Ph_2CHK		0	1	100	9.2	9.0[e]	1.72	21	73	6
Ph_2CHK	Et_2Zn	−78	20	100	10	10[e]	1.19	33	48	19
Ph_2CHK	Et_2Zn	0	1	100	9.3	8.6[e]	1.08	21	71	8

[a] $[Et_2Zn]/[\text{initiator}] = 13{\sim}25$.
[b] $M_n(\text{calcd}) = (\text{MW of monomer}) \times \text{yield}/100 \times [\text{monomer}]/[\text{initiator}] + (\text{MW of initiator})$.
[c] M_n(obsd) and M_w/M_n were determined by the SEC calibration using poly(DEA) samples.
[d] By ^{13}C NMR.
[e] M_n(obsd) was determined by the ^{1}H NMR.

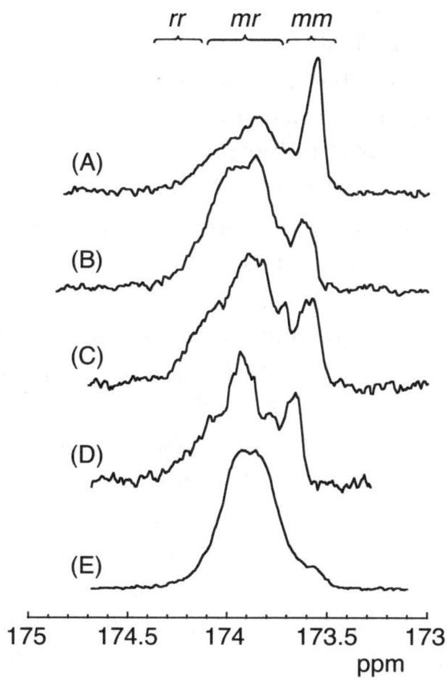

Figure 14.4 ^{13}C NMR spectra of carbonyl region of poly(DEA)s obtained with DEPrK (A, at −78 °C for 1 h; B, at 0 °C for 1 h), and with DEPrK/Et_2Zn (C, at −78 °C for 1 h; D, at −78 °C for 20 h; E, at 0 °C for 1 h).

in the presence of Et_2Zn at 0 °C were highly *mr*-rich ones (76–89%), as shown in Table 14.5. The observed triad *mr* values were apparently higher than the values of polymers produced directly with DEPrK or produced with DEPrK/Et_2Zn at −78 °C. It should also be emphasized that the *mr* values were even higher than those of the polymers produced with Ph_2CHK/Et_2Zn at 0 °C ($mr = 71\%$), indicating the synthetic advantage of the novel amide enolate initiator. The added Lewis-acidic Et_2Zn and the penultimate $CONEt_2$ moiety as a ligand may play very important roles in fixing the conformation of the main chain during the course of polymerization in the formation of highly heterotactic poly(DEA)s. The amide enolate initiator, DEPrK, effectively increases the *mr* content of the resulting polymer from the initial stage of polymerization, since the coordination effect of the $CONEt_2$ moiety of the initiator residue is superior to the effect of the Ph_2CH terminal group.

It is noteworthy that the polymerization of DEA with amide enolates leads to well-defined water-soluble poly(DEA) without hydrophobic terminal groups as well as the control of molecular weight, MWD, and stereoregularity. Those novel polymer samples are unique and advantageous for investigating the effect of terminal functionality on water solubility and thermosensitivity.

14.4 Anionic Polymerization of Protected *N*-Isopropylacrylamide

Since all the monomers shown above are *N,N*-dialkylacrylamide derivatives, they have no acidic amide protons on the nitrogen atoms. On the other hand, the acidic amide protons of acrylamide and *N*-alkylacrylamides ($pK_a = 25$–26) (Bordwell, 1988) prohibit anionic vinyl polymerization via abstraction and alternatively induce hydrogen-transfer polymerization (1,4-addition mode) under drastic basic conditions (Breslow *et al.*, 1957; Kennedy and Otsu, 1972). The protection of the amide protons of acrylamide and *N*-alkylacrylamides is evidently required for exclusive vinyl polymerization under anionic conditions. Nakahama, Hirao, and their coworkers previously succeeded in the living anionic polymerization of styrenes, 1,3-dienes, and methacrylates with protected functionalities, followed by complete deprotection of the resultant polymers to give a wide variety of functional polymers with predicted molecular weights and narrow MWD (Nakahama and Hirao, 1990; Hirao *et al.*, 2002). This strategy proved to be a powerful synthetic route for well-defined polymers bearing reactive functional groups, such as OH, NH_2, SH, COOH, CHO, $COCH_3$, and C≡CH groups, in each monomer unit. Even in the field of organic synthesis, the protection of amide is rather limited in number and still a challenging issue compared with numerous successful examples for amino, hydroxyl, and carboxyl groups.

Poly(*N*-isopropylacrylamide), poly(NIPAM), is a representative water-soluble, temperature-sensitive polymer attracting great interest in applications as intelligent materials such as hydrogels (Wu *et al.*, 1992; Kokufuta *et al.*, 1993), drug delivery devices (Bae *et al.*, 1990), polymer-bound catalysts (Bergbreiter *et al.*, 2000), dendritic hosts (Kimura *et al.*, 2000), and permeation membranes (Park *et al.*, 1998). The aqueous solution of poly(NIPAM) exhibits a reversible phase transition with a lower critical solution temperature (LCST) at around 32 °C (Heskins and Guillet, 1968; Fugishige *et al.*, 1989). In the last decade, developments in new controlled polymerization systems, such as atom-transfer radical polymerization (ATRP) (Masci *et al.*, 2004; Xia *et al.*, 2005), organotellurium-mediated radical polymerization (Goto *et al.*, 2003), reversible addition–fragmentation transfer (RAFT) polymerization (Ganachaud

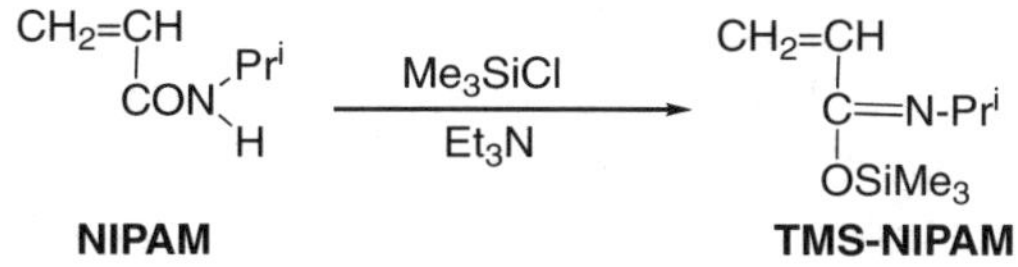

Scheme 14.3 Synthesis of TMS-NIPAM.

et al., 2000; Convertine *et al.*, 2004), and nitroxide-mediated polymerization (NMP) (Schulte *et al.*, 2005), enabled one to synthesize poly(NIPAM)s with controlled molecular weights and relatively narrow MWD ($M_w/M_n = 1.1–1.3$), but conversion was limited to less than 80% in most cases. These new radical polymerization systems gave water-soluble atactic poly(NIPAM)s ($m/r = 50/50$), similar to the conventional radical polymerization system that gave poly(NIPAM) samples with ill-controlled chain structures. Interestingly, isotactic poly(NIPAM) ($m = 92\%$) was obtained via the polymerization of NIPAM with AIBN in the presence of rare-earth-metal triflates such as $Sc(OTf)_3$ and $Y(OTf)_3$ (Isobe *et al.*, 2001). On the other hand, the free-radical polymerization of NIPAM in the presence of hexamethylphosphoramide (HMPA) afforded syndiotactic poly(NIPAM) ($r = 72\%$) (Hirano *et al.*, 2005). However, the molecular weights and their distributions were not regulated in these stereospecific polymerization systems.

Kitayama and his coworkers synthesized a new protected NIPAM monomer, *O*-trimethylsilyl-*N*-isopropylacrylamide (TMS-NIPAM) by the reaction of NIPAM and trimethylsilyl chloride in the presence of triethylamine, as shown in Scheme 14.3 (Kitayama *et al.*, 2002). Anionic polymerization of TMS-NIPAM with *tert*-BuLi in the presence of trialkylaluminum in toluene gave highly isotactic poly(NIPAM) ($m = 97\%$). The trimethylsilyl group of the resulting poly(TMS-NIPAM) was extremely labile and easily removable to afford the expected poly(NIPAM) by only treating it with methanol. Unfortunately, the molecular weights of the polymers were not regulated, plausibly because of side reactions. It should be noted that the resulting poly(NIPAM) was insoluble in water due to its high isotacticity. This negative water solubility of isotactic poly(NIPAM) is consistent with previous examples obtained by free-radical polymerization of NIPAM in the presence of rare-earth-metal triflates (Isobe *et al.*, 2001).

In our previous work, the acidic amide proton of NIPAM is purposefully masked with a more robust methoxymethyl group to prevent termination and chain-transfer reactions during the course of anionic polymerization (Ishizone and Ito, 2002; Ito and Ishizone, 2004; Ito and Ishizone, 2006). Thus, the novel monomer, *N*-methoxymethyl-*N*-isopropylacrylamide (MOM-NIPAM), was synthesized by the reaction of NIPAM and chloromethyl methyl ether in the presence of potassium *tert*-butoxide in diethyl ether, as shown in Scheme 14.4. The chemical

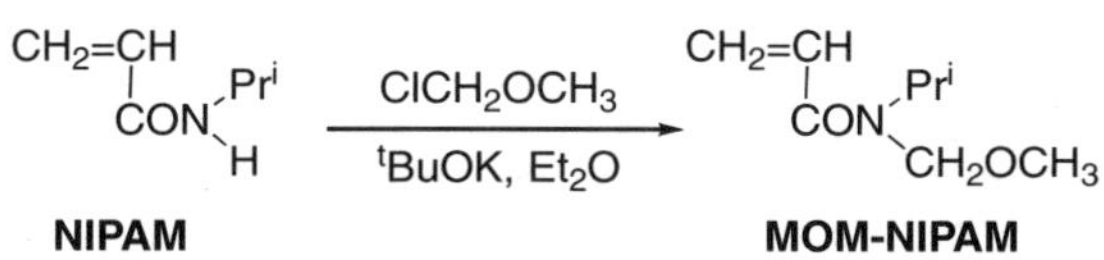

Scheme 14.4 Synthesis of MOM-NIPAM.

structure of MOM-NIPAM corresponds to an *N,N*-dialkylacrylamide derivative without an acidic amide proton. Anionic polymerization of MOM-NIPAM was carried out with DMPLi, Ph_2CHLi, Ph_2CHK, and triphenylmethylpotassium (Ph_3CK) in THF at −78 °C. From ^1H and ^{13}C NMR and IR measurements, the methoxymethyl protecting group was found to be intact during the course of anionic vinyl polymerization.

The results of the polymerization are summarized in Table 14.6. The polymerization reactions with organolithiums homogeneously proceeded, and complete conversions of MOM-NIPAM were always attained in THF at −78 °C within 3 h. The observed molecular weights of the resulting poly(MOM-NIPAM)s agreed with the calculated values. Although the MWD of the polymer obtained with DMPLi was very broad ($M_w/M_n = 2.06$), the binary initiator system of DMPLi and LiCl (8.3 molar equivalent) at −78 °C gave a polymer with $M_w/M_n = 1.31$. Increasing the amount of LiCl 26-fold, the MWD was further narrowed ($M_w/M_n = 1.19$). This is in sharp contrast to the polymerization of other *N,N*-dialkylacrylamides, where polymers having very broad MWD ($M_w/M_n = 3$–10) have been generated with DMPLi/LiCl in heterogeneous polymerization systems. The CH_3OCH_2N moiety may induce homogeneous polymerization by possible coordination to the lithium counterion. A more drastic additive effect on the narrowing of MWD was ascertained upon the addition of Et_2Zn. The M_w/M_n values were significantly reduced to around 1.1, similar to the cases of other

Table 14.6 Anionic polymerization of MOM-NIPAM in THF at −78 °C[a]

Initiator	$LiCl/Li^+$	Et_2Zn/M^+	Time h	$M_n \times 10^{-3}$		M_w/M_n[i]	Tacticity (%)[d]
				calcd.[b]	obsd.[c]		*m*/*r*
DMPLi			1	11	15	2.06	57/43
DMPLi/LiCl	8.3		1	14	13	1.31	78/22
DMPLi/LiCl	26		1	11	14	1.19	85/15
DMPLi/Et_2Zn		17	1.5	11	12	1.11	23/77
Ph_2CHLi/Et_2Zn		13	1	10	12	1.10	25/75
DMPLi/Et_2Zn		43	1	10	12	1.10	20/80
DMPLi/Et_2Zn		15	3	12	14	1.11	22/78
DMPLi/Et_2Zn[e]		54	3	16	18	1.13	17/83
$Ph_2CHLi/LiCl/Et_2Zn$	6.5	14	1	10	12	1.13	82/18
Ph_2CHK			2	10	11	1.25	69/31
Ph_2CHK/Et_2Zn		11	20	5.9	5.1	1.10	52/48
Ph_2CHK/Et_2Zn		17	2	12	13	1.12	50/50
Ph_2CHK/Et_2Zn		18	20	23	23	1.10	50/50
Ph_2CHK/Et_2Zn		18	3	32	30	1.08	47/53
Ph_2CHK/Et_2Zn		20	6	52	67	1.06	48/52
Ph_2CHK/Et_2Zn		15	5 min	12	12	1.12	52/48
Ph_3CK/Et_2Zn		16	20	11	11	1.07	51/49

[a] Yield∼100%.
[b] M_n(calcd) = (MW of monomer) × [monomer]/[initiator] + (MW of initiator).
[c] M_n(obsd) was determined by the ^1H NMR. d M_w/M_n was determined by the SEC calibration using polystyrene standards.
[d] By ^1H NMR.
[e] At −95 °C.

N,N-dialkylacrylamides, *tert*-butyl acrylate, and various methacrylates bearing functional groups. Furthermore, the novel ternary initiator system consisted of Ph_2CHLi, LiCl, and Et_2Zn quantitatively gave a polymer with the predicted M_n value and a narrow MWD. The organopotassium initiators, such as Ph_2CHK or Ph_3CK, also induce anionic polymerization of MOM-NIPAM quantitatively in THF at −78 °C. In particular, the MWD of the polymers was very narrow ($M_w/M_n = 1.1$), when polymerization was performed in the presence of Et_2Zn. The stability of the propagating carbanion of poly(MOM-NIPAM) either at 0 °C or −78 °C was confirmed by the quantitative efficiency of sequential block copolymerization with Ph_2CHK/Et_2Zn using DEA as the second monomer. It was thus substantiated that anionic polymerization of the protected monomer, MOM-NIPAM, quantitatively proceeded in a living fashion in the presence of Lewis-acidic Et_2Zn, as well as other *N,N*-dialkylacrylamides.

The methoxymethyl protecting group of poly(MOM-NIPAM) was completely hydrolyzed by treating it with aqueous hydrochloric acid in 1,4-dioxane or THF at ambient temperature for 20 h, as shown in Scheme 14.5. The ^1H and ^{13}C NMR and IR spectroscopies of the isolated polymer were identical with those of authentic poly(NIPAM) obtained by free-radical polymerization of NIPAM. The SEC trace of the deprotected poly(NIPAM) ($M_w/M_n = 1.08$) measured in THF shifted toward a lower molecular weight region while maintaining the unimodal and narrow shape of the original poly(MOM-NIPAM) ($M_w/M_n = 1.06$). Furthermore, the degree of polymerization of deprotected poly(NIPAM) estimated by ^1H NMR (DP = 68) was in good agreement with the calculated value (DP = 69). These results indicate that the acidic deprotection of poly(MOM-NIPAM) quantitatively proceeds to give poly (NIPAM) having narrow MWD and the predicted molecular weight without side reactions, such as main chain degradation and crosslinking.

The diad configuration of the deprotected poly(NIPAM) was determined by ^1H NMR measurement in DMSO-d_6 at 120 °C by comparing the signal intensity of the methylene protons of the polymer main chain, as shown in Table 14.6. Figure 14.5 shows the typical ^1H NMR spectra of a series of poly(NIPAM)s synthesized by anionic polymerization of MOM-NIPAM along with that obtained by free-radical polymerization of NIPAM with AIBN. Evidently, the initiator systems of the polymerization determine the tacticity of the obtained poly(NIPAM). The *r* diad content in poly(NIPAM)s could be widely changed from 15 to 83% by choosing the anionic initiator systems for MOM-NIPAM, while free-radical polymerization of NIPAM always induced an atactic configuration (*r* = 50%). The poly(NIPAM)s obtained with the Li^+/Et_2Zn initiator system possessed highly syndiotactic configurations (*r* = 75–83%), while either atactic (*r* = 50%) or isotactic poly(NIPAM) (*r* = 15–22%) was obtained with a K^+/Et_2Zn or $Li^+/LiCl$ initiator system, respectively. It is apparent that the addition of Et_2Zn increases the *r* content of poly(NIPAM) along with the narrowing of MWD, as described above. When a large excess amount of Et_2Zn (43 molar equiv.) was added to the reaction mixture, the *r* content slightly increased from 75% to 80%. The highest *r* content

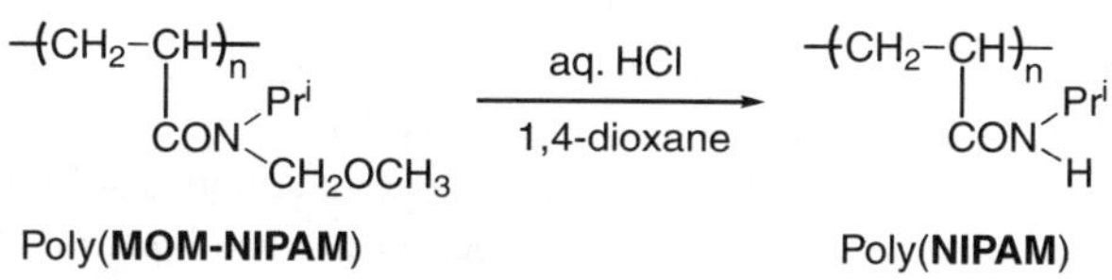

Scheme 14.5 Deprotection of poly(MOM-NIPAM).

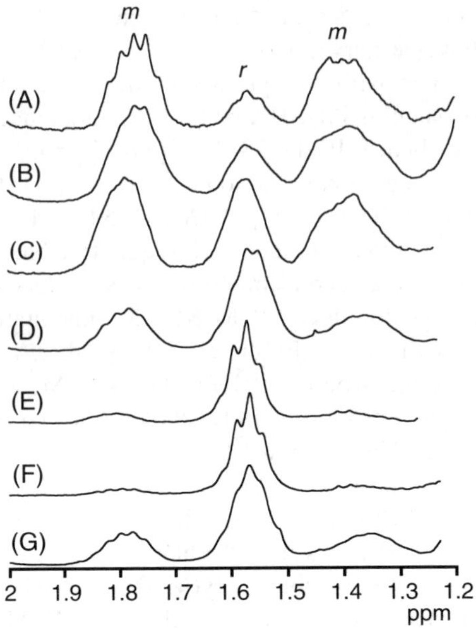

Figure 14.5 ^{1}H NMR spectra of methylene proton of poly(NIPAM)s measured in DMSO-d_6 at 120 °C. (A) m/r = 85/15, (B) m/r = 82/18, (C) m/r = 70/30, (D) m/r = 50/50, (E) m/r = 23/77, (F) m/r = 17/83, and (G) obtained with AIBN, m/r = 50/50.

(83%) of poly(NIPAM) was attained by the polymerization system with DMPLi at −95 °C in the presence of 54-fold Et_2Zn. This is the highest r content of poly(NIPAM) previously reported, whereas a syndiotactic poly(NIPAM) (r = 72%) was prepared via radical polymerization of NIPAM in the presence of HMPA (Hirano *et al.*, 2005). The ternary initiator system of $Ph_2CHLi/LiCl/Et_2Zn$ afforded isotactic poly(NIPAM) of low r content (r = 18%). This means that the added LiCl is superior to Et_2Zn in controlling stereoregularity, while the additive effect of Et_2Zn is simultaneously observed in MWD narrowing (M_w/M_n = 1.13). On the other hand, the stereoregularity of the poly(NIPAM) (r = 50%) obtained with K^+/Et_2Zn is comparable to that of the polymer obtained by free-radical polymerization of NIPAM with AIBN in benzene.

The stereoregularity of poly(NIPAM) strongly affects properties such as solubility and glass-transition temperature (Ito and Ishizone, 2004; Ito and Ishizone, 2006). For example, atactic and syndiotactic poly(NIPAM)s ($42 < r < 83\%$) were water soluble as expected, whereas isotactic one ($r < 31\%$) was insoluble in water. Similar negative water solubility for isotactic poly(NIPAM) has been reported (Isobe *et al.*, 2001; Kitayama *et al.*, 2002). The phase-transition behavior of aqueous solutions of poly(NIPAM) was analyzed by monitoring its transmittance at 500 nm in water. Figure 14.6 illustrates the cloud point traces of poly (NIPAM)s having varied stereoregularity. Atactic poly(NIPAM) (r = 50%) shows representative T_c at around 32 °C, as expected. On the other hand, an aqueous solution of poly(NIPAM) with a slightly m-rich configuration (r = 42%) becomes cloudy from 20 °C. It should be emphasized again that isotactic poly(NIPAM) (m > 69%) was insoluble in water. By contrast, syndiotactic poly(NIPAM) samples with r content of 77% and 83% presented higher T_c values

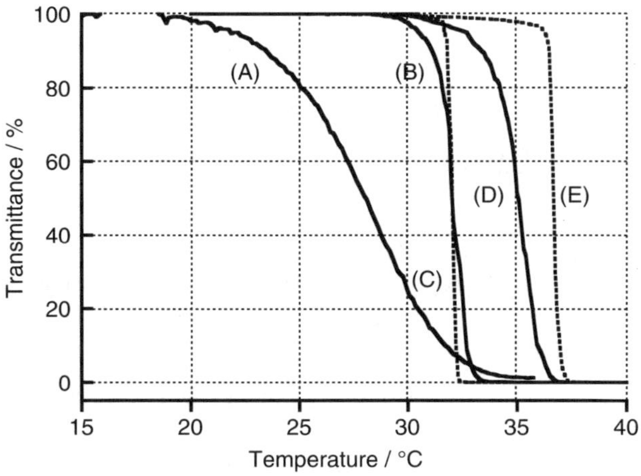

Figure 14.6 Transmittance vs. temperature curves (500 nm, 0.3 °C/min) obtained for 0.2wt% aqueous solution of poly(NIPAM)s having different stereoregularity. (curve A, $M_n = 7{,}600$, $M_w/M_n = 1.15$, $m/r = 58/42$), (curve B, $M_n = 6{,}600$, $M_w/M_n = 1.12$, $m/r = 50/50$), (curve C, $M_n = 48{,}000$, $M_w/M_n = 1.07$, $m/r = 48/52$), (curve D, $M_n = 7{,}600$, $M_w/M_n = 1.09$, $m/r = 23/77$), (curve E, $M_n = 14{,}000$, $M_w/M_n = 1.12$, $m/r = 17/83$).

at 35 and 37 °C, respectively. The cloud points of an aqueous solution of poly(NIPAM)s increased from 32 °C to 37 °C with increasing *r* content. This intriguing result indicates that the stereoregularity of poly(NIPAM) significantly influences water solubility and the cloud point in water.

14.5 Anionic Polymerization of *N,N*-Dialkylmethacrylamides

It is well known that acrylates and methacrylates, α,β-unsaturated esters, readily undergo vinyl polymerization under radical and anionic conditions. Similar to esters, radical polymerization of α,β-unsaturated amides, such as acrylamide, methacrylamide, and their *N*-monoalkyl-substituted derivatives, proceed in vinyl addition modes, while anionic vinyl polymerization is often accompanied by hydrogen-transfer polymerization due to the highly acidic amide hydrogen of these monomers (Breslow *et al.*, 1957; Kennedy and Otsu, 1972). As described above, a variety of *N,N*-dialkylacrylamides are capable of radical and anionic polymerization to afford vinyl polymers.

Contrary to the usual positive polymerizability of *N,N*-dialkylacrylamides, their methacylamide counterparts, such as *N,N*-dimethylmethacrylamide (DMMA) (Chart 14.3), showed a very strange negative polymerizability under radical and anionic conditions (Butler and Myers, 1971; Kodaira and Aoyama, 1974; Okamoto and Yuki, 1981). The only structural difference between DMMA and its polymerizable acryloyl counterpart, DMA, is the presence of an α-methyl substituent in the acryloyl framework. On the basis of ^{1}H and ^{13}C NMR study and MNDO calculations, the unique nonpolymerizability of DMMA is explained by the twisted conformation between the C=C and C=O groups due to the intramolecular steric repulsion between the α-CH_3 or CH_2= group and *N*-alkyl substituents (Kodaira *et al.*, 1990; Xie and Hogen-Esch, 1996). In addition, the partially conjugated planar structure of the O=C-NR_2 moiety in DMMA might also play an important role in reducing the π-conjugation

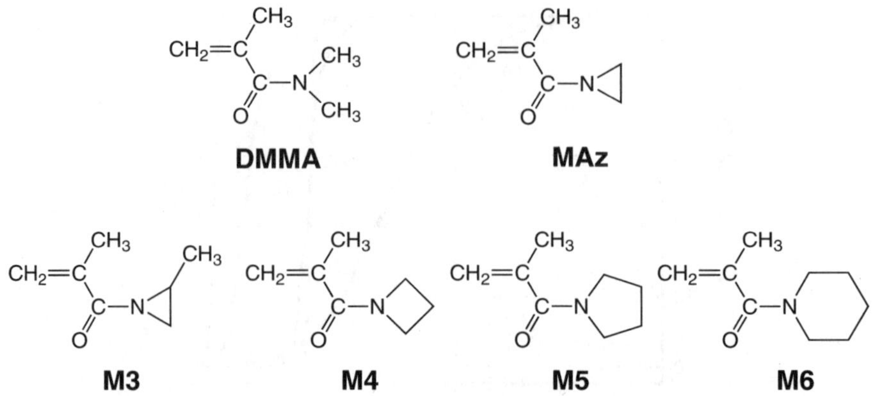

Chart 14.3 *N,N*-Dialkylmethacrylamides.

between the C=C and C=O double bonds. As an intriguing exception, vinyl polymerization of *N*-methacryloylaziridine (MAz), a cyclic analog of DMMA, has been reported (Watanabe *et al.*, 1970; Okamoto and Yuki, 1981). In fact, MAz, possessing a highly strained and small three-membered aziridine ring, can be polymerized with either AIBN or *n*-BuLi, but the detailed results of this polymerization have not been shown. It is noteworthy that plausible side reactions, such as the ring-opening reaction of the aziridine ring, are also suggested in the anionic polymerization of MAz due to the formation of insoluble products.

We have recently overcome the difficulty in the polymerizability of *N,N*-dialkylmethacrylamides by the appropriate molecular design of monomers. A series of methacrylamides carrying cyclic substituents, such as 2-methylaziridine (**M3**), azetidine (**M4**), pyrrolidine (**M5**), and piperidine (**M6**) rings, were synthesized and anionically polymerized in THF (Suzuki *et al.*, 2007; Suzuki *et al.*, 2008; Suzuki *et al.*, 2010) (Chart 14.3). Contrary to the negative polymerizability of DMMA, *N,N*-dialkylmethacrylamides with small three- and four-membered rings, such as **M3** and **M4**, quantitatively undergo anionic vinyl polymerization, as can be seen in Table 14.7. The highly strained aziridine and azetidine rings were intact during the course of anionic polymerization, but a nucleophilic ring-opening reaction of *N*-acrylaziridine was reported (Lin *et al.*, 1993; Sweeney, 2002). The bulky π-stabilized organolithium initiators, such as DMPLi and Ph_2CHLi, effectively initiated the polymerization of **M3** and **M4** to give polymers with well-controlled primary structures either in the absence or in the presence of LiCl. Even at 0 °C, tailored poly(**M3**) and poly(**M4**) with narrow MWD and predetermined molecular weights were produced with organolithiums in the presence of LiCl. The binary initiator systems of DMPLi/Et_2Zn and Ph_2CHLi/Et_2Zn also made it possible to control the polymerization of **M3** and **M4**, respectively. On the other hand, the poly(**M3**)s produced with Ph_2CHK possessed slightly broader MWD compared to the polymers obtained with organolithiums. The addition of Et_2Zn markedly retarded the polymerization of **M4** at −78 °C, and no apparent polymerization virtually proceeded with Ph_2CHK/Et_2Zn even after 72 h. On the other hand, at 0 °C, **M4** monomers were consumed by Ph_2CHK/Et_2Zn within 24 h to afford a polymer with a tailored chain length.

We have compared the relative polymerizability of **M3** and **M4** in the polymerization systems of DMPLi/LiCl in THF from a kinetic viewpoint. Figure 14.7 shows a series of SEC

Table 14.7 Anionic polymerization of **M3**, **M4**, and **M5** in THF

Monomer	Initiator	Additive/ M^+	Temp. °C	Time h	Yield %	$10^{-3}M_n$ calcd[a]	$10^{-3}M_n$ obsd[b]	M_w/M_n[c]	Tacticity (%)[d] *mm*	*mr*	*rr*
M3	DMPLi		−78	15	90	6.1	7.0	1.14			
M3	DMPLi/LiCl	3.8	−78	72	100	6.1	7.2	1.05			
M3	DMPLi/LiCl	5.2	−40	24	100	47	50	1.09			
M3	DMPLi/LiCl	4.3	0	15 min	100	15	16	1.05			
M3	DMPLi/Et_2Zn	11	−78	72	100	12	15	1.08			
M3	Ph_2CHK		−78	15	100	15	16	1.20			
M3	Ph_2CHK/Et_2Zn	12	−78	17	82	13	6.2	1.29			
M4	Ph_2CHLi		−40	16	100	18	23	1.08	60	33	7
M4	Ph_2CHLi/LiCl	4.6	−78	96	72	11	14	1.14	93	7	<1
M4	Ph_2CHLi/LiCl	4.1	−40	24	100	18	19	1.05	70	25	5
M4	Ph_2CHLi/LiCl	6.5	0	2	100	15	18	1.05	68	27	5
M4	Ph_2CHLi/Et_2Zn	17	−40	36	85	26	39	1.06	27	54	19
M4	Ph_2CHNa		−40	13	100	17	19	1.13	52	42	6
M4	Ph_2CHNa/Et_2Zn	16	−40	72	69	9.2	6.9	1.06	11	57	32
M4	Ph_2CHK		0	16	100	12	16	1.17	12	54	34
M4	Ph_2CHK/Et_2Zn	12	−78	72	0	—	—				
M4	Ph_2CHK/Et_2Zn	13	0	24	100	17	19	1.07	13	62	25
M5	Ph_2CHLi		0	72	69	6.2	8.7	1.07			
M5	DMPLi/LiCl	4.5	−40	72	44	3.8	1.6	1.06			
M5	DMPLi/LiCl	5.1	0	168	65	6.3	8.1	1.28			
M5	Ph_2CHK		0	168	77	6.5	7.8	2.51			
M5	Ph_2CHK/Et_2Zn	11	0	168	68	6.7	7.7	1.14			

[a] M_n(calcd) = (MW of monomer) × yield/100 × [monomer]/[initiator] + (MW of initiator).
[b] M_n(obsd) was determined by the ^{1}H NMR.
[c] M_w/M_n was determined by the SEC calibration using polystyrene standards.
[d] By ^{13}C NMR.

traces of poly(**M3**)s obtained with DMPLi/LiCl (4.9 equiv.) in THF at −50 °C. The SEC traces exhibit unimodal and narrow MWD and clearly shift from the lower molecular weight region to the higher side, as the conversion of **M3** increases with polymerization time. This clearly substantiates that no chain-transfer or termination reaction occurred during the polymerization reaction of **M3** at −50 °C. Figure 14.8 illustrates typical first-order plots for **M3**, showing good linearity at each temperature from −40 to −60 °C. This clearly verifies that the concentration of the propagating enolate anion derived from **M3** is almost constant during polymerization under the employed conditions. Similar results were obtained for the polymerization of **M4**, when the temperature was varied from −20 to −40 °C. The k_p^{ap} values of **M3** and **M4** were estimated from the first-order plots and strongly depended on the polymerization temperature, as listed in Table 14.8. For example, the k_p^{ap} values for the polymerization of **M3** and **M4** at −40 °C were 0.165 and 0.0157 L mol^{-1} s^{-1}, respectively. Thus, the observed k_p^{ap} value of **M3** was approximately 10 times larger than that of **M4** under similar polymerization conditions. Typical Arrhenius plots of k_p^{ap} for the anionic polymerization of **M3** are drawn in Figure 14.9, and the

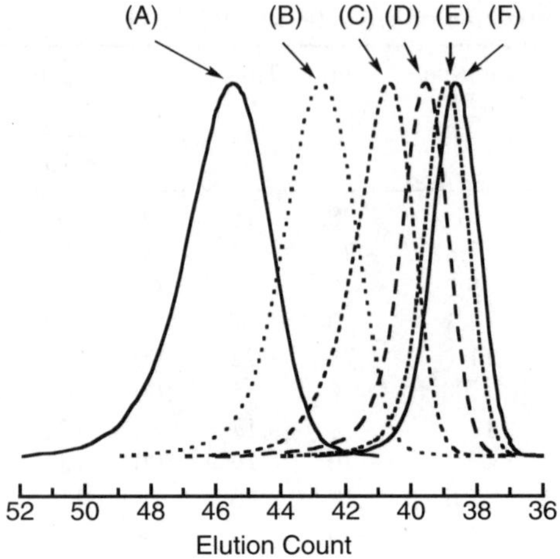

Figure 14.7 SEC curves of a series of poly(**M3**)s obtained with DMPLi and 4.9-fold LiCl in THF at −50 °C: (A) after 6 min, 21% conversion, $M_n = 5{,}300$, $M_w/M_n = 1.10$; (B) after 29 min, 41% conversion, $M_n = 7{,}500$, $M_w/M_n = 1.07$; (C) after 62 min, 62% conversion, $M_n = 10{,}000$, $M_w/M_n = 1.06$; (D) after 125 min, 80% conversion, $M_n = 12{,}000$, $M_w/M_n = 1.05$; (D) after 210 min, 91% conversion, $M_n = 14{,}000$, $M_w/M_n = 1.03$; (E) after 300 min, 96% conversion, $M_n = 15{,}000$, $M_w/M_n = 1.03$.

relationship between k_p^{ap} and the polymerization temperature is expressed in the equation shown below:

$$\ln k_p^{ap} = -5.85 \times 10^3/\mathrm{T} + 23.3\ \mathrm{L\ mol^{-1}\ s^{-1}} (\text{for } \mathbf{M3})$$

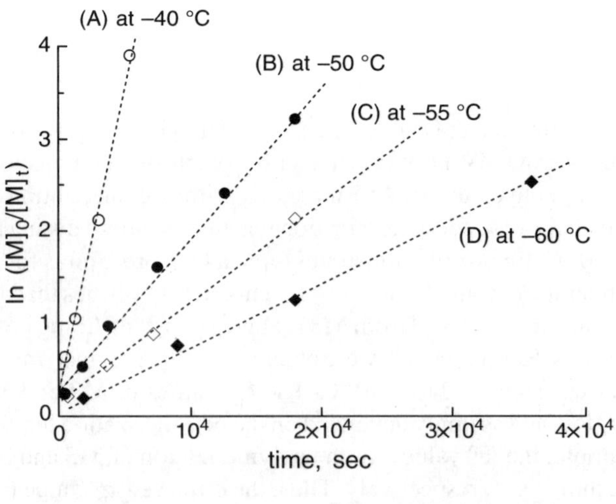

Figure 14.8 First-order plots for the polymerization of **M3** at $[M]_0 = 0.47–0.49$ M and $[I]_0 = 3.8–4.2 \times 10^{-3}$ M: (A) at −40 °C; (B) at −50 °C; (C) at −55 °C; (D) at −60 °C.

Table 14.8 Apparent rate constants of anionic polymerization of **M3** and **M4**[a]

Monomer	Temperature, °C	k_p^{ap}, L mol^{-1} s^{-1}
M3	−40	0.165 ± 0.01
M3	−50	0.0522 ± 0.005
M3	−55	0.0279 ± 0.003
M3	−60	0.0169 ± 0.002
M4	−20	0.129 ± 0.01
M4	−30	0.0540 ± 0.005
M4	−35	0.0297 ± 0.003
M4	−40	0.0157 ± 0.002

[a] in THF in the presence of 4.2–4.9-fold of LiCl.

The activation energies of the polymerization, ΔE_a^{ap}, of **M3** and **M4** were thus obtained to be 49 ± 4 and 51 ± 5 kJ mol^{-1}, respectively. Although these ΔE_a^{ap}s were similar, a slightly larger ΔE_a^{ap} value for **M4** supported lower polymerizability compared to **M3**. Furthermore, it should be emphasized that both the ΔE_a^{ap}s of **M3** and **M4** were markedly larger than the reported values of the anionic polymerization of MMA (20–25 kJ mol^{-1}) under similar conditions (Kunkel *et al.*, 1992; Baskaran *et al.*, 1999). The relative anionic polymerizability among **M3**, **M4**, and MMA is thus estimated as follows: **M4** < **M3** ≪ MMA. Besides anionic polymerization, radical polymerization of **M3** and **M4** certainly proceeded with AIBN in toluene to give polymers in 77 and 40% yields, respectively. **M3** also underwent coordination–addition polymerization with a transition-metal catalyst (Miyake *et al.*, 2009), further indicating the positive polymerizability.

Attempts at anionic polymerization of **M5** also succeeded in giving polymers with relatively narrow MWD in 60–80% yields, but polymerization completion was not achieved under all of the various reaction conditions (Table 14.7). By contrast, no polymerization of **M6**, the

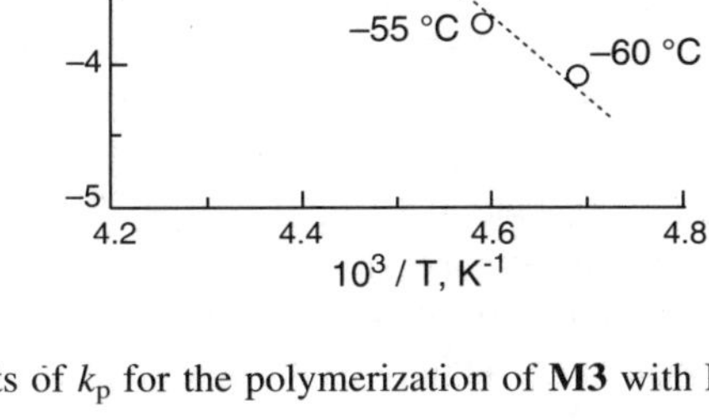

Figure 14.9 Arrhenius plots of k_p for the polymerization of **M3** with DMPLi and 4.4–4.9-fold LiCl in THF.

monomer with a strain-free six-membered piperidine ring, occurred at all, even under rather drastic conditions. Upon the addition of **M6** to DMPLi, Ph_2CHLi, or Ph_2CHK, the characteristic color of the initiator slowly disappeared, suggesting the occurrence of an initiation reaction. However, in each case, no polymer was obtained upon quenching the polymerization after 72 h, and the unreacted **M6** was recovered quantitatively. This observed polymerization behavior of **M6** is in accordance with the negative polymerizability of DMMA.

It was thus demonstrated that the polymerizability of a series of *N,N*-dialkylmethacrylamides carrying a small-membered ring decreases drastically by increasing the ring size from three to six members (**M3** > **M4** > **M5** > **M6** = DMMA). It should be noted again that all the acryloyl counterparts, **A3**, **A4**, **A5**, **A6**, and DMA, showed similar high anionic polymerizability, as usually expected for polar monomers. The varied polymerizability of methacrylamide monomers can be correlated with the β-carbon chemical shifts in the ^{13}C NMR spectra, which reasonably reflect the electron densities of the C=C bonds accepting the nucleophilic attack of the initiator and the propagating carbanion (Ishizone *et al.*, 1993). In fact, the ^{13}C NMR chemical shifts of the vinyl β-carbons of **M3**, **M4**, **M5**, **M6**, and DMMA were observed at 124.5, 119.1, 116.0, 114.6, and 115.4 ppm, respectively. The upfield chemical shifts for **M5**, **M6**, and DMMA (115–116 ppm) are far from the downfield values of typical polar monomers, such as methyl acrylate (130.6 ppm), MMA (125.3 ppm), and DMA (125.8 ppm), but rather comparable to that of styrene (113.8 ppm).

Furthermore, the conformations of the monomers are considered to be essential to determine the polymerizability of *N,N*-dialkylmethacrylamides, as previously suggested on the basis of the NMR chemical shifts and theoretical calculations (Okamoto and Yuki, 1981; Kodaira *et al.*, 1990; Xie and Hogen-Esch, 1996; Miyake *et al.*, 2009). In fact, the dihedral angles between the C=C and C=O groups of **M3**, **M4**, **M5**, **M6**, and DMMA in stable conformations were estimated from an *ab initio* calculation to be 14, 37, 48, 58, and 55°, respectively. These dihedral angles are in good accordance with the relative polymerizability (**M3** > **M4** > **M5** > **M6** = **DMMA**). The conjugation effects between C=C and C=O seem to be negligible in **M5**, **M6**, and DMMA in contrast to the effective π-conjugation system in **M3**. The planar geometry of the (O=C)-NR_2 moieties and the steric repulsions between the α-methyl or CH_2= group and two alkyl substituents on the amide nitrogen in **M5**, **M6**, and DMMA presumably induce the twisted conformations of the C=C−C=O system and result in extremely low polymerizability. On the other hand, aziridine and azetidine amides have characteristic conformation features as unusual nonplanar amides with pyramidal nitrogen (Anet and Osyany, 1967; Otani *et al.*, 2003). The highly strained rings in **M3** and **M4** prohibit conjugation between the C=O and NR_2 group to diminish the planarity, and eventually afford positive polymerizability due to the effective π-conjugation between the C=C and C=O double bonds. These findings in ^{13}C NMR and theoretical calculation suggest a suitable molecular design for *N,N*-dialkylmethacrylamides to realize positive polymerizability.

The tacticities of poly(**M3**) and poly(**M5**) were difficult to determine, since suitable signals in the ^1H and ^{13}C NMR spectra were not well split or overlapped with other signals. On the other hand, the stereoregularity of poly(**M4**) could be estimated from the relative ^{13}C NMR signal intensity of the α-methyl carbon at around 18.0–21.7 ppm, as shown in Table 14.7. Figure 14.10 illustrates a series of expanded ^{13}C NMR spectra of poly(**M4**) measured in DMSO-d_6 at 75 °C. Clearly, the tacticity of poly(**M4**) was largely changed by the polymerization conditions. The *mm* content of the poly(**M4**)s produced with Ph_2CHLi, Ph_2CHNa, and Ph_2CHK at 0 or −40 °C was 60, 52, and 12%, respectively. On the other hand, the *mr* content

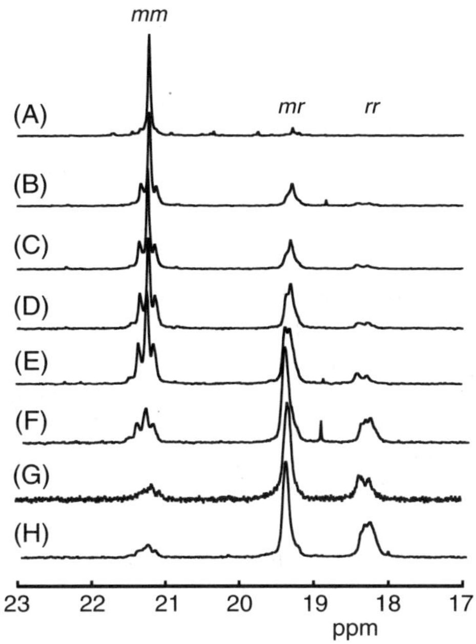

Figure 14.10 ^{13}C NMR spectra of α-methyl carbon of poly(**M4**)s measured in DMSO-d_6 at 75 °C. (A) DMPLi/LiCl at −78 °C, *mm*/*mr*/*rr* = 92/8/<1; (B) $Ph_2CHLi/LiCl$ at −40 °C, *mm*/*mr*/*rr* = 72/23/5; (C) $Ph_2CHLi/LiCl$ at 0 °C, *mm*/*mr*/*rr* = 68/27/5; (D) Ph_2CHLi at −40 °C, *mm*/*mr*/*rr* = 60/33/7; (E) Ph_2CHNa at −40 °C, *mm*/*mr*/*rr* = 52/42/6; (F) Ph_2CHLi/Et_2Zn at −40 °C, *mm*/*mr*/*rr* = 27/54/19; (G) Ph_2CHK/Et_2Zn at 0 °C, *mm*/*mr*/*rr* = 13/62/25; (H) Ph_2CHK at 0 °C, *mm*/*mr*/*rr* = 12/54/34.

was 33 (Li^+), 42 (Na^+), and 54% (K^+) and evidently increased with the radius of the counterion. The smaller cation induces the isotactic configurations of poly(**M4**), but the larger one prefers heterotactic configurations. After the addition of Lewis-acidic Et_2Zn to the initiator, the *mr* content of poly(**M4**) became predominant regardless of the counterion. In the cases of poly(DEA) and poly(NIPAM), the initiator system of Ph_2CHLi/Et_2Zn usually gave syndiotactic polymers, while the *rr* content of the poly(**M4**) obtained with Ph_2CHLi/Et_2Zn was only 19%. By contrast, in the presence of LiCl, the *mm* content tended to increase by lowering the polymerization temperature and reached 92% at −78 °C. Thus, we have found that the addition of LiCl or Et_2Zn to the organolithium initiator allows isotactic-specific or atactic polymerization of **M4**, respectively. From a synthetic viewpoint, the tacticity of poly(**M4**) can be tuned by changing the counterion of the initiator, the additive, and the polymerization temperature.

All the resulting poly(**M4**)s were readily soluble in water due to their high polarity, and no thermosensitivity was observed for aqueous solutions between 0 and 95 °C. On the other hand, an aqueous solution of poly(**M5**) showed a typical cloud point at 26–28 °C, while poly(**M3**) was insoluble in water. The exothermic crosslinking of poly(**M3**) readily started at around 120 °C due to the ring-opening reaction of the highly strained aziridine ring, indicating the potential of poly(**M3**) as a new thermosetting resin.

14.6 Conclusions

This review demonstrates that living anionic polymerization of a series of *N,N*-dialkylacrylamides, such as DMA, DEA, and MOM-NIPAM, quantitatively proceed to afford polymers possessing predictable molecular weights and narrow MWD. New initiator systems, such as organolithium/LiCl, organolithium/Et_2Zn, and organopotassium/Et_2Zn, often allow simultaneous control of the stereoregularity of the resulting polymers, as well as the molecular-weight control. Novel potassium amide enolates proved to be versatile anionic initiators and allow tailored synthesis of poly(DEA)s without hydrophobic initiator residues. The stereospecific synthesis of well-defined poly(NIPAM)s was achieved for a new protected monomer, MOM-NIPAM, by tuning the counterions of the anionic initiators and the additives. The new strategy to increase the ring strain of the amide substituent appears to be a suitable molecular design to provide anionic polymerizability for *N,N*-dialkylmethacrylamides. In fact, three-membered aziridine and four-membered azetidine methacrylamide derivatives, **M3** and **M4**, quantitatively undergo anionic polymerization in a living manner. The additive effect of Lewis-acidic compounds on stereoregularity was observed only for the anionic polymerization of a series of *N,N*-dialkyl(meth)acrylamides and not for (meth) acrylates. The polar amide moieties in the monomers and resulting polymers may play a very important role in regulating tacticity through possible coordination with the counterions and the Lewis-acidic additives.

References

Anderson, B.C., Andrews, G.D., Arthur, P. Jr *et al.* (1981) Anionic polymerization of methacrylates. Novel functional polymers and copolymers. *Macromolecules*, **14**, 1599–1601.

Anet, F.A.L. and Osyany, J.M. (1967) Nuclear magnetic resonance spectra and nitrogen inversion in 1-acylaziridines. *Journal of the American Chemical Society*, **89**, 352–356.

Bae, Y.H., Okano, T., and Kim, S.W. (1990) Temperature dependence of swelling of crosslinked poly(*N,N'*-alkyl substituted acrylamides) in water. *Journal of Polymer Science, Part B: Polymer Physics*, **28**, 923–936.

Ballard, D.G.H., Bowles, R.J., Haddleton, D.M. *et al.* (1992) Controlled polymerization of methyl methacrylate using lithium aluminum alkyls. *Macromolecules*, **25**, 5907–5913.

Baltes, T., Garret-Flaudy, F., and Freitag, R. (1999) Investigation of the LCST of polyacrylamides as a function of molecular parameters and the solvent composition. *Journal of Polymer Science, Part A: Polymer Chemistry*, **37**, 2977–2989.

Baskaran, D. (2003) Strategic developments in living anionic polymerization of alkyl (meth)acrylates. *Progress in Polymer Science*, **28**, 521–581.

Baskaran, D. and Sivaram, S. (1997) Specific salt effect of lithium perchlorate in living anionic polymerization of methyl methacrylate and *tert*-butyl acrylate. *Macromolecules*, **30**, 1550–1555.

Baskaran, D., Müller, A.H.E., and Sivaram, S. (1999) Effect of lithium perchlorate on the kinetics of the anionic polymerization of methyl methacrylate in tetrahydrofuran. *Macromolecules*, **32**, 1356–1361.

Bergbreiter, D.E., Osburn, P.L., Wilson, A., and Sink, E.M. (2000) Palladium-catalyzed C–C coupling under thermomorphic conditions. *Journal of the American Chemical Society*, **122**, 9058–9064.

Bordwell, F.G. (1988) Equilibrium acidities in dimethyl sulfoxide solution. *Accounts of Chemical Research*, **21**, 456–463.

Breslow, D.S., Hulse, G.E., and Matlack, A.S. (1957) Synthesis of poly-β-alanine from acrylamide. A novel synthesis of β-alanine. *Journal of the American Chemical Society*, **79**, 3760–3763.

Butler, G.B. and Myers, R. (1971) The fundamental basis for cyclopolymerization. IV. Radiation initiated solid-state polymerization of certain dimethacrylamides. *Journal of Macromolecular Science-Chemistry*, **A5**, 135–166.

Butler, K., Thomas, P.R., and Tyler, G.J. (1960) Stereospecific polymerization of some polar vinyl monomers. *Journal of Polymer Science*, **48**, 357–366.

Collins, S., Ward, D.G., and Suddaby, K.H. (1994) Group-transfer polymerization using metallocene catalysts: Propagation mechanisms and control of polymer stereochemistry. *Macromoleules*, **27**, 7222–7224.

Convertine, A.J., Ayres, N., Scales, C.W. *et al.* (2004) Facile, controlled, room-temperature RAFT polymerization of *N*-isopropylacrylamide. *Biomacromolecules*, **5**, 1177–1180.

Davis, T.P., Haddleton, D.M., and Richards, S.N. (1994) Controlled polymerization of acrylates and methacrylates. *Journal of Macromolecular Science Reviews in Macromolecular Chemistry*, **C34**, 243–324.

Dvořánek, L. and Vlček, P. (1994) Anionic polymerization of acrylates. 8. Kinetics of the anionic polymerization of butyl acrylate initiated with the complex initiator lithium ester enolate/lithium *tert*-butoxide. *Macromolecules*, **27**, 4881–4885.

Eggert, M. and Freitag, R. (1994) Poly-*N,N*-diethylacrylamide prepared by group-transfer polymerization: synthesis, characterization, and solution properties. *Journal of Polymer Science, Part A: Polymer Chemistry*, **32**, 803–813.

Fayt, R., Forte, R., Jacobs, C. *et al.* (1987) New initiator system for the "living" anionic polymerization of *tert*-alkyl acrylates. *Macromolecules*, **20**, 1442–1444.

Freitag, R., Baltes, T., and Eggert, M. (1994) A comparison of thermoreactive water-soluble poly-*N,N*-diethylacrylamide prepared by anionic and by group-transfer polymerization. *Journal of Polymer Science, Part A: Polymer Chemistry*, **32**, 3019–3030.

Fugishige, S., Kubota, K., and Ando, I. (1989) Phase transition of aqueous solutions of poly(*N*-isopropylacrylamide) and poly(*N*-isopropylmethacrylamide). *Journal of Physical Chemistry*, **93**, 3311–3313.

Gallei, M., Schmidt, B.V.K.J., Klein, R., and Rehahn, M. (2009) Defined poly[styrene-*block*-(ferrocenylmethyl methacrylate)] diblock copolymers via living anionic polymerization. *Macromolecular Rapid Communications*, **30**, 1463–1469.

Ganachaud, F., Monteiro, M.J., Gilbert, R.G. *et al.* (2000) Molecular weight characterization of poly(*N*-isopropylacrylamide) prepared by living free-radical polymerization. *Macromolecules*, **33**, 6738–6745.

Gassman, P.G. and Fox, B.L. (1966) Synthesis of *β*-substituted alkylamines via alkylation of *N,N*-disubstituted amides. *Journal of Organic Chemistry*, **31**, 982–983.

Gia, H.-B. and Mcgrath, J.E. (1980) High resolution NMR spectra of poly *N,N*-dimethylacrylamide in $CDCl_3$ solution. *Polymer Bulletin*, **2**, 837–840.

Goto, A., Kwak, Y., Fukuda, T. *et al.* (2003) Mechanism-based invention of high-speed living radical polymerization using organotellurium compounds and azo-initiators. *Journal of the American Chemical Society*, **125**, 8720–8721.

Han, S., Hagiwara, M., and Ishizone, T. (2003) Synthesis of thermally sensitive water-soluble polymethacrylates by living anionic polymerizations of oligo(ethylene glycol) methyl ether methacrylates. *Macromolecules*, **36**, 8312–8319.

Hatada, K. and Kitayama, T. (2000) Structurally controlled polymerizations of methacrylates and acrylates. *Polymer International*, **49**, 11–47.

Hatada, K., Ute, K., Tanaka, K. *et al.* (1986) Living and highly isotactic polymerization of methyl methacrylate by t-C_4H_9MgBr in toluene. *Polymer Journal*, **18**, 1037–1047.

Heskins, M. and Guillet, J.E. (1968) Solution properties of Poly(*N*-isopropylacrylamide). *Journal of Macromolecular Science-Chemistry*, **A2**, 1441–1455.

Hild, G., Lamps, J.P., and Rempp, P. (1993) Synthesis and characterization of anionic 2,3-epoxypropyl methacrylate polymers and of related random and block copolymers with methyl methacrylate. *Polymer*, **34**, 2875–2882.

Hirano, T., Miki, H., Seno, M., and Sato, T. (2005) Effect of polymerization conditions on the syndiotactic-specificity in radical polymerization of *N*-isopropylacrylamide and fractionation of the obtained polymer according to the stereoregularity. *Polymer*, **46**, 5501–5505.

Hirao, A., Loykulnant, S., and Ishizone, T. (2002) Recent advance in living anionic polymerization of functionalized styrene derivatives. *Progress in Polymer Science*, **27**, 1399–1471.

Hoogeveen, N.G., Cohen Stuart, M.A., and Fleer, G.J. (1996) Novel water-soluble block copolymers of dimethylaminoethyl methacrylate and dihydroxypropyl methacrylate. *Macromolecular Chemistry and Physics*, **197**, 2553–2564.

Ishizone, T. and Ito, M. (2002) Synthesis of well-defined poly(*N*-isopropylacrylamide) by the anionic polymerization of *N*-methoxymethyl-*N*-isopropylacrylamide. *Journal of Polymer Science, Part A: Polymer Chemistry*, **40**, 4328–4332.

Ishizone, T., Hirao, A., and Nakahama, S. (1993) Anionic polymerization of monomers containing functional groups. 6. Anionic block copolymerization of styrene derivatives para-substituted with electron-withdrawing groups. *Macromolecules*, **26**, 6964–6975.

Ishizone, T., Yoshimura, K., Hirao, A., and Nakahama, S. (1998a) Controlled anionic polymerization of *tert*-butyl acrylate with diphenylmethyl anions in the presence of dialkylzinc. *Macromolecules*, **31**, 8706–8712.

Ishizone, T., Uehara, G., Hirao, A. *et al.* (1998b) Anionic polymerization of monomers containing functional groups, 11. Anionic polymerizations of alkynyl methacrylates. *Macromolecular Chemistry and Physics*, **199**, 1827–1834.

Ishizone, T., Yoshimura, K., Yanase, E., and Nakahama, S. (1999) Controlled anionic polymerization of *tert*-butyl acrylate with diphenylmethylpotassium in the presence of triethylborane. *Macromolecules*, **32**, 955–957.

Ishizone, T., Yanase, E., Matsushita, T., and Nakahama, S. (2001) Controlled anionic polymerization of methacrylonitrile with diphenylmethylpotassium in the presence of triethylborane. *Macromolecules*, **34**, 6551–6553.

Ishizone, T., Tajima, H., Torimae, H., and Nakahama, S. (2002) Anionic polymerizations of 1-adamantyl methacrylate and 3-methacryloyloxy-1,1′-biadamantane. *Macromolecular Chemistry and Physics*, **203**, 2375–2384.

Ishizone, T., Han, S., Okuyama, S., and Nakahama, S. (2003) Synthesis of water-soluble polymethacrylates by living anionic polymerization of trialkylsilyl-protected oligo(ethylene glycol) methacrylates. *Macromolecules*, **36**, 42–49.

Ishizone, T., Yashiki, D., Kobayashi, M. *et al.* (2007) Potassium enolates of *N,N*-dialkylamides as initiators of anionic polymerization. *Journal of Polymer Science, Part A: Polymer Chemistry*, **45**, 1260–1271.

Ishizone, T., Seki, A., Hagiwara, M. *et al.* (2008) Anionic polymerization of oligo(ethylene glycol) alkyl ether methacrylates: effect of side chain length and ω-alkyl group of side chain on cloud point in water. *Macromolecules*, **41**, 2963–2967.

Isobe, Y., Fujioka, D., Habaue, S., and Okamoto, Y. (2001) Efficient Lewis acid-catalyzed stereocontrolled radical polymerization of acrylamides. *Journal of the American Chemical Society*, **123**, 7180–7181.

Ito, M. and Ishizone, T. (2004) Synthesis of well-defined block copolymers containing poly(*N*-isopropylacrylamide) segment by anionic block copolymerization of *N*-methoxymethyl-*N*-isopropylacrylamide. *Designed Monomers and Polymers*, **7**, 11–24.

Ito, M. and Ishizone, T. (2006) Living anionic polymerization of *N*-methoxymethyl-*N*-isopropylacrylamide: Synthesis of well-defined poly(*N*-isopropylacrylamide) having various stereoregularity. *Journal of Polymer Science, Part A: Polymer Chemistry*, **44**, 4832–4845.

Janata, M., Lochmann, L., and Müller, A.H.E. (1990) Mechanism and kinetics of the anionic polymerization of acrylates, 1. Oligomerization of *tert*-butyl acrylate and characterization of products. *Die Makromolekulare Chemie*, **191**, 2253–2260.

Janata, M., Lochmann, L., Vlček, P. *et al.* (1992) Mechanism and kinetics of the anionic polymerization of acrylates, 2. Polymerization of *tert*-butyl acrylate in a flow tube reactor and effect of lithium chloride and lithium *tert*-butoxide. *Die Makromolekulare Chemie*, **193**, 101–112.

Kennedy, J.P. and Otsu, T. (1972) Hydrogen transfer polymerization with anionic catalysts and the problem of anionic isomerization polymerization. *Journal of Macromolecular Science-Reviews in Macromolecular Chemistry*, **C6**, 237–283.

Kimura, M., Kato, M., Muto, T. *et al.* (2000) Temperature-sensitive dendritic hosts: synthesis, characterization, and control of catalytic activity. *Macromolecules*, **33**, 1117–1119.

Kitayama, T., Shinozaki, T., Masuda, E. *et al.* (1988) Highly syndiotactic poly(methyl methacrylate) with narrow molecular-weight distribution formed by *tert*-butyllithium-trialkylaluminum in toluene. *Polymer Bulletin*, **20**, 505–510.

Kitayama, T., Shibuya, W., and Katsukawa, K. (2002) Synthesis of highly isotactic poly(*N*-isopropylacrylamide) by anionic polymerization of a protected monomer. *Polymer Journal*, **34**, 405–409.

Kobayashi, M., Okuyama, S., Ishizone, T., and Nakahama, S. (1999) Stereospecific anionic polymerization of *N,N*-dialkylacrylamides. *Macromolecules*, **32**, 6466–6477.

Kobayashi, M., Ishizone, T., and Nakahama, S. (2000a) Additive effect of triethylborane on anionic polymerization of *N,N*-dimethylacrylamide and *N,N*-diethylacrylamide. *Macromolecules*, **33**, 4411–4416.

Kobayashi, M., Ishizone, T., and Nakahama, S. (2000b) Synthesis of highly isotactic poly(*N,N*-diethylacrylamide) by anionic polymerization with Grignard reagents and diethylzinc. *Journal of Polymer Science, Part A: Polymer Chemistry*, **38**, 4677–4685.

Kodaira, T. and Aoyama, F. (1974) Cyclopolymerization. II. Mechanism of the free-radical polymerization of N-n-propyldimethacrylamide. *Journal of Polymer Science: Polymer Chemistry*, **12**, 897–910.

Kodaira, T., Tanahashi, H., and Hara, K. (1990) Cyclopolymerization XVII. Anionic cyclopolymerization tendency of *N*-methyldiacrylamide and *N*-substituted dimethacrylamides. *Polymer Journal*, **22**, 649–659.

Kokufuta, E., Zhang, Y.-Q., Tanaka, T., and Mamada, A. (1993) Effects of surfactants on the phase transition of poly(*N*-isopropylacrylamide) gel. *Macromolecules*, **26**, 1053–1059.

Krom, J.A. and Streitwieser, A. (1996) Cesium ion pair acidities of some *N,N*-dialkylacetamides and aggregation of their cesium enolates in tetrahydrofuran. *Journal of Organic Chemistry*, **61**, 6354–6359.

Kunkel, D., Müller, A.H.E., Janata, M., and Lochmann, L. (1992) The role of association/complexation equilibria in the anionic polymerization of (meth)acrylates. *Die Makromolekulare Chemie., Macromolecular Symposia*, **60**, 315–326.

Lin, P.Y., Bentz, G., and Stamm, H. (1993) Regioselectivity in nucleophilic ring opening of 2-methylaziridines. Lag of bond making as model for the abnormal opening. *Journal für praktische Chemie*, **335**, 23–34.

Masci, G., Giacomelli, L., and Crescenzi, V. (2004) Atom transfer radical polymerization of *N*-isopropylacrylamide. *Macromolecular Rapid Communications*, **25**, 559–564.

Miyake, G., Caporaso, L., Cavallo, L., and Chen, E.Y.-X. (2009) Coordination–addition polymerization and kinetic resolution of methacrylamides by chiral metallocene catalyst. *Macromolecules*, **42**, 1462–1471.

Mori, H., Wakisaka, O., Hirao, A., and Nakahama, S. (1994a) Protection and polymerization of functional monomers, 23. Synthesis of well-defined poly(2-hydroxyethyl methacrylate) by means of anionic living polymerization of protected monomers. *Macromolecular Chemistry and Physics*, **195**, 3213–3224.

Mori, H., Hirao, A., and Nakahama, S. (1994b) Protection and polymerization of functional monomers. 21. Anionic living polymerization of (2,2-dimethyl-1,3-dioxolan-4-yl)methyl methacrylate. *Macromolecules*, **27**, 35–39.

Nakahama, S. and Hirao, A. (1990) Protection and polymerization of functional monomers: anionic living polymerization of protected monomers. *Progress in Polymer Science*, **15**, 299–335.

Nakahama, S., Kobayashi, M., Ishizone, T. *et al.* (1997) Polymerization of *N,N*-dialkylacrylamides with anionic initiators modified by diethylzinc. *Journal of Macromolecular Science Part A: Pure and Applied Chemistry*, **A34**, 1845–1855.

Nakhmanovich, B.I., Prudskova, T.N., Arest-Yakubovich, A.A., and Müller, A.H.E. (2001) Copolymerization of *N,N*-dimethylacrylamide with styrene and butadiene: The first example of polar growing chain end/nonpolar monomer cross-initiation. *Macromolecular Rapid Communications*, **22**, 1243–1248.

Oare, D.A., Henderson, M.A., Sanner, M.A., and Heathcock, C.H. (1990) Stereochemistry of the Michael addition of *N,N*-disubstituted amide and thioamide enolates to α,β-unsaturated ketones. *Journal of Organic Chemistry*, **55**, 132–157.

Okamoto, Y. and Yuki, H. (1981) Anionic polymerization of *N*-methacryloylaziridine. *Journal of Polymer Science: Polymer Chemistry Edition*, **19**, 2647–2650.

Otani, Y., Nagae, O., Naruse, Y. *et al.* (2003) An evaluation of amide group planarity in 7-azabicyclo[2.2.1]heptane amides. Low amide bond rotation barrier in solution. *Journal of the American Chemical Society*, **125**, 15191–15199.

Ozaki, H., Hirao, A., and Nakahama, S. (1995) Anionic polymerization of alkyl methacrylates in the presence of diethylzinc. *Macromolecular Chemistry and Physics*, **196**, 2099–2111.

Park, Y.S., Ito, Y., and Imanishi, Y. (1998) Permeation control through porous membranes immobilized with thermosensitive polymer. *Langmuir*, **14**, 910–914.

Reetz, M.T. (1988) New methods for the anionic polymerization of α-activated olefins. *Angewandte Chemie, International Edition in English.*, **27**, 994–998.

Reetz, M.T., Knauf, T., Minet, U., and Bingel, C. (1988) Metal-free carbanion salts as initiators for the anionic polymerization of acrylic and methacrylic acid esters. *Angewandte Chemie, International Edition in English*, **27**, 1373–1374.

Schulte, T., Siegenthaler, K.O., Luftmann, H. *et al.* (2005) Nitroxide-mediated polymerization of *N*-isopropylacrylamide: Electrospray ionization mass spectrometry, matrix-assisted laser desorption ionization mass spectrometry, and multiple-angle laser light scattering studies on nitroxide-terminated poly-*N*-isopropylacrylamides. *Macromolecules*, **38**, 6833–6840.

Shiohara, K., Habaue, S., and Okamoto, Y. (1996) Stereospecific polymerization of *N,N*-diphenylacrylamide. *Polymer Journal*, **28**, 682–685.

Sogah, D.Y., Hertler, W.R., Webster, O.W., and Cohen, G.M. (1987) Group-transfer polymerization. Polymerization of acrylic monomers. *Macromolecules*, **20**, 1473–1488.

Sugiyama, K., Azuma, H., Watanabe, T. *et al.* (2003) Anionic polymerization of 2-haloethyl methacrylates. *Polymer*, **44**, 4157–4164.

Suzuki, T., Kusakabe, J., and Ishizone, T. (2007) Anionic polymerization of *N*-methacryloyl-2-methylaziridine and *N*-methacryloylazetidine. *Macromolecular Symposia*, **249–250**, 412–416.

Suzuki, T., Kusakabe, J., and Ishizone, T. (2008) Living anionic polymerization of *N*-methacryloyl-2-methylaziridine: Polymerizable *N,N*-dialkylmethacrylamide. *Macromolecules*, **41**, 1929–1936.

Suzuki, T., Kusakabe, J., Kitazawa, K. *et al.* (2010) Living anionic polymerization of *N*-methacryloylazetidine: Anionic polymerizability of *N,N*-dialkylmethacrylamides. *Macromolecules*, **43**, 107–116.
Sweeney, J.B. (2002) Aziridines: epoxides' ugly cousins? *Chemical Society Reviews*, **31**, 247–258.
Tsvetanov, C.B., Müller, A.H.E., and Schulz, G.V. (1985) Dependence of the propagation rate constants on the degree of polymerization in the initial stage of the anionic polymerization of methyl methacrylate in tetrahydrofuran. *Macromolecules*, **18**, 863–868.
Varshney, S.K., Hautekeer, J.P., Fayt, R. *et al.* (1990) Anionic polymerization of (meth)acrylic monomers. 4. Effect of lithium salts as ligands on the "living" polymerization of methyl methacrylate using monofunctional initiators. *Macromolecules*, **23**, 2618–2622.
Wang, J.S., Jérôme, R., Warin, R., and Teyssié, Ph. (1993) Anionic polymerization of acrylic monomers. 10. ^{13}C and ^{7}Li NMR studies on the monomeric model of living poly(methyl methacrylate). *Macromolecules*, **26**, 1402–1406.
Watanabe, N., Sakai, M., Sakakibara, Y., and Uchino, N. (1970) Polymerization and copolymerization of aziridine derivatives possessing carbon-carbon double bond. *Kogyo Kagaku Zasshi*, **73**, 1056–1058.
Webster, O.W. (2000) The discovery and commercialization of group-transfer polymerization. *Journal of Polymer Science, Part A: Polymer Chemistry*, **38**, 2855–2860.
Webster, O.W., Hertler, W.R., Sogah, D.Y. *et al.* (1983) Group-transfer polymerization. 1. A new concept for addition polymerization with organosilicon initiators. *Journal of the American Chemical Society*, **105**, 5706–5708.
Woodbury, R.P. and Rathke, M.W. (1977) Isolation and reactions of α-lithio *N,N*-dimethylacetamide. *Journal of Organic Chemistry*, **42**, 1688–1690.
Wu, X.S., Hoffman, A.S., and Yager, P. (1992) Synthesis and characterization of thermally reversible macroporous poly (*N*-isopropylacrylamide) hydrogels. *Journal of Polymer Science, Part A: Polymer Chemistry*, **30**, 2121–2129.
Xia, Y., Yin, X., Burke, N.A.D., and Stöver, H.D.H. (2005) Thermal response of narrow-disperse poly(*N*-isopropylacrylamide) prepared by atom transfer radical polymerization. *Macromolecules*, **38**, 5937–5943.
Xie, X. and Hogen-Esch, T.E. (1996) Anionic synthesis of narrow molecular-weight distribution water-soluble poly(*N,N*-dimethylacrylamide) and poly(N-acryloyl-*N*′-methylpiperazine). *Macromolecules*, **29**, 1746–1752.
Yasuda, H., Yamamoto, H., Yokota, K. *et al.* (1992) Synthesis of monodispersed high molecular weight polymers and isolation of an organolanthanide(III) intermediate coordinated by a penultimate poly(MMA) unit. *Journal of the American Chemical Society*, **114**, 4908–4910.
Yasuda, H., Yamamoto, H., Yamashita, M. *et al.* (1993) Synthesis of high molecular weight poly(methyl methacrylate) with extremely low polydispersity by the unique function of organolanthanide(III) complexes. *Macromolecules*, **26**, 7134–7143.
Zagala, A.P. and Hogen-Esch, T.E. (1996) Living anionic polymerization of methyl methacrylate at ambient temperatures in the presence of the tetraphenylphosphonium cation. *Macromolecules*, **29**, 3038–3039.
Zhang, H. and Ruckenstein, E. (1998a) Graft copolymers by combined anionic and cationic polymerizations based on the homopolymerization of a bifunctional monomer. *Macromolecules*, **31**, 746–752.
Zhang, H. and Ruckenstein, E. (1998b) Living anionic polymerization of 1-(alkoxy)ethyl methacrylates and the preparation of well-defined poly(methacrylic acid). *Macromolecules*, **31**, 7575–7580.

15

Complex Macromolecular *Chimeras*

Hermis Iatrou, Marinos Pitsikalis, Georgios Sakellariou and Nikos Hadjichristidis
Department of Chemistry, University of Athens, Panepistimiopolis, Zografou, 15771 Athens, Greece

15.1 Introduction

The synthetic polypeptides, although far from being strictly monodisperse or constructed from a precise sequence and composition of α-amino acid residues, possess the ability, as their natural relative – proteins, to form α-helix and β-sheet motifs. In order to approach the perfection found in nature, these synthetic polypeptides must possess the lowest possible molecular weight and compositional dispersities. Thus, their self-assembly into precisely defined nanostructures, a requirement for appropriate functionality, will be favored. Such nanostructures can then serve in the construction of nanomaterials and devices for biomedical and pharmaceutical applications (Duncan, 2003).

For many years, synthetic homo/copolypeptides have been obtained by the ring-opening polymerization (ROP) of α-amino acid N-carboxyanhydrides (NCAs) (Leuchs, 1906), (Leuchs and Manasse, 1907), (Leuchs and Geiger, 1908), (Szwarc, 1965), (Sekiguchi, 1981) generally using primary amines (e.g. n-hexylamine) as initiators, under an inert atmosphere. However, attempts to synthesize well-defined living polypeptides (Scheme 15.1) with amino initiators have been plagued, for more than 50 years, by unwanted polymerization mechanisms and termination reactions (Kricheldorf, 2006). This problem has been overcome by replacing the amino-initiators with organonickel (Deming, 1997) or ammonium chloride (Dimitrov and Schlaad, 2003) and very recently by hexamethyldisilazane (Lu and Cheng, 2007) initiators.

Complex Macromolecular Architectures: Synthesis, Characterization, and Self-Assembly, First Edition.
Edited by Nikos Hadjichristidis, Akira Hirao, Yasuyuki Tezuka and Filip Du Prez.

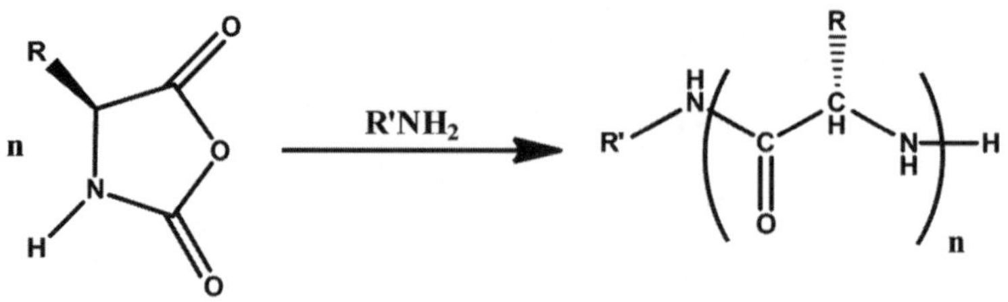

Scheme 15.1

Alternatively, high-vacuum techniques (HVT) have been employed to create living conditions for the amino-initiated ROP of NCAs. Apparently with this method, the unwanted polymerization mechanism (activated monomer) is either avoided or is insignificant, thus favoring the desired normal amine mechanism (Aliferis *et al.*, 2004).

All these living systems lead to well-defined homo/copolypeptides with high molecular weight and structural homogeneity (Aliferis *et al.*, 2005; Papadopoulos *et al.*, 2005; Hanski *et al.*, 2006; Iatrou *et al.*, 2007). Furthermore, these systems can be employed for the synthesis of polymer/polypeptide hybrid block copolymers, molecular *chimeras* (Schlaad and Antonietti, 2003), a term borrowed from the Greek mythology. The *Chimera* (in Greek *Χίμαιρα*) was an awesome fire-breathing she-monster (Homer, C8th BC) usually represented as an hybrid of a lion, goat and serpent (Figure 15.1). Biosciences started using the term *chimera* long ago to describe an organism composed of tissues that are genetically different.

Polymer–polypeptide hybrid block copolymers are able to self-assemble either in bulk (microphase separation) or in selective solvents (micellization), like conventional block copolymers, and in addition are organized within the microphases into α-helix or β-sheet

Figure 15.1 The *Chimera*, a fire-breathing she-monster of the Greek mythology (Homer, Iliad) usually represented as a hybrid of a lion, goat, and serpent.

motifs (Klok and Lecommandoux, 2006) thus leading to a rich variety of nanostructures (Hadjichristidis *et al.*, 2003).

The variety of the nanostructures could become even richer and more complex in multiblock copolymers, multiblock multicomponent (e.g. terpolymers) or in nonlinear (e.g. miktoarm stars) (Lodge, 2003) *chimeras,* approaching that found in nature. Therefore, the synthesis of well defined complex *chimeras* adds another dimension to the suprastructural hierarchy of the block copolymers with potential biomedical applications (Deming, 2007).

Although, there are many reviews on linear diblock *chimeras*, only a very few have addressed the synthesis of polypeptide-based materials with different macromolecular architectures (Hadjichristidis *et al.*, 2009; Deming, 2006; Deming, 2000). In this chapter our discussion on the synthesis of well-defined complex *chimeras*, including linear multiblock co/terpolymer, as well as nonlinear copolymer chimeras (e.g. star, graft, block-graft, dendritic), will hopefully enrich the current literature on macromolecular chimeras. This chapter will not address the self-assembly of these complex chimeras since this issue is discussed in another chapter of this volume.

15.2 Linear Multiblock Chimeras

Research on the synthesis of polypeptide hybrid diblock copolymers began in the mid-1970s (Yamashita *et al.*, 1975). Advances in polymer synthesis, culminating in the development of controlled/living polymerization techniques and click chemistry (Sumerlin and Vogt, 2010), in combination with the living ROP of NCAs, allowed for the preparation not only of simple diblock but also of a wide variety of multiblock *Chimeras*. In this section we will describe the materials synthesized according to the macroinitiators used for the ROP of the NCAs: amino- and transition-metal complexes.

15.2.1 Primary Amine Macroinitiators

ω-Functionalized polymers carrying amines on both ends can be used as macroinitiators for the ROP of NCAs. This remains the most widely applied method for the synthesis of a plethora of linear multicomponent hybrid block co/terpolymers. The major advantage of this method is the fact that several living/controlled polymerization techniques are able to prepare a variety of well-defined polymers with a high degree of amine functionalization. When combined with the recently developed ROP of NCAs under high vacuum, these methods lead to well-defined poplypeptide hybrids.

Most of these hybrid copolymers involve the use of telechelic poly(ethylene oxide), PEO, carrying amine end groups. This interest can be attributed to the fact that the amino-PEOs are commercially available in a variety of molecular weights with narrow molecular-weight distributions, are water soluble and biocompatible. Furthermore, these PEO polypeptide hybrids can form amphiphilic, double-hydrophilic and rod–coil–rod block copolymers, leading to materials with very interesting properties in solution and in bulk.

α,ω-Diamino-PEOs were employed as difunctional macroinitiators for the synthesis of ABA triblock copolymers, where A the polypeptide and B is PEO. Using this difunctional initiator, PBLG-*b*-PEO-*b*-PBLG (BLG: γ-benzyl-L-glutamate)(Cho, Kim and Komoto, 1990; Floudas *et al.*, 2003), PZLL-*b*-PEO-*b*-PZLL (ZLL: ε-carbobenzyloxy L-lysine) (Cho *et al.*, 1994;

Yang *et al.*, 2005) and poly[(DL-Val)-*co*-(DL-Leu)]-*b*-PEO-*b*-*p*oly[(DL-Val)-*co*-(DL-Leu)] (Cho *et al.*, 2003) have been prepared. Earlier studies resulted in poorly defined products, since diblocks and homopolymers were present in the final reaction products.

The better purification of the reagents (NCA, initiator, solvent) was found to offer a higher level of structural control. Thus, using a PEO-NH_2 macroinitiator and employing high-vacuum techniques, the synthesis of the PEO-*b*-PZLL-*b*-PBLG triblock terpolymers was achieved (Karatzas *et al.*, 2009) by sequential polymerization. Size-exclusion chromatography (SEC) analysis revealed that the macroinitiator was quantitatively consumed and that the final products displayed monomodal peaks of relatively narrow molecular-weight distribution. In addition, the molecular weights were very close to the stoichiometric values, indicating that when using the high-vacuum techniques the polymerization was free of termination reactions.

Hybrid block copolymers with aliphatic polyesters, such as poly(ε-caprolactone), PCL and polylactide, PLA, represent a very interesting class of polymeric materials. Due to the low immunogenicity, biocompatibility, biodegradability and excellent mechanical properties of aliphatic polyesters, these hybrids have found applications in tissue engineering and drug delivery (Langer, 2000; Tomson *et al.*, 1995; Gombotz and Petit, 1995). Moreover, the presence of the polyesters modifies the stability of the hybrids, since it reduces the enzymatic degradation of the polypeptides. Finally, the formation of crystalline polyester domains greatly influences the self-assembly behavior of the hybrid structures.

The general synthetic scheme for the polyester hybrids starts with the synthesis of the diamino polyester (A) macroinitiators, followed by the ROP of the desired NCA to give the corresponding BAB hybrids (B are the polypeptide chains).

The seven-membered cyclic initiator 2,2-dibutyl-2-stanna-1,3-dioxepane was used by Kricheldorf *et al.* (1998) to produce macrocyclic PCL (Scheme 15.2). Subsequent reaction of the two OHs with 4-nitrobenzoyl chloride yielded the corresponding α,ω-dinitrobenzoyl PCL, which was converted to the α,ω-diamine PCL by catalytic hydrogenation. These polymers were then employed as macroinitiators for the polymerization of the Gly-, Ala-, Phe- and BLG-NCAs to afford the respective triblock *chimeras* (Kricheldorf and Hauser, 2001). The low nucleophilicity of the 4-aminobenzoyl end-groups prevented the aminolytic cleavage of the PCL chains during synthesis and storage. The samples were characterized by viscometry, IR and NMR spectroscopy. The yields of the NCA polymerization were high but not quantitative (between 60 and 80%). The poor solubility of the polypeptides in organic solvents rendered impossible the determination of the molecular-weight distribution of the copolymers by SEC. However, there was evidence (NMR) that during the BLG-NCA polymerization, a mixture of di- and triblocks was obtained, due to the low initiation rate compared to that of propagation for this specific monomer.

Hybrid terpolymers have also been prepared by Deng *et al.* (2005) using poly(ethylene oxide)-*b*-poly(L-lactic acid) (PEG-*b*-PLLA) macroinitiators (Scheme 15.3). PEG-*b*-PLLA-OH was prepared by the ROP of LLA using monomethoxy PEG-OH in the presence of $Sn(Oct)_2$. The end-hydroxyl group was transformed to amino-group by reaction with *tert*-butoxycarbonyl-L-phenylalanine in the presence of dicyclohexylcarbodiimide, DCC. The deprotection of the amino group was performed under anhydrous conditions with trifluoroacetic acid at 0 °C, affording the PEG-*b*-PLLA-NH_2 macroinitiator, which was further used for the ROP of BLG-NCA leading to the synthesis of the triblock terpolymer PEG-*b*-PLA-*b*-PBLG. Characterization of the products revealed that the macroinitiators were efficiently prepared and were quantitatively consumed during the polymerization reactions. SEC analysis

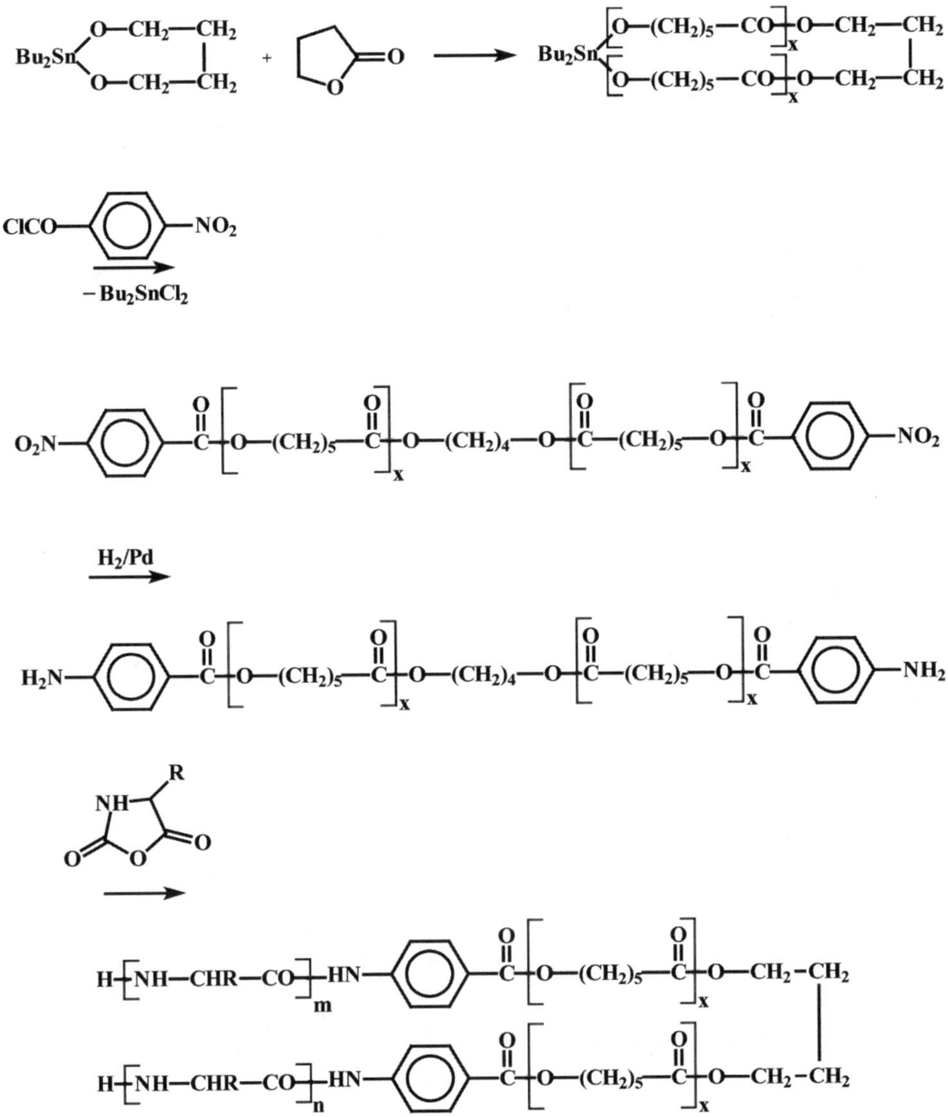

Scheme 15.2

and NMR spectroscopy confirmed that rather well-defined products of relatively narrow molecular-weight distribution were prepared.

The procedure developed by Chen *et al.* was adopted by Xia, Whittaker *et al.* for the synthesis of PEG-*b*-PLLA-*b*-PLL triblock terpolymers (Peng *et al.*, 2009), as shown in Scheme 15.4. The amine end groups of the PEG-*b*-PLA-NH_2 block copolymers served as macroinitiators for the polymerization of ZLL-NCA leading to the preparation of the desired terpolymers. The samples were characterized only by NMR and IR spectroscopy.

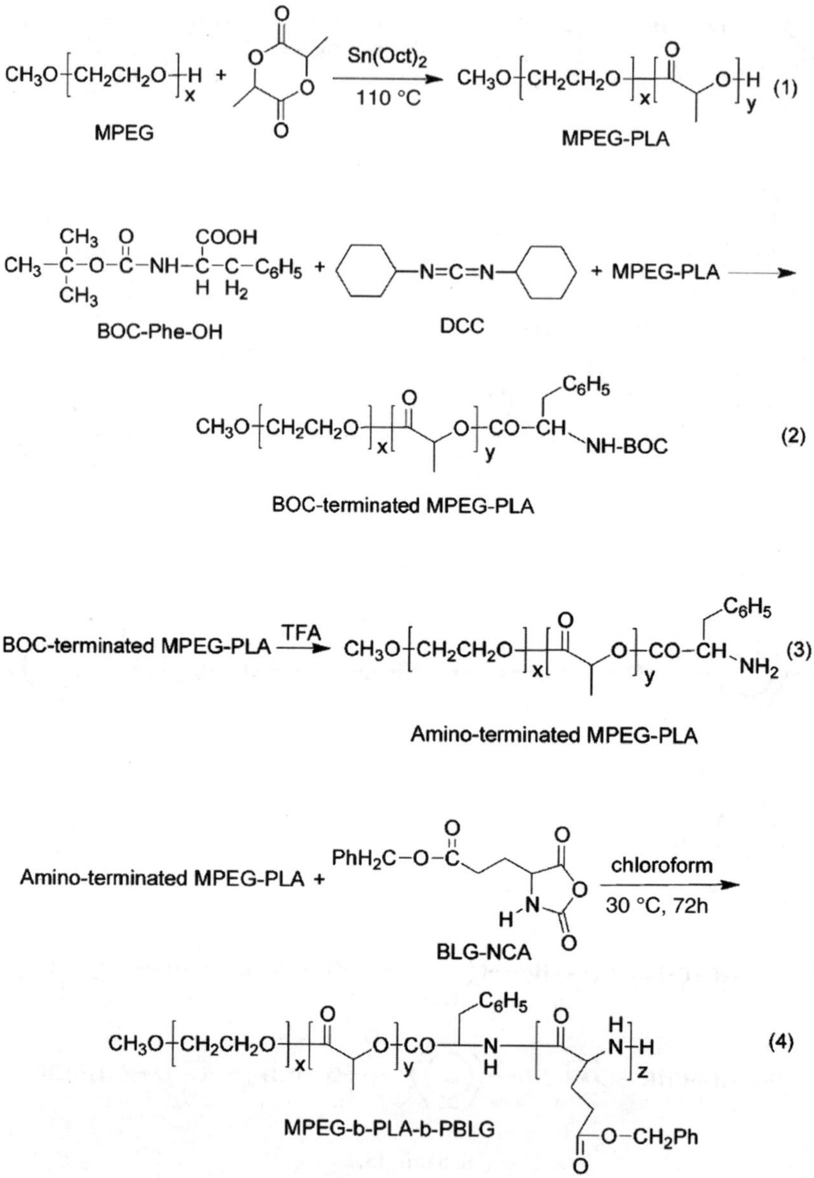

Scheme 15.3

Linear pentablock terpolymers PBLL-*b*-PBLG-*b*-PS-*b*-PBLG-*b*-PBLL (BLL: *tert*-butyloxycarbonyl-L-lysine, PS: polystyrene) were recently synthesized by Karatzas *et al.* (2008) using α,ω-diamino-PS. This polymer was prepared by anionic polymerization of styrene with sodium/naphthalene (difunctional initiator) and termination with 1-(3-bromopropyl)-2,2,5,5-tetramethyl-aza-2,5-disilacyclopentane, and of deprotection by successive

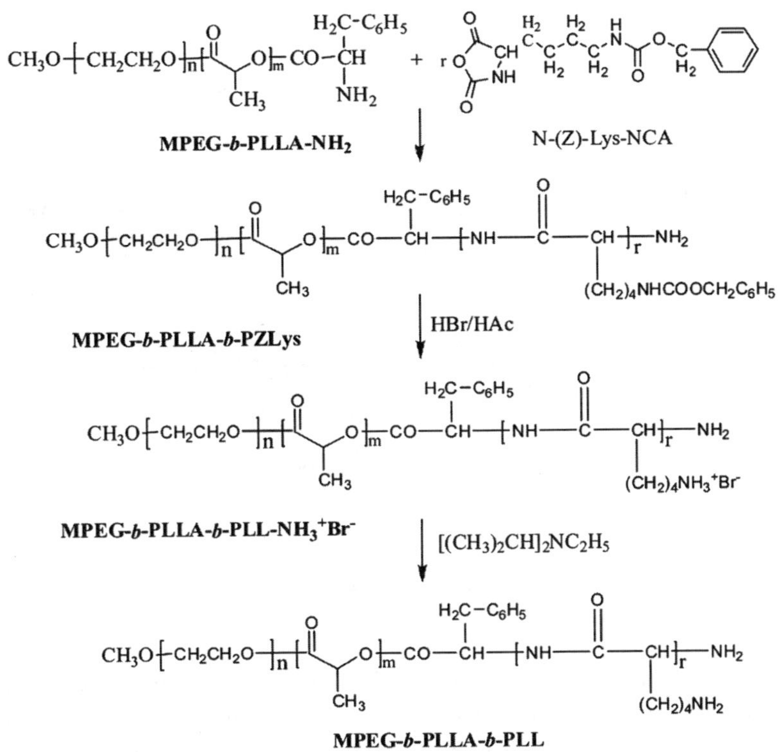

Scheme 15.4

precipitations of the polymer in methanol. The α,ω-diamino-PS was then used for the sequential polymerization of BLG-NCA and BLL-NCA under high vacuum (Scheme 15.5). The molecular weights were in very close agreement with the stoichiometric values and the molecular-weight distributions very narrow, indicating that well-defined samples were prepared with this methodology.

Step-growth polymerization is another technique leading to telechelic amine macroinitiators. The Yamamoto coupling polymerization of 2,7-dibromo-9,9-dihexylfluorene was performed by Kong *et al.* (2004) followed by the end-capping reaction with N-(p-bromobenzyl)phthalimide. Deprotection with hydrazine afforded the telechelic amine macroinitiators, which were subsequently used for the polymerization of BLG-NCA. Due to the nature of the step-growth polymerization the molecular-weight distribution, of the triblock copolymers was relatively broad (M_w/M_n values higher than 2.0) (Scheme 15.6).

An amine-functionalized TEMPO radical was employed by Karatzas *et al.* (2009) for the synthesis of poly(N-vinyl pyrrolidone), PNVP, copolymers with PBLG and PZLL blocks, PNVP-*b*-PBLG-*b*-PZLL (Scheme 15.7). The polypeptide blocks were initially prepared using the amino group as initiator of the ROP of NCAs. The TEMPO moiety did not interfere with the ROP and polypeptides with very narrow molecular-weight

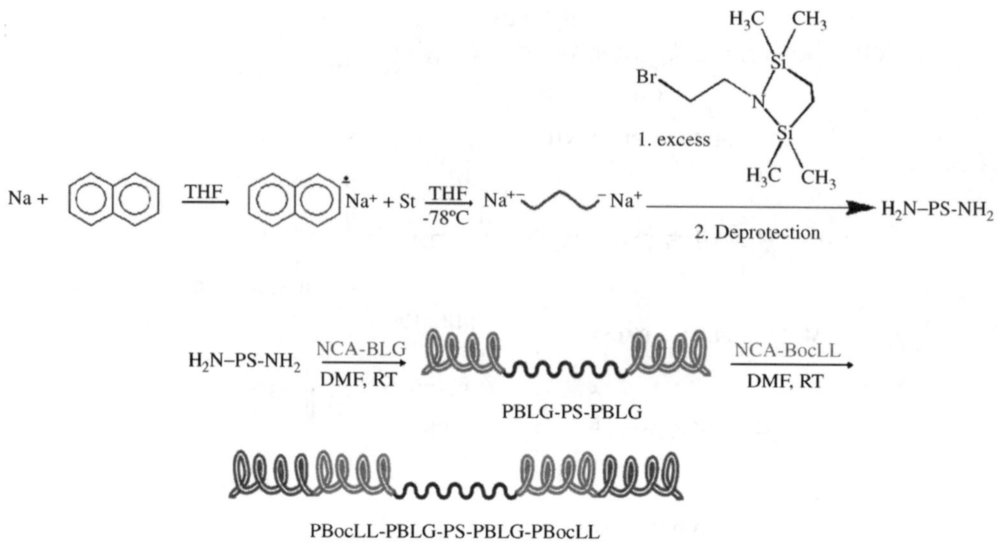

Scheme 15.5

distributions were obtained. The final step involved the nitroxide-mediated polymerization of NVP leading to *chimeras* with relatively low polydispersity (1.16–1.22) and controlled molecular weights ($22.6–96.5 \times 10^3$ g mol^{-1}).

15.2.2 Transition-Metal Complex Macroinitiators

Zero-valent metal complexes cannot be used directly for the synthesis of polypeptide hybrid block copolymers. However, N^{α}-allyloxycarbonyl-amino acid allyl amides can be used as universal precursors for the amido-amidate nickelacycle initiators (Scheme 15.8). As shown in Scheme 15.8, the N^{α}-allyloxycarbonyl-amino acid derivatives may undergo tandem oxidative additions to nickel(0) giving the nickelacycle initiators. This method was initially employed for the synthesis of block copolypeptides and was then expanded by Deming and coworkers to a variety of hybrid structures.

This methodology was applied for the synthesis of PBLG-*b*-polyoctenamer-*b*-PBLG and PBLG-*b*-polyethylene-*b*-PBLG triblock *chimeras* (Brzezinska and Deming, 2001) (Scheme 15.9). At the beginning, acyclic diene metathesis, ADMET, polymerization was employed for the synthesis of α,ω-diamino-functionalized polyoctenamer using Grubbs' catalyst $RuCl_2$-(=CHPh)$(PCy_3)_2$. This product was converted to the bisallyloxycarbonyl-L-leucine -terminated polyoctenamer, followed by reaction with [1,2-bis(diethylphosphino) ethane]Ni(COD), depeNi(COD), to produce the active sites for the ROP of BLG-NCA. After the NCA polymerization, the Wilkinson catalyst was employed to hydrogenate the polyoctenamer block, affording the corresponding polyethylene block. This methodology was also applied for the synthesis of triblock copolymers where the middle block was either poly(ethylene glycol), PEG, or poly(dimethyl siloxane), PDMS (Brzezinska *et al.*, 2002).

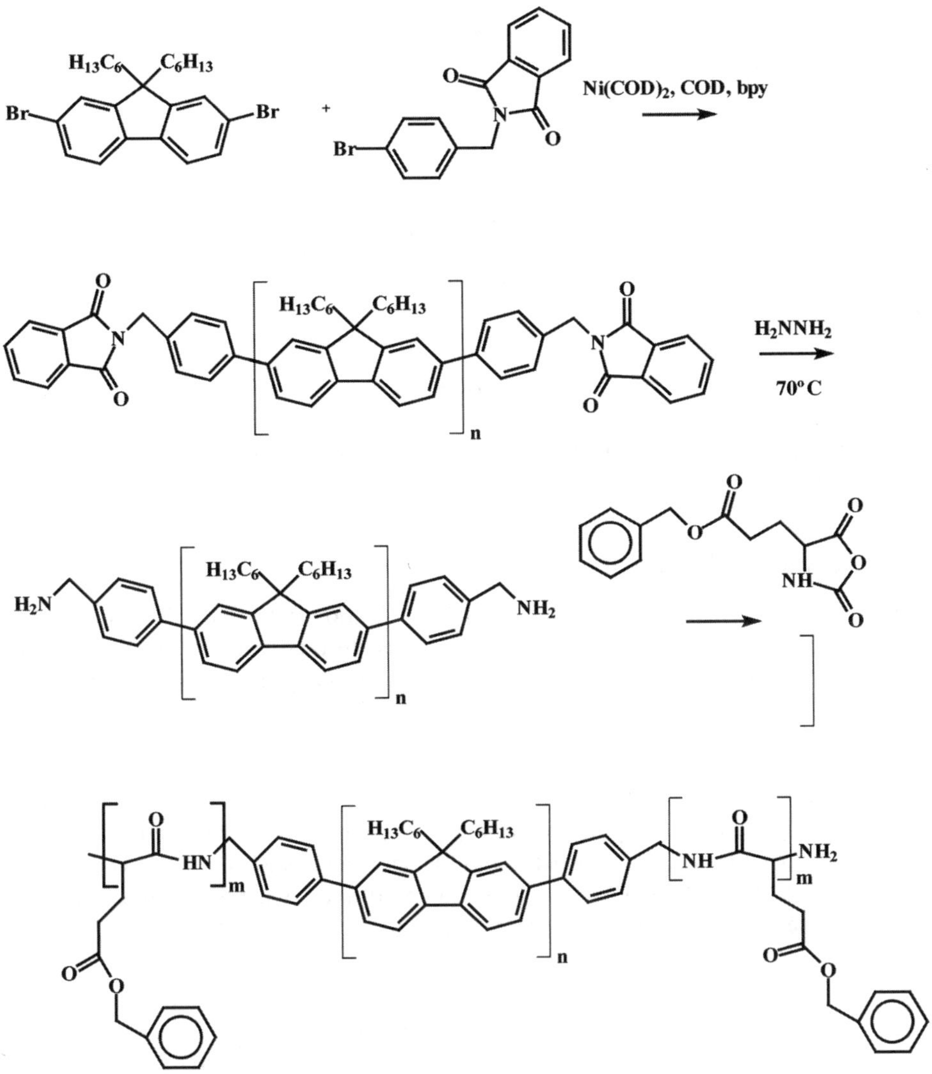

Scheme 15.6

It is well known that amido-metallacycle end groups are the active species for the synthesis of polypeptides by transition-metal complexes. Therefore, electrophilic reagents, such as isocyanates, can react with these end groups through formation of stable urea linkages. Based on this concept, excess isocyanate end-capped PEG was reacted with living PBLG to produce PBLG-*b*-PEG (Brzezinska *et al.*, 2002) (Scheme 15.10). SEC and NMR measurements revealed that the coupling reaction was near quantitative. This work was further expanded to the synthesis of CABAC pentablock terpolymers (C: PEG, A: PBLG and B: polyoctenamer, PEG or PDMS). The synthesis was achieved through the coupling reaction of the living ABA triblock copolymers with isocyanate end-capped PEG. Repeated precipitation from THF solutions into methanol was employed for the purification of the pentablock *chimeras*. SEC

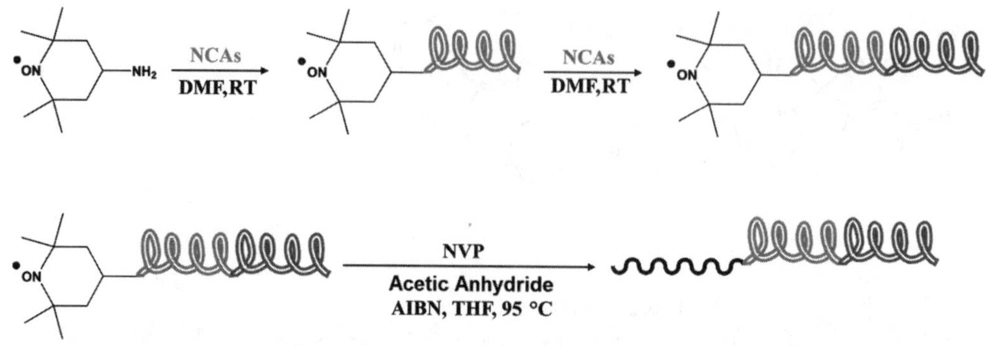

Scheme 15.7

analysis confirmed that monomodal traces of relatively narrow molecular-weight distribution were obtained.

15.3 Nonlinear Chimeras

Controlled/living polymerization techniques in combination with living ROP of NCAs can lead not only to linear but also to nonlinear *chimeras*. In this section we will describe the materials synthesized according to their structure: star, comb, brush-block, and dendritic-like *chimeras*.

15.3.1 Star Chimeras

Star polymers are branched polymers consisting of several linear chains emanating from a central core. Among the general synthetic routes (Mishra and Kobayashi, 1999; Hadjichristidis *et al.*, 2001; 2006; 2007) for star polymers, the use of multifunctional initiators has preferentially been used for the synthesis of polypeptide-based stars.

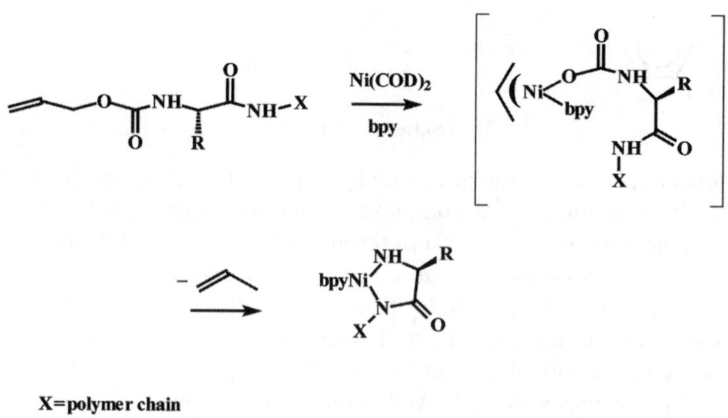

Scheme 15.8

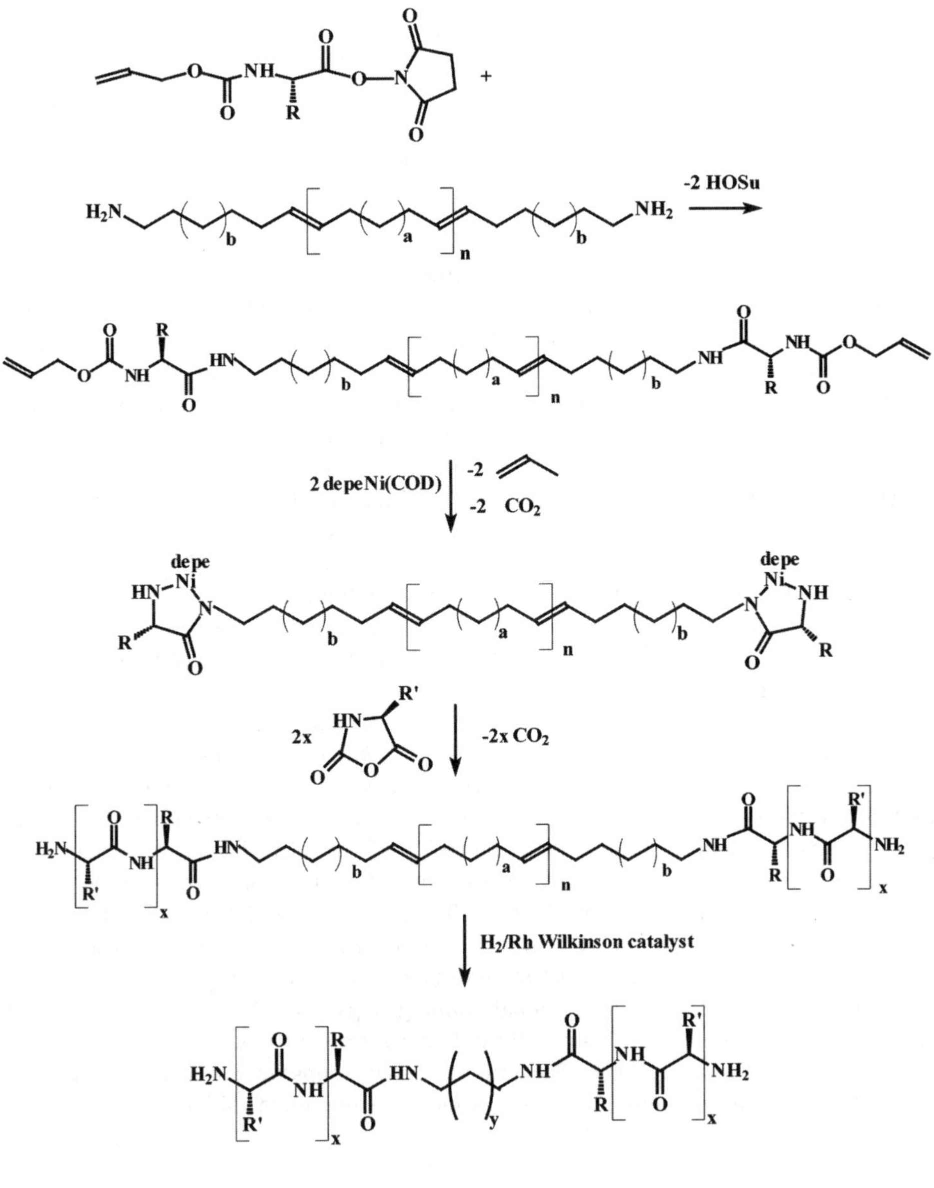

HOSu = N-hydroxysuccinimide
depe = 1,2-bis(diethylphosphino)ethane

Scheme 15.9

This method is referred to as the "core-first" or "arm-out" or divergent approach. According to this procedure, multifunctional compounds capable of simultaneously initiating the polymerization of several arms are used. There are several requirements a multifunctional initiator has to fulfill in order to produce star polymers with controllable molecular weights, uniform arm lengths, and low molecular-weight distribution. All initiating sites must be equally

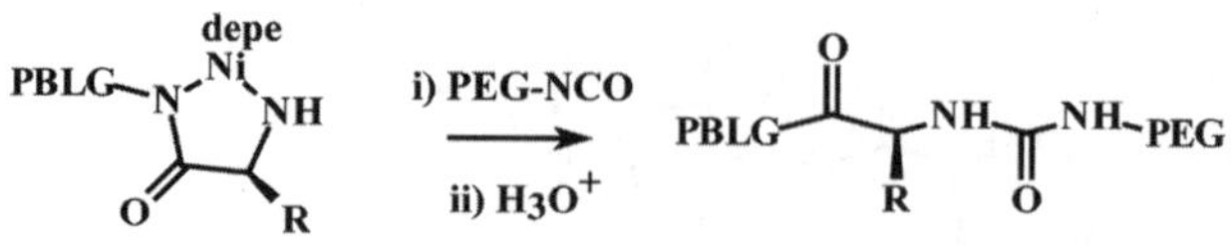

Scheme 15.10

reactive and the initiation rate must be higher than the propagation rate. The characterization of the star polymers produced by this method is difficult, since the molecular weight of the arms cannot be measured directly. The number of arms can be defined indirectly by several methods, such as end-group analysis, determination of the branching parameters, which are the ratios of either the mean square radius of gyration, or intrinsic viscosity or hydrodynamic radius of the star to the corresponding linear with the same molecular weight. Finally, the determination of the functionality can be achieved by isolation of the arms after cleavage (i.e. hydrolysis), if possible, and subsequent analysis. This approach is the most widely used for the synthesis of polypeptide stars. Using this methodology and a combination of different polymerization techniques several polypeptide-based star hybrid polymers have been prepared.

ω-Amino 4-arm PEO stars were efficiently used by Karatzas *et al.* (2009) for the ROP of BLG-NCA leading to (PEO-*b*-PBLG)$_4$ star-block copolymers (Scheme 15.11). The multifunctional macroinitiator was quantitatively consumed and thus giving a product of relatively narrow molecular-weight distribution and molecular weight very close to the stoichiometric value.

A combination of ATRP and ROP was reported by Abraham *et al.* (2006) for the synthesis of (PS-*b*-PBLG)$_3$ 3-arm star-block *chimeras*. ATRP techniques were initially adopted for the synthesis of 3-arm PS star using a suitable cynurate-based initiator. The end-bromine groups were transformed to amines by a three-step azidation route followed by conversion to nickel-amine complex macroinitiators suitable for the polymerization of BLG-NCA (Scheme 15.12). The functionality of the precursor PS star was confirmed by the measurement of the molecular weights of the stars and the corresponding arms. The arms were isolated by the hydrolysis of the ester groups at the core. The final products were obtained in moderate yields and possessed relatively narrow molecular-weight distributions ($1.1 < M_w/M_n < 1.4$).

The polyamidoamine with four terminal amine groups (G0-PAMAM) was employed by Dong *et al.* as tetrafunctional initiator for the polymerization of BLG-NCA leading to the four-arm (PBLG)$_4$ stars (Qiu *et al.*, 2009). The terminal amine groups were further functionalized by reaction with 2-bromo-2-methylpropionyl bromide to introduce initiation sites for ATRP.

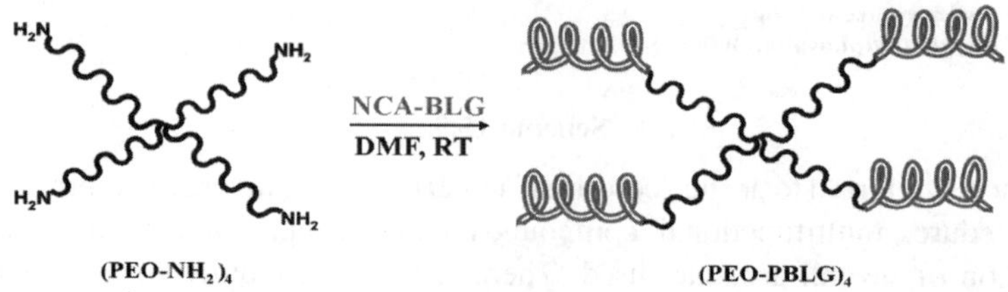

Scheme 15.11

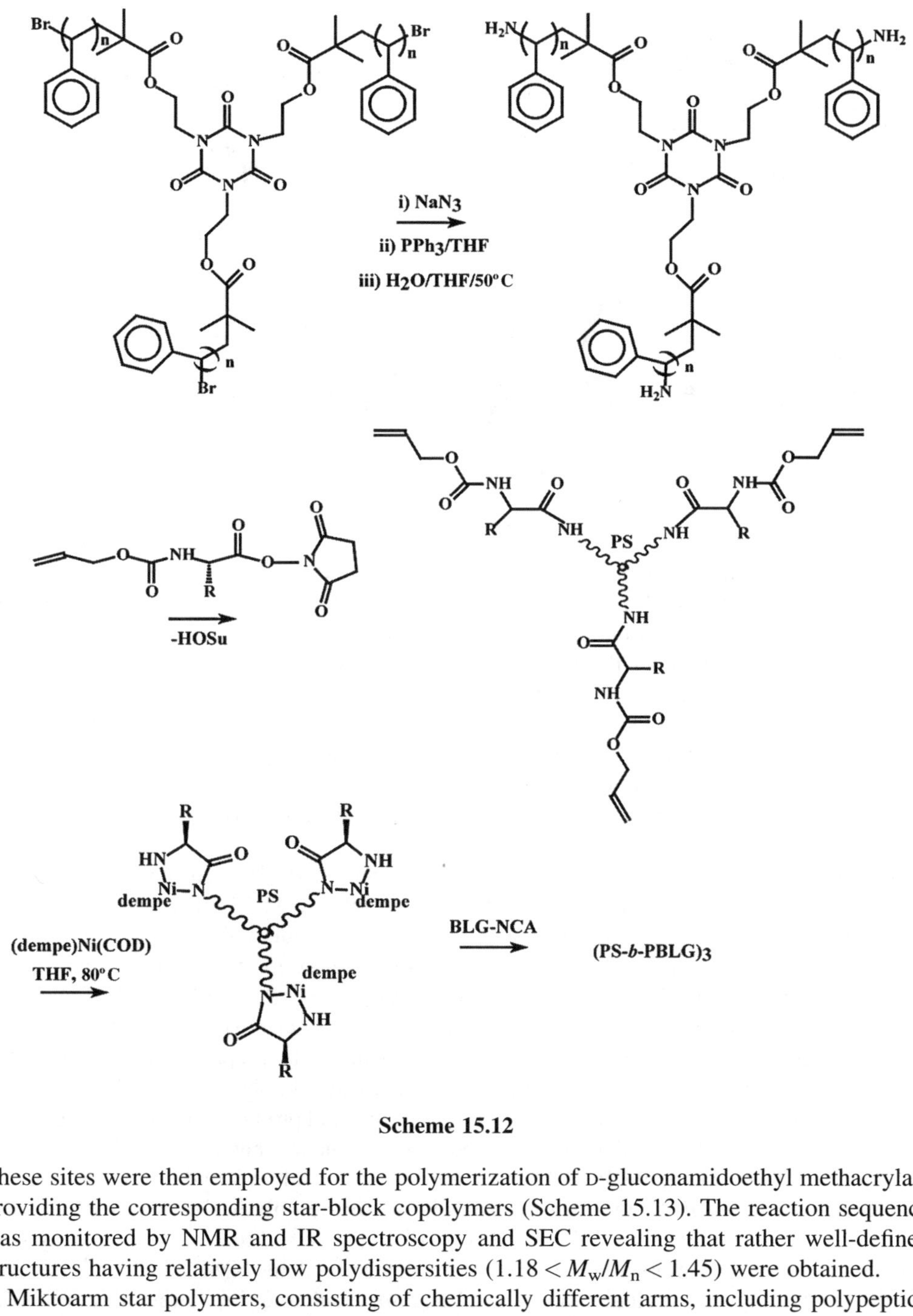

Scheme 15.12

These sites were then employed for the polymerization of D-gluconamidoethyl methacrylate providing the corresponding star-block copolymers (Scheme 15.13). The reaction sequence was monitored by NMR and IR spectroscopy and SEC revealing that rather well-defined structures having relatively low polydispersities ($1.18 < M_w/M_n < 1.45$) were obtained.

Miktoarm star polymers, consisting of chemically different arms, including polypeptide chains have also been synthesized. A combination of ATRP and ROP were employed by Babin *et al.* (2005; 2008) for the synthesis of $PS(PBLG)_2$ miktoarm star *chimeras* (Scheme 15.14). Styrene was first polymerized by ATRP, followed by reaction with a

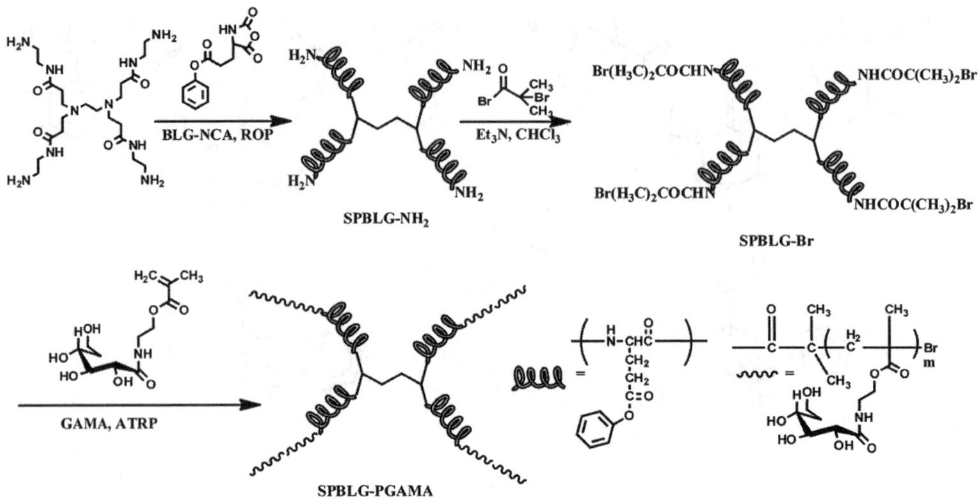

Scheme 15.13

large excess of 1-aminotriethylenetriamine for the nucleophilic substitution of the end-bromine group, leading to PS chains bearing two amine end groups. Due to the nonquantitative nature of this reaction, the unfunctionalized PS chains were removed by selective precipitation in hexane after treatment of the $PS(NH_2)_2$ with aqueous HCl. The ω-amine groups were subsequently used for the polymerization of BLG-NCA, to afford the desired structures. Hydrolysis with an excess of hydrogen bromide in trifluoroacetic acid led to the cleavage of the benzyloxycarbonyl protective groups to afford the corresponding amphiphilic miktoarm star *chimeras*. SEC analysis revealed that the $PS(NH_2)_2$ macroinitiators were quantitatively consumed and that the molecular-weight distributions of the final products were fairly narrow ($1.09 < M_w/M_n < 1.22$).

$(PLLA)_2$PBLG 3-miktoarm stars were prepared by Sun *et al.* (2009) using the trifunctional initiator 2-benzyloxycarbonylamino-1,3-propanediol (Scheme 15.15). The two available hydroxyl groups were employed for the polymerization of L-lactide in the presence of $Sn(Oct)_2$. Subsequent cleavage of the benzyloxycarbonyl protective group, with hydrogen bromide in acetic acid, afforded PLLA chains with a central amine group, $(PLLA)_2NH_2$, which served as macroinitiator for the polymerization of BLG-NCA to give the desired miktoarm. NMR analysis revealed that the $(PLLA)_2NH_2$ macroinitiator was efficiently prepared. SEC experiments confirmed the complete consumption of the macroinitiator and the absence of any homopolymer trace. However, the molecular-weight distributions of the final products were not measured due to the self-assembly of the miktoarm star copolymers in the solvent used for SEC.

A commercially available PEO bearing two central- and two end-amine groups was employed by Cho *et al.* (2000) as a macroinitiator for the polymerization of BLG-NCA, leading to the synthesis $(PEO\text{-}b\text{-}PBLG)_2(PBLG)_2$ 4-miktoarm star copolymers, (Scheme 15.16). The molecular weights were determined by NMR analysis. However, details concerning the efficiency of this synthetic approach and the characterization of the products were not provided.

Taking advantage of the unique features of anionic polymerization to produce end- or in-chain amino-functionalized polymers, as well as of the living nature of the ROP of NCAs under

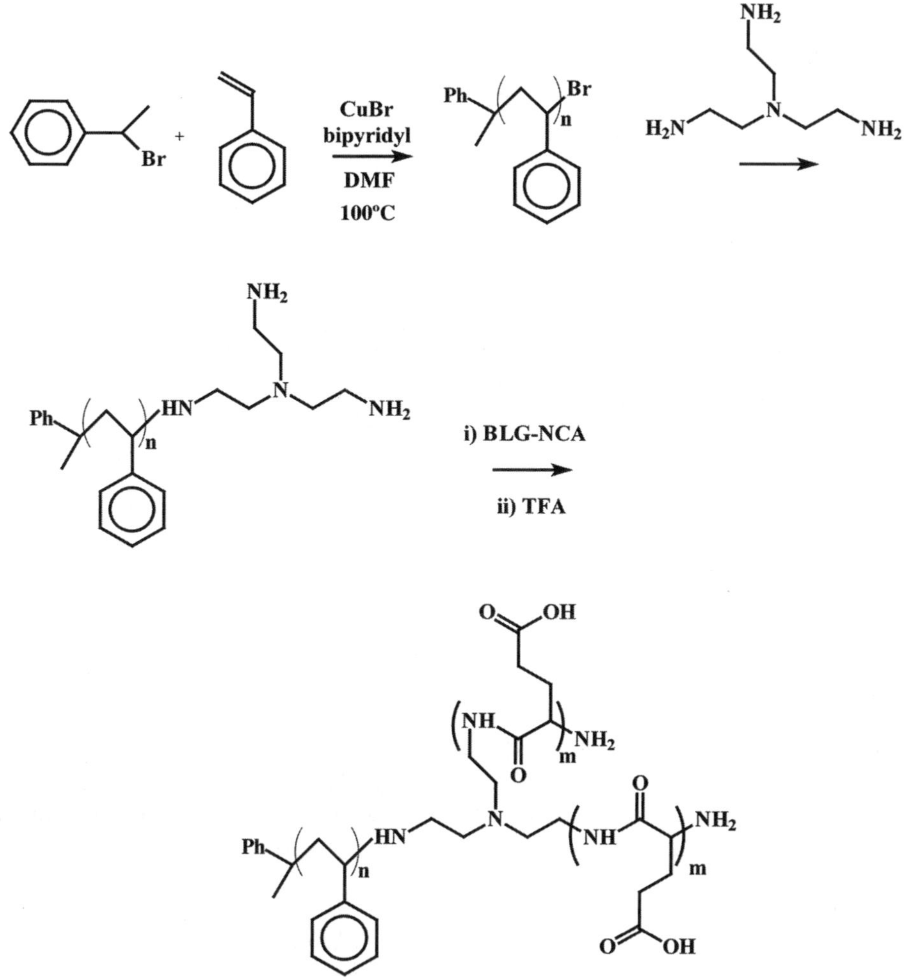

Scheme 15.14

high-vacuum conditions, a variety of well-defined miktoarm star hybrids was synthesized and fully characterized by Karatzas *et al.* (2008). These structures include: $(PS)_2$(PBLG), $(PS)_2$(PBLL), (PS)(PI)(PBLG), (PS)(PI)(PBLL), $(PS)_2$[P(α-MeS)](PBLG), $(PS)_2$[P(α-MeS)](PBLL), $(PS)_2(PBLG)_2$ and $(PS)_2(PBLL)_2$ (Scheme 15.17). The synthetic strategy involved the preparation of diphenyl ethylene, DPE, functionalized polymers [DPE-chain-end-functionalized PI (PI-D), DPE-in-chain-functionalized polystyrene (PS-D-PS) and DPE-in-chain-difunctionalized polystyrene], shown in Scheme 15.18. These structures were subsequently activated by reaction with a living polymer chain or *s*-BuLi followed by reaction with 1-(3-bromopropyl)-2,2,5,5-tetramethyl-aza-2,5-disilacyclopentane. The silyl-protected group was cleaved by treatment with HCL or *p*-toluenesulfonic acid, leading to the synthesis of the corresponding amine-functionalized polymers. These products were employed as macroinitiators for the polymerization of BLG- and BLL-NCAs to afford the desired miktoarm stars. Combined characterization by NMR, SEC and light-scattering measurements confirmed the

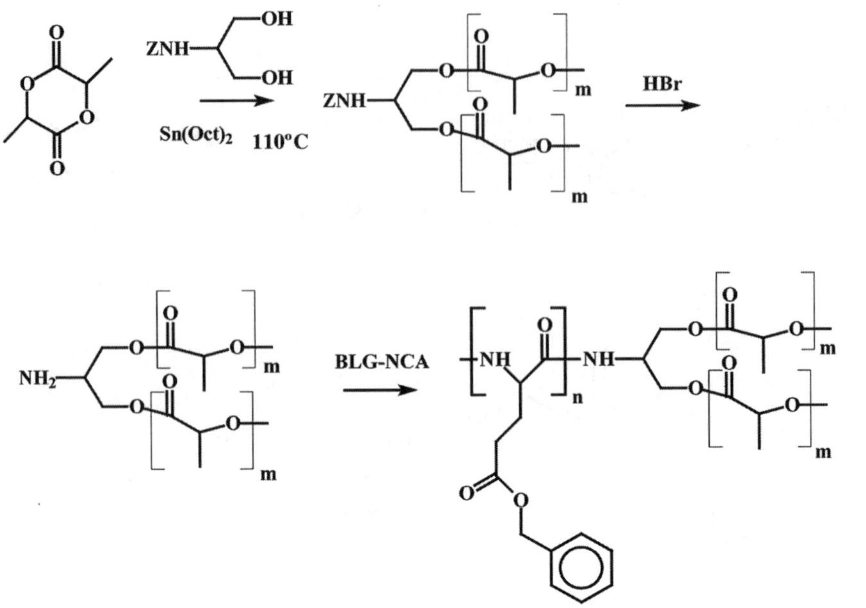

Scheme 15.15

efficiency of this synthetic scheme and the homogeneity of prepared products. The only restriction of this procedure was the steric hindrance of the in-chain functionalized macroinitiators. This became obvious from the fact that $(PS)_2(NH_2)_2$ macroinitiators led to miktoarm stars with broader molecular-weight distributions, whereas the more sterically hindered $(PS)_4(NH_2)_2$ macroinitiators afforded multimodal distributions.

15.3.2 Comb, Brush-Block, Dendritic-Like Chimeras

The limitations in achieving true living polymerization conditions of NCAs have, in the past, rendered difficult the synthesis of more complex structures containing polypeptide chains, such

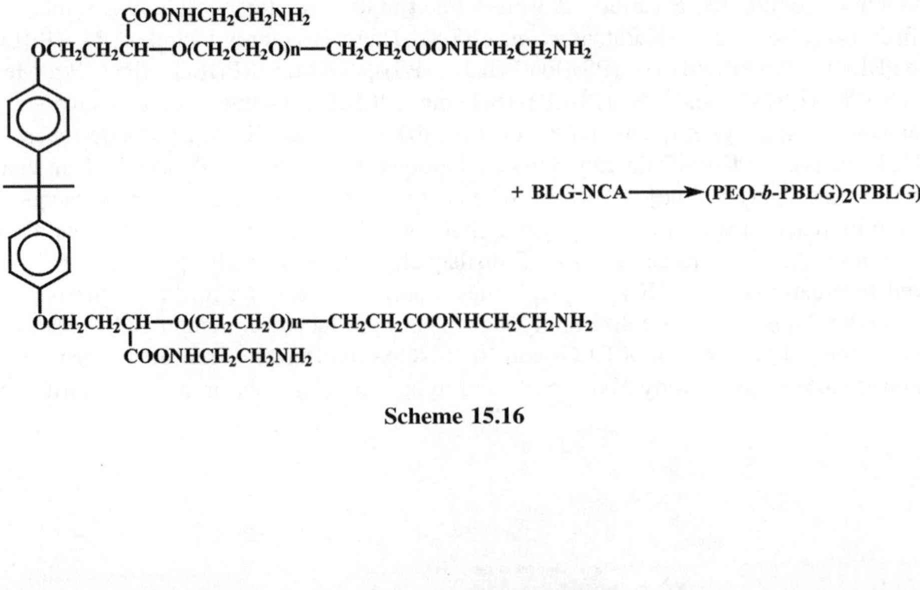

Scheme 15.16

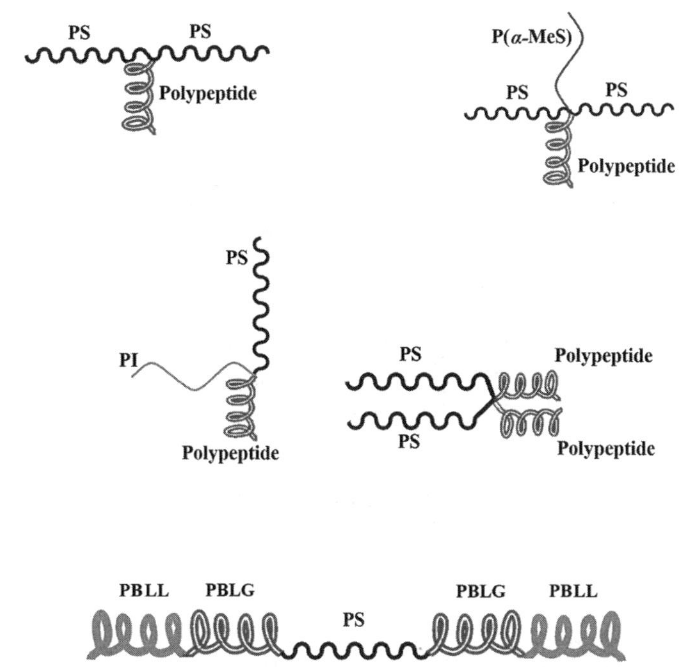

Scheme 15.17

as comb-shaped *chimeras*. Three general methods have been developed for the synthesis of randomly branched graft copolymers: (a) the "grafting-onto", (b) the "grafting-from" and (c) the macromonomer method (or "grafting-through" method) (Pitsikalis *et al.*, 1998; Hadjichristidis *et al.*, 2001; 2003b; 2006).

In the "grafting-from" method active sites are generated randomly along the backbone. These sites are capable of initiating the polymerization of a second monomer, leading to the synthesis of graft copolymers. This method is widely used for the synthesis of graft hybrids, especially with polypeptide side chains. Amine groups were generated along the backbone and then served as initiation sites for the ROP of NCAs. An early example includes the conventional radical copolymerization of N-benzylacrylamide with N-(3-aminopropyl)methacrylamide hydrochloride. The copolymer obtained was then dehydrochlorinated to afford free primary amine groups, subsequently functioning as initiation sites for the polymerization of Ala-NCA (Boulahia *et al.*, 1989). (Scheme 15.19) In another study, N-methyl-N-(4-vinylphenethyl) ethylenediamine was copolymerized with 2-hydroxyethyl methacrylate. The side amine groups were subsequently used to copolymerize the NCAs of β-benzyl L-aspartate and β-(4-phenylazobenzyl) L-aspartate to give the graft copolymer *chimera* (Aoyama *et al.*, 1990a; 1990b (Scheme 15.20) These structures were not characterized thoroughly in terms of the molecular and structural characteristics.

The "grafting-from" approach was also employed in more recent studies. N-*tert*-butoxycarbonyl-N′-(2-methacryloyl)-1,3-diaminopropane was prepared by Zhang *et al.* (2005) and polymerized by free-radical polymerization. The quantitative deprotection of the amine group was performed in a mixture of dichloromethane and trifluoroacetic acid,

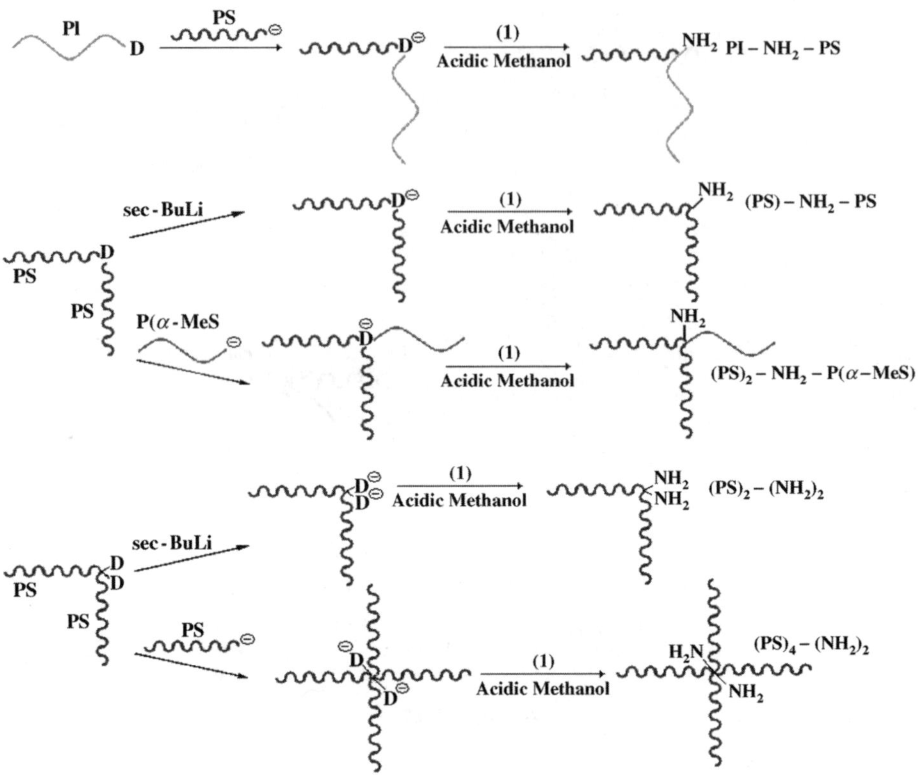

Scheme 15.18

followed by oligomerization of the BLG- or ZLL-NCAs affording the corresponding molecular brushes (Scheme 15.21). The samples were characterized by NMR and AFM techniques.

A similar methodology was applied by Xiang *et al.* (2009) for the synthesis of chitosan grafted with poly(L-tryptophan) chains (Scheme 15.22). Deacetylated chitosan (deacetylation degree of amino groups 75%) was employed as a multifunctional macroinitiator for the polymerization of the L-tryptophan-NCA in ethyl acetate solutions leading to the synthesis of chitosan-*g*-poly(L-tryptophan) graft copolymers. The products were analyzed by NMR and IR spectroscopy.

Using the grafting-from methodology amphiphilic block-brush copolymers were prepared by Cai *et al.* (2009). Amine-terminated PEO was employed as macroinitiator for the polymerization of BLG-NCA leading to the synthesis of PEO-*b*-PBLG block copolymers. Subsequent aminolysis of the benzyl ester groups of the PBLG block using 1,2-diaminoethane led to the synthesis of the corresponding PEO-*b*-poly[(2-aminoethyl)-L-glutamate)], PEO-*b*-P(ELG-NH_2), copolymers. The amine moieties of the P(ELG-NH_2) blocks served as initiators for the polymerization of BLG-NCA leading to the synthesis of the desired block-brush copolymer, PEO-b-[PELG-g-PBLG] (Scheme 15.23). The reaction sequence was monitored by SEC. It was found that the molecular-weight distribution of the initial PEO-*b*-PBLG block copolymer ($M_w/M_n = 1.2$) was substantially broadened after the grafting from

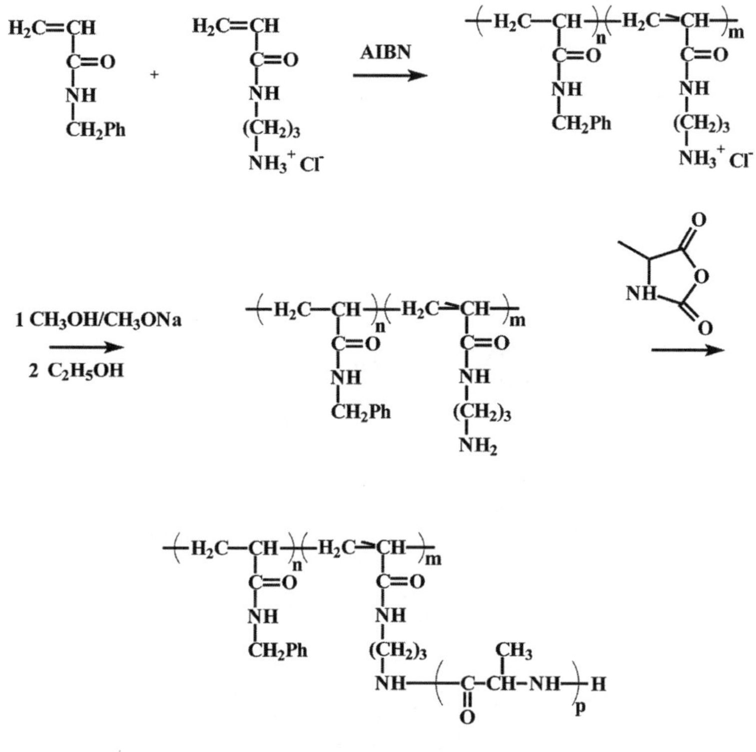

Scheme 15.19

procedure ($1.6 < M_w/M_n < 2.0$). The molecular weights of the samples were obtained by NMR spectroscopy.

The third method for the synthesis of graft copolymers is based on macromonomers, which is an oligomeric or polymeric chain bearing a polymerizable end group. Macromonomers having two polymerizable end groups have also been reported. Copolymerization of preformed macromonomers with another monomer yields graft copolymers. In an early report, *m,p*-vinylbenzylamine was used as initiator for the oligomerization of DL-phenylalanine-NCA, leading to narrow molecular-weight distribution macromonomers of low molecular weight (Scheme 15.24). These macromonomers were then copolymerized with MMA and S, to give PMMA-*g*-PPhe and PS-*g*-PPhe graft copolymers, respectively (Takaki *et al.*, 1987).

Polystyrenes dendronized with L-Lysine dendrons of different generations were prepared by the procedure given in Scheme 15.25 (Lubbert *et al.*, 2005). 4-Vinylbenzylamine was linked with the appropriate protected L-Lysine dendrons under standard peptide coupling conditions, affording the corresponding dendronized macromonomers. The macromonomers were then subjected to conventional radical polymerization to give products with broad molecular-weight distributions. Monomer conversion and polymer molecular weight were found to increase with increasing dendron size.

A polyamidoamine dendron (PAMAM) carrying a Boc-protected amine group was also synthesized by Harada *et al.* (2006; 2009). The protective group was then removed

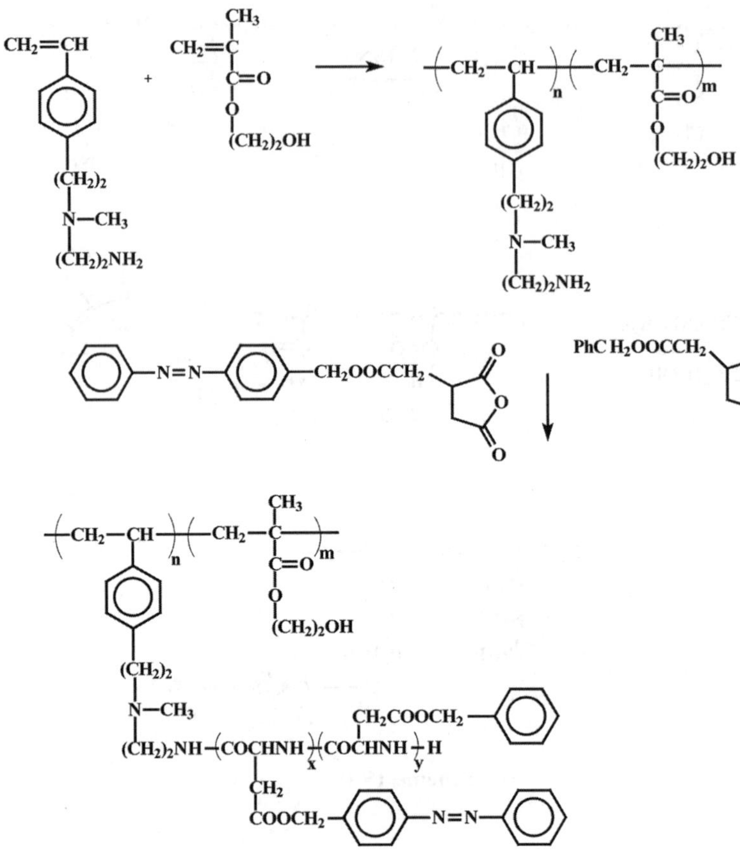

Scheme 15.20

and the primary amine group was employed as initiator for the polymerization of ZLL-NCA leading to linear-dendritic block copolymer (Scheme 15.26).

Linear/dendron-like polypeptide hybrid structures were prepared by Dong *et al.* by combination of ROP and click chemistry (Hua *et al.*, 2009). The synthesis of linear PCL/dendron-like PBLG, PCL-*b*-(Dm-PBLG), hybrid copolymers having 2^m ($m = 0,1,2,3$) PBLG branches is described in Scheme 15.27. Benzyl alcohol was used as initiator and $Sn(Oct)_2$ as the catalyst for the ROP of ε-CL. The end-hydroxyl group was then quantitatively tosylated and was further transformed into azide-terminated PCL, PCL-N_3. Propargyl focal point PAMAM-typed dendrons with 2^m ($m = 0,1,2,3$) primary amine groups, Dm, were synthesized and then click conjugated with the PCL-N_3 to produce amine-terminated PCL dendrons. The available amine groups were employed as initiation sites for the polymerization of BLG-NCA leading to the synthesis of the desired structure. SEC analysis revealed that the products had moderate polydispersities ($1.13 < M_w/M_n < 1.36$). Detailed characterization by SEC, MALDI, NMR spectroscopy, DSC and WAXD verified the synthesis of the target structures.

Using similar methodology the synthesis of asymmetric PEO-b-(Dm-PBLG) and symmetric (Dm-PBLG)-b-PEO-b-(Dm-PBLG) hybrid copolymers having 2^m ($m = 0,1,2,3$) PBLG

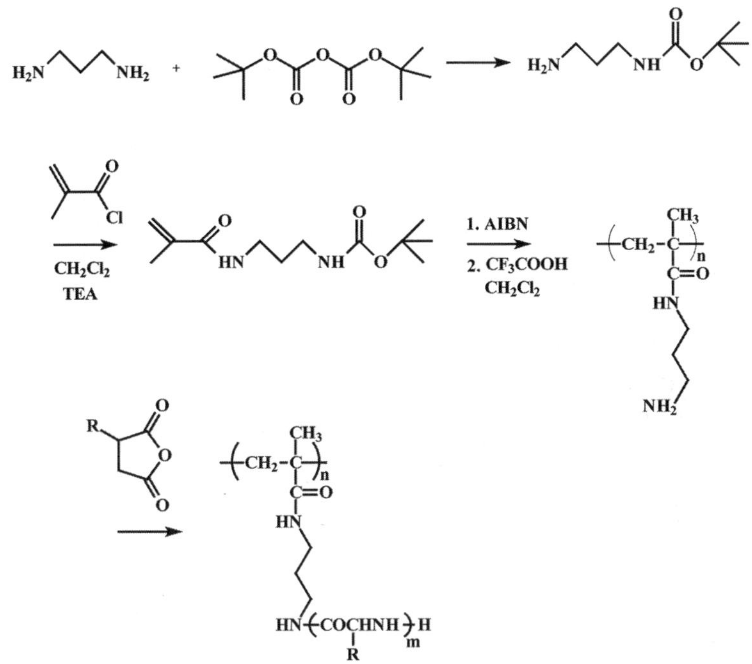

Scheme 15.21

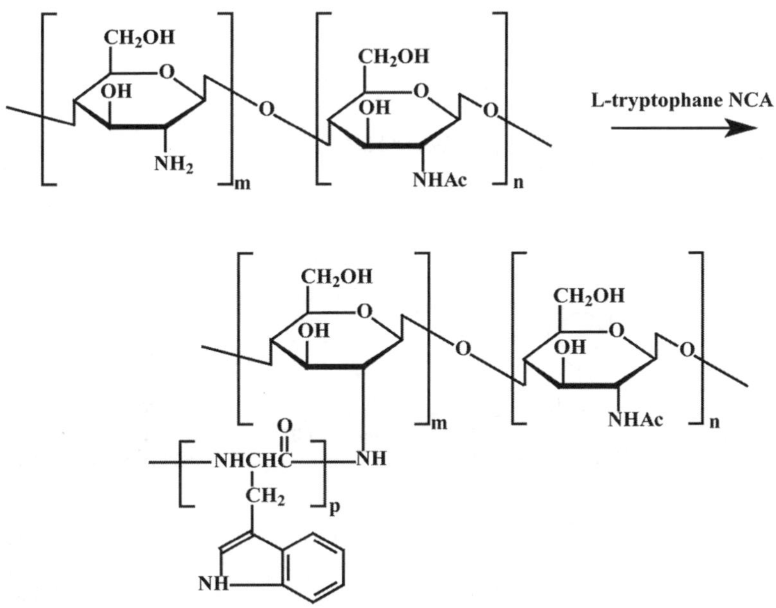

Scheme 15.22

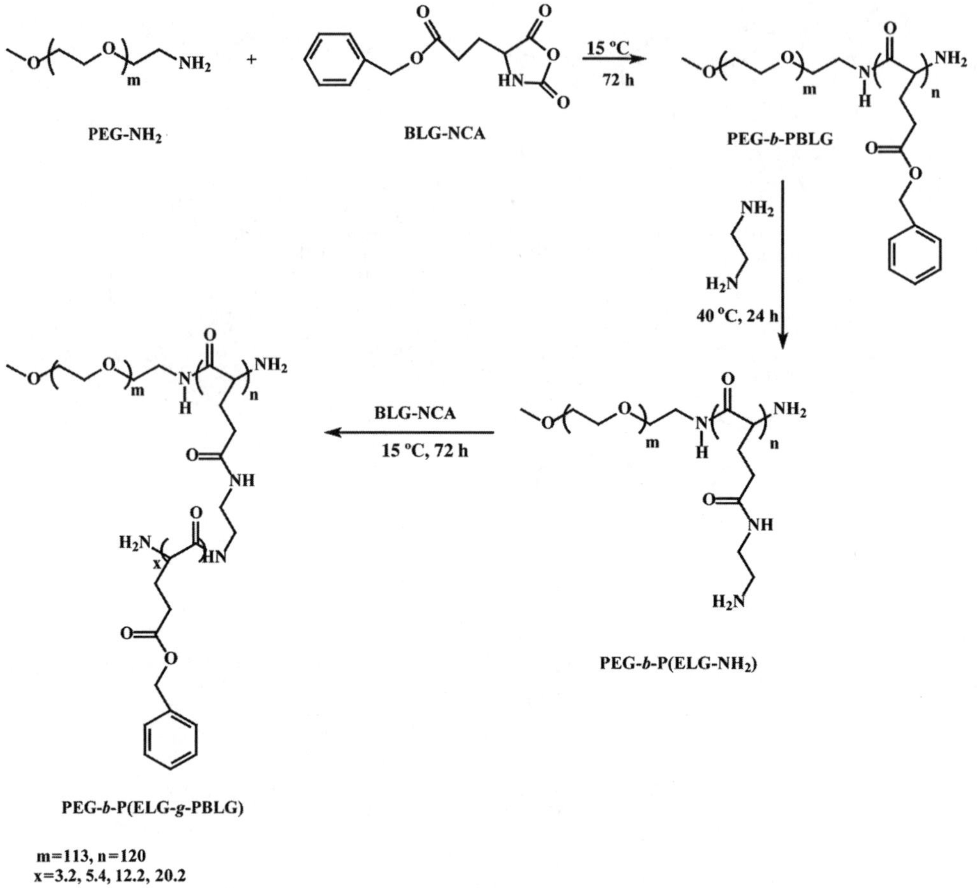

Scheme 15.23

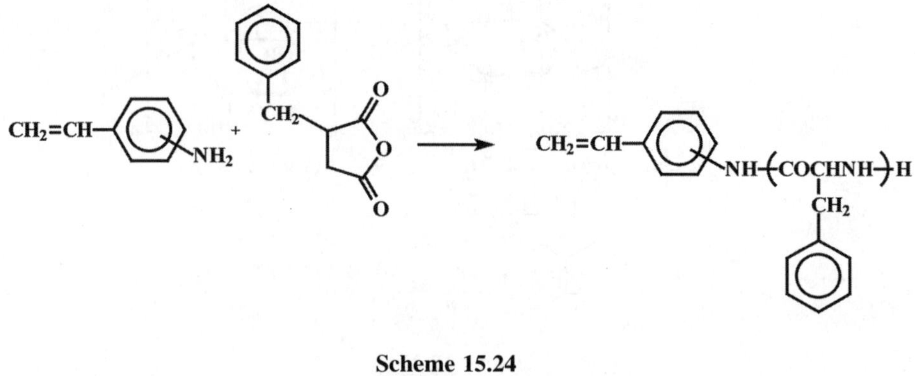

Scheme 15.24

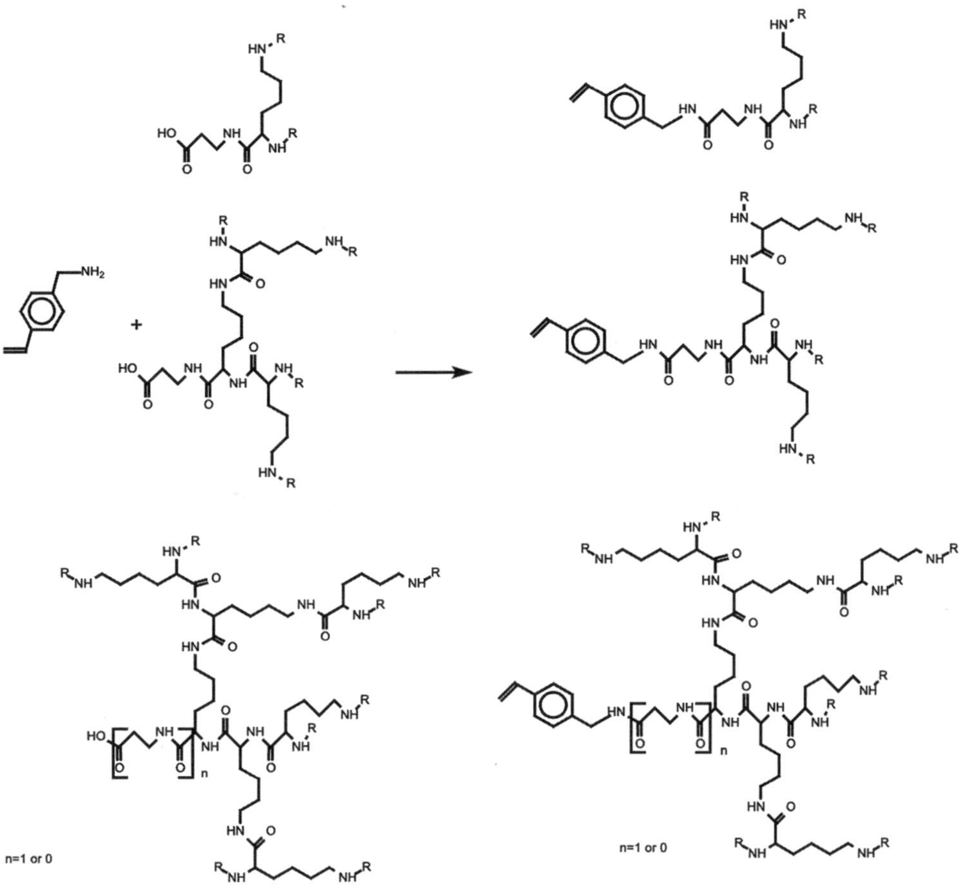

Scheme 15.25

branches was reported by Peng *et al.* (2009) (Scheme 15.28). Both the "arm-first" and "core-first" strategies were employed for the synthesis of the desired products. According to the "arm-first" approach, the propargyl focal point dendrons, Dm having 2^m ($m = 0,1,2,3$) primary amine groups were first used for the ROP of BLG-NCA, leading to the formation of "clickable" dendron-like PBLG polymers having 2^m branches. These structures were then click conjugated with azide-terminated PEO at the one or both chain ends (PEO-N_3, N_3-PEO-N_3) producing the asymmetric and the symmetric block hybrids, respectively. Alternatively, according to the "core-first" approach the propargyl focal point dendrons were first click conjugated with PEO-N_3 or N_3-PEO-N_3 generating primary amine-terminated PEO dendrons, which were subsequently used for the ROP of BLG-NCA leading to the synthesis of the target structures. SEC analysis, NMR and FT-IR spectroscopy revealed that the products were rather well defined and had relatively narrow molecular-weight distributions.

Hyperbranched amphiphilic multiarm polyethyleneimine-PBLG copolymers were prepared by Tian *et al.* (2006) The synthetic approach involved the ROP of BLG-NCA,

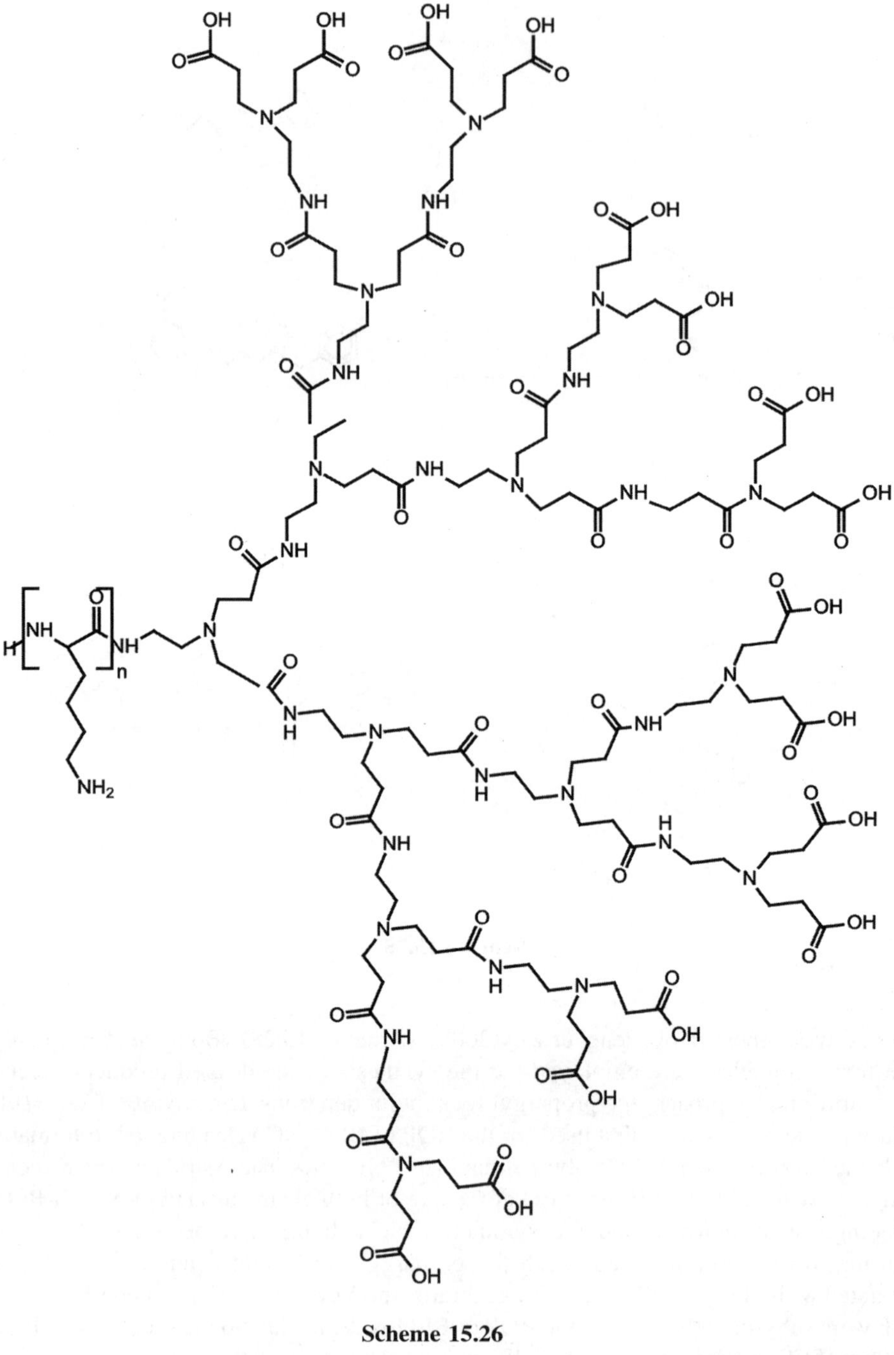

Scheme 15.26

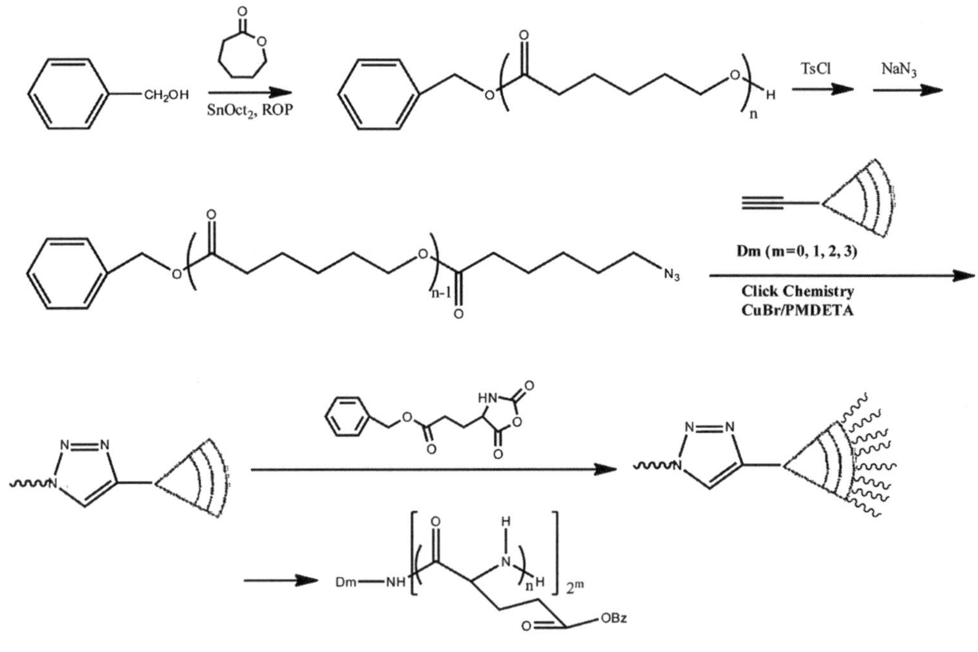

Scheme 15.27

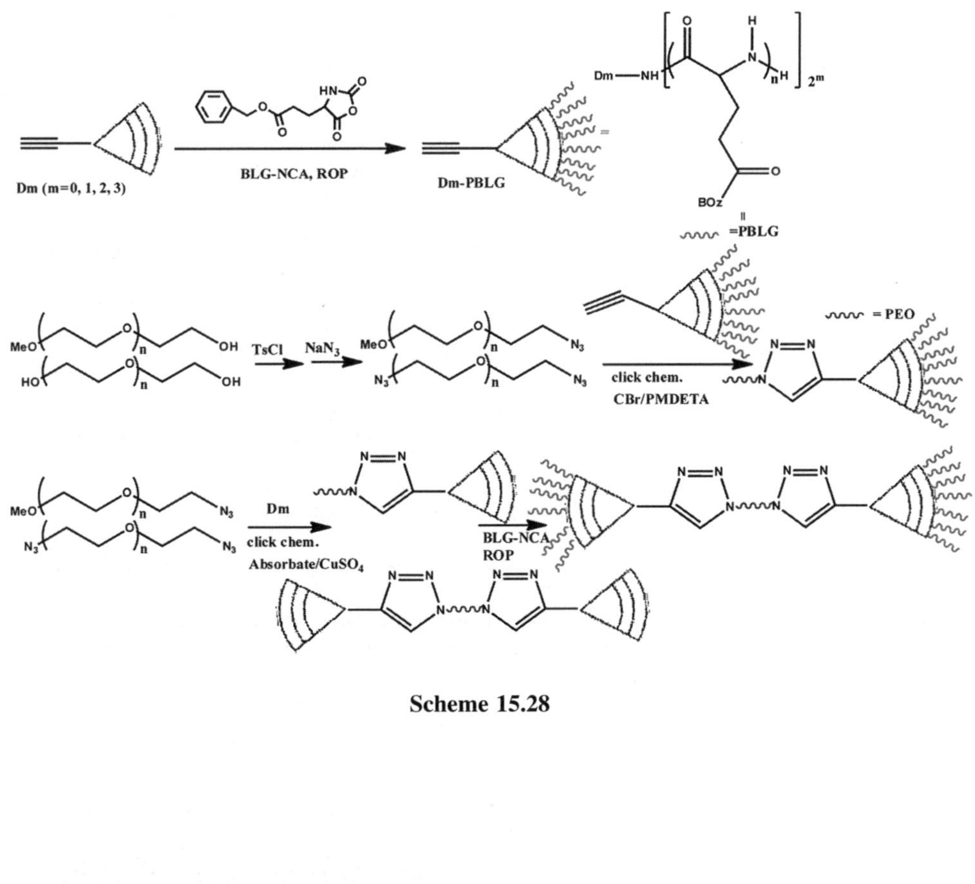

Scheme 15.28

initiated by a hyperbranched polyethyleneimine (PEI) macroinitiator. No data concerning the polydispersity indices of the PEI precursors or the final copolymers were provided. In this work, the micellar properties in aqueous solutions of the multiarm copolymers synthesized were examined.

15.4 Concluding Remarks

It is clear that the new developments in living/controlled polymerizations, as well as in the living ROP of NCAs, hold tremendous promise for the synthesis of well-defined complex *chimeras* materials with controllable molecular weight, sequence, composition and molecular-weight distribution. Although much development is still required, these complex polypeptide-based materials begin to rival their natural counterparts in terms of complexity and accuracy. Such well-defined materials can self-assemble into precisely defined nanostructures, a requirement for the development of new biomedical and pharmaceutical tools with a wide range of tunable properties. Furthermore, studies on these novel *chimeras* will shed light on the natural phenomena associated with protein function and functionality.

References

Abraham, S., Ha, C.-S., and Kim, I. (2006) Self-assembly of star-shaped polystyrene-*block*-polypeptide copolymers synthesized by the combination of atom transfer radical polymerization and ring-opening living polymerization of α-amino acid-N-carboxyanhydrides. *Journal of Polymer Science Part A: Polymer Chemistry*, **44**, 2774–2783.

Aliferis, T., Iatrou, H., and Hadjichristidis, N. (2004) Living polypeptides. *Biomacromolecules*, **5**, 1653–1656.

Aliferis, T., Iatrou, H., and Hadjichristidis, N. (2005) Well-defined linear multiblock and branched polypeptides by linking chemistry. *Journal of Polymer Science Part A: Polymer Chemistry*, **43**, 4670–4673.

Aoyama, M., Watanabe, J., and Inoue, S. (1990a) Photoregulation of permeability across a membrane from a graft copolymer containing a photoresponsive polypeptide branch. *Journal of American Chemical Society*, **112**, 5542–5545.

Aoyama, M., Youda, A., Watanabe, J. *et al.* (1990b) Synthesis and circular dichroic and photoresponsive properties of a graft copolymer containing an azoaromatic polypeptide branch and its membrane. *Macromolecules*, **23**, 1458–1463.

Babin, J., Leroy, C., Lecommandoux, S. *et al.* (2005) Towards an easy access to amphiphilic rod-coil miktoarm star copolymers. *Chemical Communications*, **15**, 1993–1995.

Babin, J., Taton, D., Brinkmann, M. *et al.* (2008) Synthesis and self-assembly in bulk of linear and mikto-arm star block copolymers based on polystyrene and poly(glutamic acid). *Macromolecules*, **41**, 1384–1392.

Boulahia, J., Carrière, F., and Sekiguchi, H. (1989) Synthesis and characterization of copolymers from acrylamide and methacrylamide carrying peptide side-chains. *Makromolekulare Chemie*, **190**, 1975–1986.

Brzezinska, K.R., Curtin, S.A., and Deming, T.J. (2002) Polypeptide end-capping using functionalized isocyanates: Preparation of pentablock copolymers. *Macromolecules*, **35**, 2970–2976.

Brzezinska, K.R. and Deming, T.J. (2001) Synthesis of ABA triblock copolymers via acyclic diene metathesis polymerization and living polymerization of alpha-amino acid-N-carboxyanhydrides. *Macromolecules*, **34**, 4348–4354.

Cai, C., Zhu, W., Chen, T. *et al.* (2009) Synthesis and self-assembly behavior of amphiphilic polypeptide-based brush-coil block copolymers. *Journal of Polymer Science Part A: Polymer Chemistry*, **47**, 5967–5978.

Cho, C.-S., Kim, S.-W., and Komoto, T. (1990) Synthesis and structural study of an ABA block-copolymer consisting of poly(gamma-benzyl L-glutamate) as the A-block and poly(ethylene oxide) as the B-block. *Makromolekulare Chemie*, **191**, 981–991.

Cho, C.-S., Jo, B.-W., Kwon, J.-K. *et al.* (1994) Synthesis and structural study of an ABA block-copolymer consisting of poly(epsilon-benzyloxycarbonyl-L-lysine) as the A-block and poly(ethylene oxide) as the B-block. *Macromolecular Chemistry and Physics*, **195**, 2195–2206.

Cho, C.-S., Jeong, Y.-I., Kim, S.-H. *et al.* (2000) Thermoplastic hydrogel based on hexablock copolymer composed of poly(gamma-benzyl L-glutamate) and poly(ethylene oxide). *Polymer*, **41**, 5185–5193.
Cho, I., Kim, J.-B., and Jung, H.-J. (2003) Synthesis and characterization of di- and triblock copolymers of poly (ethylene oxide) and poly(DL-valine-co-DL-leucine). *Polymer*, **44**, 5497–5500.
Deming, T.J. (1997) Facile synthesis of block copolypeptides of defined architecture. *Nature*, **390**, 386–389.
Deming, T.J. (2000) Living polymerization of alpha-amino acid-N-carboxyanhydrides. *Journal of Polymer Science Part A: Polymer Chemistry*, **38**, 3011–3018.
Deming, T.J. (2006) Polypeptide and polypeptide hybrid copolymer synthesis via NCA polymerization. *Advances in Polymer Science*, **202**, 1–18.
Deming, T.J. (2007) Synthetic polypeptides for biomedical applications. *Progress in Polymer Science*, **32**, 858–875.
Deng, C., Rong, G.Z., Tian, H.Y. *et al.* (2005) Synthesis and characterization of poly(ethylene glycol)-*b*-poly(L-lactide)-*b*-poly(L-glutamic acid) triblock copolymer. *Polymer*, **46**, 653–659.
Dimitrov, I. and Schlaad, H. (2003) Synthesis of nearly monodisperse polystyrene-polypeptide block copolymers via polymerisation of N-carboxyanhydrides. *Chemical Communications*, **23**, 2944–2945.
Duncan, R. (2003) The dawning era of polymer therapeutics. *Nature Reviews Drug Discovery*, **2**, 347–360.
Floudas, G., Papadopoulos, P., Klok, H.-A. *et al.* (2003) Hierarchical self-assembly of poly(gamma-benzyl-L-glutamate)-poly(ethylene glycol)-poly(gamma-benzyl-L-glutamate) rod-coil-rod triblock copolymers. *Macromolecules*, **36**, 3673–3683.
Gombotz, W.R. and Petit, D.K. (1995) Biodegradable polymers for protein and peptide drug-delivery. *Bioconjugate Chemistry*, **6**, 332–351.
Hadjichristidis, N., Pitsikalis, M., Pispas, S. *et al.* (2001a) Polymers with complex architecture by living anionic polymerization. *Chemical Reviews*, **101**, 3747–3792.
Hadjichristidis, N., Pitsikalis, M., Pispas, S. *et al.* (2001b) Polymers with complex architecture by living anionic polymerization. *Chemical Reviews*, **101**, 3747–3792.
Hadjichristidis, N., Pispas, S., and Floudas, G. (2003a) *Block Copolymers: Synthetic Strategies, Physical Properties and Applications*, Wiley Inc., New Jersey.
Hadjichristidis, N., Pitsikalis, M., Iatrou, I. *et al.* (2003b) The strength of the macromonomer strategy for complex macromolecular architecture: Molecular characterization, properties and applications of polymacromonomers. *Macromolecular Rapid Communications*, **24**, 979–1013.
Hadjichristidis, N., Iatrou, H., Pitsikalis, M. *et al.* (2006a) Macromolecular architectures by living and controlled/living polymerizations. *Progress in Polymer Science*, **31**, 1068–1132.
Hadjichristidis, N., Iatrou, H., Pitsikalis, M. *et al.* (2006b) Macromolecular architectures by living and controlled/living polymerizations. *Progress in Polymer Science*, **31**, 1068–1132.
Hadjichristidis, N., Pitsikalis, M., and Iatrou, H. (2007) Polymers with star-related structures, in *Macromolecular Engineering. Precise Synthesis, Materials Properties, Applications*, Wiley Inc., New York, pp. 909–972.
Hadjichristidis, N., Iatrou, H., Pitsikalis, M. *et al.* (2009) Synthesis of well-defined polypeptide-based materials via the ring-opening polymerization of α-amino acid N-carboxyanhydrides. *Chemical Reviews*, **109**, 5528–5578.
Hanski, S., Houbenov, N., Ruokolainen, J. *et al.* (2006) Hierarchical ionic self-assembly of rod-comb block copolypeptide-surfactant complexes. *Biomacromolecules*, **7**, 3379–3384.
Harada, A., Kawamura, M., Matsuo, T. *et al.* (2006) Synthesis and characterization of a head-tail type polycation block copolymer as a nonviral gene vector. *Bioconjugate Chemistry*, **17**, 3–5.
Harada, A., Nakanishi, K., Ichimura, S. *et al.* (2009) Spontaneous formation of narrowly-distributed self-assembly from polyamidoamine dendron-poly(L-lysine) block copolymers through helix-coil transition of poly(L-lysine) block. *Journal of Polymer Science Part A: Polymer Chemistry*, **47**, 1217–1223.
Homer (C8th BC), Iliad, Z' 179–182.
Hua, C., Dong, C.-M., and Wei, Y. (2009) Versatile strategy for the synthesis of dendronlike polypeptide/linear poly (ε-caprolactone) block copolymers via click chemistry. *Biomacromolecules*, **10**, 1140–1148.
Iatrou, H., Frielinghaus, H., Hanski, S. *et al.* (2007) Architecturally induced multiresponsive vesicles from well-defined polypeptides. Formation of gene vehicles. *Biomacromolecules*, **8**, 2173–2181.
Karatzas, A., Iatrou, H., Hadjichristidis, N. *et al.* (2008) Complex macromolecular chimeras. *Biomacromolecules*, **9**, 2072–2080.
Karatzas, A., Bilalis, P., Iatrou, H. *et al.* (2009) Synthesis of well-defined functional macromolecular *chimeras* based on poly(ethylene oxide) or poly(N-vinyl pyrrolidone). *Reactive and Functional Polymers*, **69**, 435–440.

Klok, H.-A. and Lecommandoux, S. (2006) Solid-state structure, organization and properties of peptide-synthetic hybrid block copolymers. *Advances in Polymer Science*, **202**, 75–111.

Kong, X. and Jenekhe, S.A. (2004) Block copolymers containing conjugated polymer and polypeptide sequences: Synthesis and self-assembly of electroactive and photoactive nanostructures. *Macromolecules*, **37**, 8180–8183.

Kricheldorf, H.R. (2006) Polypeptides and 100 years of chemistry of alpha-amino acid N-carboxyanhydrides. *Angewandte Chemie-International Edition*, **45**, 5752–5784.

Kricheldorf, H.R. and Hauser, K. (1998) Macrocycles. Telechelic polylactones via macrocyclic polymerization. *Macromolecules*, **31**, 614–620.

Kricheldorf, H.R. and Hauser, K. (2001) Polylactones. A-B-A triblock copolymers of various polypeptides. Syntheses involving 4-aminobenzoyl-terminated poly(epsilon caprolactone) as B block. *Biomacromolecules*, **2**, 1110–1115.

Langer, R. (2000) Biomaterials in drug delivery and tissue engineering: One laboratory's experience. *Accounts of Chemical Research*, **33**, 94–101.

Leuchs, H. (1906) On the glycine carbonic acid. *Berichte der Deutschen Chemischen Gesellschaft*, **39**, 857–861.

Leuchs, H. and Manasse, W. (1907) Isomerism of carbethoxy glycylglycine esters. *Berichte der Deutschen Chemischen Gesellschaft*, **40**, 3235–3249.

Leuchs, H. and Geiger, W. (1908) The anhydrides of alpha-amino-N-carboxylic and alpha-amino acids. *Berichte der Deutschen Chemischen Gesellschaft*, **41**, 1721–1726.

Lodge, T.P. (2003) Block copolymers: Past successes and future challenges. *Macromolecular Chemistry and Physics*, **204**, 265–273.

Lu, H. and Cheng, J.J. (2007) Hexamethyldisilazane-mediated controlled polymerization of alpha-amino acid N-carboxyanhydrides. *Journal of American Chemical Society*, **129**, 14114–14115.

Lubbert, A., Nguyen, T.Q., Sun, F. *et al.* (2005) L-lysine dendronized polystyrene. *Macromolecules*, **38**, 2064–2071.

Mishra, M.K. and Kobayashi, S. (1999) *Star and Hyperbranched Polymers*, Marcel Dekker Inc., New York.

Papadopoulos, P., Floudas, G., Schnell, I. *et al.* (2005) Nanodomain-induced chain folding in poly(gamma-benzyl-L-glutamate)-*b*-polyglycine diblock copolymers. *Biomacromolecules*, **6**, 2352–2361.

Peng, H., Xiao, Y., Mao, X. *et al.* (2009) Amphiphilic triblock copolymers of methoxy-poly(ethyleneglycol)-*b*-poly(L-lactide)-*b*-poly(L-lysine) for enhancement of osteoblast attachment and growth. *Biomacromolecules*, **10**, 95–104.

Peng, S.-M., Chen, Y., Hua, C. *et al.* (2009) Dendron-like polypeptide/linear poly(ethylene oxide) biohybrids with both asymmetrical and symmetrical topologies synthesized via the combination of click chemistry and ring-opening polymerization. *Macromolecules*, **42**, 104–113.

Pitsikalis, M., Pispas, S., Mays, J.W. *et al.* (1998) Nonlinear block copolymer architectures. *Advances in Polymer Science*, **135**, 1–137.

Qiu, S., Huang, X., Dai, X.-H. *et al.* (2009) Star-shaped polypeptide/glycopolymer biohybrids: synthesis, self-assembly, biomolecular recognition, and controlled drug release behavior. *Journal of Polymer Science Part A: Polymer Chemistry*, **47**, 2009–2023.

Schlaad, H. and Antonietti, M. (2003) Block copolymer with amino acid sequences: molecular chimeras of polypeptides and synthetic polymers. *The European Physical Journal E*, **10**, 17–23.

Sekiguchi, H. (1981) Mechanism of N-carboxy-alpha-amino acid anhydride (NCA) polymerization. *Pure and Applied Chemistry*, **53**, 1689–1714.

Sumerlin, B. and Vogt, A. (2010) Macromolecular engineering through click chemistry and other efficient transformations. *Macromolecules*, **43**, 1–13.

Sun, J., Chen, X., Guo, J. *et al.* (2009) Synthesis and self-assembly of a novel Y-shaped copolymer with a helical polypeptide arm. *Polymer*, **50**, 455–461.

Szwarc, M. (1965) The kinetics and mechanism of N-carboxy-alpha-amino-acid anhydride (NCA) polymerization to poly-amino acids. *Advances in Polymer Science*, **4**, 1–65.

Takaki, M., Asami, R., Hanada, Y. *et al.* (1987) Preparation of (vinylbenzylimino)oligo(DL-phenylalanine NCA) macromer with narrow molecular-weight distribution. *Polymer Bulletin*, **18**, 105–110.

Tian, H., Chen, X., Lin, H. *et al.* (2006) Micellization and reversible pH-sensitive phase transfer of the hyperbranched multiarm PEI-PBLG copolymer. *Chemistry-A European Journal*, **12**, 4305–4312.

Tomson, R.C., Wake, M.C., Yaszemski, M.J. *et al.* (1995) Biodegradable polymer scaffolds to regenerate organs. *Advances in Polymer Science*, **122**, 245–274.

Xiang, Y., Si, J., Zhang, Q. *et al.* (2009) Homogeneous graft copolymerization and characterization of novel artificial glycoprotein: Chitosan-poly(L-tryptophan) copolymers with secondary structural side chains. *Journal of Polymer Science Part A: Polymer Chemistry*, **47**, 925–934.

Yamashita, Y., Iwaya, Y., and Ito, K. (1975) Block copolymerization. Polymerization of NCA of methyl D-glutamate by telechelic polystyrene having glycyl groups as active chain ends. *Makromolekulare Chemie*, **176**, 1207–1216.

Yang, Z., Yuan, J., and Cheng, S. (2005) Self-assembling of biocompatible BAB amphiphilic triblock copolymers PLL(Z)-PEG-PLL(Z) in aqueous medium. *European Polymer Journal*, **41**, 267–274.

Zhang, B., Fischer, K., and Schmidt, M. (2005) Cylindrical polypeptide brushes. *Macromolecular Chemistry and Physics*, **206**, 157–162.

Part Two

Characterization and Self-Assembly

16

Self-Assembly and Applications of Polyferrocenylsilane Block Copolymers

George R. Whittell, Jessica Gwyther, David A. Rider and Ian Manners
School of Chemistry, University of Bristol, Bristol BS8 1TS, UK

16.1 Introduction

Many of the valuable properties and applications of solid-state and biological materials can be attributed to the presence of metallic elements. Examples include magnetic materials used in data storage and electrical conductors and superconductors, electrochromic materials, and catalysts (including metalloenzymes). It has long been recognized that the incorporation of metal centers into synthetic polymer chains may also lead to desirable properties and thereby generate a new and versatile class of functional materials with an enhanced ease of processing.

Synthetic problems associated with the creation of macromolecular chains containing metal centers as a key structural component have held back progress in the field since the 1950s. Over the past decade or so, these synthetic obstacles have been, in part, overcome through many creative procedures to prepare new materials (Kingsborough and Swager, 1999; Manners, 2001; Newkome *et al.*, 1999; Whittell and Manners, 2007; Schubert and Eschbaumer, 2002; Holliday and Swager, 2005; Archer, 2001; Manners, 2004; Chandrasekhar, 2005). The new approaches that are now available have led to macromolecular structures in which metals are not only incorporated via the use of traditional covalent bonds but also by potentially reversible coordination interactions to give "metallosupramolecular" systems. Of comparable importance to these synthetic developments, these new materials have shown that they can possess a diverse range of interesting and useful properties and potential applications that complement those of

Complex Macromolecular Architectures: Synthesis, Characterization, and Self-Assembly, First Edition.
Edited by Nikos Hadjichristidis, Akira Hirao, Yasuyuki Tezuka and Filip Du Prez.

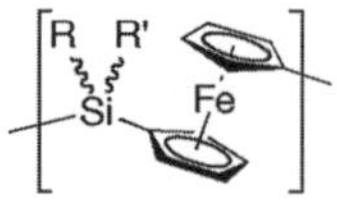

PFS

Figure 16.1 Structure of **PFS** macromolecules and melt-fabricated samples.

organic macromolecules (Kingsborough and Swager, 1999; Manners, 2001; Newkome *et al.*, 1999; Schubert and Eschbaumer, 2002; Holliday and Swager, 2005; Archer, 2001; Manners, 2004; Chandrasekhar, 2005; Nguyen *et al.*, 1999). The additional challenge of preparing well-defined metallopolymer structures has become a key goal since the early 1990s. Metallopolymers with precisely controlled chain length and end-group function are attracting rapidly expanding attention as a result of current advances in the self-assembly of block copolymers in the bulk, thin films and solution.

This chapter focuses on polyferrocenylsilanes (**PFS**s) where iron and silicon are in the main chain. Subsequently, **PFS** block copolymers will be reviewed. These materials represent an area of rapidly growing interest as a result of their self-assembly into phase-separated metal-rich nanodomain structures in thin films and micelles in block-selective solvents. The resulting nanostructured materials have a wealth of potential applications and recent breakthroughs in this area are discussed. The subject matter of the chapter is divided up into subsections covering **PFS** homopolymer and block copolymer synthesis, solution and solid-state self-assembly and applications of the latter, which have been extensively developed by ourselves and our collaborators and also by other research groups.

High molecular weight **PFS** macromolecules (Figure 16.1) were discovered in the early 1990s (Foucher *et al.*, 1992) and represent an interesting class of metal-containing polymers. Tuning of the physical properties of **PFS** is possible through the choice of substituents at the backbone silicon atom (Table 16.1). **PFS**s bearing alkyl, alkoxy, aryloxy and amino groups are typical. When symmetrically substituted, **PFS**s are often semicrystalline, whereas unsymmetrically substituted materials are generally amorphous (Rulkens *et al.*, 1996). Additionally, the presence of iron in **PFS**s imparts unique properties that are otherwise not found or difficult to acquire with traditional organic polymers. Some of these characteristics include interesting redox activity (associated with the reversible Fe(II)/Fe(III) couple), semi- and photoconductivity, and an ability to act as a magnetic ceramic or catalyst precursor, or as an etch resist to plasmas and electron beams (Manners, 2001; Manners, 2004; Kulbaba and Manners, 2001).

Table 16.1 Thermal transition and molecular weight data for representative polyferrocenylsilanes (**PFS**s) prepared by thermal or transition-metal-catalyzed ROP

Polymer	R/OR′	R′/OR	$T_g(T_m)^b$/°C	M_n^a	PDI[c]
2	H	H	16 (165)	–[e]	–[e]
2	Me	Me	33 (122–143)	3.4×10^5	1.5
2	Et	Et	22 (108)	4.8×10^5	1.6
2	*n*-Pr	*n*-Pr	24 (98)	8.5×10^5	2.7
2	*n*-Bu	*n*-Bu	3 (116, 129)	3.4×10^5	2.6
2	*n*-C_6H_{13}	*n*-C_6H_{13}	−26	7.6×10^5	1.5
2	Me	H	9 (87,102)	4.2×10^5	2.0
2	Me	Et	28	2.9×10^5	1.7
2	Me	$CH_2CH_2CF_3$	59	8.1×10^5	3.3
2	Me	$CH{=}CH_2$	28	7.7×10^5	2.1
2	Me	$CH_2CH{=}CH_2$	7	1.1×10^5	1.2
2	Me	*n*-$C_{18}H_{37}$	1 (16)	5.6×10^5	2.5
2	Me	Ph	90	1.5×10^5	2.0
2	Me	Fc[d]	99	7.1×10^5	2.3
2	Me	5-Norbornyl	81	1.1×10^5	1.5
5	OMe	OMe	19	1.5×10^5	1.9
5	OC_6H_{13}	OC_6H_{13}	−51	0.9×10^5	2.4
5	$O(CH_2)_{17}CH_3$	$O(CH_2)_{17}CH_3$	(32)	2.3×10^5	2.2
5	OPh	OPh	54	2.3×10^5	2.0
5	OC_6H_4Ph-*p*	OC_6H_4Ph-*p*	97	5.4×10^5	2.4
10	Me	$OCH_2CH{=}CH_2$	8	2.8×10^5	2.1
10	Me	$OCH_2CH_2OCOC{=}CH_2(Me)$	16	3.0×10^5	2.7

[a] Gel permeation chromatography (GPC) data obtained from the analysis of THF polymer solutions that contained 0.1% $[Bu_4N]Br$ and molecular weight values are relative to polystyrene standards. Although in this case GPC provides only molecular-weight estimates absolute determinations of M_w by static light scattering for several polymers have indicated that GPC underestimates the real values by a factor of two (Kulbaba and Manners, 2001).

[b] DSC data collected at a heating rate of at 10 °C/min.

[c] $PDI = M_w/M_n$.

[d] $Fc = (\eta^5\text{-}C_5H_4)Fe(\eta^5\text{-}C_5H_5)$.

[e] Insoluble polymer.

16.2 Synthesis of PFS Homopolymers

16.2.1 Thermal Ring-Opening Polymerization

The thermally induced ring-opening polymerization (ROP) of ring-strained sila[1]ferrocenophane monomers represents a functional-group-tolerant synthetic procedure for the preparation of **PFS**s. Typically, monomers bearing tilted cyclopentadienyl ligands are used where the release of strain (60–80 kJ/mol) thermodynamically drives the ROP. Alkyl-, aryl-, alkoxy-, aryloxy-, and amino-functionalized **PFS**s are readily available using this methodology (Scheme 16.1). Additionally, copolymerization with other ring-strained monomers, such as the tetrasilacyclobutane, $[SiMePh]_4$, or other bridged [1]ferrocenophanes is also possible

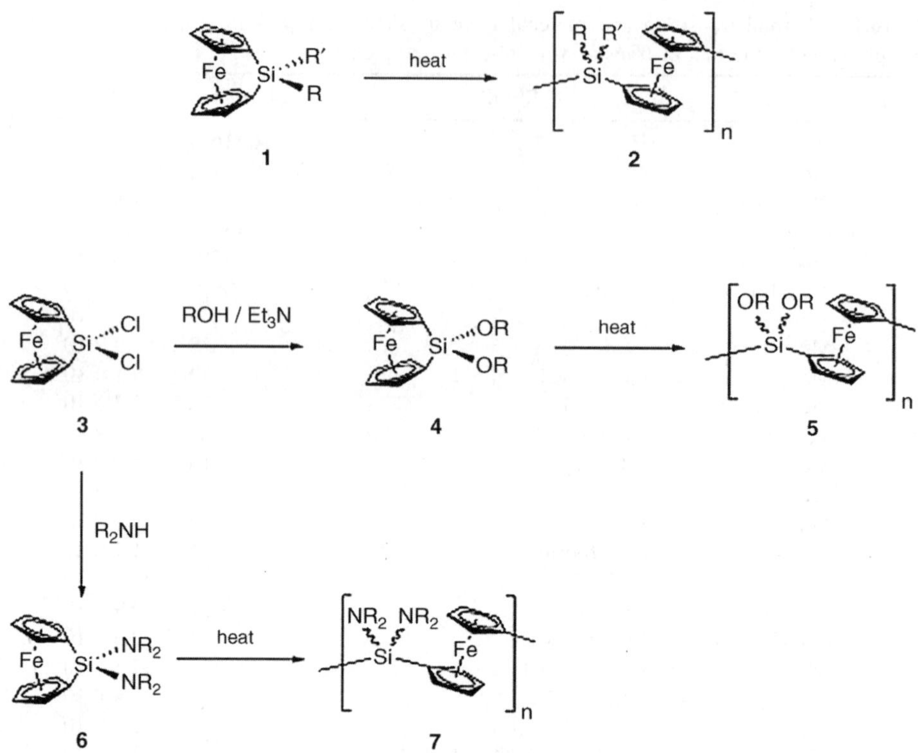

Scheme 16.1 Thermal ring-opening polymerization of ring-strained sila[1]ferrocenophanes.

(MacLachlan *et al.*, 1996; Kulbaba *et al.*, 2001; Hultzsch *et al.*, 1995; Pudelski *et al.*, 1995; Peckham *et al.*, 1996; Rulkens *et al.*, 1997). High molecular weight **PFS**s have been obtained using thermal ROP of ferrocenophane monomers in the molten state or in high boiling solvents, however, limited control over the materials' molecular weight and distribution of chain lengths is currently available (PDI > 1.5).

16.2.2 Transition-Metal-Catalyzed ROP

The transition-metal catalyzed ROP of silicon-bridged [1]ferrocenophanes proceeds in solution at room temperature using Pt^{II}, Pt^{0}, Rh^{I} as precatalysts (Ni *et al.*, 1995; Reddy *et al.*, 1995; Gómez-Elipe *et al.*, 1998). Studies have shown that the active catalysts are likely to be heterogeneous metal colloids (Temple *et al.*, 2001). When termination agents, molecules bearing a Si−H bond, are used molecular weight control is possible by controlling the initial ratio of the monomer to the silane (Scheme 16.2) (Bartole-Scott *et al.*, 2003). Molecular-weight distributions with PDIs in excess of 1.4 are typical. As an extension, the transition-metal catalyzed ROP methodology has also proven useful in the synthesis of **PFS**s bearing chlorine atoms at silicon that permits a wide range of substitution reactions along the polymer backbone (Zechel *et al.*, 1996; Cyr *et al.*, 2004; Kim *et al.*, 2006).

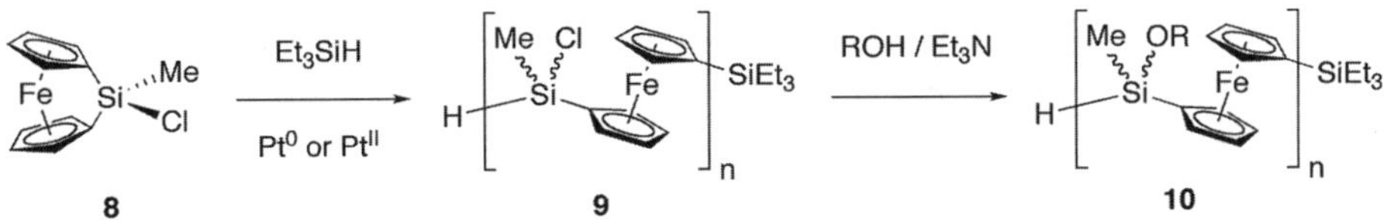

Scheme 16.2 Transition-metal catalyzed ring-opening polymerization of ring-strained sila[1] ferrocenophanes.

16.2.3 "Classical" Living Anionic ROP

Highly purified ferrocenophane monomers have been shown to undergo living anionic ROP under mild conditions using alkyllithium initiators. The polymerization mechanism involves nucleophilic attack by the anionic initiator (or during propagation, by the living anionic chain end) on silicon with concomitant cleavage of the *ipso*-carbon–silicon bond, generating cyclopentadienyl-based carbanion. Polymerization proceeds with no significant termination or side reactions and thus has allowed for the preparation of **PFS**s with predictable molecular weights and narrow molecular-weight distributions (PDI < 1.2). These characteristics, indicative of a living polymerization, have been confirmed with the symmetrically methyl-substituted (**1**, R = R′ = Me) and the unsymmetrically ethyl and methyl-substituted (**1**, R = Me, R′ = Et) ferrocenophane monomers (Figure 16.2) and also with *i*-Pr and Ph substituents

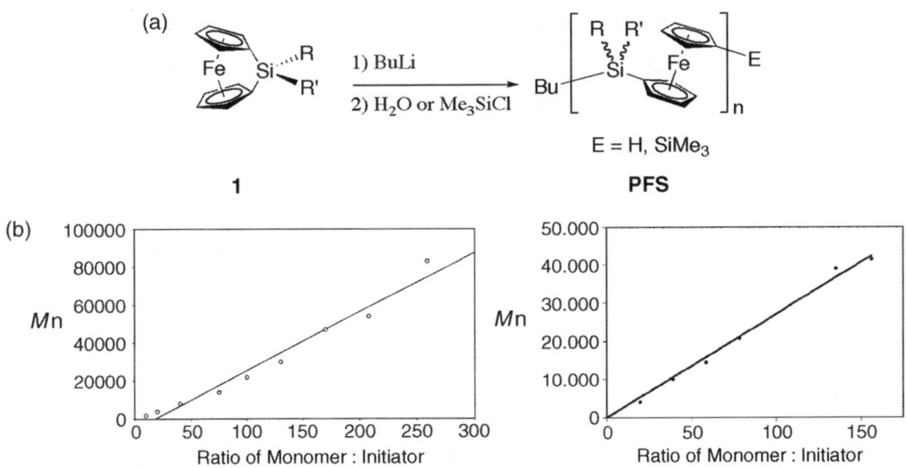

Figure 16.2 (a) Schematic and (b) molecular weight versus monomer to initiator ratio plots for the living anionic polymerization of dimethyl- and ethylmethylsila[1]ferrocenophane monomers (**1** : R = R′ = Me, ♦ trace; and R = Me, R′ = Et, ○ trace; respectively). ((a) Reprinted with permission from Y. Ni, R. Rulkens and I. Manners, "Transition metal-based polymers with controlled architectures: Well-defined poly (ferrocenylsilane) homopolymers and multiblock copolymers via the living anionic ring-opening polymerization of silicon-bridged [1]ferrocenophanes," *Journal of the Americal Chemical Society*, **118**, 4102, 1996. © 1996 American Chemical Society. (b) Reprinted with permission from D.A. Rider, K.A. Cavicchi, K.N. Power-Billard *et al.*, "Diblock copolymers with amorphous atactic polyferrocenylsilane blocks: Synthesis, characterization, and self-assembly of polystyrene-block-poly(ferrocenylethylmethylsilane) in the bulk state," *Macromolecules*, **38**, 6931, 2005. © 2005 American Chemical Society.)

(Ni *et al.*, 1996; Rider *et al.*, 2005; Gwyther and Manners, 2009). It has also been shown that random anionic copolymerization of these monomers proceeds with analogous molecular weight control (Roerdink *et al.*, 2005).

16.2.4 Photocontrolled Living Anionic ROP

Recently, the photolytic living anionic ROP of ferrocenophane monomers has been reported (Scheme 16.3). Initiation occurs by irradiation of a monomer (M) solution with light (UV-Vis) in the presence of the anionic initiator, sodium cyclopentadienylide ($Na[C_5H_5]$). Propagation proceeds via the cleavage of the Fe-η^5-cyclopentadienyl bonds in the photoexcited monomer (M^*) (Tanabe and Manners, 2004; Tanabe *et al.*, 2006). This method differs fundamentally from the living anionic ROP where the organolithium initiated anionic polymerization occurs in the absence of UV-Vis irradiation and involves Si–Cp bond cleavage. Significantly, the propagating centers for the new photolytic methodology are silyl-substituted cyclopentadienyl anions, which are less basic than iron coordinated Cp anions. Variation of the ratio of

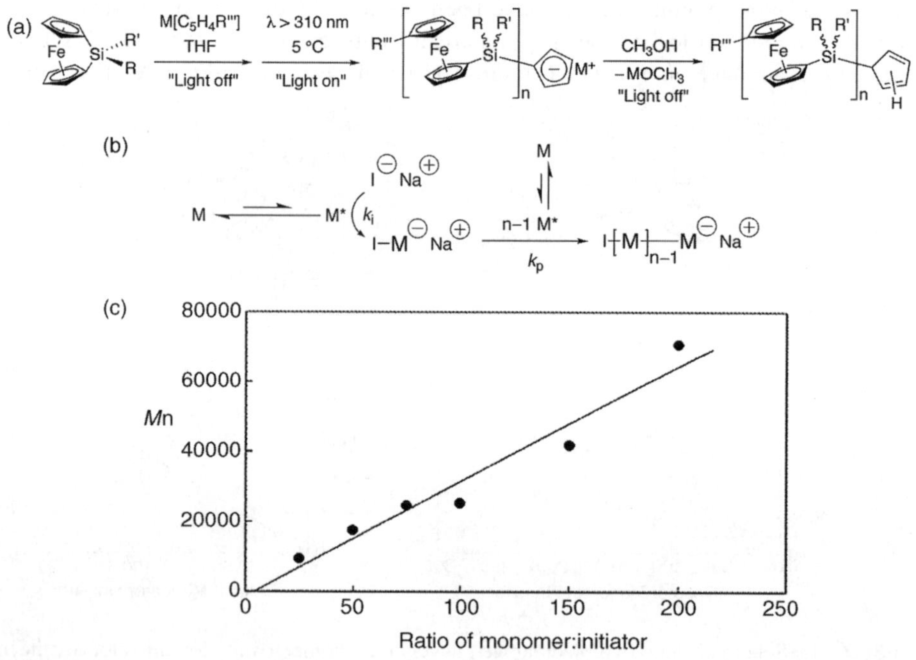

Scheme 16.3 (a) Photocontrolled living anionic ring-opening polymerization of dimethylsila[1]ferrocenophane. Using Pyrex-filtered emission from a mercury lamp, monomer **1** is polymerized with a metal Cp salt initiator. The living polymer chain is capped with MeOH to give **PFS** with a narrow molecular-weight distribution. (b) Mechanism for photolytic living ROP. On irradiation with light, some monomers, M, are promoted to a photoexcited state, M^*. The photoexcited monomer, M^*, (which is potentially solvated) then reacts with the initiator (I, with rate constant k_i) to form a ring-opened monomer, which undergoes chain propagation (with rate constant k_p) in the presence of more photoexcited monomer (reproduced from Tanabe *et al.* (2006) with permission).

monomer to initiator ratio from 25 : 1 to 200 : 1 under UV-Vis irradiation at 5 °C, followed by chain termination with water, afforded samples of polyferrocenylsilane with controlled molecular weights and narrow polydispersities (PDI = 1.04–1.21). This method not only provides a new concept for living anionic polymerization but also offers opportunities to synthesize **PFS** block copolymers under mild conditions.

Due to the availability of controlled polymerization routes for sila[1]ferrocenophane monomeric precursors, well-defined architectures with organic and inorganic coblocks are available. The incorporation of **PFS** segments into self-organizing motifs, such as block copolymers, provides further possibilities for supramolecular chemistry and the development of functional nanomaterials (Massey *et al.*, 1998a; Lammertink *et al.*, 1999; Li *et al.*, 2000; Eitouni and Balsara, 2004; Kloninger and Rehahn, 2004a; Lammertink *et al.*, 2000). The following section summarizes recent developments in the synthesis and self-assembly of **PFS** block copolymers as well as their applications in material science.

16.3 Synthesis of PFS Block Copolymers

In the mid 1990s, the living anionic ring-opening polymerization (ROP) of silicon-bridged [1] ferrocenophanes (**1**) using initiators such as *n*-BuLi was reported (Figure 16.2a) (Ni *et al.*, 1996). This permitted the synthesis of **PFS** with controlled molecular weights and narrow molecular-weight distributions (polydispersities < 1.2). The comparative reactivity of living anionic chain end groups follows the sequence of **PI** ≥ **PS** > **PFS** > **PDMS** (**PI**: polyisoprene, **PS**: polystyrene, **PDMS**: polydimethylsiloxane) (Ni *et al.*, 1996). Therefore, by sequential addition of monomers in the order of decreasing end-group reactivity, block copolymers with skeletal transition-metal atoms were successfully prepared. The first "prototypical" block copolymers that we synthesized by consecutive addition of monomers in a one-pot anionic polymerization was the organic–organometallic (**PS-*b*-PFS**) (Scheme 16.4) and the organometallic–inorganic (**PFS-*b*-PDMS**) block copolymers (Scheme 16.5) (Schubert and Eschbaumer, 2002). Extension of this method to other monomers has allowed the synthesis of **PI-*b*-PFS** (Cao *et al.*, 2002), **PFS-*b*-PMVS** (**PMVS**: polymethylvinylsiloxane) (Wang *et al.*, 2004), **PI-*b*-PS-*b*-PFS**, and **PFP-*b*-PFS-*b*-PDMS** (**PFP**: polyferrocenylphosphine) triblock copolymers and related materials (Wang *et al.*, 2002a). In principle, this sequential anionic polymerization technique can be applied to integrate a **PFS** block with any polymers compatible with the anionic polymerization mechanism. The exceptions are methacrylate-type polymers as side reactions, such as nucleophilic attack of the carbanionic chain ends at the

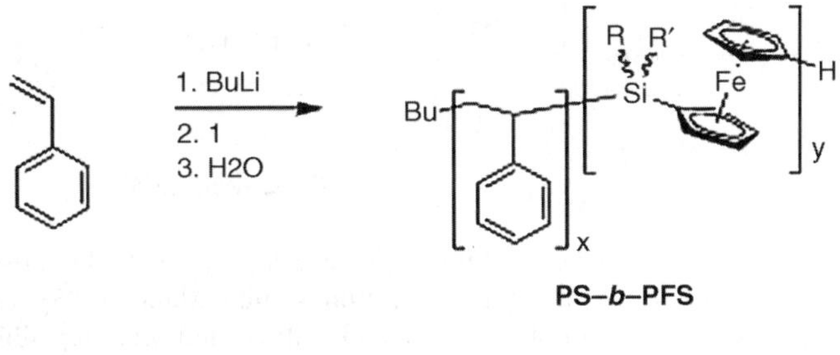

PS–*b*–PFS

Scheme 16.4 Sequential anionic polymerization of styrene and **1** (R = R′ = Me).

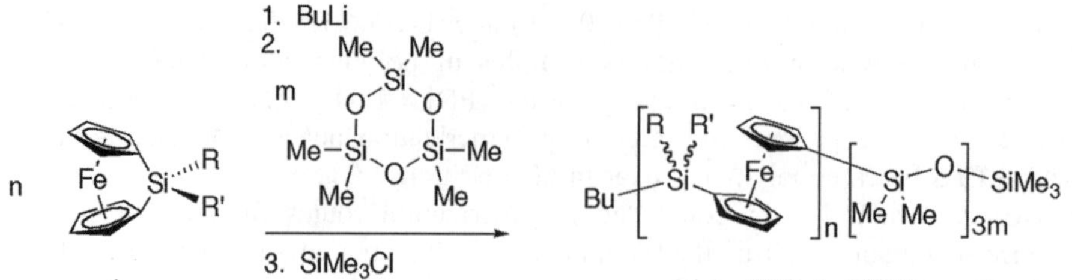

Scheme 16.5 Sequential anionic polymerization of **1** ($R = R' = Me$) and hexamethylcyclotrisiloxane.

carbonyl groups may occur if the monomers are added directly at the anionic chain end derived from **1**. Until recently, this type of block copolymer could therefore not be obtained in high yields with well-defined molecular weights and architectures.

The first **PFS-*b***-polymethacrylate copolymer was reported in 2002 and involved a two-step anionic polymerization (Scheme 16.6) (Wang *et al.*, 2002b). Hydroxyterminated polyferrocenylsilane (**PFS-OH**) was initially synthesized using *t*-butyldimethylsilyloxy-1-propyllithium as an initiator bearing a protected alcohol functionality. Once isolated, **PFS-OH** was deprotonated using potassium hydride to afford the alkoxy chain end, which then initiates dimethylaminoethylmethacrylate (**DMAEMA**) for anionic polymerization. The **PFS-*b*-PDMAEMA** block copolymer was obtained in a high yield ($M_n = 11\,000$, PDI = 1.3) and with a high **PFS** content (**PFS**: **PDMAEMA** = 1 : 5).

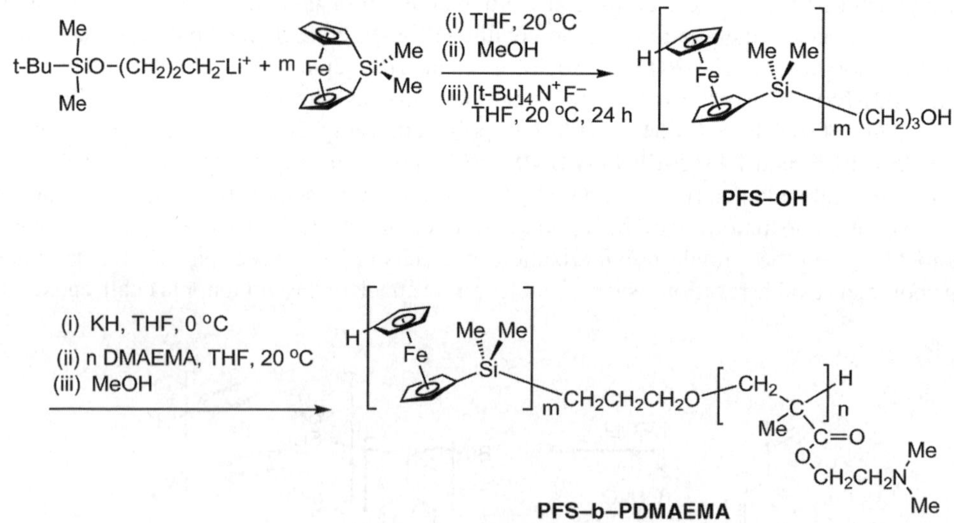

Scheme 16.6 Synthesis of polyferrocenyldimethylsilane-*b*-polydimethylaminoethylmethacylate. (Reproduced with permission from X.S. Wang, M.A. Winnik and I. Manners, "Synthesis and aqueous self-assembly of a polyferrocenylsilane-block-poly(aminoalkyl methacrylate) diblock copolymer," *Macromolecular Rapid Communications*, 2002, **23**, 210. © Wiley-VCH Verlag GmbH & Co. KGaA.).

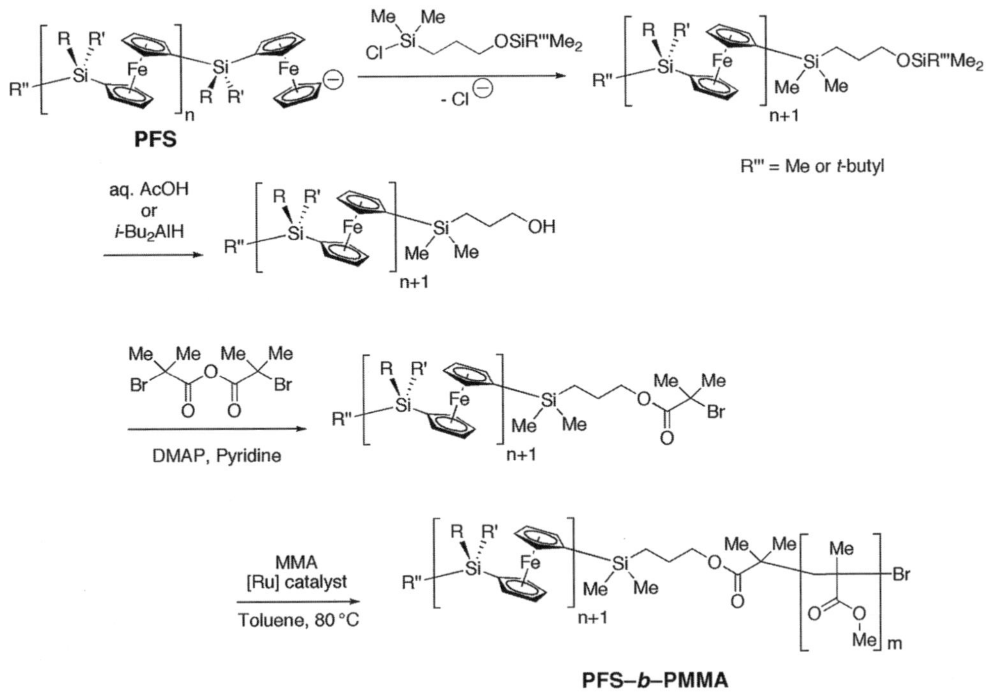

Scheme 16.7 Synthesis of polyferrocenyldimethylsilane-*b*-polymethylmethacrylate via chain-end switch from anionic to ATRP.

Alternatively, Vancso and coworkers recently investigated the anionic ROP of **1** (R = R′ = Me) followed by end-group transformations for the isolation of a **PFS**-based atom-transfer radical polymerization (ATRP) macroinitiator for methylmethacrylate (**MMA**) polymerization (Korczagin *et al.*, 2004). By quenching the living anionic **PFS** chain end with siloxypropylchlorosilane, a homopolymer bearing a protected alcohol chain end was generated (Scheme 16.7). Subsequent chain-end hydrolysis and derivatization with 2-bromoisobutyric anhydride afforded a **PFS** macroinitiator for the ATRP of **MMA**. The macroinitiators were isolated as homopolymers with narrow molecular-weight distributions ($M_n = 8000$–12 000; PDI = 1.02–1.07). Following introduction of **MMA** monomer in the presence of an ATRP Ru-based catalyst, **PFS-*b*-PMMA** diblock copolymers were produced. The metal-containing diblock copolymers were available with tunable **PFS** content (between 7–26 wt.%) via variations in the in the macroinitiator concentration with respect to **MMA** concentration. Molecular weights were also varied from 40 500–101 000 (PDI = 1.06–1.18) by maintaining constant monomer to macroinitiator ratios. **MMA** conversion was 45–60% in all cases.

In another contribution, Rehahn *et al.* successfully synthesized **PFS-*b*-PMMA** without the need for homopolymer isolation (Kloninger and Rehahn, 2004a). To circumvent the undesired reaction of anionic **PFS** chain ends with the carbonyl functionality of **MMA** that prevented sequential anionic polymerization, the reactivity of the living anionic **PFS** chain end was mediated with diphenylethylene (**DPE**). First, to ensure optimal coupling of **DPE**, living **PFS**

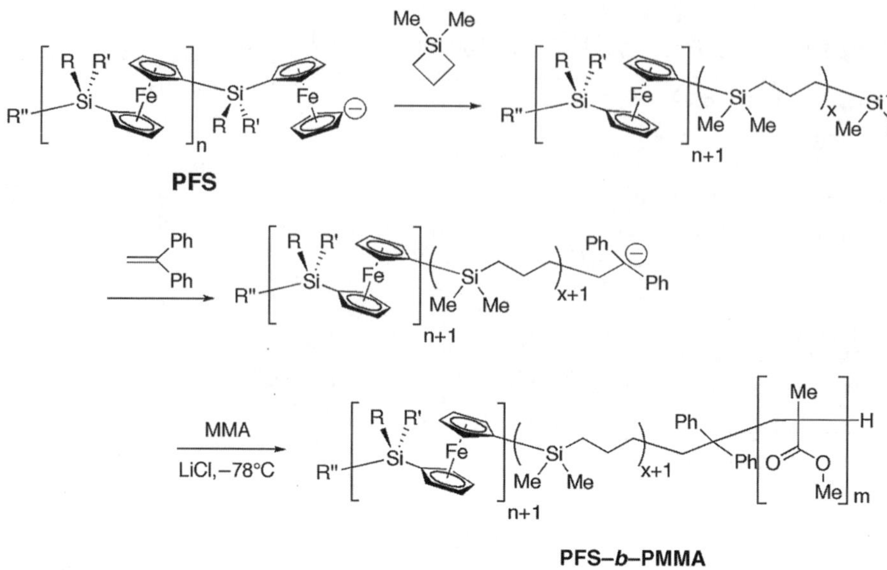

Scheme 16.8 Synthesis of polyferrocenyldimethylsilane-*b*-polymethylmethacrylate via chain end mediation using DPE.

chain ends were used for the ring-opening reaction of dimethylsilacyclobutane (Scheme 16.8). By MALDI-TOF analysis of oligomeric model homopolymers for **PFS**, it was found that incorporation of 1–3 units of the silacyclobutane were required for efficient anionic chain end transfer to **DPE** (>95% by ^{1}H-NMR spectroscopy). Having sufficiently reduced the reactivity of the chain via incorporation of **DPE**, subsequent anionic polymerization of **MMA** was completed. Unfortunately, approximately 5% of the chain ends were reported to die due to undesired spontaneous chain termination or homocoupling reactions. Nevertheless, following selective precipitation **PFS-*b*-PMMA** diblock copolymers were isolated with tunable molecular weights ($M_n = 27\,000$–84 500, PDI = 1.03–1.08) and **PFS** content (27–47 wt. %).

The Rehahn approach to **PFS-*b*-PMMA** has also allowed the preparation of the diblock copolymer **PFS-*b*-P2VP** (Wang *et al.*, 2007a) (**P2VP** = poly(2-vinylpyridine) and triblock copolymers such as **PS-*b*-PFS-*b*-P2VP** (Chuang *et al.*, 2009a).

Highly metallized **PFS** block copolymers have been prepared using the photocontrolled living anionic ROP approach (Tanabe *et al.*, 2006). A key to the success of this synthetic method is the lower reactivity propagating site. Thus, the living anionic **PFS** chain ends cleanly induce the ROP of an acetylides-functionalized monomer, previously found to have its anionic ROP complicated by alkyne-based side reactions (Scheme 16.9). First, irradiation of a mixture of the acetylide-substituted monomer and 0.08 equivalents of $Na[C_5H_5]$ in THF with light afforded the living **PFS** with acetylide side chains (**acetylidePFS**). The light was then switched off, **1** (R = R′ = Me) added, and the solution was further irradiated. After switching the light off and capping the living block copolymer using methanol, copolymer **acetylidePFS-*b*-PFS** was isolated. The pendant acetylide groups were then transformed into Ni-based clusters with $[\{NiCp(CO)\}_2]$ (Cp = η^5-C_5H_5) to yield the highly metalized block copolymer **NiPFS-*b*-PFS**. Preliminary investigations have shown that this green, air-stable block

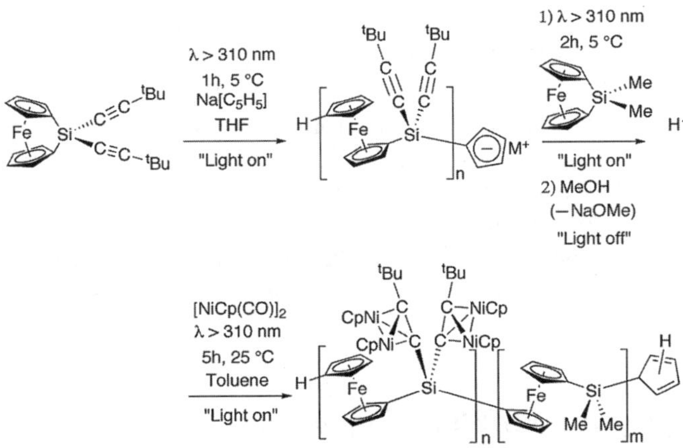

Scheme 16.9 Synthesis of **NiPFS-*b*-PFS** block copolymer with controlled architecture by sequential photocontrolled ROP.

copolymer can self-assemble in the bulk solid state into phase-separated nanodomains of **NiPFS** and Fe-only **PFS**.

The hybridization of **PFS** with a polypeptide, recently reported by Manners, Winnik *et al.*, produced a new organometallic-organic block copolymer offering opportunities for biological applications for **PFS** (Kim *et al.*, 2005a). As shown in Scheme 16.10, a **PFS-*b*-PBLG** (**PBLG**:

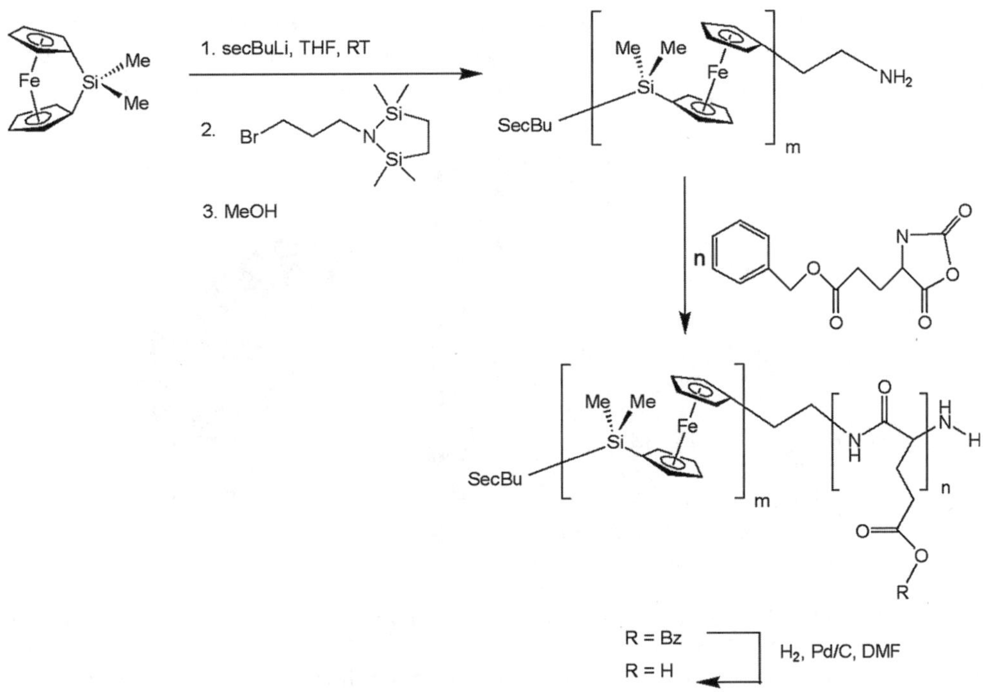

Scheme 16.10 Synthesis of **PFS-*b*-PBLG** block copolymer.

poly(γ-benzyl-L-glutamate)) block copolymer was synthesized through two steps: preparation of an amino-terminated **PFS** (**PFS-NH$_2$**, $M_n = 9330$, PDI = 1.10) through the anionic ROP of ferrocenylsilane monomer **1** (R = R′ = Me), followed by a ROP of γ-benzyl-L-glutamate-*N*-carboxyanhydrides initiated by **PFS-NH$_2$** macroinitiator (Kim *et al.*, 2005a). **PFS-*b*-PBLG** samples with controlled molecular weights (M_n = 15 300–30 200) and narrow polydispersities (PDI = 1.13–1.21) were isolated with tunable polyferrocenylsilane contents (**PFS**: **PBLG** = 1: 1.8–1: 3.5). Moreover, amphiphilic block copolymers were easily isolated through hydrogenative cleavage of the benzyl group from the **PBLG** block with H_2/Pd in DMF. The resulting **PFS**-*b*-poly(L-glutamic acid) (**PFS-*b*-PGA**) copolymers were soluble in aqueous base (0.1 M NaOH).

16.4 Solution Self-Assembly of PFS Block Copolymers

Block-selective solvents are used to induce aggregation of diblock copolymer macromolecules where there is sequestering of the incompatible blocks within a core stabilized by solvent-swollen corona-forming blocks. In most cases the solvent-swollen corona, with a dramatically increased volume, provides a significant driving force for curvature of the segment/segment interface in the diblock copolymer micelle and consequently spheres are the most common resulting morphology. However, solution self-assembly of **PFS** block copolymers with two methyl groups on silicon allows the generation of discrete supramolecular organometallic nanomaterials with a range of micellar morphologies including cylinders (Massey *et al.*, 1998a), tubes (Raez *et al.*, 2002a), fibers (Raez *et al.*, 2002b), and tapes and platelets (Cao *et al.*, 2002). For example, when a sample of **PFS$_{50}$-*b*-PDMS$_{300}$** (block ratio of **PFS**: **PDMS** = 1: 6) was heated in *n*-hexane, a selective solvent for **PDMS**, micellar aggregates formed and could be visualized by TEM (Figure 16.3a). TEM imaging of these morphologies was effective as the high electron density of iron-rich, core-forming **PFS** blocks was

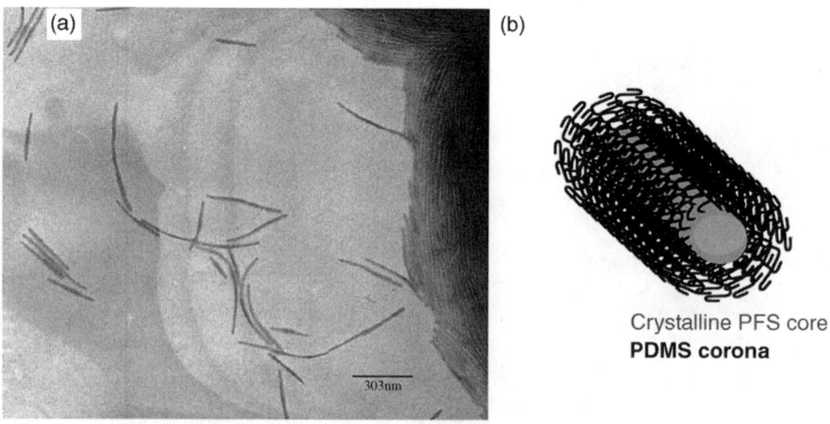

Figure 16.3 (a) TEM image and (b) scheme for **PFS** nanocylinders self-assembled from **PFS$_{50}$-*b*-PDMS$_{300}$** (**PFS** : **PDMS** = 1 : 6) in hexane. (Reproduced with permission from J.A. Massey, K.N. Power, M.A. Winnik and I. Manners, "Organometallic nanostructures: Self-assembly of poly(ferrocene) block copolymers," *Advanced Materials*, 1998, **10**, 1559. © Wiley-VCH Verlag GmbH & Co. KGaA.)

sufficient for contrast in the electron microscope (Massey *et al.*, 1998b). As schematically illustrated in Figure 16.3b, the cylinders possess an iron-rich, organometallic core of **PFS** surrounded by an insulating sheath of **PDMS**. The cylindrical morphology observed in these systems was found to be a result of semicrystallinity in the core-forming **PFS** blocks, a characteristic that was confirmed by wide-angle *X*-ray scattering (Kulbaba and Manners, 2001; Nguyen *et al.*, 1998; Pudelski *et al.*, 1996). Although other segments have been investigated, the most well-studied block is poly(ferrocenyldimethylsilane) and the designation **PFS** will represent this material in this section unless otherwise noted.

Apart from cylinders, **PFS-*b*-PDMS** block copolymers can also self-assemble into apparently hollow tubular structures when the block ratios of **PFS: PDMS** reaches 1: 12 (Raez *et al.*, 2002a). For example, $\mathbf{PFS_{40}\text{-}\mathit{b}\text{-}PDMS_{480}}$ (**PFS: PDMS** = 1: 12) self-organized in hexane to form nanotubes in which the **PFS** blocks aggregate to form a shell with a cavity in the middle of the tube, while the **PDMS** blocks form the corona (Raez *et al.*, 2002b). Several additional TEM techniques including dark-field TEM and energy-filtered TEM also supported the nanotube structures (Frankowski *et al.*, 2004). The presence of a hollow cavity was further supported by trapping tetrabutyllead in the void and performing energy-dispersive *X*-ray measurements on the resulting structure.

In 2006, Vancso *et al.* reported the solution state self-assembly of **PFS-*b*-PMMA** block copolymers in the PMMA-selective solvent acetone (Korczagin *et al.*, 2006). The resulting micellar morphology was found to depend on the length of the **PFS** block with respect to the PMMA block. For a **PFS**: **PMMA** block ratio of 1 : 23, monomolecular or small spherical micelles were the majority aggregate formed while only a fraction of the block copolymers associated into cylindrical micelles. As the **PFS** fraction increased such that the **PFS**: **PMMA** block ratios were in the range of 1 : 10–1 : 6, a larger proportion of copolymer chains aggregated into cylindrical micelles. At a block ratio of 1 : 6, the block copolymers formed well-defined cylindrical micelles exclusively, with an overall diameter of 22 nm and lengths of over 3 μm. If the **PFS** was the majority polymer segment (**PFS**: **PMMA** block ratios >1 : 1), only crew-cut spherical aggregates were observed. Interestingly, variable temperature (25–55 °C) high-resolution ^{1}H NMR studies on the cylindrical micellar solutions showed that the **PFS** core was in a mobile state, and that solvent molecules readily penetrated inside of the micellar core. It was therefore concluded that the contrast in polarity between the two blocks, supported by the polar acetone solvent, was the primary driving force for micelle formation and not the semicrystallinity associated with the **PFS** core. This was further supported by a rod-to-sphere transition temperature of 60 °C for the cylindrical micellar solutions. Shown in Figure 16.4 are AFM images of the cylindrical **PFS-*b*-PMMA** micelles exhibiting good thermal stability up to 160 °C.

The essential role of block ratio in determining the micellar morphologies of **PFS** block copolymers was also discerned in our studies on the **PI-*b*-PFS** system (Cao *et al.*, 2002). $\mathbf{PI_{30}\text{-}\mathit{b}\text{-}PFS_{60}}$ and $\mathbf{PI_{320}\text{-}\mathit{b}\text{-}PFS_{53}}$ have similar **PFS** length but differ significantly in **PI** length, with block ratios (**PI: PFS**) of 1 : 2 and 6 : 1, respectively. As shown in Figure 16.5, cylinders were observed with the block copolymers containing the longer **PI** chains, (Figure 16.5a), while the tape-like micelles, as shown in Figure 16.5b, self-assembled from $\mathbf{PI_{30}\text{-}\mathit{b}\text{-}PFS_{60}}$.

Under certain circumstances, spherical aggregates were also found. For example, when **PFS** cylindrical micelles were heated above the T_m of the **PFS** block (circa 120–145 °C) (Raez *et al.*, 2002a) or an amorphous poly(ferrocenylmethylphenylsilane) (Massey *et al.*, 2000) block was used to replace semicrystalline **PFS** in the block copolymers,

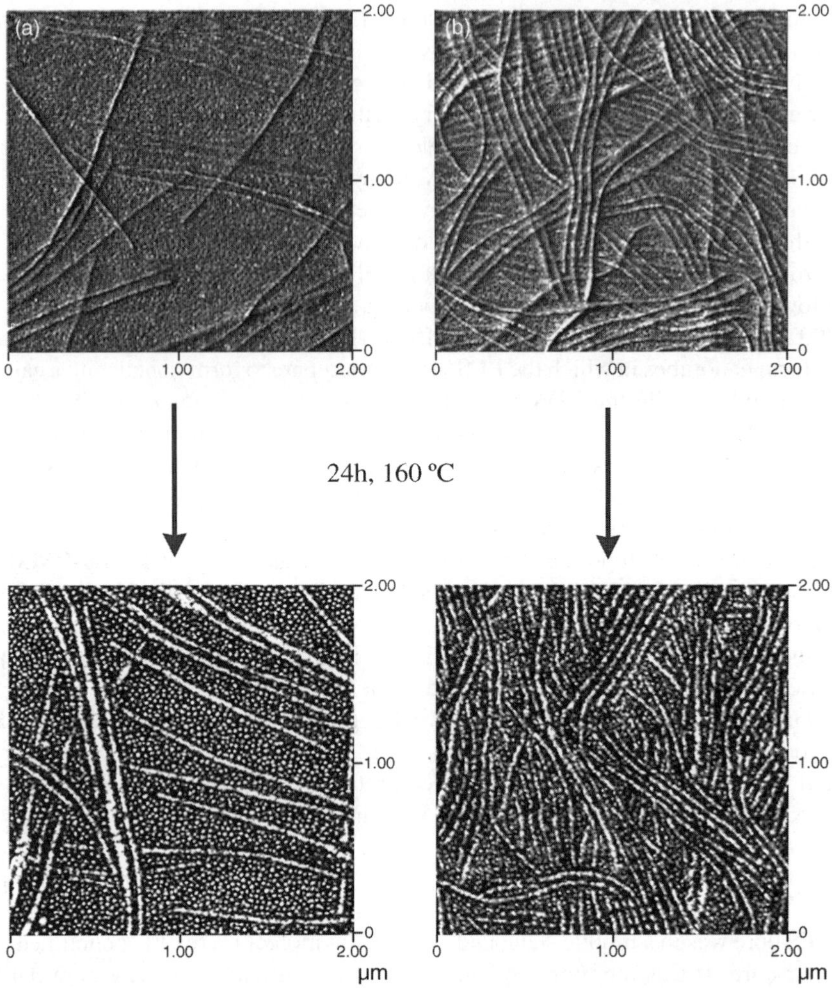

Figure 16.4 AFM phase images of silicon substrates dip coated from a solution of **PFDMS-*b*-PMMA** in acetone. Polymers: (a) $\mathbf{PFS_{50}}$**-*b*-**$\mathbf{PMMA_{1167}}$, (b) $\mathbf{PFS_{50}}$**-*b*-**$\mathbf{PMMA_{531}}$, and after annealing in a vacuum at 160 °C for 24 h. Phase scale is 10°. (Reprinted with permission from I. Korczagin, M.A. Hempenius, R.G. Fokkink *et al.*, "Self-assembly of poly(ferrocenyldimethylsilane-b-methyl methacrylate) block copolymers in a selective solvent," *Macromolecules*, **39**, 2306, 2006. © 2006 American Chemical Society.)

spherical aggregates became the only morphology observed by TEM. These experiments suggested that the crystallization of **PFS** chains could be a possible reason for the formation of observed one-dimensional supramolecular assemblies. **PFS** crystallization was indeed detected in both **PFS** nanocylinders (Massey *et al.*, 2000) and nanotubes (Raez *et al.*, 2002a) by wide-angle X-ray scattering (WAXS) experiments on the solid micellar samples dried from the corresponding solutions.

PFS block copolymers are also able to self-assemble in water when hydrophilic blocks are incorporated (Gohy *et al.*, 2004; Wang *et al.*, 2005a). Following the successful synthesis of

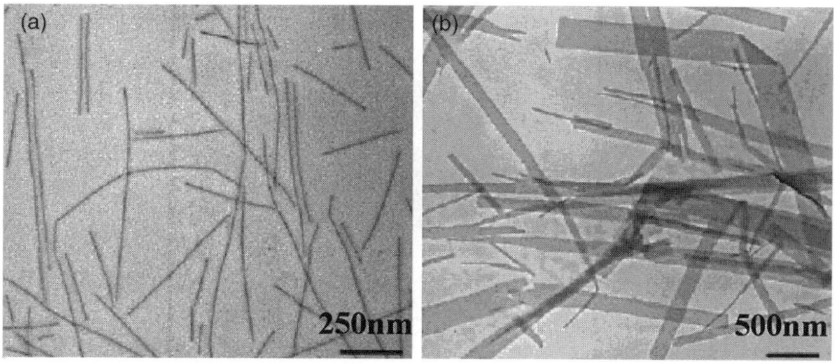

Figure 16.5 TEM images for the micelles self-assembled from (a) **PI_{320}-*b*-PFS_{53}** in THF/Hexane (2/8, v/v) and (b) **PI_{30}-*b*-PFS_{60}** in THF/Hexane (3/7 v/v). (Reprinted with permission from L. Cao, I. Manners and M.A. Winnik, "Influence of the interplay of crystallization and chain stretching on micellar morphologies: Solution self-assembly of coil–crystalline poly(isoprene-block-ferrocenylsilane)," *Macromolecules*, **32**, 8258, 2002. © 2002 American Chemical Society.)

PFS_9-*b*-$PDMAEMA_{50}$ (Scheme 16.6), quaternization of the block copolymer with methyl iodide was achieved to afford **PFS_9-*b*-$qPDMAEMA_{50}$** (Wang *et al.*, 2005a). Due to the lack of a highly soluble block, complex micellization behavior was observed for **PFS_9-*b*-$PDMAEMA_{50}$** in water. However, alcohol solvents such as EtOH, *i*-PrOH, which are good solvents for **PDMAEMA**, induce **PFS_9-*b*-$PDMAEMA_{50}$** to self-assemble into cylindrical micelles by simply dissolving the block copolymers in the solvent. Moreover, water-soluble cylindrical micelles could then be prepared via the dialysis of **PFS_9-*b*-$PDMAEMA_{50}$** in ethanol against water (Figure 16.6) (Wang *et al.*, 2005a).

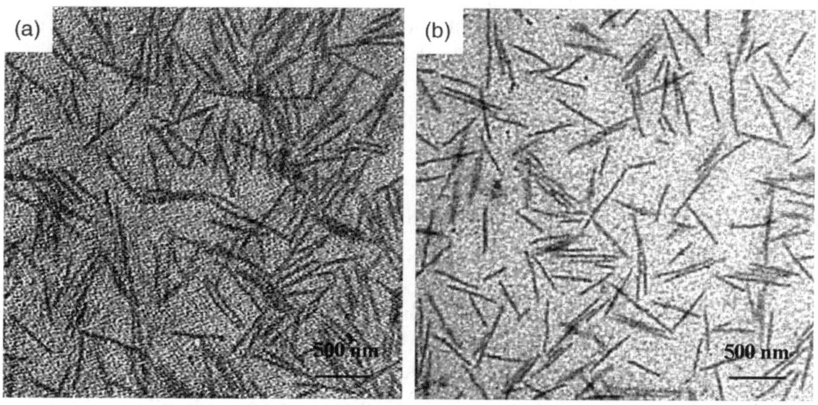

Figure 16.6 TEM for the micelles prepared by (a) mixing **PFS_9-*b*-$PDMAEMA_{50}$** with ethanol and (b) replacing ethanol with water by dialysis. (Reprinted with permission from X.S. Wang, M.A. Winnik and I. Manners, "Synthesis and self-assembly of poly(ferrocenyldimethylsilane-b-dimethylaminoethyl methacrylate): Toward water-soluble cylinders with an organometallic core," *Macromolecules*, **38**, 1928, 2005. © 2005 American Chemical Society.)

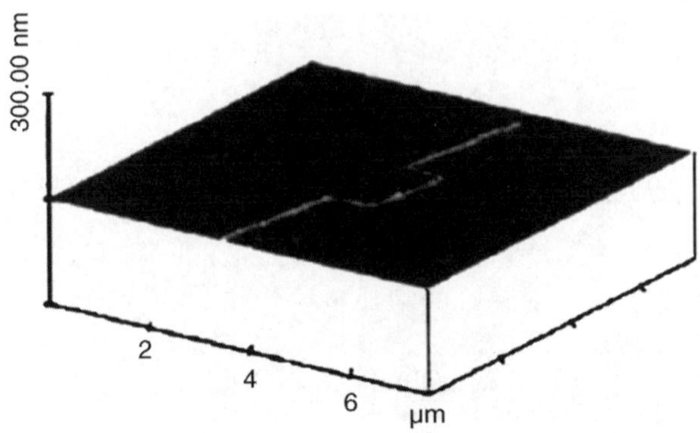

Figure 16.7 Scanning force micrograph for the ceramic line derived from aligned **PFS-*b*-PDMS** cylindrical micelles by H_2 plasma etching. (Reprinted with permission from J.A. Massey, M.A. Winnik, I. Manners *et al.*, "Fabrication of oriented nanoscopic ceramic lines from cylindrical micelles of an organometallic polyferrocene block copolymer," *Journal of the American Chemical Society*, **123**, 3147, 2001. © 2001 American Chemical Society.)

A potential application of these well-defined aggregates is as etch resists for semiconducting substrates (such as GaAs or Si) (Cao *et al.*, 2003). In a collaboration with J. Spatz and M. Möller at the University of Ulm, Germany, the **PFS** cylinders have been positioned on the surface of a GaAs resist by capillary forces along grooves, which were previously formed from electron beam etching of the surface. Subsequent reactive plasma etching generated connected ceramic lines of reduced size (Figure 16.7) (Massey *et al.*, 2001).

Despite the success in the fabrication of nanoceramic lines by plasma etching of aligned micellar precursors, our further exploitations of **PFS** nanostructures were often frustrated by the labile nature of the self-assembled micelles in that they easily undergo dissociation in a common solvent or at an elevated temperature.

Several groups have demonstrated that the stability of self-assembled micelles can be dramatically improved by performing a shell-crosslinking reaction (Thurmond *et al.*, 1996; Thurmond *et al.*, 1997; Ding and Liu, 1998; Buetuen *et al.*, 1999). The synthesis of **PFS-*b*-PMVS** and **PI-*b*-PFS** block copolymers through sequential anionic polymerizations, followed by self-assembly in hexane allowed access to either nanocylinders or nanotubes with pendent vinyl groups attached to the coronal chains. By taking advantage of pendent vinyl groups, we performed a Pt-catalyzed hydrosilylation crosslinking reaction with tetramethyldisiloxane as a crosslinker, leading to shell-crosslinked nanocylinders and nanotubes (Wang *et al.*, 2004; Wang *et al.*, 2003).

As a result of shell-crosslinking, the one-dimensional micellar structures were locked-in and preserved even if transferred from hexane to a common solvent for both blocks (Wang *et al.*, 2004; Wang *et al.*, 2003). Figure 16.8 illustrates a TEM image for $\mathbf{PI_{320}\text{-}\mathit{b}\text{-}PFS_{53}}$ shell-crosslinked micelles. The TEM sample was prepared from a micellar solution in THF, a common solvent for both **PFS** and **PI** blocks.

We have further demonstrated that the shell-crosslinking reaction is essential in making **PFS** nanoceramics. Upon heating up to 600 °C with a temperature ramp of 1 K min^{-1} under N_2, we pyrolyzed the shell-crosslinked micelles. The resulting ceramics with excellent shape retention were characterized by TEM (Figure 16.9). In a control experiment, pyrolysis of

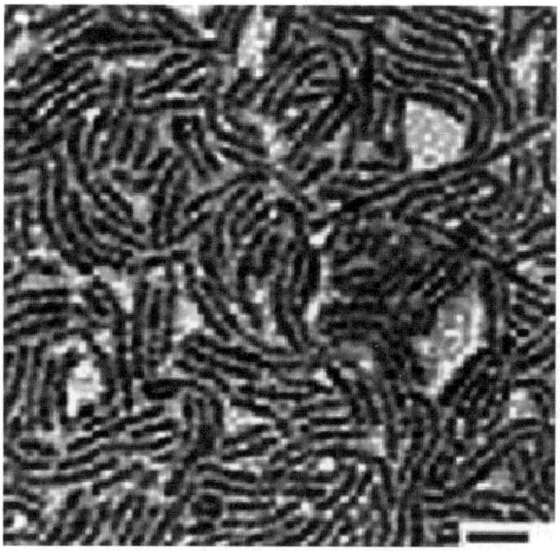

Figure 16.8 TEM image for shell-crosslinked $\mathbf{PI_{320}}$-*b*-$\mathbf{PFS_{53}}$ cylindrical micelle solid samples prepared by drying a drop of THF solution on a carbon-coated copper grid. THF is a common solvent for both blocks. Scale bar = 250 nm. (Reprinted with permission from X.S. Wang, A.C. Arsenault, G.A Ozin *et al.*, "Shell cross-linked cylinders of polyisoprene-b-ferrocenyldimethylsilane: Formation of magnetic ceramic replicas and microfluidic channel alignment and patterning," *Journal of the American Chemical Society*, **125**, 12686, 2003. © 2003 American Chemical Society.)

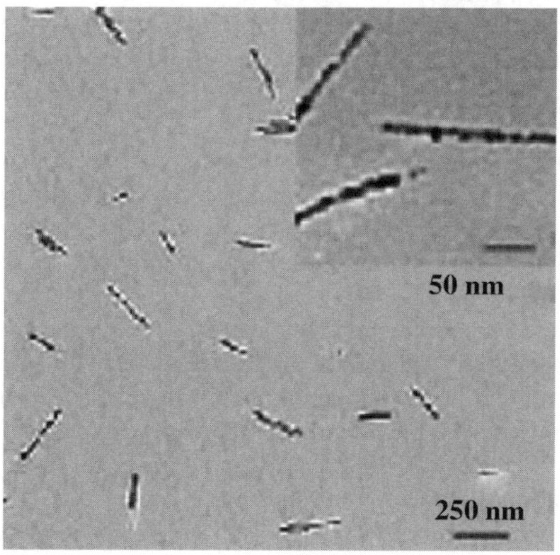

Figure 16.9 TEM image for cylindrical nanoceramic replica derived from $\mathbf{PI_{320}}$-*b*-$\mathbf{PFS_{53}}$ shell-crosslinked micelles through a pyrolysis process under N_2 with a temperature ramp of $1\,K\,min^{-1}$ up to 600 °C. (Reprinted with permission from X.S. Wang, A.C. Arsenault, G.A Ozin *et al.*, "Shell cross-linked cylinders of polyisoprene-b-ferrocenyldimethylsilane: Formation of magnetic ceramic replicas and microfluidic channel alignment and patterning," *Journal of the American Chemical Society*, **125**, 12686, 2003. © 2003 American Chemical Society.)

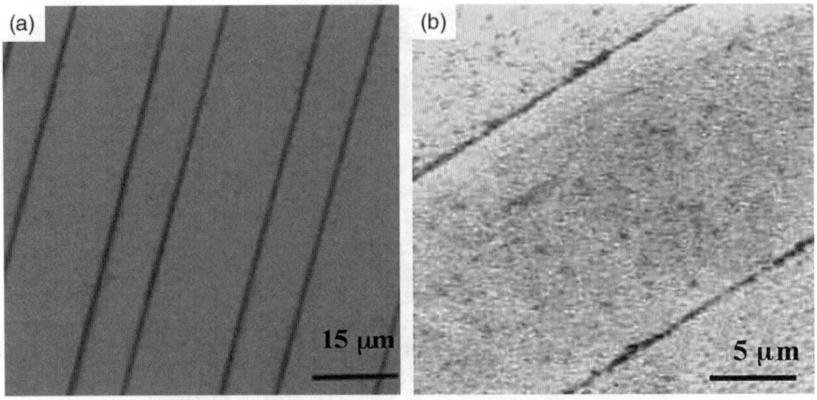

Figure 16.10 (a) Optical and (b) SEM images for microfluidically aligned shell-crosslinked $\mathbf{PI_{320}}$-***b***-$\mathbf{PFS_{53}}$ nanocylinders. (Reprinted with permission from X.S. Wang, A.C. Arsenault, G.A Ozin *et al.*, "Shell cross-linked cylinders of polyisoprene-b-ferrocenyldimethylsilane: Formation of magnetic ceramic replicas and microfluidic channel alignment and patterning," *Journal of the American Chemical Society*, **125**, 12686, 2003. © 2003 American Chemical Society.)

uncrosslinked $\mathbf{PI_{320}}$-***b***-$\mathbf{PFS_{53}}$ cylindrical micelles resulted in destruction of the structure. This comparison indicated that the shell-crosslinking plays an essential role in the shape retention and permits the formation of a ceramic replica, presumably due to the high ceramic yield that results from a crosslinked structure (Wang *et al.*, 2004; Wang *et al.*, 2003).

Shell-crosslinked micelles have also been aligned and patterned on a flat silicon substrate by a microfluidic technique (Figure 16.10). Ordered magnetic nanoceramic arrays derived from a pyrolysis process may be of interest as magnetic memory materials or as catalysts for the growth of carbon nanotubes (Rider *et al.*, 2005).

As for shell-crosslinked nanotubes and cylinders, the conservation of redox-activity due to **PFS** chains has been illustrated by performing cyclic voltammetry experiments (Wang *et al.*, 2004). This new type of nanostructure provides the chance to encapsulate guest compounds or particles in the cavities through an *in-situ* redox reaction using **PFS** as a reductant (Wang *et al.*, 2005b).

Ag nanoparticles were prepared within the nanotubes and cylinders via a reaction with Ag $[PF_6]$. We found that partial preoxidation of the **PFS** domains with an organic oxidant such as tris(4-bromophenyl)aminium hexachloroantimonate is a key step for the efficient formation of one-dimensional arrays of silver nanoparticles confined within the nanotubes (Figure 16.11) (Wang *et al.*, 2005b; Wang *et al.*, 2008). Further attempts to synthesize metal nanowires through this encapsulation method are currently in progress.

PFS-*b*-PBLG block copolymers, which are soluble in hot toluene, were found to form optically transparent gels upon cooling to ambient temperature (Kim *et al.*, 2005a). The gelation was thermally reversible in all cases. The critical concentration for gelation was found: (i) to strongly depend on the chain length of the **PBLG** block, suggesting that the interaction between the **PBLG** helices is a crucial factor in the gelation process and (ii) to be related to the chain length ratio between the **PFS** and the **PBLG** blocks. Moreover, when the chain length ratio between the **PFS** and the **PBLG** exceeded 1 : 3, incomplete gelation was observed

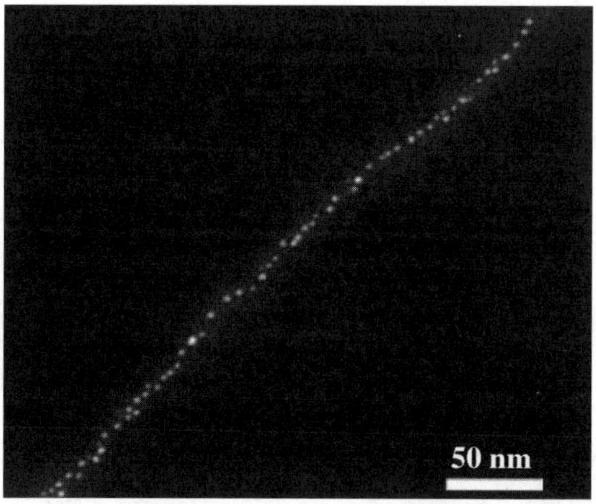

Figure 16.11 TEM image in dark-field mode for shell-crosslinked cylinders of **PFS_{53}-*b*-PI_{320}** block copolymers encapsulated with Ag nanoparticles. (Reprinted with permission from H. Wang, X. Wang, M.A. Winnik and I. Manners, "Redox-mediated synthesis and encapsulation of inorganic nanoparticles in shell-cross-linked cylindrical polyferrocenylsilane block copolymer micelles," *Journal of the American Chemical Society*, **130**, 12921, 2008. © 2008 American Chemical Society.)

(macroscopic phase separation into a dense block copolymer gel and the free solution phase). Hydrogen bonding was also shown to be a negligible factor in the gelation process. These **PFS-*b*-PBLG** block copolymer gels, with **PBLG**-based helices and a random coil of **PFS** at one end, formed fibrous nanoribbons assembled from the 1-dimensional antiparallel stacking of the building blocks in a monolayer fashion (Figure 16.12). It was proposed that strong dipolar interactions involving the phenyl groups on adjacent **PBLG** helices stabilized the stacked structure where the **PFS** blocks protruded outside of the ribbon into the toluene-rich environment, thereby preventing aggregation of nanoribbons.

In cases where cylindrical micelles can be prepared from the self-assembly of **PFS**-containing block copolymers, attention is now been drawn to attaining control over the size of the resulting micelles. As self-assembly of diblock copolymers results from segregation of the two blocks, it stands to reason that the widths of both the core and corona will be related to the molecular masses of the **PFS** and coblock, respectively. The length of the micelle, however, can range from tens of nm to a few μm and this parameter is less closely related to the chemical structure of the component block copolymer. During initial work on the stability of **PFS-*b*-PDMS** cylindrical micelles, it was observed that treatment by ultrasound resulted in the formation of shorter cylinders but that retained the original width (Massey *et al.*, 1998a). The same was true for cylindrical micelles that consisted of **PI_{264}-*b*-PFS_{68}**, as we found that these could be grown in length by addition of further unimer (Wang *et al.*, 2007a). This growth process was remarkable for a number of reasons. First, the length of the cylindrical micelles was found to increase linearly with the concentration of unimer added and this feature led us to liken the process to that of a living polymerization. Secondly, the number of block copolymer macromolecules per unit length of the micelle was calculated and this confirmed suggestions from TEM that the internal structure did not change upon micellar growth. Finally, there was no change in the distribution of micelle lengths over several months. This was consistent with the

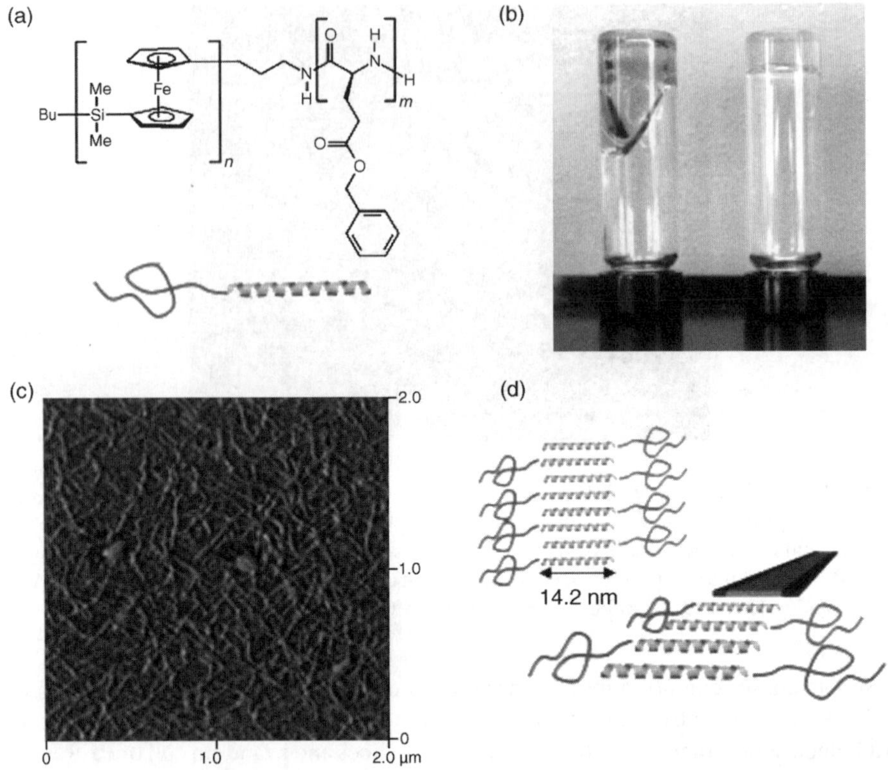

Figure 16.12 (a) Structure of **PFS-*b*-PBLG**. (b) Optical photograph of inverted vials containing a toluene swollen gel of **PFS-*b*-PBLG**. (c) AFM of **PFS-*b*-PBLG** gel and (d) schematic for nanoribbon mechanism. (Reproduced with permission from K.T. Kim, C. Park, G.W.M. Vandermeulen *et al.*, "Gelation of helical polypeptide – random coil diblock copolymers by a nanoribbon mechanism," *Angewandte Chemie International Edition*, 2009, **44**, 7964. © Wiley-VCH Verlag GmbH & Co. KGaA.)

micelles being stable to fusion and a concentration of free unimer in solution being required for growth. The absence of cylinder growth when a unimer containing an amorphous **PFS** block was utilized suggested that this process was best viewed as an example of epitaxial growth, relying on the presence of the semicrystalline **PFS** block.

It has been possible to prepare comicelles that contain compartmentalized corona by the addition of a different **PFS**-containing diblock copolymer to a preformed cylindrical micelle (Wang *et al.*, 2007b). Thus, addition of **PMVS$_{300}$-*b*-PFS$_{48}$** to cylinders of **PI$_{320}$-*b*-PFS$_{53}$** afforded micelles with a continuous semicrystalline **PFS** core, but with a **PI** corona in the center and **PMVS** coronae at the ends (Figure 16.13). By preparing similar comicelles with quarternized poly(2-vinylpyridine) (**q-P2VP**) corona in the center and neutral **P2VP** in the extremities, it has also been possible to prepare composite nanostructures by selectively decorating the former with Au nanoparticles and PbS quantum dots (Wang *et al.*, 2007c).

It has recently been shown that the semicrystalline **PFS** can nucleate epitaxial growth when present in forms other than cylindrical micelles. Reducing the volume fraction of the

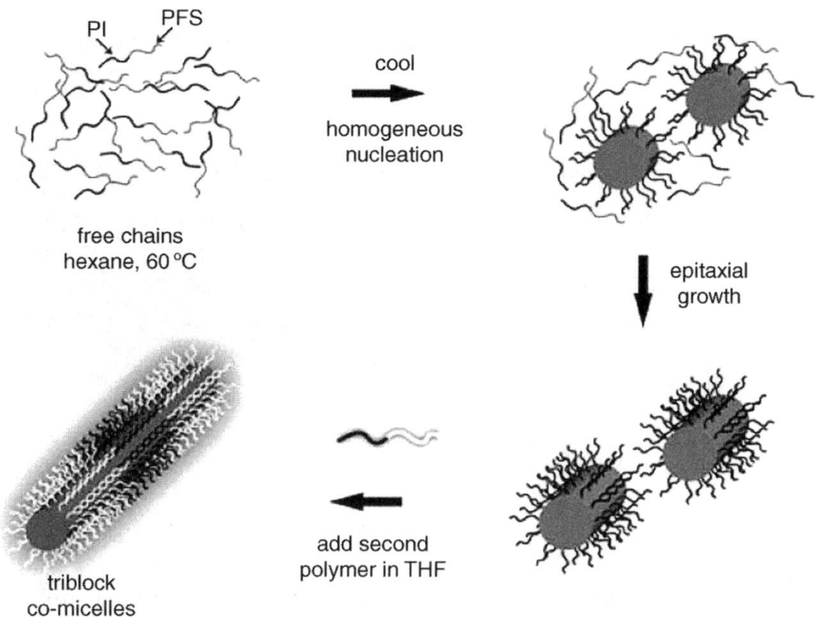

Figure 16.13 Mechanism for the formation of triblock comicelles via the self-assembly of **PFS**-containing diblock copolymers (reproduced from Wang *et al.* (2007b) with permission).

corona-forming block can lead to a formal morphology change from cylinders to platelets upon self-assembly in a block-selective solvent (Cao *et al.*, 2002). $\mathbf{PI_{76}}$-***b***-$\mathbf{PFS_{76}}$, for instance, forms such platelets and these can also be fragmented by ultrasound in much the same way as cylindrical micelles. Interestingly, addition of $\mathbf{PI_{342}}$-***b***-$\mathbf{PFS_{76}}$ unimers to a suspension of these small platelets results in epitaxial growth from only two of the available six faces (Gaedt *et al.*, 2009). Furthermore, as this block copolymer would be expected to afford cylindrical micelles on account of the **PFS** volume fraction, hybrid "scarf-like" micelles were obtained where the length of the "tassels" scaled linearly with the quantity of unimer added (Figure 16.14a). Cylindrical micelles have also been grown via epitaxial growth by addition of this same block copolymer to an annealed film of $\mathbf{PFS_{129}}$ (Figure 16.14b).

In addition to demonstrations of homoepitaxial growth, where nano-objects containing the same semicrystalline **PFS** core have been grown from one another, nucleation of a copolymer containing a compositionally different semicrystalline component has also been achieved from **PFS**. Poly(ferrocenyldimethylgermane) (**PFG**) is a semicrystalline polymer with a short lattice parameter that is 8% larger than that of the analogous **PFS** (Gaedt *et al.*, 2009). Addition of **PFG**-containing unimers to a colloidal solution of either **PFS**-containing cylindrical micelles or platelets afforded composite nano-objects via heteroepitaxial growth.

16.5 Self-Assembly of PFS Block Copolymers in the Solid State

Polyferrocenylsilane block copolymers also phase separate in the solid state to generate periodic, nanoscopic iron-rich domains that can be observed by TEM without resorting to

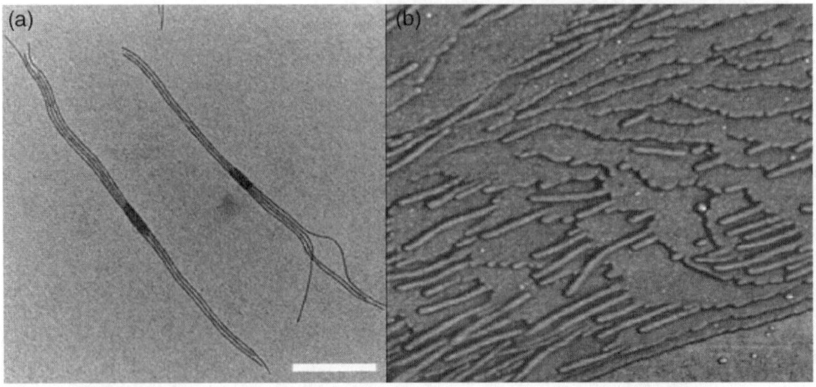

Figure 16.14 (a) TEM image of a "scarf-like" micelle and (b) AFM phase image showing cylindrical micelles grown from a PFDMS thin film. Scale bar corresponds to 500 nm (reproduced from Gaedt *et al.* (2009) with permission).

staining techniques (Ni *et al.*, 1996). This bulk self-assembly behavior has been studied on various block copolymers such as **PS-*b*-PFS** (Rider *et al.*, 2005; Lammertink *et al.*, 1999; Temple *et al.*, 2003), **PI-*b*-PFS** (Lammertink *et al.*, 2000), **PFS-*b*-PMMA** (Kloninger and Rehahn, 2004a). Additional self-assembly complications that can arise from the crystallization of **PFS** with two *n*-alkyl groups at silicon can be avoided through the use of unsymmetrically substituted **PFS**s such as poly(ferrocenylethylmethylsilane). However, the most well-studied block is poly(ferrocenyldimethylsilane) and the designation **PFS** will represent this material unless otherwise noted. This **PFS** block is normally amorphous and glassy at room temperature as annealing is required for crystallization to take place in bulk or thin films (but not in solution – Section 16.4). Shown in Figure 16.15 are TEM micrographs of solid-state bulk morphologies for **PS-*b*-PFS** (**PFS** = poly(ferrocenylethylmethylsilane)) (Rider *et al.*, 2005), **PFS-*b*-PMMA** (Kloninger and Rehahn, 2004). (blended with homopolymer), and **PMMA-*b*-PFS-*b*-PS-*b*-PFS-*b*-PMMA** block copolymers (Datta and Rehahn, 2004). Furthermore, with the **PS-*b*-PFS** system, oxidation of the **PFS** block was reported as having an influence on the order–disorder transition temperature (Eitouni and Balsara, 2004) of the phase-separated morphologies. Exploitation of the high etch resistance of **PFS** nanodomains with respect to its surrounding matrix has also been used to nanopattern surfaces (Lammertink *et al.*, 2000).

Additionally, the periodic **PFS** domains can be converted to iron-containing nanoclusters that can be used as heterogeneous catalysts (Lastella *et al.*, 2004; Lu *et al.*, 2005; Hinderling *et al.*, 2004). Spontaneous perpendicular-to-surface orientation of cylindrical **PFS** microdomains was observed throughout a thin-film sample of **PS_{374}-*b*-PFS_{45}** (**PFS** = poly(ferrocenylethylmethylsilane)). Shown in Figure 16.16a is a TEM image of the film where, following pyrolysis above 600 °C, an array of ceramic nanodomain replicas with Fe nanocrystals was produced (Figure 16.16b) (Temple *et al.*, 2003; Rider *et al.*, 2008).

The iron-rich ceramics from films **PS-*b*-PFS** function as catalysts for the growth of single-walled carbon nanotubes (SWCNT). Typically, thin films were self-assembled and treated with UV-ozone to remove organic materials and subsequently used for the simple one-step chemical vapor deposition (CVD) growth of SWCNTs. The resulting SWCNTs were characterized by

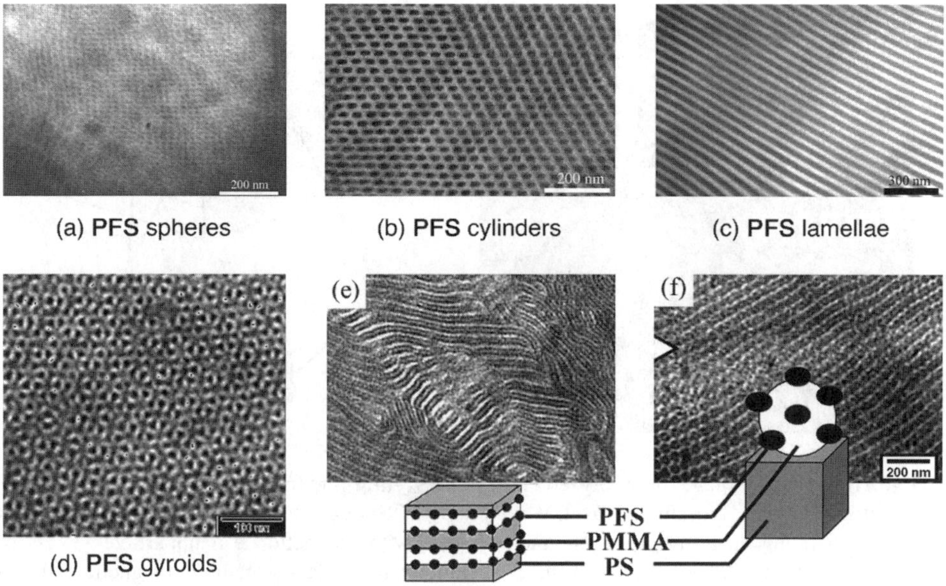

Figure 16.15 TEM images for bulk state morphologies of **PS-*b*-PFS** (a–c), **PFS-*b*-PMMA** (d) and **PMMA-*b*-PFS-*b*-PS-*b*-PFS-*b*-PMMA** (e–f) (**PFS** = poly(ferrocenylethylmethylsilane) in (a)–(c) and the dimethyl analog in (d)–(f). ((a)–(c) Reprinted with permission from D.A Rider, K.A. Cavicchi, K.N. Power-Billard *et al.*, "Diblock copolymers with amorphous atactic polyferrocenylsilane blocks: Synthesis, characterization, and self-assembly of polystyrene-block-poly(ferrocenylethylmethylsilane) in the bulk state," *Macromolecules*, **38**, 6931, 2005. © 2005 American Chemical Society. (d) Reprinted with permission from C. Kloninger and M. Rehahn, "Bicontinuous gyroidic morphologies in ferrocenyldimethylsilane-b-methyl methacrylate diblock copolymer blends," *Macromolecules*, **37**, 8319, 2004. © 2004 American Chemical Society. (e)–(f) Reproduced with permission from U. Datta and M. Rehahn, "Synthesis and self-assembly of styrene–[1]dimethylsilaferrocenophane–methyl methacrylate pentablock copolymers," *Molecular Rapid Communications*, 2004, **25**, 1615. © Wiley-VCH Verlag GmbH & Co. KGaA.)

atomic force microscopy (AFM), scanning electron microscopy (SEM), Raman spectroscopy and X-ray photoelectron spectroscopy (XPS). Shown in Figure 16.17a is an SEM of a SWCNT array generated from lithographic patterning of a **PS-*b*-PFS** film (Lastella *et al.*, 2004; Lu *et al.*, 2005). In an analogous study by Hinderling *et al.*, it was also found that multiwalled CNTs could also be grown from **PFS**-derived catalysts using a plasma treatment approach (Hinderling *et al.*, 2004). More recently, Lastella *et al.* have shown that the SWCNTs from pyrolyzed **PS-*b*-PFS** films could be used in high-throughput field effect transistors (FETs) (Lastella *et al.*, 2006). Shown in Figure 16.17b is a schematic for the high device yield (96%) preparation of individually addressable FETs. A density of up to 160 devices on a $15 \times 15\ mm^2$ chip was easily achieved (Lastella *et al.*, 2006).

In addition to the ease of fabricating SWCNT-based devices, the polymer precursor route to catalysts for SWCNT growth has proven useful in achieving high-quality SWCNT characterization where substrate effects are avoided (Lu *et al.*, 2006a). Here, an array of well-defined and highly ordered catalyst-containing nanostructures was generated from a

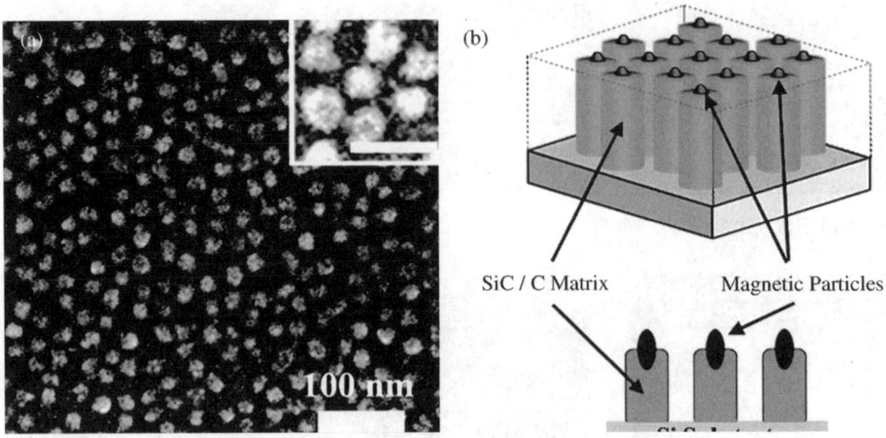

Figure 16.16 (a) SEM image for pyrolyzed $\mathbf{PS_{374}}$-***b***-$\mathbf{PFS_{45}}$ thin-film (600 °C, 2 h, N_2) thin film in which the PS matrix has been crosslinked by UV radiation (**PFS** = poly(ferrocenylethylmethylsilane). (b) Schematic illustrating an array of nanodomain replicas with magnetic Fe nanoparticles. (Reprinted with permission from D.A. Rider, K. Liu, J.-C. Eloi *et al.*, "Nanostructured magnetic thin films from organometallic block copolymers: Pyrolysis of self-assembled polystyrene-block-poly (ferrocenylethylmethylsilane)," *ACS Nano*, **2**, 263, 2008. © 2008 American Chemical Society.)

thin film of self-assembled **PS-*b*-PFS** that was then used to synthesize suspended SWCNTs. For catalyst patterning, a motif that exploits the compatibility of the block copolymer films with conventional novolac-based photoresist lithography was accomplished (Figure 16.18a–f). First, a self-assembled catalyst-containing block copolymer film was cast on a SiO_2 (blue color) substrate. Secondly, UV-patterning of a photoresist film on top of

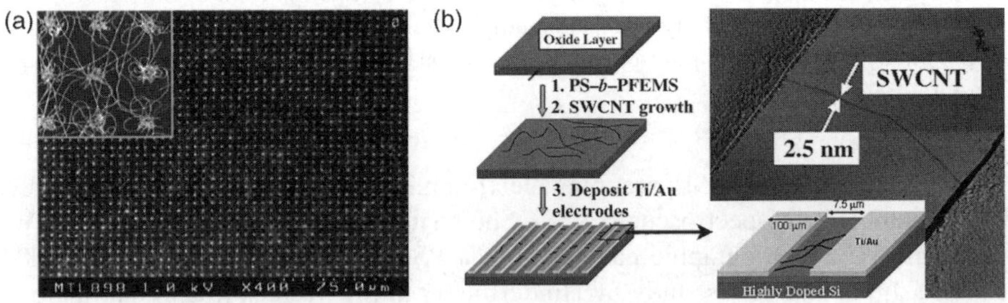

Figure 16.17 (a) SEM images for single-walled carbon nanotubes (SWCNT) synthesized by chemical vapor deposition (CVD) with iron-rich ceramic particles derived from patterned **PS-*b*-PFS** islands as catalyst. (Reprinted with permission from J.Q. Lu, T.E. Kopley, N. Moll *et al.*, "High-quality single-walled carbon nanotubes with small diameter, controlled density, and ordered locations using a polyferrocenylsilane block copolymer catalyst precursor," *Chemistry of Materials*, **17**, 2227, 2005. © 2005 American Chemical Society.) (b) Schematic and AFM height image (11 μm^2) of high-throughput field effect transistors from SWCNTs afforded from pyrolyzed **PS-*b*-PFS** films. (**PFS** = poly(ferrocenylethylmethylsilane). (Reproduced with permission from S. Lastella, G. Mallick, R. Woo *et al.*, "Parallel arrays of individually addressable single-walled carbon nanotube field-effect transistors," *Journal of Applied Physics*, **99**, 024302/1. © 2006 American Institute of Physics.)

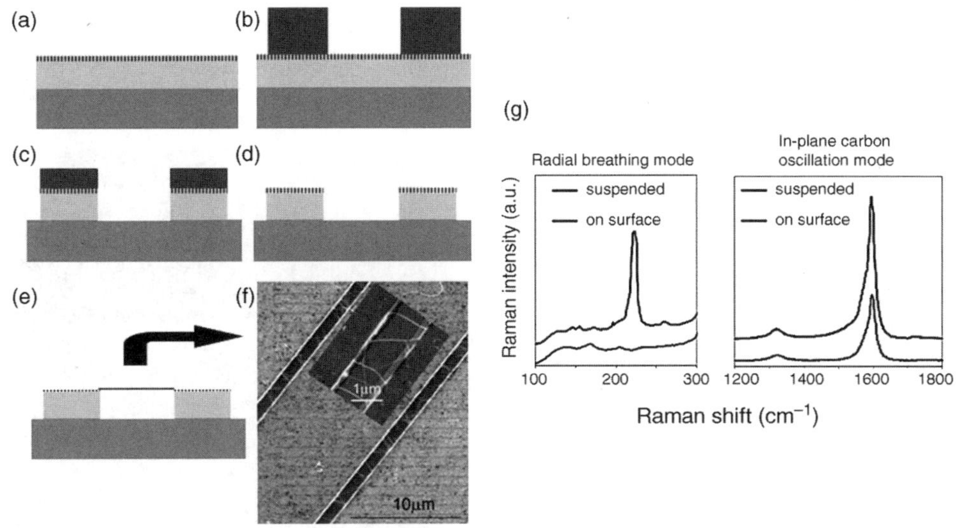

Figure 16.18 (a) Self-assembled precatalyst-containing **PS-*b*-PFS** (**PFS** = poly(ferrocenylethylmethylsilane)) block copolymer film on SiO_2 (blue color). (b) Patterned photoresist film (red color) on top of the catalyst-containing self-assembled polymer film. (c) Resist pattern transferred into the catalyst-containing polymer film and into the silicon oxide by ICP etching. (d) Catalyst-containing nanostructures after solvent strip of photoresist and UV-ozonation removing all the organic components. (e) SWCNTs suspended across a patterned gap and (f) representative SEM (inset depicting six suspended bundles of SWCNTs). (g) Comparison of Raman spectra of suspended and substrate-supported SWCNTs produced from iron-containing nanostructures. (Reprinted with permission from J. Lu, T. Kopley, D. Dutton *et al.*, "Generating suspended single-walled carbon nanotubes across a large surface area via patterning self-assembled catalyst-containing block copolymer thin films," *The Journal of Physical Chemistry B*, **110**, 10585, 2006. © 2006 American Chemical Society.)

the catalyst-containing self-assembled polymer film was completed. Thirdly, the resist pattern was transferred into the catalyst-containing polymer film and into the thermally grown silicon oxide by and inductively coupled plasma (ICP) etch. Next, a solvent strip of the photoresist was performed such that UV-ozonation of the polymer film was possible. Lastly, the chemical vapor deposition of SWCNTs was performed such that extending CNTs would bridge the patterned gaps during their growth (Figures 16.18e–f). Figure 16.18g compares the Raman spectra of suspended SWCNTs with those that were in direct contact with the surface. Enhanced Raman signals for suspended CNTs were observed and this has hence proven useful for accurate determination of CNT diameters. This method for producing suspended tubes is expected to facilitate the characterization of SWCNTs by optical means and harness suspended tubes for device applications.

As the interest in nanolithography grows, efficient routes to nanotemplates are required. An emerging method involves films of phase-separated diblock copolymers. As an interesting application of metal-containing block copolymers, we discuss the lithographic template produced using **PS-*b*-PFS**. Ross and coworkers demonstrated that by spin coating a film of a **PS-*b*-PFS,** a material that afforded plasma etch-resistant **PFS** spheres in a **PS** matrix, direct transfer of the pattern to underlying layers was possible (Cheng *et al.*, 2001). Using oxygen

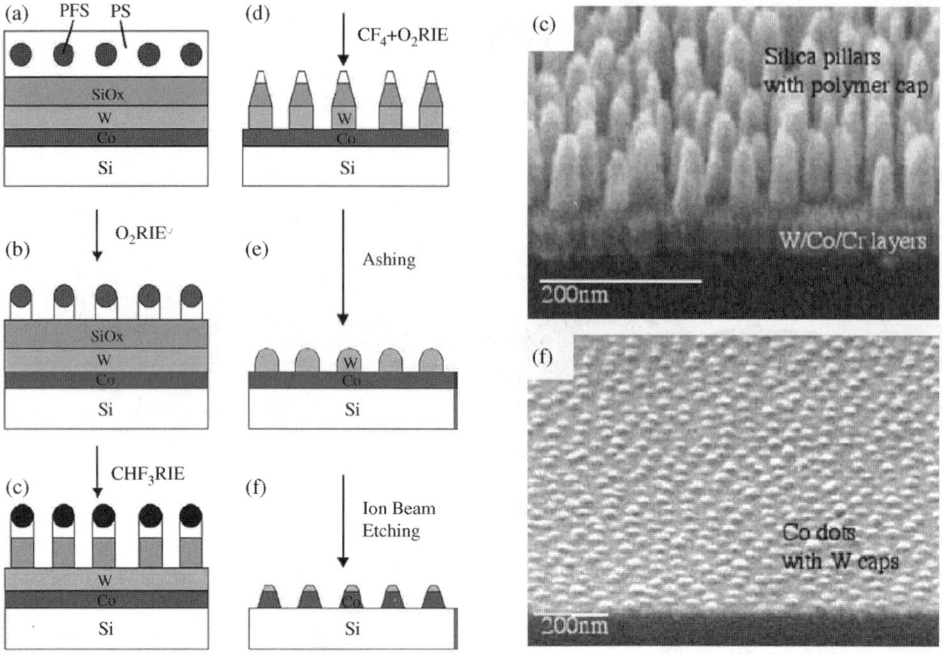

Figure 16.19 (a) Thin film of **PS-*b*-PFS** (**PFS** spheres) on silicon substrate with Co, W and SiO_x layers. (b) Oxygen reactive ion etch of **PS-*b*-PFS**. (c) CHF_3 reactive ion etch of SiO_x layer (SEM of resulting film at upper right). (d) $CF_4 + O_2$ reactive ion etch of **PS-*b*-PFS** derived lithographic mask and W layer. (e) Ashing to remove all material above W. (f) Ion-beam etching to produce cobalt dots with W caps. (Reproduced with permission from J.Y. Cheng, C.A. Ross, V.Z.-H. Chan *et al.*, "Formation of a cobalt magnetic dot array via block copolymer lithography," *Advanced Materials*, 2001, **13**, 1174. © Wiley-VCH Verlag GmbH & Co. KGaA.)

reactive ion etching (RIE), much of the **PS** matrix is eliminated so as to permit the CF_3H etch of an underlying silica layer (Figure 16.19c). Further CF_4 and O_2 RIE followed by ashing permits the patterned etching of tungsten. Lastly, an ion-beam etch was used to transfer the pattern into the cobalt layer. The end result is an array of tungsten-capped cobalt dots that have dimensions on the nanometer scale and whose pattern has been inherited from the **PS-*b*-PFS** film (Figure 16.19f).

Cheng *et al.* have studied the influence of topographically patterned substrates on the morphology and close packing of **PS-*b*-PFS** (Cheng *et al.*, 2003; Cheng *et al.*, 2006). Recently, the distortion of the spherical block copolymer morphology and a transition of the packing lattice were observed for **PFS** spheres in a **PS** matrix when confined within step-defined and V-shaped nanochannels, respectively (Cheng *et al.*, 2006; Chuang *et al.*, 2006b). Without confinement effects, the equilibrium thin-film solid-state morphology of the **PS-*b*-PFS** studied was a close-packed lattice of **PFS** spheres in a **PS** matrix (row spacing = $d_0 = 29$ nm). When a similar thin film is allowed to fill step-defined nanochannels of differing widths ($W = 30$–80 nm), interesting morphological deviations were observed. By using oxygen reactive ion etching (RIE), the researchers were able to quantify the spherical distortions that result from the channel confinement of the block copolymer. Shown in Figure 16.20a is a

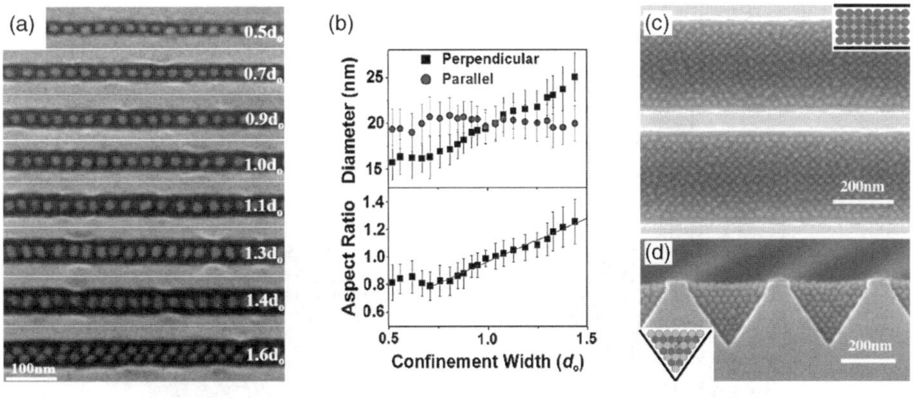

Figure 16.20 (a) SEM images of **PS-*b*-PFS** films in channels of different confinement widths (W). (Reprinted with permission from J.Y. Cheng, F. Zhang, V.P. Chuang *et al.*, "Self-assembled one-dimensional nanostructure arrays," *Nano Letters*, **6**, 2099, 2006. © 2006 American Chemical Society.) A single row of spheres forms when $0.6 < W/d_0 < 1.5$ where d_0 is the equilibrium periodicity of the block copolymer. (b) SEM images of **PS-*b*-PFS** films in V-shaped grooves that enforce a fcc packing arrangement. (Reprinted with permission from V.P. Chuang, J.Y. Cheng, T.A. Savas and C.A. Ross, "Three-dimensional self-assembly of spherical block copolymer domains into V-shaped grooves," *Nano Letters*, **6**, 2332, 2006. © 2006 American Chemical Society.)

composite image of **PFS** block copolymer domains within the channels of differing confinement widths. Each case is defined and labeled according to the confinement fraction: W/d_0, where $0.5 \leq W/d_0 \leq 1.6$. It was found that an elliptical distortion of the spheres occurred for $W/d_0 > 0.65$ (Figures 16.20b and c). When the nanochannel is sufficiently wide to accommodate an additional row of spheres (Figure 16.20a, the case of $W/d_0 = 1.6$), the ellipsoidal distortion is removed. In V-shaped grooves however, the trough shape was found to promote the formation of a well-ordered face-centered cubic (fcc) close-packed sphere array rather than the body-centered cubic (bcc) packing found in the equilibrium bulk state. Unlike the hexagonal symmetry seen in monolayers of spherical domains, the top surface shows a square symmetry (Figure 16.20c) where the (111) planes of the lattice are now parallel to the groove walls (Figure 16.20d). It is anticipated that these novel substrate determined manipulations of the block copolymer morphology and lattice will be useful as novel routes to new features and geometries for block copolymer lithography.

Nanotextured Ag surfaces generated from a **PFS** diblock copolymer were found to exhibit high and uniform surface-enhanced Raman spectroscopy (SERS) activity (Lu *et al.*, 2006b). First, the metal-containing diblock copolymer, **PS-*b*-PFS (PFS** = poly(ferrocenylethylmethylsilane)), was self-assembled in a thin film into hexagonally close-packed standing cylindrical **PFS**-based nanostructures surrounded by a **PS** matrix. UV-ozonation was then used to selectively remove the **PS** matrix and convert **PFS** into iron-containing silicon oxide cylinders. A silver nanotextured surface was subsequently generated by sputtering a thin layer of silver over these inorganic cylinders (Figure 16.21). Uniformly enhanced Raman signals on the silver nanotextured surfaces have been observed for absorbed benzenethiol, with enhancement factors up to 10^6 (Figure 16.21, bottom). This metal-containing block copolymer template approach provides a simple and straightforward method to fabricate SERS active

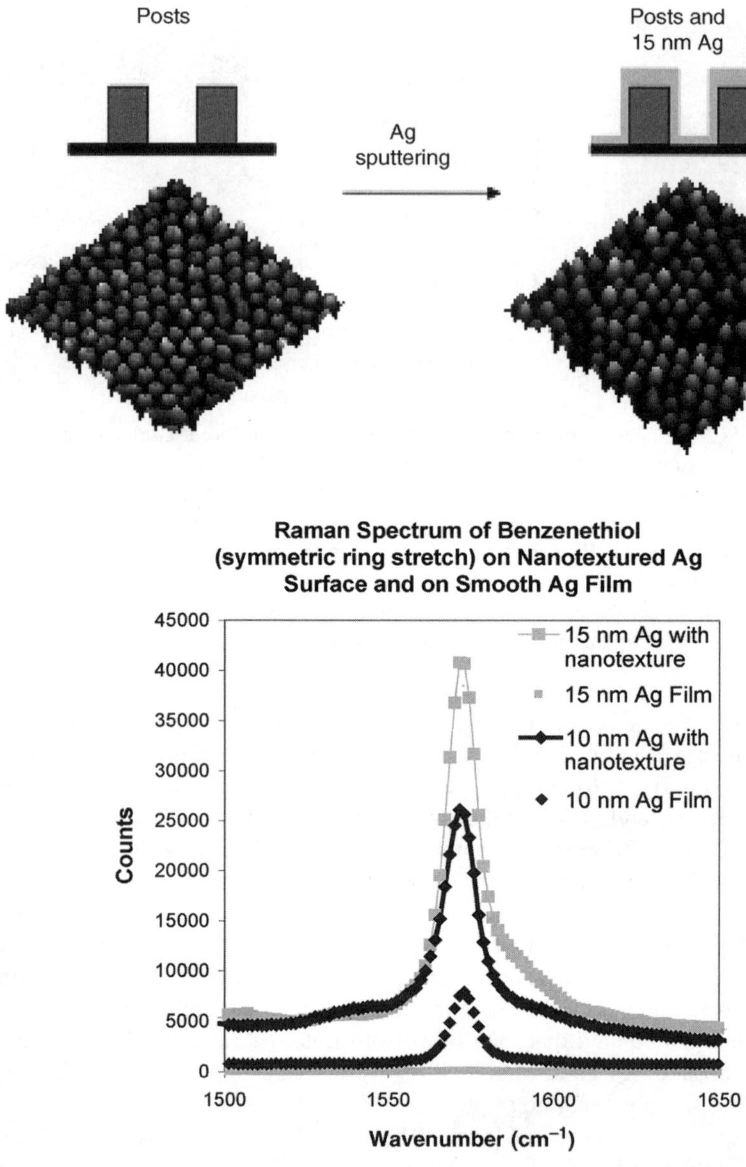

Figure 16.21 Generation of nanotextured Ag film (AFM height image top right) from sputtering on UV-ozone etched **PS-*b*-PFS** film (**PFS** = poly(ferrocenylethylmethylsilane)) (AFM height image top left). Surface-enhanced Raman spectra (bottom) of benzenethiol on nanotextured Ag films (10 and 15 nm: -◆- and -■- respectively) and control experiments with smooth Ag films. (Reproduced with permission from J.Q. Lu, D. Chamberlin, D.A. Rider *et al.*, "Using a ferrocenylsilane-based block copolymer as a template to produce nanotextured Ag surfaces: Uniformly enhanced surface enhanced Raman scattering active substrates," *Nanotechnology*, **17**, 5792, 2006. © 2006 IOP Publishing.)

substrates with great manufacturing potential. Furthermore, since the size and spacing of the cylinders can be adjusted by tailoring the polymer chain lengths, the electromagnetic field can potentially be tuned to achieve even higher SERS activity. More importantly, this block copolymer approach has established the potential for reproducible fabrication of uniformly enhancing SERS-active substrates.

To date, block copolymer self-assembly for nanolithographic applications has mainly focused on diblock copolymers, however, triblock terpolymers offer a wider range of geometries than the simple line and dot morphologies available to diblock copolymers.

Nanoscaled ring arrays have been prepared from a **PS-*b*-PFS-*b*-P2VP (PFS** = poly(ferrocenylethylmethylsilane)) triblock terpolymer (Chuang *et al.*, 2009a). Such ring-shaped structures have potential applications in the fabrication of memories or sensors (Castano *et al.*, 2005; Podbielski *et al.*, 2006; Miller *et al.*, 2002) and quantum devices (Aharonov and Bohm, 1959; Yu *et al.*, 2007; Levy *et al.*, 1990). Thin film self-assembly of the **PS-*b*-PFS-*b*-P2VP** triblock terpolymer gave a core–shell cylinder morphology of 50 nm period consisting of a **PS** core with **PFS** shell in a **P2VP** matrix perpendicular to the film surface. The **PS** and **P2VP** were selectively removed by oxygen RIE leaving partly oxidized **PFS** rings 33 nm in diameter and 11 nm wide. The presence of two organic (**PS** and **P2VP**) blocks and an organometallic (**PFS**) block results in high etch selectivity due to the resistance of **PFS** to oxygen plasma and enhanced pattern transfer compared to all organic triblock terpolymers. As the **PFS** cylinders did not span the entire film thickness, the ring patterns were transferred into a **PS** layer by imprinting (Figure 16.22A).

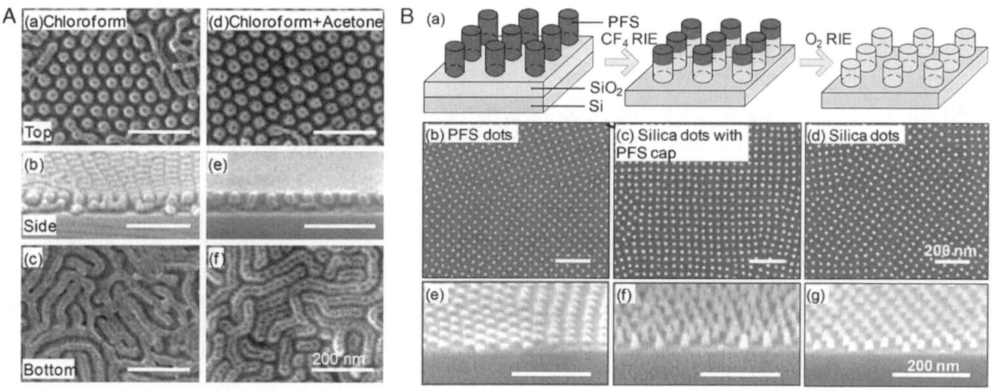

Figure 16.22 (A) SEM images of thin films of **PS-*b*-PFS-*b*-P2VP (PFS** = poly(ferrocenylethylmethylsilane)) after annealing in (a)–(c) chloroform and (d)–(f) a mixed solvent vapor of chloroform and acetone for 4 h at room temperature, followed by etching with oxygen RIE. (a) and (d): Plan view, (b) and (e): side view and (c) and (f): bottom view. (B) Fabrication of square arrays of posts from **PI-*b*-PS-*b*-PFS** (**PFS** = poly(ferrocenylethylmethylsilane)). (a) Schematic of the pattern transfer process. (b)–(g) SEM images of (b)–(d) plan view and (e)–(g) side views of (b)(e) **PFS** posts after **PI** and **PS** removal; (c)(f) silica posts with **PFS** cap; (d)(g) silica posts. (Reprinted with permission from V.P. Chuang, J. Gwyther, R.A. Mickiewicz *et al.*, "Templated self-assembly of square symmetry arrays from an ABC triblock terpolymer," *Nano Letters*, **9**, 4364, 2009. © 2009 American Chemical Society.).

Using a similar approach, thin films of a **PI-*b*-PS-*b*-PFS** triblock terpolymer and a **PI-*b*-PS-*b*-PFS/PS** homopolymer blend (**PFS** = poly(ferrocenylethylmethylsilane)) have been used to prepare nanoscaled square-symmetry arrays (Chuang *et al.*, 2009b). Access to such geometries is very difficult using the self-assemble of diblock copolymers, yet it is a device geometry considered essential for future lithography needs (Garner, Herr, and Krautschik), for example as via array structures in integrated circuits. Square-packed **PI** and **PFS** cylinders in a **PS** matrix were formed in thin films of the **PI-*b*-PS-*b*-PFS** triblock terpolymer and **PI-*b*-PS-*b*-PFS/PS** blend. Long-range order was improved by addition of the **PS** homopolymer, which is thought to stabilize the square-packed morphology by occupying volume between the cylinders allowing the highly stretched triblock terpolymer chains to relax. The organic blocks were again removed via oxygen RIE to expose square-packed **PFS** cylinders of 40.5 nm period that extended throughout the film thickness. Pattern transfer of the **PFS** domains to a 30-nm thick silica film by successive etching (Figure 16.22B) gave a square array of silica posts of 20 nm diameter and 30 nm height.

The thin-film self-assembly of the **PI-*b*-PS-*b*-PFS/PS** blend was also templated on topographical substrates with shallow groves etched into the silicon. The periodicity of the templated square arrays could be varied with trench width and orientation of the **PFS** domains controlled by treating the substrate with a **PS** brush. Preferential wetting of the middle **PS** block to the trench wall changes the lattice vector of the **PFS** array from 90° to 45° with respect to the trench edge. In thicker films, inplane cylinders were obtained where the orientation of the cylinders with respect to the trench wall could again be controlled by treating the substrate with a **PS** brush.

16.6 Summary

The study of block copolymers and their applications in nanoscience represents one of the most active areas in polymer materials science and **PFS** block copolymers are making a significant contribution to this field. **PFS** block copolymers can be prepared via living anionic and photocontrolled living anionic ring-opening polymerization routes and well-defined **PFS** block copolymer architectures are readily accessible. These materials exhibit both predicted behavior in terms of self-assembled morphologies, especially in the solid state, as well as unprecedented morphologies in solution as a result of PFS core crystallization. The presence of Fe and Si in **PFS** block copolymers offer exciting possibilities for the development of new functional nanostructured materials via simple self-assembly approaches. Applications as etch resists for the creation of nanopatterned surfaces for SERS applications, high-density magnetic data storage, and catalytic applications have already been realized. Further interesting phenomena and applications are likely to be discovered in the future.

Acknowledgments

For the relevant sections that describe our work, the authors would like to acknowledge extensive collaborations with Professor Mitch Winnik at the University of Toronto on **PFS** block copolymer assembly in solution and collaborations with Professor Tom Russell (U. Massachusetts), Dr Jennifer Lu (Agilent Technologies), and Professor Chang Ryu (RPI) and Professor Caroline Ross (MIT) on **PFS** block copolymer thin films.

References

Aharonov, Y. and Bohm, D. (1959) Significance of electromagnetic potentials in the quantum theory. *Physical Review*, **115**, 485.

Archer, R.D. (2001) *Inorganic and Organometallic Polymers*, Wiley VCH, Weinheim.

Bartole-Scott, A., Resendes, R., and Manners, I. (2003) Transition metal-catalyzed ring-opening polymerization of silicon-bridged [1]ferrocenophanes in the presence of functional silanes: molecular weight control and synthesis of telechelic poly (ferrocenyl silanes). *Macromolecular Chemistry and Physics*, **204**, 1259.

Buetuen, V., Lowe, A.B., Billingham, N.C., and Armes, S.P. (1999) Synthesis of zwitterionic shell cross-linked micelles. *Journal of the American Chemical Society*, **121**, 4288.

Cao, L., Manners, I., and Winnik, M.A. (2002) Influence of the interplay of crystallization and chain stretching on micellar morphologies: solution self-assembly of coil-crystalline poly(isoprene-block-ferrocenylsilane). *Macromolecules*, **35**, 8258.

Cao, L., Massey, J.A., Winnik, M.A. *et al.* (2003) Reactive ion etching of cylindrical polyferrocenylsilane block copolymer micelles: fabrication of ceramic nanolines on semiconducting substrates. *Advanced Functional Materials*, **13**, 271.

Castano, F.J., Morecroft, D., Jung, W., and Ross, C.A. (2005) Spin-dependent scattering in multilayered magnetic rings. *Physical Review Letters*, **95**, 137201.

Chandrasekhar, V. (2005) *Inorganic and Organometallic Polymers*, Springer, New York.

Cheng, J.Y., Ross, C.A., Chan, V.Z.-H. *et al.* (2001) Formation of a cobalt magnetic dot array via block copolymer lithography. *Advanced Materials*, **13**, 1174.

Cheng, J.Y., Ross, C.A., Thomas, E.L. *et al.* (2003) Templated self-assembly of block copolymers: effect of substrate topography. *Advanced Materials*, **15**, 1599.

Cheng, J.Y., Zhang, F., Smith, H.I. *et al.* (2006a) Pattern registration between spherical block-copolymer domains and topographical templates. *Advanced Materials*, **18**, 597.

Cheng, J.Y., Zhang, F., Chuang, V.P. *et al.* (2006b) Self-assembled one-dimensional nanostructure arrays. *Nano Letters*, **6**, 2099.

Chuang, V.P., Cheng, J.Y., Savas, T.A., and Ross, C.A. (2006) Three-dimensional self-assembly of spherical block copolymer domains into V-shaped grooves. *Nano Letters*, **6**, 2332.

Chuang, V.P., Ross, C.A., Gwyther, J., and Manners, I. (2009a) Self-assembled nanoscale ring arrays from a polystyrene-b-polyferrocenylsilane-b-poly(2-vinylpyridine) triblock terpolymer thin film. *Advanced Materials*, **21**, 3789.

Chuang, V.P., Gwyther, J., Mickiewicz, R.A. *et al.* (2009b) Templated self-assembly of square symmetry arrays from an ABC triblock terpolymer. *Nano Letters*, **9**, 4364.

Cyr, P.W., Rider, D.A., Kulbaba, K., and Manners, I. (2004) Photopatternable metallopolymers: photocrosslinking and photolithography of polyferrocenylsilane methacrylates. *Macromolecules*, **37**, 3959.

Datta, U. and Rehahn, M. (2004) Synthesis and self-assembly of styrene- [1]dimethylsilaferrocenophane-methyl methacrylate pentablock copolymers. *Macromolecular Rapid Communications*, **25**, 1615.

Ding, J. and Liu, G. (1998) Polystyrene block poly(2-cinnamoylethyl methacrylate) nanospheres with cross-linked shells. *Macromolecules*, **31**, 6554.

Eitouni, H.B. and Balsara, N.P. (2004) Effect of chemical oxidation on the self-assembly of organometallic block copolymers. *Journal of the American Chemical Society*, **126**, 7446.

Foucher, D.A., Tang, B.-Z., and Manners, I. (1992) Ring-opening polymerization of strained, ring-tilted ferrocenophanes - a route to high molecular weight poly(ferrocenylsilanes). *Journal of the American Chemical Society*, **114**, 6246.

Frankowski, D.J., Raez, J., Manners, I. *et al.* (2004) Formation of dispersed nanostructures from poly(ferrocenyldimethylsilane-b-dimethylsiloxane) nanotubes upon exposure to supercritical carbon dioxide. *Langmuir*, **20**, 9304.

Gaedt, T., Ieong, N.S., Cambridge, G. *et al.* (2009) Complex and hierarchical micelle architectures from diblock copolymers using living, crystallization-driven polymerizations. *Nature Materials*, **8**, 144.

Garner, C.M., Herr, D., and Krautschik, C.Directed self-assembly requirements for sub-20nm lithography. www.itrs.net/Links/2007ITRS/LinkedFiles/ERM/DSAR11%20082107.DOC.

Gohy, J.-F., Lohmeijer, B.G.G., Alexeev, A. *et al.* (2004) Cylindrical micelles from the aqueous self-assembly of an amphiphilic poly(ethylene oxide)-b-poly(ferrocenylsilane) (PEO-b-PFS) block copolymer with a metallo-supramolecular linker at the block junction. *Chemistry - A European Journal*, **10**, 4315.

Gómez-Elipe, P., Resendes, R., Macdonald, P.M., and Manners, I. (1998) Transition metal catalyzed ring-opening polymerization (ROP) of silicon-bridged [1]ferrocenophanes: facile molecular weight control and the remarkably convenient synthesis of poly(ferrocenes) with regioregular, comb, star, and block architectures. *Journal of the American Chemical Society*, **120**, 8348.

Gwyther, J. and Manners, I. (2009) Diblock copolymers with amorphous, high glass transition temperature, organometallic block: synthesis, characterisation and self-assembly of polystyrene-*b*-poly(ferrocenylisopropylmethylsilane) in the bulk state. *Polymer* **50**, 5384–5389

Hinderling, C., Keles, Y., Stoeckli, T. *et al.* (2004) Organometallic block copolymers as catalyst precursors for templated carbon nanotube growth. *Advanced Materials*, **16**, 876.

Review: Holliday, B.J. and Swager, T.M. (2005) Conducting metallopolymers: the roles of molecular architecture and redox matching. *Chemical Communications*, **1**, 23.

Hultzsch, K.C., Nelson, J.M., Lough, A.J., and Manners, I. (1995 Synthesis, characterization, and homopolymerization and copolymerization behavior of the silicon-bridged [1] chromarenophane $Cr(\eta6\text{-}C_6H_5)_2SiMe_2$. *Organometallics*, **14**, 5496.

Kim, K.T., Vandermeulen, G.W.M., Winnik, M.A., and Manners, I. (2005a) Organometallic-polypeptide block copolymers: synthesis and properties of poly(ferrocenyldimethylsilane)-b-poly-(gamma-benzyl-L-glutamate). *Macromolecules*, **38**, 4958.

Kim, K.T., Park, C., Vandermeulen, G.W.M. *et al.* (2005b) Gelation of helical polypeptide-random coil diblock copolymers by a nanoribbon mechanism. *Angewandte Chemie-International Edition*, **44**, 7964.

Kim, K.T., Han, J., Ryu, C.Y. *et al.* (2006) Synthesis, characterization, and AFM studies of dendronized polyferrocenylsilanes. *Macromolecules*, **39**, 7922.

Review: Kingsborough, R.P. and Swager, T.M. (1999) *Progress in Inorganic Chemistry*, **48**, 123.

Kloninger, C. and Rehahn, M. (2004a) 1,1-dimethylsilacyclobutane-mediated living anionic block copolymerization of [1]dimethylsilaferrocenophane and methyl methacrylate. *Macromolecules*, **37**, 1720.

Kloninger, C. and Rehahn, M. (2004b) Bicontinuous gyroidic morphologies in ferrocenyldimethylsilane-b-methyl methacrylate diblock copolymer blends. *Macromolecules*, **37**, 8319.

Korczagin, I., Hempenius, M.A., and Vancso, G.J. (2004) Poly(ferrocenyisilane-block-methacrylates) via sequential anionic and atom transfer radical polymerization. *Macromolecules*, **37**, 1686.

Korczagin, I., Hempenius, M.A., Fokkink, R.G. *et al.* (2006) Self-assembly of poly(ferrocenyldimethylsilane-b-methyl methacrylate) block copolymers in a selective solvent. *Macromolecules*, **39**, 2306.

Kulbaba, K. and Manners, I. (2001) Polyferrocenylsilanes: metal-containing polymers for materials science, self-assembly and nanostructure applications. *Macromolecular Rapid Communications*, **22**, 711.

Kulbaba, K., MacLachlan, M.J., Evans, C.E.B., and Manners, I. (2001) Organometallic gels: characterization and electrochemical studies of swellable, thermally crosslinked poly(ferrocenylsilane)s. *Macromolecular Chemistry and Physics*, **202**, 1768.

Lammertink, R.G.H., Hempenius, M.A., Thomas, E.L., and Vancso, G.J. (1999) Periodic organic-organometallic microdomain structures in poly(styrene-block-ferrocenyldimethylsilane) copolymers and blends with corresponding homopolymers. *Journal of Polymer Science Part B-Polymer Physics*, **37**, 1009.

Lammertink, R.G.H., Hempenius, M.A., van den Enk, J.E. *et al.* (2000) Nanostructured thin films of organic-organometallic block copolymers: one-step lithography with poly(ferrocenylsilanes) by reactive ion etching. *Advanced Materials*, **12**, 98.

Lastella, S., Jung, Y.J., Yang, H. *et al.* (2004) Density control of single-walled carbon nanotubes using patterned iron nanoparticle catalysts derived from phase-separated thin films of a polyferrocene block copolymer. *Journal of Materials Chemistry*, **14**, 1791.

Lastella, S., Mallick, G., Woo, R. *et al.* (2006) Parallel arrays of individually addressable single-walled carbon nanotube field-effect transistors. *Journal of Applied Physics*, **99**, 024302/1.

Levy, L.P., Dolan, G., Dunsmuir, J., and Bouchiat, H. (1990) Magnetization of mesoscopic copper rings - evidence for persistent currents. *Physical Review Letters*, **64**, 2074.

Li, W., Sheller, N., Forster, M.D. *et al.* (2000) Morphology and ordering behavior of a poly(styrene)-b-poly (ferrocenyldimethylsilane)diblock copolymer. *Polymer*, **41**, 719.

Lu, J.Q., Kopley, T.E., Moll, N. *et al.* (2005) High-quality single-walled carbon nanotubes with small diameter, controlled density, and ordered locations using a polyferrocenylsilane block copolymer catalyst precursor. *Chemistry of Materials*, **17** 2227.

Lu, J., Kopley, T., Dutton, D. *et al.* (2006a) Generating suspended single-walled carbon nanotubes across a large surface area via patterning self-assembled catalyst-containing block copolymer thin films. *Journal of Physical Chemistry B*, **110**, 10585.

Lu, J.Q., Chamberlin, D., Rider, D.A. *et al.* (2006b) Using a PFS block copolymer as template to produce SERS. *Nanotechnology*, **17**, 5792.

MacLachlan, M.J., Lough, A.J., and Manners, I. (1996) Spirocyclic [1]ferrocenophanes: novel cross-linking agents for ring-opened poly(ferrocenes). *Macromolecules*, **29**, 8562.

Review: Manners, I. (2001) Materials science - putting metals into polymers. *Science*, **294**, 1664.

Manners, I. (2004) *Synthetic Metal-Containing Polymers*, Wiley-VCH, Weinheim.

Massey, J.A., Power, K.N., Winnik, M.A., and Manners, I. (1998a) Organometallic nanostructures: self-assembly of poly(ferrocene) block copolymers. *Advanced Materials*, **10**, 1559.

Massey, J., Power, K.N., Manners, I., and Winnik, M.A. (1998b) Self-assembly of a novel organometallic-inorganic block copolymer in solution and the solid state: nonintrusive observation of novel wormlike poly (ferrocenyldimethylsilane)-b-poly(dimethylsiloxane) micelles. *Journal of the American Chemical Society*, **120**, 9533.

Massey, J.A., Temple, K., Cao, L. *et al.* (2000) Self-assembly of organometallic block copolymers: The role of crystallinity of the core-forming polyferrocene block in the micellar morphologies formed by poly(ferrocenylsilane-b-dimethylsiloxane) in n-alkane solvents. *Journal of the American Chemical Society*, **122**, 11577.

Massey, J.A., Winnik, M.A., Manners, I. *et al.* (2001) Fabrication of oriented nanoscopic ceramic lines from cylindrical micelles of an organometallic polyferrocene block copolymer. *Journal of the American Chemical Society*, **123**, 3147.

Miller, M.M., Prinz, G.A., Cheng, S.F., and Bounnak, S. (2002) Detection of a micron-sized magnetic sphere using a ring-shaped anisotropic magnetoresistance-based sensor: A model for a magnetoresistance-based biosensor. *Applied Physics Letters*, **81**, 2211.

Review: Newkome, G.R., He, E.F., and Moorefield, C.N. (1999) Suprasupermolecules with novel properties: metallodendrimers. *Chemical Reviews*, **99**, 1689.

Nguyen, P., Stojcevic, G., Kulbaba, K. *et al.* (1998) Synthesis, characterization, and properties of symmetrically substituted, ring-opened poly(ferrocenylalkoxy/aryloxysilanes). *Macromolecules*, **31**, 5977.

Review: Nguyen, P., Gómez-Elipe, P., and Manners, I. (1999) Organometallic polymers with transition metals in the main chain. *Chemical Reviews*, **99**, 1515.

Ni, Y., Rulkens, R., Pudelski, J.K., and Manners, I. (1995) Transition-metal-catalyzed ring-opening polymerization of silicon-bridged [1]ferrocenophanes at ambient-temperature. *Macromolecular Rapid Communications*, **16**, 637.

Ni, Y., Rulkens, R., and Manners, I. (1996) Transition metal-based polymers with controlled architectures: Well-defined poly(ferrocenylsilane) homopolymers and multiblock copolymers via the living anionic ring-opening polymerization of silicon-bridged [1]ferrocenophanes. *Journal of the American Chemical Society*, **118**, 4102.

Peckham, T.J., Massey, M., Edwards, D.A. *et al.* (1996) Synthesis, characterization, and properties of high molecular weight poly(ferrocenylgermanes) and poly(ferrocenylsilane)-poly(ferrocenylgermane) random copolymers. *Macromolecules*, **29**, 2396.

Podbielski, J., Giesen, F., and Grundler, D. (2006) Spin-wave interference in microscopic rings. *Physical Review Letters*, **96**, 167207.

Pudelski, J.K., Rulkens, R., Foucher, D.A. *et al.* (1995) Synthesis and properties of poly(ferrocenyldihyrdrosilane) homopolymer and random copolymers. *Macromolecules*, **28**, 7301.

Pudelski, J.K., Foucher, D.A., Honeyman, C.H. *et al.* (1996) Synthesis, characterization, and properties of high molecular weight poly(methylated ferrocenylsilanes) and their charge transfer polymer salts with tetracyanoethylene. *Macromolecules*, **29**, 1894.

Raez, J., Manners, I., and Winnik, M.A. (2002a) Nanotubes from the self-assembly of asymmetric crystalline-coil poly (ferrocenylsilane-siloxane) block copolymers. *Journal of the American Chemical Society*, **124**, 10381.

Raez, J., Manners, I., and Winnik, M.A. (2002b) Fiberlike structures from the self-assembly of a highly asymmetric poly(ferrocenyldimethylsilane-b-dimethylsiloxane) in dilute solution. *Langmuir*, **18**, 7229.

Reddy, N.P., Yamashita, H., and Tanaka, M. (1995) Platinum-catalyzed or palladium-catalyzed ring-opening homo-polymerization of copolymerization of silicon-bridged and germanium-bridged [1]ferrocenophanes. *Journal of the Chemical Society. Chemical Communications*, 2263.

Rider, D.A., Cavicchi, K.A., Power-Billard, K.N. *et al.* (2005) Diblock copolymers with amorphous atactic poly-ferrocenylsilane blocks: synthesis, characterization, and self-assembly of polystyrene-block-poly(ferrocenylethyl-methylsilane) in the bulk state. *Macromolecules*, **38**, 6931.

Rider, D.A., Liu, K., Eloi, J.-C. *et al.* (2008) Nanostructured magnetic thin films from organometallic block copolymers: Pyrolysis of self-assembled polystyrene-block-poly(ferrocenylethylmethylsilane). *ACS Nano*, **2**, 263.

Roerdink, M., Hempenius, M.A., and Vancso, G.J. (2005) Large area ordering at room temperature in thin films of poly (isoprene-block-ferrocenylsilane)s for nanofabrication. *Chemistry of Materials*, **17**, 1275.

Rulkens, R., Perry, R., Lough, A.J. *et al.* (1996) Linear oligo(ferrocenyldimethylsilanes) with between two and nine ferrocene units: Electrochemical and structural models for poly(ferrocenylsilane) high polymers. *Journal of the American Chemical Society*, **118**, 12683.

Rulkens, R., Resendes, R., Verma, A. *et al.* (1997) Ring-opening copolymerization of cyclotetrasilanes and silicon-bridged [1]ferrocenophanes: Synthesis and properties of polysilane-poly(ferrocenylsilane) random copolymers. *Macromolecules*, **30**, 8165.

Review: Schubert, U.S. and Eschbaumer, C. (2002) Macromolecules containing bipyridine and terpyridine metal complexes: Towards metallosupramolecular polymers. *Angewandte Chemie-International Edition*, **41**, 2893.

Tanabe, M. and Manners, I. (2004) Photolytic living anionic ring-opening polymerization (ROP) of silicon-bridged [1] ferrocenophanes via an iron-cyclopentadienyl bond cleavage mechanism. *Journal of the American Chemical Society*, **126**, 11434.

Tanabe, M., Vandermeulen, G.W.M., Chan, W.Y. *et al.* (2006) Photocontrolled living polymerizations. *Nature Materials*, **5**, 467.

Tang, C.B., Lennon, E.M., Fredrickson, G.H., Kramer, E.J., and Hawker, C.J. (2008) Evolution of block copolymer lithography to highly ordered square arrays. *Science*, **322**, 429.

Temple, K., Jäkle, F., Sheridan, J.B., and Manners, I. (2001) The nature of the active catalyst in late transition metal-mediated ring-opening polymerization (ROP) reactions: Mechanistic studies of the platinum-catalyzed ROP of silicon-bridged [1]ferrocenophanes. *Journal of the American Chemical Society*, **123**, 1355.

Temple, K., Kulbaba, K., Power-Billard, K.N. *et al.* (2003) Spontaneous vertical ordering and pyrolytic formation of nanoscopic ceramic patterns from poly(styrene-b-ferrocenylsilane). *Advanced Materials*, **15**, 297.

Thurmond, K.B. II, Kowalewski, T., and Wooley, K.L. (1996) The preparation of shell-cross-linked small particles. *Journal of the American Chemical Society*, **118**, 7239.

Thurmond, K.B. II, Kowalewski, T., and Wooley, K.L. (1997) A synthetic study of the factors affecting the dimensions and properties of amphiphilic core-shell nanospheres. *Journal of the American Chemical Society*, **119**, 6656.

Wang, X.S., Winnik, M.A., and Manners, I. (2002a) Synthesis and solution self-assembly of coil-crystalline-coil polyferrocenylphosphine-b-polyferrocenylsilane-b-polysiloxane triblock copolymers. *Macromolecules*, **35**, 9146.

Wang, X.S., Winnik, M.A., and Manners, I. (2002b) Synthesis and aqueous self-assembly of a polyferrocenylsilane-block-poly(aminoalkyl methacrylate) diblock copolymer. *Macromolecular Rapid Communications*, **23**, 210.

Wang, X.S., Arsenault, A.C., Ozin, G.A. *et al.* (2003) Shell cross-linked cylinders of polylsoprene-b-ferrocenyldimethylsilane: formation of magnetic ceramic replicas and microfluidic channel alignment and patterning. *Journal of the American Chemical Society*, **125**, 12686.

Wang, X.-S., Winnik Mitchell, A., and Manners, I. (2004) Swellable, redox-active shell-crosslinked organometallic nanotubes. *Angewandte Chemie-International Edition*, **43**, 3703.

Wang, X.S., Winnik, M.A., and Manners, I. (2005a) Synthesis and self-assembly of poly(ferrocenyldimethylsilane-b-dimethylaminoethyl methacrylate): Toward water-soluble cylinders with an organometallic core. *Macromolecules*, **38**, 1928.

Wang, X.S., Wang, H., Winnik, M.A., and Manners, I. (2005b) Redox-induced synthesis and encapsulation of metal nanoparticles in shell-cross-linked organometallic nanotubes. *Journal of the American Chemical Society*, **127**, 8924.

Wang, H., Winnik, M.A., and Manners, I. (2007a) Synthesis and self-assembly of poly(ferrocenyldimethylsilane-b-2-vinylpyridine) diblock copolymers. *Macromolecules*, **40**, 3784.

Wang, X., Guerin, G., Wang, H. *et al.* (2007b) Cylindrical block copolymer micelles and co-micelles of controlled length and architecture. *Science*, **317**, 644.

Wang, H., Lin, W., Fritz, K.P. *et al.* (2007c) Cylindrical block co-micelles with spatially selective functionalization by nanoparticles. *Journal of the American Chemical Society*, **129**, 12924.

Wang, H., Wang, X., Winnik, M.A., and Manners, I. (2008) Redox-mediated synthesis and encapsulation of inorganic nanoparticles in shell-cross-linked cylindrical polyferrocenylsilane block copolymer micelles. *Journal of the American Chemical Society*, **130**, 12921.

Whittell, G.R. and Manners, I. (2007) Metallopolymers: new multifunctional materials. *Advanced Materials*, **19**, 3439.

Yu, L.W., Chen, K.J., Song, J. *et al.* (2007) New self-limiting assembly model for Si quantum rings on Si(100). *Physical Review Letters*, **98**, 166102.

Zechel, D.L., Hultzsch, K.C., Rulkens, R. *et al.* (1996) Thermal and transition-metal-catalyzed ring-opening polymerization (ROP) of [1]silaferrocenophanes with chlorine substituents at silicon: A route to tunable poly (ferrocenylsilanes). *Organometallics*, **15**, 1972.

17

Functional Polymeric Nanostructures Prepared by Self-Assembly and Beyond

Rachel K. O'Reilly
Department of Chemistry, University of Warwick, Gibbet Hill Road, Coventry, CV4 7AL, UK

17.1 Methods for Polymer Particle Formation

The desire to control the composition, structure, and function of organic materials on the nanometer-size scale has received increasing interest due to their potential contribution in a wide range of applications, for example in the sequestration and release of small-molecule reagents and products. To access such structures with controlled large nanoscale dimensions a number of techniques have been developed and will be discussed in this review. Given the encapsulation focus of this review spherical and vesicular morphologies are proposed to be of most significant interest for the encapsulation of small molecules, given their relative ease of synthesis and established sequestration potential. A number of techniques are presented for the synthesis of functional polymer particles, however; often the range of sizes available by a technique varies from around 10 nm to 10 μm as does the stability, polydispersity and properties of the resultant materials.

17.1.1 Self-Assembly to form Micelles and Vesicles

To facilitate the synthesis of well-defined nanostructures, methodologies have been developed that utilize covalent interactions to form robust building blocks, such as block copolymers. These can be designed to undergo a second assembly step via noncovalent interactions, with a relatively small number of components, to form 3D nanostructures, such as spheres, toroids,

Complex Macromolecular Architectures: Synthesis, Characterization, and Self-Assembly, First Edition.
Edited by Nikos Hadjichristidis, Akira Hirao, Yasuyuki Tezuka and Filip Du Prez.

helices, rods, disks, vesicles and tubes (Reiss, 2003; Soo and Eisenberg, 2004; Hamley, 2005). The ability to control the self-assembly provides an exciting opportunity in the manufacture of unique materials that demonstrate properties that are not otherwise accessible by conventional techniques (Jain and Bates, 2003). This self-assembly utilizes a balance of covalent and noncovalent interactions to allow access to a wide range of nanosized morphologies for applications as delivery vehicles (Harada and Kataoka, 2006).

Aided by the recent advances in precision polymer synthesis, via developments in controlled-radical polymerization (CRP) techniques, there is great opportunity in terms of flexibility, diversity and functionality in the design of the polymeric building blocks (Kamigaito *et al.*, 2001; Matyjaszewski and Xia, 2001; Moad *et al.*, 2006). This versatility allows for control over their self-assembly into nanoscale objects of varying size and shape. In particular, the application of these techniques towards the facile synthesis of well-defined functionalized hydrophobic and hydrophilic (co)polymers represents an important advance in the field of polymer science, enabling the construction of polymers with a high degree of control over molecular weight, sequence and end functionalization, for utilization in the self-assembly of well-defined nanostructures. The desire to develop well-defined synthetic macromolecules is driven by the target of achieving a high level of control over the self-assembly process leading to materials with functional groups complexity and structural precision similar to that found in biological and natural systems.

This self-assembly process involves the association of an insoluble segment or segment(s) of the copolymer chain, when the copolymer is placed in a solvent system that solvates only a portion of the overall chain, at concentrations at or above the critical micelle concentration (cmc). A simple solution thermodynamic measure of aggregate stability or cmc can be expressed in terms of n, the degree of polymerization, by Equation 17.1,

$$\text{cmc} \sim \exp(-n\varepsilon_\text{h}/k_\text{b}T) \tag{17.1}$$

where $k_\text{b}T$ is the thermal energy.
ε_h is the monomer's effective interaction parameter with the bulk solution and is key in determining the aggregates stability.

The cmc of a nanostructure is an important property and can readily be measured in the laboratory using fluorescent probes such as pyrene (Ananthapadmanabhan *et al.*, 1985; Khougaz *et al.*, 1994). This is achieved by monitoring pyrene's characteristic strong fluorescence in nonpolar domains and weak fluorescence and polar shift of the excitation peak in polar solvents that can be used to probe the local pyrene environment. Experimentally, an increase in intensity of the pyrene corresponds to encapsulation of pyrene in a hydrophobic environment and hence nanostructure formation.

Experimentally, nanostructure formation is achieved by dissolving the block copolymer in a good solvent for all of the blocks and then gradually adding a nonsolvent for one of the blocks, followed by complete replacement of the good solvent by the nonsolvent via dialysis. This process can produce well-defined kinetically frozen aggregates, known as polymer micelles, which in aqueous solution consist of a nonsolvated core domain surrounded by a solvated corona or shell layer. This creates a favorable microenvironment within the assembled structure for the sequestration of molecules with limited affinity for the dispersing phase. These amphiphilic assemblies can consist of several hundred block copolymers and hence overall can have large micrometer sizes.

Micellar assemblies continuously undergo the formation and breakdown in structure and how quickly this occurs depends on the interaction parameter and also the cmc value. The cmc value for natural lipids in aqueous solution range from μmol to picomolar concentrations and this equates to lipid exchange rates of hours between aggregates. However, for block copolymers much more stable and kinetically trapped aggregates are formed that thus have significantly lower cmc values.

17.1.1.1 Amphiphilic Micelles

Most simply, polymer micelles are formed spontaneously via the solution-state self-assembly of amphiphilic multiblock copolymers, which consist of hydrophobic and hydrophilic chain segments (Figure 17.1) (O'Reilly *et al.*, 2006).

The major forces governing the assembly of amphiphiles into well-defined structures are;

a. hydrophobic attraction between hydrophobic moieties;
b. repulsion between hydrophilic end groups.

Both of these effects force amphiphilic molecules to be in contact with the solution and the stiff competition between these two opposing forces is mirrored in the variety of morphologies that can readily be obtained (Figure 17.2).

Spherical polymer micelles and nanoparticles are perhaps the most accessible assemblies, leading to significant potential in applications ranging from delivery vehicles for therapeutics, molecular imaging agents and as precursors to nanosized microelectronic devices. Often, this diverse spectrum of applications requires both robust stabilized structures and also the covalent attachment of functional molecules to the nanoparticle scaffold (Wooley, 2000). The ability to tailor the physical, chemical and biological properties of these various nanostructures is central in the development of their potential applications in particular in drug-delivery applications. Thus, there has been great recent interest in the development of methodologies for the synthesis, crosslinking and functionalization of polymer micelles. These functional and stabilized synthetic nanostructures can be compared to natural viral architectures, as both

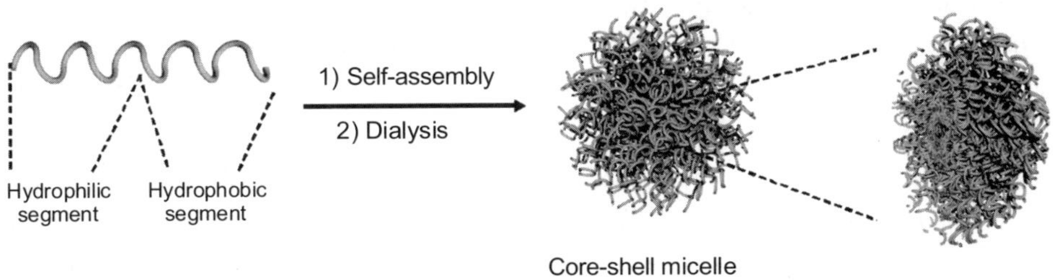

Figure 17.1 Solution-state self-assembly of amphiphilic diblock copolymers into spherical micelles. (Reproduced from R.K. O'Reilly, C.J. Hawker and K.L. Wooley, "Cross-linked block copolymer micelles: Functional nanostructures of great potential and versatility," *Chemical Society Reviews*, **35**, 1068–1083, 2006, by permission of the Royal Society of Chemistry.)

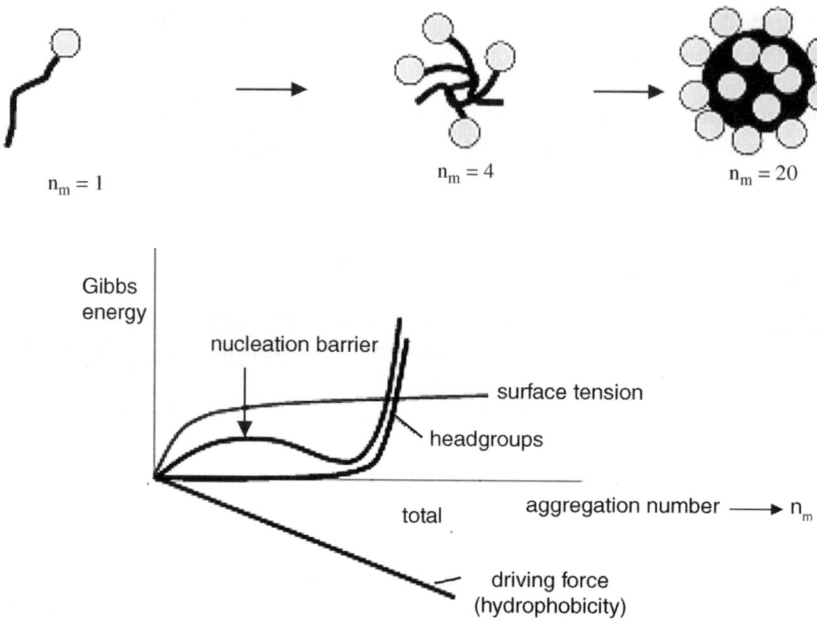

Figure 17.2 Schematic to illustrate the thermodynamics of micellization, with contributions to the Gibbs free energy as a function of aggregate size, showing the driving force (due to the hydrophobic effect), surface-tension correction and head-group term. The sum curve or nucleation barrier of these three parameters has a minimum at the optimum aggregation number. (With kind permission from Springer Science + Business Media: *Colloid & Polymer Science*, "Supramolecular perspectives in colloid science," **286**, © 2008, 855–864, M.A.C. Stuart.)

moieties possess self-assembled shells, have similar length scales and are capable of cell selective targeting (Hubbell, 2003). Given this analogy, the design of complex synthetic nanoparticles uses techniques that are based primarily on the basic construction tools found in Nature – a combination and balance of weak and strong interactions, such as hydrophobic effects, electrostatic interactions, and covalent bonds.

The versatile nature of the self-assembly process allows for significant versatility in the chemical nature of the polymer micelles that permits fine tuning of the material properties, shapes and sizes. For example, the core composition can be varied to include glassy, crystalline or fluid-like materials, can be hydrolyzed to produce an entirely hydrophilic structure, or fully degraded and removed to afford a solvent-filled nanocage. Additionally, the shell layer can be positively or negatively charged or neutral, and can contain reactive functional groups following the crosslinking reaction throughout the shell domain, to allow for the subsequent stabilization of the micelle (Figure 17.3). The ability to control particle diameters readily in the range of 10–100 nm by adjusting the block copolymer chain lengths and the ability to establish a hollow polymer nanocage by removal of the core domain (of similar dimensions as biologically functional transport nanomaterials, e.g. virus particles) represents a significant advantage over other constructs, such as linear polymers, surfactants or dendrimers. In addition, polymer micelles have been demonstrated to be more stable and also have a

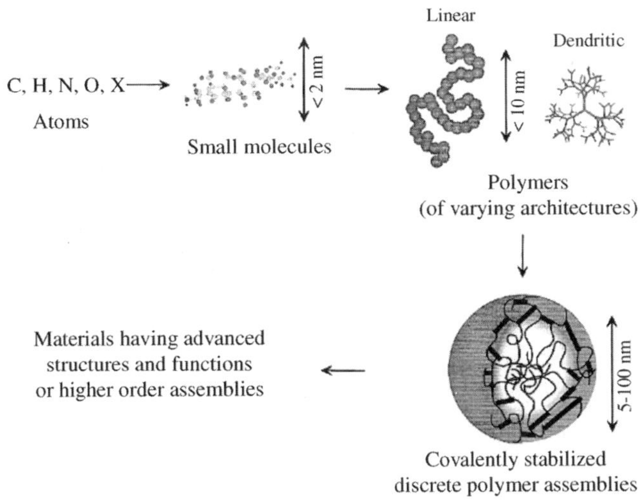

Figure 17.3 Schematic of the preparation of polymeric assemblies. (Reprinted from K.L. Wooley, "Shell crosslinked polymer assemblies: Nanoscale constructs inspired from biological systems," *Journal of Polymer Science Part A: Polymer Chemistry*, **38**, 9, 1397–1407 (Figure 1), © 2000, with the permission of John Wiley & Sons, Inc.)

significantly lower cmc than surfactant micelles. Further discussion of surfactant micelles is beyond the scope of this review and, instead, the reader is directed to a recent review article by Armes and coworker (Lui and Armes, 2001a).

17.1.1.2 Lypophilic Micelles

In fact, the polymer building blocks used in self-assembly need not be amphiphilic and phase-separated nanostructures can also be formed from completely hydrophobic or lypophilic diblock copolymers that contain two segments with differing solubilities (such as polystyrene-*b*-polyisoprene) and hence can undergo phase separation in selective solvents. One such example of completely hydrophobic phase-separated micelles are those reported by Wooley and coworkers that can be obtained from toluene and acetone solutions of a [polystyrene-*alt*-poly(maleic anhydride)]-*b*-polyisoprene triblock (Germack *et al.*, 2006). In addition, Manners and Winnik have reported the synthesis of poly(isoprene-*b*-ferrocenephenylphosphine) polymers which can form well-defined micelles (with a metal-containing core) or cylinder micelles in *n*-hexane (Figure 17.4) (Cao *et al.*, 2001; Wang *et al.*, 2003; Wang *et al.*, 2004). These inorganic polymer micelles are of great interest given their preference to assembly into rod-like materials. This is as a consequence of the high crystallinity of the core domain that favors lamellae formation over curvature that would form spheres, as a consequence of the favorable block lengths – the compromise between these two effects is the formation of stable and well-defined rod-like micelles. Of particular note in these systems is the ability to shorten the length of the rods by sonication and the ability to chain extend the rods by the addition of a second monomer feed by epitaxial growth (Gaedt *et al.*, 2009).

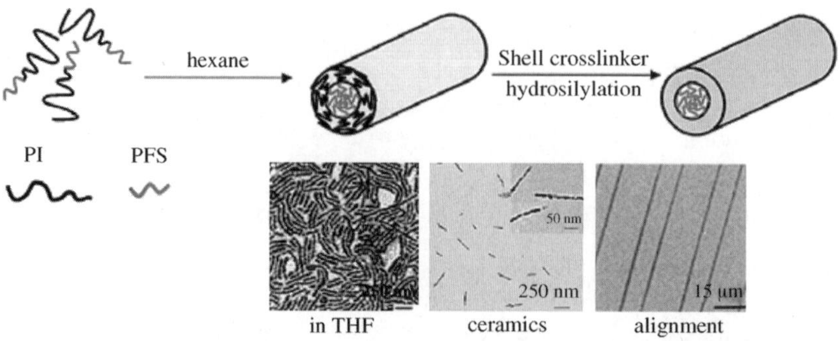

Figure 17.4 Schematic of lipophilic iron-functionalized micelles prepared by Manners. (Reprinted with permission from X.-S. Wang, A. Arsenault, G.A. Ozin *et al.*, "Shell cross-linked cylinders of polyisoprene-b-ferrocenyldimethylsilane: Formation of magnetic ceramic replicas and microfluidic channel alignment and patterning," *Journal of the American Chemical Society*, **125**, 42, 12686–12687, 2003. © 2003 American Chemical Society.)

17.1.1.3 Reverse Micelles

Conversely, inverse micellar structures are also accessible and are known as reverse micelles (Desjardins and Eisenberg, 1991). These can be formed by adding a nonsolvent for the hydrophilic block to afford the opposite of a conventional micelle, in which the hydrophilic core is surrounded by a hydrophobic shell in a hydrophobic surrounding media (Jung *et al.*, 2003; Babin *et al.*, 2008). Eisenberg showed that in a solvent of low polarity, diblock ionomers composed of a short PMMA segment and a longer PS chain could form reverse micelles composed of an ionic core surrounded by the soluble PS block (Zhong *et al.*, 1992). Reverse micelles allow for the sequestration of hydrophilic reagents and also allows for the suspension of core–shell micelles in hydrophobic solvents, which may be important in industrial applications. Indeed, Leroux and coworkers recently reported core crosslinked reverse micelles from a star-shaped polymer for the controlled release of hydrophilic guest molecules (Gao *et al.*, 2008). Of further interest is the synthesis of star-shaped polymers and their assembly into reverse micelles, given their well-defined nature and also often relative ease of synthesis compared to block copolymer technologies. In related work, Zubarev clearly demonstrated that the morphology of the solution assembly formed depended on the nature of the solvent as a consequence of the rigid-core domain of the star polymer (Teng and Zubarev, 2003). There have been a handful of reports on the application of these reverse micelles, for example as nanoreactors (Mecking *et al.*, 2000) and for the extraction of water-soluble molecules from organic solvents (Fadnavis *et al.*, 1997). However, these systems are somewhat underrepresented in the literature compared to conventional micelles and this may be as a consequence of their application in organic rather than aqueous media and hence unsuitability for many biological applications. It is proposed that these reverse micelles might be able to behave as water scavengers for applications in packaging or catalysis. In principle, a wide range of reverse micelles can be accessed as has been reported for conventional micelles, however, this has yet to be fully explored.

17.1.1.4 Other Morphologies

Many nanostructure architectures such as spheres, toriods, helices, rods, disks and vesicles (often called polymersomes) can be accessed by controlling both the conditions under which polymer solution self-assembly (solvent type, temperature, rate of assembly) is conducted and by altering the block copolymer composition (Figure 17.5). This is an interesting area of self-assembly and much work has been focused on predicting and tuning the morphologies accessible under particular assembly conditions. Of particular recent interest is the extension of these diblock copolymer systems to triblock copolymers where a much wider range of complex morphologies are now obtainable as a resultant of more complex interpolymer interactions. Of particular interest are recent reports of the controlled synthesis of hollow toriodal morphologies and also multicomponent and segmented spherical micelles (Li *et al.*, 2004; Walther *et al.*, 2009; Lutz and Laschewsky, 2005; Kim and Taton, 2007). It has been proposed that by extending the assembling copolymer system to a triblock architecture a whole new class of novel nanomaterials with unique properties and applications can now be realized. This area of solution self-assembly and the generation of a wide range of morphologies is beyond the scope of this review but the reader is directed to a recent paper in the area (Lowik and Van Hest, 2004). Overall, the ability to control the assembly product provides an exciting opportunity in the manufacture of unique materials that demonstrate properties that are not otherwise accessible by conventional techniques.

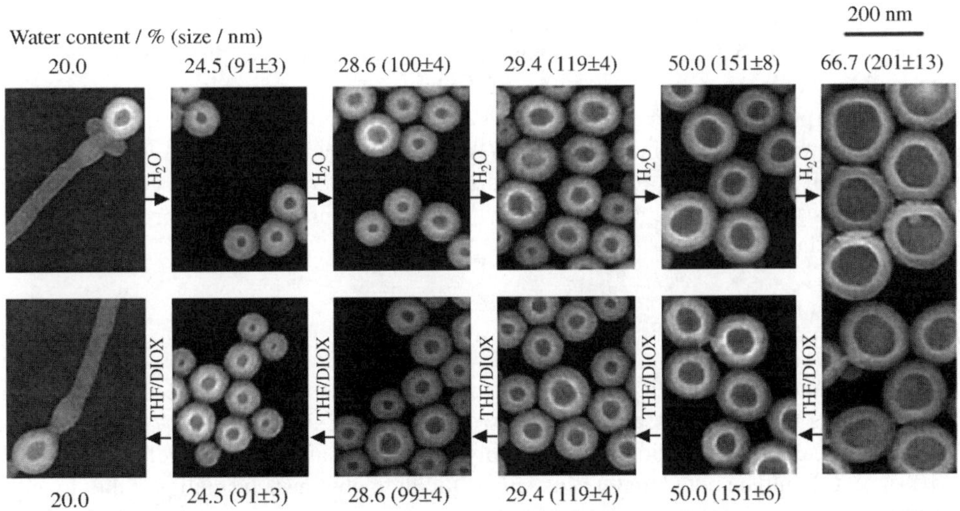

Figure 17.5 Reversibility of vesicles in responsive to increasing or decreasing water content for PAA_{44}-*b*-PS_{300} in THF/dioxane. (Reprinted with permission from L. Luo and A. Eisenberg, "Thermodynamic size control of block copolymer vesicles in solution," *Langmuir*, **17**, 22, 6804–6811, 2001. © 2001 American Chemical Society.)

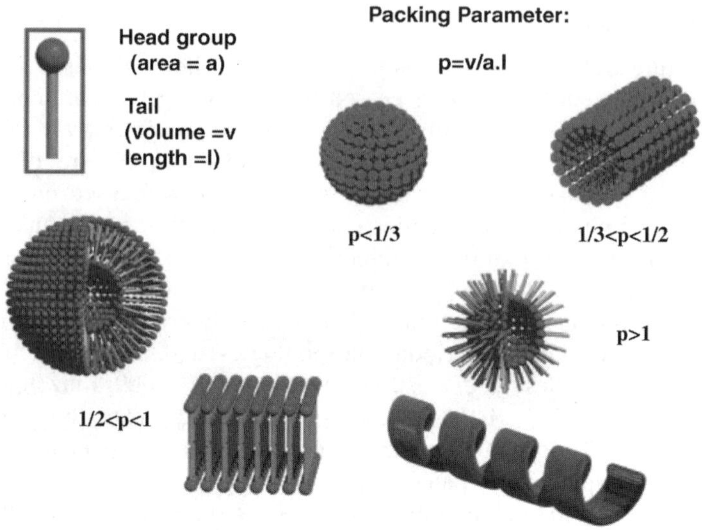

Figure 17.6 Effect of changing the packing parameter, *p*, on the morphology obtained from the assembly of an amphiphilic block copolymer. (Reprinted with permission from D.M. Vriezema, M.C. Aragonès, J.A.A.W. Elemans *et al.*, "Self-assembled nanoreactors," *Chemical Reviews*, **105**, 4, 1445–1490, 2005. © 2005 American Chemical Society.)

For example, altering the relative block copolymer compositions, changes the subsequent packing parameter p, (where $p = v/la$, in which a is the surface area of the hydrophilic or head group and v and l are the volume and length of the hydrophobic segment) and allows access to a range of morphologies (Figures 17.6 and 17.7) such as hollow cage-like or vesicle-type structures and worm-like, cylinder micelles (Isrealachvilli *et al.*, 1976). However, it should be noted that if other forces are present during the polymer assembly such as crystallization then these parameters do not always effectively predict the morphology of the resultant material.

17.1.1.5 Properties

Of key importance is the ability to tune the molecular dimensions to tune the morphology afforded upon solution self-assembly and also the length scales of the colloidal aggregates afforded. Typically, this solution self-assembly technique can afford particles in the range 10–100 nm; it should be noted that it is usually difficult to obtain such small particles by either emulsion or dispersion polymerization, although on using a microemulsion approach particles of a similar length scale can be synthesized. The covalent stabilization and tailoring of these dynamic nanostructures can be achieved using a range of chemistries, to afford robust functional nanoparticles. Many examples of the stabilization, functionalization and decoration of these polymer particles have been reported in the literature. The characterization of the size, shape and distribution of these polymer micelles can be achieved using a plethora of techniques such as dynamic light scattering (DLS), static light scattering (SLS), transmission electron

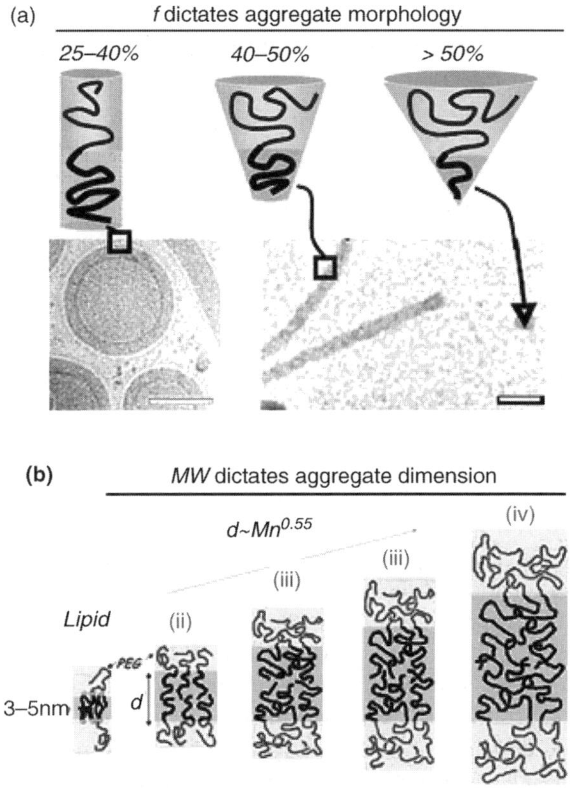

Figure 17.7 (a) Schematics of block copolymer fractions with respective cryo-TEM images showing vesicle, worm and spherical micelles. (b) Schematic scaling of polymersome membrane thickness with copolymer molecular weight. (Reproduced with permission from F. Ahmed, P.J. Photos and D.E. Discher "Polymersomes as viral capsid mimics," *Drug Development Research*, 2006, **67**, 1, 4–14. © Wiley-Liss, Inc.)

microscopy (TEM), atomic force microscopy (AFM), analytical ultracentrifugation, viscometry, size-exclusion chromatography (SEC) and small-angle neutron scattering (SANS). It is important that a range of solution and solid-state techniques are utilized to fully elucidate the morphology, distribution and size of the resultant assemblies.

This three-dimensional assembly process is driven by the hydrophobic effect that causes the two domains to selectively phase separate and form a phase-separated nanometer-sized particle. Within these two distinct domains of micelles and vesicles small molecules and reagents can be sequestered either in a nonspecific binding (hydrophobic interactions) or a more selective (i.e. metal complexation, host–guest, complimentary donor–acceptor) interaction. Nonspecific binding can be very readily utilized for the simple encapsulation of reagents within a nanostructure that exists in a surrounding media that they are not soluble in or in a stable media for the reagent. Although this strategy sometimes does not offer the specific controlled release of the encapsulates, except by nanostructure destruction, using external stimuli.

An advantage of the selective encapsulation is that the small molecule can be readily released on the application of external stimuli such as temperature, pH, stress or light without the degradation or destruction of the nanostructure. This can allow for more ready recycling and reuse.

Polymeric micelles and vesicles are unstable at elevated temperature, at concentrations below their cmc and to changes in solvent composition. Given the need for well-defined robust nanostructures for applications as controlled encapsulation and delivery vehicles there has been significant interest in the stabilization of these spherical nanostructures. The stabilization of self-assembled structures is often achieved by covalent crosslinks and by analogy, it has been demonstrated that the selective formation of such crosslinks between the polymer chains that compose a specific segment of a block copolymer micelle affords a single, nanostructured macromolecule or nanoparticle (Thurmond *et al.*, 1996). The overall effect is that the formation of covalent crosslinks throughout a specific domain offers stability to the nanostructured assemblies by providing reinforcement to the weak hydrophobic interactions that enable polymer micelle assembly and existence. There are several potential locations for crosslinking within the polymeric micelles, including at the core chain end, within the core domain, at the core/shell interface and throughout the shell layer. For a vesicle crosslinking can be introduced into the hydrophobic bilayer or within either or both of the hydrophilic coronal domains. The location of the crosslinking can dramatically affect the chemical and physical properties of the resulting materials and the extent of the crosslinking can also impact the stability, structure and thus applications of the resultant materials. In particular, the permeability of the resultant structure is altered by the crosslinking reaction and this may be utilized to further encapsulate reagents within the particle or indeed prevent ingress of reactants into the core domains of the nanostructures. Indeed, the modification of the shell domain upon crosslinking can affect the interaction of encapsulates and may either hinder or promote the release of small molecules.

Chemical fixation of either domain within the nanostructure is of interest due to the increased stability and robustness of the nanostructures relative to their precursor morphologies. It is has also been proposed that crosslinking different domains can alter the permeability of the respective domain and thus significantly affect their potential as delivery vehicles or nanosized reaction vessels (Kataoka, Harada, and Nagasaki, 2001; Rosler *et al.*, 2001). The majority of work in this area has focused on the crosslinking of the shell domain of spherical micelles through functionalities located on the backbone of corona polymer chains, to afford a central core domain that comprises of linear chains that are covalently linked to the inner surface of a permeable crosslinked shell layer. These shell crosslinked nanostructures have been demonstrated to have great potential as encapsulation and delivery agents due to their higher core mobility and more rigid form compared to core crosslinked particles (Rosler *et al.*, 2001). However, core crosslinked particles may allow for improved reagent ingress (due to the presence of a more permeable shell) and slower release characteristics given the improved encapsulation ability within the crosslinked core domain.

All of the reported crosslinking techniques are very efficient and generate robust phase-separated nanostructures (Figure 17.8) (Read and Armes, 2007). The stabilization methodologies all allow for the complete "protection" of the core or bilayer domain by the uniform stabilized shell layer, while maintaining the desired nanoscale particle diameters with narrow size distributions (with distributions of around 1–5 nm) – thus providing a confined environment within the core or bilayer domain in which reagents can be sequestered.

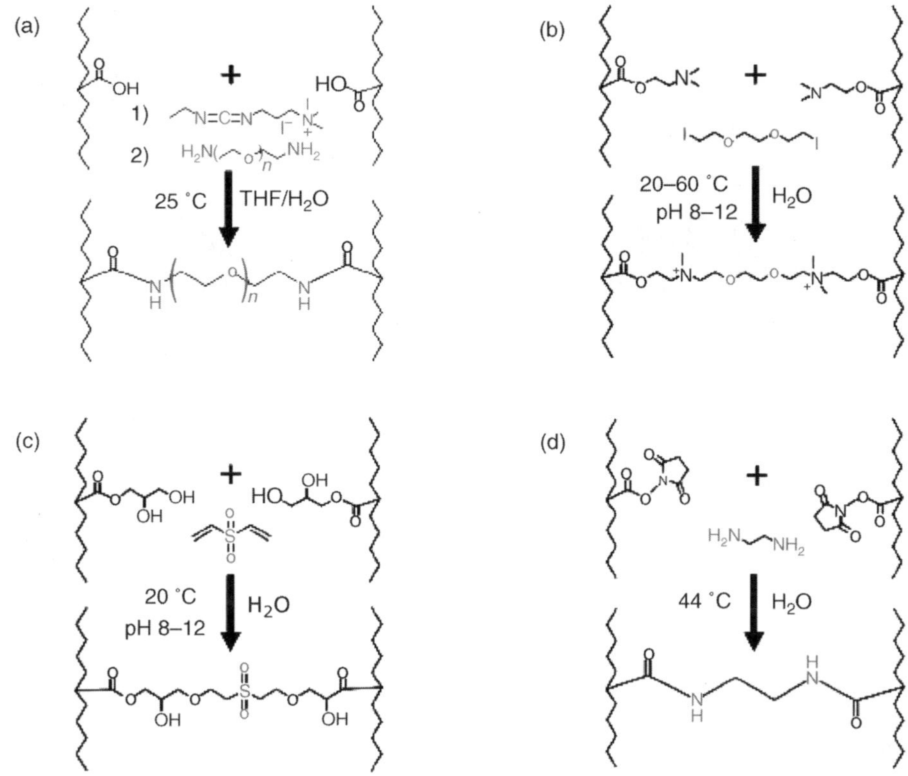

Figure 17.8 Schematic of the most commonly reported crosslinking strategies (a–d) utilized in the stabilization of polymer micelles. (Reproduced from E.S. Read and S.P. Armes, "Recent advances in shell cross-linked micelles," *Chemical Communications*, 3021–3035, 2007, by permission of the Royal Society of Chemistry.)

An important aspect of crosslinking the resultant nanoparticles is the ability to reverse this process by the application of an external trigger to allow for assembly break up and thus release of encapsulate. An elegant example has been reported by McCormick and Armes where they have utilized a disulfide bridge that can be readily cleaved upon introduction of a reducing environment either biologically or chemically (Li *et al.*, 2006). In this work, the shell crosslinked micelles were readily obtained using cystamine as a diamine crosslinker. These stabilized micelles could then be reversibly cleaved using either DTT or TCEP and the resultant micelles could once again be recrosslinked using cystamine as a thiol-exchange compound allowing for tunable release particles. The authors explored the rate of drug release from these two micellar structures and they found that the rate of release can be easily controlled given the presence or lack of crosslinks within the particle. The effect of crosslinking has been explored extensively in the literature and has been utilized for the controlled release of biomolecules, markers and also reagents.

To maintain the integrity of the aggregates it is important that the crosslinks are limited to intramicellar reactions rather than intermicellar couplings, and this is often achieved by

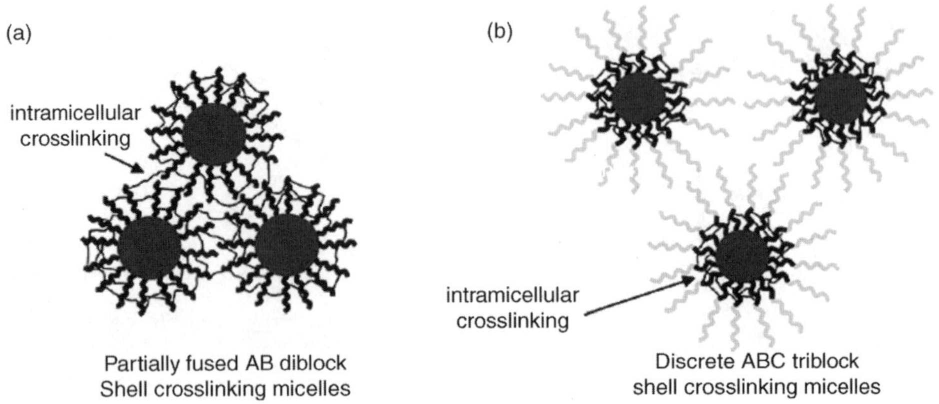

Figure 17.9 (a) AB diblock copolymer crosslinking illustrating the potential problem of micelle aggregation via intermolecular crosslinking. (b) ABC triblock copolymer in which crosslinking is confined in an inner shell layer to prevent intermolecular crosslinking. (Reproduced from E.S. Read and S.P. Armes, "Recent advances in shell cross-linked micelles," *Chemical Communications*, 3021–3035, 2007, by permission of the Royal Society of Chemistry.)

performing the crosslinking at high diblock copolymer dilution, however, this represents a significant technological disadvantage. Recent work by Armes introduced the new concept of performing the crosslinking reactions within an intermediate shell layer with steric stabilization provided by polymer chain segments extending to the nanostructure surfaces, to avoid the formation of large covalently bound aggregates (Figure 17.9) (Butun *et al.*, 2000; Lui and Armes, 2001b). The shell-crosslinked nanoparticles possess a degree of structural organization; having a well-defined external shell, an interfacial layer and a central core consisting of linear polymer chains; thus these nanoparticles consist of three distinct nanodomains, in which further chemistry can be performed.

This assembly chemistry is only limited by the preparation of an appropriate amphiphilic or hydrophobic diblock copolymer. With the structure and chemical composition of the resultant nanoparticles readily tunable using a wide range of block copolymers that are synthetically available. A broad range of polymers have been utilized as the core hydrophobic domain of shell-crosslinked nanoparticles and include styrene, isoprene, butadiene, caprolactone, poly (ethylene oxide), acrylates, methacrylate and acrylamides. Monomers that have been used to prepare the hydrophilic water-soluble domain include, among others, poly(ethylene glycol), acrylic acid, 4-vinylpyridine, (meth)acrylic acid, 2-dimethylaminoethyl methacrylate and *N*-isopropylacrylamide.

Overall, the potential uses of these crosslinked materials are improved (over naturally occurring analogs) by their increased stability towards environmental changes such as concentration and temperature, compared to their noncrosslinked micellar precursors. The subsequent selective and reversible attachment of moieties to the crosslinked self-assembled architectures further extends the potential application of these materials towards their development as well-defined polymeric supports that can behave as controlled-release vessels. However, a significant disadvantage of such materials is their sometimes difficult and time-consuming preparation, which is not ideal for large industrial scale applications.

17.1.2 *Emulsion and Miniemulsion Polymerization*

17.1.2.1 Emulsion Polymerization

There are two classes of emulsions either the oil-in-water (O/W) or water-in-oil (W/O) emulsions (Landfester, 2006; Solans *et al.*, 2005). In particular, these O/W emulsion polymerization systems are perhaps the most industrial important polymerization process for manufacturing water-based polymers, colloids and latexes. In contrast, W/O emulsions have only fully been described and explored in the last decade and are yet to find widespread industrial application. The (O/W) technique is most suitable for monomers that are semisoluble in water and whose resultant polymers are insoluble in water. The advantages of an emulsion polymerization route includes favorable polymerization kinetics (which allows for the production of high molecular weight polymers in short reaction times), favorable environmental conditions (no requirement for toxic and volatile organic solvents), good compositional control, ability to work at high solids level and high conversion (although this can lead to chain transfer to the polymer). The emulsion polymerization process can proceed by a free- or controlled-radical polymerization reaction that is carried out under heterogeneous conditions, Figure 17.10 (Chern, 2006; Lau, 2002; Nomura *et al.*, 2005).

These polymerizations normally utilize the formation of an emulsion from a mixture of water, monomer, and surfactant. The most common type of emulsion polymerization is an oil-in-water (O/W) emulsion, in which droplets of monomer (the oil) are emulsified (with surfactants) in a continuous phase of water. Water-soluble polymers, such as certain polyvinyl alcohols can also be used to act as emulsifiers/stabilizers for the monomer droplets. The polymerization takes place within the monomer latex particles, which are composed of individual polymer chains and are typically 100 nm in size. These latex particles form

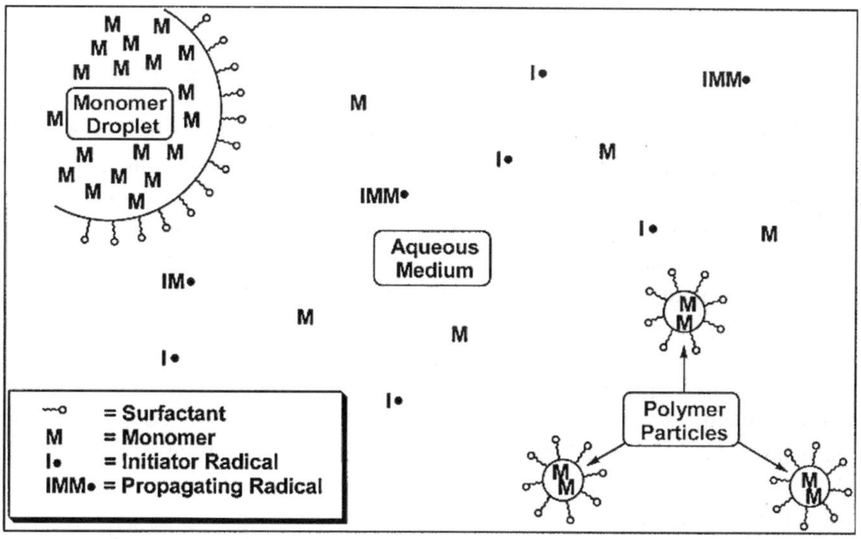

Figure 17.10 Representation of an (O/W) emulsion polymerization. (Reproduced with permission from W. Lau, "Emulsion polymerization of hydrophobia monomers," *Macromolecular Symposia*, 2002, **182**, 1, 283–289. © Wiley-VCH Verlag GmbH & Co. KGaA.)

spontaneously in the first few minutes of the process and cannot aggregate with each other because of the presence of the surfactant. That is to say, each latex particle is surrounded by the surfactant and the charged head of the surfactant repels other particles electrostatically to prevent aggregation. Hence, by changing the concentration of the surfactant in the system the size of the resultant latexes can be readily tuned. Monomers that are suitable for use in an emulsion polymerization include methacrylic and acrylic acids and esters, styrenes, butadiene, acrylonitrile and vinyl acetates. All of these monomers have a finite solubility in water that allows facile diffusion from the droplets to the growing latex particles. However, this solubility barrier does create a potential limitation for the synthesis of hydrophobic polymers by emulsion polymerization.

Of interest is the synthesis of polymer particles that have an internal microstructure that can lend itself to the loading of encapsulates. This has been achieved by the group of Suh utilizing a similar complex water-in-oil-in-water (W/O/W) system for the synthesis of hollow PMMA particles that contain water-soluble substrates (with around 90% loading rates) within the inner voids of the particle (Kim *et al.*, 2003). These three-phase systems have received relatively little attention to date due to their limited stability however, it appears that the incorporation of polymeric surfactants significantly improves their stability properties (Rosano *et al.*, 1998; Sela *et al.*, 1994). However, these systems are still intrinsically thermodynamic instable and also often have a low active entrapment efficiency. In 2003 Suh reported preparing the three-phase system and then performing a radical polymerization in the monomer oil phase. This afforded uniform polymer particles (about 100 μm) which had multihollow structures where many voids (about 100 nm) were located within the rigid polymer particles (Figure 17.11) (Kim *et al.*, 2003). These materials are proposed to be important as stable delivery nanoparticles.

17.1.2.2 Miniemulsion Polymerization

Emulsions with a droplet size on the nanometer scale (typically in the range 20–200 nm) are often referred to as miniemulsions or nanoemulsions processes (Antonietti and Landfester, 2002). The latter term is often used to highlight the size of the droplets and also it assists in avoiding misinterpretation with the term microemulsion, which is a thermodynamically stable system. Due to the small droplet size created in miniemulsions, they possess a high stability against sedimentation or creaming as the Brownian motion and diffusion rates are higher than those of sedimentation or creaming. Hence, Ostwald ripening or molecular diffusion form the main mechanisms for miniemulsion breakdown, however, this can be readily minimized by the addition of a hydrophobic agent to the dispersed phase (this lowers the Laplace pressure in the droplet) to form a metastable or fully stable latex. Often, ultrahydrophobes (such as hydrophobic dyes or chain-transfer agents) are used as the stability of the droplet increases with decreasing solubility of the hydrophobe. Compared to conventional emulsions that have a characteristic milky appearance, miniemulsions often appear transparent or translucent to the naked eye, given their smaller particle size. This can have important technological advantages especially in sensing applications.

The major advantages of miniemulsion polymerizations over related emulsion systems are:

- there is no complicated nucleation step;
- only two phases throughout the polymerization;

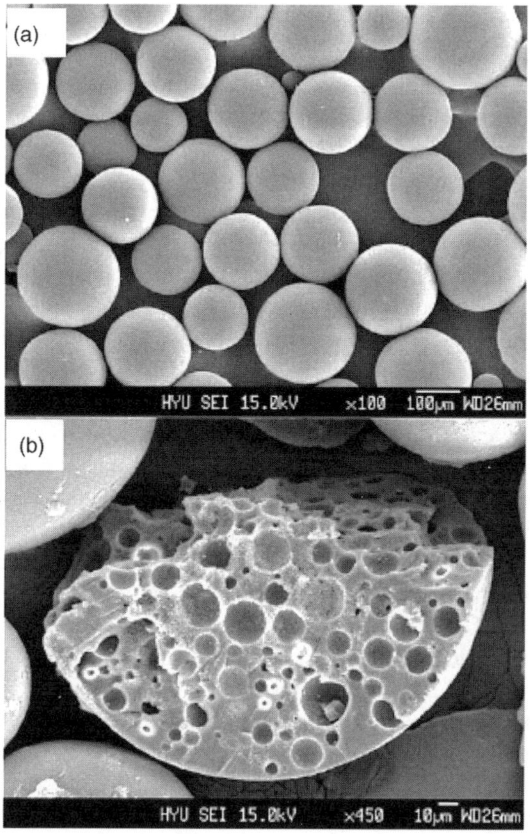

Figure 17.11 SEM image of multihollow polymer particles crosslinked with 50% EGDMA showing the (a) surface and (b) an internal image. (With kind permission from Springer Science + Business Media: *Colloid & Polymer Science*, "Multihollow polymer microcapsules by water-in-oil-in-water emulsion polymerization: Morphological study and entrapment characteristics," **281**, 2, © 2003, 157–163, J.-W. Kim, J.-Y. Ko, J.-B. Jun *et al.)*

- can readily perform O/W and W/O polymerizations;
- the final latex is a 1 : 1 copy of the initial droplets.

These features all enable more precise control over the resultant particle size and also facilitate the synthesis of a wider range of materials than conventional emulsion processes allow.

As minemulsions are nonequilibrium systems they cannot be formed spontaneously and require either formation by mechanical dispersion and/or the addition of surfactants. Mechanical dispersion is often achieved by high shear stirring, high-pressure homogenizers or ultrasound generators. It is optimal to have the shortest time of emulsification to enable the synthesis of homogenous and small droplet sizes – hence ultrasonication is the most effective and hence widely used technique. The droplet size in these systems is initially a function of the mechanical agitation of the system until the droplets reach a pseudosteady state and after this stage the droplet size does not change. The droplets' surfaces are completely covered with surfactant molecules, and no micelle formation occurs in the continuous phase. This allows for a more effective use of surfactant compared to the emulsion systems, where surfactant micelles are formed that do not form latex particles. In addition, the absence of diffusion processes in miniemulsion polymerizations makes this potentially very interesting for application in a continuous-batch process on a commercial scale.

The small droplet size (in the nm regime), high kinetic stability and optical transparency of miniemulsions enable the further application of these materials compared to conventional emulsion systems. In particular, miniemulsions have been primarily used to prepare well-defined nanometer-sized polymeric nanoparticles, using the monomer as the dispersant phase. Surfactants containing a polymerizable group (otherwise known as a surfmers) can be used to protect, stabilize and functionalize the polymer particles.

The majority of miniemulsion polymerizations utilize anionic sodium decyl sulfate (SDS) as a model surfactant, however, both cationic and nonionic surfactants have also been used. Diblock and graft copolymers can also be used as surfactants and can be designed to simultaneously also act as the hydrophobe that is industrially very favorable. The characterization of miniemulsion droplet-size distributions (DSD) has been explored by a number of researchers using a wide range of including TEM (with staining or cryo), light scattering, SAXS, CHDF and soap titration (Azad *et al.*, 1976; Ugelstad *et al.*, 1980; Choi *et al.*, 1985; Van Hamersveld *et al.*, 1999; Erdem *et al.*, 2000c; Miller *et al.*, 1994; Alduncin *et al.*, 1994; Chern and Chen, 1997; Fontenot and Schork, 1993; Landfester *et al.*, 1999). The determination of the DSD can provide a useful insight into the mechanism of formation of the droplet, however, there is currently no direct method that guarantees the accurate determination of the DSD. Instead, the droplet size is often inferred from the size of the resultant latex – however, this is not ideal.

In a similar manner to the emulsion systems it is possible to perform reverse or inverse miniemulsion polymerizations (Figure 17.12) to afford lipophobic latex particles with a central

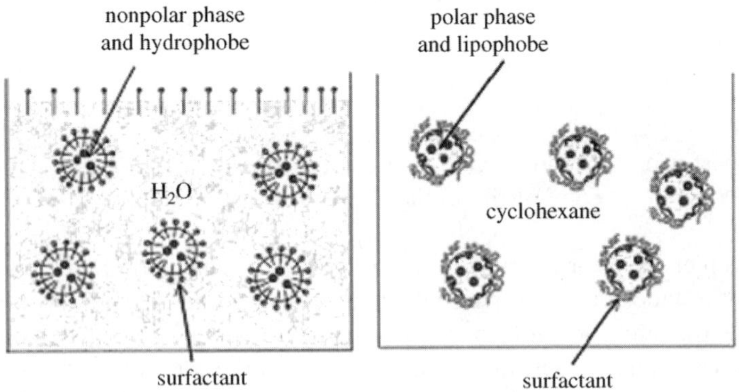

Figure 17.12 Comparison between the particles formed in a direct or (O/W) (left) and inverse (W/O) (right) miniemulsion. (Reprinted with permission from K. Landfester, "Synthesis of colloidal particles in miniemulsions," *Annual Review of Materials Research*, **36**, 231–279, 2006. © 2006 Annual Reviews, Inc.)

polar phase in an overall hydrophobic media (W/O system). In these inverse cases it was found that the use of block copolymer surfactants was highly effective due to their polymeric and sterically demanding nature that provides maximum steric stabilization for the droplet particle. The droplet stability displays a simple three-fold triangular relationship between the amount of surfactant, resulting particle size and surface coverage.

One of the key features of miniemulsion droplets is their extremely high stability that prevents any exchange reaction occurring between the payloads of different droplets. This was elegantly demonstrated by the suppression of the formation of Prussian blue upon mixing of two droplets each of which contained a reactive precursor ($K_4[Fe(CN)_6]$ and $FeCl_3$) to the blue pigment, compared to the formation of the complex (indicated by the formation of the bright blue color) when the droplets where formed by a micellar assembly or a conventional emulsion polymerization method. This is proposed to be very important in delivery applications (Landfester, 2006).

The miniemulsion process, in theory, allows for the polymerization of a wide range of monomers and for the formation of particles, as long as they are immiscible with the continuous (often water) phase. It has been proposed that each monomer droplet can be envisaged to be a small individual nanoreactor; this confined environment imparts significantly more synthetic variation and modification than can be observed in the emulsion polymerization systems. A number of polymerization techniques can be utilized in miniemulsion polymerizations including free radical, controlled radical, anionic, cationic, crosscoupling and enzymatic polymerizations (Landfester, 2006). A number of hydrophobic monomers can be polymerized via a miniemulsion route, to give particles in the size range 50–250 nm, and includes acrylates and methacrylates including those that are ultrahydrophobic and also fluorinated derivatives. Also of interest are semicrystalline polyacrylonitrile (PAN) particles, which can be only polymerized via a miniemulsion process given the inherent insolubility of the polymer in a solution of the monomer. This afford small 10-nm-sized polymer nanocrystals with interesting and complex morphologies, as evidenced by TEM analysis (Landfester and Antonietti, 2000).

In addition, the copolymerization of a range of monomers via a miniemulsion process is possible as is the polymerization to form amphiphilic copolymers to afford more complex polymer particles (Huang *et al.*, 1998; Jovanovic and Dube, 2004; Chern and Sheu, 2001; Barton, 2002). A major advantage of the miniemulsion polymerization process is the ability to tune the size of the resultant particle by tuning the amount and nature of stabilizer added to the mixture. The size of the particles formed depends on the efficiency of the steric stabilization (determined by the surfactant) as compared with the electrostatic stabilization given that the stabilization relies on a denser surfactant packing to be efficient. Hence, smaller particles (less than 100 nm) can be readily accessed by careful choice of surfactant and increasing loadings compared to monomer concentration.

It should be noted that the application of controlled-radical polymerization (CRP) techniques in emulsion polymerizations has not yet been particularly successful due to the relatively large size of the monomer droplets and insolubility of the radicals generated. However, Charleux and coworkers have reported the utilization of SG1-based alkoxyamines for the emulsion and miniemulsion polymerization of styrenes and acrylates to afford very well-defined and relatively small polymer particles (about 50 nm) (Figure 17.13) (Charleux and Nicolas, 2007). In fact, from a practical point of view very few experimental parameters had to be changed with respect to the classical solution polymerization recipe (Qi *et al.*, 2001; Cunningham, 2002). Indeed, the recent advances in miniemulsion techniques have enabled the

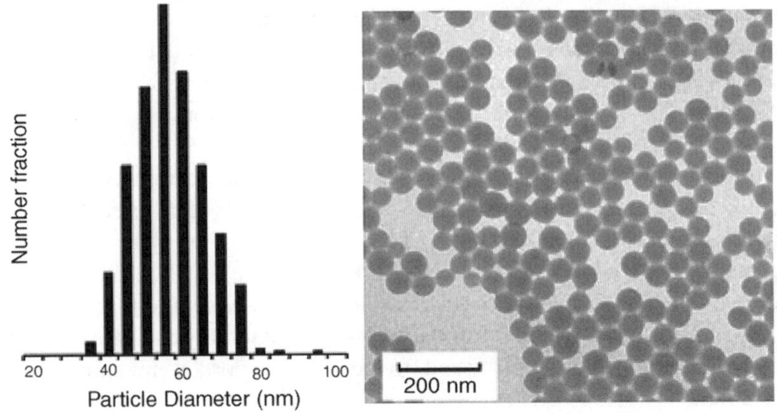

Figure 17.13 Acrylate particles afforded by the miniemulsion polymerization via NMP using SG1 initiator, size analysis as determined by TEM analysis. (Reproduced from G. Delaittre, J. Nicolas, C. Lefay *et al.*, "Surfactant-free synthesis of amphiphilic diblock copolymer nanoparticles via nitroxide-mediated emulsion polymerization," *Chemical Communications*, 614–616, 2005, by permission of the Royal Society of Chemistry.)

application of robust and versatile CRP techniques in these processes and these techniques have received increasing interest due to the smaller droplet size, which minimizes the need of transport of the stable free radicals to the polymerization loci, which is one of the major hurdles in achieving effective CRP in conventional emulsion systems (Cunningham, 2003; Li and Matjaszewski, 2003; Smulders *et al.*, 2004).

Of further interest is the *in-situ* synthesis of amphiphilic block copolymers via a miniemulsion route that allows access to surfactant-free miniemulsion systems. This can be readily achieved by utilizing an initial polymer synthesized by CRP techniques and miniemulsion polymerization as a macroinitiator for chain extension. This was shown by Charleux and coworkers to allow for the synthesis of stimuli-responsive particles by the introduction of a water-soluble monomer (*N,N*-diethylacrylamide) that exhibits a lower critical solution temperature (LCST) at about 32 °C (Delaittre *et al.*, 2007).

An important application of miniemulsion polymerizations is in the synthesis of complex particles that cannot be made by the classical polymerization or assembly methods. For example, the synthesis of magnetic or nanocomposite particles is of interest and can be achieved using miniemulsion techniques (Liu *et al.*, 2004; Zhang *et al.*, 2005). There is also the scope of encapsulating nanoparticles into the polymer latex shells to generate nanocomposites with high stability and processibility. Of further interest is the ability of the droplet to behave as a nanoreactor to create an enormous number of compartments within its structure that are separated from each other by a continuous phase, thus affording permeability and also complexity to the nanoparticle system (Meier, 1999; Landfester, 2001). Also of note, for the future application of these materials, is that it is possible to readily introduce functionality (such as amino and carboxylic acid groups) into the resultant polymer particles via the incorporation of a functional surfactant (Musyanovych *et al.*, 2007). This may facilitate their future application as selective delivery vehicles.

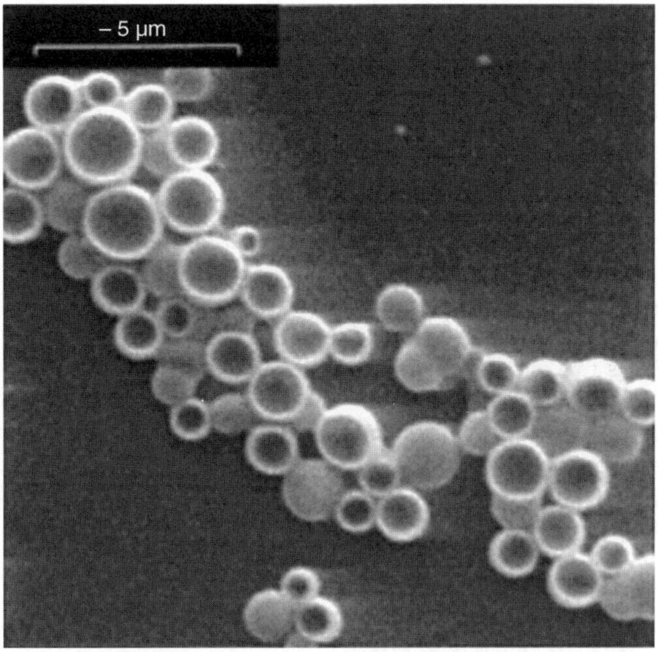

Figure 17.14 SEM image of well-defined styrene nanocapsules prepared by miniemulsion techniques and using hexadecane as the inert phase. (Reproduced with permission from K. Landfester, "The generation of nanoparticles in miniemulsions," *Advanced Materials*, 2001, **13**, 10, 765–768. © Wiley-VCH Verlag GmbH & Co. KGaA.)

Another application of emulsion and miniemulsion systems is in the synthesis of hollow particles (Jang and Ha, 2002). This can be achieved via the encapsulation of a nonsolvent within the droplet and latex particle using phase-separation techniques (McDonald *et al.*, 2000). This phase-separated polymer served as a locus for polymerization to form a crosslinked network that stabilizes the resultant hollow morphology. The solvent could then readily be removed to afford hollow structures, although in this work the resultant particles are perhaps too large (1 μm) to be widely applicable. A second example for the synthesis of hollow particles using a solution encapsulated miniemulsion route was recently reported by Zhang and coworkers (Chen *et al.*, 2007). Hollow polymeric particles (1–2 μm) can also be obtained from miniemulsion techniques by the addition of an inert oil that demixes from the polymer during the polymerization to form an oil-encapsulated polymer shell, sponge-like structures or dotted oil droplets (Figure 17.14) (Tiarks *et al.*, 2001b).

17.1.3 Suspension Polymerization

This is another important industrial polymerization technique that is suitable for water-insoluble monomer and polymers (or indeed organic insoluble monomers for reverse systems) (Trijasson *et al.*, 1990) and is especially suited to the synthesis of large 10–1000 μm polymer beads (Dowding and Vincent, 2000). The term bead or pearl polymerization is also used for the suspension polymerization process (Vivaldo-Lima *et al.*, 1997). The key difference in a suspension polymerization is that a mechanical process produces the polymer particles as a dispersed solid phase whereas emulsion polymerizations are chemical processes that require a surfactant to make the monomer emulsify. Although, a recent example by the group of Howdle demonstrated the use of nanoparticles (such as Fe_3O_4) as stabilizers to allow for the preparation of hybrid nanocomposites (Hasell *et al.*, 2007). The polymerization behavior inside the droplets is similar to that of a bulk polymerization, however, as the droplets are only 10–1000 μm in diameter, more rapid reaction rates can be tolerated.

A major advantage of suspension polymerization over bulk techniques is that the polymer separation is much more facile (Dawkins, 1989), however, the application of this technique is somewhat limited as few monomers are completely water insoluble. In addition, suspension polymerization is limited to radical polymerization systems and the nature of the agitation of the system is key, as is the viscosity in determining the size, stability and nature of the polymer particles formed. Of interest is the synthesis of porous particles via this technique that can be readily achieved by the inclusion of an inert porogen into the monomer phase, which can be readily removed after polymerization (Seidl *et al.*, 1967). Suspension polymerization has a number of advantages including the ease of heat removal, low dispersion viscosity, low levels of impurities in the resultant particle, low separation costs and the final product is in particle form. On the other hand, the disadvantages of this technique are primarily wastewater problems, lower productivity compared to bulk and a continuous process has to date not been achieved industrially.

Perhaps, of interest in the context of encapsulation, is that inorganic particulate stabilizers have been reported for application in suspension polymerization, however, the effectiveness of encapsulation over incorporation is not high enough for delivery applications (Wenning, 1956). Of greater interest is the work of Bakhshaee for the incorporation of carbon black, Vincent's report of the encapsulation of silica particles and Duguet's work on alumina encapsulation (all in PMMA particles) (Bakhshaee *et al.*, 1985; Cooper and Vincent, 1989; Duguet *et al.*, 2000). Magnetic nanoparticles were prepared recently by Ding and coworkers using a suspension polymerization method and a coprecipitation approach (Wang *et al.*, 2006). Iron oxide particles were utilized to provide unique magnetic properties that allowed for separation in a magnetic field.

17.1.4 Interfacial Polymerization

This class of polymerization type is similar to an emulsion and suspension polymerization, however, it is mostly commonly employed for polycondensation reactions. An interfacial polymerization reaction occurs at or near the interfacial boundary of two immiscible solutions. The two reagents meet at the interface and react rapidly. The basis for this method is from the Schotten–Baumann reaction between an acid chloride and a compound containing an active hydrogen atom. Under the right conditions thin flexible walls of alternating copolymer form rapidly at the interface to form liquid-filled capsules (Scott *et al.*, 2005. These capsules (about 210 nm) have been explored as direct and efficient encapsulation devices, given that their nanometer thick shells (about 10 nm) and confined liquid-core domain, offer unique sequestration potential compared to conventional polymer nanoparticles. It has been proposed that these unique vessels would have applications as high-efficiency encapsulation, and controlled-delivery vehicles.

Interfacial polymerization techniques can be used to prepare robust microcapsules (about 4 μm) that have selectively incorporated functionality such as catalytic groups in their liquid-core domains. The synthesis of these particles is achieved via an oil-in-oil emulsion in combination with interfacial polymerization methods and allows for the encapsulation of substrates not compatible with water-in-oil or oil-in-water emulsion systems (Kobalija and McQuade, 2006). Often these capsules have polyurea shells that can be formed from w/o emulsions containing reactive polyols and isocyanates. This has been very elegantly demonstrated by McQuade and coworkers for the DMAP-catalyzed acylation transformation of an alcohol within the protected environment of polyurea microcapsules formed from reaction of a polyisocyanate and a poly(vinyl alcohol) in the presence of an amine catalyst (Price *et al.*, 2006; Poe *et al.*, 2007).

It should be noted that these particles are much larger than those synthesized via mini-emulsion or self-assembly techniques, although recent work by McQuade suggests that the cores are indeed liquid like and hence can be manipulated to a greater extent than conventional solid nanoparticles (Poe *et al.*, 2008; Mason *et al.*, 2007). In addition, a very interesting feature of these microcapsules is the ability to undergo shrinking and swelling, in different solvents given the flexible nature of the shell layer. Another interesting feature of these capsules is the ability to modify the strength, permeability and size of the capsules without changing the interior polymer. This allows for the selective tailoring of the shell thickness (for example for polyurethane capsules this is achieved by altering the isocyanate ratio) and hence permeability of the organic capsules, this allows for the tuning of the catalytic activity (Figure 17.15).

There are few other examples of the application of interfacial polymerizations for the synthesis of encapsulated polymer materials. There is one notable example from the group of Tanaka, in which microcapsules were formed by interfacial polymerization using an oil-in-water emulsions with limonene oil (Mizuno *et al.*, 2004). This was achieved by dispersing

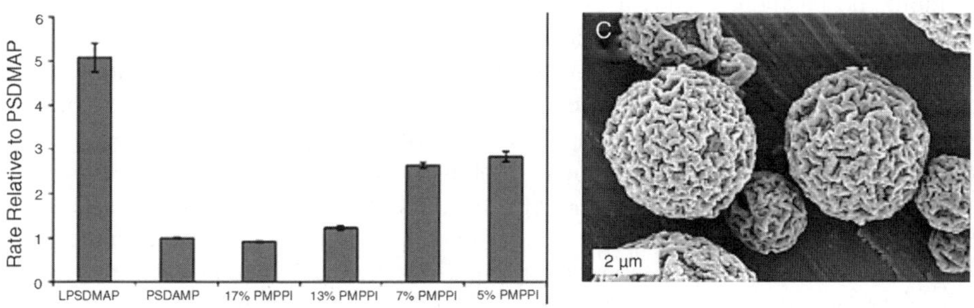

Figure 17.15 Illustration of the relative rates of the DMAP catalyzed acylation reactions by varying the nature of the catalyst (entry 1 and 2) and by changing the thickness of the shell layer (entry 3 is for the thickest shell and 6 the thinnest). SEM image of microcapsules containing DMAP functionality. The capsules have a crinkled structure due to the collapse of the relatively thick shell. (Reprinted with permission from K.E. Price, B.P. Mason, A.R. Bogdan *et al.*, "Microencapsulated linear polymers: 'Soluble' heterogeneous catalysts," *Journal of the Americal Chemical Society*, **128**, 32, 10376–10377, 2006. © 2006 American Chemical Society.)

limonene oil in an aqueous solution of poly(vinyl alcohol) with different degrees of polymerization and saponification.

17.1.5 Hyperbranched Polymer and Dendrimer Formation

The synthesis of hyperbranched polymers is of interest given their relatively large macromolecular size (albeit they are smaller than particles afforded from miniemulsion or usually smaller than those formed by micellization techniques), functionality and unimolecular properties. Hyperbranched polymers can be considered to be similar to dendritic structures but without the exact structural precision (Sunder *et al.*, 2000; Kim, 1998; Grayson and Frechet, 2001; Tomalia and Frechet, 2002). Their special architecture gives these polymers very interesting physical properties such as lower melt and solution viscosities compared to random-coil polymers, the presentation of many end groups, interesting density gradients from the core to the surface, have amorphous properties, and offer unique possibilities for delivery systems (as illustrated by dendritic box systems) (Jansen *et al.*, 1994b). In addition, a unique and useful feature of dendrimer and hyperbranched molecules is that their solution solubility is determined by the nature of the peripheral or end groups, hence relatively hydrophobic macromolecules can be readily solubilized in water by the selective attachment of polar groups at the dendrimer surface. This is important for delivery applications as this allows for the selective tailoring of the macromolecule.

The main disadvantage of dendrimers is their lack of available large-scale synthetic routes as a result of their stepwise iterative growth (either by a convergent or divergent strategy). This means that these materials often have a high price and demanding/time consuming synthetic procedures. In addition, given the synthetic difficulty in their synthesis only relatively low-generation dendrimers can be readily accessed (about 10th-generation maximum). This is due to the starburst limit or de Gennes dense packing effect, which means the number of branch ends on a dendrimer increases exponentially as a function of generation, while the surface area of the dendrimer only increases with the square of generation (and the volume with the cube of generation), hence there comes a point beyond which the dendrimer cannot grow as a consequence of a lack of space. As a consequence, the typical size of dendrimers rarely exceeds 20 nm compared to the microparticle sizes (1–10 μm) readily accessible by other, albeit less precise synthetic or assembly routes. This may be of interest if the size of the nanoparticle is key in the successful application of the particles.

It has been proposed that hyperbranched polymers can mimic some of the structural and globular properties of dendrimers, without the need for multistep synthetic strategies and also can facilitate access to larger macromolecules than accessible by dendritric synthetic routes (Hult *et al.*, 1999). The architecture of a hyperbranched polymer is similar to that of a dendrimer in that it is highly branched, however, it is less regular in structure and certainly not perfectly branched, unlike a dendrimer. The degree of branching (DB) is usually in the 35–50% range (compared to 100% in a dendrimer) and this is in fact independent of the molecular weight of the polymer, however, at higher degrees of branching gelation can sometimes occur. Hence, these materials are imperfectly branched and are polydisperse, however, can be made much more readily and on significantly larger scales than reported for dendrimers.

Hyperbranched polymers can be synthesized in one step by the polycondensation of an A_xB monomer, where $x \geq 2$. Initial examples were plagued by gelation and crosslinking reactions and hence afforded ill-defined materials (Yates and Hayes, 2004; Voit, 2005). The biggest synthetic challenge in the development of hyperbranched polymers has been how to produce these polymers that behave like ideal dendrimers, but without a multistep iterative synthetic route. In recent years, two main synthetic methods have been reported for the synthesis of hyperbranched polymers in the literature; the self-condensation vinyl polymerization(Frechet *et al.*, 1995) and the Strathclyde route (Isaure, Cormack, and Sherrington, 2003). These method of hyperbranched polymer formation has subsequently been explored by a number of groups for the synthesis of multifunctional materials (Trollas and Hedrick, 1998; Vogt *et al.*, 2007; Muller *et al.*, 1997; Bharathi and Moore, 2000).

The unimolecular nature of these branched polymeric materials is an important characteristic feature compared to multimolecular polymeric micelles. This means that these unimolecular nanosized particles are significantly more robust and stable to changes in environment than self-assembled polymer micelles (Newkome *et al.*, 1986; Frechet, 2002; Liu *et al.*, 2000). That is to say these macromolecules do not display critical micelle concentrations (cmcs) and hence do not need to be crosslinked or stabilized in a subsequent modification step after formation prior to application as stable delivery vehicles. These dendritic unimolecular micelles are static and unlike thermodynamically assembled micelles are stable and retain their cohesion regardless of their solution concentration. In addition, dendritic and hyperbranched molecules with polar interiors and nonpolar exteriors have been demonstrated to behave as unimolecular reverse micelles capable of extraction and sequestration of organics from the surrounding environment (Tully and Frechet, 2001; Frechet, 1994). The first example of a physically locked or entrapped guest molecule was reported and explored by Meijer and coworkers with the Rose Bengal and dendritic box experiment (Jansen *et al.*, 1994b). In this work, the authors used a mononuclear dendritic container (diameter about 5 nm) or so-called dendritic box to entrap guest molecules within its interior (Jansen *et al.*, 1994a).

These core–shell-type structures for dendrimers have been demonstrated to effectively encapsulate and release small molecules that are chemically incompatible with the environment external to the macromolecules (Liu *et al.*, 2000). For example, Frey and coworkers have reported the application of hyperbranched amphiphilic nanocapsules for the controlled encapsulation of hydrophilic guests and proposed that these materials may be of interest for the solubilization of inorganic compounds in organic media (Sunder *et al.*, 1999). In related work the development of dendrimer nanoreactors is of great interest and has been elegantly explored by Fréchet and coworkers (Frechet, 2002). These functional dendrimers can be used to encapsulate reactive sites for the preparation of nanoreactors that incorporate active sites within a microenvironment within the dendrimer core (Figure 17.16). The unique globular and gradient nature of the dendrimer allows for the combination of the catalytic action with rapid and directional transfer of reagents and products. This is proposed to allow for the unique application of these materials as nanosized reactors and unique delivery vehicles.

These branched macromolecules can also undergo higher-order assembly to afford larger nanometer-sized morphologies. This supramolecular self-assembly of these materials is of great interest given the ability to access a wide range of novel materials using these well-defined and multifunctional materials. For examples, the self-assembly of hyperbranched polymers can afford (over a wide range of scales) morphologies such as macroscopic tubes, micro- or nanovessel's, fibers, spherical micelles and porous films (Zhou and Yan, 2009). Generally, amphiphilic hyperbranched polymers that have a small hydrodynamic volume (about 10 nm) exist as unimolecular micelles in solvents. However, these globular molecules can then

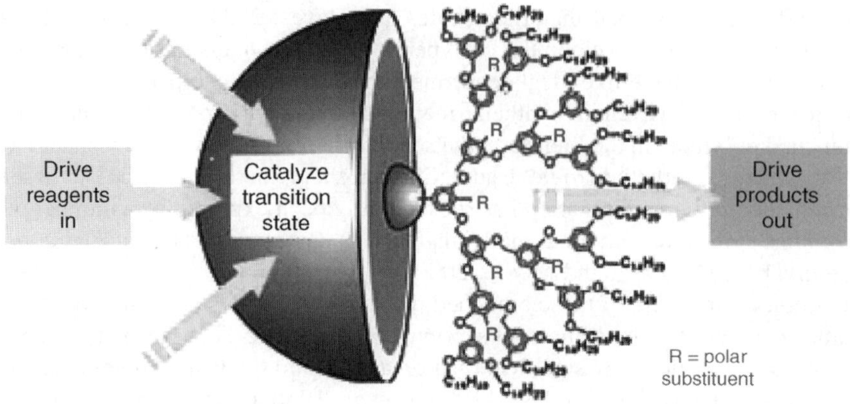

Figure 17.16 Schematic of a dendritic nanoreactor, which can use the phase-separated nature of the macromolecule to facilitate mass transport of reagents. (Reproduced with permission from J.M.J. Fréchet, "Dendrimers and supramolecular chemistry," *Proceedings of the National Academy of Sciences of the United States of America*, **99**, 8, 4782–4787. © 2002 National Academy of Sciences, USA.)

undergo a second assembly process when present in concentrations above a critical micelle concentration to further aggregate and form large (about 100 nm) micelles (Hong *et al.*, 2007; Radowski *et al.*, 2007).

17.1.6 Other Techniques for Nanoparticle Formation

There has been interesting research in the synthesis of unimolecular micellar-type structures from the selective collapse of linear polymers. This work was pioneered by Eichinger and Martin in 1983 (Martin and Eichinger, 1983a; Martin and Eichinger, 1983b). A major advantage of using the intramolecular collapse of linear polymer chains, for the preparation of nanoparticles, is that it allows access to relatively small nanoparticles in the range 5–20 nm (which can be difficult to access using conventional self-assembly techniques). This technique uses the collapse and intramolecular coupling of single polymer chains to give discrete unimolecular nanoparticles. However, a major drawback of this approach is that the competing and statistically favored intermolecular crosslinking reaction means that ultradilute conditions need to be employed to access such nanoparticles. This, for many years precluded their synthesis on useful (e.g. multigram) scales, however, more recent work in 2002 by Hawker and coworkers has overcome this difficulty with a continuous addition strategy (Harth *et al.*, 2002). This allows for the successful multigram synthesis of discrete organic soluble and stable nanoparticles by intramolecular crosslinking (Figure 17.17). These nanoparticles are thought to resemble dendrimers in morphology, however, their larger molecular size and more facile synthesis affords them greater potential as delivery vehicles.

This approach has recently been extended by the same group for the synthesis of noncovalently crosslinked nanoparticles, which can be formed reversibly upon application of external stimuli (Seo *et al.*, 2008). This group have also recently reported the application of microwave crosslinking chemistries for the synthesis of collapsed and functionalized polymer

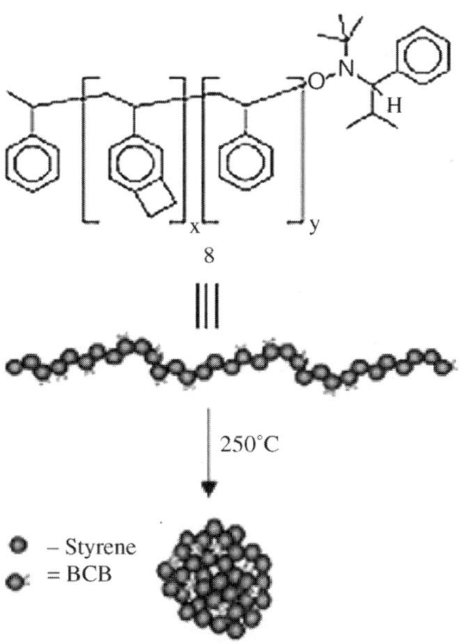

Figure 17.17 Schematic representation of the intramolecular collapse of a linear polymer to form a stable nanoparticle. (Reprinted with permission from E. Harth, B. Van Horn, V.Y. Lee *et al.*, "A facile approach to architecturally defined nanoparticles via intramolecular chain collapse," *Journal of the American Chemical Society*, **124**, 29, 8653–8660, 2002. © 2002 American Chemical Society.)

nanoparticles (An *et al.*, 2006). Recent work by Harth has utilized the intramolecular collapse of functional polymers, using thiolene click chemistry, for the formation of biocompatible and degradable nanoparticles (Van Der Ende *et al.*, 2009).

An alternative approach that has been developed very recently by Weaver and Rannard involves the synthesis of responsive branched polymer nanoparticles using a one-pot, single-step approach (Weaver *et al.*, 2008). This methodology is generic, scalable and tolerant to polymer functionality and hence offers great potential for the commercial synthesis of tailorable polymeric nanoparticles. By using a pH-responsive polymer in the preparation of the nanoparticles, the branched structures can form core–shell architectures at basic pH and under acidic pH hydrate swell to release an encapsulated payload. This current technology has only been exploited for amino methacrylate water-soluble monomers but certainly the synthetic protocol should be readily modified to allow access to similar organic collapsible particles with the ability to display tunable sequestration and release properties.

Another method for the synthesis of nanoparticles is the nanopreciptation method that has most elegantly been exploited by the group of Prud'homme (Johnson *et al.*, 2006; Prud'homme *et al.*, 2006; Gindy *et al.*, 2008; Johnson and Prud'homme, 2003) (Figure 17.18). By carefully adjusting the concentrations of the substances, as well as the mixing speed, the researchers were able to readily control the sizes of the nanoparticles in the range 80–300 nm.

This method has been primarily exploited for the entrapment of biomolecules such as proteins. The entrapment of proteins into nanoparticles still remains a difficult task, since each protein is characterized by a number of essential properties that might be somehow different

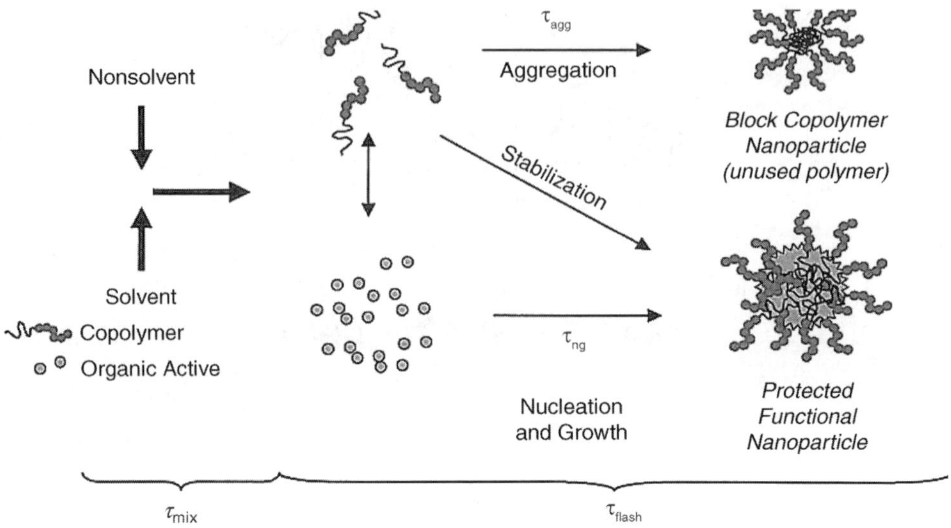

Figure 17.18 Schematic to illustrate flash nanoprecipitation as a method of nanoparticle formation. (Reproduced with permission from B.K. Johnson and R.K. Prud'homme, "Flash nanoprecipitation of organic actives and block copolymers using a confined impinging jets mixer," *Australian Journal of Chemistry*, **56**, 10, 1021–1024. © CSIRO 2003.) (http://www.publish.csiro.au/paper/CH03115.htm)

when encapsulated within a polymeric material. Recently, a modified nanoprecipitation method was optimized so as to allow hydrophilic and hydrophobic compounds to be entrapped within small nanoparticles by using miscible solvents and in the absence of water. In this method, hydrolytic degradation pathways can be minimized and other factors of instability such as interfaces and high shearing rates are avoided (Akbulut *et al.*, 2009).

17.2 Methods for Substrate Incorporation

The encapsulation of organics and inorganics in polymer micro- and nanospheres has been a long-standing research area. The methods for encapsulation can be roughly divided into two classifications. First, is the physical encapsulation, in which the polymer is absorbed (or vice versa) within a specific domain within the nanoparticle. The other is chemical encapsulation, in which there is a covalent bonding between the small molecule and the polymer chain. For example, this can be achieved by introducing a covalent bond at the interface of the polymer shell and the inorganic core either by the utilization of a coupling agent bearing a double bond in emulsion, miniemulsion or suspension polymerization (Bourgeat-Lami and Lang, 1999) or by a method that involves ATRP that starts from the surface of inorganic particles which has been selectively functionalized with initiating groups (Von Werne and Patten, 1999; Carrot *et al.*, 2001).

Encapsulation is the process in which small particles are surrounded by a coating or shell layer. The main rationale for the encapsulation of substrates is to protect the substrate from the surrounding environment and to allow controlled or triggered release at a particular time. Key considerations are the degree of loading which is possible for a given particle, which can be tailored by the size of the central domain for encapsulation and also the introduction of favorable interactions to allow for close packing of the encapsulate within the particle. In addition, the extent of leaching from the particle is crucial and should be minimized to allow for access to robust, stable and well-defined particles capable of controlled encapsulation.

It has been proposed that these nanometer- and micrometer-sized particles may have applications in biology in sensing and imaging applications or as delivery vehicles (De *et al.*, 2008). The unique properties of nanoparticles such as their similar size to biomolecules such as proteins and polynucleic acids, as well as their ability to be tuned to contain a range of functional materials such as metal centers, dye molecules, etc. make them attractive candidates for biological and materials applications. The applicable of these nanoparticle platforms for therapeutic and diagnostic activities relies on the functionalization and also encapsulation potential of these materials.

17.2.1 Encapsulation of Substrate within Polymer Particle

This can be readily achieved during the nanoparticle synthesis using self-assembly, branched macromolecular synthesis and emulsion techniques. By adding the substrate for incorporation into the solution utilized for synthesis of the nanoparticles, it can readily be incorporated into the particle using solubilizing effects. However, it should be noted that the incorporation of substrates can often affect the size, distribution or properties of the resultant nanoparticle. In addition, the mechanism of release of the substrate must be given careful consideration. Release can occur via a controlled or triggered release mechanism that is readily achieved using stimuli responsive polymers. The advantage of this encapsulation route over the conventional composite route is the more uniform dispersion of metal particles that can be achieved. This can often be key in the control of their release and also maintain a high surface area and hence activity or reactivity upon release.

This approach can be utilized for all the nanoparticle systems described above and is perhaps especially suited to the miniemulsion polymerization approach. However, the application of conventional emulsion polymerizations for the encapsulation of substrates is problematic due to the difficulty in controlling the dispersion ability of the substrates (especially inorganic) in the aqueous phase prior to polymerization. This leads to large aggregates and also means that the polymerization does not always occur at the nanoparticle surface. In addition, due to the complexity of the nucleation and growth processes inefficient encapsulation often results. It is also important to consider the nature of the substrate for encapsulation as in an emulsion polymerization the substrate must diffuse through the continuous (mostly water) phase to the reaction location, where it can polymerize. A main disadvantage of this approach is that these emulsion polymerizations are kinetically controlled. This means that encapsulation of a solid substrate can only be achieved by the diffusion of the monomer to the surface of the solid particles. Consequently, substrates that are soluble in the monomer phase but can not diffuse to the reaction locus in the micelles can not as a result be encapsulated. However, a miniemulsion process does not encounter such difficulties, especially due to the different mechanism that means the substrate does need to diffuse through the water phase and hence radical polymerizations with phase separation can occur.

There are, however, a number of examples of encapsulation of inorganics via an emulsion polymerization route. For example, dos Santos and coworkers reported the use of a semicontinuous emulsion process for the encapsulation of modified TiO_2 as an alternative approach to a batch process as higher efficiency values were achievable (De Oliveira *et al.*, 2005). In this work, the inorganic was selectively incorporated as a modified titanate using methyl methacrylate as a monomer. The authors found by TEM analysis that the encapsulated pigment exhibited a raspberry-like core–shell morphology with sizes around 400 nm.

Work by the group of Suh also reported the application of emulsion polymerization techniques (specifically water-in-oil-in-water polymerizations) for the synthesis of silver particles in multihollow porous PMMA microspheres (about 10 μm) (Lee *et al.*, 2004). The concentration of silver in the microspheres was relatively high at 0.19 wt% (as measured by atomic absorption spectroscopy). Work by Yu and coworkers for the incorporation of alumina needle like particles using emulsion polymerization was reported in 2005 (Zeng *et al.*, 2005). Well-defined core–shell alumina–polystyrene composite nanoparticles (about 50 nm) were obtained when the alumina substrate was pretreated to ensure a strong interfacial interaction between the metal and the polymer. This afforded a high yield and binding efficiency (over 90%) for the particles. TEM analysis revealed that one alumina needle per structure was ideal and higher loading afforded agglomerates of needles.

More recent work by Liu introduced a new synthetic route for the preparation of CdS nanoparticle–polymer materials using a microemulsion approach (Lui, 2008). The process involved 2 stages; the first was a γ-irradiation emulsion polymerization followed by a redox-initiated interfacial polymerization to afford hollow particles in the range 80–150 nms with shell thickness in the range 10–20 nm. The authors propose that this approach should be extendable to other inorganic materials such as Au, Ag or ZnO.

In 2009, Xue and coworkers reported the utilization of microemulsion polymerization routes for the synthesis of hollow-silica particles that are loaded with supermagnetic iron-oxide nanoparticles (SPIONS) in their shells (Li *et al.*, 2009). These particles have macroporous shells with holes in the range 50–300 nm using an oil-in-diethylene glycol microemulsion method. The loading of the SPIONS, the microsphere morphology and the size of the through holes can be readily controlled by adjusting the experimental parameters. These novel materials offer interesting magnetic potential compared to conventional hollow materials.

17.2.1.1 Miniemulsion Encapsulation

In miniemulsion techniques this can involve the encapsulation of water-insoluble materials by dissolution or dispersion into the organic phase such as fluorescent dyes or pigments for the application as tracer molecules (Figure 17.19) (Steiret and Landfester, 2007). This is achieved by first solubilizing the hydrophobic pigment and dispersing it into the monomer phase, sonicating and allowing the emulsification in the water system. It should be noted that by utilizing favorable electrostatic interactions encapsulation efficiencies can be improved. Efficient encapsulation can be proven by means of ultracentrifugation, streaming potential titration and electron microscopy methods. The authors reported extremely high loadings of polymer to substrate ratios of about 20 : 80, to afford materials that have very different thermal and physical properties to the parent polymer particles.

The introduction of species such as pigments into the monomer phase prior to miniemulsion polymerization leads to high encapsulation efficiencies. This was confirmed by sedimentation experiments in which the low-density bare polystyrene can be easily separated from

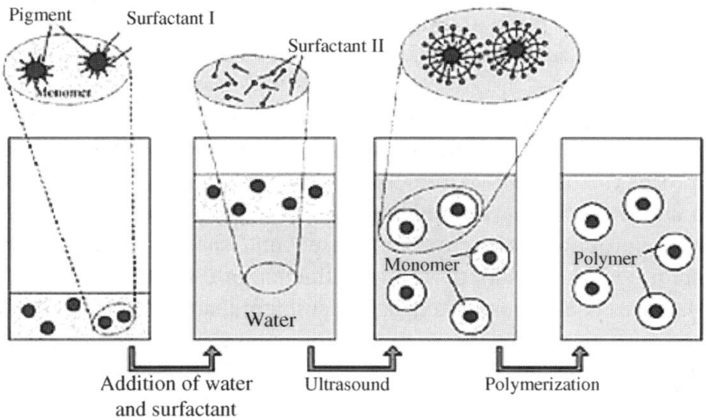

Figure 17.19 Principle of encapsulation by miniemulsion polymerization. (Reprinted with permission from K. Landfester, "Synthesis of colloidal particles in miniemulsions," *Annual Review of Materials Research*, **36**, 231–279, 2006. © 2006 Annual Reviews, Inc.)

high-density pigment particles and encapsulated species with different layer thickness and perfection. Although the existence of free uncovered pigment particles can not be excluded using this technique.

Further work by Landfester in 2007 explored the encapsulation and then explosive release of azo compounds. This was achieved by the encapsulation of a thermally unstable azo moiety via a miniemulsion technique, followed by heating to below the glass-transition temperature of the polymer and to afford decomposition of the encapsulate to cause rupture of the polymer particle and possible sudden release of the encapsulated material (Volz *et al.*, 2007). This may allow access to triggered burst-release systems and also maybe applicable to other "explosives" such as redox initiators that release a gas upon decomposition.

In addition, Erdem and coworkers have described the encapsulation of TiO_2 particles in polystyrene particles via a miniemulsion methodology in two steps (Erdem *et al.*, 2000b; Erdem *et al.*, 2000a). First, they dispersed TiO_2 in the monomer, using a stabilizer and then this phase was dispersed in aqueous solution to form stable submicrometer droplets. However, in this work they noted that the presence of TiO_2 limited the size of the droplets. In addition, all the TiO_2 was not encapsulated within the colloidal particles and the amount of material encapsulated was low at only around 3%.

Further work on the encapsulation of organics such as highly hydrophobic carbon black is also of interest because of the application of resultant materials in coatings, plastics and ink products. The direct dispersion of the pigment powder in the monomer phase prior to emulsification is again a suitable way to achieve encapsulation (Bechthold *et al.*, 2000). In this case, full encapsulation of nonagglomerated carbon black can be achieved by the suitable choice of the hydrophobe. However, this direct dispersion only allows for the incorporation of about 8 wt% carbon black because the carbon is highly agglomerated in the monomer phase de to its high hydrophobicity. In subsequent work, Antonietti utilized a novel cominiemulsion approach to increase the amount of encapsulated organic up to around 80 wt% (such as carbon black in this example) (Tiarks *et al.*, 2001a). In this work both monomer and carbon black were independently dispersed in water through the use of a surfactant and mixed afterwards in any desired ratio, cosonicated to afford hybrid particles composed of both carbon black and monomer.

Further optimization of the miniemulsion polymerization techniques allows for the synthesis of magnetic particles such as those containing magnetite. This can be achieved by modifying the process to include two miniemulsion processes and an overall three step process (Rameriz and Landfester, 2003). To achieve stable and small droplets it is key that there is a high shear rate, appropriate surfactants are used and the addition of a small-molecule hydrophobe is required as this suppresses the influence of Ostwald ripening (Landfester and Rameriz, 2003). It has been demonstrated that the miniemulsion polymerization process can be used very effectively for the synthesis of metal-encapsulated particles for applications in biomedical and also sensor applications, such as detection by magnetic resonance imaging and destruction of tumor cells by hyperthermia. For example, work by Landfester utilized miniemulsion polymerization for the synthesis of small (30–100 nm), homogeneous and stable droplets contain heterophase (magnetite) metal particles (Landfester and Rameriz, 2003). This afforded a ferrofluid that by thermogravimetric analysis revealed that the magnetite loading in the final particles was around 15 wt% and overall encapsulated particle sizes of around 100 nm. However, this reaction was somewhat heterogeneous and a distribution of particle loadings was observed by TEM analysis.

However, by moving to the three-step process that includes two miniemulsion steps (as described above) more uniform and also higher loadings at around about 40 wt% could be readily achieved. It is important to note that this three-step approach produces only magnetite-loading polymer particles with no pure polymer particles or pure magnetite particles visible by TEM analysis or ultracentifugation. In addition, magnetic measurements indicate that the magnetite aggregates consist of separate superparamagnetic particles due to the presence of the hydrophobe, this ensures that the encapsulated particles retain 60% of their magnetization compared to that observed in the bulk. These particles are in an interesting size range given their targeted application in biomedical applications as they exhibit low sedimentation rate, large specific area for biomolecule immobilization and allow for the potential integration in microfluidic technologies.

Miniemulsion polymerization has also been used for preparation of semiconductor–polystyrene core–shell microspheres. In work by Liu and coworkers they reported the one-step method by miniemulsion polymerization to synthesize ZnO–polystyrene hybrid microspheres (Zhang *et al.*, 2006). These hybrid particles offer the advantage that they retain the characteristics of ZnO nanoparticles and also the ease of processing of the polymer. The resultant hybrid microspheres had sizes around 100 nm with ZnO colloid particles present as hexagonal crystals that were about 5 nm in size. Of further interest is the encapsulation of CdSe–ZnS quantum dots in polystyrene latexes via a miniemulsion approach which was reported by Lansalot in 2006 (Joumaa *et al.*, 2006). Fluorescent latexes in the size range 100–350 nm were proposed to be of great interest in medical and biotechnological applications where their fluorescent properties are key in their application in diagnostics, imaging and optical tracking.

Wu and coworkers have explored the synthesis of SiO_2–polystyrene nanocomposite particles by a miniemulsion approach (Zhang *et al.*, 2005). Core–shell or other interesting morphologies composites were obtained depending on the size of the silica particles and the surfactant concentration. By adjusting these parameters, it was possible to control the size and morphology of the composite particles in the range 80–200 nm. In further work, the authors also utilized an acrylate–styrene copolymer shell layer and found that the presence of this second soft monomer influenced significantly the resultant morphology and also assisted in the synthesis of multicore–shell structures (Zhou *et al.*, 2006).

17.2.1.2 Self-Assembly

This introduction of substrates postpolymerization is normally utilized in polymer self-assembly to allow for the incorporation of incompatible or reactive functionalities. A number of synthetic methods can be utilized but the most facile is using the phase-separated nature of micelles or vesicles to selectively encapsulate a small molecule within a particular domain of the nanoparticle. This is often used to sequester hydrophobic small molecules within the core domain of the particle and hence provide protection from the surrounding aqueous environment. This is especially useful for the application of these nanoparticles as delivery vehicles for biological systems. Using reverse micelles hydrophilic molecules can also be selectively encapsulated within the nanoparticle. A recent example of effective encapsulation using a vesicular system was reported by Discher, who utilized poly(ethylene oxide)-*b*-poly(butadiene) for the formation of amphiphilic diblocks that self-assembled in the presence of a number of proteins to form hollow polymeric particles with specific incorporation of the protein (Aranda-Espinoza *et al.*, 2001; Bermundez *et al.*, 2002).

In more recent work by Shull and coworkers, the preparation of swollen micelle composites was investigated using self-consistent mean theory (Lefebvre and Shull, 2006). This theoretical approach concluded that nanoparticles can be encapsulated in the center of swollen micelles by including an external-field term that creates an attraction between the surface of the particle and the homopolymer. This predicts the formation of nanocomposites from a diblock copolymer (AB) and additional selectively attractive homopolymer (C). This is achieved by first preparing a layer of C homopolymer, then a layer of nanoparticles, and then a blend of A homopolymer and asymmetric AB copolymer. Then upon annealing, favorable interactions between B and C repeat units cause the copolymer to migrate to the A/C interface. Attraction between the particle and the C repeat units causes the particles to diffuse into the C phase. When the interfacial tension is lowered by the copolymer, the interface develops curvature and micelles swollen with C form around the particles in the A phase.

17.2.2 Tethering of Substrate within Polymer Particle

The introduction of functional handles to direct the encapsulation of small molecules can be readily achieved using the self-assembly of polymer di- or tri block copolymers, given the facile introduction of reactive handles into the copolymerization strategy. One advantage of this method is that the loading of the nanoparticle can be readily tailored by the designed synthesis of the parent particle.

17.2.2.1 Self-Assembly

In analogy to the crosslinking of polymer nanoparticles, the functionalization of polymer nanoparticles can be achieved at various different locations, thereby allowing for the tailoring of these materials towards specific applications such as encapsulation vehicles. Moreover, within a core–shell structure, there can be additional phase segregated domains within the core or the shell, for example the "bumpy" nanoparticles and multicompartment micelles assembled from complex mixtures of block copolymers and homopolymers (Lodge *et al.*, 2005; Zheng *et al.*, 2005). For the simplest, core–shell spherical micelle morphology, functional groups within the hydrophobic core domain or at the hydrophobic polymer chain end and may be used to introduce further functionality into the nanoparticle core to allow for improved sequestration or encapsulation. By judicious choice of the functionality introduced into the nanoparticles, these groups can be utilized as selective reactive handles and can be made to work cooperatively (e.g. acting as energy-transfer-based detectors or coupled triggered-release systems) to allow for the further tailoring of these materials towards specific applications.

The selective functionalization of the core domains of polymer micelles is an exciting area of research, due to the potential application of the resulting materials as novel sequestration systems. The covalent attachment of specific groups within the core, which can be protected from hydrolysis or degradation by the surrounding environment, allows for the delivery of specific incompatible moieties in an organic or aqueous environment. However, the functionalization of the core domain within polymer micelles and nanoparticles has received limited attention, perhaps due to the difficulty in incorporating and retaining functionality within the hydrophobic block.

Another strategy that is perhaps more synthetically versatile and straightforward is the synthesis of latent functionalized monomers, which can be incorporated selectively into the hydrophobic domain to allow for the selective and versatile functionalization of the core region. Using "click" reactive groups as the latent functionality, these click-functionalized micelles were crosslinked in an intramicellar fashion via amidation reactions within the shell layer to afford nanoparticles bearing alkynyl or azido groups within the core domains (O'Reilly *et al.*, 2005). The azido or alkynyl functionality that was embedded and dispersed throughout the hydrophobic styrenic core of the polymer micelles was demonstrated to be available for further chemical modification using "click" chemistry with complementary "click"-functionalized fluorescent dyes, thus allowing the versatile chemical modification of the core domain. Functionalization of the nanoparticle core requires that a number of issues, not present for the shell functionalization, be considered. Unlike the shell, which is a swollen hydrogel, the core is hydrophobic and for covalent attachment to occur within this domain, reagents and substrates must transverse the shell layer and core/shell interface while still being soluble in the nanoenvironment of the core. This is an area of high academic interest currently and certainly offers the potential to be a key aspect of the future application of selective encapsulation vehicles.

A very convenient method for the incorporation of metal particles into micellar systems is by a simple templating approach. This allows for the selective incorporation of metal nanoparticles into the stabilized micelle structure and offers the advantage of preventing agglomeration and precipitation and hence improves the solution processibility of the metal nanoparticles. It should be noted that to obtain stable hybrid particles there should be sufficient adhesion between the polymer chains and the metal complexes. As a result, nitrogen-containing polymers such as *N*-vinylpyridines, or those with acidic, basic or neutral complexing properties are often used. Upon micelle synthesis the polymeric material is loaded with metal and then a chemical reaction such as a reduction occurs to form the stable metal colloids within the polymeric material (Figure 17.20). As a result, these polymeric micelles are often considered to

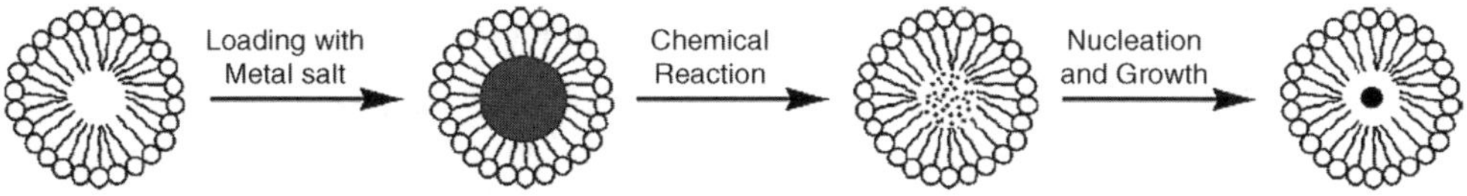

Figure 17.20 Schematic of the preparation of metal nanoparticles inside a micellar nanoreactor. (Reprinted with permission from D.M. Vriezema, M.C. Aragonès, J.A.A.W. Elemans *et al.*, "Self-assembled nanoreactors," *Chemical Reviews*, **105**, 4, 1445–1490, 2005. © 2005 American Chemical Society.)

be nanoreactors for the synthesis of stable nanometer-sized metal colloids (Forster and Antonietti, 1998; Mayer, 2001; Kastle *et al.*, 2003). This approach has been successful applied for the incorporation of metals (such as Ag, Au, Cu, Co, Ni, Pb, Pd, Pt, Rh and Zn), (Hamley, 2003; Klinghofer *et al.*, 1997) metal oxides (such as Fe_2O_3, ZnO and TiO_2) (Forster and Konrad, 2003; Spatz *et al.*, 1998) and metal sulfides (such as CdS, CoS, CuS, FeS, PbS and ZnS) (Moffit *et al.*, 1995). The latter are often prepared by the addition of hydrogen sulfide to the precursor metals encapsulated in the system prior to chemical reaction.

More specifically, the synthesis of micelles as a scaffold for the formation of metal particles has been achieved (Loginova *et al.*, 2004). This has been demonstrated very effectively by the group of Bronstein, who has utilized polystyrene as a reactive polymer to functionalize with molybdenum hexacarbonyl using UV radiation followed by reaction with H_2S to afford MoS_x encapsulated micelles. In related work Chu has reported the use of block copolymer micelles as template s for the formation of hollow molybdenum spheres whose diameter is equivalent to the size of the micellar cores (about 5 nm) (Lui *et al.*, 2000). More recently Xie and coworkers have used micelle-assisted fabrication of necklace-shaped molybdenum disulfide nanospheres (Xiong *et al.*, 2003). In this work, the authors used the aggregation-driven transformation of surfactants at low temperatures to form a necklace-shaped assembly of amorphous MoS_3 nanospheres.

17.2.2.2 Miniemulsion Polymerization

Lansalot and coworkers very recently reported a new strategy for the synthesis of iron-oxide nanoparticles using a functional monomer approach (Joumaa *et al.*, 2008). By selectively designing a functional monomer with a pendant reactive phosphate group, which can selectively interact with the iron particles, the successful incorporation of these particles into the polymer nanoparticles was achieved. This monomer was utilized as a surfmer and displayed a phosphate group at one extremity and a polymerizable methacrylic functionality at the other one that are separated by a short poly(ethylene) glycol spacer.

17.2.2.3 Suspension Polymerization

One key example of the utilization of functional monomer was that reported by Sutherland and coworkers where they utilized the suspension polymerization of a quantum-dot-functionalized monomer for the synthesis of hybrid and highly fluorescent polystyrene particles (O'Brien *et al.*, 2003).

17.2.3 Miscellaneous Encapsulation Techniques

Of recent interest is the utilization of hollow biological moieties for the encapsulation of small molecules. Of particular interest is the application of viruses as hollow cages for the sequestration and release of functional groups. This work has been explored recently by Cornelissen and coworkers who have utilized virus capsids as unique vessels for encapsulation and reactions (Comellas-Aragones *et al.*, 2007). These vessels are around about 30 nm in diameter and have very well-defined structural properties that can afford interesting higher order assembly of the particles (De La Escosura *et al.*, 2008). This virus has also been used by the Finn group as a supramolecular building block for further assembly and functionalization (Wang *et al.*, 2002). The application of ferritin-encapsulated inorganic nanoparticles has been pioneered by the group of Mann (Li, Viravaidya, and Mann, 2007). In this work they utilize an iron-storage protein apoferrtin, which is a spherical 8 nm diameter cage of polypeptides, for the encapsulation of inorganic nanoparticles such as alkaline-earth, semiconductor and transition-metal complexes (Wong and Mann, 1996).

17.3 Conclusions

Overall, the preparation of well-defined particles from tens of nanometers to tens of micrometers can be achieved using a range of techniques such as self-assembly, emulsion and miniemulsion systems, suspension and interfacial polymerization. The choice of technique is normally governed by the desired material property including key features such as its solubility, required dispersity and size. Depending on the desired application the size, morphology and function can be tuned, within each technique to allow for access to a wide range of interesting nanomaterials with specific tunable properties. The design and synthesis of nanostructures that contain a reactive functionality either tethered or sequestered in the nanostructure has also been discussed with a specific focus on the application of these materials as delivery vehicles. There is great scope and versatility in the preparation of polymeric particles and also a range of unique characteristics that can be imparted by each of the particle-preparation techniques. Overall, the wide range of techniques available to the synthetic chemist allow for access to diverse materials with well-defined properties and specific functionality, stability and solubility.

References

Ahmed, F., Photos, P.J., and Discher, D.E. (2006) Polymersomes as viral capsid mimics. *Drug Development Research*, **67**, 4–14.

Akbulut, M. *et al.* (2009) Generic method of preparing multifunctional fluorescent nanoparticles using flash nanoprecipitation. *Advanced Functional Materials*, **19**, 718–725.

Alduncin, J.A., Forcada, J., and Asua, J.M. (1994) Miniemulsion polymerization using oil-soluble initiators. *Macromolecules*, **27**, 2256–2261.

An, Z. *et al.* (2006) One-step microwave preparation of well-defined and functionalized polymeric nanoparticles. *Journal of the American Chemical Society*, **128**, 15054–15055.

Ananthapadmanabhan, K.P. *et al.* (1985) Fluorescence probes for critical micelle concentration determination. *Langmuir*, **1**, 352–355.

Antonietti, M.and Landfester, K. (2002) Polyreactions in miniemulsions. *Progress in Polymer Science*, **27**, 689–757.

Aranda-Espinoza, H. *et al.* (2001) Electromechanical limits of polymersomes. *Physical Review Letters*, **87**, 208301.

Azad, A.R.M. *et al.* (1976) Emulsification and emulsion polymerization of styrene using mixtures of cationic surfactant and long chain fatty alcohols or alkanes as emulsifiers. *ACS Symposia Series*, **24**, 1–23.
Babin, J., Lepage, M., and Zhao, Y. (2008) "Decoration" of shell cross-linked reverse polymer micelles using ATRP: a new route to stimuli-responsive nanoparticles. *Macromolecules*, **41**, 1246–1253.
Bakhshaee, M. *et al.* (1985) Encapsulation of carbon blacks into suspension polymerized copolymers. *Polymer Communications*, **26**, 185–192.
Barton, J. (2002) Inverse microemulsion polymerization of oil-soluble monomers in the presence of hydrophilic polyacrylamide nanoparticles. *Macromolecular Symposia*, **179**, 189–208.
Bechthold, N. *et al.* (2000) Miniemulsion polymerization: applications and new materials. *Macromolecular Symposia*, **151**, 549–555.
Bermundez, H. *et al.* (2002) Molecular weight dependence of polymersome membrane structure, elasticity, and stability. *Macromolecules*, **35**, 8203–8208.
Bharathi, P. and Moore, J.S. (2000) Controlled synthesis of hyperbranched polymers by slow monomer addition to a core. *Macromolecules*, **33**, 3212–3218.
Bourgeat-Lami, E. and Lang, J. (1999) Encapsulation of inorganic particles by dispersion polymerization in polar media: 2. Effect of silica size and concentration on the morphology of silica–polystyrene composite particles. *Journal of Colloid and Interface Science*, **210**, 281–289.
Butun, V. *et al.* (2000) Synthesis of shell cross-linked micelles at high solids in aqueous media, synthesis of shell cross-linked micelles at high solids in aqueous media. *Macromolecules*, **33**, 1–3.
Cao, L., Manners, I., and Winnik, M.A. (2001) Synthesis and self-assembly of the organic–organometallic diblock copolymer poly(isoprene-b-ferrocenylphenylphosphine): shell cross-linking and coordination chemistry of nanospheres with a polyferrocene core. *Macromolecules*, **34**, 3353–3360.
Carrot, G. *et al.* (2001) Atom transfer radical polymerization of n-butyl acrylate from silica nanoparticles. *Journal of Polymer Science Part A: Polymer Chemistry*, **39**, 4294–4301.
Charleux, B. and Nicolas, J. (2007) Water-soluble SG1-based alkoxyamines: a breakthrough in controlled/living free-radical polymerization in aqueous dispersed media. *Polymer*, **48**, 5813–5833.
Chen, Y. *et al.* (2007) Preparation of polymeric nanocapsules by radiation induced miniemulsion polymerization. *European Polymer Journal*, **43**, 2848–2855.
Chern, C.S. (2006) Emulsion polymerization mechanisms and kinetics. *Progress in Polymer Science*, **31**, 443–486.
Chern, C.S. and Chen, T.J. (1997) Miniemulsion polymerization of styrene using alkyl methacrylates as the reactive cosurfactant. *Colloid and Polymer Science*, **275**, 546–554.
Chern, C.S. and Sheu, J.C. (2001) Effects of carboxylic monomers on the styrene miniemulsion polymerizations stabilized by SDS/alkyl methacrylates. *Polymer*, **42**, 2349–2357.
Choi, Y.T. *et al.* (1985) Polymerization of styrene miniemulsions. *Journal of Polymer Science, Part A, Polymer Chemistry*, **23**, 2973–2987.
Cohen-Stuart, M.A. (2008) Synthesis of polyhedral particles by dispersion polymerization in supercritical carbon dioxide. *Colloid and Polymer Science*, **286**, 1435–4536.
Comellas-Aragones, M. *et al.* (2007) A virus-based single-enzyme nanoreactor. *Nature Nano*, **2**, 635–639.
Cooper, E.C. and Vincent, B. (1989) Encapsulation of filler particles in polymethylmethacrylate beads by a double-dispersion method. *Journal of Colloid and Interface Science*, **132**, 592–594.
Cunningham, M.F. (2002) Living/controlled radical polymerizations in dispersed phase systems. *Progress in Polymer Science*, **27**, 1039–1067.
Cunningham, M.F. (2003) Recent progress in nitroxide-mediated polymerizations in miniemulsion. *Comptes Rendus Chimie*, **6**, 1351–1374.
Dawkins, J.V. (1989) *Comprehensive Polymer Science. The Synthesis, Characterization and Applications of Polymers*, **4** Pergamon Press, Oxford, Chapter 14.
De La Escosura, A. *et al.* (2008) Viral capsids as templates for the production of monodisperse Prussian blue nanoparticles. *Chemical Communications*, 1542–1544.
De, M., Ghosh, P.S., and Rotello, V.M. (2008) Applications of nanoparticles in biology. *Advanced Materials*, **20**, 4225–4241.
De Oliveira, A.M. *et al.* (2005) Encapsulation of TiO2 by emulsion polymerization with methyl metacrylate (MMA). *Polymer Bulletin*, **55**, 477–484.
Delaittre, G., Save, M., and Charleux, B. (2007) Nitroxide-mediated aqueous dispersion polymerization: from water-soluble macroalkoxyamine to thermosensitive nanogels. *Macromolecular Rapid Communications*, **28**, 1528–1533.

Desjardins, A. and Eisenberg, A. (1991) Colloidal properties of block ionomers. 1. Characterization of reverse micelles of styrene-b-metal methacrylate diblocks by size-exclusion chromatography. *Macromolecules*, **24**, 5779–5790.

Discher, D.E. and Eisenberg, A. (2002) Polymer vesicles. *Science*, **297**, 967–972.

Dowding, P.J. and Vincent, B. (2000) Suspension polymerisation to form polymer beads. *Colloids Surfaces A: Physicohem Eng Aspects*, **161**, 259–269.

Duguet, E. *et al.* (2000) PMMA encapsulation of alumina particles through aqueous suspension polymerisation processes. *Macromolecular Symposia*, **151**, 365–370.

Erdem, B. *et al.* (2000a) Encapsulation of inorganic particles via miniemulsion polymerization. *Macromolecular Symposia*, **155**, 181–198.

Erdem, B. *et al.* (2000b) Encapsulation of inorganic particles via miniemulsion polymerization. I. Dispersion of titanium dioxide particles in organic media using OLOA 370 as stabilizer. *Journal of Polymer, Polymer Chemistry Edition*, **38**, 4419–4450.

Erdem, B., Sully, Y., and Ulevich, I.N. (2000c) Determination of miniemulsion droplet size via soap titration. *Langmuir*, **16**, 4890–4895.

Fadnavis, N.W., Satayavathi, B., and Desphpande, A.A. (1997) Reverse micellar extraction of antibiotics from aqueous solutions. *Biotechnology Progress*, **13**, 503–505.

Fontenot, K. and Schork, F.J. (1993) Sensitivities of droplet size and stability in monomeric emulsions. *Industrial & Engineering Chemistry Research*, **32**, 373–385.

Forster, S. and Antonietti, M. (1998) Amphiphilic block copolymers in structure-controlled nanomaterial hybrids. *Advanced Materials*, **10**, 195–217.

Forster, S. and Konrad, M. (2003) From self-organizing polymers to nano- and biomaterials. *Journal of Materials Chemistry*, **13**, 2671–2688.

Frechet, J.M.J. (1994) Functional polymers and dendrimers: reactivity, molecular architecture, and interfacial energy. *Science*, **263**, 1710–1715.

Frechet, J.M.J. (2002) Supramolecular chemistry and self-assembly special feature. *Proceedings of the National Academy of Sciences of the United States of America*, **99**, 4782–4787.

Frechet, J.M.J. *et al.* (1995) Self-condensing vinyl polymerization: an approach to dendritic materials. *Science*, **269**, 1080–1083.

Gaedt, T. *et al.* (2009) Complex and hierarchical micelle architectures from diblock copolymers using living, crystallization-driven polymerizations. *Nature Materials*, **8**, 144–150.

Gao, H. *et al.* (2008) Core cross-linked reverse micelles from star-shaped polymers. *Chemistry of Materials*, **20**, 3063–3067.

Germack, D.S. *et al.* (2006) In situ generation of nitroxide from alkoxyamines for controlled acrylate polymerization. *Journal of Polymer Science, Part A, Polymer Chemistry*, **44**, 5128–5228.

Gindy, M.E., Panagiotopoulos, A.Z., and Prud'homme, R. (2008) Composite block copolymer stabilized nanoparticles: simultaneous encapsulation of organic actives and inorganic nanostructures. *Langmuir*, **24**, 83–90.

Grayson, S.M. and Frechet, J.M.J. (2001) Convergent dendrons and dendrimers: from synthesis to applications. *Chemical Reviews*, **101**, 3819–3868.

Hamley, I. (2005) Nanoshells and nanotubes from block copolymers. *Soft Matter*, **1**, 36–43.

Hamley, I.W. (2003) Nanostructure fabrication using block copolymers. *Nanotechnology*, **14**, R39-R54.

Harada, A. and Kataoka, K. (2006) Supramolecular assemblies of block copolymers in aqueous media as nanocontainers relevant to biological applications. *Progress in Polymer Science*, **31**, 949–982.

Harth, E. *et al.* (2002) A facile approach to architecturally defined nanoparticles via intramolecular chain collapse. *Journal of the American Chemical Society*, **124**, 8653–8660.

Hasell, T. *et al.* (2007) Preparation of polymer–nanoparticle composite beads by a nanoparticle-stabilised suspension polymerisation. *Journal of Materials Chemistry*, **17**, 4382–4386.

Hong, H. *et al.* (2007) Self-assembly of large multimolecular micelles from hyperbranched star copolymers. *Macromolecular Rapid Communications*, **28**, 591–596.

Huang, H. *et al.* (1998) Miniemulsion copolymerization of styrene and butyl acrylate initiated by redox system at lower temperature-preparation and polymerization of miniemulsion. *Journal of Applied Polymer Science*, **68**, 2029–2039.

Hubbell, J.A. (2003) Enhancing drug function. *Science*, **300**, 595–596.

Hult, A., Johansson, M., and Malmstrom, E. (1999) Hyperbranched polymers. *Advances in Polymer Science*, **143**, 1–34.

Isaure, F., Cormack, P.A.G., and Sherrington, D.C. (2003) Facile synthesis of branched poly(methyl methacrylate)s. *Journal of Materials Chemistry*, **13**, 2701–2710.

Isrealachvilli, J.N., Mitchell, D.J., and Ninham, B.W. (1976) Theory of self-assembly of hydrocarbon amphiphiles into micelles and bilayers. *Journal of the Chemical Society-Faraday Transactions*, **72**, 1525–1569.

Jain, S. and Bates, F.S. (2003) On the origins of morphological complexity in block copolymer surfactants. *Science*, **300**, 460–464.

Jang, J. and Ha, H. (2002) Fabrication of hollow polystyrene nanospheres in microemulsion polymerization using triblock copolymers. *Langmuir*, **18**, 5613–5618.

Jansen, J.F.G.A., De Brabander-Van Den Berg, E.M.M., and Meijer, E.W. (1994a) The dendritic box: shape-selective liberation of encapsulated guests. *Journal of the American Chemical Society*, **117**, 4417.

Jansen, J.F.G.A., De Branbander-Van De Berg, E.M.M., and Meijer, E.W. (1994b) Encapsulation of guest molecules into a dendritic box. *Science*, **266**, 1226–1229.

Johnson, B.K. and Prud'homme, R. (2003) Flash nanoprecipitation of organic actives and block copolymers using a confined impinging jets mixer. *Australian Journal of Chemistry*, **56**, 1021–1024.

Johnson, B.K., Saad, W., and Prud'homme, R. (2006) Nanoprecipitation of pharmaceuticals using mixing and block copolymer stabilization. *ACS Symposium Series*, **924**, 278–291.

Joumaa, N. *et al.* (2006) Synthesis of quantum dot-tagged submicrometer polystyrene particles by miniemulsion polymerization. *Langmuir*, **22**, 1810–1816.

Joumaa, N. *et al.* (2008) Surface modification of iron oxide nanoparticles by a phosphate-based macromonomer and further encapsulation into submicrometer polystyrene particles by miniemulsion polymerization. *Journal of Polymer Science, Part A, Polymer Chemistry*, **46**, 327–340.

Jovanovic, R. and Dube, M.A. (2004) Butyl acrylate/vinyl acetate miniemulsion polymerization: a study on compartmentalization. *Macromolecular Chemistry and Physics*, **205**, 958–965.

Jung, H.M., Price, K.E., and Mcquade, D.T. (2003) Synthesis and characterization of cross-linked reverse micelles. *Journal of the American Chemical Society*, **125** 5351–5355.

Kamigaito, M., Ando, T., and Sawamoto, M. (2001) Metal-catalyzed living radical polymerization. *Chemical Reviews*, **101**, 3689–3745.

Kastle, G. *et al.* (2003) Micellar nanoreactors—preparation and characterization of hexagonally ordered arrays of metallic nanodots. *Advanced Functional Materials*, **13**, 853–861.

Kataoka, K., Harada, A., and Nagasaki, Y. (2001) Block copolymer micelles for drug delivery: design, characterization and biological significance. *Advanced Drug Delivery Reviews*, **47**, 113–131.

Khougaz, K., Gao, Z., and Eisenberg, A. (1994) Determination of the critical micelle concentration of block copolymer micelles by static light scattering. *Macromolecules*, **27**, 6341–6346.

Kim, B.S. and Taton, T.A. (2007) Multicomponent nanoparticles via self-assembly with cross-linked block copolymer surfactants. *Langmuir*, **23**, 2198–2202.

Kim, J.-W. *et al.* (2003) Multihollow polymer microcapsules by water-in-oil-in-water emulsion polymerization: morphological study and entrapment characteristics. *Colloid and Polymer Science*, **281**, 157–163.

Kim, Y.H. (1998) Hyperbranched polymers 10 years after. *Journal of Polymer Science, Part A, Polymer Chemistry*, **36**, 1685–1698.

Klinghofer, S. *et al.* (1997) Preparation of palladium colloids in block copolymer micelles and their use for the catalysis of the heck reaction. *Journal of the American Chemical Society*, **119**, 10116–10120.

Kobalija, M. and McQuade, D.T. (2006) Polyurea microcapsules from oil-in-oil emulsions via interfacial polymerization. *Macromolecules*, **39**, 6371–6375.

Landfester, K. (2001) The generation of nanoparticles in miniemulsions. *Advanced Materials*, **13**, 765–768.

Landfester, K. (2006) Synthesis of colloidal particle in miniemulsions. *Annual Review of Materials Research*, **36**, 231–279.

Landfester, K. and Antonietti, M. (2000) The polymerization of acrylonitrile in miniemulsions: "crumpled latex particles" or polymer nanocrystals. *Macromolecular Rapid Communications*, **21**, 820–824.

Landfester, K. *et al.* (1999) Evidence for the preservation of the particle identity in miniemulsion polymerization. *Macromolecular Rapid Communications*, **20**, 81–84.

Landfester, K. and Rameriz, L.P. (2003) Encapsulated magnetite particles for biomedical application. *Journal of Physics: Condensed Matter*, **15**, S1345-S1362.

Lau, W. (2002) Emulsion polymerization of hydrophobia monomers. *Macromolecular Symposia*, **182**, 283–289.

Lee, J.E. *et al.* (2004) Polymer/Ag composite microspheres produced by water-in-oil-in-water emulsion polymerization and their application for a preservative. *Colloid and Polymer Science*, **282**, 295–299.

Lefebvre, M.D. and Shull, K.R. (2006) Homopolymer solubilization and nanoparticle encapsulation in diblock copolymer micelles. *Macromolecules*, **39**, 3450–3457.

Li, L. *et al.* (2009) A facile one-step route to synthesize cage-like silica hollow spheres loaded with superparamagnetic iron oxide nanoparticles in their shells. *Chemical Communications*, 938–940.

Li, M. and Matjaszewski, K. (2003) Further progress in atom transfer radical polymerizations conducted in a waterborne system. *Journal of Polymer Science, Part A, Polymer Chemistry*, **41**, 3606–3614.

Li, M., Viravaidya, C., and Mann, S. (2007) Polymer-mediated synthesis of ferritin-encapsulated inorganic nanoparticles. *Small*, **3**, 1477–1481.

Li, Y. *et al.* (2006) Synthesis of reversible shell cross-linked micelles for controlled release of bioactive agents. *Macromolecules*, **39**, 2726–2728.

Li, Z. *et al.* (2004) Multicompartment micelles from ABC miktoarm stars in water. *Science*, **306**, 98–101.

Liu, M., Kono, K., and Frechet, J.M.J. (2000) Water-soluble dendritic unimolecular micelles: their potential as drug delivery agents. *Journal of Controlled Release*, **65**, 121–131.

Liu, X. *et al.* (2004) Surface modification and characterization of magnetic polymer nanospheres prepared by miniemulsion polymerization. *Langmuir*, 20, 10278–10282.

Lodge, T.P. *et al.* (2005) Simultaneous, segregated storage of two agents in a multicompartment micelle. *Journal of the American Chemical Society*, **127**, 17608–17609.

Loginova, T.P. *et al.* (2004) Molybdenum sulfide nanoparticles in block copolymer micelles: synthesis and tribiological properties. *Chemistry of Materials*, **16**, 2369–2378.

Lowik, D.W.P.M. and Van Hest, J.C.M. (2004) Peptide based amphiphiles. *Chemical Society Reviews*, **33**, 234–245.

Lui, S. and Armes, S.P. (2001a) Recent advances in the synthesis of polymeric surfactants. *Current Opinion in Colloid & Interface Science*, **6**, 249–256.

Lui, S. and Armes, S.P. (2001b) The facile one-pot synthesis of shell cross-linked micelles in aqueous solution at high solids. *Journal of the American Chemical Society*, **123**, 9910–9911.

Lui, T., Xie, Y., and Chu, B. (2000) Use of block copolymer micelles on formation of hollow MoO3 nanospheres. *Langmuir*, **16**, 9015–9022.

Lui, W. (2008) New synthetic route for preparing CdS-nanoparticle/polystyrene polymer shells hybrid materials. *Journal of Applied Polymer Science*, **109**, 3204–3209.

Luo, L. and Eisenberg, A. (2001) Thermodynamic size control of block copolymer vesicles in solution. *Langmuir*, **17**, 6804–6811.

Lutz, J.F. and Laschewsky, A. (2005) Multicomponent micelles: has the long standing dream become a reality? *Macromolecular Chemistry and Physics*, **206**, 813–817.

Martin, J.E. and Eichinger, B.E. (1983a) Dimensions of intramolecularly crosslinked polymers. 1. Theory. *Macromolecules*, **16**, 1345–1350.

Martin, J.E. and Eichinger, B.E. (1983b) Dimensions of intramolecularly crosslinked polymers. 2. Dilute solution thermodynamic parameters and photon correlation results on the polystyrene/cyclopentane system. *Macromolecules*, **16**, 1350–1358.

Mason, B.P. *et al.* (2007) A general approach to creating soluble catalytic polymers heterogenized in microcapsules. *Organic Letters*, **9**, 3449–3451.

Matyjaszewski, K. and Xia, J. (2001) Atom transfer radical polymerization. *Chemical Reviews*, **101**, 2921–2990.

Mayer, A.B.R. (2001) Colloidal metal nanoparticles dispersed in amphiphilic polymers. *Polymers for Advanced Technologies*, **12**, 96–106.

McDonald, C.J. *et al.* (2000) Emulsion polymerization of voided particles by encapsulation of a nonsolvent. *Macromolecules*, **33**, 1593–1605.

Mecking, S. *et al.* (2000) Separation of catalytically active palladium nanoclusters in compartments of amphiphilic hyperbranched polyglycerols. *Macromolecules*, **33**, 3958–3960.

Meier, W. (1999) Nanostructure synthesis using surfactants and copolymers. *Curr. Opin. Colloid Interface Sci.* **4**, 6–14.

Miller, C.M. *et al.* (1994) Characterization of miniemulsion droplet size and stability using capillary hydrodynamic fractionation. *Journal of Colloid and Interface Science*, **162**, 11–18.

Mizuno, K., Taguchi, Y., and Tanaka, M. (2004) A fundamental study on encapsulation of an oil phase ingredient by polycondensation polymerization: the effect of water-soluble polymer on stability of limonene droplets. *Journal of Chemical Engineering of Japan*, **37**, 882–888.

Moad, G., Rizzardo, E., and Thang, S.H. (2006) Living radical polymerization by the RAFT process first update. *Australian Journal of Chemistry*, **59**, 669–692.

Moffit, M. *et al.* (1995) Size control of nanoparticles in semiconductor-polymer composites. 2. Control via sizes of spherical ionic microdomains in styrene-based diblock ionomers. *Chemistry of Materials*, **7**, 1185–1192.

Muller, A.H.E., Yan, D., and Wulkow, M. (1997) Molecular parameters of hyperbranched polymers made by self-condensing vinyl polymerization. 1. Molecular weight distribution. *Macromolecules*, **30**, 7015–7023.

Musyanovych, A. *et al.* (2007) Effect of hydrophilic comonomer and surfactant type on the colloidal stability and size distribution of carboxyl- and amino-functionalized polystyrene particles prepared by miniemulsion polymerization. *Langmuir*, **23**, 5367–5376.

Newkome, G.R. *et al.* (1986) Chemistry of micelles series. Part 2. Cascade molecules. Synthesis and characterization of a benzene[9]3-arborol. *Journal of the American Chemical Society*, **108**, 849–850.

Nomura, M., Tobita, H., and Suzuki, K. (2005) Emulsion polymerization: kinetic and mechanistic aspects. *Advances in Polymer Science*, **175**, 1–128.

O'Brien, P. *et al.* (2003) Quantum dot-labelled polymer beads by suspension polymerisation. *Chemical Communications*, 2532–2533.

O'Reilly, R.K., Hawker, C.J., and Wooley, K.L. (2006) Crosslinked block copolymer micelles: functional nanostructures of great potential and versatility. *Chemical Society Reviews*, **35**, 1068–1083.

O'Reilly, R.K. *et al.* (2005) Functionalization of micelles and shell cross-linked nanoparticles using click chemistry. *Chemistry of Materials*, **17**, 5976–5988.

Poe, S.L., Kobaslija, M., and Mcquade, T. (2007) Mechanism and application of a microcapsule enabled multicatalyst reaction. *Journal of the American Chemical Society*, **129**, 9216–9221.

Poe, S.L., Kobaslija, M., and Mcquade, T. (2008) Microcapsule enabled multicatalyst system. *Journal of the American Chemical Society*, **128**, 15586–15587.

Price, K.E. *et al.* (2006) Microencapsulated linear polymers: "soluble" heterogeneous catalysts. *Journal of the American Chemical Society*, **128**, 10376–10377.

Prud'homme, R., Saad, W., and Mayer, L. (2006) Nanoparticle formation by flash nanoprecipitation using block copolymer directed self-assembly. *NSTI Nanotech*, **2**, 824.

Qi, J., Charluex, B., and Matyjaszewski, K. (2001) Controlled/living radical polymerization in aqueous media: homogeneous and heterogeneous systems. *Progress in Polymer Science*, **26**, 2083–2134.

Radowski, M.R. *et al.* (2007) Supramolecular aggregates of dendritic multishell architectures as universal nanocarriers. *Angewandte Chemie-International Edition*, **46**, 1265–1269.

Rameriz, L.P. and Landfester, K. (2003) Magnetic polystyrene nanoparticles with a high magnetite content obtained by miniemulsion processes. *Macromolecular Chemistry and Physics*, **204**, 22–31.

Read, E.S. and Armes, S.P. (2007) Recent advances in shell cross-linked micelles. *Chemical Communications*, 3021–3035.

Reiss, G. (2003) Micellization of block copolymers. *Progress in Polymer Science*, **28**, 1107–1170.

Rosano, H.L., Gandolfo, F.G., and Hidrot, J.D.P. (1998) Stability of W1/O/W2 multiple emulsions: influence of ripening and interfacial interactions. *Colloids and Surfaces A-Physicochemical and Engineering Aspects*, **138**, 109–121.

Rosler, A., Vandermeulen, G.W.M., and Klok, H.A. (2001) Advanced drug delivery devices via self-assembly of amphiphilic block copolymers. *Advanced Drug Delivery Reviews*, **53**, 95–108.

Scott, C. *et al.* (2005) Liquid-core capsules via interfacial polymerization: a free-radical analogy of the nylon rope trick. *Journal of the American Chemical Society*, **127**, 4160–4161.

Seidl, J. *et al.* (1967) Makroporöse styrol-divinylbenzol-copolymere und ihre verwendung in der chromatographie und zur darstellung von ionenaustauschern. *Advances in Polymer Science*, **5**, 113–213.

Sela, Y., Magdassi, S., and Garti, N. (1994) Newly designed polysiloxane-graft-poly (oxyethylene) copolymeric surfactants: preparation, surface activity and emulsification properties. *Colloid and Polymer Science*, **272**, 684–691.

Seo, M. *et al.* (2008) Polymeric nanoparticles via noncovalent cross-linking of linear chains. *Macromolecules*, **41**, 6413–6418.

Smulders, W.W., Jones, C.W., and Schork, F.J. (2004) Synthesis of block copolymers using RAFT miniemulsion polymerization in a train of CSTRs. *Macromolecules*, **37**, 9345–9354.

Solans, C. *et al.* (2005) Nano-emulsions. *Current Opinion in Colloid & Interface Science*, **10**, 102–110.

Soo, P.L. and Eisenberg, A. (2004) Polymer–silicate nanocomposites produced by in situ atom transfer radical polymerization. *Journal of Polymer Science, Part A, Polymer Chemistry*, **42**, 923–938.

Spatz, J.C. *et al.* (1998) Controlled mineralization and assembly of hydrolysis-based nanoparticles in organic solvents combining polymer micelles and microwave techniques. *Advanced Materials*, **10**, 473–475.

Steiret, N. and Landfester, K. (2007) Encapsulation of organic pigment particles via miniemulsion polymerization. *Macromolecular Materials and Engineering*, **292**, 1111–1125.

Sunder, A., Heinemann, J., and Frey, H. (2000) Controlling the growth of polymer trees: concepts and perspectives for hyperbranched polymers. *Chemistry – A European Journal*, **6**, 2499–2506.

Sunder, A. *et al.* (1999) Molecular nanocapsules based on amphiphilic hyperbranched polyglycerols. *Angewandte Chemie-International Edition*, **38**, 3552–3555.

Teng, J. and Zubarev, E.R. (2003) Synthesis and self-assembly of a heteroarm star amphiphile with 12 alternating arms and a well-defined core. *Journal of the American Chemical Society*, **125**, 11840–11841.

Thurmond, K.B., Kowalewski, T., and Wooley, K.L. (1996) Water-soluble knedel-like structures: the preparation of shell-cross-linked small particles. *Journal of the American Chemical Society*, **118**, 7239–7240.

Tiarks, F., Landfester, K., and Antonietti, M. (2001a) Encapsulation of carbon black by miniemulsion polymerization. *Macromolecular Chemistry and Physics*, **202**, 51–60.

Tiarks, F., Landfester, K., and Antonietti, M. (2001b) Preparation of polymeric nanocapsules by miniemulsion polymerization. *Langmuir*, **17**, 908–917.

Tomalia, D.A. and Frechet, J.M.J. (2002) Discovery of dendrimers and dendritic polymers: a brief historical perspective. *Journal of Polymer Science, Part A, Polymer Chemistry*, **40**, 2719–2728.

Trijasson, P., Frere, Y., and Gramain, P. (1990) Inverse suspension polymerization of poly(ethylene oxide) methacrylate macromonomers. *Macromolecular Rapid Communications*, **11**, 239–243.

Trollas, M. and Hedrick, J.L. (1998) Hyperbranched poly(ε-caprolactone) derived from intrinsically branched AB2 macromonomers. *Macromolecules*, **31**, 4390–4395.

Tully, D.C. and Frechet, J.M.J. (2001) Dendrimers at surfaces and interfaces: chemistry and applications. *Chemical Communications*, 1229–1239.

Ugelstad, J. *et al.* (1980) Swelling of oligomer-polymer particles. New methods of preparation. *Advances in Colloid and Interface Science*, **13**, 101–140.

Van Der Ende, A. *et al.* (2009) Tailored polyester nanoparticles: post-modification with dendritic transporter and targeting units via reductive amination and thiol-ene chemistry. *Soft Matter*, **5**, 1417–1425.

Van Hamersveld, E.M.S., Van Es, J.J.G.S., and Cupurus, F.P. (1999) Oil–acrylate hybrid emulsions, mini-emulsion polymerization and characterization. *Colloids and Surfaces A-Physicochemical and Engineering Aspects*, **153**, 285–296.

Vivaldo-Lima, E. *et al.* (1997) An updated review on suspension polymerization. *Industrial & Engineering Chemistry Research*, **36**, 939–965.

Vogt, A.P., Gondi, S.R., and Sumerlin, B.S. (2007) Hyperbranched polymers via RAFT copolymerization of an acryloyl trithiocarbonate. *Australian Journal of Chemistry*, **60**, 396–399.

Voit, B. (2005) Hyperbranched polymers—all problems solved after 15 years of research? *Journal of Polymer Science, Part A, Polymer Chemistry*, **43**, 2679–2699.

Volz, M. *et al.* (2007) Nano-explosions of nanoparticles for sudden release of substances by embedded azo-components as obtained via the miniemulsion process. *Macromolecular Materials and Engineering*, **292**, 1237–1244.

Von Werne, T. and Patten, T.E. (1999) Preparation of structurally well-defined polymer–nanoparticle hybrids with controlled/living radical polymerizations. *Journal of the American Chemical Society*, **121**, 7409–7410.

Vriezema, D.M. *et al.* (2005) Self-assembled nanoreactors. *Chemical Reviews*, **105**, 1445–1489.

Walther, A., Dreshsler, M., and Muller, A.H.E. (2009) Structures of amphiphilic Janus discs in aqueous media. *Soft Matter*, **5**, 385–390.

Wang, Q. *et al.* (2002) Natural supramolecular building blocks. *Chemistry & Biology*, **9**, 805–811.

Wang, X. *et al.* (2006) Magnetic molecularly imprinted polymer particles synthesized by suspension polymerization in silicone oil. *Macromolecular Rapid Communications*, **27**, 1180–1184.

Wang, X.S. *et al.* (2003) Shell cross-linked cylinders of polyisoprene-b-ferrocenyldimethylsilane: formation of magnetic ceramic replicas and microfluidic channel alignment and patterning. *Journal of the American Chemical Society*, **125**, 12686–12687.

Wang, X.S., Winnik, M.A., and Manners, I. (2004) Swellable, redox-active shell-crosslinked organometallic nanotubes. *Angewandte Chemie-International Edition*, **43**, 3703–3707.

Weaver, J.V.M. *et al.* (2008) pH-Responsive branched polymer nanoparticles. *Soft Matter*, **4**, 985–992.

Wenning, H. (1956) Über den Mechanismus der dreiphasigen Emulgierung bei der Perlpolymerisation. *Makromolekulare Chemie-Macromolecular Chemistry and Physics*, **20**, 196–213.

Wong, K.K.W. and Mann, S. (1996) Biomimetic synthesis of cadmium sulfide-ferritin nanocomposites. *Advanced Materials*, **8**, 928–932.

Wooley, K.L. (2000) Shell crosslinked polymer assemblies: nanoscale constructs inspired from biological systems. *Journal of Polymer Science, Part A, Polymer Chemistry*, **38**, 1397–1407.

Xiong, Y. *et al.* (2003) Micelle-assisted fabrication of necklace-shaped assembly of inorganic fullerene-like molybdenum disulfide nanospheres. *Chemical Physics Letters*, **382**, 180–185.

Yates, C.R. and Hayes, W. (2004) Synthesis and applications of hyperbranched polymers. *European Polymer Journal*, **40**, 1257–1281.

Zeng, Z., Yu, J., and Guo, Z.-X. (2005) Preparation of functionalized core-shell alumina/polystyrene composite nanoparticles. *Macromolecular Chemistry and Physics*, **206**, 1558–1567.

Zhang, J. *et al.* (2006) ZnO/PS core–shell hybrid microspheres prepared with miniemulsion polymerization. *Journal of Colloid and Interface Science*, **301**, 78–84.

Zhang, S.W. *et al.* (2005) Synthesis of SiO2/polystyrene nanocomposite particles via miniemulsion polymerization. *Langmuir*, **21**, 2124–2128.

Zheng, R., Liu, G., and Yan, X. (2005) Polymer nano- and microspheres with bumpy and chain-segregated surfaces. *Journal of the American Chemical Society*, **127**, 15358–15359.

Zhong, X.F., Varshnay, S.K., and Eisenberg, A. (1992) Critical micelle lengths for ionic blocks in solutions of polystyrene-b-poly(sodium acrylate) ionomers. *Macromolecules*, **25**, 7160–7167.

Zhou, J. *et al.* (2006) Synthesis of SiO2/poly(styrene-co-butyl acrylate) nanocomposite microspheres via miniemulsion polymerization. *Journal of Polymer Science Part A-Polymer Chemistry*, **44**, 3202–3209.

Zhou, Y. and Yan, D. (2009) Supramolecular self-assembly of amphiphilic hyperbranched polymers at all scales and dimensions: progress, characteristics and perspectives. *Chemical Communications*, 1172–1188.

18

Morphologies of Block and Star-Branched Polymers with Three Components

Hirokazu Hasegawa
Department of Polymer Chemistry, Kyoto University, Japan

18.1 Introduction

Block copolymers are a class of polymers that have molecular architectures consisting of two or more chemically distinct polymer chains connected together by covalent bonds. The simplest example is the linear AB diblock copolymer consisting of an A-block chain (a long sequence of A monomers) followed by a B-block chain (a long sequence of B monomers) as schematically shown in Figure 18.1a. The complexity increases with increasing the number of chemically distinct blocks (A, B, C, D, . . .) and the number of branching (graft-type and star-type). Simplest examples for ABC triblock terpolymers are illustrated in Figure 18.1b (linear polymer) and 1c (miktoarm (heteroarm) star polymer). Consequently, the variation of the architectures of block copolymers constructed by combination of these two parameters is limitless. In this chapter how the complexity of the block copolymers increases when the number of chemically distinct blocks increases from two to three is reviewed.

First, we consider an AB diblock copolymer for simplicity. When the segregation power between the two different block chains is weak, the block chains are mixed in the segmental level (disordered state) as illustrated in Figure 18.2a. Under the control of thermodynamics, when the segregation power between the different block chains measured by the product of the Flory–Huggins interaction parameter χ_{AB} between A and B segments and the total degree of polymerization N is strong enough, they segregate from each other within a block copolymer molecule (ordered state). (Leibler, 1980) Since the block chains cannot separate from each other due to the connectivity between them, they form a periodic domain structure of a uniform

Complex Macromolecular Architectures: Synthesis, Characterization, and Self-Assembly, First Edition.
Edited by Nikos Hadjichristidis, Akira Hirao, Yasuyuki Tezuka and Filip Du Prez.

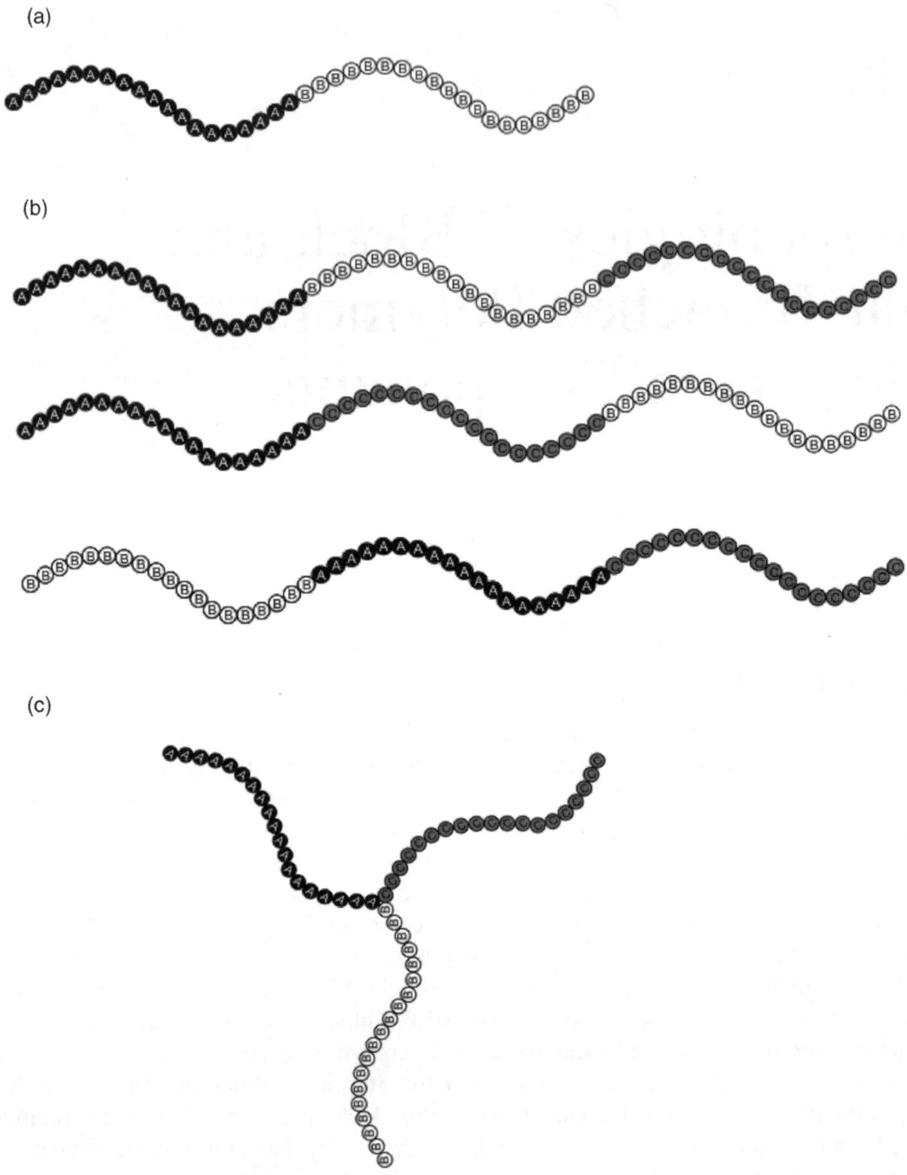

Figure 18.1 Molecular architectures: (a) AB diblock copolymer, (b) linear ABC, ACB and BAC triblock terpolymers, (c) miktoarm star ABC triblock terpolymer.

size comparable to the size of the coiled block chain, as illustrated in Figure 18.2b. Such phase separation and the resulting domains are called microphase separation and microdomains, respectively. In the case of a symmetric AB diblock copolymer in terms of volume and rigidity of the block chains, an alternating lamellar microdomain structure with planer A/B interfaces

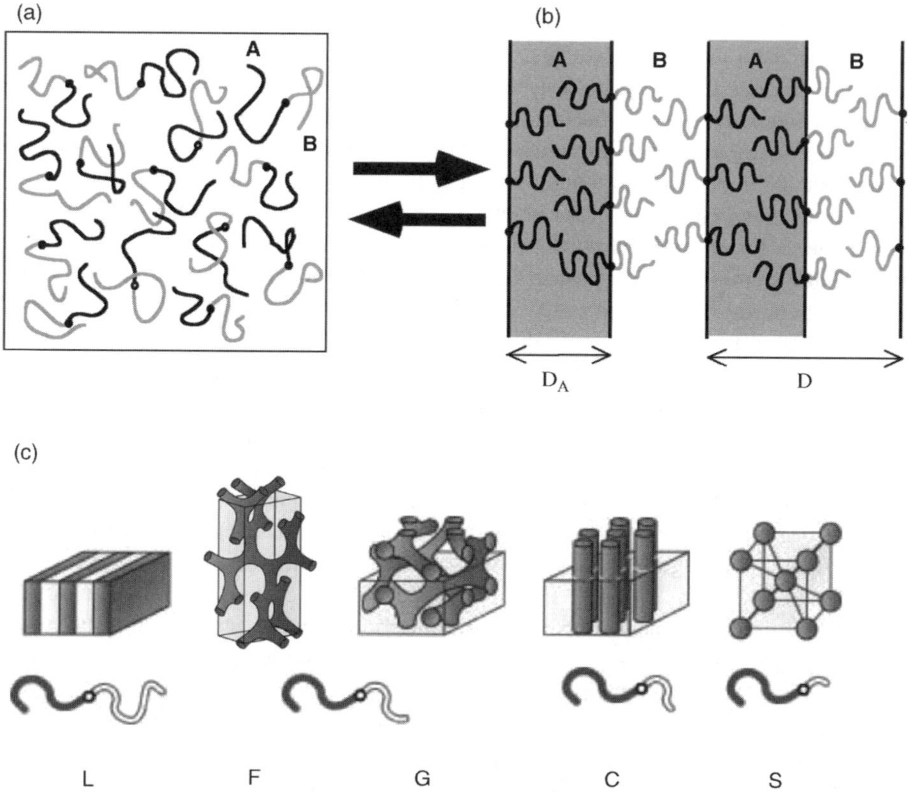

Figure 18.2 Molecular packing models for (a) disordered state and (b) ordered state, and (c) models of five equilibrium microdomain morphologies for AB diblock copolymers. L: lamellae, F: Fddd network, G: double-gyroid networks, C: hexagonally packed cylinders, S: body-centered-cubic spheres.

(L) forms. However, asymmetry in the copolymer composition (or the volume of the block chains), introduces a curvature to the A/B interfaces to alter the microdomain morphology. Five kinds of equilibrium microdomain morphologies, spherical domains (S) arranged in a body-centered cubic (bcc) lattice, cylindrical domains (C) packed in a hexagonal array, bicontinuous gyroid double-network structure (Hajduk *et al.*, 1994) characterized by gyroid minimal surface having a symmetry of I$a\bar{3}d$ space group (G), and bicontinuous F*ddd* single-network structure (Takenaka *et al.*, 2007) having a symmetry of F*ddd* space group (F) in addition to lamellar morphology as shown in Figure 18.2c, have been known for polystyrene-*block*-polyisoprene diblock copolymer, one of the most popular and most extensively studied AB diblock copolymers. When the segregation power between A and B blocks is strong (strong segregation limit: $\chi_{AB}N > 100$), microdomain morphologies that form in linear AB diblock copolymers (or linear ABA triblock copolymers as well) are basically defined by composition expressed by volume fractions ϕ_A and ϕ_B ($= 1 - \phi_A$) (Hasegawa *et al.*, 1987). On the other hand, when the segregation power between A and B blocks is weak (weak segregation limit: $20\text{–}30 \geq \chi_{AB}N$), microdomain morphologies depend on $\chi_{AB}N$ as well as volume fractions. Since χ_{AB} is usually in inverse proportion to temperature, morphological transitions as well as order–order transition may occur by changing temperature in the weak segregation limit (Khandpur *et al.*, 1995; Matsen and Bates, 1996).

Adding C block, a third component, to the end of an AB diblock copolymer produces an ABC linear triblock terpolymer. The syntheses of ABC triblock terpolymers in good quality are not so easy because each step of the sequential living polymerization of A, B and C monomers contains a chance to produce byproducts such as homopolymers and AB diblock copolymers that contaminate the system. This may be one of the reasons why the number of systems studied on ABC triblock terpolymers is much smaller than that on AB diblock copolymers. Another reason is that the number of parameters that defines the mesoscopic organizations of block copolymers increases dramatically just by adding another component to diblock copolymers. Whereas an AB diblock copolymer is defined by three parameters, ϕ_A, χ_{AB} and N, a triblock terpolymer consisting of A, B and C blocks requires six parameters, ϕ_A, ϕ_B ($\phi_C = 1 - \phi_A - \phi_B$), χ_{AB}, χ_{BC}, χ_{AC} and N in addition to the variation in block sequence (ABC, BAC and ACB, as shown in Figure 18.1b). (Zheng and Wang, 1995) Nevertheless, morphological study on ABC triblock terpolymers is much more fascinating than AB diblock copolymers because tremendous variety and complexity of microdomain morphologies can be expected for ABC triblock terpolymers. For an ABC triblock terpolymer three different states can be possible: one-phase state (disordered state of mixed ABC), two-phase state (A/mixed BC, B/mixed AC and C/mixed AB) and three-phase state (A/B/C). In this chapter, we deal with only the three-phase state of ABC triblock terpolymers because the microdomain morphologies of the two-phase state of ABC triblock terpolymers may be similar to those of diblock copolymers. When three components form their independent microdomains, mere combination of three simple geometrical shapes such as planes (lamellae), cylinders and spheres comprising L, C and S morphology of diblock copolymers, respectively, as shown in Figure 18.2c generates a far large number of morphologies, as illustrated in Figures 18.3a–f for the morphologies based on lamellae, Figures 18.4a–i based on cylinders, and Figures 18.5a–c based on spheres. The morphological variation expands further when we take into account combination with the network morphology such as F and G in Figure 18.2c. In reality, a complex morphology such as "knitting pattern" (*kp*) morphology, which is hard to imagine from our knowledge on the microdomain morphology of AB diblock copolymers, has been observed. Because of this wide

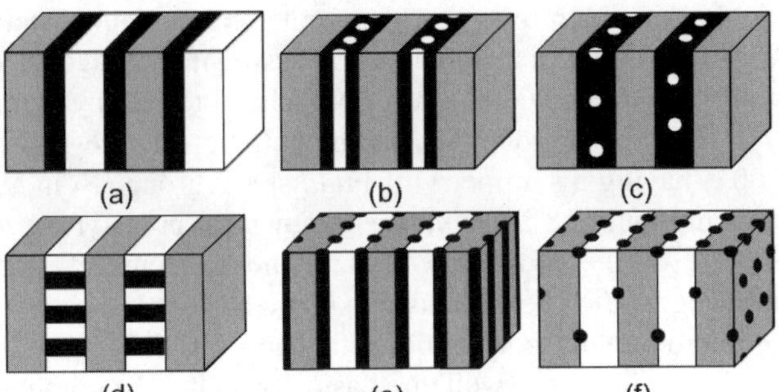

Figure 18.3 Microdomain morphology models for linear ABC triblock terpolymers based on lamellar morphology (A: gray, B: black, C: white): (a) *ll*, (b) *cil*, (c) *sil*, (d) *lil*, (e) *col*, (f) *sol*. (d), (e) and (f) are frustrated systems with A/C interfaces, which do not have any junctions.

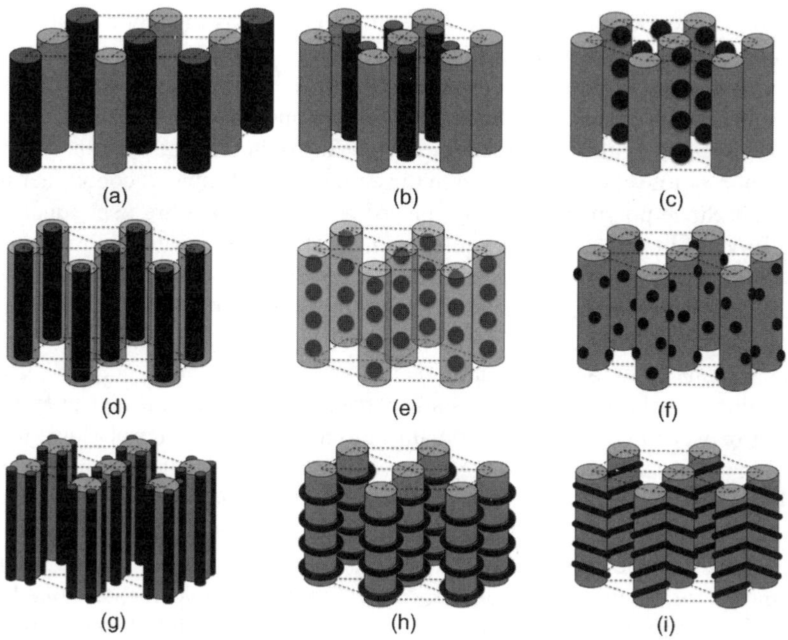

Figure 18.4 Microdomain morphology models for linear ABC triblock terpolymers based on cylindrical morphology (contrast not in regular manner): (a) *tetra-cc*, (b) *hex-cc*, (c) *cs*, (d) *cic*, (e) *sic*, (f) *soc*, (g) *coc*, (h) *roc*, (i) *hoc*.

variety of morphologies ABC triblock terpolymers should have much better capability for applications than AB diblock copolymers if their microdomain structures can be appropriately controlled and useful functions are given to each microdomain. Excellent pieces of work have shed some lights on the concepts of microdomain morphology control of ABC triblock terpolymers and are introduced below. It is also suggested to read the excellent review articles. (Bates and Fredrickson, 1999; Abetz and Simon, 2005; Bates, 2005; Meuler *et al.*, 2009).

Before we get into the details, the following problems should be pointed out. First, thermodynamic equilibrium of ABC triblock terpolymers is generally difficult to attain, in

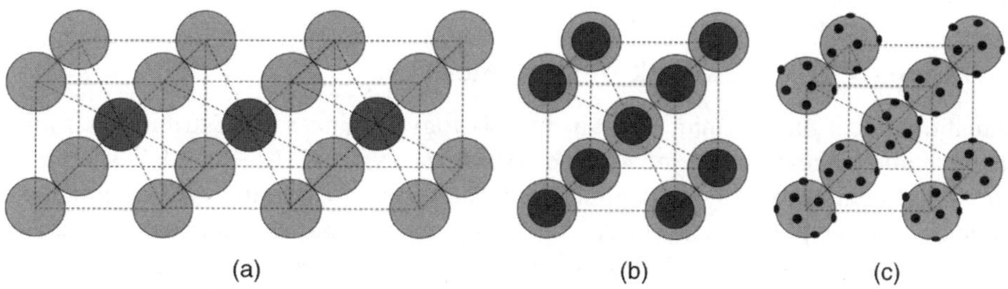

Figure 18.5 Microdomain morphology models for linear ABC triblock terpolymers based on spherical morphology (contrast not in a regular manner): (a) *ss*, (b) *sis*, (c) *sos*.

contrast to diblock copolymers for several reasons. Solution casting methods are usually used to obtain solid specimens of ABC triblock terpolymer samples except for the case of low molecular weight samples. Initial low concentration of the polymer solution ensures the disordered state, which cancels out the thermal history and the solvent enhances the mobility of the terpolymer molecules in the ordered state at higher concentration. Therefore, the equilibrium morphology can be obtained in the case of an AB diblock copolymer if a neutral solvent is used. Since polymer concentrations of A and B phases are kept equal throughout the casting process in this case, the equilibrium morphology can be maintained if solvent evaporation is slow enough. However, in the case of an ABC triblock terpolymer in general it is impossible to find a neutral solvent for all three components in principle. Consequently, the polymer concentration of one phase differs from those of the others resulting in nonequilibrium morphology in many cases. Secondly, subsequent annealing of such a specimen hardly helps in achieving equilibrium due to the high viscosity of the terpolymer and the high energy barrier for the transition between complex microdomain morphologies. The morphological transitions of the microdomains can be brought about by two kinds of molecular diffusion. One is pulling out a block chain from its microdomain and adding it into another microdomain. However, the energy barrier of such diffusion is very high in the strong segregation regime. The other involves the diffusion of junction points of the terpolymers along the interfaces. Such motion is relatively smooth for AB diblock copolymers. However, since B-block chains of ABC triblock terpolymers are bridging two interfaces with one end at the A/B interface and the other end at the B/C interface, two junction points must move cooperatively and the diffusion of B chains encounter the obstruction by chain entanglements. Therefore, the researchers who are dealing with equilibrium morphologies and phase diagrams use low molecular weight samples with which microphase separation from their disordered states is possible. On the other hand, those who are interested in finding new morphologies, which may be equilibrium or nonequilibrium, use higher molecular weight samples with which transmission electron microscope (TEM) observation is relatively easy.

Another problem in morphological study of ABC triblock terpolymers is that the structure analysis of triblock terpolymers is much more difficult than that of diblock copolymers. In addition to the complexity of microdomain morphology itself, it is necessary to differentiate three phases by giving different contrast to each microphase. Common techniques used for the structure analyses of AB diblock copolymers such as TEM observation and small-angle X-ray or neutron scattering have been employed also for ABC triblock terpolymers. However, interpretation of the data must be performed with caution.

18.2 Linear ABC Triblock Terpolymers

Systematic studies on the microdomain morphology of linear ABC triblock terpolymers (hereafter called linear ABC) are limited to several systems. Linear ABC samples were predominantly synthesized by sequential living anionic polymerizations, which limit the kinds of monomers. Thus, polystyrene (PS) and polyisoprene (PI) or polybutadiene (PB) as well as their hydrogenated polymers such as poly(ethylene-*alt*-propylene) (PEP) or poly(ethylene-*alt*-butylene) (PEB) were used as the A and B blocks, while poly(2-vinylpyridine) (P2VP), poly(4-vinylpyridine) (P4VP), poly(methyl methacrylate) (PMMA), poly(*tert*-buthyl methacrylate) (PtBMA), poly(ethylene oxide) (PEO) or polydimethylsiloxane (PDMS) was used as

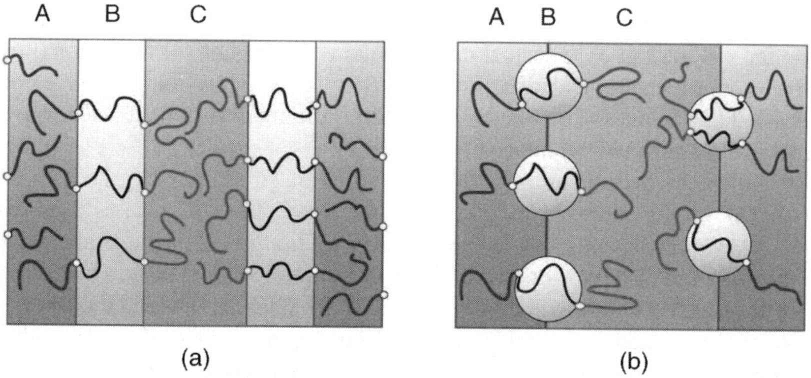

Figure 18.6 Molecular packing and microdomain interfaces in linear ABC triblock terpolymers for (a) nonfrustrated system, and (b) frustrated system.

the C block. For a systematic study to determine the morphological phase diagram of ABC triblock terpolymer, the living polymers at different conversions are taken from the reactor during the polymerization step of the third component. Then, a series of linear ABC samples with a constant A/B volume ratio but different volume fractions of C can be obtained. Usually, a polar polymer is chosen for the last sequence of the three blocks due to the high nucleophilicity of their living ends. Consequently, the resulting ABC triblock terpolymer has the segmental interaction parameters χ_{AC} and χ_{BC} both larger than χ_{AB}. Since the interfacial tension of the interfaces between two components is proportional to the square root of the interaction parameter in the strong segregation limit (Helfand and Sapse, 1975), the interfaces formed by the terpolymer are influenced by the relative magnitude of χ_{AC} and χ_{BC}. When $\chi_{AC} > \chi_{BC} \approx \chi_{AB}$, the system is "nonfrustrated" and A/B and B/C interfaces are formed as illustrated in Figure 18.6a. On the other hand, when $\chi_{AC} < \chi_{BC} \approx \chi_{AB}$, the system is "frustrated" and A/C contact is preferred to B/C contact, possibly resulting in coexistence of A/B, B/C and A/C interfaces depending on the composition, as illustrated in Figure 18.6b, despite the fact that there are no chemical junctions on the A/C interfaces. The examples of frustrated systems are PI-PS-PDMS (Shefelbine *et al.*, 1999), PS-PI-PEO (Bailey *et al.*, 2001), and PS-PB-P2VP (Hückstädt *et al.*, 2000a) and those for nonfrustrated systems are PI-PS-PEO (Epps *et al.*, 2004a; 2004b), PS-PB-PMMA (Goldacker and Abetz, 1999), and PS-PI-PDMS (Sugiyama *et al.*, 2001). Obviously, the change of the sequence order switches the frustrated system to the nonfrustrated system.

18.3 Pioneering Works

The basic idea of the three-phase microdomain morphology was first given by Riess *et al.* (1980). For a PS-PI-PMMA linear ABC system blended with PS and PMMA homopolymers, they suggested the systematic variation of the microdomain morphology (spherical – cylindrical – lamellar) analogous to AB diblock copolymer system just by considering the volume fractions of PMMA phase or (PI + PS) phase as well as PI/PS ratio. They suggested important features of the three-phase morphologies such as three-layer lamellae, core–shell cylinders or spheres, cylindrical or spherical microdomains dispersed in larger lamellar, cylindrical or spherical microdomains, and spherical microdomains attached on the surfaces of larger spherical microdomains. Some of the morphologies were presented by the impressive TEM images. The microdomain structure models of neat linear ABC given in Figures 18.3–18.5 are based on their suggestion.

Progress in the analytical techniques of the three-phase structures was made by Matsushita *et al.* (1980) and Shibayama *et al.* (1982). To differentiate three phases in PS-P4VBDMA-PI linear ABC (P4VBDMA: poly(4-vinylbenzyl)dimethylamine) they introduced two techniques, one for TEM and the other for SAXS. They differentiated three phases by giving different contrast by the double-staining technique, that is PI phase stained dark with OsO_4 and P4VBDMA phase stained less dark with phosphotungstic acid (PTA). A double-staining technique is commonly used for TEM observation of linear ABC, although just one staining agent sometimes gives different contrast to three phases. Selective staining is especially useful when identification of a particular phase is necessary. In the analysis of SAXS profiles, Shibayama *et al.* (1982) showed the importance of particle scattering. The observed SAXS intensity is the product of lattice scattering and particle scattering. The particle scattering is calculated with the step functions describing the electron-density profiles of the repeat unit and has a characteristic profiles with a series of scattering peaks. They showed that the damping of the height of those peaks is modulated for the three-phase system but decreased monotonously for the two-phase system. This suggests that caution is necessary when one indexes the scattering maxima obtained from linear ABC with a crystallographic unit cell because some reflection might be missing due to the particle scattering.

The first systematic study of the composition dependence of the microdomain morphology of linear ABC was performed by Mogi *et al.* (1992a). They prepared a PI-PS-P2VP system with equivalent PI and P2VP volume fractions but PS volume fractions ranging from 0.3 to 0.8 as shown in the phase map in Figure 18.8. This system is nonfrustrated, having roughly symmetric interfacial tensions and a large enthalpic incompatibility between the terminal blocks (i.e. $\chi_{IV} > \chi_{IS} \approx \chi_{SV}$). Therefore, no PI/P2VP interfaces were observed by TEM study. They observed systematic changes of the three-phase microdomain structures in the order of three-phase four-layer (ABCA) lamellae (Figure 18.7a, corresponding to *ll* in Figure 18.3a), ordered tricontinuous double-diamond (OTDD) network structure (Figure 18.7b) (Mogi *et al.*, 1992b), tetragonally packed A and C cylinders in B matrix (Figure 18.7c, corresponding to *tetra-cc* in Figure 18.4a), and A and C spheres packed in body-centered cubic (bcc) lattice embedded in B matrix (Figure 18.7d, corresponding to *ss* in Figure 18.5a), as expected from the morphological transition of AB diblock copolymers with composition. Although these morphologies were subsequently confirmed by SAXS (Mogi *et al.*, 1994), in their assignment of OTDD they were influenced and misled to the wrong conclusion by the preceding report of ordered-bicontinuous double-diamond (OBDD) network structures for PS-PI star block copolymers (Alward *et al.*, 1986) and diblock copolymers (Hasegawa *et al.*, 1987). OTDD contained two chemically distinct interpenetrating networks, one comprised of PI and the other of P2VP, in a PS matrix and each network consisted of cylindrical struts connected by fourfold connectors forming 6-nord ring ((6,4) nets). Later, OBDD structures were re-examined and found to be double gyroid (G) with cylindrical struts connected by threefold connectors forming 10-nord ring ((10,3) nets) as shown in Figure 18.2c. Matsushita *et al.* (1998) also re-examined the TEM images of the OTDD structures by comparing them with the simulated TEM images using minimal surface models and concluded that their network structure was gyroid. This conclusion was confirmed by SAXS (Seki *et al.*, 2000). The OTDD network was also reported in a PS-PI-P2VP sample

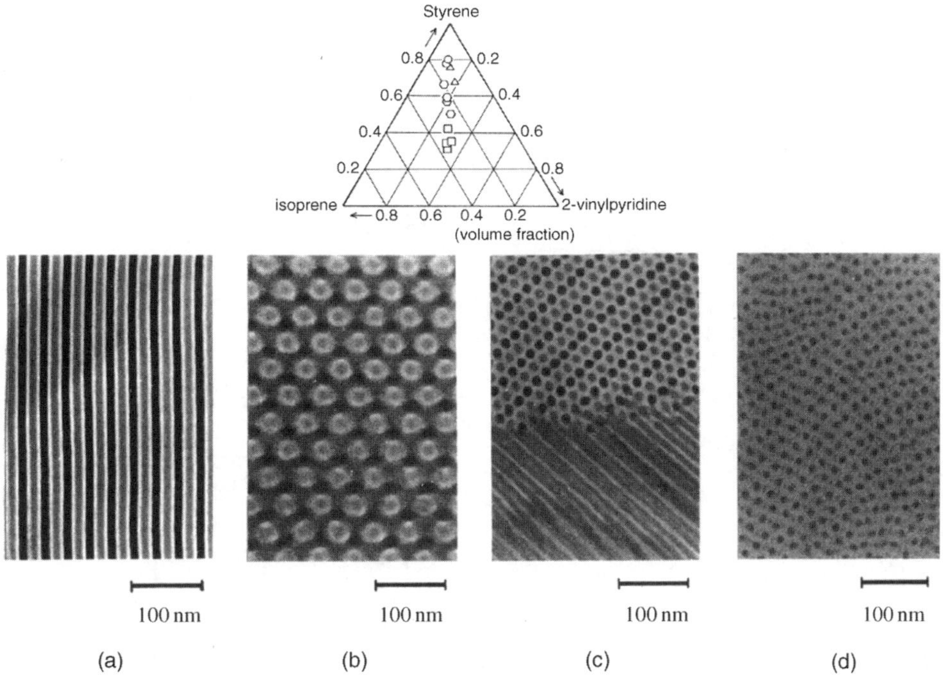

Figure 18.7 TEM images of four types of microdomain morphologies observed in PI-PS-P2VP: (a) three-phase four-layer lamellae (squares in the phase diagram), (b) assigned as OTDD first but corrected later as tricontinuous gyroid (Q^{214}) with PI and P2VP networks (hexagons), (c) tetragonally packed cylinders of PI and P2VP in PS matrix (triangles), (d) pseudo-bcc spherical microdomains of PI and P2VP in PS matrix (circles). (Reprinted with permission from Y. Mogi, H. Kotsuji, Y. Kaneko *et al.*, "Preparation and morphology of triblock copolymers of the ABC type," *Macromolecules*, **25**, 20, 5408–5411, 1992. © 1992 American Chemical Society.)

(Matsushita *et al.*, 1994). However, it was pointed out that their experimental TEM images attributed to the OTDD were consistent with the simulated images obtained from the spherical microdomains in a pseudo-bcc unit cell (Matsen, 1998).

The contribution of Mogi *et al.* (1992a) is notable because they obtained the morphological phase diagram of PI-PS-P2VP and compared it with that of PS-PI diblock copolymer, as shown in Figure 18.8. The values along the vertical axis are the volume percentages of the PS block in PI-PS-P2VP and in PS-PI. The phase boundary of lamellar morphology (*ll*) of PI-PS-P2VP is largely shifted toward the high concentration of PS. Along with this shift, the network phase was found in the samples with PS volume fractions ranging from 0.48 to 0.66. This range of ~0.18 in the PS volume fraction of PI-PS-P2VP is much broader than the gyroid phase occurring over a volume fraction range of only ~0.04 in PI-PS diblock copolymers (Khandpur *et al.*, 1995). This tendency of network formation by linear ABC was interpreted by Epps *et al.* (2004b) using the illustration in Figure 18.9. They considered a symmetric AB diblock copolymers forming L with flat interfaces (left). If a similarly sized C block is added to this diblock, an ABC triblock with equivalent block volume fractions is obtained and if the interfacial tensions are roughly symmetric ($\chi_{AB} \approx \chi_{BC}$) it would form a *ll* morphology with A/B and B/C interfaces (right) providing there is no thermodynamic driving force for the formation

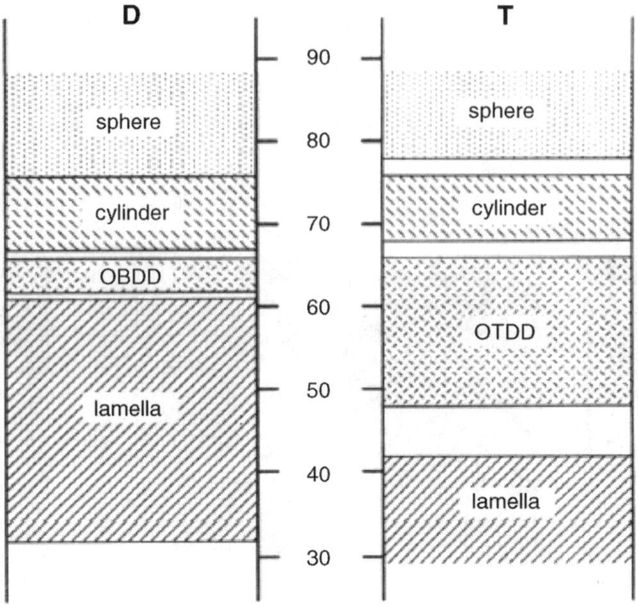

Figure 18.8 Variation of morphology with the compositions of the PI-PS-P2VP triblock copolymer (right) together with that of the PS-PI diblock copolymers (left). The values along the vertical axis are the volume percentages of the PS block in the triblock terpolymer (T) and in the diblock copolymer (D). OBDD and OTDD was corrected later as double gyroid (G) of PI networks in PS matrix and tricontinuous gyroid (Q^{214}) with PI and P2VP networks in PS matrix, respectively. (Reprinted with permission from Y. Mogi, H. Kotsuji, Y. Kaneko *et al.*, "Preparation and morphology of triblock copolymers of the ABC type," *Macromolecules*, **25**, 20, 5408–5411, 1992. © 1992 American Chemical Society.)

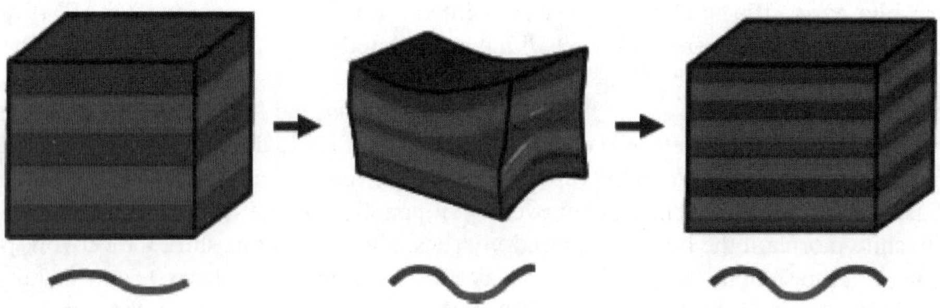

Figure 18.9 An illustration to explain how network formation in ABC triblock terpolymers is induced: the flat interfaces are preferred for symmetric compositions (left and right). Intermediate amounts of the C block destabilize the flat B/C interface, leading to a saddle surface element. (Reprinted with permission from T.H. Epps, E.W. Cochran, T.S. Bailey *et al.*, "Ordered network phases in linear poly(isoprene-b-styrene-b-ethylene oxide) triblock copolymers," *Macromolecules*, **37**, 22, 8325–8341, 2004. © 2004 American Chemical Society.)

of an A/C interface ($\chi_{AC} > \chi_{AB} \approx \chi_{BC}$). If the ABC triblocks have intermediate, asymmetric compositions they would have packing frustrations and the block chains would best be accommodated by forming the interfaces with hyperbolic surfaces such as saddle surfaces (middle) resulting in network morphologies.

18.4 Network Morphologies

Network morphologies in linear ABC are summarized in detail in the review paper by Meuler *et al.* (2009). Three types of network morphologies have been known for the equilibrium microdomain structures of linear ABC. Here, we follow their notation and call them O^{70}, Q^{214} and Q^{230}. This naming is based on the crystallographic symmetry of the repeat units. The capital letter "O" stands for orthorhombic unit cell and "Q" for cubic unit cell. The numbers in the superscripts indicate the number of the space group and O^{70}, Q^{214} and Q^{230} have F*ddd*, I4_132 and I$a\bar{3}d$ symmetry, respectively. The networks in Q^{214} and Q^{230} are both related to gyroid minimal surface and two interwoven gyroid networks are embedded in the matrix phase of the major component. The difference is that the two end blocks, A and C, form individual networks separately in the matrix phase of the middle (B) block in Q^{214} but two core–shell networks composed of A/B (cores of A encased in shells of B) or B/C blocks are embedded in the matrix phase of C or A component, respectively, in Q^{230}. Because of the spatial connectivity of each phase, the former is called a "tricontinuous gyroid" or an "alternating gyroid" and the latter a "pentacontinuous gyroid" or a "core–shell gyroid". On the other hand, in only one core–shell network composed of one of the terminal blocks and the middle block is embedded in O^{70} the matrix phase of the other terminal block that is the major component. All the networks in O^{70}, Q^{214} and Q^{230} are characterized by threefold connectors, as shown by the backbone models in the illustrations in Figure 18.10 (Epps *et al.*, 2004a). These models also show that the network consists of 6-nord rings in O^{70} ((6,3) nets) and 10-nord rings in Q^{214} and Q^{230} ((10,3) nets). These morphologies exhibit characteristic TEM images and they are shown together with the corresponding projected images calculated with the models. It should be noted that the contrast in TEM is given for the particular system, PI-PS-PEO in this case, and might change with different systems. The network morphologies are complex in nature and difficult to analyze. Therefore, it is suggested by Meuler *et al.* (2009) to use plural techniques such as TEM, SAXS, rheology, and birefringence as well as SCFT (self-consistent field theory) calculations in structure analysis of linear ABC. Strong optical anisotropy is a good indication of O^{70} because of its orthorhombic unit cells.

The Q^{214} morphology was first observed by Mogi *et al.* (1992a, 1992b) in PI-PS-P2VP nonfrustrated system as mentioned above. They proved that the Q^{214} morphology was an equilibrium morphology by observing an order–order transition (OOT) from *ll* to Q^{214} morphology upon heating using 2D-SAXS and TEM observations of the quenched samples at different annealing times (1 h to 7 days) at 423 K (Suzuki *et al.*, 2004). They used a slow-cast film from THF solution having the *ll* morphology with the lamellar planes parallel to the film surface and finally obtained the uniaxially oriented Q^{214} morphology with its (110) plane parallel to the film surface. According to the similarity of the spacings of the PI or P2VP layers in *ll* (58 nm) and (110) spacing of Q^{214} (55 nm), they proposed the mechanism of OOT as the PI or P2VP layers transformed to the (110) planes of the PI or P2VP network, the lattice planes most dense with PI or P2VP blocks. Thus, the OOT takes place so that the overall mass

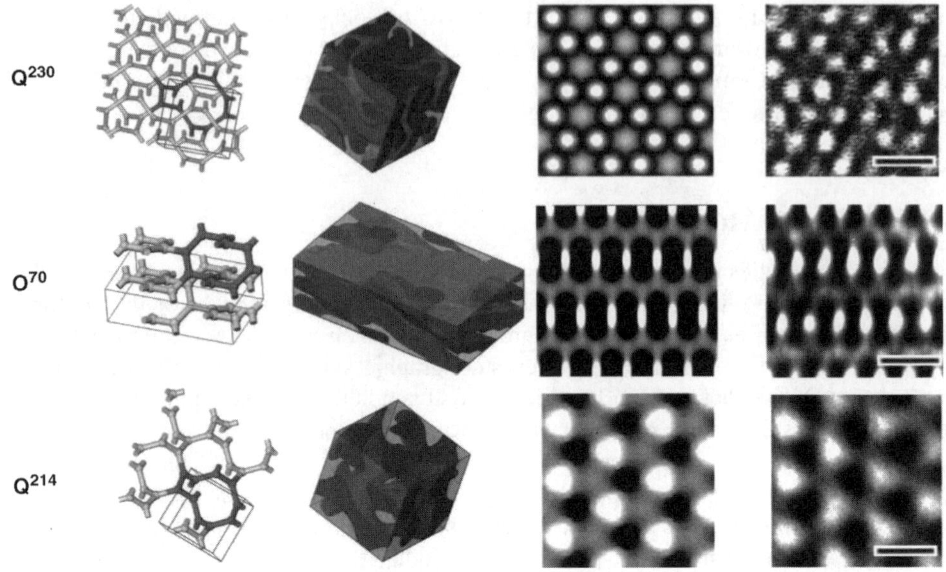

Figure 18.10 From left to right: lattice models by network backbones, 3D microdomain morphologies and 2D projections derived from the surface models and TEM images for network phases O^{70}, Q^{214} and Q^{230}. (Reprinted with permission from T.H. Epps, E.W. Cochran, C.M. Hardy *et al.*, "Network phases in ABC triblock copolymers," *Macromolecules*, **37**, 19, 7085–7088, 2004. © 2004 American Chemical Society.)

transportation is minimized. The Q^{214} morphology was also observed by Epps *et al.* (2004a, 2004b) in PI-PS-PEO nonfrustrated system.

The Q^{230} morphology in linear ABC was reported independently by two research groups in 1999. Shefelbine *et al.* reported a Q^{230} mesostructure in PI-PS-PDMS frustrated system. With only SAXS and TEM investigations they could not confirm the structure. But by supplementing their experimental analysis with SCFT calculations they could conclude that the PI-PS-PDMS triblock terpolymer had formed a pentacontinuous Q^{230} morphology with two discrete domains of PDMS cores, two separate domains of PS shells, and a PI matrix. This structure is reasonable because the PDMS occupies the cores of the networks. Then, the area of the enthalpically costly PS/PDMS interface becomes smaller than that of the less costly PI/PS interface. Their observation of the Q^{230} morphology in the PI-PS-PDMS system was limited to a relatively narrow range of compositions. Goldacker and Abetz (1999) reported a morphological study of the blends of *ll*-forming PS-PB-PMMA linear ABC (frustrated system) and L-forming PB-PMMA diblock copolymer, both having similar block lengths. They found a Q^{230} morphology (core–shell double-gyroid) with PS cores encased in PB shells embedded in PMMA matrix. They identified the morphology by comparing the TEM images with the computer simulated [110] and [112] projections of a Q^{230} morphology. They mentioned that the Q^{230} morphology was always coexisting with other morphologies such as *cic* (core–shell cylinders), *ll* (three-phase lamellae) and L (two-phase lamellae), and predominantly formed in the blend with a PB-PMMA volume fraction of 0.22. Hückstädt *et al.* (2000a) analyzed

PS-PB-P2VP frustrated system (PS wt % = 0.48, PB wt % = 0.31, $\chi_{BV} > \chi_{BS} \approx \chi_{SV}$) by TEM and identified a Q^{230} morphology containing P2VP cores, PB shells, and a PS matrix Q^{230} in a narrow range of compositions. Bailey *et al.* (2001) investigated PS-PI-PEO frustrated system having the composition $f_I \approx f_S$ using a combination of rheological measurements, SAXS and TEM, and identified a Q^{230} morphology in four samples with $0.19 < f_O < 0.23$. All those samples underwent an OOT from Q^{230} to core–shell cylinders upon heating. From the kinetic study of this transition, they suggested that the core–shell cylinder-to-Q^{230} transition likely occurred via the formation of a metastable semiperforated lamellar intermediate. Sugiyama *et al.* (2001) reported the formation of a Q^{230} morphology in the blends of nonfrustrated PS-PI-PDMS ($M_n = 42$ kg/mol, $f_S = 0.20, f_I = 0.59$) and PS and PDMS homopolymers with an overall homopolymer volume fraction ranging from 0.15 to 0.45 ($0.25 < f_S$ (f_D) < 0.34) using SAXS, SANS, and TEM measurements. Ludwigs *et al.* (2003) investigated the morphology of frustrated PS- P2VP- PtBMA system along the isopleth composition $f_S \approx f_V$ by TEM. Increasing the PtBMA block from a symmetric composition ($f_S \approx f_V \approx f_M$), the morphology changes from *ll* via Q^{230} with PS cores and P2VP shells to *cic* with PS cores and P2VP shells upon increase of the PtBMA block. These morphologies may be considered as core–shell analogs of the corresponding diblock copolymer morphologies. However, upon decrease of the PtBMA block the morphological scheme was different and a slight decrease of PtBMA lead to the formation of an inverse Q^{230} with PMMA cores and P2VP shells, but further decrease of PtBMA resulted in a cylinder-in-lamellae (*cil*) morphology.

A O^{70} morphology (Fddd) was first reported by Bailey *et al.* (2002) in a nonfrustrated PI-PS-PEO linear ABC system having the composition $f_I \approx f_S$ and $0.12 < f_O < 0.25$ between two-domain lamellae (L) of PI and mixed PS/PEO and three-domain lamellae (*ll*) of PI, PS and PEO. The thermodynamic stability of the O^{70} morphology was proved by the SCFT calculation by Tyler *et al.* (2007). Epps *et al.* (2004a, 2004b) extended the study to the ternary phase map for PI-PS-PEO system across a range of compositions between $0.1 \leq f_O \leq 0.3$, found Q^{214} and Q^{230} morphologies in addition of the O^{70} morphology and determined the compositional windows of these network morphologies as shown in Figure 18.11. It was shown that the O^{70} morphology is positioned around the isopleth composition $f_I \approx f_S$. Chatterjee *et al.* (2007) further extended those investigations to PEO-rich PI-PS-PEO system and observed an OOT from lamellae to O^{70} upon heating. While a ternary phase map of perfectly symmetric ABC triblock terpolymers should be symmetric about the $f_I = f_O$ isopleths, the comprehensive ternary phase map they obtained for PI-PS-PEO was asymmetric, which was attributed to asymmetries in the block statistical segment lengths and χ parameters. They emphasized that expansive network windows validate the strategy of using nonfrustrated linear ABC with symmetric interfacial tensions ($\chi_{AB} \approx \chi_{BC}$) to prepare multiply continuous network morphologies.

18.5 Strongly Frustrated Systems

In the strongly frustrated linear ABC system with $\chi_{AC} \ll \chi_{BC} \approx \chi_{AB}$, the A/C interfaces without junctions form and it leads to the unique microdomain morphologies. Such examples are lamellae in lamellae (*lil* in Figure 18.3d), cylinders on lamellae (*col* in Figure 18.3e), spheres on lamellae (*sol* in Figure 18.3e), spheres on cylinders (*soc* in Figure 18.4f), cylinders on cylinders (*coc* in Figure 18.4g), rings on cylinders (*roc* in Figure 18.4h), helices on cylinders (*hoc* in

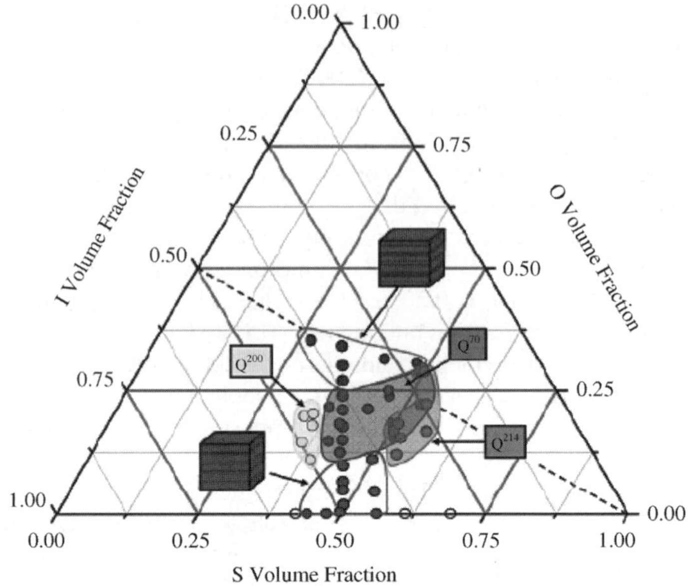

Figure 18.11 PI-PS-PEO volume fraction phase portrait in the weak segregation. Ordered and disordered states are indicated by filled and open circles, respectively. The areas of two- and three-domain lamellae identified by sketches, and three network phases, identified by Q^{230}, O^{70}, and Q^{214} are indicated. (Reprinted with permission from T.H. Epps, E.W. Cochran, C.M. Hardy *et al.*, "Network phases in ABC triblock copolymers," *Macromolecules*, **37**, 19, 7085–7088, 2004. © 2004 American Chemical Society.)

Figure 18.4i), and spheres on spheres (*sos* in Figure 18.5c). Auschra and Stadler (1993) followed by Beckmann, Auschra and Stadler (1994) and Stadler *et al.* (1995) first reported the *col*, *roc* and *sol* morphologies in PS-PEB-PMMA and PS-PB-PMMA systems with the overall molecular weight on the order of 10^2 kg/mol. In the same kind of linear ABC, Breiner *et al.* (1997, 1998) observed various kinds of cylindrical morphologies including *coc*, *soc*, and *hoc* and *sos* morphology. Figure 18.12 shows the volume-fraction–morphology portrait of PS-PB-PMMA linear ABC system (Abetz and Goldacker, 2000). When the volume fraction of the middle PB block is less than 25%, PS/PMMA interfaces are usually formed. In a PS-PER-PMMA system, a *lil* morphology ("ladder" morphology) was observed using 3D electron tomography by Kaneko *et al.* (2006).

Extremely interesting morphologies have been reported for the PS-PEB-PMMA and PS-PB-PMMA systems. They are so-called knitting pattern (*kp*) morphology (Breiner *et al.*, 1996) and barber pattern (*hoc* in Figure 18.4i) (Krappe *et al.*, 1995). Their TEM images and the suggested morphologies are shown in Figure 18.13. They both have A/C interfaces. Breiner *et al.* (1996) attributed the formation of the *kp* morphology to the asymmetric interaction between the middle PEB block and different end blocks, PS and PMMA. They considered the *kp* morphology as an intermediate of a *ll* morphology of ABC lamellae and a *col* morphology of AC lamellae with B cylinders at the A/C interfaces. They blended PS-PEB-PMMA and PS-PB-PMMA to change the volume fraction of the middle block and the χ_{AB}/χ_{AB}

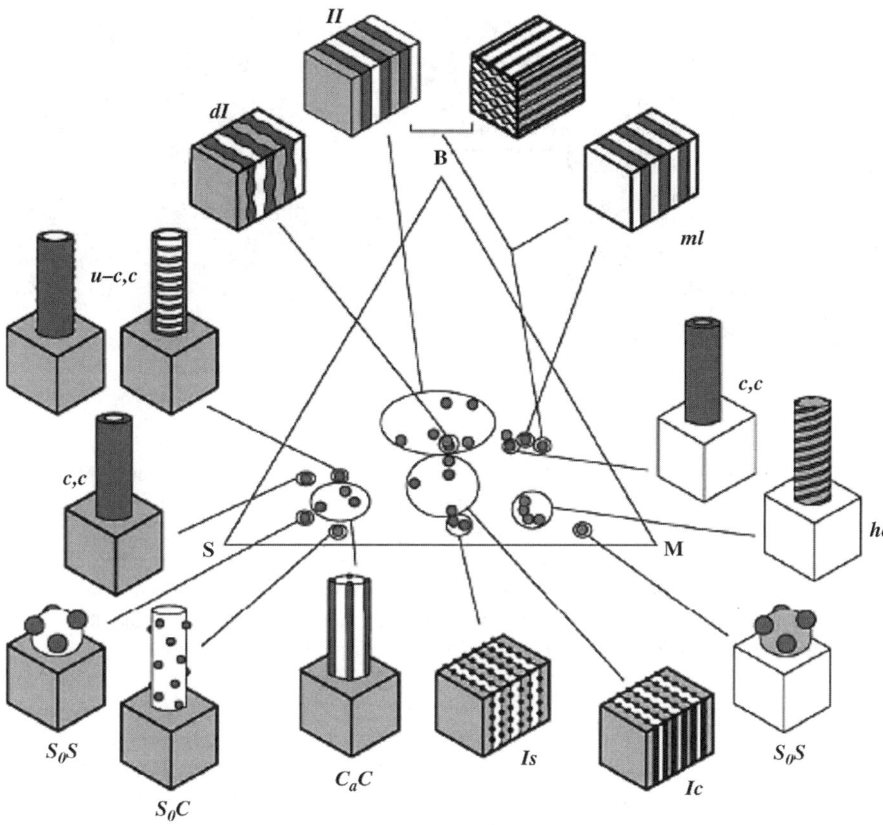

Figure 18.12 PS-PB-PMMA volume fraction morphology portrait in the strong segregation. (Reproduced with permission from V. Abetz and T. Goldacker, "Formation of superlattices via blending of block copolymers," *Macromolecular Rapid Communications*, 2000, **21**,1, 16–34. © Wiley-VCH Verlag GmbH & Co. KGaA.)

ratio systematically, and observed a systematic change of the morphology from *col* to *kp*. The stability of the *kp* morphology in the PS-PEB-PMMA system was further studied by Ott *et al.* (2001) and the solvent effect was clarified.

The barber pattern (*hoc* morphology) in PS-PB-PMMA and its blends with PS homopolymer was further studied by Jinnai *et al.* (2009) using a 3D electron tomography technique. They clarified the microdomain morphology of double helices of PB phase winding around the PS cylinders in a hexagonal array embedded in PMMA matrix. The left-handed and right-handed double helices were coexisting side-by-side to avoid the squeezed conformation of PMMA block chains. Addition of PS homopolymer increased the helical pitch and eventually broke up the helices into spheres to form *soc* morphology as shown in Figure 18.4f. The barber pattern as well as *soc* was also reported in PB-P2VP-PtBMA system (Schacher *et al.*, 2010) and PS-PFDMS (polyferrocenyldimethylsilane)-PMMA system (Kloninger and Rehahn, 2007).

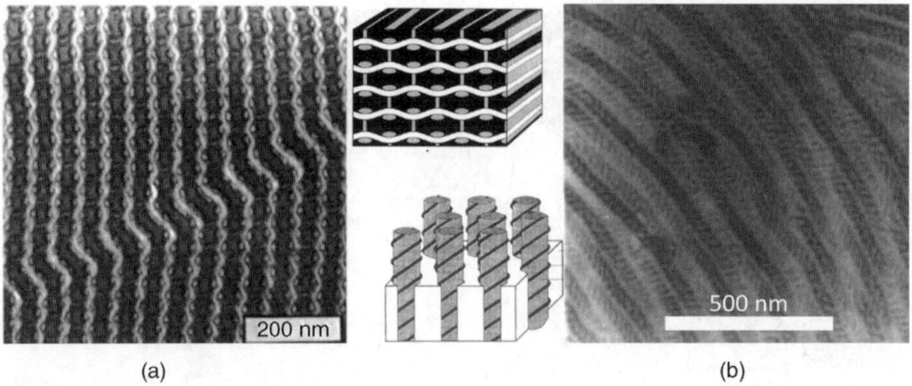

Figure 18.13 TEM images and models of (a) the knitting pattern (*kp*) and (b) barber pattern (*hoc*) in PS-PEB-PMMA. ((a) Reprinted with permission from H. Ott, V. Abetz and V. Altstädt, "Morphological studies of poly(styrene)-block-poly(ethylene-co-butylene)-block-poly(methyl methacrylate) in the composition region of the "knitting pattern" morphology," *Macromolecules*, **34**, 7, 1069–1075, 2001. © 2001 American Chemical Society. (b) Reprinted with permission from U. Krappe, R. Stadler and I. Voigt-Martin, "Chiral assembly in amorphous ABC triblock copolymers. Formation of a helical morphology in polystyrene-block-polybutadiene-block-poly(methyl methacrylate) block copolymers," *Macromolecules*, **28**, 13, 4558–4561, 1995. © 1995 American Chemical Society.)

18.6 Theoretical Approaches

Theoretical approaches to predict the microdomain morphologies of linear ABC have been reported using density-functional theory (DFT) (Nakazawa and Ohta, 1993; Zheng and Wang, 1995), strong-segregation theory (SST) (Phan and Fredrickson, 1998), Monte Carlo (MC) (Dotera and Hatano, 1996; Ko *et al.*, 2001), and self-consistent field theory (SCFT)

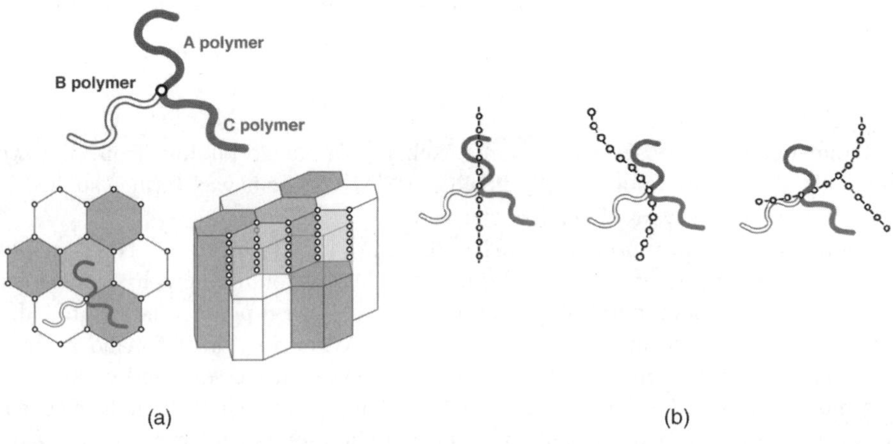

Figure 18.14 (a) The molecular packing of an ABC miktoarm star terpolymer with the junctions on linear lines forming cylindrical microdomains of a particular type (left: cross-sectional view of right). (b) Types of junction arrays: linear (left), curved or helical (middle), branched (right).

(Matsen, 1998; Tang *et al.*, 2004; Tyler *et al.*, 2007, Sun *et al.*, 2008). Theoretical description of a linear ABC system requires seven quantities: three interaction parameters, two volume fractions providing the composition, and two ratios specifying the relative segment lengths. It is necessary to consider three distinct interfaces and three distinct stretching-energy contributions from three distinct blocks instead of just one and two, respectively, for diblocks. The early theoretical works (Nakazawa and Ohta, 1993; Dotera and Hatano, 1996) were stimulated by the discovery of OTDD morphology reported for PI-PS-P2VP triblock terpolymer by Mogi *et al.* (1992b). Dotera and Hatano investigated the morphology of block copolymer melts by MC simulations with a diagonal-bond method based on the Verdier–Stockmayer model. This method has the benefit of not requiring the selection of mesostructural candidates. They examined a series of symmetric linear ABC including the OTDD composition reported by Mogi *et al.* (1992a) and predicted Q^{214} to be the equilibrium morphology in some composition but not OTDD.

After the misidentification of double gyroid (G) morphology as OBDD became clear (Hajduk *et al.*, 1995), theorists' interest shifted towards the stability of OTDD and Q^{214}. Matsen (1998) used a spectral implementation of SCFT to interrogate the phase behavior of "symmetric" ($b_A = b_B = b_C$, $\chi_{AB} = \chi_{BC} = \chi$, and $f_A = f_C$) linear ABC triblocks in the intermediate segregation regime ($\chi N < 65$). These linear ABC can be characterized by three parameters: the volume fraction of one of the terminal blocks (f), χN, and χ_{AC}/χ. He determined that, for compositions with $f \leq 0.3$, the morphology selection is not sensitive to either the χ_{AC}/χ ratio or χN and largely selected the mesostructural candidates based on the composition of the triblock terpolymer in the SCFT calculations. He predicted the sequence of equilibrium morphologies to be pseudo-bcc spheres (*ss* in Figure 18.5a), tetragonally packed cylinders (*cc* in Figure 18.4a), Q^{214} and *ll* (Figure 18.3a) with increasing f, which is analogous to the sequence of bcc spheres, hexagonally packed cylinders, Q and L in AB diblock copolymers (Figure 18.2c). The packing frustration also reduces the width of the cylinder window and, consequently, widens the network window in linear ABC triblocks relative to AB diblocks.

Phan and Fredrickson (1998) investigated symmetric linear ABC in the strong segregation limit and computed the free energies of OTDD and Q^{214} networks as well as a number of other mesostructural candidates. They predicted that the equilibrium morphology changes with composition from *ss* to *cc* to *ll* but neither of these networks was predicted to be the equilibrium mesostructure though Q^{214} always had a lower free energy than OTDD. They suggested the possibility of Q^{214} becoming stable in the intermediate segregation regime.

18.7 ABC Miktoarm Star Polymer

An ABC miktoarm star terpolymer has a configuration of a C polymer grafted at the junction point of an AB diblock copolymer as depicted in Figure 18.1c, and hereafter we call it μ-ABC. Since it is not necessary to account for the order of the sequences of A, B and C blocks in a μ-ABC, it has fewer variables to designate the system than a linear ABC. However, because of the unusual molecular architecture of a μ-ABC whose three different block chains possess the same junction, the microdomain morphologies it can take become quite different from those of a linear ABC when they microphase-separate into three phases. (In this chapter, we only consider the three-phase structures.) Three microphases of a μ-ABC must meet in space to allocate the junctions as illustrated for a particular case in Figure 18.14a (left). Consequently, all the junctions are confined on the lines where A/B, B/C and A/C interfaces meet. Therefore, there will be no junctions on the interfaces except for those lines at their intersections. When those lines are linear, as illustrated by the circles in Figure 18.14b (left), parallel cylindrical microdomains of A, B and C blocks regularly filling the space (Figure 18.14a (right)) are expected. In this case, the vertical cross sections to the cylindrical axes have the regular geometrical patterns called Archimedean tiling patterns consisting of regular polygons and there exist only 11 types, as illustrated in Figure 18.15a (Matsushita, 2007). When the junctions are allocated on curved or helical lines (Figure 18.14b (middle)) or branched lines (Figure 18.14b (right)), the expected microdomains are not cylindrical but must possess extremely complex geometry (Judas *et al.*, 2009). The experimental and theoretical investigations on the microdomain morphology of μ-ABCs were performed by several research groups. The major parts of those studies were reviewed by Matsushita (2007, 2008) and Matsushita *et al.* (2009).

A variety of patterns of cylindrical microdomain morphologies of μ-ABCs have been reported in TEM observations of PS-PI-PMMA μ-ABC (Sioula *et al.*, 1998a, 1998b), PS-PB-P2VP μ-ABC (Hückstädt *et al.*, 2000b), PS-PI-P2VP μ-ABC (Zioga *et al.*, 2000; Takano *et al.*, 2005, 2007; Hayashida *et al.*, 2006, 2007a), and PS-PI-PDMS μ-ABC (Yamauchi *et al.*, 2003). Many of the three-phase patterns can be explained by Archimedean tiling patterns, as pointed out by Gemma *et al.* (2002). An Archimedean tiling pattern is designated by a set of integers $(n_1.n_2.n_3....)$, which denotes a tiling of a vertex type in the way that n_1-gon, n_2-gon, and n_3-gon, ..., meet consecutively on each vertex, and superscripts are employed to abbreviate when possible, that is $(6^3) \equiv (6.6.6)$. They investigated the microdomain morphologies of the μ-ABCs with the arm length ratio 1:1:x with symmetric interactions between three components by MC simulation with the diagonal bond method proposed by Dotera and Hatano (1996). They found five types of polygonal cylinder morphology, all of which had the sections of the cylindrical phases exhibiting the tessellation of even-numbered polygons although the possible cylindrical morphology for a μ-ABC extends to eleven types, that is (3^6), (4^4), (6^3), $(3^4.6)$, $(3^3.4^2)$, $(3^2.4.3.4)$, $(3.4.6.4)$, $(3.6.3.6)$, (3.12^2), $(4.6.12)$, and (4.8^2), as shown in Figure 18.15a. In addition to five types of polygonal cylinder morphology, they found *sol* (Figure 18.3f), perforated layers, *col* (Figure 18.3e), columnar piled disks and lamella-in-sphere morphology (Figure 18.15b). In the three-dimensional morphologies such as *sol*, perforated layers, columnar piled disks and lamella-in-sphere morphology, the junctions of the μ-ABC molecules are located on curved lines (circles). More detailed phase diagram has been constructed from the simulation results by Ueda *et al.* (2007) as functions of arm-length ratio of μ-ABC with symmetric interactions between three components. Morphologies in the central region of their triangle phase diagram are cylindrical structures and three-dimensional structures are surrounding them. More variety of cylindrical morphologies exhibiting Archimedean tiling pattern were reported by Matsushita *et al.* (2009) in the star PI-PS-P2VP system with various compositions and some of their typical TEM images are shown in Figure 18.16. The dark, gray and bright areas correspond to PI, P2VP and PS phases stained with OsO_4, respectively. It was pointed out that the $(3^2.4.3.4)$ tiling pattern in Figures 18.16d is particularly interesting because they have a quasiperiodic structure and dodecagonal symmetry (Takano *et al.*, 2005; Hayashida *et al.*, 2007a).

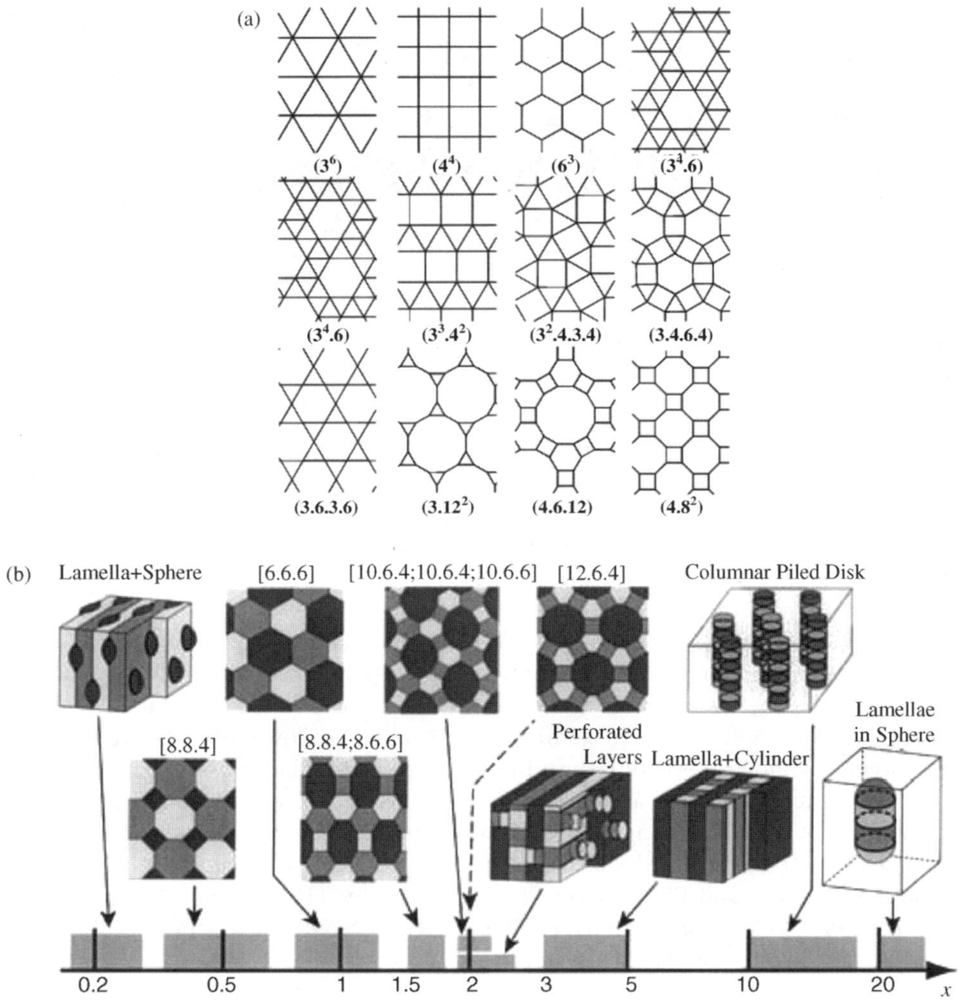

Figure 18.15 (a) The known 11 Archimedean tilings having the tessellation of regular polygons with every vertex under the same environment and the quasicrystalline tiling (3^2.4.3.4). Each tiling pattern is denoted by a set of integers ($n_1.n_2.n_3.\ldots$) in the way that n_1-gon, n_2-gon, and n_3-gon, . . . , meet consecutively on each vertex. Superscripts can be employed to abbreviate if applicable. (Reprinted with permission from Y. Matsushita, "Creation of hierarchically ordered nanophase structures in block polymers having various competing interactions," *Macromolecules*, **40**, 4, 771–776, 2007. © 2007 American Chemical Society.) (b) MC simulation results: phase diagram as functions of arm-length ratio 1:1:x of ABC star terpolymer system with symmetric interactions between three components. (Reprinted with permission from T. Gemma, A. Hatano and T. Dotera, "Monte Carlo simulations of the morphology of ABC star polymers using the diagonal bond method," *Macromolecules*, **35**, 8, 3225–3237, 2002. © 2002 American Chemical Society.)

Yamauchi *et al.* (2003, 2005) performed 3D observation of a cylindrical morphology of a PS-PI-PDMS μ-ABC using the electron tomography technique and pointed out that the cylinders are curved along the cylindrical axes resulting in the small grain size. The curving tendency of the cylinders was attributed to the frustration of the μ-ABC molecules

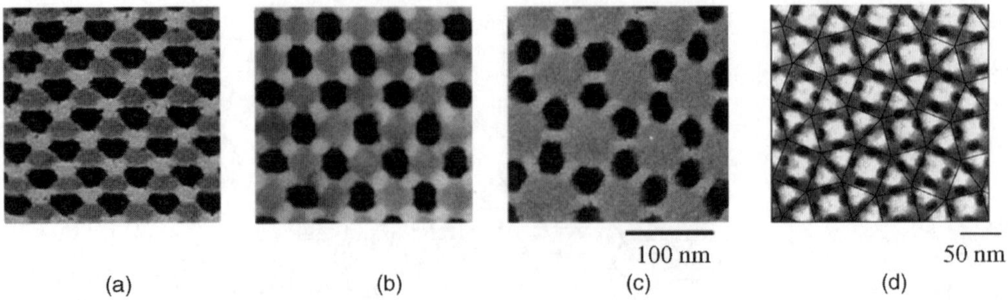

Figure 18.16 TEM images of star PI-PS-P2VP exhibiting tiling patterns. The arm length ratio is: (a) 1:1.08:0.71 (6^3), (b) 1:1.03:1.19 (4.8^2), and (c) 1:1.08:1.92 (4.6.12). (d) 1:2.3:2.0 (3^2.4.3.4). ((a)–(c) Reprinted with permission from A. Takano, S. Wada, S. Sato *et al.*, "Observation of cylinder-based microphase-separated structures from ABC star-shaped terpolymers investigated by electron computerized tomography," *Macromolecules*, **37**, 26, 9941–9946, 2004. © 2004 American Chemical Society. (d) Reprinted with permission from Y. Matsushita, "Creation of hierarchically ordered nanophase structures in block polymers having various competing interactions," *Macromolecules*, **40**, 4, 771–776, 2007. © 2007 American Chemical Society.)

confined with their junctions on the lines. They thought that the addition of a small amount of homopolymer would release the frustration of the μ-ABC molecules by filling the space and reducing the chain stretching of the μ-ABC molecules. In fact, they found the grain size dramatically increased just by adding a small amount of low molecular weight PDMS homopolymer without alternating the tiling pattern (Yamauchi *et al.*, 2005).

Under the circumstance of asymmetric interactions between three components, that is $\chi_{AC} < \chi_{AB} \approx \chi_{BC}$, partial mixing of two components might occur. Sioula *et al.* (1998a) investigated a star PS-PI-PMMA system where $\chi_{SM} < \chi_{PI} \approx \chi_{IM}$, and observed core–shell cylinders of PI cores and PS shells in PMMA matrix by TEM using OsO_4 and RuO_4 to differentiate PI and PS phase, respectively. This result suggests partial mixing of PMMA chains in the PS phase across the PS/PMMA interfaces.

18.8 Concluding Remarks

As reviewed above, many factors that affect the three-phase microdomain morphology became clear for both linear ABCs and μ-ABCs. If the functional polymers are employed instead of model polymers used for the fundamental studies, a variety of functional materials could be produced by utilizing the variety of microdomain morphology. Many ideas to control the morphology of such materials are given in the references.

References

Abetz, V. and Goldacker, T. (2000) Formation of superlattices via blending of block copolymers. *Macromolecular Rapid Communication*, **21**, 16–34.

Abetz, V. and Simon, P.F.W. (2005) Phase behaviour and morphologies of block copolymers. *Advances in Polymer Science*, **189**, 125–212.

Alward, D.B., Kinning, D.J., Thomas, E.L., and Fetters, L.J. (1986) Effect of arm number and arm molecular weight on the solid-state morphology of poly(styrene-isoprene) star block copolymers. *Macromolecules*, **19**, 215–224.

Auschra, C. and Stadler, R. (1993) New ordered morphologies in ABC triblock copolymers. *Macromolecules*, **26**, 2171–2174.

Bailey, T.S., Hardy, C.M., Epps, T.H., and Bates, F.S. (2002) A noncubic triply periodic network morphology in poly (isoprene-*b*-styrene-*b*-ethylene oxide) triblock copolymers. *Macromolecules*, **35**, 7007–7017.

Bailey, T.S., Pham, H.D., and Bates, F.S. (2001) Morphological behavior bridging the symmetric AB and ABC states in the poly(styrene-*b*-isoprene-*b*-ethylene oxide) triblock copolymer system. *Macromolecules*, **34**, 6994–7008.

Bates, F.S. (2005) Network phases in block copolymer melts. *MRS Bulletin*, **30**, 525–532.

Bates, F.S. and Fredrickson, G.L. (1999) Block copolymer – Designer soft materials. *Physics Today*, **52**(2), 32–39.

Beckmann, J., Auschra, C., and Stadler, R. (1994) "Ball at the wall" - A new lamellar multiphase morphology in a polystyrene-block-polybutadiene-block-poly(methyl methacrylate) triblock copolymer. *Macromolecular Rapid Communication*, **15**, 67–72.

Breiner, U., Krappe, U., and Stadler, R. (1996) Evolution of the "knitting pattern" morphology in ABC triblock copolymers. *Macromolecular Rapid Communication*, **17**, 567–575.

Breiner, U., Krappe, U., Abetz, V., and Stadler, R. (1997) Cylindrical morphologies in asymmetric ABC triblock copolymers. *Macromolecular Chemistry and Physics*, **198**, 1051–1083.

Breiner, U., Krappe, U., Jakob, T. *et al.* (1998) Spheres on spheres – a novel spherical multiphase morphology in polystyrene-*block*-polybutadiene-*block*-poly(methyl methacrylate) triblock copolymers. *Polymer Bulletin*, **40**, 219–226.

Chatterjee, J., Jain, S., and Bates, F.S. (2007) Comprehensive phase behavior of poly(isoprene-*b*-styrene-*b*-ethylene oxide) triblock copolymers. *Macromolecules*, **40**, 2882–2896.

Dotera, T. and Hatano, A. (1996) The diagonal bond method: A new lattice polymer model for simulation study of block copolymers. *Journal of Chemical Physics*, **105**, 8413–8427.

Epps, T.H., Cochran, E.W., Hardy, C.M. *et al.* (2004a) Network phases in ABC triblock copolymers. *Macromolecules*, **37**, 7085–7088.

Epps, T.H., Cochran, E.W., Bailey, T.S. *et al.* (2004b) Ordered network phases in linear poly(isoprene-*b*-styrene-*b*-ethylene oxide) triblock copolymers. *Macromolecules*, **37**, 8325–8341.

Gemma, T., Hatano, A., and Dotera, T. (2002) Monte Carlo simulations of the morphology of ABC star polymers using the diagonal bond method. *Macromolecules*, **35**, 3225–3237.

Goldacker, T. and Abetz, V. (1999) Core-shell cylinders and core-shell gyroid morphologies via blending of lamellar ABC triblock and BC diblock copolymers. *Macromolecules*, **32**, 5165–5167.

Hajduk, D.A., Harper, P.E., Gruner, S.M. *et al.* (1994) The gyroid: A new equilibrium morphology in weakly segregated diblock copolymers. *Macromolecules*, **27**, 4063–4075.

Hajduk, D.A., Harper, P.E., Gruner, S.M. *et al.* (1995) A re-evaluation of bicontinuous cubic phases in starblock copolymers. *Macromolecules*, **28**, 2570–2573.

Hasegawa, H., Tanaka, H., Yamasaki, K., and Hashimoto, T. (1987) Bicontinuous microdomain morphology of block copolymers. 1. Tetrapod-network structure of polystyrene-polyisoprene diblock polymers. *Macromolecules*, **20**, 1651–1662.

Hayashida, K., Kawashima, W., Takano, A. *et al.* (2006) Archimedean tiling patterns of ABC star-shaped terpolymers studied by microbeam small-angle X-ray scattering. *Macromolecules*, **39**, 4869–4872.

Hayashida, K., Dotera, T., Takano, A. *et al.* (2007a) Polymeric quasicrystal: Mesoscopic quasicrystalline tiling in ABC star polymers. *Physical Review Letters*, **98**, 195502.

Hayashida, K., Saito, N., Arai, S. *et al.* (2007b) Hierarchical morphologies formed by ABC star-shaped terpolymers. *Macromolecules*, **40**, 3695–3699.

Helfand, E. and Sapse, A.M. (1975) Theory of unsymmetric polymer-polymer interfaces. *Journal of Chemical Physics*, **62**, 1327–1331.

Hückstädt, H., Goldacker, T., Göpfert, A., and Abetz, V. (2000a) Core-shell double gyroid morphologies in ABC triblock copolymers with different chain topologies. *Macromolecules*, **33**, 3757–3761.

Hückstädt, H., Göpfert, A., and Abetz, V. (2000b) Synthesis and morphology of ABC heteroarm star terpolymers of polystyrene, polybutadiene and poly(2-vinylpyridine). *Macromolecular Chemistry and Physics*, **201**, 296–307.

Jinnai, H., Kaneko, T., Matsunaga, K. *et al.* (2009) A double helical structure formed from an amorphous, achiral ABC triblock terpolymer. *Soft Matter*, **5**, 2042–2046.

Judas, J., Kirkensgaard, K., and Hyde, S. (2009) Beyond amphiphiles: coarse-grained simulations of star-polyphile liquid crystalline assemblies. *Physical Chemistry Chemical Physics*, **11**, 2016–2022.

Kaneko, T., Suda, K., Satoh, K. *et al.* (2006) A "ladder" morphology in an ABC triblock copolymer. *Macromolecular Symposia*, **242**, 80–86.

Khandpur, A.K., Förster, S., Bates, F.S. *et al.* (1995) Polyisoprene-polystyrene diblock copolymer phase diagram near the order-disorder transition. *Macromolecules*, **28**, 8796–8806.

Kloninger, C. and Rehahn, M. (2007) Styrene-ferrocenyldimethylsilane-methyl methacrylate triblock copolymers: synthesis and phase morphology. *Macromolecular Chemistry and Physics*, **208**, 833–840.

Ko, M.J., Kim, S.H., and Jo, W.H. (2001) A molecular dynamics simulation on the self-assembly of ABC triblock copolymers, 1. Effects of block composition in symmetric triblock copolymers. *Macromolecular Theory and Simulations*, **10**, 381–388.

Krappe, U., Stadler, R., and Voigt-Martin, I. (1995) Chiral assembly in amorphous ABC triblock copolymers. Formation of a helical morphology in polystyrene-*block*-polybutadiene-*block*-poly(methyl methacrylate) block copolymers. *Macromolecules*, **28**, 4558–4561 (correction in *Macromolecules*, 28, 7583).

Leibler, L. (1980) Theory of microphase separation in block copolymers. *Macromolecules*, **13**, 1602–1617.

Ludwigs, S., Böker, A., Abetz, V. *et al.* (2003) Phase behavior of linear polystyrene-*block*-poly(2-vinylpyridine)-*block*-poly(*tert*-butyl methacrylate) triblock terpolymers. *Polymer*, **44**, 6815–6823.

Matsen, M.W. (1998) Gyroid versus double-diamond in ABC triblock copolymer melts. *Journal of Chemical Physics*, **108**, 785–796.

Matsen, M.W. and Bates, F.S. (1996) Unifying weak- and strong-segregation block copolymer theories. *Macromolecules*, **29**, 1091–1098.

Matsushita, Y. (2007) Creation of hierarchically ordered nanophase structures in block polymers having various competing interactions. *Macromolecules*, **40**, 771–776.

Matsushuta, Y. (2008) Precise molecular design of complex polymers and morphology control of their hierarchical multiphase structures. *Polymer Journal*, **40**, 177–183.

Matsushita, Y., Choshi, H., Fujimoto, T., and Nagasawa, M. (1980) Preparation and morphological properties of a triblock copolymer of the ABC type. *Macromolecules*, **13**, 1053–1058.

Matsushita, Y., Suzuki, J., and Seki, M. (1998) Surfaces of tricontinuous structure formed by an ABC triblock copolymer in bulk. *Physica B*, **248**, 238–242.

Matsushita, Y., Takano, A., Hayashida, K. *et al.* (2009) Hierarchical nanophase-separated structures created by precisely designed polymers with complexity. *Polymer*, **50**, 2191–2203.

Matsushita, Y., Tamura, M., and Noda, I. (1994) Tricontinuous double-diamond structure formed by a styrene-isoprene-2-vinylpyridine triblock copolymer. *Macromolecules*, **27**, 3680–3682.

Meuler, A.J., Hillmyer, M.A., and Bates, F.S. (2009) Ordered network mesostructures in block polymer materials. *Macromolecules*, **42**, 7221–7250.

Mogi, Y., Kotsuji, H., Kaneko, Y. *et al.* (1992a) Preparation and morphology of triblock copolymers of the ABC type. *Macromolecules*, **25**, 5408–5411.

Mogi, Y., Mori, K., Matsushita, Y., and Noda, I. (1992b) Tricontinuous morphology of triblock copolymers of the ABC type. *Macromolecules*, **25**, 5412–5415.

Mogi, Y., Nomura, M., Kotsuji, H. *et al.* (1994) Superlattice structures in morphologies of the ABC triblock copolymers. *Macromolecules*, **27**, 6755–6760.

Nakazawa, H. and Ohta, T. (1993) Microphase separation of ABC-type triblock copolymers. *Macromolecules*, **26**, 5503–5511.

Ott, H., Abetz, V., and Altstädt, V. (2001) Morphological studies of poly(styrene)-*block*-poly(ethylene-co-butylene)-*block*-poly(methyl methacrylate) in the composition region of the "knitting pattern" morphology. *Macromolecules*, **34**, 2121–2128.

Phan, S. and Fredrickson, G.H. (1998) Morphology of symmetric ABC triblock copolymers in the strong segregation limit. *Macromolecules*, **31**, 59–63.

Riess, G., Schlienger, M., and Marti, S. (1980) New morphologies in rubber-modified polymers. *Journal of Macromolecular Science – Physics*, **B17**, 355–374.

Schacher, F., Yuan, J., Schoberth, H.G., and Müller, A.H.E. (2010) Synthesis, characterization, and bulk crosslinking of polybutadiene-*block*-poly(2-vinyl pyridine)-*block*-poly(*tert*-butyl methacrylate) block terpolymers. *Polymer*, **51**, 2021–2032.

Seki, M., Suzuki, J., and Matsushita, Y. (2000) Small-angle X-ray scattering analysis of the periodic tricontinuous network structure of symmetric ABC triblock copolymers. *Journal of Applied Crystallography*, **33**, 285–290.

Shefelbine, T.A., Vigild, M.E., Matsen, M.W. *et al.* (1999) Core-shell gyroid morphology in a poly(isoprene-*block*-styrene-*block*-dimethylsiloxane) triblock copolymer. *Journal of the American Chemical Society*, **121**, 8457–8465.

Shibayama, M., Hasegawa, H., Hashimoto, T., and Kawai, H. (1982) Microdomain Structure of an ABC-type triblock polymer of polystyrene-poly[(4-vinylbenzyl)dimethylamine]-polyisoprene cast from solutions. *Macromolecules*, **15**, 274–280.

Sioula, S., Hadjichristidis, N., and Thomas, E.L. (1998a) Novel 2-dimensionally periodic non-constant mean curvature morphologies of 3-miktoarm star terpolymers of styrene, isoprene, and methyl methacrylate. *Macromolecules*, **31**, 5272–5277.

Sioula, S., Hadjichristidis, N., and Thomas, E.L. (1998b) Direct evidence for confinement of junctions to lines in an 3-miktoarm star terpolymer microdomain structure. *Macromolecules*, **31**, 8429–8432.

Stadler, R., Auschra, C., Beckmann, J. *et al.* (1995) Morphology and thermodynamics of symmetric poly(A-*block*-B-*block*-C) triblock copolymers. *Macromolecules*, **28**, 3080–3091.

Sugiyama, M., Shefelbine, T.A., Vigild, M.E., and Bates, F.S. (2001) Phase behavior of an ABC triblock copolymer blended with A and C homopolymers. *Journal of Physical Chemistry B*, **105**, 12448–12460.

Sun, M., Wang, P., Qiu, F. *et al.* (2008) Morphology and phase diagram of ABC linear triblock copolymers: Parallel real-space self-consistent-field-theory simulation. *Physical Review E*, **77**, 016701.

Suzuki, J., Nakane, K., Takano, A., and Matsushita, Y. (2004) Self-assembly template during morphological transition of a linear ABC triblock copolymer from lamellar to Gyroid structure. *Polymer*, **45**, 8989–8997.

Takano, A., Kawashima, W., Noro, A. *et al.* (2005) A mesoscopic Archimedean tiling having a new complexity in an ABC star polymer. *Journal of Polymer Science: Part B: Polymer Physics*, **43**, 2427–2432.

Takano, A., Kawashima, W., Wada, S. *et al.* (2007) Composition dependence of nanophase-separated structures formed by star-shaped terpolymers of the $A_{1.0}B_{1.0}C_X$ type. *Journal of Polymer Science: Part B: Polymer Physics*, **45**, 2277–2283.

Takano, A., Wada, S., Sato, S. *et al.* (2004) Observation of cylinder-based microphase-separated structures from ABC star-shaped terpolymers investigated by electron computerized tomography. *Macromolecules*, **37**, 9941–9946.

Takenaka, M., Wakada, T., Akasaka, S. *et al.* (2007) Orthorhombic *Fddd* Network in Diblock Copolymer Melts. *Macromolecules*, **40**, 4399–4402.

Tang, P., Qiu, F., Zhang, H., and Yang, Y. (2004) Morphology and phase diagram of complex block copolymers: ABC linear triblock copolymers. *Physical Review E*, **69**, 031803.

Tyler, C.A., Qin, J., Bates, F.S., and Morse, D.C. (2007) SCFT study of nonfrustrated ABC triblock copolymer melts. *Macromolecules*, **40**, 4654–4668.

Ueda, K., Dotera, T., and Gemma, T. (2007) Photonic band structure calculations of two-dimensional Archimedean tiling patterns. *Physical Review B*, **75**, 195122.

Yamauchi, K., Takahashi, K., Hasegawa, H. *et al.* (2003) Microdomain morphology in an ABC 3-miktoarm star terpolymer: A study by energy-filtering TEM and 3D electron tomography. *Macromolecules*, **36**, 6962–6966.

Yamauchi, K., Akasaka, S., Hasegawa, H. *et al.* (2005) Blends of a 3-miktoarm star terpolymer (3 μ-ISD) of isoprene (I), styrene (S), and dimethylsiloxane (D) with PS and PDMS. Effect on microdomain morphology and grain size. *Macromolecules*, **38**, 8022–8027.

Zheng, W. and Wang, Z.-G. (1995) Morphology of ABC triblock copolymers. *Macromolecules*, **28**, 7215–7223.

Zioga, A., Sioula, S., and Hadjichristidis, N. (2000) Synthesis and morphology of model 3-miktoarm star terpolymers of styrene, isoprene and 2-vinyl pyridine. *Macromolecular Symposia*, **157**, 239–249.

19

Morphologies and Photophysical Properties of Conjugated Rod–Coil Block Copolymers

Chi-Ching Kuo[1], Cheng-Liang Liu[2] and Wen-Chang Chen[1,2]
[1] *Institute of Polymer Science and Engineering, National Taiwan University, Taipei, Taiwan 10617*
[2] *Department of Chemical Engineering, National Taiwan University, Taipei, Taiwan 10617*

19.1 Introduction

Conjugated rod–coil block copolymers have been extensively studied recently since they provide a powerful route towards supramolecular objects with novel architectures, functions and physical properties. It is of great interest and also a challenge to explore new materials with self-assembled morphologies. This can be achieved for instance by systematically designed block copolymers containing both conjugated rod and coil blocks with linear or nonlinear architectures. Linear block copolymers containing two or three sequences of monomers can be viewed as diblock or triblock copolymers. A variety of triblock copolymers architectures with rod and coil building blocks consist of symmetric or asymmetric arrangement, such as coil–rod–coil, rod–coil–rod, and rod–coil–coil, star-branched, graft and other complex architectures, as summarized as Figure 19.1.

Several classes of π-conjugated rod–coil block copolymers were synthesized and reported in the literature, and the rod segments include fluorene (Surin *et al.*, 2004; Tung *et al.*, 2006; Yang *et al.*, 2006; Lin *et al.*, 2008b; Lin *et al.*, 2008a; Kuo *et al.*, 2008; Li *et al.*, 2009; Lin *et al.*, 2009; Tung and Chen, 2009), phenylene (Francois *et al.*, 1995; Stupp *et al.*, 1997; Zubarev *et al.*, 1999; Leclere *et al.*, 1998; Pitois and Francois, 1999; Tew *et al.*, 2000; Mori *et al.*, 2005; Sary *et al.*, 2007; Barrau *et al.*, 2008) and quinoline (Jenekhe and Chen, 1998, 1999; Chen and Jenekhe, 2000), and so on. Chemical structures of commonly used rod and coil

Complex Macromolecular Architectures: Synthesis, Characterization, and Self-Assembly, First Edition.
Edited by Nikos Hadjichristidis, Akira Hirao, Yasuyuki Tezuka and Filip Du Prez.

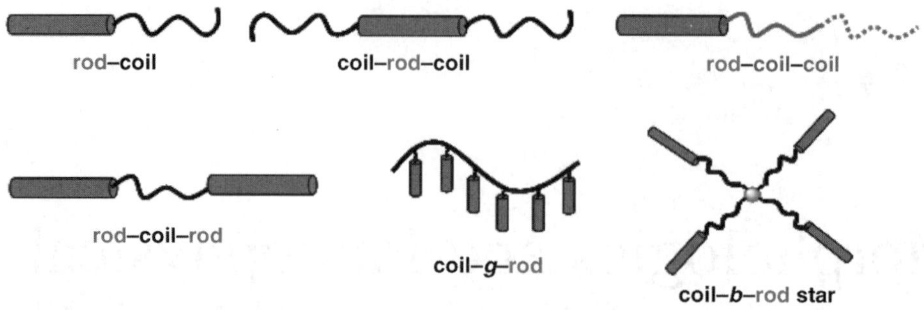

Figure 19.1 Different architectures of block copolymers containing rod and coil blocks.

blocks in the conjugated rod–coil block copolymers are shown in Figure 19.2. Aggregation and microphase separation of conjugated rod–coil block copolymers have yielded a number of nanoscale morphologies, such as lamellar, spherical, cylindrical, vesicular and microporous structures. The π–π interaction between the conjugated rods provides additional structural control factor and functionality, which differs distinctly from those of conventional coil–coil block copolymers. Furthermore, the self-assembled morphology of conjugated rod–coil block copolymers may lead to additional electronic processes such as exciton confinement and interfacial effects. Also, the combination of a stimuli-responsive coil segment with the tunable photophysical properties of the conjugated rod could produce novel multifunctional sensory materials. It is believed that conjugated rod–coil block copolymers could spark the future evolution of nanostructured polymers for multifunctional device applications.

The fabrication of the block copolymer morphologies with controlled size, shape, and functionality is a key challenge in nanotechnology. The morphology of conjugated rod–coil block copolymers has been reviewed in several recent articles (Lee *et al.*, 2001; Hadjichristidis

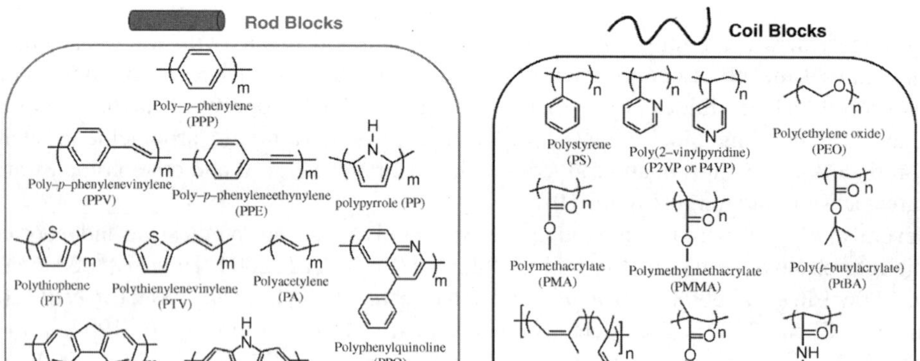

Figure 19.2 General structures of rod and coil blocks on the rod–coil block copolymers.

et al., 2003; Hoeben *et al.*, 2005; Liang *et al.*, 2007; Olsen and Segalman, 2008). In the typical coil–coil block copolymers, the microphase separated morphology is generally governed by volume fraction of each blocks (f), and the degree of incompatibility between each blocks (χN). Note that χ is the Flory–Huggins interaction parameter between two dissimilar blocks and N is the total degree of polymerization. The replacement of one block by the conjugated rod adds new structurally control parameters, including the π–π interaction and crystalline or liquid-crystalline characteristics of the conjugated rod. In addition, the stiff asymmetry between the rod and coil blocks leads an increase in the Flory–Huggins χ parameter and allows the phase separation to occur at a lower molecular weight than that of the coil–coil block copolymers. It has been suggested that two other structural factors are required to be considered for the morphology of conjugated rod coil block copolymers, the Maier–Saupe interaction strength (μN) and the ratio of coil radius of gyration and the rod length (v) (Segalman *et al.*, 2009). The controlled nanosized structure could tune the electronic, optical, and optoelectronic properties of the conjugated rod–coil block copolymers.

In this chapter, we emphasize the relationships between conjugated rod–coil block polymer morphologies and photophysical properties in different phases including solution micelles, thin films or bulk samples, electrospun nanofibers and polymer brushes. The effects of the selective solvent, rod:coil ratio, polymer architecture, concentration and geometrical confinement on the morphology and optoelectronic properties are discussed. Moreover, multifunctional sensory materials based on the semiconducting conjugated rod and stimuli-responsive coil block copolymers are also highlighted. The effects of the aggregated shape and geometrical confinements (nanofibers or brushes) on the photophysical properties are also presented and compared with those of the corresponding thin films. It is hoped that the chapter provides a comprehensive understanding on the relationship between morphology and optoelectronic properties of conjugated rod–coil block copolymers.

19.2 Solution Micelles

The solution micellar structures of block copolymers are generally driven by the following parameters of copolymer composition, selective solvent concentration, presence of additives, nature of common solvent, pH, and temperature, and so on. Segalman and coworker presented a comprehensive review (Olsen and Segalman, 2008) on the solution morphology of conjugated rod–coil block copolymers, which are classified into two categories of rod-selective solvent and coil-selective solvent systems (Figure 19.3). In the case of the rod-selective solvent, the rods pack on the shell of the assemble morphologies vesicles with different coil fractions, such as spherical and worm-like structures. It depends on the force balance of liquid-crystalline penalty between the interfacial curvature and the interfacial area per coil block. In the coil-selective solvent system, the shapes may be spherical, worm-like, elliptical, or punk-like due to the different packing geometries of the rod block with varying the rod fraction.

19.2.1 Solvent Effect

19.2.1.1 Rod-Selective Solvent System

In the rod-selective solvent system, the aggregated shape of a conjugated rod affecting the photophysical properties was well addressed by Chen and Jenekhe (Jenekhe and

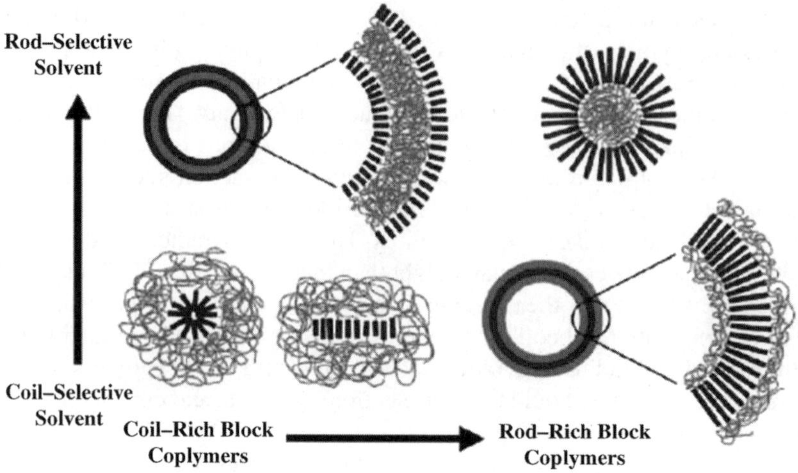

Figure 19.3 General trends on the solution micellar structures of rod–coil block copolymers. (Reprinted from *Materials Science and Engineering: R: Reports*, **62**, B.D. Olsen and R.A. Segalman, "Self-assembly of rod–coil block copolymers," 37–66, 2008, with permission from Elsevier.)

Chen, 1998, 1999). The diverse aggregated morphologies of poly(phenylquinoline)-*b*-polystyrene (PPQ-*b*-PS) were observed at different compositions of mixed trifluoroacetic acid (TFA)/dichloromethane (DCM) solvents, including spheres, lamellae, cylinders, and vesicles. Note that TFA was a good solvent for the PPQ block and DCM acted as a common solvent for both blocks. The spherical and cylindrical aggregates were further processed to form large hollow cavities with a PS inner shell and a PPQ outer shell. In addition, all of the different PPQ aggregates were observed with highly ordering crystalline features. Such different aggregation shapes led to distinct luminescence emission spectra and time-resolved photoluminescence decay dynamics. However, only the vesicular morphology was obtained in the triblock PPQ-*b*-PS-*b*-PPQ copolymer in a similar mixed TFA/DCM solvent (Chen and Jenekhe, 2000), which was quite different from the multiple morphologies observed in the diblock PPQ-*b*-PS. It implied that the vesicular morphology is thermodynamically more stable aggregate than the flat disk, microtubules, or other morphologies in such rod-selective solvent systems. Also, it demonstrated the importance of polymer architecture and the π–π interaction of the conjugated rod on the aggregated morphology.

Another reported case of the rod-selective solvent system was polyaniline-*block*-poly (ethylene glycol)-*block*-polyaniline (PANI-*b*-PEG-*b*-PANI) triblock polymers in the mixed solvent of THF/diethyl ether (Yan and Tao, 2008). In the THF solution, the micelles with a middle layer was formed by the π–π stacking of the PANI segments, while the PEG segments aggregated anisotropicly in two dimensions to form the up and down layers of the sandwich structure. By adding the rod-selective solvent of diethyl ether, the micelles changed into the spring-like helical superstructures from the short PANI block but the fibrous helical superstructures from the longer PANI block. Such helical superstructures probably resulted from the twist of the belt-like structure grown from the π–π stacking of PANI.

19.2.1.2 Coil-Selective Solvent System

In the coil-selective solvent system, the packing of the conjugated rod blocks in the micelles core tends to form spherical or cylindrical morphologies for the shorter rod block (Olsen and Segalman, 2008). But it changes to the vesicle for the longer rod block for reducing the rod bending energy. Selective structural control factors are presented below, including rod:coil ratio, solution concentration and polymer architecture.

Effect of Rod:Coil Ratio

Our group reported the morphology and photophysical properties of poly[2,7-(9,9-dihexylfluorene)]-*b*-poly(acrylic acid) (PF-*b*-PAA) in a dilute solution of dichloromethane/methanol (Figure 19.4) (Tung *et al.*, 2006). Note that dichloromethane and methanol were common and poor solvents for the PF block, respectively. Tape-like lamellar morphology was observed at a short coil length of PF-*b*-PAA because the short coil length was not enough to drive the significant π–π interaction on the PF rod. By increasing the coil length, various morphologies of the large compound micelle, sphere, or vesicle were observed with different methanol contents. It was due to the enhancement on the PAA swelling with methanol and the interfacial tension between the PF core and PAA corona. By further enhancing the long coil block, an inverted morphology of sphere or rod with the PF corona and PAA core was shown first but the core/corona was reversed at higher methanol content due to the enhanced solubility of PAA. The morphological transformation led to a significant blue-shifting on the optical absorption or fluorescence characteristics due to the possible H-aggregation formation. The significance on the π–π interaction on the polymer micellar structure in the coil-selective solvent was also demonstrated by our recent work of poly[2,7-(9,9-dihexylfluorene)]-*block*-poly[3-(trimethoxysilyl)propyl methacrylate] (PF-*b*-PTMSPMA) (Tung and Chen, 2009).

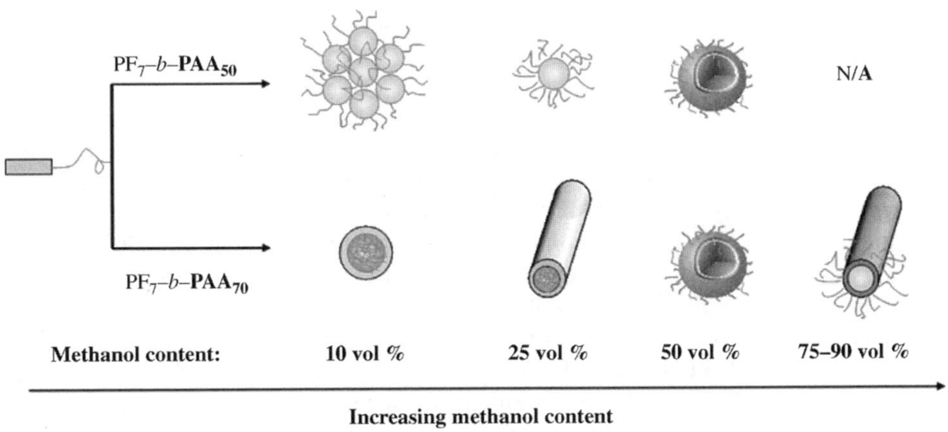

Figure 19.4 Morphological transformation on the PF-b-PAA diblock copolymers by varying the coil fraction and solvent selectivity. (Reproduced with permission from Y.-C. Tung, W.-C. Wu and W.-C. Chen, "Morphological transformation and photophysical properties of rod-coil poly[2,7-(9,9-dihexylfluorene)]-block-poly(acrylic acid) in solution," *Macromolecular Rapid Communications*, **27**, 1838–1844. © Wiley-VCH Verlag GmbH & Co. KGaA.)

Oligophenylenevinylenes (OPV)-coil diblock copolymers were shown to possess remarkable self-assembling abilities with long cylindrical micelles in dilute THF/D_2O solutions (Li *et al.*, 1999; Wang *et al.*, 2000; Wang *et al.*, 2004). The high molecular weight poly(ethylene glycol) (PEG) block drove the self-assemblied cylindrical micelles of a core diameter about 8–10 nm, which was composed of an OPV block and surrounded by a PEG corona (Wang *et al.*, 2000). A further investigation of the OPV-*block*-poly(propylene glycol) (PPG) block copolymers with different concentrations of coil-selective solvent D_2O were reported (Wang *et al.*, 2004). Bundles of fibers with a strong yellowish-green fluorescence are clearly observed at their critical water content. The packing structures for the block copolymers consisted of a hydrophobic OPV inner core and a hydrophilic PEG or PPG outer shell.

Tanaka and coworkers also studied the OPV-*block*-poly(ethylene oxide) (PEO) aggregation behavior of the block copolymers in a coil-selective solution (THF/water) by the photophysical properties of the OPV block (Mori *et al.*, 2005). With increasing water concentration, the absorption maximum was blue-shifted; the single-chain emission of molecularly dissolved OPV decreased but emission from the aggregated OPV increased. This indicated that the OPV blocks formed H-aggregates in which the OPV blocks aligned in a parallel orientation with one another. The TEM observation revealed that the block copolymers with a lower PEO weight fraction formed long cylindrical aggregates, whereas the block copolymer with a higher PEO exhibited distorted spherical aggregates. It was believed that the block copolymer preferred the spherical aggregate for the longest PEO segment in order to decrease the stretching of PEO as well as the entropy loss.

Another synthesis of coil–rod–coil block copolymers, constituted of a central conjugated segment based on 3,4-ethylenedioxythiophene and dihexylfluorene units (BEDOT-HF) with flexible PEO chains, was reported (de Cuendias *et al.*, 2008). As expected, the self-assembled morphologies depended on the volume fraction of each block. For the case of the short conjugated rod block, the spherical and cylinders micelles (J-aggregates) were observed depending on the solution concentration, whereas the system having longer conjugated block tended to generate vesicular organizations (H-aggregates).

Effect of Chemical Architecture

The tetra(p-phenylene)-*block*-poly(ethylene) (PPP-*b*-PE) rod–coil block copolymers exhibited significant structural changes from the vesicular micelles of short coil to the cylindrical micelles for a twin-rod segment, and then to sphere micelles for a long coil length. It clearly demonstrated the ability to regulate the carbohydrate-coated supramolecular structure by a systematic variation in the length of coil and the number of rods (Kim *et al.*, 2005). Recently, our group showed the polymer architecture effect on the solution morphology by comparing diblock poly[2,7-(9,9-dihexylfluorene)]-*block*-poly(2-vinylpyridine) (PF-*b*-P2VP) and triblock poly(2-vinylpyridine)-*block*-poly[2,7-(9,9-dihexylfluorene)]-*block*-poly(2-vinylpyridine) (P2VP-*b*-PF-*b*-P2VP) in the THF/methanol mixtures (Lin *et al.*, 2008a). In such a mixed solvent, while the diblock copolymers maintained a spherical micellar aggregates, triblocks were found to aggregate in elongated cylinders. The symmetric P2VP-*b*-PF-*b*-P2VP polymer chains could stack together more favorably into cylindrical arrays and provide preferential conditions for π–π interchain stacking compared with the spherical aggregates of PF-*b*-P2VP (Figure 19.5). H-aggregates were observed in the diblock PF-*b*-P2VP but neither with H-type nor J-type aggregation for the P2VP-*b*-PF-*b*-P2VP based on their photophysical properties. The importance of polymer architecture for rod–coil and coil–rod–coil block copolymers in

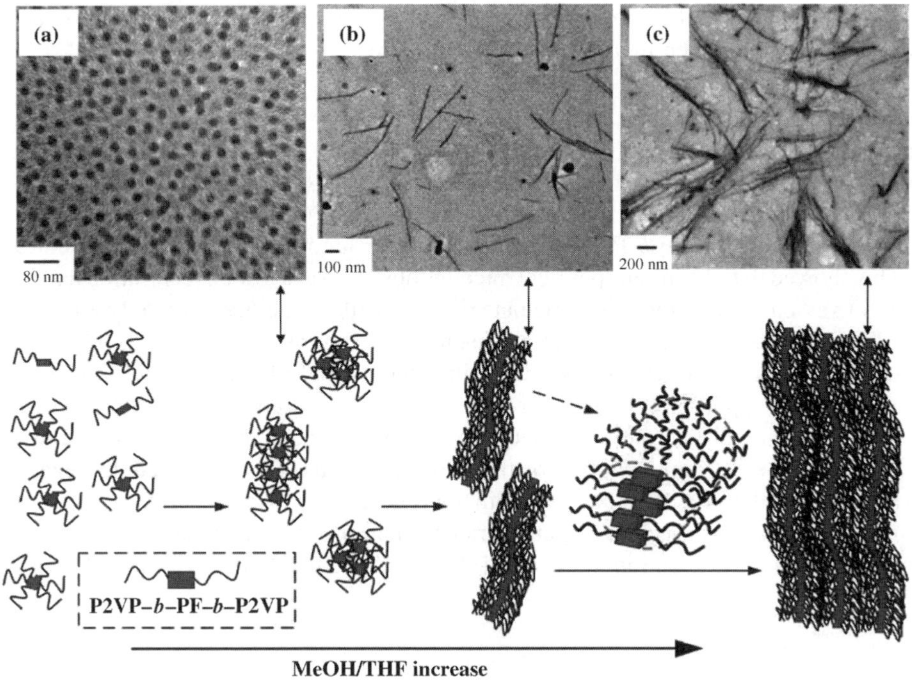

Figure 19.5 Scheme on the formation of long cylindrical and bundle of cylinders structures from the spherical micelle building blocks of P2VP-b-PF-b-P2VP. TEM images on morphological changes of P2VP-b-PF-b-P2VP in dilute solution by varying methanol (MeOH)/THF compositions (wt/wt): (a) 10/90, (b) 70/30, and (c) 90/10. (Reprinted with permission from C.-H. Lin, Y.-C. Tung, J. Ruokolainen *et al.*, "Poly[2,7-(9,9-dihexylfluorene)]-block-poly(2-vinylpyridine) rod–coil and coil–rod–coil block copolymers: Synthesis, morphology and photophysical properties in methanol/THF mixed solvents," *Macromolecules*, **41**, 8759–8769, 2008. © 2008 American Chemical Society.)

dilute solution revealed the effect of the polymer aggregated structures on the photophysical properties. A further theoretical and experiment is inevitable for analyzing the packing of conjugated rod segment.

Wu *et al.* investigated the morphologies and optical properties of OPE (oligo(p-phenyleneethynylene))-based triblock copolymers with different sequences, PS-*b*-OPE-*b*-PS (coil–rod–coil) and OPE-*b*-PS-*b*-OPE (rod–coil–rod), in different solvent mixtures of toluene and hexane (Huo *et al.*, 2007). For the PS-*b*-OPE-*b*-PS, the two PS coil end blocks could wrap the OPE middle block and made the conjugated OPE backbone stiffer and planar. However, the intrachain wrapping prevented a parallel alignment of OPE and suppressed the π–π interaction, resulting in a red-shift on its absorption and photoluminescence (PL) spectra. For the OPE-*b*-PS-*b*-OPE, the two conjugated OPE end blocks could undergo a parallel intrachain association in the dilute solution to form the π–π stacking and stabilized by the PS block. The addition of hexane could also induce interchain parallel association of OPE to form a disk-like core stabilized by a soluble PS shell. The J-type aggregation shifted the spectra of PS-*b*-OPE-*b*-PS to red, while the H-type aggregation shifted the spectra of OPE-*b*-PS-*b*-OPE to blue. As a result, they revealed that the packaging of the conjugated rod was strongly affected by the copolymer chemical architecture even in a given solvent mixture.

Concentration Effect

As described in the literature (de Cuendias *et al.*, 2008), the photophysical properties of amphiphilic coil–rod–coil block copolymer, constituted of a central conjugated segment based on 3,4-ethylenedioxythiophene and dihexylfluorene units (BEDOT-HF) with flexible PEO chains (PEO-*b*-BEDOT-HF-*b*-PEO), could be tuned through the different polymer concentration. This showed that the initial spherical micelles observed at a low concentration had a strong tendency to assemble into these intermediate rice grain-like structures, probably growing up to worm-like micelles for more concentrated solutions (Figure 19.6). Li *et al.* observed that various morphologies of OPE-*b*-PS-*b*-OPE in toluene could be correlated with the initial concentration of the polymer (Li and Wang, 2005). With the polymer concentration increasing, the relative intensity of PL confirmed that the excimers produced by the intramolecular and/or intermolecular π-electronic coupling of the OPE aggregates. The observed aggregate characteristics resulted from the adhesion, collision and fusion of worm-like micelles at high polymer concentrations. Such aggregation transformations were attributed to the cooperative effect of microphase separation in the coil-selective solvent and self-assembly driven by the π–π stacking.

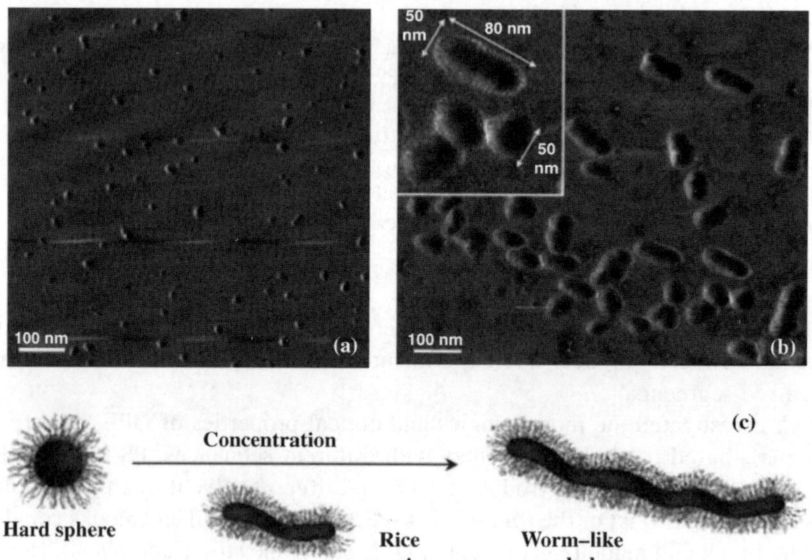

Figure 19.6 AFM topography images obtained from a drop of copolymer solutions in water deposed on a mica surface: (a) low concentration and (b) high concentration. (c) Schematic representation of the concentration effect on the morphology. (Reproduced with permission from A. de Cuendias, E. Ibarboure, S. Lecommandoux *et al.*, "Synthesis and self-assembly in water of coil-rod-coil amphiphilic block copolymers with central π-conjugated sequence," *Journal of Polymer Science Part A: Polymer Chemistry*, 2008, **46**, 4602–4616. © Wiley-VCH Verlag GmbH & Co. KGaA.)

19.2.2 Stimulus-Response Thermal and pH Effect

As discussed above, the morphology and photophysical properties of conjugated rod–coil block copolymers could be efficiently tuned through the chemical structures and solvent polarity. Such characteristics provide potential applications for environmentally stimuli sensory devices. Note that stimuli-responsive polymers are defined as polymers that undergo relatively large and abrupt, physical (such as temperature, solvents, and light) or chemical changes (such as reactants, pH, ions in solution, and chemical recognition) in response to small external changes in the environmental conditions. Here, we summarize the progress of thermo- or pH-responsive conjugated rod–coil block copolymers reported in the literature (Lu *et al.*, 2003; Balamurugan *et al.*, 2005; Yang *et al.*, 2006; Fang *et al.*, 2007; Xiao *et al.*, 2007; Lin *et al.*, 2008b; Ma *et al.*, 2008; Lin *et al.*, 2009).

Poly(N-isopropylacrylamide) (PNIPAAM) was a widely studied thermoresponsive polymers, which showed hydrophilic extended structure below LCST (32 °C) but became a compact structure above LCST. McCarley *et al.* (Balamurugan *et al.*, 2005) reported that a highly water soluble, thermally responsive poly(thiophene-*graft*-NIPAAm) (PT-*g*-PNIPAAm) underwent a reversible phase transition from a coiled structure to a collapsed globule on increasing temperature. The temperature-dependent absorption spectra of the PT-*g*-PNIPAAm in water suggested the changes in the PT effective conjugation length, which resulted from steric interactions between the collapsed PNIPAAm moieties and the PT backbone.

PF-based block copolymers consisted of a terfluorene rod segment and thermoresponsive PNIPAAm coil block (Xiao *et al.*, 2007) exhibited a reversible phase transition from a coiled structure to a collapsed globular structure at the LCST. Additionally, the energy transfer efficiency of the block copolymer doped with tetraphenylporphine tetrasulfonic acid decreased as a result of the globule-to-coil transition from PNIPAAm segments above the LCST. The result indicated that these copolymers have potential applications to be used as responsive fluorescent probes in facile detection of dye-labeled biopolymers.

Recently, we reported the synthesis and properties of new thermoresponsive) (PF-*block*-PNIPAAm-*block*-poly(N-hydroxyethylacrylamide (PHEAA) rod–coil–coil triblock copolymers) (Lin *et al.*, 2009). The LCSTs of the copolymers increased as the hydrophilic PHEAA block ratios was enhanced, since the longer PHEAA segment caused relatively stronger tendency to stretch the copolymer chains. However, the amphiphilic characteristic and strong π-electron interaction of the fluorene moiety led to the micelle morphologies of cylinders, bundles, and hollow cylinders in water, which were thermoreversible based on its LCST (Figure 19.7). The PL characteristics behaved as an on/off fluorescence indicator of temperature, resulting from the effect of the micellar aggregated morphologies on the absorption of the incident light by the inner PF block. The PL intensity of the block copolymers in water was thermoreversible based on its LCST and suggested their potential application as an on/off fluorescence indicator of temperature.

Our group reported a novel multifunctional copolymers consisting of the PNIPAAm segment and a fluorescent difluorene-containing acrylic polymer segment with pH responsiveness and/or DNA-sensing ability (Yang *et al.*, 2006). The protonation of the amino moiety at a low pH improved the solubility and prevented aggregation for fluorescence quenching. Although the aggregation of the fluorene units was significant at room temperature under the basic conditions, it was resolved at a temperature above the LCST. Moreover, after the amino groups were quaternized, the obtained polymer could be used as a biosensor because the

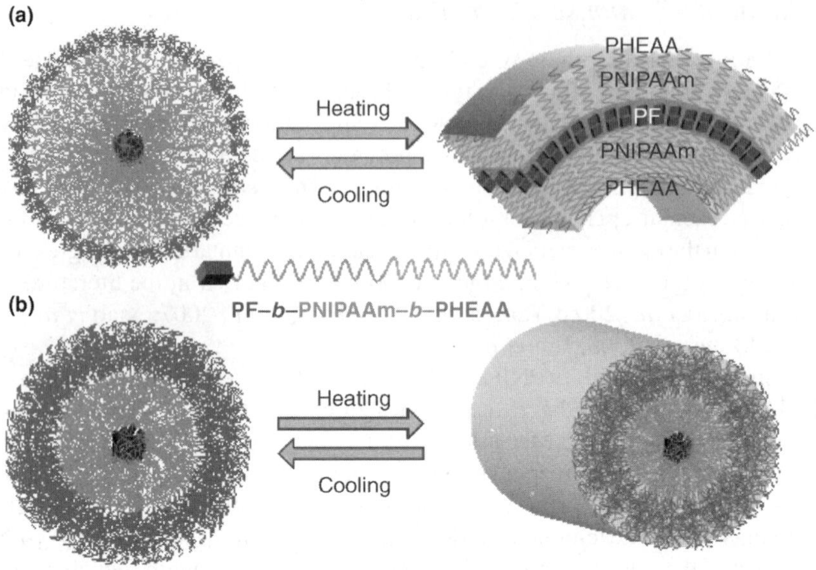

Figure 19.7 Schematic illustration of the morphological variation for (a) PF_7-*b*-$PNIPAAm_{120}$-*b*-$PHEAA_{30}$ and (b) PF_7-*b*-$PNIPAAm_{75}$-*b*-$PHEAA_{75}$ in water upon changing temperature. (Reproduced from S.-T. Lin, K. Fuchise, Y. Chen *et al.*, "Synthesis, thermomorphic characteristics, and fluorescent properties of poly[2,7-(9,9-dihexylfluorene)]-block-poly(N-isopropylacrylamide)-block-poly(N-hydroxyethylacrylamide) rod–coil–coil triblock copolymers," *Soft Matter*, **5**, 3761–3770, 2009, by permission of the Royal Society of Chemistry.)

fluorescence intensity was quenched with the addition of DNA. This demonstrated that multifunctional materials with pH- and temperature-sensing characteristics and biological molecules could be realized by the incorporation of a functional fluorene-containing moiety with PNIPAAm.

Furthermore, our group has also successfully prepared new PF-*block*-poly[2-(dimethylamino)ethylmethacrylate] (PDMAEMA) rod–coil block copolymers showing a significant variation in their surface structures and photophysical properties with respect to solvent composition (water-THF), temperature, and pH. The micellar aggregate of PF-*b*-PDMAEMA in water showed a reversible change of surface structure from cylinder bundles to spheres over a heating–cooling cycle. The LCST of the PF-*b*-PDMAEMA decreased with increasing pH, depending on the protonation to the PDMAEMA block. The PL intensity of PF-*b*-PDMAEMA in water was thermoreversible based on its LCST, which behaved as an on/off fluorescence indicator of temperature or pH (Figure 19.8) This study suggested that the PF-*b*-PDMAEMA copolymers had potential applications as multifunctional sensory materials toward solvent, temperature, and pH (Lin *et al.*, 2008b).

19.3 Thin Films or Bulk Samples

The self-assemble morphologies with nanoscale architectures in thin film of conjugated block copolymer are of current interest for device application, such as light-emitting diodes,

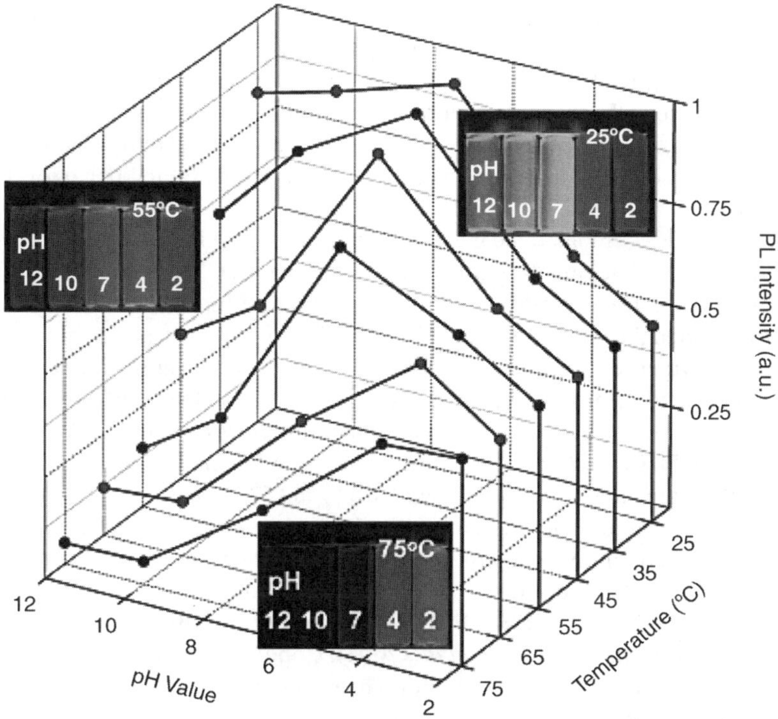

Figure 19.8 Three-dimensional plot of the variation on the PL intensity of PF_7-*b*-$PDMAEMA_{45}$ with temperature and pH. (Reproduced from S.-T. Lin, Y.-C. Tung and W.-C. Chen, "Synthesis, structures and multifunctional sensory properties of poly[2,7-(9,9-dihexylfluorene)]-block-poly[2-(dimethylamino) ethyl methacrylate] rod–coil diblock copolymers," *Journal of Materials Chemistry*, **18**, 3985–3992, 2008, by permission of the Royal Society of Chemistry.)

photodiodes, memory, solar cell, thin-film transistors, and so on. The exciton diffusion length is estimated to be around 10 nm before charge separation. Thus, control on the nanoscale morphology is a key to enhance the optoelectronic device efficiency. In general, film structures were prepared from a well-dispersed copolymer in a good solvent or solution-formed micelle in a selective solvent. Besides, compared with bulk or solution state, geometric confinement in a film of thickness that is incommensurate with the copolymer natural domain spacing. It could lead to separation of film into levels of different heights and confinement. Thus, it may also create a surface reconstruction where the thin-film nanoscale structure differs significantly from that in the solution or bulk. Here, we summarize the key factors that control the morphology of the thin film or bulk states of conjugated rod-roil block copolymers and also discuss their corresponding photophysical properties and device applications.

19.3.1 *Effect of Rod:Coil Ratio*

Hadziioannou, Mezzenga and coworkers found that the Microphase-separated morphologies of poly(paraphenylene vinylene)-*block*-poly(4-vinylpyridine) (PPV-*b*-P4VP) varied from lamellar, to hexagonal, and spherical as the volume fraction of the rod was progressively reduced (Sary *et al.*, 2007). The observed microphase morphology not only depended on the volume fraction but also on the Maier–Saupe constant, ω, which expressed as the rod–rod steric repulsion. The balance of the two contributions, ω/χ, might lead to different morphologies for a fixed χN. This means that a large value of ω/χ could drive a spherical or hexagonal phase into a lamellar smectic phase due to a predominance of the liquid-crystalline behavior over microphase separation.

Bu and the coworkers demonstrated that constraints of the packing into the unit cell outweigh the chain stretching and segment–segment repulsion energies, leading to a transition from normal (smectic A) to tilted (smectic C) rod orientation within the lamellae at a high coil fraction. They reported that the PEO coil of PEO-*b*-OFbT(oligo(fluorene-*alt*-bithiophene)s)-*b*-PEO triblock copolymers encouraged the formation of the H-type aggregation (Bu *et al.*, 2009). With the volume fraction of PEO increased, phase structures changed from the nematic to smectic, hexagonal columnar, and then lamellar nanostructures (Figure 19.9). The emission spectra of PEO-*b*-OFbT-*b*-PEO thin films were red-shifted compared to those in dilute THF solution. Also, the intensities at long wavelengths, 538 and 575 nm, were relatively enhanced as PEO length was increased. This indicated that it was possible to produce various nanostructures of rod–coil copolymers and then tune their optoelectronic properties by systematically varied their chemical structures. In addition, Leclere and coworkers described the synthesis and characterization of PF-*b*-PEO block copolymers with different block ratio and molecular architectures in THF (Surin *et al.*, 2004). Copolymers with a low volume ratio of PEO exhibited a well-defined organization into nanoribbons. A model of chain packing was proposed that the conjugated chains were π-stacked parallel to each other, edge-on over the substrate, with the PEO chains alternating on each side of the ribbon. For the copolymers with a higher average volume ratio of the PEO, the organized structures disappeared and led to untextured aggregates, probably because a long-range, regular π–π stacking of the segments could no longer take place.

Jenekhe and coworkers showed that solution-cast micellar PPQ-*b*-PS films consisted of multilayers with the hexagonally ordered arrays of spherical holes (Jenekhe and Chen, 1998, 1999). The hierarchical self-assembly approach to the microporous solids with a hollow core, a

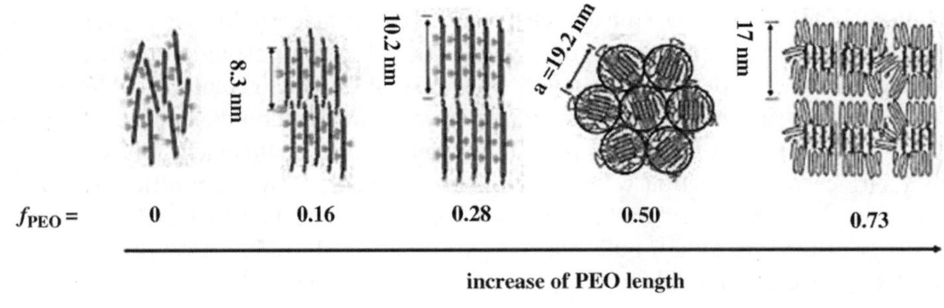

Figure 19.9 Schematic representation on the molecular arrangement of PEO-b-OFbT-b-PEO triblock copolymers with increasing the volume fraction of PEO at room temperature. (Reprinted with permission from L. Bu, Y. Qu, D. Yan *et al.*, "Synthesis and characterization of coil–rod–coil triblock copolymers comprising fluorene-based mesogenic monodisperse conjugated rod and poly(ethylene oxide) coil," *Macromolecules*, **42**, 1580–1588, 2009. © 2009 American Chemical Society.)

rod-like inner shell and flexible coil outer corona represented a nontemplate strategy. The diameter, periodicity, and wall thickness were found to depend on molecular weight and composition of the copolymers. Subsequently, Chang and coworkers (Lin *et al.*, 1999) further synthesized the rod–coil PPQ-*b*-PMMA diblock copolymers and the regularly porous honeycomb-structured film was obtained from the dichloromethane polymer solution under a flow of moist air. The diameters of the spherical pores could be controlled in the range from 0.8 to 3 μm by modifying both the relative molecular weights of the rod–coil blocks and the casting conditions. On the other hand, if the concentration of the polymer solution was too low, water droplet coalescence tended to enlarge the pore size and size distribution, and formed polygon structures. Higher relative molecular masses tended to create larger pores. The wall thickness between the pores increased linearly with the increase of the relative molecular mass. In addition, Francois and coworkers (Francois *et al.*, 1995) reported the synthesis of poly(*p*-phenylene)-*b*-polystyrene (PPP-*b*-PS) or PPP-*b*-PMMA copolymers via the combination of living anionic polymerization followed by the aromatization of poly(1,3-cyclohexadiene) moiety. Regular microporous PPP-*b*-PS membrane was also produced by using rapidly evaporating solvent, 1,2-dichloroethane, at the moist condition (Pitois and Francois, 1999). The formation of such ordered porous structure is based on a similar mechanism.

19.3.2 Architecture Effects

The grafting of rigid or semiflexible blocks, segments or mesogens as side chains or main-chain extensions led to richer and more complex morphologies. The use of the comb-type architectures with pencil flexible or rigid moieties emanate out of a main soft backbone was a common pathway to increase the morphological complexity in block copolymers. The resulting polymers usually possessed high chain rigidity and liquid-crystalline nature, depending on the length and the chemical structure of side chains.

The potential of pendant oligomers to construct functional organized materials was demonstrated by Hayakawa *et al.* (Hayakawa and Horiuchi, 2003), who created extremely regular hierarchical structures from low polydisperse copolymers of the coiled PS blocks with a semirod POTI (isoprene with the oligothiophene-modified side chains). The SEM images of the polymer film cast from CS_2 solution showed single layers of hexagonally packed micropores with a narrow size distribution (Figure 19.10). As shown in Figure 19.10, The diameter of these open pores on the surface was around 1.3 μm and a cross section showed a single layer with empty spheres with a diameter of 2.3 μm. Also, these empty spheres were separated from each other by thin walls of less 100 nm. They demonstrated the characteristics of a liquid-crystalline smectic mesophase on the π-stacked rod blocks. Annealing changed the direction of the cylinders from perpendicular to parallel at the bottom of the film, while the upper side of the film remained the same.

Our group investigated the self-assembly behavior block copolymers, P(St-Fl_n)-*b*-P2VP, in which one block is a coiled P2VP, while the other block is a stretched, semiflexible PS chain bearing a mesogen of one or two fluorene units (Li *et al.*, 2009). The fluorene chains can make effects on increasing the phase-segregation parameter χN and Khun length of the PS block. The phase behavior of P(St-Fl_2)-*b*-P2VP can be understood in terms of the conformational asymmetry of block copolymers arising from the longer fluorene side chain length as compared to the P(St-Fl_2)-*b*-P2VP. Depending on the number of pendant fluorene units, the block volume

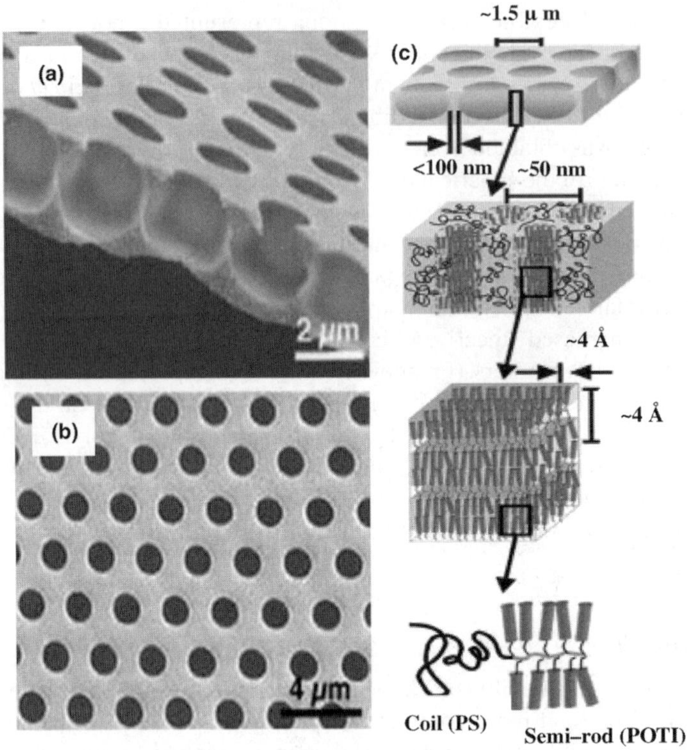

Figure 19.10 Self-organized microporous structure: (a) side view, (b) top view of the SEM images, and (c) schematic illustration of self-organized hierarchical structure. (Reproduced with permission from T. Hayakawa and S. Horiuchi, "From Angstroms to micrometers: Self-organized hierarchical structure within a polymer film," *Angewandte Chemie International Edition*, 2003, **42**, 2285–2289. © Wiley-VCH Verlag GmbH & Co. KGaA.)

ratio, the processing solvent, and thermal annealing history, spherical (S), hexagonally packed cylinders (C), lamellar (L), gyroidal (G), and hexagonally perforated lamellae (HPL) were obtained, as shown in Figure 19.11. The capability of P(St-F1$_n$)-*b*-P2VP to host COOH-modified SWCNTs through the hydrogen bonding between the COOH and P2VP moieties. The capability to encapsulate SWCNT was also found to be directly dependent on the microphase-separated structure of the block copolymers: two-dimensional (lamellae) structure being more efficient than one-dimensional morphology (cylinders).

19.3.3 Other Effects

The self-assembly thin-film morphology of rod–coil block copolymers can also be controlled by several effects, such as solvent, solution concentration, and substrate. For example, thin films of the oligofluorene-HEMA block copolymers formed either island, a string-like morphology, and honeycomb-like structures by increasing the selective solvent concentration (Chochos *et al.*, 2004). McCullough and coworkers synthesized a head-to-tail-coupled

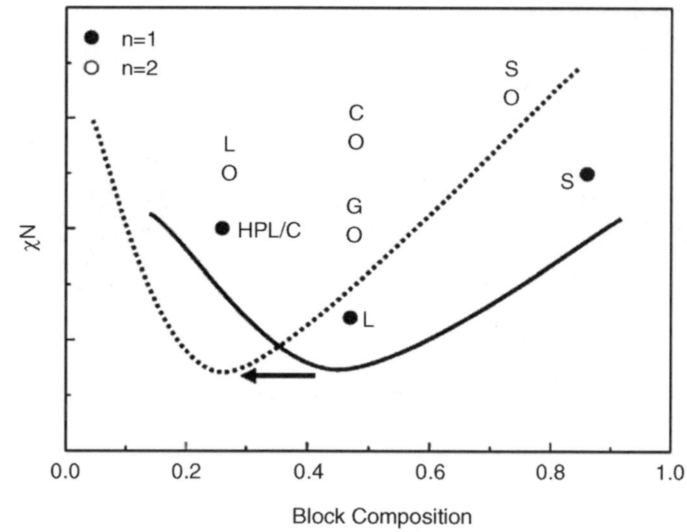

Figure 19.11 Effect of the side-chain fluorene length on phase behavior of P(St-Fl_n)-*b*-P2VP ($n = 1, 2$), solid line ($n = 1$) and dotted line ($n = 2$). The product is χN, where χ is the Flory–Huggins interaction parameter and N is the degree of polymerization. (Reprinted with permission from W. Li, H. Wang and L. Yu, "Syntheses of oligophenylenevinylenes–polyisoprene diblock copolymers and their microphase separation," *Macromolecules*, 32, 3034–3044, 1999. © 1999 American Chemical Society.)

poly(3-hexylthiophene) (P3HT)-*b*-PS and the strong tendency of regioregular P3HT to self-assemble into stacked aggregates forming highly conducting nanowires (Liu *et al.*, 2002). They also incorporated the PAA as the coil block to synthesize the amphiphilic nature of P3HT-*b*-PAA, which yielded solvatochromic behavior in several solvents of varying polarity (Craley *et al.*, 2009). UV-Vis absorption spectra revealed the presence of a vibronic structure in most polar solvents, pointing to the long-range ordering of P3HT based the self-assembly characteristic. Hillmyer and coworkers reported that regioregular P3HT-*b*-polyactide (PLA) copolymers formed a nanopitted film upon chemical etching of the PLA block (Boudouris, *et al.*, 2008).

Moreover, Segalman and coworkers found that the film structures of PPV-*b*-isoprene (PI) had a gradual reorientation from parallel to perpendicular orientation accompanied by a transition of the bulk film to a polycrystalline bimodally oriented structure with increasing film thickness (Olsen *et al.*, 2009). Hadziioannou and coworkers synthesized a series of fullerene-grafted rod poly[(2,5-di(2′-ethyl)hexyloxy)-1,4-phenylenevinylene](DEH-PPV)-coil block copolymers as a candidate active layer in photovoltaic devices (Barrau *et al.*, 2008). For the fullerene-free copolymer with a sufficiently low rod–coil weight fraction, the coil block destabilized considerably the rod–rod interactions and led to a lamellar phase with the rod block being in a liquid-crystalline state at room temperature (Figure 19.12(a)). On the other hand, microphase separation enhanced the interlamellar ordering. The results also showed the formation of either monolayer or double-layer lamellae, depending on the copolymer molecular structure. However, adding the fullerene moieties considerably affected the polymer self-assembly through the growth of fullerene nanocrystals (Figure 19.12(b)).

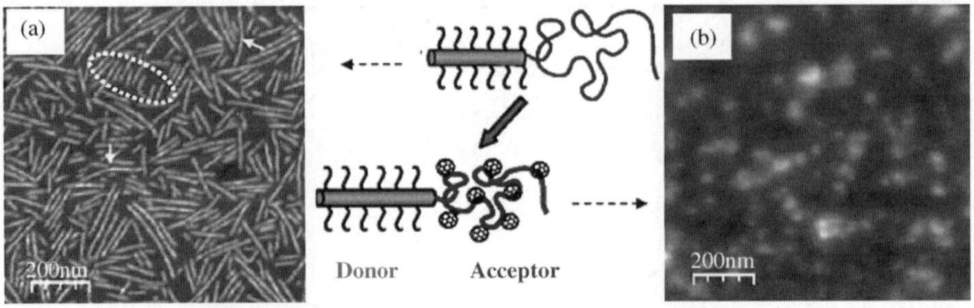

Figure 19.12 AFM phase images of poly(DEH-PPV)-coil block copolymers: (a) fullerene-free and (b) fullerene-grafted. (Reprinted with permission from S. Barrau, T. Heiser, F. Richard *et al.*, "Self-sssembling of novel fullerene-grafted donor–acceptor rod–coil block copolymers," *Macromolecules*, **41**, 2701–2710, 2008. © 2008 American Chemical Society.)

19.3.4 Optoelectronic Device Applications

Several optoelectronic applications based on conjugated rod–coil copolymers were reported, such as light-emitting diodes (LEDs) (Chochos *et al.*, 2005; Tao *et al.*, 2008; Segalman *et al.*, 2009), photovoltaic devices (PVs) (Segalman *et al.*, 2009; Tao *et al.*, 2009) and thin-film transistors (TFTs) (Sauve and McCullough, 2007). Generally, the exciton diffusion length is limited to approximately 10 nm. The resulting controlled segregation of rod–coil diblock copolymers with large donor–acceptor interface promotes exciton dissociation into free charge carriers. On the other hand, the donor and acceptor phase domain should be of this order to prevent the recombination of the photogenerated excitons. A bipolar rod–coil block copolymers consisted of a PPV rod-shaped block as the hole-transporting and -emitting material and poly(vinyloxadizaole) coil-shaped electron-transporting block (PPV-*b*-OX) lamellae oriented simultaneously both parallel to and perpendicular from the electrodes (Tao *et al.*, 2008). The LEDs using block copolymers exhibited higher efficiencies and brightness than that using either pure PPV or a blend of the two analogous homopolymers. The improved performance with microphase separation mainly resulted from the maximum effective interfacial area between the electron- and hole-transporting domains, which facilitated the exciton formation. By combining the electron-donating conjugated rod block and electron-accepting coil block also provided an opportunity to form thermodynamically optimized nanostructured films for photovoltaic applications. Active layers with organized nanodomains exhibited the higher photovoltaic efficiencies, indicating that molecular-scale segregation had a significant effect on device performances (Figure 19.13) (Segalman *et al.*, 2009). The nanofibrillar P3HT-*b*-PMA rod–coil block copolymers developed by McCullough and coworkers showed good organic semiconductor characteristics with high field effect mobilities ($\sim$0.01 cm^2/V s), which was comparable with that of regioregular P3HT, despite the presence of the nonconjugated segment (Sauve and McCullough, 2007). The fascinating self-assembly properties and active optoelectronic function of conjugated rod–coil block copolymers provide potential applications for future exploration.

The variation of nanostructures in conjugated rod–coil block copolymers holds considerable promise as the active layer in volatile organic compound (VOC) chemresistor sensors. It was reported that polythiophene rod–coil block copolymers could vary their electrical conductivity by changing rod–coil ratios when they were exposed to a variety of VOC vapors

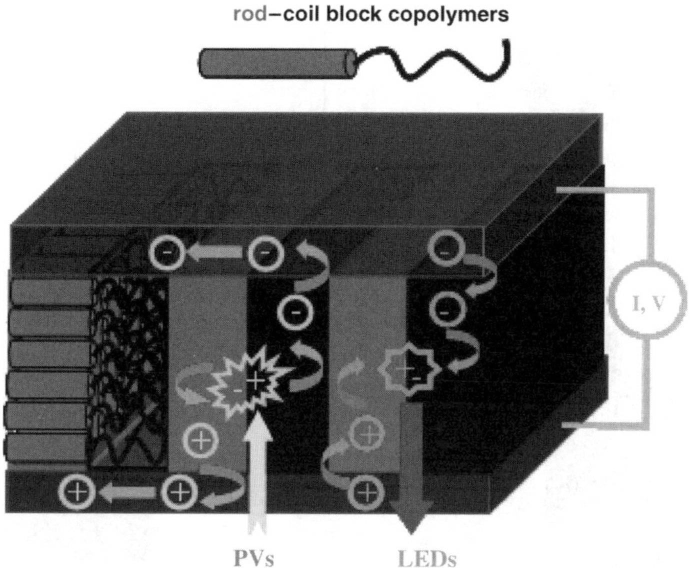

Figure 19.13 Rod–coil block copolymers for organic optoelectronics applications.

(Li *et al.*, 2006). The changing on the nanowire structures by the polymer composition and specific analytes explained the variation on the conductivity.

19.4 Electrospun Nanofibers

The intensive research of the "electrospinning (ES)" technology could fabricate one-dimensional (1D) nanostructures and nanofibers that are exceptionally long, uniform in diameter, diversified in composition and having various functions (Reneker and Chun, 1996; Li and Xia, 2004). The high surface/volume ratio of the ES fibers has attracted an extensive interest for sensory applications. Conjugated polymer ES fibers showed distinct electronic and optoelectronic properties as compared with films due to the geometrical confinement of ES process (Kuo *et al.*, 2007; Kuo *et al.*, 2008; Chen *et al.*, 2009; Kuo *et al.*, 2010; Wang *et al.*, 2009;). Although several coil–coil block copolymer based ES fibers were reported (Kalra *et al.*, 2006; Ma *et al.*, 2006; Ruotsalainen *et al.*, 2007), the morphology and properties of electrospun conjugated rod–coil polymer nanofibers was still less explored. Until now, two kinds of an electrospun conjugated rod–coil polymer, poly[2,7-(9,9-dihexyl-fluorene)]-*block*-poly(methylmethacrylate) (PF-*b*-PMMA) and the binary blends of poly (phenylquinoline)-*block*-polystyrene (PPQ-*b*-PS)/polystyrene (PS) were successfully prepared using a single-capillary spinneret. (Kuo *et al.*, 2008; Kuo, *et al.*, 2010) Since the ES technology is new to block copolymers, the basic setup is introduced in the following.

19.4.1 The Basic Setups for Electrospinning

Electrospinning (ES) represents a facile and versatile tool for manufacturing nanofibers. The ES process is involved with the polymer flow with a free surface after appearing from a

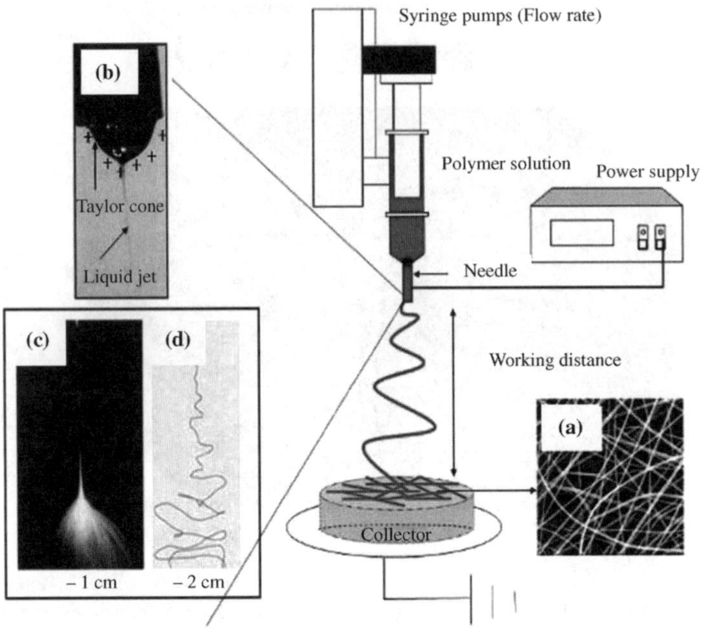

Figure 19.14 Schematic illustration of the basic setup for electrospinning. (a) The typical SEM image of nonwoven mat of polymer nanofibers. (b) CCD image of the stable con-jet spinning model. Photographs illustrating the instability region of a liquid jet and the capture time was on two different scales: (c) 1/250 s and, (d) 18 ns. (Reprinted from *Polymer*, **42**, Y.M. Shin, M.M. Hohman, M.P. Brenner and G.C. Rutledge "Experimental characterization of electrospinning: The electrically forced jet and instabilities," 9955–9967, 2001, with permission from Elsevier.)

capillary, followed by the bending and continuously stretching of the jet due to the electrostatic repulsions between the surface charges and the evaporation of solvent, as shown in the Figure 19.14.

As shown in Figure 19.14, a typical electrospinning setup consists of a high-voltage power supply, a programmable syringe pump, a syringe, a needle, and a grounded collector screen to collect nonwoven nanofibers (Figure 19.14(a)). During the process a polymer solution or composite mixture is injected from a small nozzle under the influence of a high voltage (usually in the range of 1–30 kV). The buildup of electrostatic charges on the surface of a liquid droplet induces the formation of a jet. As a result, the drop experiences two major types of electrostatic forces: the electrostatic repulsion between the surface charges; and the Coulombic force exerted by the external electric field. Under the action of these electrostatic interactions, the liquid drop is then distorted into a conical object commonly known as the Taylor cone (Figure 19.14(b)). Once the strength of the electric field has surpassed a threshold value, the electrostatic forces can overcome the surface tension of the polymer solution and thus force the ejection of a liquid jet from the nozzle. The charges carried by each element of the jet all continue to interact with each other by Coulomb's law. At the onset of bending, these off-axis forces generate a radial component of velocity that is added to the downward velocity. Figure 19.14(c) (Shin *et al.*, 2001) shows the photograph of a spinning jet. It is obvious that the jet is initially a straight line and then becomes unstable. It appears that the cone-shaped, instability region is composed of multiple jets. However, a closer examination using high-speed photography (Figure 19.14(d)) (Shin *et al.*, 2001) establishes that the conical envelope contains only a single, rapidly bending or whipping thread. The elongation of the bent segment along its own curving local path continues, generating a helical coil with many turns. As the liquid jet is continuously elongated and the solvent is evaporated, its diameter could be greatly reduced from hundreds of micrometers to as small as tens of nanometers. After that, the nonwoven nanofibers are produced, as shown in Figure 19.14(a).

19.4.2 Electrospun Nanofibers Prepared from Conjugated Rod–Coil Copolymers

19.4.2.1 Homo Rod–Coil Copolymer System

The morphology and photophysical properties of an electrospun conjugated rod–coil polymer, PF_7-*b*-$PMMA_{810}$, using a single-capillary spinneret were reported by our group (Kuo *et al.*, 2008). PMMA had a much longer coil length than PF in order to have enough molecular weight or viscosity to produce ES fibers using a single-capillary spinneret. Tetrahydrofuran (THF) (a good solvent for both blocks), THF/N, N-Dimethylformamide (DMF) with the solvent ratio of 50/50, and 100% DMF (a poor solvent for the PF block) were used to explore the solvent effects on the ES nanofibers. As shown in Figure 19.15, different fiber morphologies of PF-*b*-PMMA are observed as porous ribbon-shape (from THF solvent), smooth but less-uniform fibers (THF/DMF of 50/50 ratio), and uniform but rough fibers (from DMF). The fiber diameter decreases from 200–3000 nm of THF, to 200–700 nm of THF/DMF (50/50), and 150–300 nm of DMF, respectively. The higher dielectric constant of DMF than that of THF leads to a more evenly distributed charge on the jet and thus produces thinner fibers. In addition, the surface textures and nanoporous ES fibers probably resulted from the rapid evaporation of the THF solvent and a subsequent rapid solidification during the ES process. On the other hand, the rough fiber surface might result from the aggregation of the rigid PF segment since DMF

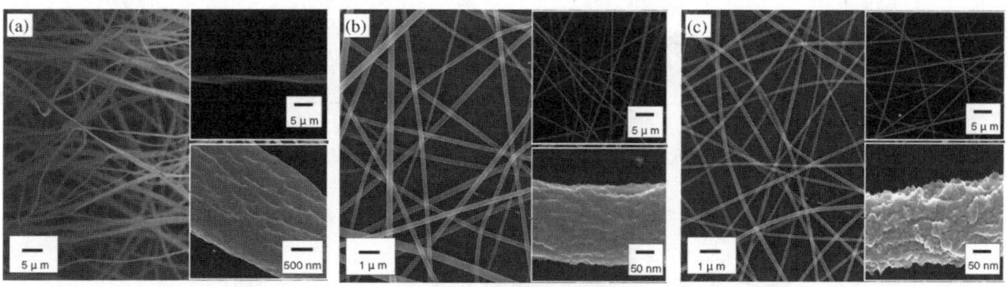

Figure 19.15 FE-SEM images of PF-b-PMMA ES fibers as a function of solution volatility: (a) THF; (b) THF/DMF (50/50); (c) DMF. The inset figures show the enlarged FE-SEM and confocal microscopy images of the above ES fibers. (Reproduced with permission from C.-C. Kuo, Y.-C. Tung, C.-H. Lin and W.-C. Chen, "Novel luminescent electrospun fibers prepared from conjugated rod–coil block copolymer of poly[2,7-(9,9-dihexylfluorene)]-block-poly(methyl methacrylate)," *Macrmolecular Rapid Communications*, 2008, **29**, 1711–1715. © Wiley-VCH Verlag GmbH & Co. KGaA.)

was a poor solvent for PF. The blue light-emitting images from the confocal microscopy in the inset figures of Figure 19.15 are consistent with those of the FE-SEM images. Moreover, PF block aggregation structures of dot-like (5–10 nm), line-like (10–20 nm), and ellipse-like structure (25–50 nm), respectively (Figure 19.16) (Kuo *et al.*, 2008), are observed for the

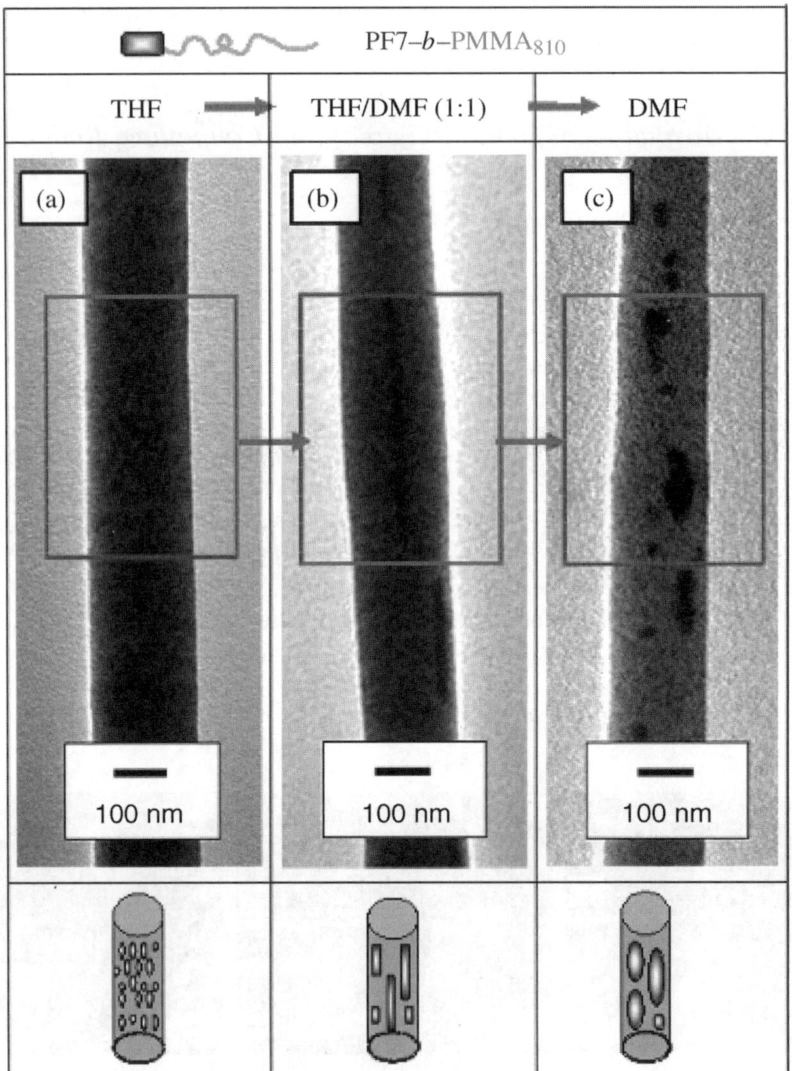

Figure 19.16 TEM images of the PF-*b*-PMMA ES fibers prepared from different solvent systems of (a) THF, (b) THF/DMF (50/50), and (c) DMF. (Reproduced with permission from C.-C. Kuo, Y.-C. Tung, C.-H. Lin and W.-C. Chen, "Novel luminescent electrospun fibers prepared from conjugated rod–coil block copolymer of poly[2,7-(9,9-dihexylfluorene)]-block-poly(methyl methacrylate)," *Macrmolecular Rapid Communications*, 2008, **29**, 1711–1715. © Wiley-VCH Verlag GmbH & Co. KGaA.)

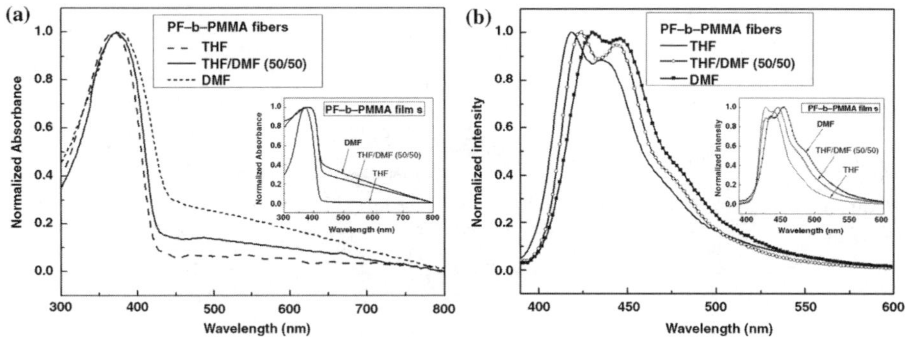

Figure 19.17 (a) UV-Vis absorption spectra and (b) photoluminescence spectra of PF-*b*-PMMA ES fibers produced from THF, THF/DMF (50/50) and DMF and those of the corresponding spin-coated films are shown in the inset figure. (Reproduced with permission from C.-C. Kuo, Y.-C. Tung, C.-H. Lin and W.-C. Chen, "Novel luminescent electrospun fibers prepared from conjugated rod–coil block copolymer of poly[2,7-(9,9-dihexylfluorene)]-block-poly(methyl methacrylate)," *Macrmolecular Rapid Communications*, 2008, **29**, 1711–1715. © Wiley-VCH Verlag GmbH & Co. KGaA.)

above three solvent systems. This suggests that the PF aggregation size increases with the enhancing DMF content and leads to the red-shifting on the absorption or luminescence peak of the prepared ES fibers (Figure 19.17). The morphological results of Figure 19.16 demonstrate the significant alteration of block copolymer microphase separation by the ES solvent. Furthermore, the aggregation size of the PF block in the ES fibers was much smaller than those in the spin-coated films due to the ES geometrical confinement and resulted in smaller absorption or luminescence maxima wavelengths in the former (Figure 19.17).

19.4.2.2 Rod–Coil Copolymer Blending System

The high surface/volume ratio and porous characteristics render the advantage of the ES nanofibers for sensory applications. We have developed new luminescent electrospun (ES) nanofibers from the binary blends of PPQ_{155}-*b*-PS_{564}/PS for pH-tunable colorimetric sensors (Kuo *et al.*, 2010). In order to have enough molecular weight or viscosity to produce ES fibers, polymer blends of PPQ-b-PS with a high molecular weight PS homopolymer was used. As shown in Figure 19.18, the PPQ-*b*-PS aggregated domain sizes in the blend ES fibers using the following solvents are 1–2 μm (CH_2Cl_2), 2–3 μm (chlorobenzene, CB), and 3–5 μm ($CHCl_3$), respectively. Note that the solubility parameters of PPQ, PS, CH_2Cl_2, CB, and $CHCl_3$ were 22.7, 18.0, 20.2, 19.6 and 18.7 ($MPa^{1/2}$), respectively. The significant difference on the solubility parameter between the processing solvents and the PPQ block probably results in a different PPQ aggregation size in the ES fibers and thus produces the PL spectra shift, changing from green, to yellow, and orange (Figures 19.18 and 19.19). The aggregated domain sizes of the PPQ-*b*-PS in the spin-coated thin films are significantly larger than those of the corresponding ES fibers and produced the red-shifting on the PL spectra, as shown in the inset of Figure 19.19. Besides, the strong stretching force driving from the electric field could also prevent or decrease the π–π stacking of PPQ rod segment during ES process (Kuo *et al.*, 2007;

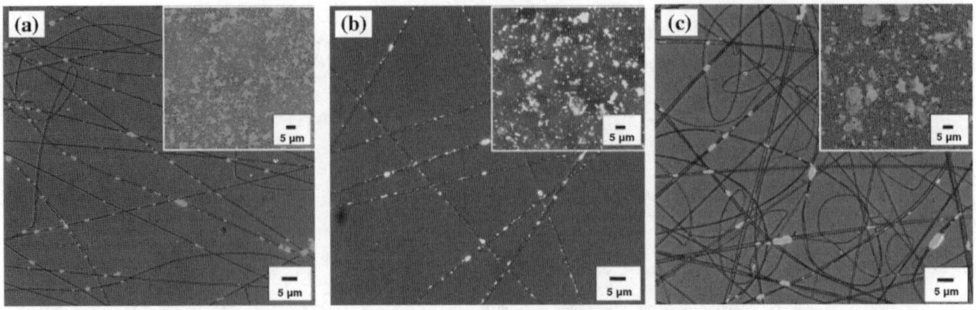

Figure 19.18 Confocal microscopy images of 5 wt% PPQ-*b*-PS/PS blend ES fibers prepared from the solvents of (a) CH_2Cl_2; (b) CB; (c) $CHCl_3$. The images of the corresponding spin-coated thin films are shown in the inset figures "Kuo, *et al.*, 2010. (Reproduced with permission from C.-C. Kuo, Y.-C. Tung and W.-C. Chen, "Morphology and pH sensing characteristics of new luminescent electrospun fibers prepared from poly(phenylquinoline)-block-polystyrene/polystyrene blends," *Macromolecular Rapid Communications*, 2010, **31**, 65–70. © Wiley-VCH Verlag GmbH & Co. KGaA.)".

Kuo *et al.*, 2008; Chen *et al.*, 2009; Kuo *et al.*, 2010). These ES fibers have a higher pH-tunable sensing ability than the spin-coated film since the former has a high surface/volume ratio. In particular, the ES fibers prepared from CH_2Cl_2 exhibited a pH-tunable photoluminescence with the emission maximum varying from 532 to 560 nm as the pH value changes from 7 to 1. The smaller PPQ aggregation and the porous surface structure probably promotes the efficient protonation (Figure 19.19(b)) (Kuo *et al.*, 2010). This demonstrates that their

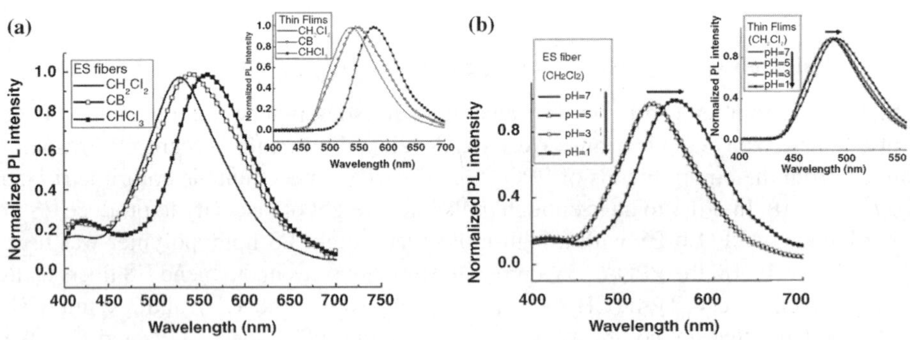

Figure 19.19 (a) Photoluminescence spectra of 5 wt% PPQ-*b*-PS/PS blend ES fibers prepared from the solvents of CH_2Cl_2; CB; $CHCl_3$. The PL spectra of the corresponding spin-coated thin films are shown in the inset figures. (b) The pH-sensing characteristics of photoluminescence spectra of 5 wt% PPQ-*b*-PS/PS blend ES fibers prepared from CH_2Cl_2 and the corresponding spin-coated thin films are shown in the inset figures. (Reproduced with permission from C.-C. Kuo, Y.-C. Tung and W.-C. Chen, "Morphology and pH sensing characteristics of new luminescent electrospun fibers prepared from poly(phenylquinoline)-block-polystyrene/polystyrene blends," *Macromolecular Rapid Communications*, 2010, **31**, 65–70. © Wiley-VCH Verlag GmbH & Co. KGaA.)

aggregate morphology and photophysical properties of the conjugated rod–coil block copolymer ES fibers are efficiently tuned through selective solvent.

19.5 Polymer Brushes

Polymer brushes have been recognized as extensively fascinating synthetic targets, because of their abilities in surface modification and potential advanced applications in switching membranes, sensors, and cell-growth control (Ito *et al.*, 1997; Zhao and Brittain, 2000; Khademhosseini *et al.*, 2003; Luzinov *et al.*, 2004). Generally, two primary approaches for production of polymer brushes are grafting-to and grafting-from methods, or commonly known as "surface-initiated polymerization". Conjugated polymer rod–coil brushes usually prepared in the former way because of the synthetic difficulty of conjugated backbones via the surface-initiated polymerization. Coil–coil-based polymer brushes have been widely reported in some literature, but using a grafting-to approach for the formation of conjugated rod–coil polymer brushes was only explored recently by our group (Wu *et al.*, 2007; Li *et al.*, 2008).

Our group reported a combined theoretical and experimental investigation of rod–coil block copolymer brushes, PF-*b*-PPEGMA-*b*-PPOPS, containing the hydrophobic fluorescent PF rod block, the hydrophilic PPEGMA coil block, and the PPOPS segment anchored onto glass or silicon substrates (Wu *et al.*, 2007). As shown in Figure 19.20, the rod blocks of polymer brushes are found to be well dispersed on the surface in their good solvent of toluene. On the other hand, they formed aggregative domains in their poor solvent of methanol. In addition, the conformations of isolated islands, worm-like structures, or network-like configuration are obtained as the grafting density of the polymer brushes is increased. However, the enhanced coil length delays the onset of the worm-like conformation. Furthermore, the aggregated domains tend to stay on top of the coil blocks for small rod-to-coil block ratio, as shown in Figure 19.21(a). However, the submergence of the aggregative domains into the coil blocks is thermodynamically preferred for a large enough rod-to-coil block ratio Figure 19.21(b). The emission peaks originated from the aggregation and/or excimer formation of PF blocks are observed after methanol treatment. For example, the PL intensity of polymer brushes shows an additional shoulder peak (Figure 19.22), which is enhanced with increasing methanol ratio in the mixed solvents. This clearly suggests that the surface structures of amphiphilic rod–coil block copolymer brushes are very sensitive to solvent stimuli.

The surface structures of rod–coil block copolymer brushes tethered onto a surface with different architectures exhibit distinct surface structures under different grafting densities and lengths of coil blocks (Li *et al.*, 2008). Figure 19.23 shows three different kinds of rod–coil block copolymer brushes, linear, Y-shaped, and comb-like structures. The comparison between the surface structures of linear and Y-shaped rod–coil block copolymer brushes with different grafting densities (ρ_s) under a selective solvent, as shown in Figure 19.24. As shown in the figure, the size of aggregates formed by the rod blocks increases with enhancing the grafting density for both the linear and Y-shaped rod–coil brushes. However, for Y-shaped polymer brushes, isolated islands of aggregates can only survive at a low surface grafting density and transform into worm-like, interconnected, honeycomb, and fully covered structures as the grafting density increases. As compared with the linear structure, the Y-shaped rod–coil brushes proceed much faster with respect to the surface grafting density because each Y-shaped brush contains two rod blocks. On the other hand, comb-like polymer brushes possess more

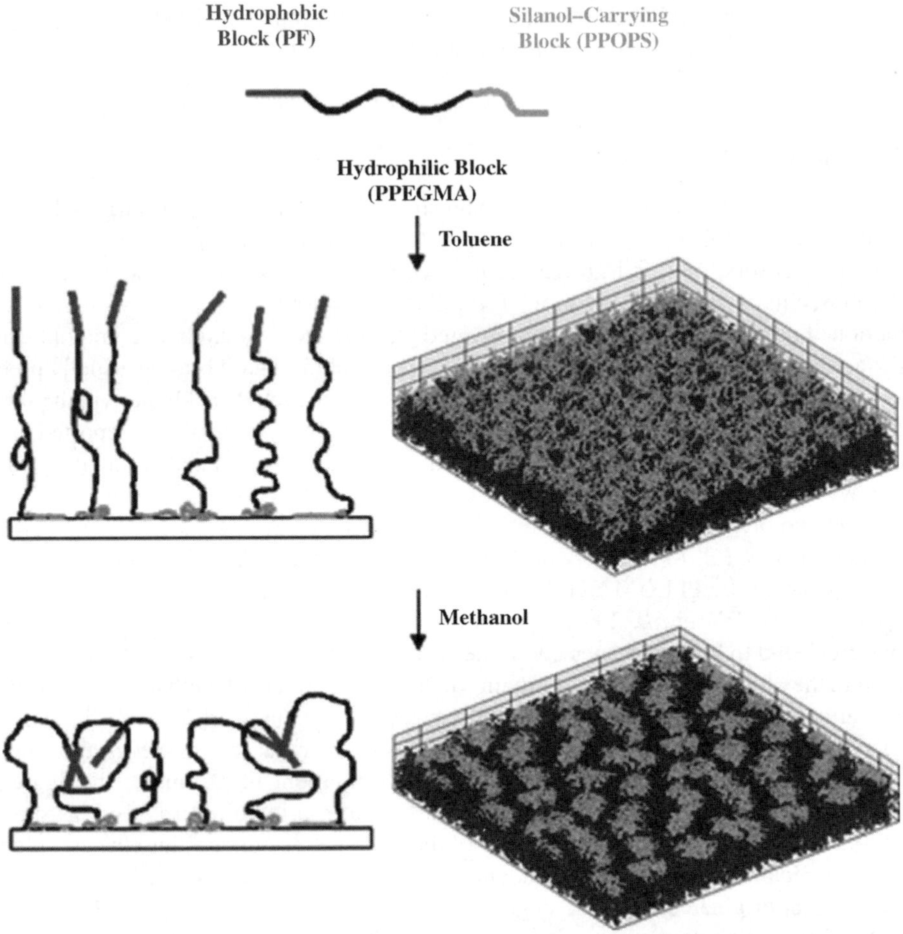

Figure 19.20 Chemical structure of PF-*b*-PPEGMA-*b*-PPOPS triblock copolymer, and the corresponding polymer brush under solvent stimuli. (Reprinted with permission from W.-C. Wu, Y. Tian, C.-Y. Chen *et al.*, "Theoretical and experimental studies on the surface structures of conjugated rod–coil block copolymer brushes," *Langmuir*, **23**, 2805–2814, 2007. © 2007 American Chemical Society.)

diverse aggregative manners than linear and Y-shaped brushes. Aggregates with the shapes of cone, cylinder, and layers of spheres are dependent on the size of the gap between each rod blocks. The above results provide a guideline for designing the polymer brushes with desired surface structures and thus desired properties.

19.6 Future Directions and Outlook

The self-assembly of conjugated rod–coil block copolymers has been demonstrated as a powerful route towards supramolecular objects with novel architectures, functions, and

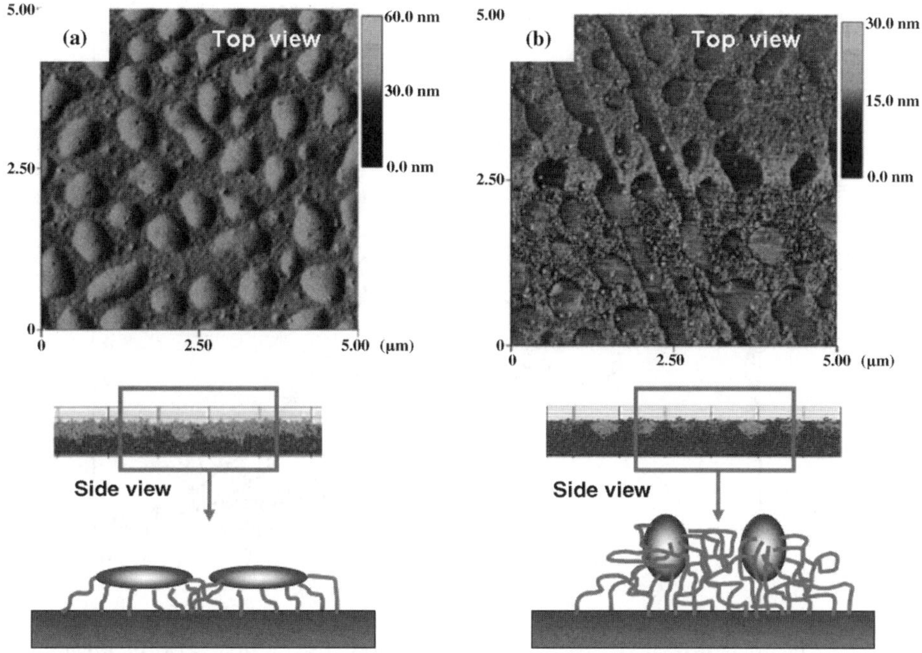

Figure 19.21 AFM images (top view) and stimulated surface structures (side view) of the rod–coil block copolymer brushes after methanol treatment: (a) PF_{16}-*b*-$PPEGMA_{44}$-*b*-$PPOPS_{16}$ and (b) PF_{16}-*b*-PPEG-MA_{92}-*b*-$PPOPS_{11}$. (Reprinted with permission from W.-C. Wu, Y. Tian, C.-Y. Chen *et al.*, "Theoretical and experimental studies on the surface structures of conjugated rod–coil block copolymer brushes," *Langmuir*, **23**, 2805–2814, 2007. © 2007 American Chemical Society.)

physical properties. This chapter provides the understanding between the morphology and photophysical properties of conjugated rod–coil block copolymers in different phases. The significant π–π interaction and the stiff asymmetry between the conjugated rod and coil blocks give them distinct morphologies in comparison with those of the coil–coil block copolymers. Controlling the molecular packaging on the conjugated rods in different phases of solutions, thin films, nanofibers or brushes remains a challenge for polymer physical scientists. Also, polymer chemists show extensive interest in exploring new synthetic schemes for preparing conjugated rod–coil block copolymers with well-defined chemical architectures. The self-organized domains of block copolymers offer switch-like control on material functionalities by the relevant order–disorder or order–order transitions. The incorporation of conjugated polymer rod moiety into block copolymers rendered various morphologies to tune the electronic and optoelectronic properties. In addition, the incorporation of the environmentally stimuli coil into conjugated block copolymers provides novel multifunctional sensory polymers with response to pH, temperature, light or metal ion-strength. It is expected that novel multifunctional sensory devices could be produced by the soft-patterning techniques with conjugated rod–coil block copolymers, such as microcontact printing and nanoimprinting. It is believed that the manipulation on the morphology

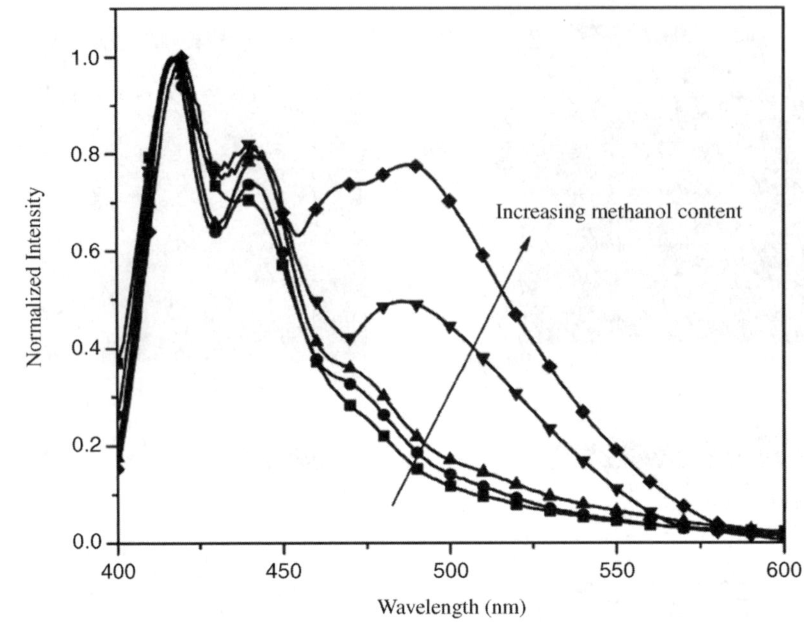

Figure 19.22 Photoluminescence spectra of polymer brush, PF_{16}-*b*-$PPEGMA_{92}$-*b*-$PPOPS_{11}$, after treatment with mixed solvent of toluene and methanol excited at the wavelength of 380 nm. (Reprinted with permission from W.-C. Wu, Y. Tian, C.-Y. Chen *et al.*, "Theoretical and experimental studies on the surface structures of conjugated rod–coil block copolymer brushes," *Langmuir*, **23**, 2805–2814, 2007. © 2007 American Chemical Society.)

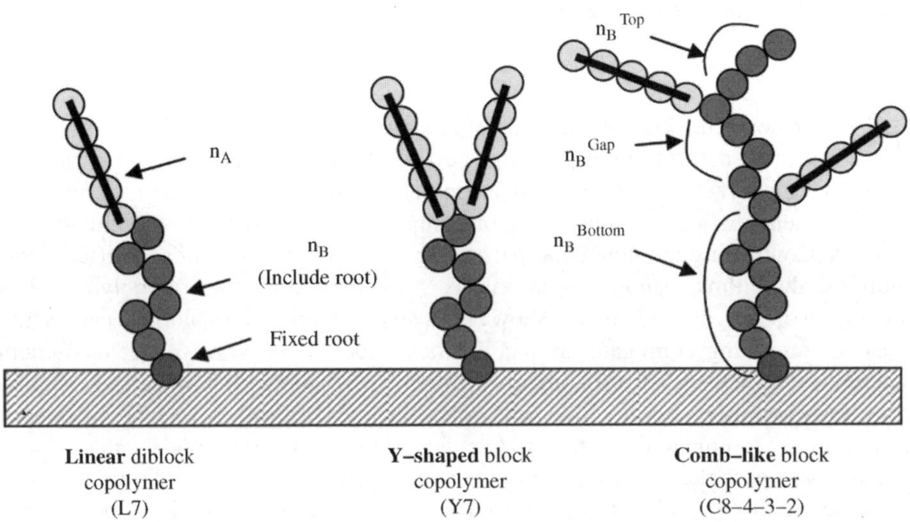

Figure 19.23 Schematic diagrams of linear, Y-shaped, and comb-like rod–coil block copolymer brushes. (Reproduced with permission from C.-S. Li, W.-C. Wu, Y.-J. Sheng and W.-C. Chen "Effects of chain architectures on the surface structures of conjugated rod-coil block copolymer," *Journal of Chemical Physics*, **128**, 154908, 2008. © 2008 American Institute of Physics.)

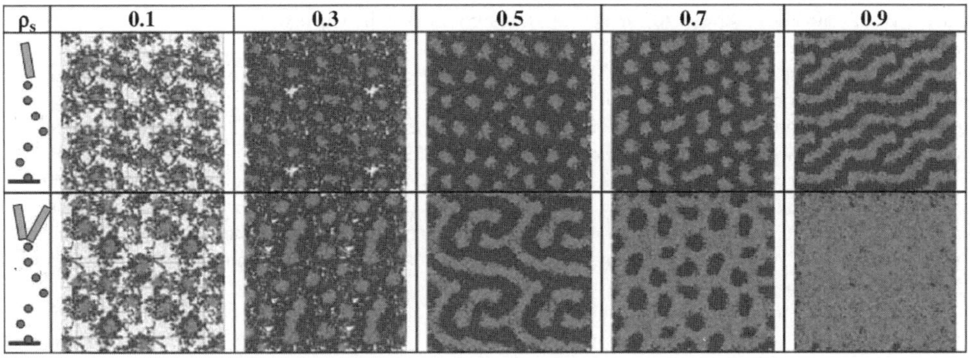

Figure 19.24 Surface structures of rod–coil block copolymer brushes. Comparison between linear brushes and Y-shaped brushes for $\rho_s = 0.1$–0.9. (Reproduced with permission from C.-S. Li, W.-C. Wu, Y.-J. Sheng and W.-C. Chen "Effects of chain architectures on the surface structures of conjugated rod-coil block copolymer," *Journal of Chemical Physics*, **128**, 154908, 2008. © 2008 American Institute of Physics.)

and properties of conjugated rod–coil block copolymers in different phases could be very important for the organic-based nanodevices.

References

Balamurugan, S.S., Bantchev, G.B., Yang, Y., and McCarley, R.L. (2005) Highly water-soluble thermally responsive poly(thiophene)-based brushes. *Angewandte Chemie-International Edition*, **44**, 4872–4876.

Barrau, S., Heiser, T., Richard, F. *et al.* (2008) Self-assembling of novel fullerene-grafted donor-acceptor rod-coil block copolymers. *Macromolecules*, **41**, 2701–2710.

Boudouris, B.W., Frisbie, C.D., and Hillmyer, M.A. (2008) Nanoporous poly(3-alkylthiophene) thin films generated from block copolymer templates. *Macromolecules*, **41**, 67–75.

Bu, L., Qu, Y., Yan, D. *et al.* (2009) Synthesis and characterization of coil-rod-coil triblock copolymers comprising fluorene-based mesogenic monodisperse conjugated rod and poly(ethylene oxide) coil. *Macromolecules*, **42**, 1580–1588.

Chen, H.C., Wang, C.T., Liu, C.L. *et al.* (2009) Full color light-emitting electrospun nanofibers prepared from PFO/MEH-PPV/PMMA ternary blends. *Journal of Polymer Science Part B-Polymer Physics*, **47**, 463–470.

Chen, X.L. and Jenekhe, S.A. (2000) Supramolecular self-assembly of three-dimensional nanostructures and microstructures: microcapsules from electroactive and photoactive rod-coil-rod triblock copolymers. *Macromolecules*, **33**, 4610–4612.

Chochos, C.L., Kallitsis, J.K., and Gregoriou, V.G. (2005) Rod-coil block copolymers incorporating terfluorene segments for stable blue light emission. *The Journal of Physical Chemistry B*, **109**, 8755–8760.

Chochos, C.L., Tsolakis, P.K., Gregoriou, V.G., and Kallitsis, JK. (2004) Influence of the coil block on the properties of rod-coil diblock copolymers with oligofluorene as the rigid segment. *Macromolecules*, **37**, 2502–2510.

Craley, C.R., Zhang, R., Kowalewski, T. *et al.* (2009) Regioregular poly(3-hexylthiophene) in a novel conducting amphiphilic block copolymer. *Macromolecular Rapid Communications*, **30**, 11–16.

de Cuendias, A., Ibarboure, E., Lecommandoux, S. *et al.* (2008) Synthesis and self-assembly in water of coil-rod-coil amphiphilic block copolymers with central π-conjugated sequence. *Journal of Polymer Science Part A-Polymer Chemistry*, **46**, 4602–4616.

Fang, C., Qi, X.Y., Fan, Q.L. *et al.* (2007) A facile route to semiconductor nanocrystal-semiconducting polymer complex using amine-functionalized rod-coil triblock copolymer as multidentate ligand. *Nanotechnology*, **18**, 035704.

Francois, B., Widawski, G., Rawiso, M., and Cesar, B. (1995) Block-copolymers with conjugated segments: synthesis and structural characterization. *Synthetic Metals*, **69**, 463–466.

Hadjichristidis, N., Pispas, S., and Floudas, G.A. (2003) *Block Copolymers: Synthesis Strategies, Physical Properties, and Applications*, John Wiley & Sons, New Jersey.

Hayakawa, T. and Horiuchi, S. (2003) From angstroms to micrometers: self-organized hierarchical structure within a polymer film. *Angewandte Chemie-International Edition*, **42**, 2285–2289.

Hoeben, F.J.M., Jonkheijm, P., Meijer, E.W., and Schenning, A.P.H.J. (2005) About supramolecular assemblies of π-conjugated systems. *Chemical Reviews*, **105**, 1491–1546.

Huo, H., Li, K., Wang, Q., and Wu, C. (2007) Self-assembly and optical property of triblock copolymers made of polystyrene and oligo(p-phenyleneethynylene) in different mixtures of toluene and hexane. *Macromolecules*, **40**, 6692–6698.

Ito, Y., Ochiai, Y., Park, Y.S., and Imanishi, Y. (1997) pH-sensitive gating by conformational change of a polypeptide brush grafted onto a porous polymer membrane. *Journal of the American Chemical Society*, **119**, 1619–1623.

Jenekhe, S.A. and Chen, X.L. (1998) Self-assembled aggregates of rod-coil block copolymers and their solubilization and encapsulation of fullerenes. *Science*, **279**, 1903–1907.

Jenekhe, S.A. and Chen, X.L. (1999) Self-assembly of ordered microporous materials from rod-coil block copolymers. *Science*, **283**, 372–375.

Kalra, V., Kakad, P.A., Mendez, S. *et al.* (2006) Self-assembled structures in electrospun poly(styrene-block-isoprene) fibers. *Macromolecules*, **39**, 5453–5457.

Khademhosseini, A., Jon, S.Y., Suh, K.Y. *et al.* (2003) Direct patterning of protein- and cell-resistant polymeric monolayers and microstructures. *Advanced Materials*, **15**, 1995–2000.

Kim, B.S., Hong, D.J., Bae, J., and Lee, M. (2005) Controlled self-assembly of carbohydrate conjugate rod-coil amphiphiles for supramolecular multivalent ligands. *Journal of the American Chemical Society*, **127**, 6333–16337.

Kuo, C.C., Lin, C.H., and Chen, W.C. (2007) Morphology and photophysical properties of light-emitting electrospun nanofibers prepared from poly(fluorene) derivative/PMMA blends. *Macromolecules*, **40**, 6959–6966.

Kuo, C.C., Tung, Y.C., and Chen, W.C. (2010) Morphology and pH sensing characteristics of new luminescent electrospun fibers prepared from poly(phenylquinoline)-block-polystyrene/polystyrene blends. *Macromolecular Rapid Communications*, **31**, 65–70.

Kuo, C.C., Tung, Y.C., Lin, C.H., and Chen, W.C. (2008) Novel luminescent electrospun fibers prepared from conjugated rod-coil block copolymer of poly[2,7-(9,9-dihexylfluorene)]-block-poly(methyl methacrylate). *Macromolecular Rapid Communications*, **29**, 1711–1715.

Leclere, Ph., Parente, V., Bredas, J.L. *et al.* (1998) Organized semiconducting nanostructures from conjugated block copolymer self-assembly. *Chemistry of Materials*, **10**, 4010–4014.

Lee, M., Cho, B.K., and Zin, W.C. (2001) Supramolecular structures from rod-coil block copolymers. *Chemical Reviews*, **101**, 3869–3892.

Li, B., Sauve, G., Iovu, M.C. *et al.* (2006) Volatile organic compound detection using nanostructures copolymers. *Nano Letters*, **6**, 1598–1602.

Li, C., Hsu, J.C., Sugiyama, K. *et al.* (2009) Synthesis and self-assembly behavior of poly(fluorenylstyrene)-block-poly (2-vinylpyridine) block copolymers and their blends with single wall carbon nanotubes. *Macromolecules*, **42**, 5793–5801.

Li, C.S., Wu, W.C., Sheng, Y.J., and Chen, W.C. (2008) Effects of chain architectures on the surface structures of conjugated rod-coil block copolymer brushes. *Journal of Chemical Physics*, **128**, 154908.

Li, D. and Xia, Y. (2004) Electrospinning of nanofibers: reinventing the wheel? *Advanced Materials*, **16**, 1151–1170.

Li, K. and Wang, Q. (2005) Multiple self-assembled nanostructures from an oligo(p-phenyleneethynylene) containing rod-coil-rod triblock copolymer. *Chemical Communications*, **38**, 4786–4788.

Li, W., Wang, H., Yu, L. *et al.* (1999) Syntheses of oligophenylenevinylenes-polyisoprene diblock copolymers and their microphase separation. *Macromolecules*, **32**, 3034–3044.

Liang, Y.Y., Wang, H.B., Yuan, S.W. *et al.* (2007) Conjugated block copolymers and co-oligomers: from supramolecular assembly to molecular electronics. *Journal of Materials Chemistry*, **17**, 2183–2194.

Lin, C.H., Tung, Y.C., Ruokolainen, J. *et al.* (2008a) Poly[2,7-(9,9-dihexylfluorene)]-*block*-poly(2-vinylpyridine) rod-coil and coil-rod-coil block copolymers: synthesis, morphology and photophysical properties in methanol/THF mixed solvents. *Macromolecules*, **41**, 8759–8769.

Lin, C.L., Tung, P.H., and Chang, F.C. (1999) Synthesis of rod-coil diblock copolymers by ATRP and their honeycomb morphologies formed by the 'breath figures' method. *Polymer*, **46**, 9304–9313.

Lin, S.T., Fuchise, K., Chen, Y. *et al.* (2009) Synthesis, thermomorphic characteristics, and fluorescent properties of poly[2,7-(9,9-dihexylfluorene)]-*block*-poly(N-isopropylacrylamide)-*block*-poly(N-hydroxyethylacrylamide) rod-coil-coil triblock copolymers. *Soft Matter*, **5**, 3761–3770.

Lin, S.T., Tung, Y.C., and Chen, W.C. (2008b) Synthesis, structures and multifunctional sensory properties of poly[2,7-(9,9-dihexylfluorene)]-*block*-poly[2-(dimethylamino) ethylmethacrylate] rod-coil diblock copolymers. *Journal of Materials Chemistry*, **18**, 3985–3992.

Liu, J., Sheina, E., Kowalewski, T., and McCullough, R.D. (2002) Tuning the electrical conductivity and self-assembly of regioregular polythiophene by block copolymerization: nanowire morphologies in new di- and triblock copolymers. *Angewandte Chemie-International Edition*, **41**, 329–332.

Lu, S., Fan, Q.L., Chua, S.J., and Huang, W. (2003) Synthesis of conjugated-ionic block copolymers by controlled radical polymerization. *Macromolecules*, **36**, 304–310.

Luzinov, I., Minko, S., and Tsukruk, V.V. (2004) Adaptive and responsive surfaces through controlled reorganization of interfacial polymer layers. *Progress in Polymer Science*, **29**, 635–698.

Ma, M., Krikorian, V., Yu, J.H. *et al.* (2006) Electrospun polymer nanofibers with internal periodic structure obtained by microphase separation of cylindrically confined block copolymers. *Nano Letters*, **6**, 2969–2972.

Ma, Z., Qiang, L., Zhang, Z. *et al.* (2008) Morphology and photophysical properties of a thermally responsive fluorescent materials based on a rod-coil tri-block copolymer. *Journal of Applied Polymer Science*, **110**, 18–22.

Mori, T., Watanabe, T., Minagawa, K., and Tanaka, M. (2005) Self-assembly of oligo(p- phenylenevinylene)-block-poly(ethylene oxide) in polar media and solubilization of an oligo(p-phenylenevinylene) homooligomer inside the assembly. *Journal of Polymer Science Part A-Polymer Chemistry*, **43**, 1569–1578.

Olsen, B.D. and Segalman, R.A. (2008) Self-assembly of rod-coil block copolymers. *Materials Science & Engineering R-Reports*, **62**, 37–66.

Olsen, B.D., Li, X., Wang, J., and Segalman, R.A. (2009) Near-surface and internal lamellar structure and orientation in thin films of rod-coil block copolymers. *Soft Matter*, **5**, 182–192.

Pitois, O. and Francois, B. (1999) Formation of ordered micro-porous membranes. *The European Physical Journal B*, **8**, 225–231.

Reneker, D.H. and Chun, I. (1996) Nanometre diameter fibres of polymer, produced by electrospinning. *Nanotechnology*, **7**, 216–223.

Ruotsalainen, T., Turku, J., Hiekkataipale, P. *et al.* (2007) Tailoring of the hierarchical structure within electrospun fibers due to supramolecular comb-coil block copolymers: polystyrene-block-poly(4-vinyl pyridine) plasticized by hydrogen bonded pentadecylphenol. *Soft Matter*, **3**, 978–985.

Sary, N., Rubatat, L., Brochon, C. *et al.* (2007) Self-assembly of poly(diethylhexyloxy-p-phenylenevinylene)-b- poly (4-vinylpyridine) rod-coil block copolymer systems. *Macromolecules*, **40**, 6990–6997.

Sauve, G. and McCullough, R.D. (2007) High field-effect mobilities for diblock copolymers of poly(3-hexylthiophene) and poly(methyl acrylate). *Advanced Materials*, **19**, 1822–1825.

Segalman, R.A., McCulloch, B., Kirmayer and, S., and Urba, J.J. (2009) Block Copolymers for Organic Optoelectronics. *Macromolecules*, **42**, 9205–9216.

Shin, Y.M., Hohman, M.M., Brenner, M.P., and Rutledge, C.C. (2001) Experimental characterization of electrospinning: the electrically forced jet and instabilities. *Polymer*, **42**, 9955–9967.

Stupp, S.I., LeBonheur, V., Walker, K. *et al.* (1997) Supramolecular materials: self-organized nanostructures. *Science*, **276**, 384–389.

Surin, M., Marsitzky, D., Grimsdale, A.C. *et al.* (2004) Microscopic morphology of polyfluorene-poly(ethylene oxide) block copolymers: influence of the block ratio. *Advanced Functional Materials*, **14**, 708–715.

Tao, Y.F., Ma, B., and Segalman, R.A. (2008) Self-Assembly rod-coil block copolymers and their application in electroluminescent devices. *Macromolecules*, **41**, 7152–7159.

Tao, Y.F., McCulloch, B., Kim, S., and Segalman, R.A. (2009) The relationship between morphology and performance of donor-acceptor rod-coil block copolymer solar cells. *Soft Matter*, **5**, 4219–4230.

Tew, G.N., Pralle, M.U., and Stupp, S.I. (2000) Supramolecular materials with electroactive chemical functions. *Angewandte Chemie-International Edition*, **39**, 517–521.

Tung, Y.C. and Chen, W.C. (2009) Poly[2,7-(9,9-dihexylfluorene)]-*block*-poly[3-(trimethoxysilyl)propyl methacrylate] (PF-b-PTMSPMA) rod-coil block copolymers: synthesis, morphology and photophysical properties in mixed solvents. *Reactive & Functional Polymers*, **69**, 507–518.

Tung, Y.C., Wu, W.C., and Chen, W.C. (2006) Morphological transformation and photophysical properties of rod-coil poly[2,7-(9,9-dihexylfluorene)]-*block*-poly(acrylic acid) in solution. *Macromolecular Rapid Communications*, **27**, 38–44.

Wang, C.T., Kuo, C.C., Chen, H.C., and Chen, W.C. (2009) Nonwoven and aligned electrospun multicomponent luminescent polymer nanofibers: effects of aggregated morphology on the photophysical properties. *Nanotechnology*, **20**, 375604.

Wang, H., Wang, H.H., Urban, V.S. *et al.* (2000) Syntheses of amphiphilic diblock copolymers containing a conjugated block and their self-assembling properties. *Journal of the American Chemical Society*, **122**, 6855–6861.

Wang, H., You, W., Jiang, P. *et al.* (2004) Supramolecular self-assembly of conjugated diblock copolymers. *Chemistry - A European Journal*, **10**, 986–993.

Wu, W.C., Tian, Y., Chen, C.Y. *et al.* (2007) Theoretical and experimental studies on the surface structures of conjugated rod-coil block copolymer brushes. *Langmuir*, **23**, 2805–2814.

Xiao, X., Fu, Y.Q., Zhou, J.J. *et al.* (2007) Reversible thermally responsive and luminescent coil-rod-coil triblock copolymers. *Macromolecular Rapid Communications*, **28**, 1003–1009.

Yan, L. and Tao, W. (2008) Synthesis of achiral PEG-PANI rod-coil block copolymers and their helical superstructures. *Journal of Polymer Science Part A-Polymer Chemistry*, **46**, 12–20.

Yang, C.C., Tian, Y., Jen, A.K.Y., and Chen, W.C. (2006) New environmentally responsive fluorescent N-isopropylacrylamide copolymer and its application to DNA sensing. *Journal of Polymer Science Part A-Polymer Chemistry*, **44**, 5495–5504.

Zhao, B. and Brittain, W.J. (2000) Polymer brushes: surface-immobilized macromolecules. *Progress in Polymer Science*, **25**, 677–710.

Zubarev, E.R., Pralle, M.U., Li, L.M., and Stupp, S.I. (1999) Conversion of supramolecular clusters to macromolecular objects. *Science*, **283**, 523–526.

20

Bulk Self-Assembly of Linear Hybrid Polypeptide-Based Diblock and Triblock Copolymers

Sébastien Lecommandoux[1,2]
[1]*Université de Bordeaux, ENSCPB, 16 avenue Pey Berland, 33607 Pessac Cedex, France*
[2]*CNRS, Laboratoire de Chimie des Polymères Organiques, UMR5629 Pessac, France*

20.1 Introduction

Considerable effort has been devoted to elucidating the process by which block copolymers self-assemble into structures (Antonietti and Förster, 2003; Bates and Fredrickson, 1999; Colfen, 2001; Forster and Antonietti, 1998; Forster and Konrad, 2003). It is widely recognized that block copolymer morphologies result from the interplay of competing forces – one that minimizes the interfacial area of the chemically incompatible copolymer blocks and another that minimizes the entropically unfavorable chain stretching resulting from the segregation of those blocks (Bates and Fredrickson, 1990; Hamley, 1998). Through variation of the volume fraction and block properties, a range of morphologies can be produced (Mao and Ober, 1997; Walther and Finkelmann, 1996).

A large portion of the research up to this point has been carried out on materials composed of a synthetic block and a peptide block. These peptide–polymer conjugates combine the advantages of synthetic polymers (such as solubility, processability, and rubber elasticity) with those of polypeptides or protein-like polymers (such as secondary structure, functionality, and biocompatibility). While synthetic polymers generally present a coil structure, peptide sequences can adopt ordered conformations such as α-helices or β-strands (Figure 20.1). The former case produces block copolymers with a rod–coil character; such rod–coil block copolymers have generated unconventional nanoscale structures not observed for purely amorphous block copolymers. Copolymers containing peptide sequences with a β-strand

Complex Macromolecular Architectures: Synthesis, Characterization, and Self-Assembly, First Edition.
Edited by Nikos Hadjichristidis, Akira Hirao, Yasuyuki Tezuka and Filip Du Prez.

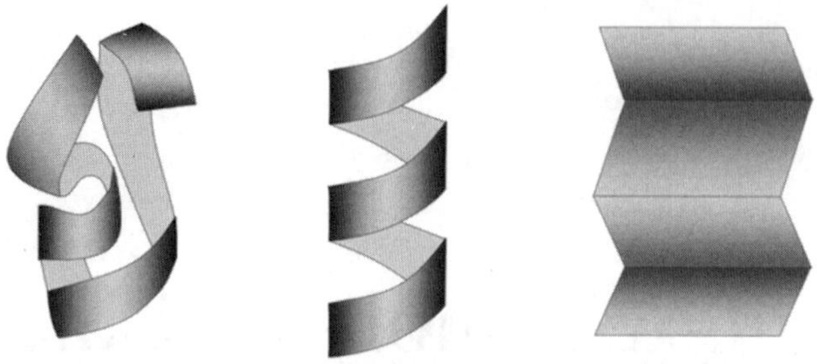

Figure 20.1 Schematic representations of secondary structures commonly adopted by polypeptides.

conformation offer an additional means to direct nanoscale structure formation through intermolecular hydrogen bonding.

Combining peptide sequences and synthetic polymers is useful not only for enhanced control over nanoscale structure formation, but also for production of biologically interactive materials. Biomimetic hybrid polymers may also produce sophisticated superstructures with new material properties. "Smart" materials based on polypeptides may reversibly change conformation and associated properties in response to an environmental stimulus, such as a shift in pH or temperature (Rodriguez-Hernandez *et al.*, 2005). Polypeptide block copolymers may also be used as model systems to study generic self-assembly processes in natural proteins. Obviously, such materials could be of significant interest for a variety of biomedical and bioanalytical applications.

This chapter will focus on linear polymer structures composed of synthetic polypeptide or polypeptide-polymer systems formed in bulk. Our aim is to present an overview of bulk self-assembled structures from polypeptide-based conjugates, specifically diblock and triblock architectures. Recent advances in theoretical developments and the first phase diagrams of such kind of polypeptide rod–coil systems will be also presented.

20.2 Diblock Copolymer Architectures

20.2.1 Polydiene-Based Diblock Copolymers

One of the first reports on the nanoscale solid-state structure of peptide–synthetic hybrid block copolymers was published by Gallot and coworkers in 1976 (Perly *et al.*, 1976). In this publication, the solid-state structure of a series of polybutadiene-*block*-poly(γ-benzyl L-glutamate) (PB-*b*-PBLG) and polybutadiene-*block*-poly(N^5-hydroxyethyl L-glutamine) (PB-*b*-PHLGln) diblock copolymers was investigated using a combination of techniques, including infrared and circular dichroism (CD) spectroscopy, X-ray scattering (SAXS), and transmission electron microscopy (TEM). The block copolymers covered a broad composition range with peptide contents ranging from 19 to 75%. Interestingly, SAXS and TEM revealed a well-ordered lamellar superstructure characterized by up to four higher-order Bragg spacings for all of the investigated samples. Wide-angle X-ray scattering (WAXS) experiments indicated that the α-helical peptide blocks were assembled in a hexagonal array. For a number of block copolymer samples it was found that the calculated length of the peptide helix was larger than the thickness of the polypeptide layer. To accommodate this difference, it was proposed that the peptide helices were folded in the peptide layer. The lamellar structure consists of planar, parallel equidistant sheets. Each sheet is obtained by superposition of two layers: (i) the PB chains in a more or less random coil conformation and (ii) the α-helical polypeptide blocks in a hexagonal array of folded chains. The hexagonal-in-lamellar structure was also found for polybutadiene-*block*-poly(ε-benzyloxycarbonyl L-lysine) (PB-*b*-PZLLys) and polybutadiene-*block*-poly(L-lysine) (PB-*b*-PLLys) block copolymers by the same authors (Billot *et al.*, 1976; Billot *et al.*, 1977; Douy and Gallot, 1977; Douy and Gallot, 1982).

Two decades later, the solid-state nanoscale structure of polyisoprene-*block*-poly(ε-benzyloxycarbonyl L-lysine) (PI-*b*-PZLLys) diblock copolymers was reported (Babin *et al.*, 2005). Diblock copolymers composed of a PI block with a number-average degree of polymerization of 49 and a PZLLys block containing 61–178 amino acid residues were investigated with dynamic mechanical analysis (DMA) and X-ray scattering (SAXS and WAXS). For the PI_{49}-*b*-$PZLLys_{35}$, PI_{49}-*b*-$PZLLys_{61}$ and PI_{49}-*b*-$PZLLys_{92}$, the X-ray scattering data were in agreement with a hexagonal-in-lamellar morphology. Interestingly, for PI_{49}-*b*-$PZLLys_{123}$ and PI_{49}-*b*-$PZLLys_{178}$ a hexagonal-in-hexagonal structure was found. This morphology is illustrated in Figure 20.2. This structure is unprecedented for polydiene-based peptide hybrid block copolymers, but has also been found for low molecular weight polystyrene-*block*-poly (γ-benzyl L-glutamate) (PS-*b*-PBLG) copolymers (Klok and Lecommandoux, 2001).

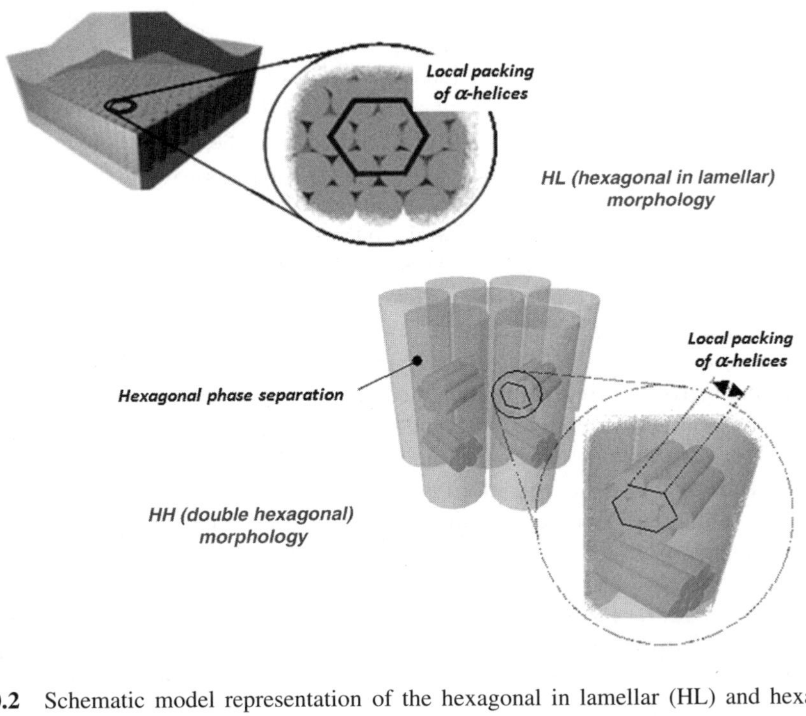

Figure 20.2 Schematic model representation of the hexagonal in lamellar (HL) and hexagonal in hexagonal (HH) morphologies.

20.2.2 Polystyrene-Based Diblock Copolymers

In an early study, Gallot and coworkers (Billot *et al.*, 1977) reported on the bulk nanoscale structure of PS-*b*-PZLLys diblock copolymers, which were based on a PS block with a number-average molecular weight of $M_n = 37$ kg/mol and had peptide contents ranging from 18 to 80 mol%. X-ray scattering patterns of dry samples following evaporation from dioxane showed two sets of signals, characteristic again of a hexagonal-in-lamellar superstructure. For several samples, the calculated length of the peptide helices was larger than the peptide layer thickness as determined from the X-ray data. In these cases, it was proposed that the helical PZLLys chains were folded in the peptide layer. Thus, the bulk nanoscale structure of the PS-*b*-PZLLys copolymers can be described in terms of the same hexagonal-in-lamellar model as it was also proposed for the PB-based block copolymer described earlier (Figure 20.2). Along the same lines, Douy and Gallot (Douy and Gallot, 1982) also studied the bulk nanoscale organization of PS-*b*-PBLG. For block copolymers composed of a PS block with $M_n = 25$ kg/mol and containing 31–94 mol% peptide, the same hexagonal-in-lamellar morphology as described above for the PS-*b*-PZLLys was found. The biocompatibility of PS-*b*-PBLG copolymers was later discussed by Mori *et al.* (1986). They studied the thrombus formation by exposing films of the diblock copolymers and the corresponding homopolypeptides to fresh canine blood. It was found that thrombus formation on the diblock copolymer films was reduced compared to the corresponding homopolymers. For the block copolymers, thrombus formation decreased with decreasing PBLG block length. Also, adsorption of plasma proteins such as bovine serum albumine, bovine γ-globulin, and bovine plasma fibrinogen was reduced on the block copolymers compared to PS homopolymer.

The characterization of the solid-state nanoscale organization of PS-polypeptide hybrid block copolymers has more recently been refined in a series of publications by Schlaad and coworkers (Losik *et al.*, 2004; Schlaad *et al.*, 2002; Schlaad, *et al.*, 2004). In a first report, three PS-*b*-PZLLys diblock copolymers were investigated (Schlaad *et al.*, 2002) by SAXS from dimethylformamide (DMF) cast films, confirming the hexagonal-in-lamellar morphology published earlier by Gallot *et al.* (Billot *et al.*, 1977). In this paper, Schlaad and coworkers went a step further and analyzed their SAXS data using the interface-distribution concept and the curvature-interface formalism. These evaluation techniques suggested that the bulk nanoscale structure of the PS-*b*-PZLLys diblock copolymers does not consist of plain but undulated lamellae. The concept of the interface distribution function and the curvature-interface formalism were also applied to compare the solid-state structure of two virtually identical PS-based diblock copolymers; PS_{52}-*b*-$PZLLys_{111}$ ($\phi_{peptide} = 0.82$) and PS_{52}-*b*-$PBLG_{104}$ ($\phi_{peptide} = 0.79$) (Losik *et al.*, 2004). Analysis of the SAXS data obtained on DMF cast films indicated a hexagonal-in-undulated (or zigzag) lamellar morphology for both block copolymers. In addition, the influence of the polydispersity of the polypeptide block on the solid-state morphology of PS-*b*-PZLLys diblock copolymers has been studied (Schlaad *et al.*, 2004). To this end, a series of five diblock copolymers was prepared from an identical ω-amino-polystyrene macroinitiator ($P_n = 52$; polydispersity index, PDI $= 1.03$). The peptide content in these diblock copolymers varied between 0.43 and 0.68 and the PDI ranged from 1.03 to 1.64. Evaluation of the SAXS data with the interface distribution function and the curvature-interface formalism confirmed, as expected, the hexagonal-in-undulated (or zigzag) lamellar solid-state morphology. Fractionation of the peptide helices according to their length leads (locally) to the formation of a nearly planar, parallel lamellar interface, which is disrupted by kinks (undulations).

Reiter and Schlaad and coworkers (Ludwigs *et al.*, 2005) used atomic force microscopy (AFM) to investigate the formation of hierarchical structures of PS_{52}-*b*-$PBLG_{104}$ in thin films. Thin films with a thickness of ~4 and 40 nm were prepared by spin coating of dilute polymer solutions on silicon substrates and were subsequently annealed in saturated tetrahydrofuran (THF) vapor to achieve a controlled crystallization of the α-helical PBLG. On the smallest length scale, the structure was found to be built of short ribbons or lamellae of interdigitated polymer chains. PBLG helices were fully stretched in thin films, in contrast to what has been observed in the 3D organized bulk mesophase (see above). Depending on the time of solvent annealing, different ordered structures on the micrometer length scale could be observed.

The examples discussed so far all involved relatively high molecular weight diblock copolymers. In this case, the molecular weight of the polypeptide block is usually sufficiently high so that it forms a stable α-helix and the common hexagonal-in-lamellar morphology is found. The situation changes, however, when the molecular weight of the block copolymers is significantly decreased. The influence of molecular weight on the solid-state organization of polystyrene-based peptide–synthetic hybrid block copolymers has been studied for a series of low-molecular weight PS-*b*-PBLG and PS-*b*-PZLLys (Klok *et al.*, 2000; Lecommandoux *et al.*, 2001). These diblock copolymers consisted of a short PS block with $P_n \sim 10$, a polypeptide block containing ~10–80 amino acid repeat units and were characterized by means of variable-temperature Fourier transform infrared (FTIR) spectroscopy and X-ray scattering. These experiments allowed the construction of "phase diagrams", which are shown in Figure 20.3. The phase diagrams reveal a number of interesting features. At temperatures below 200 °C and for sufficiently long polypeptide blocks, a hexagonal arrangement of the diblock copolymers was found, analogous to the hexagonal-in-lamellar morphology of the high-molecular weight analogues. Upon decreasing the length of the peptide block, however, several novel solid-state structures were discovered. For very short peptide block lengths (PS_{10}-*b*-$PBLG_{10}$, PS_{10}-*b*-$PZLLys_{20}$, PS_{10}-*b*-$PZLLys_{40}$ and PS_{10}-*b*-$PZLLys_{60}$) a lamellar supramolecular structure was found. This is due to the fact that, for such short peptide block lengths, a substantial fraction of the peptide blocks adopts a β-strand secondary structure.

Self-assembly of these diblock copolymers in a β-sheet type fashion results in the lamellar structures observed by SAXS. For PS_{10}-*b*-$PBLG_{20}$ a peculiar and previously unprecedented structure was found. This structure, which consisted of hexagonally packed diblock copolymer molecules that are organized in a hexagonal superlattice, has been referred to as the double-hexagonal or hexagonal-in-hexagonal morphology.

Apart from several unconventional solid-state nanoscale structures, another factor that distinguishes the phase diagrams in Figure 20.3 from those of the majority of conventional, conformationally isotropic block copolymers is the influence of temperature. For a number of diblock copolymers, increasing the temperature above 200 °C results in a change from a hexagonal-in-hexagonal (PS_{10}-*b*-$PBLG_{20}$) or hexagonal (PS_{10}-*b*-$PBLG_{40}$, PS_{10}-*b*-$PZLLys_{80}$) to a lamellar morphology. FTIR spectroscopy experiments suggested that these morphological transitions are induced by an increase in the fraction of peptide blocks that have a β-strand conformation. Interestingly, the self-assembly of the amphiphilic form of these diblock copolymers, namely polystyrene-*block*-poly(glutamic acid) (PS-*b*-PGA) revealed that the high immiscibility between PS and PGA blocks (conformational asymmetry) drove all of these rod–coil block copolymers to self-assemble in a hexagonal in lamellar (HL) morphology in the nanometer size range, regardless of their composition and architecture (Babin *et al.*, 2008) (Figure 20.4).

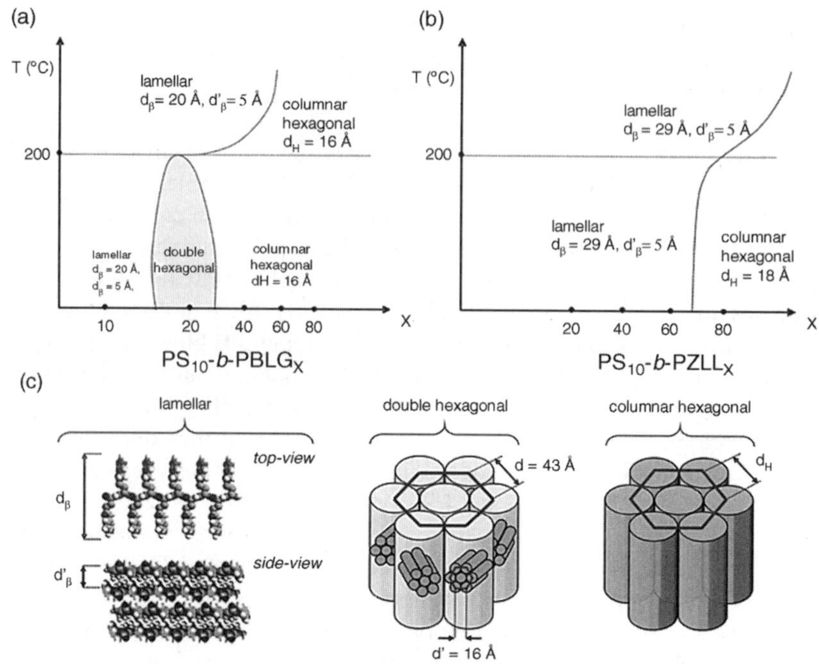

Figure 20.3 Phase diagrams describing the solid-state nanoscale structure of (a) PS-b-PBLG and (b) PS-b-PZLLys diblock copolymers. (c) Illustration of the lamellar, double hexagonal and hexagonal morphologies found for the low molecular-weight hybrid block copolymers. (Reprinted with permission from H.-A. Klok, J.F. Langenwalter and S. Lecommandoux, "Self-assembly of peptide-based diblock oligomers," *Macromolecules*, **33**, 22, 7819–7826, 2000. © 2000 American Chemical Society; S. Lecommandoux, M.-F. Achard, J.F. Langenwalter and H.-A. Klok, "Self-assembly of rod-coil diblock oligomers based on α-helical peptides," *Macromolecules*, **34**, 26, 9100–9111. © 2001 American Chemical Society.)

20.2.3 *Polyether-Based Diblock Copolymers*

Poly(ethylene glycol)–polypeptide block copolymers are of particular interest, from both a structural and a functional point of view. In addition to microphase separation and the tendency of the peptide blocks towards aggregation, crystallization of poly(ethylene glycol) (PEG) introduces an additional factor that can influence the structure formation of these hybrid block copolymers. Ma and coworkers (Zhang *et al.*, 2003) have investigated the solid-state structure and properties of three poly(ethylene glycol)-*block*-poly(L-Alanine) (PEG-*b*-PLAla) copolymers that were prepared from a PEG macroinitiator with $M_n = 2$ kg/mol. The diblock copolymers contained 39.8, 49.6 and 65.5 mol% alanine. From FTIR spectra and differential scanning calorimetry (DSC) measurements, the authors proposed a microphase-separated bulk structure.

AB diblock and ABA triblock copolymers composed of PEG as the A block and random coil segments of poly(D,L-valine-*co*-D,L-leucine) as the B block(s) were investigated by Cho and coworkers (Cho *et al.*, 2003). DSC experiments revealed PEG crystallization and showed that

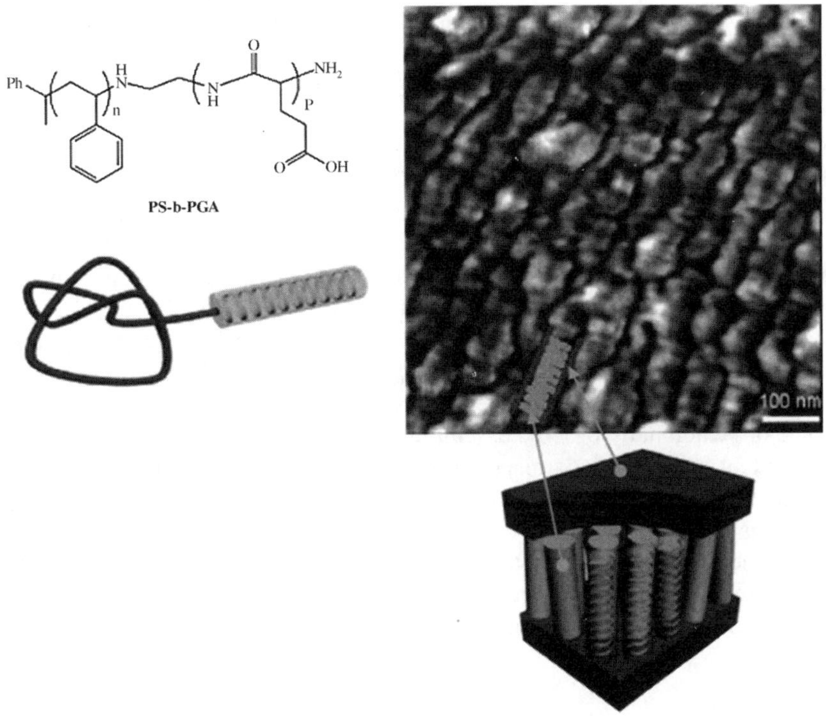

Figure 20.4 AFM image in phase mode of the self-assembled lamellar structure of PS-b-PGA block copolymers. (Reprinted with permission from J. Babin, D. Taton, M. Brinkmann and S. Lecommandoux, "Synthesis and self-assembly in bulk of linear and mikto-arm star block copolymers based on polystyrene and poly(glutamic acid)," *Macromolecules*, **41**, 4, 1384–1392, 2008. © 2008 American Chemical Society.)

the PEG melting temperature was decreased compared to that of PEG homopolymer. TEM micrographs suggested a lamellar microphase-separated structure for one of the triblock copolymer samples.

20.2.4 Polyester-Based Diblock Copolymers

One of the first studies focusing on the solid-state properties of polypeptide–polyester synthetic hybrid block copolymers was reported by Jérôme and coworkers (Degée *et al.*, 1993). Differential scanning calorimetry (DSC) experiments on a poly(ε-caprolactone)$_{50}$-*block*-poly(γ-benzyl L-glutamate)$_{40}$ (PCL$_{50}$-*b*-PBLG$_{40}$) diblock copolymer revealed two endotherms. The first endotherm was found at 60 °C and is due to the melting of the PCL. The second endotherm that was located at 110 °C, was, mistakenly, interpreted as the melting transition of PBLG. This transition, however, is not a melting transition, but instead reflects the conformational transition of the PBLG helix from a 7/2 to a 18/5 helical structure. Although no further structural investigations were carried out, the observation of two separate endotherms occurring at temperatures identical to the transitions found for the respective homopolymers is a first indication of the existence of a microphase-separated structure. Similar results were reported by Chen and coworkers (Rong *et al.*, 2003) who investigated the thermal properties of a series of PCL-*b*-PBLG copolymers composed of PCL blocks containing 13–51 repeat units peptide segments with 22–52 amino acid repeat units.

Caillol *et al.* (2003) have studied the solid-state structure and properties of a series of poly(L-lactide)-*block*-poly(γ-benzyl L-glutamate) (PLLac-*b*-PBLG) copolymers. The PLLac block in these copolymers contained 10–40 repeat units and the peptide segments were composed of 20–100 repeat units. DSC thermograms of the block copolymers revealed three transitions corresponding to the glass transition temperature (T_g) of PLLac (~50 °C), the 7/2–18/5 helix transition of PBLG (~100 °C) and the melting temperature of PLLac (~160 °C), respectively. This observation already provided a first hint towards a microphase-separated bulk morphology. WAXS experiments, which were performed at 100 °C, indicated the existence of hexagonally ordered assemblies of α-helical PBLG chains. With decreasing glutamate content, the peaks corresponding to this hexagonal organization decreased in intensity and another scattering peak appeared, which was ascribed to a lamellar assembly of PBLG chains with a β-strand secondary structure. Increasing the temperature to 200 °C resulted in the melting of PLLac, but also led to a decrease in intensity of the diffraction peaks corresponding to the hexagonally ordered α-helical PBLG segments and an increase in the fraction of PBLG segments that are ordered in a lamellar β-strand fashion.

20.2.5 Diblock Copolypeptides

A step forward in the design of hierarchically ordered structures with biofunctionality has been the recent reports on the synthesis of block copolymers based on polypeptides. Despite important synthetic efforts, the solid-state morphology of purely peptidic block copolymers remains largely unexplored. Hadjichristidis and coworkers (Papadopoulos *et al.*, 2005) recently investigated the self-assembly of a series of narrow polydispersity poly(γ-benzyl L-glutamate)-*block*-poly(L-glycine) (PBLG-*b*-PLGly) diblock copolymer within the composition range $0.67 < f_{PBLG} < 0.97$ and the temperature range $303 < T < 433$ K. SAXS, WAXS, ^{13}C NMR and DSC were used for the structure investigation coupled with dielectric spectroscopy for both the peptide secondary structure and the associated dynamics. Not only did these techniques provide insight into the nanophase morphology, but they also gave information about the type and persistence of peptide secondary structures. The main finding of the study was that the nanoscale confinement induces the chain folding of polyglycine β-sheets within their domain. Hexagonal-in-lamellar and cylinder-on-hexagonal nanostructures have been especially evidenced. In addition, nanoscale confinement proves to be important in controlling the persistence length of secondary peptide motifs. The same authors recently studied the stability, persistence and dynamics of the peptide secondary motifs in a series of poly(γ-benzyl-L-glutamate)-b-polyalanine (PBLG-*b*-PAla) polypeptides (Gitsas *et al.*, 2008b) and PBLG-*b*-PZLLys (Gitsas *et al.*, 2008a) through a combination of structural (X-rays, NMR) and dynamic (dielectric spectroscopy, NMR) probes. The unfavorable enthalpic interactions between the unlike blocks gave rise to nanophase separation that resulted in the destabilization of PAla β-sheets. Contrary to this, the overall helicity of PBLG α-helices was enhanced.

20.2.6 Miscellaneous Diblock Copolymers

Recently, the enhanced control of bulk and surface ordering of polypeptides has been employed to drive the processing and self-assembly of conductive rigid segments. Mainly triblock copolymer structures have been studied (see the following section). Manners and Winnik (Wang *et al.*, 2008) synthesized a new type of metallopolymer–polypeptide block copolymer poly(ferrocenyldimethylsilane)-*b*-poly(ε-benzyloxycarbonyl-L-lysine) (PFS-*b*-PZLLys) and studied their self-organization behavior in both the bulk state and in solution. In the bulk state, a cylindrical in-lamellar structure was observed in a compositionally asymmetric sample. Rod-like micelles with a polyferrocenylsilane core formed in a polypeptide selective solvent alone or in the presence of a common solvent.

20.3 Triblock Copolymer Architectures

20.3.1 Polydiene-Based Diblock Copolymers

Whereas Gallot *et al.* have mainly studied the solid-state organization of PB-based diblock copolymers, Nakajima *et al.* concentrated on ABA-type hybrid block copolymers containing PB as the B component ("once-broken rods"). In a first series of publications, the structure and properties of PBLG-*b*-PB-*b*-PBLG triblock copolymers containing 7.5–32.5 mol% (=3.0–14.3 vol%) PB were investigated (Nakajima *et al.*, 1979b). Infrared spectroscopy and WAXS experiments on films of the triblock copolymers indicated that the PBLG blocks were predominantly α-helical. From the WAXS experiments, it was concluded that the PBLG blocks assembled into different structures, depending on the type of solvent that was used to cast the films. In benzene-cast films, the peptide helices were relatively poorly ordered, similar to the so-called form A morphology of PBLG (Mckinnon and Tobolsky, 1968). In contrast, the PBLG segments in films cast from chloroform ($CHCl_3$) were well ordered and contained paracrystalline and mesomorphic regions. Based on TEM analysis, a cylindrical microstructure was proposed for a triblock copolymer containing 8 vol% PB. Interestingly, copolymers having the same composition but polypeptide segments made of either enantiomerically pure or racemic γ-benzyl glutamate exhibited not only different secondary structures (α-helix or random coil, respectively; FTIR and WAXS), but also different superstructures (TEM). A cylindrical or lamellar morphology was proposed in the first case and a more spherical superstructure in the second case (Hayashi *et al.*, 1984).

The bulk and surface structure of solvent cast films from a series of PBLG-*b*-PB-*b*-PBLG triblock copolymers with much higher PB contents (50–80 mol%) than the samples discussed above have been discussed by Gervais *et al.* (1988). The organization of these copolymers was compared with that of three other triblock copolymers with approximately the same PB content (~50 mol%) but that were composed of poly(N^{ε}-trifluoroacetyl L-lysine) (PTLLys), poly(N^5-hydroxyethyl L-glutamine) (PHLGln) or polysarcosine (PSar) as the peptide block. For any sample investigated, X-ray scattering experiments indicated a hexagonal-in-lamellar bulk morphology. X-ray photoelectron spectroscopy (XPS) measurements revealed that for the triblock copolymers with hydrophobic peptide blocks, that is, PBLG or PTLLys, the surface composition was identical to that in the bulk of the sample. In contrast, the surfaces of films prepared from the triblock copolymers with the more hydrophilic peptide segments, that is, PHLGln or PSar, were PB enriched. Furthermore, the XPS data suggested that the lamellar superstructures formed by the triblock copolymers were perpendicular to the air/polymer interface.

In addition, the solid-state organization and properties of PZLLys-*b*-PB-*b*-PZLLys triblock copolymers have been investigated. Kugo *et al.* (1982b) have studied copolymers composed of a central PB block with $M_n = 3.6$ kg/mol and PB contents ranging from 12 to 52 mol%. WAXS patterns obtained from solution cast triblock copolymer films were in agreement with the α-helical secondary structure of the peptide blocks. The bulk microphase-separated structure of the five different block copolymer samples could be successfully characterized by means of TEM. For the samples with the largest PB volume fraction (56 and 65 vol%), a lamellar superstructure was found. For triblock copolymers with smaller PB volume fractions, the electron micrographs suggested cylindrical and spherical microphase-separated structures.

Other polybutadiene-based ABA type triblock copolymers that have been investigated include poly(γ-methyl L-glutamate)-*block*-polybutadiene-*block*-poly(γ-methyl L-glutamate) (PMLG-*b*-PB-*b*-PMLG) and poly(γ-methyl D,L-glutamate)-*block*-polybutadiene-*block*-poly (γ-methyl D,L-glutamate) (PMGlu-*b*-PB-*b*-PMGlu) (Hayashi *et al.*, 1984; Kugo *et al.*, 1983). Infrared spectroscopy experiments on solvent-cast films indicated that the incorporation of 50% of the D-isomer disrupts the α-helical secondary structure and induces a random coil conformation in significant portions of the peptide blocks. Interestingly, different morphologies were observed when comparing the TEM images of PMLG-*b*-PB-*b*-PMLG and PMGlu-*b*-PB-*b*-PMGlu samples with the same PB content (≈30 mol%). For the first, a cylindrical morphology and for the second a spherical structure was proposed (Hayashi *et al.*, 1984). The attenuated total reflection infrared (ATR-IR) spectra further showed that adsorption of bovine serum albumine (BSA) and bovine fibrinogen (BF) did not lead to denaturation. From these observations, the authors concluded that the surfaces of the PMGlu-*b*-PB-*b*-PMGlu membranes only weakly or reversibly interact with these plasma proteins and it was predicted that this may also lead to a good overall biocompatibility.

The solid-state structure and properties of PELGlu-*b*-PB-*b*-PELGlu (PELGlu = poly(γ-ethyl L-glutamate)) triblock copolymers containing 31.5–94.5 mol% (=17–88 vol%) PELGlu have been studied using the same techniques as described above for the other triblock copolymers (Chen *et al.*, 1981; Sato *et al.*, 1985). Based on the electron micrographs, a spherical microphase-separated structure was proposed for the copolymer containing 17 vol% PB, while cylindrical and lamellar morphologies were suggested for triblock copolymers containing 28 and 44 vol%, respectively, 68 and 88 vol% PB. The biocompatibility of the PELGlu-*b*-PB-*b*-PELGlu triblock copolymers was assessed by coating the samples onto a polyester mesh fiber cloth, which was subsequently subcutaneously implanted in mongrel dogs for four weeks. It was found that the foreign-body reaction and degradation of the PELGlu-*b*-PB-*b*-PELGlu samples were less pronounced as compared to PMLG-*b*-PB-*b*-PMLG, PBLG-*b*-PB-*b*-PBLG and PZLLys-*b*-PB-*b*-PZLLys triblock copolymers.

The bulk nanoscale structure of a series of PBLG-*b*-PI-*b*-PBLG copolymers containing 37.4–81.1 mol% PBLG was studied by means of infrared spectroscopy, WAXS, dynamic mechanical analysis and electron microscopy (Yoda *et al.*, 1995a; Yoda *et al.*, 1996; Yoda *et al.*, 1995b; Yoda *et al.*, 1997). Based on the electron micrographs, a cylindrical morphology was proposed for triblock copolymers containing 74.6 and 81.1 mol% PBLG. Water-permeability measurements also supported the microphase-separated bulk morphology (Yoda *et al.*, 1996).

Treatment of a PBLG-*b*-PI-*b*-PBLG film with a mixture of 3-amino-1-propanol and 1,8-octamethylenediamine led to the formation of hydrophilic, crosslinked PHLGln-*b*-PI-*b*-PHLGln membranes (Yoda *et al.*, 1994). The swelling ratio of these membranes in pseudoextracellular fluid (PECF) was found to decrease with increasing PI content and increasing crosslink density. Enzymatic degradation experiments using papain showed that the triblock copolymer films were more resistant towards degradation than the corresponding homopolypeptide membranes. The half-times for sample degradation increased with decreasing peptide content, which was in agreement with the swelling behavior of the membranes.

20.3.2 Polystyrene-Based Triblock Copolymers

Tanaka and coworkers (Imanishi *et al.*, 1985) studied ABA-type triblock copolymers composed of a central PS block flanked by two polypeptide segments (PBLG, PZLLys or PSar). TEM of a $CHCl_3$-cast film of $PBLG_{25}$-*b*-PS_{165}-*b*-$PBLG_{25}$ revealed a lamellar phase separated structure. In contrast, no microphase separation was observed in a film of $PSar_{73}$-*b*-PS_{421}-*b*-$PSar_{73}$. The authors proposed that the different block copolymer morphologies could be related to the different secondary structure of the peptide block; while the PBLG segments are predominantly helical, the PSar may not form any regular secondary structure. Fibrinogen adsorption on the block copolymer films was studied with ATR-IR spectroscopy and compared with that on the corresponding homopolymer films (Kugo *et al.*, 1987). It was found that fibrinogen adsorption on PS and PSar homopolymer films and on PSar-*b*-PS-*b*-PSar triblock copolymer films led to denaturation of the protein. In contrast, protein adsorption on the microphase-separated PBLG-*b*-PS-*b*-PBLG surfaces was reported to stabilize the protein's secondary structure. Blood-clotting tests suggested that thrombus formation was retarded compared to the respective homopolymers.

Samyn and coworkers (Janssen *et al.*, 1990) extended the investigations of ABA triblock copolymers and studied the solid-state organization of three different PBLG-*b*-PS-*b*-PBLG copolymers containing 34, 55 and 92 wt% PBLG. TEM micrographs of ultramicrotomed and RuO_4-stained specimens and SAXS experiments indicated a lamellar morphology for the copolymers with 34 and 55 wt% PBLG. The sample containing 92 wt% PBLG did not form a lamellar structure. WAXS patterns yielded d-spacings reflecting the intermolecular distance between neighboring peptide α-helices.

20.3.3 Polysiloxane-Based Triblock Copolymers

Kumaki and coworkers (Kumaki *et al.*, 1985) have studied the structure, antithrombogenicity and oxygen permeability of ABA triblock copolymers composed of polydimethylsiloxane (PDMS) as the B block and PBLG, PBGlu (poly(γ-benzyl D,L-glutamate)), PZLLys or PSar as the A block. TEM images of DMF-cast films provided evidence for a microphase-separated morphology for $PZLLys_{49}$-*b*-$PDMS_{400}$-*b*-$PZLLys_{49}$ and $PZLLys_{91,160}$-$PDMS_{256}$-$PZLLys_{91,160}$. The images revealed a spherical morphology composed of PDMS islands in a PZLLys matrix. The formation of these spherical domains was attributed to the solvent that was used for sample preparation. While DMF is a good solvent for PZLLys, it is a poor solvent for PDMS. In a separate publication, the same authors also described nonspherical microphase-separated structures (Kang *et al.*, 1988a). In CH_2Cl_2-cast films of a triblock copolymer with a very high PDMS content ($PBLG_{48}$-*b*-$PDMS_{508}$-*b*-$PBLG_{48}$, 83 mol% PDMS) more extended, rod-like PBLG aggregates in a matrix of PDMS were observed. The TEM experiments also provided insight into the effects of peptide secondary structure and the nature of the casting solvent on the thin film morphology. $PBLG_x$-*b*-$PDMS_{30}$-*b*-$PBLG_x$ ($x = 8$, 20, 43) triblock copolymers have also been studied by Papon and coworkers (Ibarboure *et al.*, 2007). Their self-assembly of a double-hexagonal structure was addressed using X-ray scattering and AFM imaging. The thermal behavior of these structures was investigated using dynamic mechanical analysis and differential scanning calorimetry.

The structure and the associated dynamics of a series of poly(γ-benzyl-l-glutamate)-*b*-poly-(dimethyl siloxane)-*b*-poly(γ-benzyl-l-glutamate) (PBLG-*b*-PDMS-*b*-PBLG) triblock copolymers were investigated using small- and wide-angle X-ray scattering, NMR, transmission electron microscopy, and dielectric spectroscopy, respectively (Papadopoulos *et al.*, 2006). The structural analysis revealed phase separation in the case of the longer blocks with defected α-helical segments embedded within the block copolymer nanodomains. The α-helical persistence length was found to depend on the degree of segregation; thermodynamic confinement and chain stretching results in the partial annihilation of helical defects. The dynamics of polypeptides and hybrid copolymers has been recently reviewed (Floudas and Spiess, 2009).

Studies on adsorption/denaturation of proteins and oxygen permeation measurements from these triblock copolymers also describe the relationship between the film morphology and these properties. A detailed study of the gas-permeation properties of PBLG-*b*-PDMS-*b*-PBLG films (PDMS = polydimethylsiloxane) cast from dichloromethane (CH_2Cl_2) and DMF solution with PDMS contents ranging from 46 to 83 mol% was reported (Kang *et al.*, 1988b) and revealed that the oxygen permeability of the triblock copolymer films in water exponentially increased with increasing PDMS content, in agreement with a microphase-separated morphology of the membranes.

20.3.4 Polyether-Based Triblock Copolymers

Kugo *et al.* (1987) studied the solid-state conformation of the peptide segment of a series of PBLG-*b*-PEG-*b*-PBLG copolymers containing a PEG segment with a molecular weight of 4 kg/mol and 36–86 mol% PBLG. FTIR spectroscopy experiments on $CHCl_3$-cast films revealed that the PBLG blocks, which had degrees of polymerization of 25–276, had an α-helical secondary structure. The helix content of the triblock copolymer containing $PBLG_{276}$ blocks was found to be similar to that of the PBLG homopolymer. Swelling the triblock copolymer films with water resulted in a decrease in helix content, as indicated by the CD spectra. This decrease in helicity was attributed to competition of water clusters to form hydrogen bonds with the peptide backbone. The effect was even more pronounced when pseudo-extracellular fluid was used instead of water.

A first detailed study of the solid-state nanoscale structure of peptide–PEG hybrid block copolymers was published by Cho *et al.* (1990). These authors investigated thin, $CHCl_3$-cast films of PBLG-*b*-PEG-*b*-PBLG copolymers, which were composed of a PEG block of 2 kg/mol and contained 25–76 mol% PBLG. TEM micrographs of RuO_4-stained specimens revealed a lamellar morphology for triblock copolymers containing 25–64 mol% PBLG. The microphase-separated structure was proposed to consist of chain-folded, crystalline PEG domains and helical PBLG domains (IR). WAXS patterns were consistent with the ordered, crystalline-like solid-state modification of PBLG. In contrast, in films cast from benzene, the peptide blocks only formed poorly ordered arrays. The sensitivity of the organization of the PBLG blocks towards the nature of the casting solvent is identical to the behavior of the PBLG homopolymer.

While the data reported by Cho *et al.* described the structure and organization of thin solvent-cast films of PBLG-*b*-PEG-*b*-PBLG, Floudas and coworkers have extensively studied the bulk nanoscale organization of these materials (Floudas *et al.*, 2003). To this end, a series of triblock copolymers with PBLG volume fractions (f_{PBLG}) ranging from 0.07 to 0.89 was investigated using SAXS/WAXS, polarizing optical microscopy (POM), DSC and FT-IR spectroscopy. For triblock copolymers with $f_{PBLG} \leq 0.25$, PEG crystallization was observed, but with significant undercooling. Triblock copolymers with $f_{PBLG} \geq 0.43$ did not show PEG crystallization. SAXS experiments, which were carried out at 373 K, that is, above the melting point of PEG, also revealed a different behavior for triblock copolymers with small and large PBLG volume fractions. For triblock copolymers with $f_{PBLG} \geq 0.43$ only a weakly phase-separated structure was found, whereas for samples with $f_{PBLG} \leq 0.25$ the SAXS data clearly indicated a microphase-separated structure. WAXS patterns showed that in the microphase-separated state the PEG phase was semicrystalline and the peptide phase consisted of hexagonally ordered assemblies of PBLG α-helices that coexisted with β-sheet structures. For triblock copolymers with $f_{PBLG} \geq 0.43$, PEG is amorphous and interspersed with aggregates of α-helical PBLG segments and unordered peptide chains.

Additional insight into the solid-state nanoscale organization of the triblock copolymers just discussed could be obtained by combining SAXS/WAXS with different microscopic techniques (TEM and AFM) (Parras *et al.*, 2005). A "broken lamellar" morphology was observed in the TEM micrographs of $PBLG_{58}$-*b*-PEG_{90}-*b*-$PBLG_{58}$ ($f_{PBLG} = 0.58$). Annealing converted this metastable structure into a nonuniform microphase-separated pattern, which was proposed to consist of "puck-like" PEG domains in a PBLG matrix. For $PBLG_{105}$-*b*-PEG_{90}-*b*-$PBLG_{105}$ ($f_{PBLG} = 0.67$), a lamellar morphology was found in the as-cast film, which was transformed into a "broken lamellar" structure upon annealing. Based on the results, a morphology map was constructed.

Cho *et al.* (1994) reported on the solid-state structure of a series of PZLLys-*b*-PEG-*b*-PZLLys composed of a PEG block with $M_n = 2$ kg/mol and PZLLys contents of 25.2, 49.9 and 83.0 mol% (= 68, 86 and 98 vol%). Infrared spectra of $CHCl_3$-cast films were in agreement with a helical secondary structure of the peptide blocks. DSC experiments provided a first hint for the existence of a microphase-separated structure and revealed two T_gs for all samples. The higher T_g was very close to that of the PZLLys homopolymer and the lower T_g approximately 20 °C higher than that of PEG homopolymer. A PEG melting transition was not observed. These results were interpreted in terms of a microphase-separated structure with hard, crystalline PZLLys domains and soft, amorphous PEG segments. The presence of a microphase-separated structure was confirmed by TEM micrographs of RuO_4-stained thin films.

Akashi and coworkers (Tanaka *et al.*, 2004) reported on the solid-state nanoscale structure of ABA type triblock copolymers composed of a central PEG block flanked by two poly(β-benzyl L-aspartate) (PBLAsp) blocks. The molecular weight of the central PEG block was 11 or 20 kg/mol and the degrees of polymerization of the peptide blocks ranged from 12 to 32. WAXS and POM studies on CH_2Cl_2-cast films showed PEG crystallization in all samples.

DSC experiments showed that the melting temperature of the crystalline PEG domains decreased linearly with increasing PBLAsp content, reflecting the strong influence of the peptide segments on PEG crystallization. More interestingly, the authors found that heating the as-cast films above 333 K and cooling down to 303 K, converted a certain fraction of the α-helical PBLAsp chains into β-strands and was accompanied by a decrease in PEG crystallinity. On a macroscopic level, this led to increased strength and elasticity of the films.

Cho *et al.* (1988) have studied triblock copolymers composed of a middle block of poly (propylene glycol) (PPG) with a molecular weight of 2000 flanked by two PBLG segments. Three triblock copolymer samples were investigated with PPG contents of 17.0, 26.0 and 60.0 mol%, respectively. According to infrared spectra and WAXS that were recorded from $CHCl_3$-cast films, the PBLG blocks possessed an α-helical secondary structure. No further details on the solid-state nanoscale structure and the possibility of microphase separation were reported. Platelet adhesion on glass beads coated with block copolymers containing 33 or 47 mol% PPG was reduced compared to beads modified with PMLGlu homopolymer or block copolymers with 70.0 mol% PPG. These differences were attributed to differences in surface composition and morphology, which, unfortunately, were not further discussed.

Very recently, Mezzenga and coworkers (Sanchez-Ferrer and Mezzenga, 2010) studied a series of triblock, but also diblock and star-like copolymers based on PPG and PBLG. The influence on their secondary structure and solid-state organization have been studied in terms of their architecture, molecular weight of the PPG coil block (from 2 to 5 kDa), and volume fraction of the rod PBLG block (0.50 or 0.75). The degree of polymerization of the polypeptide segment into the arms of the block copolymers, strongly affects the final alpha-helical secondary structure and the corresponding self-assembling of the block copolymers. Below 20 amino acid residues, a mixture of secondary structures (alpha-helices, beta-sheets, and unordered segments) and microphase separation of the blocks is present. Above 20 repeating units, the microphase separation goes together with macrophase separation of the pure alpha-helical secondary structure, which ends up in an orthogonal lamellar phase or rods (Figure 20.5).

20.3.5 Miscellaneous Triblock Copolymers

The structure and properties of an ABA triblock copolymer composed of a polyetherurethaneurea (PEUU) B block with $M_n = 15.8$ kg/mol and two $PBLG_{29}$ B blocks were described by Ito and coworkers (Ito *et al.*, 1993). DSC thermograms showed a single endotherm located between the T_g of the PEUU and the 7/2–18/5 helix transition of PBLG, suggesting that there was no phase separation. Platelet adhesion on DMF-cast films of the triblock copolymer was significantly reduced compared to the respective homopolymers and a PEUU/PBLG blend. Thrombus formation on block copolymer films was found to be ~50% less compared to a glass surface. However, no significant difference in antithrombogenicity between PEUU, PBLG, their blend and the block copolymer was observed.

Another, very early study focused on two PBLG-*b*-PBAN-*b*-PBLG (PBAN = poly(butadiene-*co*-acrylonitrile)) triblock copolymers (Barenberg *et al.*, 1981). These block copolymers were composed of a PBAN block with $M_n = 3.4$ kg/mol and two PBLG segments containing either 80 or 160 repeat units. TEM micrographs of OsO_4-stained films cast from dioxane, which

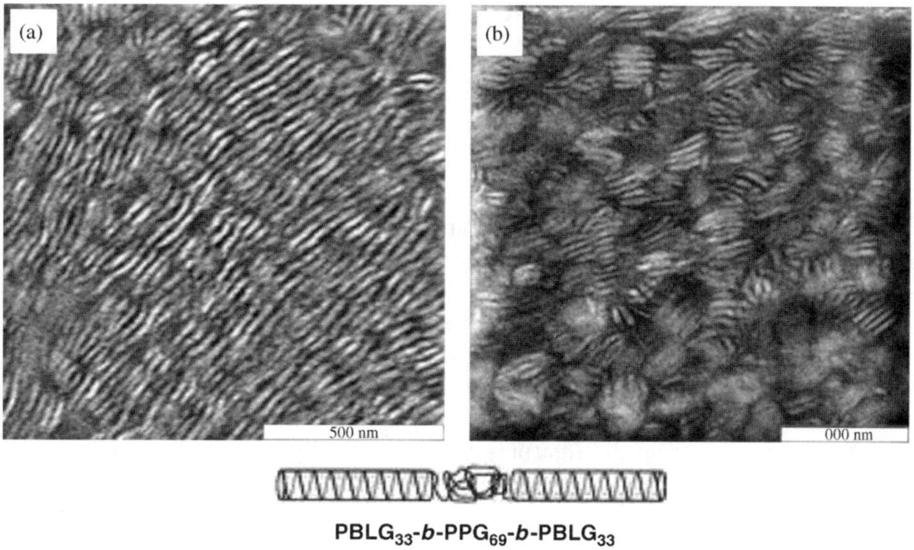

Figure 20.5 TEM images from the sample PBLG-b-PPG-b-PBLG after annealing. (a) Well-ordered domains and (b) clusters showing the phase separation between high helix content (lamellar zones) and mixture of secondary structures (gray zones). (Reprinted with permission from A. Sanchez-Ferrert and R. Mezzenga, "Secondary structure-induced micro- and macrophase separation in rod-coil polypeptide diblock, triblock and star-block copolymers," *Macromolecules*, **43**, 2, 1093–1100, 2010. © 2010 American Chemical Society.)

is a selective solvent for PBLG, revealed a lamellar morphology. When the nonselective solvent $CHCl_3$ was used for the preparation of the TEM specimens, the images were more homogeneous and phase separation was less distinct. This suggests that, depending on the solvent conditions, the PBAN block can affect PBLG secondary structure.

Electro- and photoactive peptide–synthetic hybrid triblock copolymers have been prepared using bis(benzyl amine)-terminated poly(9,9-dihexylfluorene-2,7-diyl) (PHF) as a macroinitiator for the ring-opening polymerization of BLGlu-NCA (Kong and Jenekhe, 2004). The electroactive and photoactive properties of the triblock copolymers were similar to those of the PHF homopolymer, indicating that the introduction of the PBLG segments did not interfere with charge injection and transport and other materials properties. The thin-film morphologies of the triblock copolymers were investigated with SFM using different casting solvents. Depending on solvent preparation, spherical and fibrillar nanostructures have been obtained. Later, triblock copolymer structures PBLG-*b*-PHF-*b*-PBLG with different PBLG ratio have been studied by Mezzenga and coworkers (Rubatat *et al.*, 2008). Depending on the solvent-casting history, the copolymers exhibit two different conformations: a rod–rod–rod or coil–rod–coil configuration. The observed photoluminescence spectra and dominant excited lifetimes of the poly(9,9-dihexylfluorene-2,7-diyl) block were found to differ markedly in rod–rod–rod and coil–rod–coil configurations and were correlated to the morphology of the self-assembled triblock copolymers.

20.4 Theory and Phase Diagram

While the self-assembly behavior of flexible block copolymers has been studied for over three decades now, in many practical applications, the blocks of the copolymer are characterized by some degree of structural rigidity. For instance, a flexible block may be covalently linked to a more rigid block to obtain a desirable combination of fracture and tensile strength properties. In addition, copolymers comprising conjugated polymeric units of current interest in optical and electrical applications possess a natural rigidity arising from the conjugated nature of the monomeric units (Jenekhe and Chen, 1998; Jenekhe and Chen, 2000; Leclere *et al.*, 2003; Leclere *et al.*, 1998). Synthetic polypeptides (Deming, 1997; van Hest and Tirrell, 2001), the self-assembly of which exhibits many characteristics similar to natural peptides and proteins, possess units such as α-helices and β-sheets that exhibit pronounced rigidity compared to the typical flexible polymers. Understandably, experiments on such rod–coil polymers have suggested an extremely rich morphological phase behavior arising from the interplay between the liquid-crystalline ordering of the rod-like units and the microphase separation transition between the unlike units (Lee *et al.*, 2001).

Considering the polypeptide in α-helical conformation as rigid rods, one can compare the behavior of these polypeptide-based diblock copolymers with theoretical phase diagrams calculated for rod–coil block copolymers (Halperin, 1990; Olsen and Segalman, 2008; Olsen *et al.*, 2008; Reenders and ten Brinke, 2002). The first approaches were developed in the 80's based on simple comparisons with the behavior of liquid crystals. More recently, Reenders and ten Brinke proposed a model based on the weak-segregation calculation, where they derived a Landau free-energy functional for both a compositional and a nematic order parameter. The phase behavior of a melt of monodisperse rod–coil diblocks was then studied, the excluded volume interaction between the rod blocks being modeled by an attractive Maier–Saupe interaction, typical in molecular liquid crystals. The incompatibility between rod and coil blocks was modeled by the usual Flory–Huggins interaction. The predicted theoretical phase diagram is very different from the one calculated for "classical coil–coil" block copolymers (Figure 20.6). The lamellar domain expected for rod–coil copolymers is extremely large, due to the strong intermolecular interactions between the rigid moieties (described from the attractive Maier–Saupe interaction parameter). It is only for highly asymmetric copolymers that morphologies with curved interfaces, such as hexagonal and cubic, can be described.

Based on all the experimental work on hybrid polypeptide block copolymers, one can attempt to represent a pseudophase diagram describing the different morphologies obtained as a function of the polypeptide volume fraction. Because we do not have a sufficiently large series of diblock copolymers with the same chemical blocks, we cannot define a real phase diagram and calculate the Flory–Huggins χ parameter. We decided to combine the data obtained from different systems based on polypeptides (PBLG or PZLLys) and polyvinyl segments (polystyrene, polyisoprene and polybutadiene) with a diblock architecture. All these diblock copolymers behave as rod–coils. We have thus been able to represent this pseudophase diagram by plotting the total number of monomer unit N_{total} as a function of the polypeptide (or rod) volume fraction (Figure 20.7). The first observation is that most of the phase diagram consists of the previously described hexagonal in Lamellar (HL) morphology, consisting of a local order of the polypeptide

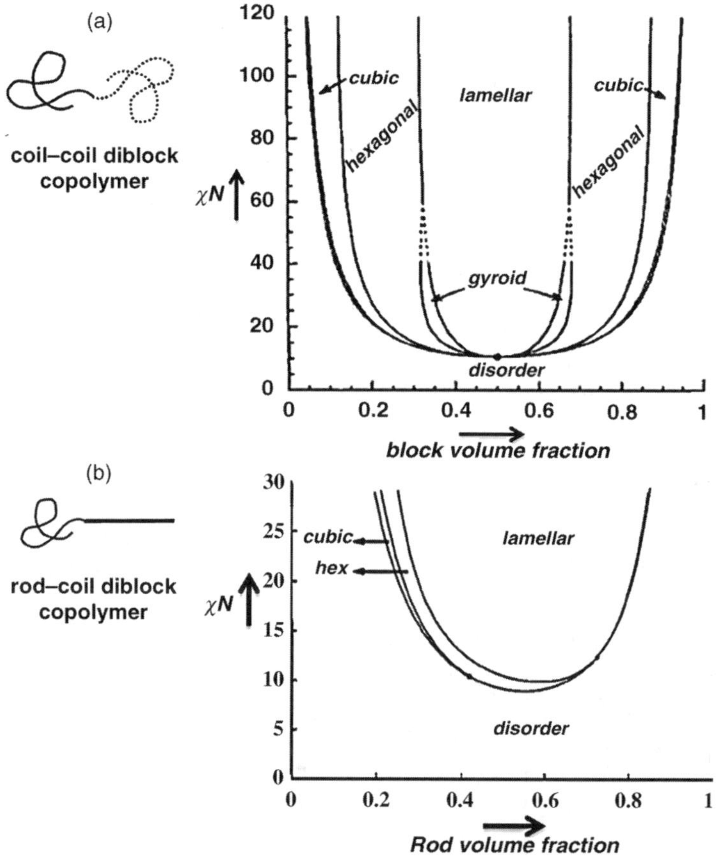

Figure 20.6 Theoretical phase diagrams developed for (a) coil–coil and (b) rod–coil block copolymers. (Reprinted with permission from M.W. Matsen and F.S. Bates, "Unifying weak- and strong-segregation block copolymer theories," *Macromolecules*, **29**, 4, 1091–1098, 1996. © 1996 American Chemical Society; M. Reenders and G. ten Brinke, "Compositional and orientational ordering in rod-coil diblock copolymer melts," *Macromolecules*, **35**, 8, 3266–3280, 2002. © 2002 American Chemical Society.)

chains and a lamellar phase separation between both blocks. In a very small part of the phase diagram, a double hexagonal (HH) structure can be observed, mainly for short and highly asymmetric block copolymers. It is worthwhile to mention that such behavior has also been nicely described in rod–coils with a conductive and rigid segment (Sary *et al.*, 2007), confirming the universality of the approach (Figure 20.7). Also interestingly, it has been described that the critical χN values for phase separation in rod–coils are lower than those expected for coil–coil analogues (Borsali *et al.*, 2001). As a result, lower molar masses can be used to induce phase separation in rod–coils, and typical organization dimensions can also be lower. Rod–coils are then interesting building blocks to afford nanostructures with high ordering.

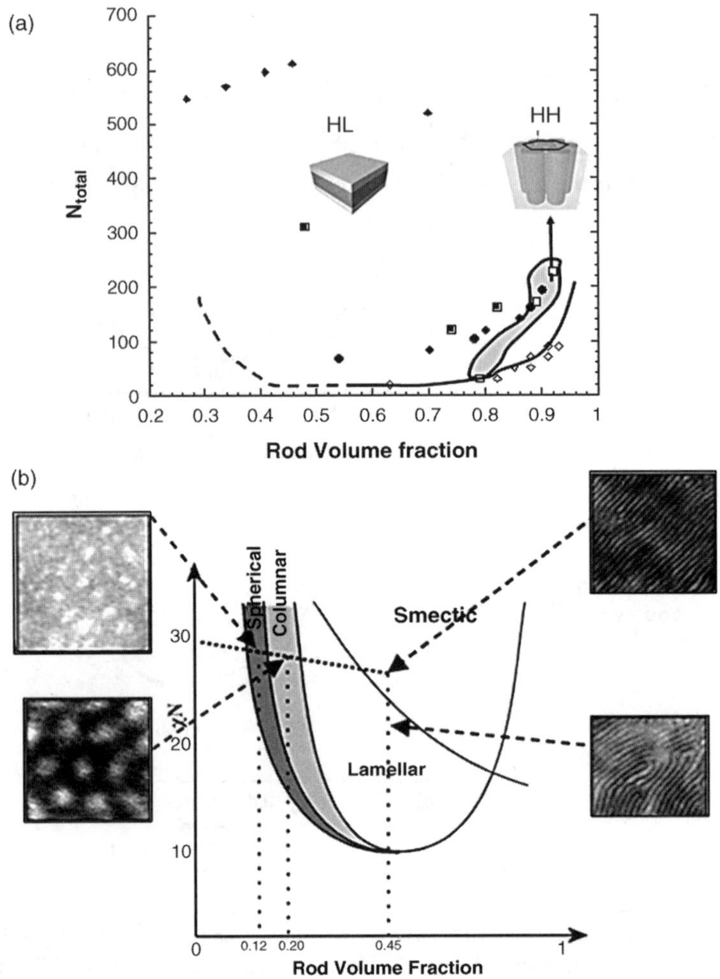

Figure 20.7 (Pseudo) experimental phase diagrams obtained for (a) polypeptide-based and (b) conductive-based block copolymers. (Reprinted with permission from N. Sary, L. Rubatat, C. Brochon *et al.*, "Self-assembly of poly(diethylhexyloxy-*p*-phenylenevinylene)-*b*-poly(4-vinylpyridine) rod-coil block copolymer systems," *Macromolecules*, **40**, 19, 6990–6997, 2007. © 2007 American Chemical Society.)

20.5 Conclusion

Polypeptide-based block copolymers are an attractive and promising class of block copolymer structures with a proven ability to mimic some of the basic properties of proteins. The most notable difference between peptide-based block copolymers and their fully synthetic and amorphous analogs is their hierarchical solid-state organization. In contrast to most synthetic amorphous block copolymers (which typically exhibit structural order only over a single length scale) peptide-based block copolymers can form hierarchically organized nanoscale structures that cover several different length scales. At the smallest length scale, peptide sequences fold into regular secondary structures, such as α-helices or β-strands. At the next level, peptide α-helices and β-strands can assemble into hexagonal superstructures and β-sheets, respectively. Finally, phase separation between the peptide and synthetic blocks leads to the formation of ordered domains with the largest characteristic length scales. This behavior is very similar to that of proteins for which the primary structure controls the secondary, tertiary and quaternary structures.

Lamellar phase-separated morphologies have been observed for a large number of peptide-based block copolymers. In comparison to conventional, fully amorphous diblock copolymers, these lamellar structures are typically observed over a much broader range of compositions. This behavior in bulk, as in solution, can be easily explained with a consideration of the flat interface generated by the rod–rod packing. In addition to these more conventional morphologies, structural investigation of peptide-based block copolymers has also revealed novel phase-separated structures, previously unknown for fully amorphous diblock copolymers. These observations underscore the fact that the solid-state structure formation of peptide-based block copolymers is not solely dictated by phase separation (as is the case for amorphous diblock copolymers), but is also influenced by other factors such as intra- and intermolecular hydrogen bonding and chain conformation. While much of the early interest in peptide–polymer conjugates was driven by their potential application as membrane materials or for the development of antithrombogenic surfaces, more recent studies reveal that these materials can also possess interesting mechanical properties and can also be used as template for nanofabrication and devices.

References

Antonietti, M. and Förster, S. (2003) Vesicles and liposomes: A self-assembly principle beyond lipids. *Advanced Materials*, **15**, 1323–1333.

Babin, J., Rodriguez-Hernandez, J., Lecommandoux, S. *et al.* (2005) Self-assembled nanostructures from peptide-synthetic hybrid block copolymers: Complex, stimuli-responsive rod-coil architectures. *Faraday Discussions*, **128**, 179–192.

Babin, J., Taton, D., Brinkmann, M., and Lecommandoux, S. (2008) Synthesis and self-assembly in bulk of linear and mikto-arm star block copolymers based on polystyrene and poly(glutamic acid). *Macromolecules*, **41**, 1384–1392.

Barenberg, S., Anderson, J.M., and Geil, P.H. (1981) Structure and properties of two plastic peptide triblock, ABA, copolymers of poly (γ-benzyl-L-glutamate), A, and poly(butadiene/acrylonitrile), B. *International Journal of Biological Macromolecules*, **3**, 82–90.

Bates, F.S. and Fredrickson, G.H. (1990) Block Copolymer Thermodynamics: Theory and Experiment. *Annual Review of Physical Chemistry*, **41**, 525–557.

Bates, F.S. and Fredrickson, G.H. (1999) Block copolymers - designer soft materials. *Physics Today*, **52**, 32–38.

Billot, J.P., Douy, A., and Gallot, B. (1976) Synthesis and structural study of block copolymers with a hydrophobic polyvinyl block and a hydrophilic polypeptide block: Copolymers polystyrene/poly(L-lysine) and polybutadiene/poly(L-lysine). *Makromolekulare Chemie-Macromolecular Chemistry and Physics*, **177**, 1889–1893.

Billot, J.P., Douy, A., and Gallot, B. (1977) Preparation, fractionation, and structure of block copolymers polystyrene-poly(carbobenzoxy-L-lysine) and polybutadiene-poly(carbobenzoxy-L-lysine). *Makromolekulare Chemie-Macromolecular Chemistry and Physics*, **178**, 1641–1650.

Borsali, R., Lecommandoux, S., Pecora, R., and Benoit, H. (2001) Scattering properties of rod-coil and once-broken rod block copolymers. *Macromolecules*, **34**, 4229–4234.

Caillol, S., Lecommandoux, S., Mingotaud, A.F. *et al.* (2003) Synthesis and self-assembly properties of peptide - Polylactide block copolymers. *Macromolecules*, **36**, 1118–1124.

Chen, G.W., Hayashi, T., and Nakajima, A. (1981) Synthesis and molecular characterization of A–B–A tri-block copolymers composed of poly(γ-ethyl L-glutamate) as the A component and polybutadiene as the B component. *Polymer Journal*, **13**, 433–442.

Cho, C.S., Jo, B.W., Kwon, J.K., and Komoto, T. (1994) Synthesis and structural study of an ABA block copolymer consisting of poly(ε-benzyloxycarbonyl-L-lysine) as the a block and poly(ethylene oxide) as the B block. *Macromolecular Chemistry and Physics*, **195**, 2195–2206.

Cho, C.S., Kim, S.W., and Komoto, T. (1990) Synthesis and structural study of an ABA block copolymer consisting of poly(γ-benzyl L-glutamate) as the a block and poly(ethylene oxide) as the B block. *Makromolekulare Chemie-Macromolecular Chemistry and Physics*, **191**, 981–991.

Cho, C.S., Kim, S.W., Sung, Y.K., and Kim, K.Y. (1988) Synthesis and structural study of an ABA block copolymer consisting of poly(γ-benzyl L-glutamate) as the A component and poly(propylene glycol) as the B component. *Makromolekulare Chemie-Macromolecular Chemistry and Physics*, **189**, 1505–1510.

Cho, I., Kim, J.B., and Jung, H.J. (2003) Synthesis and characterization of di- and triblock copolymers of poly(ethylene oxide) and poly(DL-valine-co-DL-leucine). *Polymer*, **44**, 5497–5500.

Colfen, H. (2001) Double-hydrophilic block copolymers: Synthesis and application as novel surfactants and crystal growth modifiers. *Macromolecular Rapid Communications*, **22**, 219–252.

Degée, P., Dubois, P., Jerome, R., and Teyssie, P. (1993) Synthesis and characterization of biocompatible and biodegradable poly(ε-caprolactone-b-γ-benzylglutamate) diblock copolymers. *Journal of Polymer Science Part A: Polymer Chemistry*, **31**, 275–278.

Deming, T.J. (1997) Polypeptide materials: New synthetic methods and applications. *Advanced Materials*, **9**, 299.

Douy, A. and Gallot, B. (1977) Structure of AB block copolymers with a polypeptide block - Effect of chain conformation. *Polymer Engineering and Science*, **17**, 523–526.

Douy, A. and Gallot, B. (1982) Block copolymers with a polyvinyl and a polypeptide block - Factors governing the folding of the polypeptide chains. *Polymer*, **23**, 1039–1044.

Floudas, G., Papadopoulos, P., Klok, H.A. *et al.* (2003) Hierarchical self-assembly of poly(gamma-benzyl-L-glutamate)-poly(ethylene glycol)-poly(gamma-benzyl-L-glutamate) rod-coil-rod triblock copolymers. *Macromolecules*, **36**, 3673–3683.

Floudas, G. and Spiess, H.W. (2009) Self-assembly and dynamics of polypeptides. *Macromolecular Rapid Communications*, **30**, 278–298.

Forster, S. and Antonietti, M. (1998) Amphiphilic block copolymers in structure-controlled nanomaterial hybrids. *Advanced Materials*, **10**, 195.

Forster, S. and Konrad, M. (2003) From self-organizing polymers to nano- and biomaterials. *Journal of Materials Chemistry*, **13**, 2671–2688.

Gervais, M., Douy, A., Gallot, B., and Erre, R. (1988) X-ray photoelectron spectroscopy of ABA polypeptide polybutadiene-polypeptide-polybutadiene triblock copolymers. *Polymer*, **29**, 1779–1783.

Gitsas, A., Floudas, G., Mondeshki, M. *et al.* (2008a) Effect of chain topology on the self-organization and dynamics of block copolypeptides: From diblock copolymers to stars. *Biomacromolecules*, **9**, 1959–1966.

Gitsas, A., Floudas, G., Mondeshki, M. *et al.* (2008b) Control of peptide secondary structure and dynamics in poly (gamma-benzyl-L-glutamate)-b-polyalanine peptides. *Macromolecules*, **41**, 8072–8080.

Halperin, A. (1990) Rodcoil copolymers: Their aggregation behavior. *Macromolecules*, **23**, 2724–2731.

Hamley, I. (1998) *The Physics of Block Copolymers* (ed. I.W. Hamley), Oxford University Press, Oxford.

Hayashi, T., Chen, G.W., and Nakajima, A. (1984) Synthesis and structural study of the A–B–A tri-block copolymer consisting of poly(γ-benzyl D,L-glutamate) or poly(γ-methyl D,L-glutamate) as the A component and polybutadiene as the B component. *Polymer Journal*, **16**, 739–749.

Ibarboure, E., Papon, E., and Rodriguez-Hernandez, J. (2007) Nanostructured thermotropic PBLG-PDMS-PBLG block copolymers. *Polymer*, **48**, 3717–3725.

Imanishi, Y., Tanaka, M., and Bamford, C.H. (1985) Block copolymerization of vinyl compounds by the terminal - group activation of poly(α-amino acids) and the characterization of the block copolymers. *International Journal of Biological Macromolecules*, **7**, 89–99.

Ito, Y., Miyashita, K., Kashiwagi, T., and Imanishi, Y. (1993) Synthesis and interactions with blood of polyether-urethaneurea/polypeptide block copolymers. *Biomaterials Artificial Cells and Immobilization Biotechnology*, **21**, 571–580.

Janssen, K., Vanbeylen, M., Samyn, C. *et al.* (1990) Morphology of ABA triblock copolymers consisting of poly (γ-benzyl L-glutamate) as the A component and polystyrene as the B component. *Makromolekulare Chemie-Macromolecular Chemistry and Physics*, **191**, 2777–2785.

Jenekhe, S.A. and Chen, X.L. (1998) Self-assembled aggregates of rod-coil block copolymers and their solubilization and encapsulation of fullerenes. *Science*, **279**, 1903–1907.

Jenekhe, S.A. and Chen, X.L. (2000) Supramolecular photophysics of self-assembled block copolymers containing luminescent conjugated polymers. *Journal of Physical Chemistry B*, **104**, 6332–6335.

Kang, I.K., Ito, Y., Sisido, M., and Imanishi, Y. (1988a) Adsorption of plasma proteins and platelet adhesion on to polydimethylsiloxane/poly(gamma-benzyl L-glutamate) block copolymer films. *Biomaterials*, **9**, 138–144.

Kang, I.K., Ito, Y., Sisido, M., and Imanishi, Y. (1988b) Gas permeability of the film of block and graft copolymers of polydimethylsiloxane and poly(gamma-benzyl L-glutamate). *Biomaterials*, **9**, 349–355.

Klok, H.A., Langenwalter, J.F., and Lecommandoux, S. (2000) Self-assembly of peptide-based diblock oligomers. *Macromolecules*, **33**, 7819–7826.

Klok, H.A. and Lecommandoux, S. (2001) Supramolecular materials via block copolymer self-assembly. *Advanced Materials*, **13**, 1217–1229.

Kong, X.X. and Jenekhe, S.A. (2004) Block copolymers containing conjugated polymer and polypeptide sequences: Synthesis and self-assembly of electroactive and photoactive nanostructures. *Macromolecules*, **37**, 8180–8183.

Kugo, K., Hata, Y., Hayashi, T., and Nakajima, A. (1982a) Studies on membrane surfaces of A-B-A tri-block copolymers consisting of poly(ε-N-benzyloxycarbonyl-L-lysine) as the A component and polybutadiene as the B component. *Polymer Journal*, **14**, 401–410.

Kugo, K., Hayashi, T., and Nakajima, A. (1982b) Synthesis, structure, and mechanical properties of A-B-A tri-block copolymers consisting of poly(ε-N-benzyloxy-carbonyl-L-lysine) as the A component and polybutadiene as the B component. *Polymer Journal*, **14**, 391–399.

Kugo, K., Murashima, M., Hayashi, T., and Nakajima, A. (1983) Structure and properties of membrane surfaces of A-B-A tri-block copolymers consisting of poly(γ-methyl D,L-glutamate) as the A component and polybutadiene as the B component. *Polymer Journal*, **15**, 267–277.

Kugo, K., Ohji, A., Uno, T., and Nishino, J. (1987) Synthesis and conformations of A-B-A tri-block copolymers with hydrophobic poly(γ-benzyl l-glutamate) and hydrophilic poly(ethylene oxide). *Polymer Journal*, **19**, 375–381.

Kumaki, T., Sisido, M., and Imanishi, Y. (1985) Antithrombogenicity and oxygen permeability of block and graft copolymers of polydimethylsiloxane and poly(α-amino acid). *Journal of Biomedical Materials Research*, **19**, 785–811.

Leclere, P., Hennebicq, E., Calderone, A. *et al.* (2003) Supramolecular organization in block copolymers containing a conjugated segment: A joint AFM/molecular modeling study. *Progress in Polymer Science*, **28**, 55–81.

Leclere, P., Parente, V., Bredas, J.L. *et al.* (1998) Organized semiconducting nanostructures from conjugated block copolymer self-assembly. *Chemistry of Materials*, **10**, 4010–4014.

Lecommandoux, S., Achard, M.F., Langenwalter, J.F., and Klok, H.A. (2001) Self-assembly of rod-coil diblock oligomers based on alpha-helical peptides. *Macromolecules*, **34**, 9100–9111.

Lee, M., Cho, B.K., and Zin, W.C. (2001) Supramolecular structures from rod-coil block copolymers. *Chemical Reviews*, **101**, 3869–3892.

Losik, M., Kubowicz, S., Smarsly, B., and Schlaad, H. (2004) Solid-state structure of polypeptide-based rod-coil block copolymers: Folding of helices. *European Physical Journal E*, **15**, 407–411.

Ludwigs, S., Krausch, G., Reiter, G. *et al.* (2005) Structure formation of a polystyrene-block-poly(gamma-benzyl L-glutamate) in thin films. *Macromolecules*, **38**, 7532–7535.

Mao, G. and Ober, C.K. (1997) Block copolymers containing liquid crystalline segments. *Acta Polymerica*, **48**, 405–422.

Matsen, M.W. and Bates, F.S. (1996) Unifying weak- and strong-segregation block copolymer theories. *Macromolecules*, **29**, 1091–1098.

Mckinnon, A.J. and Tobolsky, A.V. (1968) Structure and properties of poly(γ-benzyl-L-glutamate) cast from dimethylformamide. *Journal of Physical Chemistry*, **72**, 1157.

Mori, A., Ito, Y., Sisido, M., and Imanishi, Y. (1986) Interaction of polystyrene/poly(gamma-benzyl L-glutamate) and poly(methyl methacrylate)/poly(gamma-benzyl L-glutamate) block copolymers with plasma proteins and platelets. *Biomaterials*, **7**, 386–392.

Nakajima, A., Hayashi, T., Kugo, K., and Shinoda, K. (1979a) Synthesis and structural study of the A-B-A type block copolymer consisting of poly(γ-benzyl L-glutamate) as the A component and polybutadiene as the B component. *Macromolecules*, **12**, 840–843.

Nakajima, A., Kugo, K., and Hayashi, T. (1979b) Mechanical properties and water permeability of A–B–A tri-block copolymer membranes consisting of poly(γ-benzyl L-glutamate) as the A-component and polybutadiene as the B-component. *Polymer Journal*, **11**, 995–1001.

Nakajima, A., Kugo, K., and Hayashi, T. (1979c) Microheterophase structure of the A-B-A type block copolymer consisting of α-helical poly(γ-benzyl L-glutamate) as the A component and polybutadiene as the B component. *Macromolecules*, **12**, 844–848.

Olsen, B.D. and Segalman, R.A. (2008) Self-assembly of rod-coil block copolymers. *Materials Science & Engineering R-Reports*, **62**, 37–66.

Olsen, B.D., Shah, M., Ganesan, V., and Segalman, R.A. (2008) Universalization of the phase diagram for a model rod-coil diblock copolymer. *Macromolecules*, **41**, 6809–6817.

Papadopoulos, P., Floudas, G., Schnell, I. *et al.* (2005) Nanodomain-induced chain folding in poly(gamma-benzyl-L-glutamate)-b-polyglycine diblock copolymers. *Biomacromolecules*, **6**, 2352–2361.

Papadopoulos, P., Floudas, G., Schnell, I. *et al.* (2006) Thermodynamic confinement and α-helix persistence length in poly(γ-benzyl-l-glutamate)-b-poly(dimethyl siloxane)-b-poly(γ-benzyl-l-glutamate) triblock copolymers. *Biomacromolecules*, **7**, 618–626.

Parras, P., Castelletto, V., Hamley, I.W., and Klok, H.A. (2005) Nanostructure formation in poly(gamma-benzyl-L-glutamate)-poly(ethylene glycol)-poly(gamma-benzyl-L-glutamate) triblock copolymers in the solid state. *Soft Matter*, **1**, 284–291.

Perly, B., Douy, A., and Gallot, B. (1976) Block copolymers polybutadiene-poly(benzyl-L-glutamate) and polybutadiene-poly(N5-hydroxypropylglutamine) preparation and structural study by X-ray and electron microscopy. *Makromolekulare Chemie-Macromolecular Chemistry and Physics*, **177**, 2569–2589.

Reenders, M. and Ten Brinke, G. (2002) Compositional and orientational ordering in rod-coil diblock copolymer melts. *Macromolecules*, **35**, 3266–3280.

Rodriguez-Hernandez, J., Checot, F., Gnanou, Y., and Lecommandoux, S. (2005) Toward 'smart' nano-objects by self-assembly of block copolymers in solution. *Progress in Polymer Science*, **30**, 691–724.

Rong, G., Deng, M., Deng, C. *et al.* (2003) Synthesis of poly(ε-caprolactone)-b-Poly(γ-benzyl-l-glutamic acid) block copolymer using amino organic calcium catalyst. *Biomacromolecules*, **4**, 1800–1804.

Rubatat, L., Kong, X.X., Jenekhe, S.A. *et al.* (2008) Self-assembly of polypeptide/pi-conjugated polymer/polypeptide triblock copolymers in rod-rod-rod and coil-rod-coil conformations. *Macromolecules*, **41**, 1846–1852.

Sanchez-Ferrer, A. and Mezzenga, R. (2010) Secondary structure-induced micro- and macrophase separation in rod-coil polypeptide diblock, triblock, and star-block copolymers. *Macromolecules*, **43**, 1093–1100.

Sary, N., Rubatat, L., Brochon, C. *et al.* (2007) Self-assembly of poly(diethylhexyloxy-p-phenylenevinylene)-b-poly (4-vinylpyridine) rod-coil block copolymer systems. *Macromolecules*, **40**, 6990–6997.

Sato, H., Nakajima, A., Hayashi, T. *et al.* (1985) Microheterophase structure, permeability, and biocompatibility of A-B-A triblock copolymer membranes composed of poly(γ-ethyl L-glutamate) as the A component and polybutadiene as the B component. *Journal of Biomedical Materials Research*, **19**, 1135–1155.

Schlaad, H., Kukula, H., Smarsly, B. *et al.* (2002) Solid-state morphologies of linear and bottlebrush-shaped polystyrene-poly(Z-L-lysine) block copolymers. *Polymer*, **43**, 5321–5328.

Schlaad, H., Smarsly, B., and Losik, M. (2004) The role of chain-length distribution in the formation of solid-state structures of polypeptide-based rod-coil block copolymers. *Macromolecules*, **37**, 2210–2214.

Tanaka, S., Ogura, A., Kaneko, T. *et al.* (2004) Precise synthesis of ABA triblock copolymers comprised of poly (ethylene oxide) and poly(beta-benzyl-L-aspartate): A hierarchical structure inducing excellent elasticity. *Macromolecules*, **37**, 1370–1377.

van Hest, J.C.M. and Tirrell, D.A. (2001) Protein-based materials, toward a new level of structural control. *Chemical Communications*, 1897–1904.

Walther, M. and Finkelmann, H. (1996) Structure formation of liquid crystalline block copolymers. *Progress in Polymer Science*, **21**, 951–979.

Wang, Y., Zou, S., Kim, K.T. *et al.* (2008) Organometallic-polypeptide block copolymers: synthesis and self-assembly of poly(ferrocenyldimethylsilane)-b-poly(epsilon-benzyloxycarbonyl-L-Lysine). *Chemistry-A European Journal*, **14**, 8624–8631.

Yoda, R., Komatsuzaki, S., and Hayashi, T. (1995a) Surface properties and biocompatibility of A-B-A type block copolymer membranes consisting of poly(γ-benzyl-L-glutamate) as the A component and polyisoprene as the B component. *Biomaterials*, **16**, 1203–1209.

Yoda, R., Komatsuzaki, S., and Hayashi, T. (1996) Membrane properties of A-B-A type block copolymer consisting of poly(gamma-benzyl-L-glutamate) as the A component and polyisoprene as the B component. *European Polymer Journal*, **32**, 233–237.

Yoda, R., Komatsuzaki, S., Nakanishi, E., and Hayashi, T. (1995b) Microheterophase structure of A-B-A type block copolymer consisting of poly(γ-benzyl -glutamate) as the a component and polyisoprene as the B component. *European Polymer Journal*, **31**, 335–339.

Yoda, R., Komatsuzaki, S., Nakanishi, E. *et al.* (1994) Preparation and properties of A-B-A type block copolymer membranes consisting of poly(N-hydroxypropyl-L-glutamine) as the A component and polyisoprene as the B component. *Biomaterials*, **15**, 944–949.

Yoda, R., Shimoda, M., Komatsuzaki, S. *et al.* (1997) Morphology study of poly(gamma-benzyl L-glutamate) and polyisoprene copolymer by pulsed proton NMR. *European Polymer Journal*, **33**, 815–821.

Zhang, G., Ma, J., Li, Y., and Wang, Y. (2003) Synthesis and characterization of poly (L-alanine)-block-poly (ethylene glycol) monomethyl ether as amphiphilic biodegradable co-polymers. *Journal of Biomaterials Science, Polymer Edition*, **14**, 1389–1400.

21

AFM Study of Comb (Co)Polymers with Complex Chain Architecture

Michel Schappacher and Alain Deffieux
CNRS, Université de Bordeaux, Laboratoire de Chimie des Polymères Organiques, 16 avenue Pey Berland, 33607, Pessac Cedex France

21.1 Introduction

The synthesis and study of macromolecules of increasing molecular complexity is a very active research domain in close relation with the development of molecular materials for nanosciences and nanotechnologies. In recent years, highly branched (combs) and hyperbranched (dendritic) polymers have in particular gained widespread attention due to their original structure and unique bulk and solution properties, which strongly differ from their linear counterparts. This is explained for example by their large number of terminal functional groups and their intrinsic structure and shape. These characteristics lead to specific rheological properties, as well as to the possibility to accommodate guest molecules within the macromolecule. Compared to dendrimers and dendrigrafts, comb polymers characterized by a lower degree of branching still possess a highly compact architecture and a high number of chain ends (Bosman *et al.*, 1999; Frechet and Tomalia, 2002; Roovers, 1999; Teertstra and Gauthier, 2004). Since they can be prepared more readily and economically than hyperbranched polymers they are good candidates to replace hyperbranched polymers in a number of applications.

In order to precisely control the chemical composition, dimensions, chain topology and final properties of complex macromolecular structures, various synthetic strategies have been developed. They involve the combination of a series of controlled polymerization processes, polymer functionalization and coupling reactions. These multistage approaches generally involve the synthesis of reactive polymers and copolymers with precise chain architecture and topology prefiguring the final targeted structure. This includes the preparation of reactive

Complex Macromolecular Architectures: Synthesis, Characterization, and Self-Assembly, First Edition.
Edited by Nikos Hadjichristidis, Akira Hirao, Yasuyuki Tezuka and Filip Du Prez.

or potentially reactive star, dendrimer-like, cyclic, multicyclic macromolecules. These are then used as starting macromolecule for the preparation of second and eventually higher generations of densely branched and hyperbranched macromolecular objects. To recover the corresponding structures in satisfactory yield, the reactions involved in the successive building steps must be highly efficient and selective. Besides the applied construction strategy, another important aspect of complex multistage synthesis concerns the use of adequate methods to characterize the final polymer structure and determine their molecular parameters. It is also important to make a follow up of the macromolecule building up to identify at what stages side reactions eventually occur. Conventional molecular characterization techniques such as spectroscopy (NMR, IR), chromatography (SEC, HPLC) are appropriate and useful to characterize the molecular structure and dimensions of the first macromolecular block, but they become inadequate for the final and complex architecture of macromolecular object scattering methods (light, neutron), which allow the determination of the average dimensions and shape of macromolecules do give limited structural insight, and are of little help to understand when and why the obtained structures deviate from the targeted ones.

Direct visualization of isolated macromolecules by atomic force microscopy (AFM) imaging and the statistical treatment performed on a large set of these molecules is an interesting and efficient approach that was recently used to study the structure of comb polymers (Dziezok *et al.*, 1997; Sheiko *et al.*, 2003; Viville *et al.*, 2001; Viville *et al.*, 2000; Viville *et al.*, 2004). Besides the analysis of single macromolecular objects with linear and star-like architecture this method was used to investigate the selectivity of various controlled-radical polymerization processes (Beers *et al.*, 1999; Beers *et al.*, 1998; Börner *et al.*, 2002; Qin *et al.*, 2003; Min *et al.*, 2007).

A similar strategy was applied to the characterization of comb polystyrenes with various chain architectures prepared by the combination of living anionic and cationic polymerizations (Deffieux and Schappacher, 1999; Schappacher *et al.*, 1999; Schappacher and Deffieux, 1997; Schappacher and Deffieux, 2008a). This allowed us to establish a precise diagnostic of the selectivity/efficiency of the various elementary reactions used, in terms of yield, architecture control and molecular dimensions of the targeted molecules. From this analysis it was also possible to identify the main side reactions occurring during the synthesis, their extent and their impact on the final macromolecular objects.

21.2 Strategies of Comb Synthesis

Besides the macromonomer approach that mainly allows the preparation of combs with a linear backbone architecture, two main strategies have been proposed for the synthesis of comb polymers with more complex chain architectures (Scheme 21.1): (1) the grafting-from method, which requires the use of a central chain bearing reactive sites from which the growth of side chains can be initiated and (2) the grafting-onto approach, in which the functional groups on the central backbone chain are used as a grafting site to attach side chains bearing a reactive end.

In our group we have mainly focused on the preparation of comb polymers by the grafting-onto approach. It is based on the preparation as reactive backbone of poly (chloroethyl vinyl ether) (PCEVE) with various architectures ranging from linear, star, hyperbranched, to cyclic. PCEVE is used for grafting living anionic chains, typically

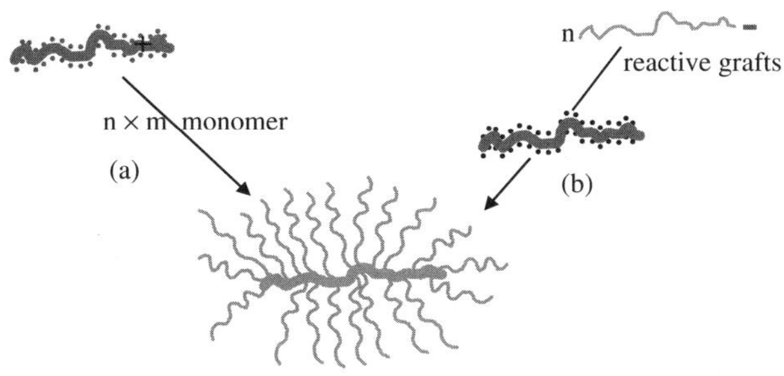

Scheme 21.1 Grafting-from (a) and grafting-onto (b) strategies for the preparation of comb polymers.

polystyryllithium or polyisoprenyllithium by functional deactivation onto the chloroethylether functions (Deffieux and Schappacher, 1999; Schappacher and Deffieux, 1997). Since both the backbone and graft building blocks are synthesized, respectively, by living cationic and living anionic polymerizations, the dimensions of the backbone and of the branches can be precisely controlled, whereas the efficiency and selectivity of the coupling reaction permits an almost quantitative substitution of the chloride groups of the PCEVE chain. This allows tuning of the dimensions and the structure of the combs giving access to molecular objects of high chain density and of various architectures and shapes, that is; cylinders, stars, rings.

As a step further, the possibility to prepare macromolecular objects with topologically organized elementary building blocks has been explored. These macromolecules associate subdomains of different chemical composition leading to amphipathic subdomains and paving the way to more sophisticated nanostructures.

Special emphasis is given to the correlation between the synthetic strategy applied, the molecular architecture and the supramolecular organization of these macromolecules through AFM imaging. This is the main focus of this review.

21.3 Linear Combs with Polystyrene Branches (Deffieux and Schappacher, 1998; Deffieux and Schappacher, 1999; Schappacher *et al.*, 1999; Schappacher and Deffieux, 1997)

The synthetic approach to generate combs with polystyrene (PS) branches is summarized in Scheme 21.2: a poly(chloroethyl vinyl ether) (PCEVE) (*a*) with precise molar mass and narrow molar-mass distribution is first synthesized by living cationic polymerization of chloroethyl vinyl ether (CEVE) using an α-iodoether adduct as chain precursor and zinc chloride as catalyst to trigger the polymerization. It allows the preparation of PCEVE with controlled molar masses, up to 100 000 g/mole ($\overline{DP}_n \sim 1000$). After purification by precipitation and drying, the PCEVEs are used as reactive backbone for the synthesis of polystyrene (PS) comb copolymers. Living polystyryllithium chains (*b*) eventually capped by a diphenylethylene (DPE) unit, with molar masses ranging from 1000 to 20 000 g/mol, are prepared in a separate step by anionic polymerization using *sec*-butyl lithium as initiator.

The PS comb synthesis was finally achieved by adding dropwise the living PS solution onto a known amount of PCEVE in THF. Discoloration of the living PS was used as an indicator of the

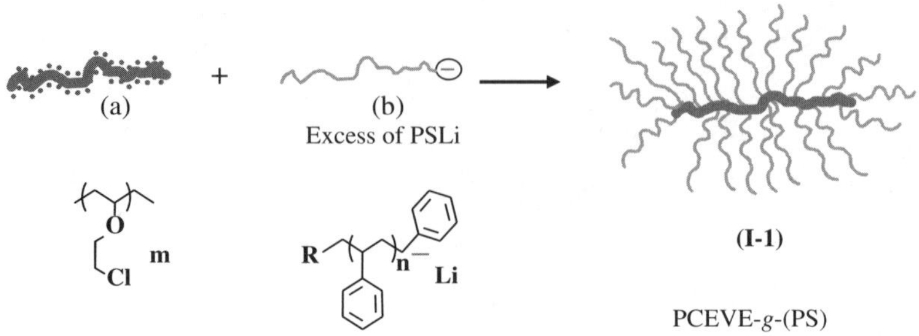

Scheme 21.2 Synthesis of polystyrene combs by grafting PSLi onto chloroether units of a linear PCEVE.

PS ends grafting. Typically, a slight excess of PSLi chains with respect to the PCEVE units (1.2 : 1 Cl) was used to insure a high grafting density and the formation of densely grafted comb polystyrenes (PCEVE-*g*-PS) (**I-1**).

Since both the PCEVE backbone and the PS branches have precisely controlled dimensions the combs exhibit tunable and precise shape and dimensions. The characteristics of some linear PS combs are collected in Table 21.1. They show a unimodal distribution and narrow dispersity (I_d). Molar masses determined using elastic light scattering are close to the theoretical values calculated, assuming one PS graft per chloroethyl vinyl ether unit, indicating very high grafting yields. They typically range between 90% and almost 100% of PCEVE unit substitution, respectively, for long and short PCEVE backbones $\overline{DP}_n$. The molecular architecture of the (PCEVE-*g*-PS) combs, in particular their highly branched structure and high density in PS segments strongly affect the dimensions, size and shape of the macromolecules, both in solution and in the solid state (Dziezok *et al.*, 1997). For example, the radius of gyration of a comb PS in a good solvent (THF) is about 4 times smaller than the corresponding linear PS of same molar mass, which means that mass density is increased by a factor of sixty within an isolated comb macromolecule.

Table 21.1 Synthesis and characteristics of (PCEVE-g-PS) combs: Influence of PCEVE backbone and PS branch lengths on the molar mass and on the grafting efficiency

PCEVE $\overline{DP}_{n\ exp}$	PSLi $\bar{M}_{n\ exp}$	$\bar{M}_{n\ th}$ [a]	$\bar{M}_{w\ exp}$ [b]	I_p	f [c]
6	10 400	62 400	67 500	1.27	6.5
36	3000	108 000	104 400	1.01	35
36	19 200	691 200	691 000	1.05	36
60	3000	180 000	179 300	1.03	60
280	3000	840 000	755 000	1.01	251
760	3000	2280 000	2 110 000	1.01	703
700	7300	5 314 000	5 420 000	1.06	742
1000	3000	3 000 000	2 680 000	1.03	897

[a] Assuming one PS graft by CEVE unit.
[b] Measured by light scattering.
[c] Number of PS branches.

Thanks to their very high molar masses (often several millions g/mol) and high chain compactness (PCEVE-*g*-PS) combs, can be readily observed as individual molecules by atomic force microscopy (AFM) in the tapping mode. This technique appears to be a powerful tool to observe the shape, dimensions, absolute molar mass, molar-mass distribution, chain conformation and internal organization of individual macromolecules (Dziezok *et al.*, 1997; Gallyamov *et al.*, 2004; Matyjaszewski *et al.*, 2003; Percec *et al.*, 1998; Potemkin *et al.*, 2004; Qin *et al.*, 2003; Sheiko *et al.*, 2003; Sheiko and Moller, 2001; Spatz *et al.*, 1997).

In practice AFM polymer samples are prepared by spin-casting their diluted organic solutions on freshly cleaved mica or highly oriented pyrolytic graphite (HOPG) substrates, selected according to their affinity for the polymer. HOPG of low polarity is the preferred substrate for polystyrene-based polymers. Samples are analyzed in the tapping mode using probes of commercially available silicon tips with a typical radius of curvature of less than 10 nm and the magnitude of the tip–sample interactions are lowered as much as possible.

The AFM images in Figure 21.1, recorded in air in the phase and topographic modes, respectively, show two sets of individual comb macromolecules, each one separated from the adjacent macromolecules. Figure 21.1a shows grain rice combs constituted of a relatively short PCEVE backbone bearing long PS branches, whereas worm-like macromolecular objects obtained by grafting short PS branches onto long PCEVE backbone are presented in Figure 21.1b.

The shorter PS combs exhibit a rigid conformation with an average size of about 100 nm length and 50 nm width, while the more elongated and narrower ones (300 nm × 20 nm) present a folded conformation, indicating that upon decreasing the molar mass of PS branches, the stiffness of the chain is decreased. When the branch density is high, the steric repulsion between the side chains leads to elongation of the backbone and then the persistence length of the molecule becomes comparable to its total length, approaching the aspect of rod-like macromolecules.

In AFM images prepared from more concentrated PS comb solutions, a succession of continuous objects forming a molecular monolayer is observed (Viville *et al.*, 2001; Viville *et al.*, 2000). Discernable interruptions between macromolecules indicate very little overlapping and interdiffusion of PS branches, in agreement with the highly compact comb architecture.

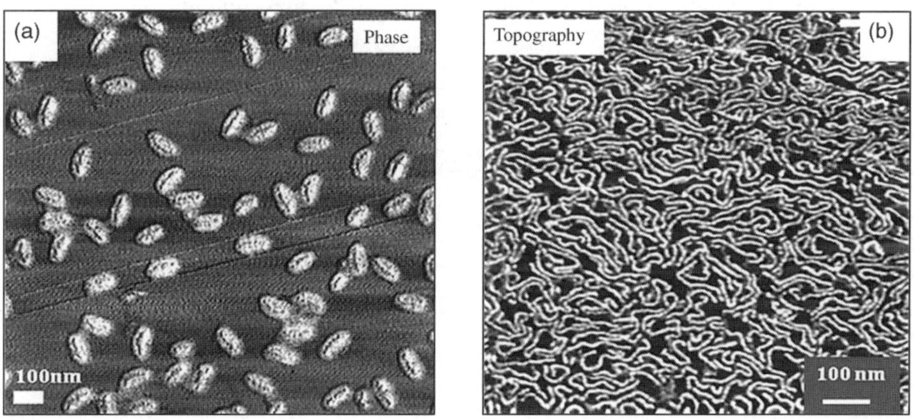

Figure 21.1 Phase and topographic AFM images of individual PCEVE-*g*-PS combs: PCEVE with $\overline{DP}_n = 300$ and 700, PS with $\overline{DP}_n = 400$ and 70 (left and right, respectively).

Since these interruptions are branch ends, the contour length (L_{tot}) distribution of the molecules can be calculated. Considering that the molecular mass is proportional to the contour length (L_{tot}) of the molecules, it is then possible to determine the molecular-length distribution, similarly to the molecular mass distribution obtained from SEC measurements. Values corresponding to the number-average molecular length (L_n) and the length-average molecular length (L_w) of sample presented Figure 21.1 (right), are 160 nm and 168 nm, respectively. It is worth noting here that the distribution ($L_w/L_n = 1.05$) is very narrow. This dispersity index of the molecular length obtained from AFM images is in very good agreement with the value determined by SEC measurements ($\bar{M}_W/\bar{M}_n = 1.06$). Considering, as an extreme case, that a fully extended PCEVE backbone with 700 monomer units yields a theoretical value of ~175 nm for the main-chain contour length, we can conclude that the objects appearing in the two images of Figure 21.1 correspond to single molecules with a fully extended backbone but with still some flexibility for combs with long backbone and short PS branches.

21.4 Star Combs with PCEVE Backbone and PS Branches (Deffieux and Schappacher, 1999)

The synthesis and study of star polymers has been the subject of numerous research studies in recent years. Two main strategies have been developed for their preparation (Scheme 21.3).

The divergent or "core-first" method: This consists of the initiation of a controlled polymerization from a multifunctional initiator, which serves finally as the star central core. This technique requires a controlled and quantitative initiation followed by a living-like polymerization process without transfer or termination. In most cases, stars obtained by this approach possess a limited number of branches typically between three to five or six. A precise determination of the number of star branches using conventional characterization techniques is not easy and generally average values can be obtained.

The convergent or "arm-first" method: This consists first in the synthesis of arms by a living/controlled polymerization. The latter are finally linked to each other by adding a multifunctional agent to the polymerization system. Addition of a small amount of a bis-monomer

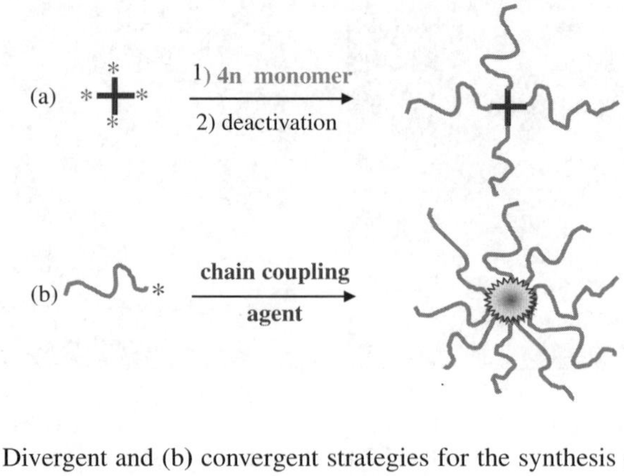

Scheme 21.3 (a) Divergent and (b) convergent strategies for the synthesis of star-like polymers.

(divinylbenzene, divinylether, etc.) at the end of the polymerization, to form a crosslinked core, has often been used.

Stars constituted of monomer units bearing reactive functions, prepared by these different strategies have been used as precursor for the construction of the corresponding comb star polymers. As explained in the previous section, different methodologies can be used: branches can be introduced on the star backbone by either grafting-from or grafting-onto approaches.

The grafting-onto approach developed for the preparation of linear combs from a PCEVE backbone was used in our group. PCEVE stars were first prepared as precursor using both the divergent and convergent methods. Besides studies of their properties the grafting-onto strategy was used as a way to magnify the initial star macromolecules, in order to allow their direct visualization by AFM and make possible a critical retroanalysis of the elementary reactions that are involved during their synthesis.

21.4.1 Comb Stars by the Divergent Approach (Schappacher and Deffieux, 2005)

The synthesis of PS comb stars using the core-first method (Boyce *et al.*, 2004) can be achieved in two distinct stages: (1) preparation of a multiarm poly(chloroethyl vinyl ether) star as reactive backbone (Schappacher and Deffieux, 2000) and (2) grafting of polystyryllithium chains onto the reactive chloroethyl ether functions of the PCEVE blocks (Deffieux and Schappacher, 1998; Schappacher and Deffieux, 1997), as indicated in Scheme 21.4. For example, a star-like 4-arm α-acetal-PS is first prepared by reacting α-acetal-PSLi oligomers of low $\overline{DP}_n$ onto pentaerythritol tetrakis(4-chlorobutylether). The obtained tetra-ended ω-acetal star PS oligomer is then used as a tetrafunctional precursor for the synthesis of the four-branch PCEVE. To that aim the four acetal-PS end groups are derived quantitatively in the presence of tetramethylsilyl iodide (TMSI) into α-iodoether termini from which the cationic growth of PCEVE branches is initiated. The CEVE-controlled polymerization, performed in conditions described in a previous section of this paper, yields a 4-arm PCEVE in which the initial short PS blocks act as a spacer between the central connecting point and the PCEVE branches. SEC analysis of the star polymer shows a good agreement with the expected structure with a ratio between the apparent Mw (refractive index detection) and the absolute M_w (light-scattering detection) of 0.86, which fit with the predicted contraction of the hydrodynamic volume of star polymers. The slight molar-mass distribution broadening (dispersity index of 1.18) and the presence of two small shoulders at high elution volume (Figure 21.2) suggests, however, both the formation of small amounts of stars with a lower number of arms and of residual homo-PCEVE is formed by side initiation. The characteristics of the PCEVE star polymer are collected in Table 21.2. An average number of 3.4 arms are obtained from the ratio of the absolute weight molar mass of the star over the molar mass of one single branch, which is equivalent to PCEVE formed by side initiation.

This lower average branch number could be explained either by incomplete initiation of the CEVE polymerization by the 4 acetal ends of the precursor and/or by some termination or transfer during the propagation reaction. It is difficult at this stage, using SEC or other conventional analytical techniques, to further identify the nature of the side reactions that predominantly occur during the star synthesis.

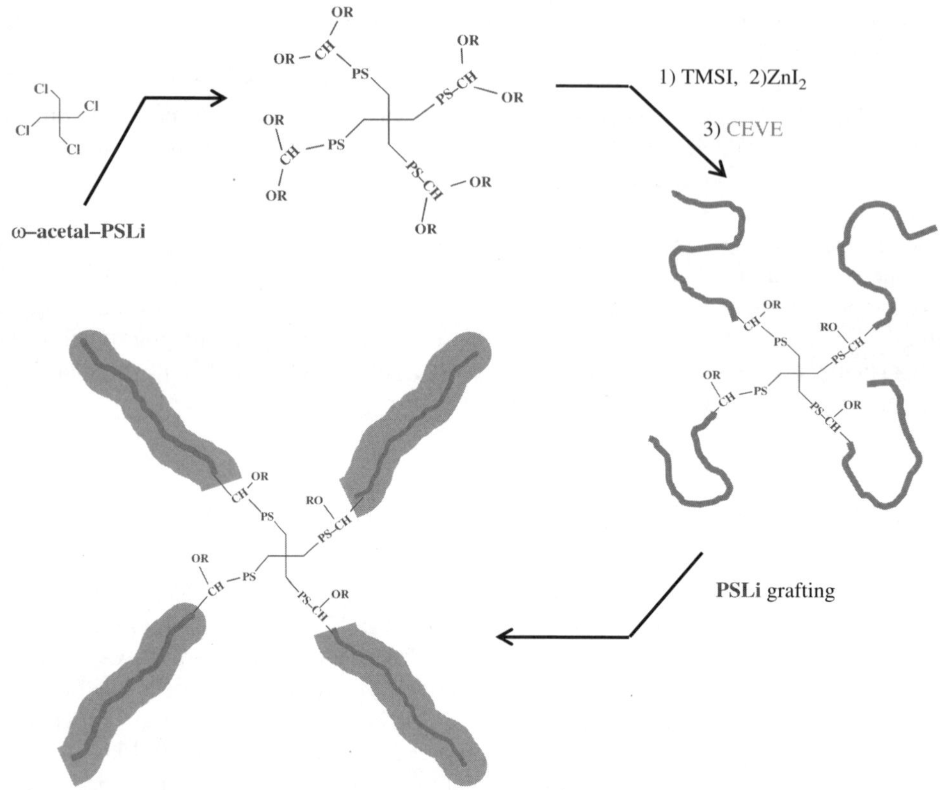

Scheme 21.4 Synthesis of 4-branch star PCEVE and of the corresponding PS combs.

The corresponding 4-arm star comb polystyrene were prepared by grafting polystyryllithium chains onto the CEVE units of the four-arm star PCEVE by adding PSLi in slight excess with respect to the CEVE units, as already described. After elimination of the ungrafted PS chains the SEC chromatogram, Figure 21.2 shows the presence of a unimodal and narrow peak

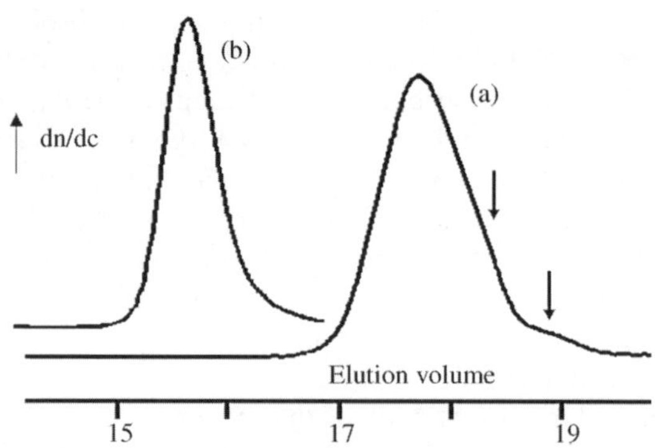

Figure 21.2 SEC chromatograms (a) of 4-arm star PCEVE and (b) of the corresponding 4-arm star PS comb.

Table 21.2 Dimensions and characteristic parameters of one-arm linear PCEVE, 4-arm PCEVE and of the corresponding linear and 4-arm comb PS (Reprinted with permission from M. Schappacher and A. Deffieux, "AFM image analysis applied to the investigation of elementary reactions in the synthesis of comb star copolymers," *Macromolecules*, **38**, 4942–4946, 2005 © 2005 American Chemical Society.)

	$\bar{M}_{n\ SEC}$	$\bar{M}_{w\ SEC}$	I_p	$\bar{M}_{w\ LS}$	f	$\bar{M}_{w\ app}/\bar{M}_{w\ star}$
Linear PCEVE arm[a]	42 000	44 000	1.06	35 400		
Star (PCEVE)$_4$[a]	96 600	114 000	1.18	132 000	3.4[b]	0.86
Linear PS comb arm[c]	—	800 000	1.09	2 700 000		0.29
Star comb PS[c]		1 700 000	1.07	9 000 000	3.34[b]	0.19

[a] Recovered by fractionation of the crude star solution.
[b] Determined from the ratio $\bar{M}_{w\ LS}$ star/$\bar{M}_{w\ SEC}$ linear arm.
[c] PS graft: $\bar{M}_{w\ SEC}$: 10 400 g/mole; $I_p = 1.03$.

distribution corresponding to the PS comb star. The grafting efficiency calculated from the ratio between the PS star molar mass and a PS graft molar mass is about 80%, which corresponds to the substitution of 4 over 5 CEVE units by a PS graft.

The main molecular characteristics of the "4-arm" comb star polystyrene are collected in Table 21.2. The $\bar{M}_{w\ RI(app)}/\bar{M}_{w\ LS}$ ratio, which is related to the volume contraction due to branching is 0.19 for the star comb versus 0.29 for the linear comb of same structure. These very low values agree with the highly branched chain architecture of the comb polymers on which is superposed the volume contraction related to the comb star architecture. The radius of gyration (R_g) and hydrodynamic radius (R_h) of the star comb on the one hand are very close to each other as are R_g and R_H of the corresponding one-branch linear comb polymer. They are equal to 45 nm, in THF at 25 °C for the star comb ($\bar{M}_w = 9\,000\,000$ g/mol) and 25 nm for the similar linear comb ($\bar{M}_w = 2\,700\,000$ g/mol) in the same conditions.

Due to their high compactness and increased mass density compared to linear polymers, the comb macromolecules can be visualized individually as a single molecular object by AFM (Boyce *et al.*, 2004; Matyjaszewski *et al.*, 2003; Schappacher and Deffieux, 2005; Sheiko, 2000; Sheiko *et al.*, 2003). Figure 21.3 shows the topographic and phase images of a series of isolated objects corresponding to star-like macromolecules (Schappacher and Deffieux, 2005). Their architecture confirms the predominant formation of 4-arm stars. The arms appear as relatively rigid and can only bend smoothly. As can be seen on the topographic image in Figure 21.3, the arms are positioned randomly around the central connecting point corresponding to the core of the stars. This can be attributed to the relative flexibility of the four short PS blocks constituting the tetrafunctional initiator backbone, at the central junction point (Scheme 21.3).

The selectivity of the synthetic pathway including the successive elementary reaction steps was examined in more detail through direct analysis of a set of several hundred individual macromolecules: each star polymer was individually characterized, counting the number of arms and measuring the arms length. Although the fraction of 4-arm comb stars largely predominates, analysis of the images shows the presence of 3-arm stars, along with a very small amount of 2-arm (long linear comb) and of one arm (Figure 21.4b). Moreover, among the 4- and 3-arm star macromolecules we can also distinguish between symmetric stars constituted

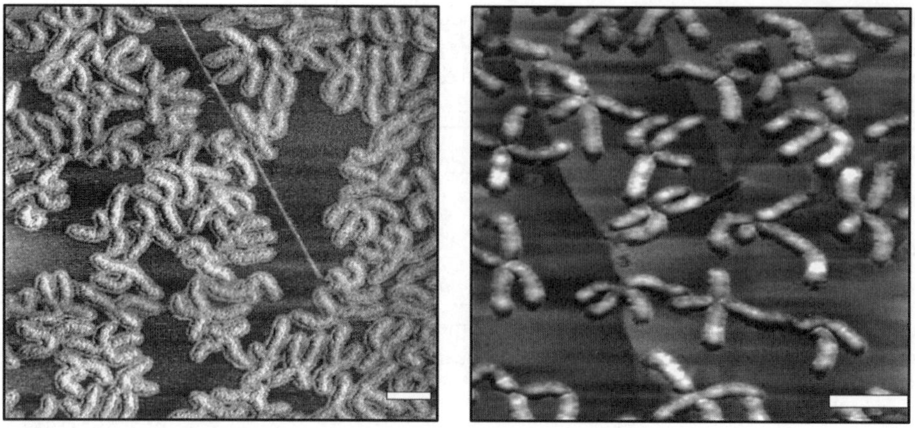

Figure 21.3 AFM images of the 4-arm comb stars (I)-(PS-*b*-PCEVE-g-PS)[4] obtained from deposition on HOPG from their diluted CH_2Cl_2 solution. (a) Phase and (b) topographical images: bar = 100 nm. (Reprinted with permission from M. Schappacher and A. Deffieux, "AFM image analysis applied to the investigation of elementary reactions in the synthesis of comb star copolymers," *Macromolecules*, **38**, 4942–4946, 2005 © 2005 American Chemical Society.)

of 4 or 3 arms of relatively close length and asymmetric stars, (Figure 21.4c), which possess one or even two much shorter arms. The different polymer structures identified and their percentages are indicated in Table 21.3.

It is assumed that the PS grafting reaction proceeds randomly and homogeneously along the PCEVE backbone, and does not affect through chain breaking or interchain coupling the primary architecture of the backbone macromolecule, that is, the number of arms and the relative initial dimension parameters (global DP_n and DP_n of each arm). Under this assumption, analysis of AFM images can give an insight into the elementary chemical processes involved in the building of the star polymers.

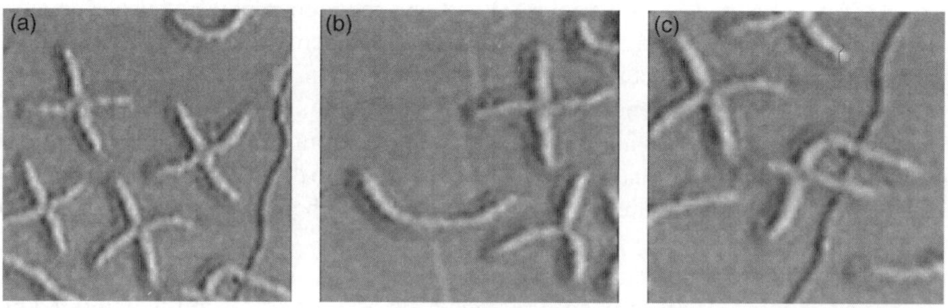

Figure 21.4 AFM images of isolated tetra-arm PS comb stars showing the main structural defects introduced during the synthesis: (i) number of branches; (ii) inhomogeneity in branch length. (Reprinted with permission from M. Schappacher and A. Deffieux, "AFM image analysis applied to the investigation of elementary reactions in the synthesis of comb star copolymers," *Macromolecules*, **38**, 4942–4946, 2005 © 2005 American Chemical Society.)

Table 21.3 Structure and percentage of the different comb stars as determined by AFM analysis (Reprinted with permission from M. Schappacher and A. Deffieux, "AFM image analysis applied to the investigation of elementary reactions in the synthesis of comb star copolymers," *Macromolecules*, **38**, 4942–4946, 2005 © 2005 American Chemical Society.)

4-arm star %			3-arm star %		linear (2 and 1 arms) %	
76			21		3	
symmetric	asymmetric		symmetric	asymmetric.	—	
%	%	%	%	%	%	%
4 arms	3 + 1 degen.	2 + 2 degen.	3 arms	2 + 1 degen.	eq. to 2 arms	one arm
47	25	4	19	2	3	<<1

degen: degenerated arm resulting from transfer or termination reactions.

In particular, the presence of macromolecules with a reduced number of arms is indicative of a nonperfect (uncompleted) initiation. The relative amount of 4-, 3- and 2-arm star fractions allow quantification of the efficiency of the initiation step of the CEVE polymerization from the acetal-PS ends (step 2, Scheme 21.4). AFM analysis yields an initiation efficiency of 91% with respect to the theoretical number of acetal functions (Table 21.4). This may arise from the presence of ill initiator structures that are not tetrafunctional, or to an incomplete formation by TMSI and/or activation of PS α-iodoether ends. The formation of degenerated branches on the other hand is indicative of the presence of side reactions. The aborted growth of some PCEVE branches can be explained by irreversible termination or transfer reactions during the propagation, and highlights the limitation in the living/controlled nature of the CEVE cationic polymerization, in particular when high molar mass PCEVE are targeted.

This suggests that beside secondary initiation resulting from hydrolyzed TMSI, a fraction of the linear PCEVE chains may come from transfer reactions. The overall impact of these different processes (incomplete initiation, transfer or termination) reduces the efficiency of the synthesis of symmetric 4-arm stars to 47%, the other 53% having one or several structural defects (lower number of branches, degenerated branch).

Table 21.4 Quantification of elementary reaction processes through statistical analysis of AFM images of isolated 4-arm polystyrene comb stars (Reprinted with permission from M. Schappacher and A. Deffieux, "AFM image analysis applied to the investigation of elementary reactions in the synthesis of comb star copolymers," *Macromolecules*, **38**, 4942–4946, 2005 © 2005 American Chemical Society.)

Efficiency of CEVE initiation from PS acetal-ends[a] %	% of size-controlled PCEVE arms[b]	Overall arm efficiency[c] %	Average number of controlled arms $\bar{F}'$ per star molecule[d]
91	91	85	3.40

[a] $I_{eff} = 4(\%\ \text{4-arm star}) + 3(\%\ \text{3-arm star}) + 2(\%\ \text{2-arm star})/4 \times 100$.
[b] $= \%4(4) + \%3(3 + 1) + \%2(2 + 2)/(\%4\ \text{arms} + \%3\ \text{arms} + \%2\ \text{arms})$.
[c] %Initiation eff x % dimension-controlled arms.
[d] $F' = 85 \times 4/100$.

This shows that AFM analysis of isolated hyperbranched macromolecules of relatively high molar mass is a powerful tool to investigate their size and architecture. It also allows investigation of the elementary chemical reactions occurring during the polymer synthesis and their selectivity, highlighting the main side reactions responsible for the formation of ill structures.

21.4.2 Comb Stars by the Convergent Approach (Ariura, 2009)

As indicated in Scheme 21.3b star polymers can be synthesized by a convergent, arm-first, approach. A series of PS comb stars was prepared following this strategy in order to investigate both the structure of the stars and the selectivity of the elementary steps of the convergent synthesis by molecular AFM analysis. To this aim star polymers with long PCEVE branches ($\overline{DP}_n > 500$) were synthesized by sequential copolymerization of CEVE and cyclohexyl divinylether (CDVE). Introduction of di-functional CDVE at the final stage of the CEVE polymerization allowed both its incorporation at the end of the growing CEVE chain accompanied by chain crosslinking, as explicated in Scheme 21.5.

In a first set of experiments the CEVE polymerization was initiated using $HCl/ZnCl_2$ as catalytic system in toluene at $-20\,°C$ and CDVE was added after complete conversion of CEVE. In these conditions the chain-coupling reaction is slow and incomplete, leading both to star-like PCEVE of broad molar mass distribution in admixture with a large amount of linear PCEVE chains (Table 21.5).

To improve the chain-end-coupling process a stronger Lewis acid, $SnCl_4$, was added at this stage to trigger the crosslinking reaction (entries 3–6). Best results were obtained using ratio CDVE/living PCEVE chains of about 90. The presence of a limited amount of linear poly (CEVE) chains was still observed in the crude polymer sample (Figure 21.5) indicating that the

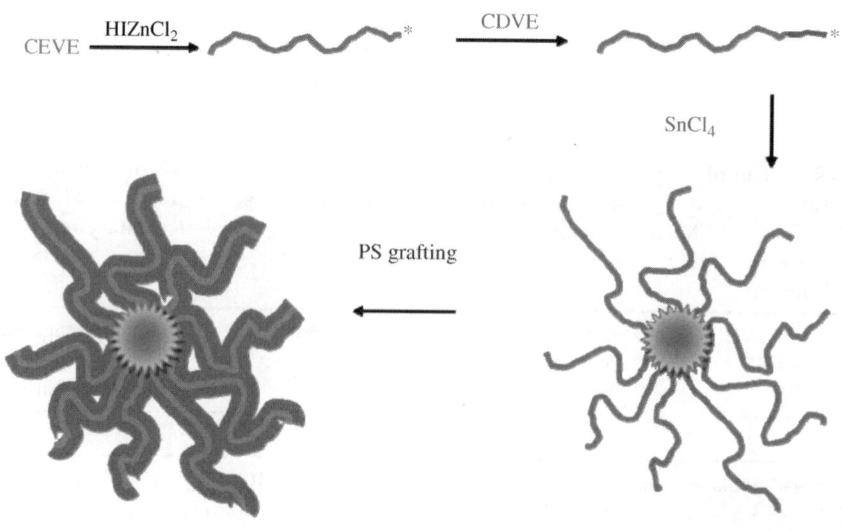

Scheme 21.5 Synthesis of a PS comb star by the convergent approach.

Table 21.5 Characteristics of PS stars obtained by the convergent strategy

Entry	$\bar{M}_w \times 10^{-4}$ PCEVE arm[a]	I_p	r[b]	Reaction temp. °C	$\bar{M}_w \times 10^{-5}$ star PCEVE[c]	I_p	N_a[d]
1	5.8	1.13	22	−10/–	—	broad	
2	4.8	1.10	35	−20/–	—	broad	
3	5.7	1.13	68	−20/−20[e]	—	broad	
4	7.7	1.23	22	−15/−5[e]	—	broad	
5	6.2	1.14	35	−15/−5[e]	7.5	1.25	12
6	12.7	1.08	90	−15/−5[e]	3.7	1.10	28

[a] SEC.
[b] [CDVE]/[P*].
[c] SEC-MALLS.
[d] $\bar{M}_w$ star/$\bar{M}_w$ arm.
[e] After addition of $SnCl_4$.

coupling was not quantitative. This was attributed to dead PCEVE chains resulting from termination/transfer reactions during the formation of the high molar mass PCEVE chains ended by CDEVE units. The SEC chromatograms of PCEVE before (linear) and after adding CDVE and $SnCl_4$ (star polymer) are shown in Figure 21.5. The linear PCEVE fraction was removed by precipitation to get, according to SEC analysis, almost pure PCEVE stars of low dispersity (<1.25).

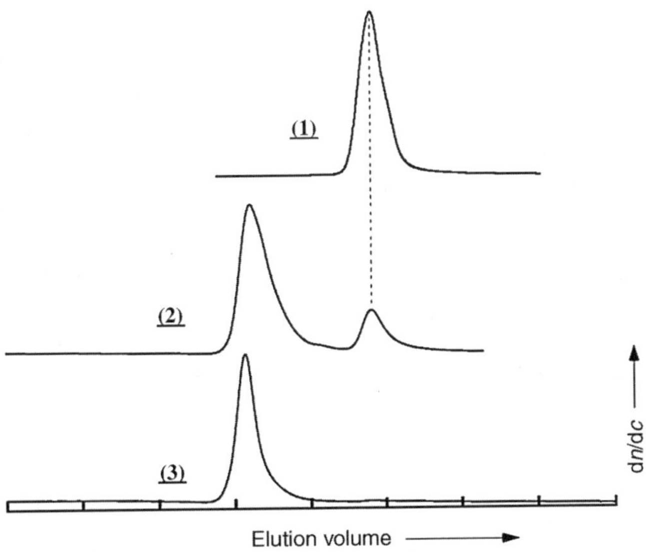

Figure 21.5 SEC chromatograms of a PCEVE arm (1) and of a crude (2) and purified (3) PCEVE star prepared by the convergent approach.

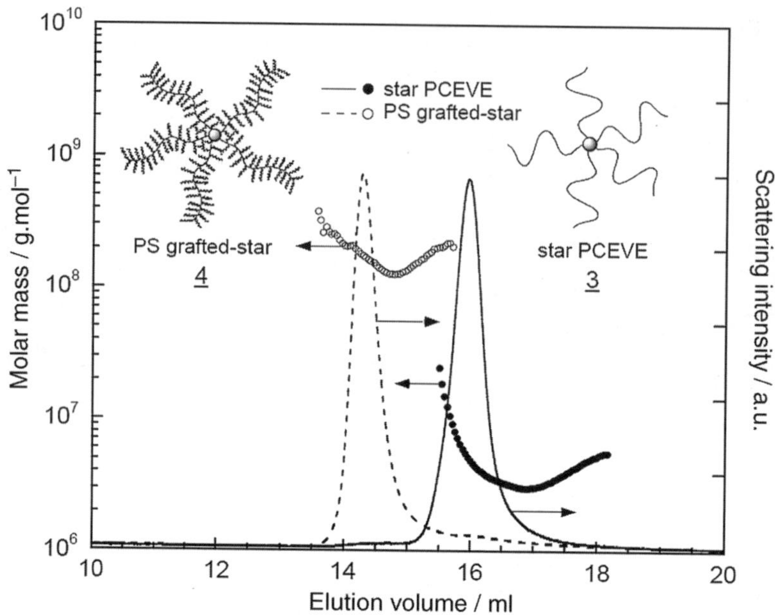

Figure 21.6 Light-scattering intensity and molar mass versus elution volume for a star PCEVE and PS grafted PCEVE star.

The average number of arms of PCEVE stars calculated from the ratio $\bar{M}_w$ star/$\bar{M}_w$ branch, determined by light scattering, was found to be strongly dependent on the amount and conditions of addition of CDVE and ranged from about 12 to 28 arms.

The PCEVE stars were used as reactive backbone to graft PSLi chains ($\bar{M}_n = 15\,000$ g/mole) and synthesize the corresponding PS comb stars, as previously reported for the synthesis of other comb architectures. The comb stars were then recovered free from ungrafted PS by selective precipitation. The SEC chromatograms of a PCEVE star and of the corresponding PS grafted comb star, obtained using light-scattering detection, are presented in Figure 21.6. They show a clean and quantitative shift of the initial peak population corresponding to entry 6 in Table 21.5 towards a new peak at much lower elution volume, corresponding to comb star polymers of much higher molar mass and of low dispersity. Table 21.6 resumes the main characteristics of the obtained PS comb star.

Table 21.6 Characteristics of PS comb star prepared by the convergent strategy after grafting of polystyrene chains with a $\overline{DP}_n$ of 160

Comb architecture		$\bar{M}_w$	I_p	R_g[a] nm	N_a[b]
single arm	$PCEVE_{1220}$-g-PS_{160}	1.5×10^7	1.28	110	1
star[c]	star-($PCEVE_{1220}$-g-PS_{160})	2.8×10^8	1.12	190	14

[a] Measured in THF at 25 °C.
[b] Number of arms determined from the molar mass ratio between comb star and single-arm comb.
[c] Entry 6 of Table 21.5.

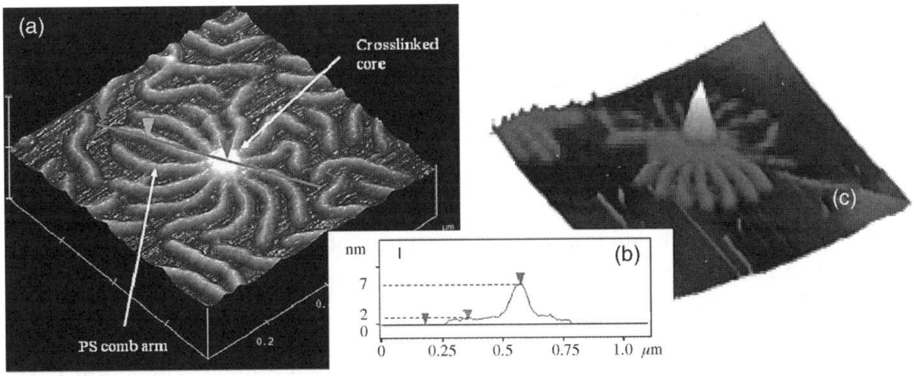

Figure 21.7 (a) Topographic AFM image on HOPG of a comb star obtained by the convergent approach, (b) its cross-sectional analysis and (c) three-dimensional representation.

The star-shape structure and the molecular characteristics of the combs were further examined by AFM in the tapping mode. Images of a series of PS comb stars obtained by the convergent method are shown in Figures 21.7 and 21.8. The magnified image of a star molecule deposited on HOPG, Figure 21.7 clearly shows the PS comb branches surrounding a central core. The comb branches exhibit a relatively uniform size (about 200 nm length, 50 nm width) with only 2 nm height, indicating that they are completed laying on the surface, thanks to their strong affinity with HOPG. The central higher (7 nm) and whiter zone corresponds to the harder densely crosslinked core.

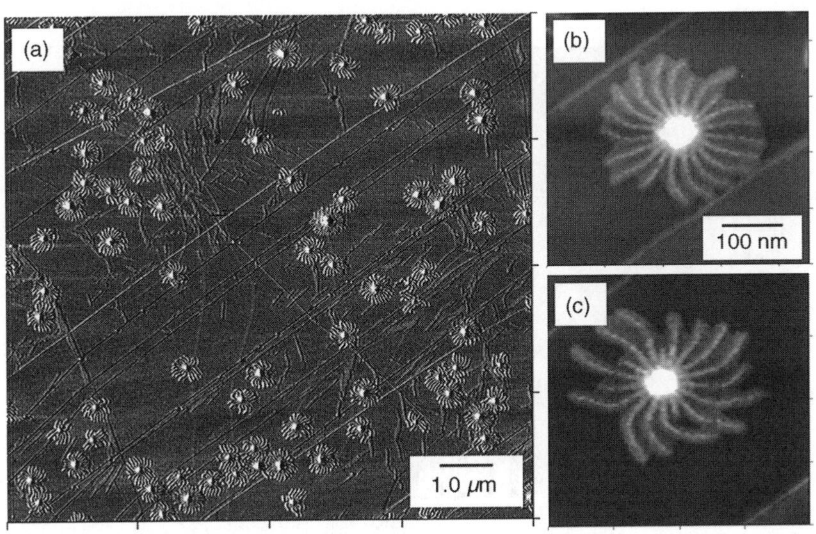

Figure 21.8 Phase (a) and topographic (b, c) AFM images of comb stars (sample of Table 21.6) obtained by the convergent method. (a) General view; (b) and (c) images of isolated macromolecules showing the hard central crosslinked core and the relatively softer attached PS comb branches.

The set of objects presented in Figure 21.8 illustrates a typical population of macromolecules obtained by the convergent strategy. Indeed, a more detailed investigation of each macromolecule shows that this approach yields stars with relatively heterogeneous arm length and number of branches, information that was hardly detectable by SEC. As can be seen in the inserts (b) and (c) of Figure 21.8 some very short branches attached to the crosslinked core are observed. They likely result from a delayed initiation in agreement with transfer reactions during the CEVE polymerization. This is inconsistent with the population of linear PCEVE chains formed during the synthesis of PCEVE stars and eliminated by fractional precipitation.

Further analysis of the star comb population performed by counting the number of arms of each macromolecule was conducted over a large set of objects in order to get the statistical number and distribution (Figure 21.9). As may be seen, the main population possesses between 15 to 19 arms, a much higher value than achievable by the divergent method. The discrepancy observed in the average number of branches between data obtained by SEC analysis (average of 14), Table 21.6 and the AFM images can be tentatively attributed to a lower PS grafting yield for the star as compared to an isolated branch. This is due to steric restrictions around the central core of the star, which yields to an overestimation of the molar mass of the comb PS branch.

In addition, a second star population with a maximum corresponding to objects with about 30–32 branches is also formed. This second set of objects in relatively low proportion can be attributed to core-to-core coupling during the synthesis of the PCEVE stars. These results illustrate the interest of AFM analysis to get a much more precise vision of both single and large population of complex macromolecules.

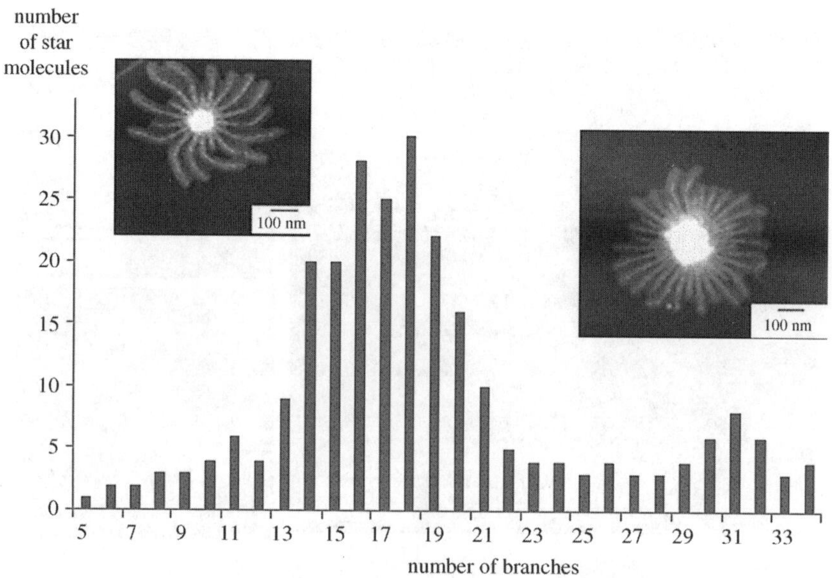

Figure 21.9 Statistical analysis of the number of branches over a set of star comb macromolecules obtained by the convergent approach.

21.5 Macrocyclic PS Combs

The interest in cyclic macromolecules began more than 50 years ago with theoretical prediction of the effect of chain cyclization on the polymer properties (Zimm and Stockmayer, 1949; Bloomfield and Zimm, 1966) and the discovery of naturally occurring macrocycles such as DNA (Dubelcco and Vogt, 1963; Chandler *et al.*, 1964; Glaubiger and Hearst, 1967; Kiger *et al.*, 1968; Mitra *et al.*, 1967). Still today, large polymer macrocycles remain a fascinating curiosity for theoreticians and physicists and a challenging area for synthetic chemists (Deffieux and Borsali, 2007).

Although these endless chains have quite specific properties, because of their still flexible chain they do not exhibit the characteristics of shape persistent objects. To improve chain stiffness the use of macrocycles bearing grafted chains has also been explored (Boydston *et al.*, 2009; Li, Jérôme, and Lecomte, 2008; Jia *et al.*, 2006; Pang *et al.*, 2007). In most cases ring sizes remained low as compared to the grafts and evidence for a ring-shape structure could be scarcely observed. Indeed, although some fascinating properties can be expected for shape-persistent macrocycles, both as isolated object and as self-assembled system, up to now few studies have been focused on their synthesis.

In order to achieve the synthesis of large polymer rings that could be used also as precursor of comb macrocycles, a new synthesis strategy has been developed in our laboratory (Schappacher and Deffieux, 2008a; Schappacher and Deffieux, 2008b; Schappacher and Deffieux, 2008c), Scheme 21.6. It is based on the preparation of linear **ABC** triblock copolymers, in which the long central **B** block is surrounded by two short A and C sequences constituted of monomer units bearing reactive antagonist functions. The A**B**C triblocks were synthesized by sequential living cationic copolymerization successively of *ter*-butyldimethylsilylethoxy vinyl ether (SiVE), chloroethyl vinyl ether (CEVE) as monomer for the central block, and cyclohexanedimethanol divinyl ether (CDVE). As previously reported, the chloroether group of the central PCEVE block could then be used for introducing PS grafts. To avoid or strongly limit reaction of the second vinyl ether function of CDVE, which may yield interchain coupling, its polymerization was stopped at very low conversion (<10%) and the A**B**C triblock was recovered by selective precipitation. The characteristics of linear A**B**C triblocks are collected in Table 21.7. The central PCEVE **B** block has $\overline{DP}_n$ value ranging from 548 to about 1000, whereas the external A and C sequences remain limited to a very small number of monomer units, ranging from 3 to 12.

Deprotection of the hydroxyl groups of the SiVE sequence can be readily achieved by reaction of the triblock copolymer with tetrabutylammonium fluoride (TBAF) in THF, thus

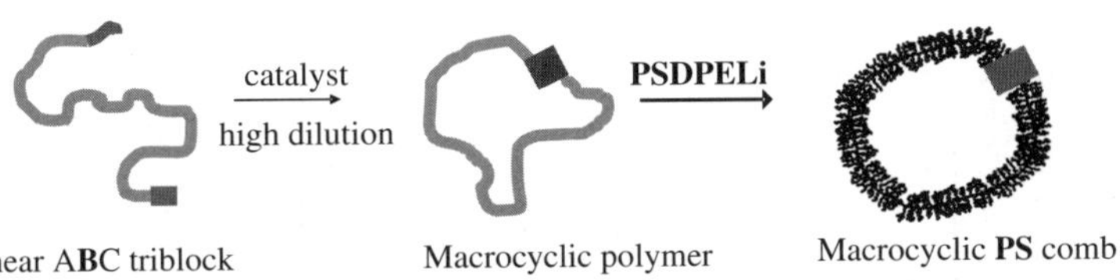

Scheme 21.6 Strategy for the synthesis of macrocyclic combs. (Reprinted with permission from M. Schappacher and A. Deffieux, "Atomic force microscopy imaging and dilute solution properties of cyclic and linear polystyrene combs," *Journal of the American Chemical Society*, **130**, 14684–14689, 2008. © 2008 American Chemical Society.)

Table 21.7 Characteristics of linear (L) A**B**C triblocks prepared by living cationic polymerization and of the corresponding cyclic (C) with poly (CEVE) as central block (Reprinted with permission from M. Schappacher and A. Deffieux, "Atomic force microscopy imaging and dilute solution properties of cyclic and linear polystyrene combs," *Journal of the American Chemical Society*, **130**, 14684–14689, 2008. © 2008 American Chemical Society.)

Triblock ABC	$M_n \times 10^{-3a}$	$M_p \times 10^{-3a}$	M_w/M_n	$M_{pC}/M_{pL.}$	$M_w \times 10^{-3b}$	$Dp_nA^c_{(SiVE)}$	$Dp_nB^d_{(CEVE)}$	$Dp_nC^e_{(CDVE)}$	Estimated cyclization yield[f] %
L-1	64.24	71.25	1.05			3	548	2	70
				0.80	59.2				
C-1	52.68	71.25	1.21						
L-2	83.80	95.42	1.05						65
						5	774	4	
C-2	67.04	67.04	1.17	0.79	83.8				
									68
L-3	103.98	117.50	1.06			12	956	3	
				0.78	104.7				
C-3	81.10	91.50	1.15						

[a] Determined SEC in THF by means of PS standards.
[b] Measured by SLS with a refractive index increment (dn/dc) of 0.09.
[c] Determined by ^1H NMR from integration ratio between CEVE protons (calculated from M_{wPCEVE}) and $(CH_3)_2Si$ protons.
[d] Calculated from $(M_{wsls} - (M_{nSiVE} + M_{nCDVE}))$.
[e] CDVE vinyl ether protons.
[f] From deconvolution of SEC peaks.

yielding hydroxyethyl vinyl ether (HEVE) units and the corresponding poly (HEVE-*b*-CEVE-*b*-CDVE) linear triblock, with unchanged total $\overline{\text{DP}}_n$ and molar mass distribution.

In a second step the A and C triblock end sequences are reacted in the presence of an acid as catalyst, under highly dilute conditions to favor macrocyclization by intramolecular coupling of the functions of A and C sequences. This is achieved by adding dropwise a solution of the copolymer into a large volume of methylene dichloride containing traces of an acid catalyst.

The intramolecular coupling between the hydroxyl functions of the HOVE units of block A and the pendant vinyl ether functions of block C was then achieved in presence of pyridinium *para*-toluene sulfonic acid salt (PTSA) as catalyst. The reaction between hydroxyl functions and vinyl ether groups yield the rapid formation of acetal links between the A and C sequences, as described in Scheme 21.7.

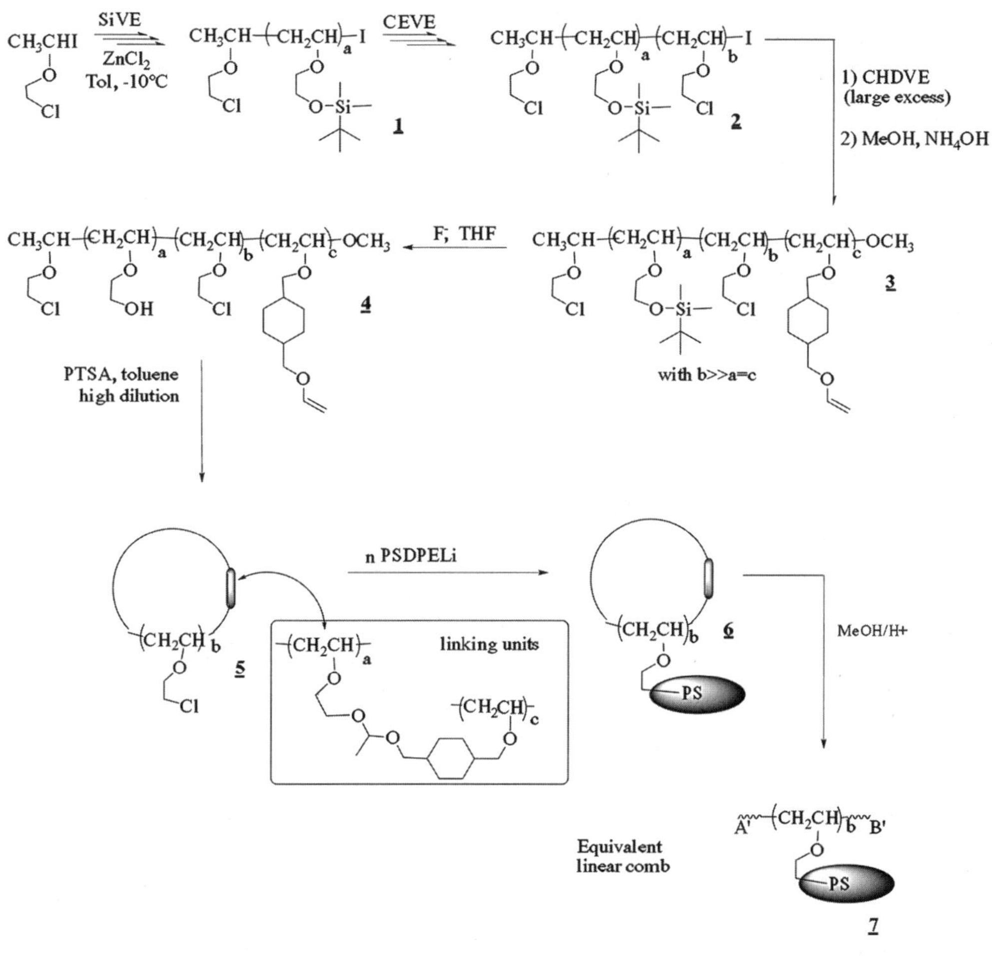

Scheme 21.7 Strategy used for the synthesis of large PCEVE rings and of corresponding macrocyclic PS combs. (Reprinted with permission from M. Schappacher and A. Deffieux, "Atomic force microscopy imaging and dilute solution properties of cyclic and linear polystyrene combs," *Journal of the American Chemical Society*, **130**, 14684–14689, 2008. © 2008 American Chemical Society.)

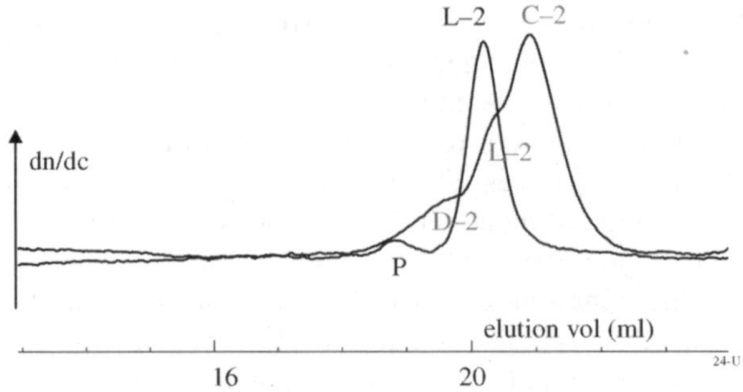

Figure 21.10 SEC of a linear A**B**C poly (HEVE-*b*-CEVE-*b*-CDVE) ***4*** linear triblock copolymer (sample L-2, Table 21.1) and of the corresponding copolymer after cyclization (C-2). P is the fraction of linears that are formed through CDVE coupling. L-2 and D-2 are remaining linears and polycondensates (mainly dimers). (Reprinted with permission from M. Schappacher and A. Deffieux, "Atomic force microscopy imaging and dilute solution properties of cyclic and linear polystyrene combs," *Journal of the American Chemical Society*, **130**, 14684–14689, 2008. © 2008 American Chemical Society.)

The SEC chromatogram of the crude polymer **5** after the cyclization step is shown Figure 21.10. A strong shift of the main SEC peak towards higher elution volume is observed, indicating a decrease in the hydrodynamic volume of the macromolecule, in agreement with chain cyclization. For all samples, the ratio between the apparent molar mass (M_{peak}) of the cyclic and of the linear A**B**C copolymers is close to 0.8, Table 21.7. However, as may be noticed on the SEC chromatogram Figure 21.10, a significant peak broadening of the cyclized polymer is observed and two peak shoulders indicate the presence both the remaining linear A**B**C precursor and the other linear or cyclic structures of higher molar masses corresponding to chain dimerization. Besides linear dimers, the peak position suggests the possible formation of bicyclic, tadpole and cyclic dimers. However, this could not be proved clearly by SEC or by other conventional analytical technique.

Deconvolution of the main SEC peak attributed to cyclic gives as a first estimate a cyclization yield in the range 65 to 70%, Table 21.10. As will be shown latter, these values are higher than those directly measured by AFM imaging on the corresponding cyclic combs. This could be ascribed to the coelution of low molar mass linears with cyclics.

The synthesis of macrocyclic PS combs was achieved as previously shown by grafting living polystyrene onto the PCEVE backbone. In order to achieve a high grafting density, the living polystyrene was added in slight excess with respect to the CEVE units (1.2 equivalents). The comb copolymers were separated from the unreacted PS by selective precipitation in a heptane/cyclohexane mixture. The SEC chromatogram of a cyclic PS comb is shown in Figure 21.11. This closely reflects the SEC profile of the cyclized A**B**C copolymer, Figure 21.10, except that all peaks are shifted towards much higher molar masses. The presence of peak shoulders attributed to linear combs as well as to various types of dimeric combs is observed again.

These results are clearly confirmed by tapping-mode AFM analysis performed on solid deposits of the PS combs obtained by casting their THF solution on oriented graphite (HOPG). The AFM picture of a crude comb sample is presented in Figure 21.12a. The image shows

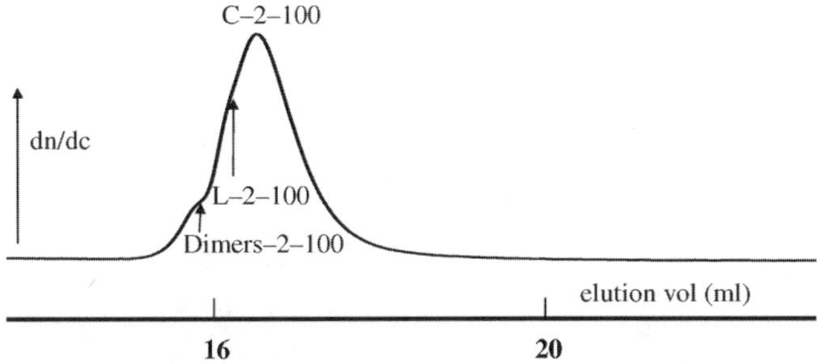

Figure 21.11 SEC chromatogram of a cyclic PS comb obtained after elimination of excess of ungrafted PS (sample C-2-100, Table 21.2). (Reprinted with permission from M. Schappacher and A. Deffieux, "Atomic force microscopy imaging and dilute solution properties of cyclic and linear polystyrene combs," *Journal of the American Chemical Society*, **130**, 14684–14689, 2008. © 2008 American Chemical Society.)

single molecules isolated on the surface with ring structures corresponding to macrocyclic combs and to a fraction of linear combs coming from uncyclized PCEVE precursor. In addition, we can also notice the formation in small proportions of comb dimers corresponding to linear (L_D) cyclic- (C_D), tadpole- (T_D) and eight-shape bicycles (B_D) (Figure 21.12b). The linear structures are believed to result from the formation of some **BC** and more likely **AB** diblock copolymers, present in admixture with the A**BC** precursor, rather than from a low efficiency of the cyclization step. Diblock copolymer formation can be attributed to transfer and/or termination reactions during the vinyl ethers cationic copolymerization. Yields in the target ring polymers can be estimated directly from AFM images by counting the different comb architectures. They are in the range of 50–60%, which is slightly lower than cyclization yields estimated by SEC (Table 21.7) around 65–70%. This may be explained by the coelution of a small fraction of low molar mass linears with the cyclics.

Further purification of the cyclic combs was partly achieved by preferential precipitation of the various dimers and linear combs by slowly decreasing the temperature of their cyclohexane solution. As may be seen on the SEC chromatograms, Figure 21.13, the shoulder and small peak of linear and dimeric structures, respectively, are suppressed when incrementally decreasing the temperature from 20 to 18 °C, whereas the SEC peak of cyclic combs gets narrower and shifts to slightly higher elution volumes. The precipitation of linear combs at higher temperature than cyclics of same molar masses can be related to the difference in their Θ temperature as already reported for linear and cyclic polystyrenes. As shown in AFM images presented in Figure 21.12c it is also possible to significantly enrich the cyclic comb fraction, up to about 80%, by SEC fractionation. The influence of both the PCEVE and the PS graft $\overline{DP}_n$s on the ring size, the ring section and the central cavity dimension is illustrated with samples in which either the backbone $\overline{DP}_n$ and/or the PS grafts length are different, Figures 21.14 and 21.15. The corresponding ring parameters are collected in Table 21.8. As may be seen, the ring diameter measured at the middle of the ring section corresponds to an almost fully stretched PCEVE backbone. The ring section increases with the $\overline{DP}_n$ of the PS grafts in agreement with a close packing of PS branches around the backbone. For the longer PS grafts the central cavity is partly or even almost completely filled (Figures 21.14b and d).

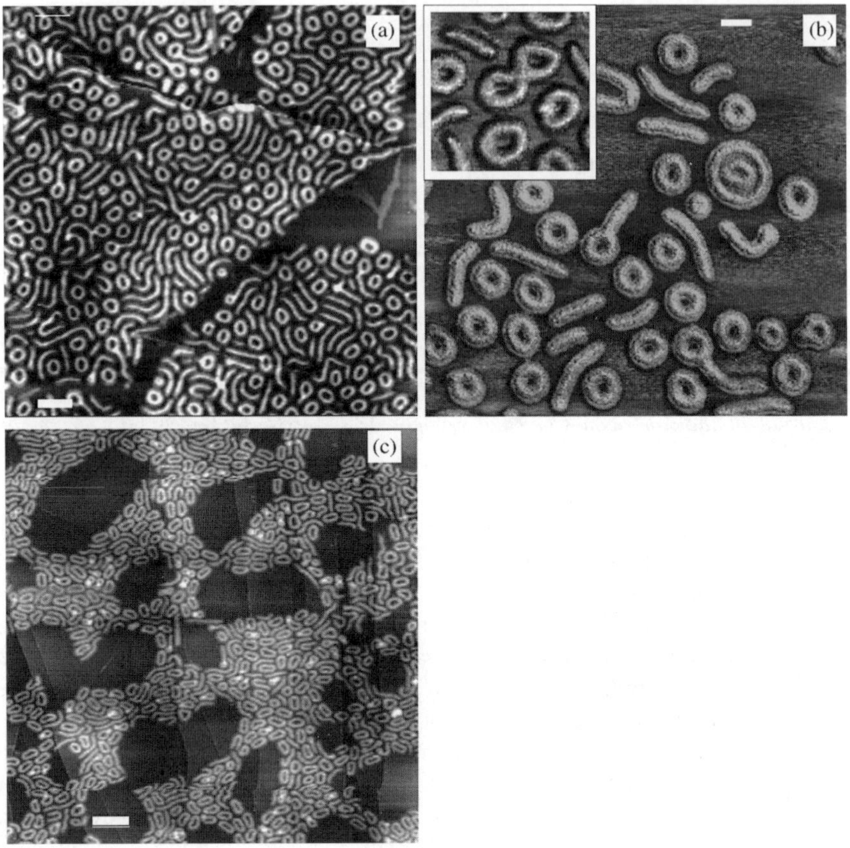

Figure 21.12 AFM images of crude and fractionated polystyrene cyclic combs (sample C-2-100, Table 21.8) obtained from deposits of their THF solution on HOPG: only the fraction of ungrafted PS has been eliminated. (a) General view of the crude sample (height image scale bar: 200 nm). (b) Magnified phase image (scale bar:100 nm) showing some of the other comb architectures formed (Scheme 21.3). (c) Enriched cyclic fraction obtained by temperature-induced selective precipitation (height image, scale bar: 200 nm). (Reprinted with permission from M. Schappacher and A. Deffieux, "Atomic force microscopy imaging and dilute solution properties of cyclic and linear polystyrene combs," *Journal of the American Chemical Society*, **130**, 14684–14689, 2008. © 2008 American Chemical Society.)

21.5.1 Reopening of the Cyclic Comb Ring (Schappacher and Deffieux, 2008a)

The cyclization process through intramolecular formation of acetal linkages (Scheme 21.2) allows reopening of the rings via quantitative acetal hydrolysis in acidic media, as confirmed by SEC and AFM imaging, Figure 21.16.

This permits cyclic and linear combs to be obtained with the same constitutional parameters (backbone $\overline{DP}_n$, number and length of PS branches) that are very useful models for direct comparison of the properties of linear and cyclic combs. The SEC chromatograms of the

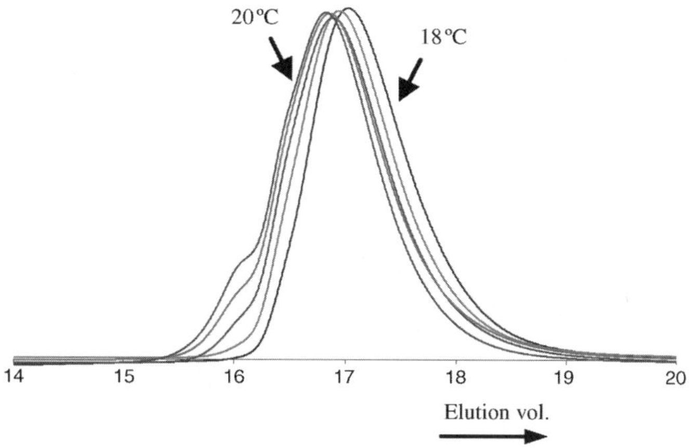

Figure 21.13 Evolution of the SEC chromatogram of the cyclohexane soluble fraction of cyclic PS comb (sample C-2-100) recovered at decreasing temperatures in the range 20–18 °C. (Reprinted with permission from M. Schappacher and A. Deffieux, "Atomic force microscopy imaging and dilute solution properties of cyclic and linear polystyrene combs," *Journal of the American Chemical Society*, **130**, 14684–14689, 2008. © 2008 American Chemical Society.)

corresponding comb polymers are shown in Figure 21.17, whereas the evolution of the comb molar mass and gyration radius as a function of the elution volume for linear and cyclic homologs is shown in Figure 21.18.

The ratio between the SEC apparent peak molar mass of the cyclic and linear combs range from 0.72 to 0.84 depending on the comb parameters (backbone and PS lengths). The more compact cyclic comb (C-1-170), with no detectable central cavity shows the smallest difference with its linear analog. This can be tentatively explained by behavior closer to a compact sphere than to a ring. The main characteristics of linear and cyclic PS combs, including apparent and absolute molar masses, hydrodynamic radius and radius of gyration in THF are collected in Table 21.9. The $<R_{gc}>^2/<R_{gl}>^2$ and R_{hc}/R_{hl} values of the PS combs are close to 0.5 and 0.9, respectively, in very good agreement with data previously determined for linear and cyclic polystyrenes. This shows that the architecture of the chain is still largely predominating over other parameters in comb structures. This can be explained by the fact that the original linear and cyclic polymers have been magnified to the same extent by the grafting process, thus preserving their original characteristics.

21.5.2 Imaging of Catenane, Eight-Shaped and Trefoil Knotted Polymer Rings: Combs as Magnified Polymer Structures (Schappacher and Deffieux, 2009)

Most of the available characterization techniques of macromolecules are based on the determination of average properties shared by an entire set of molecules of a given system (chemical composition, tacticity, dimensional parameters, melting temperature, etc.). In this context, imaging techniques offer a quite different and unique approach allowing the analysis of macromolecules as single objects. Numerous studies have been devoted over

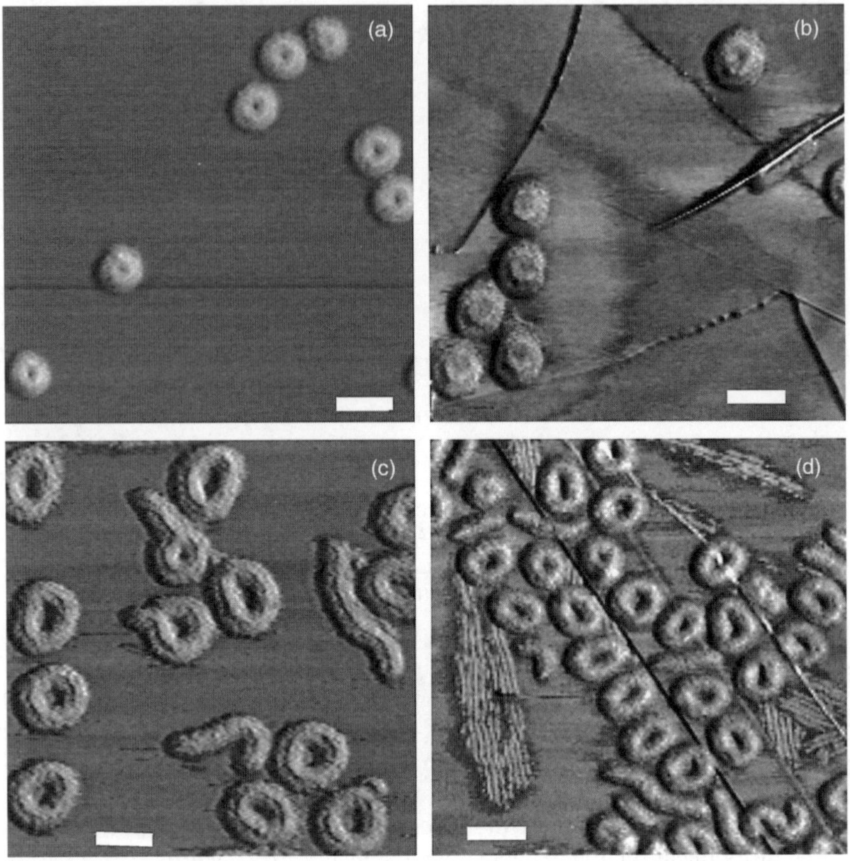

Figure 21.14 AFM phase images on HOPG of cyclic polystyrene combs obtained from A**B**C precursors with central PCEVE B block of decreasing $\overline{DP}_n$ and/or with different PS graft lengths: (a) C-1-100, (b) C-1-170, (c) C-2-100, (d) C-3-170 cyclic comb fractions recovered by SEC fractionation (scale bars: 100 nm). (Reprinted with permission from M. Schappacher and A. Deffieux, "Atomic force microscopy imaging and dilute solution properties of cyclic and linear polystyrene combs," *Journal of the American Chemical Society*, **130**, 14684–14689, 2008. © 2008 American Chemical Society.)

the past decades to molecular imaging of biomolecules using electronic microscopy and atomic force microscopy (AFM). The visualization of single DNA molecules and the discovery that they can exist in the form of single and interlocked rings of various complexities has been reported (Clayton and Vinograd, 1967; Vinograd and Lebowitz, 1966; Vinograd *et al.*, 1965). Since then, catenaned and knotted single- and double-stranded DNA rings were repeatedly observed (Cluzel *et al.*, 1996; Levene *et al.*, 1995; Yamaguchi *et al.*, 2000a; Yamaguchi *et al.*, 2000b; Lyubchenko *et al.*, 2002). Fewer studies concern the AFM investigation of neutral polymer chains at the molecular level (Kumaki *et al.*, 1996; Sheiko and Moller, 2001). Indeed, neutral linear polymer chains with sizes ranging from a few to several hundred nanometers possess a cross section of only a few atoms. They are softer than biomolecules and polyelectrolytes due to

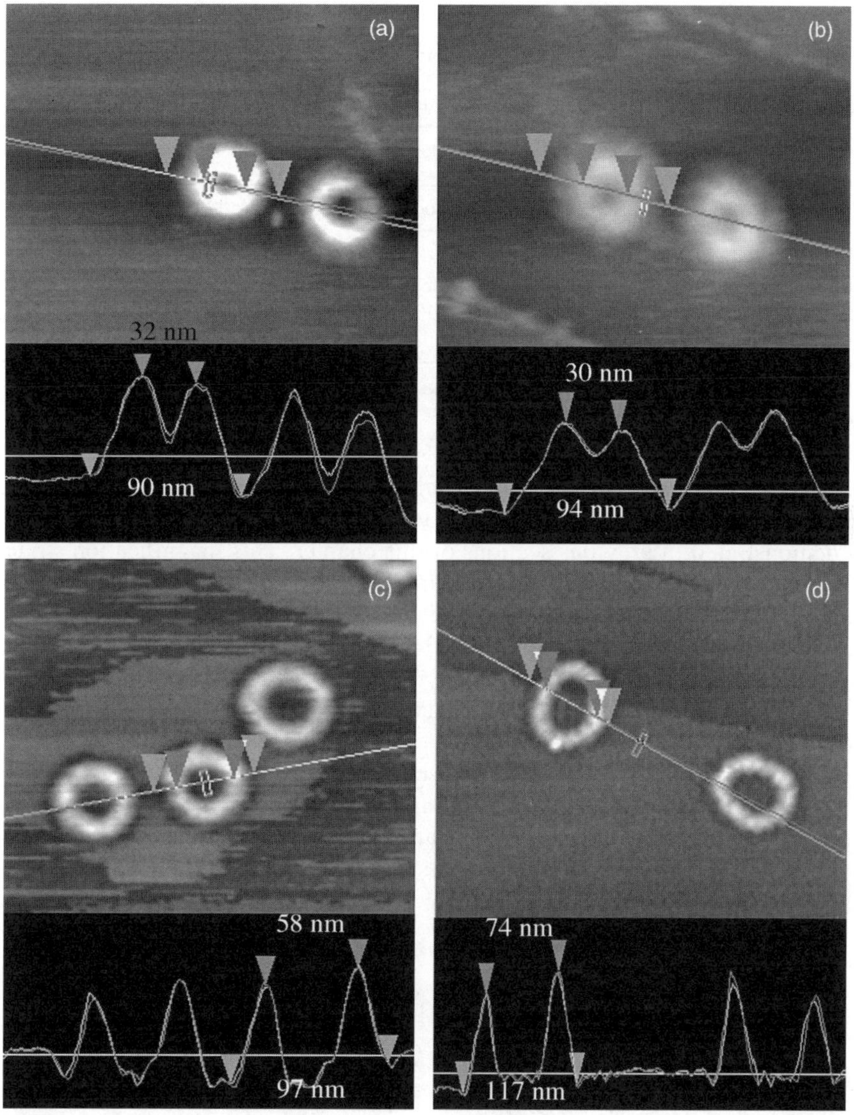

Figure 21.15 AFM section analysis of isolated cyclic PS combs corresponding to samples (a), (b), (c), and (d), of Figure 21.14. (Reprinted with permission from M. Schappacher and A. Deffieux, "Atomic force microscopy imaging and dilute solution properties of cyclic and linear polystyrene combs," *Journal of the American Chemical Society*, **130**, 14684–14689, 2008. © 2008 American Chemical Society.)

weaker interactions with substrates and their local concentration can widely fluctuate because of chain motion. This makes them highly diffuse objects that cannot be easily observed by imaging microscopy. To circumvent this problem, a magnification process of the size and density of macromolecules prior to imaging should be used to homogeneously enlarge isolated polymer chains before observation by AFM. This is what is achieved by

Table 21.8 Dimensions of cyclic PS combs as determined by measurements on AFM images (Reprinted with permission from M. Schappacher and A. Deffieux, "Atomic force microscopy imaging and dilute solution properties of cyclic and linear polystyrene combs," *Journal of the American Chemical Society*, **130**, 14684–14689, 2008. © 2008 American Chemical Society.)

PS Comb reference	z $\overline{DP}_n$ PCEVE (B block)	Circumference (nm)		Ring diameter (nm)
		Calculated[a]	Measured[b]	
C-1-100	548	137	122	94
C-1-170	548	137	116	96
C-2-100	774	193	188	105
C-3-170	956	239	250	127

[a] Calculated on the basis of a complete chain extension (C–C bond = 0.25 nm).
[b] Measured at the center of the ring section.

densely grafting oligomers on the main chain to be studied, that is, making comb-like structures with a cross section thicker and more readily detectable by the AFM tip, as already demonstrated in previous sections of this chapter. This magnification strategy was applied recently to a detailed analysis macromolecule-by-macromolecule of ring polymers obtained by end-to-end cyclization of the **ABC** linear polymer chains. Besides the expected main formation of ring-shaped macromolecules already described, AFM imaging allowed us to visualize the presence in small amounts of very unusual topological isomers including catenanes, eight-shaped and knotted rings that also form during the ring-closure process.

To obtain a representative image of the starting macromolecule, the grafting reaction should preserve the chain integrity and its architecture in order to give a homogeneous translation of its initial dimensions. This means that the grafting reaction should be highly

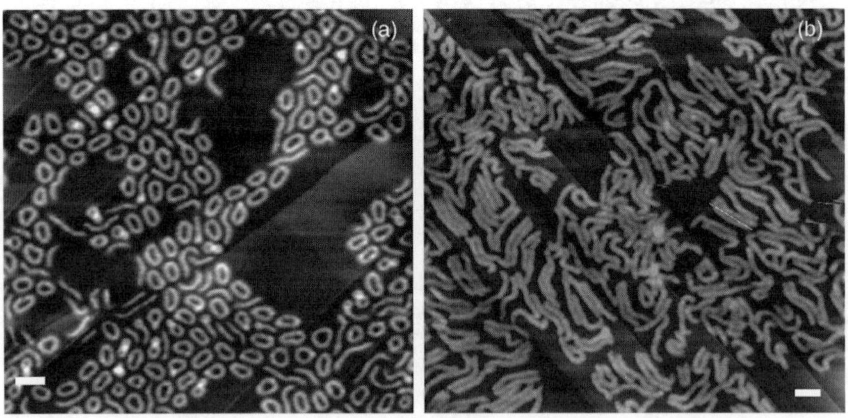

Figure 21.16 AFM image of cyclic combs after treatment in acidic media. (Reprinted with permission from M. Schappacher and A. Deffieux, "Atomic force microscopy imaging and dilute solution properties of cyclic and linear polystyrene combs," *Journal of the American Chemical Society*, **130**, 14684–14689, 2008. © 2008 American Chemical Society.)

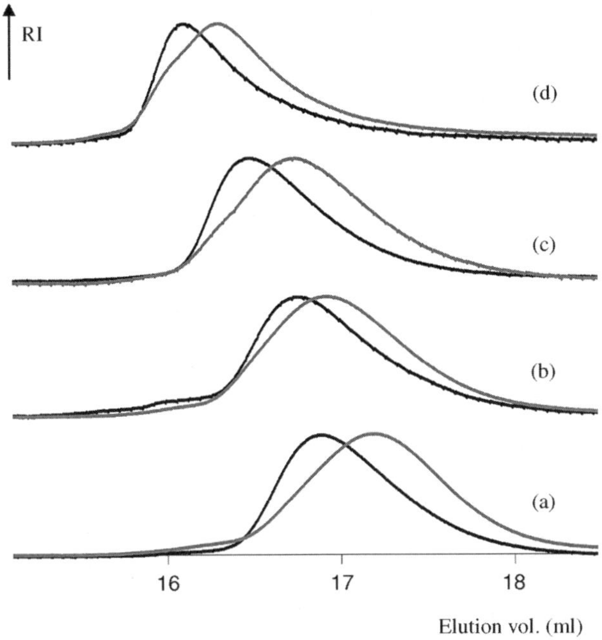

Figure 21.17 SEC chromatograms of cyclic and linear PS combs obtained by ring opening through acetal hydrolysis under acidic treatment: (a) C-1-100, (b) C-2-100, (c) C-1-170, (d) C-3-170 (see Figure 21.14). (Reproduced with permission from M. Schappacher and A. Deffieux, "Imaging of catenated, figure-of-eight, and trefoil knot polymer rings," *Angewandte Chemie International Edition*, 2009, **32**, 5930–5933. © Wiley-VCH Verlag GmbH & Co. KGaA.)

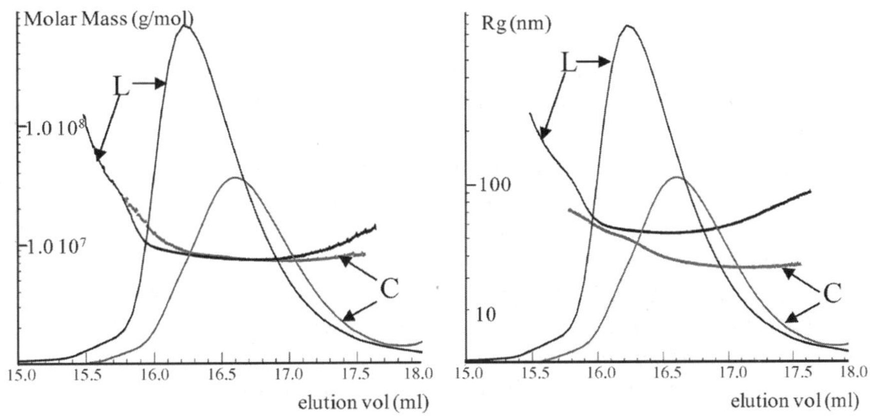

Figure 21.18 Elution volume dependence of the molar mass and of the radius of gyration for a linear PS comb "L-2-100" (L) and of its cyclic homolog "C-2-100" (C). (Reproduced with permission from M. Schappacher and A. Deffieux, "Imaging of catenated, figure-of-eight, and trefoil knot polymer rings," *Angewandte Chemie International Edition*, 2009, **32**, 5930–5933. © Wiley-VCH Verlag GmbH & Co. KGaA.)

Table 21.9 Characteristic dimensions and parameters of linear (L) and cyclic (C) PCEVE-g-PS combs (Reprinted with permission from M. Schappacher and A. Deffieux, "Atomic force microscopy imaging and dilute solution properties of cyclic and linear polystyrene combs," *Journal of the American Chemical Society*, **130**, 14684–14689, 2008. © 2008 American Chemical Society.)

Combs[a]	$\bar{M}_n \times 10^{-5b}$	$\bar{M}_P \times 10^{-5b}$	$\frac{\bar{M}_w}{\bar{M}_n}$	$\frac{M_{PC}}{M_{PL}}$	$\bar{M}_w \times 10^{-5c}$	$R_g{}^d$ (nm)	$\frac{R_{gC}{}^2}{R_{gL}{}^2}$	$R_h{}^e$ (nm)	$\frac{R_g}{R_h}$	$\frac{R_{hC}}{R_{hL}}$
L-1-100	7.90	10.01	1.17		60.05	39		34	1.04	
				0.74			0.53			0.90
C-1-100	6.70	7.40	1.13		58.30	28		31	0.93	
L-1-170	8.80	11.43	1.12		68.60	37		36	0.95	
				0.84			0.62			0.92
C-1-170	7.99	9.68	1.16		64.60	30		33	0.90	
L-2-100	11.22	15.20	1.15		79.00	54		40	1.35	
				0.72			0.48			0.90
C-2-100	9.42	10.88	1.12		75.40	37		36	1.04	
L-3-170	17.17	21.59	1.17		119.00	72		49	1.47	
				0.74			0.49			0.88
C-3-170	12.63	15.98	1.19		112.00	50		43	1.15	

[a] Refs of linear (L) and cyclic (C) samples (Table 21.1) with $\overline{DP}_n$ of the PS grafts: $\bar{M}_n = 10\,700$, PDI = 1.04; PS_{170}: $\bar{M}_n = 17\,950$, PDI = 1.03.
[b] Determined by SEC in THF on the basis of PS standards.
[c] Measured by SLS with a refractive index increment (dn/dc) of 0.18.
[d] Measured by SEC/SLS in THF.
[e] Measured by DLS in THF.
[f] Measured at 20 °C in THF.

selective and efficient. These criteria currently hamper this approach to a restricted number of macromolecules with an appropriate number of homogenously distributed grafting sites of suitable reactivity. When these conditions are fulfilled, magnified macromolecules can be prepared and individually visualized and characterized as representative models of the initial polymer chains. This approach is illustrated in Scheme 21.8 in the case of a ring-polymer synthesis.

Through this approach, the architecture, the shape and the dimensions of each macromolecule constitutive of a given polymer system can be determined. To that aim a representative set of macromolecules, constituted of several hundreds to up to a thousand entities, was taken and the macromolecules were analyzed one-by-one.

From these data, a statistical analysis of the whole set of macromolecules can be performed. This permits a measure of the yield of targeted macromolecules as well as a precise determination of the number and characteristics of supplementary macromolecular structures formed through secondary reactions. These are important data that yield information on their mechanisms of formation.

Images of trefoil knotted macromolecular rings are shown in Figure 21.19. They are constituted of a closed ring in which the chain is entangled, forming knots that cannot unfasten except by cutting the ring (Amabilino and Stoddart, 1995; Dietrich-Buchecker *et al.*, 2005; Dietrich-Buchecker and Sauvage, 1990; Frisch and Wasserman, 1961; Rapenne *et al.*, 1996). As emphasized on the schematic representation, insets Figure 21.19, the path of the knot

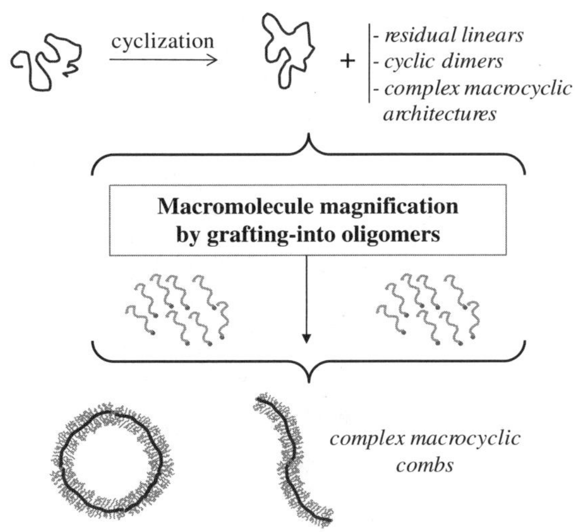

Scheme 21.8 Magnification strategy based on grafting-onto process applied to the investigation by AFM of secondary structures formed during formation of ring macromolecules. (Reproduced with permission from M. Schappacher and A. Deffieux, "Imaging of catenated, figure-of-eight, and trefoil knot polymer rings," *Angewandte Chemie International Edition*, 2009, **32**, 5930–5933. © Wiley-VCH Verlag GmbH & Co. KGaA.)

crosses three times over itself. This corresponds to the simplest knotted-ring structure (trefoil). As previously described (Dobrowolski, 2003; Perret-Aebi *et al.*, 2004) trefoil knots are topological stereoisomers of the corresponding simple rings. As shown in Figure 21.19 they are present in the form of right and left topological diastereoisomers. Their average backbone length of about 600 nm, as measured on AFM images Figure 21.19, is twice that of the corresponding single rings (300 nm).

It may be concluded that they are formed by cyclization of dimerized linear precursors, as shown in Figure 21.20b, while the formation of trefoil knots isomers of single rings is not observed.

Images of eight-shaped rings are shown in Figures 21.20a–c and for comparison a single-ring dimer is shown in Figure 21.20d. Eight-shaped rings are constituted of two simple rings covalently linked at one position and possess an overall ring-chain length corresponding to that of 2 attached single rings ($2 \times$ **ABC** copolymers). The presence of a covalent link between the two rings is confirmed by the section at the connecting point (80 nm), higher than the ring section (68 nm) and smaller than the sum of two chains (136 nm), see bottom of Figure 21.20c. It is also supported by the same height profile along the ring polymer chain including the connecting point. This should not be the case for a twisted dimer. These structures probably correspond to the cyclization of the linear moieties of a previously formed tadpole polymer observed in significant amount. Their possible mechanism of formation is shown in Scheme 21.9. Both the formation of tadpoles and of eight-shaped rings can be attributed to the presence of residual reactive functions of A and C blocks after a first cyclization process. These remaining functions can then further react to fix another chain precursor by one or two of his ends.

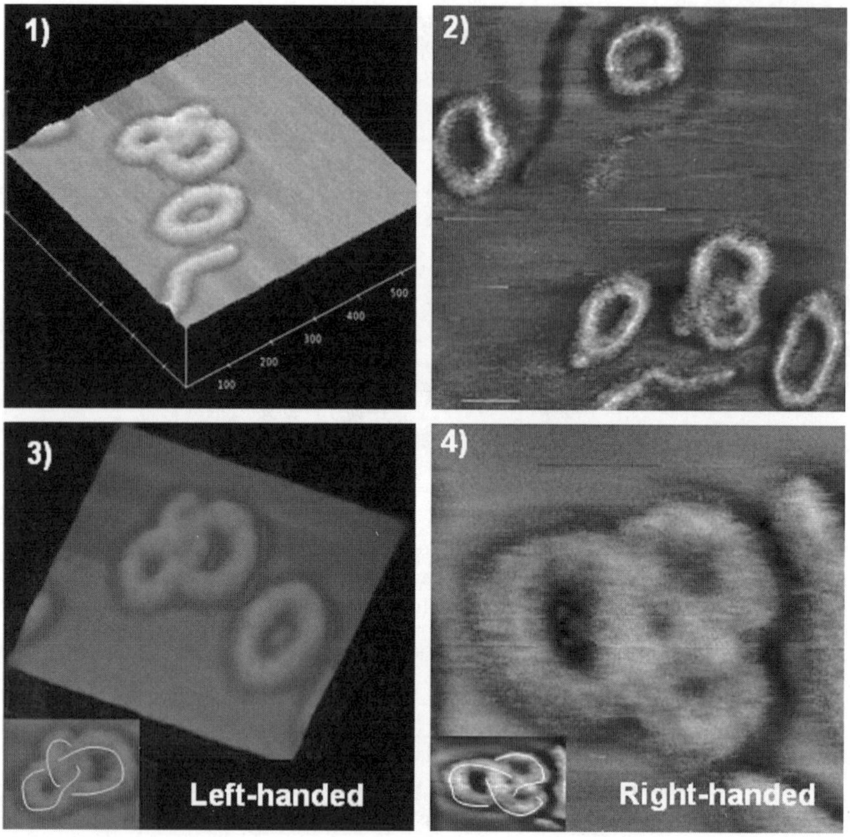

Figure 21.19 AFM images (topographic: (1) and (2); phase: (3) and (4)) of trefoil-knotted macromolecular rings formed as secondary structures in the end-to-end chain cyclization of an ABC linear precursor. (3) and (4) are topological diastereoisomers. The overall trefoil ring chain length corresponds to that of a ring dimer (2 × ABC copolymers). Macromolecules were magnified by grafting polystyrene oligomers along the PCEVE block. DP_n of the PCEVE block in the precursor: 956; DP_n of the PS grafts: 170. (Reproduced with permission from M. Schappacher and A. Deffieux, "Imaging of catenated, figure-of-eight, and trefoil knot polymer rings," *Angewandte Chemie International Edition*, 2009, **32,** 5930–5933. © Wiley-VCH Verlag GmbH & Co. KGaA.)

[2]Catenanes (Amabilino and Stoddart, 1995; Deffieux and Schappacher, 2009; Dobrowolski, 2003) constituted of two interlocked rings with two crossing points are shown in Figure 21.21; the two white spots that appear in the image correspond to the crossing points of greater heights. In contrast to eight-shaped rings, the two rings in [2]catenanes are not chemically linked to each other but are inseparable without breaking one ring, as indicated in Figures 21.21a and d. Their formation likely involves chain entanglement between a ring polymer and a linear precursor, prior to its ring closing.

Catenated and trefoil knotted rings are fascinating nontrivial topological figures that have occupied a privileged position in art and history, from the most ancient civilizations to modern art. They are the subject of basic studies and many scientific fields, from mathematics

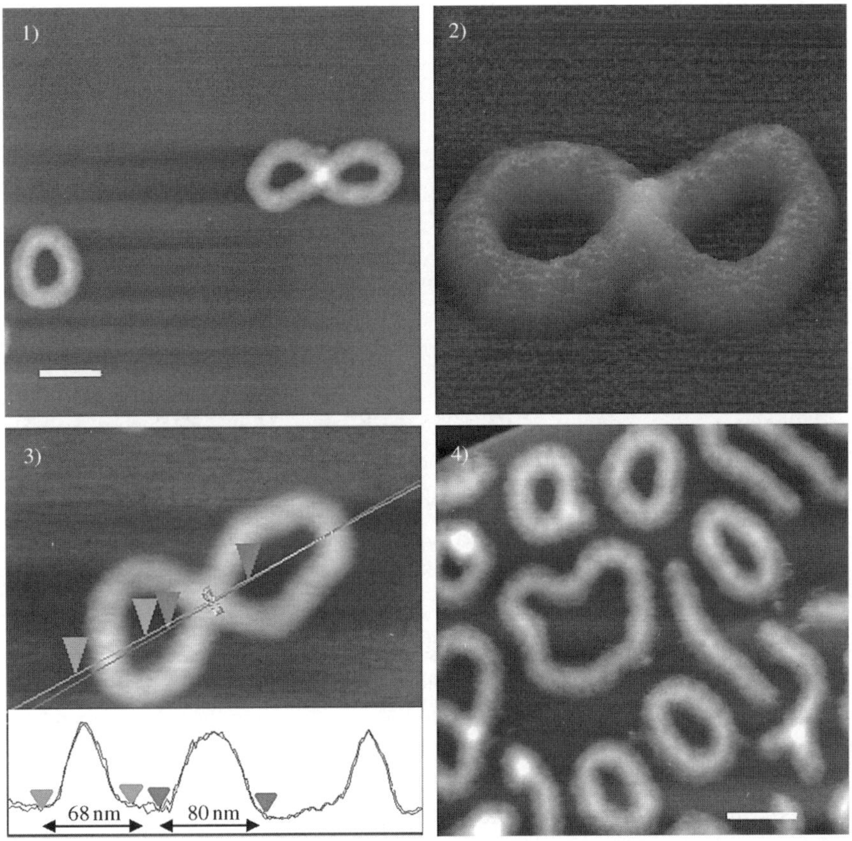

Figure 21.20 AFM images of eight-shaped and dimer rings: (a) topographic; (b) 3D image magnified; (c) topographic section; (d) single and dimer rings; Dimer rings are other secondary structures formed in the end-to-end chain cyclization. Scale bar 100 nm. The initial rings were magnified by grafting polystyrene oligomers along the PCEVE block. DP_n of the PCEVE block in the linear precursor: 956; DP_n of the PS grafts: 170. (Reproduced with permission from M. Schappacher and A. Deffieux, "Imaging of catenated, figure-of-eight, and trefoil knot polymer rings," *Angewandte Chemie International Edition*, 2009, **32**, 5930–5933. © Wiley-VCH Verlag GmbH & Co. KGaA.)

to chemistry. Their esthetic aspect is undoubtedly one major reason for which organic chemists have tried to synthesize them. So far, two different methods of syntheses of molecular catenanes and knots have been successfully developed in topological chemistry (Amabilino and Stoddart, 1995; Frisch and Wasserman, 1961). The first is based on template effects induced by a transition metal (Dietrich-Buchecker *et al.*, 2005; Dietrich-Buchecker *et al.*, 1999), which gathers and disposes molecular fragments in a predictable geometry. The second relies on the use of hydrogen bond-set on precursor molecules and is purely organic (Brüggemann *et al.*, 2007; Guidry *et al.*, 2005).

The directed synthesis of catenaned macrocyclic polymers with molar mass of about 10 000 g mol^{-1} has been attempted recently. The formation in low yields (around 5%)

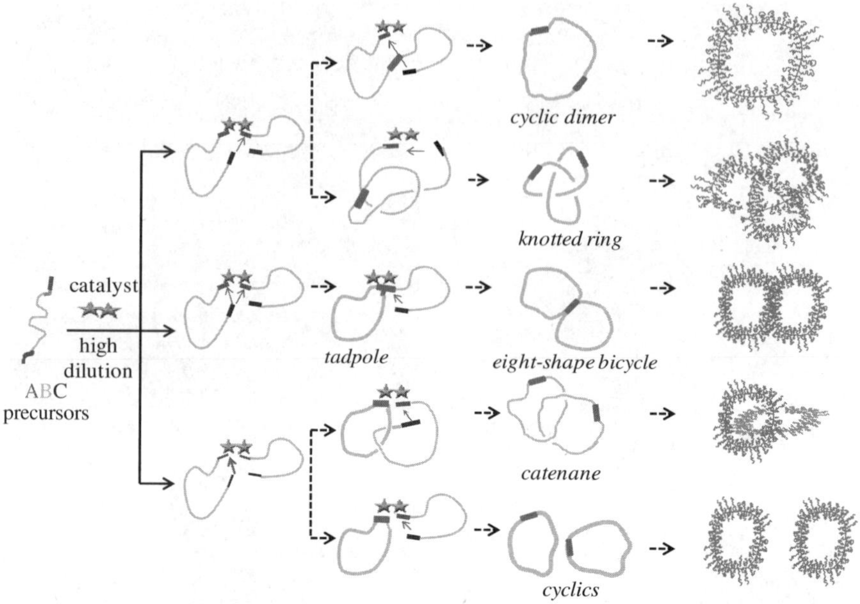

Scheme 21.9 Reaction pathways yielding to the different linear, cyclic and bicyclic structures during the end-to-end cyclization of an ABC linear triblock copolymer, bearing A and C sequences with antagonist reactive functions followed by the magnification of the obtained structures. The two stars schematize catalyst aggregates with several acidic sites.

of interlocked rings of polystyrene and poly(vinyl pyridine) (Gan *et al.*, 2002) and of polystyrene and polyisoprene rings (Ohta *et al.*, 2008) was reported by authors on the basis of size-exclusion chromatography analysis.

The formation of knotted macromolecular rings, was also claimed in polymer systems exhibiting a linear to ring-chain equilibrium or during the end-to-end ring closure of linear-chain precursors (McKenna *et al.*, 1987). However, the presence of catenaned and knotted rings could not be clearly demonstrated.

The present approach brings a direct proof of the formation of these polymer structures during the end-to-end cyclization of long polymer chains.

The relative amounts of these different chain topologies were determined on sets of more than one thousand macromolecules for **A**B**C** copolymer precursors with various degrees of polymerization (DP_n). The central PCEVE block DP_ns ranged from 548 to 956, which correspond to chains with about 1100 and 1900 carbon atoms, respectively. As detailed in Table 21.10, for a DP_n of 956 the percentages, respective to single-ring macromolecules formed, are about 0.3% in ring dimers, 1% in eight-shape bicycles, 0.3% in [2]catenanes and less than 0.1% in knotted rings. Although the fractions of ring dimers and eight-shaped bicycles are not notably influenced by the backbone DP_n, the knotted rings and catenanes fractions decrease significantly with the PCEVE backbone length, suggesting that it is an important parameter in the formation of such nontrivial chain arrangements. This is further supported by the absence of the corresponding knots with same DP_n than the main single-ring fraction. All the nontrivial cyclic structures observed, eight-shaped bicycles, catenanes and knotted

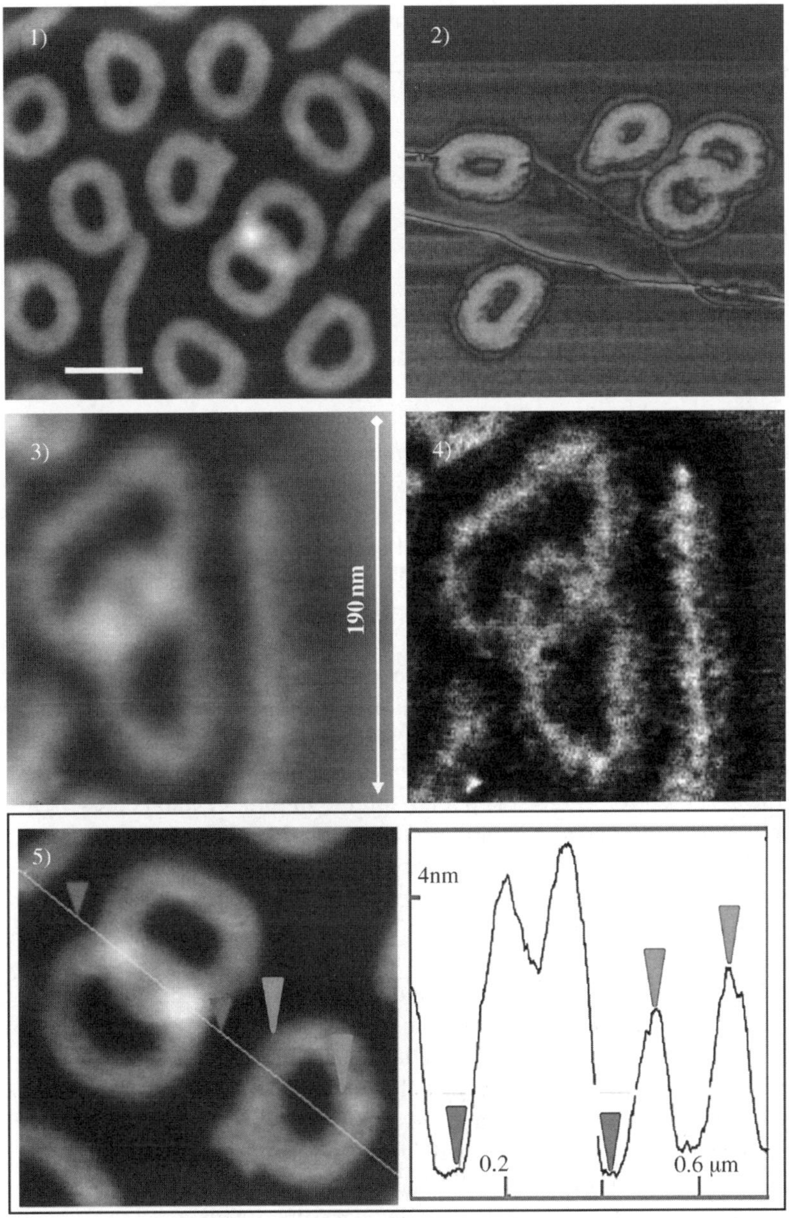

Figure 21.21 AFM images of [2]catenanes formed as secondary structures in the end-to-end chain cyclization of ABC linear precursor (scale bar 100 nm): (a) topographic image; (b) 3D image; (c) and (d) topographic and phase image showing more clearly the ring-crossing path; (e) section image of a [2] catenane and of a single ring. The initial ring macromolecules were magnified by grafting polystyrene oligomers along the PCEVE block. DP_n of the PCEVE block in the linear precursor: 956; DP_n of the PS grafts: 170. (Reproduced with permission from M. Schappacher and A. Deffieux, "Imaging of catenated, figure-of-eight, and trefoil knot polymer rings," *Angewandte Chemie International Edition*, 2009, **32**, 5930–5933. © Wiley-VCH Verlag GmbH & Co. KGaA.)

Table 21.10 Molar percentages of different ring structures as a function of the degree of polymerization of the B block of linear ABC precursors used for cyclization

DPn of B block in ABC precursor	Ring polymer[a] %	Ring dimer %[b]	Eight-shaped ring %[b]	Catenanes %[b]	Knotted rings %[b]
956	60	0.3	0.8	0.3 (100)[c]	0.05 (16)[c]
774	65	0.3	0.8	0.2 (66)[c]	<0.05 (<16)[c]
548	62	0.3	0.8	not obs.	not obs.

[a] Mole % of main cyclic fraction.
[b] Molar percentages based on ring polymer.
[c] Values in parentheses correspond to molar percentages with respect to ring dimer.

rings, possess an overall chain length corresponding to that of the dimer ring and therefore can be considered as topological isomers of the dimer ring. As indicated in Table 21.10 the proportion of the different nontrivial rings with respect to dimer rings is much higher than expected considering a unique effect of the backbone length. For instance, the molar fraction of catenanes is almost equal to that of dimer rings and the amount of knotted rings is as high as 15–20% of dimer rings.

This strongly suggests a mechanism of cyclization that involves a template effect. The end-to-end ring closing process is believed to proceed first by docking of an ABC chain precursor on catalyst acid sites through protonation of the vinyl ether block C. Once activated and under high dilution conditions the vinyl ether functions thus react with the hydroxyl groups of block A of the same polymer chain, forming acetal linkages. However, given that the acetalization catalyst is poorly soluble in toluene and is present mainly in the form of molecular aggregates it is believed that several ABC chain precursors can dock simultaneously on a same aggregate (Scheme 21.9). Interchain entanglement before cyclization as well as the crossreaction between reactive A and C sequences of two neighboring chains could explain the higher formation probability of dimer rings, tadpoles, eight-shaped bicycles, catenanes and knotted rings.

The possibility of magnifying macromolecules by grafting oligomers uniformly along their chain appears to be a very powerful method to investigate by direct imaging the structure and topology of macromolecules. This strategy allows a precise characterization macromolecule-by-macromolecule, of the architecture and dimensions of comb structures but also of the initial polymer products of a reaction. It also permits visualization and quantification of secondary molecular structures generated in the reaction and, from this information, to elucidate their specific mechanisms of formation. This strategy is clearly illustrated in the present work in which nontrivial macromolecular rings such as eight-shaped rings, [2] catenanes and trefoil knotted rings have been visualized and the probable mechanisms yielding to their formation identified.

Acknowledgments

This Research was supported by The Centre National de la Recherche Scientifique (CNRS) and the Ministère de l' Education et de la Recherche.

References

Amabilino, D.B. and Stoddart, J.F. (1995) Interlocked and intertwined structures and superstructures. *Chemical Reviews*, **95**, 2725–2828.

Ariura, F. (2009) Synthèse de polymères et copolymères en peigne et études de leurs propriétés spécifiques, PhD in *Sciences Chimiques*, University of Bordeaux, Bordeaux.

Beers, K.L., Caynor, S.G., Matyjaszewski, K. *et al.* (1999) Poly(at-butyl acrylate) brush macromolecules by atrp: Determination of contour length from fractionation and afm studies. *American Chemical Society, Polymer Preprints, Division of Polymer Chemistry*, **40**, 446–447.

Beers, K.L., Gaynor, S.G., Matyjaszewski, K. *et al.* (1998) Synthesis of densely grafted copolymers by atom transfer radical polymerization. *Macromolecules*, **31**, 9413–9415.

Bloomfield, V. and Zimm, B.H. (1966) Viscosity, sedimentation, et cetera, of ring- and straight-chain polymers in dilute solution. *The Journal of Chemical Physics*, **44**, 315–323.

Börner, H.G., Duran, D., Matyjaszewski, K. *et al.* (2002) Synthesis of molecular brushes with gradient in grafting density by atom transfer polymerization. *Macromolecules*, **35**, 3387–3394.

Bosman, A.W., Janssen, H.M., and Meijer, E.W. (1999) About dendrimers: Structure, physical properties, and applications. *Chemical Reviews*, **99**, 1665–1688.

Boyce, J.R., Shirvanyants, D., Sheiko, S.S. *et al.* (2004) Multiarm molecular brushes: Effect of the number of arms on the molecular weight polydispersity and surface ordering. *Langmuir*, **20**, 6005–6011.

Boydston, A.J., Holcombe, T.W., Unruh, D.A. *et al.* (2009) A direct route to cyclic organic nanostructures via ring-expansion metathesis polymerization of a dendronized macromonomer. *Journal of the American Chemical Society*, **131**, 5388–5389.

Brüggemann, J., Bitter, S., Müller, S. *et al.* (2007) Spontaneous knotting – from oligoamide threads to trefoil knots. *Angewandte Chemie – International Edition*, **46**, 254–259.

Chandler, B., Hayashi, M., Hayashi, M.N., and Spiegelman, S. (1964) Circularity of the replicating form of a single-stranded DNA virus. *Science*, **143**, 47–49.

Clayton, D.A. and Vinograd, J. (1967) Circular dimer and catenate forms of mitochondrial DNA in human leukaemic leucocytes. *Nature*, **216**, 652–657.

Cluzel, P., Lebrun, A., Heller, C. *et al.* (1996) DNA: An extensible molecule. *Science*, **271**, 792–794.

Deffieux, A. and Borsali, R. (2007) Controlled synthesis and Properties of macrocyclic polymers, in *Macromolecular Engineering* (eds. K. Matyjaszewski, Y. Gnanou,and L. Leibler), Wiley-VCH, Weinheim.

Deffieux, A. and Schappacher, M. (1998) Design of new polymer architectures by combination of anionic and cationic living polymerization techniques. *Macromolecular Symposia*, **132**, 45–55.

Deffieux, A. and Schappacher, M. (1999) Synthesis and characterization of star and comb polystyrenes using isometric poly(chloroethyl vinyl ether) oligomers as reactive backbone. *Macromolecules*, **32**, 1797–1802.

Dietrich-Buchecker, C., Colasson, B.X., and Sauvage, J.P. (2005) Molecular knots. *Topics in Current Chemistry*, **249**, 261–283

Dietrich-Buchecker, C., Rapenne, G., and Sauvage, J.P. (1999) Synthesis of catenanes and molecular knots by copper (I)-directed formation of the precursors followed by ruthenium(II)-catalysed ring-closing metathesis. *Coordination Chemistry Reviews*, **185–186**, 167–176.

Dietrich-Buchecker, C. and Sauvage, J.P. (1990) Templated synthesis of interlocked macrocyclic ligands, the catenands. Preparation and characterization of the prototypical bis-30 membered ring system. *Tetrahedron*, **46**, 503–512.

Dobrowolski, J.C. (2003) Classification of topological isomers: knots, links, rotaxanes, etc. *Croatica Chemica Acta*, **76**, 145–152.

Dubelcco, R. and Vogt, M. (1963) Evidence for a ring structure of polyoma virus DNA. *Proc. Natl. Acad. Sci. USA*, **50**, 236–240.

Dziezok, P., Sheiko, S.S., Fischer, K. *et al.* (1997) Cylindrical molecular brushes. *Angewandte Chemie (International Edition in English)*, **36**, 2812–2815.

Frechet, J.M. and Tomalia, D.A. (2002) *Dendrimers and Other Dendritic Polymers*, Wiley, New York.

Frisch, H.L. and Wasserman, E. (1961) Chemical topology. *Journal of the American Chemical Society*, **83**, 3789–3795.

Gallyamov, M.O., Tartsch, B., Khokhlov, A.R. *et al.* (2004) Real-time scanning force microscopy of macromolecular conformational transitions. *Macromolecular Rapid Communications*, **25**, 1703–1707.

Gan, Y., Dong, D., and Hogen-Esch, T.E. (2002) Synthesis and characterization of a catenated polystyrene-poly (2-vinylpyridine) block copolymer. *Macromolecules*, **35**, 6799–6803.

Glaubiger, D. and Hearst, J.E. (1967) Effect of superhelical structure on the secondary structure of DNA rings. *Biopolymers – Peptide Science Section*, **5**, 691–696.

Guidry, E.N., Cantrill, S.J., Stoddart, J.F., and Grubbs, R.H. (2005) Magic ring catenation by olefin metathesis. *Organic Letters*, **7**, 2129–2132.

Jia, Z., Fu, Q. and Huang, J. (2006) Synthesis of amphiphilic macrocyclic graft copolymer consisting of a poly(ethylene oxide) ring and multi-polystyrene lateral chains. *Macromolecules*, **39**, 5190–5193.

Kiger, J.A. Jr, Young, E.T. II, and Sinsheimer, R.L. (1968) Purification and properties of intracellular lambda DNA rings. *Journal of Molecular Biology*, **33**(2), 395–400.

Kumaki, J., Nishikawa, Y., and Hashimoto, T. (1996) Visualization of single-chain conformations of a synthetic polymer with atomic force microscopy. *Journal of the American Chemical Society*, **118**, 3321–3322.

Levene, S.D., Donahue, C., Boles, T.C., and Cozzarelli, N.R. (1995) Analysis of the structure of dimeric DNA catenanes by electron microscopy. *Biophysical Journal*, **69**, 1036–1045.

Li, H., Jérôme, R., and Lecomte, P. (2008) Amphiphilic sun-shaped polymers by grafting macrocyclic copolyesters with PEO. *Macromolecules*, **41**, 650–654.

Lyubchenko, Y.L., Shlyakhtenko, L.S., Binus, M. *et al.* (2002) Visualization of hemiknot DNA structure with an atomic force microscope. *Nucleic Acids Research*, **30**, 4902–4909.

Matyjaszewski, K., Qin, S., Boyce, J.R. *et al.* (2003) Effect of initiation conditions on the uniformity of three-arm star molecular brushes. *Macromolecules*, **36**, 1843–1849.

Mckenna, G.B., Hadziioannou, G., Lutz, P. *et al.* (1987) Dilute solution characterization of cyclic polystyrene molecules and their zero-shear viscosity in the melt. *Macromolecules*, **20**, 498–512.

Min, K., Yu, S., Lee, H.I. *et al.* (2007) High yield synthesis of molecular brushes via ATRP in miniemulsion. *Macromolecules*, **40**, 6557–6563.

Mitra, S., Reichard, P., Inman, R.B. *et al.* (1967) Enzymic synthesis of deoxyribonucleic acid. XXII. Replication of a circular single-stranded DNA template by DNA polymerase of *Escherichia coli*. *Journal of Molecular Biology*, **24** (3),429–442

Ohta, Y., Kushida, Y., Kawaguchi, D. *et al.* (2008) Preparation, characterization, and nanophase-separated structure of catenated polystyrene-polyisoprene. *Macromolecules*, **41**, 3957–3961.

Pang, X., Wang, G., Jia, Z. *et al.* (2007) Preparation of the amphiphilic macro-rings of poly(ethylene oxide) with multi-polystyrene lateral chains and their extraction for dyes. *Journal of Polymer Science, Part A: Polymer Chemistry*, **45**, 5824–5837.

Percec, V., Ahn, C.H., Cho, W.D. *et al.* (1998) Visualizable cylindrical macromolecules with controlled stiffness from backbones containing libraries of self-assembling dendritic side groups. *Journal of the American Chemical Society*, **120**, 8619–8631.

Perret-Aebi, L.E., Von Zelewsky, A., Dietrich-Buchecker, C., and Sauvage, J.P. (2004) Stereoselective synthesis of a topologically chiral molecule: The trefoil knot. *Angewandte Chemie – International Edition*, **43**, 4482–4485.

Potemkin, I.I., Khokhlov, A.R., Prokhorova, S. *et al.* (2004) Spontaneous curvature of comblike polymers at a flat interface. *Macromolecules*, **37**, 3918–3923.

Qin, S., Matyjaszewski, K., Xu, H., and Sheiko, S.S. (2003) Synthesis and visualization of densely grafted molecular brushes with crystallizable poly(octadecyl methacrylate) block segments. *Macromolecules*, **36**, 605–612.

Rapenne, G., Dietrich-Buchecker, C., and Sauvage, J.P. (1996) Resolution of a molecular trefoil knot. *Journal of the American Chemical Society*, **118**, 10932–10933.

Roovers, J. (1999) *Branched Polymers I and II*, Springer Verlag, Berlin.

Schappacher, M., Billaud, C., Paulo, C., and Deffieux, A. (1999) Synthesis, dimensions and solution properties of linear and macrocyclic poly(chloroethyl vinyl ether)-g-polystyrene comblike polymers. *Macromolecular Chemistry and Physics*, **200**, 2377–2386.

Schappacher, M. and Deffieux, A. (1997) New comblike copolymers of controlled structure and dimensions obtained by grafting polystyryllithium onto poly(chloroethyl vinyl ether) chains. *Macromolecular Chemistry and Physics*, **198**, 3953–3961.

Schappacher, M. and Deffieux, A. (2000) New polymer chain architecture: Synthesis and characterization of star polymers with comb polystyrene branches. *Macromolecules*, **33**, 7371–7377.

Schappacher, M. and Deffieux, A. (2005) AFM image analysis applied to the investigation of elementary reactions in the synthesis of comb star copolymers. *Macromolecules*, **38**, 4942–4946.

Schappacher, M. and Deffieux, A. (2008a) Atomic force microscopy imaging and dilute solution properties of cyclic and linear polystyrene combs. *Journal of the American Chemical Society*, **130**, 14684–14689.

Schappacher, M. and Deffieux, A. (2008b) Large macrocyclic combs: From synthesis to AFM characterization. *American Chemical Society, Polymer Preprints, Division of Polymer Chemistry*, **49**(1), 54–55 ed.

Schappacher, M. and Deffieux, A. (2008c) Synthesis of macrocyclic copolymer brushes and their self-assembly into supramolecular tubes. *Science*, **319**, 1512–1515.

Schappacher, M. and Deffieux, A. (2009) Imaging of catenated, figure-of-eight, and trefoil knot polymer rings. *Angewandte Chemie – International Edition*, **48**, 5930–5933.

Sheiko, S.S. (2000) Imaging of polymers using scanning force microscopy: From superstructures to individual molecules. *Advances in Polymer Science*, **151**, 61–174

Sheiko, S.S., da Silva, M., Shirvaniants, D. *et al.* (2003) Measuring molecular weight by atomic force microscopy. *Journal of the American Chemical Society*, **125**, 6725–6728.

Sheiko, S.S. and Moller, M. (2001) Visualization of macromolecules – A first step to manipulation and controlled response. *Chemical Reviews*, **101**, 4099–4123.

Spatz, J.P., Sheiko, S., Moller, M. *et al.* (1997) Tapping scanning force microscopy in air-theory and experiment. *Langmuir*, **13**, 4699–4703.

Teertstra, S.J. and Gauthier, M. (2004) Dendrigraft polymers: Macromolecular engineering on a mesoscopic scale. *Progress in Polymer Science (Oxford)*, **29**, 277–327.

Vinograd, J. and Lebowitz, J. (1966) Physical and topological properties of circular DNA. *Journal of General Physiology*, **49**, 103–125.

Vinograd, J., Lebowitz, J., Radloff, R. *et al.* (1965) The twisted circular form of polyoma viral DNA. *Proceedings of the National Academy of Sciences of the United States of America*, **53**, 1104–1111.

Viville, P., Deffieux, A., Schappacher, M. *et al.* (2001) Surface organization of single hyperbranched polymer molecules, as studied by atomic force microscopy. *Materials Science and Engineering C*, **15**, 311–314.

Viville, P., Deffieux, A., Schappacher, M. *et al.* (2000) Surface organization of hvperbranched polymer molecules, as studied by atomic force microscopy. *American Chemical Society, Polymer Preprints, Division of Polymer Chemistry*, **41**, 1501–1502.

Viville, P., Leclere, P., Deffieux, A. *et al.* (2004) Atomic force microscopy study of comb-like vs. arborescent graft copolymers in thin films. *Polymer*, **45**, 1833–1843.

Yamaguchi, H., Kubota, K., and Harada, A. (2000a) Direct observation of DNA catenanes by atomic force microscopy. *Chemistry Letters*, 384–385.

Yamaguchi, H., Kubota, K., and Harada, A. (2000b) Preparation of DNA catenanes and observation of their topological structures by atomic force microscopy. *Nucleic Acids Symposium Series*, **44**, 229–230.

Zimm, B.H. and Stockmayer, W.H. (1949) The dimensions of chain molecules containing branches and rings. *The Journal of Chemical Physics*, **17**, 1301–1314.

22

Tunable Thermoresponsive Polymers by Molecular Design

Richard Hoogenboom
Supramolecular Chemistry Group, Department of Organic Chemistry, Ghent University, Krijgslaan 281 S4, B-9000 Ghent, Belgium

22.1 Introduction

The continuously increasing complexity of molecular materials has driven the development of so-called "smart" materials that respond to changes in the environment. A wide variety of such smart materials are nowadays accessible regarding both triggering environmental factor as well as the response. The triggering event can be used either to create adaptive smart materials that automatically respond to changes in the environment or for triggered smart materials that can be addressed at will. In particular, the first class of adaptive materials is very appealing to scientists since adaptivity forms the basis of human and natural life. A wide range of triggers have been investigated to induce changes in smart materials including chemical or biological agents for sensors (Kim *et al.*, 2009; Qing *et al.*, 2009; Amir *et al.*, 2009), mechanical forces (Davis *et al.*, 2009; Piermattei *et al.*, 2009), UV-Vis or near-IR irradiation (Sershen *et al.*, 2000; Keurentjes *et al.*, 2009), magnetic and electrical fields (Wood *et al.*, 2008; Hergt *et al.*, 2004), as well as temperature (Schild, 1992). On the other hand, the response of smart materials to the external triggering event can be manifold too. For example, the surface properties of a material can change (Qing *et al.*, 2009; Russell, 2002), a shape change can be induced (Oosten *et al.*, 2009), the absorption or emission can vary (Davis *et al.*, 2009), or the material can undergo a phase transition (Keurentjes *et al.*, 2009; Schild, 1992).

Responsive polymeric materials that undergo a solubility phase transition by variation in temperature are particularly interesting since they allow the development of smart soluble materials or smart fluids. The majority of such responsive polymeric materials that undergo a phase transition in aqueous solution exhibit a lower critical solution temperature (LCST).

Complex Macromolecular Architectures: Synthesis, Characterization, and Self-Assembly, First Edition.
Edited by Nikos Hadjichristidis, Akira Hirao, Yasuyuki Tezuka and Filip Du Prez.

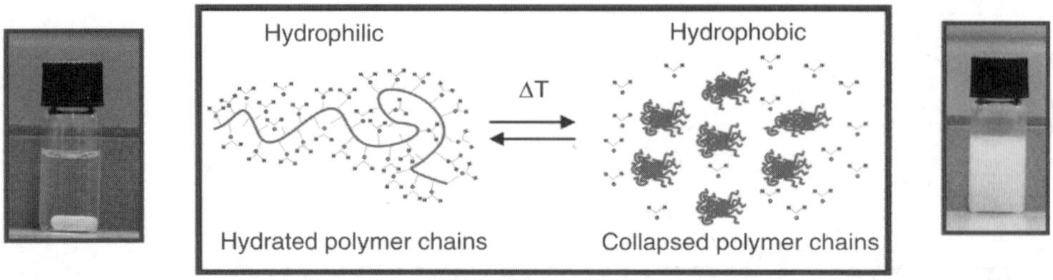

Figure 22.1 Schematic representation of the lower critical solution temperature (LCST) transition of hydrophilic polymers in water. At low temperature the polymers are hydrated, resulting in a clear solution (left) while dehydration at the LCST transition causes hydrophobic collapse and aggregation of the polymer chains leading to a cloudy solution (right).

Below the LCST transition the polymer chains are hydrated, resulting in a clear aqueous polymer solution (Figure 22.1, left). When passing the LCST transition, the hydration shell is lost, rendering the polymer chains hydrophobic causing them to precipitate and aggregate, resulting in large precipitated globules giving rise to a cloudy solution (Figure 22.1, right). Therefore, the LCST transition is often approximated by determining the cloud point of the solution upon heating. However, it is important to note that per definition the LCST is the lowest temperature at which the phase separation occurs. As such, a complete phase diagram of polymer concentration has to be measured before the LCST can be determined. In many cases, only a small concentration regime, mostly low polymer concentration, is evaluated and, thus, cloud points have to be reported rather than LCST temperatures. The aqueous LCST transition of polymers is an entropy-driven transition, that is, the release of water from the hydration layer causes an increase in entropy that drives the phase transition (Schild, 1992).

Hydrophilic polymers that exhibit a LCST transition in aqueous solution should have a delicate balance between hydrophilic and hydrophobic groups. The hydrophilic groups are hydrated at lower temperatures making the polymer soluble, while the hydrophobic groups are required to induce the phase transition. Incorporation of more hydrophilic comonomers causes a rise in LCST, while more hydrophobic comonomers lower the LCST (Schild, 1992; Aoshima and Kanaoka, 2008). The most frequently studied LCST polymer is poly(*N*-isopropylacrylamide) (PNIPAM), which has a LCST of $\sim$32 °C (Schild, 1992). The LCST being close to body temperature is just one of the reasons for the widespread use of PNIPAM. At least as important is the very flat phase diagram, that is, the cloud point is almost independent of changes in polymer concentration (Okada and Tanaka, 2005). This is of course extremely important for the development of robust smart materials based on LCST transitions since the response temperature should be constant. In addition, the cloud point shows also relatively little dependency on the pH value and ionic strength of the solution making the phase transition even more robust (Schild, 1992). However, the very high glass-transition temperature of PNIPAM ($T_g \approx 140\,°C$) leads to vitrification of the precipitated polymer phase above the LCST, which together with the formation of intramolecular hydrogen bonds causes a slight hysteresis in the phase transition (Van Durme *et al.*, 2004; Wang *et al.*, 1998). Both the formation of a glassy precipitate as well as the hysteresis are sometimes undesirable for applications.

Therefore, there has been a search for low T_g alternatives for PNIPAM that show similar beneficial properties, namely LCST close to body temperature and a robust phase transition,

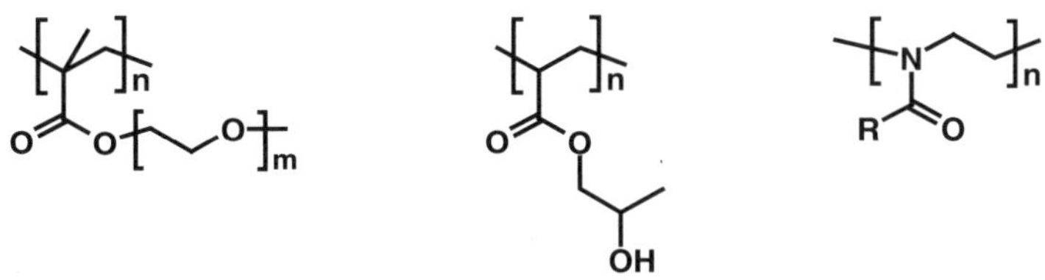

Scheme 22.1 Structures of poly(methoxyoligoethylene glycol methacrylate)s (left), poly(2-hydroxy-propyl acrylate) (middle) and poly(2-oxazoline)s (right).

without vitrification and hysteresis. In this chapter, three potentially alternative polymers to PNIPAM will be discussed in detail: poly(methoxyoligoethylene glycol methacrylate)s (POEGMAs; Section 22.3), poly(2-hydroxypropyl acrylate)s (PHPA) (Section 22.4) as well as poly(2-oxazoline)s (Section 22.5). The polymer structures are depicted in Scheme 22.1. The focus of these sections will be on the use of controlled and living polymerization techniques to make well-defined copolymers. In addition, tuning of the cloud points by controlling the molecular structure of the copolymers, for example, monomer composition and chain length, will be discussed in detail. However, before going to these classes of polymers, the importance of LCST polymers will be illustrated in the next section based on selected application examples.

22.2 Applications of Thermoresponsive Polymers

A diverse selection of applications of thermoresponsive polymers will be discussed in this section. This overview is not intended as a full or even representative review of all reported applications, but it serves to demonstrate the almost unlimited potential of LCST polymers as smart materials and hopefully inspires you, the reader, to develop additional unprecedented smart responsive materials.

22.2.1 Sensors

During the LCST transition, the polymer chains change from a dissolved hydrated state into a collapsed (partial) dehydrated state. This transition thus comprises a change from a hydrophilic to a hydrophobic microenvironment that can be exploited for the development of temperature sensors. To do so, the thermoresponsive polymer can be functionalized with a solvatochromic dye that changes absorbance or emission upon variation of the microenvironment. This concept was first applied by Winnik using PNIPAM side-chain labeled with pyrene where the pyrene emission was studied as a function of temperature to probe the local environment of the polymer (Winnik, 1990). Even though this study did not intend to develop a fluorescent thermometer, a clear variation of the ratio of the first and third emission bands was observed around the LCST, indicating a change in the polarity of the microenvironment.

This initial concept of studying the microenvironment during temperature-induced polymer phase transitions has been more recently extended to the development of fluorescent thermometers. Uchiyama and Iwai developed a polymeric sensor based on a thermoresponsive copolymer of *N,N*-dimethylaminopropylacrylamide and *N-t*-butylacrylamide with a benzofuran dye in the side chain (Uchiyama *et al.*, 2004). The chosen benzofuran dye has a higher fluorescent quantum yield in nonpolar media compared to polar media. As such, the fluorescence is enhanced 9-fold when the polymer phase-transition temperature is crossed due to the less-polar microenvironment inside the collapsed polymer globules. The use of a thermoresponsive polymer comprising basic dimethylamino groups renders the fluorescence of the polymer solution both temperature and pH responsive, which the authors proposed as inputs for the development of molecular AND logic gates. In addition, a range of PNIPAM copolymers with a variety of dyes were reported as polymeric fluorescent thermometers (Uchiyama *et al.*, 2003; Matsumura and Katoh, 2008; Wang *et al.*, 2009). More recently, Shiraishi and coworkers reported a copolymer of NIPAM with a hemicyanine dye as fluorescent thermometer of which the fluorescence is 20-fold enhanced upon increasing the temperature above the LCST transition (Shiraishi *et al.*, 2008). The same authors developed a LCST-based fluorescent sensor that shows selective emission enhancement at a specific temperature range (Shiraishi *et al.*, 2007). When crossing the LCST of the PNIPAM-based copolymer decorated with rhodamine units, the fluorescence is increased due to the more hydrophobic microenvironment of the dye. However, further increasing the temperature leads to aggregation of the hydrophobic polymer chains leading to the formation of large aggregates. It is proposed that these large aggregates scatter away the light before it reaches the rhodamine dye resulting in strong quenching. As a result, the fluorescence is only enhanced in between 25 °C and 35 °C. Practical use of polymeric fluorescent thermometers was demonstrated by Uchiyama and coworkers by using them for diagnostics (Gota *et al.*, 2009). A PNIPAM nanogel was prepared with an incorporated water-sensitive benzofuran fluorophore that is quenched in contact with water. These nanogels could be used for accurate (less than 0.5 °C) temperature sensing within living cells without interference with cellular components.

The temperature-sensing range of such LCST-based thermometers was significantly expanded by Kim and Wong using a pluronic triblock copolymer of ethylene oxide and propylene oxide end-functionalized with a Cy5.5 near-infrared fluorescent dye (Lee *et al.*, 2009). This copolymer revealed a continuous near-infrared fluorescence decrease from 0 to 80 °C, which was ascribed to the gradual transition from dissolved polymer chains, via weakly aggregated chains and micelles to large aggregates (Figure 22.2).

In contrast to the large number of fluorescent thermometers based on LCST polymers with solvatochromic dyes, only a limited number of optical sensors with a visible color change have been reported. A spiropyran-conjugated PNIPAM that undergoes a reversible color change upon increasing the temperature due to isomerization of the spiropyran was reported by Shiraishi and coworkers (Shiraishi *et al.*, 2009). However, isomerization from the colorless spiropyran form to the colored merocyanine isomer only occurs upon UV irradiation and, therefore, the temperature sensing ability of the copolymer is only evident under continuous UV irradiation of the solution. The most famous solvatochromic dye, namely Reichardts dye (Reichardt, 1992), was incorporated in PNIPAM by Ritter to develop an optical sensor (Koopmans and Ritter, 2007). At pH 10, when the Reichardts dye is deprotonated, a clear color shift was observed from orange to red when crossing the LCST due to the change in polarity around the dye. However, the protonated dye at pH 1 did not show a color shift. Inspired by this pH effect on the sensing behavior, we recently developed a dual sensor that simultaneously senses the temperature and pH value by combination of a poly(methoxydiethylene glycol methacrylate) thermoresponsive polymer with a pH-responsive

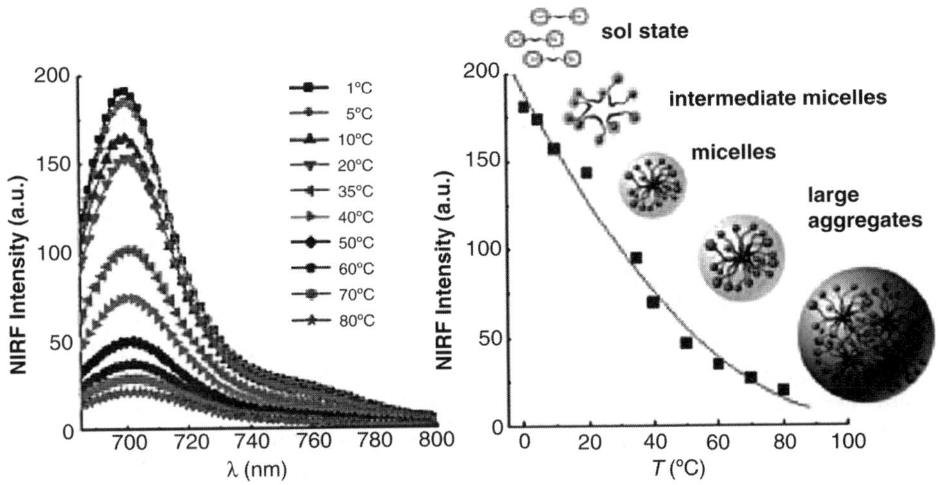

Figure 22.2 Temperature-sensing behavior of a pluronic end functionalized with a Cy5.5 near-infrared fluorescent dye. Left: Near-infrared fluorescence (NIRF) spectra at different temperatures. Right: NIRF intensity as function of temperature as well as the temperature-dependent aggregation state of the pluronic. (Reproduced with permission from S.-Y. Lee, S. Lee, I.-C. Youn *et al.*, "A near-infrared fluorescence based optical thermosensor," *Chemistry – A European Journal*, 2009, **15**, 25, 6103–6106. © Wiley-VCH Verlag GmbH & Co. KGaA.)

solvatochromic dye, namely Disperse red 1 (Pietsch *et al.*, 2009). At low pH, this polymer revealed a color change from red to orange upon increasing the temperature, while at neutral pH this color shift was lost. Nonetheless, the red intensity did vary with temperature at neutral pH and the intensity ratio of the two absorption maxima of Disperse red 1 revealed a strong temperature dependence. Together, measuring a single UV spectrum and determination of the aborption maximum as well as the intensity ratio directly provided both the temperature and pH values of the solution in between pH 1 and pH 7 and 10 and 20 °C.

22.2.2 Temperature Responsive Self-Assembly

Incorporating a temperature-responsive polymer structure in block copolymer architectures provides control over the solubility of a part of the copolymer structure. As such, a double hydrophilic block copolymer consisting of a permanently hydrophilic block and a LCST polymer block is converted into an amphiphilic block copolymer when the solution is heated above the LCST (Figure 22.3). This phenomenon has intrigued polymer scientists and the temperature-induced self-assembly of a large variety of different double hydrophilic thermoresponsive block copolymers into micellar structures has been extensively studied. These results have been captured in a large number of review articles and will not be further addressed here (Gil. and Hudson, 2004; Dimitrova *et al.*, 2007; McCormick *et al.*, 2008; Wei *et al.*, 2009; Rodriguez-Hernandez *et al.*, 2005; Smith *et al.*, 2010).

Temperature-responsive gelation was reported by Hamley for a thermoresponsive amphiphilic copolymer consisting of permanently hydrophobic poly(methyl methacrylate) (PMMA)

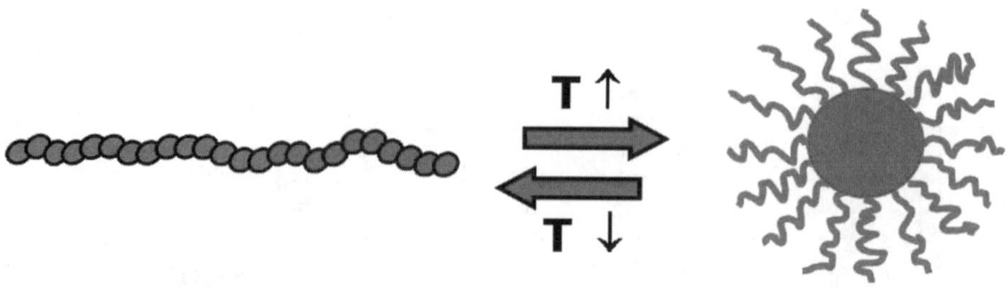

Figure 22.3 Ideal schematic representation of temperature-induced diblock copolymer self-assemble into a spherical micelle.

and PNIPAM (Tang *et al.*, 2006). At temperatures below the LCST, this copolymer forms hydrated spherical micelles that gel the solution at 10 wt% polymer concentration due to the formation of an ordered array of packed micelles, while collapse of the PNIPAM above the LCST induces phase separation in a polymer-rich phase and a water-rich phase. Similarly, Hoogenboom and Schubert recently reported the temperature-responsive gelation of double hydrophobic block copolymers based on poly(styrene) and PMMA (Hoogenboom *et al.*, 2009a). The temperature responsiveness is due to the upper critical solution temperature (UCST) of PMMA in ethanol–water mixtures, that is, the polymer is insoluble at low temperatures and dissolves upon heating. At elevated temperatures, this block copolymer forms spherical micelles in an ethanol–water (80–20 vol%) mixture with PS as core and PMMA as corona, while lowering the temperature results in the formation of a micellar gel at as low as 1 wt% polymer concentration.

Besides the use of amphiphilic diblock copolymers, temperature-induced gelation can be achieved using thermoresponsive hydrophilic triblock copolymers having outer thermoresponsive blocks. Armes reported triblock copolymers with a poly(2-methacryloyloxyethyl phosphorylcholine) (PMPC) hydrophilic middle block and PNIPAM outer blocks (Li *et al.*, 2005). Above the LCST the PNIPAM blocks collapse and the presence of collapsed PNIPAM blocks in different micellar cores results in physical crosslinking of the solution resulting in gelation. Introducing a reducible dithio-moiety in the middle of PMPC block resulted in the formation of biochemically degradable gels above the LCST of PNIPAM (Li *et al.*, 2006). Further control over the gelation process was achieved using ABC triblock copolymers with different stimuli-responsive A and C blocks (Li *et al.*, 2005). At the LCST of the poly(propylene oxide) (PPO) A-block, defined spherical micelles are obtained, while crossing the LCST of the PNIPAM C-block results in the formation of a micellar network (Figure 22.4).

The incorporation of a thermoresponsive middle block in an amphiphilic triblock copolymer provides temperature-induced control over the hydrophilic:hydrophobic ratio that governs the micellar structure that is formed. This concept has been demonstrated by Grubbs using a triblock copolymer consisting of a hydrophilic poly(ethylene oxide) (PEO) block, a thermoresponsive PNIPAM middle block and a hydrophobic poly(butadiene) block (Sundararaman *et al.*, 2008). Below the LCST of PNIPAM, the polymer is mostly hydrophilic, resulting in the formation of spherical micelles in aqueous solution. Increasing the temperature above the LCST of PNIPAM results in a block copolymer that is mostly hydrophobic inducing restructuring of the micellar aggregates into vesicular structures. The restructuring of the triblock copolymer aggregates could be reversibly switched between spherical micelles and

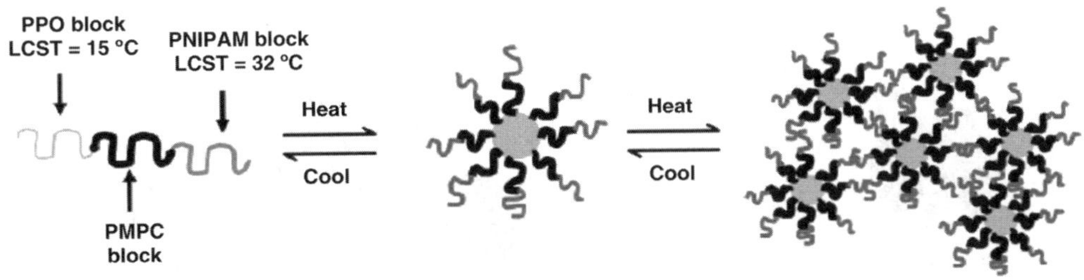

Figure 22.4 Two-step temperature-induced gelation from a hydrophilic ABC-triblock copolymers with thermoresponsive A- and C-blocks. (Reprinted with permission from C. Li, N.J. Buurma, I. Haq *et al.*, "Synthesis and characterization of biocompatible thermoresponsive ABC and ABA triblock copolymer gelators," *Langmuir*, **21**, 24, 11026–11033, 2005. © 2005 American Chemical Society.)

vesicles by variation of the temperature. Similarly, Schubert and coworkers demonstrated UCST-based switching between spherical micelles and vesicles of a supramolecular amphiphilic triblock copolymer (Ott *et al.*, 2009). A triblock copolymer of PEO, poly(*p*-trifluoromethylstyrene) (PTFMS) as middle block and polystyrene was prepared using a metallo-supramolecular block junction between the PEO and the PTFMS. It was found that the PTFMS exhibits UCST behavior in 2-propanol, which was exploited for temperature-induced control over the solvophilic: solvophobic ratio. At low temperatures where the PTFMS is insoluble, vesicular structures were formed in 2-propanol while increasing the temperature, and thus dissolution of the PTFMS block, led to restructuring of the aggregated into spherical micelles.

22.2.3 Selected Biomedical Applications

Thermoresponsive polymers have attracted major scientific interest for a range of biomedical applications (De las Heras Alercon *et al.*, 2005). The majority of studies focused on the use of PNIPAM due to its LCST being close to body temperature. Nonetheless, in recent years the focus has more and more shifted to other alternative polymers with a LCST close to body temperature and three such alternatives will be discussed in more detail in the following sections. A large number of studies focused on the formation of bioconjugates of stimuli-responsive polymers (Gil and Hudson, 2004; Rzaev, Dincer, and Piskin, 2007), biomolecule separation (Kikuchi and Okano, 2002; Tan *et al.*, 2009) as well as the use of thermoresponsive block copolymers for drug delivery (Wei *et al.*, 2009; Hoffman, 1987; Schmaljohann, 2006; Rapoport, 2007). In this section, a selection of other appealing biomedical applications of thermoresponsive polymers will be highlighted.

The formation of protein-thermoresponsive polymer conjugates is an attractive strategy to modify the activity and recovery of proteins (Hoffman, 1987; Hoffman and Stayton, 2007). Hoffman and coworkers reported the conjugation of streptavidin with preformed PNIPAM chains (Stayton *et al.*, 1995). Below the LCST of PNIPAM normal binding of biotin to streptavidin was observed, while precipitation of PNIPAM above the LCST blocked the active sites of the protein, disabling the binding of biotin (Figure 22.5, top). The switching of biotin

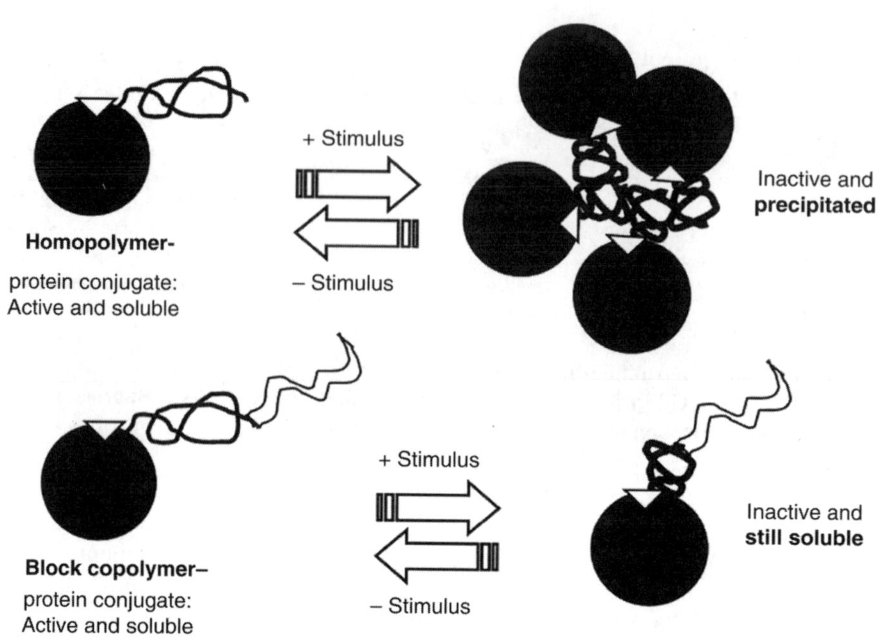

Figure 22.5 Schematic representation of modulating protein activity by conjugation with a thermoresponsive homopolymer (top) or block copolymer (bottom). (Reprinted from Progress in *Polymer Science*, **32**, A.S. Hoffman and P.S. Stayton, "Conjugates of stimuli-responsive polymers and proteins," 922–932, 2007, with permission from Elsevier.)

binding was found to be reversible by variation of the temperature. Maynard and coworkers reported the conjugation of bovine serum albumin (BSA) and lysozyme with PNIPAM (Heredia *et al.*, 2005). The proteins were functionalized with an initiator for atom-transfer radical polymerization (ATRP) and, subsequently, the PNIPAM was grown from the protein. The conjugation with PNIPAM allowed simple isolation of the protein–polymer hybrids by increasing the temperature, resulting in precipitation. Importantly, the bioactivity of lysozyme was retained after modification with PNIPAM. More recently, Sumerlin applied a similar strategy for the synthesis of BSA–PNIPAM conjugates employing BSA modified with a trithiocarbonate as chain-transfer agent for the reversible addition fragmentation chain-transfer (RAFT) polymerization of NIPAM (De *et al.*, 2008). In contrast to the previous example, it was reported that the activity of BSA decreases when increasing the temperature above the LCST of PNIPAM. The activity of the BSA–PNIPAM conjugate could be reversibility switched by changing the temperature, as was demonstrated by five heat–cool cycles. The majority of the protein-responsive polymer conjugates aggregate and precipitate upon increasing the temperature above the LCST, which is induced by the hydrophobic collapse of the thermoresponsive polymer. To avoid the aggregation and precipitation of the conjugates, Stayton and coworkers functionalized streptavidin with a double hydrophilic block copolymer comprising a thermoresponsive block in between the protein and the hydrophilic polymer (Kulkarni *et al.*, 2006). Increasing the temperature above the LCST leads to collapse of the thermoresponsive polymer and, thus, blocking of the protein active site. However, the outer hydrophilic polymer block

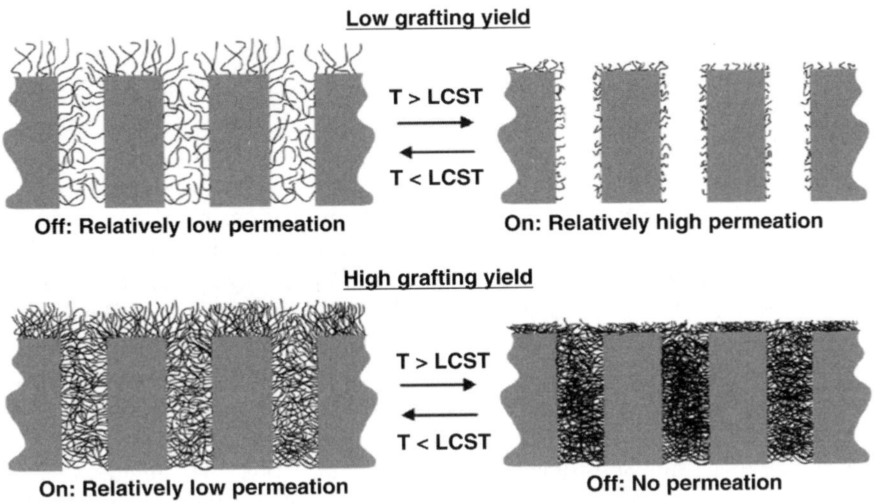

Figure 22.6 Schematic representation of controlling the permeation through microporous membranes using thermoresponsive polymeric grafts. Top: Positive response membranes at low grafting density. Bottom: Negative response membranes at high grafting density. (Reprinted from *Journal of Membrane Science*, **322**, 1, M.A.M.E. Vertommen, H.-J.L. Cornelissen, C.H.J.T. Dietz *et al.*, "Pore-covered thermoresponsive membranes for repeated on-demand drug release," 243–248, 2008, with permission from Elsevier.)

prevents further aggregation of the hydrophobic collapsed polymers and solubilizes the inactive protein–polymer conjugate (Figure 22.5, bottom).

Thermoresponsive polymers can also be used to control the permeation through microporous membranes. Grafting a thermoresponsive polymer on the surface (and in the pores) of such a membrane can be used to switch the permeability of the membrane as depicted in Figure 22.6. Iwata and coworkers demonstrated that the filtration rate through a membrane grafted with PNIPAM sharply increased when the temperature was increased above the LCST (Iwata *et al.*, 1991). This increase in permeation above the LCST is referred to as positive thermoresponsive membranes (Figure 22.6, top), whereby the hydrated chains close the pore below the LCST while collapse of the chains onto the membrane above the LCST results in opening of the pores. When the grafting density is increased, so that there are enough collapsed chains above the LCST to fill and block the pores, the permeation through the membrane is higher below the LCST, which is referred to as negative thermoresponsive membranes (Park *et al.*, 1998; Figure 22.6, bottom). This negative permeation mechanism results in relatively low permeation below the LCST since the molecules need to diffuse through the hydrated polymer chains that fill the pores. Recently, Keurentjes and coworkers reported a positive thermoresponsive membrane for controlled release of BSA (Vertommen *et al.*, 2008). A reservoir containing a solution of BSA was covered with a PNIPAM-grafted microporous membrane. Below the LCST no BSA was released since this protein is too large to diffuse through the hydrated polymer layer. Upon increasing the temperature the pores are opened and BSA release was observed. The membrane could be reversibly switched by changing the temperature, making it suitable as an on-demand controlled-delivery device.

The previous examples of biomedical applications of thermoresponsive polymers dealt with controlling protein activity and permeation. The final examples in this section will cover temperature-controlled adhesion of biomaterials to "smart" surfaces (Nath and Chilkoti, 2002). The properties of thermoresponsive polymer grafted surfaces can be switched from hydrophilic below the LCST to hydrophobic above the LCST (Chen and Hoffman, 1995). Alexander reported the synthesis of PNIPAM-based polymer grafted surfaces with cloud points ranging from 20 to 40 °C (Cunliffe *et al.*, 2003). Subsequent evaluation of protein absorption and bacterial attachment revealed a significant increase of both phenomena above the LCST due to the formation of a hydrophobic surface. The remarkable change in surface properties of thermoresponsive polymer grafts around the LCST was first explored for cell detachment by Okano and coworkers (Okano *et al.*, 1995). Hepatocytes and endothelial cells were incubated on the hydrophobic PNIPAM surface at 37 °C. Subsequent decrease of the temperature led to hydration of the PNIPAM resulting in ~80% cell detachment. More recently, Lutz and coworkers reported the use of poly(methoxyethylene glycol methacrylate) grafted surfaces as switchable cell-adhesion substrates (Wischerhoff *et al.*, 2008). The hydrophobic surface above the LCST allowed fibroblast cell adhesion and proliferation, as indicated by the spreading out of the cells (Figure 22.7a) while decreasing the temperature-enforced cell detachment due to the biorepellency of the resulting hydrated PEGylated surface, as indicated by the rounding up of the cells (Figure 22.7b).

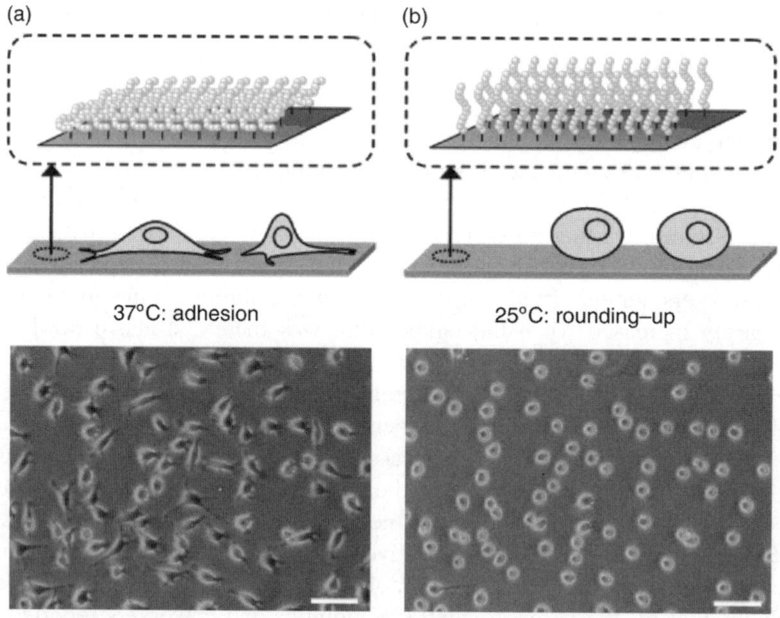

Figure 22.7 Schematic representation and microscopy images of fibroblast cells on a thermoresponsive poly(methoxyoligoethylene glycol methacrylate) grafted surface leading to cell adhesion above the LCST (a) and cell detachment upon decreasing the temperature below the LCST (b). (Reproduced with permission from E. Wischerhoff, K. Uhlig, A. Lankenau *et al.*, "Controlled cell adhesion on PEG-based switchable surfaces," *Angewandte Chemie International Edition*, 2008, **47**, 30, 5666–5668. © Wiley-VCH Verlag GmbH & Co. KGaA.)

22.3 Methoxyoligoethylene Glycol Methacrylate (OEGMA)-based Thermoresponsive (Co)polymers by RAFT

Poly(ethylene oxide) (PEO) is one of the most studied biomaterials since it is biocompatible and it is almost not recognized by the immune system when used *in vivo*, the so-called stealth behavior (Nucci *et al.*, 1991; Duncan, 2000; Harris and Veronese, 2003; Pasut and Veronese, 2007; Knop *et al.*, 2010). Even though PEO does not exhibit LCST behavior in aqueous solution under atmospheric pressure, it does undergo a phase transition when heated beyond the boiling point of water (Saeki *et al.*, 1876). The LCST strongly depends on the polymer molar mass and ranges from ~100 °C for PEO with a molar mass of one million g/mol to ~175 °C for PEO with a molar mass just over 2000 g/mol. Upon increasing the hydrophobicity of the polymer by introducing methylene side groups resulting in poly(propylene oxide), the LCST decreases below room temperature, for example, 18 °C for PPO with a molar mass of 3000 g/mol and hydrophilic hydroxyl end groups (Saito, 1986). Obviously, intermediate cloud points are accessible by statistical copolymerization of ethylene oxide and propylene oxide (Alred *et al.*, 1994; Persson *et al.*, 2000; Campese *et al.*, 2003).

To avoid the anionic (co)polymerization of ethylene oxide for the preparation of PEO-based thermoresponsive polymers, oligoethylene oxide (meth)acrylate monomers have been introduced that can be easily (co)polymerized using radical polymerization methods. The combination of relatively hydrophobic methoxy end group and the hydrophobic poly((meth) acrylate) polymer backbone with moderate oligoethylene oxide lengths results in thermoresponsive poly(methoxyethylene glycol (meth)acrylate)s (POEGMAs) without compromising the biocompatibility of PEO (Neugebauer, 2007; Lutz, 2008).

The first report on the thermoresponsive behavior of POEGMAs appeared in 2003 by Ishizone and coworkers (Han *et al.*, 2003). Well-defined POEGMAs with 1, 2 and 3 ethylene glycol units were prepared by living anionic polymerization. The POEGMAs with one ethylene glycol unit turned out to be insoluble in water, while the polymers with two and three ethylene oxide units revealed thermoresponsive properties in 0.2 wt% aqueous solutions, whereby the cloud points strongly depended on the polymer molar mass. The LCST of poly (methoxydiethylene glycol methacrylate) (PDEGMA) has been reported to be ~25 °C and a LCST of ~50 °C was reported for poly(methoxytriethylene glycol methacrylate) (PTEGMA) (Lutz, 2008). More recently, Ishizone and coworkers reported the living anionic polymerization of methoxytetraethylene glycol methacrylate (Ishizone *et al.*, 2008). The resulting polymer revealed a cloud point of 68 °C (0.2 wt%) in aqueous solution. In addition, changing the methoxy end group of the oligoethylene oxide grafts to more hydrophobic ethoxy groups significantly suppressed the cloud points by at least 20 °C, as depicted in Figure 22.8. The RAFT polymerization of OEGMA (molar mass = 430 g/mol) was reported by Laschewsky and the corresponding POEGMA revealed a cloud point of 83 °C (Mertoglu *et al.*, 2005). Besides homopolymers, block copolymers were prepared by RAFT that exhibit temperature-induced micellization as discussed in Section 22.2.2. The nitroxide-mediated controlled-radical polymerization (NMP) of methoxyoligoetylene glycol acrylates (OEGAs) resulting in well-defined thermoresponsive polymers was reported by Zhao and coworkers (Hua *et al.*, 2006a; Hua *et al.*, 2006b).

All previous examples focused on the homopolymerization of OEGMA (or OEGA) monomers resulting in POEGMAs with LCSTs that are relatively far apart. More precise tuning of the LCST by statistical copolymerization of DEGMA (cloud point (CP) = 28 °C) and

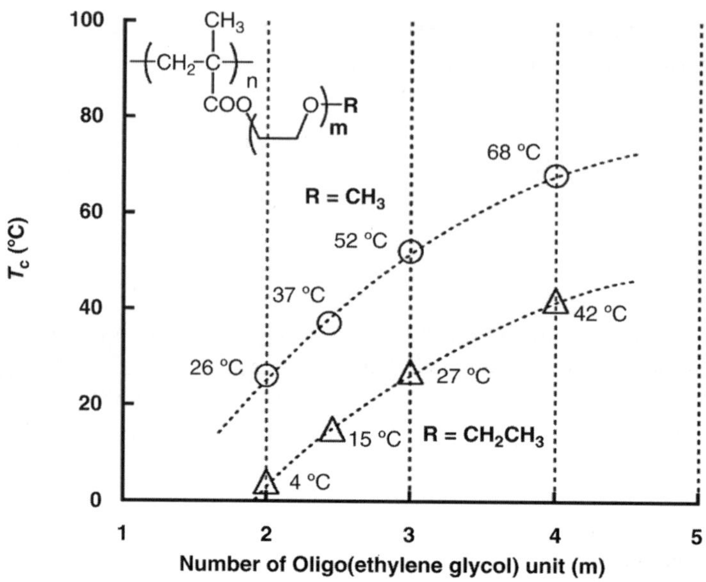

Figure 22.8 Relationship between number of oligo(ethylene oxide) units (*m*) and cloud point of polymers with methoxy and ethoxy-functionalized ethylene oxide grafts determined at 0.2 wt% polymer concentration. (Reprinted with permission from T. Ishizone, A. Seki, M. Hagiwara and S. Han, "Anionic polymerizations of oligo(ethylene glycol) alkyl ether methacrylates: Effect of side chain length and ω-alkyl group of side chain on cloud point in water," *Macromolecules*, **41**, 8, 2963–2967, 2008. © 2008 American Chemical Society.)

OEGMA (molar mass = 475 g/mol; CP = 90 °C) was first reported by Lutz and coworkers using atom-transfer radical polymerization (ATRP) (Lutz and Hoth, 2006; Lutz *et al.*, 2007). A linear increase in CP was observed when the fraction of OEGMA was increased from 0 to 30 mol% allowing accurate control over the CP (~1 °C). In fact, the CP could be predicted using the following simple linear equation: CP = 28 + 1.04 × mol% OEGMA. Even finer control over the CP (~0.5 °C) was achieved by copolymerization of DEGMA and OEGMA with a molar mass of 300 g/mol (Lutz *et al.*, 2009). In addition, Lutz and coworkers evaluated these thermoresponsive DEGMA–OEGMA copolymers as alternatives for the "gold standard" PNIPAM (Lutz *et al.*, 2006). The salt and concentration dependence of the CP were found to be similar for P(DEGMA–OEGMA) and PNIPAM while the molar mass dependence of the CP was even smaller for P(DEGMA–OEGMA) compared to PNIPAM. Therefore, P(DEGMA–OEGMA) can be regarded as alternative to PNIPAM representing an even more robust temperature-induced phase transition. More recently, Matyjaszewski reported the statistical copolymerization of DEGMA and TEGMA by ATRP revealing a linear dependence of the CP on the copolymer composition for all possible monomer compositions (Yamamoto *et al.*, 2008). Cloud-point tuning by copolymerization of OEGMA (molar mass = 475 g/mol) and ethoxyoligoethylene glycol methacrylate (EOEGMA; molar mass = 246 g/mol) using ATRP was demonstrated by Alexander (Magnusson *et al.*, 2008). The PEOEGMA had a CP of 24 °C that gradually increased to 42.5 °C with incorporation of up to 20 mol% OEGMA. The effect of Hofmeister ions on the phase transitions was explored and a controlled-release system was developed using block copolymer micelles that disassemble upon the addition of salts due to a salt-induced decrease of the CP.

The majority of studies on thermoresponsive POEGMAs with tunable cloud points is based on well-defined polymers that are prepared by ATRP. However, the most promising applications of thermoresponsive POEGMAs are biomedical applications that do not allow the presence of copper traces. Therefore, stringent purification of the polymers from ATRP is required. In recent years, we developed a research program on the synthesis of well-defined (co)polymers of OEGMA employing the RAFT polymerization technique, which excludes the use of copper(I) catalysts making the materials directly suitable for biomedical applications. It should be noted that the cytotoxicity of polymers prepared by RAFT depends on the used chain-transfer agent (CTA), that is, polymers prepared with trithiocarbonate CTAs show lower cytotoxicity compared to polymers prepared with dithiobenzoate CTAs (Chang *et al.*, 2009; Pissuwan *et al.*, 2010). Our recent efforts in the synthesis and evaluation of systematical POEGMA (co)polymer libraries will be discussed in the following.

22.3.1 Systematical Polymer Libraries

To understand the effect of molecular structure on the thermoresponsive OEGMA (co) polymers, polymer libraries were evaluated. By systematically varying the structure, for example, length or composition, of the copolymers followed by evaluation of the thermoresponsive properties, structure–property relationships are revealed. However, the preparation of such libraries of polymers can be very time consuming and is prone to human errors due to less-challenging repetition of similar experiments. Therefore, we have adapted the use of automated parallel synthesis robots for both the optimization of polymerization conditions as well as the preparation of polymer libraries (Hoogenboom *et al.*, 2003; Meier *et al.*, 2004). These synthesis robots will be discussed in this section as well as their use for the optimization of the RAFT polymerization process to demonstrate the added value of such equipment.

Versatile automated parallel synthesis robots are a prerequisite for use in an academic laboratory since they should be applicable for a wide range of polymerization methods. We have adapted the synthesis robots from Chemspeed Technologies for this purpose since they comprise a modular platform that can be easily varied to enable a wide range of polymerization scales, reaction temperatures as well as the number of parallel reactions (www.chemspeed.com). Moreover, the modular synthesis platform is covered by a glovebox to create an inert atmosphere, which is necessary for most living and controlled polymerization methods. The number and scale of the parallel polymerizations can be adjusted by placing various reactor blocks in the modular platform. Agitation is performed by vortexing the reactors. Heating and cooling is achieved by an external cryostat that is connected to the double jackets of the reactors. On top of the reactors, cold-finger reflux condensers can be placed, connected to a second cryostat for cooling, to perform reactions under reflux and to avoid evaporation of the solvent. These reactors (with or without reflux) are covered with a top reactor block that controls the atmosphere inside the reactors from open to the air, inert gas or even reduced pressure for evaporation of liquids. The robot system is further equipped with a liquid-handling system to dose the reagents in predefined amounts into the reactors. Importantly, this liquid-handling system can also withdraw samples from the polymerization mixtures during heating and vortexing to investigate the polymerization kinetics. Samples can be directly injected into gas chromatographs or size-exclusion chromatographs (SECs) for online monitoring of the progress of the polymerizations (Hoogenboom *et al.*, 2004) or the samples can be stored in sample vials that can be placed in the modular synthesizer platform as well. The entire automated parallel robot system is computer controlled and can perform the reactions in an unattended manner. For a more detailed description of the robot systems as well as some tips and tricks to perform automated parallel polymerization, the reader is referred to previous publications (Hoogenboom and Schubert, 2003; Hoogenboom and Schubert, 2005; Zhang *et al.*, 2005; Guerrero-Sanchez *et al.*, 2006). More recently, we reported a detailed experimental protocol, including programming, for the automated parallel RAFT polymerization of various monomers utilizing the Chemspeed Accelerator synthesis robot (Becer *et al.*, 2008a). This protocol was used for the optimization of the RAFT polymerization as well as the preparation of the (co)polymer libraries that are discussed in the following.

Before the preparation of systematical (co)polymer libraries to understand the effect of molecular structure on the thermoresponsive properties, the RAFT polymerization conditions were optimized using 2-cyano-2-butyl dithiobenzoate (CBDB) as RAFT agent and azobisisobutyronitrile (AIBN) as radical initiator. The RAFT polymerization mechanism and the structures of CBDB and AIBN are depicted in Figure 22.9. The control over the RAFT free-radical polymerization process strongly depends on the ratio of radical initiator to RAFT agent and temperature. Both these parameters control the number of radicals that are generated and, thus, influence the main propagation equilibrium, that is, the number of dormant and active chains in the system that controls the extent of termination reactions.

At first, the RAFT polymerization with CBDB and AIBN was optimized in the automated parallel synthesis robot for the polymerization of MMA (Fijten *et al.*, 2004). The RAFT agent: AIBN ratio was varied from 1:1 to 1:0.05 for polymerizations in toluene at 70 °C revealing uncontrolled polymerizations at a 1:1 ratio and very slow controlled polymerizations with a

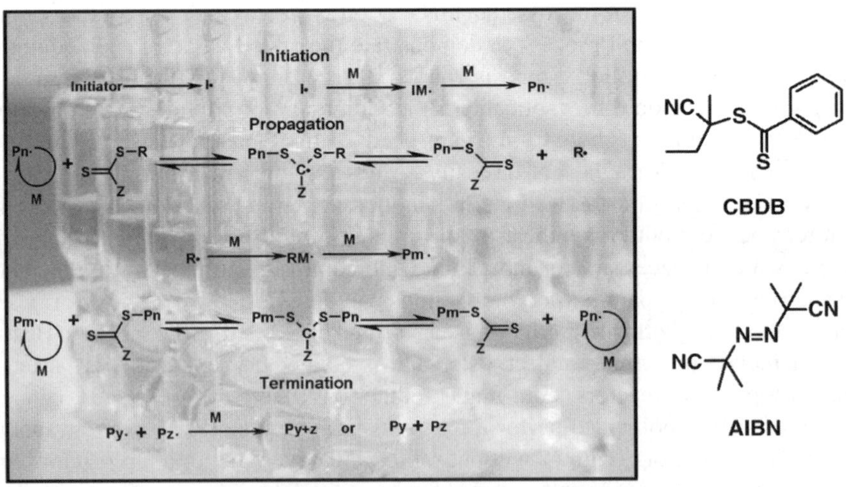

Figure 22.9 Left: Schematic representation of the main reactions in the RAFT polymerization process. The background shows a picture of the automated parallel RAFT polymerizations. Right: Chemical structure of the used 2-cyano-2-butyl dithiobenzoate (CBDB) RAFT agent and azibisisobutyronitrile (AIBN) initiator.

1:0.05 ratio. However, using a CBDB:AIBN ratio of 1:0.25 resulted in relatively fast and controlled polymerizations representing the optimal conditions. Subsequent temperature optimizations were performed by variation of the temperature from 70 to 100 °C, while the CBDB:AIBN ratio was kept constant at 1:0.25. To ensure a controlled polymerization, a first MMA block was synthesized up to 80% conversion followed by addition of more MMA. The chain extension at the different temperatures was evaluated by SEC. Only at 70 °C satisfactory chain extension was observed indicating that this it the optimal polymerization temperature. The reproducibility of the MMA RAFT polymerization in different automated parallel synthesizers as well as lab reactions was also demonstrated to ensure that our results can be transferred to classical lab experiments as well (Paulus *et al.*, 2005). Similar optimization studies for other monomers revealed that indeed 70 °C and a CBDB:AIBN ratio of 1 : 0.25 are optimal with regard to both control over the polymerization and reaction rate for the polymerization of various acrylates and methacrylates (Fijten *et al.*, 2005; Hoogenboom *et al.*, 2005).

When accelerating the synthesis of libraries of thermoresponsive polymers, the screening of solution behavior should also be accelerated to avoid the formation of bottlenecks resulting in piling up of samples. Therefore, we have adapted a parallel turbidimeter from Avantium technologies, Crystal16, that has been developed to speed up polymorph screening in the pharmaceutical industry (www.crystal16.com). This turibidimeter has four blocks of four sample positions that can be separately heated and cooled in between −15 °C and 105 °C. The transmittance of each position is continuously monitored during heating and cooling using a red laser on one side and a detector on the other. We have adapted this device to investigate polymer-solution properties such as LCST and UCST behavior, micellization and dispersion stability allowing fast screening making it ideally suited for incorporation into a high-throughput experimentation workflow (Hoogenboom *et al.*, 2008a).

22.3.2 OEGMA Homopolymers

The effect of molecular weight on the cloud points of POEGMA was investigated by systematically varying the ratio between the monomer and CBDB during the automated parallel polymerizations resulting in libraries of PDEGMA, $POEGMA_{475}$ (molar mass = 475 g/mol) and $PEOEGMA_{246}$ (molar mass = 246 g/mol) with [M]/[CBDB] ranging from 10 to 100 (Becer *et al.*, 2008b). The polymerizations were performed in ethanol at 70 °C using a [CBDB]:[AIBN] ratio of 1:0.25. The variation of molecular weight with [M]/[CBDB] ratio was confirmed by SEC and the molar-mass distributions revealed the formation of well-defined polymers with polydispersity indices (PDIs) below 1.30. In general, thermoresponsive polymers show a strong dependence of CP with molar mass, especially at low molar mass due to the enhanced influence of the end groups. Similarly, it is observed that the CPs of the PDEGMA and $POEGMA_{475}$ increase with increasing molar mass up to a [M]/[CBDB] ratio of 40 (Figure 22.10). The increase of CP with molar mass indicates that the end groups, namely 2-cyano-2-butyl and dithiobenzoate, are more hydrophobic than the polymer structure, which becomes more pronounced for shorter chains. When the oligoethylene oxide side chain is shortened to only one unit resulting in methoxyethylene glycol methacrylate, the resulting polymers are insoluble in water. On the other hand, increasing the oligoethylene glycol length to $OEGMA_{1100}$ (molar mass = 1100 g/mol) results in polymers that are water soluble in liquid

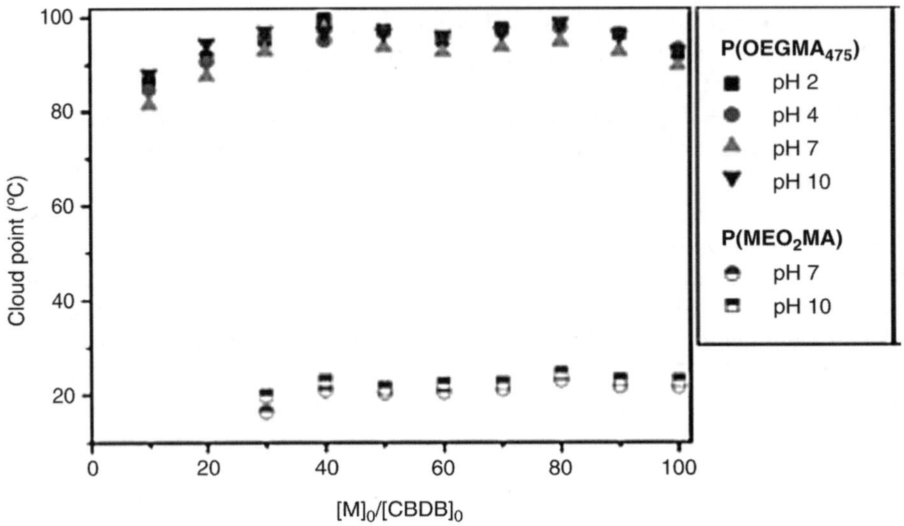

Figure 22.10 Dependence of cloud points on the molar mass (represented by [M]/[CBDB]) and pH value for $POEGMA_{475}$ and PDEGMA determined at 0.5 wt% polymer concentration. (Reprinted from C.R. Becer, S. Hahn, M.W.M. Fijten *et al.*, "Libraries of methacrylic acid and oligo(ethylene glycol) methacrylate copolymers with LCST behaviour," *Journal of Polymer Science Part A: Polymer Chemistry*, **46**, 21, 7138–7147 (Figure 2), © 2008, with the permission of John Wiley & Sons, Inc.)

water at all temperatures accessible under atmospheric pressure. When the hydrophobicity of the monomer is increased by replacing the methoxy end groups for ethoxy end groups, polymers with CPs around 20 °C were obtained for $PEOEGMA_{246}$, which is similar to the CP of PDEGMA. Furthermore, the CPs of the POEGMA homopolymers were found to be nearly independent of the solution pH value, although a minor increase of CP (1–2 °C) was observed at pH values other than 7, which was ascribed to slightly stronger interactions between the oligoethylene oxide side chains with hydroxide ions and hydronium protons compared to water.

22.3.3 OEGMA Copolymers with Dimethylaminoethyl Methacrylate (DMAEMA)

To increase the responsiveness of OEGMA-based polymers, a basic comonomer was introduced resulting in dual responsive copolymers that respond to both changes in temperature and pH value. Therefore, $OEGMA_{300}$ was copolymerized with the basic dimethylaminoethyl methacrylate (DMAEMA; Figure 22.11) using RAFT polymerization at 70 °C in toluene with CBDB as RAFT agent and AIBN as initiator ([CBDB]:[AIBN] = 1:0.25) (Fournier *et al.*, 2007). The automated parallel synthesis robot was employed for the formation of a copolymer library in which the monomer composition was varied from DMAEMA to OEGMA in steps of 10 mol%. The copolymerization parameters were determined to be $r_{DMAEMA} = 0.93$ and $r_{OEGMA} = 0.66$, indicating the formation of a minor monomer gradient through the polymer chains with a preferential incorporation of DMAEMA at the beginning of

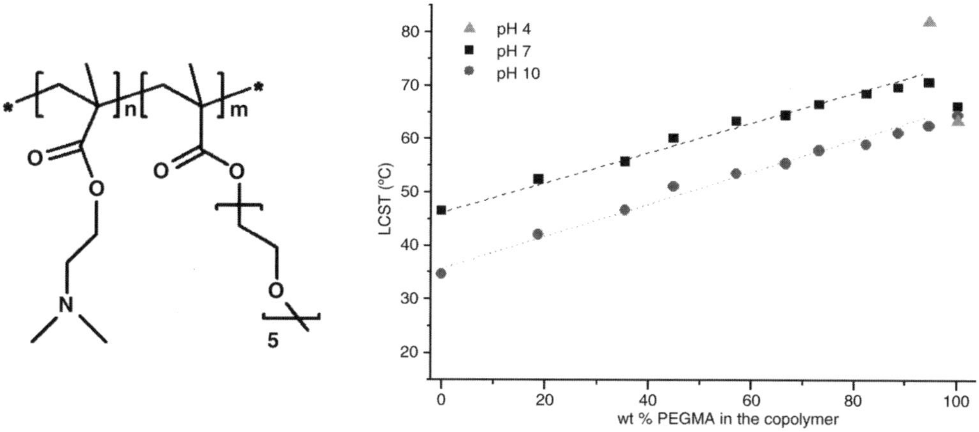

Figure 22.11 Structure of OEGMA–DMAEMA copolymers (left) and their cloud points as function of pH value and monomer composition (right) determined at 0.5 wt% polymer concentration. (Reprinted with permission from D. Fournier, R. Hoogenboom, H.M.L. Thijs *et al.*, "Tunable pH- and temperature-sensitive copolymer libraries by reversible addition-fragmentation chain transfer copolymerizations of methacrylates," *Macromolecules*, **40**, 4, 915–920, 2007. © 2007 American Chemical Society.)

the polymer chain. SEC of the resulting copolymers revealed the formation of well-defined copolymers, while ^{1}H NMR spectroscopy demonstrated that the monomer composition closely resembled the expected monomer composition.

The cloud points of the copolymers were investigated at pH 4, 7 and 10 to study both the temperature and pH-responsive phase transitions (Figure 22.11). As expected, the $POEGMA_{300}$ homopolymer revealed a CP at 60 °C nearly independent of the solution pH. On the other hand, the PDMAEMA homopolymer revealed a CP of 47 °C at pH 7 and a CP of 35 °C at pH 10 due to a lower extent of protonation at pH 10 rendering the polymer more hydrophobic. However, increasing the protonation of PDMAEMA at pH 4 resulted in a loss of thermoresponsiveness, that is, the polymer is soluble at all investigated temperatures (0–100 °C). At pH 10 where the copolymers are not protonated, the CP depended linearly on the composition, allowing accurate control over the CP. At pH 7, the CP also revealed a linear dependence on the composition, except for the POEGMA homopolymer. Apparently, the partial protonation of the DMAEMA units renders the polymer with 10 mol% DMAEMA already more hydrophilic than the POEGMA even though the PDMAEMA homopolymer is less hydrophilic than POEGMA at this pH value. All copolymers, except with 10 mol% DMAEMA, lost their thermoresponsive behavior at pH 4 due to the highly increased hydrophilicity of the copolymer upon protonation of the DMAEMA units. As such, well-defined dual-responsive copolymers were successfully prepared by RAFT copolymerization of DMAEMA and OEGMA.

22.3.4 OEGMA Copolymers with Methacrylic Acid (MAA)

In addition to the basic pH- and temperature-responsive copolymers of DMAEMA and OEGMA, acidic dual-responsive copolymers would be interesting to further broaden the scope of OEGMA-based "smart" materials. Stöver and coworkers previously reported that the

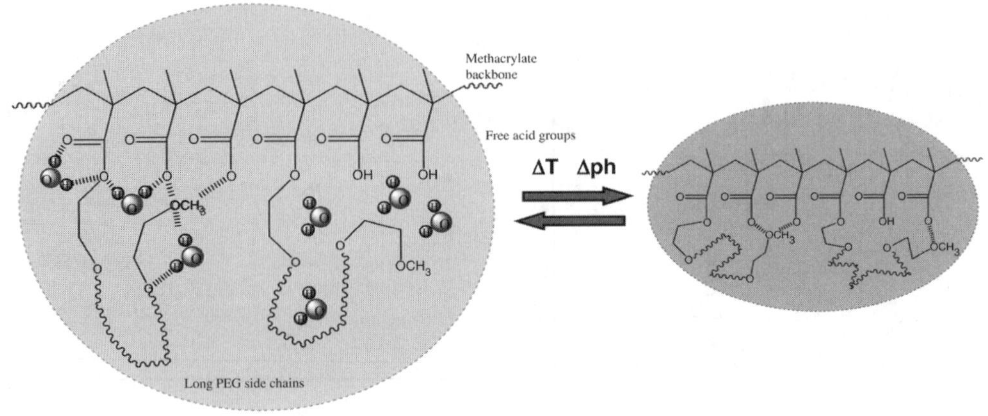

Figure 22.12 Schematic representation of the influence of intramolecular hydrogen bonding on the LCST transition of OEGMA–MAA copolymers.

CPs of ill-defined copolymers consisting of methacrylic acid (MAA) and OEGMA, prepared by free-radical polymerization, were strongly influenced by variation of pH due to the formation of intramolecular hydrogen bonds between the carboxylic acid groups and the ethylene glycol chains (Jones *et al.*, 2005a). To gain further understanding of the thermoresponsive behavior of such copolymers, we have investigated systematical libraries of well-defined copolymers based on OEGMA (molar mass = 475 and 1100 g/mol) and MAA that were prepared by the automated parallel RAFT procedure (Becer *et al.*, 2008b). PMAA is a hydrophilic polymer and was found to be fully water soluble at pH 2, 4, 7 and 10 in the range from 0 to 100 °C. $POEGMA_{475}$ is also a hydrophilic polymer and exhibits a CP at 90 °C. Incorporation of the more hydrophilic PMAA leads to a loss of thermosensitivity at pH 7 and 10 due to the increased hydrophilicity of the copolymers. Surprisingly, at pH 2 and 4 incorporation of the more hydrophilic PMAA into $POEGMA_{475}$ leads to a gradual decrease in CP from 90 °C without PMAA to ~25 °C with 80 mol% MAA. This unexpected decrease in CP upon incorporation of the more hydrophilic MAA can be ascribed to the formation of intramolecular hydrogen bonds between the protonated carboxylic acid groups and the oligoethylene oxide groups. These hydrogen bonds compete with hydration of the polymer chains and, thus, lower the CPs, as is schematically depicted in Figure 22.12 (Jones *et al.*, 2005b). The effect of intramoleular hydrogen bonding is even more pronounced in the copolymers of MAA with $OEGMA_{1100}$. Both PMAA and $POEGMA_{1100}$ homopolymers do not exhibit thermoresponsive behavior at any pH value. However, copolymers with 60–90 mol% MAA revealed thermoresponsive solubility transitions at pH 2, 4 and 7, whereby the CP decreased with increasing MAA content. These copolymers of MAA and OEGMA are interesting dual-responsive materials of which the solubility can be addressed by variation of both the temperature and pH value.

22.4 Thermoresponsive Poly(2-hydroxypropylacrylate)s by NMP

As previously discussed, poly(methoxyethylene glycol methacrylate) is insoluble in water (Han *et al.*, 2003; Becer *et al.*, 2008b). Similarly, poly(methoxyethylene glycol acrylate)

(PMEGA) that is less hydrophobic due to the absence of the methylene units attached to the polymer backbone is also reported to be water insoluble (Tanaka *et al.*, 2000). In contrast, the structural isomer poly(2-hydroxyethyl methacrylate) (PHEMA) is water soluble at low molar mass (DP < 30), exhibits LCST behavior at intermediate molar mass (30 < DP < 50) and is insoluble at higher molar mass. These different properties of PMEGA and PHEMA demonstrate that subtle differences in the hydrophilic–hydrophobic balance of the polymers as well as that the location of both hydrophilic and hydrophobic moieties can have strong effects on the solubility and thermoresponsive behavior of the polymers.

Another structural isomer of both PMEGA and PHEA is poly(hydroxypropyl acrylate) (PHPA), which consists of a mixture of 2-hydroxypropyl acrylate (~75 mol%) and 1-methyl-2-hydroxyethyl acrylate (25 mol%) resulting from the reaction of acrylic acid with propylene oxide (Scheme 22.2). Already back in 1975 it was reported by Taylor and Cerankowski that PHPA exhibits a LCST at 16 °C (Taylor and Cerankowski, 1975). After this initial report, it took 20 years before this monomer was again used as building block for thermoresponsive materials, namely hydrogels (Perera and Shanks, 1995; Christova *et al.*, 2003). In addition, one other recent report appeared on the synthesis of thermoresponsive copolymers based on HPA and the more hydrophilic aminoethyl methacrylate hydrochloride by free-radical polymerization resulting in temperature- and pH-responsive copolymers (Deng *et al.*, 2009). Besides the temperature and pH responsiveness of the copolymers, the effect of Hofmeister ions on the temperature transition was discussed.

In addition to these few examples on ill-defined thermoresponsive materials based on HPA, well-defined HPA copolymers were prepared by our group using NMP, which will be discussed in more detail in the next section. In general, HPA was copolymerized with more hydrophilic comonomers that do not show thermoresponsive behavior to be able to tune the CP in between the low CP of PHPA and 100 °C. The NMP was performed at 110 °C in *N,N*-dimethylformamide (DMF) using *N*-tert-butyl-*N*-(1′-diethylphosphono-2,2′-dimethylpropyl)-O-(2-carboxyl-prop-2-yl) alkoxyamine initiator (BlocBuilder™ initiator registered trademark of Arkema; Scheme 22.2) (Nicolas *et al.*, 2004a; Nicolas *et al.*, 2004b). Automated parallel optimization of the polymerization conditions revealed that 10–20 % of excess *N*-tertbutyl-*N*-(1′-diethylphosphono-2,2′-dimethylpropyl) free nitroxide (SG-1; Scheme 22.2) (Grimaldi *et al.*, 2000; Lacroix-Desmazes *et al.*, 2001) was required to control the polymerization of acrylates (Becer *et al.*, 2006).

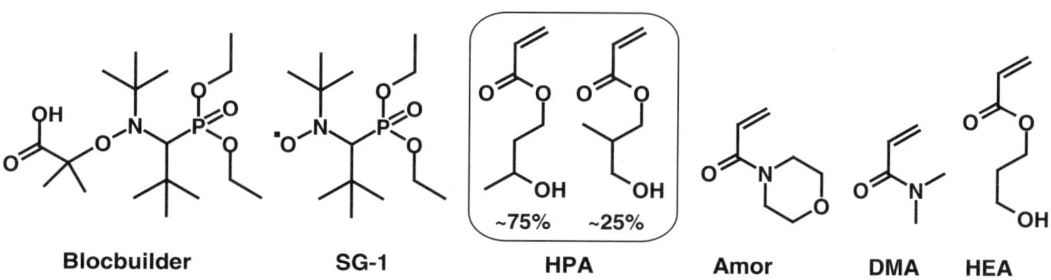

Scheme 22.2 Structures of Blocbuilder initiator, SG-1 free radical, 2-hydroxypropyl acrylate (HPA; mixture of isomers), acryloyl morpholine (Amor), *N,N*-dimethylacrylamide (DMA) and 2-hydroxyethylacrylate (HEA).

22.4.1 Libraries of HPA Copolymers

Two series of copolymers were prepared using the automated parallel synthesis robot by nitroxide-mediated copolymerization of HPA with acryloyl morpholine (Amor; Scheme 22.2) and *N,N*-dimethylacrylamide (DMA; Scheme 22.2) (Eggenhuisen *et al.*, 2008). Well-defined copolymers were prepared with the desired compositions ranging from pure HPA to pure Amor in steps of 10 mol%. The copolymerization parameters were not determined, but the polymerization rates for the homopolymerization were in a similar range and the copolymers had the desired composition at ~50% monomer conversion indicating that the copolymers have a close to ideal random monomer distribution. The Amor–HPA copolymers with less than 50 wt% HPA did not show thermoresponsive behavior due to the too high hydrophilicity of P(Amor) at 0.5 wt% polymer concentration. The DMA–HPA copolymers only revealed a CP with more than 65 wt% HPA at 0.5 wt% polymer concentration demonstrating that PDMA is more hydrophilic than P(Amor). Upon increasing the weight fraction of HPA in the copolymers, the CP gradually decreased to ~28 °C (0.5 wt%) for the HPA homopolymer (Figure 22.13). The incorporation of more hydrophilic comonomers into PHPA can thus be used to accurately tune the CP of PHPA. Comparison of the turbidity curves obtained during both heating and cooling revealed very little hysteresis at 0.5 wt% polymer concentration for the more hydrophobic copolymers with a CP below 50 °C. However, the more hydrophilic copolymers revealed tens of degrees difference between heating and cooling, whereby redissolution upon cooling occurred at higher temperatures, and the cloud points increased with repetitive heating–cooling cycles at 0.5 wt% polymer concentration. These observations were ascribed to the formation of relatively stable hydrophilic collapsed coils that have a smaller tendency to further aggregate

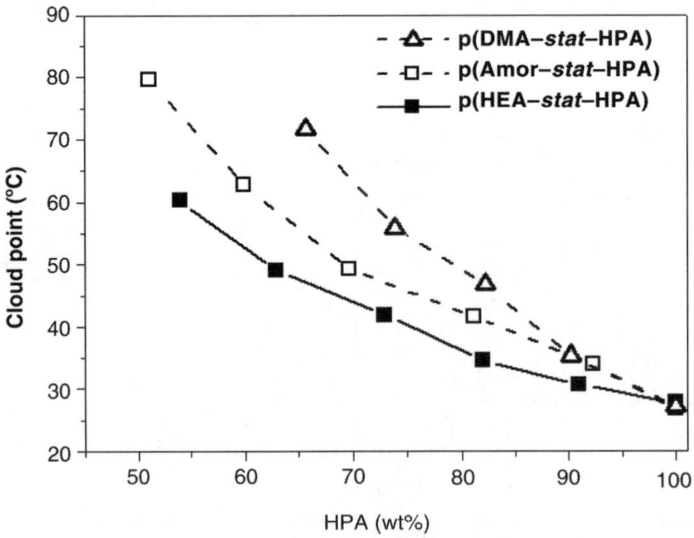

Figure 22.13 Cloud points versus 2-hydroxypropyl acrylate (HPA) weight fraction in copolymers with the more hydrophilic *N,N*-dimethylacrylamide (DMA), acryloyl morpholine (Amor) and 2-hydroxyethyl acrylate (HEA) determined at 0.5 wt% polymer concentration.

compared to the more hydrophobic copolymers. This higher stability of the collapsed polymer coils suppresses the formation of aggregates that are large enough to scatter the light and, thus, can not be detected by turbidimetry. This proposed mechanism was verified by increasing the polymer concentration, and thus the likelihood for the formation of aggregates, for the more hydrophilic copolymer. Upon increasing the polymer concentration to 2.5 wt% the CP decreased to 40 °C and both the hysteresis and increase of CP with repetitive cycles were diminished.

More recently, a copolymer library was prepared by nitroxide-mediated copolymerization of HPA with the very similar monomer 2-hydroxyethyl acrylate (HEA) (Hoogenboom *et al.*, 2009b). The copolymerization kinetics revealed the formation of well-defined ideal random copolymers with reactivity ratios both equal to one. The thermoresponsive properties of these HEA–HPA copolymers are also plotted in Figure 22.13 revealing that PHEA is slightly less hydrophilic than P(Amor) since the CPs are slightly lower at the same weight fraction of incorporated HPA. Nonetheless, also incorporation of HEA into HPA can be used to control the CP of the copolymers over a wide temperature range. The thermoresponsive properties of all three copolymer series reveal a nonlinear relationship between the weight fraction of HPA and the CPs. This nonlinearity is most likely due to the presence of a large hydrophobic SG-1 end group in combination with the moderate molar masses of the copolymers. The CPs of the HEA–HPA copolymers were also evaluated as a function of polymer concentration demonstrating that the concentration dependence is higher for the more hydrophilic copolymers as was also discussed above. An overall empirical equation was established that predicts the CP as function of both HPA content and polymer concentration.

22.5 Thermoresponsive Poly(2-oxazoline)s

Poly(2-oxazoline)s are synthetic polyamides that can be regarded as "pseudo"-polypeptides because of their structural similarity. As such, most poly(2-oxazoline)s are biocompatible and the water-soluble poly(2-methyl-2-oxazoline) and poly(2-ethyl-2-oxazoline) (PEtOx) have similar stealth behavior as PEO (Adams and Schubert, 2007; Hoogenboom, 2009). Well-defined poly(2-oxazoline)s can be prepared by living cationic ring opening polymerization of 2-oxazoline monomers (Kobayashi, 1990; Aoi and Okada, 1996; Kobayashi and Uyama, 2002). The substituent on the 2-position of the oxazoline ring forms the amidic side chains of the poly(2-oxazoline) and thus governs the polymer properties, that is, water-soluble polymers are accessible using small aliphatic side chains such as methyl or ethyl, while longer aliphatic and aromatic substituents result in hydrophobic polymers (Adams and Schubert, 2007; Hoogenboom, 2009; Hoogenboom, 2007). Since both hydrophilic and hydrophobic poly (2-oxazoline)s are readily available, the hydrophilic–hydrophobic balance can also be tuned to create thermoresponsive polymers.

Three poly(2-oxazoline)s were reported to exhibit thermoresponsive behavior in water, namely PEtOx (Christova *et al.*, 2003; Lin *et al.*, 1988), poly(2-*i*-propyl-2-oxazoline) (P*i*PropOx) (Uyama and Kobayashi, 1992; Diab *et al.*, 2004; Park *et al.*, 2004) and poly (2-*n*-propyl-2-oxazoline) (P*n*PropOx) (Park and Kataoka, 2007; Hoogenboom *et al.*, 2008b) with LCSTs of ~60 °C, ~38 °C and ~25 °C, respectively (Scheme 22.3). The cloud points of the poly(2-oxazoline)s can be controlled by introducing hydrophilic or hydrophobic end groups (Huber *et al.*, 2008) or by the copolymerization with more hydrophobic or hydrophilic

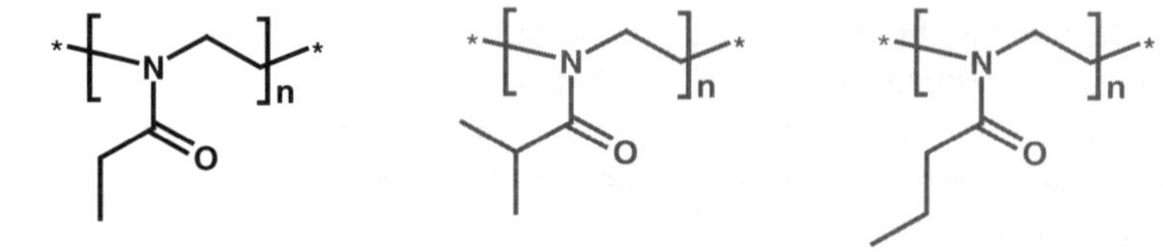

Scheme 22.3 Structures and LCSTs of poly(2-ethyl-2-oxazoline) (PEtOx), poly(2-*i*-propyl-2-oxazoline) (P*i*PropOx) and poly(2-*n*-propyl-2-oxazoline) (P*n*PropOx).

comonomers (Park and Kataoka, 2007; Park and Kataoka, 2006; Huber and Jordan, 2008; Diehl and Schlaad, 2009). Even though P*i*PropOx seems to be most suitable for biomedical applications based on its LCST close to body temperature, continued heating above the LCST of an aqueous polymer solution leads to irreversible crystallization resulting in the formation of hierarchically self-assembled fibers (Meyer *et al.*, 2007; Demirel *et al.*, 2007). Therefore, alternative poly(2-oxazoline)s are required to develop robust thermoresponsive biomaterials, which will be discussed in the following section.

22.5.1 Copolymers of 2-Ethyl-2-oxazoline and 2-N-propyl-2-oxazoline

Ideally, copolymers with a tunable phase-transition temperature should be developed that show no hysteresis between heating and cooling for use as robust thermoresponsive biomaterials. Statistical copolymers of EtOx and *n*PropOx were studied to see whether they could meet these requirements (Hoogenboom *et al.*, 2008b). At first, the copolymerization parameters were determined revealing the formation of ideal random copolymers with reactivity ratios equal to one. Subsequently, libraries of PEtOx and P*n*PropOx homopolymers were prepared with varying molar mass. Cloud-point determinations at 0.5 wt% polymer concentration revealed a strong molar-mass dependence for PEtOx and only a minor effect for the more hydrophobic P*n*PropOx, which follows the same trend for the concentration effects that was discussed above for the HPA copolymers. Based on this initial investigation, a copolymer library was designed in which the degree of polymerization of the copolymers was varied to be 50, 100 and 150 and the monomer composition was systematically varied from EtOx to *n*PropOx in steps of 10 mol% resulting in a copolymer library consisting of 33 members. These members were all prepared by the living cationic ring-opening polymerization resulting in well-defined copolymers with the desired compositions.

The thermoresponsive properties of the library of copolymers was first screened at 0.5 wt% polymer concentration, demonstrating that the CPs could be tuned from 25 °C, the CP of P*n*PropOx, up to 100 °C while some copolymers were too hydrophilic and did not show a CP. In general, the CP decreased with increasing *n*PropOx fraction as well as with increasing polymer chain length. An empirical equation was developed that accurately predicts the CPs of the copolymers at 0.5 wt% within the boundaries of the library. Detailed evaluation of the turbidimetry curves upon both heating and cooling revealed almost no hysteresis (Figure 22.14, left).

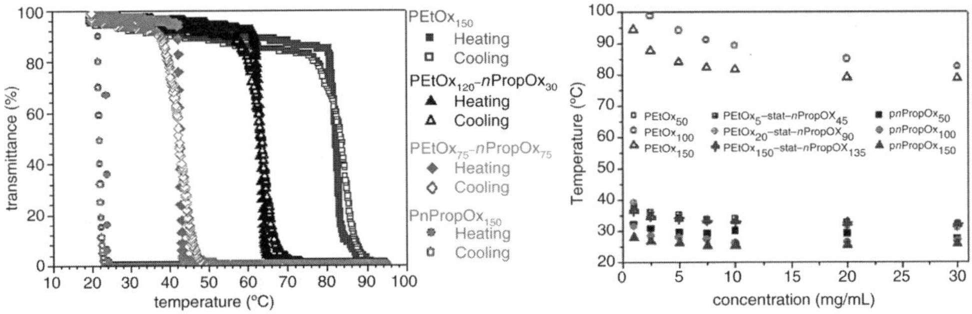

Figure 22.14 Representative turbidity curves for the copolymers of EtOx and *n*PropOx determined at 0.5 wt% polymer concentration (left) as well as the concentration dependence of the cloud points for selected copolymers (right). (Reproduced from R. Hoogenboom, H.M. Thijs, M.J. Jochems *et al.*, "Tuning the LCST of poly(2-oxazoline)s by varying composition and molecular weight: Alternatives to poly(N-isopropylacrylamide)?," *Chemical Communications*, **30**, 44, 5758–5760, 2008, by permission of the Royal Society of Chemistry.)

In addition, the copolymers with a CP close to body temperature have a robust phase transition that does not significantly vary with concentration (Figure 22.14, right). As such, these copolymers of EtOx and *n*PropOx are thermoresponsive materials with a robust and tunable CP, making them suitable candidates for biomedical applications as alternatives to PNIPAM.

22.5.2 Poly(oligo(2-ethyl-2-oxazoline) methacrylate)s

Well-defined thermoresponsive poly(2-oxazoline)s are relatively easy accessible using the living cationic ring-opening polymerization. However, only a limited number of research groups around the world are active in this class of polymers. Therefore, a more readily available poly(2-oxazoline) building block that can be polymerized using controlled-radical polymerization techniques would significantly broaden the scope of these materials. Therefore, the preparation of vinyl-functionalized oligo(2-oxazoline)s would be required, which can be achieved using functional initiators (Uyama and Kobayashi, 1991; Guinot *et al.*, 1996) or end-capping agents for the polymerization of 2-oxazolines (Miyamoto *et al.*, 1989; Kobayashi *et al.*, 1989; Christova *et al.*, 1997). Recently, we compared three different mehods for the preparation of oligo(2-ethyl-2-oxazoline) methacrylates (OEtOxMAs): (1) initiation of the living polymerization of 2-oxazolines with methacryloyl chloride; (2) end capping the polymerization with sodium carbonate to introduce a hydroxyl group followed by esterification with methacryloyl chloride and (3) direct end capping of the polymerization with triethylammonium methacrylate (Weber *et al.*, 2009a). The direct end-capping method (3) was found to be the best method leading to the highest degree of functionalization as well as the lowest amount of undesired side products such as bismethacrylates.

A series of OEtOxMA macromonomers was prepared with different oligo(2-ethyl-2-oxazoline) chain length. Subsequent controlled-radical polymerization of the macromonomers with the RAFT polymerization technique, using CBDB and AIBN, resulted in the formation of well-defined comb POEtOxMAs (Scheme 22.4) (Weber *et al.*, 2009b). The comb copolymers all exhibited a temperature-induced phase transition in aqueous solution at 0.5 wt% polymer

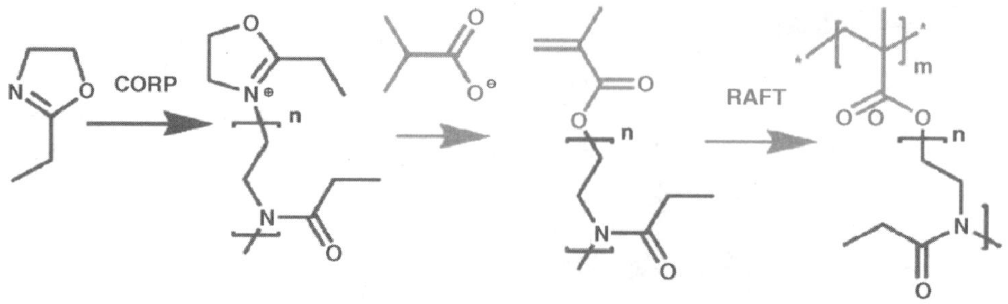

Scheme 22.4 Schematic representation of the synthesis of poly(oligo[2-ethyl-2-oxazoline] methacrylate) by subsequent cationic ring-opening polymerization of 2-ethyl-2-oxazoline followed by end capping with triethylammonium methacrylate and RAFT polymerization of the macromonomer.

concentration in the range from 66 to 95 °C. Upon increasing the methacrylate backbone length, the CP decreases, although to a lesser extent compared to the linear PEtOx. Furthermore, no clear trend was observed in the CPs upon variation of the oligo(2-ethyl-2-oxazoline) side-chain length. The CP of the POEtOxMA could be tuned by RAFT copolymerization with a more hydrophobic monomer, namely MMA. The cloud points linearly decreased from 80 °C for the pure comb POEtOxMa to 40 °C for a graft copolymer of OEtOxMA with 80 mol% MMA. Furthermore, RAFT copolymerization of OEtOxMA with the more hydrophilic MAA also resulted in the formation of well-defined graft copolymers (Weber *et al.*, 2010). Similar to the copolymers of OEGMA with MAA, the cloud points of POEtOxMA in water decreased upon incorporation of MAA due to the formation of intramolecular hydrogen bonds (Figure 22.12), which was confirmed by temperature-dependent ^{1}H NMR spectroscopic studies in deuterated water.

Copolymers of OEtOxMA and MAA with more defined methacrylate chain lengths were prepared by direct end capping of the 2-ethyl-2-oxazoline polymerization with a preformed deprotonated PMAA, whereby the OEtOxMA grafting density was varied by controlling the stiochiometry (Weber *et al.*, 2010). The thermoresponsive behavior of these copolymers was studied at 0.5 wt% aqueous solutions by turbidimetry. In water, PBS buffer and at pH 10, copolymers with 65 mol% grafting of OEtOxMA revealed CPs at 70 to 90 °C. In acidic solution at pH 4, the CPs were significantly lower due to protonation of the MAA units enabling intramolecular hydrogen bonding. In fact, polymers with 45 mol% up to 90 mol% grafting density showed a gradual increase of the CP from 30 to 70 °C. As such, the MAA-OEtOxMA copolymers are interesting dual-responsive materials that can be addressed by both temperature and pH.

22.6 Concluding Remarks

Thermoresponsive polymers have been used for a wide range of applications in "smart" materials ranging from sensors via advanced self-assembly to a variety of biomedical applications. Especially for biomedical appliations, PNIPAM is most frequently used despite the fact that it has some drawbacks such as vitrification of the collapsed polymer phase as well as the occurrence of hysteresis between heating and cooling. Therefore, continuous efforts are being made to develop alternative polymers that are superior to PNIPAM for biomedical applications. In this chapter, three such alternatives were discussed in detail, namely POEGMAs, PHPA-based copolymers and poly(2-oxazoline)s. The transition temperatures of these polymers can be easily tuned by copolymerization with more hydrophilic or hydrophobic monomers to direct the hydrophilic–hydrophobic balance of the polymers. Furthermore, the introduction of basic or acidic comonomers can be exploited to develop dual-responsive polymers that respond to changes in both temperature and pH.

In general, these alternative thermoresponsive polymer classes exhibit sharp phase transitions and, due to their low glass-transition temperatures and the absence of intramolecular hydrogen bonding, there is almost no hysteresis between heating and cooling. Therefore, these polymers are at least as good as PNIPAM and it is expected that they will be widely used for biomedical applications and will progressively replace PNIPAM.

Acknowledgment

The author would like to thank the Netherlands Scientific Organization (NWO; Veni-grant) for financial support.

References

Adams, N. and Schubert, U.S. (2007) Poly(2-oxazolines) in biological and biomedical application contexts. *Advanced Drug Delivery Reviews*, **59**, 1504–1520.

Alred, P.A., Kozlowski, A., Harris, J.M., and Tjerneld, F. (1994) Application of temperature-induced phase partitioning at ambient temperature for enzyme purification. *Journal of Chromatography A*, **659**, 289–298.

Amir, R.J., Zhong, S., Pochan, D.J., and Hawker, C.J. (2009) Enzymatically triggered self-assembly of block copolymers. *Journal of the American Chemical Society*, **131**, 13949–13951.

Aoi, K. and Okada, M. (1996) Polymerization of oxazolines. *Progress in Polymer Science*, **21**, 151–208.

Aoshima, S. and Kanaoka, S. (2008) Synthesis of stimuli-responsive polymers by living polymerization: Poly(*N*-isopropylacrylamide) and poly(vinyl ether)s. *Advances in Polymer Science*, **210**, 169–208.

Becer, C.R., Paulus, R.M., Hoogenboom, R., and Schubert, U.S. (2006) Optimization of the nitroxide-mediated radical polymerization conditions for styrene and tert-butyl acrylate in an automated parallel synthesizer. *Journal of Polymer Science, Part A: Polymer Chemistry*, **44**, 6202–6213.

Becer, C.R., Groth, A.M., Hoogenboom, R. *et al.* (2008a) Protocol for automated kinetic investigation/optimization of the RAFT polymerization of various monomers. *QSAR and Combinatorial Science*, **27**, 977–983.

Becer, C.R., Hahn, S., Fijten, M.W.M. *et al.* (2008b) Libraries of methacrylic acid and oligo(ethylene glycol) methacrylate copolymers with LCST behaviour. *Journal of Polymer Science, Part A: Polymer Chemistry*, **46**, 7138–7147.

Campese, G.M., Rodrigues, E.M.G., Tambourgi, E.B., and Pessoa, A. Jr (2003) Determination of cloud point temperatures for different copolymers. *Brazilian Journal of Chemical Engineering*, **20**, 335–337.

Chang, C.-W., Bays, E., Tao, L. *et al.* (2009) Differences in cytotoxicity of poly(PEGA)s synthesized by reversible addition–fragmentation chain transfer polymerization. *Chemical Communications*, 3580–3582.

Chen, G. and Hoffman, A.S. (1995) Graft copolymers that exhibit temperature-induced phase transitions over a wide range of pH. *Nature*, **373**, 49–52.

Christova, D., Velichkova, R., and Goethals, E.J. (1997) Bis-macromonomers of 2-alkyl-2-oxazolines – synthesis and polymerization. *Macromolecular Rapid Communications*, **18**, 1067–1073.

Christova, D., Velichkova, R., Loos, W. *et al.* (2003) New thermo-responsive polymer materials based on poly(2-ethyl-2-oxazoline) segments. *Polymer*, **44**, 2255–2261.

Cunliffe, D., De las Heras Alarcon, C., Peters, V. *et al.* (2003) Thermoresponsive surface-grafted poly(*N*-isopropylacrylamide) copolymers: Effect of phase transitions on protein and bacterial attachment. *Langmuir*, **19**, 2888–2899.

Davis, D.A., Hamilton, A., Yang, J. *et al.* (2009) Force-induced activation of covalent bonds in mechanoresponsive polymeric materials. *Nature*, **459**, 68–72.

De, P., Li, M., Gondi, S.R., and Sumerlin, B.S. (2008) Temperature-regulated activity of responsive polymer-protein conjugates prepared by grafting-from via RAFT polymerization. *Journal of the American Chemical Society*, **130**, 11288–11289.

De las Heras Alercon, C., Pennadam, S., and Alexander, C. (2005) Stimuli responsive copolymers for biomedical applications. *Chemical Society Reviews*, **34**, 276–285.

Demirel, A.L., Meyer, M., and Schlaad, H. (2007) Formation of polyamide nanofibers by directional crystallization in aqueous solution. *Angewandte Chemie International Edition*, **46**, 8622–8624.

Deng, K.L., Tian, H., Zhang, P.F. *et al.* (2009) Synthesis and characterization of a novel temperature-pH responsive copolymer of 2-hydroxypropyl acrylate and aminoethyl methacrylate hydrochloric salt. *Xpress Polymer Letters*, **3**, 97–104.

Diab, C., Akiyama, Y., Kataoka, K., and Winnik, F.M. (2004) Microcalorimetric study of the temperature-induced phase separation in aqueous solutions of poly(2-isopropyl-2-oxazolines). *Macromolecules*, **37**, 2556–2562.

Diehl, C. and Schlaad, H. (2009) Thermo-responsive polyoxazolines with widely tuneable LCST. *Macromolecular Bioscience*, **9**, 157–161.

Dimitrova, I., Trzebickab, B., Müller, A.H.E. *et al.* (2007) Thermosensitive water-soluble copolymers with doubly responsive reversibly interacting entities. *Progress in Polymer Science*, **32**, 1275–1343.

Duncan, R. (2000) The dawning era of polymer therapeutics. *Nature Reviews Drug Discovery*, **2**, 347–360.

Eggenhuisen, T.M., Becer, C.R., Fijten, M.W.M. *et al.* (2008) Libraries of statistical hydroxypropyl acrylate containing copolymers with LCST properties prepared by NMP. *Macromolecules*, **41**, 5132–5140.

Fijten, M.W.M., Meier, M.A.R., Hoogenboom, R., and Schubert, U.S. (2004) Automated parallel investigations/optimizations of the reversible addition-fragmentation chain transfer polymerization of methyl methacrylate. *Journal of Polymer Science, Part A: Polymer Chemistry*, **42**, 5775–5783.

Fijten, M.W.M., Paulus, R.M., and Schubert, U.S. (2005) Systematic parallel investigation of RAFT polymerizations for eight different (meth)acrylates: A basis for the designed synthesis of block and random copolymers. *Journal of Polymer Science, Part A: Polymer Chemistry*, **43**, 3831–3839.

Fournier, D., Hoogenboom, R., Thijs, H.M.L. *et al.* (2007) Tunable pH- and temperature-sensitive copolymer libraries by reversible addition-fragmentation chain transfer copolymerizations of methacrylates. *Macromolecules*, **40**, 915–920.

Gil., E.S. and Hudson, S.M. (2004) Stimuli-responsive polymers and their bioconjugates. *Progress in Polymer Science*, **29**, 1173–1222.

Gota, C., Okabe, K., Funatsu, T. *et al.* (2009) Hydrophilic fluorescent nanogel thermometer for intracellular thermometry. *Journal of the American Chemical Society*, **131**, 2766–2767.

Grimaldi, S., Finet, J.-P., Le Moigne, F. *et al.* (2000) Acyclic β-phosphonylated nitroxides: A new series of counter-radicals for "living"/controlled free radical polymerization. *Macromolecules*, **33**, 1141–1147.

Guerrero-Sanchez, C., Paulus, R.M., Fijten, M.W.M. *et al.* (2006) High-throughput experimentation in synthetic polymer chemistry: From RAFT and anionic polymerizations to process development. *Applied Surface Science*, **252**, 2555–2561.

Guinot, P., Bryant, L., Chow, T.Y., and Saegusa, T. (1996) Random and block 2-alkyl-2-oxazolines telechelic macromonomers. *Macromolecular Chemistry and Physics*, **197**, 1–17.

Han, S., Hagiwara, M., and Ishizone, T. (2003) Synthesis of thermally sensitive water-soluble polymethacrylates by living anionic polymerizations of oligo(ethylene glycol) methyl ether methacrylates. *Macromolecules*, **36**, 8312–8319.

Harris, J.M. and Veronese, F.M. (2003) Peptide and protein pegylation II – clinical evaluation. *Advanced Drug Delivery Reviews*, **55**, 1259–1260.

Heredia, K.L., Bontempo, D., Ly, T. *et al.* (2005) *In situ* preparation of protein-"smart" polymer conjugates with retention of bioactivity. *Journal of the American Chemical Society*, **127**, 16955–16960.

Hergt, R., Hiergeist, R., Hilger, I. *et al.* (2004) Maghemite nanoparticles with very high AC-losses for application in RF-magnetic hyperthermia. *Journal of Magnetism and Magnetic Materials*, **270**, 345–357.

Hoffman, A.S. (1987) Applications of thermally reversible polymers and hydrogels in therapeutics and diagnostics. *Journal of Controlled Release*, **6**, 297–305.

Hoffman, A.S. and Stayton, P.S. (2007) Conjugates of stimuli-responsive polymers and proteins. *Progress in Polymer Science*, **32**, 922–932.

Hoogenboom, R. (2007) Poly(2-oxazoline)s: Alive and kicking. *Macromolecular Chemistry and Physics*, **208**, 18–25.

Hoogenboom, R. (2009) Poly(2-oxazoline)s: A polymer class with numerous potential applications. *Angewandte Chemie International Edition*, **48**, 7978–7994.

Hoogenboom, R. and Schubert, U.S. (2003) The fast and the curious: High-throughput experimentation in synthetic polymer chemistry. *Journal of Polymer Science, Part A: Polymer Chemistry*, **41**, 2425–2434.

Hoogenboom, R. and Schubert, U.S. (2005) High-throughput synthesis equipment applied to polymer research. *Review of Scientific Instruments*, **76**, 062202.

Hoogenboom, R., Meier, M.A.R., and Schubert, U.S. (2003) Combinatorial methods, automated synthesis and high-throughput screening in polymer research: Past and present. *Macromolecular Rapid Communications*, **24**, 16–32.

Hoogenboom, R., Fijten, M.W.M., Abeln, C.H., and Schubert, U.S. (2004) High-throughput investigation of polymerization kinetics by online monitoring of GPC and GC. *Macromolecular Rapid Communications*, **25**, 237–242.

Hoogenboom, R., Schubert, U.S., Van Camp, W., and Du Prez, F.E. (2005) RAFT polymerization of 1-ethoxyethyl acrylate: A novel route toward near-monodisperse poly(acrylic acid) and derived block copolymer structures. *Macromolecules*, **38**, 7653–7659.

Hoogenboom, R., Thijs, H.M.L., Wouters, D. *et al.* (2008a) Tuning solution polymer properties by binary water–ethanol solvent mixtures. *Soft Matter*, **4**, 1003–1107.

Hoogenboom, R., Thijs, M.H.L., Jochems, M.J.H.C. *et al.* (2008b) Tuning the LCST of poly(2-oxazoline)s by varying composition and molecular weight: alternatives to poly(*N*-isopropylacrylamide)? *Chemical Communications*, 5758–5760.

Hoogenboom, R., Rogers, S., Can, A. *et al.* (2009a) Self-assembly of double hydrophobic block copolymers in water–ethanol mixtures: from micelles to thermoresponsive micellar gels. *Chemical Communications*, 5582–5584.

Hoogenboom, R., Popescus, D., Steinhauer, W. *et al.* (2009b) Nitroxide-mediated copolymerization of 2-hydroxyethyl acrylate and 2-hydroxypropyl acrylate: Copolymerization kinetics and thermoresponsive properties. *Macromolecular Rapid Communications*, **30**, 2042–2048.

Hua, F., Jiang, X., Li, D., and Zhao, B. (2006a) Well-defined thermosensitive, water-soluble polyacrylates and polystyrenics with short pendant oligo(ethylene glycol) groups synthesized by nitroxide-mediated radical polymerization. *Journal of Polymer Science, Part A: Polymer Chemistry*, **44**, 2454–2467.

Hua, F., Jiang, X., and Zhao, B. (2006b) Temperature-induced self-association of doubly thermosensitive diblock copolymers with pendant methoxytris(oxyethylene) groups in dilute aqueous solutions. *Macromolecules*, **39**, 3476–3479.

Huber, S. and Jordan, R. (2008) Modulation of the lower critical solution temperature of 2-alkyl-2-oxazoline copolymers. *Colloid and Polymer Science*, **286**, 395–402.

Huber, S., Hutter, N., and Jordan, R. (2008) Effect of end group polarity upon the lower critical solution temperature of poly(2-isopropyl-2-oxazoline). *Colloid and Polymer Science*, **286**, 1653–1661.

Ishizone, T., Seki, A., Hagiwara, M., and Han, S. (2008) Anionic polymerizations of oligo(ethylene glycol) alkyl ether methacrylates: Effect of side chain length and ω-alkyl group of side chain on cloud point in water. *Macromolecules*, **41**, 2963–2967.

Iwata, H., Oodate, M., Uyama, Y. *et al.* (1991) Preparation of temperature-sensitive membranes by graft polymerization onto a porous membrane. *Journal of Membrane Science*, **55**, 119–130.

Jones, J.A., Novo, N., Flagler, K. *et al.* (2005a) Thermoresponsive copolymers of methacrylic acid and poly(ethylene glycol) methyl ether methacrylate. *Journal of Polymer Science, Part A: Polymer Chemistry*, **43**, 6095–6104.

Jones, J.A., Novo, N., Flagler, K. *et al.* (2005b) Thermoresponsive copolymers of methacrylic acid and poly(ethylene glycol) methyl ether methacrylate. *Journal of Polymer Science, Part A: Polymer Chemistry*, **43**, 6095–6104.

Keurentjes, J.T.F., Kemmere, M.F., Vertommen, M.A.M.E. *et al.* (2009) Externally triggered glass transition switch for localized on-demand drug delivery. *Angewandte Chemie International Edition*, **48**, 9867–9870.

Kikuchi, A. and Okano, T. (2002) Intelligent thermoresponsive polymeric stationary phases for aqueous chromatography of biological compounds. *Progress in Polymer Science*, **27**, 1165–1193.

Kim, K.T., Cornelissen, J.J.L.M., Nolte, R.J.M., and Van Hest, J.C.M. (2009) Polymeric monosaccharide receptors responsive at neutral pH. *Journal of the American Chemical Society*, **131**, 13908–13909.

Knop, K., Hoogenboom, R., Fischer, D., and Schubert, U.S. (2010) PEG in drug delivery: Pros and cons as well as potential alternatives. *Angewandte Chemie International Edition*, **49**. doi: 10. 1002/anie.200902672

Kobayashi, S. (1990) Ethyleneimine polymers. *Progress in Polymer Science*, **15**, 751–823.

Kobayashi, S. and Uyama, H. (2002) Polymerization of cyclic imino ethers: From its discovery to the present state of the art. *Journal of Polymer Science, Part A: Polymer Chemistry*, **40**, 192–209.

Kobayashi, S., Masuda, E., Shoda, S., and Shimano, Y. (1989) Synthesis of acryl- and methacryl-type macromonomers and telechelics by utilizing living polymerization of 2-oxazolines. *Macromolecules*, **22**, 2878–2884.

Koopmans, C. and Ritter, H. (2007) Color change of N-isopropylacrylamide copolymer bearing Reichardts dye as optical sensor for lower critical solution temperature and for host-guest interaction with β-cyclodextrin. *Journal of the American Chemical Society*, **129**, 3502–3503.

Kulkarni, S., Schilli, C., Grin, B. *et al.* (2006) Controlling the aggregation of conjugates of streptavidin with smart block copolymers prepared via the RAFT polymerization technique. *Biomacromolecules*, **7**, 2736–2741.

Lacroix-Desmazes, P., Lutz, J.-F., Chauvin, F. *et al.* (2001) Living radical polymerization: Use of an excess of nitroxide as a rate moderator. *Macromolecules*, **34**, 8866–8871.

Lee, S.-Y., Lee, S., Youn, I.-C. *et al.* (2009) A near-infrared fluorescence based optical thermosensor. *Chemistry – European Journal*, **15**, 6103–6106.

Li, C., Buurma, N.J., Haq, I. *et al.* (2005) Synthesis and characterization of biocompatible thermoresponsive ABC and ABA triblock copolymer gelators. *Langmuir*, **21**, 11026–11033.

Li, C., Madsen, J., Armes, S.P., and Lewis, A.L. (2006) A new class of biochemically degradable, stimulus-responsive triblock copolymer gelators. *Angewandte Chemie International Edition*, **45**, 3510–3513.

Lin, P., Clash, C., Pearce, E.M., and Kwei, T.K. (1988) Solubility and miscibility of poly(ethyl oxazoline). *Journal of Polymer Science, Part B: Polymer Physics*, **26**, 603–619.

Lutz, J.-F. (2008) Polymerization of oligo(ethylene glycol) (meth)acrylates: Toward new generations of smart biocompatible materials. *Journal of Polymer Science, Part A: Polymer Chemistry*, **46**, 3459–3470.

Lutz, J.-F. and Hoth, A. (2006) Preparation of ideal PEG analogues with a tunable thermosensitivity by controlled radical copolymerization of 2-(2-methoxyethoxy)ethyl methacrylate and oligo(ethylene glycol) methacrylate. *Macromolecules*, **39**, 893–896.

Lutz, J.-F., Akdemir, O., and Hoth, A. (2006) Point by point comparison of two thermosensitive polymers exhibiting a similar LCST: Is the age of poly(NIPAM) over? *Journal of the American Chemical Society*, **128**, 13046–13047.

Lutz, J.-F., Weichenhan, K., Akdemir, O., and Hoth, A. (2007) About the phase transitions in aqueous solutions of thermoresponsive copolymers and hydrogels based on 2-(2-methoxyethoxy)ethyl methacrylate and oligo(ethylene glycol) methacrylate. *Macromolecules*, **40**, 2503–2508.

Lutz, J.-F., Hoth, A., and Schade, K. (2009) Design of oligo(ethylene glycol)-based thermoresponsive polymers: an optimization study. *Designed Monomers and Polymers*, **12**, 343–353.

Magnusson, J.P., Khan, A., Pasparakis, G. *et al.* (2008) Ion-sensitive "isothermal" responsive polymers prepared in water. *Journal of the American Chemical Society*, **130**, 10852–10853.

Matsumura, Y. and Katoh, A. (2008) Synthesis of 2, 3-dimorpholino-6-aminoquinoxaline derivatives and application to a new intramolecular fluorescent probe.. *Journal of Luminescence*, **128**, 625–630.

McCormick, C.L., Sumerlin, B.S., Lokitz, B.S., and Stempka, J.E. (2008) RAFT-synthesized diblock and triblock copolymers: thermally-induced supramolecular assembly in aqueous media. *Soft Matter*, **4**, 1760–1773.

Meier, M.A.R., Hoogenboom, R., and Schubert, U.S. (2004) Combinatorial methods, automated synthesis and high-throughput screening in polymer research: The evolution continues. *Macromolecular Rapid Communications*, **25**, 21–33.

Mertoglu, M., Garnier, S., Laschewsky, A. *et al.* (2005) Stimuli responsive amphiphilic block copolymers for aqueous media synthesised via reversible addition fragmentation chain transfer polymerisation (RAFT). *Polymer*, **46**, 7726–7740.

Meyer, M., Antonietti, M., and Schlaad, H. (2007) Unexpected thermal characteristics of aqueous solutions of poly(2-isopropyl-2-oxazoline). *Soft Matter*, **3**, 430–431.

Miyamoto, M., Naka, K., Tokumizu, M., and Saegusa, T. (1989) End capping of growing species of poly(2-oxazoline) with carboxylic acid: A novel and convenient route to prepare vinyl- and carboxy-terminated macromonomers. *Macromolecules*, **22**, 1604–1607.

Nath, N. and Chilkoti, A. (2002) Creating 'smart' surfaces using stimuli responsive polymers. *Advanced Materials*, **14**, 1243–1247.

Neugebauer, D. (2007) Graft copolymers with poly(ethylene oxide) segments. *Polymer International*, **56**, 1469–1498.

Nicolas, J., Charleux, B., Guerret, O., and Magnet, S. (2004a) Nitroxide-mediated controlled free-radical emulsion polymerization of styrene and n-butyl acrylate with a water-soluble alkoxyamine as initiator. *Angewandte Chemie International Edition*, **43**, 6186–6189.

Nicolas, J., Charleux, B., Guerret, O., and Magnet, S. (2004b) Novel SG1-based water-soluble alkoxyamine for nitroxide-mediated controlled free-radical polymerization of styrene and n-butyl acrylate in miniemulsion. *Macromolecules*, **37**, 4453–4463.

Nucci, M.L., Shorr, R., and Abuchowski, A. (1991) The therapeutic value of poly(ethylene glycol)-modified proteins. *Advanced Drug Delivery Reviews*, **6**, 133–151.

Okada, Y. and Tanaka, F. (2005) Cooperative hydration, chain collapse, and flat LCST behavior in aqueous poly(*N*-isopropylacrylamide) solutions. *Macromolecules*, **35**, 4465–4471.

Okano, T., Yamada, N., Okuhara, M. *et al.* (1995) Mechanism of cell detachment form temperature-modulated, hydrophilic-hydrophobic polymer surfaces. *Biomaterials*, **16**, 297–303.

Oosten, C.L., Bastiaansen, C.W.M., and Broer, D.J. (2009) Printed artificial cilia from liquid-crystal network actuators modularly driven by light. *Nature Materials*, **8**, 677–682.

Ott, C., Hoogenboom, R., Hoeppener, S. *et al.* (2009) Tuning the morphologies of amphiphilic metallo-supramolecular triblock terpolymers: from spherical micelles to switchable vesicles. *Soft Matter*, **5**, 84–91.

Park, J.-S. and Kataoka, K. (2006) Precise control of lower critical solution temperature of thermosensitive poly(2-isopropyl-2-oxazoline) via gradient copolymerization with 2-ethyl-2-oxazoline as a hydrophilic comonomer. *Macromolecules*, **39**, 6622–6630.

Park, J.-S. and Kataoka, K. (2007) Comprehensive and accurate control of thermosensitivity of poly(2-alkyl-2-oxazoline)s via well-defined gradient or random copolymerization. *Macromolecules*, **40**, 3599–3609.

Park, Y.S., Ito, Y., and Imanishi, Y. (1998) Permeation control through porous membranes immobilized with thermosensitive polymer. *Langmuir*, **14**, 910–914.

Park, J.-S., Akiyama, Y., Winnik, F.M., and Kataoka, K. (2004) Versatile synthesis of end-functionalized thermosensitive poly(2-isopropyl-2-oxazolines). *Macromolecules*, **37**, 6786–6792.

Pasut, G. and Veronese, F.W. (2007) Polymer–drug conjugation, recent achievements and general strategies. *Progress in Polymer Science*, **32**, 933–961.

Paulus, R.M., Fijten, M.W.M., De la Mar, M. *et al.* (2005) Reversible addition-fragmentation chain transfer polymerization on different synthesizer platforms. *QSAR and Combinatorial Science*, **24**, 863–867.

Perera, D.I. and Shanks, R.A. (1995) Network characteristics of homopolymer and some copolymers of poly(2-hydroxyethyl methacrylate). *Polymer International*, **37**, 133–139.

Persson, J., Kaul, A., and Tjerneld, F. (2000) Polymer recycling in aqueous two-phase extractions using thermoseparating ethylene oxide-propylene oxide copolymers. *Journal of Chromatography B*, **743**, 115–126.

Piermattei, A., Karthikeyan, S., and Sijbesma, R.P. (2009) Activating catalysts with mechanical force. *Nature Chemistry*, **1**, 133–137.

Pietsch, C., Hoogenboom, R., and Schubert, U.S. (2009) Soluble polymeric dual sensor for temperature and pH value. *Angewandte Chemie International Edition*, **48**, 5653–5656.

Pissuwan, D., Boyer, C., Gunasekaran, K. *et al.* (2010) In vitro cytotoxicity of RAFT polymers. *Biomacromolecules*, **11**, 412–420.

Qing, G., Wang, X., Fuchs, H., and Sun, T. (2009) Nucleotide-responsive wettability on a smart polymer surface. *Journal of the American Chemical Society*, **131**, 8370–8371.

Rapoport, N. (2007) Physical stimuli-responsive polymeric micelles for anti-cancer drug delivery. *Progress in Polymer Science*, **32**, 962–990.

Reichardt, C. (1992) Solvatochromism, thermochromism, piezochromism, halochromism, and chiro-solvatochromism of pyridinium N-phenoxide betaine dyes. *Chemical Society Reviews*, **21**, 147–153.

Rodriguez-Hernandez, J., Checot, F., Gnanou, Y., and Lecommandoux, S. (2005) Toward 'smart' nano-objects by self-assembly of block copolymers in solution. *Progress in Polymer Science*, **30**, 691–724.

Russell, T.P. (2002) Surface responsive materials. *Science*, **297**, 964–967.

Rzaev, Z.M.O., Dincer, S., and Piskin, E. (2007) Functional copolymers of *N*-isopropylacrylamide for bioengineering applications. *Progress in Polymer Science*, **32**, 534–595.

Saeki, S., Kuwahara, N., Nakata, M., and Kaneko, M. (1876) Upper and lower critical solution temperatures in poly (ethylene glycol) solutions. *Polymer*, **17**, 685–689.
Saito, S. (1986) Cloud points of poly(propylene oxide) and poly(vinyl methylether) in the presence of polymeric acids. *Colloids and Surfaces*, **19**, 351–357.
Schild, H.G. (1992) Poly(*N*-isopropylacrylamide): Experiment, theory and application. *Progress in Polymer Science*, **17**, 163–249.
Schmaljohann, D. (2006) Thermo- and pH-responsive polymers in drug delivery. *Advanced Drug Delivery Reviews*, **58**, 1655–1670.
Sershen, S.R., Westcott, S.L., Halas, N.J., and West, J.L. (2000) Temperature-sensitive polymer-nanoshell composites for photothermally modulated drug delivery. *Journal of Biomedical Materials Research*, **51**, 293–298.
Shiraishi, Y., Miyamoto, R., Zhang, X., and Hirai, T. (2007) Rhodamine-based fluorescent thermometer exhibiting selective emission enhancement at a specific temperature range. *Organic Letters*, **9**, 3921–3924.
Shiraishi, Y., Miyamoto, R., and Hirai, T. (2008) A hemicyanine-conjugated copolymer as highly sensitive fluorescent thermometer. *Langmuir*, **24**, 4273–4279.
Shiraishi, Y., Miyamoto, R., and Hirai, T. (2009) Spiropyran-conjugated thermoresponsive copolymer as colorimetric thermometer with linear and reversible color change. *Organic Letters*, **11**, 1571–1574.
Smith, A.E., Xua, X., and McCormick, C.L. (2010) Stimuli-responsive amphiphilic (co)polymers via RAFT polymerization. *Progress in Polymer Science*, 005. doi: 10.1016/j.progpolymsci.2009.11
Stayton, P.S., Shimoboji, T., Long, C. *et al.* (1995) Control of protein-ligand recognition using a stimuli-responsive polymer. *Nature*, **378**, 472–474.
Sundararaman, A., Stephan, T., and Grubbs, R.B. (2008) Reversible restructuring of aqueous block copolymer assemblies through stimulus-induced changes in amphiphilicity. *Journal of the American Chemical Society*, **130**, 12264–12265.
Tan, I., Zarafshani, Z., Lutz, J.-F., and Titirici, M.M. (2009) PEGylated chromatography: Efficient bioseparation on silica monoliths grafted with smart biocompatible polymers. *ACS Applied Materials and Interfaces*, **1**, 1869–1872.
Tanaka, M., Motomura, T., Kawada, M. *et al.* (2000) Blood compatible aspects of poly(2-methoxyethylacrylate) (PMEA)—relationship between protein adsorption and platelet adhesion on PMEA surface. *Biomaterials*, **21**, 1471–1481.
Tang, T., Castelletto, V., Parras, P. *et al.* (2006) Thermo-responsive poly(methyl methacrylate)-block-poly(N-isopropylacrylamide) block copolymers synthesized by RAFT polymerization: Micellization and gelation. *Macromolecular Chemistry and Physics*, **207**, 1718–1726.
Taylor, L.D. and Cerankowski, L.D. (1975) Preparation of films exhibiting a balanced temperature dependence to permeation by aqueous solutions – a study of lower consolute behaviour. *Journal of Polymer Science: Polymer Chemistry Edition*, **13**, 2551–2570.
Uchiyama, S., Matsumura, Y., Prasanna de Silva, A., and Iwai, K. (2003) Fluorescent molecular thermometers based on polymers showing temperature-induced phase transitions and labeled with polarity-responsive benzofurazans. *Analytic Chemistry*, **75**, 5936–5935.
Uchiyama, S., Kawai, N., Prasanna de Silva, A., and Iwai, K. (2004) Fluorescent polymeric AND logic gate with temperature and pH as inputs. *Journal of the American Chemical Society*, **126**, 3032–3033.
Uyama, H. and Kobayashi, S. (1991) Synthesis of poly(2-oxazoline) macromonomers having a vinyl ester group. *Macromolecules*, **24**, 614–615.
Uyama, H. and Kobayashi, S. (1992) A novel thermo-sensitive polymer. Poly(2-iso-propyl-2-oxazoline). *Chemistry Letters*, 1643–1646.
Van Durme, K., Van Assche, G., and Van Mele, B. (2004) Kinetics of demixing and remixing in poly(*N*-isopropylacrylamide)/water studied by modulated temperature DSC. *Macromolecules*, **37**, 9596–9605.
Vertommen, M.A.M.E., Cornelissen, H.-J.L., Dietz, C.H.J.T. *et al.* (2008) Pore-covered thermoresponsive membranes for repeated on-demand drug release. *Journal of Membrane Science*, **322**, 243–248.
Wang, X., Qiu, X., and Wu, C. (1998) Comparison of the coil-to-globule and the globule-to-coil transitions of a single poly(*N*-isopropylacrylamide) homopolymer chain in water. *Macromolecules*, **31**, 2972–2976.
Wang, D., Miyamoto, R., Shiraishi, Y., and Hirai, T. (2009) BODIPY-conjugated thermoresponsive copolymer as a fluorescent thermometer based on microviscosity. *Langmuir*, **25**, 13176–13182.
Weber, C., Becer, C.R., Baumgaertel, A. *et al.* (2009a) Preparation of methacrylate end-functionalized poly(2-ethyl-2-oxazoline) macromonomers. *Designed Monomers and Polymers*, **12**, 149–165.

Weber, C., Becer, C.R., Hoogenboom, R., and Schubert, U.S. (2009b) Lower critical solution temperature behavior of comb and graft shaped poly[oligo(2-ethyl-2-oxazoline)methacrylate]s. *Macromolecules*, **42**, 2865–2971.

Weber, C., Becer, C.R., Guenther, W. *et al.* (2010) Dual responsive methacrylic acid and oligo(2-ethyl-2-oxazoline) containing graft copolymers. *Macromolecules*, **43**, 160–167.

Wei, H., Cheng, S.-X., Zhang, X.-X., and Zhuo, R.-X. (2009) Thermo-sensitive polymeric micelles based on poly(*N*-isopropylacrylamide) as drug carriers. *Progress in Polymer Science*, **34**, 893–910.

Winnik, F.M. (1990) Fluorescence studies of aqueous solutions of poly(*N*-isopropylacrylamide) below and above their LCST. *Macromolecules*, **23**, 233–242.

Wischerhoff, E., Uhlig, K., Lankenau, A. *et al.* (2008) Controlled cell adhesion on PEG-based switchable surfaces. *Angewandte Chemie International Edition*, **47**, 5666–5668.

Wood, K.C., Zacharia, N.S., Schmidt, D.J. *et al.* (2008) Electroactive controlled release thin films. *Proceeding of the National Academy of Sciences of the United States of America*, 105, pp. 2280–2285.

www.chemspeed.com (last accessed January 14, 2010).

www.crystal16.com (last accessed January 14, 2010).

Yamamoto, S.-I., Pietrasik, J., and Matyjaszewski, K. (2008) The effect of structure on the thermoresponsive nature of well-defined poly(oligo(ethylene oxide) methacrylates) synthesized by ATRP. *Journal of Polymer Science, Part A: Polymer Chemistry*, **46**, 194–202.

Zhang, H., Hoogenboom, R., Meier, M.A.R., and Schubert, U.S. (2005) Combinatorial and high-throughput approaches in polymer science. *Measurement Science and Technology*, **16**, 203–211.

23

Fluorine-Containing Block Copolymers: Synthesis and Application as a Template for Nanocellular and Porous Structures Using Supercritical Carbon Dioxide

Hideaki Yokoyama[1] and Kenji Sugiyama[2]
[1] *Department of Advanced Materials Science, Graduate School of Frontier Sciences, The University of Tokyo, Japan*
[2] *Department of Chemical Science and Technology, Faculty of Bioscience and Applied Chemistry, Hosei University, Japan*

23.1 Introduction

Fluorine-containing polymers have unique properties due to their carbon–fluorine bonds that lead to a low dielectric constant, low refractive index, low surface tension and extremely large interaction with other nonfluorinated polymers. Block copolymers with fluorine-containing blocks therefore form ordered structures even with low molecular weight due to their large interaction parameters (Ren *et al.*, 2002). Such a large interaction parameter between fluorine-containing blocks and nonfluorine-containing blocks often leaves the morphologies trapped in inequilibrium ones depending on the processing history, and never leads to equilibrium morphologies on an experimental timescale. Because of difficulties in achieving equilibrium morphologies and in synthesizing fluorine-containing

Complex Macromolecular Architectures: Synthesis, Characterization, and Self-Assembly, First Edition.
Edited by Nikos Hadjichristidis, Akira Hirao, Yasuyuki Tezuka and Filip Du Prez.

monomers, which will be described later, fluorine-containing block copolymers have not been studied for understanding the general aspects of the physical properties of block copolymers. However, fluorine-containing block copolymers have intrinsic properties that are related to fluorine-containing blocks and their segmental structures, in which commonly short fluoroalkyl chains are attached as side chains. An additional interesting property of fluorine-containing polymers is that they have a higher affinity to carbon dioxide than ordinary polymers without fluorine do. Some fluorine-containing polymers even dissolve in supercritical carbon dioxide (O'Neill *et al.*, 1998). Therefore, fluorine-containing block copolymers can selectively absorb carbon dioxide in their fluorine-containing domains and selectively swell.

In this chapter, we focus on the fact that fluorine-containing polymers are in a rare class of polymers that have a high affinity to supercritical carbon dioxide (O'Neill *et al.*, 1998). We describe the polymerization of fluorine-containing block copolymers and their use to selectively localize supercritical carbon dioxide in fluorine-containing domains for introducing nanocellular and nanoporous structures.

23.2 Synthesis of Well-Defined Block Copolymers Containing Perfluoroalkylated Polymer Segments

In general, perfluoroalkylated polymers having long perfluoroalkyl chains are insoluble in most common organic solvents. In order to overcome this problem and exert their characteristic features, perfluoroalkyl chains are often incorporated into conventional polymers. Moreover, control of the primary structures of perfluorinated polymers has recently been receiving much attention from the viewpoints of both synthetic and physical studies that reveal the basic roles of the incorporated perfluoroalkyl chains. The well-defined block copolymers containing perfluoroalkyl groups employed here are typical examples that meet such requirements. A variety of well-defined block copolymers containing perfluoroalkylated polymer segments has been synthesized by methodologies based on living/controlled polymerization (Giannetti, 2001; Hillmyer and Lodge, 2002; Reisinger and Hillmyer, 2002; Imae, 2003; Krishnan *et al.*, 2004; Hansen *et al.*, 2007). They are usually synthesized by living/controlled polymerization of a conventional monomer followed by a fluorine-containing monomer based on the sequential monomer addition technique. In addition, the introduction of perfluoroalkyl groups by a modification reaction of premade functional block copolymers is an alternate way to obtain fluorine-containing block copolymers with well-defined structures. The advantage of the polymer modification method is that perfluoroalkylated block copolymers can be obtained even if it is difficult to produce them directly by living polymerization. Additional advantages are that a precursory block copolymer can be readily synthesized and modified in several ways to enable direct comparison between different polymer designs.

As described in the Introduction, perfluoroalkylated polymer segments play an important role in nanofabrication under supercritical carbon dioxide conditions. It should be mentioned that those nanocellular structures can be defined by the segment ratio and block lengths of perfluorinated polymer segments. Therefore, some well-defined architectural features of block copolymers, such as controlled composition, predictable molecular weight, and narrow molecular-weight distribution, are essential for this study.

23.2.1 Living/Controlled-Radical Polymerization

Recently, controlled-radical polymerization, such as atom-transfer radical polymerization (ATRP), nitroxide-mediated radical polymerization (NMP), and reversible addition-fragmentation chain transfer (RAFT) polymerization, has been widely developed and applied to perfluoroalkylated block copolymer synthesis, as shown in Figure 23.1 (Zhang *et al.*, 1999; Lim *et al.*, 2002; Hussain *et al.*, 2002; Li *et al.*, 2002a; Andruzzi *et al.*, 2002; Gopalan *et al.*, 2002; Radhakrishnan *et al.*, 2004; Hikita *et al.*, 2004; Shemper and Mathias, 2004; Jankova and Hvilsted, 2005; Galli *et al.*, 2005). Styrene and (meth)acrylate monomers containing perfluoroalkyl groups are used as second monomers in sequential polymerization. ABA triblock copolymers as well as AB diblock copolymers are synthesized by using a difunctional initiator. The molecular weights of these block copolymers are around 5000–50 000 and the molecular-weight distribution is relatively narrow, the M_w/M_n values being 1.1–1.5. In addition, the iniferter radical polymerization method and diazo macroinitiator method were also reported (Guan and DeSimone, 1994; Kassis *et al.*, 1996; Arnold *et al.*, 2002; Kim *et al.*, 1999; Bertolucci *et al.*, 2004). Unfortunately, the resulting block copolymers were not well controlled in chain structure.

23.2.2 Living Cationic Polymerization

The successful synthesis of AB diblock copolymers by living cationic polymerization is based on the sequential polymerization technique. Vinyl ether derivatives bearing perfluoroalkyl groups are employed as monomers. 1,1,2-Trichlorotrifluoroethane is often effective to avoid precipitates during polymerization and to produce polymers with controlled structures (Choi *et al.*, 1988; Percec and Lee, 1992; Vandooren *et al.*, 1994). Similarly, a block copolymer

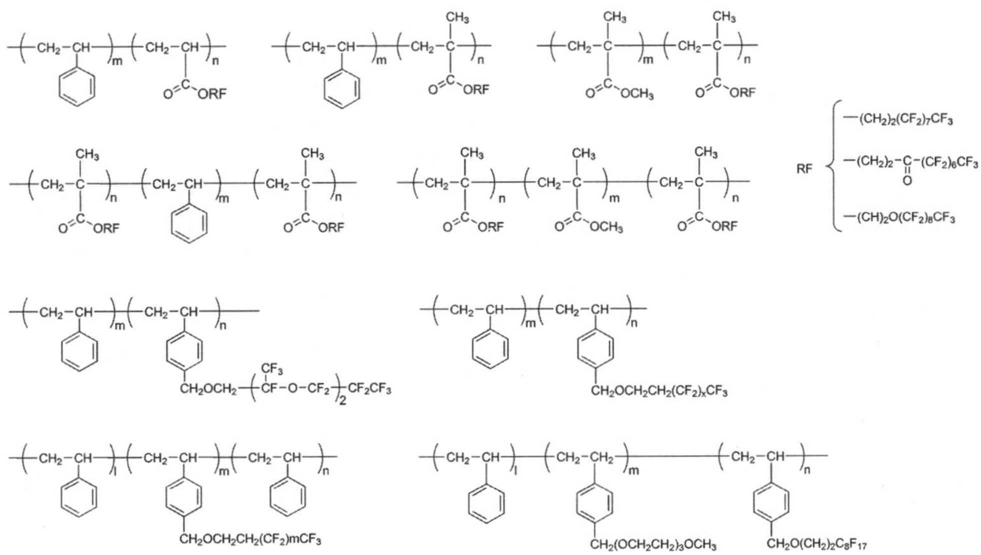

Figure 23.1 Block copolymers containing perfluoroalkylated polymer segments.

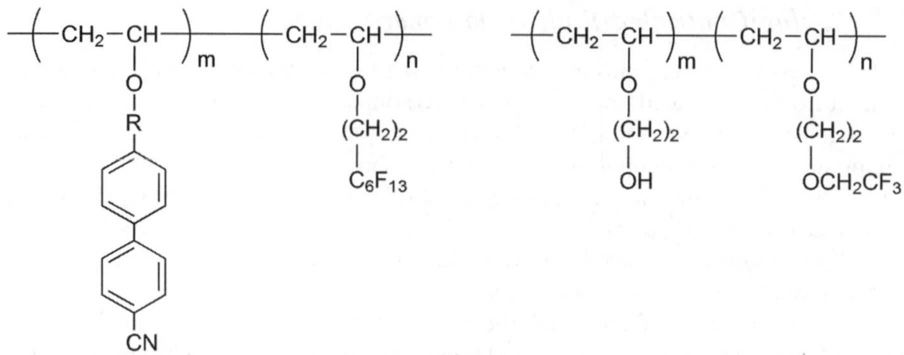

Figure 23.2 Perfluoroalkylated block copolymers obtained by living cationic polymerization (Percec and Lee, 1992; Matsumoto *et al.*, 2000).

composed of poly(2-hydroxyethyl vinyl ether) and poly(2-(2,2,2-trifluoroethoxy)ethyl vinyl ether) can be synthesized by sequential cationic polymerization of 2-(acetoxyethyl) vinyl ether and 2-(2,2,2-trifluoroethoxy)ethyl vinyl ether, followed by hydrolysis of the acetyl protecting group as shown in Figure 23.2 (Matsumoto *et al.*, 1999; Matsumoto *et al.*, 2000).

23.2.3 Living Anionic Polymerization

Although it is essentially difficult for a perfluoroalkyl group to coexist with a highly reactive anionic species, living anionic polymerization of (meth)acrylate monomers bearing perfluoroalkyl groups proceeds under suitable conditions due to the low nucleophilicity of the propagating enolate anions (Yong *et al.*, 1997; Ishizone *et al.*, 1999). Several AB block copolymers of perfluoroalkyl methacrylate with MMA (Yong *et al.*, 1997) or *tert*-butyl methacrylate (Tanaka *et al.*, 1999; Busse *et al.*, 2002) were successfully synthesized by the sequential monomer addition technique (Figure 23.3). It should be noted that the reactivities of methacrylic monomers are similar to each other, which allow block copolymer synthesis in any sequence order. For instance, well-defined ABC triblock terpolymers with three different sequences (ABC, ACB, and BAC) were successfully synthesized by changing the addition order of three methacrylic monomers substituted with either 2-perfluorobutylethyl, 2-trimethylsilyloxyethyl, or *tert*-butyl groups based on sequential anionic polymerization (Scheme 23.1) (Ishizone *et al.*, 1999; Tanaka *et al.*, 1999).

On the other hand, the monomer addition order is important and regulated when styrene is used, due to the large reactivity difference between styrene and perfluoroalkyl methacrylate. Polystyryllithium end capped with 1,1-diphenylethylene can completely initiate the

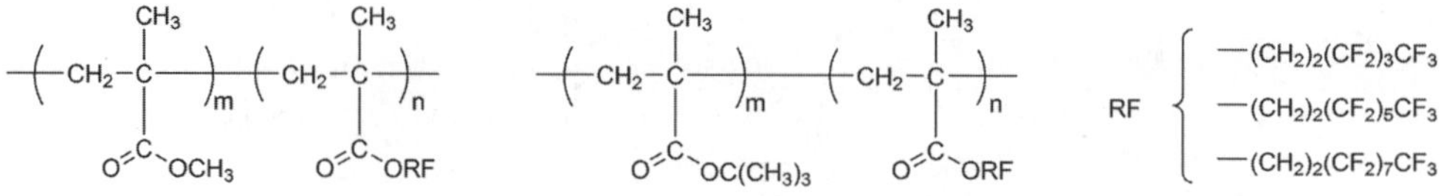

Figure 23.3 Perfluoroalkylated block copolymers obtained by living anionic polymerization.

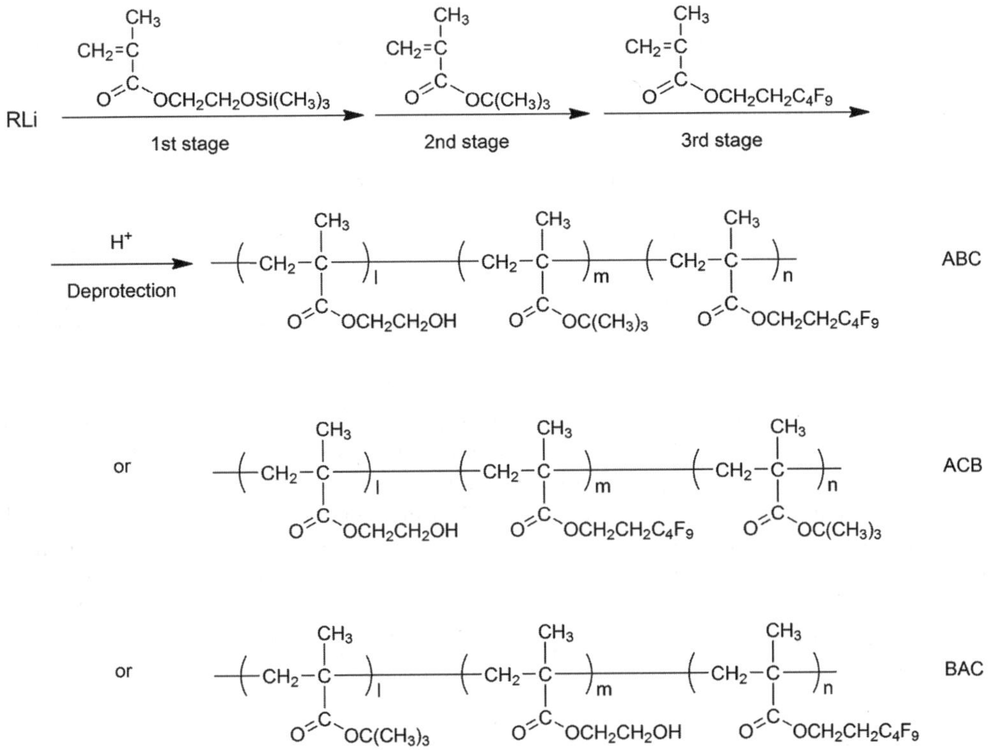

Scheme 23.1 Synthesis of ABC triblock terpolymers containing poly[2-(perfluorobutyl)ethyl methacrylate] segment by living anionic polymerization.

polymerization of perfluoroalkyl methacrylate to give a well-defined AB diblock copolymer (A: polystyrene, B: poly(perfluoroalkyl methacrylate)), as shown in Scheme 23.2. However, block copolymer (BA) is not formed in the reverse monomer addition order. In this study, polystyrene-*block*-poly[2-(perfluorooctyl)ethyl methacrylate] was synthesized by living

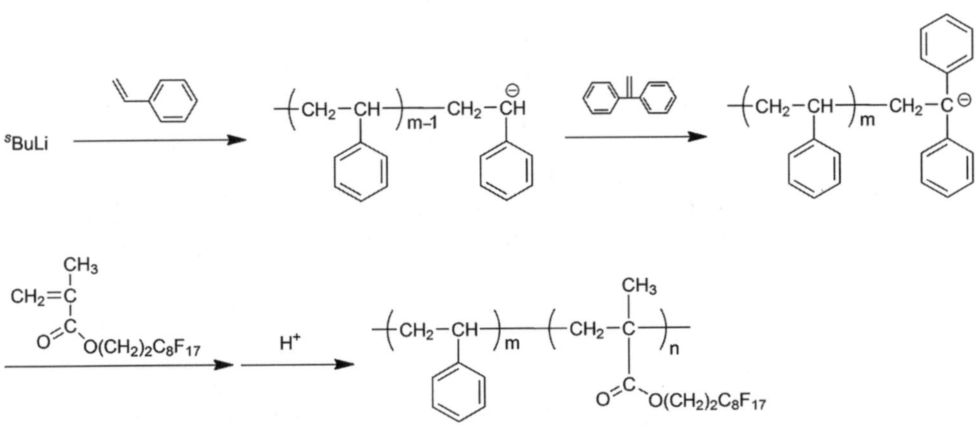

Scheme 23.2 Synthesis of polystyrene-*block*-poly[2-(perfluorooctyl)ethyl methacrylate].

Table 23.1 Synthesis of polystyrene-*block*-poly(2-perfluorooctylethyl methacrylate)

$M_n \times 10^{-3}$		M_w/M_n	Fluorine content
Calcd	Obsd (PS-PFMA)		wt-%
26.1	25.0 (21.8–3.2)	1.02	7.8$_2$
24.4	23.0 (17.7–5.3)	1.02	13.0
28.7	27.5 (20.1–7.4)	1.03	16.4
33.7	32.9 (22.7–10.2)	1.05	18.3
31.5	30.3 (19.7–10.6)	1.03	21.2
38.0	38.0 (21.0–17.0)	1.03	27.2

anionic polymerization with the sequential addition of two monomers, styrene and 2-(perfluorooctyl)ethyl methacrylate, in this order. The results are also summarized in Table 23.1. The resulting polymers always possessed predictable molecular weights and narrow molecular-weight distribution, the M_w/M_n values being 1.02–1.08 (Li *et al.*, 2004; Yokoyama and Sugiyama, 2005; Li *et al.*, 2006). Thus, the resulting block copolymers were observed to possess well-controlled structures with respect to chain length and composition. The resulting polymers are used for nanofabrication under supercritical carbon dioxide conditions.

23.2.4 Group-Transfer Polymerization

Group-transfer polymerization (GTP) is also effective to produce AB diblock copolymers of MMA with alkyl methacrylate bearing either a perfluorobutylethyl or perfluorohexylethyl group (Krupers and Möller, 1997a; Krupers and Möller, 1997b). Unfortunately, a monomer with a long perfluorooctylethyl group is difficult to polymerize due to a solubility problem, even in sequential polymerization using a second monomer. By using the same GTP system, AB block copolymers composed of tetrahydropyranyl methacrylate with some perfluoroalkyl methacrylates can be synthesized as shown in Figure 23.4 (Yang *et al.*, 2000).

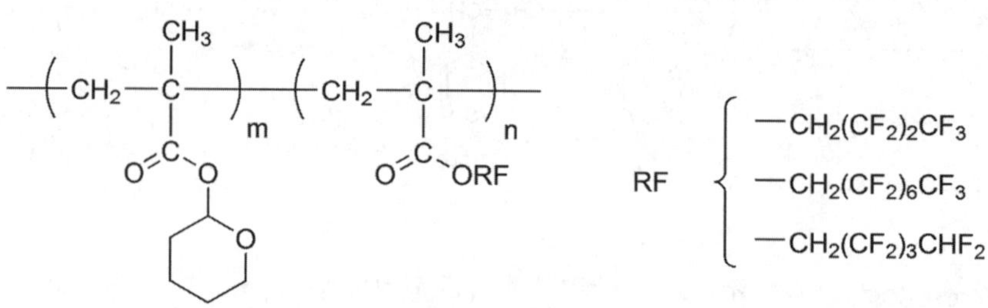

Figure 23.4 Perfluoroalkylated block copolymers obtained by group-transfer polymerization (Yang *et al.*, 2000).

23.2.5 *Introduction of Perfluoroalkyl Groups via Polystyrene-Block-Polyisoprene*

Polystyrene-*block*-polyisoprene (PS-PI) was first synthesized by living anionic polymerization. The microstructure of the PI segment was controlled with a 1,2- and 3,4-structure in more than 97 mol-%. Hydroxylation of the vinyl groups in the PI segment was carried out by hydroboration with 9-BBN, followed by esterification with perfluoroalkylacid chloride to introduce the corresponding perfluoroalkyl groups as shown in Scheme 23.3 (Iyengar *et al.*, 1996; Wang *et al.*, 1997; Genzer *et al.*, 2000a; Genzer *et al.*, 2000b; Genzer *et al.*, 2000c; Xiang *et al.*, 2000; Hayakawa *et al.*, 2000; Li *et al.*, 2002b; Hexemer *et al.*, 2004; You *et al.*, 2006).

The same precursor block copolymer, PS-PI, was treated with $C_6F_{13}(CH_2)_2Si(CH_3)_2H$ under Pt-catalyzed hydrosilylation reaction conditions. The corresponding perfluoroalkyl groups were introduced into the vinyl groups in the PI segment to directly produce the target block copolymer as shown in Scheme 23.4 (Hwang *et al.*, 1995; Jeyaprakash *et al.*, 2000).

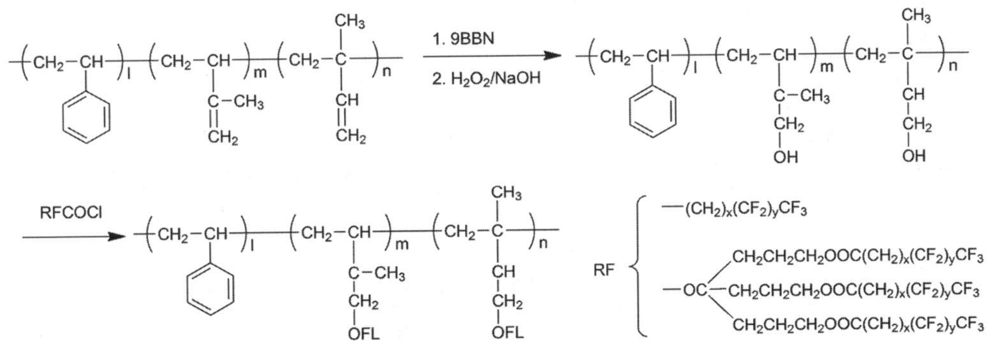

Scheme 23.3 Synthesis of perfluoroalkylated polystyrene-*block*-polyisoprene via hydroxylation followed by esterification (Iyengar *et al.*, 1996; Wang *et al.*, 1997; Genzer *et al.*, 2000a, 2000b, 2000c; Xiang *et al.*, 2000; Hayakawa *et al.*, 2000; Li *et al.*, 2002a; Hexemer *et al.*, 2004; You *et al.*, 2006).

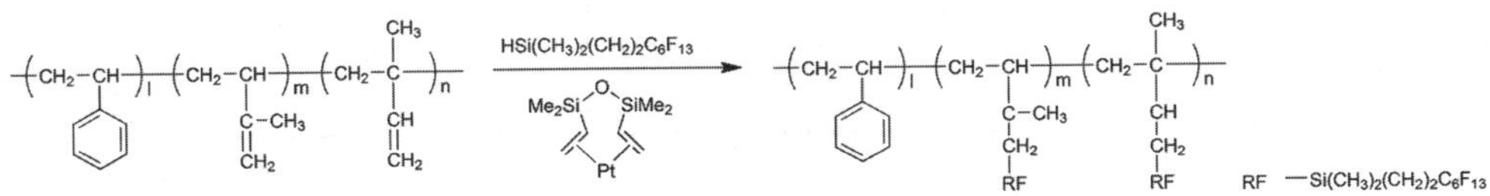

Scheme 23.4 Synthesis of perfluoroalkylated polystyrene-*block*-polyisoprene via hydrosilylation (Hwang *et al.*, 1995; Jeyaprakash *et al.*, 2000).

23.2.6 *Introduction of Perfluoroalkyl Groups via Polystyrene-Block-Poly (4-Vinylphenol)*

A precursory polystyrene-*block*-poly(4-vinylphenol) is first prepared by living anionic block copolymerization of styrene and 4-*tert*-butyldimethylsilyloxystyrene, followed by deprotection of the silyl group. The resulting polymer is reacted with 3-perfluorooctylpropyl bromide to afford the target polystyrene-*block*-poly(4-(3-perfluorooctylpropoxy)styrene)

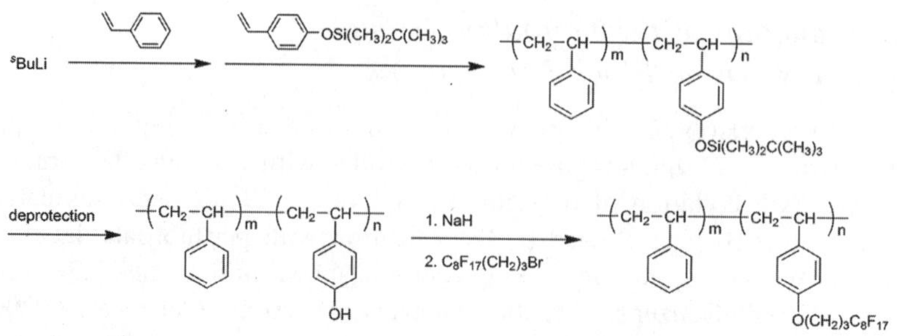

Scheme 23.5 Synthesis of polystyrene-*block*-poly[4-(3-perfluorooctylpropoxy)styrene].

under Williamson reaction conditions (Sugiyama *et al.*, 2002; Hirao *et al.*, 2004; Yokoyama *et al.*, 2004a, 2004b; Yokoyama and Sugiyama, 2004), as shown in Scheme 23.5. The resulting block copolymers showed sharp monomodal SEC peaks without any shoulders or tailings (Figure 23.5). The M_n values observed by SEC coupled with ^{1}H NMR were always in good agreement with those predicted. Their molecular-weight distribution remained very narrow after two steps of deprotection and the Williamson reaction. All M_w/M_n values were observed to be less than 1.05. These results strongly indicate that the resulting block copolymers are the expected ones with well-defined architecture, as shown in Table 23.2.

Table 23.2 Synthesis of polystyrene-block-poly(4-(2-perfluorooctylethoxy)styrene)

$M_n \times 10^{-3}$		M_w/M_n	Fluorine content
Calcd	Obsd (PS-PFSt)		wt-%
22.2	21.3 (17.0–4.29)	1.02	11.2
12.8	12.5 (8.85–3.65)	1.02	16.3
22.8	23.2 (16.0–7.19)	1.03	17.3
30.5	29.4 (19.2–10.2)	1.04	19.3
16.1	17.0 (10.9–6.15)	1.03	20.1
17.5	16.4 (7.53–8.87)	1.05	30.1

23.3 Application as a Template to Nanocellular and Porous Structures Using Supercritical Carbon Dioxide

23.3.1 Supercritical Fluids

When a fluid is pressurized and heated above its critical point, it is called a supercritical fluid. Supercritical fluids have high diffusivity like a gas even in solid and dissolution capability of materials like a liquid. In addition, near the critical point, a small change in pressure or temperature results in large changes in density and other properties, such as solubility, allowing a supercritical fluid to be fine tuned. Supercritical fluids are considered to be a substitute for organic solvents in a wide range of industrial and laboratory processes. Among them,

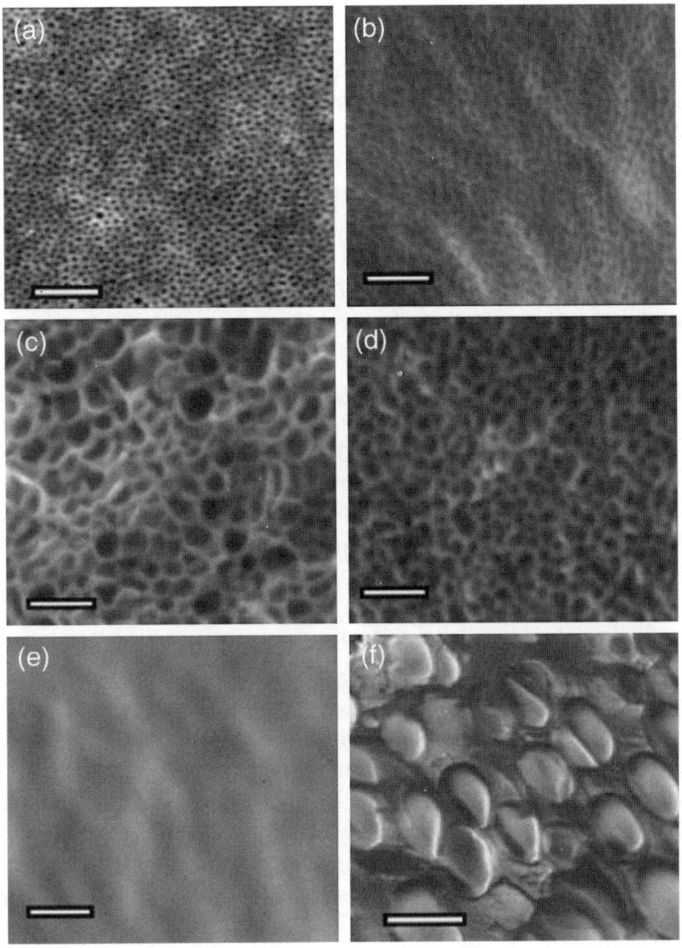

Figure 23.5 (a) An SFM topographic image of original PS-PFMA block copolymer; dark domains are spherical PFMA domains: the bar indicates 250 nm. (b–d) SEM images of the cryofractured surfaces of PS-PFMA processed at a CO_2 saturation pressure of 10 MPa (b) 20 MPa, (c) and (d) 30 MPa: the bar indicates 100 nm. (e) An SEM image of the cryofractured surface of as-cast PS-PFMA: the bar indicates 100 nm. (f) An SEM image of the cryofractured surface of PS-PFMA processed at a CO_2 saturation pressure of 20 MPa, followed by rapid depressurization at 60 °C: the bar indicates 2 μm. (Reproduced with permission from H. Yokoyama, L. Li, T. Nemoto and K. Sugiyama, "Tunable nanocellular polymeric monoliths using fluorinated block copolymer templates and supercritical carbon dioxide," *Advanced Materials*, 2004, **16**, 1542–1546. © Wiley-VCH Verlag GmbH & Co. KGaA.)

supercritical carbon dioxide is the most commonly used supercritical fluids. In particular, supercritical carbon dioxide is known to swell a wide range of polymers and can be used for a variety of applications. In the following sections we focus on the use of supercritical carbon dioxide with polymers.

23.3.2 *Supercritical Carbon Dioxide and Microcells*

Carbon dioxide in particular has been extensively studied due to its mild critical point (7.38 MPa and 304 K) and a low environmental impact. The particularly attractive properties of liquid and supercritical carbon dioxide for polymer processes are gas-like diffusivity even in solid polymers, continuously tunable solvent power and selectivity, and complete elimination at the end of the process (Cooper and Holmes, 1999; Cooper, 2003). An example of extensively studied processes using supercritical carbon dioxide and polymers is the introduction of "microcells," (Goel and Beckman, 1994a; Goel and Beckman, 1994b; Sun and Mark, 2002; Han *et al.*, 2003; Sun *et al.*, 2004) that are cells of a typical size of the order of micrometers and are frozen bubbles of carbon dioxide in polymers. The typical cell size ranges from being on the order of micrometers to millimeters, depending on the process condition, such as carbon dioxide pressure, depressurizing temperature, and the solubility of carbon dioxide in polymers (Sun and Mark, 2002; Sun *et al.*, 2004). Such microcells are much smaller than macrocells of typical polymer foams such as Styrofoam formed by using conventional blowing agents, so microcells have been expected to show improved mechanical properties related to their reduced size. In a conventional microcellular process with carbon dioxide in ordinary polymers, a polymer specimen is swollen and plasticized with carbon dioxide, and then rapidly depressurized to atmospheric pressure. Most polymeric materials do not dissolve in carbon dioxide; however, many hydrophobic polymers such as polyethylene, (Colton and Suh, 1987) polypropylene (Colton and Suh, 1987), polystyrene (Cooper, 2003), poly(ethyl methacrylate) (Goel and Beckman, 1994a), and poly(methyl methacrylate) (Goel and Beckman, 1994a; Goel and Beckman, 1994b) are significantly swollen with supercritical carbon dioxide and thus microcells can be introduced (Goel and Beckman, 1993). For a polymer having a high glass-transition temperature, the specimen must be heated to a temperature above its glass-transition temperature immediately after the rapid depressurization of carbon dioxide, but most conventional polymers are rapidly depressurized to foam due to their reduced glass-transition temperatures in supercritical carbon dioxide (Wang *et al.*, 1982; Condo and Johnston, 1994; Condo *et al.*, 1994). During rapid depressurization or heating, the solubility of the carbon dioxide remaining in the polymeric materials is reduced, and then the oversaturated carbon dioxide in the polymeric materials nucleates and grows as bubbles to form cellular structures. The minimum size of a nucleus of a carbon dioxide bubble in polymeric materials is thermodynamically limited to roughly 1 micrometer.

23.3.3 *Self-Assembly of Block Copolymers as a Template*

In order to overcome the lower limit of cell size, we need to employ artificial nuclei of cells with much smaller size. One of the possible strategies to employ small nuclei for carbon dioxide bubbles is the use of self-assembled block copolymer domains with a high affinity to carbon dioxide. We describe in this section how fluorine-containing block domains of block copolymers work as nuclei and how a nanometer-size cell (nanocell) can be introduced in each fluorine-containing block domain.

Block copolymers are a class of soft matter that self-assemble into nanoscopic periodic domain structures, as extensively described throughout this book (Bates and Fredrickson, 1990; Lazzari and Lopez-Quintela, 2003; Park *et al.*, 2003). The size of block copolymer domains typically ranges from 5 to 100 nm, primarily depending on the molecular size of the block copolymers.

The symmetry of the block fraction plays an important role in determining which morphology is selected by a particular block copolymer from a class of ordered morphologies, such as spheres, cylinders, gyroids and lamellae.

We first focus on fluorine-containing block copolymers that form spherical domains of a fluorine-containing block since they provide well-ordered and densely packed spherical nuclei for nanobubbles of carbon dioxide, which lead to closed nanocells. Such spherical domains of fluorine-containing blocks are expected to work as a template to provide nanocells, which are cells on a nanometer scale and much smaller than conventional microcells using homopolymers and supercritical carbon dioxide. In order for spherical domains to work as nuclei for nanobubbles, an affinity of fluorine-containing domains to carbon dioxide higher than that of the other block domains is essential.

23.3.4 Fabrication of Nanocells

Yokoyama *et al.* first reported a successful fabrication of cellular structures with a uniform diameter of from 10 to 30 nm and number density of 10^{16} cells/cm^3 in block copolymers with CO_2-philic fluorine-containing blocks using carbon dioxide (Li *et al.*, 2004; Yokoyama *et al.*, 2004a, 2004b; Yokoyama and Sugiyama, 2005; Li *et al.*, 2006; Yokoyama *et al.*, 2007). We first show an example of successful fabrication of nanocellular structures templated by the self-assembled fluorine-containing block copolymer domains (Yokoyama *et al.*, 2004a, 2004b). A polystyrene-*b*-poly(perfluorooctylethyl methacrylate) (PS-PFMA) diblock copolymer was synthesized via sequential anionic polymerization of styrene and perfluorooctylethyl methacrylate, as described earlier in this chapter. The PS-PFMA block copolymer, the PFMA block of which is soluble in supercritical carbon dioxide (O'Neill *et al.*, 1998), works as a template in this system for fabricating polymeric nanocellular monoliths. The specimen was pressurized with carbon dioxide in a pressure range of 7.5–30 MPa at 60 °C for 1 h to absorb carbon dioxide into the fluorine-containing domains. The temperature was then reduced to 0 °C in 10 min, while maintaining the carbon dioxide pressure. Reducing the temperature to 0 °C causes the swollen continuous PS domains to be glassy (Wang *et al.*, 1982) to prevent collapse of nanocells, otherwise the cells blow up too much upon depressurization and become microcells. SEM images of the cryofractured surfaces of the nanocellular PS-PFMA monoliths after the CO_2 process at 10, 20, and 30 MPa are shown in Figures 23.5b–d, respectively. The cryofractured surface of the as-cast PS-PFMA monolith is shown in Figure 23.5e, in which neither domains nor cells were found. In the case of the PS-PFMA specimen depressurized at a higher temperature (60 °C), we found microcells instead of nanocells, as shown in Figure 23.5f. It appears that the temperature at which the specimen is depressurized is critical for nanocellular formation.

The average cell diameters, estimated by image analyses of the SEM pictures of the cryofractured surfaces, as well as the domain spacing obtained by small-angle X-ray scattering (SAXS), are plotted in Figure 23.6a as a function of CO_2 saturation pressure, P, with bars denoting the standard deviations of the cell diameters. Both the average cell diameter and the domain spacing increase with increasing saturation pressure when $P < 20$ MPa, and decrease with increasing saturation pressure when $P < 20$ MPa. As the CO_2 pressure increases to 20 MPa or higher, hydrostatic pressure effects dominate and the

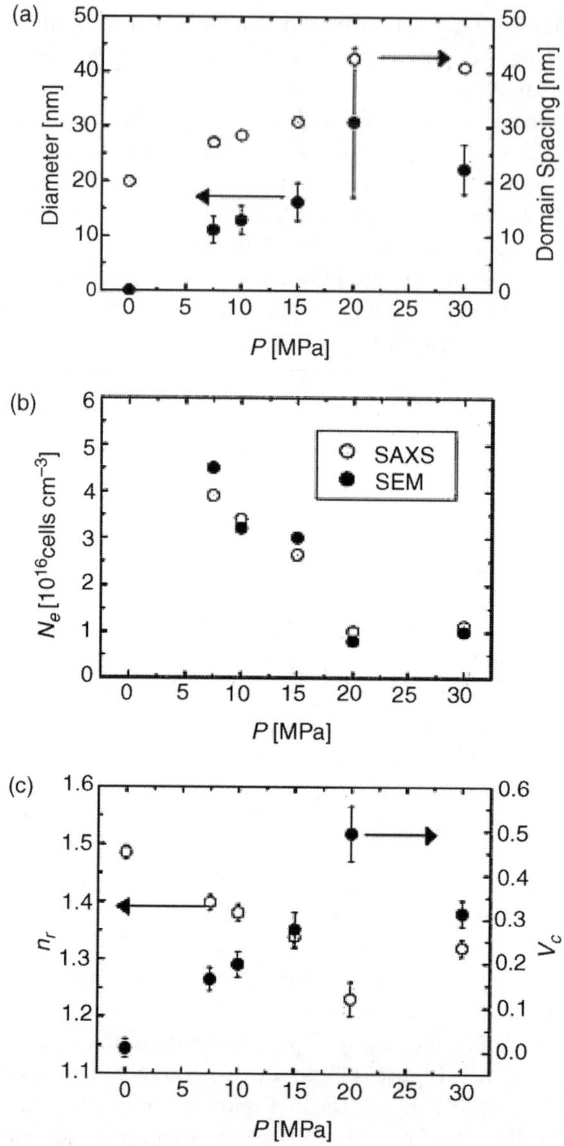

Figure 23.6 Dependence of the nanocellular structure on CO_2 saturation pressure, *P*. (a) The average diameter of the cells (solid circles) estimated from the SEM images and the domain spacing, $D\ (= 2\pi/q_{max})$, obtained by SAXS (open circles) plotted against CO_2 saturation pressure. The bars on the diameters denote the standard deviations of the cell size distribution. (b) The number densities of the cells estimated from the SEM images (solid circles) and from the domain spacing, *D*, assuming closest packing (open circles). (c) The dependence on CO_2 saturation pressure of the refractive indices (open circles) and the volume fractions of the cells (solid circles). Plots were reproduced with permission from H. Yokoyama, L. Li, T. Nemoto and K. Sugiyama, "Tunable nanocellular polymeric monoliths using fluorinated block copolymer templates and supercritical carbon dioxide," *Advanced Materials*, 2004, **16**, 1542–1546. © Wiley-VCH Verlag GmbH & Co. KGaA.

pressure dependence changes (Wang *et al.*, 1982). A similar optimum pressure has been reported for pore sizes of crosslinked polymeric monoliths (Hebb *et al.*, 2003), swelling poly(methyl methacrylate) networks (Goel and Beckman, 1993), and binodal temperatures of polystyrene/polyisoprene blends (Walker *et al.*, 1999).

The number of cells per unit volume, N_c, was estimated using the following relation,

$$N_c = N_{SEM}{}^{3/2}$$

where N_{SEM} is the number of cells per unit area in the SEM image. Another N_c was determined by small-angle X-ray scattering and a model assuming closest packing of cells. The N_c values were in the range of $1\text{–}5 \times 10^{16}$ cells/cm^3, and two different methods give approximately the same number. This result suggests that the nanocells are closely packed. The refractive index was measured and plotted in Figure 23.6c. The varying refractive index by introducing nanoscopic cells is a proof of the introduction of voids in the material with a size much smaller than the wavelength of visible light. The refractive index can be converted to a volume fraction of the void, or porosity, in nanocellular films using the Lorentz–Lorenz formula.

The refractive index, reduced down to 1.23, corresponds to 50 vol% of the empty nanocells. It is surprising that this method can provide such a high porosity since adding too much carbon dioxide to PFMA domains may induce order-to-order transition to cylinders or even to lamellae. At this point, the reason is not yet clearly understood and further study is necessary. A schematic picture of the nanocellular structure fabricated by this method is shown in Figure 23.7. The empty voids are surrounded by CO_2-philic PFMA domains with thin shells in the PS matrix. A direct proof of such a structure was later given by grazing-incidence small-angle X-ray scattering study of nanocellular thin films (Yokoyama *et al.*, 2007), in which a core–shell model of empty voids surrounded by a high electron density shell of PFMA can successfully fit the experimental scattering

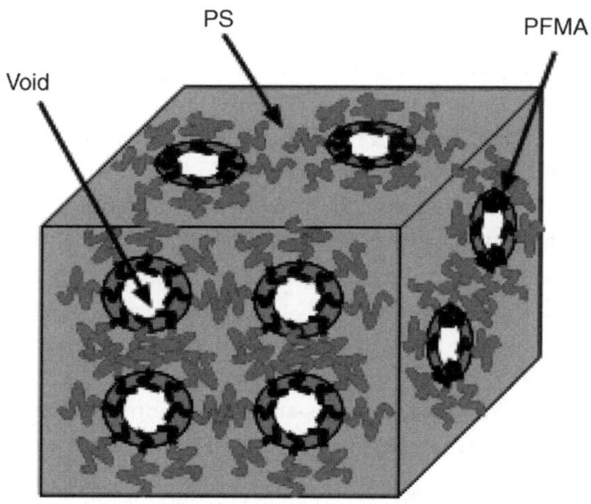

Figure 23.7 A schematic picture of the expected nanocellular structure. The domains (dark gray regions) of the PFMA block (black curves) surround the cells (white regions) while the PS continuous domain (light gray region) of the PS block (gray curves) supports the nanocellular monoliths.

function, although such a structure is expected from the mechanism of nanocellular formation using carbon dioxide.

23.3.5 Crossover from Nanocells to Microcells

A typical conventional rapid expansion at 60 °C (above the glass-transition temperature, T_g, of PS in supercritical carbon dioxide) introduced microcells, as already shown in Figure 23.5f. Thermal quenching to 0 °C ensures that the CO_2-swollen structures are fixed by the glass transition of continuous PS domains swollen with CO_2 which occurs at approximately 30 °C (Wang *et al.*, 1982; Condo *et al.*, 1994) and is well below the glass-transition temperature of PS in air. PFMA has the T_g of 77 °C, which is lower than that of PS, and is plasticized even more severely in CO_2 due to its higher CO_2-philicity. Therefore, in the presence of CO_2, the T_g of PS is higher than that of PFMA and is the determining factor for the mobility of the block copolymer chains and for the development of nanocells. By varying the depressurization temperature, crossover behavior from nanocells to microcells was found in a similar block copolymer with fluorine-containing blocks (Yokoyama and Sugiyama, 2005). Monoliths of poly[styrene-block-4-(perfluorooctylpropyloxy)styrene] (PS-PFS) with a weight-average molecular weight, M_w, of 94 000 g mol^{-1} and PFS volume fraction, f_{PFS}, of 0.23 and polystyrene (PS) with M_w of 99 000 g mol^{-1} are pressurized with carbon dioxide of 20 MPa at 80 °C for 1 h and depressurized at different temperatures. Reducing the depressurization temperature, T_d, to 0 °C followed by slow depressurization proved to be the essential step for the successful fabrication of nanocellular structures, as shown in the previous section. Reducing T_d to 0 °C prevents the growth of cells filled with CO_2 even with fairly slow depressurization, that is, 0.5 MPa/min. It is essentially a nucleation-only process with the help of fluorine-containing nanodomains, as opposed to the conventional nucleation and growth mechanism. PS continuous domains become glassy at T_d of 0 °C, and the size of the cells is successfully controlled by the saturation pressure of CO_2 within the range of 10–30 nm. The same nanocellular structures can also be found in PS-PFS, irrespective of the different backbone chemical structures and higher molecular weights. T_d was varied from 0 up to 40 °C, while maintaining the other controlling parameters. An SEM image of PS-PFS quenched to 25 °C followed by depressurization at 0.5 MPa/min is shown in Figure 23.8. The size of the cells depressurized at 25 °C is significantly larger than that at 0 °C, as shown later, but the cells are closed with a much broader size distribution. The walls separating the cells apparently become thin relative to the cell size compared to those depressurized at 0 °C. As T_d is further increased to 32 °C, larger cells (<500 nm, microcells) surrounded by the smaller cells (<100 nm, nanocells) are found in Figure 23.9; the size distribution apparently becomes bimodal. At a CO_2 pressure of 30 MPa, a T_d of 32 °C is close to the reduced glass-transition temperature of PS in CO_2. The reduced glass-transition temperature of PS is approximately 40 °C and almost constant in the pressure range of 10 to 30 MPa. Therefore, the larger microcells in Figure 23.9 may be formed by the mechanism of microcellular formation widely observed in CO_2-swollen homopolymers depressurized at around or above their glass-transition temperatures. When T_d is further increased to 40 °C, which is the proximity of the reduced glass-transition temperature of PS in the presence of CO_2 at 30 MPa, microcells begin to dominate (Yokoyama and Sugiyama, 2005). Some of the cells reach

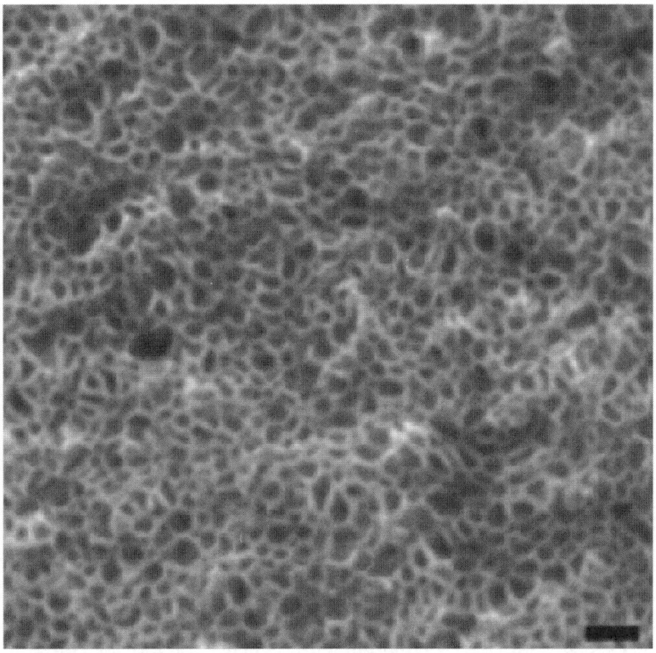

Figure 23.8 An SEM image of a microcellular PS-PFS monolith processed at 60 °C and 30 MPa and depressurized at 25 °C at a rate of 0.5 MPa/min. The bar indicates 200 nm. (Reprinted with permission from H. Yokoyama and K. Sugiyama, "Nanocellular structures in block copolymers with CO_2-philic blocks using CO_2 as a blowing agent: Crossover from nicro- to nanocellular structures with depressurization temperature," *Macromolecules*, **38**, 10516–10522, 2005. © 2005 American Chemical Society.)

1 μm in diameter; however, we still find nanocells surrounding microcells. The nano- and microcellular formations appear to be independent and to be based on completely different mechanisms.

23.3.6 Nonspherical Nanoporous Structures

In the previous section, CO_2 as a porogen was introduced into spherical fluorine-containing domains to create nanometer cells. However, block copolymers form a variety of morphologies other than the spherical domain structure that have been successfully used as templates for nanocellular structures. Nonspherical domains of fluorine-containing blocks were employed as templates for nonspherical nanoporous structures (Yokoyama *et al.*, 2008). Two batches of polystyrene-*b*-poly(perfluorooctylethyl methacrylate) (PS-PFMA) were sysnthesized by sequential living anionic polymerization. The molecular weights of PS-PFMAs are 20 000–13 000 and 10 000–7670 g mol^{-1}. Those PS-PFMAs were mixed to adjust the volume fraction of PFMA to 33 vol%. To prepare thin-film specimens, α,α,α-trifluorotoluene (TFT) and hexafluorobenzene (HFB) were used as spin-casting solvents. TFT and HFB have different selectivity for PS and PFMA and provide two different initial morphologies. Films with two different initial morphologies were pressurized with CO_2 at 8 MPa and 60 °C for 30 min, and then were thermally quenched to 0 °C, similar to the

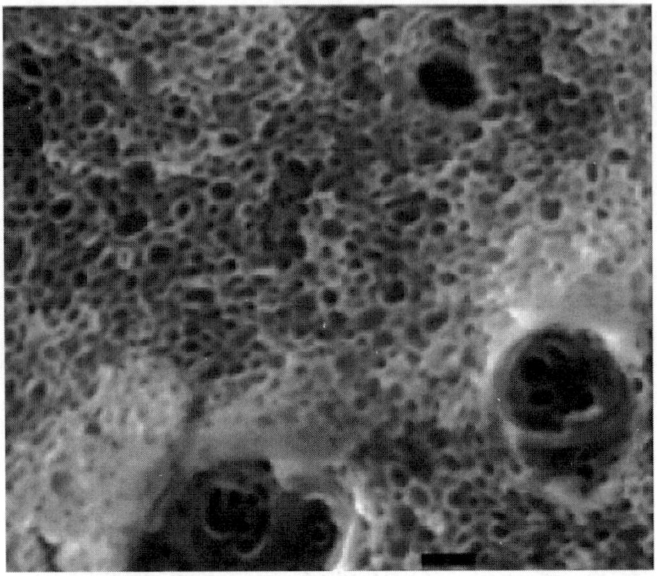

Figure 23.9 An SEM image of a nano- and microcellular PS-PFS monolith processed at 60 °C and 30 MPa and depressurized at 32 °C at a rate of 0.5 MPa/min with scale bars of 200 nm. (Reprinted with permission from H. Yokoyama and K. Sugiyama, "Nanocellular structures in block copolymers with CO_2-philic blocks using CO_2 as a blowing agent: Crossover from nicro- to nanocellular structures with depressurization temperature," *Macromolecules*, **38**, 10516–10522, 2005. © 2005 American Chemical Society.)

nanocellular process for spherical domains. Again, this thermal quenching ensures that the CO_2-swollen structures are fixed by the glass transition of PS swollen with CO_2. Finally, CO_2 pressure was released at a rate of 0.5 MPa/min. The films were removed from the high-pressure vessel and the surfaces of the films were scratched with a razor blade at the liquid nitrogen temperature to be observed by cross-sectional SEM (Figure 23.10). The film spun-cast from TFT, which shows network-like domains without any preferential orientation before the process, was converted by the CO_2 process to parallel nanosheets horizontally stacked on the substrates as shown in Figure 23.10a (Yokoyama *et al.*, 2008). Such a unique structure was also confirmed by a grazing-incidence small-angle X-ray scattering (GISAXS) study, and the spacing of nanosheets was consistently obtained to be 20.9 ± 0.4 nm (Yokoyama *et al.*, 2008). It should be noted that this structure is not a simple densely packed lamellar structure but separated block copolymer bilayer sheets, because SEM clearly shows the layered structures despite the fact that SEM is not capable of distinguishing between PS and PFMA domains due to their poor electronic contrast. Upon depressurization, the swollen lamellae would have collapsed to dense lamellae if the swollen lamellae were defect free. However, the existence of defects bridging the lamellae apparently prevents the lamellae from collapsing completely. Such defects supporting the nanosheets are occasionally found in Figure 23.10. Since the preferential segregation of PFMA domains to the surface is even more enhanced in CO_2 (Yokoyama and Sugiyama, 2004) than that in air or in a vacuum, the CO_2 interface induces parallel orientation of the swollen lamellae. Therefore, after freezing and depressurization, parallel orientation of stacked nanosheets is preferentially formed.

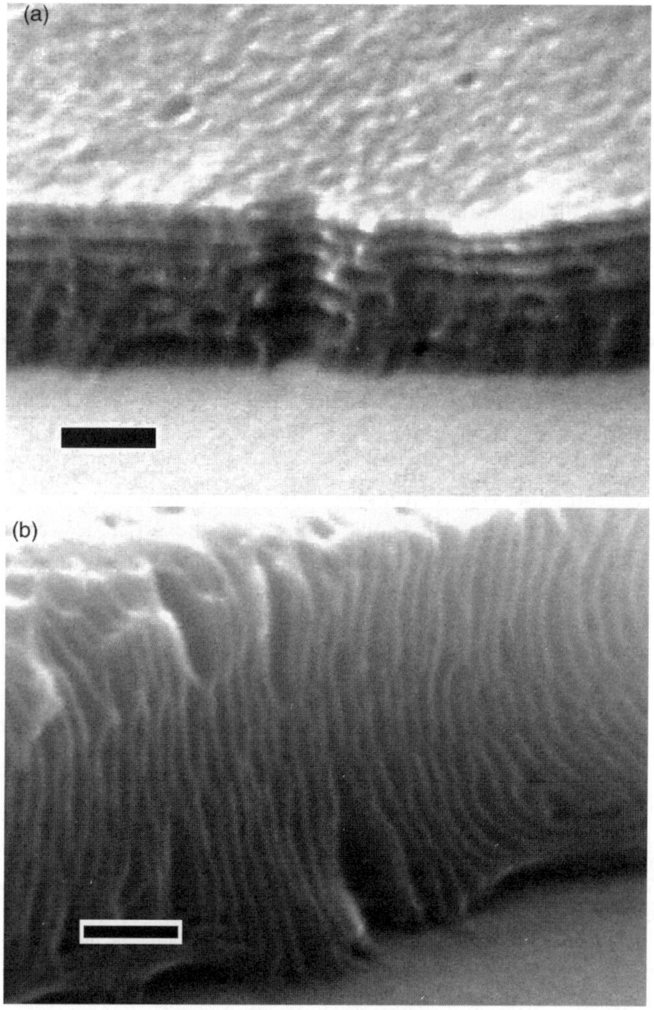

Figure 23.10 SEM images of nanosheet structures that appeared after the CO_2 process. Depending on the initial morphology, the films of the same block copolymers show different nanosheet structures. Cross-sectional views of nanosheet thin films cast from (a) TFT and (b) HFB after the CO_2 process at a saturation pressure of 8 MPa followed by scratching at liquid nitrogen temperature. The bars indicate 100 nm. (Reprinted with permission from H. Yokoyama, L. Li, C. Dutriez *et al.*, "Horizontally and vertically aligned polymeric nanosheets: CO_2-induced morphological changes of block copolymer thin films," *Macromolecules*, **41**, 8626–8631, 2008. © 2008 American Chemical Society.)

On the other hand, casting from HFB induces poorly but perpendicularly oriented PS cylinders to the surface in the PFMA matrix (Yokoyama *et al.*, 2008). Such films were processed with CO_2 in the same manner and perpendicularly standing nanosheets on substrates were found as shown in Figure 23.10b. Such perpendicularly standing nanosheets were also confirmed by GISAXS experiments.

As already discussed, the segregation of PFMA induces parallel orientation of the nanosheets; however, the vertical orientation of the PS cylinders implies orientation of the swollen bilayer nanosheets perpendicular to the surface. Although this structure may change to a horizontal orientation, which is the probable equilibrium structure, we did not observe such a morphological change in our experimental conditions.

The appearance of nanosheet structures after the CO_2 process is equivalent to the lamellar morphology in CO_2, which is different from either the as-cast morphology or that expected from the block fraction. It is reasonable to conclude that the selective swelling of the minority PFMA domains with CO_2 increases the effective volume fraction of PFMA domains and induces order-to-order transition.

By using fluorine-containing block copolymers with a high fraction of fluorine-containing block domains, nonspherical open pores can be introduced. It was also shown that block copolymer morphologies presented an order-to-order transition induced by the swelling of fluorine-containing block domains with carbon dioxide. A combination of fluorine-containing block copolymers and carbon dioxide potentially provide a wide variety of morphologies that lead to nanocellular and nanoporous structures. Such nanocellular and nanoporous structures may be used for a wide range of applications, such as low dielectric, low refractive index, low heat-conducting and separation materials.

References

Andruzzi, L., Chiellini, E., Galli, G. *et al.* (2002) Engineering low surface energy polymers through molecular design: Synthetic routes to fluorinated polystyrene-based block copolymers. *Journal of Materials Chemistry*, **12**, 1684–1692.

Arnold, M.E., Nagai, K., Spontak, R.J. *et al.* (2002) Microphase separated block copolymers comprising low surface energy fluorinated blocks and hydrophilic blocks: synthesis and characterization. *Macromolecules*, **35**, 3697–3707.

Bates, F.S. and Fredrickson, G.H. (1990) Block copolymer thermodynamics - theory and experiment. *Annual Review of Physical Chemistry*, **41**, 525–557.

Bertolucci, M., Galli, G., Chiellini, E., and Wynne, K.J. (2004) Wetting behavior of films of new fluorinated styrene-siloxane block copolymers. *Macromolecules*, **37**, 3666–3672.

Busse, K., Kressler, J., van Eck, D., and Höring, S. (2002) Synthesis of amphiphilic block copolymers based on tert-butyl methacrylate and 2-(N-methylperfluorobutanesulfonamide)ethyl methacrylate and its behavior in water. *Macromolecules*, **35**, 178–184.

Choi, W.O., Sawamoto, M., and Higashimura, T. (1988) Living cationic homo- and copolymerization of vinyl ethers bearing a perfluoroalkyl pendant. *Polymer Journal*, **20**, 201–206.

Colton, J.S. and Suh, N.P. (1987) The nucleation of microcellular thermoplastic foam with additives.1. Theoretical considerations. *Polymer Engineering and Science*, **27**, 485–492.

Condo, P.D. and Johnston, K.P. (1994) In-situ measurement of the glass-transition temperature of polymers with compressed fluid diluents. *Journal of Polymer Science Part B-Polymer Physics*, **32**, 523–533.

Condo, P.D., Paul, D.R., and Johnston, K.P. (1994) Glass transitions of polymers with compressed fluid diluents - type-ii and type-iii behavior. *Macromolecules*, **27**, 365–371.

Cooper, A.I. (2003) Porous materials and supercritical fluids. *Advanced Materials*, **15**, 1049–1059.

Cooper, A.I. and Holmes, A.B. (1999) Synthesis of molded monolithic porous polymers using supercritical carbon dioxide as the porogenic solvent. *Advanced Materials*, **11**, 1270–.

Galli, G., Martinelli, E., Chiellini, E. *et al.* (2005) Low surface energy characteristics of mesophase-forming ABC and ACB triblock copolymers with fluorinated B. blocks. *Molecular Crystals and Liquid Crystals*, **441**, 211–226.

Genzer, J., Sivaniah, E., Kramer, E.J. *et al.* (2000a) Temperature dependence of molecular orientation on the surfaces of semifluorinated polymer thin films. *Langmuir*, **16**, 1993–1997.

Genzer, J., Sivaniah, E., Kramer, E.J. *et al.* (2000b) The orientation of semifluorinated alkanes attached to polymers at the surface of polymer films. *Macromolecules*, **33**, 1882–1887.

Genzer, J., Sivaniah, E., Kramer, E.J. *et al.* (2000c) Molecular orientation of single and two armed monodendron semifluorinated chains on "soft" and "hard" surfaces studied using NEXAFS. *Macromolecules*, **33**, 6068–6077.

Giannetti, E. (2001) Semi-crystalline fluorinated polymers. *Polymer International*, **50**, 10–26.

Gopalan, P., Andruzzi, L., Li, X., and Ober, C.K. (2002) Fluorinated mesogen-jacketed liquid-crystalline polymers as surface-modifying agents: Design, synthesis, and characterization. *Macromolecular Chemistry and Physics*, **203**, 1573–1583.

Guan, Z. and DeSimone, J.M. (1994) Fluorocarbon-based heterophase polymeric materials. 1. Block copolymer surfactants for carbon dioxide applications. *Macromolecules*, **27**, 5527–5532.

Goel, S.K. and Beckman, E.J. (1993) Plasticization of poly(methyl methacrylate) (pmma) networks by supercritical carbon-dioxide. *Polymer*, **34**, 1410–1417.

Goel, S.K. and Beckman, E.J. (1994a) Generation of microcellular polymeric foams using supercritical carbon-dioxide.1. Effect of pressure and temperature on nucleation. *Polymer Engineering and Science*, **34**, 1137–1147.

Goel, S.K. and Beckman, E.J. (1994b) Generation of microcellular polymeric foams using supercritical carbon-dioxide.2. Cell-growth and skin formation. *Polymer Engineering and Science*, **34**, 1148–1156.

Han, X.M., Koelling, K.W., Tomasko, D.L., and Lee, L.J. (2003) Effect of die temperature on the morphology of microcellular foams. *Polymer Engineering and Science*, **43**, 1206–1220.

Hansen, N.M.L., Jankova, K., and Hvilsted, S. (2007) Fluoropolymer materials and architectures prepared by controlled radical polymerizations. *European Polymer Journal*, **43**, 255–293.

Hayakawa, T., Wang, J., Xiang, M. *et al.* (2000) Effect of changing molecular end groups on surface properties: synthesis and characterization of poly(styrene-*b*-semifluorinated isoprene) block copolymers with -CF_2H end groups. *Macromolecules*, **33**, 8012–8019.

Hebb, A.K., Senoo, K., Bhat, R., and Cooper, A.I. (2003) Structural control in porous cross-linked poly(methacrylate) monoliths using supercritical carbon dioxide as a "Pressure-adjustable" Porogenic solvent. *Chemistry of Materials*, **15**, 2061–2069.

Hexemer, A., Sivaniah, E., Kramer, E.J. *et al.* (2004) Managing polymer surface structure using surface active block copolymer mixtures. *Journal of Polymer Science Part B-Polymer Physics*, **42**, 411–420.

Hikita, M., Tanaka, K., Nakamura, T. *et al.* (2004) Aggregation states and surface wettability in films of poly(styrene-block-perfluorooctyl ethyl acrylate) diblock copolymers synthesized by atom transfer radical polymerization. *Langmuir*, **20**, 5304–5310.

Hillmyer, M.A. and Lodge, T.P. (2002) Synthesis and self-assembly of fluorinated block copolymers. *Journal of Polymer Science*, **40**, 1–8.

Hirao, A., Sakai, S., and Sugiyama, K. (2004) Synthesis of well-defined chain-end-functionalized polystyrenes with four, eight, and sixteen perfluorooctyl groups and their surface characterization. *Polymers for Advanced Technologies*, **15**, 15–25.

Hussain, H., Budde, H., Höring, S. *et al.* (2002) Synthesis and characterization of poly(ethylene oxide) and poly (perfluorohexylethyl methacrylate) containing triblock copolymers. *Macromolecular Chemistry and Physics*, **203**, 2103–2112.

Hwang, S.S., Ober, C.K., Perutz, S. *et al.* (1995) Block copolymers with low surface energy segments: siloxane- and perfluoroalkane-modified blocks. *Polymer*, **36**, 1321–1325.

Imae, T. (2003) Fluorinated polymers. *Current Opinion in Colloid and Interface Science*, **8**, 307–314.

Ishizone, T., Sugiyama, K., Sakano, Y. *et al.* (1999) Anionic polymerizations of perfluoroalkyl methacrylates and synthesis of well-defined ABC triblock copolymers of methacrylates containing hydrophilic, hydrophobic, and perfluoroalkyl groups. *Polymer Journal*, **31**, 983–988.

Iyengar, D.R., Perutz, S.M., Dai, C.-A. *et al.* (1996) Surface segregation studies of fluorine-containing diblock copolymers. *Macromolecules*, **29**, 1229–1234.

Jankova, K. and Hvilsted, S. (2005) Novel fluorinated block copolymer architectures fuelled by atom transfer radical polymerization. *Journal of Fluorine Chemistry*, **126**, 241–250.

Jeyaprakash, J.D., Samuel, S., Dhamodharan, R., and Ober, C.K. (2000) A solvent-free method for the synthesis of block copolymers with fluorinated pendant groups by a hydrosilylation reaction. *Journal of Polymer Science Part A-Polymer Chemistry*, **38**, 1179–1183.

Kassis, C.M., Steehler, J.K., Betts, D.E. *et al.* (1996) XPS studies of fluorinated acrylate polymers and block copolymers with polystyrene. *Macromolecules*, **29**, 3247–3254.

Kim, D.-K., Lee, S.-B., Doh, K.-S., and Nam, Y.-W. (1999) Synthesis of block copolymers having perfluoroalkyl and silicone-containing side chains using diazo macroinitiator and their surface properties. *Journal of Applied Polymer Science*, **74**, 1917–1926.

Krishnan, S., Kwark, Y.-J., and Ober, C.K. (2004) Fluorinated polymers: Liquid crystalline properties and applications in lithography. *Chemical Record*, **4**, 315–330.

Krupers, M.J. and Möller, M. (1997a) Synthesis and characterization of semifluorinated polymers via group transfer polymerization *J. Journal of Fluorine Chemistry*, **82**, 119–124.

Krupers, M.J. and Möller, M. (1997b) Semifluorinated diblock copolymers. Synthesis, characterization, and amphiphilic properties. *Macromolecular Chemistry and Physics*, **198**, 2163–2179.

Li, K., Wu, P., and Han, Z. (2002a) Preparation and surface properties of fluorine-containing diblock copolymers. *Polymer*, **43**, 4079–4086.

Li, L., Yokoyama, H., Nemoto, T., and Sugiyama, K. (2004) Facile fabrication of nanocellular block copolymer thin films using supercritical carbon dioxide. *Advanced Materials*, **16**, 1226–1229.

Li, L., Nemoto, T., Sugiyama, K., and Yokoyama, H. (2006) CO_2 foaming in thin film of block copolymer containing fluorinated blocks. *Macromolecules*, **39**, 4746–4755.

Li, X., Andruzzi, L., Chiellini, E. *et al.* (2002b) Semifluorinated aromatic side-group polystyrene-based block copolymers: Bulk structure and surface orientation studies. *Macromolecules*, **35**, 8078–8087.

Lim, K.T., Lee, M.Y., Moon, M.J. *et al.* (2002) Synthesis and properties of semifluorinated block copolymers containing poly(ethylene oxide) and poly(fluorooctyl methacrylate) via atom transfer radical polymerization. *Polymer*, **43**, 7043–7049.

Lazzari, M. and Lopez-Quintela, M.A. (2003) Block copolymers as a tool for nanomaterial fabrication. *Advanced Materials*, **15**, 1583–1594.

Matsumoto, K., Kubota, M., Matsuoka, H., and Yamaoka, H. (1999) Water-soluble fluorine-containing amphiphilic block copolymer: Synthesis and aggregation behavior in aqueous solution. *Macromolecules*, **32**, 7122–7127.

Matsumoto, K., Mazaki, H., Nishimura, R. *et al.* (2000) Perfluoroalkyl-philic character of poly(2-hydroxyethyl vinyl ether)-block-poly[2-(2,2, 2-trifluoroethoxy)ethyl vinyl ether] micelles in water: Selective solubilization of perfluorinated compounds. *Macromolecules*, **33**, 8295–8300.

O'Neill, M.L., Cao, Q., Fang, R. *et al.* (1998) Solubility of homopolymers and copolymers in carbon dioxide. *Industrial & Engineering Chemistry Research*, **37**, 3067–3079.

Park, C., Yoon, J., and Thomas, E.L. (2003) Enabling nanotechnology with self assembled block copolymer patterns. *Polymer*, **44**, 6725–6760.

Percec, V. and Lee, M. (1992) Molecular engineering of liquid crystal polymers by living polymerization XXIII: Synthesis and characterization of AB block copolymers based on ω-[(4-cyano-4′-biphenyl)-oxy]alkyl vinyl ether, *1H, 1H, 2H, 2H*-perfluorodecyl vinyl ether, and 2-(4-biphenyloxy)ethyl vinyl ether with *1H, 1H, 2H, 2H*-perfluorodecyl vinyl ether. *J M S – Pure and Applied Chemistry*, **A29**, 723–740.

Radhakrishnan, K., Switek, K.A., and Hillmyer, M.A. (2004) Synthesis of semifluorinated block copolymers by atom transfer radical polymerization. *Journal of Polymer Science Part A-Polymer Chemistry*, **42**, 853–861.

Reisinger, J.J. and Hillmyer, M.A. (2002) Synthesis of fluorinated polymers by chemical modification. *Progress in Polymer Science*, **27**, 971–1005.

Ren, Y., Lodge, T.P., and Hillmyer, M.A. (2002) Effect of selective perfluoroalkylation on the segregation strength of polystyrene - 1, 2-polybutadiene block copolymers. *Macromolecules*, **35**, 3889–3894.

Shemper, B.S. and Mathias, L.J. (2004) Synthesis and characterization of statistical and block fluorinated copolymers with linear and star-like architectures via ATRP. *European Polymer Journal*, **40**, 651–665.

Sugiyama, K., Nemoto, T., and Hirao, A. (2002) Synthesis and surface characterization of novel perfluorooctyl-functionalized polymers with well-defined architectures. *Macromolecular Symposia*, **181**, 135–153.

Sun, H.L. and Mark, E.J. (2002) Preparation, characterization, and mechanical properties of some microcellular polysulfone foams. *Journal of Applied Polymer Science*, **86**, 1692–1701.

Sun, X.H., Liu, H.J., Li, G. *et al.* (2004) Investigation on the cell nucleation and cell growth in microcellular foaming by means of temperature quenching. *Journal of Applied Polymer Science*, **93**, 163–171.

Tanaka, Y., Hasegawa, H., Hashimoto, T. *et al.* (1999) A study of three-phase structures in ABC triblock copolymers. *Polymer Journal*, **31**, 989–994.

Vandooren, C., Jérôme, R., and Teyssié, Ph. (1994) Living cationic polymerization of *1H, 1H, 2H,* 2H perfluorooctyl vinyl ether. *Polymer Bulletin*, **32**, 387–393.

Walker, T.A., Raghavan, S.R., Royer, J.R. *et al.* (1999) Enhanced miscibility of low-molecular-weight polystyrene polyisoprene blends in supercritical co2. *Journal of Physical Chemistry B*, **103**, 5472–5476.

Wang, J., Mao, G., Ober, C.K., and Kramer, E.J. (1997) Liquid crystalline, semifluorinated side group block copolymers with stable low energy surfaces: synthesis, liquid crystalline structure, and critical surface tension. *Macromolecules*, **30**, 1906–1914.

Wang, W.C.V., Kramer, E.J., and Sachse, W.H. (1982) Effects of high-pressure CO_2 on the glass-transition temperature and mechanical-properties of polystyrene. *Journal of Polymer Science Part B-Polymer Physics*, **20**, 1371–1384.

Xiang, M., Li, X., Ober, C.K. *et al.* (2000) Surface stability in liquid-crystalline block copolymers with semifluorinated monodendron side groups. *Macromolecules*, **33**, 6106–6119.

Yang, S., Wang, J., Ogino, K. *et al.* (2000) Low-surface-energy fluoromethacrylate block copolymers with patternable elements. *Chemistry of Materials*, **12**, 33–40.

Yokoyama, H. and Sugiyama, K. (2004) Surface Hydrophobicity of fluorinated block copolymers enhanced by supercritical carbon dioxide annealing. *Langmuir*, **20**, 10001–10006.

Yokoyama, H. and Sugiyama, K. (2005) Nanocellular structures in block copolymers with CO_2-philic blocks using CO_2 as a blowing agent: Crossover from micro- to nanocellular structures with depressurization temperature. *Macromolecules*, **38**, 10516–10522.

Yokoyama, H., Tanaka, K., Takahara, A. *et al.* (2004a) Surface structure of asymmetric fluorinated block copolymers. *Macromolecules*, **37**, 939–945.

Yokoyama, H., Dutriez, C., Li, L. *et al.* (2007) Grazing incident small angle x-ray scattering study of polymer thin films with embedded ordered nanometer cells. *Journal of Chemical Physics*, **127**, 014904.

Yokoyama, H., Li, L., Dutriez, C. *et al.* (2008) Horizontally and vertically aligned polymeric nanosheets: Co2-induced morphological changes of block copolymer thin films. *Macromolecules*, **41**, 8626–8631.

Yokoyama, H., Li, L., Nemoto, T., and Sugiyama, K. (2004b) Tunable nanocellular polymeric monoliths using fluorinated block copolymer templates and supercritical carbon dioxide. *Advanced Materials*, **16**, 1542–1546.

Yong, T.-M., Hems, W.P., van Nunen, J.L.M. *et al.* (1997) Synthesis of fluorinated block copolymers and their application as novel polymerization surfactants in supercritical carbon dioxide. *Chemical Communications*, 1811–1812.

You, F., Paik, M.Y., Häckel, M. *et al.* (2006) Controlled and suppression of surface relief grating in liquid-crystalline perfluoroalkyl-azobenzene polymers. *Advanced Functional Materials*, **16**, 1577–1581.

Zhang, Z.B., Ying, S.K., and Shi, Z.Q. (1999) Synthesis of fluorine-containing block copolymers via ATRP 2. Synthesis and characterization of semifluorinated di- and triblock copolymers. *Polymer*, **40**, 5439–5444.

24

Architectural Polymers, Nanostructures, and Hierarchical Structures from Block Copolymers

Ian Wyman and Guojun Liu
Department of Chemistry, Queen's University, 90 Bader Lane, Kingston, Ontario, K7L 3N6, Canada

24.1 Introduction

Structures govern function. This applies both to the macroscopic world, as well as to the molecular world. Because of this, building architectures from molecules with controlled shape and dimension has been a long passion for chemists.

Over the past 18 years, our group has been systematically advancing and developing methodologies for the preparation of architectures from block copolymers. These architectures have included architectural polymers, nanostructures, and hierarchical structures. Reviewed in this chapter are our major contributions to this field. This chapter is not meant to be a comprehensive review of the field, but instead will focus primarily on our major contributions.

Architectural polymers (Hirao *et al.*, 2005; Hadjichristidis *et al.*, 2006) refer, in this chapter, to polymers that have finite molecular weights and are architecturally more complex than linear chains. They can be cyclic, star-shaped, combed, balloon-shaped, and H-shaped. Illustrated in Figure 24.1 are the structures of cyclic, balloon-shaped, and H-shaped polymers, as well as of a miktoarm star ABC triblock copolymer.

A nanostructure should possess a regular geometric shape, which can be spherical, cylindrical, vesicular, tubular, and so forth. Also, the smallest dimension, such as the diameter of a cylinder, within a nanostructure should be less than 100 nm. A hierarchical structure is typically larger than a nanostructure. More importantly, hierarchical structures are formed through the multitier assembly of block copolymers, and have structural order at several length

Complex Macromolecular Architectures: Synthesis, Characterization, and Self-Assembly, First Edition.
Edited by Nikos Hadjichristidis, Akira Hirao, Yasuyuki Tezuka and Filip Du Prez.

Figure 24.1 Structures for cyclic, balloon-shaped, and H-shaped polymers as well as a miktoarm star ABC triblock copolymer.

scales. For example, micelles, which are typically nanostructures, can be prepared through the Tier I assembly of a diblock copolymer in a block-selective solvent. A hierarchical structure is formed via the further assembly of the micelles or the Tier II assembly of a diblock copolymer (Yan *et al.*, 2004; Hu *et al.*, 2008a).

24.2 Block Copolymer Self-Assembly

The basis for the various methodologies that we have developed for the preparation of architectures from block copolymers is self-assembly. The self-assembly process occurs in either the solid state or block-selective solvents.

In the absence of specific interactions such as hydrogen bonding or electrostatic attractions, most polymers at sufficiently high molecular weights do not undergo mixing due to the minimal entropy gain associated with mixing polymers. Likewise, the different blocks of a coil–coil diblock copolymer do not undergo mixing either. In the bulk phase under equilibrium conditions, the minor block is segregated from the major block of a diblock copolymer, forming regularly shaped and uniformly spaced nanodomains. The shapes of the segregated domains in a diblock copolymer are governed mostly by the volume fraction of the minority block, *f*, and by block incompatibility. Figure 24.2 shows the equilibrium morphologies documented for coil–coil diblock copolymers (Khandpur *et al.*, 1995; Bates and Fredrickson, 1999). At a volume fraction of $< \sim 25\%$, the minority block forms a body-centered cubic spherical phase in the matrix of the major block. Meanwhile, the morphology changes to hexagonally packed cylinders at a volume fraction of $\sim 30\%$. Alternating lamellae are formed at approximately equal volume fractions for the two blocks. At a volume fraction of $\sim 38\%$, the minor block forms gyroid or perforated layers at moderate and high incompatibility,

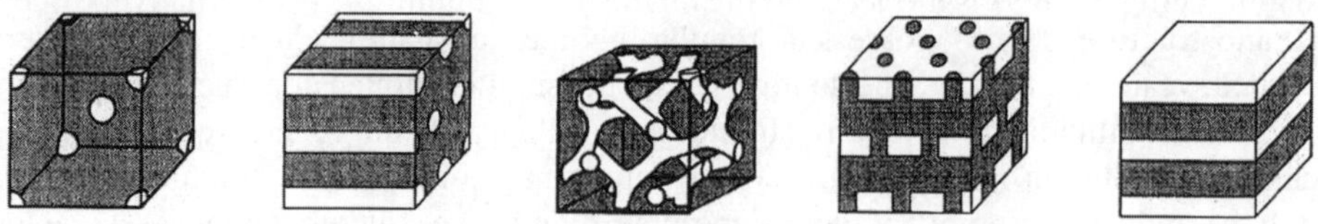

Figure 24.2 Bulk segregation patterns of coil–coil diblock copolymers. (Reprinted with permission from A.K. Khandpur, S. Foerster, F.S. Bates *et al.*, "Polyisoprene-polystyrene diblock copolymer phase diagram near the order-disorder transition," *Macromolecules*, **28**, 26, 8796–8806, 1995. © 1995 American Chemical Society.)

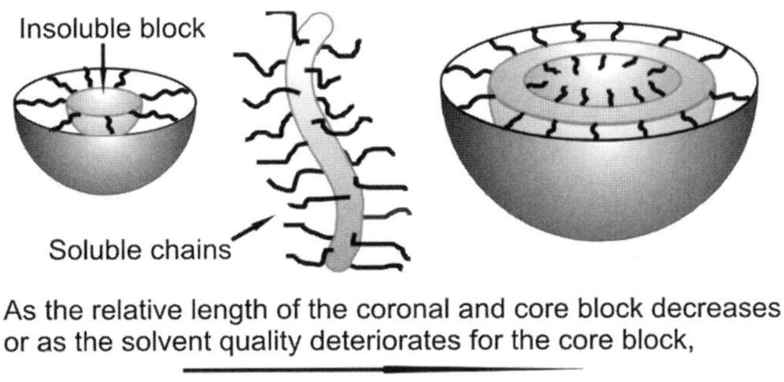

Figure 24.3 Depiction of the structures of half of a spherical micelle (left), a cylindrical micelle (middle), and half of a vesicular micelle (right) formed from a diblock copolymer. The dark block is soluble and the gray block is insoluble.

respectively. These interesting morphological transitions have been established experimentally (Hajduk *et al.*, 1994; Hashimoto *et al.*, 1994; Breiner *et al.*, 1997; Breiner *et al.*, 1998; Hamley, 1999) and can be accounted for theoretically by statistical thermodynamics (Leibler, 1980; Matsen and Schick, 1994; Matsen and Bates, 1996). Furthermore, the smallest dimension of a segregated domain, for example, the diameter of a cylinder, is proportional to the two-thirds power of the molar mass of the minor block, and can typically be tuned from ~5 to ~50 nm by changing the molar mass of the block (Bates and Fredrickson, 1999).

In analogy to their bulk behavior, diblock copolymers also self-assemble in block-selective solvents, which solubilize one, but not the other, block. This yields micelles of various shapes. Illustrated in Figure 24.3 are the shapes and chain-packing arrangements in spherical, cylindrical, and vesicular micelles of diblock copolymers (Zhang and Eisenberg, 1995; Zhang and Eisenberg, 1996; Zhang *et al.*, 1996; Ding and Liu, 1997; Ding *et al.*, 1997; Stupp *et al.*, 1997; Cornelissen *et al.*, 1998; Jenekhe and Chen, 1998; Cameron *et al.*, 1999; Engelkamp *et al.*, 1999; Jenekhe and Chen, 1999; Jain and Bates, 2003; Lazzari *et al.*, 2006; Wang *et al.*, 2007; Gadt *et al.*, 2009; Hayward and Pochan, 2010). For coil–coil diblock copolymers, the insoluble block normally aggregates, to produce spherical micelles, if the soluble block is long. As the length of the soluble block decreases relative to the insoluble block, the shapes of the micelles change to yield micelles with core–corona curvatures that are lower than those of spherical micelles. Thus, as the relative length of the soluble block is reduced, the morphologies of the micelles may change from spherical, to cylindrical, to vesicular micelles, as well as micelles having other shapes (Figure 24.3). Alternatively, one can induce changes of micelle morphologies by adjusting the quality of the selective solvent (Ding *et al.*, 1997; Cameron *et al.*, 1999; Bang *et al.*, 2006). Normally, with a greater solvent/core interfacial tension, or a poorer solvent for the core block, the formation of micelles with reduced core/corona interfacial curvatures is favored.

Aside from thermodynamic control, micelle-like aggregates (MAs) of various shapes can be produced as kinetically trapped products. Kinetic trapping occurs if chains of different MAs undergo negligible chain exchange, and fusion of different MAs or fission of MAs into smaller MAs do not take place (Jain and Bates, 2004). Chain exchange occurs mainly in a stepwise fashion. Initially, a MA loses one chain at a time to the solvent phase via chain dissociation.

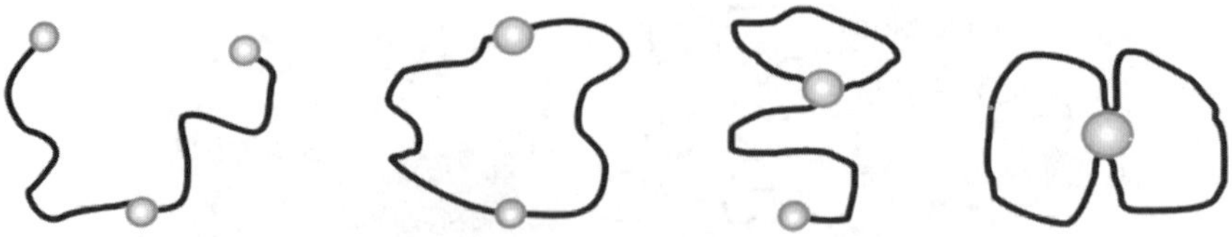

Figure 24.4 Possible structures for single-chain micelles of an ABABA pentablock copolymer in block-selective solvents for the B blocks. From left to right are a triple-beaded chain, a double-beaded loop, a tethered balloon, and a butterfly. The beads correspond to the A blocks, while the loops correspond to the B blocks.

Subsequently, another MA uptakes these chains from the solvent phase and thus becomes larger. The rate of the chain uptake process is evidently proportional to the critical micellar concentration (cmc) or the concentration of unimolecular chains that are solubilized under a given set of solvation conditions. If a solvent is sufficiently poor for the insoluble block(s), the cmc can be very low. Thus, the chain-exchange kinetics can be slow for block copolymers. Fusion or fission of micelles would not occur to a significant extent for MAs whose cores are protected from contacting one another by thick repulsive coronas.

ABC triblock copolymers can also undergo self-assembly in bulk or block-selective solvents. Triblock copolymers have many more block segregation patterns in bulk than diblock copolymers, and some of the patterns are very intricate and visually striking (Bates and Fredrickson, 1999). The shapes of micelles that are formed by triblock copolymers in block-selective solvents are also far more diverse (Li *et al.*, 2004; Pochan *et al.*, 2004; Lodge *et al.*, 2005; Zhu and Jiang, 2005; Cui *et al.*, 2007; Hu *et al.*, 2008b; Njikang *et al.*, 2008a; Schmalz *et al.*, 2008; Dupont *et al.*, 2009). The complexity of the block segregation patterns increase even further for tetra- and pentablock copolymers.

Micelle and MA formation in block-selective solvents has received much attention in the past 15 years since the seminal discovery of multiple morphologies of block copolymer MAs by Zhang and Eisenberg (Zhang and Eisenberg, 1995). Always coexisting at a concentration equal to the critical micellar concentration with MAs are unimolecular or single-chain micelles (Sadron, 1962; Tuzar and Kratochvil, 1993). Because of the low cmc values of block copolymers under normal conditions, these unimolecular micelles have largely been ignored by researchers. It is illustrative to imagine the possible shapes that can be taken by the unimolecular micelles of an ABABA pentablock copolymer in a block-selective solvent for B. These can include a triple-beaded chain (A in Figure 24.4), a double-beaded loop (B), a tethered balloon (C), and a butterfly (D). Evidently, the butterfly shape will be favored if the block-selective solvent is very poor for the A blocks. As the solvent quality improves for the A blocks, the tethered balloon, followed by the double-beaded loop, and then the triple-beaded chain shapes will be favored. Thus, the shapes of the unimolecular micelles can be diverse and fascinating as well. They are also ideal precursors for architectural polymers.

24.3 Our Approaches to Block Copolymer Architectures

Figure 24.5 depicts the approaches that we take to prepare architectures from block copolymers. Regardless of the architecture, the starting point is always polymer design. During this

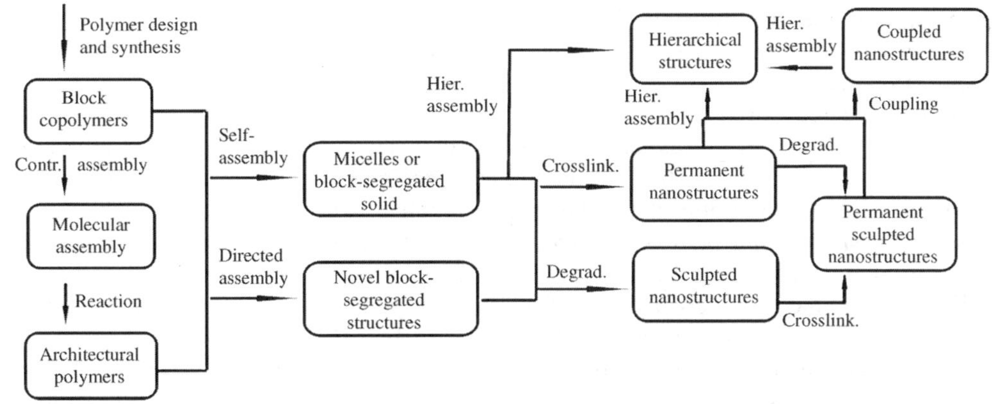

Figure 24.5 Approaches taken by our group to prepare architectures from block copolymers.

stage, we start by deciding which structure we want to prepare and devise a way to make it. A typical preparation involves the assembly of the block copolymer or copolymers and possibly performing chemical reactions on the assembled structures. Depending on the application, functionalities or functional groups might be required of the target structure. All of these aspects need to be considered while formulating a strategy and determining which polymer(s) should be prepared.

Once a target polymer has been identified, we then formulate the best way to prepare the polymers. Techniques used by our group for polymer preparation include anionic polymerization (Guo *et al.*, 1996), cationic polymerization (Liu *et al.*, 1994a), atom-transfer polymerization (Zheng *et al.*, 2007), and Grignard metathesis polymerization (Liu *et al.*, 2002c). We also modify the polymers synthesized from these primary techniques to make derivatives of these polymers. We have, for example, made diblock copolymers by clicking different homopolymers together (Fu and Liu, 2010) and made graft copolymers by the "grafting-to" method (Lu *et al.*, 2004).

Our key contribution to polymer syntheses has been in the introduction of poly(2-cinnamoyloxyethyl methacrylate), or PCEMA, into block copolymers (Figure 24.6, Liu *et al.*, 1994a; Guo *et al.*, 1996). This polymer is reasonably stable thermally, and can be stored as a solid in amber-colored bottles for years without crosslinking. In a segregated PCEMA domain, where the PCEMA block of different chains is mixed, PCEMA readily undergoes photocrosslinking when irradiated with light having wavelengths near 274 nm. This is

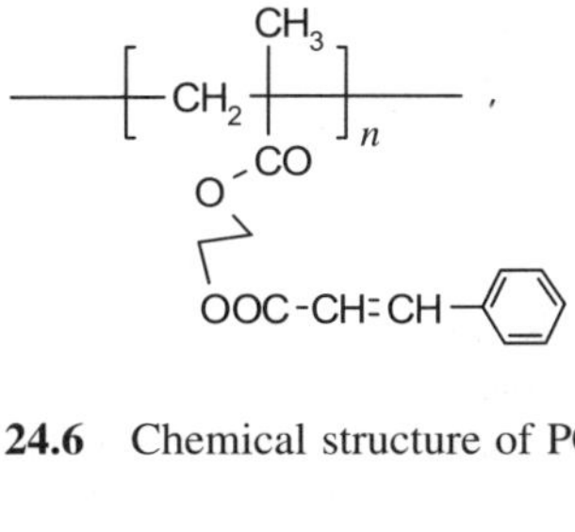

Figure 24.6 Chemical structure of PCEMA.

achieved via the 1,4-cycloaddition of two CEMA double bonds (Guillet, 1985). This photocrosslinking reaction facilitates the structural "locking in" of self-assembled block copolymers.

After a targeted polymer is synthesized, it is then used for assembly. The assembly process can include *self-assembly*, either in a block-selective solvent or in the solid state. In a block-selective solvent, the self-assembled structures may form micelles or single-chain micelles. In the solid state, the self-assembled structures will normally be the periodically spaced and regularly shaped domain structures that have been mentioned earlier in Section 24.2.

We have combined block copolymer self-assembly and emulsion formation and developed a method we refer to as *directed assembly* (Lu *et al.*, 2001a; Lu *et al.*, 2001b; Lu *et al.*, 2003). Using directed assembly, we have been able to create 2D spaces (Zheng *et al.*, 2005) or confined volumes (Lu *et al.*, 2001a; Lu *et al.*, 2001b; Lu *et al.*, 2003) for block copolymer self-assembly to take place. We were also able to direct the chain orientation and packing in an assembled structure (Zheng and Liu, 2007).

A key component in the synthetic strategies of our architectures is the *chemical processing* of the assembled structures. Chemical processing refers to the performance of selective chemistry to a domain or domains of a block-segregated copolymer in order to generate novel molecules or functional materials. The structural locking of molecular assemblies has, for example, allowed us to prepare *architectural polymers* including cyclic polymers (Hu *et al.*, 2009) and miktoarm polymers (Dou *et al.*, 2010). The chemical processing of a self-assembled copolymer in a block-selective solvent or in bulk allows the preparation of novel polymer nanostructures. The exact type of nanostructure produced depends on the type of chemical processing performed. By performing selective crosslinking a stable, or "*permanent*", block copolymer nanostructure is produced (Liu *et al.*, 1994b; Guo *et al.*, 1996; Liu *et al.*, 1996). Further degradation, including full or partial decomposition of the other block(s), yields "*permanent sculpted*" nanostructures (Liu *et al.*, 1997; Liu *et al.*, 1999; Stewart and Liu, 1999; Stewart and Liu, 2000). If chemical processing invokes only the degradation of one block of an assembled copolymer structure, while maintaining the integrity of the rest of the structure, "*sculpted*" nanostructures are obtained.

Permanent and permanent sculpted nanostructures are stable architectures. They can therefore be further chemically processed to yield more complex structures. For example, we have covalently linked different stable block copolymer nanostructures to yield *coupled nanostructures*. The end coupling of a hydrophobic nanotube with a hydrophilic nanotube yielded an amphiphilic nanotube diblock structure (Yan *et al.*, 2004). The attachment of hydrophobic block copolymer nanotubes by their ends to the surfaces of hydrophilic nanospheres bearing carboxyl groups yielded a "supersurfactant" (Liu *et al.*, 2003).

A nanotube diblock structure can be viewed as a macroscopic counterpart of a diblock copolymer. In block-selective solvents, a multiblock nanotube sample was shown to undergo *hierarchical assembly* to form "supermicelles" (Yan *et al.*, 2004). Aside from coupled nanostructures, ordinary micelles have also been used to perform "*double*", or hierarchical, assembly in order to yield hierarchical structures (Hu *et al.*, 2008a). In these cases, Tier I assembly is performed using block copolymers to form micelles. At the Tier II level, the micelles are further assembled into hierarchical structures.

Figure 24.7 shows some examples of structures that our group has prepared using the different methodologies reviewed above. The double helix was a self-assembled micelle of a

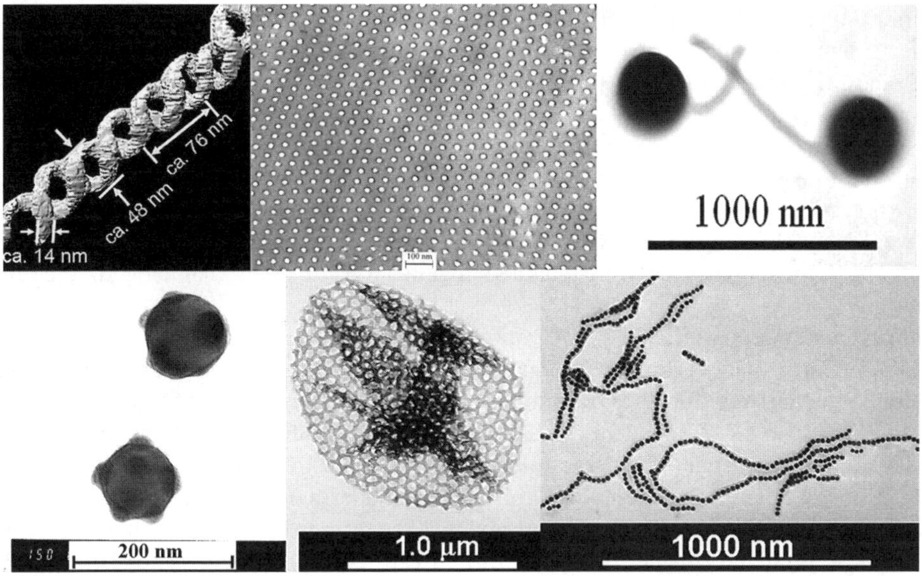

Figure 24.7 TEM and TEM topography images of some structures prepared by our group. Top left: TEM topography image of a self-assembled structure – double helix of an ABC linear triblock copolymer. (Reproduced with permission from J. Dupont, K.-I. Niihara, R. Kimoto and H. Jinnai, "Self-assembled ABC triblock copolymer double and triple helices," *Angewandte Chemie, International Edition*, **48**, 33, 6144–6147, 2009. © Wiley-VCH Verlag GmbH & Co. KGaA.) Top middle: TEM image of a permanent sculpted nanostructure – thin film containing nanochannels. Top right: TEM image of a coupled nanostructure – supersurfactant. (Reprinted with permission from G. Liu, X. Yan, Z. Li *et al.*, "End coupling of block copolymer nanotubes to nanospheres," *Journal of the American Chemical Society*, **125**, 46, 14039–14045, 2003. © 2003 American Chemical Society.) Bottom left: TEM image of a structure prepared from directed assembly – virus-like particles. (Reprinted with permission from R. Zheng, G. Liu and X. Yan, "Polymer nano- and microspheres with bumpy and chain-segregated surfaces," *Journal of the American Chemical Society*, **127**, 44, 15358–15359, 2005. © 2005 American Chemical Society.) Bottom center: TEM image of a structure from double assembly – a microcage. Bottom right: TEM image of Co nanoparticle coated by a diblock copolymer surfactant. (Reprinted with permission from Z. Zhou, G. Liu and D. Han, "Coating and structural locking of dipolar chains of cobalt nanoparticles," *ACS Nano*, **3**, 1, 165–172, 2009. © 2009 American Chemical Society.)

linear ABC triblock copolymer formed in a solvent that was marginal for the A block, poor for the B block, and good for the C block (Dupont *et al.*, 2009). The nanochannel-containing thin film was prepared from an AB diblock copolymer (Liu *et al.*, 1997; Liu and Ding, 1998; Liu *et al.*, 1999). In the solid state, the A block formed hexagonally packed cylinders that were dispersed in the B matrix. The permanent sculpted nanostructure was prepared by photo-crosslinking the B matrix and degrading the A cylinders. The supersurfactant, a coupled nanostructure, was prepared by end coupling block copolymer nanotubes to the surfaces of nanospheres (Liu *et al.*, 2003). Meanwhile, the virus-like particles were prepared from directed assembly by assembling two diblock surfactants at the interface between water and polymer-containing oil droplets (Zheng *et al.*, 2005). To prepare the microcages, an ABC triblock was assembled at Tier I into cylindrical micelles. The cylindrical micelles were then subjected to further assembly, which involved the fusion and mutation of the cylindrical micelles at the water and oil droplet interface (Hu *et al.*, 2008a). Thus, this was a product of the double-assembly approach.

Aside from the above approaches, we have frequently used diblock copolymers as surfactants to enable the preparation of various inorganic nanoparticles (Liu *et al.*, 2005a; Zhou *et al.*, 2009). Shown at the bottom right of Figure 24.7 is a TEM image of Co nanoparticles that were prepared using PCEMA-*b*-PAA as the surfactant, where PAA denotes poly(acrylic acid). The Co nanoparticles were chained together due to their magnetic dipole–dipole interactions.

Other than the preparation of block copolymer architectures, we also study the nature of the architectures that we prepare and explore their applications. We have, for example, developed techniques for the fractionation and characterization of block copolymer nanofibers. (Liu *et al.*, 2002a; Liu *et al.*, 2002b) We also studied the viscometric properties of block copolymer nanofiber fractions. (Liu *et al.*, 2002b; Yan and Liu, 2004) For applications, we have, for example, hybridized polymer nanostructures by incorporating inorganic nanoparticles or functional groups into them in order to render them functionality. We have prepared magnetic (Underhill and Liu, 2000b) and catalytic (Underhill and Liu, 2000a) nanospheres, triblock/Fe_2O_3 superparamagnetic nanofibers (Yan *et al.*, 2001), polymer-coated Co dipolar chains (Zhou *et al.*, 2009), and tetrablock/Pd hybrid fibers (Li and Liu, 2003). We have also prepared fluorescent nanospheres (Li *et al.*, 2002) and water-dispersible superparamagnetic microspheres (Liu *et al.*, 2005b). Through collaboration with industrial partners, we have immobilized bioreagents onto the surfaces of these spheres. These polymer–bioreagent hybrid structures were then explored for their applications in immunoassays and for diagnostic systems. We have also explored the use of diblock copolymer micelles as a friction-reduction additive in automobile engine oils and the results we obtained demonstrated the fantastic lubrication properties of diblock copolymer spherical micelles with PCEMA and poly(2-cinnamoyloxyethyl acrylate) (PCEA) cores.

Since our work on the preparation of nanostructures and hierarchical structures has been reviewed previously (Lazzari *et al.*, 2006; Liu, 2008), it will not be described in this chapter. Rather, we will discuss in some length, our more recent work involving the preparation of molecular architectures from block copolymer precursors.

24.4 A Block Copolymer Approach to Architectural Polymers

We have so far used two approaches to prepare architectural polymers from block copolymer precursors. Both approaches involved block copolymer assembly in the first step and chemical locking of the assembled structure in the second step. In Approach 1, the block copolymer undergoes intramolecular assembly in a block-selective solvent to yield single-chain micelles. In Approach 2, two block copolymers are used. Both of these copolymers that are used possess a short block that can associate with a corresponding short block in the other copolymer. The associated blocks are then stitched together chemically, in order to yield an architectural polymer. We will use two examples in this section to illustrate these two approaches.

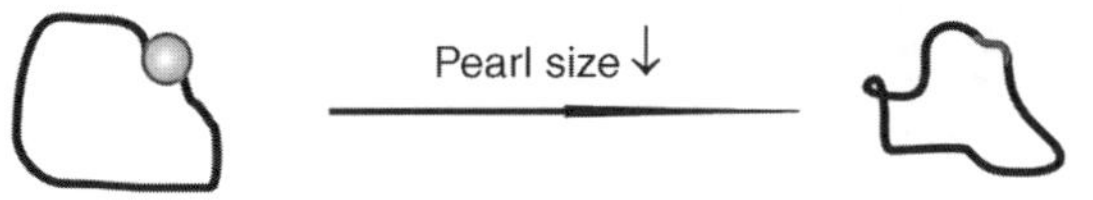

Figure 24.8 Structural comparison between a pearl-ring molecule and a macrocycle.

24.4.1 Pearl-Ring Molecules

A pearl-ring molecule is a single-beaded loop, where the loop is composed of a polymer chain and the bead is a crosslinked nanoglobule with a diameter of less than 5 nm (Figure 24.8). When the diameter of the bead approaches that of the polymer chain, a pearl-ring molecule becomes a traditional architectural polymer – a macrocycle.

Macrocycles have been of fundamental interest for quite some time. Polymer physicists are fascinated by macrocyclic polymers due to their appealing topology and interesting properties (Zimm and Stockmayer, 1949). The lack of chains ends has, for example, been predicted to lead to different chain dynamics in polymer melts and concentrated solutions (Klein, 1986; Rubinstein, 1986). This cyclic topology also leads to a reduced radius of gyration (Zimm and Stockmayer, 1949) and intrinsic viscosity (Yamakawa, 1969) for macrocycles relative to their linear counterparts of the same molecular weight.

Macrocycles have tantalized the imagination of synthetic chemists due to the immense challenge associated with the preparation of large uniform macrocycles in pure form and in large quantities (Hogen-Esch, 2006; Endo, 2008). The preparation of macrocycles from the cyclization of long polymer chains is difficult because polymer chain ends have a low probability of contacting one another, and intramolecular cyclization has to compete with interchain coupling. In order to minimize intermolecular coupling macrocycles are prepared traditionally under high dilution conditions, which limit the amount of polymer available per volume of solvent. There have been many efforts made to overcome these difficulties and to facilitate the production of large macrocycles in high concentration and high purity.

Assuming that the end cyclization is performed in a theta solvent for a polymer chain, Jacobson and Stockmayer (Jacobson and Stockmayer, 1950) determined that the probability at any given instant of finding chain ends which are within v_e and capable of reaction is:

$$P_r = \left(\frac{3}{2\pi}\right)^{3/2} \frac{v_e}{R_n^3} \tag{24.1}$$

where R_n is the root-mean-square end-to-end distance of the chain, and v_e is the reaction volume, within which the end coupling reaction occurs effectively. Thus, one can either decrease R_n or increase v_e in order to increase P_r.

ABA triblock copolymers are ideal precursors for cyclic polymers, because one can increase v_e by using a pair of reactive terminal A blocks rather than a pair of reactive end groups. More importantly, the terminal blocks can be preassociated or preassembled intramolecularly in a block-selective solvent for the B block at high dilution. This leads

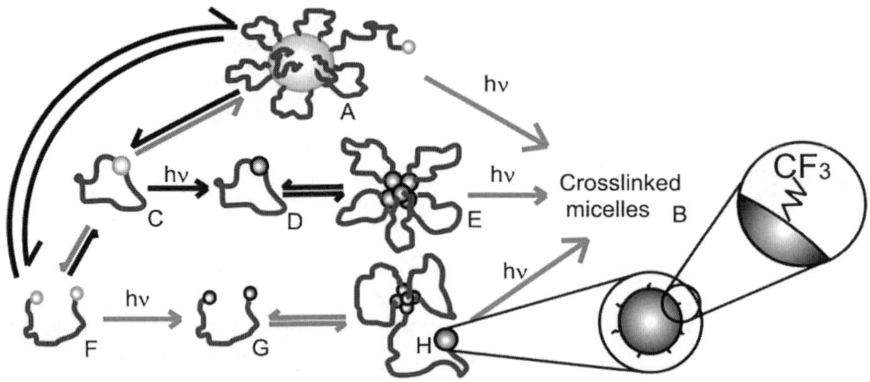

Figure 24.9 Schematic depiction of possible structures and processes occurring in an ABA triblock copolymer system in a selective solvent for B and under irradiation. The processes denoted by gray arrows should be avoided for macrocycle formation. On the right are some structural details for the pompons of P (CEMA-*r*-TFAEMA). The TFAEMA units should segregate preferentially on the surface. (Reprinted with permission from J. Hu, R. Zheng, J. Wang *et al.*, "Macrocycles from the photochemical coupling of preassociated terminal blocks of copolymers," *Macromolecules*, **42**, 13, 4638–4645, 2009. © 2009 American Chemical Society.)

to the formation of a beaded-loop precursor, as depicted by Structure C in Figure 24.9. In this way, R_n approaches zero and $P_r \rightarrow 1$.

Evidently, the triblock precursor approach has its complications. First, a beaded loop, rather than a macrocycle, is produced in this way. Thus, optimization is required in order to determine the shortest lengths of A blocks that are required to prepare the cyclic structure. Secondly, smart experimental designs are required to eliminate many of the possible side reactions. Figure 24.9 shows the different species that coexist with unimer cycles and the side reactions that they can undergo.

Always in equilibrium with the unimer cycles are pompon–coil–pompon unimers (PCPs, shown as F in Figure 24.9), which possess two end pompons. The PCP structure is favored entropically. The driving force for unimer-cycle formation from PCP is a reduction of the interfacial energy of the system when two end pompons of a PCP are fused together in a macrocycle. If the concentration of the polymer is above the cmc, micelles, for example, spherical micelles or Structure A, can also exist. As the reaction progresses, the product, the pearl-ring molecules (Structure D in Figure 24.9), should increase in concentration. Micelles can also form from the pearl rings (Structures D $\rightarrow$ E).

The triblock copolymer micelles (Structure A), micelles of the pearl rings (Structure E), and the PCPs (Structure F) can all undergo side reactions. If we use a triblock copolymer with the terminal blocks made of PCEMA, the photolysis of both the triblock copolymer micelles and the pearl-ring micelles (Structures A and E, respectively) lead to crosslinked micelles (Structure B). The photolysis of PCPs results in crosslinked PCPs (Structure F $\rightarrow$ G).

To eliminate the side reactions, we devised three more strategies. First, we used a sufficiently poor solvent for the end blocks to ensure that the single-beaded loops were the dominant species at low polymer concentrations. Secondly, we mitigated the micelles by adopting a slow reactant addition strategy, which involved the slow pumping of a concentrated block copolymer solution into a solvent reservoir under constant stirring and irradiation. Since the reactant was added slowly and reagent added earlier had been converted to macrocycles and other species, the polymer concentration remained low throughout the preparation and the micelle–unimer equilibrium favored unimer cycle formation. Thirdly, we used terminal blocks consisting of a random copolymer, rather than a PCEMA homopolymer. In these terminal blocks, the photocrosslinkable CEMA units were insoluble and thus segregated themselves from the solvent phase. The soluble units were inert and presumably projected themselves into the solvent from the surfaces of the pompons that were formed by the insoluble CEMA domains. These soluble units acted to prevent the photoactive insoluble CEMA units in the pompons of the different macrocycles from contacting one another, and thus prevented the chemical coupling of different pompons.

For our initial proof-of-concept project, we prepared and used a pentablock copolymer poly[(2-cinnamoyloxyethyl methacrylate)-*ran*-(2-trifluoroacetoxyethyl methacrylate)]-*block*-poly(solketal methacrylate)-*block*-poly(*tert*-butyl acrylate)-*block*-poly(solketal methacrylate)-*block*-poly[(2-cinnamoyloxyethyl methacrylate)-*ran*-(2-trifluoroacetoxyethyl methacrylate)], abbreviated as P(CEMA-*r*-TFAEMA)-*b*-PSMA-*b*-PtBA-*b*-PSMA-*b*-P(CEMA-*r*-TFAEMA) and shown in Figure 24.10.

The CEMA units were used to facilitate the cyclization reaction. Some TFAEMA units were introduced into the terminal blocks because they were not photoactive and were soluble in the cyclization media, which consisted of chloroform/cyclohexane with a cyclohexane (CHX) volume fraction of 90%. An intermediate PSMA-*b*-PtBA-*b*-PSMA block copolymer was also used, mainly to demonstrate that this methodology could be used to prepare cyclic block copolymers. The PSMA-*b*-PtBA-*b*-PSMA block copolymer should have behaved no differently from a homopolymer block, because all of the three middle blocks were soluble in the solvation media used for pearl-ring preparation. Seventy units were used for each P(CEMA-*r*-TFAEMA) block in this initial study, in order to ensure that unimer cycles would form. As was shown previously (Hu *et al.*, 2009), the reduction of the surface free energy accompanying the fusion of two pompons of a PCP is:

$$\Delta F_{\mathrm{A}} \propto \gamma (M_{\mathrm{e}}/\rho)^{2/3} \tag{24.2}$$

where γ is the interfacial tension between the insoluble units of the terminal blocks and the solvent, ρ is the density of the pompons, and M_{e} is the molar mass of a terminal insoluble block.

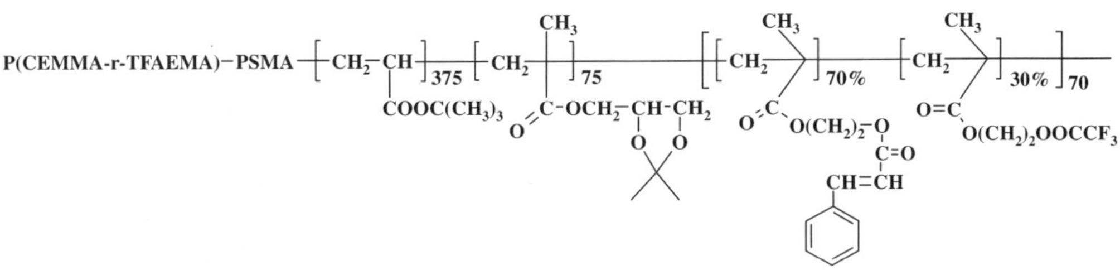

Figure 24.10 Chemical structure of P(CEMA-*r*-TFAEMA)-*b*-PSMA-*b*-PtBA-*b*-PSMA-*b*-P(CEMA-*r*-TFAEMA). This pentablock copolymer has a symmetric structure, and thus the structures of the repetitive PSMA and P(CEMA-*r*-TFAEMA) blocks on the left are omitted.

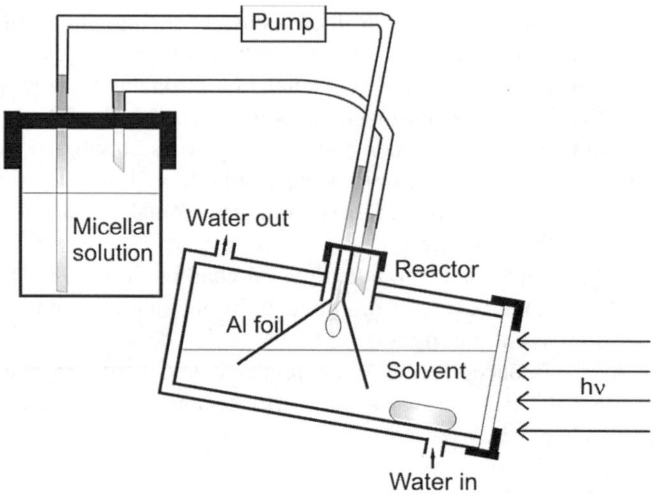

Figure 24.11 Schematic diagram of the apparatus used to prepare pearl-ring molecules.

Thus, the higher the molar mass of the terminal blocks, the greater the predicted reduction of the interfacial energy accompanying the pompon fusion process.

Figure 24.11 shows the apparatus that was used for the pearl-ring preparation. Instead of using highly diluted block copolymer solutions, we started by using a micellar solution at a polymer concentration of 4.0 mg/mL or higher. The solution was then pumped at 0.15 mL/min into a solvent reservoir at a constant temperature and under constant stirring and irradiation. The pumped solution was delivered via a syringe needle, and the tip of the needle was covered by an aluminum foil trumpet. The chloroform/cyclohexane mixture used to prepare the polymer solution and the solvent had the same cyclohexane (CHX) volume fraction, f_{CHX}, which was typically 90%. The initial polymer solution and solvent volume used was 30 mL for each of the solutions. The amount of time required to pump the entire polymer solution into the reactor was approximately 200 min. After polymer addition, the sample was irradiated for another 60–120 min to ensure a CEMA double-bond conversion of ~40%. In order to minimize solvent evaporation during photolysis, the reactor temperature was regulated to 20 °C. Also, the reactor and the pumping liquid formed a closed system to minimize solvent evaporation. After photolysis, solid samples were obtained in quantitative yield after rotary evaporation of solvent.

The aluminum trumpet was used so that the micelles were not irradiated after their addition into the reactor. Rather, they were allowed to gradually dissociate into unimers in the protected region before they became irradiated. The irradiation of the unimer cycles converted them into pearl rings.

Figure 24.12 compares the size-exclusion chromatography traces of the pentablock copolymer and the copolymer after it was irradiated at a final concentration of 2.0 mg/mL. We noticed the following features about the data. First, the peak maximum of the irradiated sample was shifted by ~0.5 min toward the low molecular weight region. Secondly, no crosslinked micellar peak was observed for the irradiated sample despite the fact that the final polymer concentration was 2.0 mg/mL, which was 49 times greater than the cmc value (0.041 mg/mL) for the

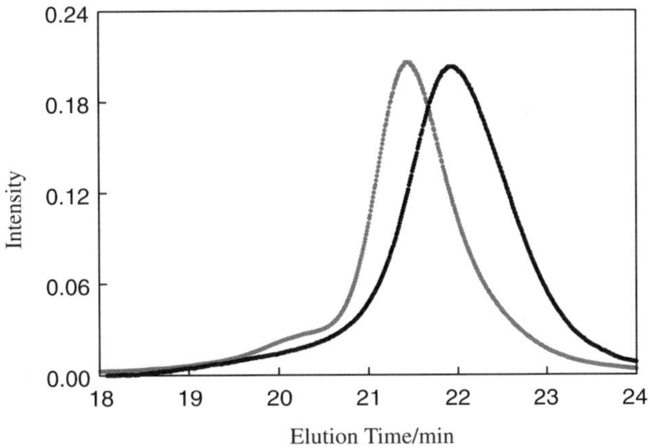

Figure 24.12 Comparison of SEC traces of the pentablock copolymer (gray) and the pentablock copolymer after irradiation at a final concentration of 2.0 mg/mL (black). (Reprinted with permission from J. Hu, R. Zheng, J. Wang *et al.*, "Macrocycles from the photochemical coupling of preassociated terminal blocks of copolymers," *Macromolecules*, **42**, 13, 4638–4645, 2009. © 2009 American Chemical Society.)

polymer. More interestingly, the shoulder on the high molecular weight side of the peak was smaller for the irradiated sample than that of the precursor.

The SEC peak shifted to the low molecular weight side, because macrocycles or crosslinked PCPs were prepared. The sample should have consisted mostly of macrocycles, because the peak shift was significant. Quantitatively, the peak molecular weights, based on polystyrene standards, of the precursor and the irradiated sample were 6.9×10^4 and 5.2×10^4 Da, respectively. Thus, a 25% reduction of the apparent molecular weight was observed for the irradiated sample. This reduction is comparable to those reported for PS macrocycles with a comparable molecular weight.

One can also calculate the volume compaction of the irradiated sample relative to its precursor. According to theories of universal SEC calibration (Young and Lovell, 1991), polymers of different families with the same retention time possess the same hydrodynamic volume. The hydrodynamic volume V_P of a monodisperse PS sample with molecular weight M_P is:

$$V_P = \frac{[\eta]_P M_P}{2.5} \tag{24.3}$$

where $[\eta]_P$ is the intrinsic viscosity of the hypothetical monodisperse sample. For PS in THF, the SEC eluant, $[\eta]_P$ can be calculated using (Spatorico and Coulter, 1973):

$$[\eta]_P = 1.10 \times 10^{-2} M_P^{0.725} \text{ mL/g} \tag{24.4}$$

The combined use of Equations 24.3 and 24.4 and the PS-equivalent peak molecular weights M_P yielded a volume reduction of 39% for the irradiated sample relative to its precursor. This volume collapse was large. It could not have been achieved by crosslinking PCPs because of the relatively short end blocks, but instead had to be due to macrocycle formation.

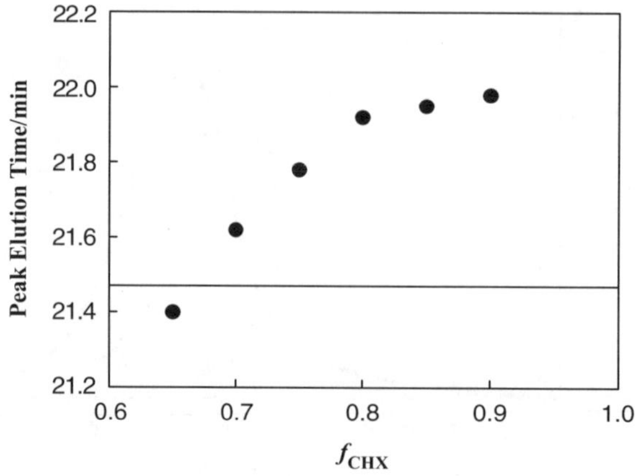

Figure 24.13 Variation of peak elution time with f_{CHX} for P(CEMA-*r*-TFAEMA)-*b*-PSMA-*b*-PtBA-*b*-PSMA-*b*-P(CEMA-*r*-TFAEMA) samples irradiated at different f_{CHX} values. (Reprinted with permission from J. Hu, R. Zheng, J. Wang *et al.*, "Macrocycles from the photochemical coupling of preassociated terminal blocks of copolymers," *Macromolecules*, **42**, 13, 4638–4645, 2009. © 2009 American Chemical Society.)

We have also determined the intrinsic viscosities $[\eta]$ of the macrocycles and their precursor, and the ratio between the two was 0.64. This compares well with the theoretical ratio of 0.66 (Yamakawa, 1969). Thus, this proves unambiguously the production of a polymer solution composed mostly of macrocycles.

Last but not the least, we have repeated the preparation described above in $CHCl_3$/CHX at CHX volume fractions (f_{CHX}) of 65%, 70%, 75%, 80%, and 85%, and performed SEC analysis of all of the resultant samples. Plotted in Figure 24.13 is the variation of the SEC elution peak time of the irradiated product as a function of f_{CHX}. Also shown in the figure is a straight line indicating the peak elution time of 21.47 min for the pentablock copolymer. Figure 24.13 shows that only the sample irradiated at $f_{CHX} = 65\%$ had a peak elution time shorter than 21.47 min. Initially the peak elution times of the other samples increased steeply with f_{CHX}. Above $f_{CHX} = 80\%$, the elution time increased only slightly with f_{CHX}.

A logical explanation for the observed peak elution time variation trend for samples with $f_{CHX} \geq 70\%$ is that these samples consisted of a mixture of pearl-ring molecules and cross-linked PCPs. As the content of the more compact pearl-ring molecules increased with increasing f_{CHX}, the peak elution time increased. Above a critical f_{CHX} value of ~80%, macrocycles became the dominant product and thus the peak elution time increased only slightly with further increases of f_{CHX}.

The sample irradiated at $f_{CHX} = 65\%$ behaved uniquely because the P(CEMA-*r*-TFAEMA) blocks were soluble and had not collapsed yet at this solvent composition. An open-ended block structure facilitated intermolecular coupling, yielding high molecular weight species. This caused the SEC peak to appear at an elution time shorter than 21.47 min.

The block copolymer approach to prepare cyclic polymers is very robust. Even at a final polymer concentration of 5.0 mg/mL, no crosslinked micelle peak was observed.

The above example shows the key steps involved in the preparation of architectures from the chemical processing of assembled block copolymers. To ensure success, the polymers have to be properly designed, and the conditions for polymer assembly have to be optimized. This is then followed by the precise execution of the chemistry.

Aside from preparing pearl-ring molecules, we have used this strategy to also make tadpole molecules (Tao and Liu, 1997; Njikang *et al.*, 2008b). Tadpole preparation involved the use of diblock copolymers containing a photocrosslinkable PCEMA block. Single-chain micelles were first made in a selectively poor solvent for the PCEMA block. The photocrosslinking of the PCEMA block yielded a diblock copolymer with a crosslinked globular PCEMA block and a random-coil second block. These molecules were called tadpoles, due to their resemblance to the shape of a tadpole. The fat head was composed of crosslinked PCEMA units and the slim tail consisted of a random-coil chain.

24.4.2 Miktoarm Star Polymer

A miktoarm star polymer is a star polymer consisting of arms that have different chemical compositions. A miktoarm ABC star polymer (μ-ABC) is a star polymer consisting of three arms made of polymers A, B, and C, respectively. Miktoarm ABC star polymers exhibit interesting bulk block segregation patterns (Vlassopoulos *et al.*, 2001; Yang *et al.*, 2001; Zhu *et al.*, 2003; Hadjichristidis *et al.*, 2005; Babin *et al.*, 2008), and form compartmentalized micelles in block-selective solvents (Pispas *et al.*, 2000; Li *et al.*, 2004; Li *et al.*,2006b; Saito *et al.*, 2008; Liu *et al.*, 2009). Various applications have been proposed for the self-assembled, or block-segregated, structures of μ-ABC. The block-segregation properties of μ-ABCD should be even more complex and interesting.

Despite these properties, miktoarm star copolymers have not been widely studied, except by Lodge, Hillmyer, and coworkers (Li *et al.*, 2004; Li *et al.*, 2006a; Li *et al.*, 2006b; Liu *et al.*, 2009) as well as by Hadjichristidis and collaborators (Pispas *et al.*, 2000; Hadjichristidis *et al.*, 2005). This is likely due to the difficulty associated with the synthesis of miktoarm star copolymers. The chemical stitching of associated block copolymers is a convenient and viable approach to the synthesis of miktoarm star polymers. This was demonstrated by us in a recent proof-of-concept study by showing the preparation of a series of μ-AB_2 samples from this method. Reviewed in this section are the key points of that study (Dou *et al.*, 2010).

In that study, the triblock copolymers used included poly(methyl methacrylate)-*block*-poly-[(2-cinnamoyloxyethyl methacrylate)-*ran*-(2-glycinoyloxyethyl methacrylate)]-*block*-poly(methyl methacrylate) (PMMA-*b*-$PGNH_2$-*b*-PMMA) or poly(methyl methacrylate)-*block*-poly-[[(2-cinnamoyloxyethyl methacrylate)-*ran*-[2-(4′-(6″-aminohexylamino)-4′-oxo)-butanoyloxyethyl methacrylate]]-*block*-poly(methyl methacrylate) (PMMA-*b*-$PHNH_2$-*b*-PMMA). Each of these triblock copolymers contained a short middle block bearing amino groups. The diblock copolymer used was poly(*tert*-butyl acrylate)-*block*-poly[(2-cinnamoyloxyethyl methacrylate)-*ran*-(2-(3′-carboxypropanoyloxy)ethyl methacrylate)] (PtBA-*b*-PSCOOH). The carboxyl-bearing block of the diblock copolymer was short and comparable in length to the amino-bearing block of the triblock copolymer, that was intended to associate with the diblock.

While the original paper (Dou *et al.*, 2010) reported the study of several pairs of diblock and triblock copolymers, we will briefly discuss, in this section, results obtained from two pairs of PMMA-*b*-PNH_2-*b*-PMMA and PtBA-*b*-PSCOOH samples (Figure 24.14). Here, PNH_2 can be either $PGNH_2$ or $PHNH_2$. The repeat unit number, n, for each of the PMMA blocks of the PMMA-*b*-PNH_2-*b*-PMMA samples was 285 and that for the middle block,

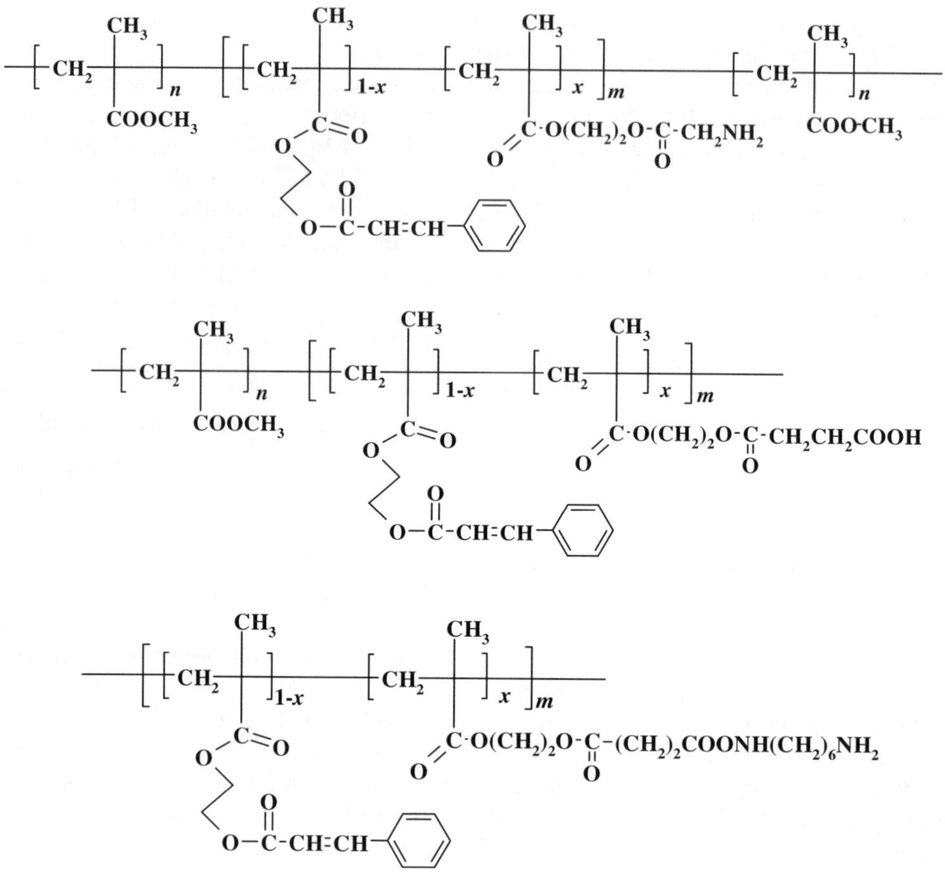

Figure 24.14 Structures of PMMA-*b*-PGNH$_2$-*b*-PMMA (top), PtBA-*b*-PSCOOH (middle), and the PHNH$_2$ block (bottom). (Reprinted with permission from H. Dou, L. Hong and G. Liu, "Miktoarm star copolymers from the chemical stitching of associating block copolymers," *Macromolecules*, **43**, 10, 4629–4637, 2010. © 2010 American Chemical Society.)

m, was 44. Of the 44 units, the fraction *x* of amino units was 30%, and the CEMA molar fraction was 70%. The repeat unit number, *n*, for the PtBA block of PtBA-PSCOOH was 410, and that for the PSCOOH block, *m*, was 37. Of the 37 units, 30% contained carboxyl groups.

Evidently, PMMA-*b*-PNH$_2$-*b*-PMMA and PtBA-PSCOOH should associate in organic solvents such as CH_2Cl_2 via the ion pairing between the ammonium cations of the PNH$_2$ block and the carboxylate anions of the PSCOOH blocks. The ions were formed as a result of proton transfer from the carboxyl groups to the amino groups. If one PtBA-*b*-PSCOOH chain was associated with one PMMA-*b*-PNH$_2$-*b*-PMMA chain, a three-armed cluster was formed (A → B, Figure 24.15). The association of two PtBA-*b*-PSCOOH chains with one PMMA-*b*-PNH$_2$-*b*-PMMA chain should result in the formation of a four-armed cluster (D → E). Amidization of the carboxyl and amino groups should covalently stitch together the three-armed and four-armed clusters, B → C and E → F, to yield μ-(PtBA)(PMMA)$_2$ and μ-(PtBA)$_2$(PMMA)$_2$, respectively.

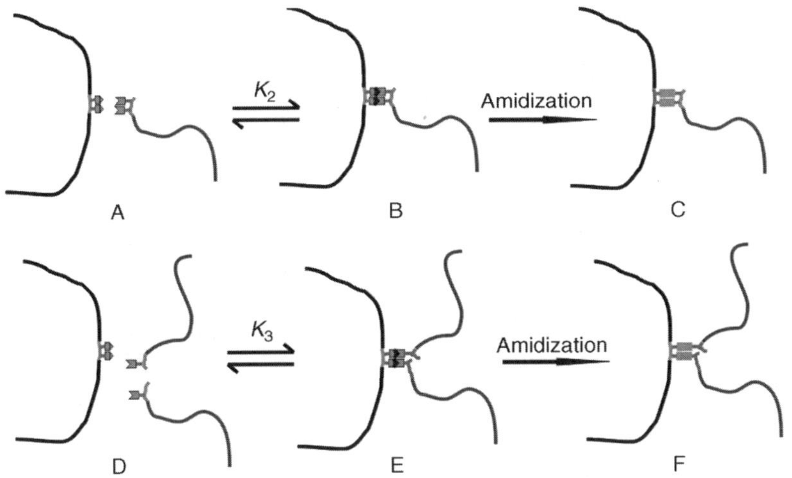

Figure 24.15 "Idealized reaction mechanism" for miktoarm preparation from PtBA-*b*-PSCOOH and PMMA-*b*-PNH_2-*b*-PMMA.

Cinnamoyl groups were introduced into the amino- and carboxyl-bearing blocks of the copolymers, because we initially intended to take advantage of the photodimerization reaction of the cinnamoyl units (Guo *et al.*, 1996; Liu *et al.*, 1999; Stewart and Liu, 2000) and thus stitch together the associated polymer chains. However, the amidization reaction was later found to be more efficient than the cinnamoyl dimerization reaction. Therefore, the cinnamoyl units performed no apparent function in this study.

To prepare the miktoarm polymers, we started by equilibrating the carboxyl- and amino-containing polymers at equimolar concentrations in CH_2Cl_2 overnight to establish the polymer association equilibrium. A coupling agent, 2-chloro-1-methyl pyridinium iodide, was then added together with triethyl amine, which was added to neutralize the HCl and HI produced from the amidization reaction. The coupling reaction was performed in refluxing CH_2Cl_2 under N_2 protection for 6 h or longer (Dou *et al.*, 2010).

Figure 24.16 compares the SEC traces from the refractive index (RI) detector for an equimolar physical mixture of PMMA-*b*-$PHNH_2$-*b*-PMMA and PtBA-PSCOOH and also the crude product obtained from coupling an equimolar mixture of two polymers at a total polymer precursor concentration, c_p, of 0.5 mg/mL (Dou *et al.*, 2010). The physical mixture showed a single narrow SEC peak derived from the superimposition of the peaks of the individual components. The SEC trace of the amidized mixture exhibited not only the peak corresponding to the precursor mixture, but also showed a high molecular weight peak. This high molecular weight peak also had a small shoulder. Our detailed analyses indicated that the main product peak belonged to the desired product μ-(PtBA)$(PMMA)_2$, and the shoulder arose from μ-$(PtBA)_2(PMMA)_2$. Thus, this strategy allowed the production of the miktoarm star polymer μ-(PtBA)$(PMMA)_2$ containing some impurity μ-$(PtBA)_2(PMMA)_2$.

We do not know the exact reason for the production of μ-$(PtBA)_2(PMMA)_2$ together with μ-(PtBA)$(PMMA)_2$. It could arise from the mismatch between the number of carboxyl groups per PtBA-*b*-PSCOOH chain and that of the amino groups per PMMA-*b*-$PHNH_2$-*b*-PMMA chain. It was difficult to produce an exact match between the average numbers of carboxyl and

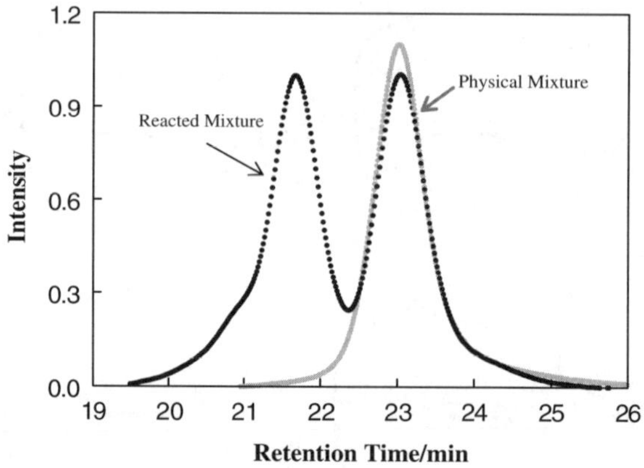

Figure 24.16 Comparison of SEC RI traces of an equimolar mixture (●) of PMMA-*b*-PGNH$_2$-*b*-PMMA and PtBA-*b*-PSCOOH and the above mixture after amidization (●). (Reprinted with permission from H. Dou, L. Hong and G. Liu, "Miktoarm star copolymers from the chemical stitching of associating block copolymers," *Macromolecules*, **43**, 10, 4629–4637, 2010. © 2010 American Chemical Society.)

amino groups, because the reactions used for attaching these groups were not quantitative. Precise quantification of the carboxyl and amino groups by ^{1}H NMR was also difficult, due to the low content of these groups. Furthermore, these numbers fluctuated from chain to chain, and thus a mismatch was very likely for any given pair of PtBA-*b*-PSCOOH and PMMA-*b*-PHNH$_2$-*b*-PMMA P1 chains, even if the average numbers were matched exactly.

The production of a mixture was unfortunate. However, the presence of the μ-(PtBA)$_2$(PMMA)$_2$ contaminant should not stop the self-assembly of μ-(PtBA)(PMMA)$_2$ in the solid state, or in block-selective solvents. It will only shift the phase diagrams of μ-(PtBA)(PMMA)$_2$ somewhat. As far as the preparation of materials from the self-assembly of miktoarm copolymers is concerned, the presence of this contaminant is therefore of no consequence.

We were able to resolve the individual peaks using a commercial program and thus obtain the integrated areas of the individual peaks. Ignoring the errors that could derive from the differences in the refractive indices of the different polymers, we should be able to equate the relative areas of μ-(PtBA)(PMMA)$_2$ and μ-(PtBA)$_2$(PMMA)$_2$ to their yields. For the above case, the SEC yields for μ-(PtBA)(PMMA)$_2$ and μ-(PtBA)$_2$(PMMA)$_2$ were 38 and 13%, respectively.

We observed that the content of the μ-(PtBA)$_2$(PMMA)$_2$ contaminant decreased under identical preparation conditions after replacing PMMA-*b*-PHNH$_2$-*b*-PMMA with PMMA-*b*-PGNH$_2$-*b*-PMMA as the amino-bearing reactant. Figure 24.17 shows plots of the yields of μ-(PtBA)(PMMA)$_2$ and μ-(PtBA)$_2$(PMMA)$_2$ production as functions of the total reactant concentration, c_p, for the amidization reaction between PMMA-*b*-PGNH$_2$-*b*-PMMA and PtBA-*b*-PSCOOH. At low c_ps, the μ-(PtBA)$_2$(PMMA)$_2$ content was negligible. The yield of μ-(PtBA)(PMMA)$_2$ production was low as well. Figure 24.17 also shows that the yield of μ-(PtBA)$_2$(PMMA)$_2$, and the total yield of μ-(PtBA)(PMMA)$_2$ and μ-(PtBA)$_2$(PMMA)$_2$, increased as c_p increased. The yield of μ-(PtBA)(PMMA)$_2$ increased initially, and subsequently decreased, with increasing c_p.

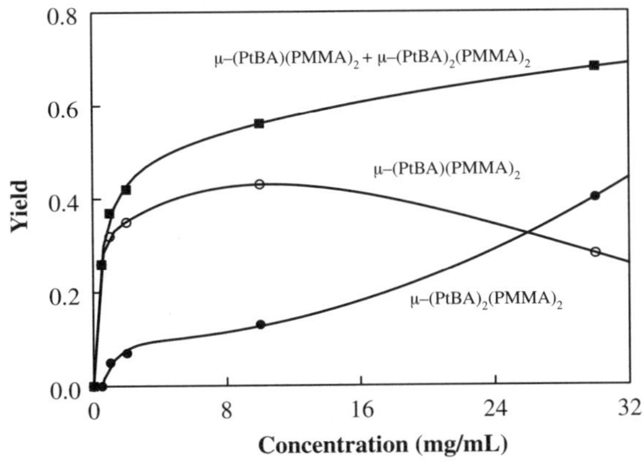

Figure 24.17 Plots of variation of the yields of μ-$(PtBA)_1(PMMA)_2$ (O) and μ-$(PtBA)_2(PMMA)_2$ (●), as well as that of μ-$(PtBA)_1(PMMA)_2$ and μ-$(PtBA)_2(PMMA)_2$ (■), as functions of the total concentration (c_p) for PMMA-*b*-$PGNH_2$-*b*-PMMA and PtBA-*b*-PSCOOH. (Reprinted with permission from H. Dou, L. Hong and G. Liu, "Miktoarm star copolymers from the chemical stitching of associating block copolymers," *Macromolecules*, **43**, 10, 4629–4637, 2010. © 2010 American Chemical Society.)

The miktoarm polymer yield increased with reactant concentration because a higher reactant concentration favored the association of the polymers. The μ-$(PtBA)_2(PMMA)_2$ yield increased at high reactant concentrations for two reasons. First, there was a mismatch between the average numbers of carboxyl and amino groups between the two types of polymers. The average number of carboxyl groups per PtBA-*b*-PSCOOH chain was 11 and the average number of amino groups per PMMA-*b*-$PGNH_2$-*b*-PMMA chain was 13. Secondly, the repulsion among the associating chains decreased substantially as the polymer concentration surpassed the overlap concentration. Above the overlap concentration, the difference between the repulsion energies of a $(PtBA)_2(PMMA)_2$ cluster and a $(PtBA)(PMMA)_2$ cluster might be small. Some $(PtBA)_2(PMMA)_2$ clusters were formed, probably to help maximize ion pairing.

Relative to PMMA-*b*-$PGNH_2$-*b*-PMMA, the use of PMMA-*b*-$PHNH_2$-*b*-PMMA helped increase the yield of miktoarm polymer formation at $c_p = 0.50$ mg/mL. This was probably due to the larger separation between the amino groups and the polymer backbone in PMMA-*b*-$PHNH_2$-*b*-PMMA than in PMMA-*b*-$PGNH_2$-*b*-PMMA. The larger separation helped alleviate the repulsion between the different chains in this low polymer concentration regime and thus increased the yield of miktoarm polymers produced.

In conclusion, we have been able to use the assembly or association and reaction strategy to prepare miktoarm copolymers. For this particular system, we could separate the pure miktoarm star polymers from their precursors by fractionation precipitation. We are now working on the preparation of μ-ABC and μ-ABCD polymers from this technique and are exploring the applications of the μ-ABC or μ-ABCD polymers produced.

24.5 Conclusions

Block copolymers are versatile precursors for the preparation of various kinds of architectures, ranging from architectural polymers to nanostructures and hierarchical structures. This chapter described the methodologies that we have used and developed to prepare the various architectures. Examined in some detail has been our recent use of the assembly and reaction strategy for the preparation of architectural polymers. In one example, we discussed the intramolecular assembly of an ABA triblock copolymer in a block-selective solvent for the central B block and the optimization of the experimental conditions to favor unimolecular cycle formation. The associated ends of the unimolecular cycles were then stitched together to yield pearl-ring molecules. In another example, two polymer chains have been allowed to undergo controlled intermolecular association. The associated chains were then chemically stitched together to yield miktoarm copolymers.

For both projects involving the preparation of architectural polymers, further experiments are underway in determining the smallest relative size that is required for the reactive blocks to facilitate the preparations. Also, methods are being developed for the visualization of the cyclic products by atomic force microscopy. Potential applications are also being explored for the products prepared. There have been many years of research on architectural polymers. The past research has been mainly focused on devising new preparation methodologies. We see the development of applications for architectural polymers as an underdeveloped area providing great research opportunities for the future.

References

Babin, J., Taton, D., Brinkmann, M., and Lecommandoux, S. (2008) Synthesis and self-assembly in bulk of linear and mikto-arm star block copolymers based on polystyrene and poly(glutamic acid). *Macromolecules*, **41**, 1384–1392.

Bang, J., Jain, S.M., Li, Z.B. *et al.* (2006) Sphere, cylinder, and vesicle nanoaggregates in poly(styrene-*b*-isoprene) diblock copolymer solutions. *Macromolecules*, **39**, 1199–1208.

Bates, F.S. and Fredrickson, G.H. (1999) Block copolymers - designer soft materials. *Physics Today*, **52**, 32–38.

Breiner, U., Krappe, U., Abetz, V., and Stadler, R. (1997) Cylindrical morphologies in asymmetric ABC triblock copolymers. *Macromolecular Chemistry and Physics*, **198**, 1051–1083.

Breiner, U., Krappe, U., Thomas, E.L., and Stadler, R. (1998) Structural characterization of the "Knitting Pattern" in Polystyrene-*block*-poly(ethylene-co-butylene)-*block*-poly(methylmethacrylate) triblock copolymers. *Macromolecules*, **31**, 135–141.

Cameron, N.S., Corbierre, M.K., and Eisenberg, A. (1999) 1998 E.W.R. Steacie award lecture asymmetric amphiphilic block copolymers in solution: a morphological wonderland. *Canadian Journal of Chemistry*, **77**, 1311–1326.

Cornelissen, J., Fischer, M., Sommerdijk, N., and Nolte, R.J.M. (1998) Helical superstructures from charged poly (styrene)-poly(isocyanodipeptide) block copolymers. *Science*, **280**, 1427–1430.

Cui, H.G., Chen, Z.Y., Zhong, S. *et al.* (2007) Block copolymer assembly via kinetic control. *Science*, **317**, 647–650.

Ding, J.F. and Liu, G.J. (1997) Polyisoprene-*block*-poly(2-cinnamoylethyl methacrylate) vesicles and their aggregates. *Macromolecules*, **30**, 655–657.

Ding, J.F., Liu, G.J., and Yang, M.L. (1997) Multiple morphologies of polyisoprene-*block*-poly(2-cinnamoylethyl methacrylate) and Polystyrene-block-poly(2-cinnamoylethyl methacrylate) micelles in organic solvents. *Polymer*, **38**, 5497–5501.

Dou, H., Hong, L., and Liu, G. (2010) Miktoarm star copolymers from the chemical stitching of associating block copolymers. *Macromolecules*, **43**, 4629–4637.

Dupont, J., Liu, G., Niihara, K.-I. *et al.* (2009) Self-assembled ABC triblock copolymer double and triple helices. *Angewandte Chemie-International Edition*, **48**, 6144–6147.

Endo, K. (2008) Synthesis and properties of cyclic polymers. *Advances in Polymer Science*, **217**, 121–183.

Engelkamp, H., Middelbeek, S., and Nolte, R.J.M. (1999) Self-assembly of disk-shaped molecules to coiled-coil aggregates with tunable helicity. *Science*, **284**, 785–788.

Fu, Y. and Liu, G. (2010) Modular synthesis of a block copolymer with a cleavable linkage via click chemistry. *Science China Chemistry*, **53**, 1128–1133.

Gadt, T., Ieong, N.S., Cambridge, G. *et al.* (2009) Complex and hierarchical micelle architectures from diblock copolymers using living, crystallization-driven polymerizations. *Nature Materials*, **8**, 144–150.

Guillet, J.E. (1985) *Polymer Photophysics and Photochemistry - An Introduction to the Study of Photoprocesses in Macromolecules*, Cambridge University Press, Cambridge.

Guo, A., Liu, G.J., and Tao, J. (1996) Star polymers and nanospheres from cross-linkable diblock copolymers. *Macromolecules*, **29**, 2487–2493.

Hadjichristidis, N., Iatrou, H., Pitsikalis, M., and Mays, J. (2006) Macromolecular architectures by living and controlled/living polymerizations. *Progress in Polymer Science*, **31**, 1068–1132.

Hadjichristidis, N., Iatrou, H., Pitsikalis, M. *et al.* (2005) Linear and non-linear triblock terpolymers. synthesis, self-assembly in selective solvents and in bulk. *Progress in Polymer Science*, **30**, 725–782.

Hajduk, D.A., Harper, P.E., Gruner, S.M. *et al.* (1994) The gyroid - a new equilibrium morphology in weakly segregated diblock copolymers. *Macromolecules*, **27**, 4063–4075.

Hamley, I.W. (1999) *Block Copolymers*, Oxford University Press, Oxford.

Hashimoto, T., Kawamura, T., Harada, M., and Tanaka, H. (1994) Small-angle scattering from hexagonally packed cylindrical particles with paracrystalline distortion. *Macromolecules*, **27**, 3063–3072.

Hayward, R.C. and Pochan, D.J. (2010) Tailored assemblies of block copolymers in solution: It is all about the process. *Macromolecules*, **43**, 3577–3584.

Hirao, A., Hayashi, M., Loykulnant, S., and Sugiyama, K. (2005) Precise syntheses of chain-multi-functionalized polymers, star-branched polymers, star-linear block polymers, densely branched polymers, and dendritic branched polymers based on iterative approach using functionalized 1,1-diphenylethylene derivatives. *Progress in Polymer Science*, **30**, 111–182.

Hogen-Esch, T.E. (2006) Synthesis and characterization of macrocyclic vinyl aromatic polymers. *Journal of Polymer Science Part A-Polymer Chemistry*, **44**, 2139–2155.

Hu, J.W., Liu, G.J., and Nijkang, G. (2008a) Hierarchical interfacial assembly of ABC triblock copolymer. *Journal of the American Chemical Society*, **130**, 3236–3237.

Hu, J.W., Njikang, G., and Liu, G.J. (2008b) Twisted ABC triblock copolymer cylinders with segregated A and C coronal chains. *Macromolecules*, **41**, 7993–7999.

Hu, J.W., Zheng, R.H., Wang, J. *et al.* (2009) Macrocycles from the photochemical coupling of preassociated terminal blocks of block copolymers. *Macromolecules*, **42**, 4638–4645.

Jacobson, H. and Stockmayer, W.H. (1950) Intramolecular reaction in polycondensations. 1. The theory of linear systems. *Journal of Chemical Physics*, **18**, 1600–1606.

Jain, S. and Bates, F.S. (2003) On the origins of morphological complexity in block copolymer surfactants. *Science*, **300**, 460–464.

Jain, S. and Bates, F.S. (2004) Consequences of nonergodicity in aqueous binary PEO-PB micellar dispersions. *Macromolecules*, **37**, 1511–1523.

Jenekhe, S.A. and Chen, X.L. (1998) Self-assembled aggregates of rod-coil block copolymers and their solubilization and encapsulation of fullerenes. *Science*, **279**, 1903–1907.

Jenekhe, S.A. and Chen, X.L. (1999) Self-assembly of ordered microporous materials from rod-coil block copolymers. *Science*, **283**, 372–375.

Khandpur, A.K., Forster, S., Bates, F.S. *et al.* (1995) Polyisoprene-polystyrene diblock copolymer phase diagram near the order-disorder transition. *Macromolecules*, **28**, 8796–8806.

Klein, J. (1986) Dynamics of entangled linear, branched, and cyclic polymers. *Macromolecules*, **19**, 105.

Lazzari, M., Liu, G.J., and Lecommandoux, S. (2006) *Block Copolymers in Nanoscience*, Wiley-VCH, Weinheim.

Leibler, L. (1980) Theory of microphase separation in block co-polymers. *Macromolecules*, **13**, 1602–1617.

Li, Z. and Liu, G.J. (2003) Water-dispersible tetrablock copolymer synthesis, aggregation, nanotube preparation, and impregnation. *Langmuir*, **19**, 10480–10486.

Li, Z., Liu, G.J., Law, S.J., and Sells, T. (2002) Water-soluble fluorescent diblock nanospheres. *Biomacromolecules*, **3**, 984–990.

Li, Z.B., Hillmyer, M.A., and Lodge, T.P. (2006a) Laterally nanostructured vesicles, polygonal bilayer sheets, and segmented wormlike micelles. *Nano Letters*, **6**, 1245–1249.

Li, Z.B., Hillmyer, M.A., and Lodge, T.P. (2006b) Morphologies of multicompartment micelles formed by ABC miktoarm star terpolymers. *Langmuir*, **22**, 9409–9417.

Li, Z.B., Kesselman, E., Talmon, Y. *et al.* (2004) Multicompartment micelles from ABC miktoarm stars in water. *Science*, **306**, 98–101.

Liu, C., Hillmyer, M.A., and Lodge, T.P. (2009) Multicompartment micelles from pH-responsive miktoarm star block terpolymers. *Langmuir*, **25**, 13718–13725.

Liu, G.J. (2008) Block copolymer nanotubes derived from self-assembly. *Advances in Polymer Science*, **220**, 29–64.

Liu, G.J. and Ding, J.F. (1998) Diblock thin films with densely hexagonally packed nanochannels. *Advanced Materials*, **10**, 69–71.

Liu, G.J., Ding, J.F., Guo, A. *et al.* (1997) Potential skin layers for membranes with tunable nanochannels. *Macromolecules*, **30**, 1851–1853.

Liu, G.J., Ding, J.F., Hashimoto, T. *et al.* (1999) Thin films with densely, regularly packed nanochannels: preparation, characterization, and applications. *Chemistry of Materials*, **11**, 2233–2240.

Liu, G.J., Hu, N.X., Xu, X.Q., and Yao, H. (1994a) Cross-linked polymer brushes. 1. Synthesis of poly[beta-(vinyloxy) ethyl cinnamate]-*b*-poly(isobutyl vinyl ether). *Macromolecules*, **27**, 3892–3895.

Liu, G.J., Qiao, L.J., and Guo, A. (1996) Diblock copolymer nanofibers. *Macromolecules*, **29**, 5508–5510.

Liu, G.J., Xu, X.Q., Skupinska, K. *et al.* (1994b) Cross-linked polymer brushes. 2. Formation and properties of poly (isobutyl vinyl ether)-b-Poly[2-(vinyloxy) ethyl cinnamate] brushes. *Journal of Applied Polymer Science*, **53**, 1699–1707.

Liu, G.J., Yan, X.H., and Duncan, S. (2002a) Polystyrene-block-polyisoprene nanofiber fractions. 1. Preparation and static light-scattering study. *Macromolecules*, **35**, 9788–9793.

Liu, G.J., Yan, X.H., Li, Z. *et al.* (2003) End coupling of block copolymer nanotubes to nanospheres. *Journal of the American Chemical Society*, **125**, 14039–14045.

Liu, G.J., Yan, X.H., Lu, Z.H. *et al.* (2005a) One-pot synthesis of block copolymer coated cobalt nanocrystals. *Chemistry of Materials*, **17**, 4985–4991.

Liu, G.J., Yan, X.H., Qiu, X.P., and Li, Z. (2002b) Fractionation and solution properties of PS-b-PCEMA-b-PtBA nanofibers. *Macromolecules*, **35**, 7742–7747.

Liu, G.J., Yang, H.S., and Zhou, J.Y. (2005b) Preparation of magnetic microspheres from water-in-oil emulsion stabilized by block copolymer dispersant. *Biomacromolecules*, **6**, 1280–1288.

Liu, J.S., Sheina, E., Kowalewski, T., and McCullough, R.D. (2002c) Tuning the electrical conductivity and self-assembly of regioregular polythiophene by block copolymerization: Nanowire morphologies in new di- and triblock copolymers. *Angewandte Chemie-International Edition*, **41**, 329–332.

Lodge, T.P., Bang, J.A., Li, Z.B. *et al.* (2005) Strategies for controlling intra- and intermicellar packing in block copolymer solutions: Illustrating the flexibility of the self-assembly toolbox. *Faraday Discussions*, **128**, 1–12.

Lu, Z.H., Liu, G.J., and Duncan, S. (2004) Polysulfone-graft-poly(tert-butyl acrylate): Synthesis, nanophase separation, poly(tert-butyl acrylate) hydrolysis, and pH-dependent iridescence. *Macromolecules*, **37**, 174–180.

Lu, Z.H., Liu, G.J., and Liu, F.T. (2001a) Block copolymer microspheres containing intricate nanometer-sized segregation patterns. *Macromolecules*, **34**, 8814–8817.

Lu, Z.H., Liu, G.J., and Liu, F.T. (2003) Water-dispersible porous polyisoprene-*block*-poly(acrylic acid) microspheres. *Journal of Applied Polymer Science*, **90**, 2785–2793.

Lu, Z.H., Liu, G.J., Phillips, H. *et al.* (2001b) Palladium nanoparticle catalyst prepared in poly(acrylic acid)-lined channels of diblock copolymer microspheres. *Nano Letters*, **1**, 683–687.

Matsen, M.W. and Bates, F.S. (1996) Unifying weak- and strong-segregation block copolymer theories. *Macromolecules*, **29**, 1091–1098.

Matsen, M.W. and Schick, M. (1994) Stable and unstable phases of a diblock copolymer melt. *Physical Review Letters*, **72**, 2660–2663.

Njikang, G., Han, D.H., Wang, J., and Liu, G.J. (2008a) ABC triblock copolymer micelle-like aggregates in selective solvents for A and C. *Macromolecules*, **41**, 9727–9735.

Njikang, G., Liu, G.J., and Curda, S.A. (2008b) Tadpoles from the intramolecular photo-cross-linking of diblock copolymers. *Macromolecules*, **41**, 5697–5702.

Pispas, S., Hadjichristidis, N., Potemkin, I., and Khokhlov, A. (2000) Effect of architecture on the micellization properties of block copolymers: A_2B miktoarm stars vs AB diblocks. *Macromolecules*, **33**, 1741–1746.

Pochan, D.J., Chen, Z.Y., Cui, H.G. *et al.* (2004) Toroidal triblock copolymer assemblies. *Science*, **306**, 94–97.

Rubinstein, M. (1986) Dynamics of ring polymers in the presence of fixed obstacles. *Physical Review Letters*, **57**, 3023–3026.

Sadron, C. (1962) Les polymeres organises. *Pure and Applied Chemistry*, **4**, 347–357.

Saito, N., Liu, C., Lodge, T.P., and Hillmyer, M.A. (2008) Multicompartment micelles from polyester-containing ABC miktoarm star terpolymers. *Macromolecules*, **41**, 8815–8822.

Schmalz, H., Schmelz, J., Drechsler, M. *et al.* (2008) Thermo-reversible formation of wormlike micelles with a microphase-separated corona from a semicrystalline triblock terpolymer. *Macromolecules*, **41**, 3235–3242.

Spatorico, A.L. and Coulter, B. (1973) Molecular-weight determinations by gel-permeation chromatography and viscometry. *Journal of Polymer Science Part B-Polymer Physics*, **11**, 1139–1150.

Stewart, S. and Liu, G. (2000) Block copolymer nanotubes. *Angewandte Chemie-International Edition*, **39**, 340–344.

Stewart, S. and Liu, G.J. (1999) Hollow nanospheres from polyisoprene-*block*-poly(2-cinnamoylethyl methacrylate)-*block*-poly(tert-butyl acrylate). *Chemistry of Materials*, **11**, 1048–1054.

Stupp, S.I., LeBonheur, V., Walker, K. *et al.* (1997) Supramolecular materials: Self-organized nanostructures. *Science*, **276**, 384–389.

Tao, J. and Liu, G.J. (1997) Polystyrene-*block*-poly(2-cinnamoylethyl methacrylate) tadpole molecules. *Macromolecules*, **30**, 2408–2411.

Tuzar, Z. and Kratochvil, P. (1993) Micelles of block and graft copolymers in solutions. *Surface and Colloid Science*, **15**, 1–83.

Underhill, R.S. and Liu, G.J. (2000a) Preparation and performance of Pd particles encapsulated in block copolymer nanospheres as a hydrogenation catalyst. *Chemistry of Materials*, **12**, 3633–3641.

Underhill, R.S. and Liu, G.J. (2000b) Triblock nanospheres and their use as templates for inorganic nanoparticle preparation. *Chemistry of Materials*, **12**, 2082–2091.

Vlassopoulos, D., Fytas, G., Pakula, T., and Roovers, J. (2001) Multiarm star polymers dynamics. *Journal of Physics-Condensed Matter*, **13**, R855–R876.

Wang, X.S., Guerin, G., Wang, H. *et al.* (2007) Cylindrical block copolymer micelles and co-micelles of controlled length and architecture. *Science*, **317**, 644–647.

Yamakawa, H. (1969) *Modern Theory of Polymer Solution*, Harper and Row, New York.

Yan, X.H. and Liu, G.J. (2004) Preparation, characterization, and solution viscosity of polystyrene-*block*-polyisoprene nanofiber fractions. *Langmuir*, **20**, 4677–4683.

Yan, X.H., Liu, G.J., and Li, Z. (2004) Preparation and phase segregation of block copolymer nanotube multiblocks. *Journal of the American Chemical Society*, **126**, 10059–10066.

Yan, X.H., Liu, G.J., Liu, F.T. *et al.* (2001) Superparamagnetic triblock copolymer/Fe_2O_3 hybrid nanofibers. *Angewandte Chemie-International Edition*, **40**, 3593–3596.

Yang, L.Z., Hong, S., Gido, S.P. *et al.* (2001) I_5S miktoarm star block copolymers: Packing constraints on morphology and discontinuous chevron tilt grain boundaries. *Macromolecules*, **34**, 9069–9073.

Young, R.J. and Lovell, P.A. (1991) *Introduction to Polymers*, Chapman & Hall, London.

Zhang, L.F. and Eisenberg, A. (1995) Multiple morphologies of "Crew-Cut" aggregates of polystyrene-*b*-poly(acrylic acid) block copolymers. *Science*, **268**, 1728–1731.

Zhang, L.F. and Eisenberg, A. (1996) Multiple morphologies and characteristics of "Crew-Cut" micelle-like aggregates of polystyrene-*b*-poly(acrylic acid) diblock copolymers in aqueous solutions. *Journal of the American Chemical Society*, **118**, 3168–3181.

Zhang, L.F., Yu, K., and Eisenberg, A. (1996) Ion-induced morphological changes in "Crew-Cut" aggregates of amphiphilic block copolymers. *Science*, **272**, 1777–1779.

Zheng, R.H. and Liu, G.J. (2007) Water-dispersible oil-filled ABC triblock copolymer vesicles and nanocapsules. *Macromolecules*, **40**, 5116–5121.

Zheng, R.H., Liu, G.J., and Jao, T.C. (2007) Poly[(2-ethylhexyl acrylate)-*ran*-(*tert*-butyl acrylate)]-*block*-poly (2-cinnamoyloxyethyl acrylate) synthesis and properties. *Polymer*, **48**, 7049–7057.

Zheng, R.H., Liu, G.J., and Yan, X.H. (2005) Polymer nano- and microspheres with bumpy and chain-segregated surfaces. *Journal of the American Chemical Society*, **127**, 15358–15359.

Zhou, Z.H., Liu, G.J., and Han, D.H. (2009) Coating and structural locking of dipolar chains of cobalt nanoparticles. *ACS Nano*, **3**, 165–172.

Zhu, J.T. and Jiang, W. (2005) Self-assembly of ABC triblock copolymer into giant segmented wormlike micelles in dilute solution. *Macromolecules*, **38**, 9315–9323.

Zhu, Y.Q., Gido, S.P., Moshakou, M. *et al.* (2003) Effect of junction point functionality on the lamellar spacing of symmetric $(PS)_n(PI)_n$ miktoarm star block copolymers. *Macromolecules*, **36**, 5719–5724.

Zimm, B.H. and Stockmayer, W.H. (1949) The dimensions of chain molecules containing branches and rings. *Journal of Chemical Physics*, **17**, 1301–1314.

25

Block Copolymer Nanostructured Thin Films for Advanced Patterning

Michelle A. Chavis, Evan L. Schwartz and Christopher K. Ober
Department of Materials Science and Engineering, Cornell University, Ithaca, NY 14853

25.1 Introduction

Moore's Law has established the benchmark pace that guides the exponential growth of transistor density on microchips (Moore, 1965). The continuation of this law has historically been achieved by the shrinking of transistor critical dimensions, but the march down to higher circuit densities and smaller feature sizes is becoming increasingly difficult with every successive technology generation. In today's electronic devices, the patterning of sub-30 nm critical dimensions and below is required in order to demonstrate performance improvements over existing brands. Currently, there are many successful techniques for microfabrication in this size regime, but each come with a very high cost of ownership and process complexity. In the race for the smallest, densest, and most powerful microchip, alternative fabrication methods that can demonstrate performance improvements without increasing cost end up receiving a lot of attention.

One idea to create nanostructured device layers is to use the natural processes of self-assembly to position organic and inorganic molecules from the "bottom-up" (Feynman, 1959). One particular approach that has received a considerable amount of attention is the use of block copolymers as a patterning material because of their inherent ability to form periodic nanostructures in the mesoscopic (10–50 nm) length scale, coupled to the facility of one of the polymers to be selectively degraded to create nanoporous thin films. These nanoporous films could act like a stencil mask for subsequent etching steps into the underlying substrate. This topic has attracted much attention and inspired several excellent reviews (Fasolka and

Complex Macromolecular Architectures: Synthesis, Characterization, and Self-Assembly, First Edition.
Edited by Nikos Hadjichristidis, Akira Hirao, Yasuyuki Tezuka and Filip Du Prez.

Mayes, 2001; Krausch and Magerle, 2002; Hamley, 2003; Lazzari and Lopez-Quintela, 2003; Park *et al.*, 2003; Hawker and Russell, 2005; Li *et al.*, 2005; Segalman, 2005; Krishnamoorty, 2006; Stoykovich and Nealey, 2006; Black, 2007a; Kim, Lee *et al.*, 2008; Cheng *et al.*, 2006a; Park *et al.*, 2008; Black *et al.*, 2007b). So-called "block copolymer lithography" would be considerably less expensive than current lithography techniques, which are described in the next section. Since the mid-1990s, researchers in this field have consistently demonstrated pattern densities greater than 10^{11} features per square centimeter, feature sizes less than 50 nm, and pattern transfer into a variety of substrates such as Si, Si_3N_4 and Ge (Park *et al.*, 2008). This chapter discusses how this emergent field might be used as a tool to extend traditional lithographic techniques down into the sub-30-nm length scale, in addition to discussing some of the technical challenges facing the field of block copolymer lithography in the future.

25.2 "Top-Down" Patterning Using Optical Photolithography

The success of the semiconductor industry is based on its ability to fabricate hundreds of millions of devices on a single microchip. Each microchip is a complex three-dimensional network of wires, vias and insulating layers, each requiring a unique set of microfabrication processes. One patterning technique called photolithography is used to write the specific patterns required for each layer. Photolithography is a process in which ultraviolet (UV) photons are used to write high-resolution spatial patterns in a radiation-sensitive polymer layer, called a photoresist. The UV photons pass through a chrome-coated photomask containing a positive or negative replica of the specific geometries required for each layer. In areas of the mask that have been cleared of chrome, UV photons pass through and strike the photoresist, which contain photoactivated acid generator (PAG) that are designed to absorb strongly at the UV wavelength used for imaging and produce a strong acid when exposed (Wallraff and Hinsberg, 1999; Reichmanis *et al.*, 2002; Ito, 2005). Modern "chemically amplified" photoresists depend on a brief post exposure bake in order to initiate a cascade of chemical reactions that spread homogenously through the exposed parts of the film (Reichmanis *et al.*, 1991). The photoactive compounds cause a chemical change of the polymer at the molecular level, which result in physical changes in the film, most importantly a change in solubility. The mechanism for the solubility switch may vary according to the specific chemistry of the resist. In the case of a positive-tone resist, the generated acid species in the exposed area results in the deprotection of an acid-labile protecting group from the polymer chain that makes the exposed area soluble in a developing solvent, usually an aqueous base. In the case of a negative-tone resist, the acid catalyzes a condensation reaction between a crosslinking resin and a hydroxyl or carboxyl-containing polymer species (Shaw *et al.*, 1997). Therefore, the part of the resist that is exposed to light becomes insoluble and remains on the substrate after immersion into the developing solvent, while the uncrosslinked areas are cleared away as shown in Figure 25.1. Once a patterned template is created on the wafer, it can be used as a stencil to scribe patterns into an underlying substrate, which is described in more detail in Section 25.5.

Traditional photolithography is highly dependent on the physics of aerial image formation, which can be quantified using known optics formulas. The ultimate resolution (R) of optical lithography is governed by the Rayleigh equation,

$$R = \frac{k\lambda}{\mathrm{NA}} \tag{25.1}$$

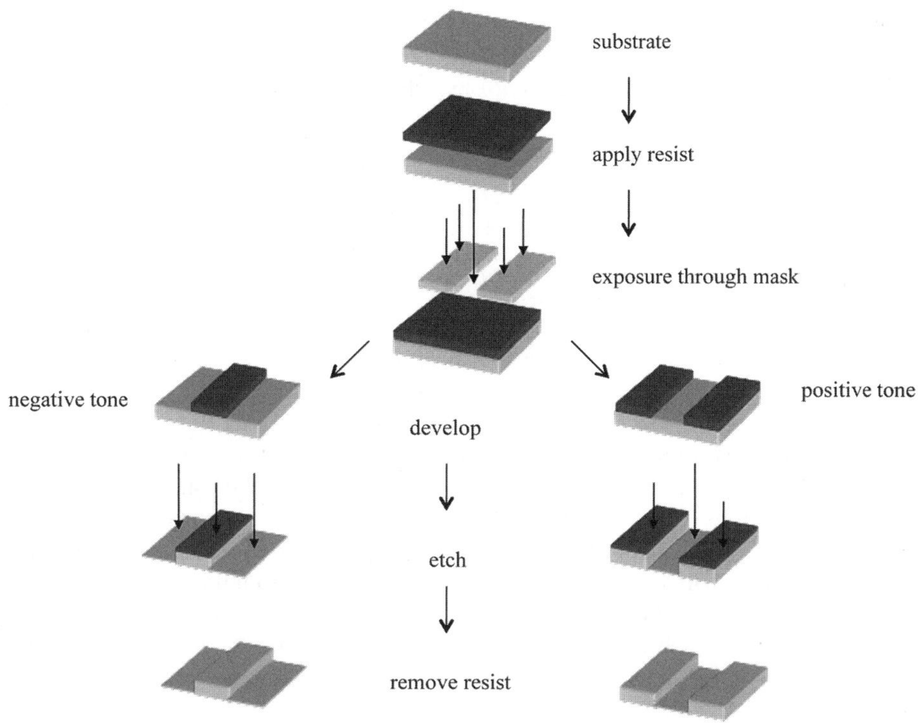

Figure 25.1 Schematic of traditional photolithography using positive-tone or negative-tone resist chemistry.

where k is a process parameter, λ is the wavelength of photons used, and NA is the numerical aperture. The current state-of-the-art processes reduces R through the use of deep-ultraviolet photons with $\lambda = 193$ nm, in addition to the use of a droplet of water on top of the resist to increase NA. So called "193-nm immersion photolithography" can efficiently produce arbitrary device pattern geometries with critical dimension (CD) control below 30 nm (Wallraff and Hinsberg, 1999; Reichmanis *et al.*, 2002; Rothschild *et al.*, 2004; Sanders, 2010). Other processing tricks, such as using two consecutive lithography steps separated by an intermediate etch step into the substrate (double patterning) can be used to extend this technology below 20 nm, at the expense of an increase in cost ownership and process complexity (Mack, 2007). Contrary to the use of expensive toolsets, self-assembly makes use of the natural laws of chemical thermodynamics to define ultimate nanolithographic resolution.

25.3 Patterning Using Block Copolymers

Block copolymers, hereby abbreviated as BCs, are a class of macromolecules in which two or more chemically distinct polymer chains are covalently linked at their respective chain ends. Due to the very small mixing entropy for large-chain polymers, each of the polymer "blocks" strive to pull apart from each other, but are restrained by the effect of the covalent bond that limits the maximum distance that the polymer "blocks" can separate. The free energy of the system is minimized by the formation of a minimum interaction volume between the polymers, and a maximum in chain conformational entropy, which results in the formation of remarkable periodic structures at mesoscopic (10–50 nm) length scales. This behavior has historically been referred to as "microphase separation" due to the formation of patterns that form in the molten and solid states (Bates and Fredrickson, 1999).

25.3.1 Block Copolymer Architectures

The simplest of BC architectures is that of the linear AB diblock, consisting of a chain of A monomers covalently bonded to a chain of B monomers. Linear diblock copolymers self-assemble into various morphologies such as spherical, cylindrical, gyroid and lamellae depending on the volume fraction f_A of one block relative to the other, each resulting in unique patterning capabilities, Figure 25.2 (Bates and Fredrickson, 1999).

The microphase separation is determined by molecular parameters such as polymer chain length, volume fractions of the component "blocks", and degree of incompatibility between the blocks (χ). Typically, the bulk structure of BCs are studied using transmission electron microscopy (TEM), or small-angle X-ray scattering (SAXS). From these techniques a rough estimate of the domain spacing, L_o, can be found. Through careful synthetic control, it is also possible to control L_o by changing the total molecular weight (N) of the BC. Theoreticians such as Semenov correctly predicted that the microdomain size (L_o) scales proportionally to $N^{2/3}$ weight for strongly segregating block copolymer pairs ($\chi N \gg 10.5$) (Hadjichristidis *et al.*, 2005). This theoretical result has also been proved experimentally (Xu *et al.*, 2001).

Advances in BC synthetic techniques have enabled the creation of well-defined block copolymers with complex, nonlinear architectures (Hadjichristidis *et al.*, 2005). For example, the incorporation of a third type of monomer enables the formation of ABC triblocks or other advanced architectures, resulting in a wide variety of exotic morphologies (Cheng *et al.*, 2002). For triblock copolymers, a plethora of new and exotic structures have been theoretically predicted and many of them have been observed experimentally, such as a helical structures (Boker *et al.*, 2002), "knitting patterns" (Breiner *et al.*, 1996), core–shell morphologies (Guo *et al.*, 2006), "Archimedean tiling" patterns (Hayashida *et al.*, 2006), and double gyroid structures that consist of two interpenetrating polymer networks (Crossland *et al.*, 2009). Additional complexity can be built into the polymer by changing the molecular topology, such as changing the block sequence (ACB instead of ABC), or by forming star-type architectures in which all three polymers are joined in the center (Hayashida *et al.*, 2007). Small changes in the

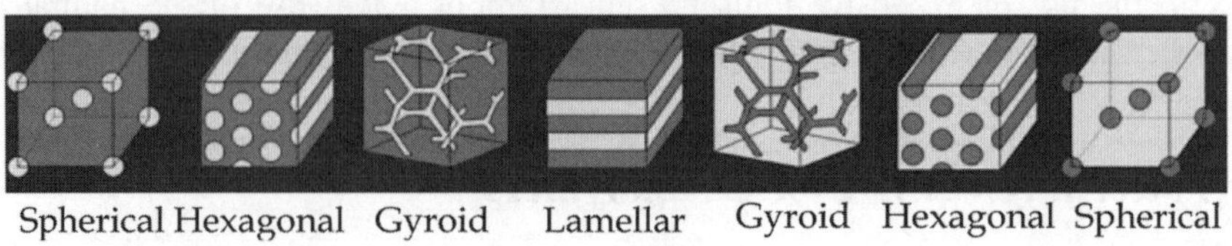

Figure 25.2 Representation of the equilibrium microdomain structure as f_A is increased for fixed χN. (Reproduced with permission from F.S. Bates and G.H. Fredrickson, "Block copolymers – Designer soft materials," *Physics Today*, **52**, 32, 1999. © 1999 American Institute of Physics.)

synthetic strategy can lead to vastly different types of morphologies, as well as differences in material properties. There also exist nonlinear architectures in which the phase behavior varies significantly from those of linear polymers. Examples include BCs with one or more blocks composed of a star, comb or dendrimer structures. Further discussion of BC phase behavior is beyond the scope of this chapter, the reader is referred to a review by Abetz and Simon (Abetz and Simon, 2005).

25.3.2 Nanostructure Formation Using Block Copolymer Thin Films

The lithography community has taken an active interest in BCs in part due to their ability to form nanoscopic features over large areas, simple processing, and low cost of ownership. A typical process flow for BC lithography is shown in Figure 25.3 and some examples of block copolymers used in BC patterning are shown in Table 25.1. Traditionally, poly(styrene-*block*-methylmethacrylate), (PS-*b*-PMMA) has been used as a model system for study

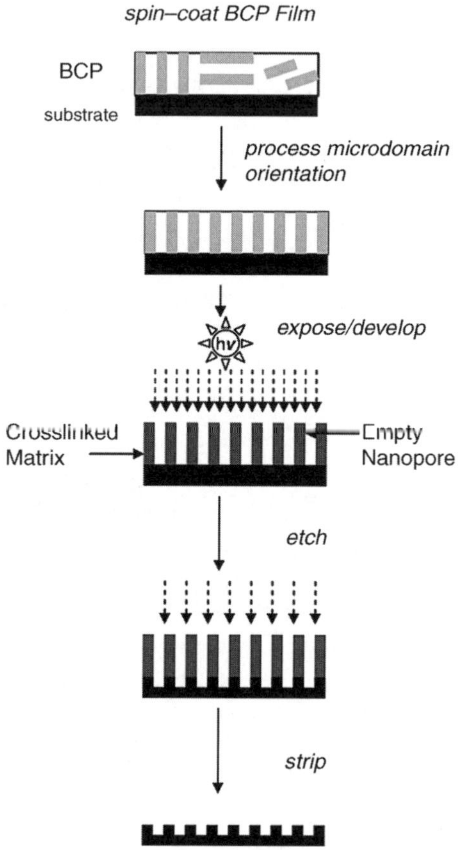

Figure 25.3 Schematic of block copolymer lithography using a combination of positive- and negative-tone resist chemistry. (Reproduced with permission from Kurt E. Geckeler and Hiroyuki Nishide (eds), *Advanced Nanomaterials*, Vol. 1, Wiley-VCH, Wienheim. © 2010.)

Table 25.1 Some commonly used block copolymers for block copolymer lithography

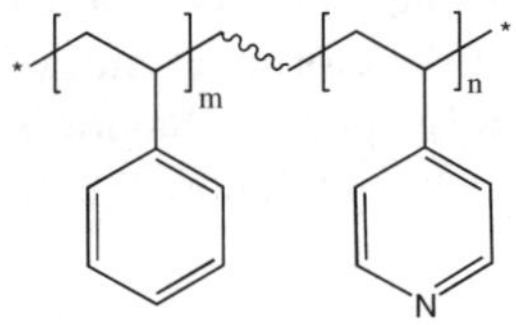	**Poly (styrene-*block*-4-vinylpyridine)** (Hammond *et al.*, 2005)
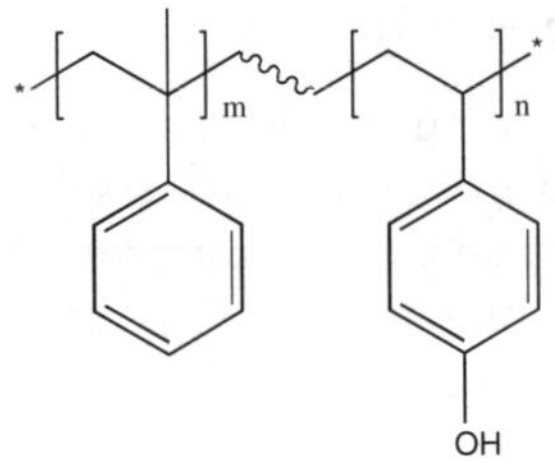	**Poly (*α*-methylstyrene-*block*-4-hydroxystyrene)** (Bosworth *et al.*, 2008; Bosworth *et al.*, 2009)
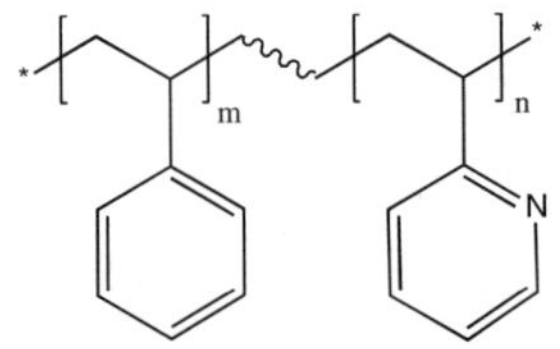	**Poly (styrene-*block*-2-vinylpyridine)** (Segalman *et al.*, 2001)
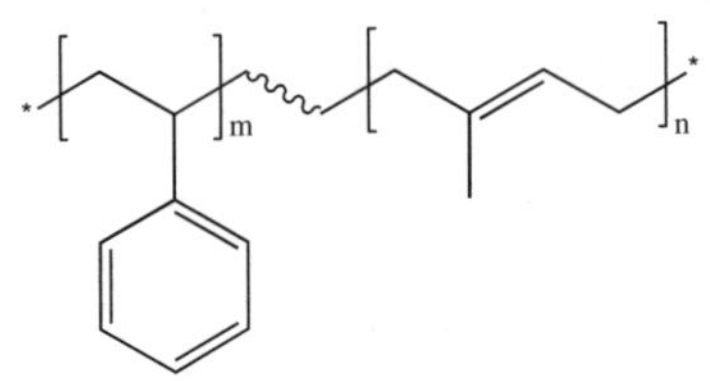	**Poly (styrene-*block*-isoprene)** (Park *et al.*, 2009)
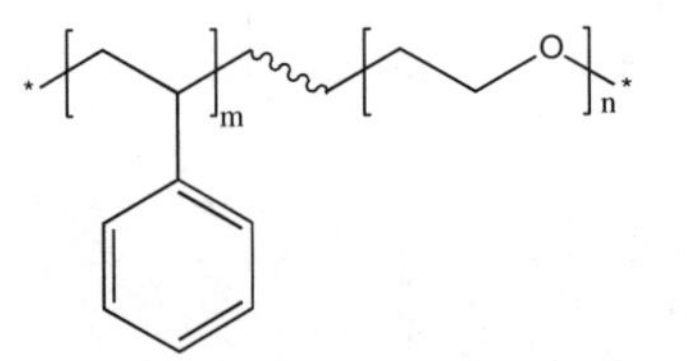	**Poly (styrene-*block*-ethylene oxide)** (Jeong *et al.*, 2004; Freer *et al.*, 2005; Morikawa *et al.*, 2006; Park *et al.*, 2009)
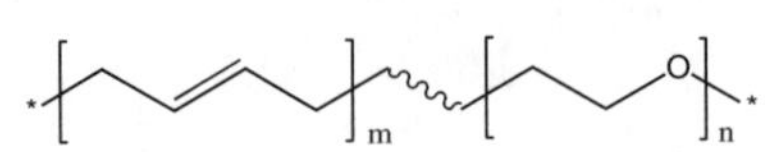	**Poly (butadiene-*block*-ethylene oxide)** (Fasolka *et al.*, 1997)
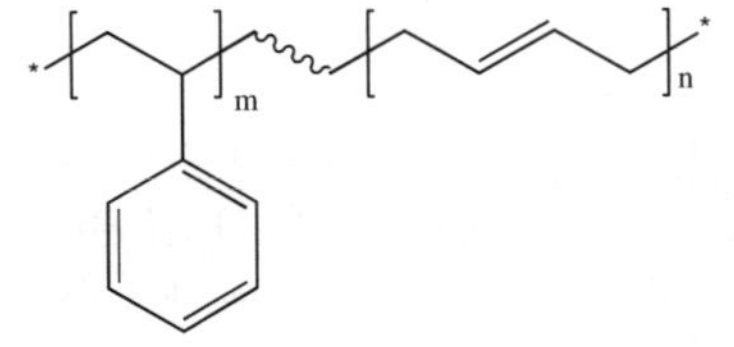	**Poly (styrene-*block*-butadiene)** (Park *et al.*, 2008)

Table 25.1 (*Continued*)

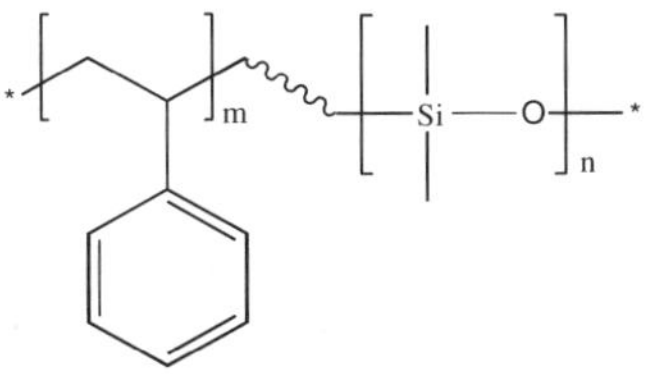	**Poly (styrene-*block*-dimethylsiloxane)** (Nose, 1995; Bita *et al.*, 2009)
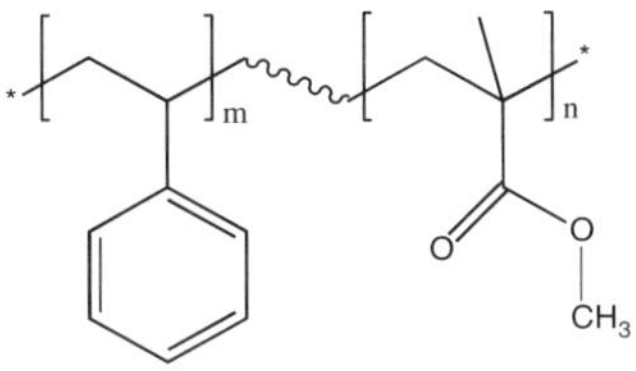	**Poly (styrene-*block*-methyl methacrylate)** (Guillet, 1985; Mansky *et al.*, 1997a, 1997b, 1997c; Huang *et al.*, 1999; Rockford *et al.*, 1999; Thurn-Albrecht *et al.*, 2000; Black, 2007a)
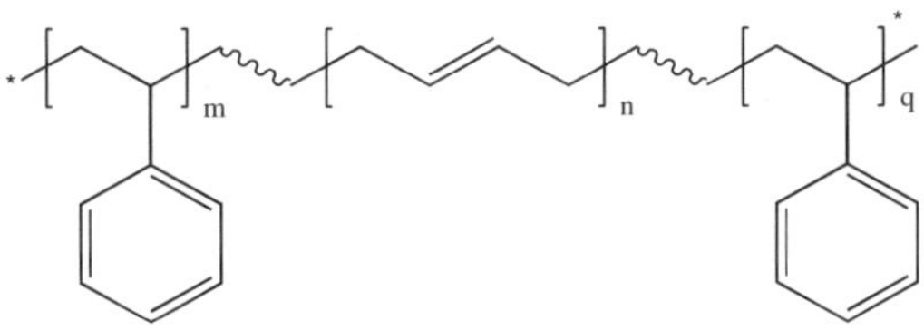	**Poly (styrene-*block*-butadiene-*block*-styrene)** (Kim and Libera, 1998)

in this area, due to its commercial availability and similarity to other common resist materials (Thurn-Albrecht *et al.*, 2000). In this review, we will use this polymer to illustrate several key concepts that will be discussed in subsequent sections.

For lithographic applications, typically the bulk BC is dissolved in a solvent that is suitable for both blocks and spin coated as a thin film on a substrate of choice, usually a cleaned silicon wafer. The wafer is then thermally annealed under vacuum in order to arrange the microdomains in either a parallel or perpendicular fashion, as required for the application of choice. The annealing step is also important to impart the polymer microdomains with a certain degree of "long-range order" (LRO), meaning that the position of the microdomains can be predicted over large areas with very few defects.[1] Upon exposure to deep-UV radiation, each one of the respective blocks acts quite differently. In PS-*b*-PMMA, the PS phase crosslinks, similar to a negative-tone photoresist in traditional photolithography, while the PMMA phase is degraded and removable via wet or dry etching, similar to a positive-tone photoresist. The mechanism of degradation is a chain-scission process, in which the polymer breaks up into its constituent monomers in

[1] The standard length scale for the phrase "good long-range order" has changed as material learning and processing techniques have advanced over the past decade. Not more than five years ago, the phrase "good long-range order" might have meant zero defects over μm^2 areas on the substrate, as measured by atomic force microscopy (AFM) or scanning electron microscopy (SEM). At the time of this writing, using chemical epitaxy techniques (Section 25.4.2), researchers have learned to reduce defectivity by over four orders of magnitude, easily producing self-assembled films with zero defects over cm^2 areas. (Bosworth, J. *et al.* (2008). *ACS Nano* **2(7)**: 1396–1402.) However, it should be noted that this is still not good enough for most semiconductor applications, which have become used to patterning with near zero defects on 8-inch silicon wafers (>300 cm^2 area) using standard optical photolithography techniques.

response to UV light (Guillet, 1985). In turn, the MMA monomers can be selectively removed by immersion in a selective solvent such as acetic acid. The crosslinking of the PS matrix phase adds mechanical strength to the template and avoids the collapse of the nanoporous structure on itself, since the creation of new surfaces is thermodynamically unfavorable. Furthermore, the high carbon content of the PS phase may act as an etch resist in order to carry out various semiconductor etching steps in the underlying substrate to create a replica of the self-assembled nanoporous template. For this reason, perpendicular orientations are more desirable, due to their connection from the top surface of the film to the bottom. In a less-common approach, the nanopores may be filled with metal via electrodeposition to form a negative image of the nanoporous template after removal of the organic phase. The example of PS-*b*-PMMA highlights many important principles in block copolymer lithography which will be discussed in more detail in subsequent sections.

25.3.3 Reaching Equilibrium Using Thermal Annealing

A two-dimensional template for lithographic applications requires a continuous thin film of periodic microdomains on a surface, typically applied to a substrate by spin coating from a dilute solution of organic solvent. However, the fast evaporation resulting from the spin-coating process leaves the microdomains trapped in a disordered, nonequilibrium state. In order to remove energetically costly defects in the film microstructure, polymer chains need mobility in order to reach an equilibrium condition. Typically, heating the film above the glass-transition temperature (T_g) of both blocks is used to enable the microdomains to reach an equilibrium morphology. A general rule of thumb is that reasonable mobility of the polymer film occurs at temperatures 60 °C above the polymer glass-transition temperature (T_g) [$(T - T_g) > 60\,°C$]. Higher annealing temperatures lead to an increase in polymer diffusivity and faster pattern formation. However, the upper limit of the annealing temperature is the polymer's decomposition temperature (T_{dec}), at which the polymer may spontaneously undergo crosslinking and/or chain scission. Kramer and coworkers have found that the best way to induce large grain sizes of a poly(styrene-*block*-4-vinylpyridine) (PS-*b*-P4VP) block copolymer is to increase the temperature of the film above the order–disorder transition temperature (T_{ODT}) where the polymer blocks mix together and remove defects, and then slowly cool the film to below T_{ODT}, where the now more strongly immiscible blocks are allowed to microphase separate to form very sharp interfaces between the blocks (Hammond *et al.*, 2005). For PS-*b*-PMMA, Black and coworkers have found a very narrow processing window between T_{ODT} and T_{dec}, meaning that it is very difficult to use this two-step thermal anneal process in order to improve microdomain ordering in this system (Black, 2007c).

25.3.4 Equilibrium Thin-Film Physics

Even after thermal annealing, the morphology of a BC thin film may be different from that of the bulk BC microstructure. In thin films, the phase behavior is complicated by different effects originating from the film thickness, and interactions between the polymer/substrate and polymer/air interface. Film thickness can be controlled by varying the concentration of the polymer solution and rotational velocity of the substrate during the spin-coating process. In the latter case, the surface energies of each interface will play a large role in the ultimate orientation of the block copolymer microdomains, which have been the subject of a number of theoretical and experimental studies (Park *et al.*, 2008). The two situations that can occur when preferential interactions exist between the blocks and the interfaces are called symmetric and antisymmetric wetting (Russell *et al.*, 1989; Ausserre *et al.*, 1990; Coulon *et al.*, 1990). In symmetric wetting, one block wets both top and bottom interfaces, while in antisymmetric wetting one of the blocks preferentially interacts exclusively with one of the interfaces in order to minimize the free energy of the system.

Most of the current theoretical research of symmetric boundary conditions has been done on lamellar diblock films. Various techniques have been used to determine the influence of film thickness on morphology. Turner used a scaling analysis to study symmetric versus antisymmetric lamellae as a function of film thickness and surface potential (Turner, 1992). Walton and coworkers used this same scaling analysis to study vertically oriented lamellae relative to the substrate (Walton *et al.*, 1994). It was found that this arrangement relaxes the entropic penalty imposed on the chains. Comparing the surface and interfacial energies associated with lamellar orientation, it was determined that domain orientation is film-thickness dependent and for neutral surface energies vertical-oriented lamellae are stable for all film thicknesses (Walton *et al.*, 1994).

Most systems, however, have antisymmetric boundary conditions, such as in the case of films that are supported on atomically flat substrates such as silica or mica. Antisymmetric systems thin-film BCs have received much less theoretical analysis due to the complexity of these systems. Matsen and coworkers have used systems with film thicknesses greater than L_o and concluded that these hybrid films are metastable (Matsen, 1997). Tang and coworkers used systems with film thicknesses equal to the period of the blocks and demonstrated stability in the morphology, which has been supported experimentally by several research groups (Russell *et al.*, 1991; Morkved and Jaeger, 1997; Tang, 2000). In other words, if the film thickness does not match with integer values of L_o, the film will form with two distinct thicknesses as a result of incomplete layer formation. For example, terraces of height L_o will form on top of a foundation film of height $n \times L_o$, with n an integer. Also called "islands and holes", they are most commonly found in spherical, and parallel lamellar morphologies (Segalman *et al.*, 2003a, 2003b; Papadakis *et al.*, 2004). There is a high energy cost for the formation of islands and holes due to the resulting increase in contact area between the polymer film and air. Because of this high energy cost, polymer surface tension usually avoids the formation of a discrete terrace step, usually forming a graded interface in which the film thickness changes over the span of several microdomains. At these transitions, where the local film thickness differs from integer multiples of the lattice constant, differences in the size, or even morphology of the microdomains compared to the bulk may result. For example, cylinder-forming diblocks have been shown to form spheres at the interface between two block copolymer layers (Böker *et al.*, 2002).

25.3.5 Microdomain Orientational Control for Lithography

As discussed previously, vertically oriented cylinders are preferred for many applications in lithography, such as the patterning of vias, contact holes, or bit-patterned media. If neither block in the block copolymer is preferentially attracted to the substrate, the microdomains will tend to orient vertically to maximize the total conformational entropy of the polymer chains. Mansky and coworkers have found that it is possible to force this vertical orientation to occur by controlling the interfacial interactions between the AB diblock copolymer and the substrate. One approach to control such interactions is to modify the surface energy of the substrate using thin films of (A-*ran*-B) random copolymers. The random copolymer surfaces are designed to have a specific binary mixture of each polymer such that the surface energy avoids preferential wetting of either block towards the substrate, resulting in an effective "neutral" substrate. Russell *et al.* demonstrated this technique with the use of a poly(styrene-*ran*-methyl methacrylate) brush layer to treat a silicon substrate before applying a PS-*b*-PMMA block copolymer leading to a perpendicular orientation of the microdomains (Mansky *et al.*, 1997a, 1997b, 1997c; Huang *et al.*, 1999; Huang *et al.*, 2000). Another recent approach to tune the surface energy of the substrate uses an organosilicate polymer called methyl silsesquioxane. Increasing the temperature during the thermal cure step causes more complete condensation of the silicate cage and an accompanying increase in the hydrophobicity of the substrate (Suh *et al.*, 2010).

It is also possible to force vertical orientations by physically roughening the surface of the substrate (Fasolka *et al.*, 1997; Sivaniah *et al.*, 2005). Sivaniah *et al.*, found that as the surface roughness increased, the free energy of parallel-oriented cylinders increased. After the substrate passes a threshold "critical roughness", a perpendicular orientation of the microdomains became the energetically favored orientation (Fasolka *et al.*, 1997; Sivaniah *et al.*, 2005). Other techniques to overcome the natural tendency for microdomains to lie parallel to the substrate include the use of homopolymer blending (Jeong *et al.*, 2004), electric fields (Thurn-Albrecht *et al.*, 2000), magnetic fields (Sakurai, 2008), and surfactant additives (Char *et al.*, 2008).

25.3.6 Lateral and Orientational Control Using Solvent Annealing

Solvent annealing is another method used to gain control over the lateral and orientational ordering of the microdomains. As mentioned previously, for optimal long-range order, sufficient mobility must be achieved in both blocks. Unlike thermal annealing, where the film is heated above the glass-transition temperature, in solvent annealing the solvent vapors act as a plasticizer, lowering the T_g of the polymer such that polymer mobility can occur even at room temperature (van Krevelen and Hoftyzer, 1976). The glass-transition temperature of a plasticized polymer can be described by:

$$T_g = \frac{T_{gP} + (KT_{gS} - T_{gP})\phi_S}{1 + (K-1)\phi_S} \tag{25.2}$$

where T_{gP} is the glass transition temperature of the polymer, T_{gS} is the glass-transition temperature of the solvent, ϕ_S is the volume fraction of the solvent (plasticizer) and K is a constant with typical values between 1 and 3. It has been demonstrated that polymer diffusion increases as the T_g decreases with the addition of solvent (van Krevelen and Hoftyzer, 1976; Zielinski and Duda, 1992; Rauch and Koehler, 2003). As a result of the mobility gained by solvent annealing, strongly segregating block pairs such as poly (styrene-*block*-ethylene oxide) (PS-*b*-PEO, $\chi = 0.08$) are capable of extremely low defect densities after exposure to solvent vapor. 5 μm^2 single grains of vertically oriented cylinders have been achieved using PS-*b*-PEO exposed to an atmosphere of benzene and water for two days (Jeong *et al.*, 2004).

There is also some degree of orientational control involved in the solvent annealing technique. Solvent evaporation from a thin film is highly directional, usually in the direction normal to the substrate. Thus, as the solvent evaporates from the top of the film towards the bottom, domains tend to orient in the same direction, inducing alignment of anisotropic microdomains normal to the direction of solvent evaporation. Kim and Libera have shown that the evaporation rate of the solvent-infused film has a strong bearing on the orientation of BC cylinders in a poly (styrene-*block*-butadiene-*block*-styrene) (PS-*b*-PB-*b*-PS) triblock copolymer (Kim and Libera, 1998). Very slow solvent-evaporation rates (0.2 nL/s) induced inplane formation of cylinders, intermediate evaporation (5 nL/s) produced vertical orientation of cylinders, while fast evaporation (200 nL/s) resulted in a vertical microphase separated structure with poor long-range order.

25.4 Combining "Top-Down" and "Bottom-Up" Patterning Techniques to Enhance Long-Range Order

With the needs of nanotechnology pushing structural dimensions to smaller and smaller limits, the ability to control and fabricate structures on the nanometer length scale is a continuous concern. The marriage of "bottom-up" patterning techniques with "top-down" patterning techniques allows for the best of both worlds, with patterning capabilities at two length scales. Using these techniques, precise control and location of the self-assembled BC films is readily achievable and offers a simple and elegant approach to patterning at small length scales.

25.4.1 Lithographically Defined Topographic Prepatterns

Lithographically patterned topography can help improve the long-range order of BC microdomains. In one of the first examples of the use of graphoepitaxy of block copolymers, Segalman and coworkers confined films of poly (styrene-*block*-2-vinylpyridine) (PS-*b*-P2VP) in microfabricated silica trenches many times larger than the domain period of the block copolymer itself, and thermally annealed the system at 180 °C for 72 h. They found the BC was able to form single crystals of up to 4.5 μm in width both in the trenches and on top of the mesas, as shown in Figure 25.4 (Segalman *et al.*, 2001). In both the trenches and mesas, it was discovered that the single grain of spheres grew "epitaxially" from the boundary edges towards the center of the patterned feature. If the mesa areas were larger than 5 μm, areas of disorder would appear in the center of the mesa. The authors also found that in order to produce good lateral ordering it is critical to control the width and depth of the topographical structures. For example, the depth of the relief structure has to be equal to or greater than L_o (Segalman *et al.*, 2001; Segalman *et al.*, 2003a, 2003b, 2003c).

Since this pioneering work, the process of using topographic prepatterns to physically constrain block copolymers has expanded dramatically (Cheng *et al.*, 2006a). Russell *et al.* investigated block copolymer morphologies confined within cylindrical pores. They found that when the diameter of the pore is comparable to the domain spacing of the block copolymer they are incommensurable (Shin *et al.*, 2004; Xiang *et al.*, 2004; Xiang *et al.*, 2005a, 2005b). Ross and coworkers have also published studies that quantified the effect of the templated self-assembly on block copolymer ordering. They found that if

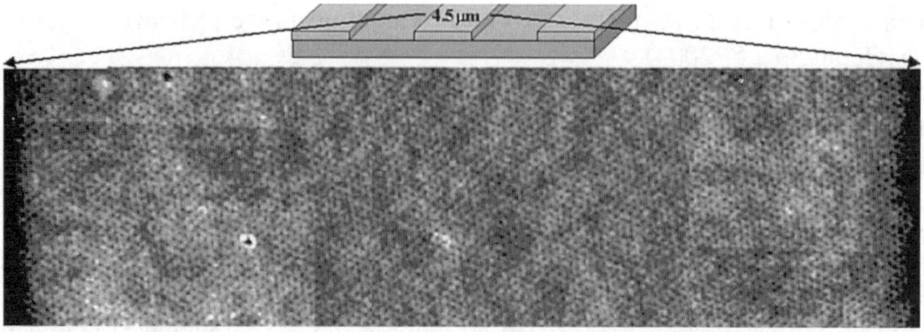

Figure 25.4 SFM of PS-PVP film on top of a mesa. When the mesa edges are less than 5 μm apart a single crystal is formed. (The horizontal lines of imperfection indicate places where 1 × 1 μm SFM images were "sewn" together in order to get a large image with high resolution.) (Reproduced with permission from R.A. Segalman, H. Yokoyama and E.J. Kramer, "Graphoepitaxy of spherical domain block copolymer films," *Advanced Materials*, 2001, **13**, 1152–1155. © Wiley-VCH Verlag GmbH & Co. KGaA.)

the trench width does not match up perfectly to an ideal integer multiple of the period of the block copolymer, the periodicity of the microdomains will compress or expand to fit the trench to minimize the free energy of pattern formation (Cheng *et al.*, 2003). However, the number of rows of block copolymer domains (N) that fit inside the trench width (W) transitions from (N) to ($N + 1$) when $W \approx (N + 0.5) \times L_o$ (Cheng *et al.*, 2004). They also quantified the increase in positional accuracy of sphere-forming poly (styrene-*block*-ferrocenyldimethylsilane) (PS-*b*-PFS) block copolymers via correlation length measurements, finding that the edge roughness of the patterned trenches and inherent fluctuations in the size of the microdomains both contribute to ~3 nm of positional error in the microdomains (Cheng *et al.*, 2006a). The authors noted that this positional error is within the tolerance for bit-patterned media (Moritz *et al.*, 2004). However, currently the field is moving away from this technique, simply because the patterned topography used to confine the BCs takes up too much expensive real estate on the wafer, which could be used for other purposes.

25.4.2 Photolithographically Defined Chemical Patterns

Another approach for lateral orientation control is the use of chemical epitaxy. Chemical epitaxy uses lithographic patterning to define hydrophobic and hydrophilic surface chemistries on a substrate in which a copolymer film is directly applied. Usually, there is a strong surface interaction between the copolymer and the substrate and at least two different areas on the substrate preferentially attract a block. Again, it is important to note that L_o and the period of the surface patterns, W, must be matched as closely as possible. Also, it is important that the top of the films are neutral or have weakly preferential interactions. If these criteria are not met, undesired morphologies and defects can be expected.

In an early work by Rockford and coworkers, alternating stripes of gold and SiO_2 were produced by grazing incidence thermal evaporation of gold on a miscut Si wafer surface. By spin coating lamellar PS-*b*-PMMA over the top of these patterns, they found the corrugated substrate acted as a guide for the perfect long-range ordering of the blocks, and the alternating chemical surface patterns created vertical orientations of the block copolymer (Rockford *et al.*, 1999). They also found that the matching between W and L_o was important in producing low defect density films. For thick films, a mismatch of greater than $\pm 10\%$ in these values caused a loss in orientational control of the microdomains (Rockford *et al.*, 2001).

Recently, this technique was revisited when a miscut sapphire crystal surface with a grazing incidence of gold evaporation was used to orient 4-cm^2-sized grains of PS-*b*-PEO (Park *et al.*, 2009). Nealey *et al.* chemically patterned surfaces and pushed block copolymer lithography into non-symmetrical structures using lithographically defined chemical patterns to direct the self-assembly of PS-*b*-PMMA, as shown in Figure 25.5 (Wang *et al.*, 2000a; Wang *et al.*, 2000b; Yang *et al.*, 2000; Wang *et al.*, 2003). In the first paper from the Nealey group in this area, they patterned a photoresist using extreme-ultraviolet interference lithography (EUV-IL) and transferred the resist pattern with an oxygen plasma etch into an underlying hydrophobic self-assembled monolayer (SAM). After clearing the resist, the resulting substrate featured nanopatterned stripes of nonpolar/polar regions that could be used as a strong surface field to attract the PS and PMMA domains, respectively, of a lamellar PS-*b*-PMMA BC (Kimura *et al.*, 2003).

Since these early studies, the Nealey group has moved towards the use of traditional UV lithography for patterning purposes. Stoykovich and coworkers have shown that vertical lamellae can be directed around the sharp bend and jog structures typical of a microelectronic circuit (Stoykovich *et al.*, 2005). In the case of sharp bends, the block copolymer pattern required the addition of PMMA homopolymer into the BC, which preferentially segregated to areas of high pattern curvature to alleviate the bending stress of the PMMA domain. However, in this work, the block copolymer merely replicates the underlying surface pattern, thereby gaining little lithographic advantage in terms of resolution.

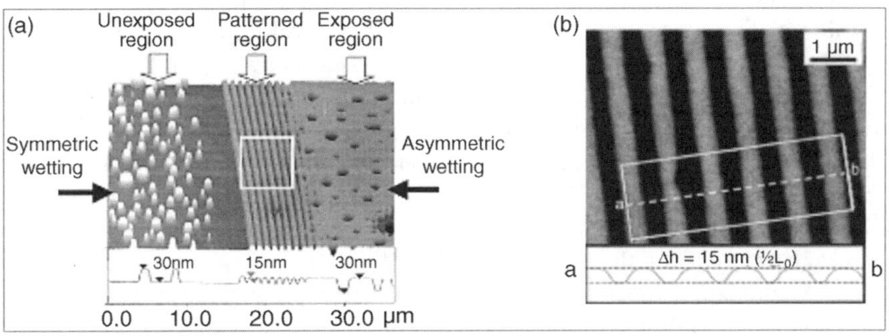

Figure 25.5 (a) AFM image of the surface of a P(S-*b*-MMA) (L_o) 30 nm) film after deposition and annealing on a chemically patterned surface with a period of 900 nm. The initial film thickness was $\sim 2.2L_o$ (66 (2 nm). Unexposed regions showed island topography with a height of ~30 nm (L_o), indicating symmetric wetting. The topography of the P(S-*b*-MMA) film on patterned regions replicated the underlying pattern of the OTS due to different wetting behavior on adjacent stripes. Exposed regions showed hole topography with a depth of ~30 nm (L_o), indicating asymmetric wetting. (b) AFM image of the topography of the film on the patterned region marked in Figure 25.1a. The difference in height between adjacent regions was ~15 nm ($1/2L_o$). (Reprinted with permission from X.M. Yang, R.D. Peters and P.F. Nealey, "Guided self-assembly of symmetric diblock copolymer films on chemically nanopatterned substrates," *Macromolecules*, **33**, 9575–9582, 2000. © 2000 American Chemical Society.)

25.4.3 Density Multiplication of Chemical Nanopatterns

Recently, the concept of "density multiplication" has been introduced into this sub-family of block copolymer orientation control. In this approach, lithographically defined chemical surface patterns with frequencies over twice that of the block copolymer period are used for templating. After printing these "sparse" patterns, consisting of alternating stripes of neutral and PMMA-selective negative-tone resist, a lamellar PS-*b*-PMMA block copolymer was spin coated on top and used to subdivide the lithographically defined patterns up to six times without defects (Cheng *et al.*, 2008). For example, for 2× density multiplication, the negative-tone patterns are placed on a regular lattice such that every other PMMA lamellae encounters the chemical pattern. Due to the strong interactions between the blocks and the sparse chemical surface pattern, the block copolymer is able to rectify any defects in the original electron-beam (e-beam) surface pattern (as illustrated by purposely writing broken lines) while still doubling the frequency of the original pattern. Soon after, Ruiz *et al.* published a paper in *Science* discussing the same frequency doubling and "pattern rectification" of cylindrical PS-*b*-PMMA oriented normal to the surface in order to create perfectly registered hexagonal arrays of holes for bit-patterned media (Ruiz *et al.*, 2008). The concept of density multiplication is a promising technique to reduce the amount of tool time required to accurately place block copolymer features within the confines of arbitrarily placed device geometries.

25.4.4 Directly Patternable Block Copolymer Systems

The use of self-assembly periodic nanostructures offers a route to achieving the desired length scales over a large area. However, for many applications it is not sufficient, as spatial control over the microdomains is difficult to achieve through spin coating and surface prepatterning alone. The Ober group has reported on the development of a block copolymer capable of behaving as a photoresist along with possessing self-assembly capabilities. They have synthesized a block copolymer containing a negative-tone photoresist, poly (4-hydroxystyrene), as the majority phase for high quantum yield, chemically amplified crosslinking reactions with blended PAGs and a resin-type crosslinker (Angelescu *et al.*, 2004; 2005). The directly patternable block copolymer, poly(α-methylstyrene-*block*-4-hydroxystyrene) (PαMS-*b*-PHOST), can also be made into a nanoporous system through removal of the minority PαMS domain via a thermal depolymerization mechanism (Madorsky, 1964). They have found that exposing these thermally sensitive films to solvent vapors allows for long-range order at room temperature, without affecting the crosslinking behavior of the PAG molecules. Therefore, by directly patterning the polymer using optical photolithography, precise placement of self-assembled structures can occur, as shown in Figure 25.6 (Bosworth *et al.*, 2009). In addition, Bosworth *et al.* demonstrated the ability to create multiple morphologies on the same resist layer by solvent annealing in acetone to create a spherical morphology, photocrosslinking a select area of the film using e-beam lithography, then exposing the uncrosslinked area of the film in tetrahydrofuran solvent vapor to produce a parallel cylindrical morphology (Bosworth *et al.*, 2008).

Gopalan and coworkers have also taken steps towards the use of chemical amplification in block copolymers. They have synthesized a block copolymer containing a photoacid generator

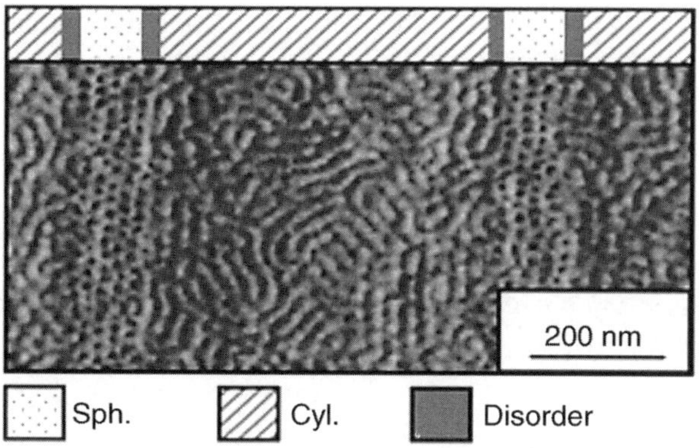

Figure 25.6 AFM phase images demonstrate a single PαMS-b-PHOST film having both regions of spherical morphology and regions of cylindrical morphology. (Reprinted with permission from J.K. Bosworth, C.T. Black and C.K. Ober, "Selective area control of self-assembled pattern architecture using a lithographically patternable block copolymer," *ACS Nano*, **3**, 1761–1766, 2009. © 2009 American Chemical Society.)

built into one block and an acid-sensitive ester linkage in the other block, essentially creating a "one-component patternable block copolymer" (Kim *et al.*, 2009). The Gopalan group also demonstrated a crosslinkable neutralizing underlayer for PS-*b*-PMMA, resulting in patterned areas of perpendicular oriented lamellae surrounded by parallel lamellae (Black *et al.*, 2007). The use of this chemically amplified resist chemistry, well known in lithography for over twenty years, in combination with UV light, could play a major role in block copolymer lithography in the future. With thoughtful choice, these techniques can be extended to other block copolymer systems, thereby widening the library of materials available for patterning on length scales below 30 nm.

25.5 Transferring Nanopatterns Using Dry Etching

For most electronic applications that are using block copolymers, the self-assembled pattern must be transferred into a semiconducting or insulating substrate. In most common practice, a "dry" plasma etching technique is used. In one such technique, called reactive ion etching (RIE), the patterned polymer "resist" is bombarded with highly charged ions that are accelerated towards the surface of the exposed semiconducting wafer. The highly energetic ions can chemically react with the substrate and the polymeric resist, and can also transfer their kinetic energy to physically remove material via a sputtering mechanism. Typically, RIE etch gases are chosen for maximum selectivity to the substrate, using fluorinated gases such as SF_6, CF_4 or CHF_3/Ar for silicon-containing materials. However, in practice 100% selectivity to the substrate does not occur, and the damage of the polymeric resist during RIE limits the amount of substrate etching that can be accomplished before completely removing the patterned resist layer.

When it comes to etching, not all polymers are created equal. In a seminal work, it was found that the etch rate of polymers in an RIE process depends critically on the carbon and oxygen content in a polymer (Gokan *et al.*, 1983). The so-called "Ohnishi parameter" quantifies the polymer etch rate (R_{etch}), as given by

$$R_{etch} \propto \frac{N}{N_c - N_o} \tag{25.3}$$

where N is equal to the total number of atoms in a monomer unit, N_c the number of carbon atoms in the monomer, and N_o the number of oxygen atoms in the monomer. Thus, etch resistance increases with increased amount of carbon atoms in a polymer, and decreases with an increased amount of oxygen in a polymer.

25.5.1 Incorporation of Inorganic Moieties for Improved Etch Resistance

Research in the block copolymer community has also focused on the incorporation of inorganic content into block copolymer structures to increase the etch resistance of the matrix phase and enable a higher aspect ratio etch into the substrate. Two methods that have been described in the literature are the direct polymerization of an inorganic-containing monomer into the block copolymer architecture, and blending inorganic molecules in solution with the block copolymer.

In one approach, block copolymers are synthesized with an inorganic-containing monomer, often containing iron or silicon. Lammertink and coworkers have specialized in the synthesis of block copolymers incorporating poly(ferrocenyldimethylsilane) (PFS), an organometallic polymer consisting of alternating dimethylsilane and ferrocene units, using ring-opening anionic polymerization (Lammertink *et al.*, 1999; Lammertink *et al.*, 2000; Lammertink *et al.*, 2001a, 2001b). By creating BCs such as PI-*b*-PFS, with a minority phase PFS, they have shown the ability to form thin BC films exhibiting organometallic "dots" packed in regular arrays. They used an O_2 RIE plasma to accomplish this, in which the iron in the PFS is much more stable to the plasma than the carbon in the BC. In fact, the PFS phase forms an etch barrier on exposure to oxygen plasma due to the creation of an oxide layer at the surface of the polymer, as determined by Auger electron spectroscopy (AES). These results are coincident with other studies on the oxygen plasma removal of thin polymer films. After the formation of the oxide layer, the etch rate is determined by the competition between the oxide growth and oxide removal by ion sputtering, leading to a 40 : 1 PI : PFS etch selectivity towards the PI block, as determined by the respective homopolymer etch rates. Thus, the inorganic dots can act as conformal masks for high aspect ratio etching into semiconductors such as silicon or gallium arsenide (Lammertink *et al.*, 2000). Poly(styrene-*b*-dimethylsiloxane), (PS-*b*-PDMS) block copolymers work in much the same fashion (Bita *et al.*, 2009).

Another approach is based on blending inorganic macromolecules into one block of the block copolymer. Through this simple approach, a highly stable template can form with an extremely high etch resistance within one of its phases. Silicon-containing polymers are good candidates for these "one-step lithography" applications using block copolymers, due also to their ability to form etch barriers upon exposure to oxygen plasma. In one example of this approach, Freer and coworkers have blended a poly (styrene-*block*-ethylene oxide) (PS-*b*-PEO) block copolymer with an inorganic silsesquioxane (MSSQ) polymer, which is a copolymer of methyl trimethoxysilane and tetraethoxy silane (Freer *et al.*, 2005). The MSSQ blended selectively into the PEO phase, and the authors generated cylindrical and spherical morphologies by changing the blend ratio of the PS-*b*-PEO and SSQ or by changing the PEO block length. Solvent annealing followed by

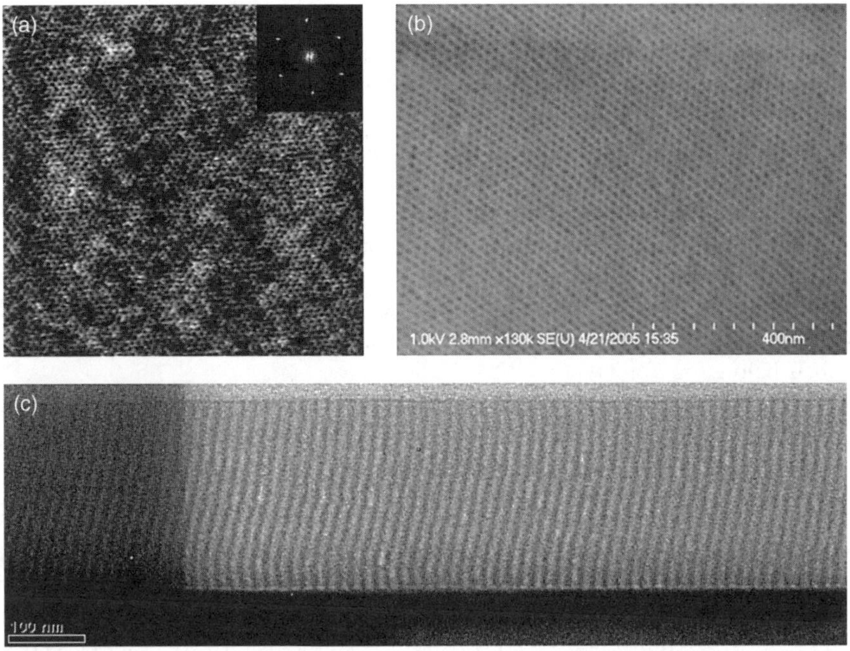

Figure 25.7 Perpendicular orientation of cylindrical pores after annealing for 48 h in a chloroform + octane vapor at 20 °C. (a and b) 1 μm × 1 μm height contrasted AFM image (10-nm height scale) and FESEM micrograph of the surface. PS-*b*-PEO with a MW of 8600 g/mol was used. The inset in (a) is the Fourier transform of the AFM image. (c) TEM cross-sectional micrograph of the porous film shown in (a) and (b). The darker and brighter regions correspond to SSQ and cylindrical pores, respectively. The film thickness is ~250 nm. (Reprinted with permission from E.M. Freer, L.E. Krupp, W.D. Hinsberg *et al.*, "Oriented mesoporous organosilicate thin films," *Nano Letters*, **5**, 2014–2018, 2005. © 2005 American Chemical Society.)

crosslinking the MSSQ phase through a high-temperature bake step allowed for the formation of highly stable, perpendicularly oriented cylinder nanotemplates that could be used for patterning applications, as shown in Figure 25.7. In a separate paper using the same principle, vertical lamellae with a 20-nm half-pitch were used to pattern lines into silicon using a low-power CF_4 plasma etch (Sundstrom *et al.*, 2006). Through a simple blending approach with amphiphilic block copolymers, high-aspect ratio nanostructures with good etch resistance were easily prepared.

25.6 Industrial Applications and Devices Using Block Copolymers

One of the first journal articles about block copolymer lithography, introduced by Park and coworkers from Princeton University in 1997, opened the door for future application of block copolymers in lithography (Park *et al.*, 2008). In this pioneering work, a monolayer of poly (styrene-*block*-butadiene) (PS-*b*-PB) was used in two different approaches to create positive-tone "holes" and negative-tone "nanopillars" on a silicon nitride substrate depending on the processing technique used on the PB block.

In the first approach, the PB block was degraded upon exposure to ozone gas. The highly reactive ozone molecules attack carbon–carbon double bonds in the diene backbone, cutting the linkages and converting the polymer into butadiene monomer, which can easily be dispersed in water. The resulting nanostructure consisted of a periodic array of 1.3×10^{11} holes per square centimeter in a PS matrix that could then be used as an etch mask for an underlying silicon nitride substrate.

The second approach demonstrated the ability to switch the durability of the PB block under RIE conditions. In the inverse process, the reactive double bonds in PB can be stained with a heavy metal vapor of osmium tetroxide, dramatically increasing the polymer's CF_4/O_2 etch resistance relative to the PS matrix ($\sim$2 : 1 etch ratio of PS: osmium-stained PB). This enabled the creation of silicon nanopillars, which foresaw later industrial applications of BCs, as described below.

25.6.1 *Metal Oxide Semiconductor (MOS) Capacitor*

Black and Guarini have demonstrated the use of a PS-*b*-PMMA-based template in order to increase the charge storage capacity of a metal oxide semiconductor (MOS) capacitor, as shown in Figure 25.8. The overall idea is to increase the device surface area using the nanoscale corrugations of the block copolymer as a mask to form the electrode topography (Black *et al.*, 2001). After processing PS-*b*-PMMA in the usual way to leave a crosslinked PS matrix interrupted by a hexagonal array of holes, an SF_6 etch is used to create etch pits in

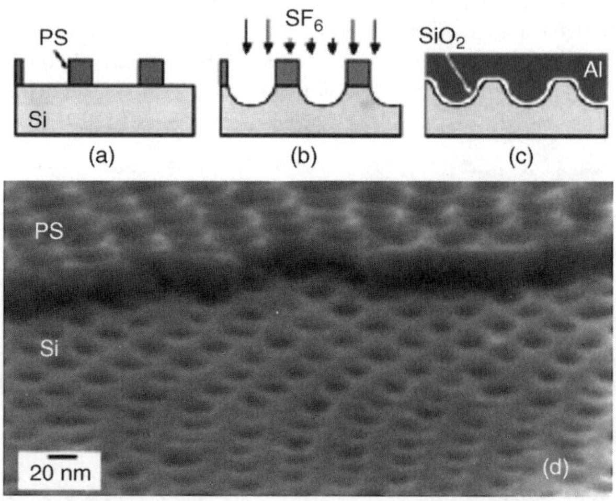

Figure 25.8 Process flow for MOS capacitors is shown. (a) Polymer template formation on silicon surface. (b) RIE pattern transfer of PS template into silicon, followed by the removal of the PS matrix. (c) SiO_2 growth followed by top A1 gate electrode deposition. (d) SEM image at a 70° tilt after RIE etch. Top image shows remaining PS template. Bottom image shows nanoscale hexagonal array has been transferred into Si counterelectrode. (Reproduced with permission from C.T. Black, K.W. Guarini, K.R. Milkove *et al.*, "Integration of self-assembled diblock copolymers for semiconductor capacitor fabrication," *Applied Physics Letter*, **79**, 409, 2001. © 2001 American Institute of Physics.)

the underlying silicon. These etch pits were later covered with aluminum as a gate electrode followed by thermally grown silicon dioxide to form the dielectric. They found that the increase in surface area gained by the block copolymer template improved the capacitance by 30%, although doubling the etch depth would have improved device capacitances up to 115%, based on theoretical calculations.

25.6.2 Bit-Patterned Media for Magnetic Hard Drives

Several hard-drive manufacturing research labs have investigated the possibility of using block copolymer microdomains as templates for high-density bit-patterned media (Cheng *et al.*, 2001; Ruiz *et al.*, 2008; Tada *et al.*, 2009; Yang *et al.*, 2010). This application requires perfect registration of the block copolymer microdomains, which, after etching of silicon nanopillars and deposition of overlaid magnetic material, translates into addressable magnetic domains. Theoretically, e-beam lithography could produce the same pattern densities, but the writing time for such a wafer would be on the order of weeks, which is obviously not industrially feasible. Researchers are currently focusing on the previously mentioned density-multiplication strategies using BCs in order to reduce the time of pattern formation. In this process, e-beam lithography is used to produce a sparse chemical template over the entire surface of the silicon wafer, which guides the perfect self-assembly of vertically oriented, cylindrical PS-*b*-PMMA. This process multiplies the pitch of the original dense lithographic patterns by a factor of at least two, which is equivalent to enhancing the density of pattern formation by a factor of at least four, as seen in Figures 25.9 and 25.10. This technique reduces the amount of writing time needed for high-density patterns, because the e-beam only needs to write every fourth dot (the BC will fill in the other three dots). Even with the savings in process time induced by the density multiplication technique, the high cost of using an e-beam on every hard-drive substrate is a nonstarter for the hard-drive industry. Rather, industrial research groups, such as those at Hitachi, are learning to use perfectly ordered BC nanostructures (low throughput) and etch them into SiO_2 as master templates for nanoimprint lithography (high throughput), as demonstrated already by several academic research groups (Li and Huck, 2004; Chuang *et al.*, 2009).

There are of course some problems associated with the use of block copolymers for bit-patterned media. The first lies with the traditionally used block copolymer for most studies to date, PS-*b*-PMMA. The relatively weak segregation strength of PS-*b*-PMMA ($\chi = 0.06$) (Russell *et al.*, 1990) means that an L_o of ~20 nm ($M_n \sim 30\,000$ g/mol) is the smallest critical dimension possible before the polymer will fail the condition $\chi N > 10.5$, meaning the polymer will no longer microphase separate (Black, 2007c). This is equivalent to a maximum memory density of ~1 Tb/in^2, which will soon be eclipsed by other technologies. Other polymers, however, are under development for sub-20 nm patterning. Pattern densities ~4 Tb/in^2 are possible with the use of other more strongly segregating block copolymers, such as PS-*b*-PDMS ($\chi = 0.26$) (Nose, 1995), which can easily produce 10 nm spacings between microdomains (Bita *et al.*, 2009).

There are also compatibility issues between the lattice geometries of block copolymers and the magnetic arrays that are typically used for bit-patterned media. Considerable amounts of money in the semiconductor industry have been spent on circuit design software and fabrication

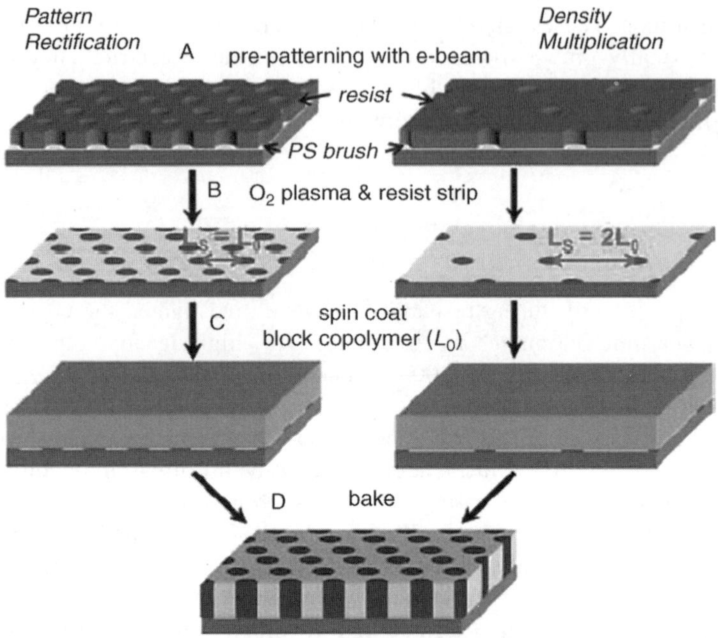

Figure 25.9 Process to create lithographically defined chemically prepatterned surfaces and subsequent directed assembly. (A) Electron-beam lithography patterns at $L_s = L_0$ (left) and $L_s = 2L_0$ (right). (B) Chemical contrast on the substrate after O_2 plasma exposure on the e-beam-defined spots above. (C) Block copolymer thin film. (D) Guided self-assembly in registration with the underlying chemical pattern. (Reproduced with permission from R. Ruiz, H. King, F.A. Detcheverry *et al.*, "Density multiplication and improved lithography by directed block copolymer assembly," *Science*, **321**, 936–939, 2008. © 2008 American Association for the Advancement of Science.)

processes that are based on a rectilinear geometry, whereas typically BC nanostructures arrange on a hexagonally close packed (HCP) lattice. E-beam chemical prepatterns with a spectrum of skew angles have demonstrated the ability to force the packing of PS-*b*-PMMA into square arrays to overcome this problem (Yang *et al.*, 2010). Recently, Park and coworkers have shown the ability to change the lattice orientation of block copolymers using supramolecular hydrogen-bonding interactions (Park *et al.*, 2008).

25.6.3 Airgap Fabrication

In industrial applications thus far, the difficulty in achieving perfect registration of the block copolymer domains over large areas has limited the industrial use of nanoporous block copolymers to applications that depend more on pore-size uniformity rather than long-range lateral ordering. IBMs "Airgap" technology, announced in the spring of 2007, is one early example of how block copolymer nanostructures can be integrated with semiconductor fabrication technology (Black *et al.*, 2007b).

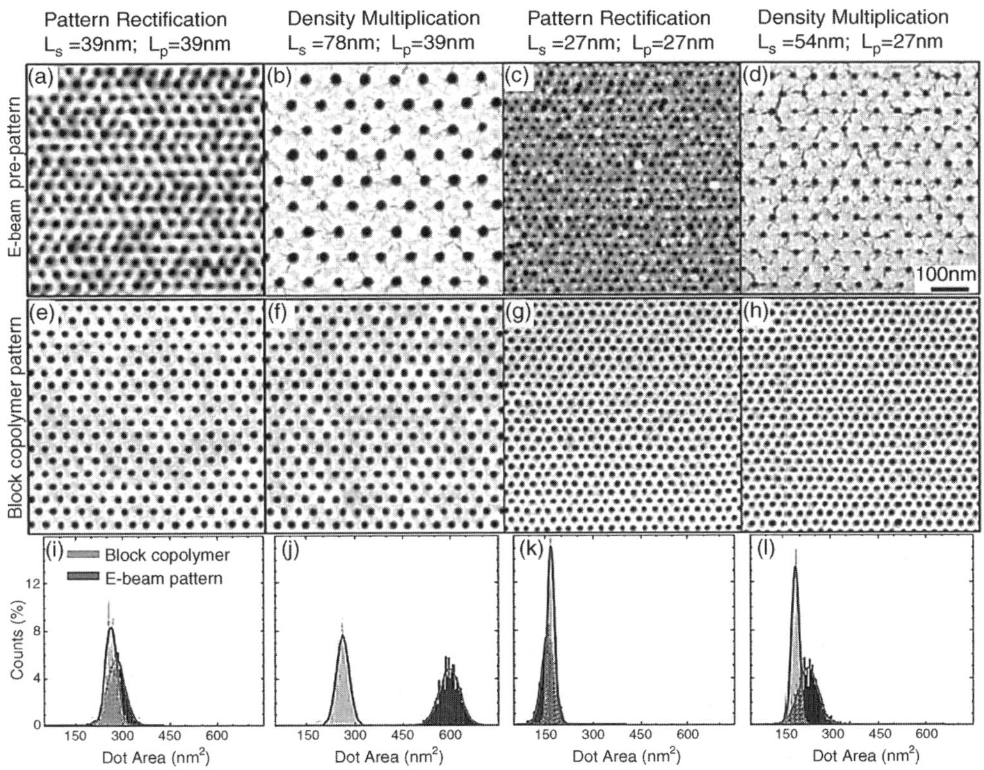

Figure 25.10 (a)–(d) SEM images of developed e-beam resist with $L_s = 39$, 78, 27, and 54 nm, respectively. (e)–(h) SEM images of the block copolymer film on top of the prepattern defined by the corresponding e-beam pattern above. The lattice pitch on the block copolymer samples is $L_p = 39$, 39, 27, and 27 nm, respectively. (i)–(l) Dot size distribution of e-beam and guided block copolymer patterns. (Reproduced with permission from R. Ruiz, H. King, F.A. Detcheverry *et al.*, "Density multiplication and improved lithography by directed block copolymer assembly," *Science*, **321**, 936–939, 2008. © 2008 American Association for the Advancement of Science.)

One of the problems facing future integrated circuit designs is the time delay associated with the addition of wiring layers that interconnect individual transistors. This is a major performance-limiting factor for modern semiconductor devices, and is only expected to get worse as device sizes are scaled down. However, the time delay can be minimized by a smart choice of the materials used in the device design. First, the resistance of current-carrying wires must be minimized by using copper wires instead of aluminum, resulting in a decrease of the RC (resistance × capacitance) time delay by 35%. Secondly, the capacitance in the dielectric insulation surrounding the wires must be minimized through the use of materials with the lowest possible dielectric constant (k). Typically, silicon dioxide ($k \sim 4$) is used as the insulation layer between the wires. The use of air ($k \sim 1$) results in a dramatic reduction in the capacitance in the dielectric layer surrounding the wires and results in an additional reduction in the RC delay by 75%. While other approaches to building "airgaps" do exist, IBM has come up with a process flow that uses the high density of a nanoporous block copolymer film to fabricate the same cheaply and efficiently (Ponoth *et al.*, 2008).

In the airgap approach, a cylinder-forming diblock copolymer (typically PS-*b*-PMMA) is spin coated on top of an organic hard mask of SiO_2. To ensure the formation of vertically oriented microdomains, a surface modification brush layer is used in order to neutralize the affinity of each of the blocks to the surface. After a brief thermal anneal and removal of the PMMA domains, the self-assembled pattern is transferred into the SiO_2 using an RIE etch. In this case, the polymer does not need to have a high degree of long-range order, the pores just need to have dimensions matching the underlying circuitry. A litho-patterned resist is used to "block-out" the areas where the airgap is not needed, which is followed by consecutive etches within the active layer to create holes in the silicon oxide insulator surrounding the wiring. The small holes from the BC template are then sealed in a final deposition step, which also results in a planar surface for subsequent device layers. They have shown that this process enables the electrical signals on the chip to travel 35% faster and reduces the power load by 15% compared to similar chips processed using conventional techniques (Black *et al.*, 2007b). Other devices using block copolymers have also been demonstrated, as discussed in detail in a recent review (Black, 2007c).

25.7 Future Challenges and Outlook

The field of block copolymer lithography has accelerated dramatically in the past ten years, and shows no signs of slowing as the industry moves towards sub-20 nm feature sizes. At this size scale, companies will become more and more interested in low-cost methods that don't require such material and labor-intensive processes as litho-etch, litho-etch double patterning (Hand, 2008) or infrastructure-intensive processes such as EUV (Rice, 2009). As we have reviewed in this chapter, there are many advantages for the use of a combined "top-down" and "bottom-up" lithography patterning technique to respond to these technical challenges.

Some of the challenges facing directed self-assembly for insertion into sub-22 nm node technology were discussed in a recent review by Daniel Herr of the Semiconductor Research Corporation (Herr, 2007). Of the challenges mentioned in this article, perhaps the largest challenge and most fundamental weakness of block copolymers is the inability to print arbitrary lines anywhere on a wafer, which means block copolymers have only been used for the simplest circuit patterns, such as lines and dots. However, pieces have been falling into place that aim to remedy this weakness with increasing amounts of control over the positional placement of the block copolymer domains, which could lead to more advanced architectures. Researchers are learning how to maneuver block copolymer morphology into simple device geometries such as jogs, bends, and t-shaped junctions using lithographic prepatterning techniques (Stoykovich *et al.*, 2005; Cheng *et al.*, 2006b; Cheng *et al.*, 2008).

Another fundamental weakness of block copolymer lithography is the inability to fabricate not only various geometries, but also geometries of multiple sizes and pitches within one resist layer. One possible solution may lie in block copolymers that can be crosslinked through the use of blended photoactive compounds, such as work shown by the Ober group on PαMS-*b*-PHOST (Bosworth *et al.*, 2008). While it has already been shown that solvent annealing using nonselective solvents can be used to enhance long-range order of this system (Bosworth *et al.*, 2009), it may also be possible to use selective solvents to swell individual microdomains to tailor the microdomain size, or if the swollen block is a majority component, the pitch of the block copolymer. Blended functional photoactive compounds in this system may allow the swollen structures to be crosslinked with UV light before moving on to the next size structure.

Other functionalities may also be possible for future generations of lithography, such as the incorporation of a conducting polymer domain that could act like a nanowire (Segalman *et al.*, 2009), or a microdomain that could be "photophysically active". In one example of this approach, photo-active azobenzene pendant groups were attached to a PS-*b*-PEO block copolymer. With the help of a hexane solvent anneal for mobility and a simultaneous *cis–trans* conformational switch of the azobenzene mesogens using UV light, the researchers were able to orient the parallel-oriented cylinders in the same direction of a linearly polarized light source (Morikawa *et al.*, 2006). Perhaps such a focused light beam could also be used to guide block copolymers into arbitrary shapes and patterns.

Block copolymer processing techniques have advanced to the point that is possible to produce zero defects over large cm^2 two-dimensional thin films (Bosworth *et al.*, 2008). However, there is plenty of room for improvement of block copolymer orientation in the third dimension. We have yet to learn how to stack block copolymer layers on top of each other to create integrated gate and contact features, although this principle may have great impact on the device community. Difficulty arises when trying to spin coat one block copolymer over the top of another due to the two resist layers intermingling after exposure to the spin-coating solvent of the upper layer. Recent work has been published in the area of organic electronics that may provide some insight into a solution for this problem, using the principle of orthogonal solvent systems (Lee *et al.*, 2008). In this work, highly chemically sensitive light-emitting organic materials were patterned into a crossbar architecture without any damage to the material, as proven by fluorescence microscopy. This technique of orthogonality requires the chemical synthesis of macromolecules that will not dissolve in typical organic solvents, such as water-soluble or fluorinated block copolymers, both of which have been demonstrated before (Hillmyer and Lodge, 2002; Nakashima and Bahadur, 2006).

Outside of the field of lithography, elegant solutions in the field of electronic materials may emerge that do not require multiple spin-coating steps to form multiple layers of patterned morphologies. For example, linear diblock copolymers may form complex 3D structures such as the gyroid and double gyroid phase, which is being rigorously studied by the energy community for solar and fuel-cell applications, but may also be interesting for the semiconductor industry for the heat dissipation of increasingly power-hungry microprocessors (Crossland *et al.*, 2009). Triblock copolymers with star-shaped architectures have been demonstrated to form interconnected cylinders between lamellar phases, which could be built-in contact and gate structures if replaced by functional polymers (Hayashida *et al.*, 2007). One could envision using these complex 3D architectures to create heterojunctions between two metals, by selective removal of each phase followed by an electrodeposition, similar to work that has already been done by Aizawa and coworkers (Aizawa and Buriak, 2006). Whether or not these exciting technologies will become widely used in semiconductor manufacturing or used only for academic exercises remains to be seen, but fundamental research in this area will undoubtedly continue in the years to come.

References

Abetz, V. and Simon, P.F.W. (2005) Phase behavior and morphologies of block copolymers. *Advances in Polymer Science*, **189**, 125–212.

Aizawa, M. and Buriak, J.M. (2006) Nanoscale patterning of two metals on silicon surfaces using an ABC triblock copolymer template. *Journal of the American Chemical Society*, **128**(17), 5877–5886.

Angelescu, D.E., Waller, J.H. *et al.* (2004) Macroscopic orientation of block copolymer cylinders in single-layer films by shearing. *Advanced Materials*, **16**(19), 1736–1740.

Angelescu, D.E., Waller, J.H. *et al.* (2005) Shear-induced alignment in thin films of spherical nanodomains. *Advanced Materials*, **17**(15), 1878–1881.

Ausserre, D., Chatenay, D. *et al.* (1990) Growth of 2-dimensional domains in copolymer thin-films. *Journal De Physique*, **51**(22), 2571–2580.

Bates, F.S. and Fredrickson, G.H. (1999) Block copolymers - designer soft materials. *Physics Today*, **52**(2), 32–38.

Bita, I., Yang, J.K.W. *et al.* (2009) Graphoepitaxy of self-assembled block copolymers on two-dimensional periodic patterned templates. *Science*, **321**, 939–943.

Black, C.T. (2007a) Block copolymers: nanowire arrays build themselves. *Nat Nano*, **2**(8), 464–465.

Black, C.T. (2007c) Polymer self assembly as a novel extension to optical lithography. *ACS Nano*, **1**(3), 147–150.

Black, C.T., Guarini, K.W. *et al.* (2001) Integration of self-assembled diblock copolymers for semiconductor capacitor fabrication. *Applied Physics Letters*, **79**(3), 409–411.

Black, C.T., Ruiz, R. *et al.* (2007b) Polymer self assembly in semiconductor microelectronics. *IBM Journal of Research and Development*, **51**(5), 605–633.

Böker, A., Elbs, H. *et al.* (2002) Microscopic mechanisms of electric-field-induced alignment of block copolymer microdomains. *Physical Review Letters*, **89**(13), 135502–135505.

Bosworth, J.K., Black, C.T. *et al.* (2009) Selective area control of self-assembled pattern architecture using a lithographically patternable block copolymer. *ACS Nano*, **3**(7), 1761–1766.

Bosworth, J.K., Ruiz, R. *et al.* (2008) Control of self assembly of lithographically patternable block copolymer films. *ACS Nano*, **2**(7), 1396–1402.

Breiner, U., Krappe, U. *et al.* (1996) Evolution of the knitting pattern morphology in ABC triblock copolymers. *Macromolecular Rapid Communications*, **17**(8), 567–575.

Char, K., Son, J.G. *et al.* (2008) Surfactant-induced orientation of thin diblock copolymer films. *PMSE Preprints*, **98**, 30.

Cheng, J.Y., Mayes, A.M. *et al.* (2004) Nanostructure engineering by templated self-assembly of block copolymers. *Nature Materials*, **3**(11), 823–828.

Cheng, J.Y., Rettner, C.T. *et al.* (2008) Dense self-assembly on sparse chemical patterns: rectifying and multiplying lithographic patterns using block copolymers. *Advanced Materials*, **20**, 3155–3158.

Cheng, J.Y., Ross, C.A. *et al.* (2001) Formation of a cobalt magnetic dot array via block copolymer lithography. *Advanced Materials (Weinheim, Germany)*, **13**(15), 1174–1178.

Cheng, J.Y., Ross, C.A. *et al.* (2006a) Templated self-assembly of block copolymers: top-down helps bottom-up. *Advanced Materials*, **18**(19), 2505–2521.

Cheng, J.Y., Ross, C.A. *et al.* (2002) Fabrication of nanostructures with long-range order using block copolymer lithography. *Applied Physics Letters*, **81**, 3657–3659.

Cheng, J.Y., Ross, C.A. *et al.* (2003) Templated self-assembly of block copolymers: effect of substrate topography. *Advanced Materials (Weinheim, Germany)*, **15**(19), 1599–1602.

Cheng, J.Y., Zhang, F. *et al.* (2006b) Pattern registration between spherical block-copolymer domains and topographical templates. *Advanced Materials*, **18**(5), 597–601.

Chuang, V.P., Ross, C.A. *et al.* (2009) Self-assembled nanoscale ring arrays from a poly(styrene-b-polyferrocenylsilane-b-poly(2-vinylpyridine) triblock terpolymer thin film. *Advanced Materials*, **21**, 3789–3793.

Coulon, G., Collin, B. *et al.* (1990). Islands and holes on the free-surface of thin diblock copolymer films. 1. Characteristics of formation and growth. *Journal De Physique*, **51**(24), 2801–2811.

Crossland, E.J.W., Kamperman, M. *et al.* (2009) A bicontinuous double gyroid hybrid solar cell. *Nano Letters*, **9**(8), 2807–2812.

Fasolka, M.J., Harris, D. *et al.* (1997) Observed substrate topography-mediated lateral patterning of diblock copolymer films. *Physics Review Letters*, **79**(16), 3018–3021.

Fasolka, M.J. and Mayes, A.M. (2001) Block copolymer thin films: physics and applications. *Annual Review of Materials Research*, **31**(1), 323–355.

Feynman, R. (1959) There's plenty of room at the bottom. Retrieved March 14, 2010, from http://www.zyvex.com/nanotech/feynman.html.

Freer, E.M., Krupp, L.E. *et al.* (2005) Oriented mesoporous organosilicate thin films. *Nano Letters*, **5**(10), 2014–2018.

Gokan, H., Esho, S. *et al.* (1983) Dry etch resistance of organic materials. *Journal of the Electrochemical Society*, **130**, 143–146.

Guillet, J. (1985) *Polymer Photophysics and Photochemistry: An Introduction to the Study of Photoprocesses in Macromolecules*, Cambridge University Press, Cambridge, UK.

Guo, S., Rzayev, J. *et al.* (2006) Nanopore and nanobushing arrays from ABC triblock thin films containing two etchable blocks. *Chemistry of Materials*, **18**, 1719–1721.

Hadjichristidis, N., Pitsikalis, M. *et al.* (2005) Synthesis of block copolymers. *Advanced Polymer Science*, **189**, 1–124.

Hamley, I.W. (2003) Nanostructure fabrication using block copolymers. *Nanotechnology*, **14**, R39–R54.

Hammond, M., Cochran, E. *et al.* (2005) Temperature dependence of order, disorder, and defects in laterally confined diblock copolymer cylinder monolayers. *Macromolecules*, **38**, 6575–6585.

Hand, A. (2008) Double patterning battles cost, complexity. *Semiconductor International*, **24**(8), 9.

Hawker, C.J. and Russell, T.P. (2005) *MRS Bulletin*, **30**, 952–966.

Hayashida, K., Saito, N. *et al.* (2007) Hierarchical morphologies formed by ABC star-shaped terpolymers. *Macromolecules*, **40**, 3695–3699.

Hayashida, K., Takano, A. *et al.* (2006) Systematic transitions of tiling patterns formed by ABC star-shaped terpolymers. *Macromolecules*, **39**, 9402–9408.

Herr, D.J.C. (2007) Challenges in directed self-assembly. *Future Fab International*, **22**, 69–71.

Hillmyer, M.A. and Lodge, T.P. (2002) Synthesis and self-assembly of fluorinated block copolymers. *Journal of Polymer Science Part A-Polymer Chemistry*, **40**, 1–8.

Huang, E., Mansky, P. *et al.* (2000) Mixed lamellar films: evolution, commensurability effects, and preferential defect formation. *Macromolecules*, **33**(1), 80–88.

Huang, E., Pruzinsky, S. *et al.* (1999) Neutrality conditions for block copolymer systems on random copolymer brush surfaces. *Macromolecules*, **32**(16), 5299–5303.

Ito, H. (2005) Chemical amplification resists for microlithography. *Advanced Polymer Science*, **172**, 37–245.

Jeong, U., Ryu, D.Y. *et al.* (2004) Enhancement in the orientation of the microdomain in block copolymer thin films upon the addition of homopolymer. *Advanced Materials (Weinheim, Germany)*, **16**(6), 533–536.

Kim, J.K., Lee, J.I. *et al.* (2008) Self assembled block copolymers: bulk to thin film. *Macromolecular Research*, **16**(4), 267–292.

Kim, G. and Libera, M. (1998) Morphological development in solvent-cast polystyrene-polybutadiene-polystyrene (SBS) triblock copolymer thin films. *Macromolecules*, **31**(8), 2569–2577.

Kim, Y.J., Kang, H. *et al.* (2009) Synthesis of photoacid generator-containing patternable diblock copolymers by reversible addition-fragmentation transfer polymerization. *Chemistry of Materials*, **21**, 3030–3032.

Kimura, M., Misner, M.J. *et al.* (2003) Long-range ordering of diblock copolymers induced by droplet pinning. *Langmuir*, **19**(23), 9910–9913.

Krausch, G. and Magerle, R. (2002) Nanostructured thin films via self-assembly of block copolymers. *Advanced Materials*, **14**(21), 1579–1583.

Krishnamoorty, S., Hinderling, C., and Heinzeimann, H. (2006) *Materials Today*, **9**(9), 40–47.

Lammertink, R.G.H., Hempenius, M.A. *et al.* (1999) Periodic organic-organometallic microdomain structures in poly (styrene-block-ferrocenyldimethylsilane) copolymers and blends with corresponding homopolymers. *Journal of Polymer Science, Part B: Polymer Physics*, **37**(10), 1009–1021.

Lammertink, R.G.H., Hempenius, M.A. *et al.* (2000) Nanostructured thin films of organic-organometallic block copolymers. One-step lithography with poly(ferrocenylsilanes) by reactive ion etching. *Advanced Materials (Weinheim, Germany)*, **12**(2), 98–103.

Lammertink, R.G.H., Hempenius, M.A. *et al.* (2001a) Poly(ferrocenyldimethylsilanes) for reactive ion etch barrier applications. *Chemistry of Materials*, **13**(2), 429–434.

Lammertink, R.G.H., Hempenius, M.A. *et al.* (2001b) Morphology and surface relief structures of asymmetric poly (styrene-block-ferrocenylsilane) thin films. *Macromolecules*, **34**(4), 942–950.

Lazzari, M. and Lopez-Quintela, M.A. (2003) Block copolymers as a tool for nanomaterial fabrication. *Advanced Materials*, **15**(19), 1583–1594.

Lee, J.-K., Chatzichristidi, M. *et al.* (2008) Acid-sensitive semiperfluoroalkyl resorcinarene: an imaging material for organic electronics. *Journal of the American Chemical Society*, **130**, 11564–11565.

Li, M., Coenjarts, C.A. *et al.* (2005) Patternable block copolymers. *Advanced Polymer Science*, **190**, 183–226.

Li, H.-W. and Huck, W.T.S. (2004) Ordered block-copolymer assembly using nanoimprint lithography. *Nano Letters*, **4** (9), 1633–1636.

Mack, C.A. (2007) The future of semiconductor lithography: after optical, what next? *Future Fab International*, **23**, 65–69.

Madorsky, S.L. (1964) *Thermal Degradation of Organic Polymers*, Interscience, New York.

Mansky, P., Liu, Y. *et al.* (1997a) Controlling polymer-surface interactions with random copolymer brushes. *Science*, **275**(5305), 1458–1460.

Mansky, P., Russell, T.P. *et al.* (1997b) Interfacial segregation in disordered block copolymers: effect of tunable surface potentials. *Physics Review Letters*, **79**(2), 237–240.

Mansky, P., Russell, T.P. *et al.* (1997c) Ordered diblock copolymer films on random copolymer brushes. *Macromolecules*, **30**(22), 6810–6813.

Matsen, M.W. (1997) Thin films of block copolymer. *Journal Chemistry Physics*, **106**(18), 7781–7792.

Moore, G. (1965) Cramming more components on integrated circuits. *Electronics*, **38**(8), 114–117.

Morikawa, Y., Nagano, S. *et al.* (2006) Optical alignment and patterning of nanoscale microdomains in a block copolymer thin film. *Advanced Materials*, **18**, 883–886.

Moritz, J., Buda, L. *et al.* (2004) Writing and reading bits on pre-patterned media. *Applied Physics Letters*, **84**(9), 1519–1521.

Morkved, T.L. and Jaeger, H.M. (1997) Thickness-induced morphology changes in lamellar diblock copolymer ultrathin films. *Europhysics Letters*, **40**(6), 643–648.

Nakashima, K. and Bahadur, P. (2006) Aggregation of water-soluble block copolymers in aqueous solutions: recent trends. *Advances in Colloid and Interface Science*, **123–126**, 75–96.

Nose, T. (1995) Coexistence curves of polystyrene/poly(dimethylsiloxane) blends. *Polymer*, **36**, 2243–2248.

Papadakis, C.M., Busch, P. *et al.* (2004) Morphological transition in thin lamellar diblock copolymer films as revealed by combined GISAXS and AFM studies. *Advances in Solid State Physics*, **44**, 327–338.

Park, H.J., Kang, M.-G. *et al.* (2009) Large area high density sub-20 nm SiO2 nanostructures fabricated by block copolymer template for nanoimprint lithography. *ACS Nano*, **9**, 2601–2608.

Park, S., Kim, B. *et al.* (2008) Fabrication of highly ordered silicon oxide dots and stripes from block copolymer thin films. *Advanced Materials*, **20**(4), 681–685.

Park, C., Yoon, J. *et al.* (2003) Enabling nanotechnology with self assembled block copolymer patterns. *Polymer*, **44** (22), 6725–6760.

Ponoth, S., Horak, D., Colburn, M.E. *et al.* (2008). IBM Airgap Process. *The Electrochemical Society, conference proceedings.*

Rauch, J. and Koehler, W. (2003) Collective and thermal diffusion in dilute, semidilute, and concentrated solutions of polystyrene in toluene. *Journal of Chemistry and Physics*, **119**(22), 11977–11989.

Reichmanis, E., Houlihan, F.M. *et al.* (1991) Chemical amplication mechanisms for microlithography. *Chemistry of Materials*, **3**(3), 394–407.

Reichmanis, E., Houlihan, F.M. *et al.* (2002) Chemical amplification mechanisms for microlithography. *Chemistry of Materials*, **3**(3), 394–407.

Rice, B. (2009) EUV Readiness- An infrastructure based perspective. *Powerpoint Slides*. Retrieved from website http://docs.google.com/viewer?a=v&q=cache:KM64HBsfQmMJ:www.semi.org/cms/groups/public/documents/web_content/ctr_030810.pdf + EUV + readiness-an + infrastructure + based + per&hl=en&gl=us&pid=bl&-srcid=ADGEESgBT43lTJEt7oI0GFBJn9qA_3qmp_RBDM6XnundmcvMZf_8SmE5GaKOvgq9vVqSpp8zS-a9DABYE_CXn8KMZifKQEkikfTCSBYMN8PAobbGCo3XxLSMcbUBRw0-Zi9zyx8GdiqlJ&sig=AHIEtbQ_cJRh_zErplSSw9o_o5pR6lAN1A&pli=1.

Rockford, L., Liu, Y. *et al.* (1999) Polymers on nanoperiodic, heterogeneous surfaces. *Physical Review Letters*, **82**(12), 2602–2605.

Rockford, L., Mochrie, S.G.J. *et al.* (2001) Propagation of nanopatterned substrate templated ordering of block copolymers in thick films. *Macromolecules*, **34**(5), 1487–1492.

Rothschild, M., Bloomstein, T.M. *et al.* (2004) Liquid immersion lithography: why, how, and when? *Journal of Vacuum Science & Technology B*, **22**, 2877–2881.

Ruiz, R., Kang, H. *et al.* (2008) Density multiplication and improved lithography by directed block copolymer assembly. *Science*, **321**, 936–939.

Russell, T.P., Coulon, G. *et al.* (1989) Characteristics of the surface-induced orientation for symmetric diblock PS/PMMA copolymers. *Macromolecules*, **22**(12), 4600–4606.

Russell, T.P., Hjelm, R.P. Jr. *et al.* (1990) Temperature dependence of the interaction parameter of polystyrene and poly (methyl methacrylate). *Macromolecules*, **23**(3), 890–893.

Russell, T.P., Menelle, A. *et al.* (1991) Unconventional morphologies of symmetric, diblock copolymers due to film thickness constraints. *Macromolecules*, **24**(23), 6263–6269.

Sakurai, S. (2008) Progress in control of microdomain orientation in block copolymers-efficiencies of various external fields. *Polymer*, **49**(12), 2781–2796.

Sanders, D.P. (2010) Advances in patterning materials for 193 nm immersion lithography. *Chemical Reviews*, **110**, 321–360.

Segalman, R.A. (2005) Patterning with block copolymer thin films. *Materials Science and Engineering: R: Reports*, **48** (6), 191–226.

Segalman, R.A., Hexemer, A. *et al.* (2003b) Edge effects on the order and freezing of a 2D array of block copolymer spheres. *Physical Review Letters*, **91**(19), 196101-1–196101-4.

Segalman, R.A., Hexemer, A. *et al.* (2003a) Effects of lateral confinement on order in spherical domain block copolymer thin films. *Macromolecules*, **36**(18), 6831–6839.

Segalman, R.A., McCulloch, B. *et al.* (2009) Block copolymers for organic optoelectronics. *Macromolecules*, **42**, 9205–9216.

Segalman, R.A., Schaefer, K.E. *et al.* (2003c) Topographic templating of islands and holes in highly asymmetric block copolymer films. *Macromolecules*, **36**(12), 4498–4506.

Segalman, R.A., Yokoyama, H. *et al.* (2001) Graphoepitaxy of spherical domain block copolymer films. *Advanced Materials*, **13**(15), 1152–1155.

Shaw, J.M., Gelorme, J.D. *et al.* (1997) Negative photoresists for optical lithography. *IBM Journal of Research and Development*, **41**(1/2), 81–94.

Shin, K., Xiang, H. *et al.* (2004) Curving and frustrating flatland. *Science*, **306**, 76.

Sivaniah, E., Hayashi, Y. *et al.* (2005) Symmetric diblock copolymer thin films on rough substrates. Kinetics and structure formation in pure block copolymer thin films. *Macromolecules*, **38**(5), 1837–1849.

Stoykovich, M.P., Mueller, M. *et al.* (2005) Directed assembly of block copolymer blends into nonregular device-oriented structures. *Science (Washington, DC, United States)*, **308**(5727), 1442–1446.

Stoykovich, M.P. and Nealey, P.F. (2006) Block copolymers and conventional lithography. *Materials Today*, **9**(9), 20–29.

Suh, H.S., Kang, H. *et al.* (2010) Orientation of block copolymer resists on interlayer dielectrics with tunable surface energy. *Macromolecules*, **43**, 461–466.

Sundstrom, L., Krupp, L.E. *et al.* (2006) Patterning ~20 nm half-pitch lines on silicon using a self-assembled organosilicate etch mask. *Applied Physics Letters*, **88**, 243107–243109.

Tada, Y., Akasaka, S. *et al.* (2009) Density multiplication by directed self-assembly of block copolymer binary blends. *Journal of Photopolymer Science and Technology*, **22**(2), 229–233.

Tang, W.H. (2000) Confinement of symmetric diblock copolymer thin films. *Macromolecules*, **33**(4), 1370–1384.

Thurn-Albrecht, T., DeRouchey, J. *et al.* (2000) Overcoming interfacial interactions with electric fields. *Macromolecules*, **33**(9), 3250–3253.

Turner, M. (1992) Equilibrium properties of a diblock copolymer lamellar phase confined between flat plates. *Physics Review Letters*, **69**(12), 1788–1791.

van Krevelen, D. and Hoftyzer, P. (1976) *Properties of Polymers: Their Estimation and Correlation Length With Chemical Structure*, 2nd edn, Elsevier Scientific Publishing Company, New York, NY.

Wallraff, G.M. and Hinsberg, W.D. (1999) Lithographic imaging techniques for the formation of nanoscopic features. *Chemical Reviews*, **99**(7), 1801–1822.

Walton, D., Kellogg, G. *et al.* (1994) A free energy model for confined diblock copolymers. *Macromolecules*, **27**(21), 6225–6228.

Wang, Q., Nath, S. *et al.* (2000a) Symmetric diblock copolymer thin films confined between homogeneous and patterned surfaces: simulations and theory. *Journal of Chemical Physics*, **112**, 9996–10011.

Wang, Q., Nealey, P.F. *et al.* (2003) Simulations of the morphology of cylinder-forming asymmetric diblock copolymer thin films on nanopatterned substrates. *Macromolecules*, **36**(5), 1731–1740.

Wang, Q., Yan, Q. *et al.* (2000b) Monte Carlo simulations of diblock copolymer thin films confined between chemically heterogeneous hard surfaces. *Macromolecules*, **33**(12), 4512–4525.

Xiang, H., Shin, K. *et al.* (2004) Block copolymers under cylindrical confinement. *Macromolecules*, **37**(15), 5660–5664.

Xiang, H., Shin, K. *et al.* (2005a) From cylinders to helices upon confinement. *Macromolecules*, **38**(4), 1055–1056.
Xiang, H., Shin, K. *et al.* (2005b) The influence of confinement and curvature on the morphology of block copolymers. *Journal Polymer Science B: Polymer Physics*, **43**(23), 3377–3383.
Xu, T., Kim, H.C. *et al.* (2001) The influence of molecular weight on nano porous polymer films. *Polymer*, **42**(21), 9091–9095.
Yang, X., Wan, L. *et al.* (2010) Directed block copolymer assembly versus electron beam lithography for bit-patterned media with areal density of 1 terabit/inch^2 and beyond. *ACS Nano*, **3**, 1844.
Yang, X.M., Peters, R.D. *et al.* (2000) Guided self-assembly of symmetric diblock copolymer films on chemically nanopatterned substrates. *Macromolecules*, **33**(26), 9575–9582.
Zielinski, J. and Duda, J. (1992) Predicting polymer/solvent diffusion coefficients using free-volume theory. *AICHE Journal*, **38**(3), 405–415.

26

Ring Polymers: Effective Isolation and Unique Properties

Haskell W. Beckham
School of Materials Science and Engineering, Georgia Institute of Technology, Atlanta, USA

Compared to the historical and current efforts focused on linear polymer architectures, relatively little has been dedicated to ring polymers. However, the existence of molecular ring topologies has been considered from the earliest days of the development of macromolecular science as a separate discipline. Prior to the emergence of Staudinger's macromolecular theory in the 1920s, the prevailing idea was that colloidal substances such as rubber and cellulose were composed of cyclic molecules held together with secondary forces (aggregate theory of polymers) (Furukawa, 1998). This was based on the lack of end groups found in chemical analysis and the widespread reluctance to accept that molecules could be very large (e.g. $> \sim 5000$ g/mol). Upon acceptance of the macromolecular concept, the cyclic topology was quickly placed aside for linear motifs, which of course became the basis for the development of the science and engineering of polymers. Cyclic structures were more often than not topologies to be avoided in quests to synthesize high molecular weight linear polymers with superior physical properties. This changed for a variety of reasons. It was discovered that peptides (Saska *et al.*, 2008), proteins (Craik, 2009) and DNA (Dulbecco and Vogt, 1963) are synthesized in nature with cyclic structures and questions were naturally asked whether the ring topology provided enhanced or unique properties over linear structures. Synthetic ring polymers were prepared but yields were typically low due to inefficient cyclization and purification schemes. This is especially true for large molecules as properties of macromolecules with different topologies are increasingly similar with increasing molecular weight, rendering isolation of ring polymers a challenge. Synthetic routes to rings commonly yield product mixtures containing a variety of topologies (Figure 26.1).

Complex Macromolecular Architectures: Synthesis, Characterization, and Self-Assembly, First Edition.
Edited by Nikos Hadjichristidis, Akira Hirao, Yasuyuki Tezuka and Filip Du Prez.

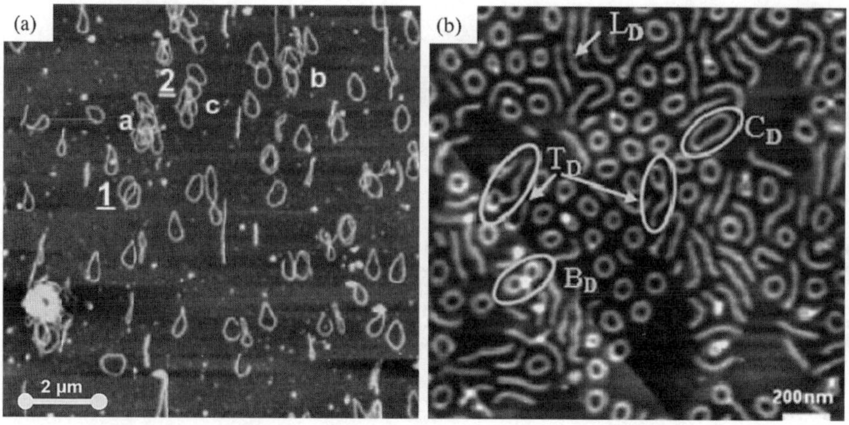

Figure 26.1 Atomic force microscopy (AFM) images of product mixtures from (a) enzymatic cyclization of DNA (Harmon *et al.*, 2003), and (b) dilute-solution cyclization of a α,ω-heterodifunctional synthetic polymer (Deffieux and Schappacher, 2009). In addition to rings, both product mixtures contain linear precursors, catenanes, and other topological byproducts that must be removed to isolate the rings. Both macromolecules were "fattened up" to facilitate visualization by AFM: DNA by coating with a protein and the synthetic polymer by grafting a high density of polystyrene side chains. ((a) Reproduced with permission from F.G. Harmon, J.P. Brockman and S.C. Kowalczykowski, "RecQ helicase stimulates both DNA catenation and changes in DNA topology by topoisomerase III," *Journal of Biological Chemistry*, **278**, 42668–42678, 2003. © 2003 The American Society for Biochemistry and Molecular Biology. (b) With kind permission from Springer Science + Business Media: *Cellular and Molecular Life Science*, "Naturally occurring and synthetic cyclic macromolecules," **66**, 15 © 2009, 2599–2602, A. Deffieux and M. Schappacher.)

Efficient routes for synthesis (Laurent and Grayson, 2009; Tezuka, 2005; Bielawski, *et al.*, 2002) and isolation (Lee *et al.*, 2000; White *et al.*, 2002; Singla *et al.*, 2003) of ring polymers have now been developed. Physical studies of natural and synthetic ring polymers have been conducted that reveal their unique properties. This chapter summarizes some of these studies, with an emphasis on synthetic ring polymers and on work published since 2000, the year that the second edition of *Cyclic Polymers*, edited by J. A. Semylen (Semlyen, 2000), was published. That book is an excellent compilation of reviews on various aspects and types of synthetic and natural ring polymers. Since its publication, research in the field of ring polymers has continued to flourish. Effective isolation and unique properties of rings are described below.

26.1 Effective Isolation

Cyclic macromolecules, or ring polymers, have been synthesized by one of three fundamental approaches: (1) backbiting of linear precursors (Jones *et al.*, 1974; Wright and Beevers, 1986; Ruddick *et al.*, 1999; Kricheldorf, 2003; Foston and Beckham, 2010a), (2) ring expansion by bond insertion (Bielawski *et al.*, 2002; Shea *et al.*, 1998; Kricheldorf *et al.*, 1998; Kricheldorf *et al.*, 2002; Bielawski *et al.*, 2003), or (3) end-to-end coupling of difunctional linear precursors in dilute solution (Geiser and Höcker, 1980; Roovers and Toporowski, 1983;

Yin and Hogen-Esch, 1993; Kubo *et al.*, 1997; Hodge and Peng, 1999; Sun *et al.*, 1995; Yan *et al.*, 1993; Yu *et al.*, 1996a; Lepoittevin *et al.*, 2000; Oike *et al.*, 2001a; Tezuka *et al.*, 2002; Oike *et al.*, 2001b; Tezuka and Komiya, 2002; Pantazis *et al.*, 2002; Schappacher and Deffieux, 2001; Cramail *et al.*, 2000; Iatrou *et al.*, 2002; Laurent and Grayson, 2006). Backbiting of linear precursors, which occurs during and therefore includes ring-chain equilibration, gives rise to a very broad distribution of cyclic and linear polymer sizes. Extensive and exhaustive purification is required to separate the rings from the linear chains and yields for any given cyclic polymer are low. Nevertheless, Semlyen *et al.* used this route to prepare a wide range of cyclic poly(dimethylsiloxane)s on which the most comprehensive set of physical data on ring polymers have been collected and reported (Semlyen, 1986; Semlyen, 2000). In principle, the ring-expansion route should not lead to linear byproducts and therefore isolation of cyclic polymers should be straightforward. However, this route is limited at the present time to a handful of ring polymers, such as cyclic poly(ethylene glycol) phthalates (Kricheldorf *et al.*, 2002), cyclic polybutadiene (Bielawski *et al.*, 2003) and cyclic polyethylene (Bielawski *et al.*, 2002). The end-to-end coupling route is the most versatile and has been used to prepare a large variety of cyclic polymers. With this route, product mixtures include linear byproducts from unreacted linear precursor, intermolecular chain extension, or end capping of the linear precursor at one or both ends by a coupling agent. Purification requires separation of cyclic products from these linear byproducts.

Of course, the first step in effectively isolating rings is simply optimization of the synthetic procedure so that linear byproducts are minimized. The least efficient method in this regard is ring-chain equilibration, while the most efficient is ring expansion. The efficiency of the versatile end-to-end coupling route is somewhere in between, depending on the details of the approach. Ring closure in dilute solution requires one of two basic types of linear precursors: α,ω-homodifunctionalized (Geiser and Höcker, 1980; Roovers and Toporowski, 1983; Lepoittevin *et al.*, 2000) or α,ω-heterodifunctionalized (Kubo *et al.*, 1997; Schappacher and Deffieux, 2001; Rique-Lurbet *et al.*, 1994). α,ω-Homodifunctionalized precursors can be cyclized with either a coupling agent or a catalyst. When a coupling agent is required, the first step is an intermolecular reaction between precursor and coupling agent. This is chain extension and must be controlled to limit its extent. Ideally, a single chain extension event is followed by rapid cyclization. One method for limiting chain extension is to add the coupling agent very slowly (e.g. using a syringe pump) to a dilute solution of linear precursor. In this manner, a molar shortage of coupling agent to linear precursor can be maintained to minimize the probability that the coupling agent reacts at both ends of a single linear precursor to create a linear byproduct. This slow-addition high-dilution method has been successfully employed to prepare cyclic poly(oxyethylene) in yields $> 75\%$ (Sun *et al.*, 1995). Even so, methods that avoid the use of coupling agents were developed and, at least in principle, might be expected to be more efficient. These include catalytically driven cyclization of α,ω-homodifunctionalized precursors and cyclization of α,ω-heterodifunctionalized precursors. In these latter two approaches, no intermolecular steps are required and ring closure can be carried out in highly dilute solution ($< 10^{-3}$ M) to maximize efficiency of cyclic formation. However, overall efficiencies depend on catalytic efficiencies and on yields for intermediate steps required to prepare precursors. For all of the dilute-solution approaches, large solvent volumes may be avoided by use of pseudo-high-dilution conditions, in which reagents are added slowly (e.g. over timescales of hours to days using multiple syringe pumps) so that transient local reagent concentrations are maintained at very low levels (Malesevic *et al.*, 2004).

Ultimately, for all of the synthetic approaches, separation of topological byproducts must be conducted to isolate ring polymers of interest, a significant challenge that becomes more daunting with increasing molecular weight as physical properties become more similar. Various strategies have been reported for isolating rings:

1. solvent fractionation (Roovers and Toporowski, 1983);
2. preparative size-exclusion chromatography (Dodgson *et al.*, 1978);
3. liquid chromatography at the critical condition (Pasch *et al.*, 1996);
4. quenching with ion-exchange resin to remove charged chains (White *et al.*, 2002);
5. inclusion complexation of linear chains with cyclodextrins (Singla *et al.*, 2003).

Solvent fractionation and size-exclusion chromatography (SEC) are the most standard methods in the list and the ones routinely employed to isolate ring polymers, including those produced biologically. Following solvent extraction of homogenized plant biomass, naturally occurring cyclic proteins (1 g per kg of plant material) are isolated by solvent fractionation and liquid chromatography. Separation of individual cyclic peptides (e.g. at least 17 from the African herb *Oldenlandia affinis*) is achieved by preparative reversed-phase liquid chromatography (Plan *et al.*, 2007).

Synthetic product mixtures of linear precursor and cyclic product often exhibit SEC chromatograms with two overlapping peaks; the peak located at longer retention times is cyclic product due to its smaller hydrodynamic volume. When size-exclusion chromatography is conducted under conditions for which retention time of linear byproducts due to size is offset by adsorptive interaction with the gel (i.e. the critical condition), rings and linear chains can be more easily resolved and separated. At the critical condition for linear polymers, retention of linear polymers is independent of size, while retention of cyclic polymers is still size dependent (Figure 26.2). This technique is referred to as liquid chromatography at the critical point of adsorption (Lepoittevin *et al.*, 2000; Pasch *et al.*, 1996), or liquid chromatography at the critical condition (LCCC) (Lee *et al.*, 2000; Cho *et al.*, 2001; Lee *et al.*, 2002; Takano *et al.*, 2007), and has been demonstrated for cyclization reaction products of dilute-solution ring closure of end-functionalized polystyrene. For the example illustrated in Figure 26.2, the critical condition for linear polystyrene is obtained on a C18-bonded column at 43 °C using a mobile phase of CH_2Cl_2/CH_3CN (57/43, v/v) and a flow rate of 0.5 mL/min (Lee *et al.*, 2000). Using LCCC requires the establishment of a critical condition through variation of temperature or solvent quality, and baseline resolution may not be possible for all linear/cyclic polymer mixtures. Characteristics that could impede separation include low molecular weights, large polydispersities, and the presence of different functional groups in the linear byproducts versus the ring polymers.

For cyclizations that proceed through charged linear precursors, ion-exchange resins may be used to remove linear byproducts from cyclic products (White *et al.*, 2002). This method was demonstrated on the cyclization reaction product obtained after deprotonating α,ω-dihydroxy-poly(dimethylsiloxane) in dilute solution (White *et al.*, 2002; Foston and Beckham, 2010a). Since residual linear species contain silanolate end groups, they can be extracted from the product mixture by quenching the reaction with a cationic exchange resin. This method has proven very effective for low molecular weight linear byproducts.

When linear byproducts are not charged, an alternative strategy is to use cyclodextrins to complex and precipitate linear chains, leaving the rings in solution. Since cyclodextrin

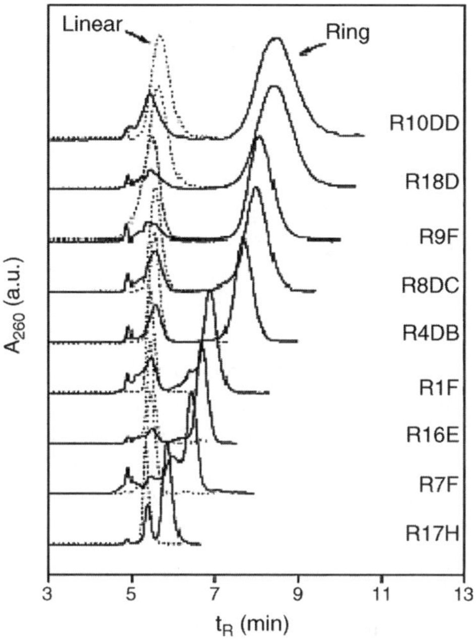

Figure 26.2 Liquid chromatograms, measured at the critical condition for linear polystyrene, of cyclization reaction products of dilute-solution ring closure of end-functionalized PS (solid lines) and their linear precursors (dotted lines) (Lee *et al.*, 2000). The linear precursors elute at nearly identical times (5.4 min), independent of molecular weight. The major peak in the cyclic reaction product, marked "Ring", elutes at times that increase with molecular weight. Note that all "Ring" products contain linear precursor impurities. (Reprinted with permission from H.C. Lee, H. Lee, W. Lee *et al.*, "Fractionation of cyclic polystyrene from linear precursor by HPLC at the chromatographic critical condition,"*Macromolecules*, **33**, 8119–8121, 2000. © 2000 American Chemical Society.)

inclusion complexes can be formed with seemingly most polymers (Huang and Gibson, 2005; Wenz *et al.*, 2006; Harada *et al.*, 2009), this strategy could have widespread utility. The method was originally demonstrated on the cyclization reaction product from the dilute-solution ring closure of α,ω-dihydroxy-poly(oxyethylene) (Singla *et al.*, 2003) but has since been shown to work for isolating cyclic PDMS (Foston and Beckham, 2010b). The inclusion complex, in which cyclodextrins thread onto and cover linear polymers, is called a polypseudorotaxane. Linear poly(oxyethylene) (POE) can be effectively complexed and precipitated from aqueous solution using α-cyclodextrin. Since POE and α-cyclodextrin can be dissolved in a common solvent (e.g. water), this method is straightforward to apply for isolating cyclic POE. The method may also be applied even without a common solvent. For example, inclusion complexes of PDMS and γ-cyclodextrin (γ-CD) can be formed by adding PDMS to a saturated aqueous solution of γ-cyclodextrin, in which the PDMS is not soluble (Okumura *et al.*, 2001; Porbeni *et al.*, 2001). The mixture is heated or sonicated, forming a cloudy suspension that is centrifuged to precipitate the complexed linear PDMS. With increasing PDMS molecular weight, the gravimetric yield of γ-CD decreases (Okumura *et al.*, 2001). Decreased yield with increasing molecular weight occurs for most polymer–cyclodextrin inclusion complexes

(Huang and Gibson, 2005; Wenz *et al.*, 2006; Harada *et al.*, 2009; Girardeau *et al.*, 2005). This method is thus less efficient for extracting large linear byproducts and must be repeated numerous times to isolate pure rings.

In many cases, multiple strategies must be employed to isolate rings. For example, Pantazis *et al.* reported the preparation of a cyclic triblock terpolymer via dilute-solution ring closure and its isolation using a two-step approach (Pantazis *et al.*, 2002). They first evaporated the solvent and continued the reaction under high-concentration conditions to convert residual linear precursor to high molecular weight linear polymer. In a second step they isolated the ring polymer by solvent fractionation. A conceptually similar two-step purification approach was described by Schulz *et al.* (2009) for the product of dilute-solution ring closure of α,ω-dialkynyl-polyisobutylene using a diazide coupling agent. Following "click" cyclization of the linear polyisobutylene (PIB) precursor, residual linear byproducts, assumed to contain azido groups at both chain ends, were further reacted with end-functionalized mono-alkynyl-poly(oxyethylene). The resulting POE-PIB-POE triblock copolymer is amphiphilic and could be separated from cyclic PIB using silica-gel chromatography.

Multiple strategies have also been employed to isolate the intended cyclic product from the mixture obtained from dilute-solution ring closure of α,ω-dihydroxy-poly(oxyethylene) (Sun *et al.*, 1995; Nam *et al.*, 2008). The reaction scheme is shown in Figure 26.3 along with size-exclusion chromatography (SEC) and ^{1}H NMR data that demonstrate how α-CD precipitation removes linear byproducts, but subsequent solvent fractionation is needed to remove large cyclic byproducts. Large cyclic byproducts are identified by the shoulder on the left (low retention time) side of the major SEC peak attributed to the product and by the absence of end-group OH peaks in the ^{1}H NMR spectrum.

As demonstrated in Figure 26.3, the effectiveness of isolation schemes can be assessed with SEC and ^{1}H NMR. Spectroscopic techniques such as NMR and IR are used to monitor the disappearance of linear precursor end groups, a task that becomes forbiddingly difficult with increasing molecular weight. Even using optimized measurement conditions, linear contaminants may be present at concentrations below detection thresholds for spectroscopic and chromatographic techniques. MALDI-TOF mass spectrometry has been extremely helpful in this regard since it can distinguish cyclic from linear chains by the presence of end groups over the entire molecular weight distribution. However, MALDI-TOF is not easy to use quantitatively since ionization or desorption efficiency can decrease with increasing molecular weight, and be different for cyclic and linear species.

Effective isolation of rings from what can be topologically complex product mixtures is important for fundamental studies of properties. Linear contaminants are impediments to interpretation of physical data (e.g. neutron scattering) (Gagliardi *et al.*, 2005) according to existing theories. Even at very low concentrations (e.g. 0.0007 g/cm^3), linear contaminants significantly influence rheological properties of rings (McKenna and Plazek, 1986). These and other properties are discussed in the next section.

26.2 Unique Properties

26.2.1 Ring Size

Tying up the two ends of a linear chain clearly reduces its pervaded volume. Reduced hydrodynamic volumes lead to increased retention times (i.e. decreased apparent molecular

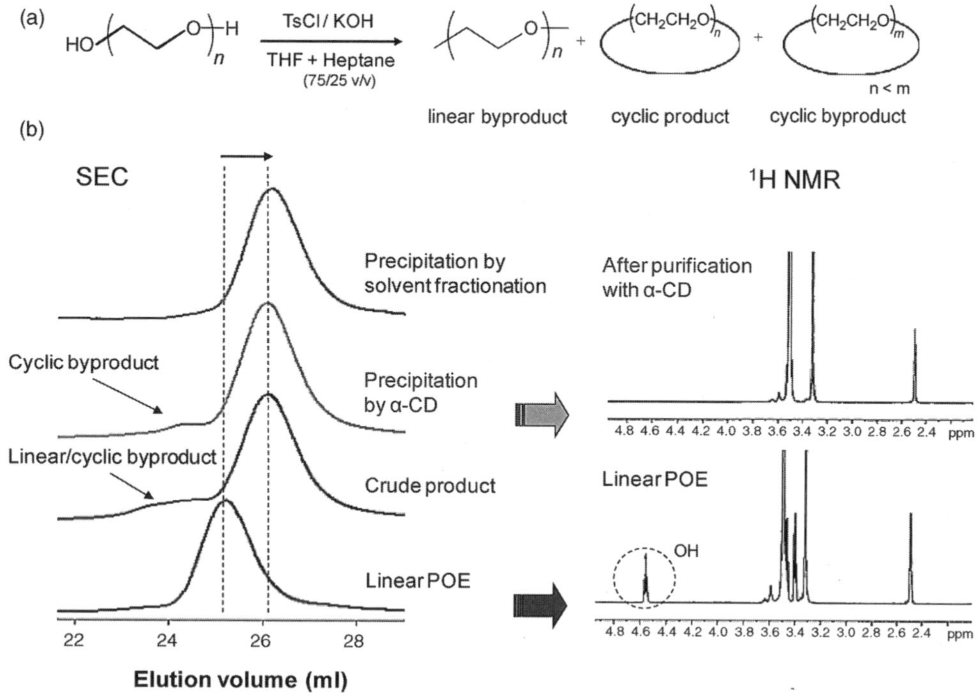

Figure 26.3 (a) Dilute-solution ring closure of α,ω-dihydroxy-poly(oxyethylene) yields a reaction product mixture that contains cyclic and linear byproducts. The cyclic product is isolated by α-CD complexation-precipitation of linear byproducts followed by solvent fractionation to remove large cyclic byproducts. (b) SEC of starting linear precursor, crude product, and product after α-CD and solvent fractionation. ^{1}H NMR spectra of linear precursor and product after purification with α-CD showing disappearance of OH end groups after α-CD precipitation of linear byproducts.

weights) in size-exclusion chromatography, which have long been used as supporting proof of cyclization (Figure 26.3). The reduction in apparent peak molecular weight is often reported as a measure of reduced hydrodynamic volume. The degree to which size is reduced upon cyclization of flexible random coils has been theoretically estimated (Bloomfield and Zimm, 1966; Bensafi *et al.*, 2000), simulated (Jang *et al.*, 2003; Zifferer and Preusser, 2001) and experimentally determined (Schappacher and Deffieux, 2008) for radii of gyration, hydrodynamic radii, and intrinsic viscosities. Ratios of these dimensional characteristics are given in Table 26.1 along with references that experimentally confirm the theoretical predictions for rings with a large variety of primary chemical structures.

The work summarized in Table 26.1 has been conducted for isolated rings in dilute solution, either at theta conditions or in good solvents. Recent small-angle neutron-scattering studies on bulk rings have shown that they adopt conformations more compact than those in solution (Arrighi *et al.*, 2004). The polymers examined were hydrogenous and deuterated linear and cyclic PDMS. For linear PDMS, $R_g \propto M_w^{0.5}$. For cyclic PDMS, $R_g \propto M_w^{0.4}$; thus, $\langle R_{gC}\rangle^2/\langle R_{gL}\rangle^2$ is not simply equal to 0.5 as in solution.

Table 26.1 Dimensional characteristics for cyclic (C) and linear (L) polymers of equivalent numbers of repeat units

Characteristic, Cyclic to Linear	Ratio	References
Radii of gyration, $\langle R_{gC}\rangle^2/\langle R_{gL}\rangle^2$	0.5	(Schappacher and Deffieux, 2008; Higgins *et al.*, 1979; Edwards *et al.*, 1983; Ishizu and Kanno, 1996)
Hydrodynamic radii, R_{hC}/R_{hL}	0.9	(Schappacher and Deffieux, 2008; Hadziioannou *et al.*, 1987; Hodgson and Amis, 1991)
SEC apparent peak molar mass, M_{pC}/M_{pL}	$\geq 0.71^a$	(Roovers and Toporowski, 1983; Oike *et al.*, 2001a; Oike *et al.*, 2001b; Schappacher and Deffieux, 2001; Iatrou *et al.*, 2002; Rique-Lurbet *et al.*, 1994; Takano *et al.*, 2007; Schappacher and Deffieux, 2008; Chen *et al.*, 2004)
Intrinsic viscosities, $[\eta]_C/[\eta]_L$	0.66	(Geiser and Höcker, 1980; Roovers and Toporowski, 1983; Lepoittevin *et al.*, 2000; Bloomfield and Zimm, 1966; Ishizu and Kanno, 1996; Hodgson and Amis, 1991; Dodgson and Semlyen, 1977; Oike *et al.*, 2000)

[a] M_{pC}/M_{pL} values increase with decreasing molecular weight (Chen *et al.*, 2004).

26.2.2 Ring Dynamics

Bulk dynamics of rings, compared to linear chains with the same number of repeat units, are affected by their reduced size, restricted conformational flexibility, and lack of chain ends. These distinctions decrease with increasing molecular weight, as is perhaps most obviously seen in the molecular-weight dependence of the glass transition (T_g). Glass transitions are higher for rings than for linear chains with the same number of repeat units, and the difference decreases with increasing molecular weight until convergence is reached. An example data set is shown in Figure 26.4 for poly(oxyethylene).

For both POE rings and linear chains, T_g decreases with decreasing molecular weight. The first comprehensive data set reported on a cyclic polymer displayed a different trend. The T_gs for cyclic PDMS *increase* with decreasing molecular weight (Clarson *et al.*, 1985; Kirst *et al.*, 1994). This led to some speculation that T_gs for all cyclic polymers should increase with decreasing molecular weight. This has not been the case. With decreasing molecular weight, a T_g increase has been reported for poly(2-vinyl-9,9-dimethylfluorene) (Chen *et al.*, 2004) but not for cyclic PS (Alberty *et al.*, 2005) or cyclic poly(2-vinylnaphthalene) (Chen *et al.*, 2004). Di Marzio and Guttman used configurational entropy considerations to predict the molecular-weight dependence of T_g for cyclic and linear polymers (Di Marzio and Guttman, 1987). Whether a given ring polymer exhibits a decrease or increase in T_g with decreasing molecular weight depends on the ratio between some structure-dependent local energies (Di Marzio and Guttman, 1987). Santangelo *et al.* noted that the T_g for cyclic PS (4.6 kg/mol, purified by LCCC) was the same as the high molecular weight limiting value for linear PS (374 K), which they attributed to "the absence of chain ends and associated configurational freedom" (Santangelo *et al.*, 2001).

Melt dynamics of rings are characterized directly by rheological and diffusion measurements. Early studies seem to have been motivated primarily by the notion that rings cannot

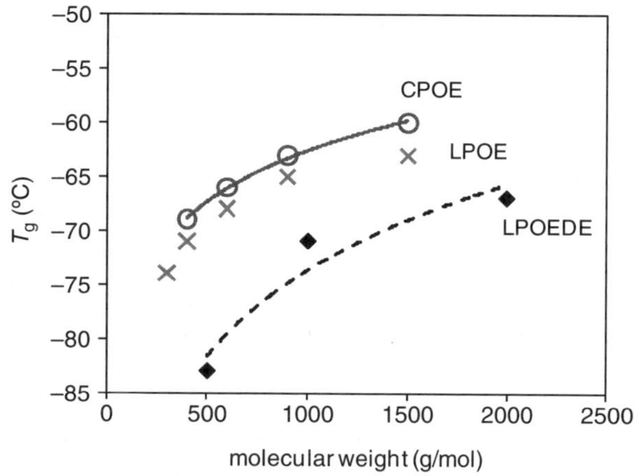

Figure 26.4 Glass-transition temperature versus molecular weight for poly(oxyethylene) (POE): cyclic (CPOE, ○), linear dihydroxy-terminated (LPOE, ×), and linear dimethoxy-terminated (LPOEDE, ◆) (Nam *et al.*, 2008). Lines are shown as guides and suggest convergence at higher molecular weight. T_g for infinitely long POE has been estimated at $-53\,°C$ (Privalko and Lobodina, 1974). (Reprinted from *Polymer*, **49**, 25, S. Nam, J. Leisen, V. Breedveld and H.W. Beckham, "Dynamics of unentangled cyclic and linear poly(oxyethylene) melts," 5467–5473, 2008, with permission from Elsevier.)

reptate in the traditional sense due to their lack of end groups. Since reptation occurs only above a critical molecular weight for entanglement (M_c), large rings were synthesized and examined. Zero-shear melt viscosities of cyclic PS (McKenna *et al.*, 1987; McKenna *et al.*, 1989; Roovers, 1985), cyclic polybutadiene (Roovers, 1988), and cyclic PDMS (Orrah *et al.*, 1988a) were consistently smaller than those of their linear analogs (i.e. same number of repeat units) at molecular weights above M_c. However, comparison of data collected on rings from different laboratories revealed some puzzling inconsistencies. It is instructive to review some of these inconsistencies and what they suggested about the samples.

McKenna *et al.* investigated cyclic PS with zero-shear viscosity (η_0) measurements across a molecular-weight (MW) range from 11 to 185 kg/mol (McKenna *et al.*, 1987). They reported that cyclic PS exhibited nearly the same critical molecular weight for entanglement, M_c, as linear PS (~40 kg/mol), and that linear and cyclic polymers exhibited a nearly identical temperature dependence of the zero-shear viscosities. Roovers collected data on cyclic polystyrene with molecular weights from 7 to 450 kg/mol (Roovers, 1985), which showed an upturn in the viscosity for cyclic PS at 340 kg/mol so that $(\eta_0)_{ring}/(\eta_0)_{linear} \approx 1$. McKenna suggested that this upturn is related to the presence of linear impurities "since small amounts of linear chains can greatly increase the viscosity of the cycles" (McKenna and Plazek, 1986). McKenna also noted that lower viscosities for the Roovers samples in the intermediate-MW range may be due to the presence of knotted rings, which are more likely as the Roovers cycles were prepared by ring closure of linear PS in a *poor* solvent (Roovers and Toporowski, 1983). The samples that McKenna examined were prepared by ring closure of linear PS in a *good* solvent. Knots and linear contaminants in rings prepared by end-to-end coupling of linear precursors are notoriously difficult to remove for high molecular weights. In later creep and recovery measurements, McKenna *et al.* (1989) reported that the M_c of cyclic

PS was roughly double that determined for linear PS (in contrast to their earlier finding that $M_{c,\text{linear}} \approx M_{c,\text{ring}}$), and that the cyclic-PS plateau modulus is half that of linear PS (consistent with Roovers' findings (Roovers, 1985)).

Roovers also examined the effect of adding linear polybutadiene (PB) to pure cyclic PB on the melt viscosity (Roovers, 1988). He attributed the smaller η_0 for cyclic PB to reduced entanglement density. He wrote, "Addition of linear polymers increases the entanglement density, as evidenced by increasing values of the plateau modulus, G_e, and thereby the melt viscosity increases" (Roovers, 1988). As an aside, he noted that synthesizing cyclic PB was "inherently more difficult" than synthesizing cyclic PS but was justified since the M_e for PB is low (2750 g/mol) so that high-MW rings do not have to be prepared and the attendant purity issues can be avoided.

Semlyen *et al.* prepared cyclic poly(dimethylsiloxane)s (PDMS) and reported on their physical properties in what is undoubtedly the most complete investigation of a cyclic polymer reported to date. Cyclic PDMS melt dynamics were investigated with viscosity measurements across a molecular-weight range from 0.5 to 32 kg/mol (Orrah *et al.*, 1988a; Cosgrove *et al.*, 1992; Dodgson *et al.*, 1980). Neat cyclic PDMS was found to exhibit an M_c of 17 kg/mol, essentially identical to that of linear PDMS. Cyclic PDMS is less viscous than equivalent-MW linear PDMS at high molecular weights, but exhibits a crossover in dynamics around 7 kg/mol and is more viscous than linear PDMS at low molecular weights (Dodgson *et al.*, 1980).

Measured using pulsed-field-gradient NMR, self-diffusion coefficients (D) of cyclic PDMS (>6 repeat units) (Cosgrove *et al.*, 1992), cyclic POE (Nam *et al.*, 2008), and cyclic alkanes (von Meerwall *et al.*, 2003) are smaller than their linear analogs (i.e. $D_{\text{ring}} < D_{\text{linear}}$). This is not true for cyclic POE when compared with linear POE end-capped with hydroxyl groups that can hydrogen bond and impede dynamics. Note that of these three polymers, only cyclic PDMS has been studied at molecular weights above M_c. In contrast to these findings, the opposite has been reported for cyclic PS (Kawaguchi *et al.*, 2006) and cyclic DNA (Robertson and Smith, 2007a; Robertson and Smith, 2007b): $D_{\text{ring}} > D_{\text{linear}}$. The latter two studies were conducted on high molecular weight polymers, thus suggesting that a crossover in diffusion dynamics may indeed occur at sufficiently high molecular weights. Larger diffusion coefficients for entangled rings are consistent with the smaller viscosities that have been reported (McKenna *et al.*, 1987; McKenna *et al.*, 1989; Roovers, 1985; Roovers, 1988; Orrah *et al.*, 1988a). Early simulation studies showed that rings diffuse either faster (Müller *et al.*, 1996; Brown and Szamel, 1998a; Brown and Szamel, 1998b) or slower (Pakula and Geyler, 1988) than linear polymers across all sizes studied. More recent simulations of cyclic and linear polyethylenes revealed that cycles diffuse faster than linear polymers only at high molecular weights, with a crossover to slower dynamics occurring near the entanglement molecular weight, M_e, of linear chains (Ozisik *et al.*, 2002; Hur *et al.*, 2006). The crossover to slower dynamics for cyclic polymers at low molecular weights (i.e. $D_{\text{ring}} < D_{\text{linear}}$) was attributed to the smaller size of cyclic chains compared to linear ones of the same length, which results in higher local polymer segment densities.

It was clear that low concentrations of linear contaminants had a large influence on the melt rheology of rings. The extent to which this is true was revealed in a recent report on the stress-relaxation behavior of PS rings (161 and 198 kg/mol) that were purified by LCCC (Section 26.1) and characterized as "pure as currently possible" (Kapnistos *et al.*, 2008). Kapnistos *et al.* conducted small-amplitude oscillatory shear measurements at different temperatures to generate a master curve via time-temperature superposition, which was Fourier transformed to yield a plot of stress-relaxation modulus (G) versus time. As opposed to linear PS, the rings did not exhibit an entanglement plateau, in contrast to the earlier studies described above. Instead, $G(t)$ for the rings continued to decrease with increasing time according to a power-law expression derived from a lattice model in which rings consist of double-folded loops (Obukhov *et al.*, 1994). In other words, the relaxation regime was extended for rings. At a given time during this relaxation regime, $G(t)_{ring} < G(t)_{linear}$ (same number of repeat units). By mixing linear chains with the "pure" rings, $G(t)$ increased, signifying a slowing down of the dynamics. This increase was characterized as significant at linear chain concentrations as low as 0.0007 g/cm^3, which is about 50 times below the overlap concentration for 198-kg/mol linear chains. At higher concentrations, the entanglement plateau was recovered. The enhanced mechanical response was attributed to the following. Trace linear chains in a sea of rings are entropically driven to thread the rings; the resulting entanglements between rings and linear chains are "longer lived than ordinary ring entanglements (double folds)" (Kapnistos *et al.*, 2008). In return, the rings bridge the linear chains forming a transient percolating network across the sample. This description is consistent with recently reported modeling studies (Iyer *et al.*, 2007; Subramanian and Shanbhag, 2008).

Systematic studies of the dynamics of ring/linear blends are particularly interesting. Melt viscosities for 50: 50 blends of equivalent-MW cyclic and linear PDMS (Orrah *et al.*, 1988b; Cosgrove *et al.*, 1996) and POE (Nam *et al.*, 2009) have been examined and are shown in Figure 26.5 as plots of viscosity ratios versus number of skeletal bonds. The viscosity ratios are ratios of experimental viscosities (η_{blend}) to viscosities calculated (η_{calc}) from the pure-component viscosities using a binary mixing law. The transition to higher-than-expected viscosities ($\eta_{blend}/\eta_{calc} > 1$) occurs for PDMS at a ring size of about 100 skeletal bonds

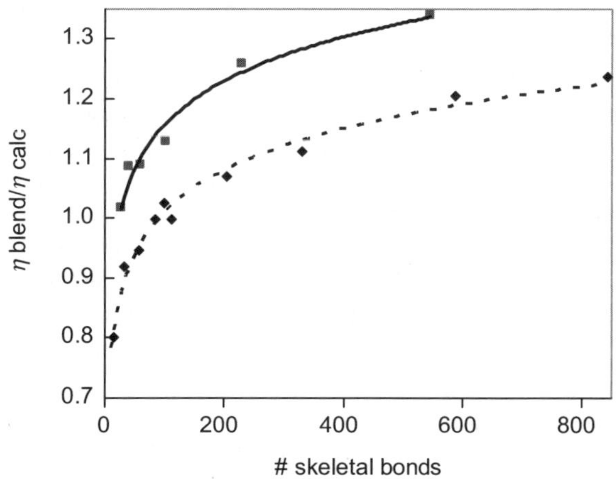

Figure 26.5 Ratios of experimental viscosities (η_{blend}) to viscosities calculated from the pure-component viscosities using a binary mixing law (η_{calc}) versus number of skeletal bonds for 50:50 blends of cyclic and linear POE (■) (Nam *et al.*, 2009) and PDMS (◆) (Orrah, *et al.*, 1988b). (Reprinted with permission from S. Nam, J. Leisen, V. Breedveld and H.W. Beckham, "Melt dynamics of blended poly (oxyethylene) chains and rings," *Macromolecules*, **42**, 3121–3128, 2009. © 2009 American Chemical Society.)

(3.7 kg/mol), and for POE at a ring size of about 27 skeletal bonds (400 g/mol), molecular weights that are well below M_c for linear PDMS (17 kg/mol) (Orrah *et al.*, 1988a) and POE (7 kg/mol) (Cheng *et al.*, 1991a), respectively. As the observed viscosity increase "cannot be due to 'conventional' molecular entanglements" (Cosgrove *et al.*, 1996), it was suggested that the increased viscosities are due to chain threading. This is consistent with earlier observations that cyclic PDMS must have at least 38 bonds to allow threading by linear PDMS (Garrido *et al.*, 1984; Garrido *et al.*, 1985a), and also with Monte Carlo simulations of cyclic/linear POE blends that reveal threading occurs for molecular weights greater than 400 g/mol (27 skeletal bonds) (Helfer *et al.*, 2003). The threading onset occurs at a higher number of skeletal bonds for PDMS since the methyl side groups require more space.

Given the importance of entanglements in physical polymer science, further studies and description of these unconventional threaded entanglements are warranted. Two recent studies (Nam *et al.*, 2009; Habuchi *et al.*, 2010) are noteworthy due to the tools and methodology employed, and also due to the cyclic polymer chosen for examination. In both studies, cyclic polyethers were studied due to the ability to synthesize and isolate these rings in high yield and purity. Characterized by a relatively low critical molecular weight for entanglement M_c (7 kg/mol) (Cheng *et al.*, 1991a), poly(oxyethylene) can be cyclized to prepare rings that traverse the Rouse to reptation regime (for linear POE), while avoiding the difficulties associated with purifying high molecular weight rings. In one study, pulsed-field-gradient (PFG) ^{1}H NMR was used and advantage taken of the ability to substitute the linear species with a perdeuterated one so that the dynamics of cyclic POE in the blend could be examined independently (Nam *et al.*, 2009). In another study, single-molecule spectroscopy on a fluorophore-labeled cyclic poly(tetrahydrofuran) allowed independent measurement of the diffusion of threaded and unthreaded species (Habuchi *et al.*, 2010).

Figure 26.6 displays self-diffusion coefficients (D) measured by pulsed-field-gradient nuclear magnetic resonance (NMR) spectroscopy (Zhao *et al.*, 2005; Zhao *et al.*, 2003; Zhao and Beckham, 2003) for binary blends of cyclic with linear POE. The PFG-NMR echo attenuation plots exhibit single-exponential decays, signifying that an averaged diffusion coefficient is measured. For each blend, the molecular weight of the cyclic and linear POE is the same, so that mole and weight fractions are equivalent. The data are plotted as the ratio of the measured average D of the blend, D_{blend}, to the D of the corresponding linear POE, D_{LPOE}, versus concentration of cyclic POE in the blend. Data are shown for three different molecular weights: 400, 900, and 1500 g/mol. For the smallest POE (400 g/mol), the D_{blend} changes linearly with concentration of cyclic POE, as predicted by a simple binary mixing law (dashed line in Figure 26.6), and consistent with observations for blends of small (i.e. 8–16 carbons) cyclic and linear alkanes (von Meerwall *et al.*, 2003). Interestingly, the D_{blend} of the higher-MW blends (900 and 1500 g/mol) drops below what would be predicted by such a simple mixing law. The translational dynamics of these higher-MW blends of cyclic with linear POE is apparently suppressed, suggesting the presence of threaded conformations or additional entanglement coupling in the melt (see schematic in Figure 26.6), which cannot form for the smallest POE. This is generally consistent with the literature reports on cyclic/linear blends of PDMS (Cosgrove *et al.*, 1996).

Assuming the D_{blend} reduction is related to the existence of threaded conformations, the fraction of threaded species across the entire CPOE concentration for the 1.5-kg/mol blend can be estimated from the magnitude of the deviation from the binary-mixing-rule prediction. For this analysis, a three-term mixing law was assumed for the average diffusion coefficient:

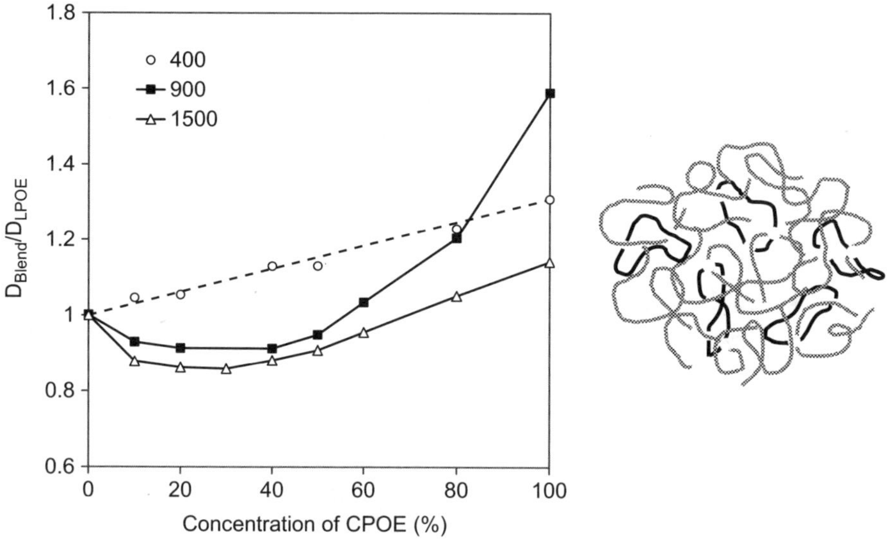

Figure 26.6 Ratio of self-diffusion coefficient of cyclic/linear blends of POE (same MW), D_{blend}, to D of the respective linear POE, versus concentration of cyclic POE in the blend. Data measured at 56 °C for three blends: 400, 900, and 1500 g/mol. Dashed line represents weighted-average D_{blend} calculated from pure-component Ds. Schematic depicts threaded conformations presumed responsible for D suppression for two higher-MW blends. (Reprinted with permission from S. Nam, J. Leisen, V. Breedveld and H.W. Beckham, "Melt dynamics of blended poly(oxyethylene) chains and rings," *Macromolecules*, **42**, 3121–3128, 2009. © 2009 American Chemical Society.)

$D_{\text{blend}} = W_{\text{fC}}D_{\text{fC}} + W_{\text{fL}}D_{\text{fL}} + W_{\text{t}}D_{\text{t}}$, where W_{fC} is the weight fraction of free (unthreaded) cyclic chains having D_{fC}, W_{fL} is the weight fraction of free linear chains having D_{fL}, and W_{t} is the weight fraction of threaded cyclic and linear chains with diffusion coefficient D_{t}. If only threaded pairs of one linear chain and one cyclic chain exist, this equation can be simplified: $D_{\text{blend}} = (W_{\text{C}} - 1/2W_{\text{t}})D_{\text{fC}} + (W_{\text{L}} - 1/2W_{\text{t}})D_{\text{fL}} + W_{\text{t}}D_{\text{t}}$, where W_{C} and W_{L} are the blend weight fractions of CPOE and LPOE. D_{t} was obtained independently by measuring a blend of CPOE and deuterated LPOE, in which PFG ^{1}H NMR only detects the hydrogenous CPOE. These data are shown in Figure 26.7(a). By extrapolating to zero CPOE concentration, D_{t} was determined to be 2.8×10^{-12} m^2/s. D_{fC} and D_{fL} are the self-diffusion coefficients for pure CPOE and LPOE and it was thus determined that the weight fraction of threaded chains, W_{t}, reaches a maximum of about 25% (Figure 26.7b).

In a recently published single-molecule study (Habuchi *et al.*, 2010), fluorophore-labeled cyclic poly(tetrahydrofuran) (3.8 kg/mol) was mixed with unlabeled linear poly(THF) (3 kg/mol) and examined at a semidilute concentration in toluene. Single-molecule trajectories were measured and diffusion coefficient distributions were constructed. Analysis of the distributions yielded two diffusion coefficients for cyclic poly(THF), 1.1 and 4.9×10^{-12} m^2/s, which were attributed to threaded and unthreaded rings, respectively. Comparison with the study discussed above using PFG NMR (Nam *et al.*, 2009) reveals intriguing similarities: low molecular weight cyclic polyethers, diffusion coefficients of 10^{-12} m^2/s, ratio of threaded to unthreaded diffusion coefficients from 3 (NMR) to 4.5

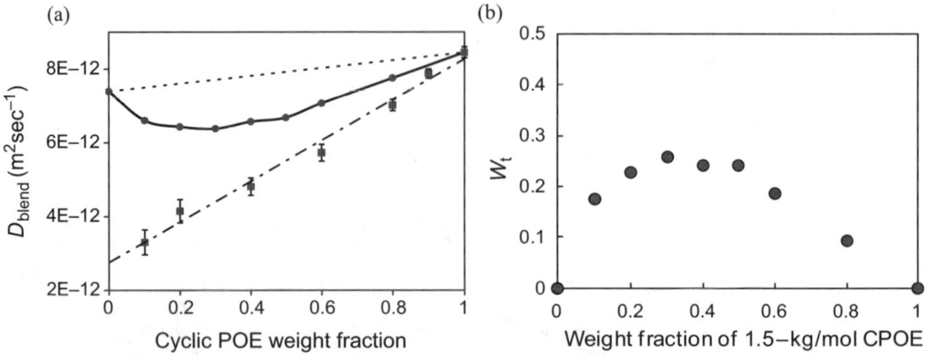

Figure 26.7 (a) Self-diffusion coefficients versus weight fraction of 1.5-kg/mol cyclic poly(oxyethylene) (CPOE) for fully hydrogenous blend with 1.5-kg/mol linear poly(oxyethylene) (LPOE) (●), and hydrogenous-cyclic-only blend with 1.6-kg/mol deuterated LPOE (■). Data were measured at 56 °C. The dashed line represents a binary-mixing-rule prediction for the fully hydrogenous blend. The dash-dot line is a linear fit to the data for the hydrogenous-cyclic-only blend: $D_{blend} = 5.5 \times 10^{-12} w_c + 2.8 \times 10^{-12}$ m^2/s. (b) Weight fraction of threaded cyclic and linear chains, w_t, versus CPOE weight fraction in the 1.5-kg/mol CPOE/LPOE blend. (Reprinted with permission from S. Nam, J. Leisen, V. Breedveld and H.W. Beckham, "Melt dynamics of blended poly(oxyethylene) chains and rings," *Macromolecules*, **42**, 3121–3128, 2009 © 2009 American Chemical Society.)

(single-molecule). Bulk methods like NMR provide ensemble-averaged diffusion coefficients, while single-molecule studies allow one to build the entire distribution. Perhaps more interesting, single-molecule techniques allow one to monitor changes in dynamics that potentially occur for individual molecules. During an experimental time window of 200 ms, Habuchi *et al.* found no evidence that a given cyclic poly(THF) became unthreaded (or threaded), as would be signified by a change in its diffusion coefficient (Habuchi *et al.*, 2010). This is remarkable as it provides a persistence time for threaded entanglements of at least 200 ms. More such studies would lend highly valuable insight.

Suppression of bulk dynamics in blends of rings and linear chains is attributed to threading. The fact that threading occurs in such blends has been unequivocally proven: crosslinking linear chains (Fyvie *et al.*, 1987; Wood *et al.*, 1993; Huang *et al.*, 1990) or attaching bulky blocking groups onto chain ends (i.e. polyrotaxanation) (Huang and Gibson, 2005; Harada *et al.*, 2009; Zhao *et al.*, 2003; Gong *et al.*, 1998) results in topological trapping. Rings that remain after exhaustive extraction, as evidenced by physical and spectroscopic data, are presumed trapped and therefore threaded. Direct and incontrovertible evidence for threading has been provided by diffusion-ordered NMR spectroscopy (DOSY), which correlates molecular size via diffusion coefficient with molecular structure via chemical shift. An example is shown in Figure 26.8 for cyclic poly(oxyethylene) (as 30-crown-10) trapped onto linear polystyrene by end capping with bulky blocking groups. In the spectrum for the polyrotaxane (Figure 26.8a), the ring (440 g/mol) exhibits the same diffusion coefficient as the much higher molecular weight polymer ($\sim 10^4$ g/mol), which can only occur if the ring is threaded onto the polymer. This is further confirmed by mixing some free ring with the polyrotaxane; the DOSY spectrum of this blend sample (Figure 26.8b) shows that unthreaded rings exhibit a separate, larger diffusion coefficient.

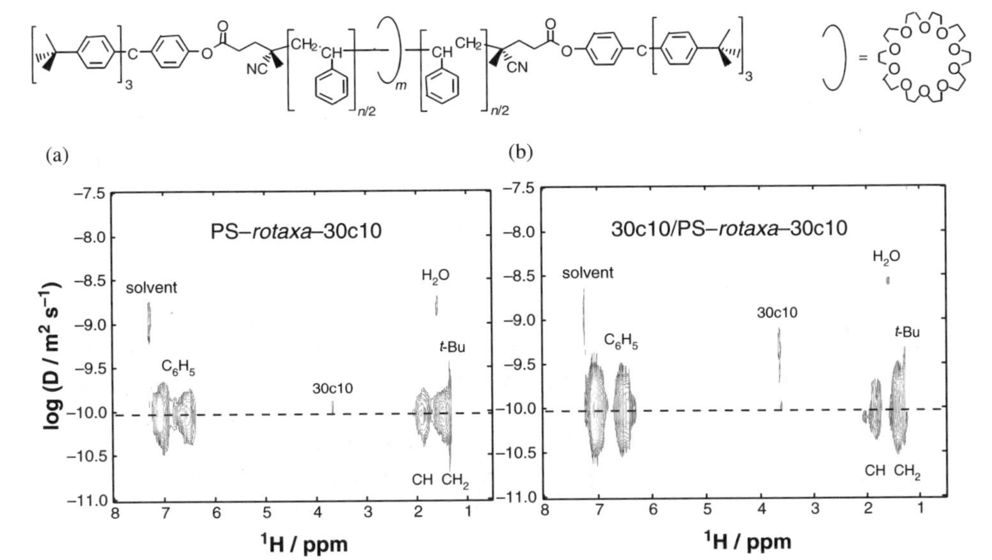

Figure 26.8 Cyclic poly(oxyethylene) (as 30-crown-10) can be threaded and trapped onto linear polystyrene by polyrotaxanation. Diffusion-ordered NMR spectra of (a) poly[(styrene)-*rotaxa*-(30-crown-10)] (PS-*rotaxa*-30c10, structure shown at top), and (b) blend of 30-crown-10 with PS-*rotaxa*-30c10 (4/96, w/w). The polystyrene sports bulky *tert*-butyl-containing end groups (labeled in spectra as *t*-Bu). The 30-crown-10 (3.7 ppm) in the polyrotaxane exhibits the same diffusion coefficient as the polystyrene (dashed horizontal line), which means it must be threaded. This is confirmed when some 30-crown-10 is added to the polyrotaxane and exhibits a separate, larger diffusion coefficient at the same chemical shift. (Reprinted with permission from T. Zhao and H.W. Beckham, "Quantitative determination of threading in rotaxanated polymers by diffusion-ordered NMR spectroscopy," *Macromolecules*, **36**, 4833–4837, 2003. © 2003 American Chemical Society.)

26.2.3 Topological Trapping to Stabilize Incompatible Blends and Introduce Localized Mobility

To broaden the palette of available polymeric materials for commercial applications, homopolymers are often blended for modifying properties and cost. Miscible polymer blends are rare and exhibit properties that are mathematical averages of the constituent components. Thus, incompatible polymer blends have been employed and exhibit properties that reflect both components, sometimes synergistically (Koning *et al.*, 1998). Blends of incompatible polymers are heterogeneous, with properties often optimal when the minor component is dispersed at μm length scales. Thus, it is important to stabilize the blend against coalescence of the dispersed phase: this is called "coarsening" and its prevention is called "blend compatibilization". Methods of blend compatibilization include (1) addition of copolymer additives (Kim *et al.*, 2005), and (2) *in situ* generation of interfacially active polymers by (a) using reactive components with complementary functional groups (Harton *et al.*, 2005), or (b) solid-state shear pulverization (Lebovitz *et al.*, 2003). These techniques require either addition of an extra component or its synthesis by chemical modification of one or more of the constituent polymers.

An alternative method for stabilizing incompatible polymer blends is topological trapping. The resulting heterogeneous ring-entrapped network or polyrotaxane can microphase separate, but cannot macrophase separate due to the threaded architecture. The maximum distance between a threaded ring and a linear chain is related to the size of the ring (Figure 26.9a). Thus, the threaded minor phase is intrinsically limited to nanoscopic dimensions. Using solid-state NMR, we found nanoscopic heterogeneous morphologies for cyclic POE threaded onto poly(acrylonitrile) (Nagapudi *et al.*, 1999a) and for cyclic POE threaded onto polystyrene (Singla, 2004). These materials are also dynamically heterogeneous as cyclic POE remains mobile even though it is threaded onto rigid linear backbones (Figure 26.9b).

Heterogeneous rotaxanated polymers are processable versions of interpenetrating polymer networks (IPNs) (Gilmer *et al.*, 1996; Jansen *et al.*, 1999a; Jansen *et al.*, 1999b; Kim and Sperling, 1997). For incompatible polymer blends, resulting morphologies and hence mechanical properties depend on processing conditions and physical properties of each component. For IPNs, when one or both components are prepared in the presence of the other, morphology control is even more complicated because it also depends on the relative rates of phase separation versus network formation. With heterogeneous rotaxanated polymers, as with conventional block copolymers, a greater degree of morphology control is allowed.

Cyclic PDMS has been topologically trapped in networks of rubbery poly(dimethylsiloxane) (Garrido *et al.*, 1985a; Garrido *et al.*, 1985b) and glassy poly(2,6-dimethyl-1,4-phenylene oxide) (PPO) (Fyvie *et al.*, 1987; Huang *et al.*, 1990) to prepare homogeneous and heterogeneous materials, respectively. The fraction of trapped rings increases with increasing ring size (Clarson *et al.*, 1986). Trapped PDMS rings, acting as diluents or plasticizers, reduce the moduli of end-linked PDMS networks (Garrido *et al.*, 1985b). As opposed to linear diluents,

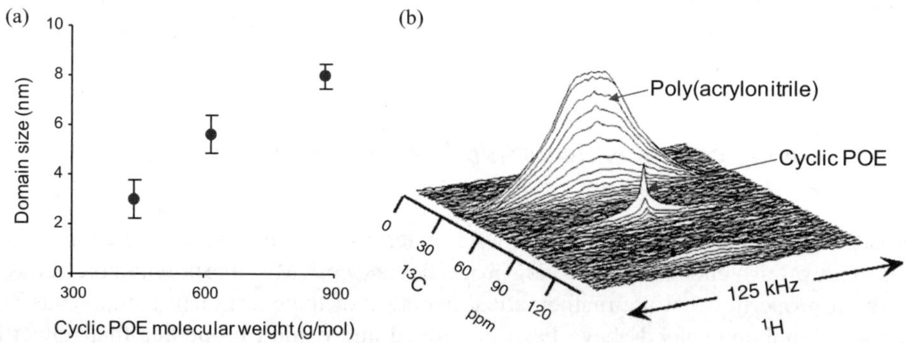

Figure 26.9 (a) Cyclic POE domain sizes in solid cyclic-POE-rotaxanated polymers, determined from NMR spin diffusion experiments, plotted as a function of cyclic POE molecular weight. Polyrotaxanes are topological copolymers in which the dispersed domain size is related to threaded ring size (Beckham *et al.*, 2003). (b) 2D wideline separation (WISE) solid-state NMR spectrum (25 °C) of cyclic POE (15 wt %) topologically trapped onto poly(acrylonitrile) (PAN) by polyrotaxanation. This NMR method correlates structure via ^{13}C chemical shifts with dynamics via ^{1}H lineshapes. A narrow ^{1}H lineshape indicates cyclic POE is mobile compared to PAN at this temperature. (Reprinted with permission from K. Nagapudi, J. Leisen, H.W. Gibson and H.W. Beckham, "Solid-state NMR investigations of poly[(acrylonitrile)-rotaxa-(60-crown-20)]," *Macromolecules*, **32**, 3025–3033, 1999. © 1999 American Chemical Society.)

topologically trapped rings impart permanent property modifications since they cannot leach out of the material. Cyclic polyesters have also been trapped in PDMS networks, but no properties were reported (Wood *et al.*, 1993). Trapping PDMS rings in glassy PPO networks decreases tensile strength, increases elongation to break, slightly decreases PPO T_g and introduces low-temperature transitions attributed to PDMS (Fyvie *et al.*, 1987; Huang *et al.*, 1990). Cyclic PDMS domains are dispersed on a 1-μm length scale (Fyvie *et al.*, 1987). Again, such property modifications are permanent with topologically trapped rings, and are consistent with those obtained traditionally by interpenetrating polymer network (IPN) formation from incompatible materials.

Trapping rings by polyrotaxanation results in similar effects on bulk properties. For example, polyether-ring-containing polyrotaxanes often exhibit reduced T_gs and increased solution viscosities (Huang and Gibson, 2005). Much less focus has been given to changes in *local* dynamics resulting from topologically trapping ring polymers. Local dynamics are important for such bulk properties as toughness and transport. Using dynamic mechanical analysis, Nagapudi *et al.* (1999b) showed that a new sub-T_g relaxation was introduced into a linear aliphatic polyurethane by trapping polyether rings onto its backbone (Figure 26.10). They used solid-state NMR to prove that the new sub-T_g relaxation was due to the onset of rapid segmental motions of the threaded rings.

26.2.4 Dissolution in Normally Incompatible Linear Polymers

Replacing one component of an incompatible polymer blend with its cyclic analog can enhance miscibility of the pair (Santore *et al.*, 1992; Kuo *et al.*, 1994; Nachlis *et al.*, 1995; Nachlis *et al.*, 1997; Singla and Beckham, 2008). This was originally predicted based on some computational studies that suggested that cyclic/linear blends may be thermodynamically favored due to the entropy gain when mixing linear chains into pure rings (Cates and Deutsch, 1986). In pure rings, entropy is restricted due to topology (e.g. no knots or catenanes). By mixing with linear chains, some of the previously unavailable configurational space becomes available since rings and linear chains can interpenetrate freely (Nachlis *et al.*, 1995; Khokhlov and Nechaev, 1996). This route to miscible polymer blends is relatively unexplored, with reports limited to cyclic polystyrene in linear poly(vinyl methyl ether) (Santore *et al.*, 1992), cyclic poly(dimethylsiloxane) in linear poly(methylphenylsiloxane) (Kuo *et al.*, 1994), cyclic polycarbonate in a number of linear amorphous polymers (Nachlis *et al.*, 1995; Nachlis *et al.*, 1997; Kambour *et al.*, 1994), and most recently cyclic poly (oxyethylene) in linear polystyrene. Data for the latter blend are shown in Figure 26.11 as POE weight fraction dissolved in PS versus overall POE weight fraction (Singla and Beckham, 2008). The weight fraction of POE dissolved in PS (w_{POE}) was determined from the DSC-observed depression in the blend T_g, the pure component T_gs and the Fox equation (Fox, 1956), $1/T_{g,blend} = w_{PS}/T_{g,PS} + w_{POE}/T_{g,POE}$, where $w_{PS} = 1 - w_{POE}$. The magnitude of the depression in the blend T_g is proportional to the amount of POE dissolved in the matrix. For cyclic POE in PS, the blend T_gs are 90, 81, 66, 35, and 20 °C for 1.5, 4.5, 10, 25, and 40 wt%, respectively. As illustrated in Figure 26.11, these T_g depressions signify that cyclic POE forms miscible blends with PS. Across the same concentration range, linear dihydroxy-terminated POE is completely immiscible and linear dimethoxy-terminated POE is partially miscible in PS. Solid-state NMR revealed that the dissolved cyclic POE domains are 1–1.5 nm and exhibit

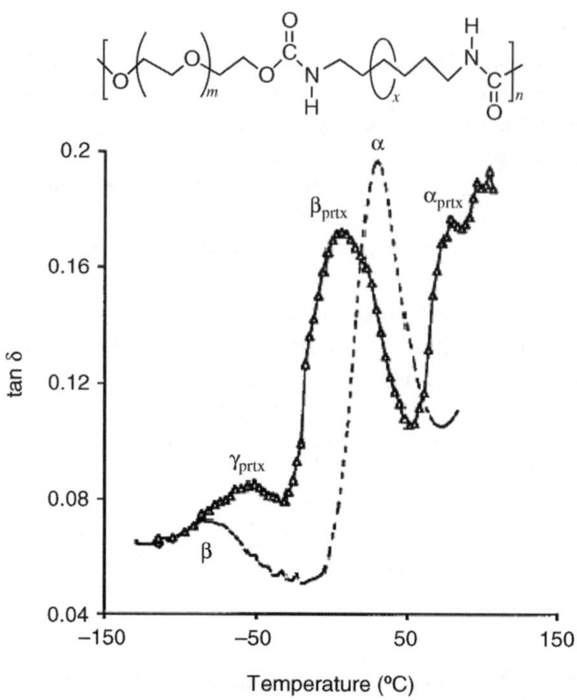

Figure 26.10 Dynamic mechanical loss tangent (1 Hz) versus temperature for a linear aliphatic polyurethane synthesized from diethylene glycol and hexamethylene diisocyanate (– – –) and the same polymer onto which cyclic poly(oxyethylene) ($X_n = 12$) is threaded (Δ, $x/n = 0.15$, 23% cyclic POE by mass) (Nagapudi *et al.*, 1999b). Two loss processes, marked α and β, are observed for the unthreaded polyurethane. The ring-threaded polymer (a polypseudorotaxane) displays an additional low-temperature relaxation γ_{prtx}, at −53 °C. The β_{prtx} peak centered around 11 °C corresponds to the glass transition of the threaded polyurethane that is slightly depressed as compared to the unthreaded backbone (α peak at 30 °C). The ring-threaded polyurethane also exhibits a high-temperature relaxation, α_{prtx}, that is not observed here for the unthreaded backbone. (Reproduced with permission from K. Nagapudi, J.P. Hunt, C. Shepherd *et al.*, "Synthesis, dynamic mechanical, and solid-state NMR spectroscopy of crown-ether-based linear aliphatic polyurethane rotaxanes," *Macromolecular Chemistry and Physics*, 1999, **200**, 2541–2550. © Wiley-VCH Verlag GmbH & Co. KGaA.)

reduced segmental mobilities compared to pure POE. Phase-separated linear POE domains exhibit higher segmental mobilities in larger domains (> ~20 nm). Blends of linear POE with PS are not stable against coarsening. By replacing linear POE with cyclic POE, stable blends of POE with PS can be prepared with T_gs that are tunable across a broad temperature range.

26.2.5 *Crystallization Rate*

Crystal structures and melt temperatures have been reported for some rings and linear chains with the same number of atoms (Yu *et al.*, 1997; Cooke *et al.*, 1998; Ungar and Zeng, 2001).

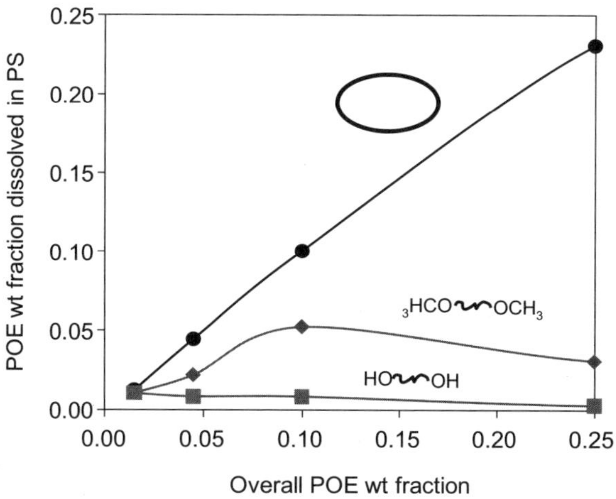

Figure 26.11 Weight fraction of POE incorporated into PS blended with cyclic POE (●, cPOE, 400 g/mol), methoxyl-terminated linear POE (◆, lPDME, 500 g/mol), or hydroxyl-terminated linear POE (■, lPOE, 400 g/mol) as a function of the overall POE weight fraction (Singla and Beckham, 2008). The weight fraction incorporated into the PS-rich phase was determined using the blend T_g from DSC and the Fox equation (Fox, 1956). Lines are drawn as guides to distinguish the behavior of the three different POE additives. (Reprinted with permission from S. Singla and H.W. Beckham, "Miscible blends of cyclic poly (oxyethylene) in linear polystyrene," *Macromolecules*, **41**, 9784–9792, 2008. © 2008 American Chemical Society.)

If lamellar thicknesses and enthalpy changes (ΔH) upon melting are similar, melt temperatures ($T_m = \Delta H/\Delta S$) for rings are higher since entropy changes (ΔS) are smaller due to restricted conformations of rings in the melt. Information on lamellar thicknesses or crystal structures do not always accompany reports on thermal properties, and it is not uncommon to see reports of T_ms for rings that are lower than T_ms for their linear counterparts (Nam *et al.*, 2008; Tezuka *et al.*, 2008).

Much less has been reported on crystallization rates of rings and comparable linear chains. Spherulitic growth was monitored for cyclic poly(tetrahydrofuran) (5 kg/mol, $T_m = 35$ °C) as a function of crystallization temperature (Tezuka *et al.*, 2008). Maximum spherulite growth rate (G_{max}) was 2.6 μm/s and occurred at −20 °C, an undercooling of 55 °C. For linear poly(THF) of the same size (5 kg/mol, $T_m = 40$ °C), $G_{max} = 6$ μm/s and occurred at −25 °C, an undercooling of 65 °C. The ring crystallizes more slowly than its comparable linear chain by a factor of 2.3. It is important to note that the linear poly(THF) is terminated with ethoxyl end groups. The importance of end groups is revealed by examining data for cyclic and linear POE terminated with hydroxyl groups that can hydrogen bond in the melt and impede transport of chain segments to growing crystal faces (Nam, 2006). In this case, rings crystallize more rapidly than comparable linear chains by a factor of 2.5 (Table 26.2). Melt-phase transport can be characterized by self-diffusion coefficients (D), which have been measured for cyclic POE and linear POEs terminated with hydroxyl and methoxyl groups (Nam *et al.*, 2008). The POE self-diffusion coefficients are ordered in the following sequence: dimethoxy-terminated linear POE > cyclic POE > dihydroxy-terminated linear POE. This is consistent with the crystallization rate data that reveal that rings crystallize more slowly than linear chains

Table 26.2 Crystallization rates and self-diffusion coefficients for rings and linear chains

Polymer	Molecular weight (kg/mol)	Crystallization rate (μm/s)	Undercooling (°C)	Self-diffusion coefficient in melt (10^{-12} m^2/s)
Cyclic PTHF (Tezuka *et al.*, 2008)	5	2.6	55	—
Linear di-OEt PTHF	5	6	65	—
Cyclic POE (Nam, 2006)	1.5	100	27	8.8
Linear di-OH POE	1.5	40	32	7.3
Linear di-OH POE (Cheng *et al.*, 1993)	3	100	24	2.5
Linear di-OMe POE	3	180	24	3

unless the chains are terminated with hydrogen-bonding hydroxyl groups. Cyclic POE crystallizes (at 22 °C) and self-diffuses (at 56 °C) more rapidly than linear dihydroxy-terminated POE by factors of 2.5 and 1.2, respectively.

Further support for the correlation of self-diffusion behavior with crystallization rate is provided by data on linear POE (3 kg/mol, $T_m = 59$ °C) terminated with hydroxyl or methoxyl groups (Table 26.2) (Cheng *et al.*, 1991b; Cheng *et al.*, 1993). The crystallization rate (at 35 °C) and self-diffusion coefficient (at 60 °C) for dimethoxy-terminated POE are greater by factors of 1.8 and 1.2, respectively. Note that use of self-diffusion coefficients measured above T_m for comparing transport rates at some crystallization temperature below T_m depends on activation energies for self-diffusion being equivalent for rings and linear chains. Activation energies are indeed the same (30 kJ/mol) for the 3-kg/mol linear POEs terminated with hydroxyl or methoxyl groups (Cheng *et al.*, 1991b).

26.2.6 Enhanced Fluorescence

Restricted dynamics and conformational freedom in cyclic polymers with pendant chromophores apparently allows some ring polymers to emit light more efficiently than linear chains of the same molecular weight. This was first reported as enhanced fluorescence for cyclic polystyrene by Gan and coworkers (Gan *et al.*, 2000; Alberty *et al.*, 2002) (cf. Figure 26.12) and has since been observed in poly(9,9-dimethyl-2-vinylfluorene) rings (Chen *et al.*, 2003). The effect diminishes with increasing molecular weight. We recently observed the same fluorescence enhancement for cyclic PS of 3 kg/mol ($X_n = 29$, $\lambda_{excite} = 253$ nm, toluene). The topology and optical properties of small (i.e. pseudoplanar) ring polymers containing pendant chromophores has led to their proposed use as artificial light-harvesting materials (Johnson *et al.*, 2008).

26.2.7 Rings on Surfaces

Since rings have no chain ends, rings at surfaces would display a very different surface from one presented by linear polymers. Surfaces devoid of chain ends are characterized by low friction and adhesion. Using a surface forces apparatus, Chen *et al.* found that adhesion and friction at

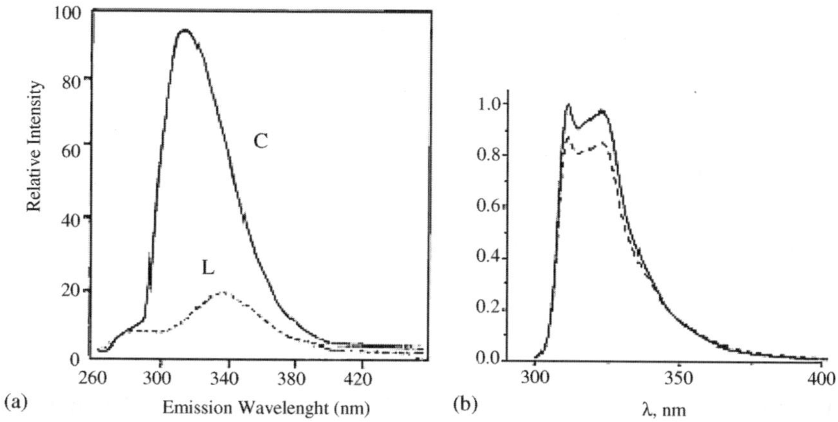

Figure 26.12 Fluorescence emission spectra of cyclic and linear (a) polystyrene ($X_n = 22$, $\lambda_{excite} = 253$ nm) (Gan *et al.*, 2000), and (b) poly(9,9-dimethyl-2-vinylfluorene) ($X_n = 18$, $\lambda_{excite} = 307$ nm) in cyclohexane (Johnson *et al.*, 2008). Dashed line is linear polymer (L) and solid line is cyclic polymer (C). ((a) Reprinted with permission from Y. Gan, D. Dong, S. Carlotti and T.E. Hogen-Esch, "Enhanced fluorescence of macrocylic polystyrene," *Journal of the Americal Chemical Society*, **122**, 2130–2131, 2000. © 2000 American Chemical Society. (b) Reprinted with permission from J.M. Johnson, R. Chen, X. Chen *et al.*, "Investigation of macrocylic polymers as artificial light harvesters: Subpicosecond energy transfer in poly(9,9-dimethyl-2-vinylfluorene)," *The Journal of Physical Chemistry B*, **112**, 16367–16381, 2008. © 2008 American Chemical Society.)

glassy polystyrene and poly(vinylbenzyl chloride) surfaces were directly proportional to the population of chain ends at the surfaces (Chen *et al.*, 2005). They did not use rings in their studies, but examined surfaces as functions of polymer molecular weight and crosslinking. Nevertheless, their findings provide a compelling reason to examine glassy surfaces composed of rings.

Soft surfaces composed of rings have been prepared using cyclic poly(dimethylsiloxane) (Patel *et al.*, 1991) and cyclic poly(oxyethylene) (Singla, 2004). Cyclic PDMS surfaces were prepared on silica by adsorption of rings from solution. Using infrared spectroscopy, Patel *et al.* found that the adsorbed amount of cyclic PDMS was greater than for equivalent-MW linear PDMS at low molar masses. Less cyclic PDMS than linear PDMS was adsorbed at high molar masses. The crossover in adsorption behavior occurred at different molecular masses depending on solvent. For hexane, the crossover from cyclic- to linear-dominated adsorption occurs around 17 kg/mol. The greater adsorption of rings at low molar masses was attributed to the smaller entropy penalty incurred when rings, which are already conformationally restricted in solution compared to linear chains, are adsorbed onto a surface. At high adsorbed amounts and for high molar masses, the adsorption entropy differences are not as significant and the ability of linear chains to form tails extended from the surface favors their adsorption. The results are in agreement with theoretical predictions (van Lent *et al.*, 1987; Sikorski, 2001).

An alternative approach to creating surfaces of rings is stimuli-induced segregation from a cast film (Figure 26.13). Blends of polystyrene and cyclic poly(oxyethylene) (1.5 kg/mol, 10 wt%) were prepared and spun cast from toluene onto glass slides (Singla, 2004). A drop

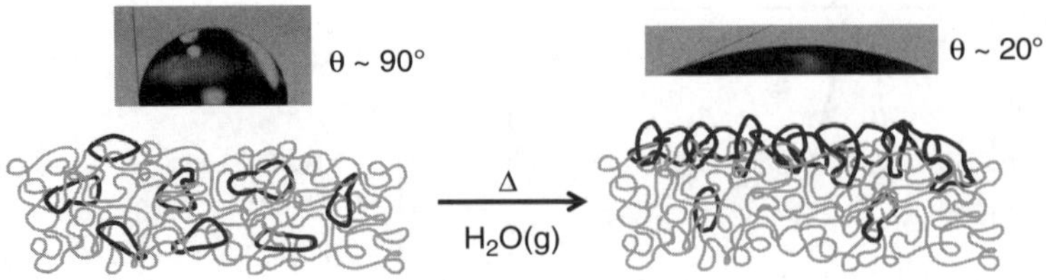

Figure 26.13 Surface segregation of cyclic poly(oxyethylene) (1.5 kg/mol) from its blend with polystyrene by annealing in a high-humidity environment. Images of water drops are shown with their contact angles above the as-spun-cast film (left) and after annealing (right).

of distilled water on the surface of this as-cast film exhibited a contact angle of 90°, consistent with that of pure polystyrene. After annealing at elevated temperatures in high humidity, the film exhibited a contact angle of 20°, indicating that cyclic POE had segregated to the surface. The hydrophilic surface was not stable to immersion in water for 3 days. Topological trapping of rings at surfaces could provide a means to stabilize such structures.

26.2.8 Medical Applications

The same property of rings that make it possible to use exclusion from nanotubular stacks of cyclodextrins to separate them from linear impurities (Section 26.1) (Singla *et al.*, 2003), that is their lack of end groups and increased cross-sectional area, provides an intriguing possibility for their application in medical therapeutics. Delivery of drugs requires persistence times in the circulation system that are longer than the time needed for elimination through nanoporous filtration tissues in the kidneys. Macromolecules have been successfully employed as carriers to increase circulation time. Exploring effects of molecular weight and chain architecture on blood-circulation time and biodistribution is currently an important research topic (Grayson and Godbey, 2008). Blood-circulation times and biodistribution were recently examined for cyclic and linear caprolactone copolymers grafted with PEG chains (Nasongkla *et al.*, 2009). For the macromolecules with molecular weights above the renal filtration threshold (i.e. > 50 kg/mol), the rings exhibited longer circulation times in mice than linear chains of the same molecular weight. This was attributed to the ability of linear chains to move more easily than rings through the nanoporous barrier in the kidneys (i.e. glomeruli) via chain-end-led reptation. As a result of longer circulation times, rings accumulated more in organs than linear chains.

Many of the macromolecules used in medical applications are biodegradable. One of these is poly(caprolactone) (PCL) which has been prepared recently in cyclic form (Hoskins and Grayson, 2009). Acid-catalyzed (*p*-toluenesulfonic acid) degradation of cyclic PCL is delayed compared to degradation of the comparable linear PCL since the first ester cleavage simply opens the ring without reducing the molecular weight. Thermal stability of cyclic and linear PCL is similar.

26.3 Outlook

Rings are the simplest nonlinear (i.e. cyclic or multicyclic) topology and the one for which the most is known. Much more complex nonlinear topologies have recently become synthetically accessible. Historically, the ability to imagine such complex topologies, and to establish a systemic nomenclature (Tezuka and Oike, 2001), has outpaced the ability to synthesize them. Nevertheless, impressive strides have been reported. Some cyclic and multicyclic polymer topologies are shown in Figure 26.14. Of the nine topologies sketched in Figure 26.14, eight have already been prepared and reported (Tezuka, 2005). With the exception of cyclic, rotaxanated and cyclic diblock topologies, most of the complex topologies shown in Figure 26.14 have been prepared using a synthetic protocol developed by Tezuka and coworkers called electrostatic assembly and covalent fixation (Tezuka, 2005; Adachi *et al.*, 2005; Tezuka *et al.*, 2004; Tezuka *et al.*, 2003; Tezuka and Fujiyama, 2005; Oike *et al.*, 2004). Linear precursors are prepared with chain-end cation functionalities that are moderately strained cyclic ammonium groups (e.g. *N*-phenylpyrrolidinium). Electrostatic assembly with multifunctional compounds containing carboxylate anions is followed by thermally induced ring opening of the cyclic ammonium groups to eliminate the charges. Such complex multicyclic topologies, particularly in blends with linear polymers, should be studied further for both practical (e.g. potentially new material performance) and theoretical (e.g. potentially new insight into entanglement coupling) reasons.

Property differences due strictly to topology have been the focus of this chapter. Thus, isolation and the unique properties of ring *homo*polymers have been described. Syntheses of ring *hetero*polymers have been reported. Cyclic diblocks, prepared primarily by dilute-solution ring closure of linear triblocks, exhibit microphase-separated morphologies in the solid state (Lescanec *et al.*, 1995; Zhu *et al.*, 2003; Takano *et al.*, 2003; Lecommandoux *et al.*, 2004) and associate in selective solvents to give micellar morphologies in solution (Iatrou *et al.*, 2002; Yu *et al.*, 1996b; Yu *et al.*, 1998; Minatti *et al.*, 2002; Minatti *et al.*, 2003). In comparing cyclic diblocks to their precursor linear copolymers, solid-state domain sizes for cyclic diblocks were found to be smaller (Lescanec *et al.*, 1995; Zhu *et al.*, 2003; Takano *et al.*, 2003; Lecommandoux *et al.*, 2004). No other general findings have been revealed as morphologies can either be the same (Lescanec *et al.*, 1995; Takano *et al.*, 2003) or different (Zhu *et al.*, 2003; Lecommandoux *et al.*, 2004; Minatti *et al.*, 2002; Minatti *et al.*, 2003), and micellar aggregation numbers can either be larger (Yu *et al.*, 1996b; Yu *et al.*, 1998) or smaller (Iatrou *et al.*, 2002) for cyclic diblocks. More studies on the combined effect of structure and topology are needed.

Large varieties of cyclic, knotted, and threaded macromolecules exist in nature (Craik, 2006). These naturally occurring nonlinear macromolecules exhibit increased stability and activity (e.g. antimicrobial, insecticidal, and so on) due to their lack of chain ends (Craik, 2009). Cyclic proteins have increased stability against proteolytic degradation, in which enzymes act on chain ends. They are also more stable against denaturation since tying up

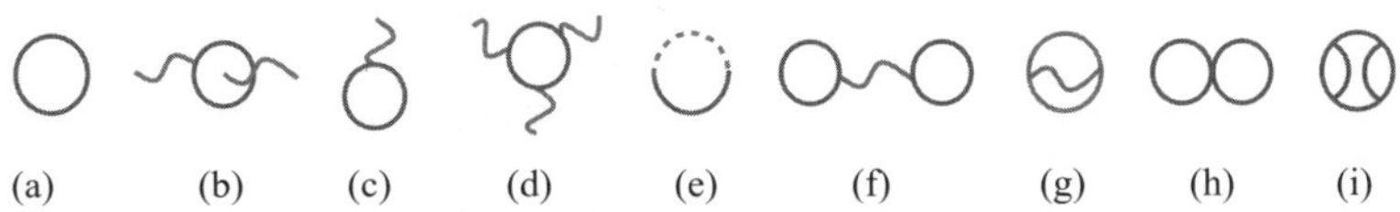

Figure 26.14 Cyclic and multicyclic macromolecular topologies: (a) cyclic, (b) rotaxanated, (c) tadpolar, (d) branched cyclic, (e) cyclic diblock, (f) manacle, (g) theta, (h) figure eight, and (i) baseball (or tennis ball, depending on your sport of choice). All but the baseball topology have been prepared.

chain ends removes high-mobility chain segments. Similarly to disulfide crosslinks, tying up chain ends helps define and stabilize folded tertiary structure. Such emerging knowledge gained from studying natural ring polymers is important for a variety of fields (e.g. protein engineering, drug design, and so on) and should prove extremely valuable for emerging medical applications of synthetic ring polymers (Grayson and Godbey, 2008).

Acknowledgments

Our work on ring polymers at Georgia Tech has greatly benefited from interactions and collaborations with Harry Gibson of Virginia Tech and Yasuyuki Tezuka of Tokyo Tech. I am grateful to them for support that has taken many forms over the years. It is my pleasure to acknowledge the contributions of colleagues and former students at Georgia Tech: Jason Hunt, Jim Baker, Karthik Nagapudi, Tiejun Zhao, Bryan White, Philip Watson, Swati Singla, Sunghyun Nam, Marcus Foston, Katie Campbell, Victor Breedveld, David Bucknall and Johannes Leisen. Funding for portions of our work in this area was provided by the National Science Foundation (DMR) and Office of Naval Research.

References

Adachi, K., Irie, H., Sato, T. *et al.* (2005) Electrostatic self-assembly and covalent fixation with cationic and anionic telechelic precursors for new loop and branch polymer topologies. *Macromolecules*, **38**, 10210–10219.

Alberty, K.A., Tillman, E., Carlotti, S. *et al.* (2002) Characterization and fluorescence of macrocyclic polystyrene by anionic end to end coupling. Role of coupling agents. *Macromolecules*, **35**, 3856–3865.

Alberty, K.A., Hogen-Esch, T.E., and Carlotti, S. (2005) Synthesis and characterization of macrocyclic vinyl-aromatic polymers. molecular weight-dependent glass transition temperatures and emission of macrocyclic polystyrene. *Macromolecular Chemistry and Physics*, **206**, 1035–1042.

Arrighi, V., Gagliardi, S., Dagger, A.C. *et al.* (2004) Conformation of cycles and linear chain polymers in bulk by SANS. *Macromolecules*, **37**, 8057–8065.

Beckham, H.W., Nagapudi, K., Girardeau, T. *et al.* (2003) NMR studies of rotaxanated polymers. *Polymer Preprints (American Chemical Society, Division of Polymer Chemistry)*, **44**, 263–264.

Bensafi, A., Maschke, U., and Benmouna, M. (2000) Cyclic polymers in good solvents. *Polymer International*, **49**, 175–183.

Bielawski, C.W., Benitez, D., and Grubbs, R.H. (2002) An "endless" route to cyclic polymers. *Science*, **297**, 2041–2044.

Bielawski, C., Benitez, D., and Grubbs, R.H. (2003) Synthesis of cyclic polybutadiene via ring-opening metathesis polymerization: The importance of removing trace linear contaminants. *Journal of the American Chemical Society*, **125**, 8424–8425.

Bloomfield, V. and Zimm, B.H. (1966) Viscosity, sedimentation, et cetera, of ring- and straight-chain polymers in dilute solution. *Journal of Chemical Physics*, **44**, 315–323.

Brown, S. and Szamel, G. (1998a) Structure and dynamics of ring polymers. *Journal of Chemical Physics*, **108**, 4705–4708.

Brown, S. and Szamel, G. (1998b) Computer simulation study of the structure and dynamics of ring polymers. *Journal of Chemical Physics*, **109**, 6184–6192.

Cates, M.E. and Deutsch, J.M. (1986) Conjectures on the statistics of ring polymers. *Journal de Physique*, **47**, 2121–2128.

Chen, R., Johnson, J.M., Bradforth, S.E., and Hogen-Esch, T.E. (2003) Ultraviolet absorption and fluorescence emission spectroscopic studies of macrocyclic and linear poly(9,9-dimethyl-2-vinylfluorene). Evidence for ground-state chromophore interactions. *Macromolecules*, **36**, 9966–9970.

Chen, R., Nossarev, G.G., and Hogen-Esch, T.E. (2004) Synthesis and thermal and spectroscopic properties of macrocyclic vinyl aromatic polymers. *Journal of Polymer Science Part A-Polymer Chemistry*, **42**, 5488–5503.

Chen, N., Maeda, N., Tirrell, M., and Israelachvili, J. (2005) Adhesion and friction of polymer surfaces: the effect of chain ends. *Macromolecules*, **38**, 3491–3503.

Cheng, S.Z.D., Barley, J.S., and von Meerwall, E.D. (1991a) Self-diffusion of poly(ethylene oxide) fractions and its influence on the crystalline texture. *Journal of Polymer Science Part B-Polymer Physics*, **29**, 515–525.

Cheng, S.Z.D., Chen, J.H., Zhang, A.Q., and Heberer, D.P. (1991b) Nonintegral and integral folding crystal growth in low-molecular mass poly(ethylene Oxide) fractions. II. End-group effect: α,ω-methoxy-poly(ethylene oxide). *Journal of Polymer Science Part B-Polymer Physics*, **29**, 299–310.

Cheng, S.Z.D., Wu, S.S., Chen, J.H. *et al.* (1993) Isothermal thickening and thinning processes in low molecular weight poly(ethylene oxide) fractions crystallized from the melt. 4. End-group dependence. *Macromolecules*, **26**, 5105–5117.

Cho, D., Park, S., Kwon, K. *et al.* (2001) Structural characterization of ring polystyrene by liquid chromatography at the critical condition and MALDI-TOF mass spectrometry. *Macromolecules*, **34**, 7570–7572.

Clarson, S.J., Dodgson, K., and Semlyen, J.A. (1985) Studies of cyclic and linear poly(dimethyl siloxanes) 19. Glass-transition temperatures and crystallization behavior. *Polymer*, **26**, 930–934.

Clarson, S.J., Mark, J.E., and Semlyen, J.A. (1986) Studies of cyclic and linear poly(dimethylsiloxanes): Effect of ring size on the trapping of cyclic polymers into network structures. *Polymer Communications*, **27**, 244–245.

Cooke, J., Viras, K., Yu, G.-E. *et al.* (1998) Large cyclic poly(oxyethylene)s: Chain folding in the crystalline state studied by Raman spectroscopy, X-ray scattering, and differential scanning calorimetry. *Macromolecules*, **31**, 3030–3039.

Cosgrove, T., Griffiths, P.C., Hollingshurst, J. *et al.* (1992) Self-diffusion and spin-spin relaxation in cyclic and linear polydimethylsiloxane melts. *Macromolecules*, **25**, 6761–6764.

Cosgrove, T., Turner, M.J., Griffiths, P.C. *et al.* (1996) Self-diffusion and spin-spin relaxation in blends of linear and cyclic polydimethylsiloxane melts. *Polymer*, **37**, 1535–1540.

Craik, D.J. (2006) Seamless proteins tie up their loose ends. *Science*, **311**, 1563–1564.

Craik, D.J. (2009) Circling the enemy: cyclic proteins in plant defence. *Trends in Plant Science*, **14**, 328–335.

Cramail, S., Schappacher, M., and Deffieux, A. (2000) Controlled synthesis and solution behavior of macrocyclic poly [(styrene)-b-(ethylene oxide)] copolymers. *Macromolecular Chemistry and Physics*, **201**, 2328–2335.

Deffieux, A. and Schappacher, M. (2009) Naturally occurring and synthetic cyclic macromolecules. *Cellular and Molecular Life Sciences*, **66**, 2599–2602.

Di Marzio, E.A. and Guttman, C.M. (1987) The glass temperature of polymer rings. *Macromolecules*, **20**, 1403–1407.

Dodgson, K. and Semlyen, J.A. (1977) Studies of cyclic and linear poly(dimethylsiloxanes): 1. Limiting viscosity number-molecular weight relationships. *Polymer*, **18**, 1265–1268.

Dodgson, K., Sympson, D., and Semlyen, J.A. (1978) Studies of cyclic and linear poly(dimethyl siloxanes): 2. Preparative gel permeation chromatography. *Polymer*, **19**, 1285–1289.

Dodgson, K., Bannister, D.J., and Semlyen, J.A. (1980) Studies of cyclic and linear poly(dimethyl siloxanes). 4. Bulk viscosities. *Polymer*, **21**, 663–667.

Dulbecco, R. and Vogt, M. (1963) Evidence for a ring structure of polyoma virus DNA. *Proceedings of the National Academy of Sciences, USA*, **50**, 236–243.

Edwards, C.J.C., Rigby, D., Stepto, R.F.T. *et al.* (1983) Studies of cyclic and linear poly(dimethylsiloxanes): 10. Calculations of radii of gyration. *Polymer*, **24**, 391–394.

Foston, M. and Beckham, H.W. (2010a) Cyclic poly(dimethylsiloxane) from kinetically controlled cyclodepolymerization of linear precursors in dilute solution. *Polymer*, **51**, 2112–2118.

Foston, M. and Beckham, H.W. (2010b) Cyclic poly(dimethylsiloxane) via ring-closing dehydrocoupling of α,ω-dihydroxy-PDMS with α,ω-dihydrido-PDMS in dilute solution. *Polymer*, **51**, 2515–2519.

Fox, T.G. (1956) Influence of diluents and of copolymer composition on the glass temperature of a polymer system. *Bulletin of the American Physical Society*, **1**, 123.

Furukawa, Y. (1998) *Inventing Polymer Science*, University of Pennsylvania Press, Philadelphia, pp. 23–34.

Fyvie, T.J., Frisch, H.L., Semylen, J.A. *et al.* (1987) Polymeric catenanes from crosslinked poly(2,6-dimethyl-1,4-phenylene oxide) and cyclic poly(dimethylsiloxane). *Journal of Polymer Science Part A-Polymer Chemistry*, **25**, 2503–2509.

Gagliardi, S., Arrighi, V., Ferguson, R. *et al.* (2005) On the difference in scattering behavior of cyclic and linear polymers. *Journal of Chemical Physics*, **122**, 0649041–0649048.

Gan, Y., Dong, D., Carlotti, S., and Hogen-Esch, T.E. (2000) Enhanced fluorescence of macrocyclic polystyrene. *Journal of the American Chemical Society*, **122**, 2130–2131.

Garrido, L., Mark, J.E., Clarson, S.J., and Semlyen, J.A. (1984) Studies of cyclic and linear poly(dimethylsiloxanes) 15. Diffusion-coefficients from network sorption measurements. *Polymer Communications*, **25**, 218–220.

Garrido, L., Mark, J.E., Clarson, S.J., and Semlyen, J.A. (1985a) Studies of cyclic and linear poly(dimethyl siloxanes) 16. Trapping of cyclics present during the end linking of linear-chains into network structures. *Polymer Communications*, **26**, 53–55.

Garrido, L., Mark, J.E., Clarson, S.J., and Semlyen, J.A. (1985b) Studies of cyclic and linear poly(dimethyl siloxanes) 17. Elastomeric properties of model PDMS networks containing either cyclics or linear chains as diluent. *Polymer Communications*, **26**, 55–57.

Geiser, D. and Höcker, H. (1980) Synthesis and investigation of macrocyclic polystyrene. *Macromolecules*, **13**, 653–656.

Gilmer, T.C., Hall, P.K., Ehrenfeld, H. *et al.* (1996) Synthesis, characterization, and mechanical properties of PMMA/poly(aromatic/aliphatic siloxane) semi-interpenetrating polymer networks. *Journal of Polymer Science Part A-Polymer Chemistry*, **34**, 1025–1037.

Girardeau, T., Zhao, T., Leisen, J. *et al.* (2005) Solid inclusion complexes of α-cyclodextrin and perdeuterated poly (oxyethylene). *Macromolecules*, **38**, 2261–2270.

Gong, C., Ji, Q., Subramaniam, C., and Gibson, H.W. (1998) Main-chain polyrotaxanes by threading crown ethers onto a preformed polyurethane: preparation and properties. *Macromolecules*, **31**, 1814–1818.

Grayson, S.M. and Godbey, W.T. (2008) The role of macromolecular architecture in passively targeted polymeric carriers for drug and gene delivery. *Journal of Drug Targeting*, **16**, 329–356.

Habuchi, S., Satoh, N., Yamamoto, T. *et al.* (2010) Multimode diffusion of ring polymer molecules revealed by a single-molecule study. *Angewandte Chemie-International Edition*, **49**, 1418–1421.

Hadziioannou, G., Cotts, P.M., ten Brinke, G. *et al.* (1987) Thermodynamic and hydrodynamic properties of dilute solutions of cyclic and linear polystyrenes. *Macromolecules*, **20**, 493–497.

Harada, A., Hashidzume, A., Yamaguchi, H., and Takashima, Y. (2009) Polymeric rotaxanes. *Chemical Reviews*, **109**, 5974–6023.

Harmon, F.G., Brockman, J.P., and Kowalczykowski, S.C. (2003) RecQ helicase stimulates both DNA catenation and changes in DNA topology by topoisomerase III. *Journal of Biological Chemistry*, **278**, 42668–42678.

Harton, S.E., Stevie, F.A., and Ade, H. (2005) Diffusion-controlled reactive coupling at polymer-polymer interfaces. *Macromolecules*, **38**, 3543–3546.

Helfer, C.A., Xu, G., Mattice, W.L., and Pugh, C. (2003) Monte Carlo simulations investigating the threading of cyclic poly(ethylene oxide) by linear chains in the melt. *Macromolecules*, **36**, 10071–10078.

Higgins, J.S., Dodgson, K., and Semylen, J.A. (1979) Studies of cyclic and linear poly(dimethylsiloxanes): 3. Neutron scattering measurements of the dimensions of ring and chain polymers. *Polymer*, **20**, 553–558.

Hodge, P. and Peng, P. (1999) Polymer-supported syntheses of some cyclic oligoamides and some cyclic alternating oligo(amide-ester)s. *Polymer*, **40**, 1871–1879.

Hodgson, D.F. and Amis, E.J. (1991) Dilute solution behavior of cyclic and linear polyelectrolytes. *Journal of Chemical Physics*, **95**, 7653–7663.

Hoskins, J.N. and Grayson, S.M. (2009) Synthesis and degradation behavior of cyclic poly(ε-caprolactone). *Macromolecules*, **42**, 6406–6413.

Huang, F. and Gibson, H.W. (2005) Polypseudorotaxanes and polyrotaxanes. *Progress in Polymer Science*, **30**, 982–1018.

Huang, W., Frisch, H.L., Hua, Y., and Semlyen, J.A. (1990) A study of the properties and topological trapping of cyclic poly(dimethylsiloxane) in poly(2,6-dimethyl-1,4-phenylene oxide) networks. *Journal of Polymer Science Part A-Polymer Chemistry*, **28**, 1807–1812.

Hur, K., Winkler, R.G., and Yoon, D.Y. (2006) Comparison of ring and linear polyethylene from molecular dynamics simulations. *Macromolecules*, **39**, 3975–3977.

Iatrou, H., Hajichristidis, N., Meier, G. *et al.* (2002) Synthesis and characterization of model cyclic block copolymers of styrene and butadiene. Comparison of the aggregation phenomena in selective solvents with linear diblock and triblock analogues. *Macromolecules*, **35**, 5426–5437.

Ishizu, K. and Kanno, H. (1996) Novel synthesis and characterization of cyclic polystyrenes. *Polymer*, **37**, 1487–1492.

Iyer, B.V.S., Lele, A.K., and Shanbhag, S. (2007) What is the size of a ring polymer in a ring-linear blend? *Macromolecules*, **40**, 5995–6000.

Jang, S.S., Cagin, T., and Goddard JIII, W.A. (2003) Effect of cyclic chain architecture on properties of dilute solutions of polyethylene from molecular dynamics simulations. *Journal of Chemical Physics*, **119**, 1843–1854.

Jansen, B.J.P., Meijer, H.E.H., and Lemstra, P.J. (1999a) Processing of (in)tractable polymers using reactive solvents part 5: morphology control during phase separation. *Polymer*, **40**, 2917–2927.

Jansen, B.J.P., Rastogi, S., Meijer, H.E.H., and Lemstra, P.J. (1999b) Rubber-modified glassy amorphous polymers prepared via chemically induced phase separation. 4. Comparison of properties of semi- and full-IPNs, and copolymers of acrylate-aliphatic epoxy systems. *Macromolecules*, **32**, 6290–6297.

Johnson, J.M., Chen, R., Chen, X. *et al.* (2008) Investigation of macrocyclic polymers as artificial light harvesters: subpicosecond energy transfer in poly(9,9-dimethyl-2-vinylfluorene). *The Journal of Physical Chemistry B*, **112**, 16367–16381.

Jones, F.R., Scales, L.E., and Semlyen, J.A. (1974) Equilibrium ring concentrations and statistical conformations of polymer-chains 13. Cyclics in 2 aliphatic polyesters. *Polymer*, **15**, 738–742.

Kambour, R.P., Nachlis, W.L., and Carbeck, J.D. (1994) A miscibility survey of bisphenol A carbonate cyclic oligomers with a spectrum of glassy polymers. *Polymer*, **35**, 209–211.

Kapnistos, M., Lang, M., Vlassopoulos, D. *et al.* (2008) Unexpected power-law stress relaxation of entangled ring polymers. *Nature Materials*, **7**, 997–1002.

Kawaguchi, D., Masuoka, K., Takano, A. *et al.* (2006) Comparison of interdiffusion behavior between cyclic and linear polystyrenes with high molecular weights. *Macromolecules*, **39**, 5180–5182.

Khokhlov, A.R. and Nechaev, S.K. (1996) Topologically driven compatibility enhancement in the mixtures of rings and linear chains. *Journal of Physics*, **6**, 1547–1555.

Kim, S.C. and Sperling, L.H. (eds) (1997) *IPNs Around the World*, John Wiley & Sons, Chichester, England.

Kim, J., Gray, M.K., Zhou, H. *et al.* (2005) Polymer blend compatibilization by gradient copolymer addition during melt processing: stabilization of dispersed phase to static coarsening. *Macromolecules*, **38**, 1037–1040.

Kirst, K.U., Kremer, F., Pakula, T., and Hollingshurst, J. (1994) Molecular dynamics of cyclic and linear poly (dimethylsiloxanes). *Colloid & Polymer Science*, **272**, 1420–1429.

Koning, C., Van Duin, M., Pagnoulle, C., and Jerome, R. (1998) Strategies for compatibilization of polymer blends. *Progress in Polymer Science*, **23**, 707–757.

Kricheldorf, H.R. (2003) Macrocycles. 21. Role of ring-ring equilibria in thermodynamically controlled polycondensations. *Macromolecules*, **36**, 2302–2308.

Kricheldorf, H.R., Lee, S.-R., and Schittenhelm, N. (1998) Macrocycles. 1. macrocyclic polymerizations of (thio) lactones - stepwise ring expansion and ring contraction. *Macromolecular Chemistry and Physics*, **199**, 273–282.

Kricheldorf, H.R., Al-Masri, M., and Schwarz, G. (2002) Macrocycles. 20. Cyclic poly(ethylene glycol) phthalates via ring-exchange substitution. *Macromolecules*, **35**, 8936–8942.

Kubo, M., Hayashi, T., Kobayashi, H. *et al.* (1997) Synthesis of α-carboxyl, ω-amino heterodifunctional polystyrene and its intramolecular cyclization. *Macromolecules*, **30**, 2805–2807.

Kuo, C.M., Clarson, S.J., and Semlyen, J.A. (1994) Studies of cyclic and linear poly(dimethylsiloxane)s: 31. Effect of molecular architecture/topology on blends of poly(methylphenylsiloxane) and poly(dimethylsiloxane). *Polymer*, **35**, 4623–4626.

Laurent, B.A. and Grayson, S.M. (2006) An efficient route to well-defined macrocyclic polymers via "Click" cyclization. *Journal of the American Chemical Society*, **128**, 4238–4239.

Laurent, B.A. and Grayson, S.M. (2009) Synthetic approaches for the preparation of cyclic polymers. *Chemical Society Reviews*, **38**, 2202–2213.

Lebovitz, A.H., Khait, K., and Torkelson, J.M. (2003) Sub-micron dispersed-phase particle size in polymer blends: overcoming the Taylor limit via solid-state shear pulverization. *Polymer*, **44**, 199–206.

Lecommandoux, S., Borsali, R., Schappacher, M. *et al.* (2004) Microphase separation of linear and cyclic block copolymers poly(styrene-*b*-isoprene): SAXS experiments. *Macromolecules*, **37**, 1843–1848.

Lee, H.C., Lee, H., Lee, W. *et al.* (2000) Fractionation of cyclic polystyrene from linear precursor by HPLC at the chromatographic critical condition. *Macromolecules*, **33**, 8119–8121.

Lee, W., Lee, H., Lee, H.C. *et al.* (2002) Retention behavior of linear and ring polystyrene at the chromatographic critical condition. *Macromolecules*, **35**, 529–538.

Lepoittevin, B., Dourges, M.-A., Masure, M. *et al.* (2000) Synthesis and characterization of ring-shaped polystyrenes. *Macromolecules*, **33**, 8218–8224.

Lescanec, R.L., Hajduk, D.A., Kim, G.Y. *et al.* (1995) Comparison of the lamellar morphology of microphase-separated cyclic block copolymers and their linear precursors. *Macromolecules*, **28**, 3485–3489.

Malesevic, M., Strijowski, U., Bächle, D., and Sewald, N. (2004) An improved method for the solution cyclization of peptides under pseudo-high dilution conditions. *Journal of Biotechnology*, **112**, 73–77.

McKenna, G.B. and Plazek, D.J. (1986) The viscosity of blends of linear and cyclic molecules of similar molecular mass. *Polymer Communications*, **27**, 304–306.

McKenna, G.B., Hadziioannou, G., Lutz, G. *et al.* (1987) Dilute solution characterization of cyclic polystyrene molecules and their zero-shear viscosity in the melt. *Macromolecules*, **20**, 498–512.

McKenna, G.B., Hostetter, B.J., Hadjichristidis, N. *et al.* (1989) A study of the linear viscoelastic properties of cyclic polystyrenes using creep and recovery measurements. *Macromolecules*, **22**, 1834–1852.

Minatti, E., Borsali, R., Schappacher, M. *et al.* (2002) Effect of cyclization of polystyrene/polyisoprene block copolymers on their micellar morphology. *Macromolecular Rapid Communications*, **23**, 978–982.

Minatti, E., Viville, P., Borsali, R. *et al.* (2003) Micellar morphological changes promoted by cyclization of PS-b-PI copolymer: DLS and AFM experiments. *Macromolecules*, **36**, 4125–4133.

Müller, M., Wittmer, J.P., and Cates, M.E. (1996) Topological effects in ring polymers: A computer simulation study. *Physical Review E*, **53**, 5063–5074.

Nachlis, W.L., Bendler, J.T., Kambour, R.P., and MacKnight, W.J. (1995) Topological effects on blend miscibility. *Macromolecules*, **28**, 7869–7878.

Nachlis, W.L., Bendler, J.T., Kambour, R.P., and MacKnight, W.J. (1997) Topological effects on blend miscibility. *Pure and Applied Chemistry*, **69**, 151–155.

Nagapudi, K., Leisen, J., Gibson, H.W., and Beckham, H.W. (1999a) Solid-state NMR investigations of poly [(acrylonitrile)-rotaxa-(60-crown-20)]. *Macromolecules*, **32**, 3025–3033.

Nagapudi, K., Hunt, J.P., Shepherd, C. *et al.* (1999b) Synthesis, dynamic mechanical, and solid-state NMR spectroscopy of crown-ether-based linear aliphatic polyurethane rotaxanes. *Macromolecular Chemistry and Physics*, **200**, 2541–2550.

Nam, S. (2006) Dynamics of Cyclic and Linear poly(oxyethylene), and Threading Conformation in their Blends. Ph.D. Thesis, *Georgia Institute of Technology*; Atlanta 147pp.

Nam, S., Leisen, J., Breedveld, V., and Beckham, H.W. (2008) Dynamics of unentangled cyclic and linear poly (oxyethylene) melts. *Polymer*, **49**, 5467–5473.

Nam, S., Leisen, J., Breedveld, V., and Beckham, H.W. (2009) Melt dynamics of blended poly(oxyethylene) chains and rings. *Macromolecules*, **42**, 3121–3128.

Nasongkla, N., Chen, B., Macaraeg, N. *et al.* (2009) Dependence of pharmokinetics and biodistribution on polymer architecture: Effect of cyclic versus linear polymers. *Journal of the American Chemical Society*, **131**, 3842–3843.

Obukhov, S.P., Rubinstein, M., and Duke, T. (1994) Dynamics of a ring polymer in a gel. *Physical Review Letters*, **73**, 1263–1266.

Oike, H., Imaizumi, H., Mouri, T. *et al.* (2000) Designing unusual polymer topologies by electrostatic self-assembly and covalent fixation. *Journal of the American Chemical Society*, **122**, 9592–9599.

Oike, H., Mouri, T., and Tezuka, Y. (2001a) Efficient polymer cyclization by electrostatic self-assembly and covalent fixation with telechelic poly(tetrahydrofuran) having cyclic ammonium salt groups. *Macromolecules*, **34**, 6592–6600.

Oike, H., Hamada, M., Eguchi, S. *et al.* (2001b) Novel synthesis of single- and double-cyclic polystyrenes by electrostatic self-assembly and covalent fixation with telechelics having cyclic ammonium salt groups. *Macromolecules*, **34**, 2776–2782.

Oike, H., Uchibori, A., Tsuchitani, A. *et al.* (2004) Designing loop and branch topology with cationic star telechelics through effective selection of mono- and difunctional counteranions. *Macromolecules*, **37**, 7595–7601.

Okumura, H., Kawaguchi, Y., and Harada, A. (2001) Preparation and characterization of inclusion complexes of poly (dimethylsiloxane)s with cyclodextrins. *Macromolecules*, **34**, 6338–6343.

Orrah, D.J., Semlyen, J.A., and Ross-Murphy, S.B. (1988a) Studies of cyclic and linear poly(dimethylsiloxanes) 27. Bulk viscosities above the critical molar mass for entanglement. *Polymer*, **29**, 1452–1454.

Orrah, D.J., Semlyen, J.A., and Ross-Murphy, S.B. (1988b) Studies of cyclic and linear poly(dimethyl siloxanes) 28. Viscosities and densities of ring and chain poly(dimethyl siloxane) blends. *Polymer*, **29**, 1455–1458.

Ozisik, R., von Meerwall, E.D., and Mattice, W.L. (2002) Comparison of the diffusion coefficients of linear and cyclic alkanes. *Polymer*, **43**, 629–635.

Pakula, T. and Geyler, S. (1988) Cooperative relaxations in condensed macromolecular systems. 3. Computer-simulated melts of cyclic polymers. *Macromolecules*, **21**, 1665–1670.

Pantazis, D., Schulz, D.N., and Hadjichristidis, N. (2002) Synthesis of a model cyclic triblock terpolymer of styrene, isoprene, and methyl methacrylate. *Journal of Polymer Science Part A-Polymer Chemistry*, **40**, 1476–1483.

Pasch, H., Deffieux, A., Henze, I. *et al.* (1996) Analysis of macrocyclic polystyrenes. 1. Liquid chromatographic investigations. *Macromolecules*, **29**, 8776–8782.

Patel, A., Cosgrove, T., and Semlyen, J.A. (1991) Studies of cyclic and linear poly(dimethylsiloxanes): 30. Adsorption studies on silica in solution. *Polymer*, **32**, 1313–1317.

Plan, M.R.R., Göransson, U., Clark, R.J. *et al.* (2007) The cyclotide fingerprint in *Oldenlandia affinis*: Elucidation of chemically modified, linear and novel macrocyclic peptides. *Chembiochem: A European Journal of Chemical Biology*, **8**, 1001–1011.

Porbeni, F.E., Edeki, E.M., Shin, I.D., and Tonelli, A.E. (2001) Formation and characterization of the inclusion complexes between poly(dimethylsiloxane) and polyacrylonitrile with γ-cyclodextrin. *Polymer*, **42**, 6907–6912.

Privalko, V.P. and Lobodina, A.P. (1974) Glass transition in the lower homologues of polyethyleneoxide. *European Polymer Journal*, **10**, 1033–1038.

Rique-Lurbet, L., Schappacher, M., and Deffieux, A. (1994) A new strategy for the synthesis of cyclic polystyrenes: Principle and application. *Macromolecules*, **27**, 6318–6324.

Robertson, R.M. and Smith, D.E. (2007a) Self-diffusion of entangled linear and circular DNA molecules: Dependence on length and concentration. *Macromolecules*, **40**, 3373–3377.

Robertson, R.M. and Smith, D.E. (2007b) Strong effects of molecular topology on diffusion of entangled DNA molecules. *Proceedings of the National Academy of Sciences, USA*, **104**, 4824–4827.

Roovers, J. (1985) Melt properties of ring polystyrenes. *Macromolecules*, **18**, 1359–1361.

Roovers, J. (1988) Viscoelastic properties of polybutadiene rings. *Macromolecules*, **21**, 1517–1521.

Roovers, J. and Toporowski, P.M. (1983) Synthesis of high molecular weight ring polystyrenes. *Macromolecules*, **16**, 843–849.

Ruddick, C.L., Hodge, P., Zhuo, Y. *et al.* (1999) Cyclo-depolymerization of polyundecanoate and related polyesters: characterization of cyclic oligoundecanoates and related cyclic oligoesters. *Journal of Materials Chemistry*, **9**, 2399–2405.

Santangelo, P.G., Roland, C.M., Chang, T. *et al.* (2001) Dynamics near the glass transition of low molecular weight cyclic polystyrene. *Macromolecules*, **34**, 9002–9005.

Santore, M.M., Han, C.C., and McKenna, G.B. (1992) A comparison of the thermodynamic stability and phase separation kinetics of polymer blends containing cyclic chains of high molecular weight. *Macromolecules*, **25**, 3416–3423.

Saska, I., Colgrave, M.L., Jones, A. *et al.* (2008) Quantitative analysis of backbone-cyclised peptides in plants. *Journal of Chromatography B*, **872**, 107–114.

Schappacher, M. and Deffieux, A. (2001) α-acetal-ω-bis(hydroxymethyl) heterodifunctional polystyrene: Synthesis, characterization, and investigation of intramolecular end-to-end ring closure. *Macromolecules*, **34**, 5827–5832.

Schappacher, M. and Deffieux, A. (2008) Atomic force microscopy imaging and dilute solution properties of cyclic and linear polystyrene combs. *Journal of the American Chemical Society*, **130**, 14684–14689.

Schulz, M., Tanner, S., Barqawi, H., and Binder, W.H. (2009) Macrocyclization of polymers via ring-closing metathesis and azide/alkyne-"Click"-reactions: An approach to cyclic polyisobutylenes. *Journal of Polymer Science Part A-Polymer Chemistry*, **48**, 671–680.

Semlyen, J.A. (ed.) (2000) *Cyclic Polymers*, 2nd edn, Kluwer Academic, Dordrecht, The Netherlands.

Semlyen, J.A. (ed.) (1986) *Cyclic Polymers*, Elsevier, New York.

Shea, K.J., Lee, S.Y., and Busch, B.B. (1998) A new strategy for the synthesis of macrocycles. The polyhomologation of boracyclanes. *Journal of Organic Chemistry*, **63**, 5746–5747.

Sikorski, A. (2001) Computer simulation of adsorbed polymer chains with a different molecular architecture. *Macromolecular Theory and Simulations*, **10**, 38–45.

Singla, S. (2004) Topological Effects on Properties of Multicomponent Polymer Systems. PhD Thesis, *Georgia Institute of Technology*; Atlanta 213pp.

Singla, S. and Beckham, H.W. (2008) Miscible blends of cyclic poly(oxyethylene) in linear polystyrene. *Macromolecules*, **41**, 9784–9792.

Singla, S., Zhao, T., and Beckham, H.W. (2003) Purification of cyclic polymers prepared from linear precursors by inclusion complexation of linear byproducts with cyclodextrins. *Macromolecules*, **36**, 6945–6948.

Subramanian, G. and Shanbhag, S. (2008) Conformational properties of blends of cyclic and linear polymer melts. *Physical Review E*, **77**, 011801–011809.

Sun, T., Yu, G.-E., Price, C. *et al.* (1995) Cyclic polyethers. *Polymer*, **36**, 3775–3778.

Takano, A., Kadoi, O., Hirahara, K. *et al.* (2003) Preparation and morphology of ring-shaped polystyrene-*block*-polyisoprenes. *Macromolecules*, **36**, 3045–3050.

Takano, A., Kushida, Y., Aoki, K. *et al.* (2007) HPLC characterization of cyclization reaction product obtained by end-to-end ring closure reaction of a telechelic polystyrene. *Macromolecules*, **40**, 679–681.

Tezuka, Y. (2005) Topological polymer chemistry by dynamic selection from electrostatic polymer self-assembly. *The Chemical Record*, **5**, 17–26.

Tezuka, Y. and Fujiyama, K. (2005) Construction of polymeric δ-graph: A doubly fused tricyclic topology. *Journal of the American Chemical Society*, **127**, 6266–6270.

Tezuka, Y. and Komiya, R. (2002) Metathesis polymer cyclization with telechelic poly(THF) having allyl groups. *Macromolecules*, **35**, 8667–8669.

Tezuka, Y. and Oike, H. (2001) Topological polymer chemistry: Systematic classification of nonlinear polymer topologies. *Journal of the American Chemical Society*, **123**, 11570–11576.

Tezuka, Y., Mori, K., and Oike, H. (2002) Efficient synthesis of cyclic poly(oxyethylene) by electrostatic self-assembly and covalent fixation with telechelic precursor having cyclic ammonium salt groups. *Macromolecules*, **35**, 5707–5711.

Tezuka, Y., Tsuchitani, A., Toshioka, Y., and Oike, H. (2003) Synthesis of θ-shaped poly(THF) by electrostatic self-assembly and covalent fixation with three-armed star telechelics having cyclic ammonium salt groups. *Macromolecules*, **36**, 65–70.

Tezuka, Y., Tsuchitani, A., and Oike, H. (2004) Synthesis of polymeric topological isomers through electrostatic self-assembly and covalent fixation with star telechelic precursors. *Macromolecular Rapid Communications*, **25**, 1531–1535.

Tezuka, Y., Ohtsuka, T., Adachi, K. *et al.* (2008) A defect-free ring polymer: Size controlled cyclic poly(tetrahydrofuran) consisting exclusively of the monomer unit. *Macromolecular Rapid Communications*, **29**, 1237–1241.

Ungar, G. and Zeng, X.-b. (2001) Learning polymer crystallization with the aid of linear, branched and cyclic model compounds. *Chemical Reviews*, **101**, 4157–4188.

van Lent, B., Scheutjens, J., and Cosgrove, T. (1987) Self-consistent field theory for the adsorption of ring polymers from solution. *Macromolecules*, **20**, 366–370.

von Meerwall, E., Ozisik, R., Mattice, W.L., and Pfister, P.M. (2003) Self-diffusion of linear and cyclic alkanes, measured with pulsed-gradient spin-echo nuclear magnetic resonance. *Journal of Chemical Physics*, **118**, 3867–3873.

Wenz, G., Han, B.-H., and Müller, A. (2006) Cyclodextrin rotaxanes and polyrotaxanes. *Chemical Reviews*, **106**, 782–817.

White, B.M., Watson, P.W., Barthelme, E.E., and Beckham, H.W. (2002) Synthesis and efficient purification of cyclic poly(dimethylsiloxane). *Macromolecules*, **35**, 5345–5348.

Wood, B.R., Joyce, S.J., Scrivens, G. *et al.* (1993) Cyclic polyesters: 2. Topological trapping experiments and theoretical studies. *Polymer*, **34**, 3059–3063.

Wright, P.V. and Beevers, M.S. (1986) Preparation of cyclic polysiloxanes, in *Cyclic Polymers* (ed. J.A. Semlyen), Elsevier, New York, pp. 85–133.

Yan, Z.-G., Yang, Z., Price, C., and Booth, C. (1993) Cyclization of poly(ethylene glycol)s and related block copolymers. *Makromolekulare Chemie-Rapid Communications*, **14**, 725–732.

Yin, R. and Hogen-Esch, T.O. (1993) Synthesis and characterization of narrow molecular weight distribution polystyrene-poly(dimethylsiloxane) macrocyclic block copolymers and their isobaric precursors. *Macromolecules*, **26**, 6952–6957.

Yu, G.-E., Sinnathamby, P., Price, C., and Booth, C. (1996a) Preparation of large cyclic poly(oxyethylene)s. *Chemical Communications*, **1**, 31–32.

Yu, G.-E., Yang, Z., Attwood, D. *et al.* (1996b) Association and surface properties of a cyclic block copolymer of ethylene oxide and butylene oxide (cyclo-B_8E_{42}) in water. *Macromolecules*, **29**, 8479–8486.

Yu, G.-E., Sun, T., Yan, Z.-G. *et al.* (1997) Low-molar-mass cyclic poly(oxyethylene)s studied by Raman spectroscopy, X-ray scattering and differential scanning calorimetry. *Polymer*, **38**, 35–42.

Yu, G.-E., Yang, Z., Attwood, D. *et al.* (1998) Effect of cyclization on the association behavior of block copolymers in aqueous solution. Comparison of oxyethylene/oxypropylene block copolymers cyclo-$P_{34}E_{104}$ and $E_{52}P_{34}E_{52}$. *Langmuir*, **14**, 2278–2285.

Zhao, T. and Beckham, H.W. (2003) Direct synthesis of cyclodextrin-rotaxanated poly(ethylene glycol)s and their self-diffusion behavior in dilute solution. *Macromolecules*, **36**, 9859–9865.

Zhao, T., Beckham, H.W., and Gibson, H.W. (2003) Quantitative determination of threading in rotaxanated polymers by diffusion-ordered NMR spectroscopy. *Macromolecules*, **36**, 4833–4837.

Zhao, T., Beckham, H.W., Ricks, H.L., and Bunz, U.H.F. (2005) Self-diffusion of poly(*p*-phenyleneethynylene)s in dilute solution determined by pulsed-field-gradient NMR. *Polymer*, **46**, 4839–4844.

Zhu, Y., Gido, S., Iatrou, H. *et al.* (2003) Microphase separation of cyclic block copolymers of styrene and butadiene and of their corresponding linear triblock copolymers. *Macromolecules*, **36**, 148–152.

Zifferer, G. and Preusser, W. (2001) Monte Carlo simulation studies of the size and shape of ring polymers. *Macromolecular Theory and Simulations*, **10**, 397–407.

Index

Complex Macromolecular Architectures: Synthesis, Characterization, and Self-Assembly, First Edition.
Edited by Nikos Hadjichristidis, Akira Hirao, Yasuyuki Tezuka and Filip Du Prez.